**CONVERSIONS BETWEEN U.S. CUSTOMARY UNITS AND SI UNITS (Continued)**

| U.S. Customary unit | | Times conversion factor | | Equals SI unit | |
|---|---|---|---|---|---|
| | | Accurate | Practical | | |
| Moment of inertia (area) | | | | | |
| inch to fourth power | in.$^4$ | 416,231 | 416,000 | millimeter to fourth power | mm$^4$ |
| inch to fourth power | in.$^4$ | $0.416231 \times 10^{-6}$ | $0.416 \times 10^{-6}$ | meter to fourth power | m$^4$ |
| Moment of inertia (mass) | | | | | |
| slug foot squared | slug-ft$^2$ | 1.35582 | 1.36 | kilogram meter squared | kg·m$^2$ |
| Power | | | | | |
| foot-pound per second | ft-lb/s | 1.35582 | 1.36 | watt (J/s or N·m/s) | W |
| foot-pound per minute | ft-lb/min | 0.0225970 | 0.0226 | watt | W |
| horsepower (550 ft-lb/s) | hp | 745.701 | 746 | watt | W |
| Pressure; stress | | | | | |
| pound per square foot | psf | 47.8803 | 47.9 | pascal (N/m$^2$) | Pa |
| pound per square inch | psi | 6894.76 | 6890 | pascal | Pa |
| kip per square foot | ksf | 47.8803 | 47.9 | kilopascal | kPa |
| kip per square inch | ksi | 6.89476 | 6.89 | megapascal | MPa |
| Section modulus | | | | | |
| inch to third power | in.$^3$ | 16,387.1 | 16,400 | millimeter to third power | mm$^3$ |
| inch to third power | in.$^3$ | $16.3871 \times 10^{-6}$ | $16.4 \times 10^{-6}$ | meter to third power | m$^3$ |
| Velocity (linear) | | | | | |
| foot per second | ft/s | 0.3048* | 0.305 | meter per second | m/s |
| inch per second | in./s | 0.0254* | 0.0254 | meter per second | m/s |
| mile per hour | mph | 0.44704* | 0.447 | meter per second | m/s |
| mile per hour | mph | 1.609344* | 1.61 | kilometer per hour | km/h |
| Volume | | | | | |
| cubic foot | ft$^3$ | 0.0283168 | 0.0283 | cubic meter | m$^3$ |
| cubic inch | in.$^3$ | $16.3871 \times 10^{-6}$ | $16.4 \times 10^{-6}$ | cubic meter | m$^3$ |
| cubic inch | in.$^3$ | 16.3871 | 16.4 | cubic centimeter (cc) | cm$^3$ |
| gallon (231 in.$^3$) | gal. | 3.78541 | 3.79 | liter | L |
| gallon (231 in.$^3$) | gal. | 0.00378541 | 0.00379 | cubic meter | m$^3$ |

*An asterisk denotes an *exact* conversion factor

*Note:* To convert from SI units to USCS units, *divide* by the conversion factor

**Temperature Conversion Formulas**

$$T(°C) = \frac{5}{9}[T(°F) - 32] = T(K) - 273.15$$

$$T(K) = \frac{5}{9}[T(°F) - 32] + 273.15 = T(°C) + 273.15$$

$$T(°F) = \frac{9}{5}T(°C) + 32 = \frac{9}{5}T(K) - 459.67$$

# Mechanics of Materials

## Eighth Edition

# Mechanics of Materials

## Eighth Edition

### James M. Gere
Professor Emeritus, Stanford University

### Barry J. Goodno
Georgia Institute of Technology

CENGAGE
Learning

Australia • Brazil • Japan • Korea • Mexico • Singapore • Spain • United Kingdom • United States

**Mechanics of Materials, Eighth Edition**
James M. Gere and Barry J. Goodno

Publisher, Global Engineering:
    Christopher M. Shortt

Senior Acquisitions Editor: Randall Adams

Senior Developmental Editor: Hilda Gowans

Editorial Assistant: Tanya Altieri

Team Assistant: Carly Rizzo

Marketing Manager: Lauren Betsos

Media Editor: Chris Valentine

Manager, Content and Media Production:
    Patricia M. Boies

Senior Content Project Manager:
    Jennifer Ziegler

Production Service: RPK Editorial Services, Inc.

Copyeditor: Shelly Gerger-Knechtl

Proofreader: Pamela Ehn

Indexer: Shelly Gerger-Knechtl

Compositor: Integra

Senior Art Director: Michelle Kunkler

Internal Designer: Juli Cook/Plan-It-Publishing

Cover Designer: Andrew Adams/4065042
    Canada Inc.

Cover Image: © Bruno Ehrs/Corbis

Rights Acquisitions Specialist: Deanna Ettinger

Text and Image Permissions Researcher:
    Kristiina Paul

Manufacturing Planner: Doug Wilke

For product information and technology assistance, contact us at
**Cengage Learning Customer & Sales Support, 1-800-354-9706.**

For permission to use material from this text or product,
submit all requests online at **www.cengage.com/permissions**.
Further permissions questions can be emailed to
**permissionrequest@cengage.com**.

Library of Congress Control Number: 2011940308

ISBN-13: 978-1-111-57773-5

ISBN-10: 1-111-57773-0

**Cengage Learning**
200 First Stamford Place, Suite 400
Stamford, CT 06902
USA

Cengage Learning is a leading provider of customized learning solutions with office locations around the globe, including Singapore, the United Kingdom, Australia, Mexico, Brazil, and Japan. Locate your local office at: **international.cengage.com/region**.

Cengage Learning products are represented in Canada by Nelson Education Ltd.

For your course and learning solutions, visit
**www.cengage.com/engineering**.

Purchase any of our products at your local college store or at our preferred online store **www.cengagebrain.com**.

Printed in Canada
2 3 4 5 14 13 12

# CONTENTS

James Monroe Gere ix

Preface xi

Symbols xviii

Greek Alphabet xx

## 1. TENSION, COMPRESSION, AND SHEAR  2

1.1 Introduction to Mechanics of Materials  4

1.2 Statics Review  6

1.3 Normal Stress and Strain  27

1.4 Mechanical Properties of Materials  37

1.5 Elasticity, Plasticity, and Creep  45

1.6 Linear Elasticity, Hooke's Law, and Poisson's Ratio  52

1.7 Shear Stress and Strain  57

1.8 Allowable Stresses and Allowable Loads  68

1.9 Design for Axial Loads and Direct Shear  74

Chapter Summary & Review  80

*Problems*  83

## 2. AXIALLY LOADED MEMBERS  118

2.1 Introduction  120

2.2 Changes in Lengths of Axially Loaded Members  120

2.3 Changes in Lengths Under Nonuniform Conditions  130

2.4 Statically Indeterminate Structures  138

2.5 Thermal Effects, Misfits, and Prestrains  149

2.6 Stresses on Inclined Sections  164

2.7 Strain Energy  176

*2.8 Impact Loading  187

*2.9 Repeated Loading and Fatigue  195

*2.10 Stress Concentrations  197

*2.11 Nonlinear Behavior  205

*2.12 Elastoplastic Analysis  210

Chapter Summary & Review  216

*Problems*  218

## 3. TORSION  254

3.1 Introduction  256

3.2 Torsional Deformations of a Circular Bar  257

3.3 Circular Bars of Linearly Elastic Materials  260

3.4 Nonuniform Torsion  272

3.5 Stresses and Strains in Pure Shear  283

3.6 Relationship Between Moduli of Elasticity $E$ and $G$  290

3.7 Transmission of Power by Circular Shafts  291

3.8 Statically Indeterminate Torsional Members  296

3.9 Strain Energy in Torsion and Pure Shear  300

3.10 Torsion of Non-Circular Prismatic Shafts  307

3.11 Thin-Walled Tubes  316

*3.12 Stress Concentrations in Torsion  324

Chapter Summary & Review  328

*Problems*  330

## 4. SHEAR FORCES AND BENDING MOMENTS  352

4.1 Introduction  354

4.2 Types of Beams, Loads, and Reactions  354

4.3 Shear Forces and Bending Moments  361

4.4 Relationships Between Loads, Shear Forces, and Bending Moments  371

4.5 Shear-Force and Bending-Moment Diagrams  375

Chapter Summary & Review  388

*Problems*  390

## 5. STRESSES IN BEAMS (BASIC TOPICS)  402

5.1 Introduction  404

5.2 Pure Bending and Nonuniform Bending  404

5.3 Curvature of a Beam  405

*Specialized and/or advanced topics

v

5.4   Longitudinal Strains in Beams   407

5.5   Normal Stresses in Beams (Linearly Elastic Materials)   412

5.6   Design of Beams for Bending Stresses   426

*5.7   Nonprismatic Beams   435

5.8   Shear Stresses in Beams of Rectangular Cross Section   439

5.9   Shear Stresses in Beams of Circular Cross Section   448

5.10   Shear Stresses in the Webs of Beams with Flanges   451

*5.11   Built-Up Beams and Shear Flow   458

*5.12   Beams with Axial Loads   462

**5.13   Stress Concentrations in Bending   468

Chapter Summary & Review   472

*Problems*   476

6. STRESSES IN BEAMS (ADVANCED TOPICS)   506

6.1   Introduction   508

6.2   Composite Beams   508

6.3   Transformed-Section Method   517

6.4   Doubly Symmetric Beams with Inclined Loads   526

6.5   Bending of Unsymmetric Beams   533

6.6   The Shear-Center Concept   541

6.7   Shear Stresses in Beams of Thin-Walled Open Cross Sections   543

6.8   Shear Stresses in Wide-Flange Beams   546

6.9   Shear Centers of Thin-Walled Open Sections   550

*6.10   Elastoplastic Bending   558

Chapter Summary & Review   566

*Problems*   569

7. ANALYSIS OF STRESS AND STRAIN   588

7.1   Introduction   590

7.2   Plane Stress   590

7.3   Principal Stresses and Maximum Shear Stresses   598

7.4   Mohr's Circle for Plane Stress   607

7.5   Hooke's Law for Plane Stress   623

7.6   Triaxial Stress   629

7.7   Plane Strain   633

Chapter Summary & Review   648

*Problems*   652

8. APPLICATIONS OF PLANE STRESS (PRESSURE VESSELS, BEAMS, AND COMBINED LOADINGS)   670

8.1   Introduction   672

8.2   Spherical Pressure Vessels   672

8.3   Cylindrical Pressure Vessels   678

8.4   Maximum Stresses in Beams   685

8.5   Combined Loadings   694

Chapter Summary & Review   712

*Problems*   714

9. DEFLECTIONS OF BEAMS   728

9.1   Introduction   730

9.2   Differential Equations of the Deflection Curve   730

9.3   Deflections by Integration of the Bending-Moment Equation   735

9.4   Deflections by Integration of the Shear-Force and Load Equations   746

9.5   Method of Superposition   752

9.6   Moment-Area Method   760

9.7   Nonprismatic Beams   769

9.8   Strain Energy of Bending   774

*9.9   Castigliano's Theorem   779

*9.10   Deflections Produced by Impact   791

*9.11   Temperature Effects   793

Chapter Summary & Review   798

*Problems*   800

10. STATICALLY INDETERMINATE BEAMS   820

10.1   Introduction   822

10.2   Types of Statically Indeterminate Beams   822

10.3   Analysis by the Differential Equations of the Deflection Curve   825

10.4   Method of Superposition   832

*10.5   Temperature Effects   845

*10.6   Longitudinal Displacements at the Ends of a Beam   853

Chapter Summary & Review   856

*Problems*   858

**11. COLUMNS**   868

11.1   Introduction   870

11.2   Buckling and Stability   870

11.3   Columns with Pinned Ends   878

11.4   Columns with Other Support Conditions   889

11.5   Columns with Eccentric Axial Loads   899

11.6   The Secant Formula for Columns   904

11.7   Elastic and Inelastic Column Behavior   909

11.8   Inelastic Buckling   911

11.9   Design Formulas for Columns   916

Chapter Summary & Review   934

*Problems*   936

**12. REVIEW OF CENTROIDS AND MOMENTS OF INERTIA**   954

12.1   Introduction   956

12.2   Centroids of Plane Areas   956

12.3   Centroids of Composite Areas   959

12.4   Moments of Inertia of Plane Areas   962

12.5   Parallel-Axis Theorem for Moments of Inertia   965

12.6   Polar Moments of Inertia   969

12.7   Products of Inertia   971

12.8   Rotation of Axes   974

12.9   Principal Axes and Principal Moments of Inertia   976

*Problems*   980

**REFERENCES AND HISTORICAL NOTES**   987

**APPENDIX A: FE EXAM REVIEW PROBLEMS**   995

**APPENDIX B: SYSTEMS OF UNITS AND CONVERSION FACTORS**   1037

**APPENDIX C: PROBLEM SOLVING**   1051

**APPENDIX D: MATHEMATICAL FORMULAS**   1057

**APPENDIX E: PROPERTIES OF PLANE AREAS**   1063

**APPENDIX F: PROPERTIES OF STRUCTURAL-STEEL SHAPES**   1069

**APPENDIX G: PROPERTIES OF STRUCTURAL LUMBER**   1081

**APPENDIX H: DEFLECTIONS AND SLOPES OF BEAMS**   1083

**APPENDIX I: PROPERTIES OF MATERIALS**   1089

**ANSWERS TO PROBLEMS**   1095

**NAME INDEX**   1123

**SUBJECT INDEX**   1124

# James Monroe Gere
## 1925–2008

(Ed Souza/Stanford News Service)

Jim Gere in the Timoshenko Library at Stanford holding a copy of the 2nd edition of this text (photo courtesy of Richard Weingardt Consultants, Inc.)

James Monroe Gere, Professor Emeritus of Civil Engineering at Stanford University, died in Portola Valley, CA, on January 30, 2008. Jim Gere was born on June 14, 1925, in Syracuse, NY. He joined the U.S. Army Air Corps at age 17 in 1942, serving in England, France and Germany. After the war, he earned undergraduate and master's degrees in Civil Engineering from the Rensselaer Polytechnic Institute in 1949 and 1951, respectively. He worked as an instructor and later as a Research Associate for Rensselaer between 1949 and 1952. He was awarded one of the first NSF Fellowships, and chose to study at Stanford. He received his Ph.D. in 1954 and was offered a faculty position in Civil Engineering, beginning a 34-year career of engaging his students in challenging topics in mechanics, and structural and earthquake engineering. He served as Department Chair and Associate Dean of Engineering and in 1974 co-founded the John A. Blume Earthquake Engineering Center at Stanford. In 1980, Jim Gere also became the founding head of the Stanford Committee on Earthquake Preparedness, which urged campus members to brace and strengthen office equipment, furniture, and other items that could pose a life safety hazard in the event of an earthquake. That same year, he was invited as one of the first foreigners to study the earthquake-devastated city of Tangshan, China. Jim retired from Stanford in 1988 but continued to be a most valuable member of the Stanford community as he gave freely of his time to advise students and to guide them on various field trips to the California earthquake country.

Jim Gere was known for his outgoing manner, his cheerful personality and wonderful smile, his athleticism, and his skill as an educator in Civil Engineering. He authored nine textbooks on various engineering subjects starting in 1972 with *Mechanics of Materials*, a text that was inspired by his teacher and mentor Stephan P. Timoshenko. His other well-known textbooks, used in engineering courses around the world, include: *Theory of Elastic Stability*, co-authored with S. Timoshenko; *Matrix Analysis of Framed Structures* and *Matrix Algebra for Engineers*, both co-authored with W. Weaver; *Moment Distribution*; *Earthquake Tables: Structural and Construction Design Manual*, co-authored with H. Krawinkler; and *Terra Non Firma: Understanding and Preparing for Earthquakes*, co-authored with H. Shah.

Respected and admired by students, faculty, and staff at Stanford University, Professor Gere always felt that the opportunity to work with and be of service to young people both inside and outside the classroom was one of his great joys. He hiked frequently and regularly visited Yosemite and the Grand Canyon national parks. He made over 20 ascents of Half Dome in Yosemite as well as "John Muir hikes" of up to 50 miles in a day. In 1986 he hiked to the base camp of Mount Everest, saving the life of a companion on the trip. James was an active runner and completed the Boston Marathon at age 48, in a time of 3:13.

James Gere will be long remembered by all who knew him as a considerate and loving man whose upbeat good humor made aspects of daily life or work easier to bear. His last project (in progress and now being continued by his daughter Susan of Palo Alto) was a book based on the written memoirs of his great-grandfather, a Colonel (122d NY) in the Civil War.

**Mechanics of materials** is a basic engineering subject that, along with statics, must be understood by anyone concerned with the strength and physical performance of structures, whether those structures are man-made or natural. At the college level, statics is usually taught during the sophomore or junior year and is a prerequisite for the follow-on course in mechanics of materials. Both courses are required for most students majoring in mechanical, structural, civil, biomedical, petroleum, nuclear, aeronautical, and aerospace engineering. Furthermore, many students from such diverse fields as materials science, industrial engineering, architecture, and agricultural engineering also find it useful to study mechanics of materials.

## A First Course in Mechanics of Materials

In many university engineering programs today, both statics and mechanics of materials are now taught in large sections comprised of students from the variety of engineering disciplines listed above. Instructors for the various parallel sections must cover the same material, and all of the major topics must be presented so that students are well prepared for the more advanced courses required by their specific degree programs. An essential prerequisite for success in a first course in mechanics of materials is a strong foundation in statics, which includes not only understanding of fundamental concepts but also proficiency in applying the laws of statical equilibrium to solution of both two and three dimensional problems. This eighth edition begins with a new section on review of statics in which the laws of equilibrium and boundary (or support) conditions are reviewed, as well as types of applied forces and internal stress resultants, all based upon and derived from a properly drawn free body diagram. Numerous examples and end of chapter problems are included to help the student review the analysis of plane and space trusses, shafts in torsion, beams and plane and space frames and to reinforce basic concepts learned in the prerequisite course.

Many instructors like to present the basic theory of say, beam bending, and then use real world examples to motivate student interest in the subject of beam flexure, beam design, etc. In many cases, structures on campus offer easy access to beams, frames, and bolted connections which can be dissected in lecture, or on homework problems, to find reactions at supports, forces and moments in members and stresses in connections. In addition, study of causes of failures in structures and components also offers the opportunity for students to begin the process of learning from actual designs and even past engineering mistakes. A number of the new example problems and also the new or revised end-of-chapter problems in this eighth edition are based upon actual components or structures and are accompanied by photographs so that the student can see the real world problem alongside the simplified mechanics model and free body diagrams to be used in its analysis.

An increasing number of universities are using rich media lecture (and/or classroom) capture software in their large undergraduate courses in mathematics, physics, and engineering and the many new photos and enhanced graphics in the eighth edition are designed to support this enhanced lecture mode.

## New to the Eighth Edition of *Mechanics of Materials*

The main topics covered in this book are the analysis and design of structural members subjected to tension, compression, torsion, and bending, including the fundamental concepts mentioned above. Other important topics are the transformations of stress and strain, combined loadings and combined stress, deflections of beams, and stability of columns. Some additional specialized topics include the following: stress concentrations, dynamic and impact loadings, non-prismatic members, shear centers, bending of beams of two materials (or composite beams), bending of unsymmetric beams, maximum stresses in beams, energy based approaches for computing deflections of beams, and statically indeterminate beams. Review material on centroids and moments of inertia is presented in Chapter 12.

As an aid to the student reader, each chapter begins with a *Chapter Overview* which highlights the major topics to be covered in that chapter, and closes with a *Chapter Summary & Review* in which the key points as well as major mathematical formulas presented in the chapter are listed for quick review (in preparation for examinations on the material). Each chapter also opens with a photograph of a component or structure which illustrates the key concepts to be discussed in that chapter.

Some of the notable features of this eighth edition, which have been added as new or updated material to meet the needs of a modern course in mechanics of materials, are as follows:

- **Statics review**—A new section entitled *Statics Review* has been added to Chapter 1. New Section 1.2 includes four example problems which illustrate calculation of support reactions and internal stress resultants for truss, beam, circular shaft and plane frame structures. Twenty six end-of-chapter problems on statics provide the student with two and three dimensional structures to be used as practice, review or homework assignment problems of varying difficulty.

- **Expanded** *Chapter Overview* and also *Chapter Summary & Review sections*– The *Chapter Overview* and *Chapter Summary* sections have been expanded and now include key equations presented in that chapter. These summary sections will serve as a convenient review for the student of key topics and equations presented in each chapter.

- **Increased emphasis** on equilibrium, constitutive, and strain-displacement/ compatibility equations in problem solutions–Example problem and end-of-chapter problem solutions have been updated to emphasize an orderly process of explicitly writing out the equilibrium, constitutive and strain-displacement/ compatibility equations before attempting a solution.

- New/**expanded topic coverage**—The following topics have been added or have received expanded coverage: stress concentrations in axially loads bars (Sec. 2.10); torsion of noncircular shafts (Sec. 3.10); stress concentrations in bending (Sec. 5.13); transformed section analysis for composite beams (Sec. 6.3); and updated code provisions for buckling of steel, aluminum and timber columns (Sec. 11.9).

- New **example** and **end-of-chapter problems**—Forty eight new example problems have been added to the eighth edition. In addition, there are 287 new and 513 revised end-of-chapter problems out of the 1230 problems presented in the eighth edition text.

- **New appendix**—New Appendix A presents 106 FE-type review problems in *Mechanics of Materials*, which cover all of the major topics presented in the text and are representative of those likely to appear on an FE exam. Each of the problems is presented in the FE Exam format. It is expected that careful review of these problems will serve as a useful guide to the student in preparing for this important examination.

## EXAMPLES

Examples are presented throughout the book to illustrate the theoretical concepts and show how those concepts may be used in practical situations. In some cases, new photographs have been added showing actual engineering structures or components to reinforce the tie between theory and application. In both lecture and text examples, it is appropriate to begin with simplified analytical models of the structure or component and the associated free-body diagram(s) to aid the student in understanding and applying the relevant theory in engineering analysis of the system. The text examples vary in length from one to four pages, depending upon the complexity of the material to be illustrated. When the emphasis is on concepts, the examples are worked out in symbolic terms so as to better illustrate the ideas, and when the emphasis is on problem-solving, the examples are numerical in character. In selected examples throughout the text, graphical display of results (e.g., stresses in beams) has been added to enhance the student's understanding of the problem results.

## PROBLEMS

In all mechanics courses, solving problems is an important part of the learning process. This textbook offers more than 1230 problems for homework assignments and classroom discussions. The problems are placed at the end of each chapter so that they are easy to find and don't break up the presentation of the main subject matter. Also, problems are generally arranged in order of increasing difficulty thus alerting students to the time necessary for solution. Answers to all problems are listed near the back of the book. An Instructor Solution Manual (ISM) is available to registered instructors at the publisher's web site.

As noted above, a new appendix has been added to this brief edition containing more than 100 FE Exam type problems. Many students take the *Fundamentals of Engineering Examination* upon graduation, the first step on their path to registration as a Professional Engineer. These problems cover all of the major topics presented in the text and are thought to be representative of those likely to appear on an FE exam. Most of these problems are in SI units which is the system of units used on the FE Exam itself, and require use of an engineering calculator to carry out the solution. Each of the problems is presented in the FE Exam format. The student must select from 4 available answers (A, B, C or D), only one of which is the correct answer. The detailed solution for each problem is available on the student website. It is expected that careful review of these problems will serve as a useful guide to the student in preparing for this important examination.

Considerable effort has been spent in checking and proofreading the text so as to eliminate errors. If you happen to find one, no matter how trivial, please notify me by e-mail (*bgoodno@ce.gatech.edu*). We will correct any errors in the next printing of the book.

## UNITS

Both the International System of Units (SI) and the U.S. Customary System (USCS) are used in the examples and problems. Discussions of both systems and a table of conversion factors are given in Appendix B. For problems involving numerical solutions, odd-numbered problems are in USCS units and even-numbered problems are in SI units. This convention makes it easy to know in advance which system of units is being used in any particular problem. In addition, tables containing properties of structural-steel shapes in both USCS and SI units may be found in Appendix F so that solution of beam analysis and design examples and end-of-chapter problems can be carried out in either USCS or SI units.

## SUPPLEMENTS

### INSTRUCTOR RESOURCES

An Instructor's Solutions Manual (ISM) is available in both print and digital versions. The digital version is accessible to registered instructors at the publisher's web site. This web site also includes both a full set of PowerPoint slides containing all graphical images in the text, and a set of LectureBuilder PowerPoint slides of all equations and Example Problems from the text, for use by instructors during lecture or review sessions.

To access additional course materials, please visit www.cengagebrain.com. At the cengagebrain.com home page, search for the ISBN of your title (from the back cover of your book) using the search box at the top of the page. This will take you to the product page where these resources can be found.

## S.P. TIMOSHENKO (1878–1972) AND J.M. GERE (1925–2008)

Many readers of this book will recognize the name of Stephen P. Timoshenko—probably the most famous name in the field of applied mechanics. Timoshenko is generally recognized as the world's most outstanding pioneer in applied mechanics. He contributed many new ideas and concepts and became famous for both his scholarship and his teaching. Through his numerous textbooks he made a profound change in the teaching of mechanics not only in this country but wherever mechanics is taught. Timoshenko was both teacher and mentor to James Gere and provided the motivation for the first edition of this text, authored by James M. Gere and published in 1972; the second and each subsequent edition of this book were written by James Gere over the course of his long and distinguished tenure as author, educator and researcher at Stanford University. James Gere started as a doctoral student at Stanford in 1952 and retired from Stanford as a professor in 1988 having authored this and eight other well known and respected text books on mechanics, and structural and earthquake engineering. He remained active at Stanford as Professor Emeritus until his death in January of 2008.

A brief biography of Timoshenko appears in the first reference in the *References and Historical Notes* section, and also in an August 2007

*STRUCTURE* magazine article entitled "Stephen P. Timoshenko: Father of Engineering Mechanics in the U.S." by Richard G. Weingardt, P.E. This article provides an excellent historical perspective on this and the many other engineering mechanics textbooks written by each of these authors.

## ACKNOWLEDGMENTS

To acknowledge everyone who contributed to this book in some manner is clearly impossible, but I owe a major debt to my former Stanford teachers, especially my mentor and friend, and lead author, James M. Gere.

I am grateful to my many colleagues teaching Mechanics of Materials at various institutions throughout the world who have provided feedback and constructive criticism about the text; for all those anonymous reviews, my thanks. With each new edition, their advice has resulted in significant improvements in both content and pedagogy.

My appreciation and thanks also go to the reviewers who provided specific comments for this eighth edition:

Jonathan Awerbuch, Drexel University

Henry N. Christiansen, Brigham Young University

Remi Dingreville, NYU—Poly

Apostolos Fafitis, Arizona State University

Paolo Gardoni, Texas A & M University

Eric Kasper, California Polytechnic State University, San Luis Obispo

Nadeem Khattak, University of Alberta

Kevin M. Lawton, University of North Carolina, Charlotte

Kenneth S. Manning, Adirondack Community College

Abulkhair Masoom, University of Wisconsin—Platteville

Craig C. Menzemer, University of Akron

Rungun Nathan, The Pennsylvania State University, Berks

Douglas P. Romilly, University of British Columbia

Edward Tezak, Alfred State College

George Tsiatis, University of Rhode Island

Xiangwu (David) Zeng, Case Western Reserve University

Mohammed Zikry, North Carolina State University

I wish to also acknowledge my Structural Engineering and Mechanics colleagues at the Georgia Institute of Technology, many of whom provided valuable advice on various aspects of the revisions and additions leading to the current edition. It is a privilege to work with all of these educators and to learn from them in almost daily interactions and discussions about structural engineering and mechanics in the context of research and higher education. I wish to extend my thanks to my many current and former students who have helped to shape this text in its various editions. Finally, I would like to acknowledge the excellent work of German Rojas, PhD., PEng. who carefully checked the solutions of many of the new examples and end of chapter problems.

The editing and production aspects of the book were always in skillful and experienced hands, thanks to the talented and knowledgeable personnel of Cengage Learning. Their goal was the same as mine—to produce the best possible new edition of this text, never compromising on any aspect of the book.

The people with whom I have had personal contact at Cengage Learning are Christopher Carson, Executive Director, Global Publishing Program, Christopher Shortt, Publisher, Global Engineering Program, Randall Adams and Swati Meherishi, Acquisitions Editors, who provided guidance throughout the project; Hilda Gowans, Senior Developmental Editor, Engineering, who was always available to provide information and encouragement; Kristiina Paul who managed all aspects of new photo selection and permissions research; Andrew Adams who created the cover design for the book; and Lauren Betsos, Global Marketing Manager, who developed promotional material in support of the text. I would like to especially acknowledge the work of Rose Kernan of RPK Editorial Services, and her staff, who edited the manuscript and managed it throughout the production process. To each of these individuals I express my heartfelt thanks for a job well done. It has been a pleasure working with you on an almost daily basis to produce this eighth edition of the text.

I am deeply appreciative of the patience and encouragement provided by my family, especially my wife, Lana, throughout this project.

Finally, I am very pleased to continue this endeavor, at the invitation of my mentor and friend, Jim Gere. This eighth edition text has now reached its 40th year of publication. I am committed to its continued excellence and welcome all comments and suggestions. Please feel free to provide me with your critical input at *bgoodno@ce.gatech.edu*.

**Barry J. Goodno**
Atlanta, Georgia

## CengageNOW—Just What You Need To Know and Do NOW!

### CengageNOW Is An Online Teaching & Learning Resource

CengageNOW offers all of your teaching and learning resources in one intuitive program organized around the essential activities you perform for class—lecturing, creating assignments, grading, quizzing, and tracking student progress and performance. CengageNOW's intuitive "tabbed" design allows you to navigate to all key functions with a single click and a unique homepage tell you just what needs to be done and when. CengageNOW provides students access to an integrated eBook, interactive tutorials, and other multimedia tools that help them get the most out of your course.

### CengageNOW Provides More Control In Less Time

CengageNOW's flexible assignment and gradebook options provide you more control while saving you valuable time in planning and managing your course assignments. With CengageNOW you can automatically grade all assignments, weigh grades, choose points or percentages and set the number of attempts and due dates per problem to best suit your overall course plan.

### CengageNOW Delivers Better Student Outcomes

CengageNOW Personalized Study is a diagnostic tool consisting of a chapter-specific Pre-Test, Study Plan and Post-test that utilizes valuable text-specific assets to empower students to master concepts, prepare for exams and be more involved in class. It's easy to assign and if you want, results will automatically post to your gradebook. Results to Personalized Study provide immediate and ongoing feedback regarding what students are mastering and why they're not—to both you and the student. In most cases, Personalized Study, in most cases links to an integrated eBook so students can easily review topics. Best of all, CengageNOW Personalized Study is designed to help your students get a better grade in your class.

To purchase or access CengageNOW™ with eBook for Gere/Goodno, *Mechanics of Materials*, Eighth Edition go to www.cengagebrain.com.

| | |
|---|---|
| $A$ | area |
| $A_f$, $A_w$ | area of flange; area of web |
| $a, b, c$ | dimensions, distances |
| $C$ | centroid, compressive force, constant of integration |
| $c$ | distance from neutral axis to outer surface of a beam |
| $D$ | diameter |
| $d$ | diameter, dimension, distance |
| $E$ | modulus of elasticity |
| $E_r$, $E_t$ | reduced modulus of elasticity; tangent modulus of elasticity |
| $e$ | eccentricity, dimension, distance, unit volume change (dilatation) |
| $F$ | force |
| $f$ | shear flow, shape factor for plastic bending, flexibility, frequency (Hz) |
| $f_T$ | torsional flexibility of a bar |
| $G$ | modulus of elasticity in shear |
| $g$ | acceleration of gravity |
| $H$ | height, distance, horizontal force or reaction, horsepower |
| $h$ | height, dimensions |
| $I$ | moment of inertia (or second moment) of a plane area |
| $I_x$, $I_y$, $I_z$ | moments of inertia with respect to $x$, $y$, and $z$ axes |
| $I_{x1}$, $I_{y1}$ | moments of inertia with respect to $x_1$ and $y_1$ axes (rotated axes) |
| $I_{xy}$ | product of inertia with respect to $xy$ axes |
| $I_{x1y1}$ | product of inertia with respect to $x_1 y_1$ axes (rotated axes) |
| $I_P$ | polar moment of inertia |
| $I_1$, $I_2$ | principal moments of inertia |
| $J$ | torsion constant |
| $K$ | stress-concentration factor, bulk modulus of elasticity, effective length factor for a column |
| $k$ | spring constant, stiffness, symbol for $\sqrt{P/EI}$ |
| $k_T$ | torsional stiffness of a bar |
| $L$ | length, distance |
| $L_E$ | effective length of a column |
| ln, log | natural logarithm (base e); common logarithm (base 10) |
| $M$ | bending moment, couple, mass |
| $M_P$, $M_Y$ | plastic moment for a beam; yield moment for a beam |
| $m$ | moment per unit length, mass per unit length |
| $N$ | axial force |
| $n$ | factor of safety, integer, revolutions per minute (rpm) |
| $O$ | origin of coordinates |
| $O'$ | center of curvature |
| $P$ | force, concentrated load, power |
| $P_{allow}$ | allowable load (or working load) |
| $P_{cr}$ | critical load for a column |

| | |
|---|---|
| $P_P$ | plastic load for a structure |
| $P_r, P_t$ | reduced-modulus load for a column; tangent-modulus load for a column |
| $P_Y$ | yield load for a structure |
| $p$ | pressure (force per unit area) |
| $Q$ | force, concentrated load, first moment of a plane area |
| $q$ | intensity of distributed load (force per unit distance) |
| $R$ | reaction, radius |
| $r$ | radius, radius of gyration ($r = \sqrt{I/A}$) |
| $S$ | section modulus of the cross section of a beam, shear center |
| $s$ | distance, distance along a curve |
| $T$ | tensile force, twisting couple or torque, temperature |
| $T_P, T_Y$ | plastic torque; yield torque |
| $t$ | thickness, time, intensity of torque (torque per unit distance) |
| $t_f, t_w$ | thickness of flange; thickness of web |
| $U$ | strain energy |
| $u$ | strain-energy density (strain energy per unit volume) |
| $u_r, u_t$ | modulus of resistance; modulus of toughness |
| $V$ | shear force, volume, vertical force or reaction |
| $v$ | deflection of a beam, velocity |
| $v', v''$, etc. | $dv/dx$, $d^2v/dx^2$, etc. |
| $W$ | force, weight, work |
| $w$ | load per unit of area (force per unit area) |
| $x, y, z$ | rectangular axes (origin at point $O$) |
| $x_c, y_c, z_c$ | rectangular axes (origin at centroid $C$) |
| $\overline{x}, \overline{y}, \overline{z}$ | coordinates of centroid |
| $Z$ | plastic modulus of the cross section of a beam |
| $\alpha$ | angle, coefficient of thermal expansion, nondimensional ratio |
| $\beta$ | angle, nondimensional ratio, spring constant, stiffness |
| $\beta_R$ | rotational stiffness of a spring |
| $\gamma$ | shear strain, weight density (weight per unit volume) |
| $\gamma_{xy}, \gamma_{yz}, \gamma_{zx}$ | shear strains in $xy$, $yz$, and $zx$ planes |
| $\gamma_{x1y1}$ | shear strain with respect to $x_1 y_1$ axes (rotated axes) |
| $\gamma_\theta$ | shear strain for inclined axes |
| $\delta$ | deflection of a beam, displacement, elongation of a bar or spring |
| $\Delta T$ | temperature differential |
| $\delta_P, \delta_Y$ | plastic displacement; yield displacement |
| $\varepsilon$ | normal strain |
| $\varepsilon_x, \varepsilon_y, \varepsilon_z$ | normal strains in $x$, $y$, and $z$ directions |
| $\varepsilon_{x1}, \varepsilon_{y1}$ | normal strains in $x_1$ and $y_1$ directions (rotated axes) |
| $\varepsilon_\theta$ | normal strain for inclined axes |
| $\varepsilon_1, \varepsilon_2, \varepsilon_3$ | principal normal strains |
| $\varepsilon'$ | lateral strain in uniaxial stress |
| $\varepsilon_T$ | thermal strain |
| $\varepsilon_Y$ | yield strain |
| $\theta$ | angle, angle of rotation of beam axis, rate of twist of a bar in torsion (angle of twist per unit length) |

| | |
|---|---|
| $\theta_p$ | angle to a principal plane or to a principal axis |
| $\theta_s$ | angle to a plane of maximum shear stress |
| $\kappa$ | curvature ($\kappa = 1/\rho$) |
| $\lambda$ | distance, curvature shortening |
| $\nu$ | Poisson's ratio |
| $\rho$ | radius, radius of curvature ($\rho = 1/\kappa$), radial distance in polar coordinates, mass density (mass per unit volume) |
| $\sigma$ | normal stress |
| $\sigma_x, \sigma_y, \sigma_z$ | normal stresses on planes perpendicular to $x$, $y$, and $z$ axes |
| $\sigma_{x1}, \sigma_{y1}$ | normal stresses on planes perpendicular to $x_1 y_1$ axes (rotated axes) |
| $\sigma_\theta$ | normal stress on an inclined plane |
| $\sigma_1, \sigma_2, \sigma_3$ | principal normal stresses |
| $\sigma_{\text{allow}}$ | allowable stress (or working stress) |
| $\sigma_{\text{cr}}$ | critical stress for a column ($\sigma_{\text{cr}} = P_{\text{cr}}/A$) |
| $\sigma_{\text{pl}}$ | proportional-limit stress |
| $\sigma_r$ | residual stress |
| $\sigma_T$ | thermal stress |
| $\sigma_U, \sigma_Y$ | ultimate stress; yield stress |
| $\tau$ | shear stress |
| $\tau_{xy}, \tau_{yz}, \tau_{zx}$ | shear stresses on planes perpendicular to the $x$, $y$, and $z$ axes and acting parallel to the $y$, $z$, and $x$ axes |
| $\tau_{x1y1}$ | shear stress on a plane perpendicular to the $x_1$ axis and acting parallel to the $y_1$ axis (rotated axes) |
| $\tau_\theta$ | shear stress on an inclined plane |
| $\tau_{\text{allow}}$ | allowable stress (or working stress) in shear |
| $\tau_U, \tau_Y$ | ultimate stress in shear; yield stress in shear |
| $\phi$ | angle, angle of twist of a bar in torsion |
| $\psi$ | angle, angle of rotation |
| $\omega$ | angular velocity, angular frequency ($\omega = 2\pi f$) |

## GREEK ALPHABET

| | | | | | |
|---|---|---|---|---|---|
| A | $\alpha$ | Alpha | N | $\nu$ | Nu |
| B | $\beta$ | Beta | $\Xi$ | $\xi$ | Xi |
| $\Gamma$ | $\gamma$ | Gamma | O | $o$ | Omicron |
| $\Delta$ | $\delta$ | Delta | $\Pi$ | $\pi$ | Pi |
| E | $\varepsilon$ | Epsilon | P | $\rho$ | Rho |
| Z | $\zeta$ | Zeta | $\Sigma$ | $\sigma$ | Sigma |
| H | $\eta$ | Eta | T | $\tau$ | Tau |
| $\Theta$ | $\theta$ | Theta | Y | $\upsilon$ | Upsilon |
| I | $\iota$ | Iota | $\Phi$ | $\phi$ | Phi |
| K | $\kappa$ | Kappa | X | $\chi$ | Chi |
| $\Lambda$ | $\lambda$ | Lambda | $\Psi$ | $\psi$ | Psi |
| M | $\mu$ | Mu | $\Omega$ | $\omega$ | Omega |

*A star attached to a section number indicates a specialized.

# Tension, Compression, and Shear

This telecommunications tower is an assemblage of many members that act primarily in tension or compression. (Péter budella/Shutterstock)

# CHAPTER OVERVIEW

In Chapter 1, we are introduced to mechanics of materials, which examines the **stresses, strains, and displacements** in bars of various materials acted on by axial loads applied at the centroids of their cross sections. After a brief review of basic concepts presented in statics, we will learn about **normal stress ($\sigma$)** and **normal strain ($\varepsilon$)** in materials used for structural applications, then identify key properties of various materials, such as the modulus of elasticity ($E$) and yield ($\sigma_y$) and ultimate ($\sigma_u$) stresses, from plots of stress ($\sigma$) versus strain ($\varepsilon$). We will also plot shear stress ($\tau$) versus shear strain ($\gamma$) and identify the shearing modulus of elasticity ($G$). If these materials perform only in the linear range, stress and strain are related by Hooke's Law for normal stress and strain ($\sigma = E \cdot \varepsilon$) and also for shear stress and strain ($\tau = G \cdot \gamma$). We will see that changes in lateral dimensions and volume depend upon Poisson's ratio ($\nu$). Material properties $E$, $G$, and $\nu$, in fact, are directly related to one another and are not independent properties of the material.

Assemblage of bars to form structures (such as trusses) leads to consideration of average shear ($\tau$) and bearing ($\sigma_b$) stresses in their connections as well as normal stresses acting on the net area of the cross section (if in tension) or on the full cross-sectional area (if in compression). If we restrict maximum stresses at any point to **allowable** values by use of factors of safety, we can identify allowable levels of axial loads for simple systems, such as cables and bars. **Factors of safety** relate actual to required strength of structural members and account for a variety of uncertainties, such as variations in material properties and probability of accidental overload. Lastly, we will consider **design**: the iterative process by which the appropriate size of structural members is determined to meet a variety of both **strength** and **stiffness requirements** for a particular structure subjected to a variety of different loadings.

## Chapter 1 is organized as follows:

**1.1** Introduction to Mechanics of Materials 4
**1.2** Statics Review 6
**1.3** Normal Stress and Strain 27
**1.4** Mechanical Properties of Materials 37
**1.5** Elasticity, Plasticity, and Creep 45
**1.6** Linear Elasticity, Hooke's Law, and Poisson's Ratio 52

**1.7** Shear Stress and Strain 57
**1.8** Allowable Stresses and Allowable Loads 68
**1.9** Design for Axial Loads and Direct Shear 74
Chapter Summary & Review 80
Problems 83

# 1.1 INTRODUCTION TO MECHANICS OF MATERIALS

**Mechanics of materials** is a branch of applied mechanics that deals with the behavior of solid bodies subjected to various types of loading. Other names for this field of study are *strength of materials* and *mechanics of deformable bodies*. The solid bodies considered in this book include bars with axial loads, shafts in torsion, beams in bending, and columns in compression.

The principal objective of mechanics of materials is to determine the stresses, strains, and displacements in structures and their components due to the loads acting on them. If we can find these quantities for all values of the loads up to the loads that cause failure, we will have a complete picture of the mechanical behavior of these structures.

An understanding of mechanical behavior is essential for the safe design of all types of structures, whether airplanes and antennas, buildings and bridges, machines and motors, or ships and spacecraft. That is why mechanics of materials is a basic subject in so many engineering fields. Statics and dynamics are also essential, but those subjects deal primarily with the forces and motions associated with particles and rigid bodies. However, most problems in mechanics of materials begin with an examination of the external and internal forces acting on a stable deformable body. We first define the loads acting on the body, along with its support conditions, then determine reaction forces at supports and internal forces in its members or elements using the basic laws of static equilibrium (provided that it is statically determinate). A well-constructed free-body diagram is an essential part of the process of carrying out a proper static analysis of a structure.

In mechanics of materials we go beyond the concepts presented in statics to study the stresses and strains inside real bodies, that is, bodies of finite dimensions that deform under loads. To determine the stresses and strains, we use the physical properties of the materials as well as numerous theoretical laws and concepts. Later, we will see that mechanics of materials provides additional essential information, based on the deformations of the body, to allow us to solve so-called statically indeterminate problems (not possible if using the laws of statics alone).

Theoretical analyses and experimental results have equally important roles in mechanics of materials. We use theories to derive formulas and equations for predicting mechanical behavior, but these expressions cannot be used in practical design unless the physical properties of the materials are known. Such properties are available only after careful experiments have been carried out in the laboratory. Furthermore, not all practical problems are amenable to theoretical analysis alone, and in such cases physical testing is a necessity.

The historical development of mechanics of materials is a fascinating blend of both theory and experiment—theory has pointed the way to useful results in some instances, and experiment has done so in others. Such famous persons as Leonardo da Vinci (1452–1519) and Galileo Galilei (1564–1642) performed experiments to determine the strength of wires, bars, and beams, although they did not develop adequate theories (by today's standards) to explain their test results. By contrast, the famous mathematician Leonhard Euler (1707–1783) developed the mathematical theory of columns and calculated the critical load of a column in 1744, long before any experimental evidence existed to show the significance of his results. Without appropriate tests to back up his theories, Euler's results remained unused for over a hundred years, although today they are the basis for the design and analysis of most columns.*

## Problems

When studying mechanics of materials, you will find that your efforts are divided naturally into two parts: first, understanding the logical development of the concepts, and second, applying those concepts to practical situations. The former is accomplished by studying the derivations, discussions, and examples that appear in each chapter, and the latter is accomplished by solving the problems at the ends of the chapters. Some of the problems are numerical in character, and others are symbolic (or algebraic).

An advantage of *numerical problems* is that the magnitudes of all quantities are evident at every stage of the calculations, thus providing an opportunity to judge whether the values are reasonable or not. The principal advantage of *symbolic problems* is that they lead to general-purpose formulas. A formula displays the variables that affect the final results; for instance, a quantity may actually cancel out of the solution, a fact that would not be evident from a numerical solution. Also, an algebraic solution shows the manner in which each variable affects the results, as when one variable appears in the numerator and another appears in the denominator. Furthermore, a symbolic solution provides the opportunity to check the dimensions at every stage of the work.

Finally, the most important reason for solving algebraically is to obtain a general formula that can be used for many different problems. In contrast, a numerical solution applies to only one set of circumstances. Because engineers must be adept at both kinds of solutions, you will find a mixture of numeric and symbolic problems throughout this book.

Numerical problems require that you work with specific units of measurement. In keeping with current engineering practice, this book utilizes both the International System of Units (SI) and the U.S. Customary System (USCS). A discussion of both systems appears in Appendix B, where you will also find many useful tables, including a table of conversion factors.

All problems appear at the ends of the chapters, with the problem numbers and subheadings identifying the sections to which they belong. In the case of problems requiring numerical solutions, odd-numbered problems are in USCS units and even-numbered problems are in SI units.

The techniques for solving problems are discussed in detail in Appendix C. In addition to a list of sound engineering procedures, Appendix C includes sections on dimensional homogeneity and significant digits. These topics are especially important, because every equation must be dimensionally homogeneous and every numerical result must be expressed with the proper number of significant digits. In this book, final numerical results are usually presented with three significant digits when a number begins with the digits 2 through 9, and with four significant digits when a number begins with the digit 1. Intermediate values are often recorded with additional digits to avoid losing numerical accuracy due to rounding of numbers.

---

*The history of mechanics of materials, beginning with Leonardo and Galileo, is given in Refs. 1-1, 1-2, and 1-3.

# 1.2 STATICS REVIEW

In your prerequisite course on statics, you studied the *equilibrium* of rigid bodies acted upon by a variety of different forces and supported or restrained in such a way that the body was stable and at rest. As a result, a properly restrained body could not undergo rigid-body motion due to the application of static forces. You drew *free-body diagrams* of the entire body, or of key parts of the body, and then applied the *equations of equilibrium* to find external reaction forces and moments or internal forces and moments at critical points. In this section, we will review the basic static equilibrium equations and apply them to the solution of example structures (both two and three dimensional) using both scalar and vector operations (both acceleration and velocity of the body will be assumed to be zero). Most problems in mechanics of materials require a static analysis as the first step, so all forces acting on the system and causing its deformation are known. Once all external and internal forces of interest have been found, we will be able to proceed with the evaluation of stresses, strains, and deformations of bars, shafts, beams, and columns in subsequent chapters.

## Equilibrium Equations

The resultant force $R$ and resultant moment $M$ of *all* forces and moments acting on either a rigid or deformable body in equilibrium are both zero. The sum of the moments may be taken about any arbitrary point. The resulting equilibrium equations can be expressed in *vector form* as:

$$R = \Sigma F = 0 \qquad\qquad \textbf{(1-1)}$$

$$M = \Sigma M = \Sigma (r \times F) = 0 \qquad\qquad \textbf{(1-2)}$$

where $F$ is one of a number of vectors of forces acting on the body and $r$ is a position vector from the point at which moments are taken to a point along the line of application of any force $F$. It is often convenient to write the equilibrium equations in *scalar form* using a rectangular Cartesian coordinate system, either in two dimensions $(x, y)$ or three dimensions $(x, y, z)$ as

$$\Sigma F_x = 0 \qquad \Sigma F_y = 0 \qquad \Sigma M_z = 0 \qquad\qquad \textbf{(1-3)}$$

Eq. (1-3) can be used for two-dimensional or planar problems, but in three dimensions, three force and three moment equations are required:

$$\Sigma F_x = 0 \qquad \Sigma F_y = 0 \qquad \Sigma F_z = 0 \qquad\qquad \textbf{(1-4)}$$

$$\Sigma M_x = 0 \qquad \Sigma M_y = 0 \qquad \Sigma M_z = 0 \qquad\qquad \textbf{(1-5)}$$

If the number of unknown forces is equal to the number of independent equilibrium equations, these equations are sufficient to solve for all unknown reaction or internal forces in the body, and the problem is referred to as *statically determinate* (provided that the body is stable). If the body or structure is constrained by additional (or redundant) supports, it is *statically indeterminate*, and a solution is not possible using the laws of static equilibrium alone. For statically indeterminate structures, we must also examine the deformations of the structure, as will be discussed in the following chapters.

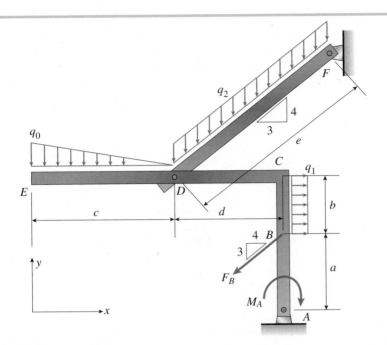

**Fig. 1-1**

Plane frame structure

## Applied Forces

External loads applied to a body or structure may be either concentrated or distributed forces or moments. For example, force $F_B$ (with units of pounds, lb, or newtons, N) in Fig. 1-1 is a point or concentrated load and is assumed to act at point $B$ on the body, while moment $M_A$ is a concentrated moment or couple (with units of lb-ft or N · m) acting at point $A$. Distributed forces may act alone or normal to a member and may have constant intensity, such as line load $q_1$ normal to member $BC$ (Fig. 1-1) or line load $q_2$ acting in the $-y$ direction on inclined member $DF$; both $q_1$ and $q_2$ have units of force intensity (lb/ft or N/m). Distributed loads also may have a linear (or other) variation with some peak intensity $q_0$ (as on member $ED$ in Fig. 1-1). Surface pressures $p$ (with units of lb/ft$^2$ or Pa), such as wind acting on a sign (Fig. 1-2), act over a designated region of a body. Finally, a body force $w$ (with units of force per unit volume, lb/ft$^3$ or N/m$^3$),

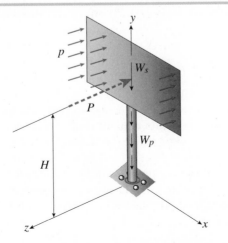

**Fig. 1-2**

Wind on sign

such as the distributed self-weight of the sign or post in Fig. 1-2, acts throughout the volume of the body and can be replaced by the component weight $W$ acting at the center of gravity (c.g.) of the sign ($W_s$) or post ($W_p$). In fact, any distributed loading (line, surface, or body force) can be replaced by a statically equivalent force at the center of gravity of the distributed loading when overall static equilibrium of the structure is evaluated using Eqs. (1-1) to (1-5).

## Free-Body Diagrams

A free-body diagram (FBD) is an essential part of a static analysis of a rigid or deformable body. All forces acting on the body, or component part of the body, must be displayed on the FBD if a correct equilibrium solution is to be obtained. This includes applied forces and moments, reaction forces and moments, and any connection forces between individual components. For example, an *overall* FBD of the plane frame in Fig. 1-1 is shown in Fig. 1-3a; all applied and reaction forces are shown on this FBD and statically equivalent concentrated loads are displayed for all distributed loads. Statically equivalent forces $F_{q0}$, $F_{q1}$, and $F_{q2}$, each acting at the c.g. of the corresponding distributed loading, are used in the equilibrium equation solution to represent distributed loads $q_0$, $q_1$, and $q_2$, respectively.

Next, the plane frame has been disassembled in Fig. 1-3b, so that *separate* FBD's can be drawn for each part of the frame, thereby exposing pin-connection forces at $D$ ($D_x$, $D_y$). Both FBD's must show all applied forces as well as reaction forces $A_x$ and $A_y$ at pin-support joint $A$ and $F_x$ and $F_y$ at pin-support joint $F$. The forces transmitted between frame elements $EDC$ and $DF$ at pin connection $D$ must be determined if the proper interaction of these two elements is to be accounted for in the static analysis.

The plane frame structure in Fig. 1-1 will be analyzed in Example 1-2 to find reaction forces at joints $A$ and $F$ and also pin-connection forces at joint $D$ using the equilibrium equations Eqs. (1-1) to (1-3). The FBD's presented in Figs. 1-3a and 1-3b are essential parts of this solution process. A *statics sign convention* is usually employed in the solution for support reactions; forces acting in the positive directions of the coordinate axes are assumed positive, and the right-hand rule is used for moment vectors.

## Reactive Forces and Support Conditions

Proper restraint of the body or structure is essential if the equilibrium equations are to be satisfied. A sufficient number and arrangement of supports must be present to prevent rigid body motion under the action of static forces. A reaction force at a support is represented by a single arrow with a slash drawn through it (see Fig. 1-3) while a moment restraint at a support is shown as a double-headed or curved arrow with a slash. Reaction forces and moments usually result from the action of applied forces of the types described above (i.e., concentrated, distributed, surface, and body forces).

A variety of different support conditions may be assumed depending on whether the problem is 2D or 3D. Supports $A$ and $F$ in the 2D plane frame structure shown in Fig. 1-1 and Fig. 1-3 are pin supports,

**Fig. 1-3**

(a) Overall FBD of plane frame structure from Fig. 1-1, and (b) Separate free-body diagrams of parts *A* through *E* and *DF* of the plane frame structure in Fig. 1-1

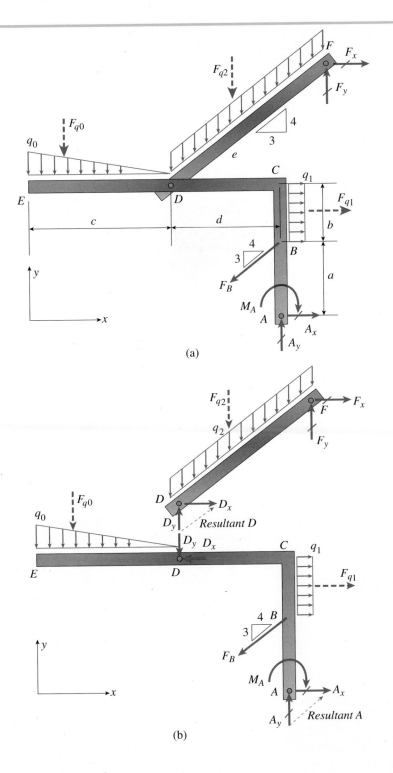

(a)

(b)

while the base of the 3D sign structure in Fig. 1-2 may be considered to be a fixed or clamped support. Some of the most commonly used idealizations for 2D and 3D supports, as well as interconnections between members or elements of a structure, are illustrated in Table 1-1. The

| Table 1-1 | Type of support or connection | Simplified sketch of support or connection | Display of restraint forces and moments, or connection forces |
|---|---|---|---|
| Reaction and Connection Forces in 2D or 3D Static Analysis | (1) Roller support—horizontal, vertical, or inclined<br><br><br>Bridge with roller support (The Earthquake Engineering Online Archive) | <br>Horizontal roller support (constrains motion in both $+y$ and $-y$ directions)<br><br><br>Vertical roller restraints<br><br><br>Rotated or inclined roller support<br><br> | (a) Two-dimensional roller support<br><br><br><br>(b) Three-dimensional roller support<br><br> |
| | (2) Pin support<br><br><br>Bridge with pin support (Courtesy of Joel Kerkhoff, P.Eng.)<br><br><br>Pin support on old truss bridge (© Barry Goodno) | <br><br><br>Pin support at $F$ in Fig. 1-1<br><br> | (a) Two-dimensional pin support<br><br><br><br>(b) Three-dimensional pin support<br><br> |

Table 1-1 (continued)

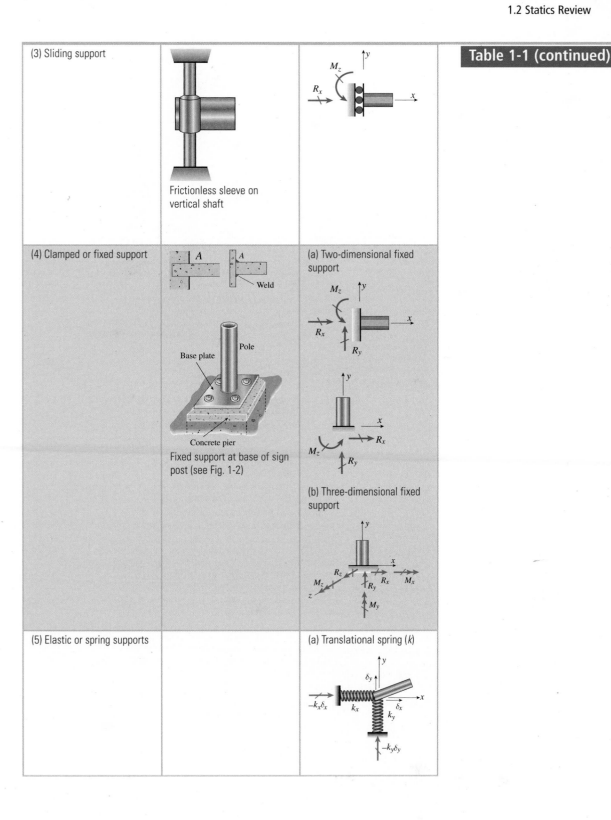

(3) Sliding support

Frictionless sleeve on vertical shaft

(4) Clamped or fixed support

Weld

Pole

Base plate

Concrete pier

Fixed support at base of sign post (see Fig. 1-2)

(a) Two-dimensional fixed support

(b) Three-dimensional fixed support

(5) Elastic or spring supports

(a) Translational spring ($k$)

**Table 1-1 (continued)**

| | | (b) Rotational spring ($k_r$) |
|---|---|---|
| | | $M_z = -k_r\,\theta_z$ |
| (6) Pinned connection (from Figs. 1-1 and 1-3) Pin connection on old bridge (© Barry Goodno) | Pinned connection at $D$ between members $EDC$ and $DF$ in plane frame (Fig. 1-1) | |
| (7) Slotted connection (modified connection from that shown in Figs. 1-1 and 1-3) | Alternate slotted connection at $D$ on plane frame (Note that the plane frame in Fig. 1-1 is *unstable* if this slotted connection is used instead of a pin at $D$.) | |
| (8) Rigid connection (internal forces and moment in members joined at $C$ of plane frame in Fig. 1-1) | Rigid connection at $C$ on plane frame | |

restraining or transmitted forces and moments associated with each type of support or connection are displayed in the third column of the table (these are not FBDs, however). The reactions forces and moments for the 3D sign structure in Fig. 1-2 are shown on the FBD in Fig. 1-4a;

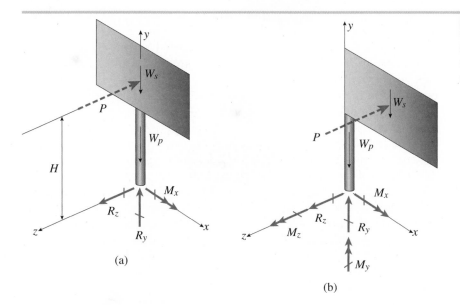

**Fig. 1-4**

(a) FBD of symmetric sign structure, and (b) FBD of eccentric sign structure

only reactions $R_y$, $R_z$, and $M_x$ are non-zero because the sign structure and wind loading are symmetric with respect to the $yz$ plane. If the sign is eccentric to the post (Fig. 1-4b), only reaction $R_x$ is zero for the case of wind loading in the $-z$ direction. (See Prob. 1.7-16 at the end of Chapter 1 for a more detailed examination of the reaction forces due to wind pressure acting on the sign structure in Fig. 1-2; forces and stresses in the base plate bolts are also computed. Several eccentric sign structures are presented for analysis as end of chapter problems in Chapter 8.)

## Internal Forces (Stress Resultants)

In our study of mechanics of materials, we will investigate the deformations of the members or elements which make up the overall deformable body. In order to compute the member deformations, we must first find the internal forces and moments (i.e., the internal stress resultants) at key points along the members of the overall structure. In fact, we will often create graphical displays of the internal axial force, torsional moment, transverse shear and bending moment along the axis of each member of the body so that we can readily identify critical points or regions within the structure. The first step is to make a section cut normal to the axis of each member so that a FBD can be drawn which displays the internal forces of interest. For example, if a cut is made at the top of member *BC* in the plane frame in Fig. 1-1, the internal *axial force* ($N_c$), *transverse shear force* ($V_c$) and *bending moment* ($M_c$) at joint *C* can be exposed as shown in the last row of Table 1-1. Fig. 1-5 shows two additional cuts made through members *ED* and *DF* in the plane frame; the resulting FBD's now can be used to find $N$, $V$, and $M$ in members $ED$ and $DF$ of the plane frame. Stress resultants $N$, $V$, and $M$ are usually taken along and normal to the member under consideration (i.e., local or member axes are used), and a *deformation sign convention* (e.g., tension is positive, compression is negative) is employed in their solution. In later chapters, we will see how these (and other) internal stress resultants are used to compute stresses in the member cross section.

The following examples are presented as a review of application of the equations of static equilibrium in the solution for external reactions and internal forces in truss, beam, circular shaft, and frame structures. First, a **truss structure** is considered and both scalar and vector solutions for reaction forces are reviewed. Then member forces are computed using the *method of joints*. Properly drawn FBD's are seen to be essential to the overall solution process. The second example involves static analysis of a **beam structure** to find reactions and internal forces at a particular section along the beam. In the third example, reactive and internal torsional moments in a **stepped shaft** are computed. And, finally, the fourth example presents the solution of the **plane frame structure** discussed here. Numerical values are assigned to applied forces and structure dimensions, and then reaction, pin connection, and selected internal forces in the frame are computed.

---

**Fig. 1-5**

FBD's for internal stress resultants in *ED* and *DF*

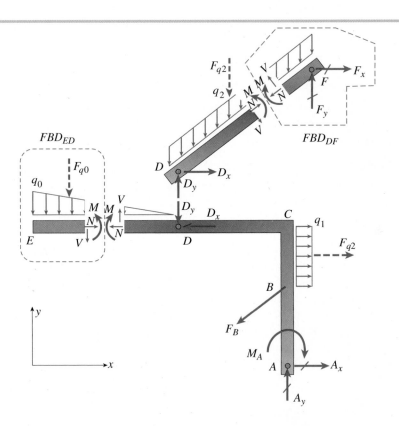

● ● ● **Example 1-1**

**Fig. 1-6**

Example 1-1: Plane truss static analysis for joint loads

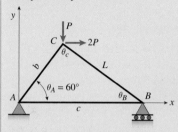

The plane truss shown in Fig. 1-6 is pin supported at $A$ and has a roller support at $B$. Joint loads $2P$ and $-P$ are applied at joint $C$. Find support reactions at joints $A$ and $B$, then solve for forces in members $AB$, $AC$ and $BC$. Use numerical properties given below.

Numerical data:

$$P = 35 \text{ kip} \quad L = 10 \text{ ft} \quad \theta_A = 60° \quad b = 0.71 \, L = 7.1 \text{ ft}$$

**Solution**

(1) Use the law of sines to find angles $\theta_B$ and $\theta_C$, then find the length ($c$) of member $AB$.
(2) Draw the FBD, then use equilibrium equations in scalar form (Eq. 1-3) to find the support reactions.
(3) Find member forces using the method of joints.
(4) Repeat solution for support reactions using a vector solution.
(5) Solve for support reactions and member forces for a 3D version of this plane (2D) truss.

**(1) Use the law of sines to find angles $\theta_B$, $\theta_C$ then find the length ($c$) of member $AB$.**

See **law of sines** in Appendix D:

$$\theta_B = \text{asin}\left(\frac{b}{L} \sin(\theta_A)\right) = 37.943° \quad \text{so} \quad \theta_C = 180° - (\theta_A + \theta_B) = 82.057°$$

and $c = L\left(\dfrac{\sin(\theta_C)}{\sin(\theta_A)}\right) = 11.436 \text{ ft}$ or $c = b \cos(\theta_A) + L \cos(\theta_B) = 11.436 \text{ ft}$

Note that the **Law of Cosines** also could be used:

$$c = \sqrt{b^2 + L^2 - 2bL \cos(\theta_C)} = 11.436 \text{ ft}$$

**(2) Draw the FBD (Fig. 1-7), then use equilibrium equations in *scalar form* (Eq. 1-3) to find the support reactions.**

Note that **the plane truss is statically determinate** since there are ($m + r = 6$) unknowns (where $m$ = number of member forces and $r$ = number of reactions), but there are ($2j = 2 \times 3 = 6$) equations of statics from the method of joints (where $j$ = number of joints).

Use equilibrium equations in scalar form to find support reactions.

Sum moments about $A$ to get reaction $B_y$:

$$B_y = \frac{[Pb \cos(\theta_A) + (2P)b \sin(\theta_A)]}{c} = 48.5 \text{ kip}$$

Sum forces in $y$ direction to get $A_y$:

$$A_y = P - B_y = -13.5 \text{ kip}$$

Sum forces in $x$ direction to get $A_x$:

$$A_x = -2P = -70 \text{ kip}$$

**Fig. 1-7**

Example 1-1: FBD of plane truss

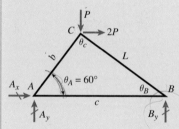

**Fig. 1-8**

Example 1-1: FBD of each joint of plane truss

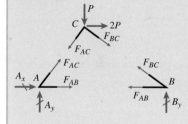

**(3) Find member forces using the method of joints.**

Draw FBDs of each joint (Fig. 1-8) then sum forces in $x$ and $y$ directions to find member forces.

Continues ➡

Sum forces in *y* direction at joint *A*:

$$F_{AC} = \frac{-A_y}{\sin(\theta_A)} = 15.59 \text{ kip}$$

*¿Pq se divide?* (handwritten)

Sum forces in *x* direction at joint *A*:

$$F_{AB} = -A_x - F_{AC}\cos(\theta_A) = 62.2 \text{ kip}$$

Sum forces in *y* direction at joint *B*:

$$F_{BC} = \frac{-B_y}{\sin(\theta_B)} \qquad F_{BC} = -78.9 \text{ kip}$$

*Se puede multiplicar por θ_C* (handwritten)

Check equilibrium at joint *C*. (First at *x* direction, then *y* direction.)

$$-F_{AC}\cos(\theta_A) + F_{BC}\cos(\theta_B) + 2P = 0 \qquad -F_{AC}\sin(\theta_A) - F_{BC}\sin(\theta_B) - P = 0$$

**(4) Repeat solution for support reactions using a *vector solution*** (show *x*, *y*, *z* components in vector format).

Position vectors to *B* and *C* from *A*:

$$r_{AB} = \begin{pmatrix} c \\ 0 \\ 0 \end{pmatrix} = \begin{pmatrix} 11.436 \\ 0 \\ 0 \end{pmatrix} \text{ft} \qquad r_{AC} = \begin{pmatrix} b\cos(\theta_A) \\ b\sin(\theta_A) \\ 0 \end{pmatrix} = \begin{pmatrix} 3.55 \\ 6.149 \\ 0 \end{pmatrix} \text{ft}$$

Force vectors at *A*, *B*, and *C*:

$$A = \begin{pmatrix} A_x \\ A_y \\ 0 \end{pmatrix} \qquad B = \begin{pmatrix} 0 \\ B_y \\ 0 \end{pmatrix} \qquad C = \begin{pmatrix} 2P \\ -P \\ 0 \end{pmatrix}$$

Sum moments about point *A*, then equate each expression to zero:

$$M_A = r_{AB} \times B + r_{AC} \times C = \begin{pmatrix} 0 \\ 0 \\ 11.436 \text{ ft } B_y - 554.66 \text{ ft}\cdot\text{kip} \end{pmatrix}$$

$$\text{so } B_y = \frac{554.66}{11.436} = 48.5 \text{ kip}$$

$$\text{or} \ldots \left| \begin{pmatrix} i & j & k \\ c & 0 & 0 \\ 0 & B_y & 0 \end{pmatrix} \right| + \left| \begin{pmatrix} i & j & k \\ \dfrac{b}{2} & b\dfrac{\sqrt{3}}{2} & 0 \\ 2P & -P & 0 \end{pmatrix} \right| = 11.436 \text{ ft } B_y \cdot k - 554.66 \text{ ft}\cdot k\cdot\text{kip}$$

Now sum forces and equate each expression to zero:

$$A + B + C \rightarrow \begin{pmatrix} A_x + 70 \text{ kip} \\ A_y + B_y - 35 \text{ kip} \\ 0 \end{pmatrix} \quad \text{so } A_x = -70 \text{ kip}$$

$$A_y = 35 - B_y = -13.5 \text{ kip}$$

Reactions $A_x$, $A_y$, and $B_y$ are the same as from the scalar solution approach.

**(5) Solve for support reactions and member forces for a 3D version of plane (2D) truss.**

To create a space truss from the plane truss, move joint *A* along the *z* axis a distance *z* while holding *B* on the *x* axis and constraining *C* to lie some

## Fig. 1-9

Example 1-1: FBD of space truss (extended version of plane truss)

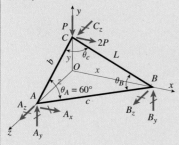

distance $y$ along the $y$ axis (see Fig. 1-9); hold member lengths ($L$, $b$, $c$) and angles ($\theta_A$, $\theta_B$, $\theta_C$) to the values used for the plane truss. Apply joint loads $2P$ and $-P$ at joint $C$. Add 3D pin support at $A$, two restraints at $B$ ($B_y$, $B_z$), and one restraint at $C$ ($C_z$).

Note that **the space truss is statically determinate** since there are ($m + r = 9$) unknowns (where $m$ = number of member forces and $r$ = number of reactions), but there are ($3j = 3 \times 3 = 9$) equations of statics from the method of joints (where $j$ = number of joints).

First, find $x$, $y$, and $z$ projections of members along coordinate axes. Then find angles $OBC$, $OBA$, and $OAC$ in each plane.

$$x = \sqrt{\frac{L^2 - b^2 + c^2}{2}} = 9.49677 \text{ ft} \qquad y = \sqrt{\frac{L^2 + b^2 - c^2}{2}} = 3.13232 \text{ ft}$$

$$z = \sqrt{\frac{-L^2 + b^2 + c^2}{2}} = 6.3717 \text{ ft}$$

$$OBC = \text{atan}\left(\frac{y}{x}\right) = 18.254° \quad OBA = \text{atan}\left(\frac{z}{x}\right) = 33.859°$$

$$OAC = \text{atan}\left(\frac{y}{z}\right) = 26.179°$$

**Draw the overall FBD (see Fig. 1-9), then use a *scalar solution* to find reactions and member forces.**

(1) Sum moments about a line through $A$, which is parallel to the $y$ axis (this will isolate reaction $B_z$, giving us one equation with one unknown):

$$B_z x + (2P)z = 0 \quad B_z = -2P\frac{z}{x} = -47 \text{ kip}$$

This is based on a *statics sign convention* so the negative sign means that force $B_z$ acts in the $-z$ direction.

(2) Sum moments about the $z$-axis to find $B_y$, then sum forces in $y$ direction to get $A_y$:

$$B_y = \frac{2P(y)}{x} = 23.1 \text{ kip} \quad \text{so} \quad A_y = P - B_y = 11.91 \text{ kip}$$

(3) Sum moments about the $x$ axis to find $C_z$:

$$C_z = \frac{A_y z}{y} = 24.2 \text{ kip}$$

(4) Sum forces in the $x$ and $z$ directions to get $A_x$ and $A_z$:

$$A_x = -2P = -70 \text{ kip} \qquad A_z = -C_z - B_z = 22.7 \text{ kip}$$

(5) Finally, use the method of joints to find member forces (a deformation sign convention is used here so positive (+) means tension and negative (−) means compression).

Sum forces in $x$ direction at joint $A$:

$$\frac{x}{c}F_{AB} + A_x = 0 \quad F_{AB} = \frac{-c}{x}A_x \quad F_{AB} = 84.3 \text{ kip}$$

Sum forces in $y$ direction at joint $A$:

$$\frac{y}{b}F_{AC} + A_y = 0 \quad F_{AC} = \frac{b}{y}(-A_y) \quad F_{AC} = -27 \text{ kip}$$

*Continues* ➡

Sum forces in $y$ direction at joint $B$:

$$\frac{y}{L}F_{BC} + B_y = 0 \quad F_{BC} = \frac{-L}{y}B_y \quad F_{BC} = -73.7 \text{ kip}$$

**Re-compute reactions for the space truss using a vector solution.**

Find position ($r$) and unit ($e$) vectors from joint $A$ to joints $B$ and $C$:

$$r_{AB} = \begin{pmatrix} x \\ 0 \\ -z \end{pmatrix} \quad e_{AB} = \frac{r_{AB}}{|r_{AB}|} = \begin{pmatrix} 0.83 \\ 0 \\ -0.557 \end{pmatrix}$$

$$r_{AC} = \begin{pmatrix} 0 \\ y \\ -z \end{pmatrix} \quad e_{AC} = \frac{r_{AC}}{|r_{AC}|} = \begin{pmatrix} 0 \\ 0.441 \\ -0.897 \end{pmatrix}$$

Sum moments about point $A$, then equate each expression to zero:

$$M_A = r_{AB} \times B + r_{AC} \times C$$

$$M_A = r_{AB} \times \begin{pmatrix} 0 \\ B_y \\ B_z \end{pmatrix} + r_{AC} \times \begin{pmatrix} 2P \\ -P \\ C_z \end{pmatrix}$$

$$= \begin{pmatrix} 6.3717 \text{ ft} \cdot B_y + 3.1323 \text{ ft} \cdot C_z - 223.01 \text{ ft} \cdot \text{kip} \\ -9.4968 \text{ ft} \cdot B_z - 446.02 \text{ ft} \cdot \text{kip} \\ 9.4968 \text{ ft} \cdot B_y - 219.26 \text{ ft} \cdot \text{kip} \end{pmatrix}$$

or $\left| \begin{pmatrix} i & j & k \\ x & 0 & -z \\ 0 & B_y & B_z \end{pmatrix} \right| = 6.3717 \text{ ft} \cdot B_y i - 9.4968 \text{ ft} \cdot B_z j + 9.4968 \text{ ft} \cdot B_y k$

and $\left| \begin{pmatrix} i & j & k \\ 0 & y & -z \\ 2P & -P & C_z \end{pmatrix} \right| = 3.1323 \text{ ft} \cdot C_z i - 223.01 \text{ ft} \cdot \text{kip } i$

$$-446.02 \text{ ft} \cdot \text{kip } j - 219.26 \text{ ft} \cdot \text{kip } k$$

Collecting coefficients of $j$ and solving:

$$B_z = \frac{446.02}{-9.4968} = -47 \text{ kip}$$

Collecting coefficients of $k$ and solving:

$$B_y = \frac{219.26}{9.4968} = 23.1 \text{ kip}$$

Collecting coefficients of $i$ and solving:

$$C_z = \frac{223.01 - 6.3717B_y}{3.1323} = 24.2 \text{ kip}$$

Complete the solution by summing forces and equating to zero:

$$\begin{pmatrix} A_x \\ A_y \\ A_z \end{pmatrix} + \begin{pmatrix} 0 \\ B_y \\ B_z \end{pmatrix} + \begin{pmatrix} 2P \\ -P \\ C_z \end{pmatrix} = \begin{pmatrix} A_x + 70.0 \text{ kip} \\ A_y - 35.0 \text{ kip} + 23.088 \text{ kip} \\ A_z - 22.733 \text{ kip} \end{pmatrix}$$

$$A_x = -70 \text{ kip} \quad A_y = 35 - 23.088 = 11.9 \text{ kip} \quad A_z = 22.7 \text{ kip}$$

Reactions $A_x$, $A_y$, $A_z$ and $B_y$, $B_z$ are the same as from the scalar solution approach.

### ● ● ● Example 1-2

The simply-supported beam structure shown in Fig. 1-10 is subjected to moment $M_A$ at pin-supported joint $A$, inclined load $F_B$ applied at joint $B$, and uniform load with intensity $q_1$ on member segment $BC$. Find support reactions at joints $A$ and $C$, then solve for internal forces at the midpoint of $BC$. Use properly drawn free-body diagrams in your solution.

### Fig. 1-10

Example 1-2: Beam static analysis for support reactions

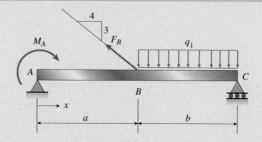

Numerical data (Newtons and meters):

$$a = 3\,\text{m} \quad b = 2\,\text{m}$$
$$M_A = 380\,\text{N}\cdot\text{m} \quad F_B = 200\,\text{N} \quad q_1 = 160\,\text{N/m}$$

**Solution**

**(1) Draw the FBD of the overall beam.** The solution for reaction forces at $A$ and $C$ must begin with a proper drawing of the FBD of the overall beam (Fig. 1-11). The FBD shows all applied and reactive forces.

### Fig. 1-11

Example 1-2: FBD of beam

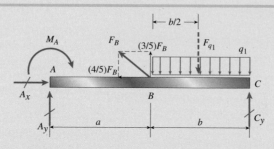

**(2) Determine statically equivalent concentrated forces.** Distributed forces are replaced by their statical equivalents ($F_{q1}$) and the components of the inclined concentrated force at $B$ may also be computed:

$$F_{q1} = q_1 b = 320\,\text{N} \quad F_{Bx} = \frac{4}{5} F_B = 160\,\text{N} \quad F_{By} = \frac{3}{5} F_B = 120\,\text{N} \quad \Leftarrow$$

**(3) Sum the moments about $A$ to find reaction force $C_y$.** This structure is statically determinate because there are three available equations from statics ($\Sigma F_x = 0$, $\Sigma F_y = 0$, and $\Sigma M = 0$) and three reaction unknowns ($A_x$, $A_y$, $C_y$). It is convenient to start the static analysis using $\Sigma M_A = 0$, because we can isolate one equation with one unknown and then easily find reaction $C_y$. A statics sign convention is used (i.e., right-hand rule or CCW is positive).

$$C_y = \frac{1}{(a+b)}\left[ M_A - F_{By}a + F_{q1}\left(a + \frac{b}{2}\right)\right] = 260\,\text{N} \quad \Leftarrow$$

Continues ⟹

**(4) Sum the forces in x and y directions to find reaction forces at A.** Now that $C_y$ is known, we can complete the equilibrium analysis to find $A_x$ and $A_y$ using $\Sigma F_x = 0$ and $\Sigma F_y = 0$. Then we can find the resultant reaction force at $A$ using components $A_x$ and $A_y$:

Sum forces in x direction:   $A_x - F_{Bx} = 0$    $A_x = F_{Bx}$    $A_x = 160 \text{ N}$

Sum forces in y direction:   $A_y + F_{By} + C_y - F_{q1} = 0$

$$A_y = -F_{By} - C_y + F_{q1}    A_y = -60 \text{ N}$$

Resultant force at $A$:   $A = \sqrt{A_x^2 + A_y^2}$    $A = 171 \text{ N}$

**(5) Find the internal forces and moment at the midpoint of member segment BC.** Now that reaction forces at $A$ and $C$ are known, we can cut a section through the beam midway between $B$ and $C$, creating left and right FBDs (Fig. 1-12). Section forces $N_c$ (axial) and $V_c$ (shear) as well as section moment ($M_c$) are exposed and may be computed using statics. Either FBD may be used to find $N_c$, $V_c$, and $M_c$; the computed internal forces and moment $N_c$, $V_c$, and $M_c$ will be the same.

Calculations based on *left FBD*:

$$\Sigma F_x = 0   N = F_{Bx} - A_x = 0 \text{ N}$$

$$\Sigma F_y = 0   V = A_y + F_{By} - q_1\left(\frac{b}{2}\right) = -100 \text{ N}$$

$$\Sigma M = 0   M = M_A + A_y\left(a + \frac{b}{2}\right) + F_{By}\left(\frac{b}{2}\right) - q_1\left(\frac{b}{2}\right)\left(\frac{b}{4}\right) = 180 \text{ N} \cdot \text{m}$$

Calculations based on *right FBD*:

$$\Sigma F_x = 0   N = 0$$

$$\Sigma F_y = 0   V = q_1\left(\frac{b}{2}\right) - C_y = -100 \text{ N}$$

$$\Sigma M = 0   M = C_y\left(\frac{b}{2}\right) - q_1\left(\frac{b}{2}\right)\left(\frac{b}{4}\right) = 180 \text{ N} \cdot \text{m}$$

The computed internal forces ($N$ and $V$) and internal moment ($M$) are the same and can be determined using either the left or right FBD. This applies for any section taken through the beam at any point along its length. Later, we will create plots or diagrams which show the variation of $N$, $V$, and $M$ over the length of the beam. These diagrams will be very useful in the design of beams, because they readily show the critical regions of the beam where $N$, $V$, and $M$ have maximum values.

## Fig. 1-12

Example 1-2: Left and right FBDs of beam

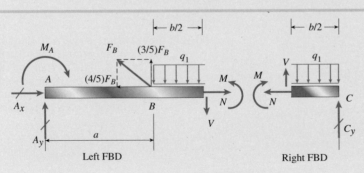

Left FBD

Right FBD

### ● ● ● Example 1-3

A stepped circular shaft is fixed at *A* and has three gears that transmit the torques shown in Fig. 1-13. Find the reaction torque at *A*, then find the internal torsional moments in segments *AB*, *BC*, and *CD*. Use properly drawn free-body diagrams in your solution.

### Fig. 1-13

Example 1-3: Stepped circular shaft in torsion

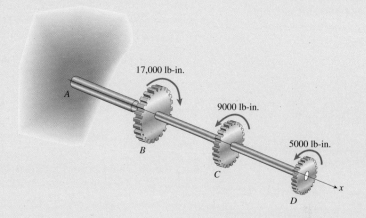

#### Solution

**(1) Draw the FBD of the overall shaft structure.** The cantilever shaft structure is statically determinate. The solution for the reaction moment at *A* must begin with a proper drawing of the FBD of the overall structure (Fig. 1-14). The FBD shows all applied and reactive torques.

### Fig. 1-14

Example 1-3: FBD of overall shaft

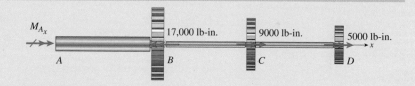

**(2) Sum the moments about the *x* axis to find the reaction moment $M_{Ax}$.** This structure is statically determinate because there is one available equation from statics ($\Sigma M_x = 0$) and one reaction unknown ($M_{Ax}$). A statics sign convention is used (i.e., right-hand rule or CCW is positive).

$$M_{Ax} - 17{,}000 \text{ lb-in.} + 9000 \text{ lb-in.} + 5000 \text{ lb-in.} = 0$$

$$M_{Ax} = -(-17000 \text{ lb-in.} + 9000 \text{ lb-in.} + 5000 \text{ lb-in.})$$

$$= 3000 \text{ lb-in.}$$

The computed result for $M_{Ax}$ is positive, so the reaction moment vector is in the positive *x* direction as assumed.

**(3) Find the internal torsional moments in each segment of the shaft.** Now that reaction moment $M_{Ax}$ is known, we can cut a section through the shaft in each segment creating left and right FBDs (Fig. 1-15). Internal

*Continues* ➡

torsional moments then may be computed using statics. Either FBD may be used; the computed internal torsional moment will be the same.

Find the internal torque $T_{AB}$ (Fig. 1-15a).

### Fig. 1-15a

Example 1-3: Left and right FBDs of shaft for each segment

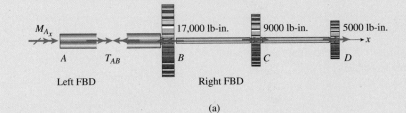

(a)

Left FBD:
$T_{AB} = -M_{Ax} = -3000$ lb-in.

Right FBD:
$T_{AB} = -17{,}000$ lb-in. $+ 9000$ lb-in.
$\quad + 5000$ lb-in. $= -3000$ lb-in.

Find the internal torque $T_{BC}$ (Fig. 1-15b).

### Fig. 1-15b

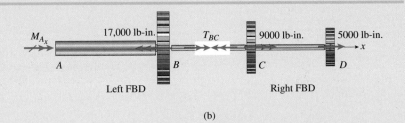

(b)

Left FBD:
$T_{BC} = -M_{Ax} + 17{,}000$ lb-in.
$\quad = 14{,}000$ lb-in.

Right FBD:
$T_{BC} = 9000$ lb-in. $+ 5000$ lb-in.
$\quad = 14{,}000$ lb-in.

Find internal torque $T_{CD}$ (Fig. 1-15c).

### Fig. 1-15c

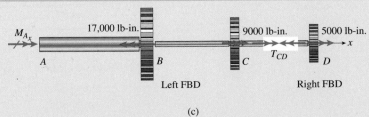

(c)

Left FBD:
$T_{CD} = -M_{Ax} + 17{,}000$ lb-in.
$\quad - 9000$ lb-in. $= 5000$ lb-in.

Right FBD:
$T_{CD} = 5000$ lb-in.

In each segment, the internal torsional moments computed using either the left or right FBDs are the same.

● ● ● **Example 1-4**

The plane frame in Fig. 1-16 is a modified version of that shown in Fig. 1-1. Initially, member *DF* has been replaced with a roller support at *D*. Moment $M_A$ is applied at pin-supported joint *A*, and load $F_B$ is applied at joint *B*. A uniform load with intensity $q_1$ acts on member *BC*, and a linearly distributed load with peak intensity $q_0$ is applied downward on member *ED*. Find the support reactions at joints *A* and *D*, then solve for internal forces at the top of member *BC*. Use numerical properties given. As a final step, remove the roller at *D*, insert member *DF* (as shown in Fig. 1-1) and reanalyze the structure to find the reaction forces at *A* and *F*.

**Fig. 1-16**

Example 1-4: Plane frame static analysis for support reactions

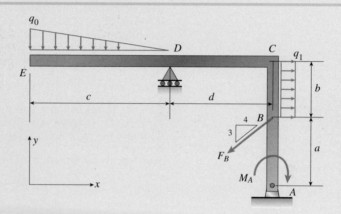

Numerical data (Newtons and meters):

$$a = 3\,\text{m} \quad b = 2\,\text{m} \quad c = 6\,\text{m} \quad d = 2.5\,\text{m}$$

$$M_A = 380\,\text{N}\cdot\text{m} \quad F_B = 200\,\text{N} \quad q_0 = 80\,\text{N/m} \quad q_1 = 160\,\text{N/m}$$

**Solution**

**(1) Draw the FBD of the overall frame.** The solution for reaction forces at *A* and *D* must begin with a proper drawing of the FBD of the overall frame (Fig. 1-17). The FBD shows all applied and reactive forces.

**Fig. 1-17**

Example 1-4: FBD of plane frame

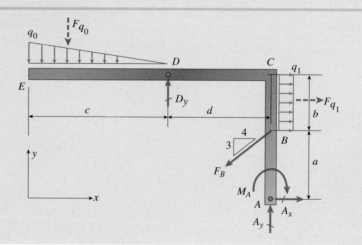

*Continues* ➡

**Example 1-4 -** *Continued*

(2) **Determine the statically equivalent concentrated forces.** Distributed forces are replaced by their statical equivalents ($F_{q0}$ and $F_{q1}$). The components of the inclined concentrated force at $B$ also may be computed:

$$F_{q0} = \frac{1}{2} \cdot q_0 c = 240 \text{ N} \qquad F_{q1} = q_1 b = 320 \text{ N}$$

$$F_{Bx} = \frac{4}{5} \cdot F_B = 160 \text{ N} \qquad F_{By} = \frac{3}{5} \cdot F_B = 120 \text{ N}$$

(3) **Sum the moments about $A$ to find reaction force $D_y$.** This structure is statically determinate because there are three available equations from statics ($\Sigma F_x = 0$, $\Sigma F_y = 0$, and $\Sigma M = 0$) and three reaction unknowns ($A_x$, $A_y$, $D_y$). It is convenient to start the static analysis using $\Sigma M_A = 0$, because we can isolate one equation with one unknown and then easily find reaction $D_y$.

$$D_y = \frac{1}{d}\left[-M_A + F_{Bx}a - F_{q1}\left(a + \frac{b}{2}\right) + F_{q0}\left(d + \frac{2}{3}c\right)\right] = 152 \text{ N}$$

(4) **Sum the forces in the $x$ and $y$ directions to find the reaction forces at $A$.** Now that $D_y$ is known, we can find $A_x$ and $A_y$ using $\Sigma F_x = 0$ and $\Sigma F_y = 0$, and then find the resultant reaction force at $A$ using components $A_x$ and $A_y$.

Sum forces in $x$ direction: $A_x - F_{Bx} + F_{q1} = 0 \quad A_x = F_{Bx} - F_{q1}$

$$A_x = -160 \text{ N}$$

Sum forces in $y$ direction: $A_y - F_{By} + D_y - F_{q0} = 0 \quad A_y = F_{By} - D_y + F_{q0}$

$$A_y = 208 \text{ N}$$

Resultant force at $A$: $A = \sqrt{A_x^2 + A_y^2} \qquad A = 262 \text{ N}$

(5) **Find the internal forces and moment at the top of member $BC$.** Now that reaction forces at $A$ and $D$ are known, we can cut a section through the frame just below joint $C$, creating upper and lower FBDs (Fig. 1-18).

### Fig. 1-18

Example 1-4: Upper and lower FBDs of plane frame

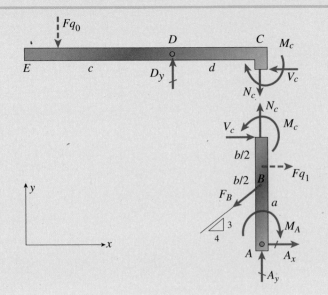

Section forces $N_c$ (axial) and $V_c$ (shear) as well as section moment ($M_c$) are exposed and may be computed using statics. Either FBD may be used to find $N_c$, $V_c$, and $M_c$; the computed stress resultants $N_c$, $V_c$, and $M_c$ will be the same.

Calculations based on **upper FBD**:

$$\Sigma F_x = 0 \qquad\qquad V_c = 0$$

$$\Sigma F_y = 0 \qquad\qquad N_c = D_y - F_{q0} = -88 \text{ N}$$

$$\Sigma M_c = 0$$

$$M_c = -D_y d + F_{q0}\left(d + \frac{2}{3}c\right) = 1180 \text{ N}\cdot\text{m}$$

Calculations based on **lower FBD**:

$$\Sigma F_x = 0 \qquad\qquad V_c = -F_{q1} + F_{Bx} - A_x = 0$$

$$\Sigma F_y = 0 \qquad\qquad N_c = F_{By} - A_y = -88 \text{ N}$$

$$\Sigma M_c = 0$$

$$M_c = -F_{q1}\frac{b}{2} + F_{Bx}b - A_x(a + b) + M_A = 1180 \text{ N}\cdot\text{m}$$

(6) **Remove the roller at D, insert member DF (as shown in Fig. 1-1) and reanalyze the structure to find reaction forces at A and F.** Member DF is pin-connected to EDC at D, has a pin support at F, and carries uniform load $q_2$ in the $-y$ direction. See Figs. 1-3a and 1-3b for the FBDs required in the solution. Note that there are now four unknown reaction forces ($A_x$, $A_y$, $F_x$ and $F_y$) but only three equilibrium equations available ($\Sigma F_x = 0$, $\Sigma F_y = 0$, $\Sigma M = 0$) for use with the overall FBD in Fig. 1-3a. To find another equation, we will have to separate the structure at pin connection D to take advantage of the fact that the moment at D is known to be zero (friction effects are assumed to be negligible); we can then use $\Sigma M_D = 0$ for either the upper FBD or the lower FBD in Fig. 1-3b to develop one more independent equation of statics. Recall that a **statics sign convention** is used here for all equilibrium equations.

Dimensions and loads for new member DF:

$$e = 5 \text{ m} \quad e_x = \frac{3}{5}e = 3 \text{ m} \quad e_y = \frac{4}{5}e = 4 \text{ m}$$

$$q_2 = 180 \text{ N/m} \quad F_{q2} = q_2 e = 900 \text{ N}$$

First, write equilibrium equations for the entire structure FBD (see Fig. 1-3a).

(a) Sum forces in x direction **for entire FBD**:

$$A_x + F_x - F_{Bx} + F_{q1} = 0 \qquad\qquad\qquad \textbf{(a)}$$

(b) Sum forces in y direction **for entire FBD**:

$$A_y + F_y - F_{q0} - F_{q2} - F_{By} = 0 \qquad\qquad\qquad \textbf{(b)}$$

*Continues* ➡

(c) Sum moments about $A$ **for entire FBD**:

$$-M_A - F_{q1}\left(a + \frac{b}{2}\right) - F_x(a + b + e_y) + F_y(e_x - d) + F_{Bx}a$$

$$+ F_{q0}\left(d + \frac{2}{3}c\right) + F_{q2}\left(d - \frac{e_x}{2}\right) = 0$$

so $-F_x(a + b + e_y) + F_y(e_x - d) = M_A + F_{q1}\left(a + \frac{b}{2}\right)$

$$-\left[F_{Bx}a + F_{q0}\left(d + \frac{2}{3}c\right) + F_{q2}\left(d - \frac{e_x}{2}\right)\right] \quad \textbf{(c)}$$

Next, write another equilibrium equation for the upper FBD in Fig. 1-3b.

(d) Sum moments about $D$ **on upper FBD**:

$$-F_x e_y + F_y e_x - F_{q2}\frac{e_x}{2} = 0 \quad \text{so} \quad -F_x e_y + F_y e_x = F_{q2}\frac{e_x}{2} \quad \textbf{(d)}$$

Solving Eqs. (c) and (d) for $F_x$ and $F_y$, we have

$$\begin{pmatrix} F_x \\ F_y \end{pmatrix} = \begin{bmatrix} -(a + b + e_y) & e_x - d \\ -e_y & e_x \end{bmatrix}^{-1}$$

$$\begin{bmatrix} M_A + F_{q1}\left(a + \frac{b}{2}\right) - \left[F_{Bx}a + F_{q0}\left(d + \frac{2}{3}c\right) + F_{q2}\left(d - \frac{e_x}{2}\right)\right] \\ F_{q2}\frac{e_x}{2} \end{bmatrix} = \begin{pmatrix} 180.6 \\ 690.8 \end{pmatrix} \text{N}$$

Now, substitute solutions for $F_x$ and $F_y$ into Eqs. (a) and (b) to find reactions $A_x$ and $A_y$:

$$A_x = -(F_x - F_{Bx} + F_{q1}) \qquad\qquad A_x = -340.6 \text{ N}$$

$$A_y = -F_y + F_{q0} + F_{q2} + F_{By} \qquad A_y = 569.2 \text{ N}$$

The resultant force at $A$ is $A = \sqrt{A_x^2 + A_y^2}$    $A = 663$ N

Sum the moments about $D$ **on lower FBD** as a check; the lower FBD is in equilibrium as required:

$$F_{q0}\left(\frac{2}{3}c\right) + F_{q1}\frac{b}{2} - F_{Bx}b - F_{By}d - M_A + A_x(a + b) + A_y d = 0 \text{ N} \cdot \text{m}$$

**(7) Finally, compute the resultant force in the pin connection at $D$.** Use the summation of forces in the upper FBD to find component forces $D_x$ and $D_y$, then find the resultant $D$ (see Fig. 1-3b).

$$\Sigma F_x = 0 \qquad\qquad D_x = -F_x = -180.6 \text{ N}$$

$$\Sigma F_y = 0 \qquad\qquad D_y = -F_y + F_{q2} = 209.2 \text{ N}$$

The resultant force at $D$ is $D = \sqrt{D_x^2 + D_y^2} = 276$ N.

# 1.3 NORMAL STRESS AND STRAIN

Now that statical equilibrium has been established and we have computed all required reaction forces and internal forces associated with the deformable body, we are ready to examine internal actions more closely. The most fundamental concepts in mechanics of materials are **stress** and **strain**. These concepts can be illustrated in their most elementary form by considering a prismatic bar subjected to axial forces. A **prismatic bar** is a straight structural member having the same cross section throughout its length, and an **axial force** is a load directed along the axis of the member, resulting in either tension or compression in the bar. Examples are shown in Fig. 1-19, where the tow bar is a prismatic member in tension and the landing gear strut is a member in compression. Other examples are the members of a bridge truss, connecting rods in automobile engines, spokes of bicycle wheels, columns in buildings, and wing struts in small airplanes.

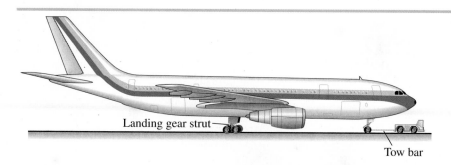

Landing gear strut

Tow bar

### Fig. 1-19

Structural members subjected to axial loads (the tow bar is in tension and the landing gear strut is in compression)

For discussion purposes, we will consider the tow bar of Fig. 1-19 and isolate a segment of it as a free body (Fig. 1-20a). When drawing this free-body diagram, we disregard the weight of the bar itself and assume that the only active forces are the axial forces $P$ at the ends. Next we consider two views of the bar, the first showing the same bar *before* the loads are applied (Fig. 1-20b) and the second showing it *after* the loads are applied (Fig. 1-20c). Note that the original length of the bar is denoted by the letter $L$, and the increase in length due to the loads is denoted by the Greek letter $\delta$ (delta).

The internal actions in the bar are exposed if we make an imaginary cut through the bar at section *mn* (Fig. 1-20c). Because this section is taken perpendicular to the longitudinal axis of the bar, it is called a **cross section**.

We now isolate the part of the bar to the left of cross section *mn* as a free body (Fig. 1-20d). At the right-hand end of this free body (section *mn*) we show the action of the removed part of the bar (that is, the part to the right of section *mn*) upon the part that remains. This action consists of continuously distributed *stresses* acting over the entire cross section, and the axial force $P$ acting at the cross section is the *resultant* of those stresses. (The resultant force is shown with a dashed line in Fig. 1-20d.)

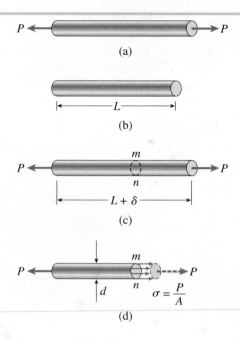

Stress has units of force per unit area and is denoted by the Greek letter $\sigma$ (sigma). In general, the stresses $\sigma$ acting on a plane surface may be uniform throughout the area or may vary in intensity from one point to another. Let us assume that the stresses acting on cross section $mn$ (Fig. 1-20d) are *uniformly distributed* over the area. Then the resultant of those stresses must be equal to the magnitude of the stress times the cross-sectional area $A$ of the bar, that is, $P = \sigma A$. Therefore, we obtain the following expression for the magnitude of the stresses:

$$\sigma = \frac{P}{A} \qquad \textbf{(1-6)}$$

This equation gives the intensity of uniform stress in an axially loaded, prismatic bar of arbitrary cross-sectional shape.

When the bar is stretched by the forces $P$, the stresses are **tensile stresses**; if the forces are reversed in direction, causing the bar to be compressed, we obtain **compressive stresses**. Inasmuch as the stresses act in a direction perpendicular to the cut surface, they are called **normal stresses**. Thus, normal stresses may be either tensile or compressive. Later, in Section 1.7, we will encounter another type of stress, called *shear stress*, that acts parallel to the surface.

When a **sign convention** for normal stresses is required, it is customary to define tensile stresses as positive and compressive stresses as negative.

Because the normal stress $\sigma$ is obtained by dividing the axial force by the cross-sectional area, it has units of force per unit of area. When USCS units are used, stress is customarily expressed in pounds per square inch (psi) or kips per square inch (ksi).* For instance, suppose that the bar of

*One kip, or kilopound, equals 1000 lb.

Fig. 1-20 has a diameter $d$ of 2.0 inches and the load $P$ has a magnitude of 6 kips. Then the stress in the bar is

$$\sigma = \frac{P}{A} = \frac{P}{\pi d^2/4} = \frac{6\,\text{k}}{\pi(2.0\ \text{in.})^2/4} = 1.91\,\text{ksi}\,(\text{or}\ 1910\ \text{psi})$$

*Area de un ⊙*
*$\pi r^2$ , $\frac{\pi d^2}{4}$*

In this example the stress is tensile, or positive.

　　When SI units are used, force is expressed in newtons (N) and area in square meters ($\text{m}^2$). Consequently, stress has units of newtons per square meter ($\text{N/m}^2$), that is, pascals (Pa). However, the pascal is such a small unit of stress that it is necessary to work with large multiples, usually the megapascal (MPa).

　　To demonstrate that a pascal is indeed small, we have only to note that it takes almost 7000 pascals to make 1 psi.* As an illustration, the stress in the bar described in the preceding example (1.91 ksi) converts to 13.2 MPa, which is $13.2 \times 10^6$ pascals. Although it is not recommended in SI, you will sometimes find stress given in newtons per square millimeter ($\text{N/mm}^2$), which is a unit equal to the megapascal (MPa).

## Limitations

The equation $\sigma = P/A$ is valid only if the stress is uniformly distributed over the cross section of the bar. This condition is realized if the axial force $P$ acts through the centroid of the cross-sectional area, as demonstrated later in this section. When the load $P$ does not act at the centroid, bending of the bar will result, and a more complicated analysis is necessary (see Sections 5.12 and 11.5). However, in this book (as in common practice) it is understood that axial forces are applied at the centroids of the cross sections unless specifically stated otherwise.

　　The uniform stress condition pictured in Fig. 1-20d exists throughout the length of the bar except near the ends. The stress distribution at the end of a bar depends upon how the load $P$ is transmitted to the bar. If the load happens to be distributed uniformly over the end, then the stress pattern at the end will be the same as everywhere else. However, it is more likely that the load is transmitted through a pin or a bolt, producing high localized stresses called stress concentrations.

　　One possibility is illustrated by the eyebar shown in Fig. 1-21. In this instance the loads $P$ are transmitted to the bar by pins that pass through the holes (or eyes) at the ends of the bar. Thus, the forces shown in the figure are actually the resultants of bearing pressures between the pins and the eyebar, and the stress distribution around the holes is quite complex. However, as we move away from the ends and toward the middle of the bar, the stress distribution gradually approaches the uniform distribution pictured in Fig. 1-20d.

　　As a practical rule, the formula $\sigma = P/A$ may be used with good accuracy at any point within a prismatic bar that is at least as far away from the stress concentration as the largest lateral dimension of the bar. In other words, the stress distribution in the steel eyebar of Fig. 1-21 is uniform at

### Fig. 1-21

Steel eyebar subjected to tensile loads $P$

---

*Conversion factors between USCS units and SI units are listed in Table B-5, Appendix B.

distances $b$ or greater from the enlarged ends, where $b$ is the width of the bar, and the stress distribution in the prismatic bar of Fig. 1-20 is uniform at distances $d$ or greater from the ends, where $d$ is the diameter of the bar (Fig. 1-20d). More detailed discussions of stress concentrations produced by axial loads are given in Section 2.10.

Of course, even when the stress is *not* uniformly distributed, the equation $\sigma = P/A$ may still be useful because it gives the *average* normal stress on the cross section.

## Normal Strain

As already observed, a straight bar will change in length when loaded axially, becoming longer when in tension and shorter when in compression. For instance, consider again the prismatic bar of Fig. 1-20. The elongation $\delta$ of this bar (Fig. 1-20c) is the cumulative result of the stretching of all elements of the material throughout the volume of the bar. Let us assume that the material is the same everywhere in the bar. Then, if we consider half of the bar (length $L/2$), it will have an elongation equal to $\delta/2$, and if we consider one-fourth of the bar, it will have an elongation equal to $\delta/4$.

In general, the elongation of a segment is equal to its length divided by the total length $L$ and multiplied by the total elongation $\delta$. Therefore, a unit length of the bar will have an elongation equal to $1/L \times \delta$. This quantity is called the *elongation per unit length*, or **strain**, and is denoted by the Greek letter $\varepsilon$ (epsilon). We see that strain is given by the equation

$$\varepsilon = \frac{\delta}{L} \qquad\qquad \textbf{(1-7)}$$

If the bar is in tension, the strain is called a **tensile strain**, representing an elongation or stretching of the material. If the bar is in compression, the strain is a **compressive strain** and the bar shortens. Tensile strain is usually taken as positive and compressive strain as negative. The strain $\varepsilon$ is called a **normal strain** because it is associated with normal stresses.

Because normal strain is the ratio of two lengths, it is a **dimensionless quantity**, that is, it has no units. Therefore, strain is expressed simply as a number, independent of any system of units. Numerical values of strain are usually very small, because bars made of structural materials undergo only small changes in length when loaded.

As an example, consider a steel bar having length $L$ equal to 2.0 m. When heavily loaded in tension, this bar might elongate by 1.4 mm, which means that the strain is

$$\varepsilon = \frac{\delta}{L} = \frac{1.4\,\text{mm}}{2.0\,\text{m}} = 0.0007 = 700 \times 10^{-6}$$

In practice, the original units of $\delta$ and $L$ are sometimes attached to the strain itself, and then the strain is recorded in forms such as mm/m, $\mu$m/m, and in./in. For instance, the strain $\varepsilon$ in the preceding illustration could be given as 700 $\mu$m/m or 700 $\times$ $10^{-6}$ in./in. Also, strain is sometimes expressed as a percent, especially when the strains are large. (In the preceding example, the strain is 0.07%.)

## Uniaxial Stress and Strain

The definitions of normal stress and normal strain are based upon purely static and geometric considerations, which means that Eqs. (1-1) and (1-2) can be used for loads of any magnitude and for any material. The principal requirement is that the deformation of the bar be uniform throughout its volume, which in turn requires that the bar be prismatic, the loads act through the centroids of the cross sections, and the material be **homogeneous** (that is, the same throughout all parts of the bar). The resulting state of stress and strain is called **uniaxial stress and strain** (although lateral strain exists as discussed in Sec. 1.6 below).

Further discussions of uniaxial stress, including stresses in directions other than the longitudinal direction of the bar, are given later in Section 2.6. We will also analyze more complicated stress states, such as biaxial stress and plane stress, in Chapter 7.

## Line of Action of the Axial Forces for a Uniform Stress Distribution

Throughout the preceding discussion of stress and strain in a prismatic bar, we assumed that the normal stress $\sigma$ was distributed uniformly over the cross section. Now we will demonstrate that this condition is met if the line of action of the axial forces is through the centroid of the cross-sectional area.

Consider a prismatic bar of arbitrary cross-sectional shape subjected to axial forces $P$ that produce uniformly distributed stresses $\sigma$ (Fig. 1-22a). Also, let $p_1$ represent the point in the cross section where the line of action of the forces intersects the cross section (Fig. 1-22b). We construct a set of $xy$ axes in the plane of the cross section and denote the coordinates of point $p_1$ by $\bar{x}$ and $\bar{y}$. To determine these coordinates, we observe that the moments $M_x$ and $M_y$ of the force $P$ about the $x$ and $y$ axes, respectively, must be equal to the corresponding moments of the uniformly distributed stresses.

The moments of the force $P$ are

$$M_x = P\bar{y} \quad M_y = -P\bar{x} \qquad \textbf{(1-8a,b)}$$

in which a moment is considered positive when its vector (using the right-hand rule) acts in the positive direction of the corresponding axis.*

The moments of the distributed stresses are obtained by integrating over the cross-sectional area $A$. The differential force acting on an element of area $dA$ (Fig. 1-22b) is equal to $\sigma dA$. The moments of this

---

*To visualize the right-hand rule, imagine that you grasp an axis of coordinates with your right hand so that your fingers fold around the axis and your thumb points in the positive direction of the axis. Then a moment is positive if it acts about the axis in the same direction as your fingers.

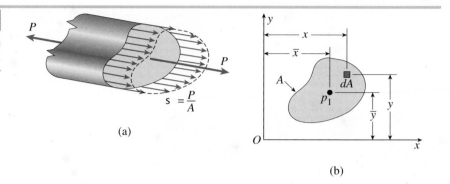

(a)

(b)

elemental force about the $x$ and $y$ axes are $\sigma y \, dA$ and $-\sigma x \, dA$, respectively, in which $x$ and $y$ denote the coordinates of the element $dA$. The total moments are obtained by integrating over the cross-sectional area:

$$M_x = \int \sigma y \, dA \quad M_y = -\int \sigma x \, dA \qquad \textbf{(1-8c,d)}$$

These expressions give the moments produced by the stresses $\sigma$.

Next, we equate the moments $M_x$ and $M_y$ as obtained from the force $P$ (Eqs. 1-8a and b) to the moments obtained from the distributed stresses (Eqs. 1-8c and d):

$$P\bar{y} = \int \sigma y \, dA \quad P\bar{x} = -\int \sigma x \, dA$$

Because the stresses $\sigma$ are uniformly distributed, we know that they are constant over the cross-sectional area $A$ and can be placed outside the integral signs. Also, we know that $\sigma$ is equal to $P/A$. Therefore, we obtain the following formulas for the coordinates of point $p_1$:

$$\bar{y} = \frac{\int y \, dA}{A} \quad \bar{x} = \frac{\int x \, dA}{A} \qquad \textbf{(1-9a,b)}$$

These equations are the same as the equations defining the coordinates of the centroid of an area (see Eqs. 12-3a and b in Chapter 12). Therefore, we have now arrived at an important conclusion: *In order to have uniform tension or compression in a prismatic bar, the axial force must act through the centroid of the cross-sectional area.* As explained previously, we always assume that these conditions are met unless it is specifically stated otherwise.

The following examples illustrate the calculation of stresses and strains in prismatic bars. In the first example we disregard the weight of the bar and in the second we include it. (It is customary when solving textbook problems to omit the weight of the structure unless specifically instructed to include it.)

### • • • Example 1-5

**Fig. 1-23**

Example 1-5: Two-tier hanging pipe stress analysis

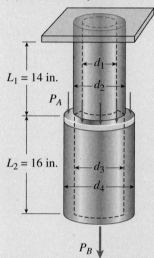

$L_1 = 14$ in.

$P_A$

$L_2 = 16$ in.

$P_B$

A hollow circular nylon pipe (see Fig 1-23) supports a load $P_A = 1750$ lb, which is uniformly distributed around a cap plate at the top of the lower pipe. A second load $P_B$ is applied at the bottom. The inner and outer diameters of the upper and lower parts of the pipe are $d_1 = 2$ in., $d_2 = 2.375$ in., $d_3 = 2.25$ in., and $d_4 = 2.5$ in., respectively. The upper pipe has a length $L_1 = 14$ in.; the lower pipe length is $L_2 = 16$ in. Neglect the self-weight of the pipes.

(a) Find $P_B$ so that the tensile stress in upper part is 2100 psi. What is the resulting stress in the lower part?
(b) If $P_A$ remains unchanged, find the new value of $P_B$ so that upper and lower parts have same tensile stress.
(c) Find the tensile strains in the upper and lower pipe segments for the loads in part (b) if the elongation of the upper pipe segment is known to be 0.138 in. and the downward displacement of the bottom of the pipe is 0.295 in.

Numerical data:

$$d_3 = 2.25 \text{ in.} \quad d_4 = 2.5 \text{ in.} \quad d_1 = 2 \text{ in.} \quad d_2 = 2.375 \text{ in.} \quad P_A = 1750 \text{ lb}$$

$$L_1 = 14 \text{ in.} \quad L_2 = 16 \text{ in.}$$

### Solution

**(a) Find $P_B$ so that the stress in the upper part is 2100 psi. What is the resulting stress in the lower part? Neglect self-weight in all calculations.**

Use the given dimensions to compute the cross-sectional areas of the upper (segment 1) and lower (segment 2) pipes (we note that $A_1$ is 1.38 times $A_2$). The stress in segment 1 is known to be 2100 psi.

$$A_1 = \frac{\pi}{4} \cdot (d_2^2 - d_1^2) = 1.289 \text{ in.}^2 \quad A_2 = \frac{\pi}{4} \cdot (d_4^2 - d_3^2) = 0.933 \text{ in.}^2$$

The axial tensile force in the upper pipe is the sum of loads $P_A$ and $P_B$. Write an expression for $\sigma_1$ in terms of both loads, then solve for $P_B$:

*[handwritten annotation: * Pq si busca (la suma de arriba) no divide entre la suma de las areas]*

$$\sigma_1 = \frac{P_A + P_B}{A_1}$$

where $\sigma_1 = 2100$ psi   so   $P_B = \sigma_1 A_1 - P_A = 957$ lb

With $P_B$ now known, the axial tensile stress in the lower segment can be computed as

$$\sigma_2 = \frac{P_B}{A_2} = 1025 \text{ psi}$$

Continues ➡

**(b) If $P_A$ remains unchanged, find the new value of $P_B$ so that upper and lower parts have same tensile stress.**

So $P_A$ = 1750 lb. Write expressions for the normal stresses in the upper and lower segments, equate these expressions, and then solve for $P_B$.

Tensile normal stress in upper segment:

$$\sigma_1 = \frac{P_A + P_B}{A_1}$$

Tensile normal stress in lower segment:

$$\sigma_2 = \frac{P_B}{A_2}$$

Equate these expressions for stresses $\sigma_1$ and $\sigma_2$ and solve for the required $P_B$:

$$P_B = \frac{\dfrac{P_A}{A_1}}{\left(\dfrac{1}{A_2} - \dfrac{1}{A_1}\right)} = 4586 \text{ lb} \qquad \Longleftarrow$$

So for the stresses to be equal in the upper and lower pipe segments, the new value of load $P_B$ is 2.62 times the value of load $P_A$.

**(c) Find the tensile strains in the upper and lower pipe segments for the loads in part (b).**

The *elongation* of the upper pipe segment is $\delta_1$ = 0.138 in. So the tensile strain in the upper pipe segment is

$$\varepsilon_1 = \frac{\delta_1}{L_1} = 9.86 \times 10^{-3} \qquad \Longleftarrow$$

The downward *displacement* of the bottom of the pipe is $\delta$ = 0.295 in. So the *net elongation* of the lower pipe segment is $\delta_2 = \delta - \delta_1$ = 0.157 in. and the tensile strain in the lower pipe segment is

$$\varepsilon_2 = \frac{\delta_2}{L_2} = 9.81 \times 10^{-3} \qquad \Longleftarrow$$

*Note:* As explained earlier, strain is a dimensionless quantity and no units are needed. For clarity, however, units are often given. In this example, $\varepsilon$ could be written as $986 \times 10^{-6}$ in./in. or 986 $\mu$in./in. for $\varepsilon_1$.

### ● ● ● Example 1-6

A semicircular bracket assembly is used to support a steel staircase in an office building. Steel rods are attached to each of two brackets using a clevis and pin; the upper end of the rod is attached to a cross beam near the roof. Photos of the bracket attachment and hanger rod supports are shown in Fig. 1-24. The weight of the staircase, and any building occupants who are using the staircase, is estimated to result in a force of 4800 N on each hanger rod.

(a) Obtain a *formula for the maximum stress* $\sigma_{max}$ in the rod, taking into account the weight of the rod itself.
(b) Calculate the *maximum stress in the rod* in MPa using numerical properties $L_r = 12$ m, $d_r = 20$ mm, $F_r = 4800$ N (note that the weight density $\gamma_r$ of steel is 77.0 kN/m³ [from Table I-1 in Appendix I]).

---

### Fig. 1-24a

Example 1-6: Hanger rods supporting steel staircase

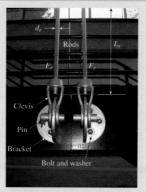

Components of hanger rod connection
(© Barry Goodno)

---

### Fig. 1-24b

Side view of hanger rod and bracket
(© Barry Goodno)

---

### Fig. 1-24c

Hanger rod attached to cross beam at roof
(© Barry Goodno)

*Continues* ➡

Numerical data:

$$L_r = 12 \text{ m} \quad d_r = 20 \text{ mm} \quad \gamma_r = 77 \text{ kN/m}^3 \quad F_r = 4800 \text{ N}$$

**Solution**

**(a) Obtain a formula for the maximum stress $\sigma_{max}$ in the rod, taking into account the weight of the rod itself.**

The maximum axial force $F_{max}$ in the rod occurs at the upper end and is equal to the force $F_r$ in the rod due to the combined staircase and occupants' weights plus the weight $W_r$ of the rod itself. The latter is equal to the weight density $\gamma_r$ of the steel times the volume $V_r$ of the rod, or

$$W_r = \gamma_r(A_r L_r) \qquad \Longleftarrow \quad (1\text{-}10)$$

in which $A_r$ is the cross-sectional area of the rod. Therefore, the *formula for the maximum stress* (from Eq. 1-6) becomes

$$\sigma_{max} = \frac{F_{max}}{A_r} \quad \text{or} \quad \sigma_{max} = \frac{F_r + W_r}{A_r} = \frac{F_r}{A_r} + \gamma_r L_r \qquad \Longleftarrow \quad (1\text{-}11)$$

**(b) Calculate the maximum stress in the rod in MPa using numerical properties.**

To calculate the maximum stress, we substitute numerical values into the preceding equation. The cross-sectional area $A_r = \pi d_r^2/4$, where $d_r = 20$ mm, and the weight density $\gamma_r$ of steel is 77.0 kN/m³ (from Table I-1 in Appendix I).

Thus,

$$A_r = \frac{\pi d_r^2}{4} = 314 \text{ mm}^2 \quad F_r = 4800 \text{ N}$$

The normal stress in the rod due to the weight of the staircase is

$$\sigma_{stair} = \frac{F_r}{A_r} = 15.3 \text{ MPa}$$

and the additional normal stress at the top of the rod due to the weight of the rod itself is

$$\sigma_{rod} = \gamma_r L_r = 0.924 \text{ MPa}$$

So the *maximum normal stress at the top of the rod* is the sum of these two normal stresses:

$$\sigma_{max} = \sigma_{stair} + \sigma_{rod} \quad \sigma_{max} = 16.2 \text{ MPa} \qquad \Longleftarrow \quad (1\text{-}12)$$

Note that

$$\frac{\sigma_{rod}}{\sigma_{stair}} = 6.05\%$$

In this example, the weight of the rod contributes about 6% to the maximum stress and should not be disregarded.

# 1.4 MECHANICAL PROPERTIES OF MATERIALS

The design of machines and structures so that they will function properly requires that we understand the **mechanical behavior** of the materials being used. Ordinarily, the only way to determine how materials behave when they are subjected to loads is to perform experiments in the laboratory. The usual procedure is to place small specimens of the material in testing machines, apply the loads, and then measure the resulting deformations (such as changes in length and changes in diameter). Most materials-testing laboratories are equipped with machines capable of loading specimens in a variety of ways, including both static and dynamic loading in tension and compression.

A typical **tensile-test machine** is shown in Fig. 1-25. The test specimen is installed between the two large grips of the testing machine and then loaded in tension. Measuring devices record the deformations, and the automatic control and data-processing systems (at the left in the photo) tabulate and graph the results.

A more detailed view of a **tensile-test specimen** is shown in Fig. 1-26 on the next page. The ends of the circular specimen are enlarged where they fit in the grips so that failure will not occur near the grips themselves. A failure at the ends would not produce the desired information about the material, because the stress distribution near the grips is not uniform, as explained in Section 1.3. In a properly designed specimen, failure will occur in the prismatic portion of the specimen where the stress distribution is uniform and the bar is subjected only to pure tension. This situation is shown in Fig. 1-26, where the steel specimen has just fractured under load. The device at the left, which is attached by two arms to the specimen, is an **extensometer** that measures the elongation during loading.

In order that test results will be comparable, the dimensions of test specimens and the methods of applying loads must be standardized. One of the major standards organizations in the United States is the American Society for Testing and Materials (ASTM), a technical society that publishes specifications and standards for materials and testing. Other standardizing organizations are the American Standards Association (ASA)

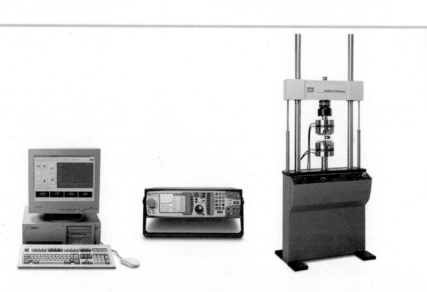

**Fig. 1-25**

Tensile-test machine with automatic data-processing system (Courtesy of MTS Systems Corporation)

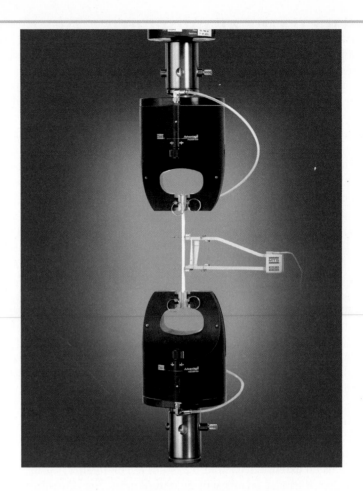

and the National Institute of Standards and Technology (NIST). Similar organizations exist in other countries.

The ASTM standard tension specimen has a diameter of 0.505 in. and a **gage length** of 2.0 in. between the gage marks, which are the points where the extensometer arms are attached to the specimen (see Fig. 1-26). As the specimen is pulled, the axial load is measured and recorded, either automatically or by reading from a dial. The elongation over the gage length is measured simultaneously, either by mechanical gages of the kind shown in Fig. 1-26 or by electrical-resistance strain gages.

In a **static test**, the load is applied slowly and the precise *rate* of loading is not of interest because it does not affect the behavior of the specimen. However, in a **dynamic test** the load is applied rapidly and sometimes in a cyclical manner. Since the nature of a dynamic load affects the properties of the materials, the rate of loading must also be measured.

**Compression tests** of metals are customarily made on small specimens in the shape of cubes or circular cylinders. For instance, cubes may be 2.0 in. on a side, and cylinders may have diameters of 1 in. and lengths from 1 to 12 in. Both the load applied by the machine and the shortening of the specimen may be measured. The shortening should be measured over a gage length that is less than the total length of the specimen in order to eliminate end effects.

Concrete is tested in compression on important construction projects to ensure that the required strength has been obtained. One type of concrete test specimen is 6 in. in diameter, 12 in. in length, and 28 days old (the age of concrete is important because concrete gains strength as it cures). Similar but somewhat smaller specimens are used when performing compression tests of rock (see Fig. 1-27).

## Stress-Strain Diagrams

Test results generally depend upon the dimensions of the specimen being tested. Since it is unlikely that we will be designing a structure having parts that are the same size as the test specimens, we need to express the test results in a form that can be applied to members of any size. A simple way to achieve this objective is to convert the test results to stresses and strains.

The axial stress $\sigma$ in a test specimen is calculated by dividing the axial load $P$ by the cross-sectional area $A$ (Eq. 1-6). When the initial area of the specimen is used in the calculation, the stress is called the **nominal stress** (other names are *conventional stress* and *engineering stress*). A more exact value of the axial stress, called the **true stress**, can be calculated by using the actual area of the bar at the cross section where failure occurs. Since the actual area in a tension test is always less than the initial area (as illustrated in Fig. 1-26), the true stress is larger than the nominal stress.

**Fig. 1-27**

Rock sample being tested in compression to obtain compressive strength, elastic modulus and Poisson's ratio (Courtesy of MTS Systems Corporation)

The average axial strain $\varepsilon$ in the test specimen is found by dividing the measured elongation $\delta$ between the gage marks by the gage length $L$ (see Fig. 1-26 and Eq. 1-7). If the initial gage length is used in the calculation (for instance, 2.0 in.), then the **nominal strain** is obtained. Since the distance between the gage marks increases as the tensile load is applied, we can calculate the **true strain** (or *natural strain*) at any value of the load by using the actual distance between the gage marks. In tension, true strain is always smaller than nominal strain. However, for most engineering purposes, nominal stress and nominal strain are adequate, as explained later in this section.

After performing a tension or compression test and determining the stress and strain at various magnitudes of the load, we can plot a diagram of stress versus strain. Such a **stress-strain diagram** is a characteristic of the particular material being tested and conveys important information about the mechanical properties and type of behavior.*

The first material we will discuss is **structural steel**, also known as *mild steel* or *low-carbon steel*. Structural steel is one of the most widely used metals and is found in buildings, bridges, cranes, ships, towers, vehicles, and many other types of construction. A stress-strain diagram for a typical structural steel in tension is shown in Fig. 1-28. Strains are plotted on the horizontal axis and stresses on the vertical axis. (In order to display all of the important features of this material, the strain axis in Fig. 1-28 is not drawn to scale.)

The diagram begins with a straight line from the origin $O$ to point $A$, which means that the relationship between stress and strain in this initial region is not only *linear* but also *proportional*.** Beyond point $A$, the proportionality between stress and strain no longer exists; hence the stress at $A$ is called the **proportional limit**. For low-carbon steels, this limit is in the range 30 to 50 ksi (210 to 350 MPa), but high-strength steels (with higher carbon content plus other alloys) can have proportional limits of more than 80 ksi (550 MPa). The slope of the straight line from $O$ to $A$ is called the **modulus of elasticity**. Because the slope has units of stress divided by strain, modulus of elasticity has the same units as stress. (Modulus of elasticity is discussed later in Section 1.6.)

With an increase in stress beyond the proportional limit, the strain begins to increase more rapidly for each increment in stress. Consequently, the stress-strain curve has a smaller and smaller slope, until, at point $B$, the curve becomes horizontal (see Fig. 1-28). Beginning at this point, considerable elongation of the test specimen occurs with no noticeable increase in the tensile force (from $B$ to $C$). This phenomenon is known as **yielding** of the material, and point $B$ is called the **yield point**. The corresponding stress is known as the **yield stress** of the steel.

In the region from $B$ to $C$ (see Fig. 1-28), the material becomes **perfectly plastic**, which means that it deforms without an increase in the applied load. The elongation of a mild-steel specimen in the perfectly plastic

---

*Stress-strain diagrams were originated by Jacob Bernoulli (1654–1705) and J. V. Poncelet (1788–1867); see Ref. 1-4.

**Two variables are said to be *proportional* if their ratio remains constant. Therefore, a proportional relationship may be represented by a straight line through the origin. However, a proportional relationship is not the same as a *linear* relationship. Although a proportional relationship is linear, the converse is not necessarily true, because a relationship represented by a straight line that does *not* pass through the origin is linear but not proportional. The often-used expression "directly proportional" is synonymous with "proportional" (Ref. 1-5).

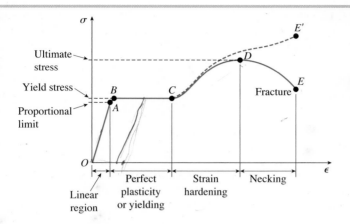

**Fig. 1-28**

Stress-strain diagram for a typical structural steel in tension (not to scale)

**Fig. 1-29**

Necking of a mild-steel bar in tension (© Barry Goodno)

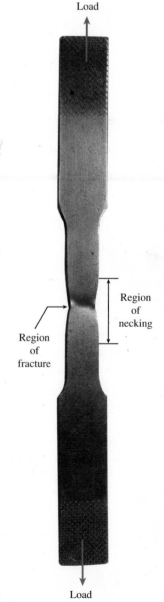

region is typically 10 to 15 times the elongation that occurs in the linear region (between the onset of loading and the proportional limit). The presence of very large strains in the plastic region (and beyond) is the reason for not plotting this diagram to scale.

After undergoing the large strains that occur during yielding in the region *BC*, the steel begins to **strain harden**. During strain hardening, the material undergoes changes in its crystalline structure, resulting in increased resistance of the material to further deformation. Elongation of the test specimen in this region requires an increase in the tensile load, and therefore the stress-strain diagram has a positive slope from *C* to *D*. The load eventually reaches its maximum value, and the corresponding stress (at point *D*) is called the **ultimate stress**. Further stretching of the bar is actually accompanied by a reduction in the load, and fracture finally occurs at a point such as *E* in Fig. 1-28.

The yield stress and ultimate stress of a material are also called the **yield strength** and **ultimate strength**, respectively. **Strength** is a general term that refers to the capacity of a structure to resist loads. For instance, the yield strength of a beam is the magnitude of the load required to cause yielding in the beam, and the ultimate strength of a truss is the maximum load it can support, that is, the failure load. However, when conducting a tension test of a particular material, we define load-carrying capacity by the stresses in the specimen rather than by the total loads acting on the specimen. As a result, the strength of a material is usually stated as a stress.

When a test specimen is stretched, **lateral contraction** occurs, as previously mentioned. The resulting decrease in cross-sectional area is too small to have a noticeable effect on the calculated values of the stresses up to about point *C* in Fig. 1-28, but beyond that point the reduction in area begins to alter the shape of the curve. In the vicinity of the ultimate stress, the reduction in area of the bar becomes clearly visible and a pronounced **necking** of the bar occurs (see Figs. 1-26 and 1-29).

If the actual cross-sectional area at the narrow part of the neck is used to calculate the stress, the **true stress-strain curve** (the dashed line *CE'* in Fig. 1-28) is obtained. The total load the bar can carry does indeed diminish after the ultimate stress is reached (as shown by curve *DE*), but this reduction is due to the decrease in area of the bar and not to a loss in strength of the material itself. In reality, the material withstands an

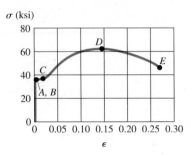

## Fig. 1-30

Stress-strain diagram for a typical structural steel in tension (drawn to scale)

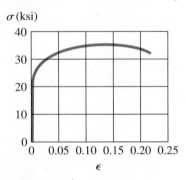

## Fig. 1-31

Typical stress-strain diagram for an aluminum alloy

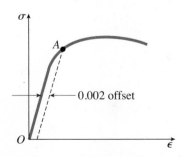

## Fig. 1-32

Arbitrary yield stress determined by the offset method

increase in true stress up to failure (point $E'$). Because most structures are expected to function at stresses below the proportional limit, the **conventional stress-strain curve** $OABCDE$, which is based upon the original cross-sectional area of the specimen and is easy to determine, provides satisfactory information for use in engineering design.

The diagram of Fig. 1-28 shows the general characteristics of the stress-strain curve for mild steel, but its proportions are not realistic because, as already mentioned, the strain that occurs from $B$ to $C$ may be more than ten times the strain occurring from $O$ to $A$. Furthermore, the strains from $C$ to $E$ are many times greater than those from $B$ to $C$. The correct relationships are portrayed in Fig. 1-30, which shows a stress-strain diagram for mild steel drawn to scale. In this figure, the strains from the zero point to point $A$ are so small in comparison to the strains from point $A$ to point $E$ that they cannot be seen, and the initial part of the diagram appears to be a vertical line.

The presence of a clearly defined yield point followed by large plastic strains is an important characteristic of structural steel that is sometimes utilized in practical design (see, for instance, the discussions of elastoplastic behavior in Sections 2.12 and 6.10). Metals such as structural steel that undergo large *permanent* strains before failure are classified as **ductile**. For instance, ductility is the property that enables a bar of steel to be bent into a circular arc or drawn into a wire without breaking. A desirable feature of ductile materials is that visible distortions occur if the loads become too large, thus providing an opportunity to take remedial action before an actual fracture occurs. Also, materials exhibiting ductile behavior are capable of absorbing large amounts of strain energy prior to fracture.

Structural steel is an alloy of iron containing about 0.2% carbon, and therefore it is classified as a low-carbon steel. With increasing carbon content, steel becomes less ductile but stronger (higher yield stress and higher ultimate stress). The physical properties of steel are also affected by heat treatment, the presence of other metals, and manufacturing processes such as rolling. Other materials that behave in a ductile manner (under certain conditions) include aluminum, copper, magnesium, lead, molybdenum, nickel, brass, bronze, monel metal, nylon, and teflon.

Although they may have considerable ductility, **aluminum alloys** typically do not have a clearly definable yield point, as shown by the stress-strain diagram of Fig. 1-31. However, they do have an initial linear region with a recognizable proportional limit. Alloys produced for structural purposes have proportional limits in the range 10 to 60 ksi (70 to 410 MPa) and ultimate stresses in the range 20 to 80 ksi (140 to 550 MPa).

When a material such as aluminum does not have an obvious yield point and yet undergoes large strains after the proportional limit is exceeded, an *arbitrary* yield stress may be determined by the **offset method**. A straight line is drawn on the stress-strain diagram parallel to the initial linear part of the curve (Fig. 1-32) but offset by some standard strain, such as 0.002 (or 0.2%). The intersection of the offset line and the stress-strain curve (point $A$ in the figure) defines the yield stress. Because this stress is determined by an arbitrary rule and is not an inherent physical property of the material, it should be distinguished from a true yield stress by referring to it as the **offset yield stress**. For a material such as aluminum, the offset yield stress is slightly above the proportional limit. In the case of structural steel, with its abrupt transition from the linear region to the region of plastic stretching, the offset stress is essentially the same as both the yield stress and the proportional limit.

Rubber maintains a linear relationship between stress and strain up to relatively large strains (as compared to metals). The strain at the proportional limit may be as high as 0.1 or 0.2 (10% or 20%). Beyond the proportional limit, the behavior depends upon the type of rubber (Fig. 1-33). Some kinds of soft rubber will stretch enormously without failure, reaching lengths several times their original lengths. The material eventually offers increasing resistance to the load, and the stress-strain curve turns markedly upward. You can easily sense this characteristic behavior by stretching a rubber band with your hands. (Note that although rubber exhibits very large strains, it is not a ductile material because the strains are not permanent. It is, of course, an elastic material; see Section 1.5.)

The ductility of a material in tension can be characterized by its elongation and by the reduction in area at the cross section where fracture occurs. The **percent elongation** is defined as

$$\text{Percent elongation} = \frac{L_1 - L_0}{L_0}(100) \qquad \textbf{(1-13)}$$

in which $L_0$ is the original gage length and $L_1$ is the distance between the gage marks at fracture. Because the elongation is not uniform over the length of the specimen but is concentrated in the region of necking, the percent elongation depends upon the gage length. Therefore, when stating the percent elongation, the gage length should always be given. For a 2-in. gage length, steel may have an elongation in the range from 3% to 40%, depending upon composition; in the case of structural steel, values of 20% or 30% are common. The elongation of aluminum alloys varies from 1% to 45%, depending upon composition and treatment.

The **percent reduction in area** measures the amount of necking that occurs and is defined as

$$\text{Percent reduction in area} = \frac{A_0 - A_1}{A_0}(100) \qquad \textbf{(1-14)}$$

in which $A_0$ is the original cross-sectional area and $A_1$ is the final area at the fracture section. For ductile steels, the reduction is about 50%.

Materials that fail in tension at relatively low values of strain are classified as **brittle**. Examples are concrete, stone, cast iron, glass, ceramics, and a variety of metallic alloys. Brittle materials fail with only little elongation after the proportional limit (the stress at point $A$ in Fig. 1-34) is exceeded. Furthermore, the reduction in area is insignificant, and so the nominal fracture stress (point $B$) is the same as the true ultimate stress. High-carbon steels have very high yield stresses—over 100 ksi (700 MPa) in some cases—but they behave in a brittle manner and fracture occurs at an elongation of only a few percent.

Stress-strain curves for two kinds of rubber in tension

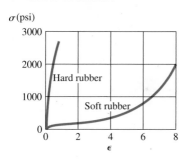

Typical stress-strain diagram for a brittle material showing the proportional limit (point $A$) and fracture stress (point $B$)

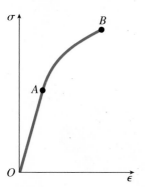

Ordinary glass is a nearly ideal brittle material, because it exhibits almost no ductility. The stress-strain curve for glass in tension is essentially a straight line, with failure occurring before any yielding takes place. The ultimate stress is about 10,000 psi (70 MPa) for certain kinds of plate glass, but great variations exist, depending upon the type of glass, the size of the specimen, and the presence of microscopic defects. **Glass fibers** can develop enormous strengths, and ultimate stresses over 1,000,000 psi (7 GPa) have been attained.

Many types of **plastics** are used for structural purposes because of their light weight, resistance to corrosion, and good electrical insulation properties. Their mechanical properties vary tremendously, with some plastics being brittle and others ductile. When designing with plastics it is important to realize that their properties are greatly affected by both temperature changes and the passage of time. For instance, the ultimate tensile stress of some plastics is cut in half merely by raising the temperature from 50° F to 120° F. Also, a loaded plastic may stretch gradually over time until it is no longer serviceable. For example, a bar of polyvinyl chloride subjected to a tensile load that initially produces a strain of 0.005 may have that strain doubled after one week, even though the load remains constant. (This phenomenon, known as *creep*, is discussed in the next section.)

Ultimate tensile stresses for plastics are generally in the range 2 to 50 ksi (14 to 350 MPa) and weight densities vary from 50 to 90 lb/ft³ (8 to 14 kN/m³). One type of nylon has an ultimate stress of 12 ksi (80 MPa) and weighs only 70 lb/ft³ (11 kN/m³), which is only 12% heavier than water. Because of its light weight, the strength-to-weight ratio for nylon is about the same as for structural steel (see Prob. 1.4-4).

A **filament-reinforced material** consists of a base material (or *matrix*) in which high-strength filaments, fibers, or whiskers are embedded. The resulting composite material has much greater strength than the base material. As an example, the use of glass fibers can more than double the strength of a plastic matrix. Composites are widely used in aircraft, boats, rockets, and space vehicles where high strength and light weight are needed.

## Compression

Stress-strain curves for materials in compression differ from those in tension. Ductile metals such as steel, aluminum, and copper have proportional limits in compression very close to those in tension, and the initial regions of their compressive and tensile stress-strain diagrams are about the same. However, after yielding begins, the behavior is quite different. In a tension test, the specimen is stretched, necking may occur, and fracture ultimately takes place. When the material is compressed, it bulges outward on the sides and becomes barrel shaped, because friction between the specimen and the end plates prevents lateral expansion. With increasing load, the specimen is flattened out and offers greatly increased resistance to further shortening (which means that the stress-strain curve becomes very steep). These characteristics are illustrated in Fig. 1-35, which shows a compressive stress-strain diagram for copper. Since the actual cross-sectional area of a specimen tested in compression is larger than the initial area, the true stress in a compression test is smaller than the nominal stress.

**Fig. 1-35**

Stress-strain diagram for copper in compression

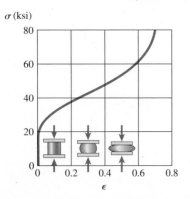

Brittle materials loaded in compression typically have an initial linear region followed by a region in which the shortening increases at a slightly higher rate than does the load. The stress-strain curves for compression and tension often have similar shapes, but the ultimate stresses in compression are much higher than those in tension. Also, unlike ductile materials, which flatten out when compressed, brittle materials actually break at the maximum load.

## Tables of Mechanical Properties

Properties of materials are listed in the tables of Appendix I at the back of the book. The data in the tables are typical of the materials and are suitable for solving problems in this book. However, properties of materials and stress-strain curves vary greatly, even for the same material, because of different manufacturing processes, chemical composition, internal defects, temperature, and many other factors.

For these reasons, data obtained from Appendix I (or other tables of a similar nature) should not be used for specific engineering or design purposes. Instead, the manufacturers or materials suppliers should be consulted for information about a particular product.

# 1.5 ELASTICITY, PLASTICITY, AND CREEP

Stress-strain diagrams portray the behavior of engineering materials when the materials are loaded in tension or compression, as described in the preceding section. To go one step further, let us now consider what happens when the load is removed and the material is *unloaded*.

Assume, for instance, that we apply a load to a tensile specimen so that the stress and strain go from the origin *O* to point *A* on the stress-strain curve of Fig. 1-36a. Suppose further that when the load is removed, the material follows exactly the same curve back to the origin *O*. This property of a material, by which it returns to its original dimensions during unloading, is called **elasticity**, and the material itself is said to be *elastic*. Note that the stress-strain curve from *O* to *A* need not be linear in order for the material to be elastic.

Now suppose that we load this same material to a higher level, so that point *B* is reached on the stress-strain curve (Fig. 1-36b). When unloading occurs from point *B*, the material follows line *BC* on the diagram. This unloading line is parallel to the initial portion of the loading curve; that is, line *BC* is parallel to a tangent to the stress-strain curve at the origin. When point *C* is reached, the load has been entirely removed, but a **residual strain**, or *permanent strain*, represented by line *OC*, remains in the material. As a consequence, the bar being tested is longer than it was before loading. This residual elongation of the bar is called the **permanent set**. Of the total strain *OD* developed during loading from *O* to *B*, the strain *CD* has been recovered elastically and the strain *OC* remains as a permanent strain. Thus, during unloading the bar returns partially to its original shape, and so the material is said to be **partially elastic**.

Between points *A* and *B* on the stress-strain curve (Fig. 1-36b), there must be a point before which the material is elastic and beyond

Fig. 1-36

Stress-strain diagrams illustrating (a) elastic behavior, and (b) partially elastic behavior

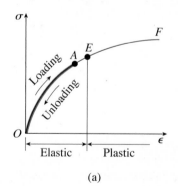

(a)

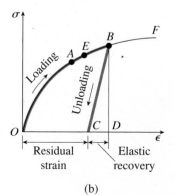

(b)

which the material is partially elastic. To find this point, we load the material to some selected value of stress and then remove the load. If there is no permanent set (that is, if the elongation of the bar returns to zero), then the material is fully elastic up to the selected value of the stress.

The process of loading and unloading can be repeated for successively higher values of stress. Eventually, a stress will be reached such that not all the strain is recovered during unloading. By this procedure, it is possible to determine the stress at the upper limit of the elastic region, for instance, the stress at point $E$ in Figs. 1-36a and b. The stress at this point is known as the **elastic limit** of the material.

Many materials, including most metals, have linear regions at the beginning of their stress-strain curves (for example, see Figs. 1-28 and 1-31). The stress at the upper limit of this linear region is the proportional limit, as explained in the preceeding section. The elastic limit is usually the same as, or slightly above, the proportional limit. Hence, for many materials the two limits are assigned the same numerical value. In the case of mild steel, the yield stress is also very close to the proportional limit, so that for practical purposes the yield stress, the elastic limit, and the proportional limit are assumed to be equal. Of course, this situation does not hold for all materials. Rubber is an outstanding example of a material that is elastic far beyond the proportional limit.

The characteristic of a material by which it undergoes inelastic strains beyond the strain at the elastic limit is known as **plasticity**. Thus, on the stress-strain curve of Fig. 1-36a, we have an elastic region followed by a plastic region. When large deformations occur in a ductile material loaded into the plastic region, the material is said to undergo **plastic flow**.

## Reloading of a Material

If the material remains within the elastic range, it can be loaded, unloaded, and loaded again without significantly changing the behavior. However, when loaded into the plastic range, the internal structure of the material is altered and its properties change. For instance, we have already observed that a permanent strain exists in the specimen after unloading from the plastic region (Fig. 1-36b). Now suppose that the material is reloaded after such an unloading (Fig. 1-37). The new loading begins at point $C$ on the diagram and continues upward to point $B$, the point at which unloading began during the first loading cycle. The material then follows the original stress-strain curve toward point $F$. Thus, for the second loading, we can imagine that we have a new stress-strain diagram with its origin at point $C$.

During the second loading, the material behaves in a linearly elastic manner from $C$ to $B$, with the slope of line $CB$ being the same as the slope of the tangent to the original loading curve at the origin $O$. The proportional limit is now at point $B$, which is at a higher stress than the original elastic limit (point $E$). Thus, by stretching a material such as steel or aluminum into the inelastic or plastic range, the *properties of the material are changed*—the linearly elastic region is increased, the proportional limit is raised, and the elastic limit is raised. However, the ductility is reduced because in the "new material" the amount of

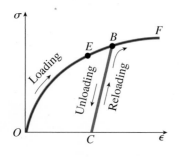

**Fig. 1-37**

Reloading of a material and raising of the elastic and proportional limits yielding beyond the elastic limit (from $B$ to $F$) is less than in the original material (from $E$ to $F$).*

## Creep

The stress-strain diagrams described previously were obtained from tension tests involving static loading and unloading of the specimens, and the passage of time did not enter our discussions. However, when loaded for long periods of time, some materials develop additional strains and are said to **creep**.

This phenomenon can manifest itself in a variety of ways. For instance, suppose that a vertical bar (Fig. 1-38a) is loaded slowly by a force $P$, producing an elongation equal to $\delta_0$. Let us assume that the loading and corresponding elongation take place during a time interval of duration $t_0$ (Fig. 1-38b). Subsequent to time $t_0$, the load remains constant. However, due to creep, the bar may gradually lengthen, as shown in Fig. 1-38b, even though the load does not change. This behavior occurs with many materials, although sometimes the change is too small to be of concern.

As another manifestation of creep, consider a wire that is stretched between two immovable supports so that it has an initial tensile stress $\sigma_0$ (Fig. 1-39). Again, we will denote the time during which the wire is initially stretched as $t_0$. With the elapse of time, the stress in the wire gradually diminishes, eventually reaching a constant value, even though the supports at the ends of the wire do not move. This process, is called **relaxation** of the material.

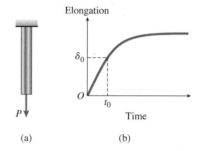

### Fig. 1-38

Creep in a bar under constant load

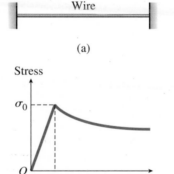

### Fig. 1-39

Relaxation of stress in a wire under constant strain

Creep is usually more important at high temperatures than at ordinary temperatures, and therefore it should always be considered in the design of engines, furnaces, and other structures that operate at elevated temperatures for long periods of time. However, materials such as steel, concrete, and wood will creep slightly even at atmospheric temperatures. For example, creep of concrete over long periods of time can create undulations in bridge decks because of sagging between the supports. (One remedy is to construct the deck with an upward **camber**, which is an initial displacement above the horizontal, so that when creep occurs, the spans lower to the level position.)

*The study of material behavior under various environmental and loading conditions is an important branch of applied mechanics. For more detailed engineering information about materials, consult a textbook devoted solely to this subject.

A machine component slides along a horizontal bar at $A$ and moves in a vertical slot at $B$. The component is represented as a rigid bar $AB$ (length $L = 5$ ft., weight $W = 985$ lb.) with roller supports at $A$ and $B$ (neglect friction). When not in use, the machine component is supported by a single wire (diameter $d = 1/8$ in.) with one end attached at $A$ and the other end supported at $C$ (see Fig. 1-40). The wire is made of a copper alloy; the stress-strain relationship for the wire is

$$\sigma(\varepsilon) = \frac{17{,}500\,\varepsilon}{1 + 240\,\varepsilon} \qquad 0 \le \varepsilon \le 0.03 \quad (\sigma \text{ in ksi})$$

**(a)** Plot a stress-strain diagram for the material; what is the modulus of elasticity $E$ (ksi)? What is the 0.2% offset yield stress (ksi)?
**(b)** Find the tensile force $T$ (lb) in the wire.
**(c)** Find the normal axial strain $\varepsilon$ and elongation $\delta$ (in.) of the wire.
**(d)** Find the permanent set of the wire if all forces are removed.

## Fig. 1-40

Example 1-7: Rigid bar supported by copper alloy wire

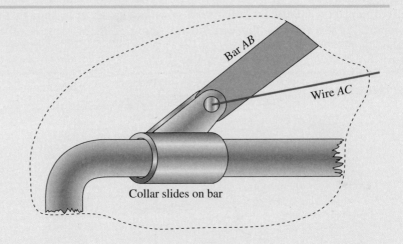

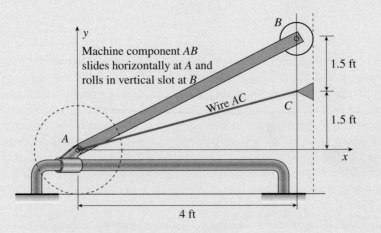

## Solution

**(a) Plot a stress-strain diagram for the material. What is the modulus of elasticity? What is the 0.2% offset yield stress (ksi)?**

Plot the function $\sigma(\varepsilon)$ for strain values between 0 and 0.03 (Fig. 1-41). The stress at strain $\varepsilon = 0.03$ is 64 ksi.

$$\sigma(\varepsilon) = \frac{17{,}500\,\varepsilon}{1 + 240\,\varepsilon} \qquad \varepsilon = 0,\, 0.001,\,\ldots,\,0.03$$

$$\sigma(0) = 0 \quad \sigma(0.03) = 64 \text{ ksi}$$

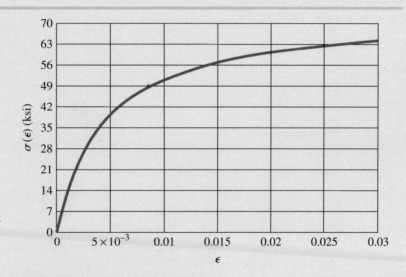

### Fig. 1-41

Stress-strain curve for copper alloy wire in Example 1-7

The slope of the tangent to the stress-strain curve at strain $\varepsilon = 0$ is the modulus of elasticity $E$ (see Fig. 1-42). Take the derivative of $\sigma(\varepsilon)$ to get the slope of the tangent to the $\sigma(\varepsilon)$ curve, and evaluate the derivative at strain $\varepsilon = 0$ to find $E$.

$$E(\varepsilon) = \frac{d}{d\varepsilon}\,\sigma(\varepsilon) \rightarrow \frac{17{,}500}{(240\,\varepsilon + 1)^2}$$

$$E = E(0) \quad E = 17{,}500 \text{ ksi}$$

Next, find an expression for the yield strain $\varepsilon_y$, the point at which the 0.2% offset line crosses the stress-strain curve (see Fig. 1-42). Substitute the expression for $\varepsilon_y$ into the $\sigma(\varepsilon)$ expression and then solve for yield stress $\sigma(\varepsilon_y) = \sigma_y$:

$$\varepsilon_y = 0.002 + \frac{\sigma_y}{E} \quad \text{and} \quad \sigma(\varepsilon_y) = \sigma_y \quad \text{or} \quad \sigma_y = \frac{17{,}500\,\varepsilon_y}{1 + 240\,\varepsilon_y}$$

Rearranging the equation in terms of $\sigma_y$ gives

$$\sigma_y^2 + \left(\frac{37E}{6000} - \frac{875}{12}\right)\sigma_y - \frac{7E}{48} = 0$$

Solving this quadratic equation for the 0.2% offset yield stress $\sigma_y$ gives $\sigma_y = 36$ ksi.

Continues

● ● ●   **Example 1-7** - *Continued*

The yield strain can be computed as

$$\varepsilon_y = 0.002 + \frac{\sigma_y}{E(\text{ksi})} = 4.057 \times 10^{-3}$$

**Fig. 1-42**

Modulus of elasticity $E$, 0.2% offset line and yield stress $\sigma_y$ and strain $\varepsilon_y$ for copper alloy wire in Example 1-7

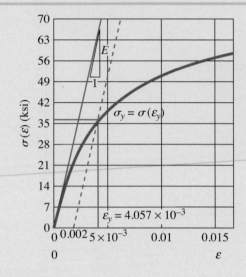

**(b) Use statics to find the tensile force $T$ (lb) in the wire; recall that bar weight $W = 985$ lb.**

Find the angle between the $x$ axis and cable attachment position at $C$:

$$\alpha_C = \text{atan}\left(\frac{1.5}{4}\right) = 20.556°$$

Sum the moments about $A$ to obtain one equation and one unknown. The reaction $B_x$ acts to the left:

$$B_x = \frac{-W(2 \text{ ft})}{3 \text{ ft}} = -656.667 \text{ lb}$$

Next, sum the forces in the $x$ direction to find the cable force $T_C$:

$$T_C = \frac{-B_x}{\cos(\alpha_C)} \quad T_C = 701 \text{ lb}$$

**(c) Find the normal axial strain $\varepsilon$ and elongation $\delta$ (in.) of the wire.**

Compute the normal stress then find the associated strain from stress-strain plot (or from the $\sigma(\varepsilon)$ equation). The wire elongation is strain times wire length.

The wire diameter, cross-sectional area, and length are

$$d = \frac{1}{8} \text{ in.} \quad A = \frac{\pi}{4}d^2 = 0.0123 \text{ in.}^2$$

$$L_C = \sqrt{(4 \text{ ft})^2 + (1.5 \text{ ft})^2} = 4.272 \text{ ft}$$

We can now compute the stress and strain in the wire and the elongation of the wire.

$$\sigma_C = \frac{T_C}{A} = 57.1 \text{ ksi}$$

Note that the stress in the wire exceeds the 0.2% offset yield stress of 36 ksi. The corresponding normal strain can be found from the $\sigma(\varepsilon)$ plot or by rearranging the $\sigma(\varepsilon)$ equation to give

$$\varepsilon(\sigma) = \frac{\sigma}{17,500 - 240\,\sigma}$$

Then,    $\varepsilon(\sigma_C) = \varepsilon_C,$    or    $\varepsilon_C = \dfrac{\sigma_C}{17,500 \text{ ksi} - 240\sigma_C} = 0.015$   ⬅

Finally, the wire elongation is

$$\delta_C = \varepsilon_C L_C = 0.774 \text{ in.}$$   ⬅

**(d) Find the permanent set of the wire if all forces are removed.**

If the load is removed from the wire, the stress in the wire will return to zero following unloading line BC in Fig. 1-43 (see also Fig. 1-36b). The **elastic recovery strain** can be computed as

$$\varepsilon_{er} = \frac{\sigma_C}{E} = 3.266 \times 10^{-3}$$

Hence, the **residual strain** is the difference between the total strain ($\varepsilon_C$) and the elastic recovery strain ($\varepsilon_{er}$) as

$$\varepsilon_{res} = \varepsilon_C - \varepsilon_{er} = 0.012$$

Finally, the **permanent set** of the wire is the product of the residual strain and the length of the wire:

$$P_{set} = \varepsilon_{res} L_C = 0.607 \text{ in.}$$   ⬅

**Fig. 1-43**

Residual strain ($\varepsilon_{res}$) and elastic recovery strain ($\varepsilon_{er}$) for copper alloy wire in Example 1-7

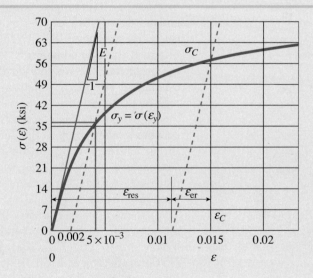

# 1.6 LINEAR ELASTICITY, HOOKE'S LAW, AND POISSON'S RATIO

Many structural materials, including most metals, wood, plastics, and ceramics, behave both elastically and linearly when first loaded. Consequently, their stress-strain curves begin with a straight line passing through the origin. An example is the stress-strain curve for structural steel (Fig. 1-28), where the region from the origin $O$ to the proportional limit (point $A$) is both linear and elastic. Other examples are the regions below *both* the proportional limits and the elastic limits on the diagrams for aluminum (Fig. 1-31), brittle materials (Fig. 1-34), and copper (Fig. 1-35).

When a material behaves elastically and also exhibits a linear relationship between stress and strain, it is said to be **linearly elastic**. This type of behavior is extremely important in engineering for an obvious reason—by designing structures and machines to function in this region, we avoid permanent deformations due to yielding.

## Hooke's Law

The linear relationship between stress and strain for a bar in simple tension or compression is expressed by the equation

$$\sigma = E\varepsilon \qquad\qquad \textbf{(1-15)}$$

in which $\sigma$ is the axial stress, $\varepsilon$ is the axial strain, and $E$ is a constant of proportionality known as the **modulus of elasticity** for the material. The modulus of elasticity is the slope of the stress-strain diagram in the linearly elastic region, as mentioned previously in Section 1.4. Since strain is dimensionless, the units of $E$ are the same as the units of stress. Typical units of $E$ are psi or ksi in USCS units and pascals (or multiples thereof) in SI units.

The equation $\sigma = E\varepsilon$ is commonly known as **Hooke's law**, named for the famous English scientist Robert Hooke (1635–1703). Hooke was the first person to investigate scientifically the elastic properties of materials, and he tested such diverse materials as metal, wood, stone, bone, and sinew. He measured the stretching of long wires supporting weights and observed that the elongations "always bear the same proportions one to the other that the weights do that made them" (Ref. 1-6). Thus, Hooke established the linear relationship between the applied loads and the resulting elongations.

Equation (1-15) is actually a very limited version of Hooke's law because it relates only to the longitudinal stresses and strains developed in simple tension or compression of a bar (*uniaxial stress*). To deal with more complicated states of stress, such as those found in most structures and machines, we must use more extensive equations of Hooke's law (see Sections 7.5 and 7.6).

The modulus of elasticity has relatively large values for materials that are very stiff, such as structural metals. Steel has a modulus of approximately 30,000 ksi (210 GPa); for aluminum, values around 10,600 ksi (73 GPa) are typical. More flexible materials have a lower modulus—values for plastics range from 100 to 2000 ksi (0.7 to 14 GPa). Some representative values of $E$ are listed in Table I-2, Appendix I. For most materials, the value of $E$ in compression is nearly the same as in tension.

Modulus of elasticity is often called **Young's modulus**, after another English scientist, Thomas Young (1773–1829). In connection with an investigation of tension and compression of prismatic bars, Young introduced the idea of a "modulus of the elasticity." However, his modulus was not the same as the one in use today, because it involved properties of the bar as well as of the material (Ref. 1-7).

## Poisson's Ratio

When a prismatic bar is loaded in tension, the axial elongation is accompanied by **lateral contraction** (that is, contraction normal to the direction of the applied load). This change in shape is pictured in Fig. 1-44, where part (a) shows the bar before loading and part (b) shows it after loading. In part (b), the dashed lines represent the shape of the bar prior to loading.

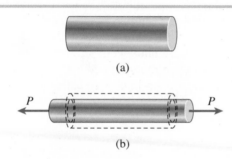

(a)

(b)

**Fig. 1-44**

Axial elongation and lateral contraction of a prismatic bar in tension: (a) bar before loading, and (b) bar after loading. (The deformations of the bar are highly exaggerated.)

Lateral contraction is easily seen by stretching a rubber band, but in metals the changes in lateral dimensions (in the linearly elastic region) are usually too small to be visible. However, they can be detected with sensitive measuring devices.

The **lateral strain** $\varepsilon'$ at any point in a bar is proportional to the axial strain $\varepsilon$ at that same point if the material is linearly elastic. The ratio of these strains is a property of the material known as **Poisson's ratio**. This dimensionless ratio, usually denoted by the Greek letter $v$ (nu), can be expressed by the equation

$$v = -\frac{\text{lateral strain}}{\text{axial strain}} = -\frac{\varepsilon'}{\varepsilon} \qquad \textbf{(1-16)}$$

The minus sign is inserted in the equation to compensate for the fact that the lateral and axial strains normally have opposite signs. For instance, the axial strain in a bar in tension is positive and the lateral strain is negative (because the width of the bar decreases). For compression we have the opposite situation, with the bar becoming shorter (negative axial strain) and wider (positive lateral strain). Therefore, for ordinary materials Poisson's ratio will have a positive value.

When Poisson's ratio for a material is known, we can obtain the lateral strain from the axial strain as follows:

$$\varepsilon' = -v\varepsilon \qquad \textbf{(1-17)}$$

When using Eqs. (1-16) and (1-17), we must always keep in mind that they apply only to a bar in uniaxial stress, that is, a bar for which the only stress is the normal stress $\sigma$ in the axial direction.

Poisson's ratio is named for the famous French mathematician Siméon Denis Poisson (1781–1840), who attempted to calculate this ratio by a molecular theory of materials (Ref. 1-8). For isotropic materials, Poisson found $v = 1/4$. More recent calculations based upon better models of atomic structure give $v = 1/3$. Both of these values are close to actual measured values, which are in the range 0.25 to 0.35 for most metals and many other materials. Materials with an extremely low value of Poisson's ratio include cork, for which $v$ is practically zero, and concrete, for which $v$ is about 0.1 or 0.2. A theoretical upper limit for Poisson's ratio is 0.5, as explained later in Section 7.5. Rubber comes close to this limiting value.

A table of Poisson's ratios for various materials in the linearly elastic range is given in Appendix I (see Table I-2). For most purposes, Poisson's ratio is assumed to be the same in both tension and compression.

When the strains in a material become large, Poisson's ratio changes. For instance, in the case of structural steel the ratio becomes almost 0.5 when plastic yielding occurs. Thus, Poisson's ratio remains constant only in the linearly elastic range. When the material behavior is nonlinear, the ratio of lateral strain to axial strain is often called the *contraction ratio*. Of course, in the special case of linearly elastic behavior, the contraction ratio is the same as Poisson's ratio.

## Limitations

For a particular material, Poisson's ratio remains constant throughout the linearly elastic range, as explained previously. Therefore, at any given point in the prismatic bar of Fig. 1-44, the lateral strain remains proportional to the axial strain as the load increases or decreases. However, for a given value of the load (which means that the axial strain is constant throughout the bar), additional conditions must be met if the lateral strains are to be the same throughout the entire bar.

First, the material must be **homogeneous**, that is, it must have the same composition (and hence the same elastic properties) at every point. However, having a homogeneous material does not mean that the elastic properties at a particular point are the same in all *directions*. For instance, the modulus of elasticity could be different in the axial and lateral directions, as in the case of a wood pole. Therefore, a second condition for uniformity in the lateral strains is that the elastic properties must be the same in all directions *perpendicular* to the longitudinal axis. When the preceding conditions are met, as is often the case with metals, the lateral strains in a prismatic bar subjected to uniform tension will be the same at every point in the bar and the same in all lateral directions.

Materials having the same properties in all directions (whether axial, lateral, or any other direction) are said to be **isotropic**. If the properties differ in various directions, the material is **anisotropic** (or **aeolotropic**).

In this book, all examples and problems are solved with the assumption that the material is linearly elastic, homogeneous, and isotropic, unless a specific statement is made to the contrary.

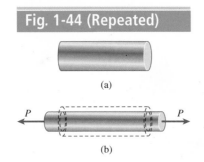

**Fig. 1-44 (Repeated)**

(a)

$P$ ←                    → $P$

(b)

### Fig. 1-45

Example 1-8: Plastic pipe compressed inside cast iron pipe

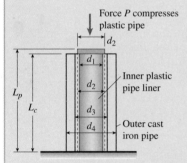

A hollow plastic circular pipe (length $L_p$, inner and outer diameters $d_1$, $d_2$, respectively; see Fig. 1-45) is inserted as a liner inside a cast iron pipe (length $L_c$, inner and outer diameters $d_3$, $d_4$, respectively).

(a) Derive a formula for the required initial length $L_p$ of the plastic pipe so that when it is compressed by some force $P$, the final length of both pipes is the same and also, at the same time, the final outer diameter of the plastic pipe is equal to the inner diameter of the cast iron pipe.

(b) Using the numerical data given below, find initial length $L_p$ (m) and final thickness $t_p$ (mm) for the plastic pipe.

(c) What is the required compressive force $P$ (N)? What are the final normal stresses (MPa) in both pipes?

(d) Compare initial and final volumes (mm³) for the plastic pipe.

Numerical data and pipe cross-section properties:

$$L_c = 0.25 \text{ m} \quad E_c = 170 \text{ GPa} \quad E_p = 2.1 \text{ GPa} \quad v_c = 0.3 \quad v_p = 0.4$$

$$d_1 = 109.8 \text{ mm} \quad d_2 = 110 \text{ mm} \quad d_3 = 110.2 \text{ mm}$$

$$d_4 = 115 \text{ mm} \quad t_p = \frac{d_2 - d_1}{2} = 0.1 \text{ mm}$$

The initial cross-sectional areas of the plastic and cast iron pipes are

$$A_p = \frac{\pi}{4}(d_2^2 - d_1^2) = 34.526 \text{ mm}^2 \quad A_c = \frac{\pi}{4} \cdot (d_4^2 - d_3^2) = 848.984 \text{ mm}^2$$

### (a) Derive a formula for the required initial length $L_p$ of the plastic pipe.

The lateral strain resulting from compression of the plastic pipe must close the gap $(d_3 - d_2)$ between the plastic pipe and the inner surface of the cast iron pipe. The required *extensional* lateral strain is positive (here, $\varepsilon_{\text{lat}} = \varepsilon'$):

$$\varepsilon_{\text{lat}} = \frac{d_3 - d_2}{d_2} = 1.818 \times 10^{-3}$$

The accompanying *compressive* normal strain in the plastic pipe is obtained using Eq. 1-17, which requires Poisson's ratio for the plastic pipe and also the required lateral strain:

$$\varepsilon_p = \frac{-\varepsilon_{\text{lat}}}{v_p} \text{ or } \varepsilon_p = \frac{-1}{v_p}\left(\frac{d_3 - d_2}{d_2}\right) = -4.545 \times 10^{-3}$$

We can now use the compressive normal strain $\varepsilon_p$ to compute the *shortening* $\delta_{p1}$ of the plastic pipe as

$$\delta_{p1} = \varepsilon_p L_p$$

At the same time, the required *shortening* of the plastic pipe (so that it will have the same final length as that of the cast iron pipe) is

$$\delta_{p2} = -(L_p - L_c)$$

*Continues* ➡

Now, equating $\delta_{p1}$ and $\delta_{p2}$ leads to a formula for the required initial length $L_p$ of the plastic pipe:

$$L_p = \frac{L_c}{1 + \varepsilon_p} \quad \text{or} \quad L_p = \frac{L_c}{1 - \dfrac{d_3 - d_2}{v_p d_2}}$$

**(b)** **Now substitute the numerical data to find the initial length $L_p$, change in thickness $\Delta t_p$, and final thickness $t_{pf}$ for the plastic pipe.**

As expected, $L_p$ is greater than the length of the cast iron pipe, $L_c = 0.25$ m, and the thickness of the compressed plastic pipe increases by $\Delta t_p$:

$$L_p = \frac{L_c}{1 - \left(\dfrac{d_3 - d_2}{v_p d_2}\right)} = 0.25114 \text{ m}$$

$$\Delta t_p = \varepsilon_{\text{lat}} t_p = 1.818 \times 10^{-4} \text{ mm} \quad \text{so} \quad t_{pf} = t_p + \Delta t_p = 0.10018 \text{ mm}$$

**(c)** **Next find the required compressive force $P$ and the final normal stresses in both pipes.**

A check on the normal compressive stress in the plastic pipe, computed using Hooke's Law (Eq. 1-15) shows that it is well below the ultimate stress for selected plastics (see Table I-3, Appendix I); this is also the final normal stress in the plastic pipe:

$$\sigma_p = E_p \varepsilon_p = -9.55 \text{ MPa}$$

The required downward force to compress the plastic pipe is

$$P_{\text{reqd}} = \sigma_p A_p = -330 \text{ N}$$

Both the initial and final stresses in the cast iron pipe are zero because no force is applied to the cast iron pipe.

**(d)** **Lastly, compare the initial and final volumes of the plastic pipe.**

The initial cross-sectional area of the plastic pipe is

$$A_p = 34.526 \text{ mm}^2$$

The final cross-sectional area of the plastic pipe is

$$A_{pf} = \frac{\pi}{4} \cdot [d_3^2 - (d_3 - 2t_{pf})^2] = 34.652 \text{ mm}^2$$

The initial volume of the plastic pipe is

$$V_{\text{pinit}} = L_p A_p = 8671 \text{ mm}^3$$

and the final volume of the plastic pipe is

$$V_{\text{pfinal}} = L_c A_{pf} \quad \text{or} \quad V_{\text{pfinal}} = 8663 \text{ mm}^3$$

The ratio of final to initial volume reveals little change:

$$\frac{V_{\text{pfinal}}}{V_{\text{pinit}}} = 0.99908$$

*Note:* The numerical results obtained in this example illustrate that the dimensional changes in structural materials under normal loading conditions are extremely small. In spite of their smallness, changes in dimensions can be important in certain kinds of analysis (such as the analysis of statically indeterminate structures) and in the experimental determination of stresses and strains.

# 1.7 SHEAR STRESS AND STRAIN

In the preceding sections we discussed the effects of normal stresses produced by axial loads acting on straight bars. These stresses are called "normal stresses" because they act in directions *perpendicular* to the surface of the material. Now we will consider another kind of stress, called a **shear stress**, that acts *tangential* to the surface of the material.

As an illustration of the action of shear stresses, consider the bolted connection shown in Fig. 1-46a. This connection consists of a flat bar $A$, a clevis $C$, and a bolt $B$ that passes through holes in the bar and clevis. Under the action of the tensile loads $P$, the bar and clevis will press against the bolt in **bearing**, and contact stresses, called **bearing stresses**, will be developed. In addition, the bar and clevis tend to *shear* the bolt, that is, cut through it, and this tendency is resisted by shear stresses in the bolt. As an example, consider the bracing for an elevated pedestrian walkway shown in the photograph.

Diagonal bracing for an elevated walkway showing a clevis and a pin in double shear (© Barry Goodno)

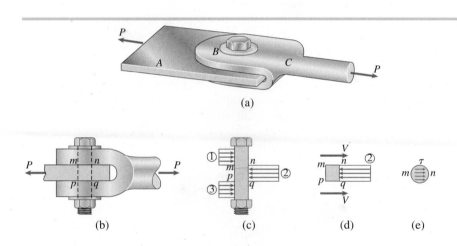

(a)

(b)        (c)        (d)        (e)

**Fig. 1-46**

Bolted connection in which the bolt is loaded in double shear

To show more clearly the actions of the bearing and shear stresses, let us look at this type of connection in a schematic side view (Fig. 1-46b). With this view in mind, we draw a free-body diagram of the bolt (Fig. 1-46c). The bearing stresses exerted by the clevis against the bolt appear on the left-hand side of the free-body diagram and are labeled 1 and 3. The stresses from the bar appear on the right-hand side and are labeled 2. The actual distribution of the bearing stresses is difficult to determine, so it is customary to assume that the stresses are uniformly distributed. Based upon the assumption of uniform distribution, we can calculate an **average bearing** stress $\sigma_b$ by dividing the total bearing force $F_b$ by the bearing area $A_b$:

$$\sigma_b = \frac{F_b}{A_b} \quad = \quad \frac{P}{B \times L} \qquad\qquad \textbf{(1-18)}$$

The **bearing area** is defined as the projected area of the curved bearing surface. For instance, consider the bearing stresses labeled 1. The projected area $A_b$ on which they act is a rectangle having a height equal to the thickness of the clevis and a width equal to the diameter of the bolt. Also, the bearing force $F_b$ represented by the stresses labeled 1 is equal to $P/2$. The same area and the same force apply to the stresses labeled 3.

Now consider the bearing stresses between the flat bar and the bolt (the stresses labeled 2). For these stresses, the bearing area $A_b$ is a rectangle with height equal to the thickness of the flat bar and width equal to the bolt diameter. The corresponding bearing force $F_b$ is equal to the load $P$.

The free-body diagram of Fig. 1-46c shows that there is a tendency to shear the bolt along cross sections $mn$ and $pq$. From a free-body diagram of the portion $mnpq$ of the bolt (see Fig. 1-46d), we see that shear forces $V$ act over the cut surfaces of the bolt. In this particular example there are two planes of shear ($mn$ and $pq$), and so the bolt is said to be in **double shear**. In double shear, each of the shear forces is equal to one-half of the total load transmitted by the bolt, that is, $V = P/2$.

The shear forces $V$ are the resultants of the shear stresses distributed over the cross-sectional area of the bolt. For instance, the shear stresses acting on cross section $mn$ are shown in Fig. 1-46e. These stresses act parallel to the cut surface. The exact distribution of the stresses is not known, but they are highest near the center and become zero at certain locations on the edges. As indicated in Fig. 1-46e, shear stresses are customarily denoted by the Greek letter $\tau$ (tau).

A bolted connection in single shear is shown in Fig. 1-47a, where the axial force $P$ in the metal bar is transmitted to the flange of the steel column through a bolt. A cross-sectional view of the column (Fig. 1-47b) shows the connection in more detail. Also, a sketch of the bolt (Fig. 1-47c) shows the assumed distribution of the bearing stresses acting on the bolt. As mentioned earlier, the actual distribution of these bearing stresses is much more complex than shown in the figure. Furthermore, bearing

**Fig. 1-47**

Bolted connection in which the bolt is loaded in single shear

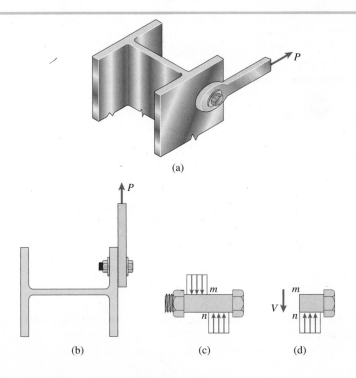

(a)

(b)        (c)        (d)

stresses are also developed against the inside surfaces of the bolt head and nut. Thus, Fig. 1-47c is *not* a free-body diagram—only the idealized bearing stresses acting on the shank of the bolt are shown in the figure.

By cutting through the bolt at section *mn* we obtain the diagram shown in Fig. 1-47d. This diagram includes the shear force *V* (equal to the load *P*) acting on the cross section of the bolt. As already pointed out, this shear force is the resultant of the shear stresses that act over the cross-sectional area of the bolt.

The deformation of a bolt loaded almost to fracture in single shear is shown in Fig. 1-48 (compare with Fig. 1-47c).

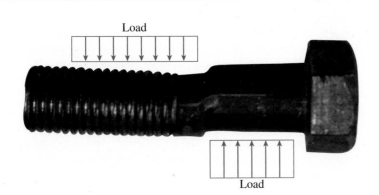

**Fig. 1-48**

Failure of a bolt in single shear (© Barry Goodno)

In the preceding discussions of bolted connections we disregarded **friction** (produced by tightening of the bolts) between the connecting elements. The presence of friction means that part of the load is carried by friction forces, thereby reducing the loads on the bolts. Since friction forces are unreliable and difficult to estimate, it is common practice to err on the conservative side and omit them from the calculations.

The **average shear stress** on the cross section of a bolt is obtained by dividing the total shear force *V* by the area *A* of the cross section on which it acts, as follows:

$$\tau_{\text{aver}} = \frac{V}{A} \qquad \textbf{(1-19)}$$

In the example of Fig. 1-47, which shows a bolt in single shear, the shear force *V* is equal to the load *P* and the area *A* is the cross-sectional area of the bolt. However, in the example of Fig. 1-46, where the bolt is in double shear, the shear force *V* equals *P*/2.

From Eq. (1-19) we see that shear stresses, like normal stresses, represent intensity of force, or force per unit of area. Thus, the **units of** shear stress are the same as those for normal stress, namely, psi or ksi in USCS units and pascals or multiples thereof in SI units.

The loading arrangements shown in Figs. 1-46 and 1-47 are examples of **direct shear** (or *simple shear*) in which the shear stresses are created by the direct action of the forces in trying to cut through the material. Direct shear arises in the design of bolts, pins, rivets, keys, welds, and glued joints.

### Fig. 1-49

Small element of material
subjected to shear stresses

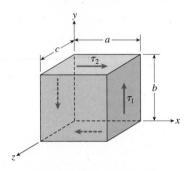

### Fig. 1-50

Element of material subjected to
shear stresses and strains

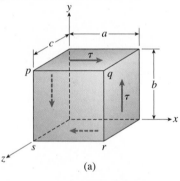

(a)

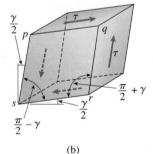

(b)

Shear stresses also arise in an indirect manner when members are subjected to tension, torsion, and bending, as discussed later in Sections 2.6, 3.3, and 5.8, respectively.

## Equality of Shear Stresses on Perpendicular Planes

To obtain a more complete picture of the action of shear stresses, let us consider a small element of material in the form of a rectangular parallelepiped having sides of lengths $a$, $b$, and $c$ in the $x$, $y$, and $z$ directions, respectively (Fig. 1-49).* The front and rear faces of this element are free of stress.

Now assume that a shear stress $\tau_1$ is distributed uniformly over the right-hand face, which has area $bc$. In order for the element to be in equilibrium in the $y$ direction, the total shear force $\tau_1 bc$ acting on the right-hand face must be balanced by an equal but oppositely directed shear force on the left-hand face. Since the areas of these two faces are equal, it follows that the shear stresses on the two faces must be equal.

The forces $\tau_1 bc$ acting on the left- and right-hand side faces (Fig. 1-49) form a couple having a moment about the $z$ axis of magnitude $\tau_1 abc$, acting counterclockwise in the figure.** Equilibrium of the element requires that this moment be balanced by an equal and opposite moment resulting from shear stresses acting on the top and bottom faces of the element. Denoting the stresses on the top and bottom faces as $\tau_2$, we see that the corresponding horizontal shear forces equal $\tau_2 ac$. These forces form a clockwise couple of moment $\tau_2 abc$. From moment equilibrium of the element about the $z$ axis, we see that $\tau_1 abc$ equals $\tau_2 abc$, or

$$\tau_1 = \tau_2 \qquad \textbf{(1-20)}$$

Therefore, the magnitudes of the four shear stresses acting on the element are equal, as shown in Fig. 1-50a.

In summary, we have arrived at the following general observations regarding shear stresses acting on a rectangular element:

1. Shear stresses on opposite (and parallel) faces of an element are equal in magnitude and opposite in direction.

2. Shear stresses on adjacent (and perpendicular) faces of an element are equal in magnitude and have directions such that both stresses point toward, or both point away from, the line of intersection of the faces.

These observations were obtained for an element subjected only to shear stresses (no normal stresses), as pictured in Figs. 1-49 and 1-50. This state of stress is called **pure shear** and is discussed later in greater detail (Section 3.5).

For most purposes, the preceding conclusions remain valid even when normal stresses act on the faces of the element. The reason is that the normal

---

*A **parallelepiped** is a prism whose bases are parallelograms; thus, a parallelepiped has six faces, each of which is a parallelogram. Opposite faces are parallel and identical parallelograms. A **rectangular parallelepiped** has all faces in the form of rectangles.

**A **couple** consists of two parallel forces that are equal in magnitude and opposite in direction.

stresses on opposite faces of a small element usually are equal in magnitude and opposite in direction; hence they do not alter the equilibrium equations used in reaching the preceding conclusions.

## Shear Strain

Shear stresses acting on an element of material (Fig. 1-50a) are accompanied by *shear strains*. As an aid in visualizing these strains, we note that the shear stresses have no tendency to elongate or shorten the element in the $x$, $y$, and $z$ directions—in other words, the lengths of the sides of the element do not change. Instead, the shear stresses produce a change in the *shape* of the element (Fig. 1-50b). The original element, which is a rectangular parallelepiped, is deformed into an oblique parallelepiped, and the front and rear faces become rhomboids.*

Because of this deformation, the angles between the side faces change. For instance, the angles at points $q$ and $s$, which were $\pi/2$ before deformation, are reduced by a small angle $\gamma$ to $\pi/2 - \gamma$ (Fig. 1-50b). At the same time, the angles at points $p$ and $r$ are increased to $\pi/2 + \gamma$. The angle $\gamma$ is a measure of the **distortion**, or change in shape, of the element and is called the **shear strain**. Because shear strain is an angle, it is usually measured in degrees or radians.

## Sign Conventions for Shear Stresses and Strains

As an aid in establishing sign conventions for shear stresses and strains, we need a scheme for identifying the various faces of a stress element (Fig. 1-50a). Henceforth, we will refer to the faces oriented toward the positive directions of the axes as the positive faces of the element. In other words, a positive face has its outward normal directed in the positive direction of a coordinate axis. The opposite faces are negative faces. Thus, in Fig. 1-50a, the right-hand, top, and front faces are the positive $x$, $y$, and $z$ faces, respectively, and the opposite faces are the negative $x$, $y$, and $z$ faces.

Using the terminology described in the preceding paragraph, we may state the sign convention for shear stresses in the following manner:

> *A shear stress acting on a positive face of an element is positive if it acts in the positive direction of one of the coordinate axes and negative if it acts in the negative direction of an axis. A shear stress acting on a negative face of an element is positive if it acts in the negative direction of an axis and negative if it acts in a positive direction.*

Thus, all shear stresses shown in Fig. 1-50a are positive.

The sign convention for shear strains is as follows:

> *Shear strain in an element is positive when the angle between two positive faces (or two negative faces) is reduced. The strain is negative when the angle between two positive (or two negative) faces is increased.*

---

*An **oblique angle** can be either acute or obtuse, but it is *not* a right angle. A **rhomboid** is a parallelogram with oblique angles and adjacent sides *not* equal. (A *rhombus* is a parallelogram with oblique angles and all four sides equal, sometimes called a *diamond-shaped figure*.)

Thus, the strains shown in Fig. 1-50b are positive, and we see that positive shear stresses are accompanied by positive shear strains.

## Hooke's Law in Shear

The properties of a material in shear can be determined experimentally from direct-shear tests or from torsion tests. The latter tests are performed by twisting hollow, circular tubes, thereby producing a state of pure shear, as explained later in Section 3.5. From the results of these tests, we can plot **shear stress-strain diagrams** (that is, diagrams of shear stress $\tau$ versus shear strain $\gamma$). These diagrams are similar in shape to tension-test diagrams ($\sigma$ versus $\varepsilon$) for the same materials, although they differ in magnitudes.

From shear stress-strain diagrams, we can obtain material properties such as the proportional limit, modulus of elasticity, yield stress, and ultimate stress. These properties in shear are usually about half as large as those in tension. For instance, the yield stress for structural steel in shear is 0.5 to 0.6 times the yield stress in tension.

For many materials, the initial part of the shear stress-strain diagram is a straight line through the origin, just as it is in tension. For this linearly elastic region, the shear stress and shear strain are proportional, and therefore we have the following equation for **Hooke's law in shear**:

$$\tau = G\gamma \tag{1-21}$$

in which $G$ is the **shear modulus of elasticity** (also called the *modulus of rigidity*).

The shear modulus $G$ has the same units as the tension modulus $E$, namely, psi or ksi in USCS units and pascals (or multiples thereof) in SI units. For mild steel, typical values of $G$ are 11,000 ksi or 75 GPa; for aluminum alloys, typical values are 4000 ksi or 28 GPa. Additional values are listed in Table I-2, Appendix I.

The moduli of elasticity in tension and shear are related by the following equation:

$$G = \frac{E}{2(1 + \nu)} \tag{1-22}$$

in which $\nu$ is Poisson's ratio. This relationship, which is derived later in Section 3.6, shows that $E$, $G$, and $\nu$ are not independent elastic properties of the material. Because the value of Poisson's ratio for ordinary materials is between zero and one-half, we see from Eq. (1-22) that $G$ must be from one-third to one-half of $E$.

The following examples illustrate some typical analyses involving the effects of shear. Example 1-5 is concerned with shear stresses in a plate, Example 1-6 deals with bearing and shear stresses in pins and bolts, and Example 1-7 involves finding shear stresses and shear strains in an elastomeric bearing pad subjected to a horizontal shear force.

● ● ● **Example 1-9**

A punch for making holes in steel plates is shown in Fig. 1-51a. Assume that a punch having diameter $d = 3/4$ in. is used to punch a hole in an 3/10 in. plate as shown in the cross-sectional view (Fig. 1-51b).

If a force $P = 24$ kips is required to create the hole, what is the average shear stress in the plate and the average compressive stress in the punch?

**Fig. 1-51**

Example 1-9: Punching a hole in a steel plate

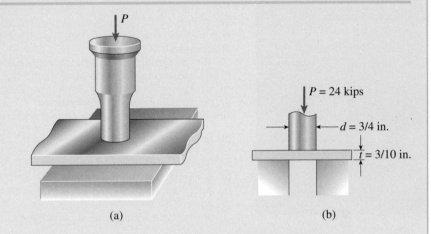

(a)    (b)

**Solution**

The average shear stress in the plate is obtained by dividing the force $P$ by the shear area of the plate. The shear area $A_s$ is equal to the circumference of the hole times the thickness of the plate, or

$$A_s = \pi dt = \pi(3/4 \text{ in.})(3/10 \text{ in.}) = 0.707 \text{ in.}^2$$

in which $d$ is the diameter of the punch and $t$ is the thickness of the plate. Therefore, the average shear stress in the plate is

$$\tau_{aver} = \frac{P}{A_s} = 24 \text{ kips}/0.707 \text{ in.}^2 = 34 \text{ ksi} \quad \Leftarrow$$

The average compressive stress in the punch is

$$\sigma_c = \frac{P}{A_{punch}} = \frac{P}{\pi d^2/4} = 24 \text{ kips}/\pi(0.75 \text{ in.})^2/4 = 54.3 \text{ ksi} \quad \Leftarrow$$

in which $A_{punch}$ is the cross-sectional area of the punch.

*Note:* This analysis is highly idealized because we are disregarding impact effects that occur when a punch is rammed through a plate. (The inclusion of such effects requires advanced methods of analysis that are beyond the scope of mechanics of materials.)

## Example 1-10

### Fig. 1-52a

Example 1-10: Semi-circular plate connection for steel staircase

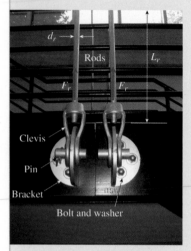

Components of eccentric hanger rod connection (from Example 1-6) (© Barry Goodno)

### Fig. 1-52b

Side view of hanger rod and bracket for eccentric connection (© Barry Goodno)

A semicircular bracket assembly is used to support a steel staircase in an office building. One design under consideration, referred to as the *eccentric design*, uses two separate L-shaped brackets, each having a clevis attached to a steel hanger rod to support the staircase (see Fig. 1-52). Photos of the bracket attachment and hanger rod in the eccentric design are shown below (see also **Example 1-6**).

The staircase designer would like to also consider a *symmetric bracket design*. The symmetric design uses a single hanger rod attached by a clevis and pin to a T-shaped bracket in which the two separate L-brackets are attached along a vertical axis (see Figs. 1-52c and d). In this design, the eccentric moment of the rod force about the $z$ axis is eliminated.

In the *symmetric design*, the weight of the staircase and the connection itself, and any building occupants who are using the staircase, is estimated to result in a force of $F_r = 9600$ N in the single hanger rod. Use numerical values for dimensions of connection components given below. Determine the following stresses in the symmetric connection.

(a) Average in-plane shear stress in bolts 1 to 6.
(b) Bearing stress between the clevis pin and the bracket.
(c) Bearing stress between the clevis and the pin.
(d) Bolt forces in the $z$ direction at bolts 1 and 4 due to moment about the $x$ axis, and resulting normal stress in bolts 1 and 4.
(e) Bearing stress between the bracket and washer at bolts 1 and 4.
(f) Shear stress through the bracket at bolts 1 and 4.

### Fig. 1-52c

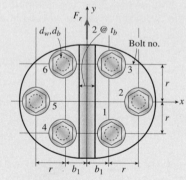

Bolts 1 through 6 in symmetric bracket design

### Fig. 1-52d

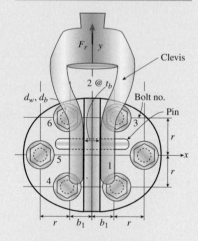

Clevis and pin in symmetric bracket design

Numerical data:

$$b_1 = 40 \text{ mm} \quad r = 50 \text{ mm} \quad t_b = 12 \text{ mm} \quad d_w = 40 \text{ mm}$$

$$d_b = 18 \text{ mm} \quad d_{pin} = 38 \text{ mm} \quad t_c = 14 \text{ mm}$$

$$e_z = 150 \text{ mm} \quad d_r = 40 \text{ mm} \quad F_r = 9600 \text{ N}$$

**Solution**

**(a) The normal tensile stress in the hanger rod** is computed as the force in the rod ($F_r$) divided by the cross-sectional area of the rod:

$$\sigma_{rod} = \frac{F_r}{A_r} = \frac{9600 \, \text{N}}{\dfrac{\pi}{4}(40 \, \text{mm})^2} = 7.64 \, \text{MPa}$$

**(b) The average in-plane shear stress in bolts 1 through 6** is equal to the force in the rod divided by the sum of the cross-sectional areas of the six bolts. This is based on the assumption that each bolt carries the same fraction of the overall rod force (see Fig. 1-52e):

$$\tau_{bolt} = \frac{F_r}{6A_{bolt}} = \frac{9600 \, \text{N}}{6\left[\dfrac{\pi}{4}(18 \, \text{mm})^2\right]} = 6.29 \, \text{MPa}$$

**Fig. 1-52e**

In-plane shear force in each bolt

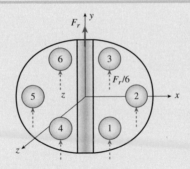

**(c)** The **bearing stress between the clevis pin and the bracket** is shown in Fig. 1-52f. The pin bears against the center portion of the bracket which has thickness of twice the bracket plate thickness ($t_b$) so the bearing stress is computed as follows:

$$\sigma_{b1} = \frac{F_r}{d_{pin} \cdot (2t_b)} = \frac{9600 \, \text{N}}{(38 \, \text{mm})(2 \times 12 \, \text{mm})} = 10.53 \, \text{MPa}$$

**Fig. 1-52f**

Bearing stresses on pin and bracket

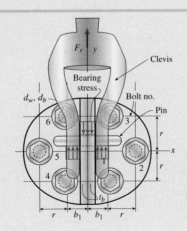

*Continues* ➡

**Example 1-10 -** *Continued*

**Fig. 1-52g**

Rod force is applied at distance $e_z$ from back plate of bracket.
(© Barry Goodno)

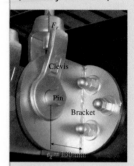

**(d)** The clevis bears against the pin in two locations (see Fig. 1-52f) so the **bearing stress between the clevis and the pin** is computed as the rod force divided by twice the clevis thickness ($t_c$) times the pin diameter:

$$\sigma_{b2} = \frac{F_r}{d_{pin} \cdot (2t_c)} = \frac{9600 \text{ N}}{(38 \text{ mm})(2)(14 \text{ mm})} = 9.02 \text{ MPa} \quad \Longleftarrow$$

**(e)** Although the connection bracket is symmetric with respect to the $yz$ plane, the rod force $F_r$ is applied at a distance $e_z = 150$ mm from the back plate (see Fig. 1-52g). This results in a moment about the $x$ axis $M_x = F_r \times e_z$ which can be converted into **two force couples**, each equal to $F_z \times 2r$, at bolt pairs 1–3 and 4–6. The resulting **tension force on bolts 1 and 4** is computed as

$$F_z = \frac{M_x}{4r} = \frac{F_r e_z}{4r} = \frac{9600 \text{ N}(150 \text{ mm})}{4(50 \text{ mm})} = 7.2 \text{ kN}$$

**Fig. 1-52h**

Moment $M_x$ can be converted into two force couples, each equal to $F_z \times 2r$, which act on bolt pairs 1-3 and 4-6.

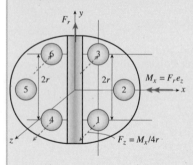

Here, we have assumed that the moment $F_r \times e_z$ acts about the $x$ axis, which eliminates bolts 2 and 5 because they lie along the $x$ axis (Fig. 1-52h). Using force $F_z$, we can now compute the normal stress in bolts 1 and 4 as

$$\sigma_1 = \sigma_4 = \frac{F_z}{\frac{\pi}{4}d_b^2} = \frac{7.2 \text{ kN}}{\frac{\pi}{4}(18 \text{ mm})^2} = 28.3 \text{ MPa} \quad \Longleftarrow$$

Note that the bolts are likely to be pretensioned so the computed stress $\sigma_1$ or $\sigma_4$ is in fact the stress increase at bolts 1 and 4, respectively, and the stress decrease at bolts 3 and 6, respectively, due to the moment $M_x$. Also, note that the symmetric bracket design eliminates the torsional moment about the $z$ axis ($M_z = F_r \times e_x$) resulting from the application of the rod force at a distance $e_x$ from the center of gravity of the bolt group in the eccentric bracket design.

**(f)** Now that we know the force $F_z$ acting on bolts 1 and 4, we can compute the **bearing stress between the bracket and washer** at bolts 1 and 4. The bearing area is the ring-shaped area of the washer so the bearing stress is

$$\sigma_{b3} = \frac{F_z}{\frac{\pi}{4}(d_w^2 - d_b^2)} = \frac{7.2 \text{ kN}}{\frac{\pi}{4}[(40 \text{ mm})^2 - (18 \text{ mm})^2]} = 7.18 \text{ MPa}$$

**(g)** Finally, the **shear stress through the bracket at bolts 1 and 4** is the force $F_z$ divided by the circumference of the washer times the thickness of the bracket:

$$\tau = \frac{F_z}{\pi d_w t_b} = \frac{7.2 \text{ kN}}{\pi(40 \text{ mm})(12 \text{ mm})} = 4.77 \text{ MPa} \quad \Longleftarrow$$

## ● ● ● Example 1-11

A bearing pad of the kind used to support machines and bridge girders consists of a linearly elastic material (usually an elastomer, such as rubber) capped by a steel plate (Fig. 1-53a). Assume that the thickness of the elastomer is $h$, the dimensions of the plate are $a \times b$, and the pad is subjected to a horizontal shear force $V$.

Obtain formulas for the average shear stress $\tau_{aver}$ in the elastomer and the horizontal displacement $d$ of the plate (Fig. 1-53b).

### Fig. 1-53

Example 1-11. Bearing pad in shear

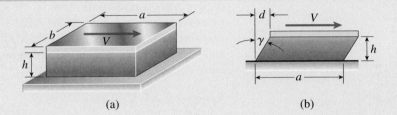

(a)                                        (b)

### Solution

Assume that the shear stresses in the elastomer are uniformly distributed throughout its entire volume. Then the shear stress on any horizontal plane through the elastomer equals the shear force $V$ divided by the area $ab$ of the plane (Fig. 1-53a):

$$\tau_{aver} = \frac{V}{ab} \quad \text{(1-23)}$$

The corresponding shear strain (from Hooke's law in shear; Eq. 1-21) is

$$\gamma = \frac{\tau_{aver}}{G_e} = \frac{V}{abG_e} \quad \text{(1-24)}$$

in which $G_e$ is the shear modulus of the elastomeric material. Finally, the horizontal displacement $d$ is equal to $h\tan\gamma$ (from Fig. 1-53b):

$$d = h\tan\gamma = h\tan\left(\frac{V}{abG_e}\right) \quad \text{(1-25)}$$

In most practical situations the shear strain $\gamma$ is a small angle, and in such cases we may replace $\tan\gamma$ by $\gamma$ and obtain

$$d = h\gamma = \frac{hV}{abG_e} \quad \text{(1-26)}$$

Equations (1-25) and (1-26) give approximate results for the horizontal displacement of the plate because they are based upon the assumption that the shear stress and strain are constant throughout the volume of the elastomeric material. In reality the shear stress is zero at the edges of the material (because there are no shear stresses on the free vertical faces), and therefore the deformation of the material is more complex than pictured in Fig. 1-53b. However, if the length $a$ of the plate is large compared with the thickness $h$ of the elastomer, the preceding results are satisfactory for design purposes.

# 1.8 ALLOWABLE STRESSES AND ALLOWABLE LOADS

Engineering has been aptly described as the *application of science to the common purposes of life*. In fulfilling that mission, engineers design a seemingly endless variety of objects to serve the basic needs of society. These needs include housing, agriculture, transportation, communication, and many other aspects of modern life. Factors to be considered in design include functionality, strength, appearance, economics, and environmental effects. However, when studying mechanics of materials, our principal design interest is **strength,** that is, *the capacity of the object to support or transmit loads*. Objects that must sustain loads include buildings, machines, containers, trucks, aircraft, ships, and the like. For simplicity, we will refer to all such objects as **structures**; thus, *a structure is any object that must support or transmit loads*.

## Factors of Safety

If structural failure is to be avoided, the loads that a structure is capable of supporting must be greater than the loads it will be subjected to when in service. Since *strength* is the ability of a structure to resist loads, the preceding criterion can be restated as follows: *The actual strength of a structure must exceed the required strength*. The ratio of the actual strength to the required strength is called the **factor of safety** $n$:

$$\text{Factor of safety} \; n = \frac{\text{Actual strength}}{\text{Required strength}} \qquad \textbf{(1-27)}$$

Of course, the factor of safety must be greater than 1.0 if failure is to be avoided. Depending upon the circumstances, factors of safety from slightly above 1.0 to as much as 10 are used.

The incorporation of factors of safety into design is not a simple matter, because both strength and failure have many different meanings. Strength may be measured by the load-carrying capacity of a structure, or it may be measured by the stress in the material. Failure may mean the fracture and complete collapse of a structure, or it may mean that the deformations have become so large that the structure can no longer perform its intended functions. The latter kind of failure may occur at loads much smaller than those that cause actual collapse.

The determination of a factor of safety must also take into account such matters as the following: probability of accidental overloading of the structure by loads that exceed the design loads; types of loads (static or dynamic); whether the loads are applied once or are repeated; how accurately the loads are known; possibilities for fatigue failure; inaccuracies in construction; variability in the quality of workmanship; variations in properties of materials; deterioration due to corrosion or other environmental effects; accuracy of the methods of analysis; whether failure is gradual (ample warning) or sudden (no warning); consequences of failure (minor damage or major catastrophe); and other such considerations. If the factor of safety is too low, the likelihood of failure will be high and the structure will be unacceptable; if the factor is too large, the structure will be wasteful of materials and perhaps unsuitable for its function (for instance, it may be too heavy).

Because of these complexities and uncertainties, factors of safety must be determined on a probabilistic basis. They usually are established by groups of experienced engineers who write the codes and specifications used by other designers, and sometimes they are even enacted into law. The provisions of codes and specifications are intended to provide reasonable levels of safety without unreasonable costs.

In aircraft design it is customary to speak of the **margin of safety** rather than the factor of safety. The margin of safety is defined as the factor of safety minus one:

$$\text{Margin of safety} = n - 1 \qquad \textbf{(1-28)}$$

Margin of safety is often expressed as a percent, in which case the value given above is multiplied by 100. Thus, a structure having an actual strength that is 1.75 times the required strength has a factor of safety of 1.75 and a margin of safety of 0.75 (or 75%). When the margin of safety is reduced to zero or less, the structure (presumably) will fail.

## Allowable Stresses

Factors of safety are defined and implemented in various ways. For many structures, it is important that the material remain within the linearly elastic range in order to avoid permanent deformations when the loads are removed. Under these conditions, the factor of safety is established with respect to yielding of the structure. Yielding begins when the yield stress is reached at *any* point within the structure. Therefore, by applying a factor of safety with respect to the yield stress (or yield strength), we obtain an **allowable stress** (or *working stress*) that must not be exceeded anywhere in the structure. Thus,

$$\text{Allowable stress} = \frac{\text{Yield strength}}{\text{Factor of safety}} \qquad \textbf{(1-29)}$$

or, for tension and shear, respectively,

$$\sigma_{\text{allow}} = \frac{\sigma_Y}{n_1} \quad \text{and} \quad \tau_{\text{allow}} = \frac{\tau_Y}{n_2} \qquad \textbf{(1-30a,b)}$$

in which $\sigma_Y$ and $\tau_Y$ are the yield stresses and $n_1$ and $n_2$ are the corresponding factors of safety. In building design, a typical factor of safety with respect to yielding in tension is 1.67; thus, a mild steel having a yield stress of 36 ksi has an allowable stress of 21.6 ksi.

Sometimes the factor of safety is applied to the **ultimate stress** instead of the yield stress. This method is suitable for brittle materials, such as concrete and some plastics, and for materials without a clearly defined yield stress, such as wood and high-strength steels. In these cases the allowable stresses in tension and shear are

$$\sigma_{\text{allow}} = \frac{\sigma_U}{n_3} \quad \text{and} \quad \tau_{\text{allow}} = \frac{\tau_U}{n_4} \qquad \textbf{(1-31a,b)}$$

in which $\sigma_U$ and $\tau_U$ are the ultimate stresses (or ultimate strengths). Factors of safety with respect to the ultimate strength of a material are usually larger than those based upon yield strength. In the case of mild steel, a factor of safety of 1.67 with respect to yielding corresponds to a factor of approximately 2.8 with respect to the ultimate strength.

## Allowable Loads

After the allowable stress has been established for a particular material and structure, the **allowable load** on that structure can be determined. The relationship between the allowable load and the allowable stress depends upon the type of structure. In this chapter we are concerned only with the most elementary kinds of structures, namely, bars in tension or compression and pins (or bolts) in direct shear and bearing.

In these kinds of structures the stresses are uniformly distributed (or at least *assumed* to be uniformly distributed) over an area. For instance, in the case of a bar in tension, the stress is uniformly distributed over the cross-sectional area provided the resultant axial force acts through the centroid of the cross section. The same is true of a bar in compression provided the bar is not subject to buckling. In the case of a pin subjected to shear, we consider only the average shear stress on the cross section, which is equivalent to assuming that the shear stress is uniformly distributed. Similarly, we consider only an average value of the bearing stress acting on the projected area of the pin.

Therefore, in all four of the preceding cases the **allowable load** (also called the *permissible load* or the *safe load*) is equal to the allowable stress times the area over which it acts:

$$\text{Allowable load} = (\text{Allowable stress})(\text{Area}) \qquad \textbf{(1-32)}$$

For bars in direct *tension* and *compression* (no buckling), this equation becomes

$$P_{\text{allow}} = \sigma_{\text{allow}} A \qquad \textbf{(1-33)}$$

in which $\sigma_{\text{allow}}$ is the permissible normal stress and $A$ is the cross-sectional area of the bar. If **the** bar has a hole through it, the *net area* is normally used when the bar is in tension. The **net area** is the gross cross-sectional area minus the area removed by the hole. For compression, the gross area may be used if the hole is filled by a bolt or pin that can transmit the compressive stresses.

For pins in *direct shear*, Eq. (1-32) becomes

$$P_{\text{allow}} = \tau_{\text{allow}} A \qquad \textbf{(1-34)}$$

in which $\tau_{\text{allow}}$ is the permissible shear stress and $A$ is the area over which the shear stresses act. If the pin is in single shear, the area is the cross-sectional area of the pin; in double shear, it is twice the cross-sectional area.

Finally, the permissible load based upon *bearing* is

$$P_{\text{allow}} = \sigma_b A_b \qquad \textbf{(1-35)}$$

in which $\sigma_b$ is the allowable bearing stress and $A_b$ is the projected area of the pin or other surface over which the bearing stresses act.

The following example illustrates how allowable loads are determined when the allowable stresses for the material are known.

## ● ● ● Example 1-12

A steel bar serving as a vertical hanger to support heavy machinery in a factory is attached to a support by the bolted connection shown in Fig. 1-54. Two clip angles (thickness $t_c = 9.5$ mm) are fastened to an upper support by bolts 1 and 2 each with a diameter of 12 mm; each bolt has a washer with a diameter of $d_w = 28$ mm. The main part of the hanger is attached to the clip angles by a single bolt (bolt 3 in Fig. 1-54a) with a diameter of $d = 25$ mm. The hanger has a rectangular cross section with a width of $b_1 = 38$ mm and thickness of $t = 13$ mm, but at the bolted connection, the hanger is enlarged to a width of $b_2 = 75$ mm. Determine the **allowable value of the tensile load P** in the hanger based upon the following considerations.

(a) The **allowable tensile stress** in the main part of the hanger is 110 MPa.
(b) The allowable tensile stress in the hanger at its **cross section through the bolt 3 hole** is 75 MPa. (The permissible stress at this section is lower because of the stress concentrations around the hole.)
(c) The **allowable bearing stress** between the hanger and the shank of bolt 3 is 180 MPa.
(d) The **allowable shear stress** in bolt 3 is 45 MPa.
(e) The **allowable normal stress** in bolts 1 and 2 is 160 MPa.
(f) The **allowable bearing stress between the washer** and the clip angle at either bolt 1 or 2 is 65 MPa.
(g) The **allowable shear stress** through the clip angle at bolts 1 and 2 is 35 MPa.

### Fig. 1-54

Example 1-12: Vertical hanger subjected to a tensile load $P$: (a) front view of bolted connection, and (b) side view of connection

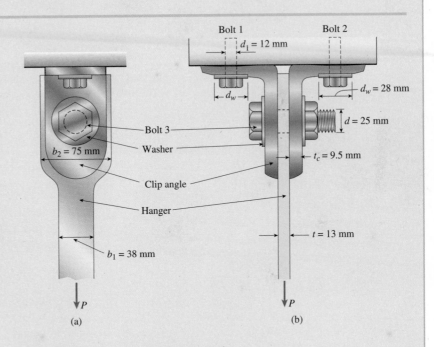

Numerical properties:

$t_c = 9.5$ mm   $t = 13$ mm   $b_1 = 38$ mm   $b_2 = 75$ mm

$d_1 = 12$ mm   $d = 25$ mm   $d_w = 28$ mm

*Continues* ➡

**Example 1-12 -** *Continued*

$$\sigma_a = 110 \text{ MPa} \quad \sigma_{a3} = 75 \text{ MPa} \quad \sigma_{ba3} = 180 \text{ MPa} \quad \tau_{a3} = 45 \text{ MPa}$$

$$\tau_{a1} = 35 \text{ MPa} \quad \sigma_{a1} = 160 \text{ MPa} \quad \sigma_{ba1} = 65 \text{ MPa}$$

### Solution

(a) The **allowable load $P_a$**, based upon the **stress in the main part of the hanger** (Fig. 1-54c) is equal to the allowable stress in tension times the cross-sectional area of the hanger (Eq. 1-33):

$$P_a = \sigma_a b_1 t = (110 \text{ MPa})(38 \text{ mm} \times 13 \text{ mm}) = 54.3 \text{ kN}$$

A load greater than this value will overstress the main part of the hanger (that is, the actual stress will exceed the allowable stress), thereby reducing the factor of safety.

(b) At the **cross section of the hanger through the bolt hole** (Fig 1-54d), we must make a similar calculation but with a different allowable stress and a different area. The net cross-sectional area (that is, the area that remains after the hole is drilled through the bar) is equal to the net width times the thickness. The net width is equal to the gross width $b_2$ minus the diameter $d$ of the hole. Thus, the equation for the **allowable load $P_b$** at this section is

$$P_b = \sigma_{a3}(b_2 - d)t = (75 \text{ MPa})(75 \text{ mm} - 25 \text{ mm})(13 \text{ mm}) = 48.8 \text{ kN}$$

(c) The allowable load based upon **bearing between the hanger and the bolt** (Fig. 1-54e) is equal to the allowable bearing stress times the bearing area. The bearing area is the projection of the actual contact area,

**Fig. 1-54c**

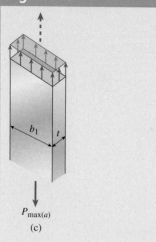

(c)

**Fig. 1-54d**

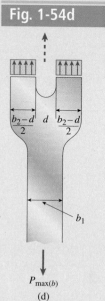

(d)

**Fig. 1-54e**

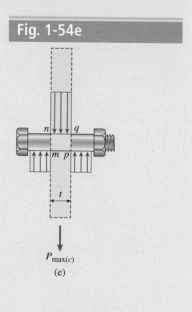

(e)

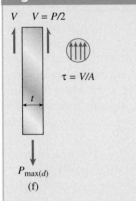

$$V \quad V = P/2$$

$$\tau = V/A$$

$t$

$$P_{\max(d)}$$
(f)

which is equal to the bolt diameter times the thickness of the hanger. Therefore, the allowable load (Eq. 1-35) is

$$P_c = \sigma_{ba3}dt = 58.5 \text{ kN} = (180 \text{ MPa})(25 \text{ mm})(13 \text{ mm}) = 58.5 \text{ kN}$$

**(d)** The allowable load $P_d$ based upon shear in the bolt (Fig. 1-54f) is equal to the allowable shear stress times the shear area (Eq. 1-34). The shear area is twice the area of the bolt because the bolt is in double shear; thus,

$$P_d = 2\tau_{a3}\left(\frac{\pi}{4} d^2\right) = 2(45 \text{ MPa})\left[\frac{\pi}{4}(25 \text{ mm})^2\right] = 44.2 \text{ kN}$$

**(e)** The **allowable normal stress** in bolts 1 and 2 is 160 MPa. Each bolt carries one half of the applied load $P$ (see Fig. 1-54g). The allowable total load $P_e$ is the product of the allowable normal stress in the bolt and the sum of the cross-sectional areas of bolts 1 and 2:

$$P_e = \sigma_{a1}(2)\left(\frac{\pi}{4} d_1^2\right) = (160 \text{ MPa})(2)\left[\frac{\pi}{4}(12 \text{ mm})^2\right] = 36.2 \text{ kN} \quad \Longleftarrow$$

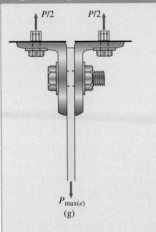

$$P/2 \qquad P/2$$

$$P_{\max(e)}$$
(g)

**(f)** The **allowable bearing stress between the washer** and the clip angle at either bolt 1 or 2 is 65 MPa. Each bolt (1 or 2) carries one half of the applied load $P$ (see Fig. 1-54h). The bearing area here is the ring-shaped circular area of the washer (the washer is assumed to fit snugly against the bolt). The allowable total load $P_f$ is the allowable bearing stress on the washer times twice the area of the washer:

$$P_f = \sigma_{ba1}(2)\left[\frac{\pi}{4}\cdot(d_w^2 - d_1^2)\right] = (65 \text{ MPa})(2)\left\{\frac{\pi}{4}[(28 \text{ mm})^2 - (12 \text{ mm})^2]\right\}$$

$$= 65.3 \text{ kN}$$

**(g)** The **allowable shear stress** through the clip angle at bolts 1 and 2 is 35 MPa. Each bolt (1 or 2) carries one half of the applied load $P$ (see Fig. 1-54i). The shear area at each bolt is equal to the circumference of the hole ($\pi \times d_w$) times the thickness of the clip angle ($t_c$).

The allowable total load $P_g$ is the the allowable shear stress times twice the shear area:

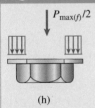

$$P_{\max(f)}/2$$

(h)

$$P_g = \tau_{a1}(2)(\pi d_w t_c) = (35 \text{ MPa})(2)(\pi \times 28 \text{ mm} \times 9.5 \text{ mm}) = 58.5 \text{ kN}$$

We have now found the allowable tensile loads in the hanger based upon all seven of the given conditions. Comparing the seven preceding results, we see that the *smallest value of the load is* $P_{\text{allow}} = 36.2$ kN. This load, which is *based upon normal stress in bolts 1 and 2* (see part (e) above), is the allowable tensile load in the hanger.

We could refine the analysis by next considering the self-weight of the entire hanger assembly (see Example 1-6).

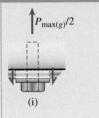

$$P_{\max(g)}/2$$

(i)

# 1.9 DESIGN FOR AXIAL LOADS AND DIRECT SHEAR

In the preceding section we discussed the determination of allowable loads for simple structures, and in earlier sections we saw how to find the stresses, strains, and deformations of bars. The determination of such quantities is known as **analysis**. In the context of mechanics of materials, analysis consists of determining the *response* of a structure to loads, temperature changes, and other physical actions. By the response of a structure, we mean the stresses, strains, and deformations produced by the loads.

Response also refers to the load-carrying capacity of a structure; for instance, the allowable load on a structure is a form of response.

A structure is said to be *known* (or *given*) when we have a complete physical description of the structure, that is, when we know all of its *properties*. The properties of a structure include the types of members and how they are arranged, the dimensions of all members, the types of supports and where they are located, the materials used, and the properties of the materials. Thus, when analyzing a structure, *the properties are given and the response is to be determined.*

The inverse process is called **design**. When designing a structure, *we must determine the properties of the structure in order that the structure will support the loads and perform its intended functions.* For instance, a common design problem in engineering is to determine the size of a member to support given loads. Designing a structure is usually a much lengthier and more difficult process than analyzing it—indeed, analyzing a structure, often more than once, is typically part of the design process.

In this section we will deal with design in its most elementary form by calculating the required sizes of simple tension and compression members as well as pins and bolts loaded in shear. In these cases the design process is quite straightforward. Knowing the loads to be transmitted and the allowable stresses in the materials, we can calculate the required areas of members from the following general relationship (compare with Eq. 1-32):

$$\text{Required area} = \frac{\text{Load to be transmitted}}{\text{Allowable stress}} \qquad \textbf{(1-36)}$$

This equation can be applied to any structure in which the stresses are uniformly distributed over the area. (The use of this equation for finding the size of a bar in tension and the size of a pin in shear is illustrated in Example 1-13, which follows.)

In addition to **strength** considerations, as exemplified by Eq. (1-36), the design of a structure is likely to involve **stiffness** and **stability**. Stiffness refers to the ability of the structure to resist changes in shape (for instance, to resist stretching, bending, or twisting), and stability refers to the ability of the structure to resist buckling under compressive stresses. Limitations on stiffness are sometimes necessary to prevent excessive deformations, such as large deflections of a beam that might interfere with its performance. Buckling is the principal consideration in the design of columns, which are slender compression members (Chapter 11).

Another part of the design process is **optimization**, which is the task of designing the best structure to meet a particular goal, such as minimum weight. For instance, there may be many structures that will support a given load, but in some circumstances the best structure will be the lightest one. Of course, a goal such as minimum weight usually must be balanced against more general considerations, including the aesthetic, economic, environmental, political, and technical aspects of the particular design project.

When analyzing or designing a structure, we refer to the forces that act on it as either **loads** or **reactions**. Loads are *active forces* that are applied to the structure by some external cause, such as gravity, water pressure, wind, and earthquake ground motion. Reactions are *passive forces* that are induced at the supports of the structure—their magnitudes and directions are determined by the nature of the structure itself. Thus, reactions must be calculated as part of the analysis, whereas loads are known in advance.

Example 1-13, on the following pages, begins with a review of **free-body diagrams** and elementary statics and concludes with the design of a bar in tension and a pin in direct shear.

When drawing free-body diagrams, it is helpful to distinguish reactions from loads or other applied forces. A common scheme is to place a slash, or slanted line, across the arrow when it represents a reactive force, as illustrated in Fig. 1-55 of the following example.

## ● ● ● Example 1-13

The cable-pipe structure *ABCD* shown in Fig. 1-55 has pin supports at points *A* and *D*, which are 6 ft apart. Member *ABC* is a steel pipe, and member *BDC* is a *continuous cable* which passes over a frictionless pulley at *D*. A sign weighing 1500 lb is suspended from bar *ABC* at points *E* and *F*.

**Determine the required diameter of the pins** at *A*, *B*, *C*, and *D* if the allowable stress in shear is 6.5 ksi. Also, **find the required cross-sectional areas of bar *ABC* and cable *BDC*** if the **allowable stresses in tension and compression** are 18 ksi and 10 ksi, respectively. (The allowable compression stress is lower because of the possibility of buckling instability.)

(**Note:** The pins at the supports are in double shear. Also, consider only the weight of the sign; disregard the weights of members *BDC* and *ABC*.)

### Fig. 1-55

Example 1-13: Cable-supported pipe *ABC* carrying sign of weight *W*

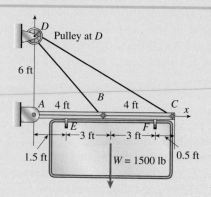

### Solution

The first step in the overall solution is to find reaction forces at supports and the tensile force in the continuous cable *BDC*. These quantities are found by applying the laws of statics to free-body diagrams (see Section 1.2). Once reaction and cable forces are known, we can find the axial forces in member *ABC* and the shear forces in pins at *A*, *B*, *C*, and *D*. We can then find the required sizes of member *ABC* and the pins at *A*, *B*, *C*, and *D*.

*Reactions:* We begin with a *free-body diagram of the entire structure* (Fig. 1-56), which shows all of the applied and reaction forces. A statics sign convention is commonly used so all reaction components are initially shown in the positive coordinate directions.

### Fig. 1-56

Example 1-13: Free-body diagram of entire structure

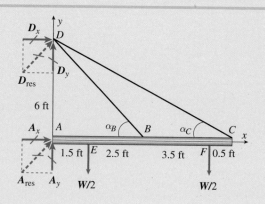

**Summing moments about point *D*** (counterclockwise moments are positive) gives

$$\Sigma M_D = 0 \quad A_x(6 \text{ ft}) - \frac{W}{2}(1.5 \text{ ft} + 7.5 \text{ ft}) = 0 \quad \text{or} \quad A_x = \frac{1500 \text{ lb}}{2}\left(\frac{9 \text{ ft}}{6 \text{ ft}}\right)$$
$$= 1125 \text{ lb}$$

Next, **sum forces in the *x* direction**:

$$\Sigma F_x = 0 \quad A_x + D_x = 0 \quad \text{or} \quad D_x = -A_x = -1125 \text{ lb}$$

The minus sign means that $D_x$ acts in the negative *x* direction.

**Summing forces in the *x* direction at joint *D*** will give us the force in the continuous cable *BDC*.

First compute angles $\alpha_B$ and $\alpha_C$ (see Fig. 1-56):

$$\alpha_B = \text{atan}\left(\frac{6}{4}\right) = 56.31° \quad \alpha_C = \text{atan}\left(\frac{6}{8}\right) = 36.87°$$

Now $\Sigma F_x = 0$ at joint *D*:

$$D_x + T(\cos(\alpha_B) + \cos(\alpha_C)) = 0 \quad \text{so} \quad T = \frac{-D_x}{(\cos(\alpha_B) + \cos(\alpha_C))}$$

$$\text{or} \quad T = \frac{-(-1125 \text{ lb})}{(\cos(\alpha_B) + \cos(\alpha_C))} = 830.4 \text{ lb}$$

Now find the vertical reaction at joint *D* where $T_B = T_C$ because the cable is one continuous cable (see Fig. 1-57).

$$D_y = T(\sin(\alpha_B) + \sin(\alpha_C)) = 1189.23 \text{ lb}$$

The plus sign means that $D_y$ acts in the positive *y* direction.

So, the resultant at *D* is

$$D_{\text{res}} = \sqrt{D_x^2 + D_y^2} = 1637.04 \text{ lb}$$

Next, sum the forces in the *y* direction for the entire free-body diagram to get $A_y$:

$$\Sigma F_y = 0 \quad A_y + D_y - W = 0 \quad \text{so} \quad A_y = -D_y + W = 310.77 \text{ lb}$$

The resultant at *A* is

$$A_{\text{res}} = \sqrt{A_x^2 + A_y^2} = 1167.13 \text{ lb}$$

### Fig. 1-57

Free-body diagram of joint D

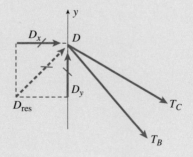

Continues ➡

**Example 1-13 -** *Continued*

**As a final check, confirm the equilibrium using the free-body diagram of pipe *ABC*** (see Fig. 1-58, where $T_B = T_C$ because the cable is one continuous cable). The forces will sum to zero, and the sum of the moments about $A$ is zero:

$$\Sigma F_x: \quad A_x - T\cos(\alpha_B) - T\cos(\alpha_C) = 0$$

$$\Sigma F_y: \quad A_y - W + T\sin(\alpha_B) + T\sin(\alpha_C) = 0$$

$$\Sigma M_A: T\sin(\alpha_B)(1.5\text{ ft} + 2.5\text{ ft}) + T\sin(\alpha_C)(8\text{ ft}) - \frac{W}{2}(1.5\text{ ft} + 7.5\text{ ft}) = 0$$

### Fig. 1-58

Free-body diagram of member ABC

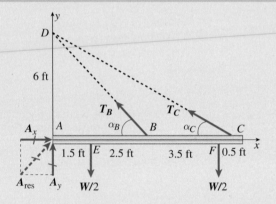

**Determine the shear forces in pins** (all are in double shear) and *required diameter of each pin.* Now that the reaction and cable forces are known, we can identify the shear forces in pins at *A, B, C,* and *D* and then find the required size of each pin.

Pin at *A*:

$$A_{pinA} = \frac{A_{res}}{2\,\tau_{allow}} = \frac{1167.13\text{ lb}}{2(6.5\text{ ksi})} = 0.0898\text{ in.}^2$$

$$\text{so} \quad d_{pinA} = \sqrt{\frac{4}{\pi}(0.0898\text{ in.}^2)} = 0.338\text{ in.}$$

Pin at *B*:

$$A_{pinB} = \frac{T}{2\,\tau_{allow}} = \frac{830.4\text{ lb}}{2(6.5\text{ ksi})} = 0.0639\text{ in.}^2$$

$$\text{so} \quad d_{pinB} = \sqrt{\frac{4}{\pi}(0.0639\text{ in.}^2)} = 0.285\text{ in.}$$

Pin at *C*:

The resultant force at *C* is the same as that at *B*, so the pin at *C* will have the same diameter as that at *B*.

Pin in pulley at *D*:

$$A_{pinD} = \frac{D_{res}}{2\tau_{allow}} = \frac{1637.04 \text{ lb}}{2(6.5 \text{ ksi})} = 0.1259 \text{ in.}^2$$

$$\text{so} \quad d_{pinD} = \sqrt{\frac{4}{\pi}(0.1259 \text{ in.}^2)} = 0.4 \text{ in.}$$

**Determine the axial force in cable *BDC* and *required cross-sectional area of the cable*.** Here we use the computed tension for *T* and higher allowable axial stress for members in tension:

$$A_{cable} = \frac{T}{\sigma_{allowT}} = 830.4 \text{ lb}/18,000 \text{ psi} = 0.046 \text{ in.}^2$$

**Determine the axial force in pipe *ABC* and *required cross-sectional area of the pipe*.** We can use free-body diagrams of portions of member *ABC* to compute the axial compressive force *N* in segments *AB* and *BC* (as discussed in Section 1.2). For segment *AB*, force $N_{AB} = -1125$ lb, while in segment *BC*, $N_{BC} = -664$ lb. The larger force $N_{AB}$ controls. Now we must use the reduced allowable axial stress for compression, so the required area is

$$A_{pipe} = \frac{1125 \text{ lb}}{\sigma_{allowC}} = 1125 \text{ lb}/10,000 \text{ psi} = 0.112 \text{ in.}^2$$

*Notes:* In this example, we intentionally omitted the self-weight of the cable–pipe structure from the calculations. However, once the sizes of the members are known, their weights can be calculated and included in the free-body diagrams given.

When the weights of the bars are included, the design of member *ABC* becomes more complicated. Not only because of its own weight but also because of the weight of the sign, member *ABC* is a ***beam*** subjected to bending and transverse shear as well as axial compression. The design of such members must wait until we study stresses in beams (Chapter 5), where a beam with axial loads is discussed as a separate topic (see Section 5.12). As the compressive forces increase, the possibility of lateral instability (or buckling) of *ABC* also becomes a concern; we will investigate this in Chapter 11.

In practice, other loads besides the weights of the structure and sign would have to be considered before making a final decision about the sizes of the pipes, cables, and pins. Loads that could be important include wind loads, earthquake loads, and the weights of objects that might have to be supported temporarily by the structure.

Finally, if cables *BD* and *CD* are *separate cables* (instead of one continuous cable as in the previous analysis where $T_B = T_C$), the forces $T_B$ and $T_C$ in the two cables are not equal in magnitude. The structure is now *statically indeterminate*, and the cable forces and the reactions at *A* cannot be determined using the equations of static equilibrium alone. We will consider problems of this type in Chapter 2, Section 2.4 (see Example 2-5).

In Chapter 1 we learned about mechanical properties of construction materials. After a brief review of **statics**, we computed normal stresses and strains in bars loaded by centroidal axial loads, and also shear stresses and strains (as well as bearing stresses) in pin connections used to assemble simple structures, such as trusses. We also defined allowable levels of stress from appropriate factors of safety and used these values to set allowable loads that could be applied to the structure.

Some of the major concepts presented in this chapter are as follows.

1. The principal objective of mechanics of materials is to determine the **stresses, strains, and displacements** in structures and their components due to the loads acting on them. These components include bars with axial loads, shafts in torsion, beams in bending, and columns in compression.

2. Prismatic bars subjected to tensile or compressive loads acting through the centroid of their cross section (to avoid bending) experience **normal stress ($\sigma$) and strain ($\varepsilon$)**

$$\sigma = \frac{P}{A}$$

$$\varepsilon = \frac{\delta}{L}$$

and either extension or contraction proportional to their lengths. These stresses and strains are **uniform** except near points of load application where high localized stresses, or **stress-concentrations**, occur.

3. We investigated the **mechanical behavior** of various materials and plotted the resulting stress-strain diagram, which conveys important information about the material. **Ductile** materials (such as mild steel) have an initial linear relationship between normal stress and strain (up to the **proportional limit**) and are said to be **linearly elastic** with stress and strain related by **Hooke's law**

$$\sigma = E\varepsilon$$

They also have a well-defined yield point. Other ductile materials (such as aluminum alloys) typically do not have a clearly definable yield point, so an arbitrary yield stress may be determined by using the **offset method**.

4. Materials that fail in tension at relatively low values of strain (such as concrete, stone, cast iron, glass ceramics and a variety of metallic alloys) are classified as **brittle**. Brittle materials fail with only little elongation after the proportional limit.

5. If the material remains within the elastic range, it can be loaded, unloaded, and loaded again without significantly changing the

behavior. However when loaded into the plastic range, the internal structure of the material is altered and its properties change. Loading and unloading behavior of materials depends on the **elasticity** and **plasticity** properties of the material, such as the **elastic limit** and possibility of **permanent set** (residual strain) in the material. Sustained loading over time may lead to **creep** and **relaxation**.

6. Axial elongation of bars loaded in tension is accompanied by lateral contraction; the ratio of lateral strain to normal strain is known as **Poisson's ratio** ($v$).

$$v = -\frac{\text{lateral strain}}{\text{axial strain}} = -\frac{\varepsilon'}{\varepsilon}$$

Poisson's ratio remains constant throughout the linearly elastic range, provided the material is homogeneous and isotropic. Most of the examples and problems in the text are solved with the assumption that the material is linearly elastic, homogeneous, and isotropic.

7. **Normal** stresses ($\sigma$) act perpendicular to the surface of the material and **shear stresses** ($\tau$) act tangential to the surface. We investigated bolted connections between plates in which the bolts were subjected to either single or double shear ($\tau_{\text{aver}}$) where

$$\tau_{\text{aver}} = \frac{V}{A}$$

as well as average **bearing** stresses ($\sigma_b$). The bearing stresses act on the rectangular projected area ($A_b$) of the actual curved contact surface between a bolt and plate.

$$\sigma_b = \frac{F_b}{A_b}$$

8. We looked at an element of material acted on by shear stresses and strains to study a state of stress referred to as **pure shear**. We saw that shear strain ($\gamma$) is a measure of the distortion or change in shape of the element in pure shear. We looked at Hooke's law in shear in which shear stress ($\tau$) is related to shear strain by the shearing modulus of elasticity $G$.

$$\tau = G\gamma$$

We noted that $E$ and $G$ are related and therefore are not independent elastic properties of the material.

$$G = \frac{E}{2(1 + v)}$$

9. **Strength** is the capacity of a structure or component to support or transmit loads. **Factors of safety** relate actual to required strength of structural members and account for a variety of uncertainties, such as variations in material properties, uncertain magnitudes or distributions of loadings, probability of accidental overload, and so on. Because of these uncertainties, factors of safety ($n_1$, $n_2$, $n_3$, $n_4$) must be determined using probabilistic methods.

10. Yield or ultimate level stresses can be divided by factors of safety to produce allowable values for use in design. For **ductile** materials,

$$\sigma_{\text{allow}} = \frac{\sigma_Y}{n_1}, \qquad \tau_{\text{allow}} = \frac{\tau_Y}{n_2}$$

while for **brittle** materials,

$$\sigma_{\text{allow}} = \frac{\sigma_U}{n_3}, \qquad \tau_{\text{allow}} = \frac{\tau_U}{n_4}.$$

A typical value of $n_1$ and $n_2$ is 1.67 while $n_3$ and $n_4$ might be 2.8.

For a pin-connected member in axial **tension**, the allowable load depends on the allowable stress times the appropriate area (e.g., net cross-sectional area for bars acted on by centroidal tensile loads, cross-sectional area of pin for pins in shear, and projected area for bolts in bearing). If the bar is in **compression**, net cross-sectional area need not be used, but buckling may be an important consideration.

11. Lastly, we considered **design**, the iterative process by which the appropriate size of structural members is determined to meet a variety of both **strength and stiffness requirements** for a particular structure subjected to a variety of different loadings. However, incorporation of factors of safety into design is not a simple matter, because both strength and failure have many different meanings.

# PROBLEMS CHAPTER 1

## Statics Review

**1.2-1** Segments *AB* and *BC* of beam *ABC* are pin connected a small distance to the right of joint *B* (see figure). Axial loads act at *A* and at the mid-span of *AB*. A concentrated moment is applied at joint *B*.

    (a) Find reactions at supports *A*, *B*, and *C*.

    (b) Find internal stress resultants *N*, *V*, and *M* at $x = 15$ ft.

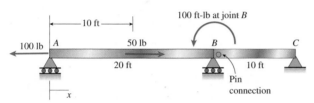

**PROB. 1.2-1**

**1.2-2** Segments *AB* and *BCD* of beam *ABCD* are pin connected at $x = 4$ m. The beam is supported by a sliding support at *A* and roller supports at *C* and *D* (see figure). A triangularly distributed load with peak intensity of 80 N/m acts on *BC*. A concentrated moment is applied at joint *B*.

    (a) Find reactions at supports *A*, *C*, and *D*.

    (b) Find internal stress resultants *N*, *V*, and *M* at $x = 5$ m.

    (c) Repeat parts (a) and (b) for the case of the roller support at *C* replaced by a linear spring of stiffness $k_y = 200$ kN/m (see figure).

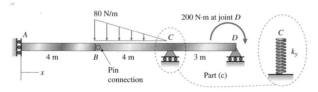

**PROB. 1.2-2**

**1.2-3** Segments *AB* and *BCD* of beam *ABCD* are pin connected at $x = 10$ ft. The beam is supported by a pin support at *A* and roller supports at *C* and *D*; the roller at *D* is rotated by 30° from the *x* axis (see figure). A trapezoidal distributed load on *BC* varies in intensity from 5 lb/ft at *B* to

2.5 lb/ft at *C*. A concentrated moment is applied at joint *A*, and a 40-lb inclined load is applied at the mid-span of *CD*.

    (a) Find reactions at supports *A*, *C*, and *D*.

    (b) Find the resultant force in the pin connection at *B*.

    (c) Repeat parts (a) and (b) if a rotational spring ($k_r = 50$ ft-lb/radian) is added at *A and* the roller at *C* is removed.

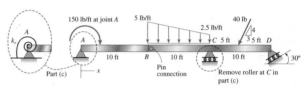

**PROB. 1.2-3**

**1.2-4** Consider the plane truss with a pin support at joint 3 and a vertical roller support at joint 5 (see figure).

    (a) Find reactions at support joints 3 and 5.

    (b) Find axial forces in truss members 11 and 13.

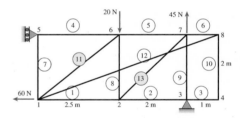

**PROB. 1.2-4**

**1.2-5** A plane truss has a pin support at *A* and a roller support at *E* (see figure).

    (a) Find reactions at all supports.

    (b) Find the axial force in truss member *FE*.

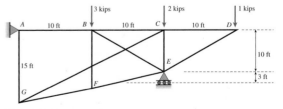

**PROB. 1.2-5**

**1.2-6** A plane truss has a pin support at *F* and a roller support at *D* (see figure).

    (a) Find reactions at both supports.

    (b) Find the axial force in truss member *FE*.

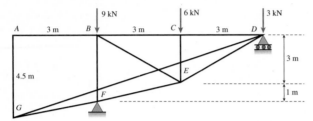

**PROB. 1.2-6**

**1.2-7** A space truss has three-dimensional pin supports at joints *O*, *B*, and *C*. Load *P* is applied at joint *A* and acts toward point *Q*. Coordinates of all joints are given in feet (see figure).

    (a) Find reaction force components $B_x$, $B_z$, and $O_z$.

    (b) Find the axial force in truss member *AC*.

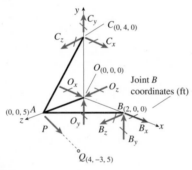

**PROB. 1.2-7**

**1.2-8** A space truss is restrained at joints *O*, *A*, *B*, and *C*, as shown in the figure. Load *P* is applied at joint *A* and load 2*P* acts downward at joint *C*.

    (a) Find reaction force components $A_x$, $B_y$, and $B_z$ in terms of load variable *P*.

    (b) Find the axial force in truss member *AB* in terms of load variable *P*.

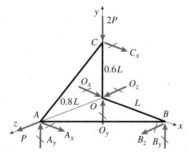

**PROB. 1.2-8**

**1.2-9** A space truss is restrained at joints *A*, *B*, and *C*, as shown in the figure. Load 2*P* is applied at in the −*x* direction at joint *A*, load 3*P* acts in the +*z* direction at joint *B* and load *P* is applied in the +*z* direction at joint *C*. Coordinates of all joints are given in terms of dimension variable *L* (see figure).

    (a) Find reaction force components $A_y$ and $A_z$ in terms of load variable *P*.

    (b) Find the axial force in truss member *AB* in terms of load variable *P*.

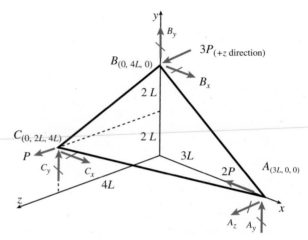

**PROB. 1.2-9**

**1.2-10** A space truss is restrained at joints *A*, *B*, and *C*, as shown in the figure. Load *P* acts in the +z direction at joint *B* and in the −z direction at joint *C*. Coordinates of all joints are given in terms of dimension variable *L* (see figure). Let *P* = 5 kN and *L* = 2 m.

    (a) Find the reaction force components $A_z$ and $B_x$.

    (b) Find the axial force in truss member *AB*.

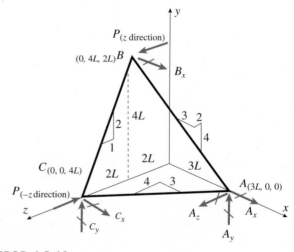

**PROB. 1.2-10**

**1.2-11** A stepped shaft $ABC$ consisting of two solid, circular segments is subjected to torques $T_1$ and $T_2$ acting in opposite directions, as shown in the figure. The larger segment of the shaft has a diameter of $d_1 = 2.25$ in. and a length $L_1 = 30$ in.; the smaller segment has a diameter $d_2 = 1.75$ in. and a length $L_2 = 20$ in. The torques are $T_1 = 21,000$ lb-in. and $T_2 = 10,000$ lb-in.

(a) Find reaction torque $T_A$ at support $A$.

(b) Find the internal torque $T(x)$ at two locations: $x = L_1/2$ and $x = L_1 + L_2/2$. Show these internal torques on properly drawn free-body diagrams (FBDs).

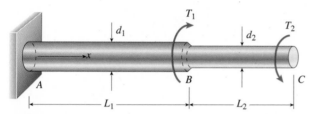

PROB. 1.2-11

**1.2-12** A stepped shaft $ABC$ consisting of two solid, circular segments is subjected to uniformly distributed torque $t_1$ acting over segment 1 and concentrated torque $T_2$ applied at $C$, as shown in the figure. Segment 1 of the shaft has a diameter of $d_1 = 57$ mm and length of $L_1 = 0.75$ m; segment 2 has a diameter $d_2 = 44$ mm and length $L_2 = 0.5$ m. Torque intensity $t_1 = 3100$ N·m/m and $T_2 = 1100$ N·m.

(a) Find reaction torque $T_A$ at support $A$.

(b) Find the internal torque $T(x)$ at two locations: $x = L_1/2$ and at $x = L_1 + L_2/2$. Show these internal torques on properly drawn free-body diagrams (FBDs).

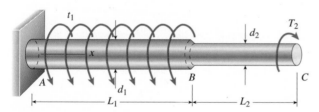

PROB. 1.2-12

**1.2-13** A plane frame is restrained at joints $A$ and $C$, as shown in the figure. Members $AB$ and $BC$ are pin connected at $B$. A triangularly distributed lateral load with a peak intensity of 90 lb/ft acts on $AB$. A concentrated moment is applied at joint $C$.

(a) Find reactions at supports $A$ and $C$.

(b) Find internal stress resultants $N$, $V$, and $M$ at $x = 3$ ft on column $AB$.

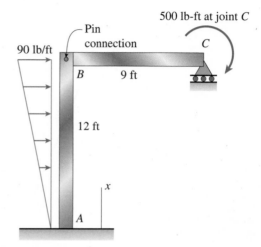

PROB. 1.2-13

**1.2-14** A plane frame is restrained at joints $A$ and $D$, as shown in the figure. Members $AB$ and $BCD$ are pin connected at $B$. A triangularly distributed lateral load with peak intensity of 80 N/m acts on $CD$. An inclined concentrated force of 200 N acts at the mid-span of $BC$.

(a) Find reactions at supports $A$ and $D$.

(b) Find resultant forces in the pins at $B$ and $C$.

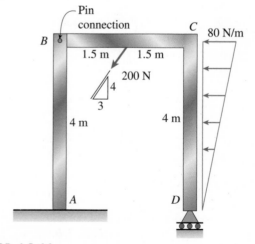

PROB. 1.2-14

**1.2-15** A 200-lb trap door (*AB*) is supported by a strut (*BC*) which is pin connected to the door at *B* (see figure).

(a) Find reactions at supports *A* and *C*.

(b) Find internal stress resultants *N*, *V*, and *M* on the trap door at 20 in. from *A*.

(a) Find reactions at supports *A* and *E*.

(b) Find internal stress resultants *N*, *V*, and *M* at point *H*.

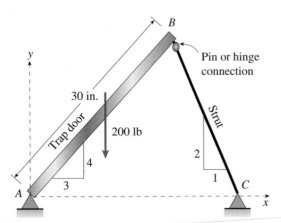

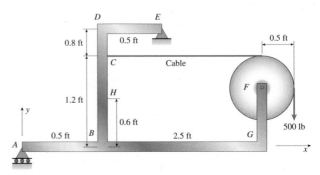

PROB. 1.2-17

PROB. 1.2-15

**1.2-16** A plane frame is constructed by using a pin connection between segments *ABC* and *CDE*. The frame has pin supports at *A* and *E* and joint loads at *B* and *D* (see figure).

(a) Find reactions at supports *A* and *E*.

(b) Find the resultant force in the pin at *C*.

**1.2-18** A plane frame with a pin support at *A* and roller supports at *C* and *E* has a cable attached at *E*, which runs over frictionless pulleys at *D* and *B* (see figure). The cable force is known to be 400 N. There is a pin connection just to the left of joint *C*.

(a) Find reactions at supports *A*, *C*, and *E*.

(b) Find internal stress resultants *N*, *V*, and *M* just to the right of joint *C*.

(c) Find resultant force in the pin near *C*.

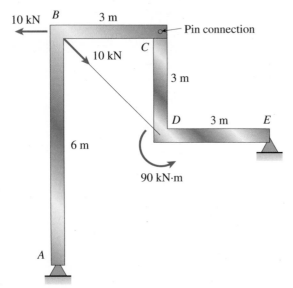

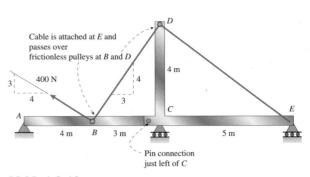

PROB. 1.2-18

PROB. 1.2-16

**1.2-17** A plane frame with pin supports at *A* and *E* has a cable attached at *C*, which runs over a frictionless pulley at *F* (see figure). The cable force is known to be 500 lb.

**1.2-19** A 150-lb rigid bar *AB*, with frictionless rollers at each end, is held in the position shown in the figure by a continuous cable *CAD*. The cable is pinned at *C* and *D* and runs over a pulley at *A*.

(a) Find reactions at supports *A* and *B*.

(b) Find the force in the cable.

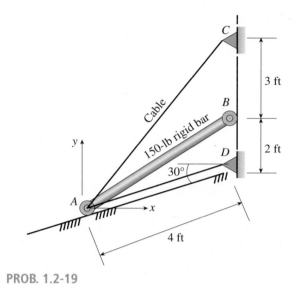

PROB. 1.2-19

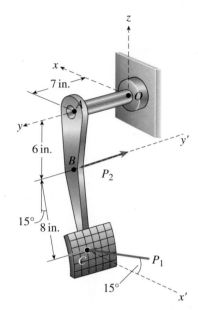

PROB. 1.2-21

**1.2-20** A plane frame has a pin support at $A$ and roller supports at $C$ and $E$ (see figure). Frame segments $ABD$ and $CDEF$ are joined just left of joint $D$ by a pin connection.
(a) Find reactions at supports $A$, $C$, and $E$.
(b) Find the resultant force in the pin just left of $D$.

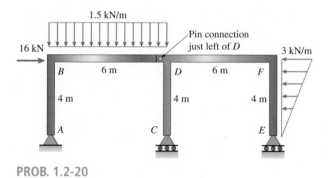

PROB. 1.2-20

**1.2-21** A special vehicle brake is clamped at $O$, (when the brake force $P_1$ is applied—see figure). Force $P_1 = 50$ lb and lies in a plane which is parallel to the $xz$ plane and is applied at C normal to line $BC$. Force $P_2 = 40$ lb and is applied at $B$ in the $-y$ direction.
(a) Find reactions at support $O$.
(b) Find internal stress resultants $N$, $V$, $T$, and $M$ at the mid-point of segment $OA$.

**1.2-22** Space frame $ABCD$ is clamped at $A$, except it is free to translate in the $x$ direction. There is also a roller support at $D$, which is normal to line $CDE$. A triangularly distributed force with peak intensity $q_0 = 75$ N/m acts along $AB$ in the positive $z$ direction. Forces $P_x = 60$ N and $P_z = -45$ N are applied at joint $C$, and a concentrated moment $M_y = 120$ N · m acts at the mid-span of member $BC$.
(a) Find reactions at supports $A$ and $D$.
(b) Find internal stress resultants $N$, $V$, $T$, and $M$ at the mid-height of segment $AB$.

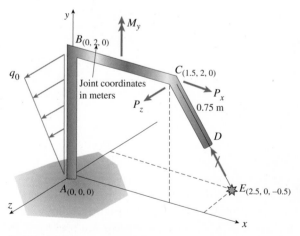

PROB. 1.2-22

**1.2-23** Space frame $ABC$ is clamped at $A$, except it is free to rotate at $A$ about the $x$ and $y$ axes. Cables $DC$ and $EC$ support the frame at $C$. Force $P_y = -50$ lb is applied at the mid-span of $AB$, and a concentrated moment $M_x = -20$ in.-lb acts at joint $B$.

(a) Find reactions at support $A$.

(b) Find cable tension forces.

moment releases at the base of member 2 so that member 2 can lengthen and shorten as the roller support at $B$ moves along the 30° incline. (These releases indicate that the internal axial force $N$ and moment $M$ must be zero at this location.)

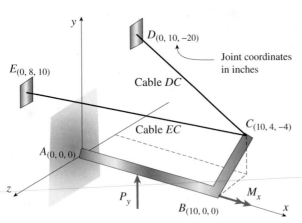

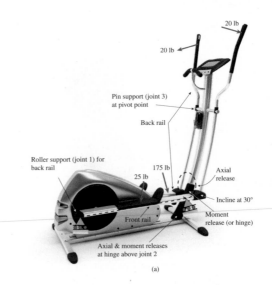

(a)

PROB. 1.2-23

**1.2-24** A soccer goal is subjected to gravity loads (in the $-z$ direction, $w = 73$ N/m for $DG$, $BG$, and $BC$; $w = 29$ N/m for all other members; see figure) and a force $F = 200$ N applied eccentrically at the mid-height of member $DG$. Find reactions at supports $C$, $D$, and $H$.

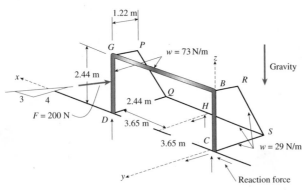

PROB. 1.2-24

**1.2-25** An elliptical exerciser machine (see figure part a) is composed of front and back rails. A simplified plane-frame model of the back rail is shown in figure part b. Analyze the plane-frame model to find reaction forces at supports $A$, $B$, and $C$ for the position and applied loads given in figure part b. Note that there are axial and

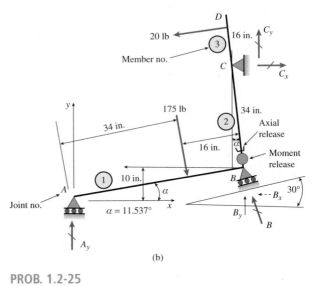

(b)

PROB. 1.2-25

**1.2-26** A mountain bike is moving along a flat path at constant velocity. At some instant, the rider (weight = 670 N) applies pedal and hand forces, as shown in the figure part a.

(a) Find reaction forces at the front and rear hubs. (Assume that the bike is pin supported at the rear hub and roller supported at the front hub.)

(b) Find internal stress resultants $N$, $V$, and $M$ in the inclined seat post (see figure part b).

(c) If $P_1$ remains at 1700 lb and $P_2$ is now set at 2260 lb, what new thickness of $BC$ will result in the same compressive stress in both parts?

(hamurishi/Shutterstock)

(b)

(hamurishi/Shutterstock)

PROB. 1.2-26

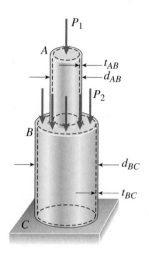

PROB. 1.3-1

**1.3-2** A force $P$ of 70 N is applied by a rider to the front hand brake of a bicycle ($P$ is the resultant of an evenly distributed pressure). As the hand brake pivots at $A$, a tension $T$ develops in the 460-mm long brake cable ($A_e = 1.075$ mm²) which elongates by $\delta = 0.214$ mm. Find normal stress $\sigma$ and strain $\varepsilon$ in the brake cable.

# Normal Stress and Strain

**1.3-1** A hollow circular post $ABC$ (see figure) supports a load $P_1 = 1700$ lb acting at the top. A second load $P_2$ is uniformly distributed around the cap plate at $B$. The diameters and thicknesses of the upper and lower parts of the post are $d_{AB} = 1.25$ in., $t_{AB} = 0.5$ in., $d_{BC} = 2.25$ in., and $t_{BC} = 0.375$ in., respectively.

(a) Calculate the normal stress $\sigma_{AB}$ in the upper part of the post.

(b) If it is desired that the lower part of the post have the same compressive stress as the upper part, what should be the magnitude of the load $P_2$?

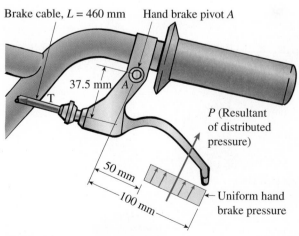

PROB. 1.3-2

**1.3-3** A bicycle rider would like to compare the effectiveness of cantilever hand brakes [see figure part (a)] versus V brakes [figure part (b)].

(a) Calculate the braking force $R_B$ at the wheel rims for each of the bicycle brake systems shown. Assume that all forces act in the plane of the figure and that cable tension $T = 45$ lbs. Also, what is the average compressive normal stress $\sigma_c$ on the brake pad ($A = 0.625$ in.²)?

(b) For each braking system, what is the stress in the brake cable (assume effective cross-sectional area of 0.00167 in.²)?

(*HINT*: Because of symmetry, you only need to use the right half of each figure in your analysis.)

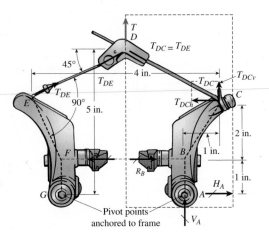

(a) Cantilever brakes

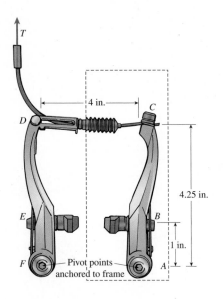

(b) V brakes

PROB. 1.3-3

**1.3-4** A circular aluminum tube with a length of $L = 420$ mm is loaded in compression by forces $P$ (see figure). The hollow segment of length $L/3$ had outside and inside diameters of 60 mm and 35 mm, respectively. The solid segment of length $2L/3$ has a diameter of 60 mm. A strain gage is placed on the outside of the hollow segment of the bar to measure normal strains in the longitudinal direction.

(a) If the measured strain in the hollow segment is $\varepsilon_h = 470 \times 10^{-6}$, what is the strain $\varepsilon_s$ in the solid part? (*Hint*: The strain in the solid segment is equal to that in the hollow segment multiplied by the ratio of the area of the hollow to that of the solid segment.)

(b) What is the overall shortening $\delta$ of the bar?

(c) If the compressive stress in the bar cannot exceed 48 MPa, what is the maximum permissible value of load $P$?

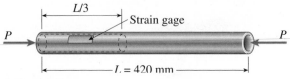

PROB. 1.3-4

**1.3-5** The cross section of a concrete corner column that is loaded uniformly in compression is shown in the figure. A circular pipe chase cut-out of 10 in. in diameter runs the height of the column (see figure).

(a) Determine the average compressive stress $\sigma_c$ in the concrete if the load is equal to 3500 kips.

(b) Determine the coordinates $x_c$ and $y_c$ of the point where the resultant load must act in order to produce uniform normal stress in the column.

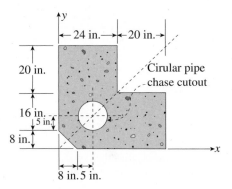

PROB. 1.3-5

**1.3-6** A car weighing 130 kN when fully loaded is pulled slowly up a steep inclined track by a steel cable (see figure). The cable has an effective cross-sectional area of 490 mm², and the angle $\alpha$ of the incline is 30°.

(a) Calculate the tensile stress $\sigma_t$ in the cable.

(b) If the allowable stress in the cable is 150 MPa, what is the maximum acceptable angle of the incline for a fully loaded car?

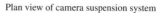

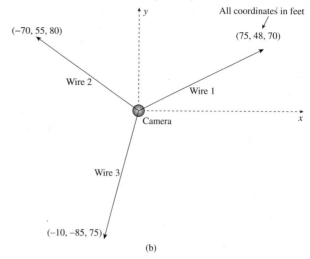

Plan view of camera suspension system

All coordinates in feet

$(-70, 55, 80)$

$(75, 48, 70)$

Wire 2

Wire 1

Camera

$y$

$x$

Wire 3

$(-10, -85, 75)$

(b)

**PROB. 1.3-7**

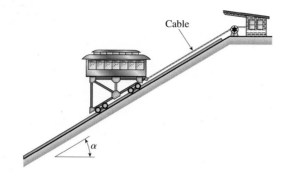

Cable

$\alpha$

**PROB. 1.3-6**

**1.3-7** Two steel wires support a moveable overhead camera weighing $W = 28$ lb (see figure part a) used for close-up viewing of field action at sporting events. At some instant, wire 1 is at an angle $\alpha = 22°$ to the horizontal and wire 2 is at an angle $\beta = 40°$. Wires 1 and 2 have diameters of 30 and 35 mils, respectively. (Wire diameters are often expressed in mils; one mil equals 0.001 in.)

(a) Determine the tensile stresses $\sigma_1$ and $\sigma_2$ in the two wires.

(b) If the stresses in wires 1 and 2 must be the same, what is the required diameter of wire 1?

(c) Now, to stabilize the camera for windy outdoor conditions, a third wire is added (see figure part b). Assume the three wires meet at a common point (coordinates (0, 0, 0) above the camera at the instant shown in figure part b). Wire 1 is attached to a support at coordinates (75 ft, 48 ft, 70 ft). Wire 2 is supported at (−70 ft, 55 ft, 80 ft). Wire 3 is supported at (10 ft, −85 ft, 75 ft). Assume that all three wires have a diameter of 30 mils. Find the tensile stresses in wires 1 to 3.

**1.3-8** A long retaining wall is braced by wood shores set at an angle of 30° and supported by concrete thrust blocks, as shown in the first part of the figure. The shores are evenly spaced, 3 m apart.

For analysis purposes, the wall and shores are idealized as shown in the second part of the figure. Note that the base of the wall and both ends of the shores are assumed to be pinned. The pressure of the soil against the wall is assumed to be triangularly distributed, and the resultant force acting on a 3-meter length of the wall is $F = 190$ kN.

If each shore has a 150 mm × 150 mm square cross section, what is the compressive stress $\sigma_c$ in the shores?

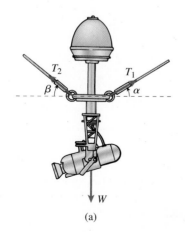

$T_2$    $T_1$

$\beta$    $\alpha$

$W$

(a)

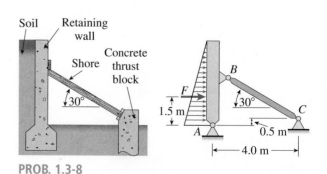

Soil    Retaining wall

Concrete thrust block

Shore

$30°$

$F$

1.5 m

$30°$

$B$

$C$

$A$

0.5 m

4.0 m

**PROB. 1.3-8**

**1.3-9** A pickup truck tailgate supports a crate ($W_C = 150$ lb), as shown in the figure. The tailgate weighs $W_T = 60$ lb and is supported by two cables (only one is shown in the figure). Each cable has an effective cross-sectional area $A_e = 0.017$ in.$^2$.

(a) Find the tensile force $T$ and normal stress $\sigma$ in each cable.

(b) If each cable elongates $\delta = 0.01$ in. due to the weight of both the crate and the tailgate, what is the average strain in the cable?

(b) If each cable elongates $\delta = 0.25$ mm due to the weight of both the crate and the tailgate, what is the average strain in the cable?

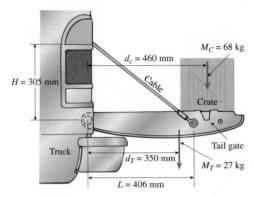

**PROB. 1.3-10**

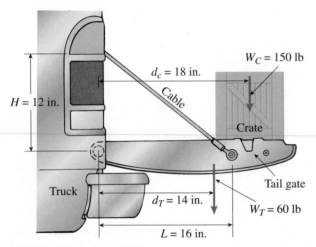

**PROBS. 1.3-9 and 1.3-10**

(© Barry Goodno)

**1.3-10** Solve the preceding problem if the mass of the tail gate is $M_T = 27$ kg and that of the crate is $M_C = 68$ kg. Use dimensions $H = 305$ mm, $L = 406$ mm, $d_C = 460$ mm, and $d_T = 350$ mm. The cable cross-sectional area is $A_e = 11.0$ mm$^2$.

(a) Find the tensile force $T$ and normal stress $\sigma$ in each cable.

**1.3-11** An L-shaped reinforced concrete slab 12 ft × 12 ft (but with a 6 ft × 6 ft cut-out) and thickness $t = 9.0$ in. is lifted by three cables attached at $O$, $B$ and $D$, as shown in the figure. The cables are combined at point $Q$, which is 7.0 ft above the top of the slab and directly above the center of mass at $C$. Each cable has an effective cross-sectional area of $A_e = 0.12$ in.$^2$.

(a) Find the tensile force $T_1$ ($i = 1, 2, 3$) in each cable due to the weight $W$ of the concrete slab (ignore weight of cables).

(b) Find the average stress $\sigma_i$ in each cable. (See Table I-1 in Appendix I for the weight density of reinforced concrete.)

(c) Add cable $AQ$ so that $OQA$ is one continuous cable, with each segment having force $T_1$, which is connected to cables $BQ$ and $DQ$ at point $Q$. Repeat parts (a) and (b). (*Hint*: There are now three force equilibrium equations and one *constraint equation*, $T_1 = T_4$.)

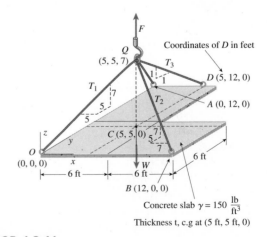

**PROB. 1.3-11**

**1.3-12** A round bar $ACB$ of length $2L$ (see figure) rotates about an axis through the midpoint $C$ with constant angular speed $\omega$ (radians per second). The material of the bar has weight density $\gamma$.

(a) Derive a formula for the tensile stress $\sigma_x$ in the bar as a function of the distance $x$ from the midpoint $C$.

(b) What is the maximum tensile stress $\sigma_{max}$?

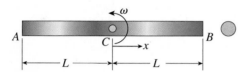

PROB. 1.3-12

**1.3-13** Two gondolas on a ski lift are locked in the position shown in the figure while repairs are being made elsewhere. The distance between support towers is $L = 100$ ft. The length of each cable segment under gondola weights $W_B = 450$ lb and $W_C = 650$ lb are $D_{AB} = 12$ ft, $D_{BC} = 70$ ft, and $D_{CD} = 20$ ft. The cable sag at $B$ is $\Delta_B = 3.9$ ft and that at $C(\Delta_C)$ is 7.1 ft. The effective cross-sectional area of the cable is $A_e = 0.12$ in.$^2$.

(a) Find the tension force in each cable segment; neglect the mass of the cable.

(b) Find the average stress ($\sigma$) in each cable segment.

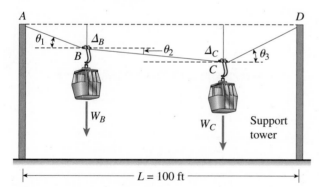

PROB. 1.3-13

**1.3-14** A crane boom of mass 450 kg with its center of mass at $C$ is stabilized by two cables $AQ$ and $BQ$ ($A_e = 304$ mm$^2$ for each cable) as shown in the figure. A load $P = 20$ kN is supported at point $D$. The crane boom lies in the $y$–$z$ plane.

(a) Find the tension forces in each cable: $T_{AQ}$ and $T_{BQ}$ (kN). Neglect the mass of the cables, but include the mass of the boom in addition to load $P$.

(b) Find the average stress ($\sigma$) in each cable.

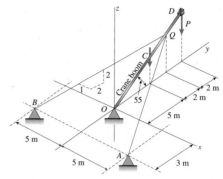

PROB. 1.3-14

# Mechanical Properties and Stress-Strain Diagrams

**1.4-1** Imagine that a long steel wire hangs vertically from a high-altitude balloon.

(a) What is the greatest length (feet) it can have without yielding if the steel yields at 40 ksi?

(b) If the same wire hangs from a ship at sea, what is the greatest length? (Obtain the weight densities of steel and sea water from Table I-1, Appendix I.)

**1.4-2** Steel riser pipe hangs from a drill rig located offshore in deep water (see figure).

(a) What is the greatest length (meters) it can have without breaking if the pipe is suspended in the air and the ultimate strength (or breaking strength) is 550 MPa?

(b) If the same riser pipe hangs from a drill rig at sea, what is the greatest length? (Obtain the weight densities of steel and sea water from Table I-1, Appendix I. Neglect the effect of buoyant foam casings on the pipe).

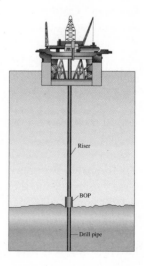

PROB. 1.4-2

**1.4-3** Three different materials, designated $A$, $B$, and $C$, are tested in tension using test specimens having diameters of 0.505 in. and gage lengths of 2.0 in. (see figure). At failure, the distances between the gage marks are found to be 2.13, 2.48, and 2.78 in., respectively. Also, at the failure cross sections the diameters are found to be 0.484, 0.398, and 0.253 in., respectively.

Determine the percent elongation and percent reduction in area of each specimen, and then, using your own judgment, classify each material as brittle or ductile.

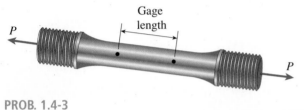

**Gage length**

**PROB. 1.4-3**

**1.4-4** The *strength-to-weight ratio* of a structural material is defined as its load-carrying capacity divided by its weight. For materials in tension, we may use a characteristic tensile stress (as obtained from a stress-strain curve) as a measure of strength. For instance, either the yield stress or the ultimate stress could be used, depending upon the particular application. Thus, the strength-to-weight ratio $R_{S/W}$ for a material in tension is defined as

$$R_{s/w} = \frac{\sigma}{\gamma}$$

in which $\sigma$ is the characteristic stress and $\gamma$ is the weight density. Note that the ratio has units of length.

Using the ultimate stress $\sigma_U$ as the strength parameter, calculate the strength-to-weight ratio (in units of meters) for each of the following materials: aluminum alloy 6061-T6, Douglas fir (in bending), nylon, structural steel ASTM-A572, and a titanium alloy. (Obtain the material properties from Tables I-1 and I-3 of Appendix I. When a range of values is given in a table, use the average value.)

**1.4-5** A symmetrical framework consisting of three pin-connected bars is loaded by a force $P$ (see figure). The angle between the inclined bars and the horizontal is $\alpha = 52°$. The axial strain in the middle bar is measured as 0.036.

Determine the tensile stress in the outer bars if they are constructed of a copper alloy having the following stress-strain relationship:

$$\sigma = \frac{18{,}000\varepsilon}{1 + 300\varepsilon} \qquad 0 \le \varepsilon \le 0.03 \qquad (\sigma = \text{ksi})$$

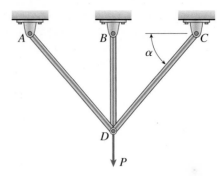

**PROB. 1.4-5**

**1.4-6** A specimen of a methacrylate plastic is tested in tension at room temperature (see figure), producing the stress-strain data listed in the accompanying table (see the next page).

Plot the stress-strain curve and determine the proportional limit, modulus of elasticity (i.e., the slope of the initial part of the stress-strain curve), and yield stress at 0.2% offset. Is the material ductile or brittle?

**PROB. 1.4-6**

| STRESS-STRAIN DATA FOR PROB. 1.4-6 | |
| --- | --- |
| **Stress (MPa)** | **Strain** |
| 8.0 | 0.0032 |
| 17.5 | 0.0073 |
| 25.6 | 0.0111 |
| 31.1 | 0.0129 |
| 39.8 | 0.0163 |
| 44.0 | 0.0184 |
| 48.2 | 0.0209 |
| 53.9 | 0.0260 |
| 58.1 | 0.0331 |
| 62.0 | 0.0429 |
| 62.1 | Fracture |

**1.4-7** The data shown in the accompanying table were obtained from a tensile test of high-strength steel. The test specimen had a diameter of 0.505 in. and a gage length of 2.00 in. (see figure for Prob. 1.4-3). At fracture, the elongation between the gage marks was 0.12 in. and the minimum diameter was 0.42 in.

Plot the conventional stress-strain curve for the steel and determine the proportional limit, modulus of elasticity (i.e., the slope of the initial part of the stress-strain curve), yield stress at 0.1% offset, ultimate stress, percent elongation in 2.00 in., and percent reduction in area.

| TENSILE-TEST DATA FOR PROB. 1.4-7 | |
|---|---|
| Load (lb) | Elongation (in.) |
| 1000 | 0.0002 |
| 2000 | 0.0006 |
| 6000 | 0.0019 |
| 10,000 | 0.0033 |
| 12,000 | 0.0039 |
| 12,900 | 0.0043 |
| 13,400 | 0.0047 |
| 13,600 | 0.0054 |
| 13,800 | 0.0063 |
| 14,000 | 0.0090 |
| 14,400 | 0.0102 |
| 15,200 | 0.0130 |
| 16,800 | 0.0230 |
| 18,400 | 0.0336 |
| 20,000 | 0.0507 |
| 22,400 | 0.1108 |
| 22,600 | Fracture |

## Elasticity and Plasticity

**1.5-1** A bar made of structural steel having the stress-strain diagram shown in the figure has a length of 48 in. The yield stress of the steel is 42 ksi and the slope of the initial linear part of the stress-strain curve (modulus of elasticity) is $30 \times 10^3$ ksi. The bar is loaded axially until it elongates 0.20 in., and then the load is removed.

How does the final length of the bar compare with its original length? (*Hint:* Use the concepts illustrated in Fig. 1-36b.)

**1.5-2** A bar of length 2.0 m is made of a structural steel having the stress-strain diagram shown in the figure. The yield stress of the steel is 250 MPa and the slope of the initial linear part of the stress-strain curve (modulus of elasticity) is 200 GPa. The bar is loaded axially until it elongates 6.5 mm, and then the load is removed.

How does the final length of the bar compare with its original length? (*Hint:* Use the concepts illustrated in Fig. 1-36b.)

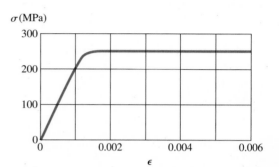

PROB. 1.5-2

**1.5-3** An aluminum bar has length $L = 6$ ft and diameter $d = 1.375$ in. The stress-strain curve for the aluminum is shown in Fig. 1-31 of Section 1.4. The initial straight-line part of the curve has a slope (modulus of elasticity) of $10.6 \times 10^6$ psi. The bar is loaded by tensile forces $P = 44.6$ k and then unloaded.

(a) What is the permanent set of the bar?
(b) If the bar is reloaded, what is the proportional limit? (*Hint:* Use the concepts illustrated in Figs. 1-36b and 1-37.)

**1.5-4** A circular bar of magnesium alloy is 750 mm long. The stress-strain diagram for the material is shown in the figure. The bar is loaded in tension to an elongation of 6.0 mm, and then the load is removed.

(a) What is the permanent set of the bar?
(b) If the bar is reloaded, what is the proportional limit? (*Hint:* Use the concepts illustrated in Figs. 1-36b and 1-37.)

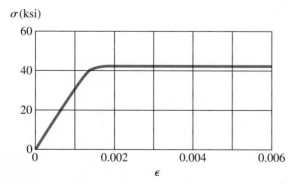

PROB. 1.5-1

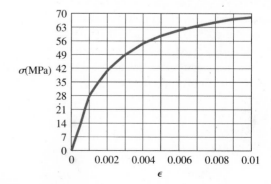

PROB. 1.5-4

**1.5-5** A wire of length $L = 4$ ft and diameter $d = 0.125$ in. is stretched by tensile forces $P = 600$ lb. The wire is made of a copper alloy having a stress-strain relationship that may be described mathematically by the following equation:

$$\sigma = \frac{18,000\varepsilon}{1 + 300\varepsilon} \quad 0 \le \varepsilon \le 0.03 \quad (\sigma = \text{ksi})$$

in which $\varepsilon$ is nondimensional and $\sigma$ has units of kips per square inch (ksi).

(a) Construct a stress-strain diagram for the material.

(b) Determine the elongation of the wire due to the forces $P$.

(c) If the forces are removed, what is the permanent set of the bar?

(d) If the forces are applied again, what is the proportional limit?

## Hooke's Law and Poisson's Ratio

*When solving the problems for Section 1.6, assume that the material behaves linearly elastically.*

**1.6-1** A high-strength steel bar used in a large crane has diameter $d = 2.00$ in. (see figure). The steel has modulus of elasticity $E = 29 \times 10^6$ psi and Poisson's ratio $v = 0.29$. Because of clearance requirements, the diameter of the bar is limited to 2.001 in. when it is compressed by axial forces.

What is the largest compressive load $P_{max}$ that is permitted?

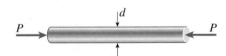

PROB. 1.6-1

**1.6-2** A round bar of 10 mm diameter is made of aluminum alloy 7075-T6 (see figure). When the bar is stretched by axial forces $P$, its diameter decreases by 0.016 mm.

Find the magnitude of the load $P$. (Obtain the material properties from Appendix I.)

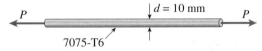

PROB. 1.6-2

**1.6-3** A polyethylene bar having diameter $d_1 = 4.0$ in. is placed inside a steel tube having inner diameter $d_2 = 4.01$ in. (see figure). The polyethylene bar is then compressed by an axial force $P$.

At what value of the force $P$ will the space between the polyethylene bar and the steel tube be closed? (For polyethylene, assume $E = 200$ ksi and $v = 0.4$.)

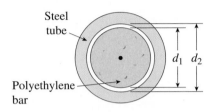

PROB. 1.6-3

**1.6-4** A circular aluminum tube of length $L = 600$ mm is loaded in compression by forces $P$ (see figure). The outside and inside diameters are $d_2 = 75$ mm and $d_1 = 63$ mm, respectively. A strain gage is placed on the outside of the tube to measure normal strains in the longitudinal direction. Assume that $E = 73$ GPa and Poisson's ration $v = 0.33$.

(a) If the compressive stress in the tube is 57 MPa, what is the load $P$?

(b) If the measured strain is $\varepsilon = 781 \times 10^{-6}$, what is the shortening $\delta$ of the tube? What is the percent decrease in its cross-sectional area? What is the volume change of the tube?

(c) If the tube has a constant outer diameter of $d_2 = 75$ mm along its entire length $L$ but *now* has increased *inner* diameter $d_3$ with a normal stress of 70 MPa over the middle third (see figure, part c) while the rest of the bar remains at normal stress of 57 MPa, what is the diameter $d_3$?

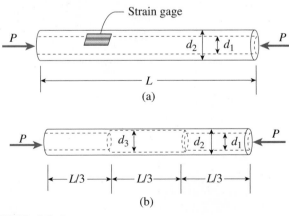

PROB. 1.6-4

**1.6-5** A bar of monel metal as in the figure (length $L = 9$ in., diameter $d = 0.225$ in.) is loaded axially by a tensile force $P$. If the bar elongates by 0.0195 in., what is the decrease in diameter $d$? What is the magnitude of the load $P$? Use the data in Table I-2, Appendix I.

**PROB. 1.6-5**

**1.6-6** A tensile test is peformed on a brass specimen 10 mm in diameter using a gage length of 50 mm (see figure). When the tensile load $P$ reaches a value of 20 kN, the distance between the gage marks has increased by 0.122 mm.

(a) What is the modulus of elasticity $E$ of the brass?

(b) If the diameter decreases by 0.00830 mm, what is Poisson's ratio?

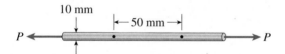

**PROB. 1.6-6**

**1.6-7** A hollow, brass circular pipe $ABC$ (see figure) supports a load $P_1 = 26.5$ kips acting at the top. A second load $P_2 = 22.0$ kips is uniformly distributed around the cap plate at $B$. The diameters and thicknesses of the upper and lower parts of the pipe are $d_{AB} = 1.25$ in., $t_{AB} = 0.5$ in., $d_{BC} = 2.25$ in., and $t_{BC} = 0.375$ in., respectively. The modulus of elasticity is 14,000 ksi. When both loads are fully applied, the wall thickness of pipe $BC$ increases by $200 \times 10^{-6}$ in.

(a) Find the increase in the inner diameter of pipe segment $BC$.

(b) Find Poisson's ratio for the brass.

(c) Find the increase in the wall thickness of pipe segment $AB$ and the increase in the inner diameter of $AB$.

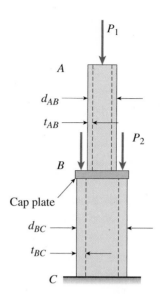

**PROB. 1.6-7**

**1.6-8** Three round, copper alloy bars having the same length $L$ but different shapes are shown in the figure. The first bar has a diameter $d$ over its entire length, the second has a diameter $d$ over one-fifth of its length, and the third has a diameter $d$ over one-fifteenth of its length. Elsewhere, the second and third bars have diameter $2d$. All three bars are subjected to the same axial load $P$.

Use the following numerical date: $P = 1400$ kN, $L = 5$ m, $d = 80$ mm, $E = 110$ GPa, and $v = 0.33$.

(a) Find the change in length of each bar.

(b) Find the change in volume of each bar.

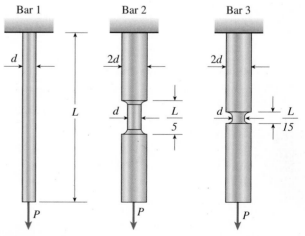

**PROB. 1.6-8**

# Shear Stress and Strain

**1.7-1** An angle bracket having thickness $t = 0.75$ in. is attached to the flange of a column by two 5/8-inch diameter bolts (see figure). A uniformly distributed load from a floor joist acts on the top face of the bracket with a pressure $p = 275$ psi. The top face of the bracket has length $L = 8$ in. and width $b = 3.0$ in.

Determine the average bearing pressure $\sigma_b$ between the angle bracket and the bolts and the average shear stress $\tau_{aver}$ in the bolts. (Disregard friction between the bracket and the column.)

**1.7-2** Truss members supporting a roof are connected to a 26-mm-thick gusset plate by a 22-mm diameter pin as shown in the figure and photo. The two end plates on the truss members are each 14 mm thick.

(a) If the load $P = 80$ kN, what is the largest bearing stress acting on the pin?

(b) If the ultimate shear stress for the pin is 190 MPa, what force $P_{ult}$ is required to cause the pin to fail in shear?

(Disregard friction between the plates.)

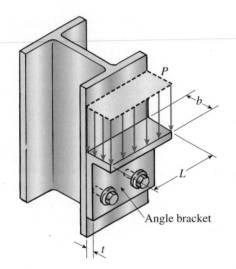

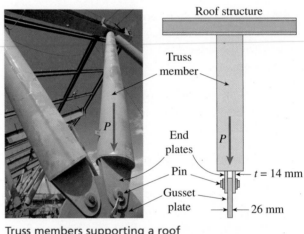

Truss members supporting a roof
(Vince Streano/Getty Images)

PROB. 1.7-2

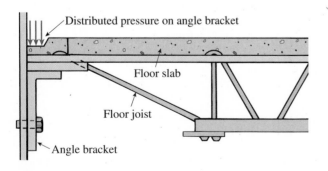

PROB. 1.7-1

**1.7-3** The upper deck of a football stadium is supported by braces, each of which transfers a load $P = 160$ kips to the base of a column (see figure part a). A cap plate at the bottom of the brace distributes the load $P$ to four flange plates ($t_f = 1$ in.) through a pin ($d_p = 2$ in.) to two gusset plates ($t_g = 1.5$ in.) (see figure parts b and c).

Determine the following quantities.

(a) The average shear stress $\tau_{aver}$ in the pin.

(b) The average bearing stress between the flange plates and the pin ($\sigma_{bf}$), and also between the gusset plates and the pin ($\sigma_{bg}$).

(Disregard friction between the plates.)

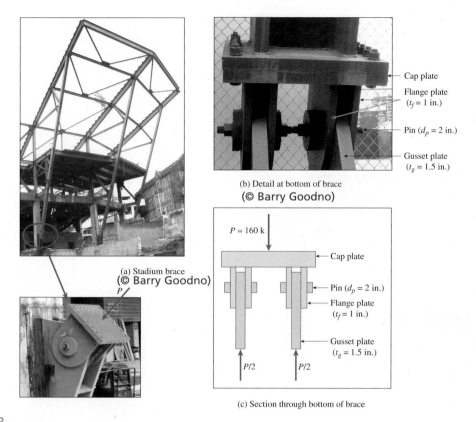

Cap plate

Flange plate
$(t_f = 1$ in.$)$

Pin $(d_p = 2$ in.$)$

Gusset plate
$(t_g = 1.5$ in.$)$

(b) Detail at bottom of brace
(© Barry Goodno)

$P = 160$ k

Cap plate

Pin $(d_p = 2$ in.$)$

Flange plate
$(t_f = 1$ in.$)$

Gusset plate
$(t_g = 1.5$ in.$)$

$P/2$     $P/2$

(c) Section through bottom of brace

(a) Stadium brace
(© Barry Goodno)

**PROB. 1.7-3**

**1.7-4** The inclined ladder $AB$ supports a house painter (85 kg) at $C$ and the self weight ($q = 40$ N/m) of the ladder itself. Each ladder rail ($t_r = 4$ mm) is supported by a shoe ($t_s = 5$ mm) which is attached to the ladder rail by a bolt of diameter $d_p = 8$ mm.

(a) Find support reactions at $A$ and $B$.
(b) Find the resultant force in the shoe bolt at $A$.
(c) Find maximum average shear ($\tau$) and bearing ($\sigma_b$) stresses in the shoe bolt at $A$.

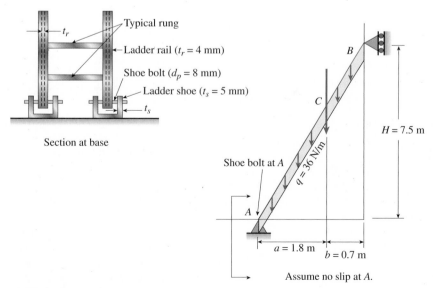

Typical rung

Ladder rail ($t_r = 4$ mm)

Shoe bolt ($d_p = 8$ mm)

Ladder shoe ($t_s = 5$ mm)

$t_r$

$t_s$

Section at base

$B$

$C$

$H = 7.5$ m

Shoe bolt at $A$

$q = 36$ N/m

$A$

$a = 1.8$ m

$b = 0.7$ m

Assume no slip at $A$.

**PROB. 1.7-4**

**1.7-5** The force in the brake cable of the V-brake system shown in the figure is $T = 45$ lb. The pivot pin at $A$ has diameter $d_p = 0.25$ in. and length $L_p = 5/8$ in.

Use dimensions show in the figure. Neglect the weight of the brake system.

(a) Find the average shear stress $\tau_{aver}$ in the pivot pin where it is anchored to the bicycle frame at $B$.

(b) Find the average bearing stress $\sigma_{b,aver}$ in the pivot pin over segment $AB$.

(© Barry Goodno)

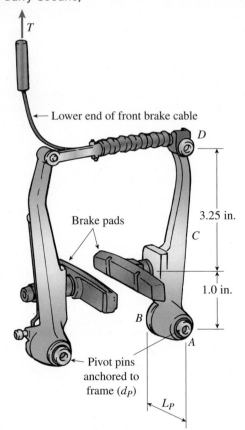

PROB. 1.7-5

**1.7-6** A steel plate of dimensions $2.5 \times 1.5 \times 0.08$ m and weighing 23.1 kN is hoisted by steel cables with lengths $L_1 = 3.2$ m and $L_2 = 3.9$ m that are each attached to the plate by a clevis and pin (see figure). The pins through the clevises are 18 mm in diameter and are located 2.0 m apart. The orientation angles are measured to be $\theta = 94.4°$ and $\alpha = 54.9°$.

For these conditions, first determine the cable forces $T_1$ and $T_2$, then find the average shear stress $\tau_{aver}$ in both pin 1 and pin 2, and then the average bearing stress $\sigma_b$ between the steel plate and each pin. Ignore the mass of the cables.

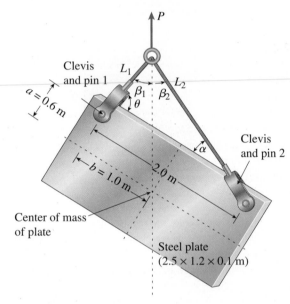

PROB. 1.7-6

**1.7-7** A special-purpose eye bolt of shank diameter $d = 0.50$ in. passes through a hole in a steel plate of thickness $t_p = 0.75$ in. (see figure) and is secured by a nut with thickness $t = 0.25$ in. The hexagonal nut bears directly against the steel plate. The radius of the circumscribed circle for the hexagon is $r = 0.40$ in. (which means that each side of the hexagon has length 0.40 in.). The tensile forces in three cables attached to the eye bolt are $T_1 = 800$ lb., $T_2 = 550$ lb., and $T_3 = 1241$ lb.

(a) Find the resultant force acting on the eye bolt.

(b) Determine the average bearing stress $\sigma_b$ between the hexagonal nut on the eye bolt and the plate.

(c) Determine the average shear stress $\tau_{aver}$ in the nut and also in the steel plate.

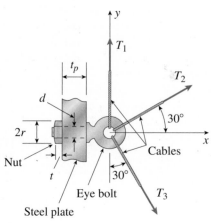

PROB. 1.7-7

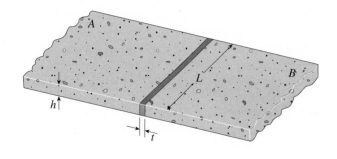

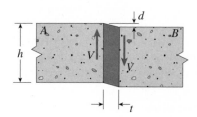

PROB. 1.7-9

**1.7-8** An elastomeric bearing pad consisting of two steel plates bonded to a chloroprene elastomer (an artificial rubber) is subjected to a shear force $V$ during a static loading test (see figure). The pad has dimensions $a = 125$ mm and $b = 240$ mm, and the elastomer has thickness $t = 50$ mm. When the force $V$ equals 12 kN, the top plate is found to have displaced laterally 8.0 mm with respect to the bottom plate.

What is the shear modulus of elasticity $G$ of the chloroprene?

**1.7-10** A flexible connection consisting of rubber pads (thickness $t = 9$ mm) bonded to steel plates is shown in the figure. The pads are 160 mm long and 80 mm wide.

(a) Find the average shear strain $\gamma_{aver}$ in the rubber if the force $P = 16$ kN and the shear modulus for the rubber is $G = 1250$ kPa.

(b) Find the relative horizontal displacement $\delta$ between the interior plate and the outer plates.

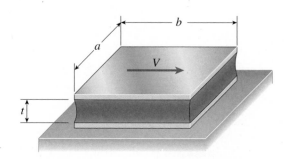

PROB. 1.7-8

**1.7-9** A joint between two concrete slabs $A$ and $B$ is filled with a flexible epoxy that bonds securely to the concrete (see figure). The height of the joint is $h = 4.0$ in., its length is $L = 40$ in., and its thickness is $t = 0.5$ in. Under the action of shear forces $V$, the slabs displace vertically through the distance $d = 0.002$ in. relative to each other.

(a) What is the average shear strain $\gamma_{aver}$ in the epoxy?

(b) What is the magnitude of the forces $V$ if the shear modulus of elasticity $G$ for the epoxy is 140 ksi?

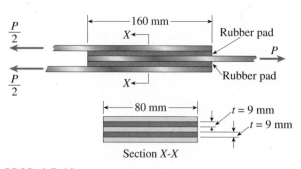

PROB. 1.7-10

**1.7-11** Steel riser pipe hangs from a drill rig located offshore in deep water (see figure). Separate segments are joined using bolted flange plages (see figure part b and photo). Assume that there are six bolts at each pipe segment connection. Assume that the total length of riser pipe is $L = 5000$ ft; outer and inner diameters are $d_2 = 16$ in. and $d_1 = 15$ in.; flange plate thickness $t_f = 1.75$ in.; and bolt and washer diameters are $d_b = 1.125$ in., and $d_w = 1.875$ in.

(a) If the entire length of the riser pipe were suspended in air, find the average normal stress $\sigma$ in each bolt, the average bearing stress $\sigma_b$ beneath each washer, and the average shear stress $\tau$ through the flange plate at each bolt location for the topmost bolted connection.

(b) If the same riser pipe hangs from a drill rig at sea, what are the normal, bearing, and shear stresses in the connection? (Obtain the weight densities of steel and sea water from Table I-1, Appendix I. Neglect the effect of buoyant foam casings on the riser pipe.)

(c)

(Courtesy of Transocean)
**PROB. 1.7-11**

**1.7-12** The clamp shown in the figure is used to support a load hanging from the lower flange of a steel beam. The clamp consists of two arms ($A$ and $B$) joined by a pin at $C$. The pin has diameter $d = 12$ mm. Because arm $B$ straddles arm $A$, the pin is in double shear.

Line 1 in the figure defines the line of action of the resultant horizontal force $H$ acting between the lower flange of the beam and arm $B$. The vertical distance from this line to the pin is $h = 250$ mm. Line 2 defines the line of action of the resultant vertical force $V$ acting between the flange and arm $B$. The horizontal distance from this line to the centerline of the beam is $c = 100$ mm. The force conditions between arm $A$ and the lower flange are symmetrical with those given for arm $B$.

Determine the average shear stress in the pin at $C$ when the load $P = 18$ kN.

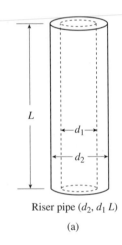

$L$

$d_1$

$d_2$

Riser pipe ($d_2$, $d_1$ $L$)

(a)

Flange plate ($t_f$), typical
bolt ($d_b$), and washer ($d_w$)

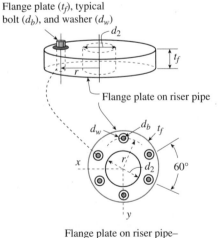

$d_2$

$t_f$

$r$

Flange plate on riser pipe

$d_w$    $d_b$  $t_f$

$x$    $r$    $d_2$    $60°$

$y$

Flange plate on riser pipe–
*plan view (n = 6 bolts shown)*

(b)

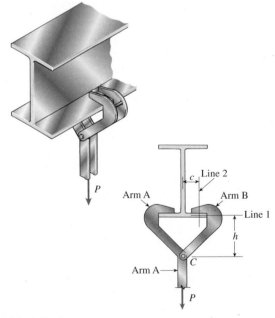

$P$    Arm A    $c$  Line 2    Arm B

Line 1

$h$

$C$

Arm A

$P$

**PROB. 1.7-12**

1.7-13 A hitch-mounted bicycle rack is designed to carry up to four 30-lb. bikes mounted on and strapped to two arms *GH* (see bike loads in the figure part a). The rack is attached to the vehicle at *A* and is assumed to be like a cantilever beam *ABCDGH* (figure part b). The weight of fixed segment *AB* is $W_1 = 10$ lb, centered 9 in. from *A* (see the figure part b) and the rest of the rack weighs $W_2 = 40$ lb, centered 19 in. from *A*. Segment *ABCDG* is a steel tube, $2 \times 2$ in., of thickness $t = 1/8$ in. Segment *BCDGH* pivots about a bolt at *B* of diameter $d_B = 0.25$ in. to allow access to the rear of the vehicle without removing the hitch rack.

When in use, the rack is secured in an upright position by a pin at *C* (diameter of pin $d_p = 5/16$ in.) (see photo and figure part c). The overturning effect of the bikes on the rack is resisted by a force couple $F \cdot h$ at *BC*.

(a) Find the support reactions at *A* for the fully loaded rack.

(b) Find forces in the bolt at *B* and the pin at *C*.

(c) Find average shear stresses $\tau_{aver}$ in both the bolt at *B* and the pin at *C*.

(d) Find average bearing stresses $\sigma_b$ in the bolt at **B** and the pin at *C*.

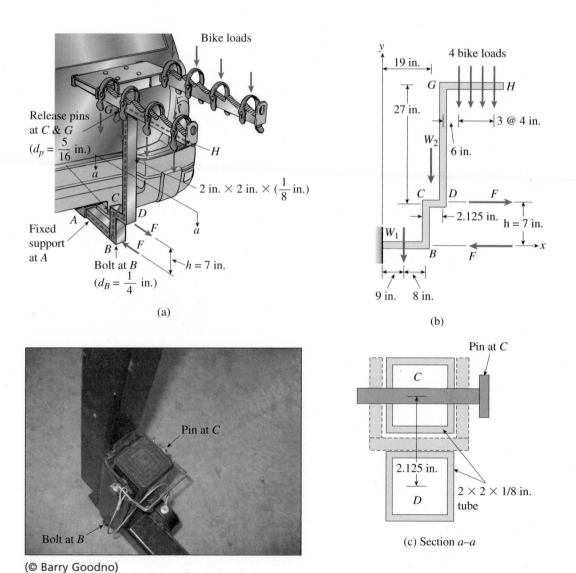

(a)

(b)

(© Barry Goodno)

(c) Section *a–a*

PROB. 1.7-13

**1.7-14** A bicycle chain consists of a series of small links, each 12 mm long between the centers of the pins (see figure). You might wish to examine a bicycle chain and observe its construction. Note particularly the pins, which we will assume to have a diameter of 2.5 mm.

In order to solve this problem, you must now make two measurements on a bicycle (see figure): (1) the length $L$ of the crank arm from main axle to pedal axle, and (2) the radius $R$ of the sprocket (the toothed wheel, sometimes called the chainring).

(a) Using your measured dimensions, calculate the tensile force $T$ in the chain due to a force $F = 800$ N applied to one of the pedals.

(b) Calculate the average shear stress $\tau_{aver}$ in the pins.

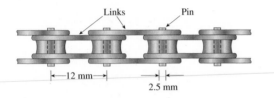

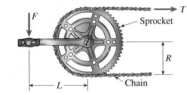

PROB. 1.7-14

**1.7-15** A shock mount constructed as shown in the figure is used to support a delicate instrument. The mount consists of an outer steel tube with inside diameter $b$, a central steel bar of diameter $d$ that supports the load $P$, and a hollow rubber cylinder (height $h$) bonded to the tube and bar.

(a) Obtain a formula for the shear stress $\tau$ in the rubber at a radial distance $r$ from the center of the shock mount.

(b) Obtain a formula for the downward displacement $\delta$ of the central bar due to the load $P$, assuming that $G$ is the shear modulus of elasticity of the rubber and that the steel tube and bar are rigid.

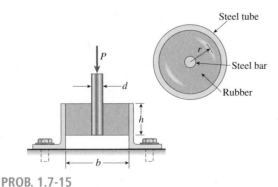

PROB. 1.7-15

**1.7-16** A removable sign post on a hurricane evacuation route has a square base plate with four slots (or cut-outs) at bolts 1 to 4 (see figure part a and photo) for ease of installation and removal. The upper portion of the post has a separate base plate which is bolted to an anchored base (see photo). Each of the four bolts has a diameter of $d_b$ and a washer with diameter of $d_w$. The bolts are arranged in a rectangular pattern ($b \times h$). Consider only **wind force $W_y$** applied in the $y$ direction at the center of pressure of the sign structure at a height $z = L$ above the base. Neglect the weight of the sign and post, and also neglect friction between the upper and lower base plates. Assume that the lower base plate and short anchored post are rigid.

(a) Find the average **shear stress** $\tau$ (MPa) at bolt 1 due to the wind force $W_y$; repeat for bolt 4

(b) Find the average **bearing stress** $\sigma_b$ (MPa) between the bolt and the base plate (thickness $t$) at bolt 1; repeat for bolt 4.

(c) Find the average **bearing stress** $\sigma_b$ (MPa) between base plate and washer at bolt 4 due to the wind force $W_y$ (assume the initial bolt pretension is zero).

(d) Find the average **shear stress** $\tau$ (MPa) through the base plate at bolt 4 due to the wind force $W_y$.

(e) Find an expression for the **normal stress** $\sigma$ in bolt 3 due to the wind force $W_y$.

See Prob. 1.8-15 for additional discussion of wind on a sign, and the resulting forces acting on a *conventional* base plate.

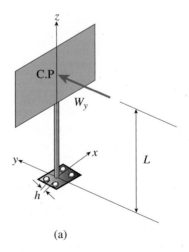

(a)

| Numerical data | |
|---|---|
| $H = 150$ mm, | $b = 96$ mm, |
| $h = 108$ mm, | $t = 14$ mm, |
| $d_b = 12$ mm, | $d_w = 22$ mm, |
| $L = 2.75$ m | $W_y = 667$ N |

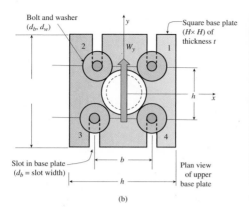

(© Barry Goodno)

**PROB. 1.7-16**

**1.7-17** A spray nozzle for a garden hose requires a force $F = 5$ lb. to open the spring-loaded spray chamber $AB$. The nozzle hand grip pivots about a pin through a flange at $O$. Each of the two flanges has thickness $t = 1/16$ in., and the pin has diameter $d_p = 1/8$ in. (see figure part a). The spray nozzle is attached to the garden hose with a quick release fitting at $B$ (see figure part b). Three brass balls (diameter $d_b = 3/16$ in.) hold the spray head in place

under water pressure force $f_p = 30$ lb. at $C$ (see figure part c). Use dimensions given in figure part a.

(a) Find the force in the pin at $O$ due to applied force F.

(b) Find average shear stress $\tau_{aver}$ and bearing stress $\sigma_b$ in the pin at $O$.

(c) Find the average shear stress $\tau_{aver}$ in the brass retaining balls at $C$ due to water pressure force $f_p$.

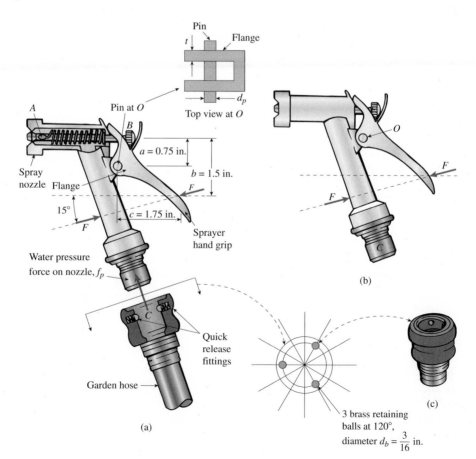

**PROB. 1.7-17**

**1.7-18** A single steel strut $AB$ with diameter $d_s = 8$ mm. supports the vehicle engine hood of mass 20 kg which pivots about hinges at $C$ and $D$ (see figure parts a and b). The strut is bent into a loop at its end and then attached to a bolt at $A$ with diameter $d_b = 10$ mm. Strut $AB$ lies in a vertical plane.

(a) Find the strut force $F_s$ and average normal stress $\sigma$ in the strut.

(b) Find the average shear stress $\tau_{aver}$ in the bolt at $A$.

(c) Find the average bearing stress $\sigma_b$ on the bolt at $A$.

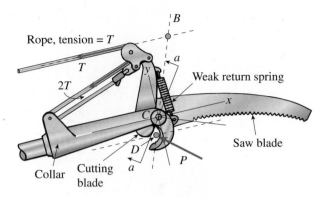

(a) Top part of pole saw

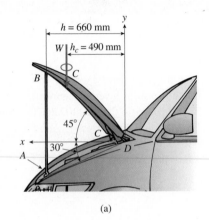

(a)

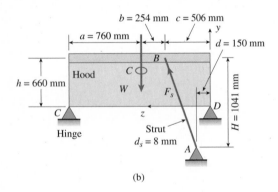

(b)

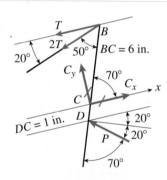

(b) Free-body diagram

**PROB. 1.7-18**

**1.7-19** The top portion of a pole saw used to trim small branches from trees is shown in the figure part a. The cutting blade $BCD$ (see figure parts a and c) applies a force $P$ at point $D$. Ignore the effect of the weak return spring attached to the cutting blade below $B$. Use properties and dimensions given in the figure.

(a) Find the force $P$ on the cutting blade at $D$ if the tension force in the rope is $T = 25$ lb (see free-body diagram in figure part b).

(b) Find force in the pin at $C$.

(c) Find average shear stress $\tau_{aver}$ and bearing stress $\sigma_b$ in the support pin at $C$ (see section $a$–$a$ through cutting blade in figure part c).

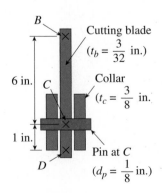

(c) Section $a$–$a$

**PROB. 1.7-19**

# Allowable Loads

**1.8-1** A bar of solid circular cross section is loaded in tension by forces $P$ (see figure). The bar has length $L = 16.0$ in. and diameter $d = 0.50$ in. The material is a magnesium alloy having modulus of elasticity $E = 6.4 \times 10^6$ psi. The allowable stress in tension is $\sigma_{allow} = 17,000$ psi, and the elongation of the bar must not exceed 0.04 in.

What is the allowable value of the forces $P$?

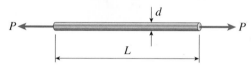

**PROB. 1.8-1**

**1.8-2** A torque $T_0$ is transmitted between two flanged shafts by means of ten 20-mm bolts (see figure and photo). The diameter of the bolt circle is $d = 250$ mm.

If the allowable shear stress in the bolts is 85 MPa, what is the maximum permissible torque? (Disregard friction between the flanges.)

Drive shaft coupling on a ship propulsion motor (Courtesy of American Superconductor)

**PROB. 1.8-2**

**1.8-3** A tie-down on the deck of a sailboat consists of a bent bar bolted at both ends, as shown in the figure. The diameter $d_B$ of the bar is 1/4 in., the diameter $d_W$ of the washers is 7/8 in., and the thickness $t$ of the fiberglass deck is 3/8 in.

If the allowable shear stress in the fiberglass is 300 psi, and the allowable bearing pressure between the washer and the fiberglass is 550 psi, what is the allowable load $P_{allow}$ on the tie-down?

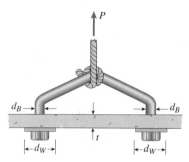

**PROB. 1.8-3**

**1.8-4** Two steel tubes are joined at $B$ by four pins ($d_p = 11$ mm), as shown in the cross section $a$–$a$ in the figure. The outer diameters of the tubes are $d_{AB} = 41$ mm and $d_{BC} = 28$ mm. The wall thicknesses are $t_{AB} = 6.5$ mm and $t_{BC} = 7.5$ mm. The yield stress in tension for the steel is $\sigma_Y = 200$ MPa and the ultimate stress in *tension* is $\sigma_U = 340$ MPa. The corresponding yield and ultimate values in *shear* for the pin are 80 MPa and 140 MPa, respectively. Finally, the yield and ultimate values in *bearing* between the pins and the tubes are 260 MPa and 450 MPa, respectively. Assume that the factors of safety with respect to yield stress and ultimate stress are 3.5 and 4.5, respectively.

(a) Calculate the allowable tensile force $P_{allow}$ considering tension in the tubes.

(b) Recompute $P_{allow}$ for shear in the pins.

(c) Finally, recompute $P_{allow}$ for bearing between the pins and the tubes. Which is the controlling value of P?

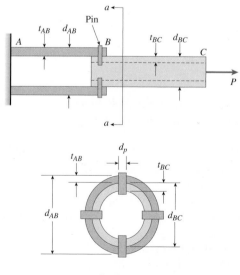

Section $a$–$a$

**PROB. 1.8-4**

**1.8-5** A steel pad supporting heavy machinery rests on four short, hollow, cast iron piers (see figure). The ultimate strength of the cast iron in compression is 50 ksi. The outer diameter of the piers is $d = 4.5$ in. and the wall thickness is $t = 0.40$ in.

Using a factor of safety of 3.5 with respect to the ultimate strength, determine the total load $P$ that may be supported by the pad.

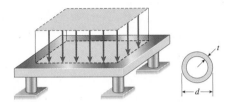

**PROB. 1.8-5**

**1.8-6** The rear hatch of a van ($BDCF$ in figure part a) is supported by two hinges at $B_1$ and $B_2$ and by two struts $A_1B_1$ and $A_2B_2$ (diameter $d_s = 10$ mm) as shown in figure part b. The struts are supported at $A_1$ and $A_2$ by pins, each with diameter $d_p = 9$ mm and passing through an eyelet of thickness $t = 8$ mm at the end of the strut (figure part b). If a closing force $P = 50$ N is applied at $G$ and the mass of the hatch $M_h = 43$ kg is concentrated at $C$:

(a) What is the force $F$ in each strut? (Use the free-body diagram of one half of the hatch in the figure part c.)

(b) What is the maximum permissible force in the strut, $F_{allow}$, if the allowable stresses are as follows: compressive stress in the strut, 70 MPa; shear stress in the pin, 45 MPa; and bearing stress between the pin and the end of the strut, 110 MPa.

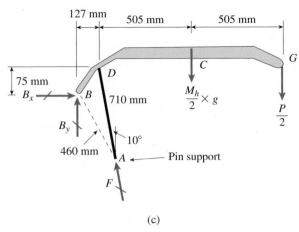

(c)

**PROB. 1.8-6**

**1.8-7** A lifeboat hangs from two ship's davits, as shown in the figure. A pin of diameter $d = 0.80$ in. passes through each davit and supports two pulleys, one on each side of the davit.

Cables attached to the lifeboat pass over the pulleys and wind around winches that raise and lower the lifeboat. The lower parts of the cables are vertical and the upper parts make an angle $\alpha = 15°$ with the horizontal. The allowable tensile force in each cable is 1800 lb, and the allowable shear stress in the pins is 4000 psi.

If the lifeboat weighs 1500 lb, what is the maximum weight that should be carried in the lifeboat?

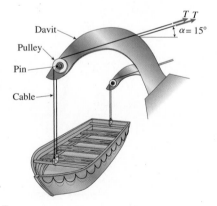

**PROB. 1.8-7**

**1.8-8** A cable and pulley system in figure part a supports a cage of mass 300 kg at $B$. Assume that this includes the mass of the cables as well. The thickness of each the three steel pulleys is $t = 40$ mm. The pin diameters are $d_{pA} = 25$ mm, $d_{pB} = 30$ mm and $d_{pC} = 22$ mm (see figure part a and part b).

(a) Find expressions for the resultant forces acting on the pulleys at $A$, $B$, and $C$ in terms of cable tension $T$.

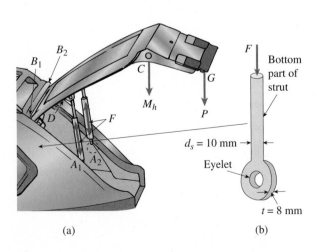

(a)

(b)

**(b)** What is the maximum weight $W$ that can be added to the cage at $B$ based on the following allowable stresses? Shear stress in the pins is 50 MPa; bearing stress between the pin and the pulley is 110 MPa.

Determine the allowable compressive force $P_{allow}$ in the spar.

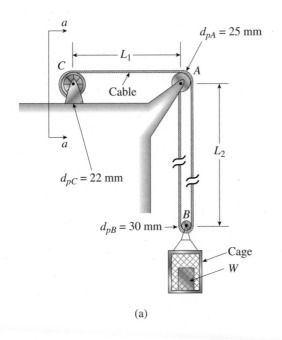

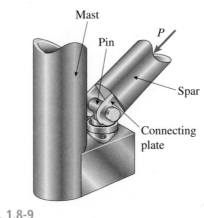

PROB. 1.8-9

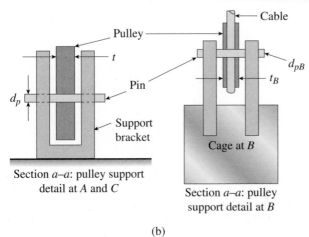

PROB. 1.8-8

**1.8-9** A ship's spar is attached at the base of a mast by a pin connection (see figure). The spar is a steel tube of outer diameter $d_2 = 3.5$ in. and inner diameter $d_1 = 2.8$ in. The steel pin has diameter $d = 1$ in., and the two plates connecting the spar to the pin have thickness $t = 0.5$ in. The allowable stresses are as follows: compressive stress in the spar, 10 ksi; shear stress in the pin, 6.5 ksi; and bearing stress between the pin and the connecting plates, 16 ksi.

**1.8-10** What is the maximum possible value of the clamping force $C$ in the jaws of the pliers shown in the figure if the ultimate shear stress in the 5-mm diameter pin is 340 MPa?

What is the maximum permissible value of the applied load $P$ if a factor of safety of 3.0 with respect to failure of the pin is to be maintained?

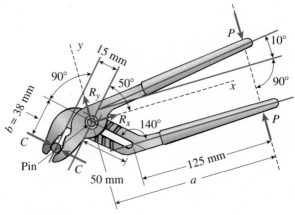

PROB. 1.8-10

**1.8-11** A metal bar *AB* of weight *W* is suspended by a system of steel wires arranged as shown in the figure. The diameter of the wires is 5/64 in., and the yield stress of the steel is 65 ksi.

Determine the maximum permissible weight $W_{max}$ for a factor of safety of 1.9 with respect to yielding.

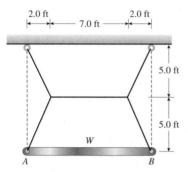

PROB. 1.8-11

**1.8-12** A plane truss is subjected to loads 2*P* and *P* at joints *B* and *C*, respectively, as shown in the figure part a. The truss bars are made of two L102 × 76 × 6.4 steel angles [see Table F-5(b): cross-sectional area of the two angles, *A* = 2180 mm², figure part b] having an ultimate stress in tension equal to 390 MPa. The angles are connected to an 12 mm-thick gusset plate at *C* (figure part c) with 16-mm diameter rivets; assume each rivet transfers an equal share of the member force to the gusset plate. The ultimate stresses in shear and bearing for the rivet steel are 190 MPa and 550 MPa, respectively.

Determine the allowable load $P_{allow}$ if a safety factor of 2.5 is desired with respect to the ultimate load that can be carried. (Consider tension in the bars, shear in the rivets, bearing between the rivets and the bars, and also bearing between the rivets and the gusset plate. Disregard friction between the plates and the weight of the truss itself.)

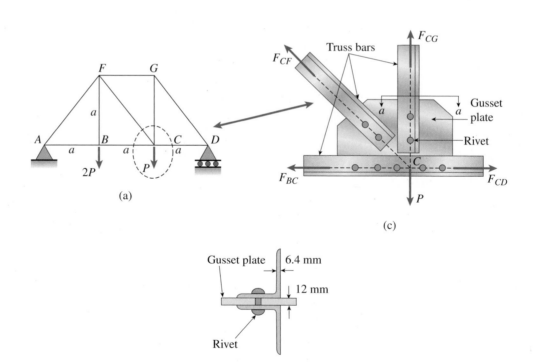

PROB. 1.8-12

**1.8-13** A solid bar of circular cross section (diameter $d$) has a hole of diameter $d/5$ drilled laterally through the center of the bar (see figure). The allowable average tensile stress on the net cross section of the bar is $\sigma_{allow}$.

(a) Obtain a formula for the allowable load $P_{allow}$ that the bar can carry in tension.

(b) Calculate the value of $P_{allow}$ if the bar is made of brass with diameter $d = 1.75$ in. and $\sigma_{allow} = 12$ ksi.

(*Hint*: Use the formulas of Case 15, Appendix E.)

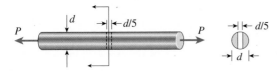

PROB. 1.8-13

**1.8-14** A solid steel bar of diameter $d_1 = 60$ mm has a hole of diameter $d_2 = 32$ mm drilled through it (see figure). A steel pin of diameter $d_2$ passes through the hole and is attached to supports.

Determine the maximum permissible tensile load $P_{allow}$ in the bar if the yield stress for shear in the pin is $\tau_Y = 120$ MPa, the yield stress for tension in the bar is $\sigma_Y = 250$ MPa and a factor of safety of 2.0 with respect to yielding is required. (*Hint*: Use the formulas of Case 15, Appendix E.)

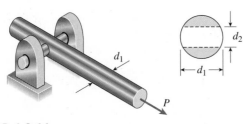

PROB. 1.8-14

**1.8-15** A sign of weight $W$ is supported at its base by four bolts anchored in a concrete footing. Wind pressure $p$ acts normal to the surface of the sign; the resultant of the uniform wind pressure is force $F$ at the center of pressure. The wind force is assumed to create equal shear forces $F/4$ in the $y$ direction at each bolt (see figure parts a and c). The overturning effect of the wind force also causes an uplift force $R$ at bolts $A$ and $C$ and a downward force $(-R)$ at bolts $B$ and $D$ (see figure part b). The resulting effects of the wind, and the associated ultimate stresses for each stress condition, are: normal stress in each bolt ($\sigma_u = 60$ ksi); shear through the base plate ($\tau_u = 17$ ksi); horizontal shear and bearing on each bolt ($\tau_{hu} = 25$ ksi and $\sigma_{bu} = 75$ ksi); and bearing on the bottom washer at $B$ (or $D$) ($\sigma_{bw} = 50$ ksi).

Find the maximum wind pressure $p_{max}$ (psf) that can be carried by the bolted support system for the sign if a safety factor of 2.5 is desired with respect to the ultimate wind load that can be carried.

Use the following numerical data: bolt $d_b = \frac{3}{4}$ in.; washer $d_w = 1.5$ in.; base plate $t_{bp} = 1$ in.; base plate dimensions $h = 14$ in. and $b = 12$ in.; $W = 500$ lb; $H = 17$ ft; sign dimensions ($L_v = 10$ ft. $\times$ $L_h = 12$ ft.); pipe column diameter $d = 6$ in., and pipe column thickness $t = 3/8$ in.

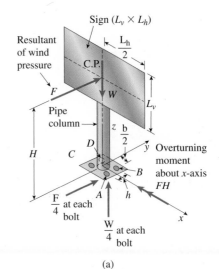

(a)

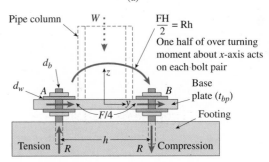

(b)

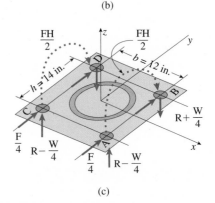

(c)

PROB. 1.8-15

**1.8-16** The piston in an engine is attached to a connecting rod $AB$, which in turn is connected to a crank arm $BC$ (see figure). The piston slides without friction in a cylinder and is subjected to a force $P$ (assumed to be constant) while moving to the right in the figure. The connecting rod, which has diameter $d$ and length $L$, is attached at both ends by pins. The crank arm rotates about the axle at $C$ with the pin at $B$ moving in a circle of radius $R$. The axle at $C$, which is supported by bearings, exerts a resisting moment $M$ against the crank arm.

(a) Obtain a formula for the maximum permissible force $P_{allow}$ based upon an allowable compressive stress $\sigma_c$ in the connecting rod.

(b) Calculate the force $P_{allow}$ for the following data: $\sigma_c = 160$ MPa, $d = 9.00$ mm, and $R = 0.28L$.

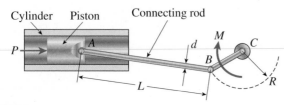

Cylinder    Piston    Connecting rod

PROB. 1.8-16

# Design for Axial Loads and Direct Shear

**1.9-1** An aluminum tube is required to transmit an axial tensile force $P = 33$ k (see figure part a). The thickness of the wall of the tube is to be 0.25 in.

(a) What is the minimum required outer diameter $d_{min}$ if the allowable tensile stress is 12,000 psi?

(b) Repeat part (a) if the tube will have a hole of diameter $d/10$ at mid-length (see figure parts b and c).

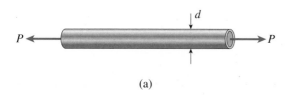

(a)

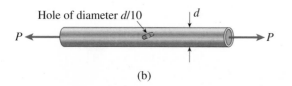

Hole of diameter $d/10$

(b)

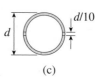

(c)

PROB. 1.9-1

**1.9-2** A copper alloy pipe having yield stress $\sigma_Y = 290$ MPa is to carry an axial tensile load $P = 1500$ kN (see figure part a). A factor of safety of 1.8 against yielding is to be used.

(a) If the thickness t of the pipe is to be one-eighth of its outer diameter, what is the minimum required outer diameter $d_{min}$?

(b) Repeat part (a) if the tube has a hole of diameter $d/10$ drilled through the entire tube as shown in the figure part b.

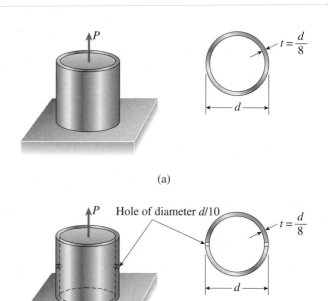

(a)

Hole of diameter $d/10$

(b)

PROB. 1.9-2

**1.9-3** A horizontal beam $AB$ with cross-sectional dimensions ($b = 0.75$ in.) $\times$ ($h = 8.0$ in.) is supported by an inclined strut $CD$ and carries a load $P = 2700$ lb at joint $B$ (see figure part a). The strut, which consists of two bars each of thickness $5b/8$, is connected to the beam by a bolt passing through the three bars meeting at joint $C$ (see figure part b).

(a) If the allowable shear stress in the bolt is 13,000 psi, what is the minimum required diameter $d_{min}$ of the bolt at $C$?

(b) If the allowable bearing stress in the bolt is 19,000 psi, what is the minimum required diameter $d_{min}$ of the bolt at $C$?

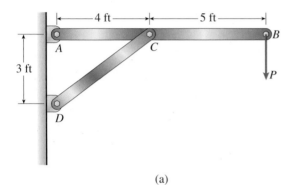

(a)

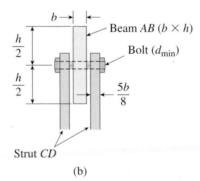

(b)

**PROB. 1.9-3**

**1.9-4** Lateral bracing for an elevated pedestrian walkway is shown in the figure part a. The thickness of the clevis plate $t_c = 16$ mm and the thickness of the gusset plate $t_g = 20$ mm (see figure part b). The maximum force in the diagonal bracing is expected to be $F = 190$ kN.

If the allowable shear stress in the pin is 90 MPa and the allowable bearing stress between the pin and both the clevis and gusset plates is 150 MPa, what is the minimum required diameter $d_{min}$ of the pin?

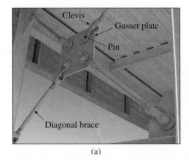

(a)

(© Barry Goodno)

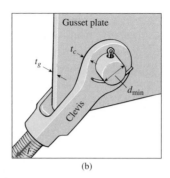

(b)

**PROB. 1.9-4**

**1.9-5** A plane truss has joint loads $P$, $2P$, and $3P$ at joints $D$, $C$, and $B$, respectively (see figure) where load variable $P = 5200$ lb. All members have two end plates (see figure for Prob. 1.7-2) which are pin-connected to gusset plates (see also figure for Prob. 1.8-12). Each end plate has thickness $t_p = 0.625$ in., and all gusset plates have thickness $t_g = 1.125$ in. If the allowable shear stress in each pin is 12,000 psi and the allowable bearing stress in each pin is 18,000 psi, what is the minimum required diameter $d_{min}$ of the pins used at either end of member $BE$?

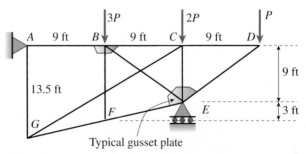

Typical gusset plate

**PROB. 1.9-5**

**1.9-6** A suspender on a suspension bridge consists of a cable that passes over the main cable (see figure) and supports the bridge deck, which is far below. The suspender is held in position by a metal tie that is prevented from sliding downward by clamps around the suspender cable.

Let $P$ represent the load in each part of the suspender cable, and let $\theta$ represent the angle of the suspender cable just above the tie. Finally, let $\sigma_{allow}$ represent the allowable tensile stress in the metal tie.

(a) Obtain a formula for the minimum required cross-sectional area of the tie.

(b) Calculate the minimum area if $P = 130$ kN, $\theta = 75°$, and $\sigma_{allow} = 80$ MPa.

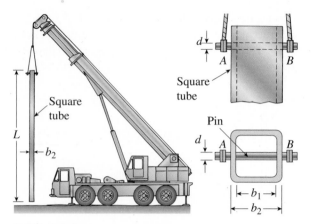

PROB. 1.9-7

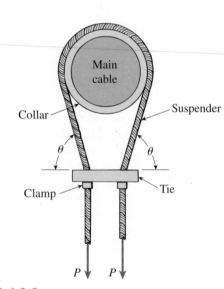

PROB. 1.9-6

**1.9-7** A square steel tube of length $L = 20$ ft and width $b_2 = 10.0$ in. is hoisted by a crane (see figure). The tube hangs from a pin of diameter $d$ that is held by the cables at points $A$ and $B$. The cross section is a hollow square with inner dimension $b_1 = 8.5$ in. and outer dimension $b_2 = 10.0$ in. The allowable shear stress in the pin is 8,700 psi, and the allowable bearing stress between the pin and the tube is 13,000 psi.

Determine the minimum diameter of the pin in order to support the weight of the tube. (*Note:* Disregard the rounded corners of the tube when calculating its weight.)

**1.9-8** A cable and pulley system at $D$ is used to bring a 230-kg pole ($ACB$) to a vertical position as shown in the figure part a. The cable has tensile force $T$ and is attached at $C$. The length $L$ of the pole is 6.0 m, the outer diameter is $d = 140$ mm, and the wall thickness $t = 12$ mm. The pole pivots about a pin at $A$ in figure part b. The allowable shear stress in the pin is 60 MPa and the allowable bearing stress is 90 MPa.

Find the minimum diameter of the pin at A in order to support the weight of the pole in the position shown in the figure part a.

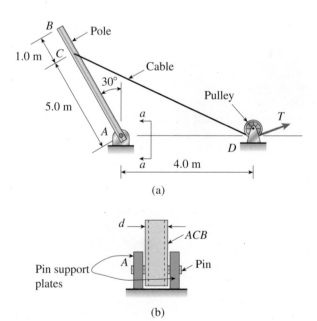

PROB. 1.9-8

**1.9-9** A pressurized circular cylinder has a sealed cover plate fastened with steel bolts (see figure). The pressure $p$ of the gas in the cylinder is 290 psi, the inside diameter $D$ of the cylinder is 10.0 in., and the diameter $d_B$ of the bolts is 0.50 in.

If the allowable tensile stress in the bolts is 10,000 psi, find the number $n$ of bolts needed to fasten the cover.

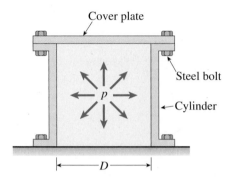

Cover plate

Steel bolt

$p$

Cylinder

$\longleftarrow$ $D$ $\longrightarrow$

**PROB. 1.9-9**

**1.9-10** A tubular post of outer diameter $d_2$ is guyed by two cables fitted with turnbuckles (see figure). The cables are tightened by rotating the turnbuckles, thus producing tension in the cables and compression in the post. Both cables are tightened to a tensile force of 110 kN. Also, the angle between the cables and the ground is 60°, and the allowable compressive stress in the post is $\sigma_c = 35$ MPa.

If the wall thickness of the post is 15 mm, what is the minimum permissible value of the outer diameter $d_2$?

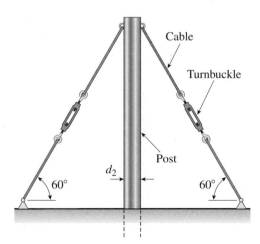

Cable

Turnbuckle

Post

$d_2$

60°    60°

**PROB. 1.9-10**

**1.9-11** A large precast concrete panel for a warehouse is being raised to a vertical position using two sets of cables

at two lift lines as shown in the figure part a. Cable 1 has length $L_1 = 22$ ft and distances along the panel (see figure part b) are $a = L_1/2$ and $b = L_1/4$. The cables are attached at lift points $B$ and $D$ and the panel is rotated about its base at $A$. However, as a worst case, assume that the panel is momentarily lifted off the ground and its total weight must be supported by the cables. Assuming the cable lift forces $F$ at each lift line are about equal, use the simplified model of one half of the panel in figure part b to perform your analysis for the lift position shown. The total weight of the panel is $W = 85$ kips. The orientation of the panel is defined by the following angles: $\gamma = 20°$ and $\theta = 10°$.

Find the required cross-sectional area $A_C$ of the cable if its breaking stress is 91 ksi and a factor of safety of 4 with respect to failure is desired.

(a)

(Courtesy Tilt-Up Concrete Association)

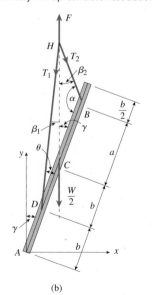

(b)

**PROB. 1.9-11**

**1.9-12** A steel column of hollow circular cross section is supported on a circular steel base plate and a concrete pedestal (see figure). The column has outside diameter $d = 250$ mm and supports a load $P = 750$ kN.

(a) If the allowable stress in the column is 55 MPa, what is the minimum required thickness $t$? Based upon your result, select a thickness for the column. (Select a thickness that is an even integer, such as 10, 12, 14, . . . , in units of millimeters.)

(b) If the allowable bearing stress on the concrete pedestal is 11.5 MPa, what is the minimum required diameter $D$ of the base plate if it is designed for the allowable load $P_{allow}$ that the column with the selected thickness can support?

(b) Also re-check the normal tensile stress in rod $BC$ and bearing stress at $B$; if either is inadequate under the additional load from platform $HF$, redesign them to meet the original design criteria.

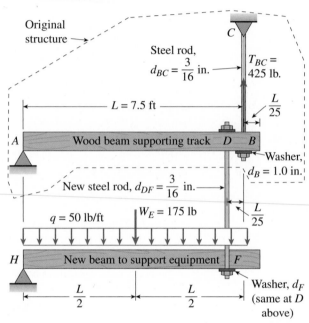

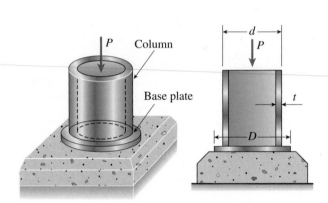

PROB. 1.9-12

PROB. 1.9-13

**1.9-13** An elevated jogging track is supported at intervals by a wood beam $AB$ ($L = 7.5$ ft) which is pinned at $A$ and supported by steel rod $BC$ and a steel washer at $B$. Both the rod ($d_{BC} = 3/16$ in.) and the washer ($d_B = 1.0$ in.) were designed using a rod tension force of $T_{BC} = 425$ lb. The rod was sized using a factor of safety of 3 against reaching the ultimate stress $\sigma_u = 60$ ksi. An allowable bearing stress $\sigma_{ba} = 565$ psi was used to size the washer at $B$.

Now, a small platform $HF$ is to be suspended below a section of the elevated track to support some mechanical and electrical equipment. The equipment load is uniform load $q = 50$ lb/ft and concentrated load $W_E = 175$ lb at mid-span of beam $HF$. The plan is to drill a hole through beam $AB$ at $D$ and install the same rod ($d_{BC}$) and washer ($d_B$) at both $D$ and $F$ to support beam $HF$.

(a) Use $\sigma_u$ and $\sigma_{ba}$ to check the proposed design for rod $DF$ and washer $d_F$; are they acceptable?

**1.9-14** A flat bar of width $b = 60$ mm and thickness $t = 10$ mm is loaded in tension by a force $P$ (see figure). The bar is attached to a support by a pin of diameter $d$ that passes through a hole of the same size in the bar. The allowable tensile stress on the net cross section of the bar is $\sigma_T = 140$ MPa, the allowable shear stress in the pin is $\tau_S = 80$ MPa, and the allowable bearing stress between the pin and the bar is $\sigma_B = 200$ MPa.

(a) Determine the pin diameter $d_m$ for which the load $P$ will be a maximum.

(b) Determine the corresponding value $P_{max}$ of the load.

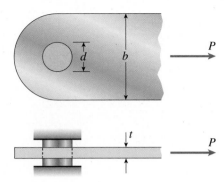

**1.9-15**  Two bars *AB* and *BC* of the same material support a vertical load *P* (see figure). The length *L* of the horizontal bar is fixed, but the angle $\theta$ can be varied by moving support *A* vertically and changing the length of bar *AC* to correspond with the new position of support *A*. The allowable stresses in the bars are the same in tension and compression.

We observe that when the angle $\theta$ is reduced, bar *AC* becomes shorter but the cross-sectional areas of both bars increase (because the axial forces are larger). The opposite effects occur if the angle $\theta$ is increased. Thus, we see that the weight of the structure (which is proportional to the volume) depends upon the angle $\theta$.

Determine the angle $\theta$ so that the structure has minimum weight without exceeding the allowable stresses in the bars. (*Note*: The weights of the bars are very small compared to the force *P* and may be disregarded.)

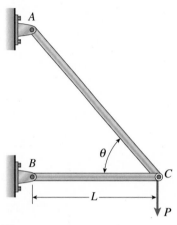

PROB. 1.9-15

# Axially Loaded Members

An oil drilling rig is comprised of axially loaded members that must be designed for a variety of loading conditions, including self-weight, impact, and temperature effects. (Joe Raedle/Reportage/ Getty Images)

# CHAPTER OVERVIEW

In Chapter 2, we consider several other aspects of axially loaded members, beginning with the determination of changes in lengths caused by loads (Sections 2.2 and 2.3). The calculation of **changes in lengths** is an essential ingredient in the analysis of statically indeterminate structures, a topic we introduce in Section 2.4. If the member is statically indeterminate, we must augment the equations of statical equilibrium with compatibility equations (which rely on **force-displacement relations**) to solve for any unknowns of interest, such as support reactions or internal axial forces in members. Changes in lengths also must be calculated whenever it is necessary to control the displacements of a structure, whether for aesthetic or functional reasons. In Section 2.5, we discuss the **effects of temperature** on the length of a bar, and we introduce the concepts of **thermal stress** and **thermal strain**. Also included in this section is a discussion of the effects of **misfits and prestrains**.

A generalized view of the stresses in axially loaded bars is presented in Section 2.6, where we discuss the **stresses on inclined sections** (as distinct from **cross sections**) of bars. Although only normal stresses act on cross sections of axially loaded bars, both normal and shear stresses act on inclined sections. Stresses on inclined sections of axially loaded members are investigated as a first step toward a more complete consideration of **plane stress states** in later chapters. We then introduce several additional topics of importance in mechanics of materials, namely, strain energy (Section 2.7), impact loading (Section 2.8), fatigue (Section 2.9), stress concentrations (Section 2.10), and nonlinear behavior (Sections 2.11 and 2.12). Although these subjects are discussed in the context of members with axial loads, the discussions provide the foundation for applying the same concepts to other structural elements, such as bars in torsion and beams in bending.

## Chapter 2 is organized as follows:

2.1 Introduction 120
2.2 Changes in Lengths of Axially Loaded Members 120
2.3 Changes in Lengths Under Nonuniform Conditions 130
2.4 Statically Indeterminate Structures 138
2.5 Thermal Effects, Misfits, and Prestrains 149
2.6 Stresses on Inclined Sections 164
2.7 Strain Energy 176

*2.8 Impact Loading 187
*2.9 Repeated Loading and Fatigue 195
*2.10 Stress Concentrations 197
*2.11 Nonlinear Behavior 205
*2.12 Elastoplastic Analysis 210
Chapter Summary & Review 216
Problems 218

*Specialized and/or advanced topics

## 2.1 INTRODUCTION

Structural components subjected only to tension or compression are known as **axially loaded members**. Solid bars with straight longitudinal axes are the most common type, although cables and coil springs also carry axial loads. Examples of axially loaded bars are truss members, connecting rods in engines, spokes in bicycle wheels, columns in buildings, and struts in aircraft engine mounts. The stress-strain behavior of such members was discussed in Chapter 1, where we also obtained equations for the stresses acting on cross sections ($\sigma = P/A$) and the strains in longitudinal directions ($\varepsilon = \delta/L$).

## 2.2 CHANGES IN LENGTHS OF AXIALLY LOADED MEMBERS

When determining the changes in lengths of axially loaded members, it is convenient to begin with a **coil spring** (Fig. 2-1). Springs of this type are used in large numbers in many kinds of machines and devices—for instance, there are dozens of them in every automobile.

When a load is applied along the axis of a spring, as shown in Fig. 2-1, the spring gets longer or shorter depending upon the direction of the load. If the load acts away from the spring, the spring elongates and we say that the spring is loaded in *tension*. If the load acts toward the spring, the spring shortens and we say it is in *compression*. However, it should not be inferred from this terminology that the individual coils of a spring are subjected to direct tensile or compressive stresses; rather, the coils act primarily in direct shear and torsion (or twisting). Nevertheless, the overall stretching or shortening of a spring is analogous to the behavior of a bar in tension or compression, and so the same terminology is used.

### Springs

The elongation of a spring is pictured in Fig. 2-2, where the upper part of the figure shows a spring in its **natural length** $L$ (also called its *unstressed length*, *relaxed length*, or *free length*), and the lower part of the figure shows the effects of applying a tensile load. Under the action of the force $P$, the spring lengthens by an amount $\delta$ and its final length becomes $L + \delta$. If the material of the spring is **linearly elastic**, the load and elongation will be proportional:

$$P = k\delta \qquad \delta = fP \qquad \textbf{(2-1a,b)}$$

in which $k$ and $f$ are constants of proportionality.

The constant $k$ is called the **stiffness** of the spring and is defined as the force required to produce a unit elongation, that is, $k = P/\delta$. Similarly, the constant $f$ is known as the **flexibility** and is defined as the elongation produced by a load of unit value, that is, $f = \delta/P$. Although we used a spring in tension for this discussion, it should be obvious that Eqs. (2-1a) and (2-1b) also apply to springs in compression.

**Fig. 2-1**

Spring subjected to an axial load $P$

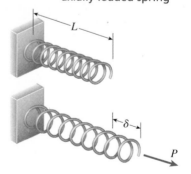

**Fig. 2-2**

Elongation of an axially loaded spring

From the preceding discussion it is apparent that the stiffness and flexibility of a spring are the reciprocal of each other:

$$k = \frac{1}{f} \qquad f = \frac{1}{k} \qquad \textbf{(2-2a,b)}$$

The flexibility of a spring can easily be determined by measuring the elongation produced by a known load, and then the stiffness can be calculated from Eq. (2-2a). Other terms for the stiffness and flexibility of a spring are the **spring constant** and **compliance**, respectively.

The spring properties given by Eqs. (2-1) and (2-2) can be used in the analysis and design of various mechanical devices involving springs, as illustrated later in Example 2-1.

## Prismatic Bars

Axially loaded bars elongate under tensile loads and shorten under compressive loads, just as springs do. To analyze this behavior, let us consider the prismatic bar shown in Fig. 2-3. A **prismatic bar** is a structural member having a straight longitudinal axis and constant cross section throughout its length. Although we often use circular bars in our illustrations, we should bear in mind that structural members may have a variety of cross-sectional shapes, such as those shown in Fig. 2-4.

The **elongation** $\delta$ of a prismatic bar subjected to a tensile load $P$ is shown in Fig. 2-5. If the load acts through the centroid of the end cross section, the uniform normal stress at cross sections away from the ends is given by the formula $\sigma = P/A$, where $A$ is the cross-sectional area. Furthermore, if the bar is made of a homogeneous material, the axial strain is $\varepsilon = \delta/L$, where $\delta$ is the elongation and $L$ is the length of the bar.

**Fig. 2-3**

Prismatic bar of circular cross section

**Fig. 2-4**

Typical cross sections of structural members

Solid cross sections

Hollow or tubular cross sections

Thin-walled open cross sections

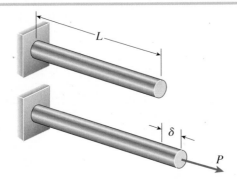

**Fig. 2-5**

Elongation of a prismatic bar in tension

Let us also assume that the material is **linearly elastic**, which means that it follows Hooke's law. Then the longitudinal stress and strain are related by the equation $\sigma = E\varepsilon$, where $E$ is the modulus of elasticity. Combining these basic relationships, we get the following equation for the elongation of the bar:

$$\delta = \frac{PL}{EA} \qquad \textbf{(2-3)}$$

This equation shows that the elongation is directly proportional to the load $P$ and the length $L$ and inversely proportional to the modulus of elasticity $E$ and the cross-sectional area $A$. The product $EA$ is known as the **axial rigidity** of the bar.

Although Eq. (2-3) was derived for a member in tension, it applies equally well to a member in compression, in which case $\delta$ represents the shortening of the bar. Usually we know by inspection whether a member gets longer or shorter; however, there are occasions when a **sign convention** is needed (for instance, when analyzing a statically indeterminate bar). When that happens, elongation is usually taken as positive and shortening as negative.

The change in length of a bar is normally very small in comparison to its length, especially when the material is a structural metal, such as steel or aluminum. As an example, consider an aluminum strut that is 75.0 in. long and subjected to a moderate compressive stress of 7000 psi. If the modulus of elasticity is 10,500 ksi, the shortening of the strut (from Eq. 2-3 with $P/A$ replaced by $\sigma$) is $\delta = 0.050$ in. Consequently, the ratio of the change in length to the original length is 0.05/75, or 1/1500, and the final length is 0.999 times the original length. Under ordinary conditions similar to these, we can use the original length of a bar (instead of the final length) in calculations.

The stiffness and flexibility of a prismatic bar are defined in the same way as for a spring. The stiffness is the force required to produce a unit elongation, or $P/\delta$, and the flexibility is the elongation due to a unit load, or $\delta/P$. Thus, from Eq. (2-3) we see that the **stiffness** and **flexibility** of a prismatic bar are, respectively,

$$k = \frac{EA}{L} \qquad f = \frac{L}{EA} \qquad\qquad \textbf{(2-4a,b)}$$

Stiffnesses and flexibilities of structural members, including those given by Eqs. (2-4a) and (2-4b), have a special role in the analysis of large structures by computer-oriented methods.

## Cables

Cables are used to transmit large tensile forces, for example, when lifting and pulling heavy objects, raising elevators, guying towers, and supporting suspension bridges. Unlike springs and prismatic bars, cables cannot resist compression. Furthermore, they have little resistance to bending and therefore may be curved as well as straight. Nevertheless, a cable is considered to be an axially loaded member because it is subjected only to tensile forces. Because the tensile forces in a cable are directed along the axis, the forces may vary in both direction and magnitude, depending upon the configuration of the cable.

Cables are constructed from a large number of wires wound in some particular manner. While many arrangements are available depending upon how the cable will be used, a common type of cable, shown in Fig. 2-6, is formed by six *strands* wound helically around a central strand. Each strand is in turn constructed of many wires, also wound helically. For this reason, cables are often referred to as **wire rope**.

Steel cables on a pulley
(© Barsik/Dreamstime.com)

**Fig. 2-6**

Typical arrangement of strands
and wires in a steel cable
(Tom Grundy/Shutterstock)

The cross-sectional area of a cable is equal to the total cross-sectional area of the individual wires, called the **effective area** or **metallic area**. This area is less than the area of a circle having the same diameter as the cable because there are spaces between the individual wires. For example, the actual cross-sectional area (effective area) of a particular 1.0-in. diameter cable is only 0.471 in.$^2$, whereas the area of a 1.0-in. diameter circle is 0.785 in.$^2$.

Under the same tensile load, the elongation of a cable is greater than the elongation of a solid bar of the same material and same metallic cross-sectional area, because the wires in a cable "tighten up" in the same manner as the fibers in a rope. Thus, the modulus of elasticity (called the **effective modulus**) of a cable is less than the modulus of the material of which it is made. The effective modulus of steel cables is about 20,000 ksi (140 GPa), whereas the steel itself has a modulus of about 30,000 ksi (210 GPa).

When determining the **elongation** of a cable from Eq. (2-3), the effective modulus should be used for $E$ and the effective area should be used for $A$.

In practice, the cross-sectional dimensions and other properties of cables are obtained from the manufacturers. However, for use in solving problems in this book (and definitely *not* for use in engineering applications), we list in Table 2-1 the properties of a particular type of cable. Note that the last column contains the *ultimate load*, which is the load that would cause the cable to break. The *allowable load* is obtained from the ultimate load by applying a safety factor that may range from 3 to 10, depending upon how the cable is to be used. The individual wires in a cable are usually made of high-strength steel, and the calculated tensile stress at the breaking load can be as high as 200,000 psi (1400 MPa).

The following examples illustrate techniques for analyzing simple devices containing springs and bars. The solutions require the use of free-body diagrams, equations of equilibrium, and equations for changes in length. The problems at the end of the chapter provide many additional examples.

| Nominal diameter | | Approximate weight | | Effective area | | Ultimate load | |
|---|---|---|---|---|---|---|---|
| in. | (mm) | lb/ft | (N/m) | in.$^2$ | (mm$^2$) | lb | (kN) |
| 0.50 | (12) | 0.42 | (6.1) | 0.119 | (76.7) | 23,100 | (102) |
| 0.75 | (20) | 0.95 | (13.9) | 0.268 | (173) | 51,900 | (231) |
| 1.00 | (25) | 1.67 | (24.4) | 0.471 | (304) | 91,300 | (406) |
| 1.25 | (32) | 2.64 | (38.5) | 0.745 | (481) | 144,000 | (641) |
| 1.50 | (38) | 3.83 | (55.9) | 1.08 | (697) | 209,000 | (930) |
| 1.75 | (44) | 5.24 | (76.4) | 1.47 | (948) | 285,000 | (1260) |
| 2.00 | (50) | 6.84 | (99.8) | 1.92 | (1230) | 372,000 | (1650) |

**Table 2-1**

Properties of steel cables*

* To be used solely for solving problems in this book.

• • •   Example 2-1

A small lab scale has a rigid L-shaped frame *ABC* consisting of a horizontal arm *AB* (length $b = 11$ in.) and a vertical arm *BC* (length $c = 7.2$ in.) pivoted at point *B*, as shown in Fig. 2-7a. The pivot is attached to the outer frame *BCD*, which stands on a laboratory bench. The position of the pointer at *C* is controlled by a spring (stiffness $k = 8.15$ lb/in.) that is attached to a threaded rod. The position of the threaded rod is adjusted by turning the knurled nut.

The *pitch* of the threads (that is, the distance from one thread to the next) is $p = 1/16$ in., which means that one full revolution of the nut will move the rod by that same amount. Initially, when there is no weight on the hanger, the nut is turned until the pointer at the end of arm *BC* is directly over the reference mark on the outer frame.

**(a)** If a weight $W = 3$ lb is placed on the hanger at *A*, **how many revolutions of the nut are required** to bring the pointer back to the mark? (Deformations of the metal parts of the device may be disregarded, because they are negligible compared to the change in length of the spring.)

**(b)** If a *rotational* spring of stiffness $k_r = kb^2/4$ is added at *B*, how many revolutions of the nut are now required? (The rotational spring provides resisting moment $k_r$ (lb-in./radian) for unit angular rotations of *ABC*.)

---

**Fig. 2-7**

Example 2-1: (a) Small lab scale, and (b) free-body diagram of L-frame *ABC* on small lab scale

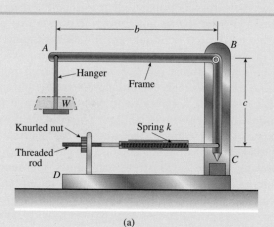

(a)

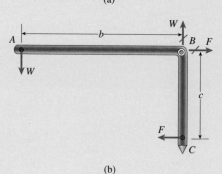

(b)

## Solution

### (a) Use an L-frame with translational spring $k$ only.

Inspection of the device (Fig. 2-7a) shows that the weight $W$ acting downward will cause the pointer at $C$ to move to the right. When the pointer moves to the right, the spring stretches by an additional amount that we can determine from the force in the spring.

To determine the force in the spring, we construct a free-body diagram of frame $ABC$ (Fig. 2-7b). In this diagram, $W$ represents the force applied by the hanger, and $F$ represents the force applied by the spring. The reactions at the pivot are indicated with slashes across the arrows (see the discussion of reactions in Sections 1.2 and 1.9).

Taking moments about point $B$ gives

$$F = \frac{Wb}{c} \tag{a}$$

The corresponding elongation of the spring (from Eq. 2-1a) is

$$\delta = \frac{F}{k} = \frac{Wb}{ck} \tag{b}$$

To bring the pointer back to the mark, we must turn the nut through enough revolutions to move the threaded rod to the left an amount equal to the elongation of the spring. Since each complete turn of the nut moves the rod a distance equal to the pitch $p$, the total movement of the rod is equal to $np$, where $n$ is the number of turns. Therefore, we obtain a formula for the required turns of the nut $n$ as

$$np = \delta = \frac{Wb}{ck} \quad \text{so} \quad n = \frac{Wb}{ckp} \tag{c, d}$$

*Numerical results*: Substitute numerical data into Eq. (d) to find the required number of turns of the nut $n$:

$$n = \frac{Wb}{ckp} = \frac{(3 \text{ lb})(11 \text{ in.})}{(7.2 \text{ in.})(8.15 \text{ lb/in.})\left(\dfrac{1}{16} \text{ in.}\right)} = 9 \text{ revolutions} \quad \Leftarrow$$

This result shows that if we rotate the nut through nine revolutions, the threaded rod will move to the left an amount equal to the elongation of the spring caused by the 3-lb load, thus returning the pointer to the reference mark.

### (b) Use an L-frame with translational spring $k$ and rotational spring $k_r$.

The rotational spring at $B$ (Fig. 2-7c) provides additional resistance in the form of a moment at $B$ to movement of the pointer at $C$. A new free-body diagram is required. If we apply a *small* rotation $\theta$ at joint $B$, the resulting forces and moments are shown in Fig. 2-7d. Summing moments about $B$, we obtain

$$Fc + k_r\theta = Wb \quad \text{or} \quad kc^2\theta + k_r\theta = Wb \tag{e, f}$$

Solving for rotation $\theta$, we obtain

$$\theta = \frac{Wb}{kc^2 + k_r} = \frac{Wb}{k\left(c^2 + \dfrac{b^2}{4}\right)} \tag{g}$$

*Continues* ➡

**Example 2-1** - *Continued*

## Fig. 2-7 (Continued)

Example 2-1: (c) Small lab scale with rotational spring added at B, and (d) free-body diagram of frame ABC with rotational spring added at B and a *small* rotation θ applied at B

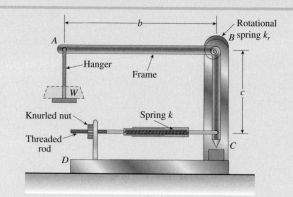

(c)

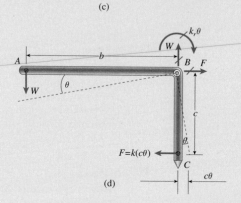

(d)

The force $F$ in the translational spring is now

$$F = k(c\theta) = kc\left[\frac{Wb}{k\left(c^2 + \dfrac{b^2}{4}\right)}\right] \quad \text{or} \quad F = \frac{Wbc}{\left(c^2 + \dfrac{b^2}{4}\right)} \quad \textbf{(h)}$$

So, the elongation of the translational spring is

$$\delta = np = \frac{F}{k} = \frac{Wbc}{k\left(c^2 + \dfrac{b^2}{4}\right)}$$

and the required number of revolutions of the nut is

$$n = \frac{Wbc}{kp\left(c^2 + \dfrac{b^2}{4}\right)}$$

so $n = \dfrac{(3\,\text{lb})(11\,\text{in.})(7.2\,\text{in.})}{(8.15\,\text{lb/in.})\left(\dfrac{1}{16}\,\text{in.}\right)\left[(7.2\,\text{in.})^2 + \dfrac{(11\,\text{in.})^2}{4}\right]} = 5.7$ revolutions

The combination of a translational spring attached to the threaded rod and a rotational spring at B results in a system with greater stiffness than that having a translational spring only. Hence, the pointer at C moves less under the load W, so fewer revolutions are required to re-center the pointer.

## ● ● ● Example 2-2

The device shown in Fig. 2-8a consists of a horizontal beam *ABC* supported by two vertical bars *BD* and *CE*. Bar *CE* is pinned at both ends but bar *BD* is fixed to the foundation at its lower end. The distance from *A* to *B* is 450 mm and from *B* to *C* is 225 mm. Bars *BD* and *CE* have lengths of 480 mm and 600 mm, respectively, and their cross-sectional areas are 1020 mm² and 520 mm², respectively. The bars are made of steel having a modulus of elasticity $E = 205$ GPa.

Assuming that beam *ABC* is rigid determine the following.

**(a)** Find the maximum allowable load $P_{max}$ if the displacement of point *A* is limited to 1.0 mm.

**(b)** If $P = 25$ kN, what is the required cross-sectional area of bar *CE* so that the displacement at point *A* is equal to 1.0 mm?

### Fig. 2-8

Example 2-2: Horizontal beam *ABC* supported by two vertical bars

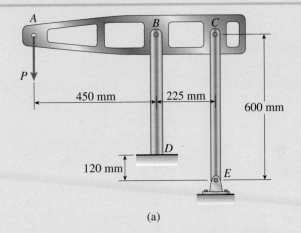

(a)

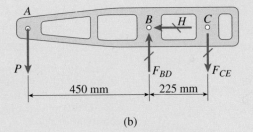

(b)

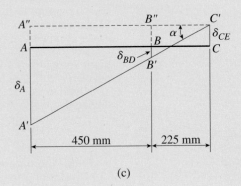

(c)

*Continues* ➡

**Example 2-2 -** *Continued*

## Solution

### (a) Determine the maximum allowable load $P_{max}$.

To find the displacement of point A, we need to know the displacements of points B and C. Therefore, we must find the changes in lengths of bars BD and CE, using the general equation $\delta = PL/EA$ (Eq. 2-3).

We begin by finding the forces in the bars from a free-body diagram of the beam (Fig. 2-8b). Because bar CE is pinned at both ends, it is a "two-force" member and transmits only a vertical force $F_{CE}$ to the beam. However, bar BD can transmit both a vertical force $F_{BD}$ and a horizontal force H. From equilibrium of beam ABC in the horizontal direction, we see that the horizontal force vanishes.

Two additional equations of equilibrium enable us to express the forces $F_{BD}$ and $F_{CE}$ in terms of the load P. Thus, by taking moments about point B and then summing forces in the vertical direction, we find

$$F_{CE} = 2P \qquad F_{BD} = 3P \qquad \text{(a)}$$

Note that the force $F_{CE}$ acts downward on bar ABC and the force $F_{BD}$ acts upward. Therefore, member CE is in tension and member BD is in compression.

The shortening of member BD is

$$\delta_{BD} = \frac{F_{BD}L_{BD}}{EA_{BD}}$$

$$= \frac{(3P)(480 \text{ mm})}{(205 \text{ GPa})(1020 \text{ mm}^2)} = 6.887P \times 10^{-6} \text{ mm} \ (P = \text{newtons}) \quad \text{(b)}$$

Note that the shortening $\delta_{BD}$ is expressed in millimeters provided the load P is expressed in newtons.

Similarly, the lengthening of member CE is

$$\delta_{CE} = \frac{F_{CE}L_{CE}}{EA_{CE}}$$

$$= \frac{(2P)(600 \text{ mm})}{(205 \text{ GPa})(520 \text{ mm}^2)} = 11.26P \times 10^{-6} \text{ mm} \ (P = \text{newtons}) \quad \text{(c)}$$

Again, the displacement is expressed in millimeters provided the load P is expressed in newtons. Knowing the changes in lengths of the two bars, we can now find the displacement of point A.

*Displacement diagram.* A displacement diagram showing the relative positions of points A, B, and C is sketched in Fig. 2-8c. Line ABC represents the original alignment of the three points. After the load P is applied, member BD shortens by the amount $\delta_{BD}$ and point B moves to B'. Also, member CE elongates by the amount $\delta_{CE}$ and point C moves to C'. Because the beam ABC is assumed to be rigid, points A', B', and C' lie on a straight line.

For clarity, the displacements are highly exaggerated in the diagram. In reality, line ABC rotates through a very small angle to its new position A'B'C' (see Note 2 at the end of this example).

Using similar triangles, we can now find the relationships between the displacements at points A, B, and C. From triangles A'A"C' and B'B"C' we get

$$\frac{A'A''}{A''C'} = \frac{B'B''}{B''C'} \quad \text{or} \quad \frac{\delta_A + \delta_{CE}}{450 + 225} = \frac{\delta_{BD} + \delta_{CE}}{225} \qquad \text{(d)}$$

in which all terms are expressed in millimeters.

Substituting for $\delta_{BD}$ and $\delta_{CE}$ from Eqs. (f) and (g) gives

$$\frac{\delta_A + 11.26P \times 10^{-6}}{450 + 225} = \frac{6.887P \times 10^{-6} + 11.26P \times 10^{-6}}{225}$$

### Fig. 2-8 (Repeated)

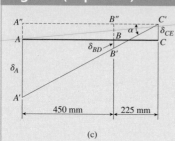

(c)

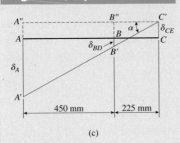

**Fig. 2-8 (Repeated)**

(c)

Finally, we substitute for $\delta_A$ its limiting value of 1.0 mm and solve the equation for the load $P$. The result is

$$P = P_{max} = 23{,}200 \text{ N (or } 23.2 \text{ kN)} \quad \Leftarrow$$

When the load reaches this value, the downward displacement at point $A$ is 1.0 mm.

*Note 1:* Since the structure behaves in a linearly elastic manner, the displacements are proportional to the magnitude of the load. For instance, if the load is one-half of $P_{max}$, that is, if $P = 11.6$ kN, the downward displacement of point $A$ is 0.5 mm.

*Note 2:* To verify our premise that line $ABC$ rotates through a very small angle, we can calculate the angle of rotation $\alpha$ from the displacement diagram (Fig. 2-8c), as follows:

$$\tan \alpha = \frac{A'A''}{A''C'} = \frac{\delta_A + \delta_{CE}}{675 \text{ mm}} \quad \text{(e)}$$

The displacement $\delta_A$ of point $A$ is 1.0 mm, and the elongation $\delta_{CE}$ of bar $CE$ is found from Eq. (g) by substituting $P = 23{,}200$ N; the result is $\delta_{CE} = 0.261$ mm. Therefore, from Eq. (i) we get

$$\tan \alpha = \frac{1.0 \text{ mm} + 0.261 \text{ mm}}{675 \text{ mm}} = \frac{1.261 \text{ mm}}{675 \text{ mm}} = 0.001868$$

from which $\alpha = 0.11°$. This angle is so small that if we tried to draw the displacement diagram to scale, we would not be able to distinguish between the original line $ABC$ and the rotated line $A'B'C'$.

Thus, when working with displacement diagrams, we usually can consider the displacements to be very small quantities, thereby simplifying the geometry. In this example we were able to assume that points $A$, $B$, and $C$ moved only vertically, whereas if the displacements were large, we would have to consider that they moved along curved paths.

**(b) Determine the required cross-sectional area of bar *CE*.**

If $P = 25$ kN, what is the required cross-sectional area of bar $CE$ so that the displacement at point $A$ is equal to 1.0 mm?

We start with the displacements relationship in Eq. (d):

$$\frac{\delta_A + \delta_{CE}}{450 \text{ mm} + 225 \text{ mm}} = \frac{\delta_{BD} + \delta_{CE}}{225 \text{ mm}} \quad \text{(d repeated)}$$

then substitute the required numerical value for $\delta_A$ and the force-displacement expressions from Eqs. (b) and (c), giving

$$\frac{\delta_A + \left(\dfrac{F_{CE}L_{CE}}{EA_{CE}}\right)}{450 \text{ mm} + 225 \text{ mm}} = \frac{\left(\dfrac{F_{BD}L_{BD}}{EA_{BD}}\right) + \left(\dfrac{F_{CE}L_{CE}}{EA_{CE}}\right)}{225 \text{ mm}} \quad \text{(f)}$$

After substituting expressions for $F_{BD}$ and $F_{CE}$ from Eq. (a), we can solve Eq. (f) for $A_{CE}$ to get the expression:

$$A_{CE} = \frac{4A_{BD}L_{CE}P}{A_{BD}E\delta_A - 9L_{BD}P} \quad \text{(g)}$$

Substituting numerical values, we find that the required cross-sectional area of bar $CE$ to ensure that point $A$ displaces 1.0 mm under an applied load $P = 25$ kN is

$$A_{CE} = \frac{4(1020 \text{ mm}^2)(600 \text{ mm})(25 \text{ kN})}{(1020 \text{ mm}^2)(205 \text{ GPa})(1.0 \text{ mm}) - 9(480 \text{ mm})(25 \text{ kN})} = 605 \text{ mm}^2 \quad \Leftarrow$$

The applied load in part (b) of $P = 25$ kN is greater than $P_{max} = 23.2$ kN in part (a), so the cross-sectional area of $CE$ is larger (as expected).

# 2.3 CHANGES IN LENGTHS UNDER NONUNIFORM CONDITIONS

When a prismatic bar of linearly elastic material is loaded only at the ends, we can obtain its change in length from the equation $\delta = PL/EA$, as described in the preceding section. In this section we will see how this same equation can be used in more general situations.

## Bars with Intermediate Axial Loads

Suppose, for instance, that a prismatic bar is loaded by one or more axial loads acting at intermediate points along the axis (Fig. 2-9a). We can determine the change in length of this bar by adding algebraically the elongations and shortenings of the individual segments. The procedure is as follows.

1.  Identify the segments of the bar (segments $AB$, $BC$, and $CD$) as segments 1, 2, and 3, respectively.

2.  Determine the internal axial forces $N_1$, $N_2$, and $N_3$ in segments 1, 2, and 3, respectively, from the free-body diagrams of Figs. 2-9b, c, and d. Note that the internal axial forces are denoted by the letter $N$ to distinguish them from the external loads $P$. By summing forces in the vertical direction, we obtain the following expressions for the axial forces:

$$N_1 = -P_B + P_C + P_D \quad N_2 = P_C + P_D \quad N_3 = P_D$$

In writing these equations we used the sign convention given in the preceding section (internal axial forces are positive when in tension and negative when in compression).

---

### Fig. 2-9

(a) Bar with external loads acting at intermediate points; (b), (c), and (d) free-body diagrams showing the internal axial forces $N_1$, $N_2$, and $N_3$

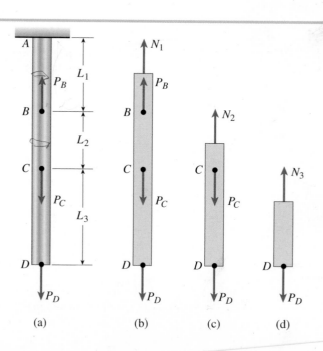

(a)  (b)  (c)  (d)

3. Determine the changes in the lengths of the segments from Eq. (2-3):

$$\delta_1 = \frac{N_1 L_1}{EA} \qquad \delta_2 = \frac{N_2 L_2}{EA} \qquad \delta_3 = \frac{N_3 L_3}{EA}$$

in which $L_1$, $L_2$, and $L_3$ are the lengths of the segments and $EA$ is the axial rigidity of the bar.

4. Add $\delta_1$, $\delta_2$, and $\delta_3$ to obtain $\delta$, the change in length of the entire bar:

$$\delta = \sum_{i=1}^{3} \delta_i = \delta_1 + \delta_2 + \delta_3$$

As already explained, the changes in lengths must be added algebraically, with elongations being positive and shortenings negative.

## Bars Consisting of Prismatic Segments

This same general approach can be used when the bar consists of several prismatic segments, each having different axial forces, different dimensions, and different materials (Fig. 2-10). The change in length may be obtained from the equation

$$\delta = \sum_{i=1}^{n} \frac{N_i L_i}{E_i A_i} \tag{2-5}$$

in which the subscript $i$ is a numbering index for the various segments of the bar and $n$ is the total number of segments. Note especially that $N_i$ is not an external load but is the internal axial force in segment $i$.

## Bars with Continuously Varying Loads or Dimensions

Sometimes the axial force $N$ and the cross-sectional area $A$ vary continuously along the axis of a bar, as illustrated by the tapered bar of Fig. 2-11a. This bar not only has a continuously varying cross-sectional area but also a continuously varying axial force. In this illustration, the load consists of two parts, a single force $P_B$ acting at end $B$ of the bar and distributed forces $p(x)$ acting along the axis. (A distributed force has units of force per unit distance, such as pounds per inch or newtons per meter.) A distributed axial load may be produced by such factors as centrifugal forces, friction forces, or the weight of a bar hanging in a vertical position.

Under these conditions we can no longer use Eq. (2-5) to obtain the change in length. Instead, we must determine the change in length of a differential element of the bar and then integrate over the length of the bar.

We select a differential element at distance $x$ from the left-hand end of the bar (Fig. 2-11a). The internal axial force $N(x)$ acting at this cross section (Fig. 2-11b) may be determined from equilibrium using either segment $AC$ or segment $CB$ as a free body. In general, this force is a function of $x$. Also, knowing the dimensions of the bar, we can express the cross-sectional area $A(x)$ as a function of $x$.

Bar consisting of prismatic segments having different axial forces, different dimensions, and different materials

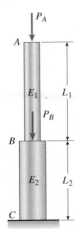

**Fig. 2-11**

Bar with varying cross-sectional area and varying axial force

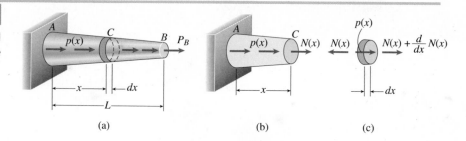

(a)          (b)          (c)

The elongation $d\delta$ of the differential element (Fig. 2-11c) may be obtained from the equation $\delta = PL/EA$ by substituting $N(x)$ for $P$, $dx$ for $L$, and $A(x)$ for $A$, as follows:

$$d\delta = \frac{N(x)dx}{EA(x)} \qquad (2\text{-}6)$$

The elongation of the entire bar is obtained by integrating over the length:

$$\delta = \int_0^L d\delta = \int_0^L \frac{N(x)dx}{EA(x)} \qquad (2\text{-}7)$$

If the expressions for $N(x)$ and $A(x)$ are not too complicated, the integral can be evaluated analytically and a formula for $\delta$ can be obtained, as illustrated later in Example 2-4. However, if formal integration is either difficult or impossible, a numerical method for evaluating the integral should be used.

## Limitations

Equations (2-5) and (2-7) apply only to bars made of linearly elastic materials, as shown by the presence of the modulus of elasticity $E$ in the formulas. Also, the formula $\delta = PL/EA$ was derived using the assumption that the stress distribution is uniform over every cross section (because it is based on the formula $\sigma = P/A$). This assumption is valid for prismatic bars but not for tapered bars, and therefore Eq. (2-7) gives satisfactory results for a tapered bar only if the angle between the sides of the bar is small.

As an illustration, if the angle between the sides of a bar is 20°, the stress calculated from the expression $\sigma = P/A$ (at an arbitrarily selected cross section) is 3% less than the exact stress for that same cross section (calculated by more advanced methods). For smaller angles, the error is even less. Consequently, we can say that Eq. (2-7) is satisfactory if the angle of taper is small. If the taper is large, more accurate methods of analysis are needed (Ref. 2-1).

The following examples illustrate the determination of changes in lengths of nonuniform bars.

● ● ● **Example 2-3**

A vertical steel bar *ABC* is pin-supported at its upper end and loaded by a force $P_1$ at its lower end (Fig. 2-12a). A horizontal beam *BDE* is pinned to the vertical bar at joint *B* and supported at point *D*. The beam carries a load $P_2$ at end *E*.

The upper part of the vertical bar (segment *AB*) has length $L_1 = 20.0$ in. and cross-sectional area $A_1 = 0.25$ in.$^2$; the lower part (segment *BC*) has length $L_2 = 34.8$ in. and area $A_2 = 0.15$ in.$^2$. The modulus of elasticity *E* of the steel is $29.0 \times 10^6$ psi. The left- and right-hand parts of beam *BDE* have lengths $a = 28$ in. and $b = 25$ in., respectively.

**(a)** Calculate the vertical displacement $\delta_C$ at point *C* if the load $P_1 = 2100$ lb and the load $P_2 = 5600$ lb. (Disregard the weights of the bar and the beam.)
**(b)** Where on segment *DE* should load $P_2$ be applied if vertical displacement $\delta_C$ must equal 0.010 in.?
**(c)** If load $P_2$ is once again applied at *E*, what new value of cross sectional area $A_2$ is required so that vertical displacement $\delta_C$ is equal to 0.0075 in.?

**Fig. 2-12**

Example 2-3: Change in length of a nonuniform bar (bar *ABC*)

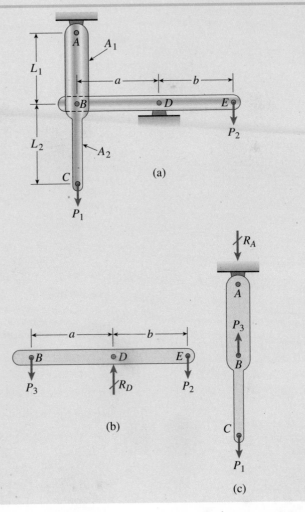

Continues

**Example 2-3 -** *Continued*

### Solution

**(a) Determine the vertical displacement at point C.**

*Axial forces in bar ABC.* From Fig. 2-12a, we see that the vertical displacement of point C is equal to the change in length of bar ABC. Therefore, we must find the axial forces in both segments of this bar.

The axial force $N_2$ in the lower segment is equal to the load $P_1$. The axial force $N_1$ in the upper segment can be found if we know either the vertical reaction at A or the force applied to the bar by the beam. The latter force can be obtained from a free-body diagram of the beam (Fig. 2-12b), in which the force acting on the beam (from the vertical bar) is denoted $P_3$ and the vertical reaction at support D is denoted $R_D$. No horizontal force acts between the bar and the beam, as can be seen from a free-body diagram of the vertical bar itself (Fig. 2-12c). Therefore, there is no horizontal reaction at support D of the beam.

Taking moments about point D for the free-body diagram of the beam (Fig. 2-12b) gives

$$P_3 = \frac{P_2 b}{a} = \frac{(5600 \text{ lb})(25.0 \text{ in.})}{28.0 \text{ in.}} = 5000 \text{ lb} \qquad \text{(a)}$$

This force acts downward on the beam (Fig. 2-12b) and upward on the vertical bar (Fig. 2-12c).

Now we can determine the downward reaction at support A (Fig. 2-12c):

$$R_A = P_3 - P_1 = 5000 \text{ lb} - 2100 \text{ lb} = 2900 \text{ lb} \qquad \text{(b)}$$

The upper part of the vertical bar (segment AB) is subjected to an axial compressive force $N_1$ equal to $R_A$, or 2900 lb. The lower part (segment BC) carries an axial tensile force $N_2$ equal to $P_1$, or 2100 lb.

*Note:* As an alternative to the preceding calculations, we can obtain the reaction $R_A$ from a free-body diagram of the entire structure (instead of from the free-body diagram of beam BDE).

*Changes in length.* With tension considered positive, Eq. (2-5) yields

$$\delta = \sum_{i=1}^{n} \frac{N_i L_i}{E_i A_i} = \frac{N_1 L_1}{EA_1} + \frac{N_2 L_2}{EA_2} \qquad \text{(c)}$$

$$= \frac{(-2900 \text{ lb})(20.0 \text{ in})}{(29.0 \times 10^6 \text{ psi})(0.25 \text{ in.}^2)} + \frac{(2100 \text{ lb})(34.8 \text{ in.})}{(29.0 \times 10^6 \text{ psi})(0.15 \text{ in.}^2)}$$

$$= -0.0080 \text{ in.} + 0.0168 \text{ in.} = 0.0088 \text{ in.}$$

in which $\delta$ is the change in length of bar ABC. Since $\delta$ is positive, the bar elongates. The displacement of point C is equal to the change in length of the bar:

$$\delta_C = 0.0088 \text{ in.}$$

This displacement is downward.

**(b) Determine the location of load $P_2$ on segment DE.**

Load $P_2$ is now positioned at some distance x to the right of point D (see Fig. 2-12d):

$$\Sigma M_D = 0 \qquad P_3 = \frac{P_2 x}{a} \qquad \text{(d)}$$

From Eq. (b), we have $R_A = P_3 - P_1$.

---

**Fig. 2-12 (Continued)**

Example 2-3: (d) Load $P_2$ at some distance x to the right of point D

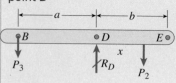

(d)

Axial compressive force in $AB$ is $R_A$ and axial tensile force in $BC$ is $P_1$, so from Eq. (c), the downward displacement at joint $C$ is

$$\delta_C = \frac{-(P_3 - P_1)L_1}{EA_1} + \frac{P_1L_2}{EA_2} \tag{e}$$

Substituting the expression for $P_3$ from Eq. (d) and solving for $x$ gives

$$\delta_C = \frac{-\left[\left(\dfrac{P_2x}{a}\right) - P_1\right]L_1}{EA_1} + \frac{P_1L_2}{EA_2} \tag{f}$$

and

$$x = \frac{a(A_1L_2P_1 + A_2L_1P_1 - A_1A_2E\delta_C)}{A_2L_1P_2} \tag{g}$$

Finally, insert numerical values and solve for distance $x$:

$$x = \frac{\begin{array}{c}28 \text{ in.}[(0.25 \text{ in.}^2)(34.8 \text{ in.})(2100 \text{ lb}) + \\ (0.15 \text{ in.}^2)(20 \text{ in.})(2100 \text{ lb}) - \\ (0.25 \text{ in.}^2)(0.15 \text{ in.}^2)(29{,}000{,}000 \text{ psi})(0.010 \text{ in.})]\end{array}}{(0.15 \text{ in.}^2)(20 \text{ in.})(5600 \text{ lb})} = 22.8 \text{ in.}$$

### (c) Determine the new cross-sectional area $A_2$.

Load $P_2$ is now once again applied at joint $E$, so force $P_3$ is obtained from Eq. (a), and the revised Eq. (f) is

$$\delta_C = \frac{-\left[\left(\dfrac{P_2b}{a}\right) - P_1\right]L_1}{EA_1} + \frac{P_1L_2}{EA_2} \tag{h}$$

We can now solve for the cross-sectional area $A_2$ in terms of other structure variables:

$$A_2 = \frac{L_2P_1}{E\left[\delta_C - \dfrac{L_1\left(P_1 - \dfrac{P_2b}{a}\right)}{A_1E}\right]} \tag{i}$$

Substitution of numerical values into Eq. (i) provides the required area of bar $BC$ if $\delta_C = 0.0075$ in. under applied loads $P_1$ and $P_2$:

$$A_2 = \frac{(34.8 \text{ in.})(2100 \text{ lb})}{(29{,}000{,}000 \text{ psi}) \cdot \left[(0.0075 \text{ in.}) - \dfrac{(20 \text{ in.})\left[2100 \text{ lb} - \dfrac{(5600 \text{ lb})(25 \text{ in.})}{(28 \text{ in.})}\right]}{(0.25 \text{ in.}^2)(29{,}000{,}000 \text{ psi})}\right]}$$

$$= 0.1626 \text{ in.}^2$$

As expected, cross-sectional area $A_2$ must increase, so that the vertical displacement at $C$ is reduced from that computed in Eq. (c).

A tapered bar *AB* of solid circular cross section and length *L* (Fig. 2-13a) is supported at end *B* and subjected to a tensile load *P* at the free end *A*. The diameters of the bar at ends *A* and *B* are $d_A$ and $d_B$, respectively.

Determine the elongation of the bar due to the load *P*, assuming that the angle of taper is small.

**Fig. 2-13**

Example 2-4: Change in length of a tapered bar of solid circular cross section

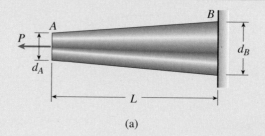

(a)

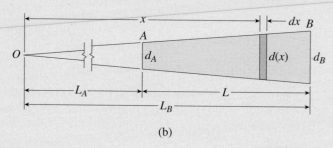

(b)

### Solution

The bar being analyzed in this example has a constant axial force (equal to the load *P*) throughout its length. However, the cross-sectional area varies continuously from one end to the other. Therefore, we must use integration (see Eq. 2-7) to determine the change in length.

*Cross-sectional area.* The first step in the solution is to obtain an expression for the cross-sectional area *A(x)* at any cross section of the bar. For this purpose, we must establish an origin for the coordinate *x*. One possibility is to place the origin of coordinates at the free end *A* of the bar. However, the integrations to be performed will be slightly simplified if we locate the origin of coordinates by extending the sides of the tapered bar until they meet at point *O*, as shown in Fig. 2-13b.

The distances $L_A$ and $L_B$ from the origin *O* to ends *A* and *B*, respectively, are in the ratio

$$\frac{L_A}{L_B} = \frac{d_A}{d_B} \tag{a}$$

as obtained from similar triangles in Fig. 2-13b. From similar triangles we also get the ratio of the diameter *d(x)* at distance *x* from the origin to the diameter $d_A$ at the small end of the bar:

$$\frac{d(x)}{d_A} = \frac{x}{L_A} \quad \text{or} \quad d(x) = \frac{d_A x}{L_A} \tag{b}$$

Therefore, the cross-sectional area at distance $x$ from the origin is

$$A(x) = \frac{\pi[d(x)]^2}{4} = \frac{\pi d_A^2 x^2}{4L_A^2} \qquad \text{(c)}$$

*Change in length.* We now substitute the expression for $A(x)$ into Eq. (2-7) and obtain the elongation $\delta$:

$$\delta = \int \frac{N(x)dx}{EA(x)} = \int_{L_A}^{L_B} \frac{Pdx(4L_A^2)}{E(\pi d_A^2 x^2)} = \frac{4PL_A^2}{\pi E d_A^2} \int_{L_A}^{L_B} \frac{dx}{x^2} \qquad \text{(d)}$$

By performing the integration (see Appendix D for integration formulas) and substituting the limits, we get

$$\delta = \frac{4PL_A^2}{\pi E d_A^2} \left[ -\frac{1}{x} \right]_{L_A}^{L_B} = \frac{4PL_A^2}{\pi E d_A^2} \left( \frac{1}{L_A} - \frac{1}{L_B} \right) \qquad \text{(e)}$$

This expression for $\delta$ can be simplified by noting that

$$\frac{1}{L_A} - \frac{1}{L_B} = \frac{L_B - L_A}{L_A L_B} = \frac{L}{L_A L_B} \qquad \text{(f)}$$

Thus, the equation for $\delta$ becomes

$$\delta = \frac{4PL}{\pi E d_A^2} \left( \frac{L_A}{L_B} \right) \qquad \text{(g)}$$

Finally, we substitute $L_A/L_B = d_A/d_B$ (see Eq. a) and obtain

$$\delta = \frac{4PL}{\pi E d_A d_B} \qquad \Longleftarrow \quad \text{(2-8)}$$

This formula gives the elongation of a tapered bar of solid circular cross section. By substituting numerical values, we can determine the change in length for any particular bar.

*Note 1:* A common mistake is to assume that the elongation of a tapered bar can be determined by calculating the elongation of a prismatic bar that has the same cross-sectional area as the midsection of the tapered bar. Examination of Eq. (2-8) shows that this idea is not valid.

*Note 2:* The preceding formula for a tapered bar (Eq. 2-8) can be reduced to the special case of a prismatic bar by substituting $d_A = d_B = d$. The result is

$$\delta = \frac{4PL}{\pi E d^2} = \frac{PL}{EA}$$

which we know to be correct.

A general formula such as Eq. (2-8) should be checked whenever possible by verifying that it reduces to known results for *special cases*. If the reduction does not produce a correct result, the original formula is in error. If a correct result is obtained, the original formula may still be incorrect but our confidence in it increases. In other words, this type of check is a necessary but not sufficient condition for the correctness of the original formula.

# 2.4 STATICALLY INDETERMINATE STRUCTURES

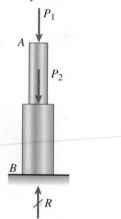

**Fig. 2-14**

Statically determinate bar

The springs, bars, and cables that we discussed in the preceding sections have one important feature in common—their reactions and internal forces can be determined solely from free-body diagrams and equations of equilibrium. Structures of this type are classified as **statically determinate**. We should note especially that the forces in a statically determinate structure can be found without knowing the properties of the materials. Consider, for instance, the bar $AB$ shown in Fig. 2-14. The calculations for the internal axial forces in both parts of the bar, as well as for the reaction $R$ at the base, are independent of the material of which the bar is made.

Most structures are more complex than the bar of Fig. 2-14, and their reactions and internal forces cannot be found by statics alone. This situation is illustrated in Fig. 2-15, which shows a bar $AB$ fixed at *both* ends. There are now two vertical reactions ($R_A$ and $R_B$) but only one useful equation of equilibrium—the equation for summing forces in the vertical direction. Since this equation contains two unknowns, it is not sufficient for finding the reactions. Structures of this kind are classified as **statically indeterminate**. To analyze such structures we must supplement the equilibrium equations with additional equations pertaining to the displacements of the structure.

To see how a statically indeterminate structure is analyzed, consider the example of Fig. 2-16a. The prismatic bar $AB$ is attached to rigid supports at both ends and is axially loaded by a force $P$ at an intermediate point $C$. As already discussed, the reactions $R_A$ and $R_B$ cannot be found by statics alone, because only one **equation of equilibrium** is available:

**Fig. 2-15**

Statically indeterminate bar

$$\Sigma F_{vert} = 0 \quad R_A - P + P_B = 0 \qquad \textbf{(2-9)}$$

An additional equation is needed in order to solve for the two unknown reactions.

The additional equation is based upon the observation that a bar with both ends fixed does not change in length. If we separate the bar from its supports (Fig. 2-16b), we obtain a bar that is free at both ends and loaded by the three forces, $R_A$, $R_B$, and $P$. These forces cause the bar to change in length by an amount $\delta_{AB}$, which must be equal to zero:

$$\delta_{AB} = 0 \qquad \textbf{(2-10)}$$

This equation, called an **equation of compatibility**, expresses the fact that the change in length of the bar must be compatible with the conditions at the supports.

In order to solve Eqs. (2-9) and (2-10), we must now express the compatibility equation in terms of the unknown forces $R_A$ and $R_B$. The relationships between the forces acting on a bar and its changes in length are known as **force-displacement relations**. These relations have various forms depending upon the properties of the material. If the material is linearly

**Fig. 2-16**

Analysis of a statically indeterminate bar

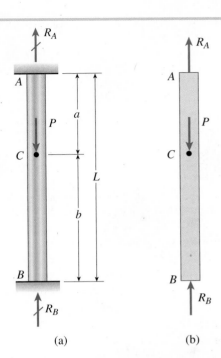

(a)                    (b)

elastic, the equation $\delta = PL/EA$ can be used to obtain the force-displacement relations.

Let us assume that the bar of Fig. 2-16 has cross-sectional area $A$ and is made of a material with modulus $E$. Then the changes in lengths of the upper and lower segments of the bar are, respectively,

$$\delta_{AC} = \frac{R_A a}{EA} \qquad \delta_{CB} = -\frac{R_B b}{EA} \qquad \textbf{(2-11a,b)}$$

where the minus sign indicates a shortening of the bar. Equations (2-11a) and (2-11b) are the force-displacement relations.

We are now ready to solve simultaneously the three sets of equations (the equation of equilibrium, the equation of compatibility, and the force-displacement relations). In this illustration, we begin by combining the force-displacement relations with the equation of compatibility:

$$\delta_{AB} = \delta_{AC} = \delta_{CB} = \frac{R_A a}{EA} - \frac{R_B b}{EA} = 0 \qquad \textbf{(2-12)}$$

Note that this equation contains the two reactions as unknowns.

The next step is to solve simultaneously the equation of equilibrium (Eq. 2-9) and the preceding equation (Eq. 2-12). The results are

$$R_A = \frac{Pb}{L} \qquad R_B = \frac{Pa}{L} \qquad \textbf{(2-13a,b)}$$

With the reactions known, all other force and displacement quantities can be determined. Suppose, for instance, that we wish to find the downward displacement $\delta_C$ of point $C$. This displacement is equal to the elongation of segment $AC$:

$$\delta_C = \delta_{AC} = \frac{R_A a}{EA} = \frac{Pab}{LEA} \qquad \textbf{(2-14)}$$

Also, we can find the stresses in the two segments of the bar directly from the internal axial forces (e.g., $\sigma_{AC} = R_A/A = Pb/AL$).

## General Comments

From the preceding discussion we see that the analysis of a statically indeterminate structure involves setting up and solving equations of equilibrium, equations of compatibility, and force-displacement relations. The equilibrium equations relate the loads acting on the structure to the unknown forces (which may be reactions or internal forces), and the compatibility equations express conditions on the displacements of the structure. The force-displacement relations are expressions that use the dimensions and properties of the structural members to relate the forces and displacements of those members. In the case of axially loaded bars that behave in a linearly elastic manner, the relations are based upon the equation $\delta = PL/EA$. Finally, all three sets of equations may be solved simultaneously for the unknown forces and displacements.

In the engineering literature, various terms are used for the conditions expressed by the equilibrium, compatibility, and force-displacement equations. The equilibrium equations are also known as *static* or *kinetic* equations; the compatibility equations are sometimes called *geometric* equations, *kinematic* equations, or equations of *consistent deformations*; and the force-displacement relations are often referred to as *constitutive relations* (because they deal with the *constitution*, or physical properties, of the materials).

For the relatively simple structures discussed in this chapter, the preceding method of analysis is adequate. However, more formalized approaches are needed for complicated structures. Two commonly used methods, the *flexibility method* (also called the *force method*) and the *stiffness method* (also called the *displacement method*), are described in detail in textbooks on structural analysis. Even though these methods are normally used for large and complex structures requiring the solution of hundreds and sometimes thousands of simultaneous equations, they still are based upon the concepts described previously, that is, equilibrium equations, compatibility equations, and force-displacement relations.*

The following two examples illustrate the methodology for analyzing statically indeterminate structures consisting of axially loaded members.

---

*From a historical viewpoint, it appears that Euler in 1774 was the first to analyze a statically indeterminate system; he considered the problem of a rigid table with four legs supported on an elastic foundation (Refs. 2-2 and 2-3). The next work was done by the French mathematician and engineer L. M. H. Navier, who in 1825 pointed out that statically indeterminate reactions could be found only by taking into account the elasticity of the structure (Ref. 2-4). Navier solved statically indeterminate trusses and beams.

● ● ● **Example 2-5**

A horizontal *rigid* bar *ABC* is pinned at end *A* and supported by two wires (*BD* and *CD*) at points *B* and *C* (Fig. 2-17). A vertical load *P* acts at end *C* of the bar. The bar has a length of 2*b* and wires *BD* and *CD* have lengths of $L_1$ and $L_2$, respectively. Also, wire *BD* has a diameter of $d_1$ and modulus of elasticity $E_1$; wire *CD* has a diameter of $d_2$ and modulus $E_2$.

**(a)** Obtain **formulas for the allowable load P** if the allowable stresses in wires *BD* and *CD*, respectively, are $\sigma_1$ and $\sigma_2$. (Disregard the weight of the bar and cables.)

**(b)** **Calculate the allowable load P** for the following conditions: Wire *BD* is made of aluminum with a modulus $E_1$ = 10,400 ksi and a diameter of $d_1$ = 1/6 in. Wire *CD* is made of magnesium with a modulus $E_2$ = 6500 ksi and a diameter of $d_2$ = 1/8 in. The allowable stresses in the aluminum and magnesium wires are $\sigma_1$ = 29 ksi and $\sigma_2$ = 25 ksi, respectively. Dimensions are *a* = 6 ft and *b* = 4 ft in Fig. 2-17.

**Fig. 2-17**

Example 2-5: (a) Analysis of a statically indeterminate cable-bar structure, (b) free-body diagram of bar *ABC*, and (c) elongation of wire *BD*

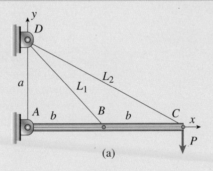

(a)

## Solution

**(a) Determine the formulas for allowable load P.**

*Equation of equilibrium.* We begin the analysis by drawing a free-body diagram of bar *ABC* (Fig. 2-17b). In this diagram, $T_1$ and $T_2$ are the unknown tensile forces in the wires and $A_x$ and $A_y$ are the horizontal and vertical components of the reaction at support *A*. We see immediately that the structure is statically indeterminate, because there are four unknown forces ($T_1$, $T_2$, $A_x$, and $A_y$) but only three independent equations of equilibrium.

Taking moments about point *A* (with counterclockwise moments being positive) yields

$$\Sigma M_A = 0 \qquad T_1(b) \sin (\alpha_B) + T_2(2b) \sin (\alpha_C) - P(2b) = 0 \qquad \textbf{(a)}$$
$$\text{or} \quad T_1 \sin (\alpha_B) + 2T_2 \sin (\alpha_C) = 2P$$

The other two equations, obtained by summing forces in the horizontal direction and summing forces in the vertical direction, are of no immediate benefit in finding $T_1$ and $T_2$, because they introduce additional unknowns $A_x$ and $A_y$.

*Equation of compatibility.* To obtain an equation pertaining to the displacements, we observe that the load *P* causes bar *ABC* to rotate about the pin support at *A*, thereby stretching the wires. The resulting displacements are shown in the displacement diagram of Fig. 2-17b, where line *ABC* represents the original position of the rigid bar and line

*Continues* ➡

● ● ●   **Example 2-5** - *Continued*

**Fig. 2-17 (Continued)**

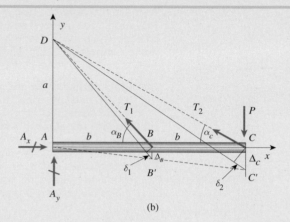

(b)

*AB'C'* represents the rotated position. The vertical downward displacements $\Delta_B$ and $\Delta_C$ are used to find the elongations $\delta_1$ and $\delta_2$ of the wires. Because these displacements are very small, the bar rotates through a very small angle (shown highly exaggerated in the figure), and we can make calculations on the assumption that points *B* and *C* move vertically downward (instead of moving along the arcs of circles).

Because the horizontal distances *AB* and *BC* are equal, we obtain the geometric relationship between the vertical displacements:

$$\Delta_C = 2\Delta_B \tag{b}$$

This is the *compatibility equation* that will allow us to find another relationship between the two cable forces once we have substituted the force-displacement relations.

First, using geometry* (see Fig. 2-17c), we can relate the vertical displacements to the wire elongations as

**Fig. 2-17 (Continued)**

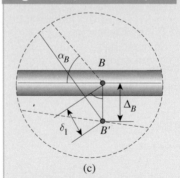

(c)

$$\delta_1 = \Delta_B \sin(\alpha_B) \quad \text{and} \quad \delta_2 = \Delta_C \sin(\alpha_C) \tag{c,d}$$

$$\Delta_C = \frac{\delta_2}{\sin(\alpha_C)} \quad \text{and} \quad 2\Delta_B = 2\frac{\delta_1}{\sin(\alpha_B)} \tag{e,f}$$

or

$$\delta_2 = 2\left(\frac{\sin(\alpha_C)}{\sin(\alpha_B)}\right)\delta_1 \tag{g}$$

*Force-displacement relations.* Since the wires behave in a linearly elastic manner, their elongations can be expressed in terms of the unknown forces $T_1$ and $T_2$ by means of the expressions:

$$\delta_1 = \left(\frac{L_1}{E_1 A_1}\right)T_1 = f_1 T_1 \quad \text{and} \quad \delta_2 = \left(\frac{L_2}{E_2 A_2}\right)T_2 = f_2 T_2 \tag{h,i}$$

---

*Note:* We also could use the dot product of $\Delta_B$ $(-j)$ and a unit vector $n = (\cos(\alpha_B)i - \sin(\alpha_B)j)$ along line *DB* to find $\delta_1$ and also the dot product of $\Delta_C$ $(-j)$ and a unit vector $n = (\cos(\alpha_C)i - \sin(\alpha_C)j)$ along line *DC* to find $\delta_2$.

where $f_1$ and $f_2$ are the *flexibilities* of wires *BD* and *CD*, respectively, and wire cross-sectional areas are

$$A_1 = \frac{\pi d_1^2}{4} \qquad A_2 = \frac{\pi d_2^2}{4} \qquad \text{(j,k)}$$

*Solution of equations.* We now solve simultaneously the three sets of equations (equilibrium, compatibility, and force-displacement) to find wire forces $T_1$ and $T_2$. First, we insert the force-displacement equations [Eqs. (h, i)] into the compatibility equation [Eq. (g)]:

$$\delta_2 = 2\left(\frac{\sin(\alpha_C)}{\sin(\alpha_B)}\right)\delta_1 = f_2 T_2 = 2\left(\frac{\sin(\alpha_C)}{\sin(\alpha_B)}\right)f_1 T_1 \qquad \text{(l)}$$

Solving for $T_2$ gives 
$$T_2 = 2\left(\frac{\sin(\alpha_C)}{\sin(\alpha_B)}\right)\left(\frac{f_1}{f_2}\right)T_1 \qquad \text{(m)}$$

Inserting this expression for $T_2$ into the equilibrium equation [Eq. (a)] and solving for $T_1$ yields

$$T_1 = \left(\frac{f_2 \sin(\alpha_B)}{f_2 \sin(\alpha_B)^2 + 4f_1 \sin(\alpha_C)^2}\right)(2P) \qquad \text{(n)}$$

and 
$$T_2 = \left(\frac{f_1 \sin(\alpha_C)}{f_2 \sin(\alpha_B)^2 + 4f_1 \sin(\alpha_C)^2}\right)(4P) \qquad \text{(o)}$$

*Allowable load P.* Now that the statically indeterminate analysis is completed and the forces in the wires are known, we can determine the permissible value of the load *P*. The stress $\sigma_1$ in wire *BD* and the stress $\sigma_2$ in wire *CD* are readily obtained from the forces [Eqs. (n) and (o)]:

$$\sigma_1 = \frac{T_1}{A_1} = \left(\frac{f_2 \sin(\alpha_B)}{f_2 \sin(\alpha_B)^2 + 4f_1 \sin(\alpha_C)^2}\right)\left(2\frac{P}{A_1}\right)$$

$$\sigma_2 = \frac{T_2}{A_2} = \left(\frac{f_1 \sin(\alpha_C)}{f_2 \sin(\alpha_B)^2 + 4f_1 \sin(\alpha_C)^2}\right)\left(4\frac{P}{A_2}\right)$$

We can solve each of these equations for a permissible value of load *P* which depends on either allowable stress $\sigma_1$ or $\sigma_2$; the smaller value of load *P* controls:

$$P_1 = \frac{\sigma_1 A_1}{2}\left(\frac{f_2 \sin(\alpha_B)^2 + 4f_1 \sin(\alpha_C)^2}{f_2 \sin(\alpha_B)}\right) \qquad \Longleftarrow \text{(2-15a)}$$

$$P_2 = \frac{\sigma_2 A_2}{4}\left(\frac{f_2 \sin(\alpha_B)^2 + 4f_1 \sin(\alpha_C)^2}{f_1 \sin(\alpha_C)}\right) \qquad \Longleftarrow \text{(2-15b)}$$

*Continues* ➡

**(b) Determine the numerical calculations for allowable load P.**

Using the numerical data and the preceding equations, we obtain the numerical values:

$$A_1 = \frac{\pi}{4}d_1^2 = \frac{\pi}{4}\left(\frac{1}{6} \text{ in.}\right)^2 = 0.02182 \text{ in.}^2$$

$$A_2 = \frac{\pi}{4}d_2^2 = \frac{\pi}{4}\left(\frac{1}{8} \text{ in.}\right)^2 = 0.01227 \text{ in.}^2$$

$$L_1 = \sqrt{a^2 + b^2} = \sqrt{(6 \text{ ft})^2 + (4 \text{ ft})^2} = 7.2111 \text{ ft}$$

$$L_2 = \sqrt{a^2 + (2b)^2} = \sqrt{(6 \text{ ft})^2 + (8 \text{ ft})^2} = 10 \text{ ft}$$

$$\alpha_B = \tan^{-1}\left(\frac{a}{b}\right) = 56.31°$$

$$\alpha_C = \tan^{-1}\left(\frac{a}{2b}\right) = 36.87°$$

$$f_1 = \frac{L_1}{E_1 A_1} = \frac{7.2111 \text{ ft}}{(10,400 \text{ ksi})(0.02182 \text{ in.}^2)} = 3.81324 \times 10^{-4} \frac{\text{in.}}{\text{lb}}$$

$$f_2 = \frac{L_2}{E_2 A_2} = \frac{10 \text{ ft}}{(6500 \text{ ksi})(0.01227 \text{ in.}^2)} = 1.50461 \times 10^{-3} \frac{\text{in.}}{\text{lb}}$$

Substituting these numerical values into Eqs. (2-15a, b) gives

$$P_1 = \frac{\sigma_1 A_1}{2}\left(\frac{f_2 \sin(\alpha_B)^2 + 4f_1 \sin(\alpha_C)^2}{f_2 \sin(\alpha_B)}\right) = 402 \text{ lb}$$

$$P_2 = \frac{\sigma_2 A_2}{4}\left(\frac{f_2 \sin(\alpha_B)^2 + 4f_1 \sin(\alpha_C)^2}{f_1 \sin(\alpha_C)}\right) = 533 \text{ lb}$$

The first result is based upon the allowable stress $\sigma_1$ in the aluminum wire, and the second is based upon the allowable stress $\sigma_2$ in the magnesium wire. The allowable load is the smaller of the two values:

$$P_{\text{allow}} = 402 \text{ lb}$$

At this load, the force $T_1$ in the aluminum wire is 633 lb, and the stress in the aluminum wire is 29 ksi (the allowable stress $\sigma_1$), while the force $T_2$ in the magnesium wire is 231 lb, and the stress in the magnesium wire is (402/533)(25 ksi) = 18.86 ksi. As expected, this stress is less than the allowable stress of $\sigma_2 = 25$ ksi.

● ● ● **Example 2-6**

A solid circular steel cylinder *S* is encased in a hollow circular copper tube *C* (Figs. 2-18a and b). The cylinder and tube are compressed between the rigid plates of a testing machine by compressive forces *P*. The steel cylinder has cross-sectional area $A_s$ and modulus of elasticity $E_s$, the copper tube has area $A_c$ and modulus $E_c$, and both parts have length *L*.

Determine the following quantities: (a) the compressive forces $P_s$ in the steel cylinder and $P_c$ in the copper tube; (b) the corresponding compressive stresses $\sigma_s$ and $\sigma_c$; and (c) the shortening $\delta$ of the assembly.

**Fig. 2-18**

Example 2-6: Analysis of a statically indeterminate structure

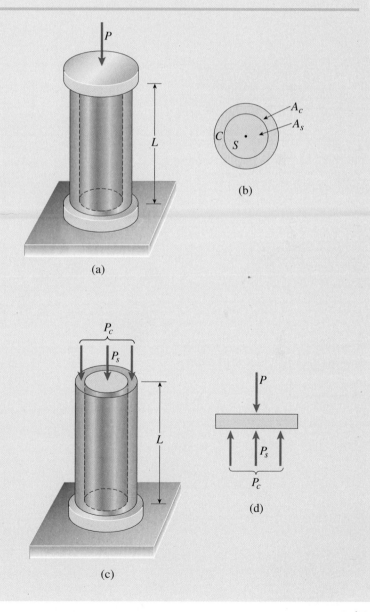

*Continues* ➡

### Solution

**(a)** *Compressive forces in the steel cylinder and copper tube.* We begin by removing the upper plate of the assembly in order to expose the compressive forces $P_s$ and $P_c$ acting on the steel cylinder and copper tube, respectively (Fig. 2-18c). The force $P_s$ is the resultant of the uniformly distributed stresses acting over the cross section of the steel cylinder, and the force $P_c$ is the resultant of the stresses acting over the cross section of the copper tube.

 *Equation of equilibrium.* A free-body diagram of the upper plate is shown in Fig. 2-18d. This plate is subjected to the force $P$ and to the unknown compressive forces $P_s$ and $P_c$; thus, the equation of equilibrium is

$$\Sigma F_{\text{vert}} = 0 \qquad P_s + P_c - P = 0 \qquad \text{(f)}$$

**Fig. 2-18 (Repeated)**

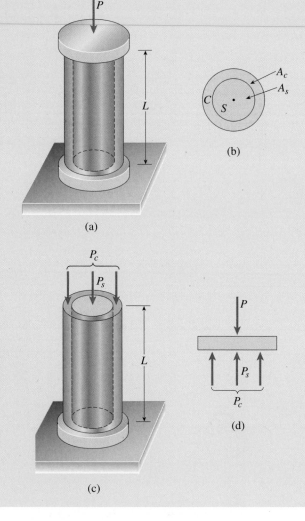

(a)

(b)

(c)

(d)

This equation, which is the only nontrivial equilibrium equation available, contains two unknowns. Therefore, we conclude that the structure is statically indeterminate.

*Equation of compatibility.* Because the end plates are rigid, the steel cylinder and copper tube must shorten by the same amount. Denoting the shortenings of the steel and copper parts by $\delta_s$ and $\delta_c$, respectively, we obtain the following equation of compatibility:

$$\delta_s = \delta_c \tag{g}$$

*Force-displacement relations.* The changes in lengths of the cylinder and tube can be obtained from the general equation $\delta = PL/EA$. Therefore, in this example the force-displacement relations are

$$\delta_s = \frac{P_s L}{E_s A_s} \qquad \delta_c = \frac{P_c L}{E_c A_c} \tag{h,i}$$

*Solution of equations.* We now solve simultaneously the three sets of equations. First, we substitute the force-displacement relations in the equation of compatibility, which gives

$$\frac{P_s L}{E_s A_s} = \frac{P_c L}{E_c A_c} \tag{j}$$

This equation expresses the compatibility condition in terms of the unknown forces.

Next, we solve simultaneously the equation of equilibrium (Eq. f) and the preceding equation of compatibility (Eq. j) and obtain the axial forces in the steel cylinder and copper tube:

$$P_s = P\left(\frac{E_s A_s}{E_s A_s + E_c A_c}\right) \qquad P_c = P\left(\frac{E_c A_c}{E_s A_s + E_c A_c}\right) \impliedby \tag{2-16a,b}$$

These equations show that the compressive forces in the steel and copper parts are directly proportional to their respective axial rigidities and inversely proportional to the sum of their rigidities.

**(b)** *Compressive stresses in the steel cylinder and copper tube.* Knowing the axial forces, we can now obtain the compressive stresses in the two materials:

$$\sigma_s = \frac{P_s}{A_s} = \frac{PE_s}{E_s A_s + E_c A_c} \qquad \sigma_c = \frac{P_c}{A_c} = \frac{PE_c}{E_s A_s + E_c A_c} \impliedby \tag{2-17a,b}$$

*Continues* ⮞

Note that the ratio $\sigma_s/\sigma_c$ of the stresses is equal to the ratio $E_s/E_c$ of the moduli of elasticity, showing that in general the "stiffer" material always has the larger stress.

**(c)** *Shortening of the assembly.* The shortening $\delta$ of the entire assembly can be obtained from either Eq. (h) or Eq. (i). Thus, upon substituting the forces (from Eqs. 2-16a and b), we get

$$\delta = \frac{P_s L}{E_s A_s} = \frac{P_c L}{E_c A_c} = \frac{PL}{E_s A_s + E_c A_c} \qquad \Longleftarrow \text{(2-18)}$$

This result shows that the shortening of the assembly is equal to the total load divided by the sum of the stiffnesses of the two parts (recall from Eq. 2-4a that the stiffness of an axially loaded bar is $k = EA/L$).

   *Alternative solution of the equations.* Instead of substituting the force-displacement relations (Eqs. h and i) into the equation of compatibility, we could rewrite those relations in the form

$$P_s = \frac{E_s A_s}{L}\,\delta_s \qquad P_c = \frac{E_c A_c}{L}\,\delta_c \qquad \text{(k,l)}$$

and substitute them into the equation of equilibrium (Eq. f):

$$\frac{E_s A_s}{L}\,\delta_s + \frac{E_c A_c}{L}\,\delta_c = P \qquad \text{·(m)}$$

This equation expresses the equilibrium condition in terms of the unknown displacements. Then we solve simultaneously the equation of compatibility (Eq. g) and the preceding equation, thus obtaining the displacements:

$$\delta_s = \delta_c = \frac{PL}{E_s A_s + E_c A_c} \qquad \text{(n)}$$

which agrees with Eq. (2-18). Finally, we substitute expression (n) into Eqs. (k) and (l) and obtain the compressive forces $P_s$ and $P_c$ (see Eqs. 2-16a and b).

   *Note:* The alternative method of solving the equations is a simplified version of the stiffness (or displacement) method of analysis, and the first method of solving the equations is a simplified version of the flexibility (or force) method. The names of these two methods arise from the fact that Eq. (m) has displacements as unknowns and stiffnesses as coefficients (see Eq. 2-4a), whereas Eq. (j) has forces as unknowns and flexibilities as coefficients (see Eq. 2-4b).

# 2.5 THERMAL EFFECTS, MISFITS, AND PRESTRAINS

External loads are not the only sources of stresses and strains in a structure. Other sources include *thermal effects* arising from temperature changes, *misfits* resulting from imperfections in construction, and *prestrains* that are produced by initial deformations. Still other causes are settlements (or movements) of supports, inertial loads resulting from accelerating motion, and natural phenomenon such as earthquakes.

Thermal effects, misfits, and prestrains are commonly found in both mechanical and structural systems and are described in this section. As a general rule, they are much more important in the design of statically indeterminate structures that in statically determinate ones.

## Thermal Effects

Changes in temperature produce expansion or contraction of structural materials, resulting in **thermal strains** and **thermal stresses**. A simple illustration of thermal expansion is shown in Fig. 2-19, where the block of material is unrestrained and therefore free to expand. When the block is heated, every element of the material undergoes thermal strains in all directions, and consequently the dimensions of the block increase. If we take corner $A$ as a fixed reference point and let side $AB$ maintain its original alignment, the block will have the shape shown by the dashed lines.

For most structural materials, thermal strain $\varepsilon_T$ is proportional to the temperature change $\Delta T$; that is,

$$\varepsilon_T = \alpha(\Delta T) \qquad \textbf{(2-19)}$$

### Fig. 2-19

Block of material subjected to an increase in temperature

in which $\alpha$ is a property of the material called the **coefficient of thermal expansion**. Since strain is a dimensionless quantity, the coefficient of thermal expansion has units equal to the reciprocal of temperature change. In SI units the dimensions of $\alpha$ can be expressed as either 1/K (the reciprocal of kelvins) or 1/°C (the reciprocal of degrees Celsius). The value of $\alpha$ is the same in both cases because a *change* in temperature is numerically the same in both kelvins and degrees Celsius. In USCS units, the dimensions of $\alpha$ are 1/°F (the reciprocal of degrees Fahrenheit).* Typical values of $\alpha$ are listed in Table I-4 of Appendix I.

When a **sign convention** is needed for thermal strains, we usually assume that expansion is positive and contraction is negative.

To demonstrate the relative importance of thermal strains, we will compare thermal strains with load-induced strains in the following manner. Suppose we have an axially loaded bar with longitudinal strains given by the equation $\varepsilon = \sigma/E$, where $\sigma$ is the stress and $E$ is the modulus of elasticity. Then suppose we have an identical bar subjected to a temperature change $\Delta T$, which means that the bar has thermal strains given by Eq. (2-19). Equating the two strains gives the equation

$$\sigma = E\alpha(\Delta T)$$

*For a discussion of temperature units and scales, see Section B.4 of Appendix B.

From this equation we can calculate the axial stress $\sigma$ that produces the same strain as does the temperature change $\Delta T$. For instance, consider a stainless steel bar with $E = 30 \times 10^6$ psi and $\alpha = 9.6 \times 10^{-6}/°F$. A quick calculation from the preceding equation for $\sigma$ shows that a change in temperature of $100°F$ produces the same strain as a stress of 29,000 psi. This stress is in the range of typical allowable stresses for stainless steel. Thus, a relatively modest change in temperature produces strains of the same magnitude as the strains caused by ordinary loads, which shows that temperature effects can be important in engineering design.

Ordinary structural materials expand when heated and contract when cooled, and therefore an increase in temperature produces a positive thermal strain. Thermal strains usually are reversible, in the sense that the member returns to its original shape when its temperature returns to the original value. However, a few special metallic alloys have recently been developed that do not behave in the customary manner. Instead, over certain temperature ranges their dimensions decrease when heated and increase when cooled.

Water is also an unusual material from a thermal standpoint—it expands when heated at temperatures above 4°C and also expands when cooled below 4°C. Thus, water has its maximum density at 4°C.

Now let us return to the block of material shown in Fig. 2-19. We assume that the material is homogeneous and isotropic and that the temperature increase $\Delta T$ is uniform throughout the block. We can calculate the increase in *any* dimension of the block by multiplying the original dimension by the thermal strain. For instance, if one of the dimensions is $L$, then that dimension will increase by the amount

$$\delta_T = \varepsilon_T L = \alpha(\Delta T)L \tag{2-20}$$

Equation (2-20) is a **temperature-displacement relation**, analogous to the force-displacement relations described in the preceding section. It can be used to calculate changes in lengths of structural members subjected to uniform temperature changes, such as the elongation $\delta_T$ of the prismatic bar shown in Fig. 2-20. (The transverse dimensions of the bar also change, but these changes are not shown in the figure since they usually have no effect on the axial forces being transmitted by the bar.)

In the preceding discussions of thermal strains, we assumed that the structure had no restraints and was able to expand or contract freely. These conditions exist when an object rests on a frictionless surface or hangs in open space. In such cases no stresses are produced by a uniform temperature change throughout the object, although nonuniform temperature

---

**Fig. 2-20**

Increase in length of a prismatic bar due to a uniform increase in temperature (Eq. 2-20)

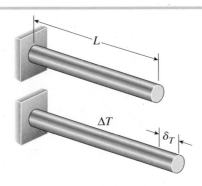

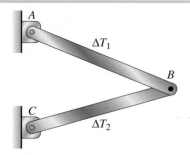

**Fig. 2-21**

Statically determinate truss with a uniform temperature change in each member

changes may produce internal stresses. However, many structures have supports that prevent free expansion and contraction, in which case **thermal stresses** will develop even when the temperature change is uniform throughout the structure.

To illustrate some of these ideas about thermal effects, consider the two-bar truss $ABC$ of Fig. 2-21 and assume that the temperature of bar $AB$ is changed by $\Delta T_1$ and the temperature of bar $BC$ is changed by $\Delta T_2$. Because the truss is statically determinate, both bars are free to lengthen or shorten, resulting in a displacement of joint $B$. However, there are no stresses in either bar and no reactions at the supports. This conclusion applies generally to **statically determinate structures**; that is, uniform temperature changes in the members produce thermal strains (and the corresponding changes in lengths) without producing any corresponding stresses.

A **statically indeterminate structure** may or may not develop temperature stresses, depending upon the character of the structure and the nature of the temperature changes. To illustrate some of the possibilities, consider the statically indeterminate truss shown in Fig. 2-22. Because the supports of this structure permit joint $D$ to move horizontally, no stresses are developed when the *entire* truss is heated uniformly. All members increase in length in proportion to their original lengths, and the truss becomes slightly larger in size.

However, if some bars are heated and others are not, thermal stresses will develop because the statically indeterminate arrangement of the bars prevents free expansion. To visualize this condition, imagine that just one bar is heated. As this bar becomes longer, it meets resistance from the other bars, and therefore stresses develop in all members.

The analysis of a statically indeterminate structure with temperature changes is based upon the concepts discussed in the preceding section, namely equilibrium equations, compatibility equations, and force-displacement relations. The principal difference is that we now use temperature-displacement relations (Eq. 2-20) in addition to force-displacement relations (such as $\delta = PL/EA$) when performing the analysis. The following two examples illustrate the procedures in detail.

Forces can develop in statically indeterminate trusses due to temperature and prestrain (Barros & Barros/Getty Images)

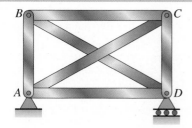

**Fig. 2-22**

Statically indeterminate truss subjected to temperature changes

**Example 2-7**

A prismatic bar $AB$ of length $L$, made of linearly elastic material, is held between immovable supports (Fig. 2-23a). The bar has a modulus of elasticity $E$ and a coefficient of thermal expansion $\alpha$.

**(a)** If the temperature of the bar is raised uniformly by an amount $\Delta T$, derive a formula for the thermal stress $\sigma_T$ developed in the bar.

**(b)** Modify the formula in part (a) if the rigid support at $B$ is replaced by an elastic support having a spring constant $k$ (Fig. 2-23b); assume that only bar $AB$ is subject to the uniform temperature increase $\Delta T$.

**(c)** Repeat part (b) but now assume that the bar is heated nonuniformly such that the temperature increase at distance $x$ from $A$ is given by $\Delta T(x) = \Delta T_0(1-x^2/L^2)$ (see Fig. 2-23c).

**Fig. 2-23**

(a) Example 2-7: Statically indeterminate bar with uniform temperature increase $\Delta T$, (b) statically indeterminate bar with elastic support and uniform temperature increase $\Delta T$, and (c) statically indeterminate bar with elastic support and nonuniform temperature increase $\Delta T(x)$

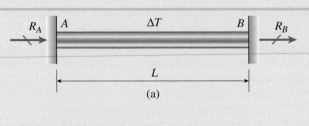

(a)

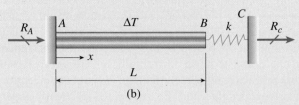

(b)

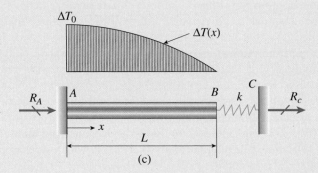

(c)

**Solution**

**(a)** Determine thermal stress in the bar fixed at $A$ and $B$ subjected to uniform temperature increase $\Delta T$.

Because the temperature increases, the bar tends to elongate but is restrained by the rigid supports at $A$ and $B$. Therefore, reactions $R_A$ and $R_B$ are developed at the supports, and the bar is subjected to uniform compressive stresses.

*Equation of Equilibrium.* The only nontrivial equation of static equilibrium is that reactions $R_A$ and $R_B$ must sum to zero. Thus, we have one equation but two unknowns, which is a *one-degree statically indeterminate problem*:

$$\Sigma F_x = 0 \qquad R_A + R_B = 0 \tag{a}$$

We will select reaction $R_B$ as the *redundant* and use the *superposition* of two statically determinate "released" structures (Fig. 2-23d) to develop an additional equation: an equation of *compatibility*. The first released structure is subjected to the temperature increase $\Delta T$ and hence elongates by amount $\delta_T$. The second elongates $\delta_B$ under redundant $R_B$, which is applied as a load. We are using a *statics sign convention*, so forces and displacements in the *x* direction are assumed to be positive.

*Equation of compatibility.* The equation of compatibility expresses the fact that the net change in length of the bar is zero, because supports *A* and *B* are fully restrained:

$$\delta_T + \delta_B = 0 \tag{b}$$

### Fig. 2-23 (Continued)

Example 2-7: (d) Statically determinate bars with support *B* removed (i.e., *released* structures)

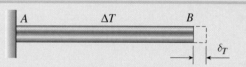

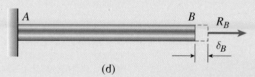

(d)

*Temperature-displacement and force-displacement relations.* The increase in length of the bar due to temperature is (Eq. 2-20)

$$\delta_T = \alpha(\Delta T)L \tag{c}$$

where $\alpha$ is the coefficient of thermal expansion of the material. The increase in bar length due to unknown applied force $R_B$ is obtained from the force-displacement relation:

$$\delta_B = R_B\left(\frac{L}{EA}\right) = R_B f_{AB} \tag{d}$$

in which *E* is the modulus of elasticity, *A* is the bar cross-sectional area, and $f_{AB}$ is the flexibility of the bar.

*Solution of equations.* Substituting Eqs. (c) and (d) into compatibility equation (b) and solving for redundant $R_B$ gives

$$R_B = \frac{-\alpha(\Delta T)L}{f_{AB}} = -EA\alpha(\Delta T) \tag{e}$$

*Continues* ➡

and from the equilibrium equation Eq. (a), we get

$$R_A = -R_B = EA\alpha(\Delta T) \qquad \text{(f)}$$

By using a statics sign convention, we now know that $R_B$ is in the negative $x$ direction, while $R_A$ is in the positive $x$ direction. As a final step, we compute the compressive stress in the bar (assuming that $\Delta T$ is positive and therefore an increase in temperature) to be

$$\sigma_T = \frac{R_A}{A} = E\alpha(\Delta T) \qquad \text{(g)}$$

*Note 1:* In this example, the reactions are independent of the length of the bar, and the stress is independent of both the length and the cross-sectional area [see Eqs. (f) and (g)]. Thus, once again, we see the usefulness of a symbolic solution, because these important features of the bar's behavior might not be noticed in a purely numerical solution.

   *Note 2:* When determining the thermal elongation of the bar [Eq. (c)], we assumed that the material was homogeneous and that the increase in temperature was uniform throughout the volume of the bar. Also, when determining the increase in length due to the reactive force [Eq. (d)], we assumed linearly elastic behavior of the material. These limitations always should be kept in mind when writing equations, such as Eqs.(c) and (d).

   *Note 3:* The bar in this example has zero longitudinal displacements, not only at the fixed ends but also at every cross section. Thus, there are **no axial strains in this bar**, and we have the special situation of *longitudinal stresses without longitudinal strains*. Of course, there are transverse strains in the bar, from both the temperature change and the axial compression.

(b) **Determine thermal stress in the bar fixed at A with elastic support at B and subjected to uniform temperature change ΔT.**

The structure in Fig. 2-23b is one-degree statically indeterminate, so we select reaction $R_C$ as the redundant and once again use the superposition of two released structures to solve the problem.

   First, **statical equilibrium** of the original indeterminate structure requires that

$$R_A + R_C = 0 \qquad \text{(h)}$$

while **compatibility of displacements** at joint C for the two released structures is expressed as

$$\delta_T + \delta_C = 0 \qquad \text{(i)}$$

In the first released structure, we apply uniform temperature change $\Delta T$ to bar $AB$ only, so

$$\delta_T = \alpha(\Delta T)L \qquad \text{(c, repeated)}$$

Note that the spring displaces in the positive $x$ direction but is not deformed by the temperature change. Next, redundant $R_C$ is applied to the end of the spring in the second released structure, resulting in displacement in the positive $x$ direction. Both bar $AB$ and the spring are subject to force $R_C$, so the total displacement at $C$ is the sum of the elongations of bar and spring:

$$\delta_C = R_C\left(\frac{L}{EA}\right) + \frac{R_C}{k} = R_C(f_{AB} + f) \tag{j}$$

where $f = 1/k$ is the flexibility of the spring. Substituting the temperature-displacement equation [Eq. (c)] and force-displacement equation [Eq. (j)] into the compatibility equation [Eq. (i)], then solving for redundant $R_C$ gives

$$R_C = \frac{-\alpha(\Delta T)L}{f_{AB} + f} = \frac{-\alpha(\Delta T)L}{\dfrac{L}{EA} + \dfrac{1}{k}} \quad \text{or} \quad R_C = -\left[\frac{EA\alpha(\Delta T)}{1 + \dfrac{EA}{kL}}\right] \tag{k}$$

Then from the equilibrium equation [Eq. (h)], we see that

$$R_A = -R_C = \frac{EA\alpha(\Delta T)}{1 + \dfrac{EA}{kL}} \tag{l}$$

Recall that we are using a *statics sign convention*, so reaction force $R_A$ is in the positive $x$ direction, while reaction force $R_C$ is in the negative $x$ direction. Finally, the compressive stress in the bar is

$$\sigma_T = \frac{R_A}{A} = \frac{E\alpha(\Delta T)}{1 + \dfrac{EA}{kL}} \quad \longleftarrow \tag{m}$$

Note that *if the spring stiffness k goes to infinity*, Eq. (l) becomes Eq. (f) and Eq. (m) becomes Eq. (g). In effect, use of an infinitely stiff spring moves the rigid support from $C$ back to $B$.

**(c) Determine thermal stress in the bar fixed at A with elastic support at B and subjected to *nonuniform* temperature change.**

The structure in Fig. 2-23c is one-degree statically indeterminate. So, once again, we select reaction $R_C$ as the redundant and, as in parts (a) and (b), use superposition of two released structures to solve the one-degree statically indeterminate problem (Fig. 2-23e,f).

| **Fig. 2-23 (Continued)** |
| --- |

Example 2-7: (e) Statically determinate bar with support C removed (i.e., *released* structure) under nonuniform temperature increase

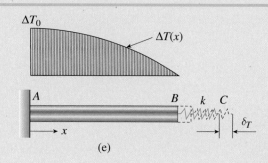

(e)

Continues ➡

● ● ●    **Example 2-7 -** *Continued*

**Fig. 2-23 (Continued)**

(f) Statically determinate bar with support *C* removed (i.e., *released* structure) under applied force $R_C$

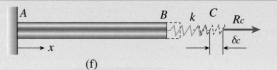

(f)

The equation of *static equilibrium* for the overall structure is Eq. (h), and the equation of *compatibility* is Eq. (i). First, we solve for displacement $\delta_T$ in the first released structure (Fig. 2-23e) as

$$\delta_T = \int_0^L \alpha[\Delta T(x)]dx = \int_0^L \alpha\left\{ \Delta T_0\left[ 1 - \left(\frac{x}{L}\right)^2 \right] \right\}dx = \frac{2}{3}\alpha(\Delta T_0)L \quad \textbf{(n)}$$

and $\delta_C$ for the second released structure (Fig. 2-23f) is the same as Eq. (j), giving

$$\delta_C = R_C(f_{AB} + f) \qquad \textbf{(j, repeated)}$$

Substituting the temperature-displacement equation [Eq. (n)] and the force-displacement equation [Eq. (j)] into the compatibility equation [Eq. (i)] gives

$$R_C = \frac{\dfrac{-2}{3}\alpha(\Delta T_0)L}{f_{AB} + f} = \frac{-2\alpha(\Delta T_0)L}{3\left(\dfrac{L}{EA} + \dfrac{1}{k}\right)} \quad \text{or} \quad R_C = -\left(\frac{2}{3}\right)\left[\frac{EA\alpha(\Delta T_0)}{1 + \dfrac{EA}{kL}}\right] \quad \textbf{(o)}$$

From the statical equilibrium equation [Eq. (h)], we get

$$R_A = -R_C = \left(\frac{2}{3}\right)\left[\frac{EA\alpha(\Delta T_0)}{1 + \dfrac{EA}{kL}}\right] \qquad \textbf{(p)}$$

Finally, the compressive stress in the bar under nonuniform temperature change $\Delta T(x) = \Delta T_0 (1 - (x/L)^2)$ is

$$\sigma_T = \frac{R_A}{A} = \left(\frac{2}{3}\right)\left[\frac{E\alpha(\Delta T_0)}{1 + \dfrac{EA}{kL}}\right] \qquad \Longleftarrow \textbf{(q)}$$

We note once again that use of an infinitely stiff spring eliminates the *EA/kL* term from Eq. (q) and provides the solution for a prismatic bar fixed at *A* and *B* under nonuniform temperature change $\Delta T(x) = \Delta T_0 (1 - (x/L)^2)$.

## ● ● ● Example 2-8

A sleeve in the form of a circular tube of length $L$ is placed around a bolt and fitted between washers at each end (Fig. 2-24a). The nut is then turned until it is just snug. The sleeve and bolt are made of different materials and have different cross-sectional areas. (Assume that the coefficient of thermal expansion $\alpha_S$ of the sleeve is greater than the coefficient $\alpha_B$ of the bolt.)

(a) If the temperature of the entire assembly is raised by an amount $\Delta T$, what stresses $\sigma_S$ and $\sigma_B$ are developed in the sleeve and bolt, respectively?
(b) What is the increase $\delta$ in the length $L$ of the sleeve and bolt?

### Fig. 2-24

Example 2-8: Sleeve and bolt assembly with uniform temperature increase $\Delta T$

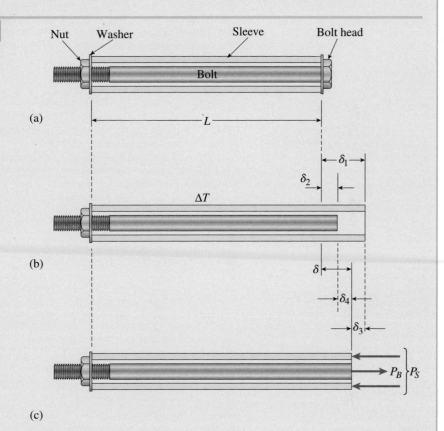

### Solution

Because the sleeve and bolt are of different materials, they will elongate by different amounts when heated and allowed to expand freely. However, when they are held together by the assembly, free expansion cannot occur and thermal stresses are developed in both materials. To find these stresses, we use the same concepts as in any statically indeterminate analysis—equilibrium equations, compatibility equations, and displacement relations. However, we cannot formulate these equations until we disassemble the structure.

A simple way to cut the structure is to remove the head of the bolt, thereby allowing the sleeve and bolt to expand freely under the temperature change $\Delta T$ (Fig. 2-24b). The resulting elongations of the sleeve and bolt

*Continues* ➡

are denoted $\delta_1$ and $\delta_2$, respectively, and the corresponding *temperature-displacement relations* are

$$\delta_1 = \alpha_S(\Delta T)L \qquad \delta_2 = \alpha_B(\Delta T)L \qquad \text{(g,h)}$$

Since $\alpha_S$ is greater than $\alpha_B$, the elongation $\delta_1$ is greater than $\delta_2$, as shown in Fig. 2-24b.

The axial forces in the sleeve and bolt must be such that they shorten the sleeve and stretch the bolt until the final lengths of the sleeve and bolt are the same. These forces are shown in Fig. 2-24c, where $P_S$ denotes the compressive force in the sleeve and $P_B$ denotes the tensile force in the bolt. The corresponding shortening $\delta_3$ of the sleeve and elongation $\delta_4$ of the bolt are

$$\delta_3 = \frac{P_S L}{E_S A_S} \qquad \delta_4 = \frac{P_B L}{E_B A_B} \qquad \text{(i,j)}$$

in which $E_S A_S$ and $E_B A_B$ are the respective axial rigidities. Equations (i) and (j) are the *load-displacement relations*.

Now we can write an *equation of compatibility* expressing the fact that the final elongation $\delta$ is the same for both the sleeve and bolt. The elongation of the sleeve is $\delta_1 - \delta_3$ and of the bolt is $\delta_2 + \delta_4$; therefore,

$$\delta = \delta_1 - \delta_3 = \delta_2 + \delta_4 \qquad \text{(k)}$$

Substituting the temperature-displacement and load-displacement relations [Eqs. (g) to (j)] into this equation gives

$$\delta = \alpha_S(\Delta T)L - \frac{P_S L}{E_S A_S} = \alpha_B(\Delta T)L + \frac{P_B L}{E_B A_B} \qquad \text{(l)}$$

from which we get

$$\frac{P_S L}{E_S A_S} + \frac{P_B L}{E_B A_B} = \alpha_S(\Delta T)L - \alpha_B(\Delta T)L \qquad \text{(m)}$$

which is a modified form of the compatibility equation. Note that it contains the forces $P_S$ and $P_B$ as unknowns.

An *equation of equilibrium* is obtained from Fig. 2-24c, which is a free-body diagram of the part of the assembly remaining after the head of the bolt is removed. Summing forces in the horizontal direction gives

$$P_S = P_B \qquad \text{(n)}$$

which expresses the obvious fact that the compressive force in the sleeve is equal to the tensile force in the bolt.

We now solve simultaneously Eqs. (m) and (n) and obtain the axial forces in the sleeve and bolt:

$$P_S = P_B = \frac{(\alpha_S - \alpha_B)(\Delta T)E_S A_S E_B A_B}{E_S A_S + E_B A_B} \qquad \text{(2-21)}$$

When deriving this equation, we assumed that the temperature increased and that the coefficient $\alpha_S$ was greater than the coefficient $\alpha_B$. Under these conditions, $P_S$ is the compressive force in the sleeve and $P_B$ is the tensile force in the bolt.

The results will be quite different if the temperature increases but the coefficient $\alpha_S$ is less than the coefficient $\alpha_B$. Under these conditions, a gap will open between the bolt head and the sleeve and there will be no stresses in either part of the assembly.

(a) *Stresses in the sleeve and bolt.* Expressions for the stresses $\sigma_S$ and $\sigma_B$ in the sleeve and bolt, respectively, are obtained by dividing the corresponding forces by the appropriate areas:

$$\sigma_S = \frac{P_S}{A_S} = \frac{(\alpha_S - \alpha_B)(\Delta T)E_S A_S E_B}{E_S A_S + E_B A_B} \qquad \Longleftarrow \text{(2-22a)}$$

$$\sigma_B = \frac{P_B}{A_B} = \frac{(\alpha_S - \alpha_B)(\Delta T)E_S A_S E_B}{E_S A_S + E_B A_B} \qquad \Longleftarrow \text{(2-22b)}$$

Under the assumed conditions, the stress $\sigma_S$ in the sleeve is compressive and the stress $\sigma_B$ in the bolt is tensile. It is interesting to note that these stresses are independent of the length of the assembly and their magnitudes are inversely proportional to their respective areas (that is, $\sigma_S/\sigma_B = A_B/A_S$).

(b) *Increase in length of the sleeve and bolt.* The elongation $\delta$ of the assembly can be found by substituting either $P_S$ or $P_B$ from Eq. (2-21) into Eq. (I), yielding

$$\delta = \frac{(\alpha_S E_S A_S + \alpha_B E_B A_B)(\Delta T)L}{E_S A_S + E_B A_B} \qquad \Longleftarrow \text{(2-23)}$$

With the preceding formulas available, we can readily calculate the forces, stresses, and displacements of the assembly for any given set of numerical data.

*Note:* As a partial check on the results, we can see if Eqs. (2-21), (2-22), and (2-23) reduce to known values in simplified cases. For instance, suppose that the bolt is rigid and therefore unaffected by temperature changes. We can represent this situation by setting $\alpha_B = 0$ and letting $E_B$ become infinitely large, thereby creating an assembly in which the sleeve is held between rigid supports. Substituting these values into Eqs. (2-21), (2-22), and (2-23), we find

$$P_S = E_S A_S \alpha_S (\Delta T) \qquad \sigma_S = E_S \alpha_S (\Delta T) \qquad \delta = 0$$

These results agree with those of Example 2-7 for a bar held between rigid supports.

As a second special case, suppose that the sleeve and bolt are made of the same material. Then both parts will expand freely and will lengthen the same amount when the temperature changes. No forces or stresses will be developed. To see if the derived equations predict this behavior, we substitute $\alpha_S = \alpha_B = \alpha$ into Eqs. (2-21), (2-22), and (2-23) and obtain

$$P_S = P_B = 0 \qquad \sigma_S = \sigma_B = 0 \qquad \delta = \alpha(\Delta T)L$$

which are the expected results.

## Misfits and Prestrains

Suppose that a member of a structure is manufactured with its length slightly different from its prescribed length. Then the member will not fit into the structure in its intended manner, and the geometry of the structure will be different from what was planned. We refer to situations of this kind as **misfits**. Sometimes misfits are intentionally created in order to introduce strains into the structure at the time it is built. Because these strains exist before any loads are applied to the structure, they are called **prestrains**. Accompanying the prestrains are prestresses, and the structure is said to be **prestressed**. Common examples of prestressing are spokes in bicycle wheels (which would collapse if not prestressed), the pretensioned faces of tennis racquets, shrink-fitted machine parts, and prestressed concrete beams.

If a structure is **statically determinate**, small misfits in one or more members will not produce strains or stresses, although there will be departures from the theoretical configuration of the structure. To illustrate this statement, consider a simple structure consisting of a horizontal beam $AB$ supported by a vertical bar $CD$ (Fig. 2-25a). If bar $CD$ has exactly the correct length $L$, the beam will be horizontal at the time the structure is built. However, if the bar is slightly longer than intended, the beam will make a small angle with the horizontal. Nevertheless, there will be no strains or stresses in either the bar or the beam attributable to the incorrect length of the bar. Furthermore, if a load $P$ acts at the end of the beam (Fig. 2-25b), the stresses in the structure due to that load will be unaffected by the incorrect length of bar $CD$.

In general, if a structure is statically determinate, the presence of small misfits will produce small changes in geometry but no strains or stresses. Thus, the effects of a misfit are similar to those of a temperature change.

The situation is quite different if the structure is **statically indeterminate**, because then the structure is not free to adjust to misfits (just as it is not free to adjust to certain kinds of temperature changes). To show this, consider a beam supported by two vertical bars (Fig. 2-26a). If both bars have exactly the correct length $L$, the structure can be assembled with no strains or stresses and the beam will be horizontal.

Suppose, however, that bar $CD$ is slightly longer than the prescribed length. Then, in order to assemble the structure, bar $CD$ must be compressed by external forces (or bar $EF$ stretched by external forces), the bars must be fitted into place, and then the external forces must be released. As a result, the beam will deform and rotate, bar $CD$ will be in compression, and bar $EF$ will be in tension. In other words, prestrains will exist in all members and the structure will be prestressed, even though no external

---

**Fig. 2-25**

Statically determinate structure with a small misfit

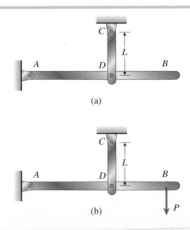

(a)

(b)

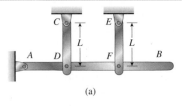

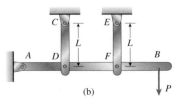

**Fig. 2-26**

Statically indeterminate structure
with a small misfit

loads are acting. If a load $P$ is now added (Fig. 2-26b), additional strains
and stresses will be produced.

The analysis of a statically indeterminate structure with misfits and
prestrains proceeds in the same general manner as described previously for
loads and temperature changes. The basic ingredients of the analysis are
equations of equilibrium, equations of compatibility, force-displacement
relations, and (if appropriate) temperature-displacement relations. The
methodology is illustrated in Example 2-9.

## Bolts and Turnbuckles

Prestressing a structure requires that one or more parts of the structure be
stretched or compressed from their theoretical lengths. A simple way to
produce a change in length is to tighten a bolt or a turnbuckle. In the case
of a **bolt** (Fig. 2-27) each turn of the nut will cause the nut to travel along
the bolt a distance equal to the spacing $p$ of the threads (called the *pitch
of the threads*). Thus, the distance $\delta$ traveled by the nut is

$$\delta = np \tag{2-24}$$

in which $n$ is the number of revolutions of the nut (not necessarily an integer). Depending upon how the structure is arranged, turning the nut can
stretch or compress a member.

In the case of a **double-acting turnbuckle** (Fig. 2-28), there are two end
screws. Because a right-hand thread is used at one end and a left-hand
thread at the other, the device either lengthens or shortens when the buckle
is rotated. Each full turn of the buckle causes it to travel a distance $p$ along
each screw, where again $p$ is the pitch of the threads. Therefore, if the turnbuckle is tightened by one turn, the screws are drawn closer together by a
distance $2p$ and the effect is to shorten the device by $2p$. For $n$ turns, we have

$$\delta = 2np \tag{2-25}$$

Turnbuckles are often inserted in cables and then tightened, thus creating
initial tension in the cables, as illustrated in the following example.

**Fig. 2-27**

The *pitch* of the threads is the
distance from one thread to the
next

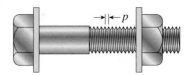

**Fig. 2-28**

Double-acting turnbuckle. (Each
full turn of the turnbuckle
shortens or lengthens the cable
by 2*p*, where *p* is the pitch of
the screw threads.)

The mechanical assembly shown in Fig. 2-29a consists of a copper tube, a rigid end plate, and two steel cables with turnbuckles. The slack is removed from the cables by rotating the turnbuckles until the assembly is snug but with no initial stresses. (Further tightening of the turnbuckles will produce a prestressed condition in which the cables are in tension and the tube is in compression.)

**(a)** Determine the forces in the tube and cables (Fig. 2-29a) when the turnbuckles are tightened by $n$ turns.
**(b)** Determine the shortening of the tube.

### Fig. 2-29

Example 2-9: Statically indeterminate assembly with a copper tube in compression and two steel cables in tension

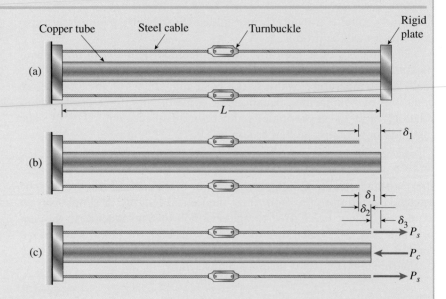

### Solution

We begin the analysis by removing the plate at the right-hand end of the assembly so that the tube and cables are free to change in length (Fig. 2-29b). Rotating the turnbuckles through $n$ turns will shorten the cables by a distance

$$\delta_1 = 2np \tag{o}$$

as shown in Fig. 2-29b.

The tensile forces in the cables and the compressive force in the tube must be such that they elongate the cables and shorten the tube until their final lengths are the same. These forces are shown in Fig. 2-29c, where $P_s$ denotes the tensile force in one of the steel cables and $P_c$ denotes the compressive force in the copper tube. The elongation of a cable due to the force $P_s$ is

$$\delta_2 = \frac{P_s L}{E_s A_s} \tag{p}$$

in which $E_sA_s$ is the axial rigidity and $L$ is the length of a cable. Also, the compressive force $P_c$ in the copper tube causes it to shorten by

$$\delta_3 = \frac{P_cL}{E_cA_c} \tag{q}$$

in which $E_cA_c$ is the axial rigidity of the tube. Equations (p) and (q) are the *load-displacement relations.*

The final shortening of one of the cables is equal to the shortening $\delta_1$ caused by rotating the turnbuckle minus the elongation $\delta_2$ caused by the force $P_s$. This final shortening of the cable must equal the shortening $\delta_3$ of the tube:

$$\delta_1 - \delta_2 = \delta_3 \tag{r}$$

which is the *equation of compatibility.*

Substituting the turnbuckle relation (Eq. o) and the load-displacement relations (Eqs. p and q) into the preceding equation yields

$$2np - \frac{P_sL}{E_sA_s} = \frac{P_cL}{E_cA_c} \tag{s}$$

or

$$\frac{P_sL}{E_sA_s} + \frac{P_cL}{E_cA_c} = 2np \tag{t}$$

which is a modified form of the compatibility equation. Note that it contains $P_s$ and $P_c$ as unknowns.

From Fig. 2-29c, which is a free-body diagram of the assembly with the end plate removed, we obtain the following equation of equilibrium:

$$2P_s = P_c \tag{u}$$

**(a)** *Forces in the cables and tube.* Now we solve simultaneously Eqs. (t) and (u) and obtain the axial forces in the steel cables and copper tube, respectively:

$$P_s = \frac{2npE_cA_cE_sA_s}{L(E_cA_c + 2E_sA_s)} \qquad P_c = \frac{4npE_cA_cE_sA_s}{L(E_cA_c + 2E_sA_s)} \quad \Longleftarrow \tag{2-26a,b}$$

Recall that the forces $P_s$ are tensile forces and the force $P_c$ is compressive. If desired, the stresses $\sigma_s$ and $\sigma_c$ in the steel and copper can now be obtained by dividing the forces $P_s$ and $P_c$ by the cross-sectional areas $A_s$ and $A_c$, respectively.

**(b)** *Shortening of the tube.* The decrease in length of the tube is the quantity $\delta_3$ (see Fig. 2-29 and Eq. q):

$$\delta_3 = \frac{P_cL}{E_cA_c} = \frac{4npE_sA_s}{E_cA_c + 2E_sA_s} \quad \Longleftarrow \tag{2-27}$$

With the preceding formulas available, we can readily calculate the forces, stresses, and displacements of the assembly for any given set of numerical data.

# 2.6 STRESSES ON INCLINED SECTIONS

In our previous discussions of tension and compression in axially loaded members, the only stresses we considered were the normal stresses acting on cross sections. These stresses are pictured in Fig. 2-30, where we consider a bar $AB$ subjected to axial loads $P$.

When the bar is cut at an intermediate cross section by a plane $mn$ (perpendicular to the $x$ axis), we obtain the free-body diagram shown in Fig. 2-30b. The normal stresses acting over the cut section may be calculated from the formula $\sigma_x = P/A$ provided that the stress distribution is uniform over the entire cross-sectional area $A$. As explained in Chapter 1, this condition exists if the bar is prismatic, the material is homogeneous, the axial force $P$ acts at the centroid of the cross-sectional area, and the cross section is away from any localized stress concentrations. Of course, there are no shear stresses acting on the cut section, because it is perpendicular to the longitudinal axis of the bar.

For convenience, we usually show the stresses in a two-dimensional view of the bar (Fig. 2-30c) rather than the more complex three-dimensional view (Fig. 2-30b). However, when working with two-dimensional figures we must not forget that the bar has a thickness perpendicular to the plane of the figure. This third dimension must be considered when making derivations and calculations.

---

**Fig. 2-30**

Prismatic bar in tension showing the stresses acting on cross section $mn$: (a) bar with axial forces $P$, (b) three-dimensional view of the cut bar showing the normal stresses, and (c) two-dimensional view

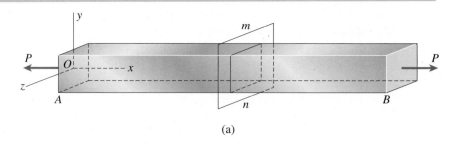

(a)

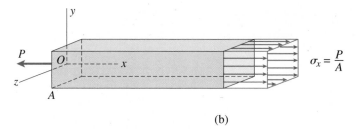

(b)

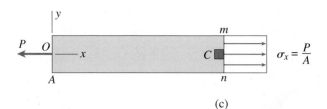

(c)

## Stress Elements

The most useful way of representing the stresses in the bar of Fig. 2-30 is to isolate a small element of material, such as the element labeled $C$ in Fig. 2-30c, and then show the stresses acting on all faces of this element. An element of this kind is called a **stress element**. The stress element at point $C$ is a small rectangular block (it doesn't matter whether it is a cube or a rectangular parallelepiped) with its right-hand face lying in cross section $mn$.

The dimensions of a stress element are assumed to be infinitesimally small, but for clarity we draw the element to a large scale, as in Fig. 2-31a. In this case, the edges of the element are parallel to the $x$, $y$, and $z$ axes, and the only stresses are the normal stresses $\sigma_x$ acting on the $x$ faces (recall that the $x$ faces have their normals parallel to the $x$ axis). Because it is more convenient, we usually draw a two-dimensional view of the element (Fig. 2-31b) instead of a three-dimensional view.

## Stresses on Inclined Sections

The stress element of Fig. 2-31 provides only a limited view of the stresses in an axially loaded bar. To obtain a more complete picture, we need to investigate the stresses acting on **inclined sections**, such as the section cut by the inclined plane $pq$ in Fig. 2-32a. Because the stresses are the same throughout the entire bar, the stresses acting over the inclined section must be uniformly distributed, as pictured in the free-body diagrams of Fig. 2-32b (three-dimensional view) and Fig. 2-32c (two-dimensional view). From the equilibrium of the free body we know that the resultant of the stresses must be a horizontal force $P$. (The resultant is drawn with a dashed line in Figs. 2-32b and 2-32c.)

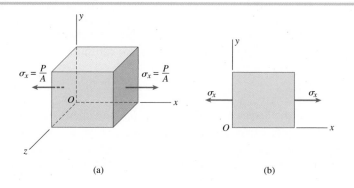

**Fig. 2-31**

Stress element at point $C$ of the axially loaded bar shown in Fig. 2-30c: (a) three-dimensional view of the element, and (b) two-dimensional view of the element

As a preliminary matter, we need a scheme for specifying the **orientation** of the inclined section $pq$. A standard method is to specify the angle $\theta$ between the $x$ axis and the normal $n$ to the section (see Fig. 2-33a). Thus, the angle $\theta$ for the inclined section shown in the figure is approximately 30°. By contrast, cross section $mn$ (Fig. 2-30a) has an angle $\theta$ equal to zero (because the normal to the section is the $x$ axis). For additional examples, consider the stress element of Fig. 2-31. The angle $\theta$ for the right-hand face is 0, for the top face is 90° (a longitudinal section of the bar), for the left-hand face is 180°, and for the bottom face is 270° (or $-90°$).

**Fig. 2-32**

Prismatic bar in tension showing the stresses acting on an inclined section *pq*: (a) bar with axial forces *P*, (b) three-dimensional view of the cut bar showing the stresses, and (c) two-dimensional view

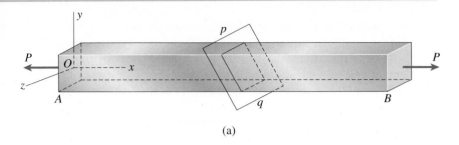

(a)

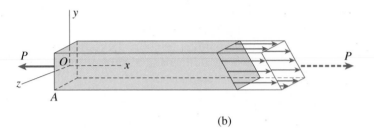

(b)

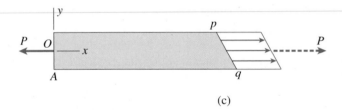

(c)

Let us now return to the task of finding the stresses acting on section *pq* (Fig. 2-33b). As already mentioned, the resultant of these stresses is a force *P* acting in the *x* direction. This resultant may be resolved into two components, a normal force *N* that is perpendicular to the inclined plane *pq* and a shear force *V* that is tangential to it. These force components are

$$N = P \cos \theta \qquad V = P \sin \theta \qquad \textbf{(2-28a,b)}$$

Associated with the forces *N* and *V* are normal and shear stresses that are uniformly distributed over the inclined section (Figs. 2-33c and d). The normal stress is equal to the normal force *N* divided by the area of the section, and the shear stress is equal to the shear force *V* divided by the area of the section. Thus, the stresses are

$$\sigma = \frac{N}{A_1} \qquad \tau = \frac{V}{A_1} \qquad \textbf{(2-29a,b)}$$

in which $A_1$ is the area of the inclined section, as follows:

$$A_1 = \frac{A}{\cos \theta} \qquad \textbf{(2-30)}$$

As usual, *A* represents the cross-sectional area of the bar. The stresses $\sigma$ and $\tau$ act in the directions shown in Figs. 2-33c and d, that is, in the same directions as the normal force *N* and shear force *V*, respectively.

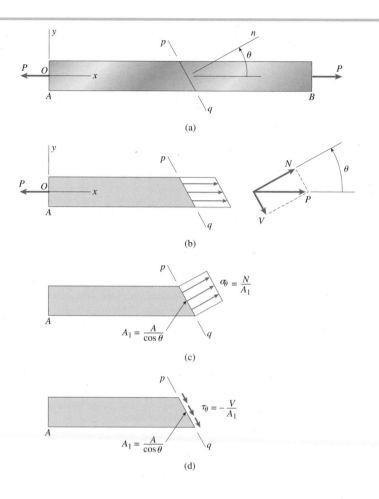

### Fig. 2-33

Prismatic bar in tension showing the stresses acting on an inclined section $pq$

At this point we need to establish a standardized **notation and sign convention** for stresses acting on inclined sections. We will use a subscript $\theta$ to indicate that the stresses act on a section inclined at an angle $\theta$ (Fig. 2-34), just as we use a subscript $x$ to indicate that the stresses act on a section perpendicular to the $x$ axis (see Fig. 2-30). Normal stresses $\sigma_\theta$ are positive in tension and shear stresses $\tau_\theta$ are positive when they tend to produce counterclockwise rotation of the material, as shown in Fig. 2-34.

For a bar in tension, the normal force $N$ produces positive normal stresses $\sigma_\theta$ (see Fig. 2-33c) and the shear force $V$ produces negative shear

### Fig. 2-34

Sign convention for stresses acting on an inclined section. (Normal stresses are positive when in tension and shear stresses are positive when they tend to produce counterclockwise rotation.)

stresses $\tau_\theta$ (see Fig. 2-33d). These stresses are given by the following equations (see Eqs. 2-28, 2-29, and 2-30):

$$\sigma_\theta = \frac{N}{A_1} = \frac{P}{A}\cos^2\theta \qquad \tau_\theta = -\frac{V}{A_1} = -\frac{P}{A}\sin\theta\cos\theta$$

Introducing the notation $\sigma_x = P/A$, in which $\sigma_x$ is the normal stress on a cross section, and also using the trigonometric relations

$$\cos^2\theta = \frac{1}{2}(1 + \cos 2\theta) \qquad \sin\theta\cos\theta = \frac{1}{2}(\sin 2\theta)$$

we get the following expressions for the **normal and shear stresses**:

$$\sigma_\theta = \sigma_x \cos^2\theta = \frac{\sigma_x}{2}(1 + \cos 2\theta) \qquad \textbf{(2-31a)}$$

$$\tau_\theta = -\sigma_x \sin\theta\cos\theta = -\frac{\sigma_x}{2}(\sin 2\theta) \qquad \textbf{(2-31b)}$$

These equations give the stresses acting on an inclined section oriented at an angle $\theta$ to the $x$ axis (Fig. 2-34).

It is important to recognize that Eqs. (2-31a) and (2-31b) were derived only from statics, and therefore they are independent of the material. Thus, these equations are valid for any material, whether it behaves linearly or nonlinearly, elastically or inelastically.

## Maximum Normal and Shear Stresses

The manner in which the stresses vary as the inclined section is cut at various angles is shown in Fig. 2-35. The horizontal axis gives the angle $\theta$ as it varies from $-90°$ to $+90°$, and the vertical axis gives the stresses $\sigma_\theta$ and $\tau_\theta$. Note that a positive angle $\theta$ is measured counterclockwise from the $x$ axis (Fig. 2-34) and a negative angle is measured clockwise.

As shown on the graph, the normal stress $\sigma_\theta$ equals $\sigma_x$ when $\theta = 0$. Then, as $\theta$ increases or decreases, the normal stress diminishes until at $\theta = \pm90°$ it becomes zero, because there are no normal stresses on sections cut parallel to the longitudinal axis. The **maximum normal stress** occurs at $\theta = 0$ and is

$$\sigma_{max} = \sigma_x \qquad \textbf{(2-32)}$$

---

### Fig. 2-35

Graph of normal stress $\sigma_\theta$ and shear stress $\tau_\theta$ versus angle $\theta$ of the inclined section (see Fig. 2-34 and Eqs. 2-31a and b)

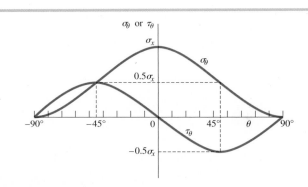

Also, we note that when $\theta = \pm45°$, the normal stress is one-half the maximum value.

The shear stress $\tau_\theta$ is zero on cross sections of the bar ($\theta = 0$) as well as on longitudinal sections ($\theta = \pm90°$). Between these extremes, the stress varies as shown on the graph, reaching the largest positive value when $\theta = -45°$ and the largest negative value when $\theta = +45°$. These **maximum shear stresses** have the same magnitude:

$$\tau_{max} = \frac{\sigma_x}{2} \tag{2-33}$$

but they tend to rotate the element in opposite directions.

The maximum stresses in a **bar in tension** are shown in Fig. 2-36. Two stress elements are selected—element $A$ is oriented at $\theta = 0°$ and element $B$ is oriented at $\theta = 45°$. Element $A$ has the maximum normal stresses (Eq. 2-32) and element $B$ has the maximum shear stresses (Eq. 2-33). In the case of element $A$ (Fig. 2-36b), the only stresses are the maximum normal stresses (no shear stresses exist on any of the faces).

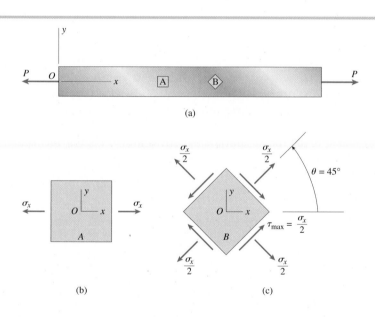

(a)

(b)          (c)

**Fig. 2-36**

Normal and shear stresses acting on stress elements oriented at $\theta = 0°$ and $\theta = 45°$ for a bar in tension

In the case of element $B$ (Fig. 2-36c), both normal and shear stresses act on all faces (except, of course, the front and rear faces of the element). Consider, for instance, the face at 45° (the upper right-hand face). On this face the normal and shear stresses (from Eqs. 2-31a and b) are $\sigma_x/2$ and $-\sigma_x/2$, respectively. Hence, the normal stress is tension (positive) and the shear stress acts clockwise (negative) against the element. The stresses on the remaining faces are obtained in a similar manner by substituting $\theta = 135°$, $-45°$, and $-135°$ into Eqs. (2-31a and b).

Thus, in this special case of an element oriented at $\theta = 45°$, the normal stresses on all four faces are the same (equal to $\sigma_x/2$) and all four shear stresses have the maximum magnitude (equal to $\sigma_x/2$). Also, note that the shear stresses acting on perpendicular planes are equal in magnitude and have directions either toward, or away from, the line of intersection of the planes, as discussed in detail in Section 1.7.

If a bar is loaded in compression instead of tension, the stress $\sigma_x$ will be compression and will have a negative value. Consequently, all stresses acting on stress elements will have directions opposite to those for a bar in tension. Of course, Eqs. (2-31a and b) can still be used for the calculations simply by substituting $\sigma_x$ as a negative quantity.

Even though the maximum shear stress in an axially loaded bar is only one-half the maximum normal stress, the shear stress may cause failure if the material is much weaker in shear than in tension. An example of a shear failure is pictured in Fig. 2-37, which shows a block of wood that was loaded in compression and failed by shearing along a 45° plane.

A similar type of behavior occurs in mild steel loaded in tension. During a tensile test of a flat bar of low-carbon steel with polished surfaces, visible *slip bands* appear on the sides of the bar at approximately 45° to the axis (Fig. 2-38). These bands indicate that the material is failing in shear along the planes on which the shear stress is maximum. Such bands were first observed by G. Piobert in 1842 and W. Lüders in 1860 (see Refs. 2-5 and 2-6), and today they are called either *Lüders' bands* or *Piobert's bands*. They begin to appear when the yield stress is reached in the bar (point *B* in Fig. 1-10 of Section 1.4).

## Uniaxial Stress

The state of stress described throughout this section is called **uniaxial stress**, for the obvious reason that the bar is subjected to simple tension or compression in just one direction. The most important orientations of stress elements for uniaxial stress are $\theta = 0$ and $\theta = 45°$ (Fig. 2-36b and c); the former has the maximum normal stress and the latter has the maximum shear stress. If sections are cut through the bar at other angles, the stresses acting on the faces of the corresponding stress elements can be determined from Eqs. (2-31a and b), as illustrated in Examples 2-10 and 2-11 that follow.

Uniaxial stress is a special case of a more general stress state known as *plane stress*, which is described in detail in Chapter 7.

**Fig. 2-37**

Shear failure along a 45° plane of a wood block loaded in compression (Jim Gere)

Load

Load

**Fig. 2-38**

Slip bands (or Lüders' bands) in a polished steel specimen loaded in tension (Jim Gere)

Load

Load

## • • • Example 2-10

A prismatic brass bar with a length of $L = 0.5$ m and having a cross-sectional area $A = 1200$ mm$^2$ is compressed by an axial load $P = 90$ kN (Fig. 2-39a).

**(a)** Determine the complete state of stress acting on an inclined section $pq$ cut through the bar at an angle $\theta = 25°$, and show the stresses on a properly oriented stress element.

**(b)** If the bar is now fixed between supports $A$ and $B$ (Fig. 2-39b) and then subjected to a temperature increase of $\Delta T = 33°$C, the compressive stress on plane $rs$ is known to be 65 MPa. Find the shear stress $\tau_\theta$ on plane $rs$. What is angle $\theta$? (Assume that modulus of elasticity $E = 110$ GPa and coefficient of thermal expansion $\alpha = 20 \times 10^{-6}$ /°C.)

**(c)** If the allowable normal stress is $\pm 82$ MPa and the allowable shear stress is $\pm 40$ MPa, find the maximum permissible temperature *increase* ($\Delta T$) in the bar if allowable stress values in the bar are not to be exceeded.

### Fig. 2-39

Example 2-10: (a) Stresses on inclined section $pq$ through bar, and (b) Stresses on inclined section $rs$ through bar

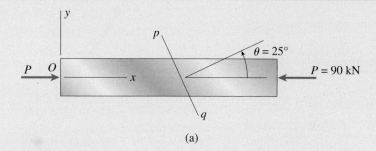

(a)

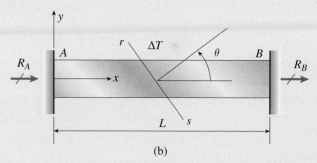

(b)

### Solution

**(a) Determine the complete state of stress on stress element aligned with inclined section $pq$.**

To find the state of stress on inclined section $pq$, we start by finding compressive normal stress $\sigma_x$ due to applied load $P$:

$$\sigma_x = \frac{-P}{A} = \frac{-90 \text{ kN}}{1200 \text{ mm}^2} = -75 \text{ MPa}$$

Next, we find normal and shear stresses from Eqs. (2-31a) and (2-31b) with $\theta = 25°$ as

$$\sigma_\theta = \sigma_x \cos(\theta)^2 = (-75 \text{ MPa}) \cos(25°)^2 = -61.6 \text{ MPa}$$

$$\tau_\theta = -\sigma_x \sin(\theta) \times \cos(\theta) = -(-75 \text{ MPa}) \sin(25°) \times \cos(25°)$$

$$= 28.7 \text{ MPa}$$

*Continues*

These stresses are shown acting on the inclined section *pq* in Fig. 2-39c. Stress element **face ab** (Fig. 2-39d) is aligned with section *pq*. Note that the normal stress $\sigma_\theta$ is *negative* (compressive), and the shear stress $\tau_\theta$ is *positive* (counterclockwise). We must now use Eqs. (2-31a) and (2-31b) to find normal and shear stresses on the remaining three faces of the stress element (see Fig. 2-39d).

**Fig. 2-39 (Continued)**

Example 2-10: (c) Stresses on element inclined section *pq* through bar, and (d) complete state of stress on element at inclined section *pq* through bar

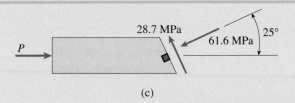

(c)

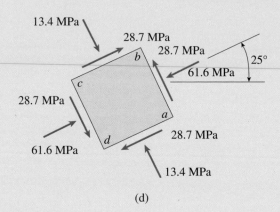

(d)

The normal and shear stresses on **face cb** are computed using angle $\theta + 90° = 115°$ in Eqs. (2-31a) and (2-31b):

$$\sigma_{cb} = \sigma_x \cos(115°)^2 = (-75 \text{ MPa}) \cos(115°)^2 = -13.4 \text{ MPa}$$

$$\tau_{cb} = -\sigma_x \sin(115°) \cos(115°) = -(-75 \text{ MPa})[\sin(115°)\cos(115°)]$$
$$= -28.7 \text{ MPa}$$

The stresses on the opposite **face cd** are the same as those on **face ab**, which can be verified by substituting $\theta = 25° + 180° = 205°$ into Eqs. (2-31a) and (2-31b). For **face ad** we substitute $\theta = 25° - 90° = -65°$ into Eqs. (2-31a) and (2-31b). The complete state of stress is shown in Fig. 2-39d.

**(b) Determine the normal and shear stresses due to temperature increase on the stress element aligned with inclined section *rs*.**

From Example 2-7, we know that reactions $R_A$ and $R_B$ (Fig. 2-39b) due to temperature increase $\Delta T = 33°$ are

$$R_A = -R_B = EA\alpha(\Delta T) \qquad \text{(a)}$$

and the resulting axial *compressive* thermal stress is

$$\sigma_T = \frac{R_A}{A} = E\alpha(\Delta T) \qquad \text{(b)}$$

So

$$\sigma_x = -(110 \text{ GPa})[20 \times 10^{-6}/^\circ\text{C}](33^\circ\text{C}) = -72.6 \text{ MPa}$$

Since the *compressive* stress on plane *rs* is known to be 65 MPa, we can find angle $\theta$ for inclined plane *rs* from Eq. (2-31a) as

$$\theta_{rs} = \cos^{-1}\left(\sqrt{\frac{\sigma_\theta}{\sigma_x}}\right) = \cos^{-1}\left(\sqrt{\frac{-65 \text{ MPa}}{-72.6 \text{ MPa}}}\right) = 18.878^\circ \quad \Longleftarrow$$

and from Eq. (2-31b), we find shear stress $\tau_\theta$ on inclined plane *rs* to be

$$\tau_\theta = -\sigma_x(\sin(\theta_{rs})\cos(\theta_{rs})) = -(-72.6 \text{ MPa})\sin(18.878^\circ)\cos(18.878^\circ)$$

$$= 22.2 \text{ MPa} \quad \Longleftarrow$$

**Fig. 2-39 (Continued)**

Example 2-10: (e) Normal and shear stresses on element at inclined section *rs* through bar

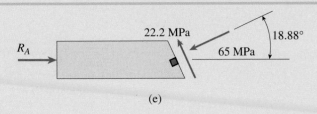

(e)

(c) **Determine the maximum permissible temperature *increase* ($\Delta$T) in the bar based on allowable stress values.**

The *maximum normal stress* $\sigma_{max}$ occurs on a stress element inclined at $\theta = 0$ (Eq. 2-32) so $\sigma_{max} = \sigma_x$. If we equate thermal stress from Eq. (b) to *allowable normal stress* $\sigma_a = 82$ MPa, we can find the value of $\Delta T_{max}$ based on allowable normal stress:

$$\Delta T_{max1} = \frac{\sigma_a}{E\alpha} = \frac{82 \text{ MPa}}{(110 \text{ GPa})[20 \times 10^{-6}/^\circ\text{C}]} = 37.3^\circ\text{C} \quad \textbf{(c)}$$

From Eq. 2-33, we see that *maximum shear stress* $\tau_{max}$ occurs at a section inclination of 45° for which $\tau_{max} = \sigma_x/2$. Using the given allowable shear stress value, $\tau_a = 40$ MPa and the relationship between maximum normal and shear stresses in Eq. 2-33, we can compute a second value for $\Delta T_{max}$ as

$$\Delta T_{max2} = \frac{2\tau_a}{E\alpha} = \frac{2(40 \text{ MPa})}{(110 \text{ GPa})[20 \times 10^{-6}/^\circ\text{C}]} = 36.4^\circ\text{C} \quad \Longleftarrow$$

The lower temperature increase value, based on not exceeding allowable shear stress $\tau_a$, controls. We could have anticipated this because $\tau_{allow} < \sigma_{allow}/2$.

## ● ● ● Example 2-11

A compression bar having a square cross section of width $b$ must support a load $P = 8000$ lb (Fig. 2-40a). The bar is constructed from two pieces of material that are connected by a glued joint (known as a *scarf joint*) along plane $pq$, which is at an angle $\alpha = 40°$ to the vertical. The material is a structural plastic for which the allowable stresses in compression and shear are 1100 psi and 600 psi, respectively. Also, the allowable stresses in the glued joint are 750 psi in compression and 500 psi in shear.

Determine the minimum width $b$ of the bar.

### Solution

For convenience, let us rotate a segment of the bar to a horizontal position (Fig. 2-40b) that matches the figures used in deriving the equations for the stresses on an inclined section (see Figs. 2-33 and 2-34). With the bar in this position, we see that the normal $n$ to the plane of the glued joint (plane $pq$) makes an angle $\beta = 90° - \alpha$, or 50°, with the axis of the bar. Since the angle $\theta$ is defined as positive when counterclockwise (Fig. 2-34), we conclude that $\theta = -50°$ for the glued joint.

The cross-sectional area of the bar is related to the load $P$ and the stress $\sigma_x$ acting on the cross sections by the equation

$$A = \frac{P}{\sigma_x} \tag{a}$$

Therefore, to find the required area, we must determine the value of $\sigma_x$ corresponding to each of the four allowable stresses. Then the smallest value of $\sigma_x$ will determine the required area. The values of $\sigma_x$ are obtained by rearranging Eqs. (2-31a and b) as

$$\sigma_x = \frac{\sigma_\theta}{\cos^2\theta} \qquad \sigma_x = -\frac{\tau_\theta}{\sin\theta\cos\theta} \tag{2-34a,b}$$

We will now apply these equations to the glued joint and to the plastic.

(a) *Values of $\sigma_x$ based upon the allowable stresses in the glued joint.* For compression in the glued joint we have $\sigma_\theta = -750$ psi and $\theta = -50°$. Substituting into Eq. (2-34a), we get

$$\sigma_x = \frac{-750 \text{ psi}}{(\cos(-50°))^2} = -1815 \text{ psi} \tag{b}$$

For shear in the glued joint we have an allowable stress of 500 psi. However, it is not immediately evident whether $\tau_\theta$ is $+500$ psi or $-500$ psi. One approach is to substitute both $+500$ psi and $-500$ psi into Eq. (2-34b) and then select the value of $\sigma_x$ that is negative. The other value of $\sigma_x$ will be positive (tension) and does not apply to this bar. Another approach is to inspect the bar itself (Fig. 2-40b) and observe from the directions of the loads that the shear stress will act clockwise against plane $pq$, which means that the shear stress is negative. Therefore, we substitute $\tau_\theta = -500$ psi and $\theta = -50°$ into Eq. (2-34b) and obtain

$$\sigma_x = -\frac{-500 \text{ psi}}{(\sin(-50°))(\cos(-50°))} = -1015 \text{ psi} \tag{c}$$

**Fig. 2-40**

Example 2-11: Stresses on an inclined section

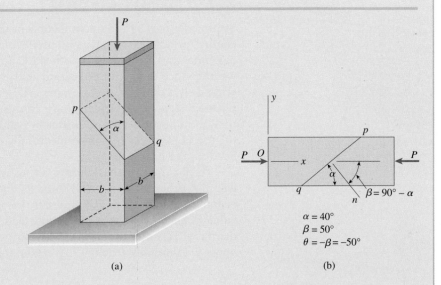

$$\alpha = 40°$$
$$\beta = 50°$$
$$\theta = -\beta = -50°$$

(a)

(b)

(b) *Values of $\sigma_x$ based upon the allowable stresses in the plastic.* The maximum compressive stress in the plastic occurs on a cross section. Therefore, since the allowable stress in compression is 1100 psi, we know immediately that

$$\sigma_x = -1100 \text{ psi} \tag{d}$$

The maximum shear stress occurs on a plane at 45° and is numerically equal to $\sigma_x/2$ (see Eq. 2-33). Since the allowable stress in shear is 600 psi, we obtain

$$\sigma_x = -1200 \text{ psi} \tag{e}$$

This same result can be obtained from Eq. (2-34b) by substituting $\tau_\theta = 600$ psi and $\theta = 45°$.

(c) *Minimum width of the bar.* Comparing the four values of $\sigma_x$ (Eqs. b, c, d, and e), we see that the smallest is $\sigma_x = -1015$ psi. Therefore, this value governs the design. Substituting into Eq. (a), and using only numerical values, we obtain the required area:

$$A = \frac{8000 \text{ lb}}{1015 \text{ psi}} = 7.88 \text{ in.}^2$$

Since the bar has a square cross section ($A = b^2$), the minimum width is

$$b_{\min} = \sqrt{A} = \sqrt{7.88 \text{ in.}^2} = 2.81 \text{ in.}$$

Any width larger than $b_{\min}$ will ensure that the allowable stresses are not exceeded.

## 2.7 STRAIN ENERGY

Strain energy is a fundamental concept in applied mechanics, and strain-energy principles are widely used for determining the response of machines and structures to both static and dynamic loads. In this section we introduce the subject of strain energy in its simplest form by considering only axially loaded members subjected to static loads. More complicated structural elements are discussed in later chapters—bars in torsion in Section 3.9 and beams in bending in Section 9.8. In addition, the use of strain energy in connection with dynamic loads is described in Sections 2.8 and 9.10.

To illustrate the basic ideas, let us again consider a prismatic bar of length $L$ subjected to a tensile force $P$ (Fig. 2-41). We assume that the load is applied slowly, so that it gradually increases from zero to its maximum value $P$. Such a load is called a **static load** because there are no dynamic or inertial effects due to motion. The bar gradually elongates as the load is applied, eventually reaching its maximum elongation $\delta$ at the same time

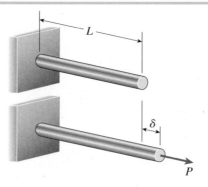

**Fig. 2-41**

Prismatic bar subjected to a statically applied load

that the load reaches its full value $P$. Thereafter, the load and elongation remain unchanged.

During the loading process, the load $P$ moves slowly through the distance $\delta$ and does a certain amount of **work**. To evaluate this work, we recall from elementary mechanics that a constant force does work equal to the product of the force and the distance through which it moves. However, in our case the force varies in magnitude from zero to its maximum value $P$. To find the work done by the load under these conditions, we need to know the manner in which the force varies. This information is supplied by a **load-displacement diagram**, such as the one plotted in Fig. 2-42. On this diagram the vertical axis represents the axial load and the horizontal axis represents the corresponding elongation of the bar. The shape of the curve depends upon the properties of the material.

Let us denote by $P_1$ any value of the load between zero and the maximum value $P$, and let us denote the corresponding elongation of the bar by $\delta_1$. Then an increment $dP_1$ in the load will produce an increment $d\delta_1$ in the elongation. The work done by the load during this incremental elongation is the product of the load and the distance through which it moves, that is, the work equals $P_1 d\delta_1$. This work is represented in the figure by the

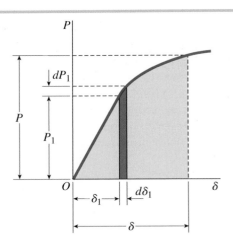

**Fig. 2-42**

Load-displacement diagram

area of the shaded strip below the load-displacement curve. The total work done by the load as it increases from zero to the maximum value $P$ is the summation of all such elemental strips:

$$W = \int_0^\delta P_1 \, d\delta_1 \qquad \textbf{(2-35)}$$

In geometric terms, *the work done by the load is equal to the area below the load-displacement curve.*

When the load stretches the bar, strains are produced. The presence of these strains increases the energy level of the bar itself. Therefore, a new quantity, called **strain energy**, is defined as the energy absorbed by the bar during the loading process. From the principle of conservation of energy, we know that this strain energy is equal to the work done by the load provided no energy is added or subtracted in the form of heat. Therefore,

$$U = W = \int_0^\delta P_1 \, d\delta_1 \qquad \textbf{(2-36)}$$

in which $U$ is the symbol for strain energy. Sometimes strain energy is referred to as **internal work** to distinguish it from the external work done by the load.

Work and energy are expressed in the same **units**. In SI, the unit of work and energy is the joule (J), which is equal to one newton meter (1 J = 1 N·m). In USCS units, work and energy are expressed in foot-pounds (ft-lb), foot-kips (ft-k), inch-pounds (in.-lb), and inch-kips (in.-k).*

## Elastic and Inelastic Strain Energy

If the force $P$ (Fig. 2-41) is slowly removed from the bar, the bar will shorten. If the elastic limit of the material is not exceeded, the bar will return to its original length. If the limit is exceeded, *a permanent set will remain* (see Section 1.5). Thus, either all or part of the strain

*Conversion factors for work and energy are given in Appendix B, Table B-5.

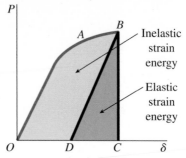

### Fig. 2-43

Elastic and inelastic strain energy

energy will be recovered in the form of work. This behavior is shown on the load-displacement diagram of Fig. 2-43. During loading, the work done by the load is equal to the area below the curve (area *OABCDO*). When the load is removed, the load-displacement diagram follows line *BD* if point *B* is beyond the elastic limit, and a permanent elongation *OD* remains. Thus, the strain energy recovered during unloading, called the **elastic strain energy**, is represented by the shaded triangle *BCD*. Area *OABDO* represents energy that is lost in the process of permanently deforming the bar. This energy is known as the **inelastic strain energy**.

Most structures are designed with the expectation that the material will remain within the elastic range under ordinary conditions of service. Let us assume that the load at which the stress in the material reaches the elastic limit is represented by point *A* on the load-displacement curve (Fig. 2-43). As long as the load is below this value, all of the strain energy is recovered during unloading and no permanent elongation remains. Thus, the bar acts as an elastic spring, storing and releasing energy as the load is applied and removed.

## Linearly Elastic Behavior

Let us now assume that the material of the bar follows Hooke's law, so that the load-displacement curve is a straight line (Fig. 2-44). Then the strain energy *U* stored in the bar (equal to the work *W* done by the load) is

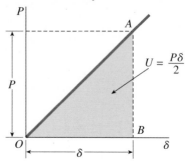

### Fig. 2-44

Load-displacement diagram for a bar of linearly elastic material

$$U = W = \frac{P\delta}{2} \qquad \textbf{(2-37)}$$

which is the area of the shaded triangle *OAB* in the figure.*

The relationship between the load *P* and the elongation $\delta$ for a bar of linearly elastic material is given by the equation

$$\delta = \frac{PL}{EA} \qquad \textbf{(2-38)}$$

Combining this equation with Eq. (2-37) enables us to express the strain energy of a **linearly elastic bar** in either of the following forms:

$$U = \frac{P^2 L}{2EA} \qquad U = \frac{EA\delta^2}{2L} \qquad \textbf{(2-39a,b)}$$

The first equation expresses the strain energy as a function of the load and the second expresses it as a function of the elongation.

From the first equation we see that increasing the length of a bar increases the amount of strain energy even though the load is unchanged (because more material is being strained by the load). On the other hand, increasing either the modulus of elasticity or the cross-sectional area decreases the strain energy because the strains in the bar are reduced. These ideas are illustrated in Examples 2-12 and 2-15.

---

*The principle that the work of the external loads is equal to the strain energy (for the case of linearly elastic behavior) was first stated by the French engineer B. P. E. Clapeyron (1799–1864) and is known as *Clapeyron's theorem* (Ref. 2-7).

Strain-energy equations analogous to Eqs. (2-39a) and (2-39b) can be written for a **linearly elastic spring** by replacing the stiffness $EA/L$ of the prismatic bar by the stiffness $k$ of the spring. Thus,

$$U = \frac{P^2}{2k} \qquad U = \frac{k\delta^2}{2} \qquad \textbf{(2-40a,b)}$$

Other forms of these equations can be obtained by replacing $k$ by $1/f$, where $f$ is the flexibility.

## Nonuniform Bars

The total strain energy $U$ of a bar consisting of several segments is equal to the sum of the strain energies of the individual segments. For instance, the strain energy of the bar pictured in Fig. 2-45 equals the strain energy of segment $AB$ plus the strain energy of segment $BC$. This concept is expressed in general terms by the following equation:

$$U = \sum_{i=1}^{n} U_i \qquad \textbf{(2-41)}$$

**Fig. 2-45**

Bar consisting of prismatic segments having different cross-sectional areas and different axial forces

in which $U_i$ is the strain energy of segment $i$ of the bar and $n$ is the number of segments. (This relation holds whether the material behaves in a linear or nonlinear manner.)

Now assume that the material of the bar is linearly elastic and that the internal axial force is constant within each segment. We can then use Eq. (2-39a) to obtain the strain energies of the segments, and Eq. (2-41) becomes

$$U = \sum_{i=1}^{n} \frac{N_i^2 L_i}{2E_i A_i} \qquad \textbf{(2-42)}$$

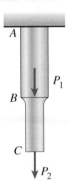

in which $N_i$ is the axial force acting in segment $i$ and $L_i$, $E_i$, and $A_i$ are properties of segment $i$. (The use of this equation is illustrated in Examples 2-12 and 2-15 at the end of the section.)

We can obtain the strain energy of a nonprismatic bar with continuously varying axial force (Fig. 2-46) by applying Eq. (2-39a) to a differential element (shown shaded in the figure) and then integrating along the length of the bar:

$$U = \int_0^L \frac{[N(x)]^2 dx}{2EA(x)} \qquad \textbf{(2-43)}$$

**Fig. 2-46**

Nonprismatic bar with varying axial force

In this equation, $N(x)$ and $A(x)$ are the axial force and cross-sectional area at distance $x$ from the end of the bar. (Example 2-13 illustrates the use of this equation.)

## Comments

The preceding expressions for strain energy (Eqs. 2-39 through 2-43) show that strain energy is *not* a linear function of the loads, not even when the material is linearly elastic. Thus, it is important to realize that *we cannot*

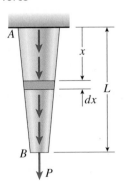

*obtain the strain energy of a structure supporting more than one load by combining the strain energies obtained from the individual loads acting separately.*

In the case of the nonprismatic bar shown in Fig. 2-45, the total strain energy is *not* the sum of the strain energy due to load $P_1$ acting alone and the strain energy due to load $P_2$ acting alone. Instead, we must evaluate the strain energy with all of the loads acting simultaneously, as demonstrated later in Example 2-13.

Although we considered only tension members in the preceding discussions of strain energy, all of the concepts and equations apply equally well to members in **compression**. Since the work done by an axial load is positive regardless of whether the load causes tension or compression, it follows that strain energy is always a positive quantity. This fact is also evident in the expressions for strain energy of linearly elastic bars (such as Eqs. 2-39a and 2-39b). These expressions are always positive because the load and elongation terms are squared.

Strain energy is a form of **potential energy** (or "energy of position") because it depends upon the relative locations of the particles or elements that make up the member. When a bar or a spring is compressed, its particles are crowded more closely together; when it is stretched, the distances between particles increase. In both cases the strain energy of the member increases as compared to its strain energy in the unloaded position.

## Displacements Caused by a Single Load

The displacement of a linearly elastic structure supporting only one load can be determined from its strain energy. To illustrate the method, consider a two-bar truss (Fig. 2-47) loaded by a vertical force $P$. Our objective is to determine the vertical displacement $\delta$ at joint $B$ where the load is applied.

When applied slowly to the truss, the load $P$ does work as it moves through the vertical displacement $\delta$. However, it does no work as it moves laterally, that is, sideways. Therefore, since the load-displacement diagram is linear (see Fig. 2-44 and Eq. 2-37), the strain energy $U$ stored in the structure, equal to the work done by the load, is

$$U = W = \frac{P\delta}{2}$$

from which we get

$$\delta = \frac{2U}{P} \tag{2-44}$$

This equation shows that under certain special conditions, as outlined in the following paragraph, the displacement of a structure can be determined directly from the strain energy.

The conditions that must be met in order to use Eq. (2-44) are as follows: (1) the structure must behave in a linearly elastic manner, and (2) only one load may act on the structure. Furthermore, the only displacement that can be determined is the displacement corresponding to the load itself (that is, the displacement must be in the direction of the load and must be at the point where the load is applied). Therefore, this method for finding displacements is extremely limited in its application and is not a good indicator of the great importance of strain-energy principles in structural mechanics. However, the method does provide an introduction to the use of strain energy. (The method is illustrated later in Example 2-14.)

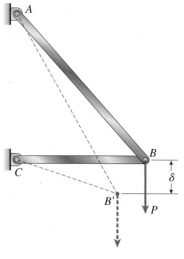

**Fig. 2-47**

Structure supporting a single load $P$

## Strain-Energy Density

In many situations it is convenient to use a quantity called **strain-energy density**, defined as the strain energy per unit volume of material. Expressions for strain-energy density in the case of linearly elastic materials can be obtained from the formulas for strain energy of a prismatic bar (Eqs. 2-39a and b). Since the strain energy of the bar is distributed uniformly throughout its volume, we can determine the strain-energy density by dividing the total strain energy $U$ by the volume $AL$ of the bar. Thus, the strain-energy density, denoted by the symbol $u$, can be expressed in either of these forms:

$$u = \frac{P^2}{2EA^2} \qquad u = \frac{E\delta^2}{2L^2} \qquad \textbf{(2-45a,b)}$$

If we replace $P/A$ by the stress $\sigma$ and $\delta/L$ by the strain $\varepsilon$, we get

$$u = \frac{\sigma^2}{2E} \qquad u = \frac{E\varepsilon^2}{2} \qquad \textbf{(2-46a,b)}$$

These equations give the strain-energy density in a linearly elastic material in terms of either the normal stress $\sigma$ or the normal strain $\varepsilon$.

The expressions in Eqs. (2-46a and b) have a simple geometric interpretation. They are equal to the area $\sigma\varepsilon/2$ of the triangle below the stress-strain diagram for a material that follows Hooke's law ($\sigma = E\varepsilon$). In more general situations where the material does not follow Hooke's law, the strain-energy density is still equal to the area below the stress-strain curve, but the area must be evaluated for each particular material.

Strain-energy density has **units** of energy divided by volume. The SI units are joules per cubic meter ($J/m^3$) and the USCS units are foot-pounds per cubic foot, inch-pounds per cubic inch, and other similar units. Since all of these units reduce to units of stress (recall that $1\ J = 1\ N{\cdot}m$), we can also use units such as pascals (Pa) and pounds per square inch (psi) for strain-energy density.

The strain-energy density of the material when it is stressed to the proportional limit is called the **modulus of resilience** $u_r$. It is found by substituting the proportional limit $\sigma_{pl}$ into Eq. (2-46a):

$$u_r = \frac{\sigma_{pl}^2}{2E} \qquad \textbf{(2-47)}$$

For example, a mild steel having $\sigma_{pl} = 36{,}000$ psi and $E = 30 \times 10^6$ psi has a modulus of resilience $u_r = 21.6$ psi (or 149 kPa). Note that the modulus of resilience is equal to the area below the stress-strain curve up to the proportional limit. *Resilience* represents the ability of a material to absorb and release energy within the elastic range.

Another quantity, called *toughness*, refers to the ability of a material to absorb energy without fracturing. The corresponding modulus, called the **modulus of toughness** $u_t$, is the strain-energy density when the material is stressed to the point of failure. It is equal to the area below the entire stress-strain curve. The higher the modulus of toughness, the greater the ability of the material to absorb energy without failing. A high modulus of toughness is therefore important when the material is subject to impact loads (see Section 2.8).

The preceding expressions for strain-energy density (Eqs. 2-45 through 2-47) were derived for *uniaxial stress*, that is, for materials subjected only to tension or compression. Formulas for strain-energy density in other stress states are presented in Chapters 3 and 7.

**Fig. 2-48**

Example 2-12: Calculation of strain energy

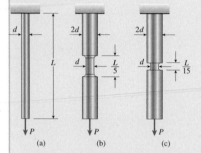

Three round bars having the same length $L$ but different shapes are shown in Fig. 2-48. The first bar has diameter $d$ over its entire length, the second has diameter $d$ over one-fifth of its length, and the third has diameter $d$ over one-fifteenth of its length. Elsewhere, the second and third bars have diameter $2d$. All three bars are subjected to the same axial load $P$.

Compare the amounts of strain energy stored in the bars, assuming linearly elastic behavior. (Disregard the effects of stress concentrations and the weights of the bars.)

**Solution**

(a) *Strain energy $U_1$ of the first bar*. The strain energy of the first bar is found directly from Eq. (2-39a):

$$U_1 = \frac{P^2L}{2EA} \qquad \Longleftarrow \text{(a)}$$

in which $A = \pi d^2/4$.

(b) *Strain energy $U_2$ of the second bar*. The strain energy is found by summing the strain energies in the three segments of the bar (see Eq. 2-42). Thus,

$$U_2 = \sum_{i=1}^{n} \frac{N_i^2 L_i}{2E_i A_i} = \frac{P^2(L/5)}{2EA} + \frac{P^2(4L/5)}{2E(4A)} = \frac{P^2L}{5EA} = \frac{2U_1}{5} \qquad \Longleftarrow \text{(b)}$$

which is only 40% of the strain energy of the first bar. Thus, increasing the cross-sectional area over part of the length has greatly reduced the amount of strain energy that can be stored in the bar.

(c) *Strain energy $U_3$ of the third bar*. Again using Eq. (2-42), we get

$$U_3 = \sum_{i=1}^{n} \frac{N_i^2 L_i}{2E_i A_i} = \frac{P^2(L/15)}{2EA} + \frac{P^2(14L/15)}{2E(4A)} = \frac{3P^2L}{20EA} = \frac{3U_1}{10} \qquad \Longleftarrow \text{(c)}$$

The strain energy has now decreased to 30% of the strain energy of the first bar.

*Note:* Comparing these results, we see that the strain energy decreases as the part of the bar with the larger area increases. If the same amount of work is applied to all three bars, the highest stress will be in the third bar, because the third bar has the least energy-absorbing capacity. If the region having diameter $d$ is made even smaller, the energy-absorbing capacity will decrease further.

We therefore conclude that it takes only a small amount of work to bring the tensile stress to a high value in a bar with a groove, and the narrower the groove, the more severe the condition. When the loads are dynamic and the ability to absorb energy is important, the presence of grooves is very damaging.

In the case of static loads, the maximum stresses are more important than the ability to absorb energy. In this example, all three bars have the same maximum stress $P/A$ (provided stress concentrations are alleviated), and therefore all three bars have the same load-carrying capacity when the load is applied statically.

● ● ● **Example 2-13**

Determine the strain energy of a prismatic bar suspended from its upper end (Fig. 2-49). Consider the following loads: (a) the weight of the bar itself, and (b) the weight of the bar plus a load $P$ at the lower end. (Assume linearly elastic behavior.)

**Fig. 2-49**

Example 2-13: (a) Bar hanging under its own weight, and (b) bar hanging under its own weight and also supporting a load $P$

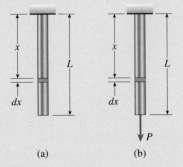

(a)          (b)

**Solution**

(a) *Strain energy due to the weight of the bar itself (Fig. 2-49a).* The bar is subjected to a varying axial force, the internal force being zero at the lower end and maximum at the upper end. To determine the axial force, we consider an element of length $dx$ (shown shaded in the figure) at distance $x$ from the upper end. The internal axial force $N(x)$ acting on this element is equal to the weight of the bar below the element:

$$N(x) = \gamma A(L - x) \tag{d}$$

in which $\gamma$ is the weight density of the material and $A$ is the cross-sectional area of the bar. Substituting into Eq. (2-43) and integrating gives the total strain energy:

$$U = \int_0^L \frac{[N(x)]^2\, dx}{2EA(x)} = \int_0^L \frac{[\gamma A(L - x)]^2\, dx}{2EA} = \frac{\gamma^2 A L^3}{6E} \tag{2-48}$$

(b) *Strain energy due to the weight of the bar plus the load P (Fig. 2-49b).* In this case the axial force $N(x)$ acting on the element is

$$N(x) = \gamma A(L - x) + P \tag{e}$$

[compare with Eq. (d)]. From Eq. (2-43) we now obtain

$$U = \int_0^L \frac{[\gamma A(L - x) + P]^2\, dx}{2EA} = \frac{\gamma^2 A L^3}{6E} + \frac{\gamma P L^2}{2E} + \frac{P^2 L}{2EA} \tag{2-49}$$

*Note:* The first term in this expression is the same as the strain energy of a bar hanging under its own weight (Eq. 2-48), and the last term is the same as the strain energy of a bar subjected only to an axial force $P$ (Eq. 2-39a). However, the middle term contains both $\gamma$ and $P$, showing that it depends upon both the weight of the bar and the magnitude of the applied load.

Thus, this example illustrates that the strain energy of a bar subjected to two loads is *not* equal to the sum of the strain energies produced by the individual loads acting separately.

● ● ●  **Example 2-14**

Determine the vertical displacement $\delta_B$ of joint $B$ of the truss shown in Fig. 2-50. Note that the only load acting on the truss is a vertical load $P$ at joint $B$. Assume that both members of the truss have the same axial rigidity $EA$.

**Fig. 2-50**

Example 2-14: Displacement of a truss supporting a single load $P$

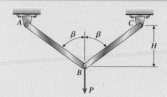

### Solution

Since there is only one load acting on the truss, we can find the displacement corresponding to that load by equating the work of the load to the strain energy of the members. However, to find the strain energy we must know the forces in the members (see Eq. 2-39a).

From the equilibrium of forces acting at joint $B$ we see that the axial force $F$ in either bar is

$$F = \frac{P}{2 \cos \beta} \tag{f}$$

in which $\beta$ is the angle shown in the figure.

Also, from the geometry of the truss we see that the length of each bar is

$$L_1 = \frac{H}{\cos \beta} \tag{g}$$

in which $H$ is the height of the truss.

We can now obtain the strain energy of the two bars from Eq. (2-39a):

$$U = (2)\frac{F^2 L_1}{2EA} = \frac{P^2 H}{4EA \cos^3 \beta} \tag{h}$$

Also, the work of the load $P$ (from Eq. 2-37) is

$$W = \frac{P\delta_B}{2} \tag{i}$$

where $\delta_B$ is the downward displacement of joint $B$. Equating $U$ and $W$ and solving for $\delta_B$, we obtain

$$\delta_B = \frac{PH}{2EA \cos^3 \beta} \qquad \Longleftarrow \tag{2-50}$$

*Note*: We found this displacement using only equilibrium and strain energy—we did not need to draw a displacement diagram at joint $B$.

### ● ● ● Example 2-15

The cylinder for a compressed air machine is clamped by bolts that pass through the flanges of the cylinder (Fig. 2-51a). A detail of one of the bolts is shown in part (b) of the figure. The diameter $d$ of the shank is 0.500 in. and the root diameter $d_r$ of the threaded portion is 0.406 in. The grip $g$ of the bolts is 1.50 in. and the threads extend a distance $t = 0.25$ in. into the grip. Under the action of repeated cycles of high and low pressure in the chamber, the bolts may eventually break.

To reduce the likelihood of the bolts failing, the designers suggest two possible modifications: (1) Machine down the shanks of the bolts so that the shank diameter is the same as the thread diameter $d_r$, as shown in Fig. 2-52a. (2) Replace each pair of bolts by a single long bolt, as shown in Fig. 2-52b. The long bolts are similar to the original bolts (Fig. 2-51b) except that the grip is increased to the distance $L = 13.5$ in.

Compare the energy-absorbing capacity of the three bolt configurations: (a) original bolts, (b) bolts with reduced shank diameter, and (c) long bolts. (Assume linearly elastic behavior and disregard the effects of stress concentrations.)

### Fig. 2-51

Example 2-15: (a) Cylinder with piston and clamping bolts, and (b) detail of one bolt

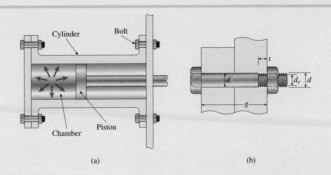

(a)                    (b)

### Solution

(a) *Original bolts.* The original bolts can be idealized as bars consisting of two segments (Fig. 2-51b). The left-hand segment has length $g - t$ and diameter $d$, and the right-hand segment has length $t$ and diameter $d_r$. The strain energy of one bolt under a tensile load $P$ can be obtained by adding the strain energies of the two segments (Eq. 2-42):

$$U_1 = \sum_{i=1}^{n} \frac{N_i^2 L_i}{2 E_i A_i} = \frac{P^2(g - t)}{2EA_s} + \frac{P^2 t}{2EA_r} \qquad \text{(j)}$$

in which $A_s$ is the cross-sectional area of the shank and $A_r$ is the cross-sectional area at the root of the threads; thus,

$$A_s = \frac{\pi d^2}{4} \qquad A_r = \frac{\pi d_r^2}{4} \qquad \text{(k)}$$

Substituting these expressions into Eq. (j), we get the following formula for the strain energy of one of the original bolts:

$$U_1 = \frac{2P^2(g - t)}{\pi E d^2} + \frac{2P^2 t}{\pi E d_r^2} \qquad \text{(l)}$$

*Continues* ➡

**Example 2-15 -** *Continued*

(b) *Bolts with reduced shank diameter.* These bolts can be idealized as prismatic bars having length $g$ and diameter $d_r$ (Fig. 2-52a). Therefore, the strain energy of one bolt (see Eq. 2-39a) is

$$U_2 = \frac{P^2 g}{2EA_r} = \frac{2P^2 g}{\pi E d_r^2} \tag{m}$$

The ratio of the strain energies for cases (1) and (2) is

$$\frac{U_2}{U_1} = \frac{g d^2}{(g - t)d_r^2 + t d^2} \tag{n}$$

or, upon substituting numerical values,

$$\frac{U_2}{U_1} = \frac{(1.50 \text{ in.})(0.500 \text{ in.})^2}{(1.50 \text{ in.} - 0.25 \text{ in.})(0.406 \text{ in.})^2 + (0.25 \text{ in.})(0.500 \text{ in.})^2} = 1.40 \quad \Longleftarrow$$

Thus, using bolts with reduced shank diameters results in a 40% increase in the amount of strain energy that can be absorbed by the bolts. If implemented, this scheme should reduce the number of failures caused by the impact loads.

(c) *Long bolts.* The calculations for the long bolts (Fig. 2-52b) are the same as for the original bolts except the grip $g$ is changed to the grip $L$. Therefore, the strain energy of one long bolt [compare with Eq. (l)] is

$$U_3 = \frac{2P^2(L - t)}{\pi E d^2} + \frac{2P^2 t}{\pi E d_r^2} \tag{o}$$

Since one long bolt replaces two of the original bolts, we must compare the strain energies by taking the ratio of $U_3$ to $2U_1$, as

$$\frac{U_3}{2U_1} = \frac{(L - t)d_r^2 + t d^2}{2(g - t)d_r^2 + 2t d^2} \tag{p}$$

Substituting numerical values gives

$$\frac{U_3}{2U_1} = \frac{(13.5 \text{ in.} - 0.25 \text{ in.})(0.406 \text{ in.})^2 + (0.25 \text{ in.})(0.500 \text{ in.})^2}{2(1.50 \text{ in.} - 0.25 \text{ in.})(0.406 \text{ in.})^2 + 2(0.25 \text{ in.})(0.500 \text{ in.})^2} = 4.18 \quad \Longleftarrow$$

Thus, using long bolts increases the energy-absorbing capacity by 318% and achieves the greatest safety from the standpoint of strain energy.

*Note:* When designing bolts, designers must also consider the maximum tensile stresses, maximum bearing stresses, stress concentrations, and many other matters.

## Fig. 2-52

Example 2-15: Proposed modifications to the bolts: (a) Bolts with reduced shank diameter, and (b) bolts with increased length

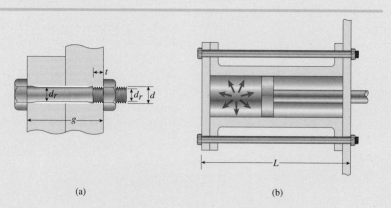

(a)                              (b)

# *2.8 IMPACT LOADING

Loads can be classified as static or dynamic depending upon whether they remain constant or vary with time. A **static load** is applied slowly, so that it causes no vibrational or dynamic effects in the structure. The load increases gradually from zero to its maximum value, and thereafter it remains constant.

A **dynamic load** may take many forms—some loads are applied and removed suddenly (*impact loads*), others persist for long periods of time and continuously vary in intensity (*fluctuating loads*). Impact loads are produced when two objects collide or when a falling object strikes a structure. Fluctuating loads are produced by rotating machinery, traffic, wind gusts, water waves, earthquakes, and manufacturing processes.

As an example of how structures respond to dynamic loads, we will discuss the impact of an object falling onto the lower end of a prismatic bar (Fig. 2-53). A collar of mass $M$, initially at rest, falls from a height $h$ onto a flange at the end of bar $AB$. When the collar strikes the flange, the bar begins to elongate, creating axial stresses within the bar. In a very short interval of time, such as a few milliseconds, the flange will move downward and reach its position of maximum displacement. Thereafter, the bar shortens, then lengthens, then shortens again as the bar vibrates longitudinally and the end of the bar moves up and down. The vibrations are analogous to those that occur when a spring is stretched and then released, or when a person makes a bungee jump. The vibrations of the bar soon cease because of various damping effects, and then the bar comes to rest with the mass $M$ supported on the flange.

The response of the bar to the falling collar is obviously very complicated, and a complete and accurate analysis requires the use of advanced mathematical techniques. However, we can make an approximate analysis by using the concept of strain energy (Section 2.7) and making several simplifying assumptions.

Let us begin by considering the energy of the system just before the collar is released (Fig. 2-53a). The potential energy of the collar with respect to the elevation of the flange is $Mgh$, where $g$ is the acceleration of gravity.* This potential energy is converted into kinetic energy as the collar falls. At the instant the collar strikes the flange, its potential energy with respect to the elevation of the flange is zero and its kinetic energy is $Mv^2/2$, where $v = \sqrt{2gh}$ is its velocity.**

During the ensuing impact, the kinetic energy of the collar is transformed into other forms of energy. Part of the kinetic energy is transformed into the strain energy of the stretched bar. Some of the energy is dissipated in the production of heat and in causing localized plastic deformations of the collar and flange. A small part remains as the kinetic energy of the collar, which either moves further downward (while in contact with the flange) or else bounces upward.

To make a simplified analysis of this very complex situation, we will idealize the behavior by making the following assumptions. (1) We assume

## Fig. 2-53

Impact load on a prismatic bar $AB$ due to a falling object of mass $M$

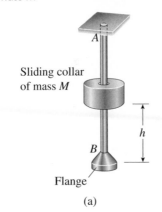

Sliding collar of mass $M$

$h$

Flange

(a)

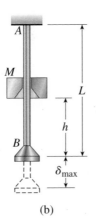

$A$

$M$

$L$

$B$

$h$

$\delta_{max}$

(b)

---

*In SI units, the acceleration of gravity $g$ = 9.81 m/s²; in USCS units, $g$ = 32.2 ft/s². For more precise values of $g$, or for a discussion of mass and weight, see Appendix B.

**In engineering work, velocity is usually treated as a vector quantity. However, since kinetic energy is a scalar, we will use the word "velocity" to mean the *magnitude* of the velocity, or the *speed*.

that the collar and flange are so constructed that the collar "sticks" to the flange and moves downward with it (that is, the collar does not rebound). This behavior is more likely to prevail when the mass of the collar is large compared to the mass of the bar. (2) We disregard all energy losses and assume that the kinetic energy of the falling mass is transformed entirely into strain energy of the bar. This assumption predicts larger stresses in the bar than would be predicted if we took energy losses into account. (3) We disregard any change in the potential energy of the bar itself (due to the vertical movement of elements of the bar), and we ignore the existence of strain energy in the bar due to its own weight. Both of these effects are extremely small. (4) We assume that the stresses in the bar remain within the linearly elastic range. (5) We assume that the stress distribution throughout the bar is the same as when the bar is loaded statically by a force at the lower end, that is, we assume the stresses are uniform throughout the volume of the bar. (In reality longitudinal stress waves will travel through the bar, thereby causing variations in the stress distribution.)

On the basis of the preceding assumptions, we can calculate the maximum elongation and the maximum tensile stresses produced by the impact load. (Recall that we are disregarding the weight of the bar itself and finding the stresses due solely to the falling collar.)

## Maximum Elongation of the Bar

The maximum elongation $\delta_{max}$ (Fig. 2-53b) can be obtained from the principle of *conservation of energy* by equating the potential energy lost by the falling mass to the maximum strain energy acquired by the bar. The potential energy lost is $W(h + \delta_{max})$, where $W = Mg$ is the weight of the collar and $h + \delta_{max}$ is the distance through which it moves. The strain energy of the bar is $EA\delta_{max}^2/2L$, where $EA$ is the axial rigidity and $L$ is the length of the bar (see Eq. 2-39b). Thus, we obtain the following equation:

$$W(h + \delta_{max}) = \frac{EA\delta_{max}^2}{2L} \qquad \textbf{(2-51)}$$

This equation is quadratic in $\delta_{max}$ and can be solved for the positive root; the result is

$$\delta_{max} = \frac{WL}{EA} + \left[ \left( \frac{WL}{EA} \right)^2 + 2h\left( \frac{WL}{EA} \right) \right]^{1/2} \qquad \textbf{(2-52)}$$

Note that the maximum elongation of the bar increases if either the weight of the collar or the height of fall is increased. The elongation diminishes if the stiffness $EA/L$ is increased.

The preceding equation can be written in simpler form by introducing the notation

$$\delta_{st} = \frac{WL}{EA} = \frac{MgL}{EA} \qquad \textbf{(2-53)}$$

in which $\delta_{st}$ is the elongation of the bar due to the weight of the collar under static loading conditions. Equation (2-52) now becomes

$$\delta_{max} = \delta_{st} + (\delta_{st}^2 + 2h\delta_{st})^{1/2} \qquad \textbf{(2-54)}$$

or

$$\delta_{max} = \delta_{st}\left[1 + \left(1 + \frac{2h}{\delta_{st}}\right)^{1/2}\right] \qquad \textbf{(2-55)}$$

From this equation we see that the elongation of the bar under the impact load is much larger than it would be if the same load were applied statically. Suppose, for instance, that the height $h$ is 40 times the static displacement $\delta_{st}$; the maximum elongation would then be 10 times the static elongation.

When the height $h$ is large compared to the static elongation, we can disregard the "ones" on the right-hand side of Eq. (2-55) and obtain

$$\delta_{max} = \sqrt{2h\delta_{st}} = \sqrt{\frac{Mv^2L}{EA}} \qquad \textbf{(2-56)}$$

in which $M = W/g$ and $v = \sqrt{2gh}$ is the velocity of the falling mass when it strikes the flange. This equation can also be obtained directly from Eq. (2-51) by omitting $\delta_{max}$ on the left-hand side of the equation and then solving for $\delta_{max}$. Because of the omitted terms, values of $\delta_{max}$ calculated from Eq. (2-56) are always less than those obtained from Eq. (2-55).

## Maximum Stress in the Bar

The maximum stress can be calculated easily from the maximum elongation because we are assuming that the stress distribution is uniform throughout the length of the bar. From the general equation $\delta = PL/EA = \sigma L/E$, we know that

$$\sigma_{max} = \frac{E\delta_{max}}{L} \qquad \textbf{(2-57)}$$

Substituting from Eq. (2-52), we obtain the following equation for the maximum tensile stress:

$$\sigma_{max} = \frac{W}{A} + \left[\left(\frac{W}{A}\right)^2 + \frac{2WhE}{AL}\right]^{1/2} \qquad \textbf{(2-58)}$$

Introducing the notation

$$\sigma_{st} = \frac{W}{A} = \frac{Mg}{A} = \frac{E\delta_{st}}{L} \qquad \textbf{(2-59)}$$

in which $\sigma_{st}$ is the stress when the load acts statically, we can write Eq. (2-58) in the form

$$\sigma_{max} = \sigma_{st} + \left(\sigma_{st}^2 + \frac{2hE}{L}\sigma_{st}\right)^{1/2} \qquad \textbf{(2-60)}$$

or

$$\sigma_{max} = \sigma_{st}\left[1 + \left(1 + \frac{2hE}{L\sigma_{st}}\right)^{1/2}\right] \qquad \textbf{(2-61)}$$

This equation is analogous to Eq. (2-55) and again shows that an impact load produces much larger effects than when the same load is applied statically.

Again considering the case where the height $h$ is large compared to the elongation of the bar (compare with Eq. 2-56), we obtain

$$\sigma_{max} = \sqrt{\frac{2hE\sigma_{st}}{L}} = \sqrt{\frac{Mv^2E}{AL}} \qquad \textbf{(2-62)}$$

From this result we see that an increase in the kinetic energy $Mv^2/2$ of the falling mass will increase the stress, whereas an increase in the volume $AL$ of the bar will reduce the stress. This situation is quite different from static tension of the bar, where the stress is independent of the length $L$ and the modulus of elasticity $E$.

The preceding equations for the maximum elongation and maximum stress apply only at the instant when the flange of the bar is at its lowest position. After the maximum elongation is reached in the bar, the bar will vibrate axially until it comes to rest at the static elongation. From then on, the elongation and stress have the values given by Eqs. (2-53) and (2-59).

Although the preceding equations were derived for the case of a prismatic bar, they can be used for any linearly elastic structure subjected to a falling load, provided we know the appropriate stiffness of the structure. In particular, the equations can be used for a spring by substituting the stiffness $k$ of the spring (see Section 2.2) for the stiffness $EA/L$ of the prismatic bar.

## Impact Factor

The ratio of the dynamic response of a structure to the static response (for the same load) is known as an **impact factor**. For instance, the impact factor for the elongation of the bar of Fig. 2-53 is the ratio of the maximum elongation to the static elongation:

$$\text{Impact factor} = \frac{\delta_{max}}{\delta_{st}} \qquad \textbf{(2-63)}$$

This factor represents the amount by which the static elongation is amplified due to the dynamic effects of the impact.

Equations analogous to Eq. (2-63) can be written for other impact factors, such as the impact factor for the stress in the bar (the ratio of $\sigma_{max}$ to $\sigma_{st}$). When the collar falls through a considerable height, the impact factor can be very large, such as 100 or more.

## Suddenly Applied Load

A special case of impact occurs when a load is applied suddenly with no initial velocity. To explain this kind of loading, consider again the prismatic bar shown in Fig. 2-53 and assume that the sliding collar is lowered gently until it just touches the flange. Then the collar is suddenly released. Although in this instance no kinetic energy exists at the beginning of extension of the bar, the behavior is quite different from that of static loading of the bar. Under static loading conditions, the load is released gradually and equilibrium always exists between the applied load and the resisting force of the bar.

However, consider what happens when the collar is released suddenly from its point of contact with the flange. Initially the elongation of the bar and the stress in the bar are zero, but then the collar moves downward under the action of its own weight. During this motion the bar elongates and its resisting force gradually increases. The motion continues until at some instant the resisting force just equals $W$, the weight of the collar. At this particular instant the elongation of the bar is $\delta_{st}$. However, the collar now has a certain kinetic energy, which it acquired during the downward displacement $\delta_{st}$. Therefore, the collar continues to move downward until its velocity is brought to zero by the resisting force in the bar. The maximum elongation for this condition is obtained from Eq. (2-55) by setting $h$ equal to zero; thus,

$$\delta_{max} = 2\delta_{st} \qquad\qquad \textbf{(2-64)}$$

From this equation we see that a suddenly applied load produces an elongation twice as large as the elongation caused by the same load applied statically. Thus, the impact factor is 2.

After the maximum elongation $2\delta_{st}$ has been reached, the end of the bar will move upward and begin a series of up and down vibrations, eventually coming to rest at the static elongation produced by the weight of the collar.[*]

## Limitations

The preceding analyses were based upon the assumption that no energy losses occur during impact. In reality, energy losses always occur, with most of the lost energy being dissipated in the form of heat and localized deformation of the materials. Because of these losses, the kinetic energy of a system immediately after an impact is less than it was before the impact. Consequently, less energy is converted into strain energy of the bar than we previously assumed. As a result, the actual displacement of the end of the bar of Fig. 2-53 is less than that predicted by our simplified analysis.

We also assumed that the stresses in the bar remain within the proportional limit. If the maximum stress exceeds this limit, the analysis becomes more complicated because the elongation of the bar is no longer proportional to the axial force. Other factors to consider are the effects of stress waves, damping, and imperfections at the contact surfaces. Therefore, we must remember that all of the formulas in this section are based upon highly idealized conditions and give only a rough approximation of the true conditions (usually overestimating the elongation).

Materials that exhibit considerable ductility beyond the proportional limit generally offer much greater resistance to impact loads than do brittle materials. Also, bars with grooves, holes, and other forms of stress concentrations (see Sections 2.9 and 2.10) are very weak against impact—a slight shock may produce fracture, even when the material itself is ductile under static loading.

---

[*]Equation (2-64) was first obtained by the French mathematician and scientist J. V. Poncelet (1788–1867); see Ref. 2-8.

A round, prismatic steel bar ($E = 210$ GPa) of length $L = 2.0$ m and diameter $d = 15$ mm hangs vertically from a support at its upper end (Fig. 2-54). A sliding collar of mass $M = 20$ kg drops from a height $h = 150$ mm onto the flange at the lower end of the bar without rebounding.

### Fig. 2-54

Example 2-16: Impact load on a vertical bar

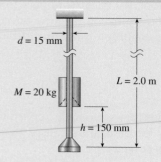

$d = 15$ mm

$M = 20$ kg

$L = 2.0$ m

$h = 150$ mm

(a) Calculate the maximum elongation of the bar due to the impact and determine the corresponding impact factor.
(b) Calculate the maximum tensile stress in the bar and determine the corresponding impact factor.

### Solution

Because the arrangement of the bar and collar in this example matches the arrangement shown in Fig. 2-53, we can use the equations derived previously (Eqs. 2-49 to 2-60).

(a) *Maximum elongation.* The elongation of the bar produced by the falling collar can be determined from Eq. (2-55). The first step is to determine the static elongation of the bar due to the weight of the collar. Since the weight of the collar is $Mg$, we calculate

$$\delta_{st} = \frac{MgL}{EA} = \frac{(20.0 \text{ kg})(9.81 \text{ m/s}^2)(2.0 \text{ m})}{(210 \text{ GPa})(\pi/4)(15 \text{ mm})^2} = 0.0106 \text{ mm}$$

From this result we see that

$$\frac{h}{\delta_{st}} = \frac{150 \text{ mm}}{0.0106 \text{ mm}} = 14{,}150$$

The preceding numerical values may now be substituted into Eq. (2-55) to obtain the maximum elongation:

$$\delta_{max} = \delta_{st}\left[1 + 1\left(1 + \frac{2h}{\delta_{st}}\right)^{1/2}\right]$$

$$= (0.0106 \text{ mm})[1 + \sqrt{1 + 2(14,150)}]$$

$$= 1.79 \text{ mm}$$

Since the height of fall is very large compared to the static elongation, we obtain nearly the same result by calculating the maximum elongation from Eq. (2-56):

$$\delta_{max} = \sqrt{2h\delta_{st}} = [2(150 \text{ mm})(0.0106 \text{ mm})]^{1/2} = 1.78 \text{ mm}$$

The impact factor is equal to the ratio of the maximum elongation to the static elongation:

$$\text{Impact factor} = \frac{\delta_{max}}{\delta_{st}} = \frac{1.79 \text{ mm}}{0.0106 \text{ mm}} = 169$$

This result shows that the effects of a dynamically applied load can be very large as compared to the effects of the same load acting statically.

(b) *Maximum tensile stress.* The maximum stress produced by the falling collar is obtained from Eq. (2-57), as follows:

$$\sigma_{max} = \frac{E\delta_{max}}{L} = \frac{(210 \text{ GPa})(1.79 \text{ mm})}{2.0 \text{ m}} = 188 \text{ MPa}$$

This stress may be compared with the static stress (see Eq. 2-59), which is

$$\sigma_{st} = \frac{W}{A} = \frac{Mg}{A} = \frac{(20 \text{ kg})(9.81 \text{ m/s}^2)}{(\pi/4)(15 \text{ mm})^2} = 1.11 \text{ MPa}$$

The ratio of $\sigma_{max}$ to $\sigma_{st}$ is 188/1.11 = 169, which is the same impact factor as for the elongations. This result is expected, because the stresses are directly proportional to the corresponding elongations (see Eqs. 2-57 and 2-59).

## Example 2-17

### Fig. 2-55

Example 2-17: Impact load on a horizontal bar

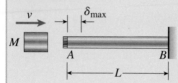

A horizontal bar $AB$ of length $L$ is struck at its free end by a heavy block of mass $M$ moving horizontally with velocity $v$ (Fig. 2-55).

(a) Determine the maximum shortening $\delta_{max}$ of the bar due to the impact and determine the corresponding impact factor.
(b) Determine the maximum compressive stress $\sigma_{max}$ and the corresponding impact factor. (Let $EA$ represent the axial rigidity of the bar.)

### Solution

The loading on the bar in this example is quite different from the loads on the bars pictured in Figs. 2-53 and 2-54. Therefore, we must make a new analysis based upon conservation of energy.

(a) *Maximum shortening of the bar.* For this analysis we adopt the same assumptions as those described previously. Thus, we disregard all energy losses and assume that the kinetic energy of the moving block is transformed entirely into strain energy of the bar.

The kinetic energy of the block at the instant of impact is $Mv^2/2$. The strain energy of the bar when the block comes to rest at the instant of maximum shortening is $EA\delta_{max}^2/2L$, as given by Eq. (2-39b). Therefore, we can write the following equation of conservation of energy:

$$\frac{Mv^2}{2} = \frac{EA\delta_{max}^2}{2L} \qquad \text{(2-65)}$$

Solving for $\delta_{max}$, we get

$$\delta_{max} = \sqrt{\frac{Mv^2L}{EA}} \qquad \longleftarrow \text{(2-66)}$$

This equation is the same as Eq. (2-56), which we might have anticipated.

To find the impact factor, we need to know the static displacement of the end of the bar. In this case the static displacement is the shortening of the bar due to the weight of the block applied as a compressive load on the bar (see Eq. 2-53):

$$\delta_{st} = \frac{WL}{EA} = \frac{MgL}{EA}$$

Thus, the impact factor is

$$\text{Impact factor} = \frac{\delta_{max}}{\delta_{st}} = \sqrt{\frac{EAv^2}{Mg^2L}} \qquad \longleftarrow \text{(2-67)}$$

The value determined from this equation may be much larger than 1.

(b) *Maximum compressive stress in the bar.* The maximum stress in the bar is found from the maximum shortening by means of Eq. (2-57):

$$\sigma_{max} = \frac{E\delta_{max}}{L} = \frac{E}{L}\sqrt{\frac{Mv^2L}{EA}} = \sqrt{\frac{Mv^2E}{AL}} \qquad \longleftarrow \text{(2-68)}$$

This equation is the same as Eq. (2-62).

The static stress $\sigma_{st}$ in the bar is equal to $W/A$ or $Mg/A$, which (in combination with Eq. 2-68) leads to the same impact factor as before (Eq. 2-67).

# *2.9 REPEATED LOADING AND FATIGUE

The behavior of a structure depends not only upon the nature of the material but also upon the character of the loads. In some situations the loads are static—they are applied gradually, act for long periods of time, and change slowly. Other loads are dynamic in character—examples are impact loads acting suddenly (Section 2.8) and repeated loads recurring for large numbers of cycles.

Some typical patterns for **repeated loads** are sketched in Fig. 2-56. The first graph (a) shows a load that is applied, removed, and applied again, always acting in the same direction. The second graph (b) shows an alternating load that reverses direction during every cycle of loading, and the third graph (c) illustrates a fluctuating load that varies about an average value. Repeated loads are commonly associated with machinery, engines, turbines, generators, shafts, propellers, airplane parts, automobile parts, and the like. Some of these structures are subjected to millions (and even billions) of loading cycles during their useful life.

A structure subjected to dynamic loads is likely to fail at a lower stress than when the same loads are applied statically, especially when the loads are repeated for a large number of cycles. In such cases failure is usually caused by **fatigue**, or **progressive fracture**. A familiar example of a fatigue failure is stressing a metal paper clip to the breaking point by repeatedly bending it back and forth. If the clip is bent only once, it does not break. But if the load is reversed by bending the clip in the opposite direction, and if the entire loading cycle is repeated several times, the clip will finally break. *Fatigue* may be defined as the deterioration of a material under repeated cycles of stress and strain, resulting in progressive cracking that eventually produces fracture.

In a typical fatigue failure, a microscopic crack forms at a point of high stress (usually at a *stress concentration*, discussed in the next section) and gradually enlarges as the loads are applied repeatedly. When the crack becomes so large that the remaining material cannot resist the loads, a sudden fracture of the material occurs (Fig. 2-57). Depending upon the nature of the material, it may take anywhere from a few cycles of loading to hundreds of millions of cycles to produce a fatigue failure.

### Fig. 2-56

Types of repeated loads: (a) load acting in one direction only, (b) alternating or reversed load, and (c) fluctuating load that varies about an average value

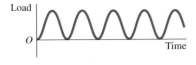

(a)

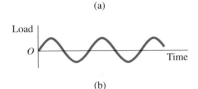

(b)

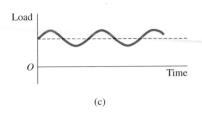

(c)

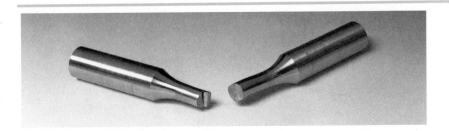

### Fig. 2-57

Fatigue failure of a bar loaded repeatedly in tension; the crack spread gradually over the cross section until fracture occurred suddenly. (Courtesy of MTS Systems Corporation)

The magnitude of the load causing a fatigue failure is less than the load that can be sustained statically, as already pointed out. To determine the failure load, tests of the material must be performed. In the case of repeated loading, the material is tested at various stress levels and the number of cycles to failure is counted. For instance, a specimen of material is placed in a fatigue-testing machine and loaded repeatedly to a certain stress, say $\sigma_1$. The loading cycles are continued until failure occurs, and the number $n$ of loading cycles to failure is noted. The test is then repeated for a different stress, say $\sigma_2$. If $\sigma_2$ is greater than $\sigma_1$, the number of cycles to failure will be smaller. If $\sigma_2$ is less than $\sigma_1$, the number will be larger. Eventually, enough data are accumulated to plot an **endurance curve**, or **S-N diagram**, in which failure stress ($S$) is plotted versus the number ($N$) of cycles to failure (Fig. 2-58). The vertical axis is usually a linear scale and the horizontal axis is usually a logarithmic scale.

The endurance curve of Fig. 2-58 shows that the smaller the stress, the larger the number of cycles to produce failure. For some materials the curve has a horizontal asymptote known as the **fatigue limit** or **endurance limit**. When it exists, this limit is the stress below which a fatigue failure will not occur regardless of how many times the load is repeated. The precise shape of an endurance curve depends upon many factors, including properties of the material, geometry of the test specimen, speed of testing, pattern of loading, and surface condition of the specimen. The results of numerous fatigue tests, made on a great variety of materials and structural components, have been reported in the engineering literature.

Typical S-N diagrams for steel and aluminum are shown in Fig. 2-59. The ordinate is the failure stress, expressed as a percentage of the ultimate stress for the material, and the abscissa is the number of cycles at which failure occurred. Note that the number of cycles is plotted on a logarithmic scale. The curve for steel becomes horizontal at about $10^7$ cycles, and the fatigue limit is about 50% of the ultimate tensile stress for ordinary static loading. The fatigue limit for aluminum is not as clearly defined as that for steel, but a typical value of the fatigue limit is the stress at $5 \times 10^8$ cycles, or about 25% of the ultimate stress.

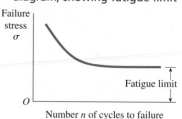

**Fig. 2-58**

Endurance curve, or S-N diagram, showing fatigue limit

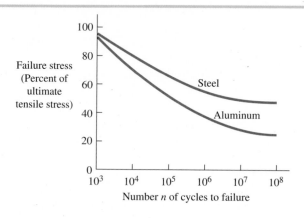

**Fig. 2-59**

Typical endurance curves for steel and aluminum in alternating (reversed) loading

Since fatigue failures usually begin with a microscopic crack at a point of high localized stress (that is, at a stress concentration), the condition of the surface of the material is extremely important. Highly polished specimens have higher endurance limits. Rough surfaces, especially those at stress concentrations around holes or grooves, greatly lower the endurance limit. Corrosion, which creates tiny surface irregularities, has a similar effect. For steel, ordinary corrosion may reduce the fatigue limit by more than 50%.

# *2.10 STRESS CONCENTRATIONS

When determining the stresses in axially loaded bars, we customarily use the basic formula $\sigma = P/A$, in which $P$ is the axial force in the bar and $A$ is its cross-sectional area. This formula is based upon the assumption that the stress distribution is uniform throughout the cross section. In reality, bars often have holes, grooves, notches, keyways, shoulders, threads, or other abrupt changes in geometry that create a disruption in the otherwise uniform stress pattern. These discontinuities in geometry cause high stresses in very small regions of the bar, and these high stresses are known as **stress concentrations**. The discontinuities themselves are known as **stress raisers**.

Stress concentrations also appear at points of loading. For instance, a load may act over a very small area and produce high stresses in the region around its point of application. An example is a load applied through a pin connection, in which case the load is applied over the bearing area of the pin.

The stresses existing at stress concentrations can be determined either by experimental methods or by advanced methods of analysis, including the finite-element method. The results of such research for many cases of practical interest are readily available in the engineering literature (for example, Ref. 2-9). Some typical stress-concentration data are given later in this section and also in Chapters 3 and 5.

## Saint-Venant's Principle

To illustrate the nature of stress concentrations, consider the stresses in a bar of rectangular cross section (width $b$, thickness $t$) subjected to a concentrated load $P$ at the end (Fig. 2-60). The peak stress directly under the load may be several times the average stress $P/bt$, depending upon the area over which the load is applied. However, the maximum stress diminishes rapidly as we move away from the point of load application, as shown by the stress diagrams in the figure. At a distance from the end of the bar equal to the width $b$ of the bar, the stress distribution is nearly uniform, and the maximum stress is only a few percent larger than the average stress. This observation is true for most stress concentrations, such as holes and grooves.

Thus, we can make a general statement that the equation $\sigma = P/A$ gives the axial stresses on a cross section only when the cross section is at least a distance $b$ away from any concentrated load or discontinuity in shape, where $b$ is the largest lateral dimension of the bar (such as the width or diameter).

The preceding statement about the stresses in a prismatic bar is part of a more general observation known as **Saint-Venant's principle**. With rare exceptions, this principle applies to linearly elastic bodies of all types. To understand Saint-Venant's principle, imagine that we have a body with a system of loads acting over a small part of its surface. For instance, suppose we have a prismatic bar of width $b$ subjected to a system of several concentrated loads acting at the end (Fig. 2-61a). For simplicity, assume that the loads are symmetrical and have only a vertical resultant.

Next, consider a different but statically equivalent load system acting over the same small region of the bar. ("Statically equivalent" means the two load systems have the same force resultant and same moment resultant.) For instance, the uniformly distributed load shown in Fig. 2-61b is statically equivalent to the system of concentrated loads shown in Fig. 2-61a. Saint-Venant's principle states that the stresses in the body caused by either of the two systems of loading are the same, provided we move away from the

## Fig. 2-60

Stress distributions near the end of a bar of rectangular cross section (width $b$, thickness $t$) subjected to a concentrated load $P$ acting over a small area

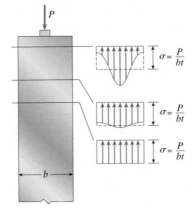

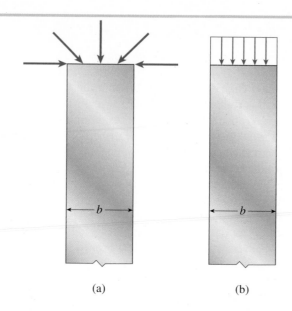

(a)          (b)

**Fig. 2-60 (Repeated)**

$\sigma = \dfrac{P}{bt}$

$\sigma = \dfrac{P}{bt}$

$\sigma = \dfrac{P}{bt}$

loaded region a distance at least equal to the largest dimension of the loaded region (distance $b$ in our example). Thus, the stress distributions shown in Fig. 2-60 are an illustration of Saint-Venant's principle. Of course, this "principle" is not a rigorous law of mechanics but is a common-sense observation based upon theoretical and practical experience.

Saint-Venant's principle has great practical significance in the design and analysis of bars, beams, shafts, and other structures encountered in mechanics of materials. Because the effects of stress concentrations are localized, we can use all of the standard stress formulas (such as $\sigma = P/A$) at cross sections a sufficient distance away from the source of the concentration. Close to the source, the stresses depend upon the details of the loading and the shape of the member. Furthermore, formulas that are applicable to entire members, such as formulas for elongations, displacements, and strain energy, give satisfactory results even when stress concentrations are present. The explanation lies in the fact that stress concentrations are localized and have little effect on the overall behavior of a member.*

## Stress-Concentration Factors

Now let us consider some particular cases of stress concentrations caused by discontinuities in the shape of a bar. We begin with a bar of rectangular cross section having a circular hole and subjected to a tensile force $P$ (Fig. 2-62a). The bar is relatively thin, with its width $b$ being much larger than its thickness $t$. Also, the hole has diameter $d$.

The normal stress acting on the cross section through the center of the hole has the distribution shown in Fig. 2-62b. The maximum stress $\sigma_{\max}$ occurs at the edges of the hole and may be significantly larger than the *nominal stress* $\sigma = P/ct$ at the same cross section. (Note that $ct$ is the net area at the cross section through the hole.) The intensity of a stress

---

*Saint-Venant's principle is named for Barré de Saint-Venant (1797–1886), a famous French mathematician and elastician (Ref. 2-10). It appears that the principle applies generally to solid bars and beams but not to all thin-walled open sections. For a discussion of the limitations of Saint-Venant's principle, see Ref. 2-11.

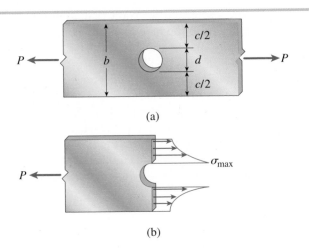

**Fig. 2-62**

Stress distribution in a flat bar with a circular hole

concentration is usually expressed by the ratio of the maximum stress to the nominal stress, called the **stress-concentration factor** $K$:

$$K = \frac{\sigma_{max}}{\sigma_{nom}} \qquad \textbf{(2-69)}$$

For a bar in tension, the nominal stress is the average stress based upon the net cross-sectional area. In other cases, a variety of stresses may be used. Thus, whenever a stress concentration factor is used, it is important to note carefully how the nominal stress is defined.

A graph of the stress-concentration factor $K$ for a bar with a hole is given in Fig. 2-63. If the hole is tiny, the factor $K$ equals 3, which means that the maximum stress is three times the nominal stress. As the hole becomes larger in proportion to the width of the bar, $K$ becomes smaller and the effect of the concentration is not as severe.

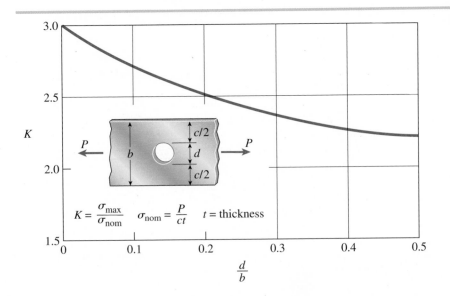

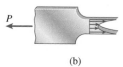

### Fig. 2-64

Stress distribution in a flat bar with shoulder fillets

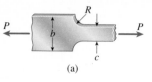

From Saint-Venant's principle, we know that, at distances equal to the width $b$ of the bar *away* from the hole in either axial direction, the stress distribution is practically uniform and equal to $P$ divided by the gross cross-sectional area ($\sigma = P/bt$).

To reduce the stress-concentration effects (see Fig. 2-64), *fillets* are used to round off the re-entrant corners.* Stress-concentration factors for two other cases of practical interest are given in Figs. 2-65 and 2-66. These graphs are for flat bars and circular bars, respectively, that are stepped down in size, forming a *shoulder*. Without the fillets, the stress-concentration factors would be extremely large, as indicated at the left-hand side of each graph where $K$ approaches infinity as the fillet radius $R$ approaches

### Fig. 2-65

Stress-concentration factor $K$ for flat bars with shoulder fillets. The dashed line is for a full quarter-circular fillet.

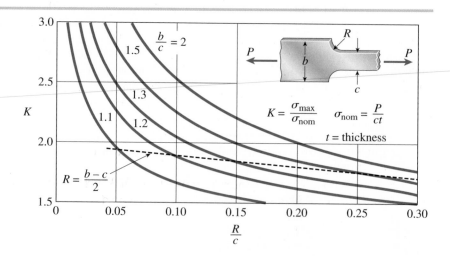

### Fig. 2-66

Stress-concentration factor $K$ for round bars with shoulder fillets. The dashed line is for a full quarter-circular fillet.

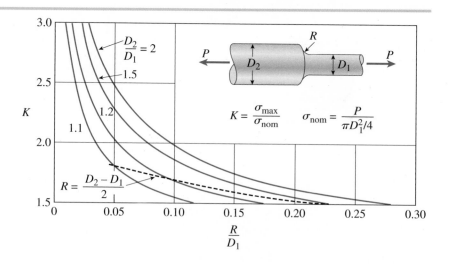

---

*A *fillet* is a curved concave surface formed where two other surfaces meet. Its purpose is to round off what would otherwise be a sharp re-entrant corner.

zero. In both cases, the maximum stress occurs in the smaller part of the bar in the region of the fillet.*

## Designing for Stress Concentrations

Because of the possibility of fatigue failures, stress concentrations are especially important when the member is subjected to repeated loading. As explained in the preceding section, cracks begin at the point of highest stress and then spread gradually through the material as the load is repeated. In practical design, the fatigue limit (Fig. 2-58) is considered to be the ultimate stress for the material when the number of cycles is extremely large. The allowable stress is obtained by applying a factor of safety with respect to this ultimate stress. Then the peak stress at the stress concentration is compared with the allowable stress.

In many situations the use of the full theoretical value of the stress-concentration factor is too severe. Fatigue tests usually produce failure at higher levels of the nominal stress than those obtained by dividing the fatigue limit by $K$. In other words, a structural member under repeated loading is not as sensitive to a stress concentration as the value of $K$ indicates, and a reduced stress-concentration factor is often used.

Other kinds of dynamic loads, such as impact loads, also require that stress-concentration effects be taken into account. Unless better information is available, the full stress-concentration factor should be used. Members subjected to low temperatures also are highly susceptible to failures at stress concentrations, and therefore special precautions should be taken in such cases.

The significance of stress concentrations when a member is subjected to static loading depends upon the kind of material. With ductile materials, such as structural steel, a stress concentration can often be ignored. The reason is that the material at the point of maximum stress (such as around a hole) will yield and plastic flow will occur, thus reducing the intensity of the stress concentration and making the stress distribution more nearly uniform. On the other hand, with brittle materials (such as glass) a stress concentration will remain up to the point of fracture. Therefore, we can make the general observation that with static loads and a ductile material the stress-concentration effect is not likely to be important, but with static loads and a brittle material the full stress-concentration factor should be considered.

Stress concentrations can be reduced in intensity by properly proportioning the parts. Generous fillets reduce stress concentrations at re-entrant corners. Smooth surfaces at points of high stress, such as on the inside of a hole, inhibit the formation of cracks. Proper reinforcing around holes can also be beneficial. There are many other techniques for smoothing out the stress distribution in a structural member and thereby reducing the stress-concentration factor. These techniques, which are usually studied in engineering design courses, are extremely important in the design of aircraft, ships, and machines. Many unnecessary structural failures have occurred because designers failed to recognize the effects of stress concentrations and fatigue.

---

*The stress-concentration factors given in the graphs are theoretical factors for bars of linearly elastic material. The graphs are plotted from the formulas given in Ref. 2-9.

**Example 2-18**

A stepped brass bar with a hole (Fig. 2-67a) has widths of $b = 9.0$ cm and $c = 6.0$ cm and a thickness of $t = 1.0$ cm. The fillets have radii equal to 0.5 cm and the hole has a diameter of $d = 1.8$ cm. The ultimate strength of the brass is 200 MPa.

(a) If a factor of safety of 2.8 is required, what is the maximum allowable tensile load $P_{max}$?
(b) Find the hole diameter $d_{max}$ at which the two segments of the bar have tensile load carrying capacity equal to that for the fillet region of the stepped bar.

### Fig. 2-67

Example 2-18: (a) Stress concentrations in stepped bar with a hole, and (b) selection of factor $K_h$ using Fig. 2-63

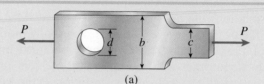

(a)

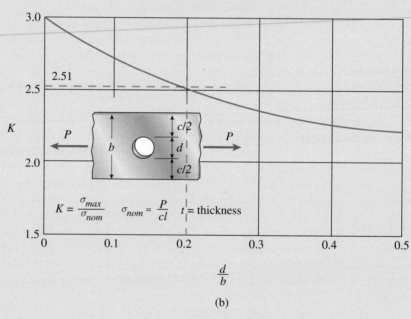

$\dfrac{d}{b}$

(b)

### Solution

**(a) Determine the maximum allowable tensile load.**

The maximum allowable tensile load is determined by comparing the product of the nominal stress times the net area in each segment of the stepped bar (i.e., segment with hole and segment with fillets).

For the segment of the bar of width $b$ and thickness $t$ and having a hole of diameter $d$, the net cross-sectional area is $(b - d)(t)$, and the nominal axial stress may be computed as

$$\sigma_1 = \frac{P}{(b - d)t} \quad \text{and} \quad \sigma_1 = \frac{\sigma_{allow}}{K_{hole}} = \frac{\left(\dfrac{\sigma_U}{FS_U}\right)}{K_{hole}} \qquad \text{(a,b)}$$

where the maximum stress has been set equal to the allowable stress and the stress concentration factor $K_{hole}$ is obtained from Fig. 2-63. Next, we equate the nominal stress expressions in Eqs. (a) and (b) and solve for $P_{max}$:

$$\sigma_1 = \frac{P}{(b-d)t} = \frac{\left(\dfrac{\sigma_U}{FS_U}\right)}{K_{hole}} \quad \text{so} \quad P_{max\,1} = \frac{\left(\dfrac{\sigma_U}{FS_U}\right)}{K_{hole}}(b-d)t \quad \text{(c)}$$

Using the given numerical properties of the bar, we see that $d/b = 1.8/9.0 = 0.2$, so $K_{hole}$ is approximately 2.51 from Fig. 2-63 (see Fig. 2-67b). We now can compute the allowable tensile load on the stepped bar *accounting for stress concentrations at the hole* using equation (c) as

$$P_{max1} = \frac{\left(\dfrac{200 \text{ MPa}}{2.8}\right)}{2.51}(9.0 \text{ cm} - 1.8 \text{ cm})(1.0 \text{ cm}) = 20.5 \text{ kN} \quad \text{(d)}$$

Next, we must investigate the tensile load carrying capacity of the stepped bar in the segment having fillets of radius $R = 0.5$ cm. Following the procedure we used to find $P_{max1}$ in Eqs. (a),(b), and (c), we now find

$$P_{max2} = \frac{\left(\dfrac{\sigma_U}{FS_U}\right)}{K_{fillet}}(ct) \quad \text{(e)}$$

The stress concentration factor $K_{fillet}$ is obtained from Fig. 2-65 using two parameters: the ratio of fillet radius to reduced width $c$ ($R/c = 0.1$) and the ratio of the bar's full width to reduced width ($b/c = 1.5$). The stress concentration factor is approximately 2.35 (see Fig. 2-67c), and so the maximum allowable tensile load based on stress concentrations in the fillet region of the bar is

$$P_{max2} = \frac{\left(\dfrac{200 \text{ MPa}}{2.8}\right)}{2.35}(6 \text{ cm})(1 \text{ cm}) = 18.24 \text{ kN} \quad \Longleftarrow \quad \text{(f)}$$

Comparing Eqs. (d) and (f), we see that the lesser value of the maximum allowable tensile load $P_{max2}$, which is based on stress concentrations in the fillet region of the stepped bar, controls here.

*Continues* ⮕

● ● ●  **Example 2-18 -** *Continued*

### Fig. 2-67 (Continued)

Example 2-18: (c) Selection of factor $K_f$ using Fig. 2-64

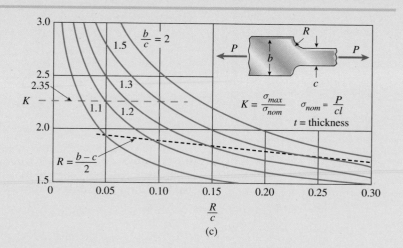

(c)

**(b) Determine the maximum hole diameter.**

Comparing Eqs. (d) and (f), we see that the segment of the stepped bar with a hole has greater tensile load capacity $P_{max}$ than the segment with the fillets. If we enlarge the hole, we will reduce the net cross sectional area, $A_{net} = (b - d)(t)$, (note that width $b$ and thickness $t$ remain unchanged), but at the same time, we will reduce the stress concentration factor $K_{hole}$ (see Fig. 2-63) because ratio $d/b$ increases.

If we use the right-hand side of Eq. (f) to $P_{max2}$ and simplify the resulting expression, we get

$$\frac{\left(\dfrac{\sigma_U}{FS_U}\right)}{K_{hole}}(b - d)t = P_{max2} \quad \text{or} \quad \frac{\left(1 - \dfrac{d}{b}\right)}{K_{hole}} = \frac{P_{max2}}{\left(\dfrac{\sigma_U}{FS_U}\right)}\left(\frac{1}{bt}\right) \quad \textbf{(g)}$$

Substituting numerical values produces the following

$$\text{Ratio} = \frac{\left(1 - \dfrac{d}{b}\right)}{K_{hole}} = \frac{18.24 \text{ kN}}{\left(\dfrac{200 \text{ MPa}}{2.8}\right)}\left[\frac{1}{9 \text{ cm}(1 \text{ cm})}\right] = 0.284 \quad \textbf{(h)}$$

### Table 2-2

Stress Concentration Factors and *d/b* values from Fig. 2-63

| d/b | $K_{hole}$ | Ratio |
|-----|-----------|-------|
| 0.2 | 2.51 | 0.32 |
| 0.3 | 2.39 | 0.29 |
| 0.4 | 2.32 | 0.26 |
| 0.5 | 2.25 | 0.22 |

We can tabulate a few values of stress concentration factor $K_{hole}$ for corresponding values of the *d/b* ratio from Fig. 2-63 as follows (see Table 2-2), showing that the required *d/b* ratio lies between 0.3 and 0.4. Several iterations using trial-and-error and values from Fig. 2-63 reveal that

$$\frac{d}{b} = \frac{2.97}{9} = 0.33 \quad \text{so} \quad \frac{(1 - 0.33)}{2.36} = 0.284 \quad \text{so} \quad d_{max} = 2.97 \text{ cm} \quad \Longleftarrow \quad \textbf{(i)}$$

Hence, the maximum hole diameter $d_{max}$ is approximately 3.0 cm if the two segments of the stepped bar are to have equal tensile load capacities.

# *2.11 NONLINEAR BEHAVIOR

Up to this point, our discussions have dealt primarily with members and structures composed of materials that follow Hooke's law. Now we will consider the behavior of axially loaded members when the stresses exceed the proportional limit. In such cases the stresses, strains, and displacements depend upon the shape of the stress-strain curve in the region beyond the proportional limit (see Section 1.4 for some typical stress-strain diagrams).

## Nonlinear Stress-Strain Curves

For purposes of analysis and design, we often represent the actual stress-strain curve of a material by an **idealized stress-strain curve** that can be expressed as a mathematical function. Some examples are shown in Fig. 2-68. The first diagram (Fig. 2-68a) consists of two parts, an initial linearly elastic region followed by a nonlinear region defined by an appropriate mathematical expression. The behavior of aluminum alloys can sometimes be represented quite accurately by a curve of this type, at least in the region before the strains become excessively large (compare Fig. 2-68a with Fig. 1-13).

In the second example (Fig. 2-68b), a single mathematical expression is used for the entire stress-strain curve. The best known expression of this kind is the Ramberg-Osgood stress-strain law, which is described later in more detail.

The stress-strain diagram frequently used for structural steel is shown in Fig. 2-68c. Because steel has a linearly elastic region followed

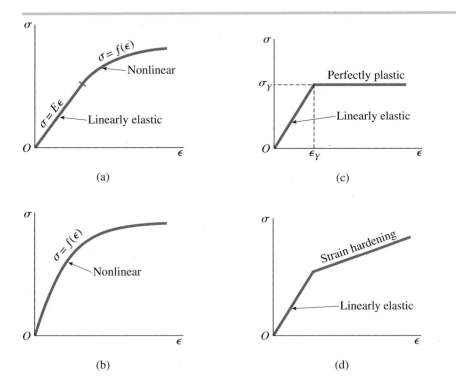

**Fig. 2-68**

Types of idealized material behavior: (a) elastic-nonlinear stress-strain curve, (b) general nonlinear stress-strain curve, (c) elastoplastic stress-strain curve, and (d) bilinear stress-strain curve

by a region of considerable yielding (see the stress-strain diagrams of Figs. 1-10 and 1-12), its behavior can be represented by two straight lines. The material is assumed to follow Hooke's law up to the yield stress $\sigma_Y$, after which it yields under constant stress, the latter behavior being known as **perfect plasticity**. The perfectly plastic region continues until the strains are 10 or 20 times larger than the yield strain. A material having a stress-strain diagram of this kind is called an **elastoplastic material** (or *elastic-plastic material*).

Eventually, as the strain becomes extremely large, the stress-strain curve for steel rises above the yield stress due to strain hardening, as explained in Section 1.4. However, by the time strain hardening begins, the displacements are so large that the structure will have lost its usefulness. Consequently, it is common practice to analyze steel structures on the basis of the elastoplastic diagram shown in Fig. 2-68c, with the same diagram being used for both tension and compression. An analysis made with these assumptions is called an **elastoplastic analysis**, or simply, **plastic analysis**, and is described in the next section.

Figure 2-68d shows a stress-strain diagram consisting of two lines having different slopes, called a **bilinear stress-strain diagram**. Note that in both parts of the diagram the relationship between stress and strain is linear, but only in the first part is the stress proportional to the strain (Hooke's law). This idealized diagram may be used to represent materials with strain hardening or it may be used as an approximation to diagrams of the general nonlinear shapes shown in Figs. 2-68a and b.

## Changes in Lengths of Bars

The elongation or shortening of a bar can be determined if the stress-strain curve of the material is known. To illustrate the general procedure, we will consider the tapered bar $AB$ shown in Fig. 2-69a. Both the cross-sectional area and the axial force vary along the length of the bar, and the material has a general nonlinear stress-strain curve (Fig. 2-69b). Because the bar is statically determinate, we can determine the internal axial forces at all cross sections from static equilibrium alone. Then we can find the stresses by dividing the forces by the cross-sectional areas, and we can find the strains from the stress-strain curve. Lastly, we can determine the change in length from the strains, as described in the following paragraph.

**Fig. 2-69**

Change in length of a tapered bar consisting of a material having a nonlinear stress-strain curve

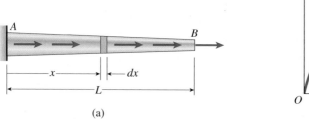

(a)

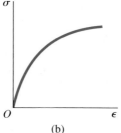

(b)

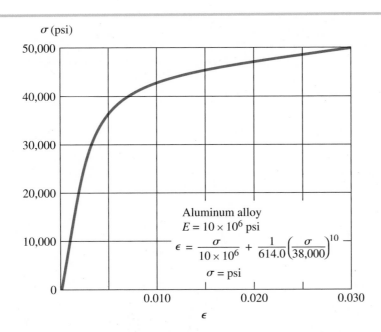

**Fig. 2-70**

Stress-strain curve for an aluminum alloy using the Ramberg-Osgood equation (Eq. 2-74)

The change in length of an element $dx$ of the bar (Fig. 2-69a) is $\varepsilon dx$, where $\varepsilon$ is the strain at distance $x$ from the end. By integrating this expression from one end of the bar to the other, we obtain the change in length of the entire bar:

$$\delta = \int_0^L \varepsilon dx \qquad \textbf{(2-70)}$$

where $L$ is the length of the bar. If the strains are expressed analytically, that is, by algebraic formulas, it may be possible to integrate Eq. (2-70) by formal mathematical means and thus obtain an expression for the change in length. If the stresses and strains are expressed numerically, that is, by a series of numerical values, we can proceed as follows. We can divide the bar into small segments of length $\Delta x$, determine the average stress and strain for each segment, and then calculate the elongation of the entire bar by summing the elongations for the individual segments. This process is equivalent to evaluating the integral in Eq. (2-70) by numerical methods instead of by formal integration.

If the strains are uniform throughout the length of the bar, as in the case of a prismatic bar with constant axial force, the integration of Eq. (2-70) is trivial and the change in length is

$$\delta = \varepsilon L \qquad \textbf{(2-71)}$$

as expected (compare with Eq. 1-2 in Section 1.3).

## Ramberg-Osgood Stress-Strain Law

Stress-strain curves for several metals, including aluminum and magnesium, can be accurately represented by the **Ramberg-Osgood equation**:

$$\frac{\varepsilon}{\varepsilon_0} = \frac{\sigma}{\sigma_0} + \alpha\left(\frac{\sigma}{\sigma}\right)^m \tag{2-72}$$

In this equation, $\sigma$ and $\varepsilon$ are the stress and strain, respectively, and $\varepsilon_0$, $\sigma_0$, $\alpha$, and $m$ are constants of the material (obtained from tension tests). An alternative form of the equation is

$$\varepsilon = \frac{\sigma}{E} + \frac{\sigma_0\alpha}{E}\left(\frac{\sigma}{\sigma_0}\right)^m \tag{2-73}$$

in which $E = \sigma_0/\varepsilon_0$ is the modulus of elasticity in the initial part of the stress-strain curve.*

A graph of Eq. (2-73) is given in Fig. 2-70 for an aluminum alloy for which the constants are as follows: $E = 10 \times 10^6$ psi, $\sigma_0 = 38{,}000$ psi, $\alpha = 3/7$, and $m = 10$. The equation of this particular stress-strain curve is

$$\varepsilon = \frac{\sigma}{10 \times 10^6} + \frac{1}{614.0}\left(\frac{\sigma}{38{,}000}\right)^{10} \tag{2-74}$$

where $\sigma$ has units of pounds per square inch (psi).

A similar equation for an aluminum alloy, but in SI units ($E = 70$ GPa, $\sigma_0 = 260$ MPa, $\alpha = 3/7$, and $m = 10$), is as follows:

$$\varepsilon = \frac{\sigma}{70{,}000} + \frac{1}{628.2}\left(\frac{\sigma}{260}\right)^{10} \tag{2-75}$$

where $\sigma$ has units of megapascals (MPa). The calculation of the change in length of a bar, using Eq. (2-73) for the stress-strain relationship, is illustrated in Example 2-19.

## Statically Indeterminate Structures

If a structure is statically indeterminate and the material behaves nonlinearly, the stresses, strains, and displacements can be found by solving the same general equations as those described in Section 2.4 for linearly elastic structures, namely, equations of equilibrium, equations of compatibility, and force-displacement relations (or equivalent stress-strain relations). The principal difference is that the force-displacement relations are now nonlinear, which means that analytical solutions cannot be obtained except in very simple situations. Instead, the equations must be solved numerically, using a suitable computer program.

---

*The Ramberg-Osgood stress-strain law is presented in Ref. 2-12.

● ● ● **Example 2-19**

A prismatic bar $AB$ of length $L = 2.2$ m and cross-sectional area $A = 480$ mm$^2$ supports two concentrated loads $P_1 = 108$ kN and $P_2 = 27$ kN, as shown in Fig. 2-71. The material of the bar is an aluminum alloy having a nonlinear stress-strain curve described by the following Ramberg-Osgood equation (Eq. 2-75):

$$\varepsilon = \frac{\sigma}{70{,}000} + \frac{1}{628.2}\left(\frac{\sigma}{260}\right)^{10}$$

in which $\sigma$ has units of MPa. (The general shape of this stress-strain curve is shown in Fig. 2-70.)

Determine the displacement $\delta_B$ of the lower end of the bar under each of the following conditions: (a) the load $P_1$ acts alone, (b) the load $P_2$ acts alone, and (c) the loads $P_1$ and $P_2$ act simultaneously.

### Fig. 2-71

Example 2-19: Elongation of a bar of nonlinear material using the Ramberg-Osgood equation

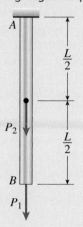

### Solution

(a) *Displacement due to the load $P_1$ acting alone.* The load $P_1$ produces a uniform tensile stress throughout the length of the bar equal to $P_1/A$, or 225 MPa. Substituting this value into the stress-strain relation gives $\varepsilon = 0.003589$. Therefore, the elongation of the bar, equal to the displacement at point $B$, is (see Eq. 2-71)

$$\delta_B = \varepsilon L = (0.003589)(2.2 \text{ m}) = 7.90 \text{ mm} \quad \Longleftarrow$$

(b) *Displacement due to the load $P_2$ acting alone.* The stress in the upper half of the bar is $P_2/A$ or 56.25 MPa, and there is no stress in the lower half. Proceeding as in part (a), we obtain the following elongation:

$$\delta_B = \varepsilon L/2 = (0.0008036)(1.1 \text{ m}) = 0.884 \text{ mm} \quad \Longleftarrow$$

(c) *Displacement due to both loads acting simultaneously.* The stress in the lower half of the bar is $P_1/A$ and in the upper half is $(P_1 + P_2)/A$. The corresponding stresses are 225 MPa and 281.25 MPa, and the corresponding strains are 0.003589 and 0.007510 (from the Ramberg-Osgood equation). Therefore, the elongation of the bar is

$$\delta_B = (0.003589)(1.1 \text{ m}) + (0.007510)(1.1 \text{ m})$$

$$= 3.95 \text{ mm} + 8.26 \text{ mm} = 12.2 \text{ mm} \quad \Longleftarrow$$

The three calculated values of $\delta_B$ illustrate an important principle pertaining to a structure made of a material that behaves nonlinearly:

*In a nonlinear structure, the displacement produced by two (or more) loads acting simultaneously is not equal to the sum of the displacements produced by the loads acting separately.*

## Fig. 2-72

Idealized stress-strain diagram for an elastoplastic material, such as structural steel

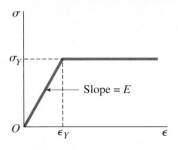

## *2.12 Elastoplastic Analysis

In the preceding section we discussed the behavior of structures when the stresses in the material exceed the proportional limit. Now we will consider a material of considerable importance in engineering design—steel, the most widely used structural metal. Mild steel (or structural steel) can be modeled as an *elastoplastic* material with a stress-strain diagram as shown in Fig. 2-72. An elastoplastic material initially behaves in a linearly elastic manner with a modulus of elasticity $E$. After plastic yielding begins, the strains increase at a more-or-less constant stress, called the **yield stress** $\sigma_Y$. The strain at the onset of yielding is known as the **yield strain** $\varepsilon_Y$.

The load-displacement diagram for a prismatic bar of elastoplastic material subjected to a tensile load (Fig. 2-73) has the same shape as the stress-strain diagram. Initially, the bar elongates in a linearly elastic manner and Hooke's law is valid. Therefore, in this region of loading we can find the change in length from the familiar formula $\delta = PL/EA$. Once the yield stress is reached, the bar may elongate without an increase in load, and the elongation has no specific magnitude. The load at which yielding begins is called the **yield load** $P_Y$ and the corresponding elongation of the bar is called the **yield displacement** $\delta_Y$. Note that for a single prismatic bar, the yield load $P_Y$ equals $\sigma_Y A$ and the yield displacement $\delta_Y$ equals $P_Y L/EA$, or $\sigma_Y L/E$. (Similar comments apply to a bar in compression, provided buckling does not occur.)

If a structure consisting only of axially loaded members is **statically determinate** (Fig. 2-74), its overall behavior follows the same pattern. The structure behaves in a linearly elastic manner until one of its members reaches the yield stress. Then that member will begin to elongate (or shorten) with no further change in the axial load in that member. Thus, the entire structure will yield, and its load-displacement diagram has the same shape as that for a single bar (Fig. 2-73).

## Fig. 2-73

Load-displacement diagram for a prismatic bar of elastoplastic material

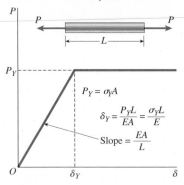

## Statically Indeterminate Structures

The situation is more complex if an elastoplastic structure is statically indeterminate. If one member yields, other members will continue to resist any increase in the load. However, eventually enough members will yield to cause the entire structure to yield.

To illustrate the behavior of a statically indeterminate structure, we will use the simple arrangement shown in Fig. 2-75 on the next page. This structure consists of three steel bars supporting a load $P$ applied through a rigid plate. The two outer bars have length $L_1$, the inner bar has length $L_2$, and all three bars have the same cross-sectional area $A$. The stress-strain diagram for the steel is idealized as shown in Fig. 2-72, and the modulus of elasticity in the linearly elastic region is $E = \sigma_Y/\varepsilon_Y$.

As is normally the case with a statically indeterminate structure, we begin the analysis with the equations of *equilibrium* and *compatibility*. From equilibrium of the rigid plate in the vertical direction we obtain

$$2F_1 + F_2 = P \qquad \textbf{(2-76)}$$

## Fig. 2-74

Statically determinate structure consisting of axially loaded members

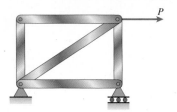

where $F_1$ and $F_2$ are the axial forces in the outer and inner bars, respectively. Because the plate moves downward as a rigid body when the load is applied, the compatibility equation is

$$\delta_1 = \delta_2 \qquad \text{(2-77)}$$

where $\delta_1$ and $\delta_2$ are the elongations of the outer and inner bars, respectively. Because they depend only upon equilibrium and geometry, the two preceding equations are valid at all levels of the load $P$; it does not matter whether the strains fall in the linearly elastic region or in the plastic region.

When the load $P$ is small, the stresses in the bars are less than the yield stress $\sigma_Y$ and the material is stressed within the linearly elastic region. Therefore, the *force-displacement relations* between the bar forces and their elongations are

$$\delta_1 = \frac{F_1 L_1}{EA} \qquad \delta_2 = \frac{F_2 L_2}{EA} \qquad \text{(2-78a,b)}$$

Substituting in the compatibility equation (Eq. 2-77), we get

$$F_1 L_1 = F_2 L_2 \qquad \text{(2-79)}$$

Solving simultaneously Eqs. (2-76) and (2-79), we obtain

$$F_1 = \frac{PL_2}{L_1 + 2L_2} \qquad F_2 = \frac{PL_1}{L_1 + 2L_2} \qquad \text{(2-80a,b)}$$

Thus, we have now found the forces in the bars in the linearly elastic region. The corresponding stresses are

$$\sigma_1 = \frac{F_1}{A} = \frac{PL_2}{A(L_1 + 2L_2)} \qquad \sigma_2 = \frac{F_2}{A} = \frac{PL_1}{A(L_1 + 2L_2)} \qquad \text{(2-81a,b)}$$

These equations for the forces and stresses are valid provided the stresses in all three bars remain below the yield stress $\sigma_Y$.

As the load $P$ gradually increases, the stresses in the bars increase until the yield stress is reached in either the inner bar or the outer bars. Let us assume that the outer bars are longer than the inner bar, as sketched in Fig. 2-75:

$$L_1 > L_2 \qquad \text{(2-82)}$$

Then the inner bar is more highly stressed than the outer bars (see Eqs. 2-81a and b) and will reach the yield stress first. When that happens, the force in the inner bar is $F_2 = \sigma_Y A$. The magnitude of the load $P$ when the yield stress is first reached in any one of the bars is called the **yield load** $P_Y$. We can determine $P_Y$ by setting $F_2$ equal to $\sigma_Y A$ in Eq. (2-80b) and solving for the load:

$$P_Y = \sigma_Y A\left(1 + \frac{2L_2}{L_1}\right) \qquad \text{(2-83)}$$

**Fig. 2-75**

Elastoplastic analysis of a statically indeterminate structure

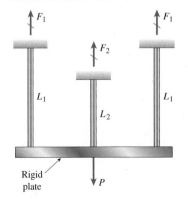

As long as the load $P$ is less than $P_Y$, the structure behaves in a linearly elastic manner and the forces in the bars can be determined from Eqs. (2-80a and b).

The downward displacement of the rigid bar at the yield load, called the **yield displacement** $\delta_Y$, is equal to the elongation of the inner bar when its stress first reaches the yield stress $\sigma_Y$:

$$\delta_Y = \frac{F_2 L_2}{EA} = \frac{\sigma_2 L_2}{E} = \frac{\sigma_Y L_2}{E} \qquad \textbf{(2-84)}$$

The relationship between the applied load $P$ and the downward displacement $\delta$ of the rigid bar is portrayed in the load-displacement diagram of Fig. 2-76. The behavior of the structure up to the yield load $P_Y$ is represented by line $OA$.

**Fig. 2-76**

Load-displacement diagram for the statically indeterminate structure shown in Fig. 2-75

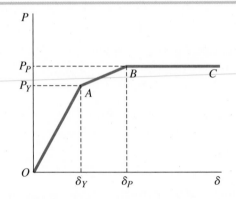

With a further increase in the load, the forces $F_1$ in the outer bars increase but the force $F_2$ in the inner bar remains constant at the value $\sigma_Y A$ because this bar is now perfectly plastic (see Fig. 2-73). When the forces $F_1$ reach the value $\sigma_Y A$, the outer bars also yield and therefore the structure cannot support any additional load. Instead, all three bars will elongate plastically under this constant load, called the **plastic load** $P_P$. The plastic load is represented by point $B$ on the load-displacement diagram (Fig. 2-76), and the horizontal line $BC$ represents the region of continuous plastic deformation without any increase in the load.

The plastic load $P_P$ can be calculated from static equilibrium [Eq. (2-76)] knowing that

$$F_1 = \sigma_Y A \quad F_2 = \sigma_Y A \qquad \textbf{(2-85 a,b)}$$

Thus, from equilibrium we find

$$P_P = 3\sigma_Y A \qquad \textbf{(2-86)}$$

The **plastic displacement** $\delta_P$ at the instant the load just reaches the plastic load $P_P$ is equal to the elongation of the outer bars at the instant they reach the yield stress. Therefore,

$$\delta_P = \frac{F_1 L_1}{EA} = \frac{\sigma_1 L_1}{E} = \frac{\sigma_Y L_1}{E} \qquad \textbf{(2-87)}$$

Comparing $\delta_P$ with $\delta_Y$, we see that in this example the ratio of the plastic displacement to the yield displacement is

$$\frac{\delta_P}{\delta_Y} = \frac{L_1}{L_2} \qquad \textbf{(2-88)}$$

Also, the ratio of the plastic load to the yield load is

$$\frac{P_P}{P_Y} = \frac{3L_1}{L_1 + 2L_2} \qquad \textbf{(2-89)}$$

For example, if $L_1 = 1.5L_2$, the ratios are $\delta_P/\delta_Y = 1.5$ and $P_P/P_Y = 9/7 = 1.29$. In general, the ratio of the displacements is always larger than the ratio of the corresponding loads, and the partially plastic region $AB$ on the load-displacement diagram (Fig. 2-76) always has a smaller slope than does the elastic region $OA$. Of course, the fully plastic region $BC$ has the smallest slope (zero).

## General Comments

To understand why the load-displacement graph is linear in the partially plastic region (line $AB$ in Fig. 2-76) and has a slope that is less than in the linearly elastic region, consider the following. In the partially plastic region of the structure, the outer bars still behave in a linearly elastic manner. Therefore, their elongation is a linear function of the load. Since their elongation is the same as the downward displacement of the rigid plate, the displacement of the rigid plate must also be a linear function of the load. Consequently, we have a straight line between points $A$ and $B$. However, the slope of the load-displacement diagram in this region is less than in the initial linear region because the inner bar yields plastically and only the outer bars offer increasing resistance to the increasing load. In effect, the stiffness of the structure has diminished.

From the discussion associated with Eq. (2-86) we see that the calculation of the plastic load $P_P$ requires only the use of statics, because all members have yielded and their axial forces are known. In contrast, the calculation of the yield load $P_Y$ requires a statically indeterminate analysis, which means that equilibrium, compatibility, and force-displacement equations must be solved.

After the plastic load $P_P$ is reached, the structure continues to deform as shown by line $BC$ on the load-displacement diagram (Fig. 2-76). Strain hardening occurs eventually, and then the structure is able to support additional loads. However, the presence of very large displacements usually means that the structure is no longer of use, and so the plastic load $P_P$ is usually considered to be the failure load.

The preceding discussion has dealt with the behavior of a structure when the load is applied for the first time. If the load is removed before the yield load is reached, the structure will behave elastically and return to its original unstressed condition. However, if the yield load is exceeded, some members of the structure will retain a permanent set when the load is removed, thus creating a prestressed condition. Consequently, the structure will have *residual stresses* in it even though no external loads are acting. If the load is applied a second time, the structure will behave in a different manner.

## Example 2-20

The structure shown in Fig. 2-77a consists of a horizontal beam $AB$ (assumed to be rigid) supported by two identical bars (bars 1 and 2) made of an elastoplastic material. The bars have length $L$ and cross-sectional area $A$, and the material has yield stress $\sigma_Y$, yield strain $\varepsilon_Y$, and modulus of elasticity $E = \sigma_Y/\varepsilon_Y$. The beam has length $3b$ and supports a load $P$ at end $B$.

(a) Determine the yield load $P_Y$ and the corresponding yield displacement $\delta_Y$ at the end of the bar (point $B$).
(b) Determine the plastic load $P_P$ and the corresponding plastic displacement $\delta_P$ at point $B$.
(c) Construct a load-displacement diagram relating the load $P$ to the displacement $\delta_B$ of point $B$.

### Fig. 2-77

Example 2-20: Elastoplastic analysis of a statically indeterminate structure

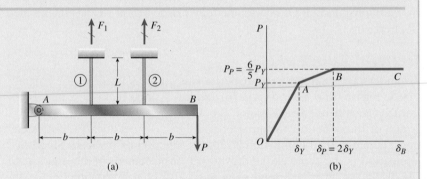

(a)                     (b)

### Solution

*Equation of equilibrium.* Because the structure is statically indeterminate, we begin with the equilibrium and compatibility equations. Considering the equilibrium of beam $AB$, we take moments about point $A$ and obtain

$$\Sigma M_A = 0 \qquad F_1(b) + F_2(2b) - P(3b) = 0$$

in which $F_1$ and $F_2$ are the axial forces in bars 1 and 2, respectively. This equation simplifies to

$$F_1 + 2F_2 = 3P \tag{g}$$

*Equation of compatibility.* The compatibility equation is based upon the geometry of the structure. Under the action of the load $P$ the rigid beam rotates about point $A$, and therefore the downward displacement at every point along the beam is proportional to its distance from point $A$. Thus, the compatibility equation is

$$\delta_2 = 2\delta_1 \tag{h}$$

where $\delta_2$ is the elongation of bar 2 and $\delta_1$ is the elongation of bar 1.

(a) *Yield load and yield displacement.* When the load $P$ is small and the stresses in the material are in the linearly elastic region, the force-displacement relations for the two bars are

$$\delta_1 = \frac{F_1 L}{EA} \qquad \delta_2 = \frac{F_2 L}{EA} \qquad \text{(i,j)}$$

Combining these equations with the compatibility condition [Eq. (h)] gives

$$\frac{F_2 L}{EA} = 2\frac{F_1 L}{EA} \quad \text{or} \quad F_2 = 2F_1 \qquad \text{(k)}$$

Now substituting into the equilibrium equation [Eq. (g)], we find

$$F_1 = \frac{3P}{5} \qquad F_2 = \frac{6P}{5} \qquad \text{(l,m)}$$

Bar 2, which has the larger force, will be the first to reach the yield stress. At that instant the force in bar 2 will be $F_2 = \sigma_Y A$. Substituting that value into Eq. (m) gives the yield load $P_Y$, as follows:

$$P_Y = \frac{5\sigma_Y A}{6} \qquad \qquad \text{(2-90)}$$

The corresponding elongation of bar 2 [from Eq. (j)] is $\delta_2 = \sigma_Y L/E$, and therefore the yield displacement at point $B$ is

$$\delta_Y = \frac{3\delta_2}{2} = \frac{3\sigma_Y L}{2E} \qquad \qquad \text{(2-91)}$$

Both $P_Y$ and $\delta_Y$ are indicated on the load-displacement diagram (Fig. 2-77b).

(b) *Plastic load and plastic displacement.* When the plastic load $P_P$ is reached, both bars will be stretched to the yield stress and both forces $F_1$ and $F_2$ will be equal to $\sigma_Y A$. It follows from equilibrium [Eq. (g)] that the plastic load is

$$P_P = \sigma_Y A \qquad \qquad \text{(2-92)}$$

At this load, the left-hand bar (bar 1) has just reached the yield stress; therefore, its elongation [from Eq. (i)] is $\delta_1 = \sigma_Y L/E$, and the plastic displacement of point $B$ is

$$\delta_P = 3\delta_1 = \frac{3\sigma_Y L}{E} \qquad \qquad \text{(2-93)}$$

The ratio of the plastic load $P_P$ to the yield load $P_Y$ is 6/5, and the ratio of the plastic displacement $\delta_P$ to the yield displacement $\delta_Y$ is 2. These values are also shown on the load-displacement diagram.

(c) *Load-displacement diagram.* The complete load-displacement behavior of the structure is pictured in Fig. 2-77b. The behavior is linearly elastic in the region from $O$ to $A$, partially plastic from $A$ to $B$, and fully plastic from $B$ to $C$.

In **Chapter 2,** we investigated the behavior of axially loaded bars acted on by distributed loads, such as self-weight, and also temperature changes and prestrains. We developed force-displacement relations for use in computing changes in lengths of bars under both uniform (i.e., constant force over its entire length) and nonuniform conditions (i.e., axial forces, and perhaps also cross-sectional area, vary over the length of the bar). Then, equilibrium and compatibility equations were developed for statically indeterminate structures in a superposition procedure leading to solution for all unknown forces, stresses, etc. We developed equations for normal and shear stresses on inclined sections and, from these equations, found maximum normal and shear stresses along the bar. The major concepts presented in this chapter are as follows:

1. The elongation or shortening ($\delta$) of prismatic bars subjected to tensile or compressive centroidal loads is proportional to both the load ($P$) and the length ($L$) of the bar, and inversely proportional to the axial rigidity ($EA$) of the bar; this relationship is called a **force-displacement relation**.

$$\delta = \frac{PL}{EA}$$

2. Cables are **tension-only elements**, and an effective modulus of elasticity ($E_e$) and effective cross-sectional area ($A_e$) should be used to account for the tightening effect that occurs when cables are placed under load.

3. The axial rigidity per unit length of a bar is referred to as its **stiffness** ($k$), and the inverse relationship is the **flexibility** ($f$) of the bar.

$$\delta = Pf = \frac{P}{k} \qquad f = \frac{L}{EA} = \frac{1}{k}$$

4. The summation of the displacements of the individual segments of a nonprismatic bar equals the elongation or shortening of the entire bar ($\delta$).

$$\delta = \sum_{i=1}^{n} \frac{N_i L_i}{E_i A_i}$$

Free-body diagrams are used to find the axial force ($N_i$) in each segment $i$; if axial forces and/or cross-sectional areas vary continuously, an integral expression is required.

$$\delta = \int_0^L d\delta = \int_0^L \frac{N(x)dx}{EA(x)}$$

5. If the bar structure is **statically indeterminate**, additional equations (beyond those available from statics) are required to solve for unknown forces. **Compatibility equations** are used to relate bar displacements to support conditions and thereby generate additional relationships among the unknowns. It is convenient to use a **superposition** of "released" (or statically determinate) structures to represent the actual statically indeterminate bar structure.

6. **Thermal effects** result in displacements proportional to the temperature change ($\Delta T$) and the length ($L$) of the bar but not stresses in statically determinate structures. The coefficient of thermal expansion ($\alpha$) of the material also is required to compute axial strains ($\varepsilon_T$) and axial displacements ($\delta_T$) due to thermal effects.

$$\varepsilon_T = \alpha(\Delta T) \quad \delta_T = \varepsilon_T L = \alpha(\Delta T)L$$

7. **Misfits** and **prestrains** induce axial forces only in statically indeterminate bars.

8. **Maximum normal** ($\sigma_{max}$) and **shear stresses** ($\tau_{max}$) can be obtained by considering an inclined stress element for a bar loaded by axial forces. The maximum normal stress occurs along the axis of the bar, but the maximum shear stress occurs at an inclination of 45° to the axis of the bar, and the maximum shear stress is one-half of the maximum normal stress.

$$\sigma_{max} = \sigma_x \quad \tau_{max} = \frac{\sigma_x}{2}$$

9. A number of advanced topics were also discussed in Chapter 2 but are not discussed in this summary. These advanced topics include: strain energy, impact loading, fatigue, stress concentrations, nonlinear behavior, and elastoplastic analysis.

# PROBLEMS CHAPTER 2

## Changes in Lengths of Axially Loaded Members

**2.2-1** The L-shaped arm $ABCD$ shown in the figure lies in a vertical plane and pivots about a horizontal pin at $A$. The arm has constant cross-sectional area and total weight $W$. A vertical spring of stiffness $k$ supports the arm at point $B$.

(a) Obtain a formula for the elongation of the spring due to the weight of the arm.

(b) Repeat part (a) if the pin support at $A$ is moved to $D$.

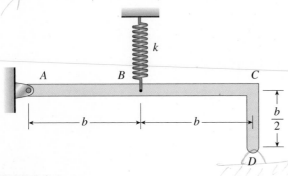

**PROB. 2.2-1**

**2.2-2** A steel cable with nominal diameter 25 mm (see Table 2-1) is used in a construction yard to lift a bridge section weighing 38 kN, as shown in the figure. The cable has an effective modulus of elasticity $E = 140$ GPa.

(a) If the cable is 14 m long, how much will it stretch when the load is picked up?

(b) If the cable is rated for a maximum load of 70 kN, what is the factor of safety with respect to failure of the cable?

**PROB. 2.2-2**

**218**    Chapter 2   Axially Loaded Members

**2.2-3** A steel wire and an aluminum alloy wire have equal lengths and support equal loads $P$ (see figure). The moduli of elasticity for the steel and aluminum alloy are $E_s = 30,000$ ksi and $E_a = 11,000$ ksi, respectively.

(a) If the wires have the same diameters, what is the ratio of the elongation of the aluminum alloy wire to the elongation of the steel wire?

(b) If the wires stretch the same amount, what is the ratio of the diameter of the aluminum alloy wire to the diameter of the steel wire?

(c) If the wires have the same diameters and same load $P$, what is the ratio of the initial length of the aluminum alloy wire to that of the steel wire if the aluminum alloy wire stretches 1.5 times that of the steel wire?

(d) If the wires have the same diameters, same initial length, and same load $P$, what is the material of the upper wire if it elongates 1.7 times that of the steel wire?

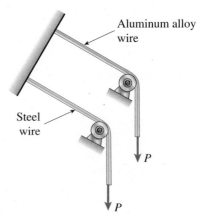

**PROB. 2.2-3**

**2.2-4** By what distance $h$ does the cage shown in the figure move downward when the weight $W$ is placed inside it? (See the figure.)

Consider only the effects of the stretching of the cable, which has axial rigidity $EA = 10,700$ kN. The pulley at $A$ has diameter $d_A = 300$ mm and the pulley at $B$ has diameter $d_B = 150$ mm. Also, the distance $L_1 = 4.6$ m, the distance $L_2 = 10.5$ m, and the weight $W = 22$ kN. (*Note:* When calculating the length of the cable, include the parts of the cable that go around the pulleys at $A$ and $B$.)

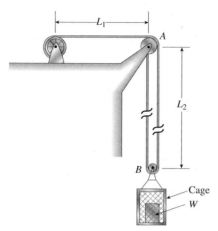

**PROB. 2.2-4**

**2.2-5** A safety valve on the top of a tank containing steam under pressure $p$ has a discharge hole of diameter $d$ (see figure). The valve is designed to release the steam when the pressure reaches the value $p_{max}$.

If the natural length of the spring is $L$ and its stiffness is $k$, what should be the dimension $h$ of the valve? (Express your result as a formula for $h$.)

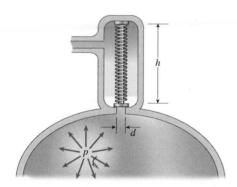

**PROB. 2.2-5**

**2.2-6** The device shown in the figure consists of a prismatic rigid pointer $ABC$ supported by a uniform translational spring of stiffness $k = 950$ N/m. The spring is positioned at distance $b = 165$ mm from the pinned end $A$ of the pointer. The device is adjusted so that when there is no load $P$, the pointer reads zero on the angular scale.

(a) If the load $P = 11$ N, at what distance $x$ should the load be placed so that the pointer will read $\theta = 2.5°$ on the scale (see figure part a)?

(b) Repeat part (a) if a rotational spring $k_r = kb^2$ is added at $A$ (see figure part b).

(c) Let $x = 7b/8$. What is $P_{max}$ (N) if $\theta$ cannot exceed $2°$? Include spring $k_r$ in your analysis.

(d) Now, if the weight of the pointer $ABC$ is known to be $W_p = 3$ N and the weight of the spring is $W_s = 2.75$ N, what initial angular position (i.e., $\theta$ in degrees) of the pointer will result in a zero reading on the angular scale once the pointer is released from rest? Assume $P = k_r = 0$.

(e) If the pointer is rotated to a vertical position (see figure part c), find the required load $P$, applied at midheight of the pointer, that will result in a pointer reading of $\theta = 2.5°$ on the scale. Consider the weight of the pointer $W_p$ in your analysis.

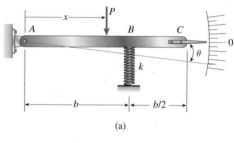

(a)

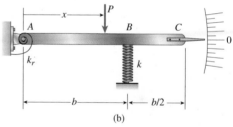

(b)

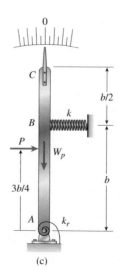

(c)

**PROB. 2.2-6**

**2.2-7** Two rigid bars are connected to each other by two linearly elastic springs. Before loads are applied, the lengths of the springs are such that the bars are parallel and the springs are without stress.

(a) Derive a formula for the displacement $\delta_4$ at point 4 when the load $P$ is applied at joint 3 and moment $PL$ is applied at joint 1, as shown in the figure part a. (Assume that the bars rotate through very small angles under the action of the load $P$.)

(b) Repeat part (a) if a rotational spring, $k_r = kL^2$, is now added at joint 6. What is the ratio of the deflection $\delta_4$ in the figure part a to that in the figure part b ?

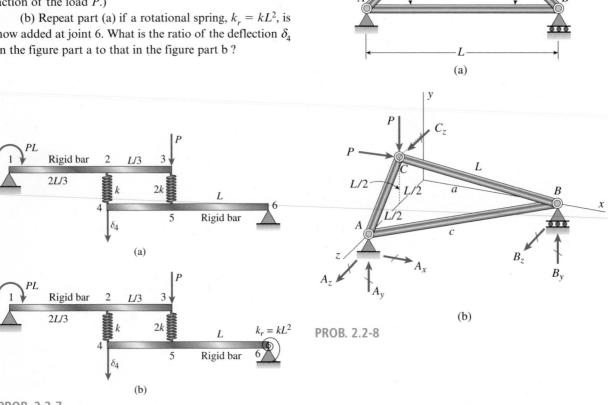

(a)

(b)

PROB. 2.2-7

PROB. 2.2-8

**2.2-8** The three-bar truss $ABC$ shown in the figure part a has a span $L = 3$ m and is constructed of steel pipes having cross-sectional area $A = 3900$ mm² and modulus of elasticity $E = 200$ GPa. Identical loads $P$ act both vertically and horizontally at joint $C$, as shown.

(a) If $P = 475$ kN, what is the horizontal displacement of joint $B$?

(b) What is the maximum permissible load value $P_{max}$ if the displacement of joint $B$ is limited to 1.5 mm?

(c) Repeat parts (a) and (b) if the plane truss is replaced by a space truss (see figure part b).

**2.2-9** An aluminum wire having a diameter $d = 1/10$ in. and length $L = 12$ ft is subjected to a tensile load $P$ (see figure). The aluminum has modulus of elasticity $E = 10,600$ ksi

If the maximum permissible elongation of the wire is 1/8 in. and the allowable stress in tension is 10 ksi, what is the allowable load $P_{max}$?

PROB. 2.2-9

**2.2-10** A uniform bar $AB$ of weight $W = 25$ N is supported by two springs, as shown in the figure. The spring on the left has stiffness $k_1 = 300$ N/m and natural length $L_1 = 250$ mm. The corresponding quantities for the spring on the right are $k_2 = 400$ N/m and $L_2 = 200$ mm. The distance between the springs is $L = 350$ mm, and the spring on the right is suspended from a support that is distance $h = 80$ mm below the point of support for the spring on the left. Neglect the weight of the springs.

(a) At what distance $x$ from the left-hand spring (figure part a) should a load $P = 18$ N be placed in order to bring the bar to a horizontal position?

(b) If $P$ is now removed, what new value of $k_1$ is required so that the bar (figure part a) will hang in a horizontal position under weight $W$?

(c) If $P$ is removed and $k_1 = 300$ N/m, what distance $b$ should spring $k_1$ be moved to the right so that the bar (figure part a) will hang in a horizontal position under weight $W$?

(d) If the spring on the left is now replaced by two springs in series ($k_1 = 300$N/m, $k_3$) with overall natural length $L_1 = 250$ mm (see figure part b), what value of $k_3$ is required so that the bar will hang in a horizontal position under weight $W$?

**2.2-11** A hollow, circular, cast-iron pipe ($E_c = 12{,}000$ ksi) supports a brass rod ($E_b = 14{,}000$ ksi) and weight $W = 2$ kips, as shown. The outside diameter of the pipe is $d_c = 6$ in.

(a) If the allowable compressive stress in the pipe is 5000 psi and the allowable shortening of the pipe is 0.02 in., what is the minimum required wall thickness $t_{c,min}$? (Include the weights of the rod and steel cap in your calculations.)

(b) What is the elongation of the brass rod $\delta_r$ due to both load $W$ and its own weight?

(c) What is the minimum required clearance $h$?

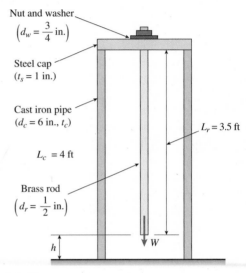

**PROB. 2.2-11**

**2.2-12** The horizontal rigid beam $ABCD$ is supported by vertical bars $BE$ and $CF$ and is loaded by vertical forces $P_1 = 400$ kN and $P_2 = 360$ kN acting at points $A$ and $D$, respectively (see figure). Bars $BE$ and $CF$ are made of steel ($E = 200$ GPa) and have cross-sectional areas $A_{BE} = 11{,}100$ mm$^2$ and $A_{CF} = 9280$ mm$^2$. The distances between various points on the bars are shown in the figure.

Determine the vertical displacements $\delta_A$ and $\delta_D$ of points $A$ and $D$, respectively.

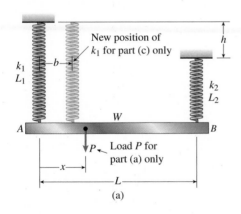

(a)

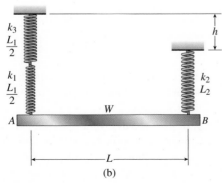

(b)

**PROB. 2.2-10**

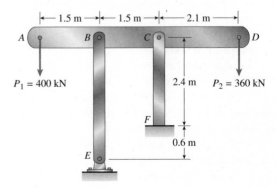

**PROB. 2.2-12**

**2.2-13** Two pipe columns ($AB$, $FC$) are pin-connected to a *rigid* beam ($BCD$), as shown in the figure. Each pipe column has a modulus of $E$, but heights ($L_1$ or $L_2$) and outer diameters ($d_1$ or $d_2$) are different for each column. Assume the inner diameter of each column is 3/4 of outer diameter. Uniformly distributed downward load $q = 2P/L$ is applied over a distance of $3L/4$ along $BC$, and concentrated load $P/4$ is applied downward at $D$.

(a) Derive a formula for the displacement $\delta_D$ at point $D$ in terms of $P$ and column flexibilities $f_1$ and $f_2$.

(b) If $d_1 = (9/8)d_2$, find the $L_1/L_2$ ratio so that beam $BCD$ displaces downward to a horizontal position under the load system in part (a).

(c) If $L_1 = 2L_2$, find the $d_1/d_2$ ratio so that beam $BCD$ displaces downward to a horizontal position under the load system in part (a).

(d) If $d_1 = (9/8)d_2$ and $L_1/L_2 = 1.5$, at what horizontal distance $x$ from $B$ should load $P/4$ be placed so that beam $BCD$ displaces downward to a horizontal position under the load system in part (a)?

(b) Repeat part (a) if a translational spring $k_1 = k/2$ is added at $C$ and a rotational spring $k_r = kb^2/2$ is added at $A$ (see figure part b).

**2.2-15** Solve the preceding problem for the following data: $b = 8.0$ in., $k = 16$ lb/in., $\alpha = 45°$, and $P = 10$ lb.

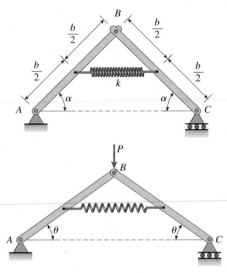

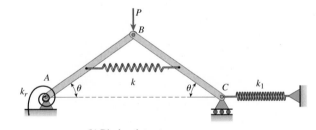

(a) Initial and displaced structures

(b) Displaced structures

PROB. 2.2-14 and 2.2-15

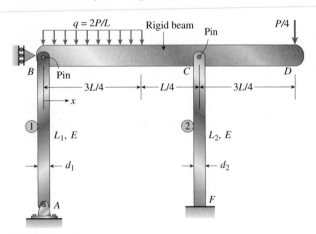

PROB. 2.2-13

**2.2-14** A framework $ABC$ consists of two rigid bars $AB$ and $BC$, each having a length $b$ (see the first part of the figure part a). The bars have pin connections at $A$, $B$, and $C$, and are joined by a spring of stiffness $k$. The spring is attached at the midpoints of the bars. The framework has a pin support at $A$ and a roller support at $C$, and the bars are at an angle $\alpha$ to the horizontal.

When a vertical load $P$ is applied at joint $B$ (see the second part of the figure part a) the roller support $C$ moves to the right, the spring is stretched, and the angle of the bars decreases from $\alpha$ to the angle $\theta$.

(a) Determine the angle $\theta$ and the increase $\delta$ in the distance between points $A$ and $C$. Also find reactions at $A$ and $C$. (Use the following data: $b = 200$ mm, $k = 3.2$ kN/m, $\alpha = 45°$, and $P = 50$ N.)

# Changes in Lengths Under Nonuniform Conditions

**2.3-1** (a) Calculate the elongation of a copper bar of solid circular cross section with tapered ends when it is stretched by axial loads of magnitude 3.0 k (see figure).

The length of the end segments is 20 in. and the length of the prismatic middle segment is 50 in. Also, the diameters at cross sections $A$, $B$, $C$, and $D$ are 0.5, 1.0, 1.0, and 0.5 in., respectively, and the modulus of elasticity is 18,000 ksi. (*Hint:* Use the result of Example 2-4.)

(b) If the total elongation of the bar cannot exceed 0.025 in., what are the required diameters at $B$ and $C$? Assume that diameters at $A$ and $D$ remain at 0.5 in.

3.0 k

20 in.

50 in.

20 in.

3.0 k

PROB. 2.3-1

**2.3-2** A long, rectangular copper bar under a tensile load $P$ hangs from a pin that is supported by two steel posts (see figure). The copper bar has a length of 2.0 m, a cross-sectional area of 4800 mm², and a modulus of elasticity $E_c = 120$ GPa. Each steel post has a height of 0.5 m, a cross-sectional area of 4500 mm², and a modulus of elasticity $E_s = 200$ GPa.

(a) Determine the downward displacement $\delta$ of the lower end of the copper bar due to a load $P = 180$ kN.

(b) What is the maximum permissible load $P_{max}$ if the displacement $\delta$ is limited to 1.0 mm?

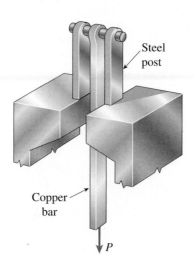

Steel
post

Copper
bar

$P$

PROB. 2.3-2

**2.3-3** An aluminum bar $AD$ (see figure) has a cross-sectional area of 0.40 in.² and is loaded by forces $P_1 = 1700$ lb, $P_2 = 1200$ lb, and $P_3 = 1300$ lb. The lengths of the segments of the bar are $a = 60$ in., $b = 24$ in., and $c = 36$ in.

(a) Assuming that the modulus of elasticity $E = 10.4 \times 10^6$ psi, calculate the change in length of the bar. Does the bar elongate or shorten?

(b) By what amount $P$ should the load $P_3$ be increased so that the bar does not change in length when the three loads are applied?

(c) If $P_3$ remains at 1300 lb, what revised cross-sectional area for segment $AB$ will result in no change of length when all three loads are applied?

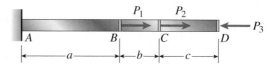

$P_1$    $P_2$

$A$    $B$    $C$    $D$    $P_3$

$a$    $b$    $c$

PROB. 2.3-3

**2.3-4** A rectangular bar of length $L$ has a slot in the middle half of its length (see figure). The bar has width $b$, thickness $t$, and modulus of elasticity $E$. The slot has width $b/4$.

(a) Obtain a formula for the elongation $\delta$ of the bar due to the axial loads $P$.

(b) Calculate the elongation of the bar if the material is high-strength steel, the axial stress in the middle region is 160 MPa, the length is 750 mm, and the modulus of elasticity is 210 GPa.

(c) If the total elongation of the bar is limited to $\delta_{max} = 0.475$ mm, what is the maximum length of the slotted region? Assume that the axial stress in the middle region remains at 160 MPa.

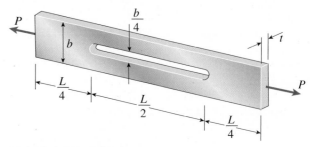

$P$    $\dfrac{b}{4}$    $t$

$b$

$\dfrac{L}{4}$    $\dfrac{L}{2}$    $\dfrac{L}{4}$    $P$

PROBS. 2.3-4 and 2.3-5

**2.3-5** Solve the preceding problem if the axial stress in the middle region is 24,000 psi, the length is 30 in., and the modulus of elasticity is $30 \times 10^6$ psi. In part (c), assume that $\delta_{max} = 0.02$ in.

**2.3-6** A two-story building has steel columns $AB$ in the first floor and $BC$ in the second floor, as shown in the figure. The roof load $P_1$ equals 400 kN and the second-floor load $P_2$ equals 720 kN. Each column has length $L = 3.75$ m. The cross-sectional areas of the first- and second-floor columns are 11,000 mm² and 3900 mm², respectively.

(a) Assuming that $E = 206$ GPa, determine the total shortening $\delta_{AC}$ of the two columns due to the combined action of the loads $P_1$ and $P_2$.

(b) How much additional load $P_0$ can be placed at the top of the column (point $C$) if the total shortening $\delta_{AC}$ is not to exceed 4.0 mm?

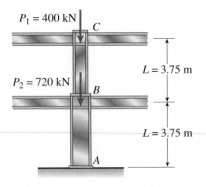

**PROB. 2.3-6**

**2.3-7** A steel bar 8.0 ft long has a circular cross section of diameter $d_1 = 0.75$ in. over one-half of its length and diameter $d_2 = 0.5$ in. over the other half (see figure part a). The modulus of elasticity is $E = 30 \times 10^6$ psi.

(a) How much will the bar elongate under a tensile load $P = 5000$ lb?

(b) If the same volume of material is made into a bar of constant diameter $d$ and length 8.0 ft, what will be the elongation under the same load $P$?

(c) If the uniform axial centroidal load $q = 1250$ lb/ft is applied to the left over segment 1 (see figure part b), find the ratio of the total elongation of the bar to that in parts (a) and (b).

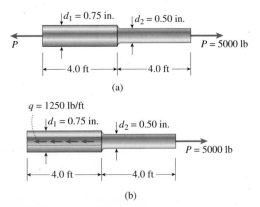

**PROB. 2.3-7**

**2.3-8** A bar $ABC$ of length $L$ consists of two parts of equal lengths but different diameters. Segment $AB$ has diameter $d_1 = 100$ mm, and segment $BC$ has diameter $d_2 = 60$ mm. Both segments have length $L/2 = 0.6$ m. A longitudinal hole of diameter $d$ is drilled through segment $AB$ for one-half of its length (distance $L/4 = 0.3$ m). The bar is made of plastic having modulus of elasticity $E = 4.0$ GPa. Compressive loads $P = 110$ kN act at the ends of the bar.

(a) If the shortening of the bar is limited to 8.0 mm, what is the maximum allowable diameter $d_{max}$ of the hole? (See figure part a.)

(b) Now, if $d_{max}$ is instead set at $d_2/2$, at what distance $b$ from end $C$ should load $P$ be applied to limit the bar shortening to 8.0 mm? (See figure part b.)

(c) Finally, if loads $P$ are applied at the ends and $d_{max} = d_2/2$, what is the permissible length $x$ of the hole if shortening is to be limited to 8.0 mm? (See figure part c.)

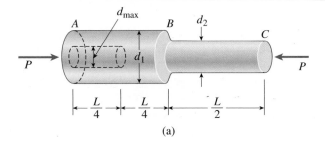

(a)

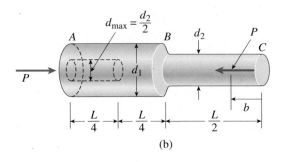

(b)

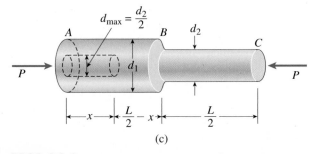

(c)

**PROB. 2.3-8**

**2.3-9** A wood pile, driven into the earth, supports a load $P$ entirely by friction along its sides (see figure part a). The friction force $f$ per unit length of pile is assumed to be uniformly distributed over the surface of the pile. The pile has length $L$, cross-sectional area $A$, and modulus of elasticity $E$.

(a) Derive a formula for the shortening $\delta$ of the pile in terms of $P$, $L$, $E$, and $A$.

(b) Draw a diagram showing how the compressive stress $\sigma_c$ varies throughout the length of the pile.

(c) Repeat parts (a) and (b) if skin friction $f$ varies linearly with depth (see figure part b).

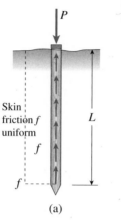

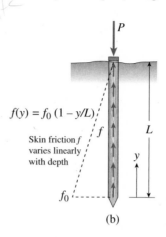

(a)

(b)

PROB. 2.3-9

**2.3-10** Consider the copper tubes joined below using a "sweated" joint. Use the properties and dimensions given.

(a) Find the total elongation of segment 2-3-4 ($\delta_{2-4}$) for an applied tensile force of $P = 5$ kN. Use $E_c = 120$ GPa.

(b) If the yield strength in shear of the tin-lead solder is $\tau_y = 30$ MPa and the tensile yield strength of the copper is $\sigma_y = 200$ MPa, what is the maximum load $P_{max}$ that can be applied to the joint if the desired factor of safety in shear is $FS_\tau = 2$ and in tension is $FS_\sigma = 1.7$?

(c) Find the value of $L_2$ at which tube and solder capacities are equal.

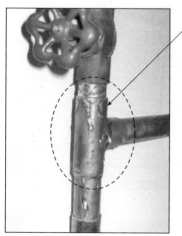

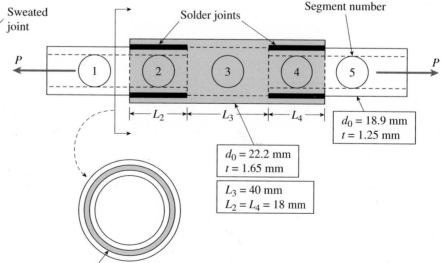

$d_0 = 22.2$ mm
$t = 1.65$ mm

$L_3 = 40$ mm
$L_2 = L_4 = 18$ mm

$d_0 = 18.9$ mm
$t = 1.25$ mm

Tin-lead solder in space between copper tubes; assume thickness of solder equal zero

(© Barry Goodno)

PROB. 2.3-10

**2.3-11** The nonprismatic cantilever circular bar shown has an internal cylindrical hole of diameter $d/2$ from 0 to $x$, so the net area of the cross section for segment 1 is $(3/4)A$. Load $P$ is applied at $x$, and load $P/2$ is applied at $x = L$. Assume that $E$ is constant.

(a) Find reaction force $R_1$.

(b) Find internal axial forces $N_i$ in segments 1 and 2.

(c) Find $x$ required to obtain axial displacement at joint 3 of $\delta_3 = PL/EA$.

(d) In part (c), what is the displacement at joint 2, $\delta_2$?

(e) If $P$ acts at $x = 2L/3$ and $P/2$ at joint 3 is replaced by $\beta P$, find $\beta$ so that $\delta_3 = PL/EA$.

(f) Draw the *axial force* (AFD: $N(x)$, $0 \le x \le L$) and *axial displacement diagrams* (ADD: $\delta(x)$, $0 \le x \le L$) using results from parts (b) through (d).

(b) What is the elongation $\delta_B$ of the entire bar?

(c) What is the ratio $\beta$ of the elongation of the upper half of the bar to the elongation of the lower half of the bar?

(d) If bar $AB$ is a riser pipe hanging from a drill rig at sea, what is the total elongation of the pipe? Let $L = 1500$ m, $A = 0.0157$ m$^2$, $E = 210$ GPa. See Appendix I for weight densities of steel and sea water. (See Probs. 1.4-2 and 1.7-11 for additional figures.)

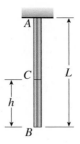

PROB. 2.3-12

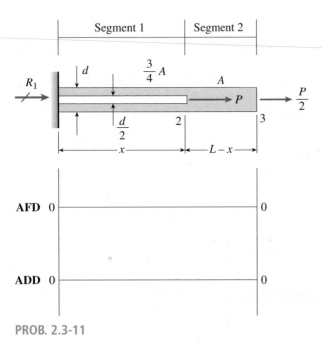

PROB. 2.3-11

**2.3-12** A prismatic bar $AB$ of length $L$, cross-sectional area $A$, modulus of elasticity $E$, and weight $W$ hangs vertically under its own weight (see figure).

(a) Derive a formula for the downward displacement $\delta_C$ of point $C$, located at distance $h$ from the lower end of the bar.

**2.3-13** A flat bar of rectangular cross section, length $L$, and constant thickness $t$ is subjected to tension by forces $P$ (see figure). The width of the bar varies linearly from $b_1$ at the smaller end to $b_2$ at the larger end. Assume that the angle of taper is small.

(a) Derive the following formula for the elongation of the bar:

$$\delta = \frac{PL}{Et(b_2 - b_1)} \ln \frac{b_2}{b_1}$$

(b) Calculate the elongation, assuming $L = 5$ ft, $t = 1.0$ in., $P = 25$ k, $b_1 = 4.0$ in., $b_2 = 6.0$ in., and $E = 30 \times 10^6$ psi.

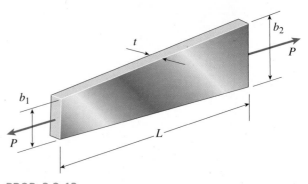

PROB. 2.3-13

**2.3-14** A post *AB* supporting equipment in a laboratory is tapered uniformly throughout its height *H* (see figure). The cross sections of the post are square, with dimensions *b* × *b* at the top and 1.5*b* × 1.5*b* at the base.

Derive a formula for the shortening $\delta$ of the post due to the compressive load *P* acting at the top. (Assume that the angle of taper is small and disregard the weight of the post itself.)

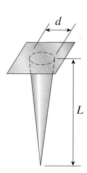

PROB. 2.3-15

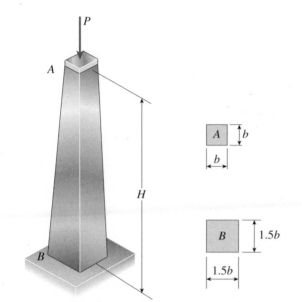

PROB. 2.3-14

**2.3-15** A long, slender bar in the shape of a right circular cone with length *L* and base diameter *d* hangs vertically under the action of its own weight (see figure). The weight of the cone is *W* and the modulus of elasticity of the material is *E*.

Derive a formula for the increase $\delta$ in the length of the bar due to its own weight. (Assume that the angle of taper of the cone is small.)

**2.3-16** A uniformly tapered tube *AB* of circular cross section and length *L* is shown in the figure. The average diameters at the ends are $d_A$ and $d_B = 2d_A$. Assume *E* is constant. Find the elongation $\delta$ of the tube when it is subjected to loads *P* acting at the ends. Use the following numerical data: $d_A = 35$ mm, *L* = 300 mm, *E* = 2.1 GPa, *P* = 25 kN. Consider the following cases:

(a) A hole of *constant* diameter $d_A$ is drilled from *B* toward *A* to form a hollow section of length *x* = *L*/2;

(b) A hole of *variable* diameter *d*(*x*) is drilled from *B* toward *A* to form a hollow section of length *x* = *L*/2 and constant thickness $t = d_A/20$.

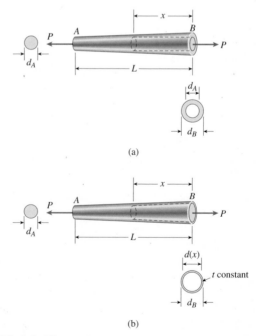

PROB. 2.3-16

**2.3-17** The main cables of a suspension bridge (see figure part a) follow a curve that is nearly parabolic because the primary load on the cables is the weight of the bridge deck, which is uniform in intensity along the horizontal. Therefore, let us represent the central region *AOB* of one of the main cables (see part b of the figure) as a parabolic cable supported at points *A* and *B* and carrying a uniform load of intensity *q* along the horizontal. The span of the cable is *L*, the sag is *h*, the axial rigidity is *EA*, and the origin of coordinates is at midspan.

(a) Derive the following formula for the elongation of cable *AOB* shown in part b of the figure:

$$\delta = \frac{qL^3}{8hEA}\left(1 + \frac{16h^2}{3L^2}\right)$$

(b) Calculate the elongation $\delta$ of the central span of one of the main cables of the Golden Gate Bridge, for which the dimensions and properties are $L = 4200$ ft, $h = 470$ ft, $q = 12,700$ lb/ft, and $E = 28,800,000$ psi. The cable consists of 27,572 parallel wires of diameter 0.196 in.

*Hint:* Determine the tensile force *T* at any point in the cable from a free-body diagram of part of the cable; then determine the elongation of an element of the cable of length *ds*; finally, integrate along the curve of the cable to obtain an equation for the elongation $\delta$.

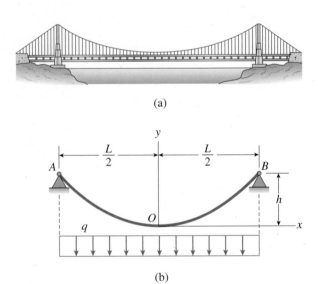

(a)

(b)

PROB. 2.3-17

**2.3-18** A bar *ABC* revolves in a horizontal plane about a vertical axis at the midpoint *C* (see figure). The bar, which has length 2*L* and cross-sectional area *A*, revolves at con-

stant angular speed $\omega$. Each half of the bar (*AC* and *BC*) has weight $W_1$ and supports a weight $W_2$ at its end.

Derive the following formula for the elongation of one-half of the bar (that is, the elongation of either *AC* or *BC*):

$$\delta = \frac{L^2\omega^2}{3gEA}(W_1 + 3W_2)$$

in which *E* is the modulus of elasticity of the material of the bar and *g* is the acceleration of gravity.

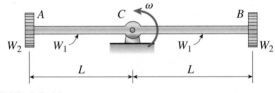

PROB. 2.3-18

## Statically Indeterminate Structures

**2.4-1** The assembly shown in the figure consists of a brass core (diameter $d_1 = 0.25$ in.) surrounded by a steel shell (inner diameter $d_2 = 0.28$ in., outer diameter $d_3 = 0.35$ in.). A load *P* compresses the core and shell, which have length $L = 4.0$ in. The moduli of elasticity of the brass and steel are $E_b = 15 \times 10^6$ psi and $E_s = 30 \times 10^6$ psi, respectively.

(a) What load *P* will compress the assembly by 0.003 in.?

(b) If the allowable stress in the steel is 22 ksi and the allowable stress in the brass is 16 ksi, what is the allowable compressive load $P_{allow}$? (*Suggestion:* Use the equations derived in Example 2-5.)

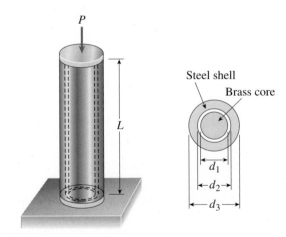

PROB. 2.4-1

**2.4-2** A cylindrical assembly consisting of a brass core and an aluminum collar is compressed by a load $P$ (see figure). The length of the aluminum collar and brass core is 350 mm, the diameter of the core is 25 mm, and the outside diameter of the collar is 40 mm. Also, the moduli of elasticity of the aluminum and brass are 72 GPa and 100 GPa, respectively.

(a) If the length of the assembly decreases by 0.1% when the load $P$ is applied, what is the magnitude of the load?

(b) What is the maximum permissible load $P_{max}$ if the allowable stresses in the àluminum and brass are 80 MPa and 120 MPa, respectively? (*Suggestion:* Use the equations derived in Example 2-5.)

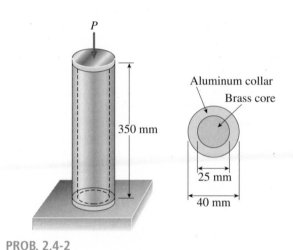

PROB. 2.4-2

**2.4-3** Three prismatic bars, two of material $A$ and one of material $B$, transmit a tensile load $P$ (see figure). The two outer bars (material $A$) are identical. The cross-sectional area of the middle bar (material $B$) is 50% larger than the cross-sectional area of one of the outer bars. Also, the modulus of elasticity of material $A$ is twice that of material $B$.

(a) What fraction of the load $P$ is transmitted by the middle bar?

(b) What is the ratio of the stress in the middle bar to the stress in the outer bars?

(c) What is the ratio of the strain in the middle bar to the strain in the outer bars?

PROB. 2.4-3

**2.4-4** A circular bar $ACB$ of diameter $d$ having a cylindrical hole of length $x$ and diameter $d/2$ from $A$ to $C$ is held between rigid supports at $A$ and $B$. A load $P$ acts at $L/2$ from ends $A$ and $B$. Assume $E$ is constant.

(a) Obtain formulas for the reactions $R_A$ and $R_B$ at supports $A$ and $B$, respectively, due to the load $P$ (see figure part a).

(b) Obtain a formula for the displacement $\delta$ at the point of load application (see figure part a).

(c) For what value of $x$ is $R_B = (6/5)R_A$? (See figure part a.)

(d) Repeat part (a) if the bar is now rotated to a vertical position, load $P$ is removed, and the bar is hanging under its own weight (assume mass density $= \rho$). (See figure part b.) Assume that $x = L/2$.

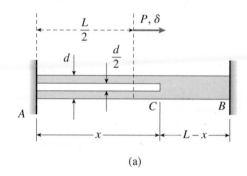

(a)

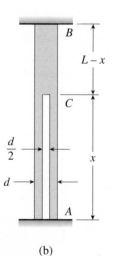

(b)

PROB. 2.4-4

**2.4-5** Three steel cables jointly support a load of 12 k (see figure). The diameter of the middle cable is 3/4 in. and the diameter of each outer cable is 1/2 in. The tensions in the cables are adjusted so that each cable carries one-third of the load (i.e., 4 k). Later, the load is increased by 9 k to a total load of 21 k.

(a) What percent of the total load is now carried by the middle cable?

(b) What are the stresses $\sigma_M$ and $\sigma_O$ in the middle and outer cables, respectively? (*Note:* See Table 2-1 in Section 2.2 for properties of cables.)

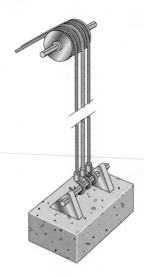

PROB. 2.4-5

**2.4-6** A plastic rod $AB$ of length $L = 0.5$ m has a diameter $d_1 = 30$ mm (see figure). A plastic sleeve $CD$ of length $c = 0.3$ m and outer diameter $d_2 = 45$ mm is securely bonded to the rod so that no slippage can occur between the rod and the sleeve. The rod is made of an acrylic with modulus of elasticity $E_1 = 3.1$ GPa and the sleeve is made of a polyamide with $E_2 = 2.5$ GPa.

(a) Calculate the elongation $\delta$ of the rod when it is pulled by axial forces $P = 12$ kN.

(b) If the sleeve is extended for the full length of the rod, what is the elongation?

(c) If the sleeve is removed, what is the elongation?

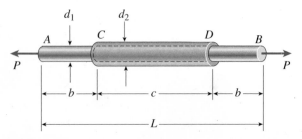

PROB. 2.4-6

**2.4-7** A tube structure is acted on by loads at $B$ and $D$, as shown in the figure. The tubes are joined using two flange plates at $C$, which are bolted together using six 0.5-in. diameter bolts.

(a) Derive formulas for the reactions $R_A$ and $R_E$ at the ends of the bar.

(b) Determine the axial displacements $\delta_B$, $\delta_c$, and $\delta_D$ at points $B$, $C$, and $D$, respectively.

(c) Draw an axial-displacement diagram (ADD) in which the abscissa is the distance $x$ from support $A$ to any point on the bar and the ordinate is the horizontal displacement $\delta$ at that point.

(d) Find the maximum value of the load variable $P$ if allowable normal stress in the bolts is 14 ksi.

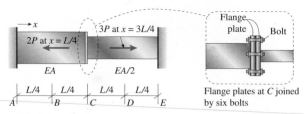

PROB. 2.4-7

**2.4-8** The fixed-end bar $ABCD$ consists of three prismatic segments, as shown in the figure. The end segments have cross-sectional area $A_1 = 840$ mm² and length $L_1 = 200$ mm. The middle segment has cross-sectional area $A_2 = 1260$ mm² and length $L_2 = 250$ mm. Loads $P_B$ and $P_C$ are equal to 25.5 kN and 17.0 kN, respectively.

(a) Determine the reactions $R_A$ and $R_D$ at the fixed supports.

(b) Determine the compressive axial force $F_{BC}$ in the middle segment of the bar.

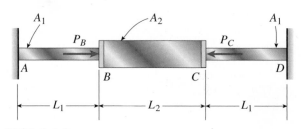

PROB. 2.4-8

**2.4-9** The aluminum and steel pipes shown in the figure are fastened to rigid supports at ends $A$ and $B$ and to a rigid plate at $C$ at their junction. The aluminum pipe is twice as long as the steel pipe. Two equal and symmetrically placed loads $P$ act on the plate at $C$.

(a) Obtain formulas for the axial stresses $\sigma_a$ and $\sigma_s$ in the aluminum and steel pipes, respectively.

(b) Calculate the stresses for the following data: $P = 12$ k, cross-sectional area of aluminum pipe $A_a = 8.92$ in.$^2$, cross-sectional area of steel pipe $A_s = 1.03$ in.$^2$, modulus of elasticity of aluminum $E_a = 10 \times 10^6$ psi, and modulus of elasticity of steel $E_s = 29 \times 10^6$ psi.

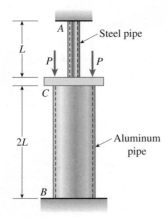

**PROB. 2.4-9**

**2.4-10** A hollow circular pipe (see figure) supports a load $P$ which is uniformly distributed around a cap plate at the top of the lower pipe. The inner and outer diameters of the upper and lower parts of the pipe are $d_1 = 50$ mm, $d_2 = 60$ mm, $d_3 = 57$ mm, and $d_4 = 64$ mm, respectively. Pipe lengths are $L_1 = 2$ m and $L_2 = 3$ m. Neglect the self-weight of the pipes. Assume that cap plate thickness is small compared to $L_1$ and $L_2$. Let $E = 110$ MPa.

(a) If the tensile stress in the upper part is $\sigma_1 = 10.5$ MPa, what is load $P$? Also, what are reactions $R_1$ at the upper support and $R_2$ at the lower support? What is the stress $\sigma_2$ (MPa) in the lower part?

(b) Find displacement $\delta$ (mm) at the cap plate. Plot the Axial Force Diagram, AFD $[N(x)]$ and Axial Displacement Diagram, ADD $[\delta(x)]$.

(c) Add the uniformly distributed load $q$ along the centroidal axis of pipe segment 2. Find $q$ (kN/m) so that $R_2 = 0$. Assume that load $P$ from part (a) is also applied.

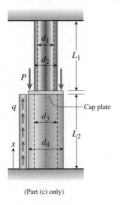

(Part (c) only)

**PROB. 2.4-10**

**2.4-11** A *bimetallic* bar (or composite bar) of square cross section with dimensions $2b \times 2b$ is constructed of two different metals having moduli of elasticity $E_1$ and $E_2$ (see figure). The two parts of the bar have the same cross-sectional dimensions. The bar is compressed by forces $P$ acting through rigid end plates. The line of action of the loads has an eccentricity $e$ of such magnitude that each part of the bar is stressed uniformly in compression.

(a) Determine the axial forces $P_1$ and $P_2$ in the two parts of the bar.

(b) Determine the eccentricity $e$ of the loads.

(c) Determine the ratio $\sigma_1/\sigma_2$ of the stresses in the two parts of the bar.

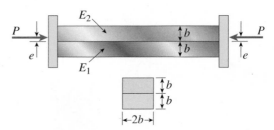

**PROB. 2.4-11**

**2.4-12** A rigid bar of weight $W = 800$ N hangs from three equally spaced vertical wires (length $L = 150$ mm, spacing $a = 50$ mm): two of steel and one of aluminum. The wires also support a load $P$ acting on the bar. The diameter of the steel wires is $d_s = 2$ mm, and the diameter of the aluminum wire is $d_a = 4$ mm. Assume $E_s = 210$ GPa and $E_a = 70$ GPa.

(a) What load $P_{allow}$ can be supported *at the midpoint of the bar* ($x = a$) if the allowable stress in the steel wires is 220 MPa and in the aluminum wire is 80 MPa? (See figure part a.)

(b) What is $P_{allow}$ if the load is positioned at $x = a/2$? (See figure part a.)

(c) Repeat part (b) if the second and third wires are *switched* as shown in figure part b.

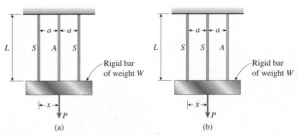

(a)          (b)

**PROB. 2.4-12**

**2.4-13** A horizontal rigid bar of weight $W = 7200$ lb is supported by three slender circular rods that are equally spaced (see figure). The two outer rods are made of aluminum ($E_1 = 10 \times 10^6$ psi) with diameter $d_1 = 0.4$ in. and length $L_1 = 40$ in. The inner rod is magnesium ($E_2 = 6.5 \times 10^6$ psi) with diameter $d_2$ and length $L_2$. The allowable stresses in the aluminum and magnesium are 24,000 psi and 13,000 psi, respectively.

If it is desired to have all three rods loaded to their maximum allowable values, what should be the diameter $d_2$ and length $L_2$ of the middle rod?

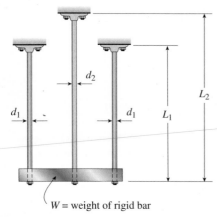

$W$ = weight of rigid bar

**PROB. 2.4-13**

**2.4-14** Three-bar truss $ABC$ (see figure) is constructed of steel pipes having a cross-sectional area $A = 3500$ mm$^2$ and a modulus of elasticity $E = 210$ GPa. Member $BC$ is of length $L = 2.5$ m, and the angle between members $AC$ and $AB$ is known to be 60°. Member $AC$ length is $b = 0.71L$. Loads $P = 185$ kN and $2P = 370$ kN act vertically and horizontally at joint $C$, as shown. Joints $A$ and $B$ are pinned supports. (Use the law of sines and law of cosines to find missing dimensions and angles in the figure.)

(a) Find the support reactions at joints $A$ and $B$. Use horizontal reaction $B_x$ as the redundant.

(b) What is the maximum permissible value of load variable $P$ if the allowable normal stress in each truss member is 150 MPa?

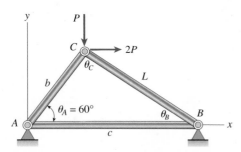

**PROB. 2.4-14**

**2.4-15** A rigid bar $AB$ of length $L = 66$ in. is hinged to a support at $A$ and supported by two vertical wires attached at points $C$ and $D$ (see figure). Both wires have the same cross-sectional area ($A = 0.0272$ in.$^2$) and are made of the same material (modulus $E = 30 \times 10^6$ psi). The wire at $C$ has length $h = 18$ in. and the wire at $D$ has length twice that amount. The horizontal distances are $c = 20$ in. and $d = 50$ in.

(a) Determine the tensile stresses $\sigma_C$ and $\sigma_D$ in the wires due to the load $P = 340$ lb acting at end $B$ of the bar.

(b) Find the downward displacement $\delta_B$ at end $B$ of the bar.

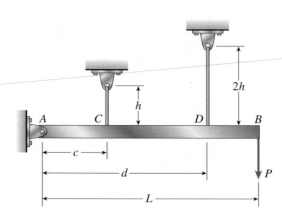

**PROB. 2.4-15**

**2.4-16** A rigid bar $ABCD$ is pinned at point $B$ and supported by springs at $A$ and $D$ (see figure). The springs at $A$ and $D$ have stiffnesses $k_1 = 10$ kN/m and $k_2 = 25$ kN/m, respectively, and the dimensions $a$, $b$, and $c$ are 250 mm, 500 mm, and 200 mm, respectively. A load $P$ acts at point $C$.

If the angle of rotation of the bar due to the action of the load $P$ is limited to 3°, what is the maximum permissible load $P_{max}$?

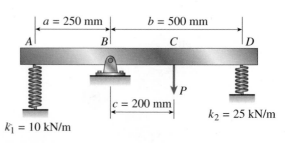

**PROB. 2.4-16**

**2.4-17** A trimetallic bar is uniformly compressed by an axial force $P = 9$ kips applied through a rigid end plate (see figure). The bar consists of a circular steel core surrounded by brass and copper tubes. The steel core has diameter 1.25 in., the brass tube has outer diameter 1.75 in., and the copper tube has outer diameter 2.25 in. The corresponding moduli of elasticity are $E_s = 30{,}000$ ksi, $E_b = 16{,}000$ ksi, and $E_c = 18{,}000$ ksi.

Calculate the compressive stresses $\sigma_s$, $\sigma_b$, and $\sigma_c$ in the steel, brass, and copper, respectively, due to the force $P$.

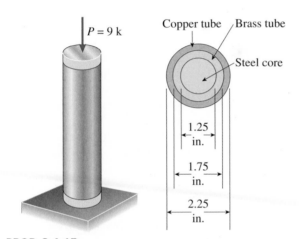

PROB. 2.4-17

# Thermal Effects

**2.5-1** The rails of a railroad track are welded together at their ends (to form continuous rails and thus eliminate the clacking sound of the wheels) when the temperature is 60°F.

What compressive stress $\sigma$ is produced in the rails when they are heated by the sun to 120°F if the coefficient of thermal expansion $\alpha = 6.5 \times 10^{-6}/°F$ and the modulus of elasticity $E = 30 \times 10^6$ psi?

**2.5-2** An aluminum pipe has a length of 60 m at a temperature of 10°C. An adjacent steel pipe at the same temperature is 5 mm longer than the aluminum pipe.

At what temperature (degrees Celsius) will the aluminum pipe be 15 mm longer than the steel pipe? (Assume that the coefficients of thermal expansion of aluminum and steel are $\alpha_a = 23 \times 10^{-6}/°C$ and $\alpha_s = 12 \times 10^{-6}/°C$, respectively.)

**2.5-3** A rigid bar of weight $W = 750$ lb hangs from three equally spaced wires, two of steel and one of aluminum (see figure). The diameter of the wires is 1/8 in. Before they were loaded, all three wires had the same length.

What temperature increase $\Delta T$ in all three wires will result in the entire load being carried by the steel wires? (Assume $E_s = 30 \times 10^6$ psi, $\alpha_s = 6.5 \times 10^{-6}/°F$, and $\alpha_a = 12 \times 10^{-6}/°F$.)

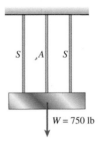

PROB. 2.5-3

**2.5-4** A steel rod of 15-mm diameter is held snugly (but without any initial stresses) between rigid walls by the arrangement shown in figure part a. (For the steel rod, use $\alpha = 12 \times 10^{-6}/°C$ and $E = 200$ GPa.)

(a) Calculate the temperature drop $\Delta T$ (degrees Celsius) at which the average shear stress in the 12-mm diameter bolt becomes 45 MPa. Also, what is the normal stress in the rod?

(b) What are the average bearing stresses in the bolt and clevis at $A$ and between the washer ($d_w = 20$ mm) and wall ($t = 18$ mm) at $B$?

(c) If the connection to the wall at $B$ is changed to an end plate with two bolts (see figure part b), what is the required diameter $d_b$ of each bolt if the temperature drop is $\Delta T = 38°C$ and the allowable bolt stress is 90 MPa?

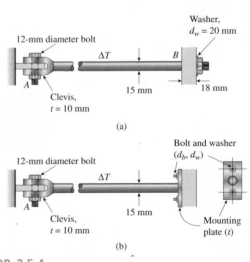

PROB. 2.5-4

**2.5-5** A bar *AB* of length *L* is held between rigid supports and heated nonuniformly in such a manner that the temperature increase $\Delta T$ at distance *x* from end *A* is given by the expression $\Delta T = \Delta T_B x^3/L^3$, where $\Delta T_B$ is the increase in temperature at end *B* of the bar (see figure part a).

(a) Derive a formula for the compressive stress $\sigma_c$ in the bar. (Assume that the material has modulus of elasticity *E* and coefficient of thermal expansion $\alpha$).

(b) Now modify the formula in part (a) if the rigid support at *A* is replaced by an elastic support at *A* having a spring constant *k* (see figure part b). Assume that only bar *AB* is subject to the temperature increase.

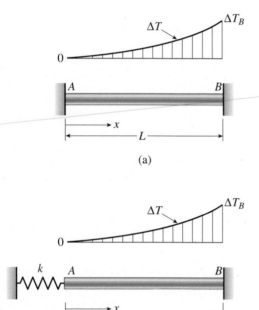

(a)

(b)

PROB. 2.5-5

**2.5-6** A plastic bar *ACB* having two different solid circular cross sections is held between rigid supports as shown in the figure. The diameters in the left- and right-hand parts are 50 mm and 75 mm, respectively. The corresponding lengths are 225 mm and 300 mm. Also, the modulus of elasticity *E* is 6.0 GPa, and the coefficient of thermal expansion $\alpha$ is $100 \times 10^{-6}/°C$. The bar is subjected to a uniform temperature increase of 30°C.

(a) Calculate the following quantities: (1) the compressive force *N* in the bar; (2) the maximum compressive stress $\sigma_c$; and (3) the displacement $\delta_C$ of point *C*.

(b) Repeat part (a) if the rigid support at *A* is replaced by an elastic support having spring constant $k = 50$ MN/m (see figure part b; assume that only the bar *ACB* is subject to the temperature increase).

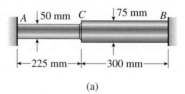

(a)

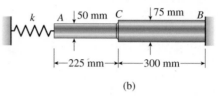

(b)

PROB. 2.5-6

**2.5-7** A circular steel rod *AB* (diameter $d_1 = 1.0$ in., length $L_1 = 3.0$ ft) has a bronze sleeve (outer diameter $d_2 = 1.25$ in., length $L_2 = 1.0$ ft) shrunk onto it so that the two parts are securely bonded (see figure).

Calculate the total elongation $\delta$ of the steel bar due to a temperature rise $\Delta T = 500°F$. (Material properties are as follows: for steel, $E_s = 30 \times 10^6$ psi and $\alpha_s = 6.5 \times 10^{-6}/°F$; for bronze, $E_b = 15 \times 10^6$ psi and $\alpha_b = 11 \times 10^{-6}/°F$.)

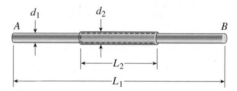

PROB. 2.5-7

**2.5-8** A brass sleeve *S* is fitted over a steel bolt *B* (see figure), and the nut is tightened until it is just snug. The bolt has a diameter $d_B = 25$ mm, and the sleeve has inside and outside diameters $d_1 = 26$ mm and $d_2 = 36$ mm, respectively.

Calculate the temperature rise $\Delta T$ that is required to produce a compressive stress of 25 MPa in the sleeve. (Use material properties as follows: for the sleeve, $\alpha_S = 21 \times 10^{-6}/°C$ and $E_S = 100$ GPa; for the bolt, $\alpha_B = 10 \times 10^{-6}/°C$ and $E_B = 200$ GPa.) (*Suggestion:* Use the results of Example 2-8.)

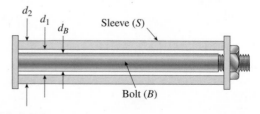

PROB. 2.5-8

**2.5-9** Rectangular bars of copper and aluminum are held by pins at their ends, as shown in the figure. Thin spacers provide a separation between the bars. The copper bars have cross-sectional dimensions 0.5 in. × 2.0 in., and the aluminum bar has dimensions 1.0 in. × 2.0 in.

Determine the shear stress in the 7/16-in. diameter pins if the temperature is raised by 100°F. (For copper, $E_c = 18,000$ ksi and $\alpha_c = 9.5 \times 10^{-6}/°F$; for aluminum, $E_a = 10,000$ ksi and $\alpha_a = 13 \times 10^{-6}/°F$.) *Suggestion:* Use the results of Example 2-8.

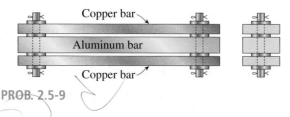

Copper bar

Aluminum bar

Copper bar

PROB. 2.5-9

**2.5-10** A rigid bar *ABCD* is pinned at end *A* and supported by two cables at points *B* and *C* (see figure). The cable at *B* has nominal diameter $d_B = 12$ mm and the cable at *C* has nominal diameter $d_C = 20$ mm. A load *P* acts at end *D* of the bar.

What is the allowable load *P* if the temperature rises by 60°C and each cable is required to have a factor of safety of at least 5 against its ultimate load?

(*Note:* The cables have effective modulus of elasticity $E = 140$ GPa and coefficient of thermal expansion $\alpha = 12 \times 10^{-6}/°C$. Other properties of the cables can be found in Table 2-1, Section 2.2.)

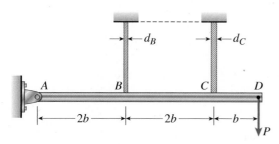

PROB. 2.5-10

**2.5-11** A rigid triangular frame is pivoted at *C* and held by two identical horizontal wires at points *A* and *B* (see figure). Each wire has axial rigidity $EA = 120$ k and coefficient of thermal expansion $\alpha = 12.5 \times 10^{-6}/°F$.

(a) If a vertical load $P = 500$ lb acts at point *D*, what are the tensile forces $T_A$ and $T_B$ in the wires at *A* and *B*, respectively?

(b) If, while the load *P* is acting, both wires have their temperatures raised by 180°F, what are the forces $T_A$ and $T_B$?

(c) What further increase in temperature will cause the wire at *B* to become slack?

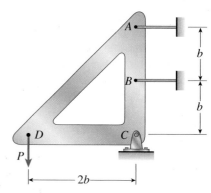

PROB. 2.5-11

# Misfits and Prestrains

**2.5-12** A steel wire *AB* is stretched between rigid supports (see figure). The initial prestress in the wire is 42 MPa when the temperature is 20°C.

(a) What is the stress $\sigma$ in the wire when the temperature drops to 0°C?

(b) At what temperature *T* will the stress in the wire become zero? (Assume $\alpha = 14 \times 10^{-6}/°C$ and $E = 200$ GPa.)

Steel wire

PROB. 2.5-12

**2.5-13** A copper bar *AB* of length 25 in. and diameter 2 in. is placed in position at room temperature with a gap of 0.008 in. between end *A* and a rigid restraint (see figure). The bar is supported at end *B* by an elastic spring with spring constant $k = 1.2 \times 10^6$ lb/in.

(a) Calculate the axial compressive stress $\sigma_c$ in the bar if the temperature *of the bar only* rises 50°F. (For copper, use $\alpha = 9.6 \times 10^{-6}/°F$ and $E = 16 \times 10^6$ psi.)

(b) What is the force in the spring? (Neglect gravity effects.)

(c) Repeat part (a) if $k \to \infty$.

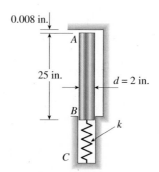

0.008 in.

25 in.

$d = 2$ in.

$k$

PROB. 2.5-13

**2.5-14** A bar $AB$ having length $L$ and axial rigidity $EA$ is fixed at end $A$ (see figure). At the other end a small gap of dimension $s$ exists between the end of the bar and a rigid surface. A load $P$ acts on the bar at point $C$, which is two-thirds of the length from the fixed end.

If the support reactions produced by the load $P$ are to be equal in magnitude, what should be the size $s$ of the gap?

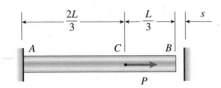

PROB. 2.5-14

**2.5-15** Pipe 2 has been inserted snugly into Pipe 1, but the holes for a connecting pin do not line up: there is a gap $s$. The user decides to apply *either* force $P_1$ to Pipe 1 *or* force $P_2$ to Pipe 2, whichever is smaller. Determine the following using the numerical properties in the box.

(a) If only $P_1$ is applied, find $P_1$ (kips) required to close gap $s$; if a pin is then inserted and $P_1$ removed, what are reaction forces $R_A$ and $R_B$ for this load case?

(b) If only $P_2$ is applied, find $P_2$ (kips) required to close gap $s$; if a pin is inserted and $P_2$ removed, what are reaction forces $R_A$ and $R_B$ for this load case?

(c) What is the maximum *shear* stress in the pipes, for the loads in parts (a) and (b)?

(d) If a temperature increase $\Delta T$ is to be applied to the entire structure to close gap $s$ (*instead of applying forces $P_1$ and $P_2$*), find the $\Delta T$ required to close the gap. If a pin is inserted after the gap has closed, what are reaction forces $R_A$ and $R_B$ for this case?

(e) Finally, if the structure (with pin inserted) then cools to the *original* ambient temperature, what are reaction forces $R_A$ and $R_B$?

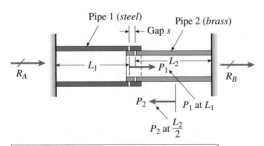

Numerical properties
$E_1 = 30{,}000$ ksi, $E_2 = 14{,}000$ ksi
$\alpha_1 = 6.5 \times 10^{-6}/°F$, $\alpha_2 = 11 \times 10^{-6}/°F$
Gap $s = 0.05$ in.
$L_1 = 56$ in., $d_1 = 6$ in., $t_1 = 0.5$ in., $A_1 = 8.64$ in.$^2$
$L_2 = 36$ in., $d_2 = 5$ in., $t_2 = 0.25$ in., $A_2 = 3.73$ in.$^2$

PROB. 2.5-15

**2.5-16** A nonprismatic bar $ABC$ made up of segments $AB$ (length $L_1$, cross-sectional area $A_1$) and $BC$ (length $L_2$, cross-sectional area $A_2$) is fixed at end $A$ and free at end $C$ (see figure). The modulus of elasticity of the bar is $E$. A small gap of dimension $s$ exists between the end of the bar and an elastic spring of length $L_3$ and spring constant $k_3$. If bar $ABC$ only (*not the spring*) is subjected to temperature increase $\Delta T$, determine the following.

(a) Write an expression for reaction forces $R_A$ and $R_D$ if the elongation of $ABC$ exceeds gap length $s$.

(b) Find expressions for the displacements of points $B$ and $C$ if the elongation of $ABC$ exceeds gap length $s$.

PROB. 2.5-16

**2.5-17** Wires $B$ and $C$ are attached to a support at the left-hand end and to a pin-supported rigid bar at the right-hand end (see figure). Each wire has cross-sectional area $A = 0.03$ in.$^2$ and modulus of elasticity $E = 30 \times 10^6$ psi. When the bar is in a vertical position, the length of each wire is $L = 80$ in. However, before being attached to the bar, the length of wire $B$ was 79.98 in. and wire $C$ was 79.95 in.

Find the tensile forces $T_B$ and $T_C$ in the wires under the action of a force $P = 700$ lb acting at the upper end of the bar.

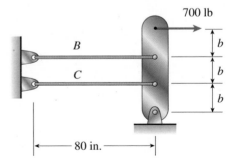

PROB. 2.5-17

**2.5-18** A rigid steel plate is supported by three posts of high-strength concrete each having an effective cross-sectional area $A = 40{,}000$ mm$^2$ and length $L = 2$ m (see figure). Before the load $P$ is applied, the middle post is shorter than the others by an amount $s = 1.0$ mm.

Determine the maximum allowable load $P_{allow}$ if the allowable compressive stress in the concrete is $\sigma_{allow} = 20$ MPa. (Use $E = 30$ GPa for concrete.)

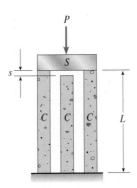

**PROB. 2.5-18**

**2.5-19** A capped cast-iron pipe is compressed by a brass rod, as shown. The nut is turned until it is just snug, then add an additional quarter turn to pre-compress the CI pipe. The pitch of the threads of the bolt is $p = 52$ mils (a mil is one-thousandth of an inch). Use the numerical properties provided.

(a) What stresses $\sigma_p$ and $\sigma_r$ will be produced in the cast-iron pipe and brass rod, respectively, by the additional quarter turn of the nut?

(b) Find the bearing stress $\sigma_b$ beneath the washer and the shear stress $\tau_c$ in the steel cap.

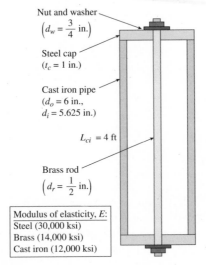

**PROB. 2.5-19**

**2.5-20** A plastic cylinder is held snugly between a rigid plate and a foundation by two steel bolts (see figure).

Determine the compressive stress $\sigma_p$ in the plastic when the nuts on the steel bolts are tightened by one complete turn.

Data for the assembly are as follows: length $L = 200$ mm, pitch of the bolt threads $p = 1.0$ mm, modulus of elasticity for steel $E_s = 200$ GPa, modulus of

elasticity for the plastic $E_p = 7.5$ GPa, cross-sectional area of one bolt $A_s = 36.0$ mm², and cross-sectional area of the plastic cylinder $A_p = 960$ mm².

**PROBS. 2.5-20 and 2.5-21**

**2.5-21** Solve the preceding problem if the data for the assembly are as follows: length $L = 10$ in., pitch of the bolt threads $p = 0.058$ in., modulus of elasticity for steel $E_s = 30 \times 10^6$ psi, modulus of elasticity for the plastic $E_p = 500$ ksi, cross-sectional area of one bolt $A_s = 0.06$ in.², and cross-sectional area of the plastic cylinder $A_p = 1.5$ in.²

**2.5-22** Consider the sleeve made from two copper tubes joined by tin-lead solder over distance $s$. The sleeve has brass caps at both ends, which are held in place by a steel bolt and washer with the nut turned just snug at the outset. Then, two "loadings" are applied: $n = 1/2$ turn applied to the nut; at the same time the internal temperature is raised by $\Delta T = 30°C$.

(a) Find the forces in the sleeve and bolt, $P_s$ and $P_B$, due to both the prestress in the bolt and the temperature increase. For copper, use $E_c = 120$ GPa and $\alpha_c = 17 \times 10^{-6}/°C$; for steel, use $E_s = 200$ GPa and $\alpha_s = 12 \times 10^{-6}/°C$. The pitch of the bolt threads is $p = 1.0$ mm. Assume $s = 26$ mm and bolt diameter $d_b = 5$ mm.

(b) Find the required length of the solder joint, $s$, if shear stress in the sweated joint cannot exceed the allowable shear stress $\tau_{aj} = 18.5$ MPa.

(c) What is the final elongation of the entire assemblage due to both temperature change $\Delta T$ and the initial prestress in the bolt?

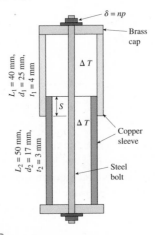

**PROB. 2.5-22**

**2.5-23** A polyethylene tube (length $L$) has a cap which when installed compresses a spring (with undeformed length $L_1 > L$) by amount $\delta = (L_1 - L)$. Ignore deformations of the cap and base. Use the force at the base of the spring as the redundant. Use numerical properties in the boxes given.

(a) What is the resulting force in the spring, $F_k$?
(b) What is the resulting force in the tube, $F_t$?
(c) What is the final length of the tube, $L_f$?
(d) What temperature change $\Delta T$ inside the tube will result in zero force in the spring?

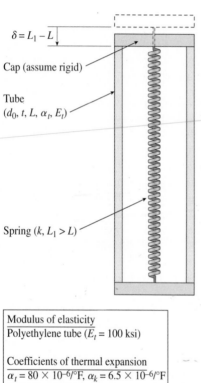

| Modulus of elasticity |
|---|
| Polyethylene tube ($E_t = 100$ ksi) |

| Coefficients of thermal expansion |
|---|
| $\alpha_t = 80 \times 10^{-6}/°F$, $\alpha_k = 6.5 \times 10^{-6}/°F$ |

| Properties and dimensions |
|---|
| $d_0 = 6$ in.  $t = \dfrac{1}{8}$ in. |
| $L_1 = 12.125$ in. $> L = 12$ in.  $k = 1.5\ \dfrac{\text{kip}}{\text{in.}}$ |

**PROB. 2.5-23**

**2.5-24** Prestressed concrete beams are sometimes manufactured in the following manner. High-strength steel wires are stretched by a jacking mechanism that applies a force $Q$, as represented schematically in part a of the figure.

Concrete is then poured around the wires to form a beam, as shown in figure part b.

After the concrete sets properly, the jacks are released and the force $Q$ is removed (see part c of the figure). Thus, the beam is left in a prestressed condition, with the wires in tension and the concrete in compression.

Let us assume that the prestressing force $Q$ produces in the steel wires an initial stress $\sigma_0 = 620$ MPa. If the moduli of elasticity of the steel and concrete are in the ratio 12:1 and the cross-sectional areas are in the ratio 1:50, what are the final stresses $\sigma_s$ and $\sigma_c$ in the two materials?

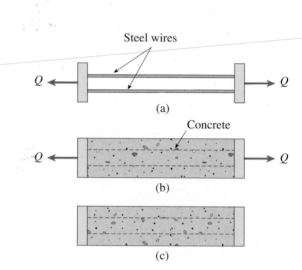

**PROB. 2.5-24**

**2.5-25** A polyethylene tube (length $L$) has a cap which is held in place by a spring (with undeformed length $L_1 < L$). After installing the cap, the spring is post-tensioned by turning an adjustment screw by amount $\delta$. Ignore deformations of the cap and base. Use the force at the base of the spring as the redundant. Use numerical properties in the boxes below.

(a) What is the resulting force in the spring, $F_k$?
(b) What is the resulting force in the tube, $F_t$?
(c) What is the final length of the tube, $L_f$?
(d) What temperature change $\Delta T$ inside the tube will result in zero force in the spring?

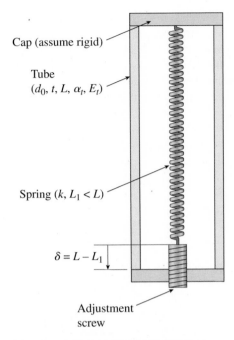

Cap (assume rigid)

Tube
$(d_0, t, L, \alpha_t, E_t)$

Spring $(k, L_1 < L)$

$\delta = L - L_1$

Adjustment
screw

Modulus of elasticity
Polyethylene tube ($E_t = 100$ ksi)

Coefficients of thermal expansion
$\alpha_t = 80 \times 10^{-6}/°\text{F}, \ \alpha_k = 6.5 \times 10^{-6}/°\text{F}$

Properties and dimensions
$d_0 = 6$ in.  $t = \dfrac{1}{8}$ in.

$L = 12$ in.  $L_1 = 11.875$ in.  $k = 1.5 \dfrac{\text{kip}}{\text{in.}}$

PROB. 2.5-25

## Stresses on Inclined Sections

**2.6-1** A steel bar of rectangular cross section (1.5 in. × 2.0 in.) carries a tensile load $P$ (see figure). The allowable stresses in tension and shear are 14,500 psi and 7,100 psi, respectively. Determine the maximum permissible load $P_{max}$.

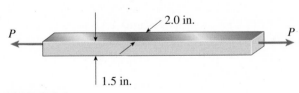

2.0 in.

$P$          $P$

1.5 in.

PROB. 2.6-1

**2.6-2** A circular steel rod of diameter $d$ is subjected to a tensile force $P = 3.5$ kN (see figure). The allowable stresses in tension and shear are 118 MPa and 48 MPa, respectively. What is the minimum permissible diameter $d_{min}$ of the rod?

$P$          $d$          $P = 3.5$ kN

PROB. 2.6-2

**2.6-3** A standard brick (dimensions 8 in. × 4 in. × 2.5 in.) is compressed lengthwise by a force $P$, as shown in the figure. If the ultimate shear stress for brick is 1200 psi and the ultimate compressive stress is 3600 psi, what force $P_{max}$ is required to break the brick?

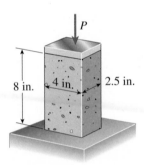

$P$

8 in.    4 in.    2.5 in.

PROB. 2.6-3

**2.6-4** A brass wire of diameter $d = 2.42$ mm is stretched tightly between rigid supports so that the tensile force is $T = 98$ N (see figure). The coefficient of thermal expansion for the wire is $19.5 \times 10^{-6}/°\text{C}$ and the modulus of elasticity is $E = 110$ GPa.

(a) What is the maximum permissible temperature drop $\Delta T$ if the allowable shear stress in the wire is 60 MPa?

(b) At what temperature change does the wire go slack?

$T$          $d$          $T$

PROBS. 2.6-4 and 2.6-5

**2.6-5** A brass wire of diameter $d = 1/16$ in. is stretched between rigid supports with an initial tension $T$ of 37 lb (see figure). Assume that the coefficient of thermal expansion is $10.6 \times 10^{-6}/°F$ and the modulus of elasticity is $15 \times 10^6$ psi.)

(a) If the temperature is lowered by 60°F, what is the maximum shear stress $\tau_{max}$ in the wire?

(b) If the allowable shear stress is 10,000 psi, what is the maximum permissible temperature drop?

(c) At what temperature change $\Delta T$ does the wire go slack?

**2.6-6** A steel bar with diameter $d = 12$ mm is subjected to a tensile load $P = 9.5$ kN (see figure).

(a) What is the maximum normal stress $\sigma_{max}$ in the bar?

(b) What is the maximum shear stress $\tau_{max}$?

(c) Draw a stress element oriented at 45° to the axis of the bar and show all stresses acting on the faces of this element.

(d) Repeat part (c) for a stress element oriented at 22.5° to the axis of the bar.

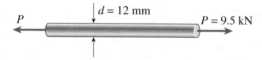

PROB. 2.6-6

**2.6-7** During a tension test of a mild-steel specimen (see figure), the extensometer shows an elongation of 0.00120 in. with a gage length of 2 in. Assume that the steel is stressed below the proportional limit and that the modulus of elasticity $E = 30 \times 10^6$ psi.

(a) What is the maximum normal stress $\sigma_{max}$ in the specimen?

(b) What is the maximum shear stress $\tau_{max}$?

(c) Draw a stress element oriented at an angle of 45° to the axis of the bar and show all stresses acting on the faces of this element.

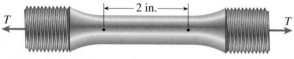

PROB. 2.6-7

**2.6-8** A copper bar with a rectangular cross section is held without stress between rigid supports (see figure). Subsequently, the temperature of the bar is raised 50°C.

(a) Determine the stresses on all faces of the elements $A$ and $B$, and show these stresses on sketches of the elements. (Assume $\alpha = 17.5 \times 10^{-6}/°C$ and $E = 120$ GPa.)

(b) If the shear stress at $B$ is known to be 48 MPa at some inclination $\theta$, find angle $\theta$ and show the stresses on a sketch of a properly oriented element.

PROB. 2.6-8

**2.6-9** The plane truss below is assembled from steel C10 × 20 shapes (see Table 3(a) in Appendix F). Assume that $L = 10$ ft and $b = 0.71L$.

(a) If load variable $P = 49$ kips, what is the maximum shear stress $\tau_{max}$ in each truss member?

(b) What is the maximum permissible value of load variable $P$ if the allowable normal stress is 14 ksi and the allowable shear stress is 7.5 ksi?

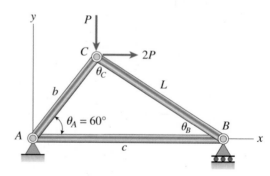

PROB. 2.6-9

**2.6-10** A plastic bar of diameter $d = 32$ mm is compressed in a testing device by a force $P = 190$ N applied as shown in the figure.

(a) Determine the normal and shear stresses acting on all faces of stress elements oriented at (1) an angle $\theta = 0°$, (2) an angle $\theta = 22.5°$, and (3) an angle $\theta = 45°$. In each case, show the stresses on a sketch of a properly oriented element. What are $\sigma_{max}$ and $\tau_{max}$?

(b) Find $\sigma_{max}$ and $\tau_{max}$ in the plastic bar if a re-centering spring of stiffness $k$ is inserted into the testing device, as shown in the figure. The spring stiffness is 1/6 of the axial stiffness of the plastic bar.

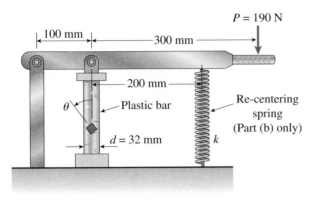

*P* = 190 N

100 mm | 300 mm

200 mm

θ    Plastic bar

*d* = 32 mm    *k*

Re-centering spring
(Part (b) only)

PROB. 2.6-10

**2.6-11** A plastic bar of rectangular cross section ($b = 1.5$ in. and $h = 3$ in.) fits snugly between rigid supports at room temperature (68°F) but with no initial stress (see figure). When the temperature of the bar is raised to 160°F, the compressive stress on an inclined plane *pq* at midspan becomes 1700 psi.

(a) What is the shear stress on plane *pq*? (Assume $\alpha = 60 \times 10^{-6}/°F$ and $E = 450 \times 10^3$ psi.)

(b) Draw a stress element oriented to plane *pq* and show the stresses acting on all faces of this element.

(c) If the allowable normal stress is 3400 psi and the allowable shear stress is 1650 psi, what is the maximum load *P* (*in +x direction*) which can be added at the quarter point (in addition to thermal effects given) without exceeding allowable stress values in the bar?

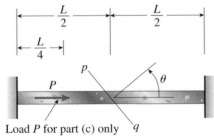

$\frac{L}{2}$    $\frac{L}{2}$

$\frac{L}{4}$

*p*    *P*    θ

Load *P* for part (c) only    *q*

PROBS. 2.6-11

**2.6-12** A copper bar of rectangular cross section ($b = 18$ mm and $h = 40$ mm) is held snugly (but without any initial stress) between rigid supports (see figure). The allowable stresses on the inclined plane *pq* at midspan, for which $\theta = 55°$, are specified as 60 MPa in compression and 30 MPa in shear.

(a) What is the maximum permissible temperature rise $\Delta T$ if the allowable stresses on plane *pq* are not to be exceeded? (Assume $\alpha = 17 \times 10^{-6}/°C$ and $E = 120$ GPa.)

(b) If the temperature increases by the maximum permissible amount, what are the stresses on plane *pq*?

(c) If the temperature rise $\Delta T = 28°C$, how far to the right of end *A* (distance $\beta L$, expressed as a fraction of length *L*) can load $P = 15$ kN be applied without exceeding allowable stress values in the bar? Assume that $\sigma_a = 75$ MPa and $\tau_a = 35$ MPa.

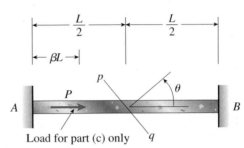

$\frac{L}{2}$    $\frac{L}{2}$

$\beta L$

*p*    θ

*A*    *P*    *B*

Load for part (c) only    *q*

PROBS. 2.6-12

**2.6-13** A circular brass bar of diameter *d* is member *AC* in truss *ABC* which has load $P = 5000$ lb applied at joint *C*. Bar *AC* is composed of two segments brazed together on a plane *pq* making an angle $\alpha = 36°$ with the axis of the bar (see figure). The allowable stresses in the brass are 13,500 psi in tension and 6500 psi in shear. On the brazed joint, the allowable stresses are 6000 psi in tension and 3000 psi in shear. What is the tensile force $N_{AC}$ in bar *AC*? What is the minimum required diameter $d_{min}$ of bar *AC*?

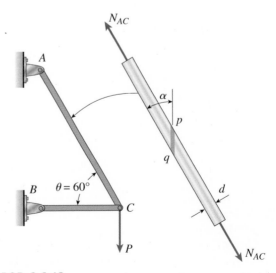

$N_{AC}$

*A*

α

*p*

*q*

*B*    $\theta = 60°$    *d*

*C*

*P*    $N_{AC}$

PROB. 2.6-13

**2.6-14** Two boards are joined by gluing along a scarf joint, as shown in the figure. For purposes of cutting and gluing, the angle $\alpha$ between the plane of the joint and the faces of the boards must be between 10° and 40°. Under a tensile load $P$, the normal stress in the boards is 4.9 MPa.

(a) What are the normal and shear stresses acting on the glued joint if $\alpha = 20°$?

(b) If the allowable shear stress on the joint is 2.25 MPa, what is the largest permissible value of the angle $\alpha$?

(c) For what angle a will the shear stress on the glued joint be numerically equal to twice the normal stress on the joint?

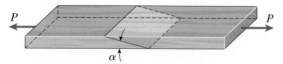

PROB. 2.6-14

**2.6-15** Acting on the sides of a stress element cut from a bar in uniaxial stress are tensile stresses of 10,000 psi and 5000 psi, as shown in the figure.

(a) Determine the angle $\theta$ and the shear stress $\tau_\theta$ and show all stresses on a sketch of the element.

(b) Determine the maximum normal stress $\sigma_{max}$ and the maximum shear stress $\tau_{max}$ in the material.

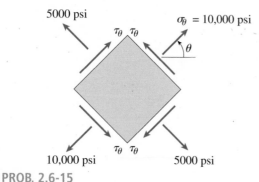

PROB. 2.6-15

**2.6-16** A prismatic bar is subjected to an axial force that produces a tensile stress $\sigma_\theta = 65$ MPa and a shear stress $\tau_\theta = 23$ MPa on a certain inclined plane (see figure). Determine the stresses acting on all faces of a stress element oriented at $\theta = 30°$ and show the stresses on a sketch of the element.

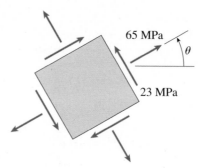

PROB. 2.6-16

**2.6-17** The normal stress on plane $pq$ of a prismatic bar in tension (see figure) is found to be 7500 psi. On plane $rs$, which makes an angle $\beta = 30°$ with plane $pq$, the stress is found to be 2500 psi.

Determine the maximum normal stress $\sigma_{max}$ and maximum shear stress $\tau_{max}$ in the bar.

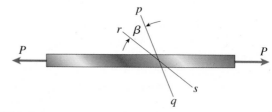

PROB. 2.6-17

**2.6-18** A tension member is to be constructed of two pieces of plastic glued along plane $pq$ (see figure). For purposes of cutting and gluing, the angle $\theta$ must be between 25° and 45°. The allowable stresses on the glued joint in tension and shear are 5.0 MPa and 3.0 MPa, respectively.

(a) Determine the angle $\theta$ so that the bar will carry the largest load $P$. (Assume that the strength of the glued joint controls the design.)

(b) Determine the maximum allowable load $P_{max}$ if the cross-sectional area of the bar is 225 mm².

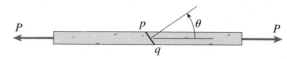

PROB. 2.6-18

**2.6-19** Plastic bar $AB$ of rectangular cross section ($b = 0.75$ in. and $h = 1.5$ in.) and length $L = 2$ ft is fixed at $A$ and has a spring support ($k = 18$ k/in.) at $C$ (see figure). Initially, the bar and spring have no stress. When the temperature of the bar is *raised* by 100°F, the *compressive* stress on an inclined plane $pq$ at $L_\theta = 1.5$ ft becomes 950 psi. Assume the spring is massless and is unaffected by the temperature change. Let $\alpha = 55 \times 10^{-6}/°F$ and $E = 400$ ksi.

(a) What is the shear stress $\tau_\theta$ on plane $pq$? What is angle $\theta$?

(b) Draw a stress element oriented to plane $pq$, and show the stresses acting on all faces of this element.

(c) If the allowable normal stress is $\pm 1000$ psi and the allowable shear stress is $\pm 560$ psi, what is the maximum permissible value of spring constant $k$ if allowable stress values in the bar are not to be exceeded?

(d) What is the maximum permissible length $L$ of the bar if allowable stress values in the bar are not to be exceeded? (Assume $k = 18$ k/in.)

(e) What is the maximum permissible temperature *increase* ($\Delta T$) in the bar if allowable stress values in the bar are not to be exceeded? (Assume $L = 2$ ft and $k = 18$ k/in.)

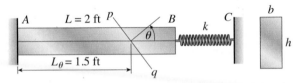

PROB. 2.6-19

## Strain Energy

*When solving the problems for Section 2.7, assume that the material behaves linearly elastically.*

**2.7-1** A prismatic bar $AD$ of length $L$, cross-sectional area $A$, and modulus of elasticity $E$ is subjected to loads $5P$, $3P$, and $P$ acting at points $B$, $C$, and $D$, respectively (see figure). Segments $AB$, $BC$, and $CD$ have lengths $L/6$, $L/2$, and $L/3$, respectively.

(a) Obtain a formula for the strain energy $U$ of the bar.

(b) Calculate the strain energy if $P = 6$ k, $L = 52$ in., $A = 2.76$ in.$^2$, and the material is aluminum with $E = 10.4 \times 10^6$ psi.

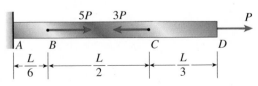

PROB. 2.7-1

**2.7-2** A bar of circular cross section having two different diameters $d$ and $2d$ is shown in the figure. The length of each segment of the bar is $L/2$ and the modulus of elasticity of the material is $E$.

(a) Obtain a formula for the strain energy $U$ of the bar due to the load $P$.

(b) Calculate the strain energy if the load $P = 27$ kN, the length $L = 600$ mm, the diameter $d = 40$ mm, and the material is brass with $E = 105$ GPa.

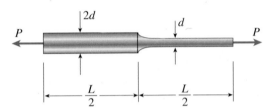

PROB. 2.7-2

**2.7-3** A three-story steel column in a building supports roof and floor loads as shown in the figure. The story height $H$ is 10.5 ft, the cross-sectional area $A$ of the column is 15.5 in.$^2$, and the modulus of elasticity $E$ of the steel is $30 \times 10^6$ psi.

Calculate the strain energy $U$ of the column assuming $P_1 = 40$ k and $P_2 = P_3 = 60$ k.

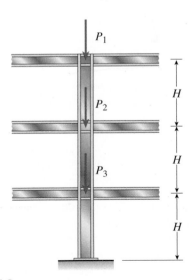

PROB. 2.7-3

**2.7-4** The bar $ABC$ shown in the figure is loaded by a force $P$ acting at end $C$ and by a force $Q$ acting at the midpoint $B$. The bar has constant axial rigidity $EA$.

(a) Determine the strain energy $U_1$ of the bar when the force $P$ acts alone ($Q = 0$).

(b) Determine the strain energy $U_2$ when the force $Q$ acts alone ($P = 0$).

(c) Determine the strain energy $U_3$ when the forces $P$ and $Q$ act simultaneously upon the bar.

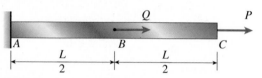

PROB. 2.7-4

**2.7-5** Determine the strain energy per unit volume (units of psi) and the strain energy per unit weight (units of in.) that can be stored in each of the materials listed in the accompanying table, assuming that the material is stressed to the proportional limit.

| DATA FOR PROBLEM 2.7-5 | | | |
|---|---|---|---|
| Material | Weight Density (lb/in.³) | Modules of elasticity (ksi) | Proportional limit (psi) |
| Mild steel | 0.284 | 30,000 | 36,000 |
| Tool steel | 0.284 | 30,000 | 75,000 |
| Aluminum | 0.0984 | 10,500 | 60,000 |
| Rubber (soft) | 0.0405 | 0.300 | 300 |

**2.7-6** The truss $ABC$ shown in the figure is subjected to a horizontal load $P$ at joint $B$. The two bars are identical with cross-sectional area $A$ and modulus of elasticity $E$.

(a) Determine the strain energy $U$ of the truss if the angle $\beta = 60°$.

(b) Determine the horizontal displacement $\delta_B$ of joint $B$ by equating the strain energy of the truss to the work done by the load.

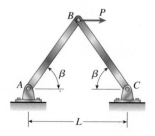

PROB. 2.7-6

**2.7-7** The truss $ABC$ shown in the figure supports a horizontal load $P_1 = 300$ lb and a vertical load $P_2 = 900$ lb. Both bars have cross-sectional area $A = 2.4$ in.² and are made of steel with $E = 30 \times 10^6$ psi.

(a) Determine the strain energy $U_1$ of the truss when the load $P_1$ acts alone ($P_2 = 0$).

(b) Determine the strain energy $U_2$ when the load $P_2$ acts alone ($P_1 = 0$).

(c) Determine the strain energy $U_3$ when both loads act simultaneously.

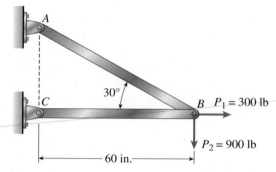

PROB. 2.7-7

**2.7-8** The statically indeterminate structure shown in the figure consists of a horizontal rigid bar $AB$ supported by five equally spaced springs. Springs 1, 2, and 3 have stiffnesses $3k$, $1.5k$, and $k$, respectively. When unstressed, the lower ends of all five springs lie along a horizontal line. Bar $AB$, which has weight $W$, causes the springs to elongate by an amount $\delta$.

(a) Obtain a formula for the total strain energy $U$ of the springs in terms of the downward displacement $\delta$ of the bar.

(b) Obtain a formula for the displacement $\delta$ by equating the strain energy of the springs to the work done by the weight $W$.

(c) Determine the forces $F_1$, $F_2$, and $F_3$ in the springs.

(d) Evaluate the strain energy $U$, the displacement $\delta$, and the forces in the springs if $W = 600$ N and $k = 7.5$ N/mm.

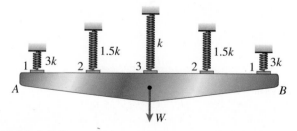

PROB. 2.7-8

**2.7-9** A slightly tapered bar $AB$ of rectangular cross section and length $L$ is acted upon by a force $P$ (see figure). The width of the bar varies uniformly from $b_2$ at end $A$ to $b_1$ at end $B$. The thickness $t$ is constant.

(a) Determine the strain energy $U$ of the bar.

(b) Determine the elongation $\delta$ of the bar by equating the strain energy to the work done by the force $P$.

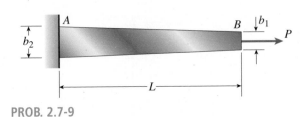

PROB. 2.7-9

**2.7-10** A compressive load $P$ is transmitted through a rigid plate to three magnesium-alloy bars that are identical except that initially the middle bar is slightly shorter than the other bars (see figure). The dimensions and properties of the assembly are as follows: length $L = 1.0$ m, cross-sectional area of each bar $A = 3000$ mm$^2$, modulus of elasticity $E = 45$ GPa, and the gap $s = 1.0$ mm.

(a) Calculate the load $P_1$ required to close the gap.

(b) Calculate the downward displacement $\delta$ of the rigid plate when $P = 400$ kN.

(c) Calculate the total strain energy $U$ of the three bars when $P = 400$ kN.

(d) Explain why the strain energy $U$ is *not* equal to $P\delta/2$. (*Hint:* Draw a load-displacement diagram.)

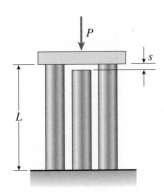

PROB. 2.7-10

**2.7-11** A block $B$ is pushed against three springs by a force $P$ (see figure). The middle spring has stiffness $k_1$ and the outer springs each have stiffness $k_2$. Initially, the springs are unstressed and the middle spring is longer than the outer springs (the difference in length is denoted $s$).

(a) Draw a force-displacement diagram with the force $P$ as ordinate and the displacement $x$ of the block as abscissa.

(b) From the diagram, determine the strain energy $U_1$ of the springs when $x = 2s$.

(c) Explain why the strain energy $U_1$ is not equal to $P\delta/2$, where $\delta = 2s$.

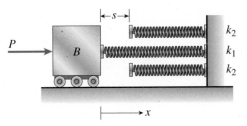

PROB. 2.7-11

**2.7-12** A bungee cord that behaves linearly elastically has an unstressed length $L_0 = 760$ mm and a stiffness $k = 140$ N/m. The cord is attached to two pegs, distance $b = 380$ mm apart, and pulled at its midpoint by a force $P = 80$ N (see figure).

(a) How much strain energy $U$ is stored in the cord?

(b) What is the displacement $\delta_C$ of the point where the load is applied?

(c) Compare the strain energy $U$ with the quantity $P\delta_C/2$.

(*Note:* The elongation of the cord is *not* small compared to its original length.)

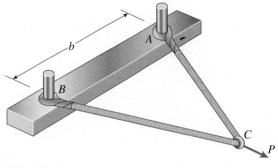

PROB. 2.7-12

## Impact Loading

*The problems for Section 2.8 are to be solved on the basis of the assumptions and idealizations described in the text. In particular, assume that the material behaves linearly elastically and no energy is lost during the impact.*

**2.8-1** A sliding collar of weight $W = 150$ lb falls from a height $h = 2.0$ in. onto a flange at the bottom of a slender vertical rod (see figure). The rod has length $L = 4.0$ ft, cross-sectional area $A = 0.75$ in.², and modulus of elasticity $E = 30 \times 10^6$ psi.

Calculate the following quantities: (a) the maximum downward displacement of the flange, (b) the maximum tensile stress in the rod, and (c) the impact factor.

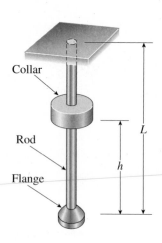

PROB. 2.8-1

**2.8-2** Solve the preceding problem if the collar has mass $M = 80$ kg, the height $h = 0.5$ m, the length $L = 3.0$ m, the cross-sectional area $A = 350$ mm², and the modulus of elasticity $E = 170$ GPa.

**2.8-3** Solve Prob. 2.8-1 if the collar has weight $W = 50$ lb, the height $h = 2.0$ in., the length $L = 3.0$ ft, the cross-sectional area $A = 0.25$ in.², and the modulus of elasticity $E = 30,000$ ksi.

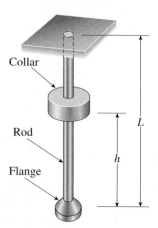

PROBS. 2.8-2 and 2.8-3

**2.8-4** A block weighing $W = 5.0$ N drops inside a cylinder from a height $h = 200$ mm onto a spring having stiffness $k = 90$ N/m (see figure). (a) Determine the maximum shortening of the spring due to the impact, and (b) determine the impact factor.

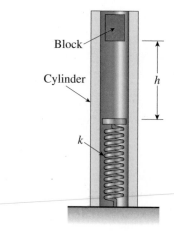

PROBS. 2.8-4 and 2.8-5

**2.8-5** Solve the preceding problem if the block weighs $W = 1.0$ lb, $h = 12$ in., and $k = 0.5$ lb/in.

**2.8-6** A small rubber ball (weight $W = 450$ mN) is attached by a rubber cord to a wood paddle (see figure). The natural length of the cord is $L_0 = 200$ mm, its cross-sectional area is $A = 1.6$ mm², and its modulus of elasticity is $E = 2.0$ MPa. After being struck by the paddle, the ball stretches the cord to a total length $L_1 = 900$ mm.

What was the velocity $v$ of the ball when it left the paddle? (Assume linearly elastic behavior of the rubber cord, and disregard the potential energy due to any change in elevation of the ball.)

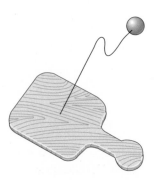

PROB. 2.8-6

**2.8-7** A weight $W = 4500$ lb falls from a height $h$ onto a vertical wood pole having length $L = 15$ ft, diameter $d = 12$ in., and modulus of elasticity $E = 1.6 \times 10^6$ psi (see figure).

If the allowable stress in the wood under an impact load is 2500 psi, what is the maximum permissible height $h$?

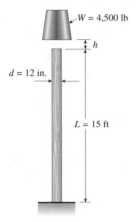

$W = 4,500$ lb
$h$
$d = 12$ in.
$L = 15$ ft

PROB. 2.8-7

**2.8-8** A cable with a restrainer at the bottom hangs vertically from its upper end (see figure). The cable has an effective cross-sectional area $A = 40$ mm² and an effective modulus of elasticity $E = 130$ GPa. A slider of mass $M = 35$ kg drops from a height $h = 1.0$ m onto the restrainer.

If the allowable stress in the cable under an impact load is 500 MPa, what is the minimum permissible length $L$ of the cable?

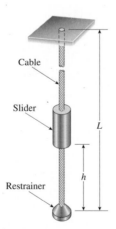

Cable
Slider
$L$
$h$
Restrainer

PROBS. 2.8-8 and 2.8-9

**2.8-9** Solve the preceding problem if the slider has weight $W = 100$ lb, $h = 45$ in., $A = 0.080$ in.², $E = 21 \times 10^6$ psi, and the allowable stress is 70 ksi.

**2.8-10** A bumping post at the end of a track in a railway yard has a spring constant $k = 8.0$ MN/m (see figure). The maximum possible displacement $d$ of the end of the striking plate is 450 mm.

What is the maximum velocity $v_{max}$ that a railway car of weight $W = 545$ kN can have without damaging the bumping post when it strikes it?

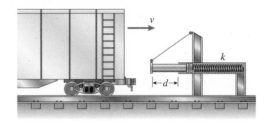

$v$
$k$
$d$

PROB. 2.8-10

**2.8-11** A bumper for a mine car is constructed with a spring of stiffness $k = 1120$ lb/in. (see figure). If a car weighing 3450 lb is traveling at velocity $v = 7$ mph when it strikes the spring, what is the maximum shortening of the spring?

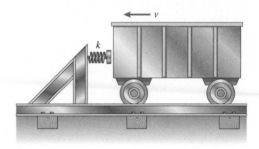

$v$
$k$

PROB. 2.8-11

**2.8-12** A bungee jumper having a mass of 55 kg leaps from a bridge, braking her fall with a long elastic shock cord having axial rigidity $EA = 2.3$ kN (see figure).

If the jumpoff point is 60 m above the water, and if it is desired to maintain a clearance of 10 m between the jumper and the water, what length $L$ of cord should be used?

PROB. 2.8-12

**2.8-13** A weight $W$ rests on top of a wall and is attached to one end of a very flexible cord having cross-sectional area $A$ and modulus of elasticity $E$ (see figure). The other end of the cord is attached securely to the wall. The weight is then pushed off the wall and falls freely the full length of the cord.

(a) Derive a formula for the impact factor.

(b) Evaluate the impact factor if the weight, when hanging statically, elongates the band by 2.5% of its original length.

PROB. 2.8-13

**2.8-14** A rigid bar $AB$ having mass $M = 1.0$ kg and length $L = 0.5$ m is hinged at end $A$ and supported at end $B$ by a nylon cord $BC$ (see figure). The cord has cross-sectional area $A = 30$ mm², length $b = 0.25$ m, and modulus of elasticity $E = 2.1$ GPa.

If the bar is raised to its maximum height and then released, what is the maximum stress in the cord?

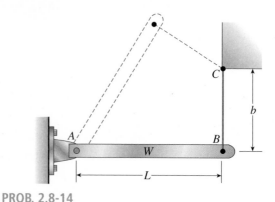

PROB. 2.8-14

# Stress Concentrations

*The problems for Section 2.10 are to be solved by considering the stress-concentration factors and assuming linearly elastic behavior.*

**2.10-1** The flat bars shown in parts a and b of the figure are subjected to tensile forces $P = 3.0$ k. Each bar has thickness $t = 0.25$ in.

(a) For the bar with a circular hole, determine the maximum stresses for hole diameters $d = 1$ in. and $d = 2$ in. if the width $b = 6.0$ in.

(b) For the stepped bar with shoulder fillets, determine the maximum stresses for fillet radii $R = 0.25$ in. and $R = 0.5$ in. if the bar widths are $b = 4.0$ in. and $c = 2.5$ in.

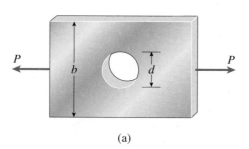

(a)

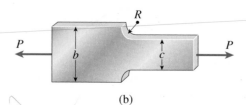

(b)

PROBS. 2.10-1 and 2.10-2

**2.10-2** The flat bars shown in parts a and b of the figure are subjected to tensile forces $P = 2.5$ kN. Each bar has thickness $t = 5.0$ mm.

(a) For the bar with a circular hole, determine the maximum stresses for hole diameters $d = 12$ mm and $d = 20$ mm if the width $b = 60$ mm.

(b) For the stepped bar with shoulder fillets, determine the maximum stresses for fillet radii $R = 6$ mm and $R = 10$ mm if the bar widths are $b = 60$ mm and $c = 40$ mm.

**2.10-3** A flat bar of width $b$ and thickness $t$ has a hole of diameter $d$ drilled through it (see figure). The hole may have any diameter that will fit within the bar.

What is the maximum permissible tensile load $P_{max}$ if the allowable tensile stress in the material is $\sigma_t$?

PROB. 2.10-3

**2.10-4** A round brass bar of diameter $d_1 = 20$ mm has upset ends of diameter $d_2 = 26$ mm (see figure). The lengths of the segments of the bar are $L_1 = 0.3$ m and $L_2 = 0.1$ m. Quarter-circular fillets are used at the shoulders of the bar, and the modulus of elasticity of the brass is $E = 100$ GPa.

If the bar lengthens by 0.12 mm under a tensile load $P$, what is the maximum stress $\sigma_{max}$ in the bar?

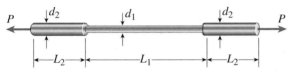

PROBS. 2.10-4 and 2.10-5

**2.10-5** Solve the preceding problem for a bar of monel metal having the following properties: $d_1 = 1.0$ in., $d_2 = 1.4$ in., $L_1 = 20.0$ in., $L_2 = 5.0$ in., and $E = 25 \times 10^6$ psi. Also, the bar lengthens by 0.0040 in. when the tensile load is applied.

**2.10-6** A prismatic bar of diameter $d_0 = 20$ mm is being compared with a stepped bar of the same diameter ($d_1 = 20$ mm) that is enlarged in the middle region to a diameter $d_2 = 25$ mm (see figure). The radius of the fillets in the stepped bar is 2.0 mm.

(a) Does enlarging the bar in the middle region make it stronger than the prismatic bar? Demonstrate your answer by determining the maximum permissible load $P_1$ for the prismatic bar and the maximum permissible load $P_2$ for the enlarged bar, assuming that the allowable stress for the material is 80 MPa.

(b) What should be the diameter $d_0$ of the prismatic bar if it is to have the same maximum permissible load as does the stepped bar?

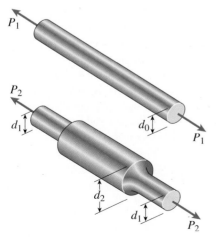

PROB. 2.10-6

**2.10-7** A stepped bar with a hole (see figure) has widths $b = 2.4$ in. and $c = 1.6$ in. The fillets have radii equal to 0.2 in.

What is the diameter $d_{max}$ of the largest hole that can be drilled through the bar without reducing the load-carrying capacity?

PROB. 2.10-7

# Nonlinear Behavior (Changes in Lengths of Bars)

**2.11-1** A bar $AB$ of length $L$ and weight density $\gamma$ hangs vertically under its own weight (see figure). The stress-strain relation for the material is given by the Ramberg-Osgood equation (Eq. 2-73):

$$\varepsilon = \frac{\sigma}{E} + \frac{\sigma_0 \alpha}{E}\left(\frac{\sigma}{\sigma_0}\right)^m$$

Derive the following formula

$$\delta = \frac{\gamma L^2}{2E} + \frac{\sigma_0 \alpha L}{(m+1)E}\left(\frac{\gamma L}{\sigma_0}\right)^m$$

for the elongation of the bar.

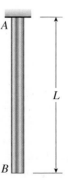

PROB. 2.11-1

**2.11-2** A prismatic bar of length $L = 1.8$ m and cross-sectional area $A = 480$ mm$^2$ is loaded by forces $P_1 = 30$ kN and $P_2 = 60$ kN (see figure). The bar is constructed of magnesium alloy having a stress-strain curve described by the following Ramberg-Osgood equation:

$$\varepsilon = \frac{\sigma}{45,000} + \frac{1}{618}\left(\frac{\sigma}{170}\right)^{10} \quad (\sigma = \text{MPa})$$

in which $\sigma$ has units of megapascals.

(a) Calculate the displacement $\delta_C$ of the end of the bar when the load $P_1$ acts alone.

(b) Calculate the displacement when the load $P_2$ acts alone.

(c) Calculate the displacement when both loads act simultaneously.

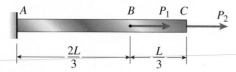

**PROB. 2.11-2**

**2.11-3** A circular bar of length $L = 32$ in. and diameter $d = 0.75$ in. is subjected to tension by forces $P$ (see figure). The wire is made of a copper alloy having the following *hyperbolic stress-strain relationship*:

$$\sigma = \frac{18,000\varepsilon}{1 + 300\varepsilon} \quad 0 \le \varepsilon \le 0.03 \quad (\sigma = \text{ksi})$$

(a) Draw a stress-strain diagram for the material.

(b) If the elongation of the wire is limited to 0.25 in. and the maximum stress is limited to 40 ksi, what is the allowable load $P$?

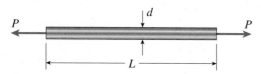

**PROB. 2.11-3**

**2.11-4** A prismatic bar in tension has length $L = 2.0$ m and cross-sectional area $A = 249$ mm$^2$. The material of the bar has the stress-strain curve shown in the figure.

Determine the elongation $\delta$ of the bar for each of the following axial loads: $P = 10$ kN, 20 kN, 30 kN, 40 kN, and 45 kN. From these results, plot a diagram of load $P$ versus elongation $\delta$ (load-displacement diagram).

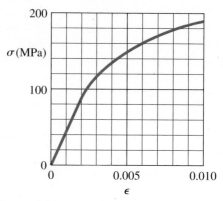

**PROB. 2.11-4**

**2.11-5** An aluminum bar subjected to tensile forces $P$ has length $L = 150$ in. and cross-sectional area $A = 2.0$ in.$^2$ The stress-strain behavior of the aluminum may be represented approximately by the bilinear stress-strain diagram shown in the figure.

Calculate the elongation $\delta$ of the bar for each of the following axial loads: $P = 8$ k, 16 k, 24 k, 32 k, and 40 k. From these results, plot a diagram of load $P$ versus elongation $\delta$ (load-displacement diagram).

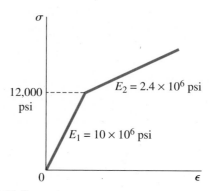

**PROB. 2.11-5**

**2.11-6** A rigid bar $AB$, pinned at end $A$, is supported by a wire $CD$ and loaded by a force $P$ at end $B$ (see figure). The wire is made of high-strength steel having modulus of elasticity $E = 210$ GPa and yield stress $\sigma_Y = 820$ MPa. The length of the wire is $L = 1.0$ m and its diameter is $d = 3$ mm. The stress-strain diagram for the steel is defined by the *modified power law*, as follows:

$$\sigma = E\varepsilon \quad 0 \le \sigma \le \sigma_Y$$

$$\sigma = \sigma_Y\left(\frac{E\varepsilon}{\sigma_Y}\right)^n \quad \sigma \ge \sigma_Y$$

(a) Assuming $n = 0.2$, calculate the displacement $\delta_B$ at the end of the bar due to the load $P$. Take values of $P$ from 2.4 kN to 5.6 kN in increments of 0.8 kN.

(b) Plot a load-displacement diagram showing $P$ versus $\delta B$.

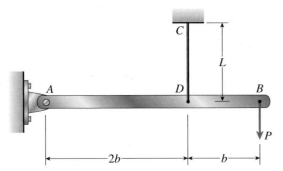

PROB. 2.11-6

# Elastoplastic Analysis

*The problems for Section 2.12 are to be solved assuming that the material is elastoplastic with yield stress $\sigma_Y$, yield strain $\varepsilon_Y$, and modulus of elasticity E in the linearly elastic region (see Fig. 2-72).*

**2.12-1** Two identical bars $AB$ and $BC$ support a vertical load $P$ (see figure). The bars are made of steel having a stress-strain curve that may be idealized as elastoplastic with yield stress $\sigma_Y$. Each bar has cross-sectional area $A$.

Determine the yield load $P_Y$ and the plastic load $P_P$.

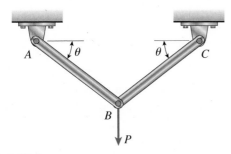

PROB. 2.12-1

**2.12-2** A stepped bar $ACB$ with circular cross sections is held between rigid supports and loaded by an axial force $P$ at midlength (see figure). The diameters for the two parts of the bar are $d_1 = 20$ mm and $d_2 = 25$ mm, and the material is elastoplastic with yield stress $\sigma_Y = 250$ MPa.

Determine the plastic load $P_P$.

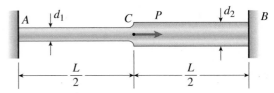

PROB. 2.12-2

**2.12-3** A horizontal rigid bar $AB$ supporting a load $P$ is hung from five symmetrically placed wires, each of cross-sectional area $A$ (see figure). The wires are fastened to a curved surface of radius $R$.

(a) Determine the plastic load $P_P$ if the material of the wires is elastoplastic with yield stress $\sigma_Y$.

(b) How is $P_P$ changed if bar $AB$ is flexible instead of rigid?

(c) How is $P_P$ changed if the radius $R$ is increased?

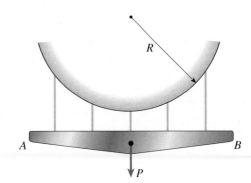

PROB. 2.12-3

**2.12-4** A load $P$ acts on a horizontal beam that is supported by four rods arranged in the symmetrical pattern shown in the figure. Each rod has cross-sectional area $A$ and the material is elastoplastic with yield stress $\sigma_Y$. Determine the plastic load $P_P$.

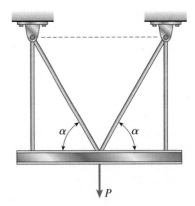

PROB. 2.12-4

**2.12-5** The symmetric truss $ABCDE$ shown in the figure is constructed of four bars and supports a load $P$ at joint $E$. Each of the two outer bars has a cross-sectional area of 0.307 in.$^2$, and each of the two inner bars has an area of 0.601 in.$^2$ The material is elastoplastic with yield stress $\sigma_Y = 36$ ksi.

Determine the plastic load $P_P$.

**PROB. 2.12-7**

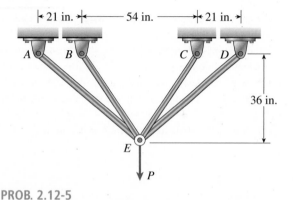

**PROB. 2.12-5**

**2.12-6** Five bars, each having a diameter of 10 mm, support a load $P$ as shown in the figure. Determine the plastic load $P_P$ if the material is elastoplastic with yield stress $\sigma_Y = 250$ MPa.

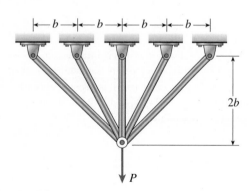

**PROB. 2.12-6**

**2.12-7** A circular steel rod $AB$ of diameter $d = 0.60$ in. is stretched tightly between two supports so that initially the tensile stress in the rod is 10 ksi (see figure). An axial force $P$ is then applied to the rod at an intermediate location $C$.

(a) Determine the plastic load $P_P$ if the material is elastoplastic with yield stress $\sigma_Y = 36$ ksi.

(b) How is $P_P$ changed if the initial tensile stress is doubled to 20 ksi?

**2.12-8** A rigid bar $ACB$ is supported on a fulcrum at $C$ and loaded by a force $P$ at end $B$ (see figure). Three identical wires made of an elastoplastic material (yield stress $\sigma_Y$ and modulus of elasticity $E$) resist the load $P$. Each wire has cross-sectional area $A$ and length $L$.

(a) Determine the yield load $P_Y$ and the corresponding yield displacement $\delta_Y$ at point $B$.

(b) Determine the plastic load $P_P$ and the corresponding displacement $\delta_P$ at point $B$ when the load just reaches the value $P_P$.

(c) Draw a load-displacement diagram with the load $P$ as ordinate and the displacement $\delta_B$ of point $B$ as abscissa.

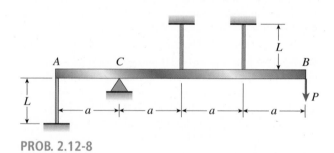

**PROB. 2.12-8**

**2.12-9** The structure shown in the figure consists of a horizontal rigid bar $ABCD$ supported by two steel wires, one of length $L$ and the other of length $3L/4$. Both wires have cross-sectional area $A$ and are made of elastoplastic material with yield stress $\sigma_Y$ and modulus of elasticity $E$. A vertical load $P$ acts at end $D$ of the bar.

(a) Determine the yield load $P_Y$ and the corresponding yield displacement $\delta_Y$ at point $D$.

(b) Determine the plastic load $P_P$ and the corresponding displacement $\delta_P$ at point $D$ when the load just reaches the value $P_P$.

(c) Draw a load-displacement diagram with the load $P$ as ordinate and the displacement $\delta_D$ of point $D$ as abscissa.

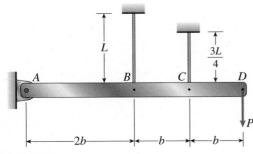

**PROB. 2.12-9**

**2.12-10** Two cables, each having a length $L$ of approximately 40 m, support a loaded container of weight $W$ (see figure). The cables, which have effective cross-sectional area $A = 48.0$ mm² and effective modulus of elasticity $E = 160$ GPa, are identical except that one cable is longer than the other when they are hanging separately and unloaded. The difference in lengths is $d = 100$ mm. The cables are made of steel having an elastoplastic stress-strain diagram with $\sigma_Y = 500$ MPa. Assume that the weight $W$ is initially zero and is slowly increased by the addition of material to the container.

(a) Determine the weight $W_Y$ that first produces yielding of the shorter cable. Also, determine the corresponding elongation $\delta_Y$ of the shorter cable.

(b) Determine the weight $W_P$ that produces yielding of both cables. Also, determine the elongation $\delta_P$ of the shorter cable when the weight $W$ just reaches the value $W_P$.

(c) Construct a load-displacement diagram showing the weight $W$ as ordinate and the elongation $\delta$ of the shorter cable as abscissa. (*Hint*: The load displacement diagram is not a single straight line in the region $0 \le W \le W_Y$.)

**2.12-11** A hollow circular tube $T$ of length $L = 15$ in. is uniformly compressed by a force $P$ acting through a rigid plate (see figure). The outside and inside diameters of the tube are 3.0 and 2.75 in., respectively. A concentric solid circular bar $B$ of 1.5 in. diameter is mounted inside the tube. When no load is present, there is a clearance $c = 0.010$ in. between the bar $B$ and the rigid plate. Both bar and tube are made of steel having an elastoplastic stress-strain diagram with $E = 29 \times 10^3$ ksi and $\sigma_Y = 36$ ksi.

(a) Determine the yield load $P_Y$ and the corresponding shortening $\delta_Y$ of the tube.

(b) Determine the plastic load $P_P$ and the corresponding shortening $\delta_P$ of the tube.

(c) Construct a load-displacement diagram showing the load $P$ as ordinate and the shortening $\delta$ of the tube as abscissa. (*Hint:* The load-displacement diagram is not a single straight line in the region $0 \le P \le P_Y$.)

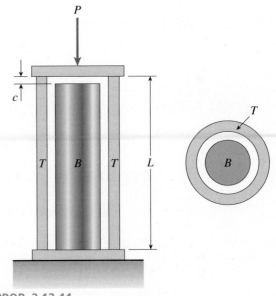

**PROB. 2.12-11**

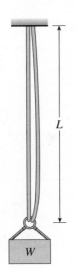

**PROB. 2.12-10**

# Torsion

Circular shafts are essential components in machines and devices for power generation and transmission.
(R. Scott Lewis at ACCO Engineered Systems)

# CHAPTER OVERVIEW

Chapter 3 is concerned with the twisting of circular bars and hollow shafts acted upon by torsional moments. First, we consider **uniform torsion** which refers to the case in which torque is constant over the length of a prismatic shaft, while **nonuniform torsion** describes cases in which the torsional moment and/or the torsional rigidity of the cross section varies over the length. As for the case of axial deformations, we must relate stress and strain and also applied loading and deformation. For torsion, recall that **Hooke's Law for shear** states that shearing stresses, $\tau$, are proportional to shearing strains, $\gamma$, with the constant of proportionality being $G$, the shearing modulus of elasticity. Both shearing stresses and shearing strains vary linearly with increasing radial distance in the cross section, as described by the **torsion formula**. The angle of twist, $\phi$, is proportional to the internal torsional moment and the torsional flexibility of the circular bar. Most of the discussion in this chapter is devoted to linear elastic behavior and small rotations of statically determinate members. However, if the bar is **statically indeterminate**, we must augment the equations of statical equilibrium with compatibility equations (which rely on **torque-displacement relations**) to solve for any unknowns of interest, such as support moments or internal torsional moments in members. **Stresses on inclined sections** also are investigated as a first step toward a more complete consideration of plane stress states in later chapters. Finally, a number of specialized and advanced topics (such as strain energy, torsion of noncircular shafts, shear flow in thin-walled tubes, and stress concentrations in torsion) are introduced at the end of this chapter.

## The topics in Chapter 3 are organized as follows:

**3.1** Introduction 256
**3.2** Torsional Deformations of a Circular Bar 257
**3.3** Circular Bars of Linearly Elastic Materials 260
**3.4** Nonuniform Torsion 272
**3.5** Stresses and Strains in Pure Shear 283
**3.6** Relationship Between Moduli of Elasticity $E$ and $G$ 290
**3.7** Transmission of Power by Circular Shafts 291
**3.8** Statically Indeterminate Torsional Members 296

**3.9** Strain Energy in Torsion and Pure Shear 300
**3.10** Torsion of Noncircular Prismatic Shafts 307
**3.11** Thin-Walled Tubes 316
**\*3.12** Stress Concentrations in Torsion 324
Chapter Summary & Review 328
Problems 330

\*Specialized and/or advanced topics

# 3.1 INTRODUCTION

In Chapters 1 and 2, we discussed the behavior of the simplest type of structural member—namely, a straight bar subjected to axial loads. Now we consider a slightly more complex type of behavior known as **torsion**. Torsion refers to the twisting of a straight bar when it is loaded by moments (or torques) that tend to produce rotation about the longitudinal axis of the bar. For instance, when you turn a screwdriver (Fig. 3-1a), your hand applies a torque $T$ to the handle (Fig. 3-1b) and twists the shank of the screwdriver. Other examples of bars in torsion are drive shafts in automobiles, axles, propeller shafts, steering rods, and drill bits.

An idealized case of torsional loading is pictured in Fig. 3-2a, which shows a straight bar supported at one end and loaded by two pairs of equal and opposite forces. The first pair consists of the forces $P_1$ acting near the midpoint of the bar and the second pair consists of the forces $P_2$ acting at the end. Each pair of forces forms a **couple** that tends to twist the bar about its longitudinal axis. As we know from statics, the **moment of a couple** is equal to the product of one of the forces and the perpendicular distance between the lines of action of the forces; thus, the first couple has a moment $T_1 = P_1 d_1$ and the second has a moment $T_2 = P_2 d_2$.

Typical USCS **units** for moment are the pound-foot (lb-ft) and the pound-inch (lb-in.). The SI unit for moment is the newton meter (N·m).

The moment of a couple may be represented by a **vector** in the form of a double-headed arrow (Fig. 3-2b). The arrow is perpendicular to the plane containing the couple, and therefore in this case both arrows are parallel to the axis of the bar. The direction (or *sense*) of the moment is indicated by the *right-hand rule* for moment vectors—namely, using your right hand, let your fingers curl in the direction of the moment, and then your thumb will point in the direction of the vector.

An alternative representation of a moment is a curved arrow acting in the direction of rotation (Fig. 3-2c). Both the curved arrow and vector representations are in common use, and both are used in this book. The choice depends upon convenience and personal preference.

Moments that produce twisting of a bar, such as the moments $T_1$ and $T_2$ in Fig. 3-2, are called **torques** or **twisting moments**. Cylindrical members that are subjected to torques and transmit power through rotation are called **shafts**; for instance, the drive shaft of an automobile or the propeller shaft of a ship. Most shafts have circular cross sections, either solid or tubular.

In this chapter we begin by developing formulas for the deformations and stresses in circular bars subjected to torsion. We then analyze the state of stress known as *pure shear* and obtain the relationship between the moduli of elasticity $E$ and $G$ in tension and shear, respectively. Next, we analyze rotating shafts and determine the power they transmit. Finally, we cover several additional topics related to torsion, namely, statically indeterminate members, strain energy, thin-walled tubes of noncircular cross section, and stress concentrations.

### Fig. 3-1

Torsion of a screwdriver due to a torque $T$ applied to the handle

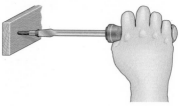

(a)

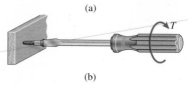

(b)

### Fig. 3-2

Circular bar subjected to torsion by torques $T_1$ and $T_2$

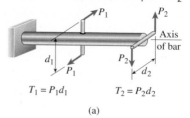

$T_1 = P_1 d_1 \qquad T_2 = P_2 d_2$

(a)

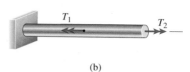

(b)

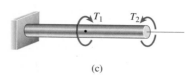

(c)

# 3.2 TORSIONAL DEFORMATIONS OF A CIRCULAR BAR

We begin our discussion of torsion by considering a prismatic bar of circular cross section twisted by torques $T$ acting at the ends (Fig. 3-3a). Since every cross section of the bar is identical, and since every cross section is subjected to the same internal torque $T$, we say that the bar is in **pure torsion**. From considerations of symmetry, it can be proved that cross sections of the bar do not change in shape as they rotate about the longitudinal axis. In other words, all cross sections remain plane and circular and all radii remain straight. Furthermore, if the angle of rotation between one end of the bar and the other is small, neither the length of the bar nor its radius will change.

To aid in visualizing the deformation of the bar, imagine that the left-hand end of the bar (Fig. 3-3a) is fixed in position. Then, under the action of the torque $T$, the right-hand end will rotate (with respect to the left-hand end) through a small angle $\phi$, known as the **angle of twist** (or *angle of rotation*). Because of this rotation, a straight longitudinal line $pq$ on the surface of the bar will become a helical curve $pq'$, where $q'$ is the position of point $q$ after the end cross section has rotated through the angle $\phi$ (Fig. 3-3b).

The angle of twist changes along the axis of the bar, and at intermediate cross sections it will have a value $\phi(x)$ that is between zero at the left-hand end and $\phi$ at the right-hand end. If every cross section of the bar has the same radius and is subjected to the same torque (pure torsion), the angle $\phi(x)$ will vary linearly between the ends.

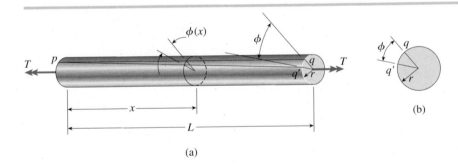

**Fig. 3-3**

Deformations of a circular bar in pure torsion

(a)

(b)

## Shear Strains at the Outer Surface

Now consider an element of the bar between two cross sections distance $dx$ apart (see Fig. 3-4a on the next page). This element is shown enlarged in Fig. 3-4b. On its outer surface we identify a small element $abcd$, with sides $ab$ and $cd$ that initially are parallel to the longitudinal axis. During twisting of the bar, the right-hand cross section rotates with respect to the left-hand cross section through a small angle of twist $d\phi$, so that points $b$ and $c$ move to $b'$ and $c'$, respectively. The lengths of the sides of the element, which is now element $ab'c'd$, do not change during this small rotation.

**Fig. 3-4**

Deformation of an
element of length *dx* cut
from a bar in torsion

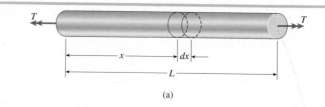

(a)

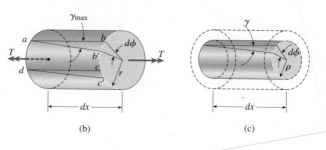

(b)                                    (c)

**Fig. 3-4**

Deformation of an
element of length *dx* cut
from a bar in torsion

However, the angles at the corners of the element (Fig. 3-4b) are no longer equal to 90°. The element is therefore in a state of **pure shear**, which means that the element is subjected to shear strains but no normal strains (see Fig. 1-28 of Section 1.4). The magnitude of the shear strain at the outer surface of the bar, denoted $\gamma_{max}$, is equal to the decrease in the angle at point $a$, that is, the decrease in angle *bad*. From Fig. 3-4b we see that the decrease in this angle is

$$\gamma_{max} = \frac{bb'}{ab} \tag{3-1}$$

where $\gamma_{max}$ is measured in radians, $bb'$ is the distance through which point $b$ moves, and $ab$ is the length of the element (equal to $dx$). With $r$ denoting the radius of the bar, we can express the distance $bb'$ as $rd\phi$, where $d\phi$ also is measured in radians. Thus, the preceding equation becomes

$$\gamma_{max} = \frac{rd\phi}{dx} \tag{3-2}$$

This equation relates the shear strain at the outer surface of the bar to the angle of twist.

The quantity $d\phi/dx$ is the rate of change of the angle of twist $\phi$ with respect to the distance $x$ measured along the axis of the bar. We will denote $d\phi/dx$ by the symbol $\theta$ and refer to it as the **rate of twist**, or the **angle of twist per unit length**:

$$\theta = \frac{d\phi}{dx} \tag{3-3}$$

**Fig. 3-3b (Repeated)**

With this notation, we can now write the equation for the shear strain at the outer surface (Eq. 3-2) as

(b)

$$\gamma_{max} = \frac{rd\phi}{dx} = r\theta \tag{3-4}$$

For convenience, we discussed a bar in pure torsion when deriving Eqs. (3-3) and (3-4). However, both equations are valid in more general cases of torsion, such as when the rate of twist $\theta$ is not constant but varies with the distance $x$ along the axis of the bar.

In the special case of pure torsion, the rate of twist is equal to the total angle of twist $\phi$ divided by the length $L$, that is, $\theta = \phi/L$. Therefore, *for pure torsion only*, we obtain

$$\gamma_{max} = r\theta = \frac{r\phi}{L} \qquad \textbf{(3-5)}$$

This equation can be obtained directly from the geometry of Fig. 3-3a by noting that $\gamma_{max}$ is the angle between lines $pq$ and $pq'$, that is, $\gamma_{max}$ is the angle $qpq'$. Therefore, $\gamma_{max}L$ is equal to the distance $qq'$ at the end of the bar. But since the distance $qq'$ also equals $r\phi$ (Fig. 3-3b), we obtain $r\phi = \gamma_{max}L$, which agrees with Eq. (3-5).

## Shear Strains Within the Bar

The shear strains within the interior of the bar can be found by the same method used to find the shear strain $\gamma_{max}$ at the surface. Because radii in the cross sections of a bar remain straight and undistorted during twisting, we see that the preceding discussion for an element $abcd$ at the outer surface (Fig. 3-4b) will also hold for a similar element situated on the surface of an interior cylinder of radius $\rho$ (Fig. 3-4c). Thus, interior elements are also in pure shear with the corresponding shear strains given by the equation (compare with Eq. 3-4):

$$\gamma = \rho\theta = \frac{\rho}{r}\gamma_{max} \qquad \textbf{(3-6)}$$

This equation shows that the shear strains in a circular bar vary linearly with the radial distance $\rho$ from the center, with the strain being zero at the center and reaching a maximum value $\gamma_{max}$ at the outer surface.

## Circular Tubes

A review of the preceding discussions will show that the equations for the shear strains (Eqs. 3-2 to 3-4) apply to **circular tubes** (Fig. 3-5) as well as to solid circular bars. Figure 3-5 shows the linear variation in shear strain between the maximum strain at the outer surface and the minimum strain at the interior surface. The equations for these strains are as follows:

$$\gamma_{max} = \frac{r_2\phi}{L} \qquad \gamma_{min} = \frac{r_1}{r_2}\gamma_{max} = \frac{r_1\phi}{L} \qquad \textbf{(3-7a,b)}$$

in which $r_1$ and $r_2$ are the inner and outer radii, respectively, of the tube.

All of the preceding equations for the strains in a circular bar are based upon geometric concepts and do not involve the material properties. Therefore, the equations are valid for any material, whether it behaves elastically or inelastically, linearly or nonlinearly. However, the equations are limited to bars having small angles of twist and small strains.

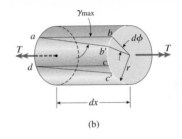

**Fig. 3-4b (Repeated)**

(b)

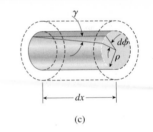

**Fig. 3-4c (Repeated)**

(c)

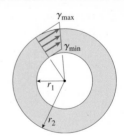

**Fig. 3-5**

Shear strains in a circular tube

# 3.3 CIRCULAR BARS OF LINEARLY ELASTIC MATERIALS

Now that we have investigated the shear strains in a circular bar in torsion (see Figs. 3-3 to 3-5), we are ready to determine the directions and magnitudes of the corresponding shear stresses. The directions of the stresses can be determined by inspection, as illustrated in Fig. 3-6a. We observe that the torque $T$ tends to rotate the right-hand end of the bar counterclockwise when viewed from the right. Therefore the shear stresses $\tau$ acting on a stress element located on the surface of the bar will have the directions shown in the figure.

For clarity, the stress element shown in Fig. 3-6a is enlarged in Fig. 3-6b, where both the shear strain and the shear stresses are shown. As explained previously in Section 2.6, we customarily draw stress elements in two dimensions, as in Fig. 3-6b, but we must always remember that stress elements are actually three-dimensional objects with a thickness perpendicular to the plane of the figure.

The magnitudes of the shear stresses can be determined from the strains by using the stress-strain relation for the material of the bar. If the material is linearly elastic, we can use **Hooke's law in shear** (Eq. 1-21):

$$\tau = G\gamma \qquad \textbf{(3-8)}$$

in which $G$ is the shear modulus of elasticity and $\gamma$ is the shear strain in radians. Combining this equation with the equations for the shear strains (Eqs. 3-2 and 3-4), we get

$$\tau_{max} = Gr\theta \qquad \tau = G\rho\theta = \frac{\rho}{r}\tau_{max} \qquad \textbf{(3-9a,b)}$$

in which $\tau_{max}$ is the shear stress at the outer surface of the bar (radius $r$), $\tau$ is the shear stress at an interior point (radius $\rho$), and $\theta$ is the rate of twist. (In these equations, $\theta$ has units of radians per unit of length.)

**Fig. 3-6**

Shear stresses in a circular bar in torsion

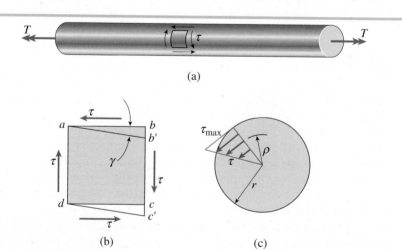

(a)

(b)

(c)

Equations (3-9a) and (3-9b) show that the shear stresses vary linearly with the distance from the center of the bar, as illustrated by the triangular stress diagram in Fig. 3-6c. This linear variation of stress is a consequence of Hooke's law. If the stress-strain relation is nonlinear, the stresses will vary nonlinearly and other methods of analysis will be needed.

The shear stresses acting on a cross-sectional plane are accompanied by shear stresses of the same magnitude acting on longitudinal planes (Fig. 3-7). This conclusion follows from the fact that equal shear stresses always exist on mutually perpendicular planes, as explained in Section 1.7. If the material of the bar is weaker in shear on longitudinal planes than on cross-sectional planes, as is typical of wood when the grain runs parallel to the axis of the bar, the first cracks due to torsion will appear on the surface in the longitudinal direction.

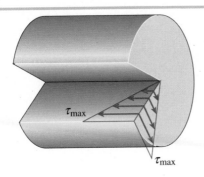

### Fig. 3-7

Longitudinal and transverse shear stresses in a circular bar subjected to torsion

The state of pure shear at the surface of a bar (Fig. 3-6b) is equivalent to equal tensile and compressive stresses acting on an element oriented at an angle of 45°, as explained later in Section 3.5. Therefore, a rectangular element with sides at 45° to the axis of the shaft will be subjected to tensile and compressive stresses, as shown in Fig. 3-8. If a torsion bar is made of a material that is weaker in tension than in shear, failure will occur in tension along a helix inclined at 45° to the axis, as you can demonstrate by twisting a piece of classroom chalk.

### Fig. 3-8

Tensile and compressive stresses acting on a stress element oriented at 45° to the longitudinal axis

## The Torsion Formula

The next step in our analysis is to determine the relationship between the shear stresses and the torque $T$. Once this is accomplished, we will be able to calculate the stresses and strains in a bar due to any set of applied torques.

The distribution of the shear stresses acting on a cross section is pictured in Figs. 3-6c and 3-7. Because these stresses act continuously around the cross section, they have a resultant in the form of a moment—a moment equal to the torque $T$ acting on the bar. To determine this resultant, we consider an element of area $dA$ located at radial distance $\rho$ from the axis of the bar (Fig. 3-9). The shear force acting on this element is equal to $\tau\, dA$, where $\tau$ is the shear stress at radius $\rho$. The moment of this

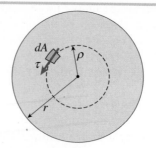

force about the axis of the bar is equal to the force times its distance from the center, or $\tau \rho dA$. Substituting for the shear stress $\tau$ from Eq. (3-9b), we can express this elemental moment as

$$dM = \tau \rho dA = \frac{\tau_{max}}{r}\rho^2 dA$$

The resultant moment (equal to the torque $T$) is the summation over the entire cross-sectional area of all such elemental moments:

$$T = \int_A dM = \frac{\tau_{max}}{r}\int_A \rho^2 dA = \frac{\tau_{max}}{r}I_P \qquad \textbf{(3-10)}$$

in which

$$I_P = \int_A \rho^2 dA \qquad \textbf{(3-11)}$$

is the **polar moment of inertia** of the circular cross section.

For a circle of radius $r$ and diameter $d$, the polar moment of inertia is

$$I_P = \frac{\pi r^4}{2} = \frac{\pi d^4}{32} \qquad \textbf{(3-12)}$$

as given in Appendix E, Case 9. Note that moments of inertia have units of length to the fourth power.*

An expression for the maximum shear stress can be obtained by rearranging Eq. (3-10), as follows:

$$\tau_{max} = \frac{Tr}{I_P} \qquad \textbf{(3-13)}$$

This equation, known as the **torsion formula**, shows that the maximum shear stress is proportional to the applied torque $T$ and inversely proportional to the polar moment of inertia $I_P$.

Typical **units** used with the torsion formula are as follows. In SI, the torque $T$ is usually expressed in newton meters (N·m), the radius $r$ in meters (m), the polar moment of inertia $I_P$ in meters to the fourth power (m⁴), and the shear stress $\tau$ in pascals (Pa). If USCS units are used, $T$ is often expressed in pound-feet (lb-ft) or pound-inches (lb-in.),

---

*Polar moments of inertia are discussed in Section 12.6 of Chapter 12.

$r$ in inches (in.), $I_P$ in inches to the fourth power (in.$^4$), and $\tau$ in pounds per square inch (psi).

Substituting $r = d/2$ and $I_P = \pi d^4/32$ into the torsion formula, we get the following equation for the maximum stress:

$$\tau_{max} = \frac{16T}{\pi d^3} \tag{3-14}$$

This equation applies only to bars of *solid circular cross section*, whereas the torsion formula itself (Eq. 3-13) applies to both solid bars and circular tubes, as explained later. Equation (3-14) shows that the shear stress is inversely proportional to the cube of the diameter. Thus, if the diameter is doubled, the stress is reduced by a factor of eight.

The shear stress at distance $\rho$ from the center of the bar is

$$\tau = \frac{\rho}{r} \tau_{max} = \frac{T\rho}{I_P} \tag{3-15}$$

which is obtained by combining Eq. (3-9b) with the torsion formula (Eq. 3-13). Equation (3-15) is a *generalized torsion formula*, and we see once again that the shear stresses vary linearly with the radial distance from the center of the bar.

## Angle of Twist

The angle of twist of a bar of linearly elastic material can now be related to the applied torque $T$. Combining Eq. (3-9a) with the torsion formula, we get

$$\theta = \frac{T}{GI_P} \tag{3-16}$$

in which $\theta$ has units of radians per unit of length. This equation shows that the rate of twist $\theta$ is directly proportional to the torque $T$ and inversely proportional to the product $GI_P$, known as the **torsional rigidity** of the bar.

For a bar in **pure torsion**, the total angle of twist $\phi$, equal to the rate of twist times the length of the bar (that is, $\phi = \theta L$), is

$$\phi = \frac{TL}{GI_P} \tag{3-17}$$

in which $\phi$ is measured in radians. The use of the preceding equations in both analysis and design is illustrated later in Examples 3-1 and 3-2.

The quantity $GI_P/L$, called the **torsional stiffness** of the bar, is the torque required to produce a unit angle of rotation. The **torsional flexibility** is the reciprocal of the stiffness, or $L/GI_P$, and is defined as the angle of rotation produced by a unit torque. Thus, we have the following expressions:

$$k_T = \frac{GI_P}{L} \qquad f_T = \frac{L}{GI_P} \tag{3-18a,b}$$

These quantities are analogous to the axial stiffness $k = EA/L$ and axial flexibility $f = L/EA$ of a bar in tension or compression (compare with Eqs. 2-4a and 2-4b). Stiffnesses and flexibilities have important roles in structural analysis.

The equation for the angle of twist (Eq. 3-17) provides a convenient way to determine the shear modulus of elasticity $G$ for a material. By conducting a torsion test on a circular bar, we can measure the angle of twist $\phi$ produced by a known torque $T$. Then the value of $G$ can be calculated from Eq. (3-17).

## Circular Tubes

Circular tubes are more efficient than solid bars in resisting torsional loads. As we know, the shear stresses in a solid circular bar are maximum at the outer boundary of the cross section and zero at the center. Therefore, most of the material in a solid shaft is stressed significantly below the maximum shear stress. Furthermore, the stresses near the center of the cross section have a smaller moment arm $\rho$ for use in determining the torque (see Fig. 3-9 and Eq. 3-10).

By contrast, in a typical hollow tube most of the material is near the outer boundary of the cross section where both the shear stresses and the moment arms are highest (Fig. 3-10). Thus, if weight reduction and savings of material are important, it is advisable to use a circular tube. For instance, large drive shafts, propeller shafts, and generator shafts usually have hollow circular cross sections.

The analysis of the torsion of a circular tube is almost identical to that for a solid bar. The same basic expressions for the shear stresses may be used (for instance, Eqs. 3-9a and 3-9b). Of course, the radial distance $\rho$ is limited to the range $r_1$ to $r_2$, where $r_1$ is the inner radius and $r_2$ is the outer radius of the bar (Fig. 3-10).

The relationship between the torque $T$ and the maximum stress is given by Eq. (3-10), but the limits on the integral for the polar moment of inertia (Eq. 3-11) are $\rho = r_1$ and $\rho = r_2$. Therefore, the polar moment of inertia of the cross-sectional area of a tube is

**Fig. 3-10**

Circular tube in torsion

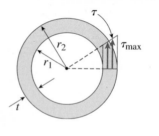

$$I_P = \frac{\pi}{2}(r_2^4 - r_1^4) = \frac{\pi}{32}(d_2^4 - d_1^4) \tag{3-19}$$

The preceding expressions can also be written in the following forms:

$$I_P = \frac{\pi r t}{2}(4r^2 + t^2) = \frac{\pi d t}{4}(d^2 + t^2) \tag{3-20}$$

in which $r$ is the *average radius* of the tube, equal to $(r_1 + r_2)/2$; $d$ is the *average diameter*, equal to $(d_1 + d_2)/2$; and $t$ is the *wall thickness* (Fig. 3-10), equal to $r_2 - r_1$. Of course, Eqs. (3-19) and (3-20) give the same results, but sometimes the latter is more convenient.

If the tube is relatively thin so that the wall thickness $t$ is small compared to the average radius $r$, we may disregard the terms $t^2$ in Eq. (3-20).

With this simplification, we obtain the following *approximate formulas* for the polar moment of inertia:

$$I_P \approx 2\pi r^3 t = \frac{\pi d^3 t}{4} \qquad (3\text{-}21)$$

These expressions are given in Case 22 of Appendix E.

*Reminders*: In Eqs. 3-20 and 3-21, the quantities $r$ and $d$ are the average radius and diameter, not the maximums. Also, Eqs. 3-19 and 3-20 are exact; Eq. 3-21 is approximate.

The torsion formula (Eq. 3-13) may be used for a circular tube of linearly elastic material provided $I_p$ is evaluated according to Eq. (3-19), Eq. (3-20), or, if appropriate, Eq. (3-21). The same comment applies to the general equation for shear stress (Eq. 3-15), the equations for rate of twist and angle of twist (Eqs. 3-16 and 3-17), and the equations for stiffness and flexibility (Eqs. 3-18a and b).

The shear stress distribution in a tube is pictured in Fig. 3-10. From the figure, we see that the average stress in a thin tube is nearly as great as the maximum stress. This means that a hollow bar is more efficient in the use of material than is a solid bar, as explained previously and as demonstrated later in Examples 3-2 and 3-3.

When designing a circular tube to transmit a torque, we must be sure that the thickness $t$ is large enough to prevent wrinkling or buckling of the wall of the tube. For instance, a maximum value of the radius to thickness ratio, such as $(r_2/t)_{max} = 12$, may be specified. Other design considerations include environmental and durability factors, which also may impose requirements for minimum wall thickness. These topics are discussed in courses and textbooks on mechanical design.

## Limitations

The equations derived in this section are limited to bars of circular cross section (either solid or hollow) that behave in a linearly elastic manner. In other words, the loads must be such that the stresses do not exceed the proportional limit of the material. Furthermore, the equations for stresses are valid only in parts of the bars away from stress concentrations (such as holes and other abrupt changes in shape) and away from cross sections where loads are applied. (Stress concentrations in torsion are discussed later in Section 3.12.)

Finally, it is important to emphasize that the equations for the torsion of circular bars and tubes cannot be used for bars of other shapes. Noncircular bars, such as rectangular bars and bars having I-shaped cross sections, behave quite differently than do circular bars. For instance, their cross sections do *not* remain plane and their maximum stresses are *not* located at the farthest distances from the midpoints of the cross sections. Thus, these bars require more advanced methods of analysis, such as those presented in books on theory of elasticity and advanced mechanics of materials.* (A brief overview of torsion of noncircular prismatic shafts is presented in Section 3.10.)

---

*The torsion theory for circular bars originated with the work of the famous French scientist
C. A. de Coulomb (1736–1806); further developments were due to Thomas Young and A. Duleau (Ref. 3-1).
The general theory of torsion (for bars of any shape) is due to the most famous elastician of all time,
Barré de Saint-Venant (1797–1886); see Ref. 2-10.

A solid steel bar of circular cross section (Fig. 3-11) has diameter $d = 1.5$ in., length $L = 54$ in., and shear modulus of elasticity $G = 11.5 \times 10^6$ psi. The bar is subjected to torques $T$ acting at the ends.

(a) If the torques have magnitude $T = 250$ lb-ft, what is the maximum shear stress in the bar? What is the angle of twist between the ends?

(b) If the allowable shear stress is 6000 psi and the allowable angle of twist is 2.5°, what is the maximum permissible torque?

### Fig. 3-11

Example 3-1: Bar in pure torsion

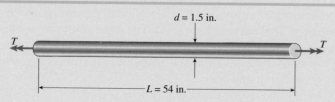

$d = 1.5$ in.

$T$ ⟵         ⟶ $T$

$L = 54$ in.

### Solution

(a) *Maximum shear stress and angle of twist.* Because the bar has a solid circular cross section, we can find the maximum shear stress from Eq. (3-14), as

$$\tau_{max} = \frac{16T}{\pi d^3} = \frac{16(250 \text{ lb-ft})(12 \text{ in./ft})}{\pi(1.5 \text{ in.})^3} = 4530 \text{ psi} \quad \Longleftarrow$$

In a similar manner, the angle of twist is obtained from Eq. (3-17) with the polar moment of inertia given by Eq. (3-12):

$$I_P = \frac{\pi d^4}{32} = \frac{\pi(1.5 \text{ in.})^4}{32} = 0.4970 \text{ in.}^4$$

$$\phi = \frac{TL}{GI_P} = \frac{(250 \text{ lb-ft})(12 \text{ in./ft})(54 \text{ in.})}{(11.5 \times 10^6 \text{ psi})(0.4970 \text{ in.}^4)} = 0.02834 \text{ rad} = 1.62° \quad \Longleftarrow$$

Thus, the analysis of the bar under the action of the given torque is completed.

(b) *Maximum permissible torque.* The maximum permissible torque is determined either by the allowable shear stress or by the allowable angle of twist. Beginning with the shear stress, we rearrange Eq. (3-14) and calculate as

$$T_1 = \frac{\pi d^3 \tau_{allow}}{16} = \frac{\pi}{16}(1.5 \text{ in.})^3(6000 \text{ psi}) = 3980 \text{ lb-in.} = 331 \text{ lb-ft}$$

Any torque larger than this value will result in a shear stress that exceeds the allowable stress of 6000 psi.

Using a rearranged Eq. (3-17), we now calculate the torque based upon the angle of twist:

$$T_2 = \frac{GI_P \phi_{allow}}{L} = \frac{(11.5 \times 10^6 \text{ psi})(0.4970 \text{ in.}^4)(2.5°)(\pi \text{rad}/180°)}{54 \text{ in.}}$$

$$= 4618 \text{ lb-in.} = 385 \text{ lb-ft}$$

Any torque larger than $T_2$ will result in the allowable angle of twist being exceeded.

The maximum permissible torque is the smaller of $T_1$ and $T_2$:

$$T_{max} = 331 \text{ lb-ft} \quad \Longleftarrow$$

In this example, the allowable shear stress provides the limiting condition.

## ● ● ● **Example 3-2**

(© culture-images GmbH/Alamy)

A steel shaft is to be manufactured either as a solid circular bar or as a circular tube (Fig. 3-12). The shaft is required to transmit a torque of 1200 N·m without exceeding an allowable shear stress of 40 MPa nor an allowable rate of twist of 0.75°/m. (The shear modulus of elasticity of the steel is 78 GPa.)

(a) Determine the required diameter $d_0$ of the solid shaft.
(b) Determine the required outer diameter $d_2$ of the hollow shaft if the thickness $t$ of the shaft is specified as one-tenth of the outer diameter.
(c) Determine the ratio of diameters (that is, the ratio $d_2/d_0$) and the ratio of weights of the hollow and solid shafts.

### **Fig. 3-12**

Example 3-2: Torsion of a steel shaft

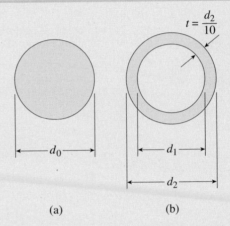

(a)                    (b)

**Solution**

(a) *Solid shaft.* The required diameter $d_0$ is determined either from the allowable shear stress or from the allowable rate of twist. In the case of the allowable shear stress we rearrange Eq. (3-14) and obtain

$$d_0^3 = \frac{16T}{\pi\tau_{allow}} = \frac{16(1200 \text{ N·m})}{\pi(4 \text{ MPa})} = 152.8 \times 10^{-6} \text{ m}^3$$

from which we get

$$d_0 = 0.0535 \text{ m} = 53.5 \text{ mm}$$

In the case of the allowable rate of twist, we start by finding the required polar moment of inertia (see Eq. 3-16):

$$I_P = \frac{T}{G\theta_{allow}} = \frac{1200 \text{ N·m}}{(78 \text{ GPa})(0.75°/\text{m})(\pi\text{rad}/180°)} = 1175 \times 10^{-9} \text{ m}^4$$

*Continues* ➡

**Example 3-2 -** *Continued*

**Fig. 3-12 (Repeated)**

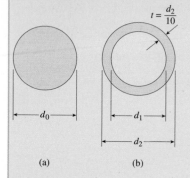

(a)          (b)

Since the polar moment of inertia is equal to $\pi d^4/32$, the required diameter is

$$d_0^4 = \frac{32 I_P}{\pi} = \frac{32(1175 \times 10^{-9} \text{ m}^4)}{\pi} = 11.97 \times 10^{-6} \text{ m}^4$$

or

$$d_0 = 0.0588 \text{ m} = 58.8 \text{ mm}$$

Comparing the two values of $d_0$, we see that the rate of twist governs the design and the required diameter of the solid shaft is

$$d_0 = 58.8 \text{ mm}$$

In a practical design, we would select a diameter slightly larger than the calculated value of $d_0$; for instance, 60 mm.

(b) *Hollow shaft*. Again, the required diameter is based upon either the allowable shear stress or the allowable rate of twist. We begin by noting that the outer diameter of the bar is $d_2$ and the inner diameter is

$$d_1 = d_2 - 2t = d_2 - 2(0.1d_2) = 0.8d_2$$

Thus, the polar moment of inertia (Eq. 3-19) is

$$I_P = \frac{\pi}{32}(d_2^4 - d_1^4) = \frac{\pi}{32}\left[d_2^4 - (0.8d_2)^4\right] = \frac{\pi}{32}(0.5904d_2^4) = 0.05796d_2^4$$

In the case of the allowable shear stress, we use the torsion formula (Eq. 3-13) as

$$\tau_{allow} = \frac{Tr}{I_P} = \frac{T(d_2/2)}{0.05796d_2^4} = \frac{T}{0.1159d_2^3}$$

Rearranging, we get

$$d_2^3 = \frac{T}{0.1159\tau_{allow}} = \frac{1200 \text{ N} \cdot \text{m}}{0.1159(40 \text{ MPa})} = 258.8 \times 10^{-6} \text{ m}^3$$

Solving for $d_2$ gives

$$d_2 = 0.0637 \text{ m} = 63.7 \text{ mm}$$

which is the required outer diameter based upon the shear stress.

In the case of the allowable rate of twist, we use Eq. (3-16) with $\theta$ replaced by $\theta_{allow}$ and $I_P$ replaced by the previously obtained expression; thus,

$$\theta_{allow} = \frac{T}{G(0.05796d_2^4)}$$

from which

$$d_2^4 = \frac{T}{0.05796G\theta_{allow}}$$

$$= \frac{1200 \text{ N}\cdot\text{m}}{0.05796(78 \text{ GPa})(0.75°/\text{m})(\pi\text{rad}/180°)} = 20.28 \times 10^{-6} \text{ m}^4$$

Solving for $d_2$ gives

$$d_2 = 0.0671 \text{ m} = 67.1 \text{ mm}$$

which is the required diameter based upon the rate of twist.

Comparing the two values of $d_2$, we see that the rate of twist governs the design and the required outer diameter of the hollow shaft is

$$d_2 = 67.1 \text{ mm}$$

The inner diameter $d_1$ is equal to $0.8d_2$, or 53.7 mm. (As practical values, we might select $d_2 = 70$ mm and $d_1 = 0.8d_2 = 56$ mm.)

(c) *Ratios of diameters and weights.* The ratio of the outer diameter of the hollow shaft to the diameter of the solid shaft (using the calculated values) is

$$\frac{d_2}{d_0} = \frac{67.1 \text{ mm}}{58.8 \text{ mm}} = 1.14$$

Since the weights of the shafts are proportional to their cross-sectional areas, we can express the ratio of the weight of the hollow shaft to the weight of the solid shaft as follows:

$$\frac{W_H}{W_S} = \frac{A_H}{A_S} = \frac{\pi(d_2^2 - d_1^2)/4}{\pi d_0^2/4} = \frac{d_2^2 - d_1^2}{d_0^2}$$

$$= \frac{(67.1 \text{ mm})^2 - (53.7 \text{ mm})^2}{(58.8 \text{ mm})^2} = 0.47$$

These results show that the hollow shaft uses only 47% as much material as does the solid shaft, while its outer diameter is only 14% larger.

*Note:* This example illustrates how to determine the required sizes of both solid bars and circular tubes when allowable stresses and allowable rates of twist are known. It also illustrates the fact that circular tubes are more efficient in the use of materials than are solid circular bars.

## Example 3-3

A hollow shaft and a solid shaft constructed of the same material have the same length and the same outer radius $R$ (Fig. 3-13). The inner radius of the hollow shaft is $0.6R$.

(a) Assuming that both shafts are subjected to the same torque, compare their shear stresses, angles of twist, and weights.
(b) Determine the strength-to-weight ratios for both shafts.

### Fig. 3-13

Example 3-3: Comparison of hollow and solid shafts

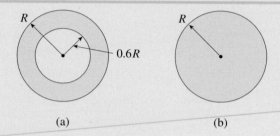

(a)                    (b)

### Solution

(a) *Comparison or shear stresses*. The maximum shear stresses, given by the torsion formula (Eq. 3-13), are proportional to $1/I_p$ inasmuch as the torques and radii are the same. For the hollow shaft, we get

$$I_P = \frac{\pi R^4}{2} - \frac{\pi(0.6R)^4}{2} = 0.4352\pi R^4$$

and for the solid shaft,

$$I_P = \frac{\pi R^4}{2} = 0.5\pi R^4$$

Therefore, the ratio $\beta_1$ of the maximum shear stress in the hollow shaft to that in the solid shaft is

$$\beta_1 = \frac{\tau_H}{\tau_S} = \frac{0.5\pi R^4}{0.4352\pi R^4} = 1.15$$

where the subscripts $H$ and $S$ refer to the hollow shaft and the solid shaft, respectively.

   *Comparison of angles of twist*. The angles of twist (Eq. 3-17) are also proportional to $1/I_p$, because the torques $T$, lengths $L$, and moduli of elasticity $G$ are the same for both shafts. Therefore, their ratio is the same as for the shear stresses:

$$\beta_2 = \frac{\phi_H}{\phi_S} = \frac{0.5\pi R^4}{0.4352\pi R^4} = 1.15$$

*Comparison of weights.* The weights of the shafts are proportional to their cross-sectional areas; consequently, the weight of the solid shaft is proportional to $\pi R^2$ and the weight of the hollow shaft is proportional to

$$\pi R^2 - \pi(0.6R)^2 = 0.64\pi R^2$$

Therefore, the ratio of the weight of the hollow shaft to the weight of the solid shaft is

$$\beta_3 = \frac{W_H}{W_S} = \frac{0.64\pi R^2}{\pi R^2} = 0.64 \qquad \Longleftarrow$$

From the preceding ratios we again see the inherent advantage of hollow shafts. In this example, the hollow shaft has 15% greater stress and 15% greater angle of rotation than the solid shaft but 36% less weight.

(b) *Strength-to-weight ratios.* The relative efficiency of a structure is sometimes measured by its *strength-to-weight ratio*, which is defined for a bar in torsion as the allowable torque divided by the weight. The allowable torque for the hollow shaft of Fig. 3-13a (from the torsion formula) is

$$T_H = \frac{\tau_{max} I_P}{R} = \frac{\tau_{max}(0.4352\pi R^4)}{R} = 0.4352\pi R^3 \tau_{max}$$

and for the solid shaft is

$$T_S = \frac{\tau_{max} I_P}{R} = \frac{\tau_{max}(0.5\pi R^4)}{R} = 0.5\pi R^3 \tau_{max}$$

The weights of the shafts are equal to the cross-sectional areas times the length $L$ times the weight density $\gamma$ of the material:

$$W_H = 0.64\pi R^2 L\gamma \qquad W_S = \pi R^2 L\gamma$$

Thus, the strength-to-weight ratios $S_H$ and $S_S$ for the hollow and solid bars, respectively, are

$$S_H = \frac{T_H}{W_H} = 0.68\frac{\tau_{max}R}{\gamma L} \qquad S_S = \frac{T_S}{W_S} = 0.5\frac{\tau_{max}R}{\gamma L} \qquad \Longleftarrow$$

In this example, the strength-to-weight ratio of the hollow shaft is 36% greater than the strength-to-weight ratio for the solid shaft, demonstrating once again the relative efficiency of hollow shafts. For a thinner shaft, the percentage will increase; for a thicker shaft, it will decrease.

# 3.4 NONUNIFORM TORSION

As explained in Section 3.2, *pure torsion* refers to torsion of a prismatic bar subjected to torques acting only at the ends. **Nonuniform torsion** differs from pure torsion in that the bar need not be prismatic and the applied torques may act anywhere along the axis of the bar. Bars in nonuniform torsion can be analyzed by applying the formulas of pure torsion to finite segments of the bar and then adding the results, or by applying the formulas to differential elements of the bar and then integrating.

To illustrate these procedures, we will consider three cases of nonuniform torsion. Other cases can be handled by techniques similar to those described here.

**Case 1.** *Bar consisting of prismatic segments with constant torque throughout each segment* (Fig. 3-14). The bar shown in part (a) of the figure has two different diameters and is loaded by torques acting at points $A$, $B$, $C$, and $D$. Consequently, we divide the bar into segments in such a way that each segment is prismatic and subjected to a constant torque. In this example, there are three such segments, $AB$, $BC$, and $CD$. Each segment is in pure torsion, and therefore all of the formulas derived in the preceding section may be applied to each part separately.

The first step in the analysis is to determine the magnitude and direction of the internal torque in each segment. Usually the torques can be determined by inspection, but if necessary they can be found by cutting sections through the bar, drawing free-body diagrams, and solving equations of equilibrium. This process is illustrated in parts (b), (c), and (d) of the figure. The first cut is made anywhere in segment $CD$, thereby exposing the internal torque $T_{CD}$. From the free-body diagram (Fig. 3-14b), we see that $T_{CD}$ is equal to $-T_1 - T_2 + T_3$. From the next diagram we see that $T_{BC}$ equals $-T_1 - T_2$, and from the last we find that $T_{AB}$ equals $-T_1$. Thus,

$$T_{CD} = -T_1 - T_2 + T_3 \quad T_{BC} = -T_1 - T_2 \quad T_{AB} = -T_1 \quad \textbf{(3-22a,b,c)}$$

Each of these torques is constant throughout the length of its segment.

When finding the shear stresses in each segment, we need only the magnitudes of these internal torques, since the directions of the stresses are not of interest. However, when finding the angle of twist for the entire bar, we need to know the direction of twist in each segment in order to combine the angles of twist correctly. Therefore, we need to establish a *sign convention* for the internal torques. A convenient rule in many cases is the following: *An internal torque is positive when its vector points away from the cut section and negative when its vector points toward the section.* Thus, all of the internal torques shown in Figs. 3-14b, c, and d are pictured in their positive directions. If the calculated torque [from Eq. (3-22a, b, or c)] turns out to have a positive sign, it means that the torque acts in the assumed direction; if the torque has a negative sign, it acts in the opposite direction.

The maximum shear stress in each segment of the bar is readily obtained from the torsion formula (Eq. 3-13) using the appropriate cross-sectional dimensions and internal torque. For instance, the maximum

**Fig. 3-14**

Bar in nonuniform torsion (Case 1)

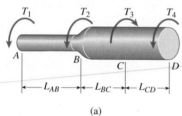

(a)

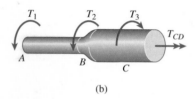

(b)

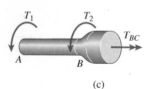

(c)

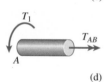

(d)

stress in segment *BC* (Fig. 3-14) is found using the diameter of that segment and the torque $T_{BC}$ calculated from Eq. (3-22b). The maximum stress in the entire bar is the largest stress from among the stresses calculated for each of the three segments.

The angle of twist for each segment is found from Eq. (3-17), again using the appropriate dimensions and torque. The total angle of twist of one end of the bar with respect to the other is then obtained by algebraic summation, as follows:

$$\phi = \phi_1 + \phi_2 + \ldots + \phi_n \qquad \textbf{(3-23)}$$

where $\phi_1$ is the angle of twist for segment 1, $\phi_2$ is the angle for segment 2, and so on, and *n* is the total number of segments. Since each angle of twist is found from Eq. (3-17), we can write the general formula

$$\phi = \sum_{i=1}^{n} \phi_i = \sum_{i=1}^{n} \frac{T_i L_i}{G_i (I_P)_i} \qquad \textbf{(3-24)}$$

in which the subscript *i* is a numbering index for the various segments. For segment *i* of the bar, $T_i$ is the internal torque (found from equilibrium, as illustrated in Fig. 3-14), $L_i$ is the length, $G_i$ is the shear modulus, and $(I_P)_i$ is the polar moment of inertia. Some of the torques (and the corresponding angles of twist) may be positive and some may be negative. By summing *algebraically* the angles of twist for all segments, we obtain the total angle of twist $\phi$ between the ends of the bar. The process is illustrated later in Example 3-4.

**Case 2.** *Bar with continuously varying cross sections and constant torque* (Fig. 3-15). When the torque is constant, the maximum shear stress in a solid bar always occurs at the cross section having the smallest diameter, as shown by Eq. (3-14). Furthermore, this observation usually holds for tubular bars. If this is the case, we only need to investigate the smallest cross section in order to calculate the maximum shear stress. Otherwise, it may be necessary to evaluate the stresses at more than one location in order to obtain the maximum.

To find the angle of twist, we consider an element of length *dx* at distance *x* from one end of the bar (Fig. 3-17). The differential angle of rotation $d\phi$ for this element is

$$d\phi = \frac{T dx}{G I_P(x)} \qquad \textbf{(3-25)}$$

in which $I_P(x)$ is the polar moment of inertia of the cross section at distance *x* from the end. The angle of twist for the entire bar is the summation of the differential angles of rotation:

$$\phi = \int_0^L d\phi = \int_0^L \frac{T dx}{G I_P(x)} \qquad \textbf{(3-26)}$$

**Fig. 3-15**

Bar in nonuniform torsion (Case 2)

If the expression for the polar moment of inertia $I_p(x)$ is not too complex, this integral can be evaluated analytically. In other cases, it must be evaluated numerically.

**Case 3.** *Bar with continuously varying cross sections and continuously varying torque* (Fig. 3-16). The bar shown in part (a) of the figure is subjected to a *distributed torque* of intensity $t$ per unit distance along the axis of the bar. As a result, the internal torque $T(x)$ varies continuously along the axis (Fig. 3-16b). The internal torque can be evaluated with the aid of a free-body diagram and an equation of equilibrium. As in Case 2, the polar moment of inertia $I_p(x)$ can be evaluated from the cross-sectional dimensions of the bar.

Knowing both the torque and polar moment of inertia as functions of $x$, we can use the torsion formula to determine how the shear stress varies along the axis of the bar. The cross section of maximum shear stress can then be identified, and the maximum shear stress can be determined.

The angle of twist for the bar of Fig. 3-16a can be found in the same manner as described for Case 2. The only difference is that the torque, like the polar moment of inertia, also varies along the axis. Consequently, the equation for the angle of twist becomes

$$\phi = \int_0^L d\phi = \int_0^L \frac{T(x)dx}{GI_p(x)}$$

(3-27)

This integral can be evaluated analytically in some cases, but usually it must be evaluated numerically.

## Limitations

The analyses described in this section are valid for bars made of linearly elastic materials with circular cross sections (either solid or hollow). Also, the stresses determined from the torsion formula are valid in regions of the bar *away* from stress concentrations, which are high localized stresses that occur wherever the diameter changes abruptly and wherever concentrated torques are applied (see Section 3.12). However, stress concentrations have relatively little effect on the angle of twist, and therefore the equations for $\phi$ are generally valid.

Finally, we must keep in mind that the torsion formula and the formulas for angles of twist were derived for prismatic bars with circular cross sections (see Section 3.10 for a brief discussion of noncircular bars in torsion). We can safely apply them to bars with varying cross sections only when the changes in diameter are small and gradual. As a rule of thumb, the formulas given here are satisfactory as long as the angle of taper (the angle between the sides of the bar) is less than 10°.

### Fig. 3-16

Bar in nonuniform torsion
(Case 3)

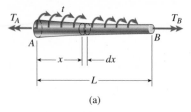

(a)

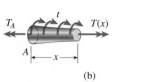

(b)

● ● ● ● **Example 3-4**

(© Bigjoker/Alamy)

A solid steel shaft *ABCDE* (Fig. 3-17) having diameter $d = 30$ mm turns freely in bearings at points *A* and *E*. The shaft is driven by a gear at *C*, which applies a torque $T_2 = 450$ N·m in the direction shown in the figure. Gears at *B* and *D* are driven by the shaft and have resisting torques $T_1 = 275$ N·m and $T_3 = 175$ N·m, respectively, acting in the opposite direction to the torque $T_2$. Segments *BC* and *CD* have lengths $L_{BC} = 500$ mm and $L_{CD} = 400$ mm, respectively, and the shear modulus $G = 80$ GPa.

Determine the maximum shear stress in each part of the shaft and the angle of twist between gears *B* and *D*.

**Fig. 3-17**

Example 3-4: Steel shaft in torsion

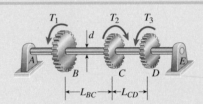

### Solution

Each segment of the bar is prismatic and subjected to a constant torque (Case 1). Therefore, the first step in the analysis is to determine the torques acting in the segments, after which we can find the shear stresses and angles of twist. (Recall that we drew free-body diagrams and then applied the laws of statics to find the reactive and internal torsional moments in a geared shaft in Example 1-3 in Section 1.2.)

*Torques acting in the segments.* The torques in the end segments (*AB* and *DE*) are zero since we are disregarding any friction in the bearings at the supports. Therefore, the end segments have no stresses and no angles of twist.

The torque $T_{CD}$ in segment *CD* is found by cutting a section through the segment and constructing a free-body diagram, as in Fig. 3-18a. The torque is assumed to be positive, and therefore its vector points away from the cut section. From equilibrium of the free body, we obtain

$$T_{CD} = T_2 - T_1 = 450 \text{ N} \cdot \text{m} - 275 \text{ N} \cdot \text{m} = 175 \text{ N} \cdot \text{m}$$

**Fig. 3-18**

Free-body diagrams for Example 3-4

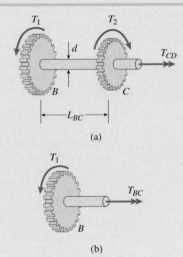

(a)

(b)

*Continues* ➡

**Example 3-4 -** *Continued*

The positive sign in the result means that $T_{CD}$ acts in the assumed positive direction.

The torque in segment $BC$ is found in a similar manner, using the free-body diagram of Fig. 3-18b:

$$T_{BC} = -T_1 = -275 \text{ N} \cdot \text{m}$$

Note that this torque has a negative sign, which means that its direction is opposite to the direction shown in the figure.

*Shear stresses.* The maximum shear stresses in segments $BC$ and $CD$ are found from the modified form of the torsion formula (Eq. 3-14); thus,

$$\tau_{BC} = \frac{16 T_{BC}}{\pi d^3} = \frac{16(275 \text{ N} \cdot \text{m})}{\pi (30 \text{ mm})^3} = 51.9 \text{ MPa} \qquad \Longleftarrow$$

$$\tau_{CD} = \frac{16 T_{CD}}{\pi d^3} = \frac{16(175 \text{ N} \cdot \text{m})}{\pi (30 \text{ mm})^3} = 33.0 \text{ MPa} \qquad \Longleftarrow$$

Since the directions of the shear stresses are not of interest in this example, only absolute values of the torques are used in the preceding calculations.

*Angles of twist.* The angle of twist $\phi_{BD}$ between gears $B$ and $D$ is the algebraic sum of the angles of twist for the intervening segments of the bar, as given by Eq. (3-23); thus,

$$\phi_{BD} = \phi_{BC} + \phi_{CD}$$

When calculating the individual angles of twist, we need the moment of inertia of the cross section:

$$I_P = \frac{\pi d^4}{32} = \frac{\pi (30 \text{ mm})^4}{32} = 79{,}520 \text{ mm}^4$$

Now we can determine the angles of twist, as

$$\phi_{BC} = \frac{T_{BC} L_{BC}}{G I_P} = \frac{(-275 \text{ N} \cdot \text{m})(500 \text{ mm})}{(80 \text{ GPa})(79{,}520 \text{ mm}^4)} = -0.0216 \text{ rad}$$

and

$$\phi_{CD} = \frac{T_{CD} L_{CD}}{G I_P} = \frac{(175 \text{ N} \cdot \text{m})(400 \text{ mm})}{(80 \text{ GPa})(79{,}520 \text{ mm}^4)} = 0.0110 \text{ rad}$$

Note that in this example the angles of twist have opposite directions. Adding algebraically, we obtain the total angle of twist:

$$\phi_{BD} = \phi_{BC} + \phi_{CD} = -0.0216 + 0.0110 = -0.0106 \text{ rad} = -0.61° \qquad \Longleftarrow$$

The minus sign means that gear $D$ rotates clockwise (when viewed from the right-hand end of the shaft) with respect to gear $B$. However, for most purposes only the absolute value of the angle of twist is needed, and therefore it is sufficient to say that the angle of twist between gears $B$ and $D$ is 0.61°. The angle of twist between the two ends of a shaft is sometimes called the *wind-up*.

*Notes:* The procedures illustrated in this example can be used for shafts having segments of different diameters or of different materials, as long as the dimensions and properties remain constant within each segment.

Only the effects of torsion are considered in this example and in the problems at the end of the chapter. Bending effects are considered later, beginning with Chapter 4.

Two sections (*AB*, *BC*) of steel drill pipe, joined by bolted flange plates at *B*, are being tested to assess the adequacy of both the pipe and the bolted connection (see Fig. 3-19). In the test, the pipe structure is fixed at *A* and a concentrated torque $2T_0$ is applied at $x = 2L/5$ and uniformly distributed torque intensity $t_0 = 3T_0/L$ is applied on pipe BC.

(a) Find expressions for internal torques *T(x)* over the length of the pipe structure.

(b) Find the maximum shear stress $\tau_{max}$ in the pipes and its location. Assume that load variable $T_0 = 2000$ kip-in. Let $G = 11,800$ ksi, and assume that both pipes have the same inner diameter, $d = 10$ in. Pipe *AB* has a thickness of $t_{AB} = 3/4$ in., while pipe *BC* has a thickness of $t_{BC} = 5/8$ in.

(c) Find expressions for twist rotations $\phi(x)$ over the length of the pipe structure. If the maximum allowable twist of the pipe structure is $\phi_{allow} = 0.5°$, find the maximum permissible value of *load variable* $T_0$ (kip-in.). Let $L = 10$ ft.

(d) Use $T_0$ from part (c) to find the number of $d_b = 7/8$-in. diameter bolts at radius $r = 15$ in. required in the flange plate connection at *B*. Assume that the allowable shear stress for the bolts is $\tau_a = 28$ ksi.

**Fig. 3-19**

Example 3-5: Two pipes in nonuniform torsion

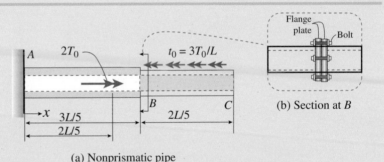

(a) Nonprismatic pipe

(b) Section at B

(Courtesy of Subsea Riser Products)

**Solution**

(a) *Internal torques T(x)*. First, we must find the reactive torque at *A* using statics (see Section 1.2, Example 1-3). Summing torsional moments about the *x* axis of the structure, we find

$$\Sigma M_x = 0 \quad R_A + 2T_0 - t_0\left(\frac{2L}{5}\right) = 0$$

so $$R_A = -2T_0 + \left(\frac{3T_0}{L}\right)\left(\frac{2L}{5}\right) = \frac{-4T_0}{5} \qquad \text{(a)}$$

Reaction $R_A$ is negative, which means that the reactive torsional moment vector is in the (−*x*) direction based on a *statics sign convention*. We now can draw *free-body diagrams* (FBD) of segments of the pipe to find internal torsional moments *T(x)* over the length of the pipe.

*Continues* ➡

**Fig. 3-20**

Example 3-5: (a) FBD of segment 1, (b) FBD of segment 2, and (c) FBD of segment 3

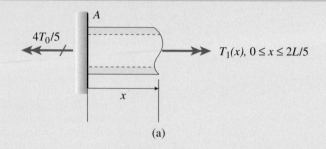

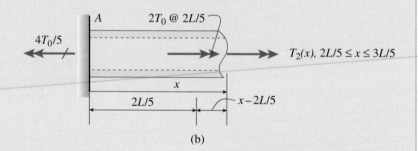

From the FBD of segment 1 (Fig. 3-20a), we see that the internal torsional moment is constant and is equal to reactive torque $R_A$. Torque $T_1(x)$ is positive, because the torsional moment vector points away from the cut section of the pipe; we refer to this as a *deformation sign convention*:

$$T_1(x) = \frac{4}{5}T_0 \quad 0 \le x \le \frac{2}{5}L \tag{b}$$

Next, a FBD of segment 2 of the pipe structure (Fig. 3-20b) gives

$$T_2(x) = \frac{4}{5}T_0 - 2T_0 = \frac{-6}{5}T_0 \quad \frac{2}{5}L \le x \le \frac{3}{5}L \tag{c}$$

where $T_2(x)$ is also constant and the minus sign means that $T_2(x)$ actually points in the negative $x$ direction.

Finally, the FBD of segment 3 of the pipe structure (Fig. 3-20c) provides the following expression for internal torsional moment $T_3(x)$:

$$T_3(x) = \frac{4}{5}T_0 - 2T_0 + t_0\left(x - \frac{3}{5}L\right) = 3T_0\left(\frac{x}{L} - 1\right) \quad \frac{3}{5}L \le x \le L \tag{d}$$

**Fig. 3-20 (Continued)**

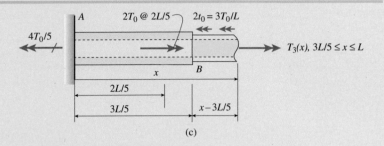

(c)

Evaluating Eq. (d) at B and C, we see that at B we have

$$T_3\left(\frac{3}{5}L\right) = 3T_0\left(\frac{3}{5} - 1\right) = \frac{-6}{5}T_0$$

and at C, we have

$$T_3(L) = 3T_0(1 - 1) = 0$$

We can now plot Eqs. (b), (c), and (d) to get a **torsional moment diagram** (Fig. 3-21) (TMD) which displays the variation of internal torsional moment over the length of the pipe structure ($x = 0$ to $x = L$).

**Fig. 3-21**

Example 3-5: Torsional moment diagram (TMD)

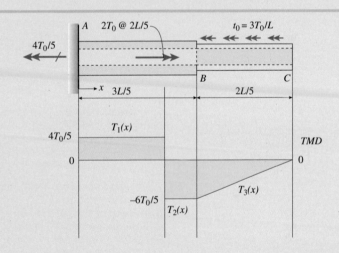

(b) *Maximum shear stress in pipe* $\tau_{max}$. We will use the torsion formula (Eq. 3-13) to compute the shear stress in the pipe. The maximum shear stress is on the surface of the pipe. The polar moment of inertia of each pipe is computed as

$$I_{pAB} = \frac{\pi}{32}\left[(d + 2t_{AB})^4 - (d)^4\right]$$

$$= \frac{\pi}{32}\left[\left[10 \text{ in.} + 2\left(\frac{3}{4}\text{ in.}\right)\right]^4 - (10 \text{ in.})^4\right] = 735.335 \text{ in.}^4$$

and

$$I_{pBC} = \frac{\pi}{32}\left[(d + 2t_{BC})^4 - (d)^4\right]$$

$$= \frac{\pi}{32}\left[\left[10 \text{ in.} + 2\left(\frac{5}{8}\text{ in.}\right)\right]^4 - (10 \text{ in.})^4\right] = 590.822 \text{ in.}^4$$

Continues ➡

Example 3-5 - *Continued*

The shear modulus $G$ is constant, so the torsional rigidity of $AB$ is 1.245 times that of $BC$. From the TMD (Fig. 3-21), we see that the maximum torsional moments in both $AB$ and $BC$ (each equal to $6T_0/5$) are near joint $B$. Applying the torsion formula to pipes $AB$ and $BC$ near $B$ gives

$$\tau_{max\,AB} = \frac{\left(\frac{6}{5}T_0\right)\left(\frac{d + 2t_{AB}}{2}\right)}{I_{pAB}}$$

$$= \frac{\left(\frac{6}{5}2000\ \text{k-in.}\right) \cdot \left[\frac{10\ \text{in.} + 2\left(\frac{3}{4}\ \text{in.}\right)}{2}\right]}{735.335\ \text{in.}^4} = 18.8\ \text{ksi}$$

$$\tau_{max\,BC} = \frac{\left(\frac{6}{5}T_0\right)\left(\frac{d + 2t_{BC}}{2}\right)}{I_{pBC}}$$

$$= \frac{\left(\frac{6}{5}2000\ \text{k-in.}\right) \cdot \left[\frac{10\ \text{in.} + 2\left(\frac{5}{8}\ \text{in.}\right)}{2}\right]}{590.822\ \text{in.}^4} = 22.8\ \text{ksi} \quad \Leftarrow$$

So the maximum shear stress in the pipe is *just to the right of* the flange plate connection at joint $B$. "Just to the right of" means that we must move an appropriate distance away from the connection to avoid any stress concentration effects at the point of attachment of the two pipes in accordance with St. Venant's principle (see Section 3.12).

(c) *Twist rotations* $\phi(x)$. Next, we use the *torque-displacement relation*, Eqs. (3-24) through (3-27), to find the variation of twist rotation $\phi$ over the length of the pipe structure. Support $A$ is fixed, so $\phi_A = \phi(0) = 0$. The internal torque from $x = 0$ to $x = 2L/5$ (segment 1) is constant, so we use Eq. (3-24) to find twist rotation $\phi_1(x)$ which varies linearly from $x = 0$ to $x = 2L/5$:

$$\phi_1(x) = \frac{T_1(x)(x)}{GI_{pAB}} = \frac{\left(\frac{4T_0}{5}\right)(x)}{GI_{pAB}} = \frac{4T_0 x}{5GI_{pAB}} \qquad 0 \le x \le \frac{2L}{5} \qquad \text{(e)}$$

Evaluating Eq. (e) at $x = 2L/5$, we find the twist rotation at the point of application of torque $2T_0$ to be

$$\phi_1\left(\frac{2L}{5}\right) = \frac{T_1\left(\frac{2L}{5}\right)\left(\frac{2L}{5}\right)}{GI_{pAB}} = \frac{\left(\frac{4T_0}{5}\right)\left(\frac{2L}{5}\right)}{GI_{pAB}} = \frac{8T_0 L}{25GI_{pAB}} = \frac{0.32T_0 L}{GI_{pAB}} \qquad \text{(f)}$$

Next, we find an expression for the variation of twist angle $\phi_2(x)$ from $x = 2L/5$ to $x = 3L/5$ (point $B$). As with $\phi_1(x)$, twist $\phi_2(x)$ varies linearly over segment 2, because torque $T_2(x)$ is constant (Fig. 3-21). Using Eq. 3-24, we get

$$\phi_2(x) = \phi_1\left(\frac{2L}{5}\right) + \frac{T_2(x)\left(x - \frac{2L}{5}\right)}{GI_{pAB}} = \frac{8T_0 L}{25 GI_{pAB}} + \frac{\left(\frac{-6}{5}T_0\right)\left(x - \frac{2L}{5}\right)}{GI_{pAB}}$$

$$= \frac{2T_0(2L - 3x)}{5 GI_{pAB}} \qquad \frac{2L}{5} \le x \le \frac{3L}{5} \quad \text{(g)}$$

Finally, we develop an expression for twist over segment 3 (or pipe $BC$). We see that the internal torsional moment now has a linear variation (Fig. 3-21), so an integral form of the torque-displacement relation [Eq. (3-27)] is required. We insert the expression for $T_3(x)$ from Eq. (d) and add the torsional displacement at $B$ to get a formula for the variation of twist in $BC$:

$$\phi_3(x) = \phi_2\left(\frac{3L}{5}\right) + \int_{\frac{3L}{5}}^{x} \frac{\left[3T_0\left(\frac{\zeta}{L} - 1\right)\right]}{GI_{pBC}}\, d\zeta$$

$$= \frac{2T_0\left[2L - 3\left(\frac{3L}{5}\right)\right]}{5 GI_{pAB}} + \int_{\frac{3L}{5}}^{x} \frac{\left[3T_0\left(\frac{\zeta}{L} - 1\right)\right]}{GI_{pBC}}\, d\zeta$$

Torque $T_3(x)$ has a linear variation, so evaluating the integral yields a quadratic expression for variation of twist in $BC$:

$$\phi_3(x) = \frac{2LT_0}{25 GI_{pAB}} + \frac{3T_0(21L^2 - 50Lx + 25x^2)}{50 GI_{pBC}L} \qquad \frac{3L}{5} \le x \le L \quad \text{(h)}$$

Substituting $x = 3L/5$, we obtain the twist at $B$:

$$\phi_3\left(\frac{3L}{5}\right) = \frac{2LT_0}{25 GI_{pAB}}$$

At $x = L$, we get the twist at $C$:

$$\phi_3(L) = \frac{2LT_0}{25 GI_{pAB}} - \frac{6LT_0}{25 GI_{pBC}} = -0.219 \frac{T_0 L}{GI_{pAB}}$$

If we assume that $I_{pAB} = 1.245 I_{pBC}$ (based on the numerical properties here), we can plot the variation of twist over the length of the pipe structure (Fig. 3-22), noting that $\phi_{max}$ occurs at $x = 2L/5$ [see Eq. (f)].

Continues ➡

**Example 3-5** - *Continued*

**Fig 3-22**

Example 3-5: Torsional displacement diagram (TDD)

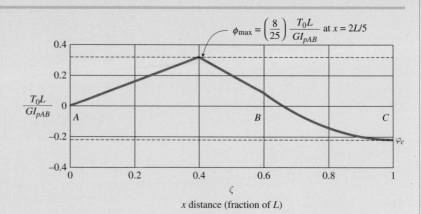

$$\phi_{\max} = \left(\frac{8}{25}\right)\frac{T_0 L}{GI_{pAB}} \text{ at } x = 2L/5$$

x distance (fraction of *L*)

(© Can Stock Photo Inc./ Nostalgie)

Finally, if we restrict $\phi_{\max}$ to the allowable value of 0.5°, we can solve for the maximum permissible value of *load variable* $T_0$ (kip-in.) using the numerical properties given previously:

$$T_{0\,\max} = \frac{GI_{pAB}}{0.32L}(\phi_{allow}) = \frac{(11800 \text{ ksi})(735.335 \text{ in.}^4)}{0.32\left[10 \text{ ft}\left(12\,\dfrac{\text{in.}}{\text{ft}}\right)\right]}(0.5°)$$

$$= 1972 \text{ k-in.} \qquad \Longleftarrow \quad \textbf{(i)}$$

(d) *Number of bolts required in flange plate.* We now use $T_{0,\max}$ from Eq. (i) to find the required number of $d_b$ = 7/8-in. diameter bolts at radius $r$ = 15 in. in the flange plate connection at *B*. The allowable shear stress in the bolts is $\tau_a$ = 28 ksi. We assume that each bolt carries an equal share of the torque at *B*, so each of *n* bolts carries shear force $F_b$ at distance *r* from the centroid of the cross section (Fig. 3-23).

**Fig. 3-23**

Example 3-5: Flange plate bolts at *B*

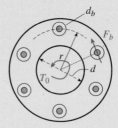

The maximum shear force $F_b$ per bolt is $\tau_a$ times the bolt cross-sectional area $A_b$, and the total torque at *B* is $6T_{0,\max}/5$ (see TMD in Fig. 3-21), so we find

$$nF_b r = \frac{6}{5}T_{0\max} \quad \text{or} \quad n = \frac{\dfrac{6}{5}T_{0\max}}{\tau_a A_b r} = \frac{\dfrac{6}{5}(1972 \text{ k-in.})}{(28 \text{ ksi})\left[\dfrac{\pi}{4}\left(\dfrac{7}{8}\text{ in.}\right)^2\right](15 \text{ in.})} = 9.37$$

Use **ten** 7/8-in. diameter bolts at a radius of 15 in. in the flange plate connection at *B*.

# 3.5 STRESSES AND STRAINS IN PURE SHEAR

When a circular bar, either solid or hollow, is subjected to torsion, shear stresses act over the cross sections and on longitudinal planes, as illustrated previously in Fig. 3-7. We will now examine in more detail the stresses and strains produced during twisting of a bar.

We begin by considering a stress element *abcd* cut between two cross sections of a bar in torsion (Figs. 3-24a and b). This element is in a state of **pure shear**, because the only stresses acting on it are the shear stresses $\tau$ on the four side faces (see the discussion of shear stresses in Section 1.7.)

The directions of these shear stresses depend upon the directions of the applied torques $T$. In this discussion, we assume that the torques rotate the right-hand end of the bar clockwise when viewed from the right (Fig. 3-24a); hence the shear stresses acting on the element have the directions shown in the figure. This same state of stress exists for a similar element cut from the interior of the bar, except that the magnitudes of the shear stresses are smaller because the radial distance to the element is smaller.

The directions of the torques shown in Fig. 3-24a are intentionally chosen so that the resulting shear stresses (Fig. 3-24b) are positive according to the sign convention for shear stresses described previously in Section 1.7. This **sign convention** is repeated here:

A shear stress acting on a positive face of an element is positive if it acts in the positive direction of one of the coordinate axes and negative if it acts in the negative direction of an axis. Conversely, a shear stress acting on a negative face of an element is positive if it acts in the negative direction of one of the coordinate axes and negative if it acts in the positive direction of an axis.

Applying this sign convention to the shear stresses acting on the stress element of Fig. 3-24b, we see that all four shear stresses are positive. For instance, the stress on the right-hand face (which is a positive face because the $x$ axis is directed to the right) acts in the positive direction of the $y$ axis; therefore, it is a positive shear stress. Also, the stress on the left-hand face (which is a negative face) acts in the negative direction of the $y$ axis; therefore, it is a positive shear stress. Analogous comments apply to the remaining stresses.

## Stresses on Inclined Planes

We are now ready to determine the stresses acting on *inclined planes* cut through the stress element in pure shear. We will follow the same approach as the one we used in Section 2.6 for investigating the stresses in uniaxial stress.

A two-dimensional view of the stress element is shown in Fig. 3-25a. As explained previously in Section 2.6, we usually draw a two-dimensional view for convenience, but we must always be aware that the element has a third dimension (thickness) perpendicular to the plane of the figure.

We now cut from the element a wedge-shaped (or "triangular") stress element having one face oriented at an angle $\theta$ to the $x$ axis (Fig. 3-25b). Normal stresses $\sigma_\theta$ and shear stresses $\tau_\theta$ act on this inclined face and are shown in their positive directions in the figure. The **sign convention** for

**Fig. 3-24**

Stresses acting on a stress element cut from a bar in torsion (pure shear)

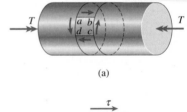

(a)

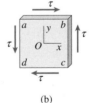

(b)

stresses $\sigma_\theta$ and $\tau_\theta$ was described previously in Section 2.6 and is repeated here:

Normal stresses $\sigma_\theta$ are positive in tension and shear stresses $\tau_\theta$ are positive when they tend to produce counterclockwise rotation of the material. (Note that this sign convention for the shear stress $\tau_\theta$ acting on an inclined plane is different from the sign convention for ordinary shear stresses $\tau$ that act on the sides of rectangular elements oriented to a set of $xy$ axes.)

The horizontal and vertical faces of the triangular element (Fig. 3-25b) have positive shear stresses $\tau$ acting on them, and the front and rear faces of the element are free of stress. Therefore, all stresses acting on the element are visible in this figure.

The stresses $\sigma_\theta$ and $\tau_\theta$ may now be determined from the equilibrium of the triangular element. The *forces* acting on its three side faces can be obtained by multiplying the stresses by the areas over which they act. For instance, the force on the left-hand face is equal to $\tau A_0$, where $A_0$ is the area of the vertical face. This force acts in the negative $y$ direction and is shown in the *free-body diagram* of Fig. 3-25c. Because the thickness of the element in the $z$ direction is constant, we see that the area of the bottom face is $A_0 \tan \theta$ and the area of the inclined face is $A_0 \sec \theta$. Multiplying the stresses acting on these faces by the corresponding areas enables us to obtain the remaining forces and thereby complete the free-body diagram (Fig. 3-25c).

## Fig. 3-25

Analysis of stresses on inclined planes: (a) element in pure shear, (b) stresses acting on a triangular stress element, and (c) forces acting on the triangular stress element (free-body diagram)

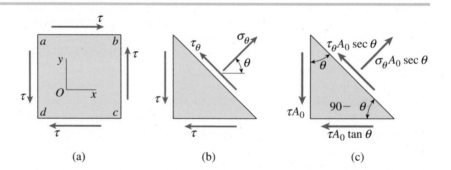

(a)    (b)    (c)

We are now ready to write two equations of equilibrium for the triangular element, one in the direction of $\sigma_\theta$ and the other in the direction of $\tau_\theta$. When writing these equations, the forces acting on the left-hand and bottom faces must be resolved into components in the directions of $\sigma_\theta$ and $\tau_\theta$. Thus, the first equation, obtained by summing forces in the direction of $\sigma_\theta$, is

$$\sigma_\theta A_0 \sec \theta = \tau A_0 \sin \theta + \tau A_0 \tan \theta \cos \theta$$

or

$$\sigma_\theta = 2\tau \sin \theta \cos \theta \tag{3-28a}$$

The second equation is obtained by summing forces in the direction of $\tau_\theta$:

$$\tau_\theta A_0 \sec \theta = \tau A_0 \cos \theta - \tau A_0 \tan \theta \sin \theta$$

or

$$\tau_\theta = \tau(\cos^2\theta - \sin^2\theta) \qquad \textbf{(3-28b)}$$

These equations can be expressed in simpler forms by introducing the following trigonometric identities (see Appendix D):

$$\sin 2\theta = 2\sin\theta\cos\theta \qquad \cos 2\theta = \cos^2\theta - \sin^2\theta$$

Then the equations for $\sigma_\theta$ and $\tau_\theta$ become

$$\sigma_\theta = \tau\sin 2\theta \qquad \tau_\theta = \tau\cos 2\theta \qquad \textbf{(3-29a,b)}$$

Equations (3-29a and b) give the normal and shear stresses acting on any inclined plane in terms of the shear stresses $\tau$ acting on the $x$ and $y$ planes (Fig. 3-25a) and the angle $\theta$ defining the orientation of the inclined plane (Fig. 3-25b).

The manner in which the stresses $\sigma_\theta$ and $\tau_\theta$ vary with the orientation of the inclined plane is shown by the graph in Fig. 3-26, which is a plot of Eqs. (3-29a and b). We see that for $\theta = 0$, which is the right-hand face of the stress element in Fig. 3-25a, the graph gives $\sigma_\theta = 0$ and $\tau_\theta = \tau$. This latter result is expected, because the shear stress $\tau$ acts counterclockwise against the element and therefore produces a positive shear stress $\tau_\theta$.

For the top face of the element ($\theta = 90°$), we obtain $\sigma_\theta = 0$ and $\tau_\theta = -\tau$. The minus sign for $\tau_\theta$ means that it acts clockwise against the element, that is, to the right on face $ab$ (Fig. 3-25a), which is consistent with the direction of the shear stress $\tau$. Note that the numerically largest shear stresses occur on the planes for which $\theta = 0$ and $90°$, as well as on the opposite faces ($\theta = 180°$ and $270°$).

From the graph we see that the normal stress $\sigma_\theta$ reaches a maximum value at $\theta = 45°$. At that angle, the stress is positive (tension) and equal numerically to the shear stress $\tau$. Similarly, $\sigma_\theta$ has its minimum value (which is compressive) at $\theta = -45°$. At both of these 45° angles, the shear stress $\tau_\theta$ is equal to zero. These conditions are pictured in Fig. 3-27 which shows stress elements oriented at $\theta = 0$ and $\theta = 45°$. The element at 45° is acted upon by equal tensile and compressive stresses in perpendicular directions, with no shear stresses.

Note that the normal stresses acting on the 45° element (Fig. 3-27b) correspond to an element subjected to shear stresses $\tau$ acting in the directions shown in Fig. 3-27a. If the shear stresses acting on the element of Fig. 3-27a are reversed in direction, the normal stresses acting on the 45° planes also will change directions.

If a stress element is oriented at an angle other than 45°, both normal and shear stresses will act on the inclined faces [see Eqs. (3-29a and b) and Fig. 3-26]. Stress elements subjected to these more general conditions are discussed in detail in Chapter 7.

The equations derived in this section are valid for a stress element in pure shear regardless of whether the element is cut from a bar in torsion or from some other structural element. Also, since Eqs. (3-29) were derived from equilibrium only, they are valid for any material, whether or not it behaves in a linearly elastic manner.

**Fig. 3-26**

Graph of normal stresses $\sigma_\theta$ and shear stresses $\tau_\theta$ versus angle $\theta$ of the inclined plane

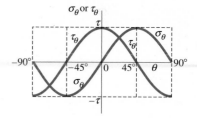

## Fig. 3-27

Stress elements oriented at $\theta = 0$ and $\theta = 45°$ for pure shear

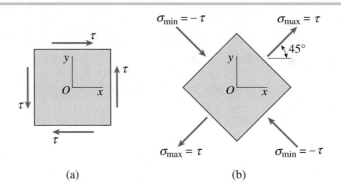

(a)    (b)

## Fig. 3-28

Torsion failure of a brittle material by tension cracking along a 45° helical surface

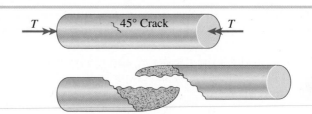

The existence of maximum tensile stresses on planes at 45° to the $x$ axis (Fig. 3-27b) explains why bars in torsion that are made of materials that are brittle and weak in tension fail by cracking along a 45° helical surface (Fig. 3-28). As mentioned in Section 3.3, this type of failure is readily demonstrated by twisting a piece of classroom chalk.

## Strains in Pure Shear

Let us now consider the strains that exist in an element in pure shear. For instance, consider the element in pure shear shown in Fig. 3-27a. The corresponding shear strains are shown in Fig. 3-29a, where the deformations are highly exaggerated. The shear strain $\gamma$ is the change in angle between two lines that were originally perpendicular to each other, as discussed previously in Section 1.7. Thus, the decrease in the angle at the lower left-hand corner of the element is the shear strain $\gamma$ (measured in radians). This same change in angle occurs at the upper right-hand corner, where the angle decreases, and at the other two corners, where the angles increase. However, the lengths of the sides of the element, including the thickness perpendicular to the plane of the paper, do not change when these shear deformations occur. Therefore, the element changes its shape from a rectangular parallelepiped (Fig. 3-27a) to an oblique parallelepiped (Fig. 3-29a). This change in shape is called a **shear distortion**.

If the material is linearly elastic, the shear strain for the element oriented at $\theta = 0$ (Fig. 3-29a) is related to the shear stress by Hooke's law in shear:

$$\gamma = \frac{\tau}{G} \tag{3-30}$$

where, as usual, the symbol $G$ represents the shear modulus of elasticity.

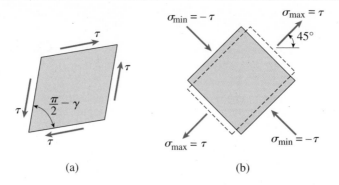

**Fig. 3-29**

Strains in pure shear: (a) shear distortion of an element oriented at $\theta = 0$, and (b) distortion of an element oriented at $\theta = 45°$

(a)                    (b)

Next, consider the strains that occur in an element oriented at $\theta = 45°$ (Fig. 3-29b). The tensile stresses acting at 45° tend to elongate the element in that direction. Because of the Poisson effect, they also tend to shorten it in the perpendicular direction (the direction where $\theta = 135°$ or $-45°$). Similarly, the compressive stresses acting at 135° tend to shorten the element in that direction and elongate it in the 45° direction. These dimensional changes are shown in Fig. 3-29b, where the dashed lines show the deformed element. Since there are no shear distortions, the element remains a rectangular parallelepiped even though its dimensions have changed.

If the material is linearly elastic and follows Hooke's law, we can obtain an equation relating strain to stress for the element at $\theta = 45°$ (Fig. 3-29b). The tensile stress $\sigma_{max}$ acting at $\theta = 45°$ produces a positive normal strain in that direction equal to $\sigma_{max}/E$. Since $\sigma_{max} = \tau$, we can also express this strain as $\tau/E$. The stress $\sigma_{max}$ also produces a negative strain in the perpendicular direction equal to $-v\tau/E$, where $v$ is Poisson's ratio. Similarly, the stress $\sigma_{min} = -\tau$ (at $\theta = 135°$) produces a negative strain equal to $-\tau/E$ in that direction and a positive strain in the perpendicular direction (the 45° direction) equal to $v\tau/E$. Therefore, the normal strain in the 45° direction is

$$\varepsilon_{max} = \frac{\tau}{E} + \frac{v\tau}{E} = \frac{\tau}{E}(1 + v) \qquad \textbf{(3-31)}$$

which is positive, representing elongation. The strain in the perpendicular direction is a negative strain of the same amount. In other words, pure shear produces elongation in the 45° direction and shortening in the 135° direction. These strains are consistent with the shape of the deformed element of Fig. 3-29a, because the 45° diagonal has lengthened and the 135° diagonal has shortened.

In the next section we will use the geometry of the deformed element to relate the shear strain $\gamma$ (Fig. 3-29a) to the normal strain $\varepsilon_{max}$ in the 45° direction (Fig. 3-29b). In so doing, we will derive the following relationship:

$$\varepsilon_{max} = \frac{\gamma}{2} \qquad \textbf{(3-32)}$$

This equation, in conjunction with Eq. (3-30), can be used to calculate the maximum shear strains and maximum normal strains in pure torsion when the shear stress $\tau$ is known.

**• • • Example 3-6**

A circular tube with an outside diameter of 80 mm and an inside diameter of 60 mm is subjected to a torque $T = 4.0$ kN·m (Fig. 3-30). The tube is made of aluminum alloy 7075-T6.

(a) Determine the maximum shear, tensile, and compressive stresses in the tube and show these stresses on sketches of properly oriented stress elements.
(b) Determine the corresponding maximum strains in the tube and show these strains on sketches of the deformed elements.
(c) What is the maximum permissible torque $T_{max}$ if the allowable normal strain is $\varepsilon_a = 0.9 \times 10^{-3}$?
(d) If $T = 4.0$ kN·m and $\varepsilon_a = 0.9 \times 10^{-3}$, what new outer diameter is required so that the tube can carry the required torque $T$ (assume that the inner diameter of the tube remains at 60 mm)?

**Fig. 3-30**

Example 3-6: Circular tube in torsion

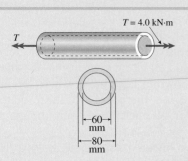

$T = 4.0$ kN·m

$T$

60 mm
80 mm

**Solution**

(a) *Maximum stresses.* The maximum values of all three stresses (shear, tensile, and compressive) are equal numerically, although they act on different planes. Their magnitudes are found from the torsion formula:

$$\tau_{max} = \frac{Tr}{I_P} = \frac{(4000\ \text{N} \cdot \text{m})(0.040\ \text{m})}{\dfrac{\pi}{32}\left[(0.080\ \text{m})^4 - (0.060\ \text{m})^4\right]} = 58.2\ \text{MPa}$$  ◀

The maximum shear stresses act on cross-sectional and longitudinal planes, as shown by the stress element in Fig. 3-31a, where the $x$ axis is parallel to the longitudinal axis of the tube.
  The maximum tensile and compressive stresses are

$$\sigma_t = 58.2\ \text{MPa} \qquad \sigma_c = -58.2\ \text{MPa}$$  ◀

These stresses act on planes at 45° to the axis (Fig. 3-31b).

(b) *Maximum strains.* The maximum shear strain in the tube is obtained from Eq. (3-30). The shear modulus of elasticity is obtained from Table I-2, Appendix I, as $G = 27$ GPa. Therefore, the maximum shear strain is

$$\gamma_{max} = \frac{\tau_{max}}{G} = \frac{58.2\ \text{MPa}}{27\ \text{GPa}} = 0.0022\ \text{rad}$$  ◀

The deformed element is shown by the dashed lines in Fig. 3-28c.
  The magnitude of the maximum normal strains (from Eq. 3-32) is

$$\varepsilon_{max} = \frac{\gamma_{max}}{2} = 0.0011$$

Thus, the maximum tensile and compressive strains are

$$\varepsilon_t = 0.0011 \qquad \varepsilon_c = -0.0011$$  ◀

The deformed element is shown by the dashed lines in Fig. 3-31d for an element with sides of unit length.

## Fig. 3-31

Stress and strain elements for the tube of Example 3-6: (a) maximum shear stresses, (b) maximum tensile and compressive stresses, (c) maximum shear strains, and (d) maximum tensile and compressive strains

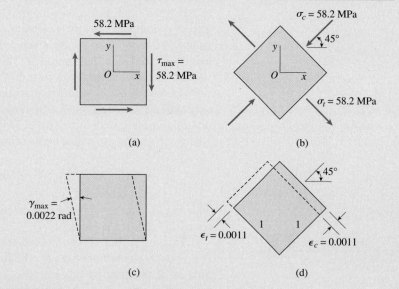

(a)

(b)

(c)

(d)

(c) *Maximum permissible torque.* The tube is in *pure shear*, so the allowable shear strain is twice the allowable normal strain (see Eq. 3-32):

$$\gamma_a = 2\varepsilon_a = 2(0.9 \times 10^{-3}) = 1.8 \times 10^{-3}$$

From the shear formula (Eq. 3-13), we get

$$\tau_{max} = \frac{T\left(\dfrac{d_2}{2}\right)}{I_p} \quad \text{so} \quad T_{max} = \frac{\tau_a I_p}{\left(\dfrac{d_2}{2}\right)} = \frac{2(G\gamma_a)I_p}{d_2}$$

where $d_2$ is the outer diameter. Substituting numerical values gives

$$T_{max} = \frac{2(27 \text{ GPa})(1.8 \times 10^{-3})\left[\dfrac{\pi}{32}\left[(0.08 \text{ m})^4 - (0.06 \text{ m})^4\right]\right]}{0.08 \text{ m}}$$

$$= 3.34 \text{ kN} \cdot \text{m}$$

(d) *New outer diameter of tube.* We can use the previous equation but with $T = 4.0$ kN·m to find the required outer diameter $d_2$:

$$\frac{I_p}{d_2} = \frac{T}{2G\gamma_a} \quad \text{or} \quad \frac{d_2^4 - (0.06 \text{ m})^4}{d_2} = \frac{\left(\dfrac{32}{\pi}\right)4 \text{ kN} \cdot \text{m}}{2(27 \text{ GPa})(1.8 \times 10^{-6})} = 0.41917 \text{ m}^3$$

Solving for the required outer diameter $d_2$ numerically gives

$$d_2 = 83.2 \text{ mm}$$

# 3.6 RELATIONSHIP BETWEEN MODULI OF ELASTICITY *E* and *G*

An important relationship between the moduli of elasticity $E$ and $G$ can be obtained from the equations derived in the preceding section. For this purpose, consider the stress element $abcd$ shown in Fig. 3-32a on the next page. The front face of the element is assumed to be square, with the length of each side denoted as $h$. When this element is subjected to pure shear by stresses $\tau$, the front face distorts into a rhombus (Fig. 3-32b) with sides of length $h$ and with shear strain $\gamma = \tau/G$. Because of the distortion, diagonal $bd$ is lengthened and diagonal $ac$ is shortened. The length of diagonal $bd$ is equal to its initial length $\sqrt{2}h$ times the factor $1 + \varepsilon_{max}$, where $\varepsilon_{max}$ is the normal strain in the 45° direction; thus,

$$L_{bd} = \sqrt{2}h(1 + \varepsilon_{max}) \tag{3-33}$$

This length can be related to the shear strain $\gamma$ by considering the geometry of the deformed element.

To obtain the required geometric relationships, consider triangle $abd$ (Fig. 3-32c) which represents one-half of the rhombus pictured in Fig. 3-32b. Side $bd$ of this triangle has length $L_{bd}$ [Eq. (3-33)], and the other sides have length $h$. Angle $adb$ of the triangle is equal to one-half of angle $adc$ of the rhombus, or $\pi/4 - \gamma/2$. The angle $abd$ in the triangle is the same. Therefore, angle $dab$ of the triangle equals $\pi/2 + \gamma$. Now using the law of cosines (see Appendix D) for triangle $abd$, we get

$$L_{bd}^2 = h^2 + h^2 - 2h^2 \cos\left(\frac{\pi}{2} + \gamma\right)$$

---

**Fig. 3-32**

Geometry of deformed element in pure shear

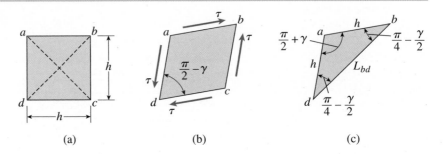

(a)                    (b)                    (c)

---

Substituting for $L_{bd}$ from Eq. (3-33) and simplifying, we get

$$(1 + \varepsilon_{max})^2 = 1 - \cos\left(\frac{\pi}{2} + \gamma\right)$$

By expanding the term on the left-hand side, and also observing that $\cos(\pi/2 + \gamma) = -\sin\gamma$, we obtain

$$1 + 2\varepsilon_{max} + \varepsilon_{max}^2 = 1 + \sin\gamma$$

Because $\varepsilon_{max}$ and $\gamma$ are very small strains, we can disregard $\varepsilon_{max}^2$ in comparison with $2\varepsilon_{max}$ and we can replace $\sin \gamma$ by $\gamma$. The resulting expression is

$$\varepsilon_{max} = \frac{\gamma}{2} \tag{3-34}$$

which establishes the relationship already presented in Section 3.5 as Eq. (3-32).

The shear strain $\gamma$ appearing in Eq. (3-34) is equal to $\tau/G$ by Hooke's law (Eq. 3-30) and the normal strain $\varepsilon_{max}$ is equal to $\tau(1 + v)/E$ by Eq. (3-31). Making both of these substitutions in Eq. (3-34) yields

$$G = \frac{E}{2(1 + v)} \tag{3-35}$$

We see that $E$, $G$, and $v$ are not independent properties of a linearly elastic material. Instead, if any two of them are known, the third can be calculated from Eq. (3-35).

Typical values of $E$, $G$, and $v$ are listed in Table I-2, Appendix I.

# 3.7 TRANSMISSION OF POWER BY CIRCULAR SHAFTS

The most important use of circular shafts is to transmit mechanical power from one device or machine to another, as in the drive shaft of an automobile, the propeller shaft of a ship, or the axle of a bicycle. The power is transmitted through the rotary motion of the shaft, and the amount of power transmitted depends upon the magnitude of the torque and the speed of rotation. A common design problem is to determine the required size of a shaft so that it will transmit a specified amount of power at a specified rotational speed without exceeding the allowable stresses for the material.

Let us suppose that a motor-driven shaft (Fig. 3-33) is rotating at an angular speed $\omega$, measured in radians per second (rad/s). The shaft transmits a torque $T$ to a device (not shown in the figure) that is performing useful work. The torque applied by the shaft to the external device has the same sense as the angular speed $\omega$, that is, its vector points to the left. However, the torque shown in the figure is the torque exerted *on the shaft* by the device, and so its vector points in the opposite direction.

In general, the work $W$ done by a torque of constant magnitude is equal to the product of the torque and the angle through which it rotates; that is,

$$W = T\psi \tag{3-36}$$

where $\psi$ is the angle of rotation in radians.

**Power** is the *rate* at which work is done, or

$$P = \frac{dW}{dt} = T\frac{d\Psi}{dt} \tag{3-37}$$

**Fig. 3-33**

Shaft transmitting a constant torque $T$ at an angular speed $\omega$

Motor

in which $P$ is the symbol for power and $t$ represents time. The rate of change $d\psi/dt$ of the angular displacement $\psi$ is the angular speed $\omega$, and therefore the preceding equation becomes

$$P = T\omega \quad (\omega = \text{rad/s}) \tag{3-38}$$

This formula, which is familiar from elementary physics, gives the power transmitted by a rotating shaft transmitting a constant torque $T$.

The **units** to be used in Eq. (3-38) are as follows. If the torque $T$ is expressed in newton meters, then the power is expressed in watts (W). One watt is equal to one newton meter per second (or one joule per second). If $T$ is expressed in pound-feet, then the power is expressed in foot-pounds per second.*

Angular speed is often expressed as the frequency $f$ of rotation, which is the number of revolutions per unit of time. The unit of frequency is the hertz (Hz), equal to one revolution per second ($s^{-1}$). Inasmuch as one revolution equals $2\pi$ radians, we obtain

$$\omega = 2\pi f \quad (\omega = \text{rad/s}, f = \text{Hz} = s^{-1}) \tag{3-39}$$

The expression for power (Eq. 3-38) then becomes

$$P = 2\pi f T \quad (f = \text{Hz} = s^{-1}) \tag{3-40}$$

Another commonly used unit is the number of revolutions per minute (rpm), denoted by the letter $n$. Therefore, we also have the following relationships:

$$n = 60f \tag{3-41}$$

and

$$P = \frac{2\pi n T}{60} \quad (n = \text{rpm}) \tag{3-42}$$

In Eqs. (3-40) and (3-42), the quantities $P$ and $T$ have the same units as in Eq. (3-38); that is, $P$ has units of watts if $T$ has units of newton meters, and $P$ has units of foot-pounds per second if $T$ has units of pound-feet.

In U.S. engineering practice, power is sometimes expressed in horsepower (hp), a unit equal to 550 ft-lb/s. Therefore, the horsepower $H$ being transmitted by a rotating shaft is

$$H = \frac{2\pi n T}{60(550)} = \frac{2\pi n T}{33{,}000} \quad (n = \text{rpm}, T = \text{lb-ft}, H = \text{hp}) \tag{3-43}$$

One horsepower is approximately 746 watts.

The preceding equations relate the torque acting in a shaft to the power transmitted by the shaft. Once the torque is known, we can determine the shear stresses, shear strains, angles of twist, and other desired quantities by the methods described in Sections 3.2 through 3.5.

The following examples illustrate some of the procedures for analyzing rotating shafts.

---

*See Table B-1, Appendix B, for units of work and power.

● ● ● ● **Example 3-7**

A motor driving a solid circular steel shaft transmits 40 hp to a gear at *B* (Fig. 3-34). The allowable shear stress in the steel is 6000 psi.

(a) What is the required diameter *d* of the shaft if it is operated at 500 rpm?
(b) What is the required diameter *d* if it is operated at 3000 rpm?

**Fig. 3-34**

Example 3-7: Steel shaft in torsion

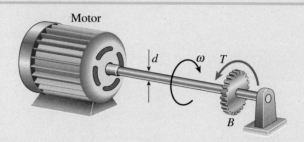

Motor

**Solution**

(a) *Motor operating at 500 rpm.* Knowing the horsepower and the speed of rotation, we can find the torque *T* acting on the shaft by using Eq. (3-43). Solving that equation for *T*, we get

$$T = \frac{33,000\,H}{2\pi n} = \frac{33,000(40\ \text{hp})}{2\pi(500\ \text{rpm})} = 420.2\ \text{lb-ft} = 5042\ \text{lb-in.}$$

This torque is transmitted by the shaft from the motor to the gear.

The maximum shear stress in the shaft can be obtained from the modified torsion formula (Eq. 3-14):

$$\tau_{max} = \frac{16T}{\pi d^3}$$

Solving that equation for the diameter *d*, and also substituting $\tau_{allow}$ for $\tau_{max}$, we get

$$d^3 = \frac{16T}{\pi \tau_{allow}} = \frac{16(5042\ \text{lb-in.})}{\pi(6000\ \text{psi})} = 4.280\ \text{in.}^3$$

from which

$$d = 1.62\ \text{in.} \qquad \blacktriangleleft$$

The diameter of the shaft must be at least this large if the allowable shear stress is not to be exceeded.

(b) *Motor operating at 3000 rpm.* Following the same procedure as in part (a), we obtain

$$T = \frac{33,000H}{2\pi n} = \frac{33,000(40\ \text{hp})}{2\pi(3000\ \text{rpm})} = 70.03\ \text{lb-ft} = 840.3\ \text{lb-in.}$$

$$d^3 = \frac{16T}{\pi \tau_{allow}} = \frac{16(840.3\ \text{lb-in.})}{\pi(6000\ \text{psi})} = 0.7133\ \text{in.}^3$$

$$d = 0.89\ \text{in.} \qquad \blacktriangleleft$$

which is less than the diameter found in part (a).

This example illustrates that the higher the speed of rotation, the smaller the required size of the shaft (for the same power and the same allowable stress).

• • • **Example 3-8**

A solid steel shaft *ABC* of 50 mm diameter (Fig. 3-35a) is driven at *A* by a motor that transmits 50 kW to the shaft at 10 Hz. The gears at *B* and *C* drive machinery requiring power equal to 35 kW and 15 kW, respectively.

Compute the maximum shear stress $\tau_{max}$ in the shaft and the angle of twist $\phi_{AC}$ between the motor at *A* and the gear at *C*. (Use *G* = 80 GPa.)

### Fig. 3-35

Example 3-8. Steel shaft in torsion

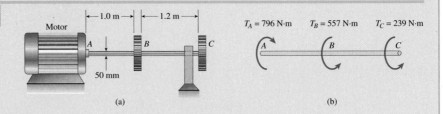

(a)                    (b)

### Solution

*Torques acting on the shaft.* We begin the analysis by determining the torques applied to the shaft by the motor and the two gears. Since the motor supplies 50 kW at 10 Hz, it creates a torque $T_A$ at end *A* of the shaft (Fig. 3-35b) that we can calculate from Eq. (3-40):

$$T_A = \frac{P}{2\pi f} = \frac{50 \text{ kW}}{2\pi(10 \text{ Hz})} = 796 \text{ N} \cdot \text{m}$$

In a similar manner, we can calculate the torques $T_B$ and $T_C$ applied by the gears to the shaft:

$$T_B = \frac{P}{2\pi f} = \frac{35 \text{ kW}}{2\pi(10 \text{ Hz})} = 557 \text{ N} \cdot \text{m}$$

$$T_C = \frac{P}{2\pi f} = \frac{15 \text{ kW}}{2\pi(10 \text{ Hz})} = 239 \text{ N} \cdot \text{m}$$

These torques are shown in the free-body diagram of the shaft (Fig. 3-35b). Note that the torques applied by the gears are opposite in direction to the torque applied by the motor. (If we think of $T_A$ as the "load" applied to the shaft by the motor, then the torques $T_B$ and $T_C$ are the "reactions" of the gears.)

The internal torques in the two segments of the shaft are now found (by inspection) from the free-body diagram of Fig. 3-35b:

$$T_{AB} = 796 \text{ N} \cdot \text{m} \qquad T_{BC} = 239 \text{ N} \cdot \text{m}$$

Both internal torques act in the same direction, and therefore the angles of twist in segments AB and BC are additive when finding the total angle of twist. (To be specific, both torques are positive according to the sign convention adopted in Section 3.4.)

*Shear stresses and angles of twist.* The shear stress and angle of twist in segment AB of the shaft are found in the usual manner from Eqs. (3-14) and (3-17):

$$\tau_{AB} = \frac{16T_{AB}}{\pi d^3} = \frac{16(796 \text{ N} \cdot \text{m})}{\pi (50 \text{ mm})^3} = 32.4 \text{ MPa}$$

$$\phi_{AB} = \frac{T_{AB}L_{AB}}{GI_P} = \frac{(796 \text{ N} \cdot \text{m})(1.0 \text{ m})}{(80 \text{ GPa})\left(\dfrac{\pi}{32}\right)(50 \text{ mm})^4} = 0.0162 \text{ rad}$$

The corresponding quantities for segment BC are

$$\tau_{BC} = \frac{16T_{BC}}{\pi d^3} = \frac{16(239 \text{ N} \cdot \text{m})}{\pi (50 \text{ mm})^3} = 9.7 \text{ MPa}$$

$$\phi_{BC} = \frac{T_{BC}L_{BC}}{GI_P} = \frac{(239 \text{ N} \cdot \text{m})(1.2 \text{ m})}{(80 \text{ GPa})\left(\dfrac{\pi}{32}\right)(50 \text{ mm})^4} = 0.0058 \text{ rad}$$

Thus, the maximum shear stress in the shaft occurs in segment AB and is

$$\tau_{\text{max}} = 32.4 \text{ MPa} \qquad \leftarrow$$

Also, the total angle of twist between the motor at A and the gear at C is

$$\phi_{AC} = \phi_{AB} + \phi_{BC} = 0.0162 \text{ rad} + 0.0058 \text{ rad} = 0.0220 \text{ rad} = 1.26° \qquad \leftarrow$$

As explained previously, both parts of the shaft twist in the same direction, and therefore the angles of twist are added.

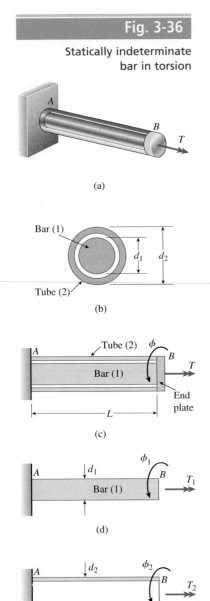

**Fig. 3-36**

Statically indeterminate
bar in torsion

# 3.8 STATICALLY INDETERMINATE TORSIONAL MEMBERS

The bars and shafts described in the preceding sections of this chapter are *statically determinate* because all internal torques and all reactions can be obtained from free-body diagrams and equations of equilibrium. However, if additional restraints, such as fixed supports, are added to the bars, the equations of equilibrium will no longer be adequate for determining the torques. The bars are then classified as **statically indeterminate**. Torsional members of this kind can be analyzed by supplementing the equilibrium equations with compatibility equations pertaining to the rotational displacements. Thus, the general method for analyzing statically indeterminate torsional members is the same as described in Section 2.4 for statically indeterminate bars with axial loads.

The first step in the analysis is to write **equations of equilibrium**, obtained from free-body diagrams of the given physical situation. The unknown quantities in the equilibrium equations are torques, either internal torques or reaction torques.

The second step in the analysis is to formulate **equations of compatibility**, based upon physical conditions pertaining to the angles of twist. As a consequence, the compatibility equations contain angles of twist as unknowns.

The third step is to relate the angles of twist to the torques by **torque-displacement relations**, such as $\phi = TL/GI_P$. After introducing these relations into the compatibility equations, they too become equations containing torques as unknowns. Therefore, the last step is to obtain the unknown torques by solving simultaneously the equations of equilibrium and compatibility.

To illustrate the method of solution, we will analyze the composite bar *AB* shown in Fig. 3-36a. The bar is attached to a fixed support at end *A* and loaded by a torque *T* at end *B*. Furthermore, the bar consists of two parts: a solid bar and a tube (Figs. 3-36b and c), with both the solid bar and the tube joined to a rigid end plate at *B*.

For convenience, we will identify the solid bar and tube (and their properties) by the numerals 1 and 2, respectively. For instance, the diameter of the solid bar is denoted $d_1$ and the outer diameter of the tube is denoted $d_2$. A small gap exists between the bar and the tube, and therefore the inner diameter of the tube is slightly larger than the diameter $d_1$ of the bar.

When the torque *T* is applied to the composite bar, the end plate rotates through a small angle $\phi$ (Fig. 3-36c) and torques $T_1$ and $T_2$ are developed in the solid bar and the tube, respectively (Figs. 3-36d and e). From equilibrium we know that the sum of these torques equals the applied load, and so the *equation of equilibrium* is

$$T_1 + T_2 = T \qquad \textbf{(3-44)}$$

Because this equation contains two unknowns ($T_1$ and $T_2$), we recognize that the composite bar is statically indeterminate.

To obtain a second equation, we must consider the rotational displacements of both the solid bar and the tube. Let us denote the angle of twist of the solid bar (Fig. 3-36d) by $\phi_1$ and the angle of twist of the tube by $\phi_2$ (Fig. 3-36e). These angles of twist must be equal because the bar and tube are securely joined to the end plate and rotate with it; consequently, the *equation of compatibility* is

$$\phi_1 = \phi_2 \tag{3-45}$$

The angles $\phi_1$ and $\phi_2$ are related to the torques $T_1$ and $T_2$ by the *torque-displacement relations*, which in the case of linearly elastic materials are obtained from the equation $\phi = TL/GI_P$. Thus,

$$\phi_1 = \frac{T_1 L}{G_1 I_{P1}} \qquad \phi_2 = \frac{T_2 L}{G_2 I_{P2}} \tag{3-46a,b}$$

in which $G_1$ and $G_2$ are the shear moduli of elasticity of the materials and $I_{P1}$ and $I_{P2}$ are the polar moments of inertia of the cross sections.

When the preceding expressions for $\phi_1$ and $\phi_2$ are substituted into Eq. (3-45), the equation of compatibility becomes

$$\frac{T_1 L}{G_1 I_{P1}} = \frac{T_2 L}{G_2 I_{P2}} \tag{3-47}$$

We now have two equations [Eqs. (3-44) and (3-47)] with two unknowns, so we can solve them for the torques $T_1$ and $T_2$. The results are

$$T_1 = T\left(\frac{G_1 I_{P1}}{G_1 I_{P1} + G_2 I_{P2}}\right) \qquad T_2 = T\left(\frac{G_2 I_{P2}}{G_1 I_{P1} + G_2 I_{P2}}\right) \tag{3-48a,b}$$

With these torques known, the essential part of the statically indeterminate analysis is completed. All other quantities, such as stresses and angles of twist, can now be found from the torques.

The preceding discussion illustrates the general methodology for analyzing a statically indeterminate system in torsion. In the following example, this same approach is used to analyze a bar that is fixed against rotation at both ends. In the example and in the problems, we assume that the bars are made of linearly elastic materials. However, the general methodology is also applicable to bars of nonlinear materials—the only change is in the torque-displacement relations.

## Example 3-9

**Fig. 3-37**

Example 3-9: Statically
indeterminate bar in torsion

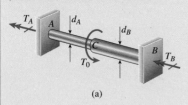

(a)

(b)

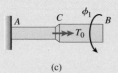

(c)

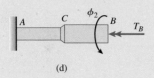

(d)

The bar *ACB* shown in Figs. 3-37a and b is fixed at both ends and loaded by
a torque $T_0$ at point *C*. Segments *AC* and *CB* of the bar have diameters $d_A$
and $d_B$, lengths $L_A$ and $L_B$, and polar moments of inertia $I_{PA}$ and $I_{PB}$, respectively. The material of the bar is the same throughout both segments.

   Obtain formulas for (a) the reactive torques $T_A$ and $T_B$ at the ends,
(b) the maximum shear stresses $\tau_{AC}$ and $\tau_{CB}$ in each segment of the bar, and
(c) the angle of rotation $\phi_C$ at the cross section where the load $T_0$ is applied.

### Solution

*Equation of equilibrium.* The load $T_0$ produces reactions $T_A$ and $T_B$ at the
fixed ends of the bar, as shown in Figs. 3-37a and b. Thus, from the equilibrium of the bar we obtain

$$T_A + T_B = T_0 \tag{f}$$

Because there are two unknowns in this equation (and no other useful
equations of equilibrium), the bar is statically indeterminate.

   *Equation of compatibility.* We now separate the bar from its support at
end *B* and obtain a bar that is fixed at end *A* and free at end *B* (Figs. 3-37c
and d). When the load $T_0$ acts alone (Fig. 3-37c), it produces an angle of twist
at end *B* that we denote as $\phi_1$. Similarly, when the reactive torque $T_B$ acts
alone, it produces an angle $\phi_2$ (Fig. 3-37d). The angle of twist at end *B* in the
original bar, equal to the sum of $\phi_1$ and $\phi_2$, is zero. Therefore, the equation
of compatibility is

$$\phi_1 + \phi_2 = 0 \tag{g}$$

Note that $\phi_1$ and $\phi_2$ are assumed to be positive in the direction shown in the
figure.

   *Torque-displacement equations.* The angles of twist $\phi_1$ and $\phi_2$ can be
expressed in terms of the torques $T_0$ and $T_B$ by referring to Figs. 3-37c and d
and using the equation $\phi = TL/GI_P$. The equations are as follows:

$$\phi_1 = \frac{T_0 L_A}{GI_{pA}} \qquad \phi_2 = -\frac{T_B L_A}{GI_{pA}} - \frac{T_B L_B}{GI_{pB}} \tag{h,i}$$

The minus signs appear in Eq. (i) because $T_B$ produces a rotation that is
opposite in direction to the positive direction of $\phi_2$ (Fig. 3-37d).

   We now substitute the angles of twist (Eqs. h and i) into the compatibility equation (Eq. g) and obtain

$$\frac{T_0 L_A}{GI_{pA}} - \frac{T_B L_A}{GI_{pA}} - \frac{T_B L_B}{GI_{pB}} = 0$$

or

$$\frac{T_B L_A}{I_{pA}} + \frac{T_B L_B}{I_{pB}} = \frac{T_0 L_A}{I_{pA}} \tag{j}$$

*Solution of equations.* The preceding equation can be solved for the torque $T_B$, which then can be substituted into the equation of equilibrium (Eq. f) to obtain the torque $T_A$. The results are

$$T_A = T_0\left(\frac{L_B I_{pA}}{L_B I_{pA} + L_A I_{pB}}\right) \quad T_B = T_0\left(\frac{L_A I_{pB}}{L_B I_{pA} + L_A I_{pB}}\right) \quad \Longleftarrow \quad (3\text{-}49\text{a,b})$$

Thus, the reactive torques at the ends of the bar have been found, and the statically indeterminate part of the analysis is completed.

As a special case, note that if the bar is prismatic ($I_{PA} = I_{PB} = I_p$) the preceding results simplify to

$$T_A = \frac{T_0 L_B}{L} \quad T_B = \frac{T_0 L_A}{L} \qquad (3\text{-}50\text{a,b})$$

where $L$ is the total length of the bar. These equations are analogous to those for the reactions of an axially loaded bar with fixed ends (see Eqs. 2-13a and 2-13b).

*Maximum shear stresses.* The maximum shear stresses in each part of the bar are obtained directly from the torsion formula:

$$\tau_{AC} = \frac{T_A d_A}{2I_{PA}} \quad \tau_{CB} = \frac{T_B d_B}{2I_{PB}}$$

Substituting from Eqs. (3-49a) and (3-49b) gives

$$\tau_{AC} = \frac{T_0 L_B d_A}{2(L_B I_{PA} + L_A I_{PB})} \quad \tau_{CB} = \frac{T_0 L_A d_B}{2(L_B I_{PA} + L_A I_{PB})} \quad \Longleftarrow \quad (3\text{-}51\text{a,b})$$

By comparing the product $L_B d_A$ with the product $L_A d_B$, we can immediately determine which segment of the bar has the larger stress.

*Angle of rotation.* The angle of rotation $\phi_C$ at section $C$ is equal to the angle of twist of either segment of the bar, since both segments rotate through the same angle at section $C$. Therefore, we obtain

$$\phi_C = \frac{T_A L_A}{GI_{PA}} = \frac{T_B L_B}{GI_{PB}} = \frac{T_0 L_A L_B}{G(L_B I_{PA} + L_A I_{PB})} \quad \Longleftarrow \quad (3\text{-}52)$$

In the special case of a prismatic bar ($I_{PA} = I_{PB} = I_p$), the angle of rotation at the section where the load is applied is

$$\phi_C = \frac{T_0 L_A L_B}{GLI_p} \qquad (3\text{-}53)$$

This example illustrates not only the analysis of a statically indeterminate bar but also the techniques for finding stresses and angles of rotation. In addition, note that the results obtained in this example are valid for a bar consisting of either solid or tubular segments.

# 3.9 STRAIN ENERGY IN TORSION AND PURE SHEAR

When a load is applied to a structure, work is performed by the load and strain energy is developed in the structure, as described in detail in Section 2.7 for a bar subjected to axial loads. In this section we will use the same basic concepts to determine the strain energy of a bar in torsion.

Consider a prismatic bar $AB$ in **pure torsion** under the action of a torque $T$ (Fig. 3-38). When the load is applied statically, the bar twists and the free end rotates through an angle $\phi$. If we assume that the material of the bar is linearly elastic and follows Hooke's law, then the relationship between the applied torque and the angle of twist will also be linear, as shown by the torque-rotation diagram of Fig. 3-39 and as given by the equation $\phi = TL/GI_P$.

**Fig. 3-38**

Prismatic bar in pure torsion

**Fig. 3-39**

Torque-rotation diagram for a bar in pure torsion (linearly elastic material)

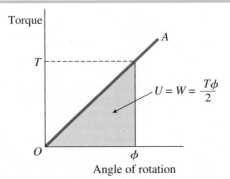

Angle of rotation

The work $W$ done by the torque as it rotates through the angle $\phi$ is equal to the area below the torque-rotation line $OA$, that is, it is equal to the area of the shaded triangle in Fig. 3-39. Furthermore, from the principle of conservation of energy we know that the strain energy of the bar is equal to the work done by the load, provided no energy is gained or lost in the form of heat. Therefore, we obtain the following equation for the strain energy $U$ of the bar:

$$U = W = \frac{T\phi}{2} \tag{3-54}$$

This equation is analogous to the equation $U = W = P\delta/2$ for a bar subjected to an axial load (see Eq. 2-37).

Using the equation $\phi = TL/GI_P$, we can express the strain energy in the following forms:

$$U = \frac{T^2 L}{2GI_P} \qquad U = \frac{GI_P \phi^2}{2L} \qquad \text{(3-55a,b)}$$

The first expression is in terms of the load and the second is in terms of the angle of twist. Again, note the analogy with the corresponding equations for a bar with an axial load (see Eqs. 2-39a and b).

The SI unit for both work and energy is the joule (J), which is equal to one newton meter (1 J = 1 N · m). The basic USCS unit is the foot-pound (ft-lb), but other similar units, such as inch-pound (in.-lb) and inch-kip (in.-k), are commonly used.

## Nonuniform Torsion

If a bar is subjected to nonuniform torsion (described in Section 3.4), we need additional formulas for the strain energy. In those cases where the bar consists of prismatic segments with constant torque in each segment (see Fig. 3-14a of Section 3.4), we can determine the strain energy of each segment and then add to obtain the total energy of the bar:

$$U = \sum_{i=1}^{n} U_i \qquad \text{(3-56)}$$

in which $U_i$ is the strain energy of segment $i$ and $n$ is the number of segments. For instance, if we use Eq. (3-55a) to obtain the individual strain energies, the preceding equation becomes

$$U = \sum_{i=1}^{n} \frac{T_i^2 L_i}{2G_i (I_P)_i} \qquad \text{(3-57)}$$

in which $T_i$ is the internal torque in segment $i$ and $L_i$, $G_i$, and $(I_P)_i$ are the torsional properties of the segment.

If either the cross section of the bar or the internal torque varies along the axis, as illustrated in Figs. 3-15 and 3-16 of Section 3.4, we can obtain the total strain energy by first determining the strain energy of an element and then integrating along the axis. For an element of length $dx$, the strain energy is (see Eq. 3-55a)

$$dU = \frac{[T(x)]^2 dx}{2GI_P(x)}$$

in which $T(x)$ is the internal torque acting on the element and $I_P(x)$ is the polar moment of inertia of the cross section at the element. Therefore, the total strain energy of the bar is

$$U = \int_0^L \frac{[T(x)]^2 dx}{2GI_P(x)} \qquad \text{(3-58)}$$

Once again, the similarities of the expressions for strain energy in torsion and axial load should be noted (compare Eqs. 3-57 and 3-58 with Eqs. 2-42 and 2-43 of Section 2.7).

The use of the preceding equations for nonuniform torsion is illustrated in the examples that follow. In Example 3-10 the strain energy is found for a bar in pure torsion with prismatic segments, and in Examples 3-11 and 3-12 the strain energy is found for bars with varying torques and varying cross-sectional dimensions.

In addition, Example 3-12 shows how, under very limited conditions, the angle of twist of a bar can be determined from its strain energy. (For a more detailed discussion of this method, including its limitations, see the subsection "Displacements Caused by a Single Load" in Section 2.7.)

## Limitations

When evaluating strain energy we must keep in mind that the equations derived in this section apply only to bars of linearly elastic materials with small angles of twist. Also, we must remember the important observation stated previously in Section 2.7, namely, *the strain energy of a structure supporting more than one load cannot be obtained by adding the strain energies obtained for the individual loads acting separately*. This observation is demonstrated in Example 3-10.

## Strain-Energy Density in Pure Shear

Because the individual elements of a bar in torsion are stressed in pure shear, it is useful to obtain expressions for the strain energy associated with the shear stresses. We begin the analysis by considering a small element of material subjected to shear stresses $\tau$ on its side faces (Fig. 3-40a). For convenience, we will assume that the front face of the element is square, with each side having length $h$. Although the figure shows only a two-dimensional view of the element, we recognize that the element is actually three dimensional with thickness $t$ perpendicular to the plane of the figure.

Under the action of the shear stresses, the element is distorted so that the front face becomes a rhombus, as shown in Fig. 3-40b. The change in angle at each corner of the element is the shear strain $\gamma$.

**FIG. 3-40**

Element in pure shear

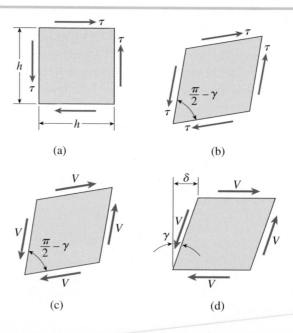

(a)  (b)

(c)  (d)

The shear forces $V$ acting on the side faces of the element (Fig. 3-40c) are found by multiplying the stresses by the areas $ht$ over which they act:

$$V = \tau h t \qquad \textbf{(3-59)}$$

These forces produce work as the element deforms from its initial shape (Fig. 3-40a) to its distorted shape (Fig. 3-40b). To calculate this work we need to determine the relative distances through which the shear forces move. This task is made easier if the element in Fig. 3-40c is rotated as a rigid body until two of its faces are horizontal, as in Fig. 3-40d. During the rigid-body rotation, the net work done by the forces $V$ is zero because the forces occur in pairs that form two equal and opposite couples.

As can be seen in Fig. 3-40d, the top face of the element is displaced horizontally through a distance $\delta$ (relative to the bottom face) as the shear force is gradually increased from zero to its final value $V$. The displacement $\delta$ is equal to the product of the shear strain $\gamma$ (which is a small angle) and the vertical dimension of the element:

$$\delta = \gamma h \qquad \textbf{(3-60)}$$

If we assume that the material is linearly elastic and follows Hooke's law, then the work done by the forces $V$ is equal to $V\delta/2$, which is also the strain energy stored in the element:

$$U = W = \frac{V\delta}{2} \qquad \textbf{(3-61)}$$

Note that the forces acting on the side faces of the element (Fig. 3-40d) do not move along their lines of action—hence they do no work.

Substituting from Eqs. (3-59) and (3-60) into Eq. (3-61), we get the total strain energy of the element:

$$U = \frac{\tau \gamma h^2 t}{2}$$

Because the volume of the element is $h^2 t$, the **strain-energy density** $u$ (that is, the strain energy per unit volume) is

$$u = \frac{\tau \gamma}{2} \qquad \textbf{(3-62)}$$

Finally, we substitute Hooke's law in shear ($\tau = G\gamma$) and obtain the following equations for the strain-energy density in pure shear:

$$u = \frac{\tau^2}{2G} \qquad u = \frac{G\gamma^2}{2} \qquad \textbf{(3-63a,b)}$$

These equations are similar in form to those for uniaxial stress (see Eqs. 2-46a and b of Section 2.7).

The SI unit for strain-energy density is joule per cubic meter (J/m$^3$), and the USCS unit is inch-pound per cubic inch (or other similar units). Since these units are the same as those for stress, we may also express strain-energy density in pascals (Pa) or pounds per square inch (psi).

In Section 3.11, we will use the equation for strain-energy density in terms of the shear stress (Eq. 3-63a) to determine the angle of twist of a thin-walled tube of arbitrary cross-sectional shape.

● ● ● **Example 3-10**

### Fig. 3-41

Example 3-10: Strain energy produced by two loads

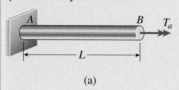

(a)

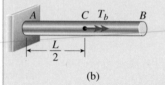

(b)

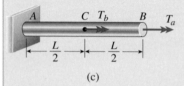

(c)

A solid circular bar $AB$ of length $L$ is fixed at one end and free at the other (Fig. 3-41). Three different loading conditions are to be considered: (a) torque $T_a$ acting at the free end; (b) torque $T_b$ acting at the midpoint of the bar; and (c) torques $T_a$ and $T_b$ acting simultaneously.

For each case of loading, obtain a formula for the strain energy stored in the bar. Then evaluate the strain energy for the following data: $T_a = 100$ N · m, $T_b = 150$ N · m, $L = 1.6$ m, $G = 80$ GPa, and $I_p = 79.52 \times 10^3$ mm⁴.

**Solution**

(a) *Torque $T_a$ acting at the free end (Fig. 3-41a).* In this case, the strain energy is obtained directly from Eq. (3-55a):

$$U_a = \frac{T_a^2 L}{2GI_P} \qquad \longleftarrow \textbf{(a)}$$

(b) *Torque $T_b$ acting at the midpoint (Fig. 3-41b).* When the torque acts at the midpoint, we apply Eq. (3-55a) to segment $AC$ of the bar:

$$U_b = \frac{T_b^2(L/2)}{2GI_P} = \frac{T_b^2 L}{4GI_P} \qquad \longleftarrow \textbf{(b)}$$

(c) *Torques $T_a$ and $T_b$ acting simultaneously (Fig. 3-41c).* When both loads act on the bar, the torque in segment $CB$ is $T_a$ and the torque in segment $AC$ is $T_a + T_b$. Thus, the strain energy (from Eq. 3-57) is

$$U_c = \sum_{i=1}^{n} \frac{T_i^2 L_i}{2G(I_P)_i} = \frac{T_a^2(L/2)}{2GI_P} + \frac{(T_a + T_b)^2(L/2)}{2GI_P}$$

$$= \frac{T_a^2 L}{2GI_P} + \frac{T_a T_b L}{2GI_P} + \frac{T_b^2 L}{4GI_P} \qquad \longleftarrow \textbf{(c)}$$

A comparison of Eqs. (a), (b), and (c) shows that the strain energy produced by the two loads acting simultaneously is *not* equal to the sum of the strain energies produced by the loads acting separately. As pointed out in Section 2.7, the reason is that strain energy is a quadratic function of the loads, not a linear function.

(d) *Numerical results.* Substituting the given data into Eq. (a), we obtain

$$U_a = \frac{T_a^2 L}{2GI_P} = \frac{(100 \text{ N} \cdot \text{m})^2(1.6 \text{ m})}{2(80 \text{ GPa})(79.52 \times 10^3 \text{ mm}^4)} = 1.26 \text{ J} \qquad \longleftarrow$$

Recall that one joule is equal to one newton-meter (1 J = 1 N · m). Proceeding in the same manner for Eqs. (b) and (c), we find

$$U_b = 1.41 \text{ J} \qquad \longleftarrow$$

$$U_c = 1.26 \text{ J} + 1.89 \text{ J} + 1.41 \text{ J} = 4.56 \text{ J} \qquad \longleftarrow$$

Note that the middle term, involving the product of the two loads, contributes significantly to the strain energy and cannot be disregarded.

● ● ● **Example 3-11**

A prismatic bar *AB*, fixed at one end and free at the other, is loaded by a distributed torque of constant intensity $t$ per unit distance along the axis of the bar (Fig. 3-42).

(a) Derive a formula for the strain energy of the bar.
(b) Evaluate the strain energy of a hollow shaft used for drilling into the earth if the data are as follows:

$$t = 480 \text{ lb-in./in.}, \quad L = 12 \text{ ft}, \quad G = 11.5 \times 10^6 \text{ psi}, \quad \text{and} \quad I_p = 17.18 \text{ in.}^4$$

**Fig. 3-42**

Example 3-11: Strain energy produced by a distributed torque

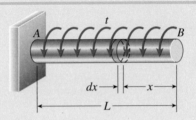

**Solution**

(a) *Strain energy of the bar.* The first step in the solution is to determine the internal torque $T(x)$ acting at distance $x$ from the free end of the bar (Fig. 3-42). This internal torque is equal to the total torque acting on the part of the bar between $x = 0$ and $x = x$. This latter torque is equal to the intensity $t$ of torque times the distance $x$ over which it acts:

$$T(x) = tx \tag{a}$$

Substituting into Eq. (3-58), we obtain

$$U = \int_0^L \frac{[T(x)]^2 dx}{2GI_p} = \frac{1}{2GI_p} \int_0^L (tx)^2 dx = \frac{t^2 L^3}{6GI_p} \quad \Longleftarrow \quad (3\text{-}64)$$

This expression gives the total strain energy stored in the bar.

(b) *Numerical results.* To evaluate the strain energy of the hollow shaft, we substitute the given data into Eq. (3-64):

$$U = \frac{t^2 L^3}{6GI_p} = \frac{(480 \text{ lb-in./in.})^2 (144 \text{ in.})^3}{6(11.5 \times 10^6 \text{ psi})(17.18 \text{ in.}^4)} = 580 \text{ in.-lb} \quad \Longleftarrow$$

This example illustrates the use of integration to evaluate the strain energy of a bar subjected to a distributed torque.

**Fig. 3-43**

Example 3-12: Tapered bar in torsion

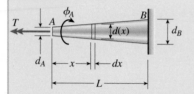

A tapered bar $AB$ of solid circular cross section is supported at the right-hand end and loaded by a torque $T$ at the other end (Fig. 3-43). The diameter of the bar varies linearly from $d_A$ at the left-hand end to $d_B$ at the right-hand end.

Determine the angle of rotation $\phi_A$ at end $A$ of the bar by equating the strain energy to the work done by the load.

**Solution**

From the principle of conservation of energy we know that the work done by the applied torque equals the strain energy of the bar; thus, $W = U$. The work is given by the equation

$$W = \frac{T\phi_A}{2} \tag{a}$$

and the strain energy $U$ can be found from Eq. (3-58).

To use Eq. (3-58), we need expressions for the torque $T(x)$ and the polar moment of inertia $I_p(x)$. The torque is constant along the axis of the bar and equal to the load $T$, and the polar moment of inertia is

$$I_p(x) = \frac{\pi}{32}\left[d(x)\right]^4$$

in which $d(x)$ is the diameter of the bar at distance $x$ from end $A$. From the geometry of the figure, we see that

$$d(x) = d_A + \frac{d_B - d_A}{L}x \tag{b}$$

and therefore

$$I_p(x) = \frac{\pi}{32}\left(d_A + \frac{d_B - d_A}{L}x\right)^4 \tag{c}$$

Now we can substitute into Eq. (3-58), as follows:

$$U = \int_0^L \frac{[T(x)]^2 dx}{2GI_p(x)} = \frac{16T^2}{\pi G}\int_0^L \frac{dx}{\left(d_A + \dfrac{d_B - d_A}{L}x\right)^4}$$

The integral in this expression can be integrated with the aid of a table of integrals (see Appendix D) with the result:

$$\int_0^L \frac{dx}{\left(d_A + \dfrac{d_B - d_A}{L}x\right)^4} = \frac{L}{3(d_B - d_A)}\left(\frac{1}{d_A^3} - \frac{1}{d_B^3}\right)$$

Therefore, the strain energy of the tapered bar is

$$U = \frac{16T^2 L}{3\pi G(d_B - d_A)}\left(\frac{1}{d_A^3} - \frac{1}{d_B^3}\right) \tag{3-65}$$

Equating the strain energy to the work of the torque (Eq. a) and solving for $\phi_A$, we get

$$\phi_A = \frac{32TL}{3\pi G(d_B - d_A)}\left(\frac{1}{d_A^3} - \frac{1}{d_B^3}\right) \quad \Longleftarrow \tag{3-66}$$

This equation gives the angle of rotation at end $A$ of the tapered bar. [*Note:* This is the same angle of twist expression obtained in the solution of Prob. 3.4-8(a).]

Note especially that the method used in this example for finding the angle of rotation is suitable only when the bar is subjected to a single load, and then only when the desired angle corresponds to that load. Otherwise, we must find angular displacements by the usual methods described in Sections 3.3, 3.4, and 3.8.

# 3.10 TORSION OF NONCIRCULAR PRISMATIC SHAFTS

In Sections 3.1 through 3.9 of this chapter, we restricted our attention to the twisting of circular shafts. Shafts with circular cross sections (either solid or hollow) do not *warp* when torsional moments are applied. Plane cross sections remain plane (as shown in Fig. 3-4), and shearing stresses and strains vary linearly with distance $\rho$ from the longitudinal axis of the shaft to the outer surface ($\rho = r$). Now we consider prismatic shafts of length $L$ acted upon by torsional moments $T$ at either end but having *noncircular* cross sections. These cross sections could be solid (such as the elliptical, triangular, and rectangular shapes shown in Fig. 3-44), or they could be thin-walled *open* cross sections such as the I-beam, channel, and Z-shaped cross sections depicted in Fig. 3-45.

These noncircular cross sections *warp* under the action of torsional moments, and this warping alters the shear stress and strain distributions in the cross section. We can no longer use the simple *torsion formula* of

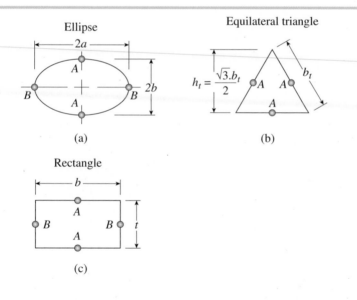

**Fig. 3-44**

Solid elliptical, triangular, and rectangular cross-sectional shapes

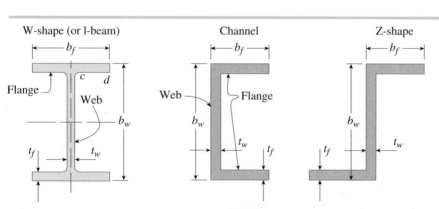

**Fig. 3-45**

Thin-walled *open* cross sections: I-beam, channel, and Z-shaped

Eq. 3-13 to compute shear stresses, and the *torque-displacement relation* of Eq. 3-17 cannot be used to find the angle of twist of the shaft. For example, warping distortions of a rectangular bar of length $L$ acted upon by torques $T$ at either end are shown in Fig. 3-46a; the cross sections remain rectangular, but a grid on the surface of the bar is distorted as shown and plus/minus $x$ displacements represent the out-of-plane warping of the cross sections. The torsional shear stress distribution in the rectangular cross section is shown in Fig. 3-46b. The shear stresses at the corners are zero, and the maximum shear stress occurs at the midpoint of the longer side (point $A$ in Figs. 3-44 and 3-46b). A more advanced theory, developed by Saint-Venant, is required to develop expressions for torsion of shafts of noncircular cross section (see Refs. 1-1, 2-1, and 2-10). Simple formulas for maximum shear stress and angle of twist for the cross-sectional shapes shown in Figs. 3-44 and 3-45 will be presented next and then used in calculations in Examples 3-13 and 3-14. However, derivation of these formulas is beyond the scope of this text; future coursework on the *theory of elasticity* and, perhaps, the *finite element method* will provide more detailed analysis relating applied torque and the resulting stress distributions and angle of twist of noncircular prismatic shafts.

**Fig. 3-46**

(a) Torsion of bar of rectangular cross section, and (b) shear stress distribution for bar of rectangular cross section acted upon by torsional moment $T$

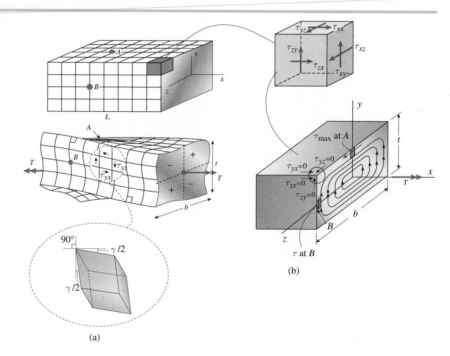

(a)

(b)

## Shear Stress Distribution and Angle of Twist

In the following discussion, we will present only the basic relations between applied torsional moment $T$ and three key items of interest for a variety of noncircular cross sections:

(1)  The location and value of the maximum shear stress. $\tau_{max}$ in the cross section

(2)  The torsional rigidity GJ

(3)  The angle of twist $\phi$ of a prismatic bar of length $L$

Constant $G$ is the shearing modulus of elasticity of the material, and variable $J$ is the *torsion constant* for the cross section. Note that only for a circular cross section does torsion constant $J$ become the polar moment of inertia $I_p$.

## Elliptical, Triangular, and Rectangular Cross Sections

The shear stress distribution for a bar with an *elliptical cross section* (2a along major axis, 2b along minor axis, area $A = \pi ab$) is shown in Fig. 3-47. The maximum shear stress is at the ends of the *minor axis* and may be computed using the expression:

$$\tau_{max} = \frac{2T}{\pi ab^2} \qquad \textbf{(3-67)}$$

where $a$ is greater than or equal to $b$. The angle of twist $\phi$ of a prismatic shaft of length $L$ with an elliptical cross section can be expressed as

$$\phi = \frac{TL}{GJ_e}$$

where torsion constant $J_e$ is

$$J_e = \frac{\pi a^3 b^3}{a^2 + b^2} \qquad \textbf{(3-68a,b)}$$

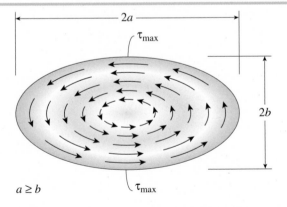

$2a$

$\tau_{max}$

$2b$

$a \geq b$

$\tau_{max}$

### Fig. 3-47

Shear stress distribution in an elliptical cross section

Note that if $a = b$, we have a *solid circular cross section* instead of an ellipse, and the expression for $J_e$ becomes the polar moment of inertia $I_P$ [Eq. (3-12)] and Eqs. (3-67) and (3-68a) reduce to Eqs. (3-13) and (3-17), respectively.

Next we consider an *equilateral triangular cross section* (Fig. 3-44b) for the shaft of length $L$ acted upon by torques $T$ at each end. Each side has dimension $b_t$, and the triangle height is $h_t$. The torsion constant $J_t$ is

$$J_t = \frac{h_t^4}{15\sqrt{3}} \qquad (3\text{-}69)$$

The maximum shear stress occurs on the surface at the *midpoint of each side* (points $A$ in Fig. 3-44b). The maximum shear stress and the angle of twist $\phi$ of a prismatic shaft of length $L$ with an *equilateral triangular cross section* can be expressed as

$$\tau_{max} = \frac{T\left(\dfrac{h_t}{2}\right)}{J_t} = \frac{15\sqrt{3}\,T}{2h_t^3} \qquad (3\text{-}70)$$

where $J_t = \dfrac{h_t^4}{15\sqrt{3}}$

and

$$\phi = \frac{TL}{GJ_t} = \frac{15\sqrt{3}\,TL}{Gh_t^4} \qquad (3\text{-}71)$$

Finally, we consider a *rectangular cross section* ($b \times t$, $b/t \geq 1$) (see Figs. 3-44c and 3-46). Theory of elasticity solutions provide expressions for *maximum shear stress at point $A$* in the cross section and the angle of twist for a variety of aspect ratios $b/t$ as

$$\tau_{max} = \frac{T}{k_1 b t^2} \qquad (3\text{-}72)$$

$$\phi = \frac{TL}{(k_2 b t^3)G} = \frac{TL}{GJ_r} \qquad (3\text{-}73)$$

where

$$J_r = k_2 b t^3$$

and dimensionless coefficients $k_1$ and $k_2$ are listed in Table 3-1.

| Table 3-1 | $b/t$ | 1.00 | 1.50 | 1.75 | 2.00 | 2.50 | 3.00 | 4 | 6 | 8 | 10 | $\infty$ |
|---|---|---|---|---|---|---|---|---|---|---|---|---|
| Dimensionless Coefficients | $k_1$ | 0.208 | 0.231 | 0.239 | 0.246 | 0.258 | 0.267 | 0.282 | 0.298 | 0.307 | 0.312 | 0.333 |
| for Rectangular Bars | $k_2$ | 0.141 | 0.196 | 0.214 | 0.229 | 0.249 | 0.263 | 0.281 | 0.298 | 0.307 | 0.312 | 0.333 |

It is especially important to note that, for the elliptical, triangular, and rectangular sections considered here, *maximum shear stress does not occur at the largest distance from the axis of the shaft like it does for circular sections.* Instead, maximum shear strain and stress occur at the *midpoints* of the sides for each section. In fact, the shear stresses are zero in the corners of the triangular and rectangular sections (as indicated by the appearance of zero shear strain at the corners of the recangular section in Fig. 3-46a, for example).

## Thin-Walled Open Cross Sections: I-beam, Angle, Channel, and Z-shape

Metal structural shapes of open cross section (see Fig. 3-45) can be represented as assemblages of rectangles for purposes of computing their torsional properties and response to applied torsional moments. Torsion constants for typical structural steel shapes are tabulated in the AISC manual (see Ref. 5-4) and may be up to 10% higher than properties based on use of rectangles to represent flanges and web. Hence, maximum shear stress values and twist angles computed using the formulas presented here may be somewhat conservative.

The total torque is assumed to be equal to the sum of the torques carried by the flanges and web. We first compute the flange $b_f/t_f$ ratio (see Fig. 3-45 for cross-sectional dimensions). Then find constant $k_1$ from Table 3-1 (interpolation between values may be necessary). For the web, we use the ratio $(b_w - 2t_f)/t_w$ in Table 3-1 to find a new constant $k_1$ for the web. The separate torsion constants for both flanges and the web are expressed as

$$J_f = k_1 b_f t_f^3 \qquad J_w = k_1 (b_w - 2t_f)\left(t_w^3\right) \qquad \textbf{(3-74a,b)}$$

The total torsion constant for the thin, open cross section is obtained (assuming two flanges) as

$$J = J_w + 2J_f \qquad\qquad \textbf{(3-75)}$$

The maximum shear stress and angle of twist then can be computed as

$$\tau_{max} = \frac{2T\left(\dfrac{t}{2}\right)}{J} \quad \text{and} \quad \phi = \frac{TL}{GJ} \qquad \textbf{(3-76a,b)}$$

where the *larger* of $t_f$ and $t_w$ is used in the formula for $\tau_{max}$.

Examples 3-13 and 3-14 illustrate the application of these formulas to obtain maximum shear stress and angle of twist values for prismatic bars with noncircular cross sections, such as those presented in Figs. 3-44 and 3-45.

● ● ● **Example 3-13**

A shaft of length $L = 6$ ft is subjected to torques $T = 45$ kip-in. at either end (Fig. 3-48). Segment $AB$ ($L_1 = 3$ ft) is made of brass ($G_b = 6000$ ksi) and has a square cross section ($a = 3$ in.). Segment $BC$ ($L_2 = 3$ ft) is made of steel ($G_s = 10,800$ ksi) and has a circular cross section ($d = a = 3$ in.). Ignore stress concentrations near $B$.

(a) Find the maximum shear stress and angle of twist for each segment of the shaft.
(b) Find a new value for the dimension $a$ of bar $AB$ if the maximum shear stress in $AB$ and $BC$ are to be equal.
(c) Repeat part (b) if the angles of twist of segments $AB$ and $BC$ are to be equal.
(d) If dimension $a$ is reset to $a = 3$ in. and bar $BC$ is now a hollow pipe with outer diameter $d_2 = a$, find the inner diameter $d_1$ so that the the angles of twist of segments $AB$ and $BC$ are equal.

| **Fig. 3-48** |
| --- |

Example 3-13: Torsion of shaft with noncircular cross section

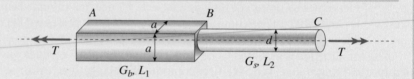

**Solution**

(a) *Maximum shear stress and angles of twist for each segment.* Both segments of the shaft have internal torque equal to applied torque $T$. For square segment $AB$, we obtain torsion coefficients $k_1$ and $k_2$ from Table 3-1, then use Eqs. 3-72 and 3-73 to compute maximum shear stress and angle of twist as

$$\tau_{max1} = \frac{T}{k_1 bt^2} = \frac{T}{k_1 a^3} = \frac{(45 \text{ k-in.})}{0.208(3 \text{ in.})^3} = 8.01 \text{ ksi} \quad \Longleftarrow \textbf{(a)}$$

$$\phi_1 = \frac{TL_1}{(k_2 bt^3)G_b} = \frac{TL_1}{k_2 a^4 G_b} = \frac{(45 \text{ k-in.})(36 \text{ in.})}{0.141(3 \text{ in.})^4(6000 \text{ ksi})}$$

$$= 2.36 \times 10^{-2} \text{ radians} \quad \Longleftarrow \textbf{(b)}$$

The maximum shear stress in $AB$ occurs at the midpoint of each side of the square cross section.

Segment $BC$ is a solid, circular cross section, so we use Eqs. 3-14 and 3-17 to compute the maximum shear stress and angle of twist for segment $BC$:

$$\tau_{max2} = \frac{16T}{\pi d^3} = \frac{16(45 \text{ k-in.})}{\pi (3 \text{ in.})^3} = 8.49 \text{ ksi} \quad \Longleftarrow \textbf{(c)}$$

$$\phi_2 = \frac{TL_2}{G_s J_p} = \frac{(45 \text{ k-in.})(36 \text{ in.})}{10,800 \text{ ksi}\left[\dfrac{\pi}{32}(3 \text{ in.})^4\right]} = 1.886 \times 10^{-2} \quad \Longleftarrow \textbf{(d)}$$

Comparing the shear stress and angle of twist values for square segment $AB$ and circular segment $BC$, we see that the steel pipe $BC$ has 6% greater maximum shear stress but 20% less twist rotation than the brass bar $AB$.

(b) *New value for dimension a of bar AB so that maximum shear stress in AB and BC are equal.* We equate expressions for $\tau_{max1}$ and $\tau_{max2}$ in Eqs. (a) and (c) and solve for the required new value of dimension $a$ of bar $AB$:

$$\tau_{max\,1} = \tau_{max2} \quad \text{so} \quad \frac{16}{\pi d^3} = \frac{1}{k_1 a_{new}^3} \quad \text{or} \quad a_{new} = \left(\frac{\pi d^3}{16 k_1}\right)^{\frac{1}{3}} = 2.94 \text{ in.} \quad \blacktriangleleft \quad \text{(e)}$$

The diameter of bar $BC$ remains unchanged at $d = 3$ in., so a slight reduction in dimension $a$ for bar $AB$ leads to the same maximum shear stress of 8.49 ksi [Eq.(c)] in the two bar segments.

(c) *New value for dimension a of bar AB so that twist rotations in AB and BC are equal.* Now, we equate expressions for $\phi_1$ and $\phi_2$ in Eqs. (b) and (d) and solve for the required new value of dimension $a$ of bar $AB$:

$$\phi_1 = \phi_2$$

$$\text{so} \quad \frac{L_1}{k_2 a_{new}^4 G_b} = \frac{L_2}{G_s I_p} \quad \blacktriangleleft \quad \text{(f)}$$

$$\text{or} \quad a_{new} = \left[\frac{L_1}{L_2}\left(\frac{G_s I_p}{k_2 G_b}\right)\right]^{\frac{1}{4}} = 3.17 \text{ in.}$$

The diameter of bar $BC$ remains unchanged at $d = 3$ in., so a slight increase in dimension $a$ for brass bar $AB$ leads to the same twist rotation of 0.01886 radians, as in Eq. (d) in each of the two bar segments.

(d) *Change segment BC to hollow pipe; find inner diameter $d_1$ so that twist rotations in AB and BC are equal.* Side dimension $a$ of square segment $AB$ is equal to 3 in., and outer diameter $d_2 = 3$ in. (Fig. 3-49). Using Eq. (3-19) for the polar moment of inertia of segment $BC$, we get for twist angle $\phi_2$:

$$\phi_2 = \frac{T L_2}{G_s\left[\dfrac{\pi}{32}\left(d_2^4 - d_1^4\right)\right]} \quad \text{(g)}$$

Once again, we equate expressions for $\phi_1$ and $\phi_2$ but now using Eqs. (b) and (g). Solving for $d_1$, we get the expression:

$$d_1 = \left[d_2^4 - 32\left(\frac{L_2}{L_1}\right)\left(\frac{G_b}{G_s}\right)\left(\frac{a^4 k_2}{\pi}\right)\right]^{\frac{1}{4}} \quad \text{(h)}$$

$$= \left[(3 \text{ in.})^4 - 32\left(\frac{3 \text{ ft}}{3 \text{ ft}}\right)\left(\frac{6000 \text{ ksi}}{10,800 \text{ ksi}}\right)\left[\frac{(3 \text{ in.})^4 (0.141)}{\pi}\right]\right]^{\frac{1}{4}} \quad \blacktriangleleft$$

$$= 2.01 \text{ in.}$$

So the square, solid brass pipe $AB$ ($a \times a$, $a = 3$ in.) and hollow steel pipe $BC$ ($d_2 = 3$ in., $d_1 = 2.01$ in.) are each 3 ft. in length and have the same twist rotation (0.0236 radians) due to applied torque $T$. However, additional calculations will show that the maximum shear stress in segment $BC$ is now increased from 8.49 ksi [Eq.(c)] to 10.64 ksi by using a hollow rather than solid bar for $BC$.

Note that by deriving the formula for inner diameter $d_1$ in Eq. (h) (rather than finding a numerical solution alone), we can also investigate other solutions of possible interest using different values of the key variables. For example, if bar $AB$ is increased in length to $L_1 = 3.5$ ft, inner diameter $d_1$ for $BC$ can be increased to 2.43 in. and the angles of twist for $AB$ and $BC$ will be the same.

**Fig. 3-49**

Hollow pipe cross section for segment BC

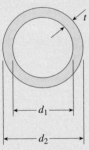

**Example 3-14**

A steel angle, **L178 × 102 × 19**, and a steel wide-flange beam, **W360 × 39**, each of length $L = 3.5$ m, are subjected to torque $T$ (see Fig. 3-50). The allowable shear stress is 45 MPa and the maximum permissible twist rotation is 5°. *Find* the value of the maximum torque $T$ than can be applied to each section. Assume that $G = 80$ GPa and ignore stress concentration effects. (See Tables F-1(b) and F-5(b) for cross-sectional properties and dimensions).

**Fig. 3-50**

Example 3-14: W-shape and angle steel sections in torsion

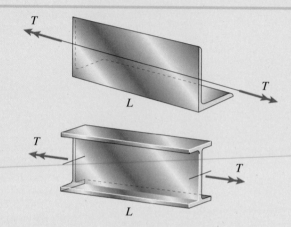

**Solution**

The angle and wide-flange steel shapes have the same cross-sectional area ($A = 4960$ mm²; see Tables F-1(b) and F-5(b)) but thicknesses of flange and web components of each section are quite different. First, we consider the angle section.

(a) *Steel angle section.* We can approximate the unequal leg angle as one long *rectangle* with length $b_L = 280$ mm and constant thickness $t_L = 19$ mm, so $b_L/t_L = 14.7$. From Table 3-1, we estimate coefficients $k_1 = k_2$ to be approximately 0.319. The *maximum allowable torques* can be obtained from Eqs. 3-72 and 3-73 based on the given allowable shear stress and allowable twist rotation, respectively, as

$$T_{max1} = \tau_a k_1 b_L t_L^2 = 45 \text{ MPa}(0.319)(280 \text{ mm})[(19 \text{ mm})^2] = 1451 \text{ N} \cdot \text{m} \quad \textbf{(a)}$$

$$T_{max2} = \phi_a(k_2 b_L t_L^3)\frac{G}{L} = \left(\frac{5\pi}{180} \text{ rad}\right)(0.319)(280 \text{ mm})[(19 \text{ mm})^3]\frac{80 \text{ GPa}}{3500 \text{ mm}}$$

$$= 1222 \text{ N} \cdot \text{m} \quad \textbf{(b)}$$

Alternatively, we can compute the torsion constant for the angle $J_L$,

$$J_L = K_1 b_L t_L^3 = 6.128 \times 10^5 \text{ mm}^4 \quad \textbf{(c)}$$

then use Eqs. 3-74 and 3-76 to find maximum allowable torque values. From Eq. 3-76a, we find $T_{max1}$ and from Eq. 3-76b, we obtain $T_{max2}$:

$$T_{max1} = \frac{\tau_a J_L}{t_L} = 1451 \text{ N} \cdot \text{m} \quad \text{and} \quad T_{max2} = \frac{GJ_L}{L}\phi_a = 1222 \text{ N} \cdot \text{m}$$

*For the angle,* the lesser value controls, so $T_{max} = 1222$ N·m

(b) *Steel W-shape.* The two flanges and the web are separate rectangles which together resist the applied torsional moment. However, the dimensions ($b$, $t$) of each of these rectangles are different: for a W 360 $\times$ 39, each flange has a width of $b_f = 128$ mm and a thickness of $t_f = 10.7$ mm [see Table F-1(b)]. The web has thickness $t_w = 6.48$ mm [Table F-5(b)] and, conservatively, $b_w = (d_w - 2t_f) = (353$ mm $- 2(10.7$ mm$)) = 331.6$ mm. Based on the $b/t$ ratios, we can find separate coefficients $k_1$ for the flanges and web from Table 3-1, then compute the torsion constants $J$ for each component using Eqs. 3-74 as:

*For the flanges:*

$$\frac{b_f}{t_f} = 11.963$$

so an estimated value for $k_{1f} = 0.316$. Then we have

$$J_f = k_{1f}b_f t_f^3 = 0.316(128 \text{ mm})[(10.7 \text{ mm})^3] = 4.955 \times 10^4 \text{ mm}^4 \quad \textbf{(d)}$$

*For the web:*

$$\frac{d_w - 2t_f}{t_w} = 51.173$$

and $k_{1w}$ is estimated as $k_{1w} = 0.329$, so

$$J_w = k_{1w}(d_w - 2t_f)(t_w^3) = 0.329[353 \text{ mm} - 2(10.7 \text{ mm})][(6.48 \text{ mm})^3]$$

$$= 2.968 \times 10^4 \text{ mm} \quad \textbf{(e)}$$

The torsion constant for the entire W360 $\times$ 39 section is obtained by adding web and flange contributions [Eqs. (d) and (e)]:

$$J_W = 2J_f + J_w = [2(4.955) + 2.968](10^4) \text{ mm}^4 = 1.288 \times 10^5 \text{ mm}^4 \quad \textbf{(f)}$$

Now, we use Eq. 3-76a and the allowable shear stress $\tau_a$ to compute the maximum allowable torque based on both flange and web maximum shear stresses:

$$T_{\max f} = \tau_a \frac{J_W}{t_f} = 45 \text{ MPa}\left(\frac{1.288 \times 10^5 \text{ mm}^4}{10.7 \text{ mm}}\right) = 542 \text{ N} \cdot \text{m} \quad \textbf{(g)}$$

$$T_{\max w} = \tau_a \frac{J_W}{t_w} = 45 \text{ MPa}\left(\frac{1.288 \times 10^5 \text{ mm}^4}{6.48 \text{ mm}}\right) = 894 \text{ N} \cdot \text{m} \quad \textbf{(h)}$$

Note that since the flanges have greater thickness than the web, the maximum shear stress will be in the flanges. So, a calculation of $T_{\max}$ based on the maximum web shear stress using Eq. (h) is not necessary.

Finally, we use Eq. 3-76b to compute $T_{\max}$ based on the allowable angle of twist:

$$T_{\max \phi} = \frac{GJ_W}{L}\phi_a = \frac{80 \text{ GPa}(1.288 \times 10^5 \text{ mm}^4)}{3500 \text{ mm}}\left(\frac{5\pi}{180} \text{ rad}\right)$$

$$= 257 \text{ N} \cdot \text{m} \quad \Longleftarrow \quad \textbf{(i)}$$

*For the W-shape,* the most restrictive requirement is the allowable twist rotation so $T_{\max} = 257$ N $\cdot$ m governs [Eq. (i)].

It is interesting to note that, even though both angle and W-shapes have the same cross-sectional area, the W-shape is considerably weaker in torsion, because its component rectangles are much thinner ($t_w = 6.48$ mm, $t_f = 10.7$ mm) than the angle section ($t_L = 19$ mm). However, we will see in Chapter 5 that, although weak in torsion, the W-shape has a considerable advantage in resisting bending and *transverse* shear stresses.

# 3.11 THIN-WALLED TUBES

With the exception of Section 3.10, in which we considered torsion of bars of noncircular cross section, the torsion theory described in the preceding sections is applicable only to solid or hollow bars of circular cross section. Circular shapes are the most efficient shapes for resisting torsion and consequently are the most commonly used. However, in lightweight structures, such as aircraft and spacecraft, thin-walled tubular members with noncircular closed cross sections are often required to resist torsion. In this section, we will analyze structural members of this kind.

To obtain formulas that are applicable to a variety of shapes, let us consider a thin-walled tube of arbitrary cross section (Fig. 3-51a). The tube is cylindrical in shape—that is, all cross sections are identical and the longitudinal axis is a straight line. The thickness $t$ of the wall is not necessarily constant but may vary around the cross section. However, the thickness must be small in comparison with the total width of the tube. The tube is subjected to pure torsion by torques $T$ acting at the ends.

## Shear Stresses and Shear Flow

The shear stresses $\tau$ acting on a cross section of the tube are pictured in Fig. 3-51b, which shows an element of the tube cut out between two cross sections that are distance $dx$ apart. The stresses act parallel to the boundaries of the cross section and "flow" around the cross section. Also, the intensity of the stresses varies so slightly *across* the thickness of the tube (because the tube is assumed to be thin) that we may assume $\tau$ to be constant in that direction. However, if the thickness $t$ is not constant, the stresses will vary in intensity as we go *around* the cross section, and the manner in which they vary must be determined from equilibrium.

To determine the magnitude of the shear stresses, we will consider a rectangular element *abcd* obtained by making two longitudinal cuts *ab* and *cd* (Figs. 3-51a and b). This element is isolated as a free body in Fig. 3-51c.

---

**Fig. 3-51**

Thin-walled tube of arbitrary cross-sectional shape

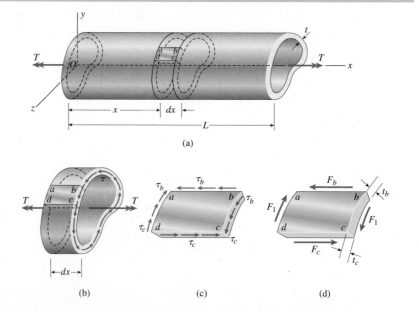

(a)

(b)   (c)   (d)

Acting on the cross-sectional face *bc* are the shear stresses $\tau$ shown in Fig. 3-51b. We assume that these stresses vary in intensity as we move along the cross section from *b* to *c*; therefore, the shear stress at *b* is denoted $\tau_b$ and the stress at *c* is denoted $\tau_c$ (see Fig. 3-51c).

As we know from equilibrium, identical shear stresses act in the opposite direction on the opposite cross-sectional face *ad*, and shear stresses of the same magnitude also act on the longitudinal faces *ab* and *cd*. Thus, the constant shear stresses acting on faces *ab* and *cd* are equal to $\tau_b$ and $\tau_c$, respectively.

The stresses acting on the longitudinal faces *ab* and *cd* produce forces $F_b$ and $F_c$ (Fig. 3-51d). These forces are obtained by multiplying the stresses by the areas on which they act:

$$F_b = \tau_b t_b dx \qquad F_c = \tau_c t_c dx$$

in which $t_b$ and $t_c$ represent the thicknesses of the tube at points *b* and *c*, respectively (Fig. 3-51d).

In addition, forces $F_1$ are produced by the stresses acting on faces *bc* and *ad*. From the equilibrium of the element in the longitudinal direction (the *x* direction), we see that $F_b = F_c$, or

$$\tau_b t_b = \tau_c t_c$$

Because the locations of the longitudinal cuts *ab* and *cd* were selected arbitrarily, it follows from the preceding equation that the product of the shear stress $\tau$ and the thickness *t* of the tube is the same at every point in the cross section. This product is known as the **shear flow** and is denoted by the letter *f*:

$$f = \tau t = \text{constant} \qquad \textbf{(3-77)}$$

This relationship shows that the largest shear stress occurs where the thickness of the tube is smallest, and vice versa. In regions where the thickness is constant, the shear stress is constant. Note that shear flow is the shear force per unit distance along the cross section.

## Torsion Formula for Thin-Walled Tubes

The next step in the analysis is to relate the shear flow *f* (and hence the shear stress $\tau$) to the torque *T* acting on the tube. For that purpose, let us examine the cross section of the tube, as pictured in Fig. 3-52. The **median line** (also called the *centerline* or the *midline*) of the wall of the tube is shown as a dashed line in the figure. We consider an element of area of length *ds* (measured along the median line) and thickness *t*. The distance *s* defining the location of the element is measured along the median line from some arbitrarily chosen reference point.

The total shear force acting on the element of area is *fds*, and the moment of this force about any point *O* within the tube is

$$dT = rfds$$

**Fig. 3-52**

Cross section of thin-walled tube

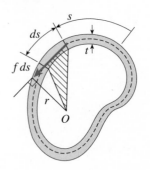

in which $r$ is the perpendicular distance from point $O$ to the line of action of the force $f\,ds$. (Note that the line of action of the force $f\,ds$ is tangent to the median line of the cross section at the element $ds$.) The total torque $T$ produced by the shear stresses is obtained by integrating along the median line of the cross section:

$$T = f\int_0^{L_m} r\,ds \qquad \text{(3-78)}$$

in which $L_m$ denotes the length of the median line.

The integral in Eq. (3-78) can be difficult to integrate by formal mathematical means, but fortunately it can be evaluated easily by giving it a simple geometric interpretation. The quantity $r\,ds$ represents twice the area of the shaded triangle shown in Fig. 3-52. (Note that the triangle has base length $ds$ and height equal to $r$.) Therefore, the integral represents twice the area $A_m$ enclosed by the median line of the cross section:

$$\int_0^{L_m} r\,ds = 2A_m \qquad \text{(3-79)}$$

It follows from Eq. (3-78) that $T = 2fA_m$, and therefore the **shear flow** is

$$f = \frac{T}{2A_m} \qquad \text{(3-80)}$$

Now we can eliminate the shear flow $f$ between Eqs. (3-77) and (3-80) and obtain a **torsion formula for thin-walled tubes**:

$$\tau = \frac{T}{2tA_m} \qquad \text{(3-81)}$$

Since $t$ and $A_m$ are properties of the cross section, the shear stresses $\tau$ can be calculated from Eq. (3-81) for any thin-walled tube subjected to a known torque $T$. (*Reminder:* The area $A_m$ is the area *enclosed* by the median line—it is *not* the cross-sectional area of the tube.)

To illustrate the use of the torsion formula, consider a thin-walled **circular tube** (Fig. 3-53) of thickness $t$ and radius $r$ to the median line. The area enclosed by the median line is

$$A_m = \pi r^2 \qquad \text{(3-82)}$$

and therefore the shear stress (constant around the cross section) is

$$\tau = \frac{T}{2\pi r^2 t} \qquad \text{(3-83)}$$

This formula agrees with the stress obtained from the standard torsion formula (Eq. 3-13) when the standard formula is applied to a circular tube with thin walls using the approximate expression $I_P \approx 2\pi r^3 t$ for the polar moment of inertia (Eq. 3-21).

As a second illustration, consider a thin-walled **rectangular tube** (Fig. 3-54) having thickness $t_1$ on the sides and thickness $t_2$ on the top and bottom. Also, the height and width (measured to the median line of

**Fig. 3-53**

Thin-walled circular tube

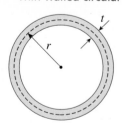

the cross section) are $h$ and $b$, respectively. The area within the median line is

$$A_m = bh \qquad \textbf{(3-84)}$$

and thus the shear stresses in the vertical and horizontal sides, respectively, are

$$\tau_{\text{vert}} = \frac{T}{2t_1 bh} \qquad \tau_{\text{horiz}} = \frac{T}{2t_2 bh} \qquad \textbf{(3-85a,b)}$$

If $t_2$ is larger than $t_1$, the maximum shear stress will occur in the vertical sides of the cross section.

## Strain Energy and Torsion Constant

The strain energy of a thin-walled tube can be determined by first finding the strain energy of an element and then integrating throughout the volume of the bar. Consider an element of the tube having area $t\,ds$ in the cross section (see the element in Fig. 3-52) and length $dx$ (see the element in Fig. 3-51). The volume of such an element, which is similar in shape to the element $abcd$ shown in Fig. 3-51a, is $t\,ds\,dx$. Because elements of the tube are in pure shear, the strain-energy density of the element is $\tau^2/2G$, as given by Eq. (3-63a). The total strain energy of the element is equal to the strain-energy density times the volume:

$$dU = \frac{\tau^2}{2G} t\,ds\,dx = \frac{\tau^2 t^2}{2G}\frac{ds}{t}\,dx = \frac{f^2}{2G}\frac{ds}{t}\,dx \qquad \textbf{(3-86)}$$

in which we have replaced $\tau t$ by the shear flow $f$ (a constant).

The total strain energy of the tube is obtained by integrating $dU$ throughout the volume of the tube, that is, $ds$ is integrated from 0 to $L_m$ around the median line and $dx$ is integrated along the axis of the tube from 0 to $L$, where $L$ is the length. Thus,

$$U = \int dU = \frac{f^2}{2G}\int_0^{L_m} \frac{ds}{t}\int_0^L dx \qquad \textbf{(3-87)}$$

Note that the thickness $t$ may vary around the median line and must remain with $ds$ under the integral sign. Since the last integral is equal to the length $L$ of the tube, the equation for the strain energy becomes

$$U = \frac{f^2 L}{2G}\int_0^{L_m} \frac{ds}{t} \qquad \textbf{(3-88)}$$

Substituting for the shear flow from Eq. (3-80), we obtain

$$U = \frac{T^2 L}{8GA_m^2}\int_0^{L_m} \frac{ds}{t} \qquad \textbf{(3-89)}$$

**Fig. 3-54**

Thin-walled rectangular tube

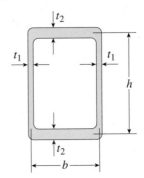

as the equation for the strain energy of the tube in terms of the torque $T$.

The preceding expression for strain energy can be written in simpler form by introducing a new property of the cross section, called the **torsion constant**. For a thin-walled tube, the torsion constant (denoted by the letter $J$) is defined as follows:

$$J = \frac{4A_m^2}{\displaystyle\int_0^{L_m} \frac{ds}{t}} \tag{3-90}$$

With this notation, the equation for **strain energy** (Eq. 3-89) becomes

$$U = \frac{T^2 L}{2GJ} \tag{3-91}$$

which has the same form as the equation for strain energy in a circular bar (see Eq. 3-55a). The only difference is that the torsion constant $J$ has replaced the polar moment of inertia $I_p$. Note that the torsion constant has units of length to the fourth power.

In the special case of a cross section having constant thickness $t$, the expression for $J$ (Eq. 3-90) simplifies to

$$J = \frac{4t A_m^2}{L_m} \tag{3-92}$$

For each shape of cross section, we can evaluate $J$ from either Eq. (3-90) or Eq. (3-92).

As an illustration, consider again the thin-walled **circular tube** of Fig. 3-53. Since the thickness is constant we use Eq. (3-92) and substitute $L_m = 2\pi r$ and $A_m = \pi r^2$; the result is

$$J = 2\pi r^3 t \tag{3-93}$$

which is the approximate expression for the polar moment of inertia (Eq. 3-21). Thus, in the case of a thin-walled circular tube, the polar moment of inertia is the same as the torsion constant.

As a second illustration, we will use the **rectangular tube** of Fig. 3-54. For this cross section we have $A_m = bh$. Also, the integral in Eq. (3-90) is

$$\int_0^{L_m} \frac{ds}{t} = 2\int_0^h \frac{ds}{t_1} + 2\int_0^b \frac{ds}{t_2} = 2\left(\frac{h}{t_1} + \frac{b}{t_2}\right)$$

Thus, the torsion constant (Eq. 3-90) is

$$J = \frac{2b^2 h^2 t_1 t_2}{b t_1 + h t_2} \tag{3-94}$$

Torsion constants for other thin-walled cross sections can be found in a similar manner.

**Fig. 3-53 (Repeated)**

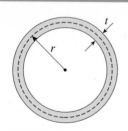

**Fig. 3-54 (Repeated)**

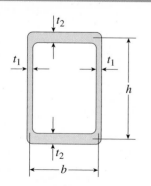

## Angle of Twist

The angle of twist $\phi$ for a thin-walled tube of arbitrary cross-sectional shape (Fig. 3-55) may be determined by equating the work $W$ done by the applied torque $T$ to the strain energy $U$ of the tube. Thus,

$$W = U \quad \text{or} \quad \frac{T\phi}{2} = \frac{T^2 L}{2GJ}$$

from which we get the equation for the angle of twist:

$$\phi = \frac{TL}{GJ} \qquad \qquad \textbf{(3-95)}$$

**Fig. 3-55**

Angle of twist $\phi$ for a thin-walled tube

Again we observe that the equation has the same form as the corresponding equation for a circular bar (Eq. 3-17) but with the polar moment of inertia replaced by the torsion constant. The quantity $GJ$ is called the **torsional rigidity** of the tube.

## Limitations

The formulas developed in this section apply to prismatic members having *closed* tubular shapes with thin walls. If the cross section is thin walled but *open*, as in the case of I-beams and channel sections, the theory given here does not apply. To emphasize this point, imagine that we take a thin-walled tube and slit it lengthwise—then the cross section becomes an open section, the shear stresses and angles of twist increase, the torsional resistance decreases, and the formulas given in this section cannot be used. Recall that the torsion of prismatic bars with noncircular cross sections was reviewed briefly in Section 3.10. This included solid rectangular, triangular, and elliptical cross sections, as well as thin-walled open sections (such as I-beams and channels). An advanced theory is required to derive formulas for shear stress and the angle of twist of such bars, so only key formulas and their application were presented.

Some of the formulas given in this section on thin-walled tubes are restricted to linearly elastic materials—for instance, any equation containing the shear modulus of elasticity $G$ is in this category. However, the equations for shear flow and shear stress (Eqs. 3-80 and 3-81) are based only upon equilibrium and are valid regardless of the material properties. The entire theory is approximate because it is based upon centerline dimensions, and the results become less accurate as the wall thickness $t$ increases.*

An important consideration in the design of any thin-walled member is the possibility that the walls will buckle. The thinner the walls and the longer the tube, the more likely it is that buckling will occur. In the case of noncircular tubes, stiffeners and diaphragms are often used to maintain the shape of the tube and prevent localized buckling. In all of our discussions and problems, we assume that buckling is prevented.

---

*The torsion theory for thin-walled tubes described in this section was developed by R. Bredt, a German engineer who presented it in 1896 (Ref. 3-2). It is often called *Bredt's theory of torsion*.

## ● ● ● Example 3-15

Compare the maximum shear stress in a circular tube (Fig. 3-56) as calculated by the approximate theory for a thin-walled tube with the stress calculated by the exact torsion theory. (Note that the tube has constant thickness $t$ and radius $r$ to the median line of the cross section.)

### Fig. 3-56

Example 3-15: Comparison of approximate and exact theories of torsion

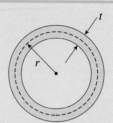

### Solution

*Approximate theory.* The shear stress obtained from the approximate theory for a thin-walled tube (Eq. 3-83) is

$$\tau_1 = \frac{T}{2\pi r^2 t} = \frac{T}{2\pi t^3 \beta^2} \tag{3-96}$$

in which the relation

$$\beta = \frac{r}{t} \tag{3-97}$$

is introduced.

*Torsion formula.* The maximum stress obtained from the more accurate torsion formula (Eq. 3-13) is

$$\tau_2 = \frac{T(r + t/2)}{I_P} \tag{a}$$

where

$$I_P = \frac{\pi}{2}\left[\left(r + \frac{t}{2}\right)^4 - \left(r - \frac{t}{2}\right)^4\right] \tag{b}$$

After expansion, this expression simplifies to

$$I_P = \frac{\pi r t}{2}(4r^2 + r^2) \tag{3-98}$$

and the expression for the shear stress (Eq. a) becomes

$$\tau_2 = \frac{T(2r + t)}{\pi r t(4r^2 + t^2)} = \frac{T(2\beta + 1)}{\pi t^3 \beta(4\beta^2 + 1)} \tag{3-99}$$

*Ratio.* The ratio $\tau_1/\tau_2$ of the shear stresses is

$$\frac{\tau_1}{\tau_2} = \frac{4\beta^2 + 1}{2\beta(2\beta + 1)} \tag{3-100}$$

which depends only on the ratio $\beta$.

For values of $\beta$ equal to 5, 10, and 20, we obtain from Eq. (3-100) the values $\tau_1/\tau_2 = 0.92$, 0.95, and 0.98, respectively. Thus, we see that the approximate formula for the shear stresses gives results that are slightly less than those obtained from the exact formula. The accuracy of the approximate formula increases as the wall of the tube becomes thinner. In the limit, as the thickness approaches zero and $\beta$ approaches infinity, the ratio $\tau_1/\tau_2$ becomes 1.

### ● ● ● Example 3-16

A circular tube and a square tube (Fig. 3-57) are constructed of the same material and subjected to the same torque. Both tubes have the same length, same wall thickness, and same cross-sectional area.

What are the ratios of their shear stresses and angles of twist? (Disregard the effects of stress concentrations at the corners of the square tube.)

**Fig. 3-57**

Example 3-16: Comparison of circular and square tubes

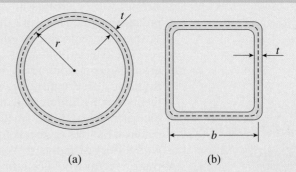

(a)                              (b)

**Solution**

*Circular tube.* For the circular tube, the area $A_{m1}$ enclosed by the median line of the cross section is

$$A_{m1} = \pi r^2 \qquad \text{(c)}$$

where $r$ is the radius to the median line. Also, the torsion constant (Eq. 3-93) and cross-sectional area are

$$J_1 = 2\pi r^3 t \qquad A_1 = 2\pi r t \qquad \text{(d,e)}$$

*Square tube.* For the square tube, the cross-sectional area is

$$A_2 = 4bt \qquad \text{(f)}$$

where $b$ is the length of one side, measured along the median line. Inasmuch as the areas of the tubes are the same, we obtain $b = \pi r/2$. Also, the torsion constant (Eq. 3-94) and area enclosed by the median line of the cross section are

$$J_2 = b^3 t = \frac{\pi^3 r^3 t}{8} \qquad A_{m2} = b^2 = \frac{\pi^2 r^2}{4} \qquad \text{(g,h)}$$

*Ratios.* The ratio $\tau_1/\tau_2$ of the shear stress in the circular tube to the shear stress in the square tube (from Eq. 3-81) is

$$\frac{\tau_1}{\tau_2} = \frac{A_{m2}}{A_{m1}} = \frac{\pi^2 r^2/4}{\pi r^2} = \frac{\pi}{4} = 0.79 \qquad \Longleftarrow \text{(i)}$$

The ratio of the angles of twist (from Eq. 3-95) is

$$\frac{\phi_1}{\phi_2} = \frac{J_2}{J_1} = \frac{\pi^3 r^3 t/8}{2\pi r^3 t} = \frac{\pi^2}{16} = 0.62 \qquad \Longleftarrow \text{(j)}$$

These results show that the circular tube not only has a 21% lower shear stress than does the square tube but also a greater stiffness against rotation.

# *3.12 STRESS CONCENTRATIONS IN TORSION

In the previous sections of this chapter we discussed the stresses in torsional members assuming that the stress distribution varied in a smooth and continuous manner. This assumption is valid provided that there are no abrupt changes in the shape of the bar (no holes, grooves, abrupt steps, and the like) and provided that the region under consideration is away from any points of loading. If such disruptive conditions do exist, then high localized stresses will be created in the regions surrounding the discontinuities. In practical engineering work these **stress concentrations** are handled by means of **stress-concentration factors**, as explained previously in Section 2.10.

The effects of a stress concentration are confined to a small region around the discontinuity, in accord with Saint-Venant's principle (see Section 2.10). For instance, consider a stepped shaft consisting of two segments having different diameters (Fig. 3-58). The larger segment has diameter $D_2$ and the smaller segment has diameter $D_1$. The junction between the two segments forms a "step" or "shoulder" that is machined with a fillet of radius $R$. Without the fillet, the theoretical stress concentration factor would be infinitely large because of the abrupt 90° reentrant corner. Of course, infinite stresses cannot occur. Instead, the material at the reentrant corner would deform and partially relieve the high stress concentration. However, such a situation is very dangerous under dynamic loads, and in good design a fillet is always used. The larger the radius of the fillet, the lower the stresses.

At a distance from the shoulder approximately equal to the diameter $D_2$ (for instance, at cross section $A$-$A$ in Fig. 3-58a) the torsional shear stresses are practically unaffected by the discontinuity. Therefore, the maximum stress $\tau_2$ at a sufficient distance to the left of the shoulder can be found from the torsion formula using $D_2$ as the diameter (Fig. 3-58b). The same general comments apply at section $C$-$C$, which is distance $D_1$ (or greater) from the toe of the fillet. Because the diameter $D_1$ is less than the diameter $D_2$, the maximum stress $\tau_1$ at section $C$-$C$ (Fig. 3-58d) is larger than the stress $\tau_2$.

The stress-concentration effect is greatest at section $B$-$B$, which cuts through the toe of the fillet. At this section the maximum stress is

$$\tau_{\max} = K\tau_{\text{nom}} = K\frac{Tr}{I_P} = K\left(\frac{16T}{\pi D_1^3}\right) \qquad \textbf{(3-101)}$$

In this equation, $K$ is the stress-concentration factor and $\tau_{\text{nom}}$ (equal to $\tau_1$) is the nominal shear stress, that is, the shear stress in the smaller part of the shaft.

Values of the factor $K$ are plotted in Fig. 3-59 as a function of the ratio $R/D_1$. Curves are plotted for various values of the ratio $D_2/D_1$. Note that when the fillet radius $R$ becomes very small and the transition from one diameter to the other is abrupt, the value of $K$ becomes quite large. Conversely, when $R$ is large, the value of $K$ approaches 1.0 and the effect of the stress concentration disappears. The dashed curve in Fig. 3-59 is for the special case of a full quarter-circular fillet, which means that $D_2 = D_1 + 2R$. (*Note:* Probs. 3.12-1 through 3.12-5 provide practice in obtaining values of $K$ from Fig. 3-59.)

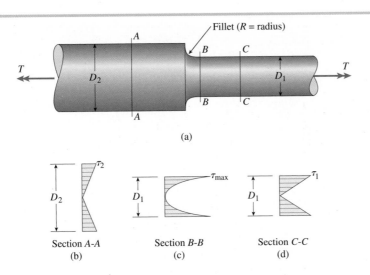

**Fig. 3-58**

Stepped shaft in torsion

Many other cases of stress concentrations for circular shafts, such as a shaft with a keyway and a shaft with a hole, are available in the engineering literature (see, for example, Ref. 2-9).

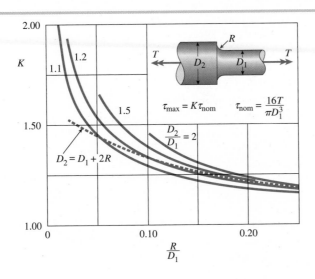

**Fig. 3-59**

Stress-concentration factor $K$ for a stepped shaft in torsion. (The dashed line is for a full quarter-circular fillet.)

$$\tau_{max} = K\tau_{nom} \qquad \tau_{nom} = \frac{16T}{\pi D_1^3}$$

As explained in Section 2.10, stress concentrations are important for brittle materials under static loads and for most materials under dynamic loads. As a case in point, fatigue failures are of major concern in the design of rotating shafts and axles (see Section 2.9 for a brief discussion of fatigue). The theoretical stress-concentration factors $K$ given in this section are based upon linearly elastic behavior of the material. However, fatigue experiments show that these factors are conservative, and failures in ductile materials usually occur at larger loads than those predicted by the theoretical factors.

### • • • Example 3-17

A stepped shaft consisting of solid circular segments ($D_1$ = 1.75 in. and $D_2$ = 2.1 in., see Fig. 3-60) has a fillet of radius $R$ = 0.2 in.

(a) Find the *maximum permissible torque $T_{max}$*, assuming that the allowable shear stress at the stress concentration is 9200 psi.
(b) If the shaft is to be replaced with a shaft with allowable shear stress of 12,500 psi, $D_2$ = 2.1 in. with a full quarter-circular fillet, carrying a torque of $T$ = 8500 lb-in., find the *smallest acceptable value of diameter $D_1$*.

### Fig. 3-60

Example 3-17: Stepped circular shaft in torsion

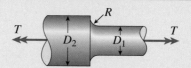

**Solution**

(a) *Maximum permissible torque.* If we compute the ratio of the shaft diameters ($D_2/D_1$ = 1.2) and the ratio of the fillet radius $R$ to diameter $D_1$ ($R/D_1$ = 0.114), we can find the stress concentration factor $K$ to be approximately 1.3 from Fig. 3-59 (repeated here). Then, equating the maximum shear stress in the smaller shaft to the allowable shear stress $\tau_a$, we get

$$\tau_{max} = K\left(\frac{16T}{\pi D_1^3}\right) = \tau_a \tag{a}$$

Solving Eq. (a) for $T_{max}$, we get

$$T_{max} = \tau_a\left(\frac{\pi D_1^3}{16K}\right) \tag{b}$$

Substituting numerical values gives

$$T_{max} = (9200 \text{ psi}) \cdot \left[\frac{\pi(1.75 \text{ in.})^3}{16(1.3)}\right] = 7447 \text{ lb-in.}$$

### Fig. 3-59 (Repeated)

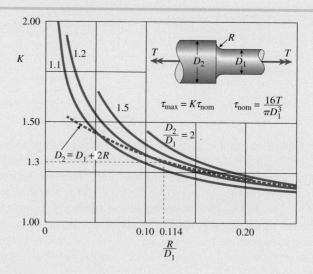

(b) *Smallest acceptable value of diameter $D_1$.* In the shaft redesign, a full quarter-circular fillet is being used so

$$D_2 = D_1 + 2R \quad \text{or} \quad R = \frac{D_2 - D_1}{2} = \frac{2.1 \text{ in.} - D_1}{2} = 1.05 \text{ in.} - \frac{D_1}{2} \quad \text{(c)}$$

Next, solve Eq. (a) for diameter $D_1$ in terms of the unknown stress concentration factor $K$:

$$D_1 = \left[ K \left( \frac{16T}{\pi \tau_a} \right) \right]^{\frac{1}{3}} = \left[ K \left[ \frac{16(8500 \text{ lb-in.})}{\pi(12{,}500 \text{ psi})} \right] \right]^{\frac{1}{3}} = \left( \frac{272K \text{ lb-in.}}{25\pi \text{ psi}} \right)^{\frac{1}{3}} \quad \text{(d)}$$

Solving Eqs. (c) and (d) *using trial and error* and using Fig. 3-59 to obtain $K$, we have the following results:

*Trial #1.*

$$D_{1a} = 1.50 \text{ in.} \quad R = 1.05 \text{ in.} - \frac{D_{1a}}{2} = 0.3 \text{ in.} \quad \frac{R}{D_{1a}} = 0.2$$

From Fig. 3-59: $K = 1.24$:

$$D_{1b} = \left( \frac{272K \text{ lb-in.}}{25\pi \text{ psi}} \right)^{\frac{1}{3}} = 1.625 \text{ in.}$$

*Trial #2.*

$$D_{1a} = 1.625 \text{ in.} \quad R = 1.05 \text{ in.} - \frac{D_{1a}}{2} = 0.238 \text{ in.} \quad \frac{R}{D_{1a}} = 0.146$$

From Fig. 3-59: $K = 1.26$:

$$D_{1b} = \left( \frac{272K \text{ lb-in.}}{25\pi \text{ psi}} \right)^{\frac{1}{3}} = 1.634 \text{ in.}$$

*Trial #3.*

$$D_{1a} = 1.64 \text{ in.} \quad R = 1.05 \text{ in.} - \frac{D_{1a}}{2} = 0.23 \text{ in.} \quad \frac{R}{D_{1a}} = 0.14$$

From Fig. 3-59: $K = 1.265$:

$$D_{1b} = \left( \frac{272K \text{ lb-in.}}{25\pi \text{ psi}} \right)^{\frac{1}{3}} = 1.636 \text{ in.}$$

**Use $D_1 = 1.64$ in.** Check the maximum shear stress:

$$\tau_{max} = K \left( \frac{16T}{\pi D_1^3} \right) = (1.265) \left[ \frac{16(8500 \text{ lb-in.})}{\pi(1.64 \text{ in.})^3} \right] = 12{,}415 \text{ psi}$$

A stepped shaft with $D_2 = 2.1$ in., $D_1 = 1.64$ in., and a full quarter-circular fillet of radius $R = 0.23$ in. will carry the required torque $T$ without exceeding the allowable shear stress in the fillet region.

In **Chapter 3**, we investigated the behavior of bars and hollow tubes acted on by concentrated torques or distributed torsional moments as well as pre-strain effects. We developed torque-displacement relations for use in computing angles of twist of bars under both uniform (i.e., constant torsional moment over its entire length) and nonuniform conditions (i.e., torques, and perhaps also polar moment of inertia, vary over the length of the bar). Then, equilibrium and compatibility equations were developed for statically indeterminate structures in a superposition procedure leading to solution for all unknown torques, rotational displacements, stresses, etc. Starting with a state of pure shear on stress elements aligned with the axis of the bar, we then developed equations for normal and shear stresses on inclined sections. A number of advanced topics were presented in the last parts of the chapter. The major concepts presented in this chapter are as follows:

1. For circular bars and tubes, the **shearing stress** ($\tau$) and **strain** ($\gamma$) vary linearly with radial distance from the center of the cross-section.

$$\tau = (\rho/r)\tau_{max} \qquad \gamma = (\rho/r)\gamma_{max}$$

2. The **torsion formula** defines the relation between shear stress and torsional moment. Maximum shear stress $\tau_{max}$ occurs on the outer surface of the bar or tube and depends on torsional moment $T$, radial distance $r$, and second moment of inertia of the cross section $I_p$, known as polar moment of inertia for circular cross sections. Thin-walled tubes are seen to be more efficient in torsion, because the available material is more uniformly stressed than solid circular bars.

$$\tau_{max} = \frac{Tr}{I_P}$$

3. The angle of twist $\phi$ of prismatic circular bars subjected to torsional moment(s) is proportional to both the torque $T$ and the length of the bar $L$, and inversely proportional to the torsional rigidity ($GI_p$) of the bar; this relationship is called the **torque-displacement relation**.

$$\phi = \frac{TL}{GI_P}$$

4. The angle of twist per unit length of a bar is referred to as its **torsional flexibility** ($f_T$), and the inverse relationship is the torsional **stiffness** ($k_T = 1/f_T$) of the bar or shaft.

$$k_T = \frac{GI_P}{L} \qquad f_T = \frac{L}{GI_P}$$

5. The summation of the twisting deformations of the individual segments of a nonprismatic shaft equals the twist of the entire bar ($\phi$). Free-body diagrams are used to find the torsional moments ($T_i$) in each segment $i$.

$$\phi = \sum_{i=1}^{n} \phi_i = \sum_{i=1}^{n} \frac{T_i L_i}{G_i (I_P)_i}$$

If torsional moments and/or cross sectional properties ($I_p$) vary continuously, an integral expression is required.

$$\phi = \int_0^L d\phi = \int_0^L \frac{T(x)dx}{GI_p(x)}$$

6. If the bar structure is **statically indeterminate**, additional equations are required to solve for unknown moments. **Compatibility equations** are used to relate bar rotations to support conditions and thereby generate additional relationships among the unknowns. It is convenient to use a **superposition** of "released" (or statically determinate) structures to represent the actual statically indeterminate bar structure.

7. **Misfits** and **prestrains** induce torsional moments only in statically indeterminate bars or shafts.

8. A circular shaft is subjected to **pure shear** due to torsional moments. **Maximum normal** and **shear stresses** can be obtained by considering an inclined stress element. The maximum shear stress occurs on an element aligned with the axis of the bar, but the maximum normal stress occurs at an inclination of 45° to the axis of the bar, and the maximum normal stress is equal to the maximum shear stress

$$\sigma_{max} = \tau$$

We can also find a relationship between the maximum shear and normal strains for the case of pure shear:

$$\varepsilon_{max} = \gamma_{max}/2$$

9. Circular shafts are commonly used to transmit mechanical power from one device or machine to another. If the torque $T$ is expressed in newton meters and $n$ is the shaft rpm, the power $P$ is expressed in watts as

$$P = \frac{2\pi n T}{60}$$

In U.S. customary units, torque $T$ is given in ft-lb and power may be given in horsepower (hp), $H$, as

$$H = \frac{2\pi n T}{33,000}$$

10. A number of advanced topics were also discussed in Chapter 3 but are not reviewed in this summary. These advanced topics include: strain energy in torsion, noncircular cross sections, thin-walled tubes, and stress concentrations.

# PROBLEMS CHAPTER 3

## Torsional Deformations

**3.2-1** A copper rod of length $L = 18.0$ in. is to be twisted by torques $T$ (see figure) until the angle of rotation between the ends of the rod is 3.0°.

(a) If the allowable shear strain in the copper is 0.0006 rad, what is the maximum permissible diameter of the rod?

(b) If the rod diameter is 0.5 in., what is the minimum permissible length of the rod?

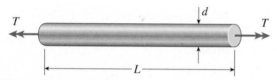

PROBS. 3.2-1 and 3.2-2

**3.2-2** A plastic bar of diameter $d = 56$ mm is to be twisted by torques $T$ (see figure) until the angle of rotation between the ends of the bar is 4.0°.

(a) If the allowable shear strain in the plastic is 0.012 rad, what is the minimum permissible length of the bar?

(b) If the length of the bar is 200 mm, what is the maximum permissible diameter of the bar?

**3.2-3** A circular aluminum tube subjected to pure torsion by torques $T$ (see figure) has an outer radius $r_2$ equal to 1.5 times the inner radius $r_1$.

(a) If the maximum shear strain in the tube is measured as $400 \times 10^{-6}$ rad, what is the shear strain $\gamma_1$ at the inner surface?

(b) If the maximum allowable rate of twist is 0.125 degrees per foot and the maximum shear strain is to be kept at $400 \times 10^{-6}$ rad by adjusting the torque $T$, what is the minimum required outer radius $(r_2)_{min}$?

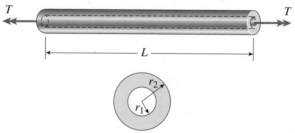

PROBS. 3.2-3, 3.2-4, and 3.2-5

**3.2-4** A circular steel tube of length $L = 1.0$ m is loaded in torsion by torques $T$ (see figure).

(a) If the inner radius of the tube is $r_1 = 45$ mm and the measured angle of twist between the ends is 0.5°, what is the shear strain $\gamma_1$ (in radians) at the inner surface?

(b) If the maximum allowable shear strain is 0.0004 rad and the angle of twist is to be kept at 0.45° by adjusting the torque $T$, what is the maximum permissible outer radius $(r_2)_{max}$?

**3.2-5** Solve the preceding problem if the length $L = 56$ in., the inner radius $r_1 = 1.25$ in., the angle of twist is 0.5°, and the allowable shear strain is 0.0004 rad.

## Circular Bars and Tubes

**3.3-1** A prospector uses a hand-powered winch (see figure) to raise a bucket of ore in his mine shaft. The axle of the winch is a steel rod of diameter $d = 0.625$ in. Also, the distance from the center of the axle to the center of the lifting rope is $b = 4.0$ in.

(a) If the weight of the loaded bucket is $W = 100$ lb, what is the maximum shear stress in the axle due to torsion?

(b) If the maximum bucket load is 125 lb and the allowable shear stress in the axle is 9250 psi, what is the minimum permissible axle diameter?

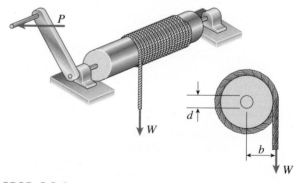

PROB. 3.3-1

**3.3-2** When drilling a hole in a table leg, a furniture maker uses a hand-operated drill (see figure) with a bit of diameter $d = 4.0$ mm.

(a) If the resisting torque supplied by the table leg is equal to 0.3 N·m, what is the maximum shear stress in the drill bit?

(b) If the allowable shear stress in the drill bit is 32 MPa, what is the maximum resisting torque before the drill binds up?

(c) If the shear modulus of elasticity of the steel is $G = 75$ GPa, what is the rate of twist of the drill bit (degrees per meter)?

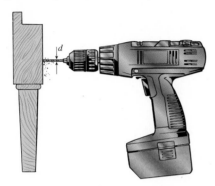

PROB. 3.3-2

**3.3-3** While removing a wheel to change a tire, a driver applies forces $P = 25$ lb at the ends of two of the arms of a lug wrench (see figure). The wrench is made of steel with shear modulus of elasticity $G = 11.4 \times 10^6$ psi. Each arm of the wrench is 9.0 in. long and has a solid circular cross section of diameter $d = 0.5$ in.

(a) Determine the maximum shear stress in the arm that is turning the lug nut (arm $A$).

(b) Determine the angle of twist (in degrees) of this same arm.

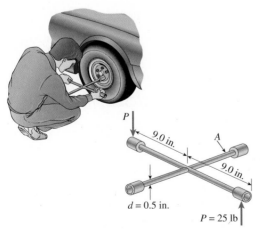

PROB. 3.3-3

**3.3-4** An aluminum bar of solid circular cross section is twisted by torques $T$ acting at the ends (see figure). The dimensions and shear modulus of elasticity are as follows: $L = 1.4$ m, $d = 32$ mm, and $G = 28$ GPa.

(a) Determine the torsional stiffness of the bar.

(b) If the angle of twist of the bar is 5°, what is the maximum shear stress? What is the maximum shear strain (in radians)?

(c) If a hole of diameter $d/2$ is drilled longitudinally through the bar, what is the ratio of the torsional stiffnesses of the hollow and solid bars? What is the ratio of their maximum shear stresses if both are acted on by the same torque?

(d) If the hole diameter remains at $d/2$, what new outside diameter $d_2$ will result in equal stiffnesses of the hollow and solid bars?

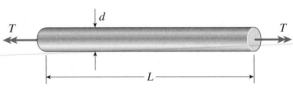

PROB. 3.3-4

**3.3-5** A high-strength steel drill rod used for boring a hole in the earth has a diameter of 0.5 in. (see figure). The allowable shear stress in the steel is 40 ksi and the shear modulus of elasticity is 11,600 ksi.

(a) What is the minimum required length of the rod so that one end of the rod can be twisted 30° with respect to the other end without exceeding the allowable stress?

(b) If the shear strain in part (a) is limited to $3.2 \times 10^{-3}$, what is the minimum required length of the drill rod?

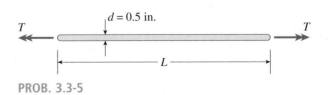

PROB. 3.3-5

**3.3-6** The steel shaft of a socket wrench has a diameter of 8.0 mm. and a length of 200 mm (see figure).

If the allowable stress in shear is 60 MPa, what is the maximum permissible torque $T_{max}$ that may be exerted with the wrench?

Through what angle $\phi$ (in degrees) will the shaft twist under the action of the maximum torque? (Assume $G = 78$ GPa and disregard any bending of the shaft.)

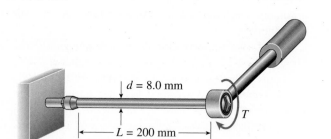

**PROB. 3.3-6**

**3.3-7** A circular tube of aluminum is subjected to torsion by torques $T$ applied at the ends (see figure). The bar is 24 in. long, and the inside and outside diameters are 1.25 in. and 1.75 in., respectively. It is determined by measurement that the angle of twist is 4° when the torque is 6200 lb-in.

(a) Calculate the maximum shear stress $\tau_{max}$ in the tube, the shear modulus of elasticity $G$, and the maximum shear strain $\gamma_{max}$ (in radians).

(b) If the maximum shear strain in the tube is limited to $2.5 \times 10^{-3}$ and the inside diameter is increased to 1.375 in., what is the maximum permissible torque?

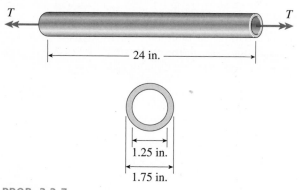

**PROB. 3.3-7**

**3.3-8** A propeller shaft for a small yacht is made of a solid steel bar 104 mm in diameter. The allowable stress in shear is 48 MPa, and the allowable rate of twist is 2.0° in 3.5 meters.

(a) Assuming that the shear modulus of elasticity is $G = 80$ GPa, determine the maximum torque $T_{max}$ that can be applied to the shaft.

(b) Repeat part (a) if the shaft is now hollow with an inner diameter of $5d/8$. Compare $T_{max}$ values to corresponding values from part (a)

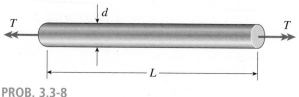

**PROB. 3.3-8**

**3.3-9** Three identical circular disks $A$, $B$, and $C$ are welded to the ends of three identical solid circular bars (see figure). The bars lie in a common plane and the disks lie in planes perpendicular to the axes of the bars. The bars are welded at their intersection $D$ to form a rigid connection. Each bar has diameter $d_1 = 0.5$ in. and each disk has diameter $d_2 = 3.0$ in.

Forces $P_1$, $P_2$, and $P_3$ act on disks $A$, $B$, and $C$, respectively, thus subjecting the bars to torsion. If $P_1 = 28$ lb, what is the maximum shear stress $\tau_{max}$ in any of the three bars?

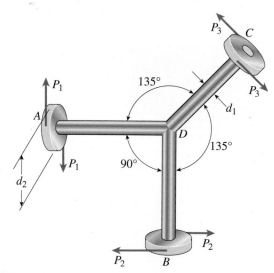

**PROB. 3.3-9**

**3.3-10** The steel axle of a large winch on an ocean liner is subjected to a torque of 1.65 kN·m (see figure).

(a) What is the minimum required diameter $d_{min}$ if the allowable shear stress is 48 MPa and the allowable rate of twist is 0.75°/m? (Assume that the shear modulus of elasticity is 80 GPa.)

(b) Repeat part (a) if the shaft is now hollow with an inner diameter of $5d/8$. Compare $d_{min}$ values to corresponding values from part (a).

**PROB. 3.3-10**

**3.3-11** A hollow steel shaft used in a construction auger has outer diameter $d_2 = 6.0$ in. and inner diameter $d_1 = 4.5$ in. (see figure). The steel has shear modulus of elasticity $G = 11.0 \times 10^6$ psi.

For an applied torque of 150 k-in., determine the following quantities:

(a) shear stress $\tau_2$ at the outer surface of the shaft,
(b) shear stress $\tau_1$ at the inner surface, and
(c) rate of twist $\theta$ (degrees per unit of length).

Also, draw a diagram showing how the shear stresses vary in magnitude along a radial line in the cross section.

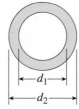

**PROBS. 3.3-11 and 3.3-12**

**3.3-12** Solve the preceding problem if the shaft has outer diameter $d_2 = 150$ mm and inner diameter $d_1 = 100$ mm. Also, the steel has shear modulus of elasticity $G = 75$ GPa and the applied torque is 16 kN·m.

**3.3-13** A vertical pole of solid, circular cross section is twisted by horizontal forces $P = 1100$ lb acting at the ends of a rigid horizontal arm $AB$ (see figure part a). The distance from the outside of the pole to the line of action of each force is $c = 5.0$ in. (see figure part b) and the pole height is $L = 14$ in.

(a) If the allowable shear stress in the pole is 4500 psi, what is the minimum required diameter $d_{min}$ of the pole?

(b) Find the torsional stiffness of the pole (kip-in./rad). Assume that $G = 10,800$ ksi.

(c) If two translational springs, each with stiffness $k = 33$ kips/in., are added at $2c/5$ from $A$ and $B$ (see figure part c), repeat part (a) to find $d_{min}$. (Hint: Consider the pole and pair of springs as "springs in parallel." )

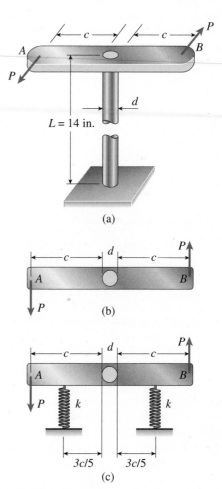

**PROBS. 3.3-13 and 3.3-14**

**3.3-14** A vertical pole of solid, circular cross section is twisted by horizontal forces $P = 5$ kN acting at the ends of a rigid horizontal arm $AB$ (see figure part a). The distance from the outside of the pole to the line of action of each force is $c = 125$ mm (see figure part b) and the pole height $L = 350$ mm.

(a) If the allowable shear stress in the pole is 30 MPa, what is the minimum required diameter $d_{min}$ of the pole?

(b) What is the torsional stiffness of the pole (kN · m/rad)?

(c) If two translational springs, each with stiffness $k = 2550$ kN/m, are added at $2c/5$ from $A$ and $B$ (see figure part c), repeat part (a) to find $d_{min}$. (*Hint*: Consider the pole and pair of springs as "springs in parallel.")

**3.3-15** A solid brass bar of diameter $d = 1.25$ in. is subjected to torques $T_1$, as shown in part a of the figure. The allowable shear stress in the brass is 12 ksi.

(a) What is the maximum permissible value of the torques $T_1$?

(b) If a hole of diameter 0.625 in. is drilled longitudinally through the bar, as shown in part b of the figure, what is the maximum permissible value of the torques $T_2$?

(c) What is the percent decrease in torque and the percent decrease in weight due to the hole?

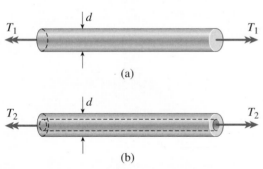

(a)

(b)

PROB. 3.3-15

**3.3-16** A hollow aluminum tube used in a roof structure has an outside diameter $d_2 = 104$ mm and an inside diameter $d_1 = 82$ mm (see figure). The tube is 2.75 m long, and the aluminum has shear modulus $G = 28$ GPa.

(a) If the tube is twisted in pure torsion by torques acting at the ends, what is the angle of twist (in degrees) when the maximum shear stress is 48 MPa?

(b) What diameter $d$ is required for a solid shaft (see figure) to resist the same torque with the same maximum stress?

(c) What is the ratio of the weight of the hollow tube to the weight of the solid shaft?

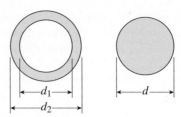

PROB. 3.3-16

**3.3-17** A circular tube of inner radius $r_1$ and outer radius $r_2$ is subjected to a torque produced by forces $P = 900$ lb (see figure part a). The forces have their lines of action at a distance $b = 5.5$ in. from the outside of the tube.

(a) If the allowable shear stress in the tube is 6300 psi and the inner radius $r_1 = 1.2$ in., what is the minimum permissible outer radius $r_2$?

(b) If a torsional spring of stiffness $k_R = 450$ kip-in./rad is added at the end of the tube (see figure part b), what is the maximum value of forces $P$ if the allowable shear stress is not to be exceeded? Assume that the tube has a length of $L = 18$ in., outer radius of $r_2 = 1.45$ in., and shear modulus $G = 10,800$ ksi. (*Hint*: Consider the tube and torsional spring as "springs in parallel.")

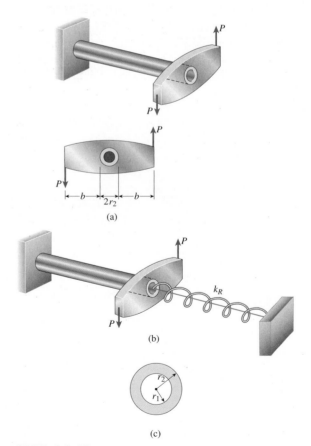

(a)

(b)

(c)

PROB. 3.3-17

# Nonuniform Torsion

**3.4-1** A stepped shaft *ABC* consisting of two solid circular segments is subjected to torques $T_1$ and $T_2$ acting in opposite directions, as shown in the figure. The larger segment of the shaft has a diameter of $d_1 = 2.25$ in. and length $L_1 = 30$ in.; the smaller segment has a diameter of $d_2 = 1.75$ in. and length of $L_2 = 20$ in. The material is steel with shear modulus $G = 11 \times 10^6$ psi, and the torques are $T_1 = 20,000$ lb-in. and $T_2 = 8000$ lb-in.

(a) Calculate the maximum shear stress $\tau_{max}$ in the shaft and the angle of twist $\phi_c$ (in degrees) at end *C*.

(b) If the maximum shear stress in *BC* must be the same as that in *AB*, what is the required diameter of segment *BC*? What is the resulting twist at end *C*?

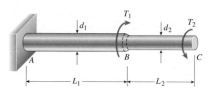

PROB. 3.4-1

**3.4-2** A circular tube of outer diameter $d_3 = 70$ mm and inner diameter $d_2 = 60$ mm is welded at the right-hand end to a fixed plate and at the left-hand end to a rigid end plate (see figure). A solid, circular bar with a diameter of $d_1 = 40$ mm is inside of, and concentric with, the tube. The bar passes through a hole in the fixed plate and is welded to the rigid end plate.

The bar is 1.0 m long and the tube is half as long as the bar. A torque $T = 1000$ N·m acts at end *A* of the bar. Also, both the bar and tube are made of an aluminum alloy with shear modulus of elasticity $G = 27$ GPa.

(a) Determine the maximum shear stresses in both the bar and tube.

(b) Determine the angle of twist (in degrees) at end *A* of the bar.

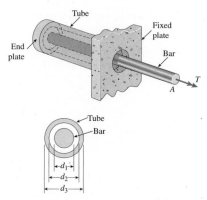

PROB. 3.4-2

**3.4-3** A stepped shaft *ABCD* consisting of solid circular segments is subjected to three torques, as shown in the figure. The torques have magnitudes of 12.5 k-in., 9.8 k-in., and 9.2 k-in. The length of each segment is 25 in. and the diameters of the segments are 3.5 in., 2.75 in., and 2.5 in. The material is steel with shear modulus of elasticity $G = 11.6 \times 10^3$ ksi.

(a) Calculate the maximum shear stress $\tau_{max}$ in the shaft and the angle of twist $\phi_D$ (in degrees) at end *D*.

(b) If each segment must have the same shear stress, find the required diameter of each segment in part (a) so that all three segments have shear stress $\tau_{max}$ from part (a). What is the resulting angle of twist at *D*?

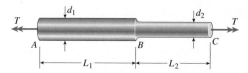

PROB. 3.4-3

**3.4-4** A solid, circular bar *ABC* consists of two segments, as shown in the figure. One segment has a diameter of $d_1 = 56$ mm and length of $L_1 = 1.45$ m; the other segment has a diameter of $d_2 = 48$ mm and length of $L_2 = 1.2$ m.

What is the allowable torque $T_{allow}$ if the shear stress is not to exceed 30 MPa and the angle of twist between the ends of the bar is not to exceed 1.25°? (Assume $G = 80$ GPa.)

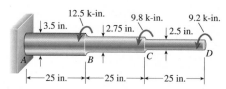

PROB. 3.4-4

**3.4-5** A hollow tube *ABCDE* constructed of monel metal is subjected to five torques acting in the directions shown in the figure. The magnitudes of the torques are $T_1 = 1000$ lb-in., $T_2 = T_4 = 500$ lb-in., and $T_3 = T_5 = 800$ lb-in. The tube has an outside diameter of $d_2 = 1.0$ in. The allowable shear stress is 12,000 psi and the allowable rate of twist is 2.0°/ft.

Determine the maximum permissible inside diameter $d_1$ of the tube.

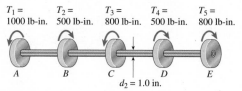

PROB. 3.4-5

**3.4-6** A shaft of solid, circular cross section consisting of two segments is shown in the first part of the figure. The left-hand segment has a diameter of 80 mm and length of 1.2 m; the right-hand segment has a diameter of 60 mm and length of 0.9 m.

Shown in the second part of the figure is a hollow shaft made of the same material and having the same length. The thickness $t$ of the hollow shaft is $d/10$, where $d$ is the outer diameter. Both shafts are subjected to the same torque.

(a) If the hollow shaft is to have the same torsional stiffness as the solid shaft, what should be its outer diameter $d$?

(b) If torque $T$ is applied at either end of both shafts, and the hollow shaft is to have the same maximum shear stress as the solid shaft, what should be its outer diamteter $d$?

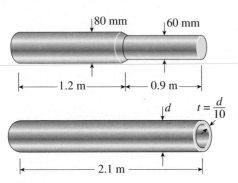

PROB. 3.4-6

**3.4-7** Four gears are attached to a circular shaft and transmit the torques shown in the figure. The allowable shear stress in the shaft is 10,000 psi.

(a) What is the required diameter $d$ of the shaft if it has a solid cross section?

(b) What is the required outside diameter $d$ if the shaft is hollow with an inside diameter of 1.0 in.?

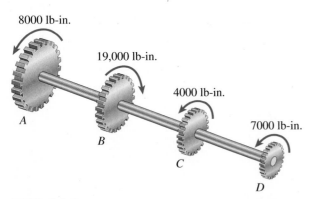

PROB. 3.4-7

**3.4-8** A tapered bar $AB$ of solid circular cross section is twisted by torques $T$ (see figure). The diameter of the bar varies linearly from $d_A$ at the left-hand end to $d_B$ at the right-hand end.

(a) Confirm that the angle of twist of the tapered bar is

$$\phi = \frac{32TL}{3\pi G(d_B - d_A)}\left(\frac{1}{d_A^3} - \frac{1}{d_B^3}\right)$$

(b) For what ratio $d_B/d_A$ will the angle of twist of the tapered bar be one-half the angle of twist of a prismatic bar of diameter $d_A$? (The prismatic bar is made of the same material, has the same length, and is subjected to the same torque as the tapered bar.)

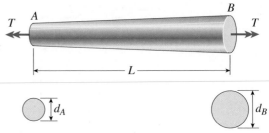

PROBS. 3.4-8, 3.4-9, and 3.4-10

**3.4-9** A tapered bar $AB$ of solid circular cross section is twisted by torques $T = 36,000$ lb-in. (see figure). The diameter of the bar varies linearly from $d_A$ at the left-hand end to $d_B$ at the right-hand end. The bar has length $L = 4.0$ ft and is made of an aluminum alloy having shear modulus of elasticity $G = 3.9 \times 10^6$ psi. The allowable shear stress in the bar is 15,000 psi and the allowable angle of twist is 3.0°.

If the diameter at end $B$ is 1.5 times the diameter at end $A$, what is the minimum required diameter $d_A$ at end $A$? [*Hint:* Use the angle of twist expression from Prob. 3.4-8(a)].

**3.4-10** The bar shown in the figure is tapered linearly from end $A$ to end $B$ and has a solid circular cross section. The diameter at the smaller end of the bar is $d_A = 25$ mm and the length is $L = 300$ mm. The bar is made of steel with shear modulus of elasticity $G = 82$ GPa.

If the torque $T = 180$ N·m and the allowable angle of twist is 0.3°, what is the minimum allowable diameter $d_B$ at the larger end of the bar? [*Hint:* Use the angle of twist expression from Prob. 3.4-8(a)]

**3.4-11** The nonprismatic cantilever circular bar shown has an internal cylindrical hole from 0 to $x$, so the net polar moment of inertia of the cross section for segment 1 is $(7/8)I_p$. Torque $T$ is applied at $x$ and torque $T/2$ is applied at $x = L$. Assume that $G$ is constant.

(a) Find reaction moment $R_1$.

(b) Find internal torsional moments $T_i$ in segments 1 and 2.

(c) Find $x$ required to obtain twist at joint 3 of $\phi_3 = TL/GI_p$.

(d) What is the rotation at joint 2, $\phi_2$?

(e) Draw the torsional moment (TMD: $T(x), 0 \le x \le L$) and displacement (TDD: $\phi(x), 0 \le x \le L$) diagrams.

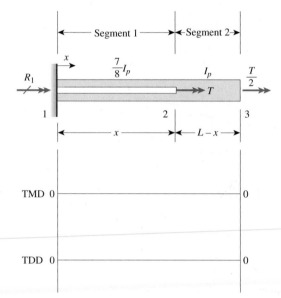

**PROB. 3.4-11**

**3.4-12** A uniformly tapered tube $AB$ of hollow circular cross section is shown in the figure. The tube has constant wall thickness $t$ and length $L$. The average diameters at the ends are $d_A$ and $d_B = 2d_A$. The polar moment of inertia may be represented by the approximate formula $I_p \approx \pi d^3 t/4$ (see Eq. 3-21).

Derive a formula for the angle of twist $\phi$ of the tube when it is subjected to torques $T$ acting at the ends.

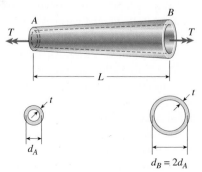

**PROB. 3.4-12**

**3.4-13** A uniformly tapered aluminum-alloy tube $AB$ of circular cross section and length $L$ is shown in the figure. The outside diameters at the ends are $d_A$ and $d_B = 2d_A$. A hollow section of length $L/2$ and constant thickness $t = d_A/10$ is cast into the tube and extends from $B$ halfway toward $A$.

(a) Find the angle of twist $\phi$ of the tube when it is subjected to torques $T$ acting at the ends. Use numerical values as follows: $d_A = 2.5$ in., $L = 48$ in., $G = 3.9 \times 10^6$ psi, and $T = 40,000$ in.-lb.

(b) Repeat part (a) if the hollow section has constant diameter $d_A$ (see figure part b).

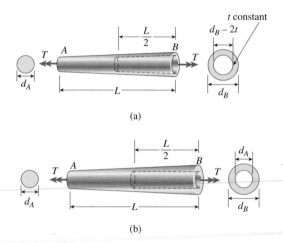

(a)

(b)

**PROB. 3.4-13**

**3.4-14** For the *thin* nonprismatic steel pipe of constant thickness $t$ and variable diameter $d$ shown with applied torques at joints 2 and 3, determine the following.

(a) Find reaction moment $R_1$.

(b) Find an expression for twist rotation $\phi_3$ at joint 3. Assume that $G$ is constant.

(c) Draw the torsional moment diagram (TMD: $T(x)$, $0 \le x \le L$).

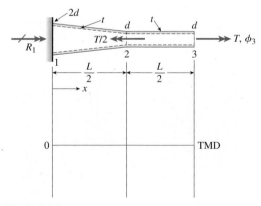

**PROB. 3.4-14**

**3.4-15** A mountain-bike rider going uphill applies torque $T = Fd$ ($F = 15$ lb, $d = 4$ in.) to the end of the handlebars $ABCD$ (by pulling on the handlebar extenders $DE$). Consider the right half of the handlebar assembly only (assume the bars are fixed at the fork at $A$). Segments $AB$ and $CD$ are prismatic with lengths $L_1 = 2$ in. and $L_3 = 8.5$ in., and with outer diameters and thicknesses $d_{01} = 1.25$ in., $t_{01} = 0.125$ in., and $d_{03} = 0.87$ in., $t_{03} = 0.115$ in., respectively as shown. Segment $BC$ of length $L_2 = 1.2$ in., however, is tapered, and outer diameter and thickness vary linearly between dimensions at $B$ and $C$.

*Consider torsion effects only.* Assume $G = 4000$ ksi is constant.

Derive an integral expression for the angle of twist $\phi_D$ of half of the handlebar tube when it is subjected to torque $T = Fd$ acting at the end. Evaluate $\phi_D$ for the given numerical values.

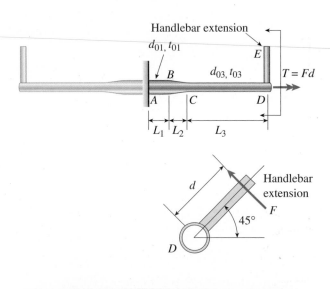

(© Barry Goodno)
PROB. 3.4-15

**3.4-16** A prismatic bar $AB$ of length $L$ and solid circular cross section (diameter $d$) is loaded by a distributed torque of constant intensity $t$ per unit distance (see figure).

(a) Determine the maximum shear stress $\tau_{max}$ in the bar.

(b) Determine the angle of twist $\phi$ between the ends of the bar.

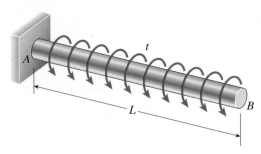

PROB. 3.4-16

**3.4-17** A prismatic bar $AB$ of solid circular cross section (diameter $d$) is loaded by a distributed torque (see figure). The intensity of the torque, that is, the torque per unit distance, is denoted $t(x)$ and varies linearly from a maximum value $t_A$ at end $A$ to zero at end $B$. Also, the length of the bar is $L$ and the shear modulus of elasticity of the material is $G$.

(a) Determine the maximum shear stress $\tau_{max}$ in the bar.

(b) Determine the angle of twist $\phi$ between the ends of the bar.

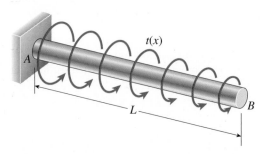

PROB. 3.4-17

**3.4-18** A nonprismatic bar $ABC$ of solid circular cross section is loaded by distributed torques (see figure). The intensity of the torques, that is, the torque per unit distance, is denoted $t(x)$ and varies linearly from zero at $A$ to a maximum value $T_0/L$ at $B$. Segment $BC$ has linearly distributed torque of intensity $t(x) = T_0/3L$ of opposite sign to that applied along $AB$. Also, the polar moment of inertia of $AB$ is twice that of $BC$, and the shear modulus of elasticity of the material is $G$.

(a) Find reaction torque $R_A$.

(b) Find internal torsional moments $T(x)$ in segments $AB$ and $BC$.

(c) Find rotation $\phi_C$.

(d) Find the maximum shear stress $\tau_{max}$ and its location along the bar.

(e) Draw the torsional moment diagram (TMD: $T(x)$), $0 \le x \le L$).

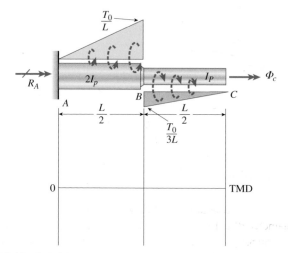

**PROB. 3.4-18**

**3.4-19** A magnesium-alloy wire of diameter $d = 4$ mm and length $L$ rotates inside a flexible tube in order to open or close a switch from a remote location (see figure). A torque $T$ is applied manually (either clockwise or counterclockwise) at end $B$, thus twisting the wire inside the tube. At the other end $A$, the rotation of the wire operates a handle that opens or closes the switch.

A torque $T_0 = 0.2$ N·m is required to operate the switch. The torsional stiffness of the tube, combined with friction between the tube and the wire, induces a distributed torque of constant intensity $t = 0.04$ N·m/m (torque per unit distance) acting along the entire length of the wire.

(a) If the allowable shear stress in the wire is $\tau_{allow} = 30$ MPa, what is the longest permissible length $L_{max}$ of the wire?

(b) If the wire has length $L = 4.0$ m and the shear modulus of elasticity for the wire is $G = 15$ GPa, what is the angle of twist $\phi$ (in degrees) between the ends of the wire?

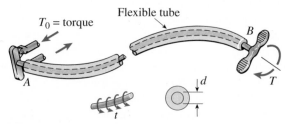

**PROB. 3.4-19**

**3.4-20** Two tubes ($AB$, $BC$) of the same material are connected by three pins (pin diameter $= d_p$) just left of $B$ as shown in the figure. Properties and dimensions for each tube are given in the figure. Torque $2T$ is applied at $x = 2L/5$ and uniformly distributed torque intensity $t_0 = 3T/L$ is applied on tube $BC$.

[*Hint*: See Eg 3-5 for torsional moment and displacement diagrams.]

(a) Find the maximum value of load variable $T$ (N · m) based on allowable shear ($\tau_a$) and bearing ($\sigma_{ba}$) stresses in the three pins which connect the two tubes at $B$. Use the following numerical properties: $L = 1.5$ m, $E = 74$ GPa, $v = 0.33$, $d_p = 18$ mm, $\tau_a = 45$ MPa, $\sigma_{ba} = 90$ MPa, $d_1 = 85$ mm, $d_2 = 73$ mm, and $d_3 = 60$ mm.

(b) What is the maximum shear stress in the tubes for the applied torque in part (a)?

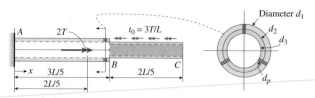

**PROB. 3.4-20**

## Pure Shear

**3.5-1** A hollow aluminum shaft (see figure) has outside diameter $d_2 = 4.0$ in. and inside diameter $d_1 = 2.0$ in. When twisted by torques $T$, the shaft has an angle of twist per unit distance equal to 0.54°/ft. The shear modulus of elasticity of the aluminum is $G = 4.0 \times 10^6$ psi.

(a) Determine the maximum tensile stress $\sigma_{max}$ in the shaft.

(b) Determine the magnitude of the applied torques $T$.

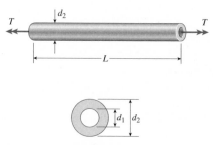

**PROBS. 3.5-1, 3.5-2, and 3.5-3**

**3.5-2** A hollow steel bar ($G = 80$ GPa) is twisted by torques $T$ (see figure). The twisting of the bar produces a maximum shear strain $\gamma_{max} = 640 \times 10^{-6}$ rad. The bar has outside and inside diameters of 150 mm and 120 mm, respectively.

(a) Determine the maximum tensile strain in the bar.

(b) Determine the maximum tensile stress in the bar.

(c) What is the magnitude of the applied torques $T$?

**3.5-3** A tubular bar with outside diameter $d_2 = 4.0$ in. is twisted by torques $T = 70.0$ k-in. (see figure). Under the action of these torques, the maximum tensile stress in the bar is found to be 6400 psi.

(a) Determine the inside diameter $d_1$ of the bar.

(b) If the bar has length $L = 48.0$ in. and is made of aluminum with shear modulus $G = 4.0 \times 10^6$ psi, what is the angle of twist $\phi$ (in degrees) between the ends of the bar?

(c) Determine the maximum shear strain $\gamma_{max}$ (in radians)?

**3.5-4** A solid circular bar of diameter $d = 50$ mm (see figure) is twisted in a testing machine until the applied torque reaches the value $T = 500$ N·m. At this value of torque, a strain gage oriented at 45° to the axis of the bar gives a reading $\varepsilon = 339 \times 10^{-6}$.

What is the shear modulus $G$ of the material?

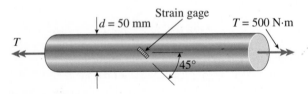

PROB. 3.5-4

**3.5-5** A steel tube ($G = 11.5 \times 10^6$ psi) has an outer diameter $d_2 = 2.0$ in. and an inner diameter $d_1 = 1.5$ in. When twisted by a torque $T$, the tube develops a maximum normal strain of $170 \times 10^{-6}$.

What is the magnitude of the applied torque $T$?

**3.5-6** A solid circular bar of steel ($G = 78$ GPa) transmits a torque $T = 360$ N·m. The allowable stresses in tension, compression, and shear are 90 MPa, 70 MPa, and 40 MPa, respectively. Also, the allowable tensile strain is $220 \times 10^{-6}$.

(a) Determine the minimum required diameter $d$ of the bar.

(b) If the bar diameter d = 40 mm, what is $T_{max}$?

**3.5-7** The normal strain in the 45° direction on the surface of a circular tube (see figure) is $880 \times 10^{-6}$ when the torque $T = 750$ lb-in. The tube is made of copper alloy with $G = 6.2 \times 10^6$ psi and $v = 0.35$.

(a) If the outside diameter $d_2$ of the tube is 0.8 in., what is the inside diameter $d_1$?

(b) If the allowable normal stress in the tube is 14 ksi, what is the maximum permissible inside diameter $d_1$?

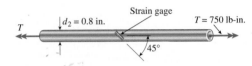

PROB. 3.5-7

**3.5-8** An aluminum tube has inside diameter $d_1 = 50$ mm, shear modulus of elasticity $G = 27$ GPa, $v = 0.33$, and torque $T = 4.0$ kN·m. The allowable shear stress in the aluminum is 50 MPa and the allowable normal strain is $900 \times 10^{-6}$.

(a) Determine the required outside diameter $d_2$.

(b) Re-compute the required outside diameter $d_2$ if allowable normal stress is 62 MPa and allowable shear strain is $1.7 \times 10^{-3}$.

**3.5-9** A solid steel bar ($G = 11.8 \times 10^6$ psi) of diameter $d = 2.0$ in. is subjected to torques $T = 8.0$ k-in. acting in the directions shown in the figure.

(a) Determine the maximum shear, tensile, and compressive stresses in the bar and show these stresses on sketches of properly oriented stress elements.

(b) Determine the corresponding maximum strains (shear, tensile, and compressive) in the bar and show these strains on sketches of the deformed elements.

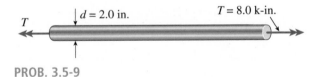

PROB. 3.5-9

**3.5-10** A solid aluminum bar ($G = 27$ GPa) of diameter $d = 40$ mm is subjected to torques $T = 300$ N·m acting in the directions shown in the figure.

(a) Determine the maximum shear, tensile, and compressive stresses in the bar and show these stresses on sketches of properly oriented stress elements.

(b) Determine the corresponding maximum strains (shear, tensile, and compressive) in the bar and show these strains on sketches of the deformed elements.

**PROB. 3.5-10**

**3.5-11** Two circular aluminum pipes of equal length $L = 24$ in. are loaded by torsional moments $T$ (see figure). Pipe 1 has outside and inside diameters $d_2 = 3$ in. and $d_1 = 2.5$ in., respectively. Pipe 2 has a constant outer diameter of $d_2$ along its entire length $L$ and an inner diameter of $d_1$ but has an increased inner diameter of $d_3 = 2.65$ in. over the middle third.

Assume that $E = 10,400$ ksi, $v = 0.33$, and allowable shear stress $\tau_a = 6500$ psi.

(a) Find the maximum acceptable torques that can be applied to Pipe 1; repeat for Pipe 2.

(b) If the maximum twist $\phi$ of Pipe 2 cannot exceed 5/4 of that of Pipe 1, what is the maximum acceptable length of the middle segment? Assume both pipes have total length $L$ and the same applied torque $T$.

(c) Find the new value of inner diameter $d_3$ of Pipe 2 if the maximum torque carried by Pipe 2 is to be 7/8 of that for Pipe 1.

(d) If the maximum normal strain in each pipe is known to be $\varepsilon_{max} = 811 \times 10^{-6}$, what is the applied torque on each pipe? Also, what is the maximum twist of each pipe? Use original properties and dimensions.

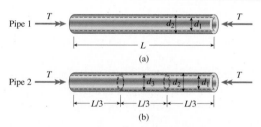

**PROB. 3.5-11**

# Transmission of Power

**3.7-1** A generator shaft in a small hydroelectric plant turns at 120 rpm and delivers 50 hp (see figure).

(a) If the diameter of the shaft is $d = 3.0$ in., what is the maximum shear stress $\tau_{max}$ in the shaft?

(b) If the shear stress is limited to 4000 psi, what is the minimum permissible diameter $d_{min}$ of the shaft?

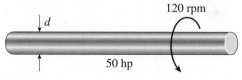

**PROB. 3.7-1**

**3.7-2** A motor drives a shaft at 12 Hz and delivers 20 kW of power (see figure).

(a) If the shaft has a diameter of 30 mm, what is the maximum shear stress $\tau_{max}$ in the shaft?

(b) If the maximum allowable shear stress is 40 MPa, what is the minimum permissible diameter $d_{min}$ of the shaft?

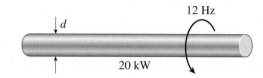

**PROB. 3.7-2**

**3.7-3** The propeller shaft of a large ship has outside diameter 18 in. and inside diameter 12 in., as shown in the figure. The shaft is rated for a maximum shear stress of 4500 psi.

(a) If the shaft is turning at 100 rpm, what is the maximum horsepower that can be transmitted without exceeding the allowable stress?

(b) If the rotational speed of the shaft is doubled but the power requirements remain unchanged, what happens to the shear stress in the shaft?

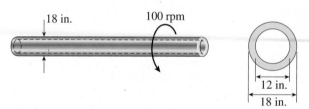

**Prob. 3.7-3**

**3.7-4** The drive shaft for a truck (outer diameter 60 mm and inner diameter 40 mm) is running at 2500 rpm (see figure).

(a) If the shaft transmits 150 kW, what is the maximum shear stress in the shaft?

(b) If the allowable shear stress is 30 MPa, what is the maximum power that can be transmitted?

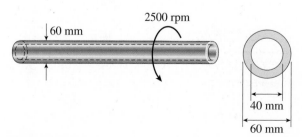

**PROB. 3.7-4**

**3.7-5** A hollow circular shaft for use in a pumping station is being designed with an inside diameter equal to 0.75 times the outside diameter. The shaft must transmit 400 hp at 400 rpm without exceeding the allowable shear stress of 6000 psi.

Determine the minimum required outside diameter *d*.

**3.7-6** A tubular shaft being designed for use on a construction site must transmit 120 kW at 1.75 Hz. The inside diameter of the shaft is to be one-half of the outside diameter.

If the allowable shear stress in the shaft is 45 MPa, what is the minimum required outside diameter *d*?

**3.7-7** A propeller shaft of solid circular cross section and diameter *d* is spliced by a collar of the same material (see figure). The collar is securely bonded to both parts of the shaft.

What should be the minimum outer diameter $d_1$ of the collar in order that the splice can transmit the same power as the solid shaft?

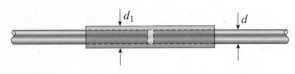

**PROB. 3.7-7**

**3.7-8** What is the maximum power that can be delivered by a hollow propeller shaft (outside diameter 50 mm, inside diameter 40 mm, and shear modulus of elasticity 80 GPa) turning at 600 rpm if the allowable shear stress is 100 MPa and the allowable rate of twist is 3.0°/m?

**3.7-9** A motor delivers 275 hp at 1000 rpm to the end of a shaft (see figure). The gears at *B* and *C* take out 125 and 150 hp, respectively.

Determine the required diameter *d* of the shaft if the allowable shear stress is 7500 psi and the angle of twist between the motor and gear *C* is limited to 1.5°. (Assume $G = 11.5 \times 10^6$ psi, $L_1 = 6$ ft, and $L_2 = 4$ ft.)

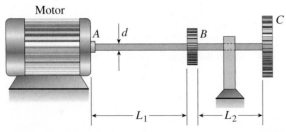

**PROBS. 3.7-9 and 3.7-10**

**3.7-10** The shaft *ABC* shown in the figure is driven by a motor that delivers 300 kW at a rotational speed of 32 Hz. The gears at *B* and *C* take out 120 and 180 kW, respectively. The lengths of the two parts of the shaft are $L_1 = 1.5$ m and $L_2 = 0.9$ m.

Determine the required diameter *d* of the shaft if the allowable shear stress is 50 MPa, the allowable angle of twist between points *A* and *C* is 4.0°, and $G = 75$ GPa.

## Statically Indeterminate Torsional Members

**3.8-1** A solid circular bar *ABCD* with fixed supports is acted upon by torques $T_0$ and $2T_0$ at the locations shown in the figure.

(a) Obtain a formula for the maximum angle of twist $\phi_{max}$ of the bar. [*Hint:* Use Eqs. (3-50a and b) of Example 3-9 to obtain the reactive torques.]

(b) What is $\phi_{max}$ if the applied torque $T_0$ at *B* is reversed in direction?

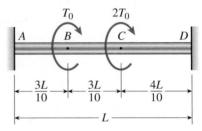

**PROB. 3.8-1**

**3.8-2** A solid circular bar *ABCD* with fixed supports at ends *A* and *D* is acted upon by two equal and oppositely directed torques $T_0$, as shown in the figure. The torques are applied at points *B* and *C*, each of which is located at distance *x* from one end of the bar. (The distance *x* may vary from zero to *L*/2.)

(a) For what distance *x* will the angle of twist at points *B* and *C* be a maximum?

(b) What is the corresponding angle of twist $\phi_{max}$? [*Hint:* Use Eqs. (3-50a and b) of Example 3-9 to obtain the reactive torques.]

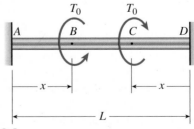

**PROB. 3.8-2**

**3.8-3** A solid circular shaft $AB$ of diameter $d$ is fixed against rotation at both ends (see figure). A circular disk is attached to the shaft at the location shown.

What is the largest permissible angle of rotation $\phi_{max}$ of the disk if the allowable shear stress in the shaft is $\tau_{allow}$? [Assume that $a > b$. Also, use Eqs. (3-50a and b) of Example 3-9 to obtain the reactive torques.]

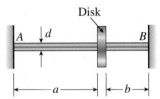

Disk

**PROB. 3.8-3**

**3.8-4** A hollow steel shaft $ACB$ of outside diameter 50 mm and inside diameter 40 mm is held against rotation at ends $A$ and $B$ (see figure). Horizontal forces $P$ are applied at the ends of a vertical arm that is welded to the shaft at point $C$.

Determine the allowable value of the forces $P$ if the maximum permissible shear stress in the shaft is 45 MPa. [Hint: Use Eqs. (3-50a and b) of Example 3-9 to obtain the reactive torques.]

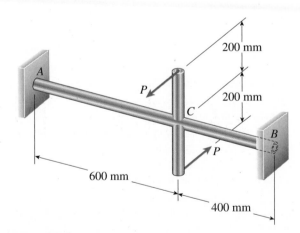

200 mm

200 mm

600 mm

400 mm

**PROB. 3.8-4**

**3.8-5** A stepped shaft $ACB$ having solid circular cross sections with two different diameters is held against rotation at the ends (see figure).

(a) If the allowable shear stress in the shaft is 6000 psi, what is the maximum torque $(T_0)_{max}$ that may be applied at section $C$? [Hint: Use Eqs. (3-49a and b) of Example 3-9 to obtain the reactive torques.]

(b) Find $(T_0)_{max}$ if the maximum angle of twist is limited to 0.55°. Let $G = 10,600$ ksi.

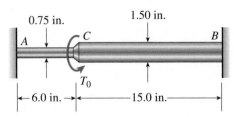

0.75 in.    1.50 in.

$A$    $C$    $B$

$T_0$

6.0 in.    15.0 in.

**PROB. 3.8-5**

**3.8-6** A stepped shaft $ACB$ having solid circular cross sections with two different diameters is held against rotation at the ends (see figure).

(a) If the allowable shear stress in the shaft is 43 MPa, what is the maximum torque $(T_0)_{max}$ that may be applied at section $C$? [Hint: Use Eqs. (3-49a and b) of Example 3-9 to obtain the reactive torques.]

(b) Find $(T_0)_{max}$ if the maximum angle of twist is limited to 1.85°. Let $G = 28$ GPa.

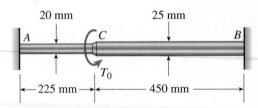

20 mm    25 mm

$A$    $C$    $B$

$T_0$

225 mm    450 mm

**PROB. 3.8-6**

**3.8-7** A stepped shaft $ACB$ is held against rotation at ends $A$ and $B$ and subjected to a torque $T_0$ acting at section $C$ (see figure). The two segments of the shaft ($AC$ and $CB$) have diameters $d_A$ and $d_B$, respectively, and polar moments of inertia $I_{PA}$ and $I_{PB}$, respectively. The shaft has length $L$ and segment $AC$ has length $a$.

(a) For what ratio $a/L$ will the maximum shear stresses be the same in both segments of the shaft?

(b) For what ratio $a/L$ will the internal torques be the same in both segments of the shaft? [Hint: Use Eqs. (3-49a and b) of Example 3-9 to obtain the reactive torques.]

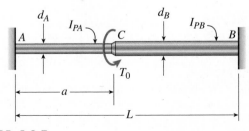

$d_A$    $I_{PA}$    $C$    $d_B$    $I_{PB}$    $B$

$A$    $T_0$

$a$

$L$

**PROB. 3.8-7**

**3.8-8** A circular bar $AB$ of length $L$ is fixed against rotation at the ends and loaded by a distributed torque $t(x)$ that varies linearly in intensity from zero at end $A$ to $t_0$ at end $B$ (see figure).

(a) Obtain formulas for the fixed-end torques $T_A$ and $T_B$.

(b) Find an expression for the angle of twist $\phi\,(x)$. What is $\phi$ max, and where does it occur along the bar?

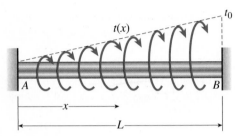

PROB. 3.8-8

**3.8-9** A circular bar $AB$ with ends fixed against rotation has a hole extending for half of its length (see figure). The outer diameter of the bar is $d_2 = 3.0$ in. and the diameter of the hole is $d_1 = 2.4$ in. The total length of the bar is $L = 50$ in.

(a) At what distance $x$ from the left-hand end of the bar should a torque $T_0$ be applied so that the reactive torques at the supports will be equal?

(b) Based on the solution for $x$ in part (a), what is $\phi_{max}$, and where does it occur? Assume that $T_0 = 87.4$ kip-in. and $G = 10{,}600$ ksi.

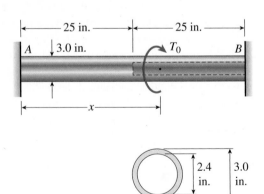

PROB. 3.8-9

**3.8-10** A solid steel bar of diameter $d_1 = 25.0$ mm is enclosed by a steel tube of outer diameter $d_3 = 37.5$ mm and inner diameter $d_2 = 30.0$ mm (see figure). Both bar and tube are held rigidly by a support at end $A$ and joined

securely to a rigid plate at end $B$. The composite bar, which has a length $L = 550$ mm, is twisted by a torque $T = 400$ N·m acting on the end plate.

(a) Determine the maximum shear stresses $\tau_1$ and $\tau_2$ in the bar and tube, respectively.

(b) Determine the angle of rotation $\phi$ (in degrees) of the end plate, assuming that the shear modulus of the steel is $G = 80$ GPa.

(c) Determine the torsional stiffness $k_T$ of the composite bar. [*Hint:* Use Eqs. (3-48a and b) to find the torques in the bar and tube.]

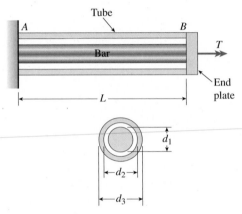

PROBS. 3.8-10 and 3.8-11

**3.8-11** A solid steel bar of diameter $d_1 = 1.50$ in. is enclosed by a steel tube of outer diameter $d_3 = 2.25$ in. and inner diameter $d_2 = 1.75$ in. (see figure). Both bar and tube are held rigidly by a support at end $A$ and joined securely to a rigid plate at end $B$. The composite bar, which has length $L = 30.0$ in., is twisted by a torque $T = 5000$ lb-in. acting on the end plate.

(a) Determine the maximum shear stresses $\tau_1$ and $\tau_2$ in the bar and tube, respectively.

(b) Determine the angle of rotation $\phi$ (in degrees) of the end plate, assuming that the shear modulus of the steel is $G = 11.6 \times 10^6$ psi.

(c) Determine the torsional stiffness $k_T$ of the composite bar. [*Hint:* Use Eqs. (3-48a and b) to find the torques in the bar and tube.]

**3.8-12** The composite shaft shown in the figure is manufactured by shrink-fitting a steel sleeve over a brass core so that the two parts act as a single solid bar in torsion. The outer diameters of the two parts are $d_1 = 40$ mm for the brass core and $d_2 = 50$ mm for the steel sleeve. The shear moduli of elasticity are $G_b = 36$ GPa for the brass and $G_s = 80$ GPa for the steel.

(a) Assuming that the allowable shear stresses in the brass and steel are $\tau_b = 48$ MPa and $\tau_s = 80$ MPa, respectively, determine the maximum permissible torque $T_{max}$ that may be applied to the shaft. [*Hint:* Use Eqs. (3-48a and b) to find the torques.]

(b) If the applied torque $T = 2500$ kN·m, find the required diameter $d_2$ so that allowable shear stress $\tau_s$ is reached in the steel.

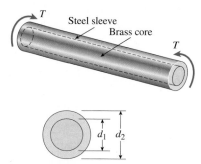

PROBS. 3.8-12 and 3.8-13

**3.8-13** The composite shaft shown in the figure is manufactured by shrink-fitting a steel sleeve over a brass core so that the two parts act as a single solid bar in torsion. The outer diameters of the two parts are $d_1 = 1.6$ in. for the brass core and $d_2 = 2.0$ in. for the steel sleeve. The shear moduli of elasticity are $G_b = 5400$ ksi for the brass and $G_s = 12,000$ ksi for the steel.

(a) Assuming that the allowable shear stresses in the brass and steel are $\tau_b = 4500$ psi and $\tau_s = 7500$ psi, respectively, determine the maximum permissible torque $T_{max}$ that may be applied to the shaft. [*Hint:* Use Eqs. (3-48a and b) to find the torques.]

(b) If the applied torque $T = 15$ kip-in., find the required diameter $d_2$ so that allowable shear stress $\tau_s$ is reached in the steel.

**3.8-14** A steel shaft ($G_s = 80$ GPa) of total length $L = 3.0$ m is encased for one-third of its length by a brass sleeve ($G_b = 40$ GPa) that is securely bonded to the steel

(see figure). The outer diameters of the shaft and sleeve are $d_1 = 70$ mm and $d_2 = 90$ mm, respectively.

(a) Determine the allowable torque $T_1$ that may be applied to the ends of the shaft if the angle of twist between the ends is limited to 8.0°.

(b) Determine the allowable torque $T_2$ if the shear stress in the brass is limited to $\tau_b = 70$ MPa.

(c) Determine the allowable torque $T_3$ if the shear stress in the steel is limited to $\tau_s = 110$ MPa.

(d) What is the maximum allowable torque $T_{max}$ if all three of the preceding conditions must be satisfied?

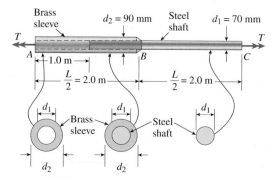

PROB. 3.8-14

**3.8-15** A uniformly tapered aluminum-alloy tube $AB$ of circular cross section and length $L$ is fixed against rotation at $A$ and $B$, as shown in the figure. The outside diameters at the ends are $d_A$ and $d_B = 2d_A$. A hollow section of length $L/2$ and constant thickness $t = d_A/10$ is cast into the tube and extends from $B$ halfway toward $A$. Torque $T_0$ is applied at $L/2$.

(a) Find the reactive torques at the supports, $T_A$ and $T_B$. Use numerical values as follows: $d_A = 2.5$ in., $L = 48$ in., $G = 3.9 \times 10^6$ psi, $T_0 = 40,000$ in.-lb.

(b) Repeat part (a) if the hollow section has constant diameter $d_A$.

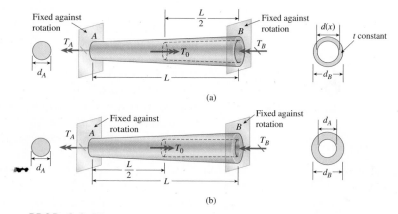

(a)

(b)

PROB. 3.8-15

**3.8-16** Two pipes ($L_1 = 2.5$ m and $L_2 = 1.5$ m) are joined at $B$ by flange plates (thickness $t_f = 14$ mm) with five bolts ($d_{bf} = 13$ mm) arranged in a circular pattern (see figure). Also, each pipe segment is attached to a wall (at $A$ and $C$, see figure) using a base plate ($t_b = 15$ mm) and four bolts ($d_{bb} = 16$ mm). All bolts are tightened until just snug. Assume $E_1 = 110$ GPa, $E_2 = 73$ GPa, $v_1 = 0.33$, and $v_2 = 0.25$. Neglect the self-weight of the pipes, and assume the pipes are in a stress-free state initially. The cross-sectional areas of the pipes are $A_1 = 1500$ mm² and $A_2 = (3/5)A_1$. The outer diameter of Pipe 1 is 60 mm. The outer diameter of Pipe 2 is equal to the inner diameter of Pipe 1. The bolt radius $r = 64$ mm for both base and flange plates.

(a) If torque $T$ is applied at $x = L_1$, find an *expression* for reactive torques $R_1$ and $R_2$ in terms of $T$.

(b) Find the maximum load variable $T$ (i.e., $T_{max}$) if allowable torsional stress in the two pipes is $\tau_{allow} = 65$ MPa.

(c) Draw torsional moment (TMD) and torsional displacement (TDD) diagrams. Label all key ordinates. What is $\phi_{max}$?

(d) Find $T_{max}$ if allowable shear and bearing stresses in the base plate and flange bolts cannot be exceeded. Assume allowable stresses in shear and bearing for all bolts are $\tau_{allow} = 45$ MPa and $\sigma_{allow} = 90$ MPa.

(e) Remove torque $T$ at $x = L_1$. Now assume the flange-plate bolt holes are misaligned by some angle $\beta$ (see figure). Find the expressions for reactive torques $R_1$ and $R_2$ if the pipes are twisted to align the flange-plate bolt holes, bolts are then inserted, and the pipes released.

(f) What is the maximum permissible misalignment angle $\beta_{max}$ if allowable stresses in shear and bearing for all bolts (from part (d) above) are not to be exceeded?

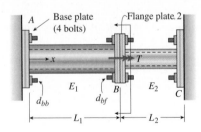

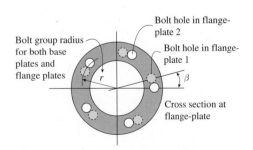

PROB. 3.8-16

# Strain Energy in Torsion

**3.9-1** A solid circular bar of steel ($G = 11.4 \times 10^6$ psi) with length $L = 30$ in. and diameter $d = 1.75$ in. is subjected to pure torsion by torques $T$ acting at the ends (see figure).

(a) Calculate the amount of strain energy $U$ stored in the bar when the maximum shear stress is 4500 psi.

(b) From the strain energy, calculate the angle of twist $\phi$ (in degrees).

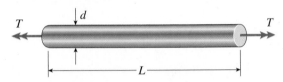

PROBS. 3.9-1 and 3.9-2

**3.9-2** A solid circular bar of copper ($G = 45$ GPa) with length $L = 0.75$ m and diameter $d = 40$ mm is subjected to pure torsion by torques $T$ acting at the ends (see figure).

(a) Calculate the amount of strain energy $U$ stored in the bar when the maximum shear stress is 32 MPa.

(b) From the strain energy, calculate the angle of twist $\phi$ (in degrees).

**3.9-3** A stepped shaft of solid circular cross sections (see figure) has length $L = 45$ in., diameter $d_2 = 1.2$ in., and diameter $d_1 = 1.0$ in. The material is brass with $G = 5.6 \times 10^6$ psi.

Determine the strain energy $U$ of the shaft if the angle of twist is 3.0°.

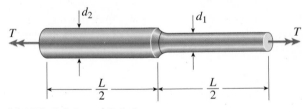

PROBS. 3.9-3 and 3.9-4

**3.9-4** A stepped shaft of solid circular cross sections (see figure) has length $L = 0.80$ m, diameter $d_2 = 40$ mm, and diameter $d_1 = 30$ mm. The material is steel with $G = 80$ GPa.

Determine the strain energy $U$ of the shaft if the angle of twist is 1.0°.

**3.9-5** A cantilever bar of circular cross section and length $L$ is fixed at one end and free at the other (see figure). The bar is loaded by a torque $T$ at the free end and by a distributed torque of constant intensity $t$ per unit distance along the length of the bar.

(a) What is the strain energy $U_1$ of the bar when the load $T$ acts alone?

(b) What is the strain energy $U_2$ when the load $t$ acts alone?

(c) What is the strain energy $U_3$ when both loads act simultaneously?

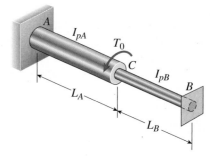

**PROB. 3.9-7**

**3.9-8** Derive a formula for the strain energy $U$ of the cantilever bar shown in the figure.

The bar has circular cross sections and length $L$. It is subjected to a distributed torque of intensity $t$ per unit distance. The intensity varies linearly from $t = 0$ at the free end to a maximum value $t = t_0$ at the support.

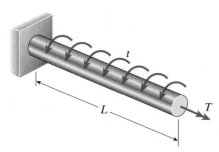

**PROB. 3.9-5**

**3.9-6** Obtain a formula for the strain energy $U$ of the statically indeterminate circular bar shown in the figure. The bar has fixed supports at ends $A$ and $B$ and is loaded by torques $2T_0$ and $T_0$ at points $C$ and $D$, respectively.

[*Hint:* Use Eqs. (3-50a and b) of Example 3-9, Section 3.8 to obtain the reactive torques.]

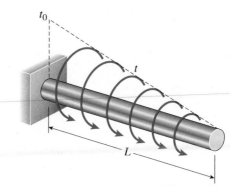

**PROB. 3.9-8**

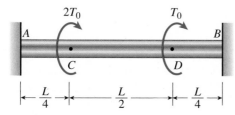

**PROB. 3.9-6**

**3.9-7** A statically indeterminate stepped shaft $ACB$ is fixed at ends $A$ and $B$ and loaded by a torque $T_0$ at point $C$ (see figure). The two segments of the bar are made of the same material, have lengths $L_A$ and $L_B$, and have polar moments of inertia $I_{PA}$ and $I_{PB}$.

Determine the angle of rotation $\phi$ of the cross section at $C$ by using strain energy.

[*Hint:* Use Eq. 3-55b to determine the strain energy $U$ in terms of the angle $\phi$. Then equate the strain energy to the work done by the torque $T_0$. Compare your result with Eq. (3-52) of Example 3-9, Section 3.8.]

**3.9-9** A thin-walled hollow tube $AB$ of conical shape has constant thickness $t$ and average diameters $d_A$ and $d_B$ at the ends (see figure).

(a) Determine the strain energy $U$ of the tube when it is subjected to pure torsion by torques $T$.

(b) Determine the angle of twist $\phi$ of the tube.

(*Note:* Use the approximate formula $I_P \approx \pi d^3 t/4$ for a thin circular ring; see Case 22 of Appendix E.)

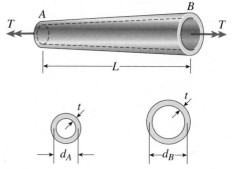

**PROB. 3.9-9**

**3.9-10** A hollow circular tube $A$ fits over the end of a solid circular bar $B$, as shown in the figure. The far ends of both bars are fixed. Initially, a hole through bar $B$ makes an angle $\beta$ with a line through two holes in tube $A$. Then bar $B$ is twisted until the holes are aligned, and a pin is placed through the holes.

When bar $B$ is released and the system returns to equilibrium, what is the total strain energy $U$ of the two bars? (Let $I_{PA}$ and $I_{PB}$ represent the polar moments of inertia of bars $A$ and $B$, respectively. The length $L$ and shear modulus of elasticity $G$ are the same for both bars.)

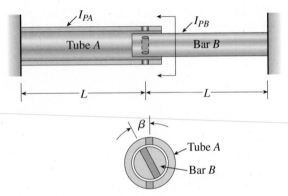

PROB. 3.9-10

**3.9-11** A heavy flywheel rotating at $n$ revolutions per minute is rigidly attached to the end of a shaft of diameter $d$ (see figure). If the bearing at $A$ suddenly freezes, what will be the maximum angle of twist $\phi$ of the shaft? What is the corresponding maximum shear stress in the shaft?

(Let $L$ = length of the shaft, $G$ = shear modulus of elasticity, and $I_m$ = mass moment of inertia of the flywheel about the axis of the shaft. Also, disregard friction in the bearings at $B$ and $C$ and disregard the mass of the shaft.)

(*Hint:* Equate the kinetic energy of the rotating flywheel to the strain energy of the shaft.)

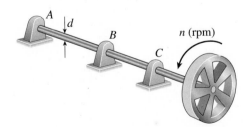

PROB. 3.9-11

# Thin-Walled Tubes

**3.11-1** A hollow circular tube having an inside diameter of 10.0 in. and a wall thickness of 1.0 in. (see figure) is subjected to a torque $T = 1200$ k-in.

Determine the maximum shear stress in the tube using (a) the approximate theory of thin-walled tubes, and (b) the exact torsion theory. Does the approximate theory give conservative or nonconservative results?

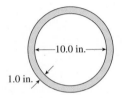

PROB. 3.11-1

**3.11-2** A solid circular bar having diameter $d$ is to be replaced by a rectangular tube having cross-sectional dimensions $d \times 2d$ to the median line of the cross section (see figure).

Determine the required thickness $t_{min}$ of the tube so that the maximum shear stress in the tube will not exceed the maximum shear stress in the solid bar.

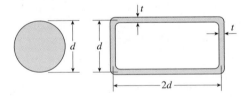

PROB. 3.11-2

**3.11-3** A thin-walled aluminum tube of rectangular cross section (see figure) has a centerline dimensions $b = 6.0$ in. and $h = 4.0$ in. The wall thickness $t$ is constant and equal to 0.25 in.

(a) Determine the shear stress in the tube due to a torque $T = 15$ k-in.

(b) Determine the angle of twist (in degrees) if the length $L$ of the tube is 50 in. and the shear modulus $G$ is $4.0 \times 10^6$ psi.

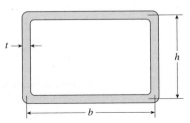

PROBS. 3.11-3 and 3.11-4

**3.11-4** A thin-walled steel tube of rectangular cross section (see figure) has centerline dimensions $b = 150$ mm and $h = 100$ mm. The wall thickness $t$ is constant and equal to 6.0 mm.

(a) Determine the shear stress in the tube due to a torque $T = 1650$ N·m.

(b) Determine the angle of twist (in degrees) if the length $L$ of the tube is 1.2 m and the shear modulus $G$ is 75 GPa.

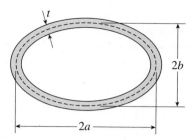

PROB. 3.11-7

**3.11-5** A thin-walled circular tube and a solid circular bar of the same material (see figure) are subjected to torsion. The tube and bar have the same cross-sectional area and the same length.

What is the ratio of the strain energy $U_1$ in the tube to the strain energy $U_2$ in the solid bar if the maximum shear stresses are the same in both cases? (For the tube, use the approximate theory for thin-walled bars.)

**3.11-8** A torque $T$ is applied to a thin-walled tube having a cross section in the shape of a regular hexagon with constant wall thickness $t$ and side length $b$ (see figure). Obtain formulas for the shear stress $\tau$ and the rate of twist $\theta$.

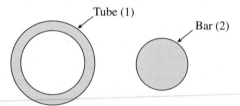

PROB. 3.11-5

**3.11-6** Calculate the shear stress $\tau$ and the angle of twist $\phi$ (in degrees) for a steel tube ($G = 76$ GPa) having the cross section shown in the figure. The tube has length $L = 1.5$ m and is subjected to a torque $T = 10$ kN·m.

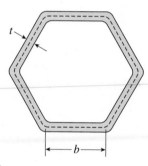

PROB. 3.11-8

**3.11-9** Compare the angle of twist $\phi_1$ for a thin-walled circular tube (see figure) calculated from the approximate theory for thin-walled bars with the angle of twist $\phi_2$ calculated from the exact theory of torsion for circular bars.

(a) Express the ratio $\phi_1/\phi_2$ in terms of the nondimensional ratio $\beta = r/t$.

(b) Calculate the ratio of angles of twist for $\beta = 5, 10,$ and 20. What conclusion about the accuracy of the approximate theory do you draw from these results?

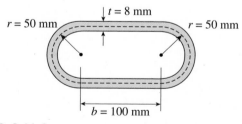

PROB. 3.11-6

**3.11-7** A thin-walled steel tube having an elliptical cross section with constant thickness $t$ (see figure) is subjected to a torque $T = 18$ k-in.

Determine the shear stress $\tau$ and the rate of twist $\theta$ (in degrees per inch) if $G = 12 \times 10^6$ psi, $t = 0.2$ in., $a = 3$ in., and $b = 2$ in. (*Note:* See Appendix E, Case 16, for the properties of an ellipse.)

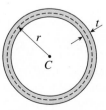

PROB. 3.11-9

**3.11-10** A thin-walled rectangular tube has uniform thickness $t$ and dimensions $a \times b$ to the median line of the cross section (see figure).

How does the shear stress in the tube vary with the ratio $\beta = a/b$ if the total length $L_m$ of the median line of the cross section and the torque $T$ remain constant?

From your results, show that the shear stress is smallest when the tube is square ($\beta = 1$).

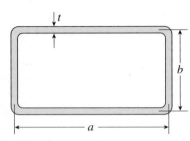

PROB. 3.11-10

**3.11-11** A tubular aluminum bar ($G = 4 \times 10^6$ psi) of square cross section (see figure) with outer dimensions 2 in. $\times$ 2 in. must resist a torque $T = 3000$ lb-in.

Calculate the minimum required wall thickness $t_{min}$ if the allowable shear stress is 4500 psi and the allowable rate of twist is 0.01 rad/ft.

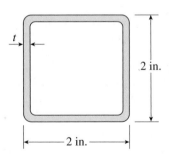

PROB. 3.11-11

**3.11-12** A thin tubular shaft of circular cross section (see figure) with inside diameter 100 mm is subjected to a torque of 5000 N·m.

If the allowable shear stress is 42 MPa, determine the required wall thickness $t$ by using (a) the approximate theory for a thin-walled tube, and (b) the exact torsion theory for a circular bar.

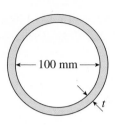

PROB. 3.11-12

**3.11-13** A long, thin-walled tapered tube $AB$ of circular cross section (see figure) is subjected to a torque $T$. The tube has length $L$ and constant wall thickness $t$. The diameter to the median lines of the cross sections at the ends $A$ and $B$ are $d_A$ and $d_B$, respectively.

Derive the following formula for the angle of twist of the tube:

$$\phi = \frac{2TL}{\pi G t}\left(\frac{d_A + d_B}{d_A^2 d_B^2}\right)$$

[*Hint:* If the angle of taper is small, we may obtain approximate results by applying the formulas for a thin-walled prismatic tube to a differential element of the tapered tube and then integrating along the axis of the tube.]

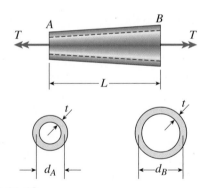

PROB. 3.11-13

## Stress Concentrations in Torsion

*The problems for Section 3.12 are to be solved by considering the stress-concentration factors.*

**3.12-1** A stepped shaft consisting of solid circular segments having diameters $D_1 = 2.0$ in. and $D_2 = 2.4$ in. (see figure) is subjected to torques $T$. The radius of the fillet is $R = 0.1$ in.

If the allowable shear stress at the stress concentration is 6000 psi, what is the maximum permissible torque $T_{max}$?

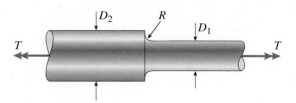

**PROBS. 3.12-1 through 3.12-5**

**3.12-2** A stepped shaft with diameters $D_1 = 40$ mm and $D_2 = 60$ mm is loaded by torques $T = 1100$ N·m (see figure).

If the allowable shear stress at the stress concentration is 120 MPa, what is the smallest radius $R_{min}$ that may be used for the fillet?

**3.12-3** A full quarter-circular fillet is used at the shoulder of a stepped shaft having diameter $D_2 = 1.0$ in. (see figure). A torque $T = 500$ lb-in. acts on the shaft.

Determine the shear stress $\tau_{max}$ at the stress concentration for values as follows: $D_1 = 0.7, 0.8$, and $0.9$ in. Plot a graph showing $\tau_{max}$ versus $D_1$.

**3.12-4** The stepped shaft shown in the figure is required to transmit 600 kW of power at 400 rpm. The shaft has a full quarter-circular fillet, and the smaller diameter $D_1 = 100$ mm.

If the allowable shear stress at the stress concentration is 100 MPa, at what diameter $D_2$ will this stress be reached? Is this diameter an upper or a lower limit on the value of $D_2$?

**3.12-5** A stepped shaft (see figure) has diameter $D_2 = 1.5$ in. and a full quarter-circular fillet. The allowable shear stress is 15,000 psi and the load $T = 4800$ lb-in.

What is the smallest permissible diameter $D_1$?

# Shear Forces and Bending Moments

Shear forces and bending moments govern the design of beams in a variety of structures such as this support structure for seating in a new stadium. (Photo provided by Brasfield & Gorrie. University of Alabama Bryant-Denny Stadium Expansion)

# CHAPTER OVERVIEW

Chapter 4 begins with a review of two-dimensional beam and frame analysis which you learned in your first course in mechanics, **Statics**. First, various types of beams, loadings, and support conditions are defined for typical structures, such as cantilever and simple beams. **Applied loads** may be concentrated (either a force or moment) or distributed. **Support conditions** include clamped, roller, pinned, and sliding supports. The number and arrangement of supports must produce a stable structure model that is either statically determinate or statically indeterminate. We will study statically determinate beam structures in this chapter, and later we will consider statically indeterminate beams in Chapter 10.

The focus in this chapter are the **internal stress resultants** (axial $N$, shear $V$, and moment $M$) at any point in the structure. In some structures, internal **"releases"** are introduced into the structure at specified points to control the magnitude of $N$, $V$, or $M$ in certain members, and must be included in the analytical model. At these release points, $N$, $V$, or $M$ may be considered to have a value of zero. Graphical displays or **diagrams** showing the variation of $N$, $V$, and $M$ over the entire structure are very useful in beam and frame design (as we will see in Chapter 5), because these diagrams quickly identify locations and values of maximum axial force, shear, and moment needed for design.

## The above topics on beams and frames are discussed in Chapter 4 as follows:

4.1 Introduction 354
4.2 Types of Beams, Loads, and Reactions 354
4.3 Shear Forces and Bending Moments 361
4.4 Relationships Between Loads, Shear Forces, and Bending Moments 371

4.5 Shear-Force and Bending-Moment Diagrams 375
Chapter Summary & Review 388
Problems 390

# 4.1 INTRODUCTION

Structural members are usually classified according to the types of loads that they support. For instance, an *axially loaded bar* supports forces having their vectors directed along the axis of the bar, and a *bar in torsion* supports torques (or couples) having their moment vectors directed along the axis. In this chapter, we begin our study of **beams** (Fig. 4-1), which are structural members subjected to lateral loads, that is, forces or moments having their vectors perpendicular to the axis of the bar.

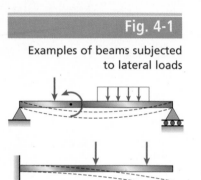

**Fig. 4-1**

Examples of beams subjected to lateral loads

The beams shown in Fig. 4-1 are classified as *planar structures* because they lie in a single plane. If all loads act in that same plane, and if all deflections (shown by the dashed lines) occur in that plane, then we refer to that plane as the **plane of bending**.

In this chapter we discuss shear forces and bending moments in beams, and we will show how these quantities are related to each other and to the loads. Finding the shear forces and bending moments is an essential step in the design of any beam. We usually need to know not only the maximum values of these quantities, but also the manner in which they vary along the axis. Once the shear forces and bending moments are known, we can find the stresses, strains, and deflections, as discussed later in Chapters 5, 6, and 9.

# 4.2 TYPES OF BEAMS, LOADS, AND REACTIONS

Beams are usually described by the manner in which they are supported. For instance, a beam with a pin support at one end and a roller support at the other (Fig. 4-2a) is called a **simply supported beam** or a **simple beam**. The essential feature of a **pin support** is that it prevents translation at the end of a beam but does not prevent rotation. Thus, end *A* of the beam of Fig. 4-2a cannot move horizontally or vertically but the axis of the beam can rotate in the plane of the figure. Consequently, a pin support is capable of developing a force reaction with both horizontal and vertical components ($H_A$ and $R_A$), but it cannot develop a moment reaction.

At end *B* of the beam (Fig. 4-2a) the **roller support** prevents translation in the vertical direction but not in the horizontal direction; hence this support can resist a vertical force ($R_B$) but not a horizontal force. Of course, the axis of the beam is free to rotate at *B* just as it is at *A*. The vertical reactions at roller supports and pin supports may act *either* upward or downward, and the horizontal reaction at a pin support may act either to the left or to the right. In the figures, reactions are indicated by slashes across the arrows in order to distinguish them from loads, as explained previously in Section 1.9.

The beam shown in Fig. 4-2b, which is fixed at one end and free at the other, is called a **cantilever beam**. At the **fixed support** (or *clamped support*) the beam can neither translate nor rotate, whereas at the free end it may do both. Consequently, both force and moment reactions may exist at the fixed support.

The third example in the figure is a **beam with an overhang** (Fig. 4-2c). This beam is simply supported at points *A* and *B* (that is, it has a pin support

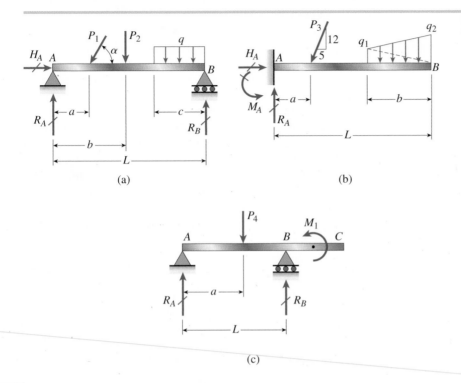

Types of beams: (a) simple beam, (b) cantilever beam, and (c) beam with an overhang

at $A$ and a roller support at $B$) but it also projects beyond the support at $B$. The overhanging segment $BC$ is similar to a cantilever beam except that the beam axis may rotate at point $B$.

When drawing sketches of beams, we identify the supports by **conventional symbols**, such as those shown in Fig. 4-2. These symbols indicate the manner in which the beam is restrained, and therefore they also indicate the nature of the reactive forces and moments. However, *the symbols do not represent the actual physical construction*. For instance, consider the examples shown in Fig. 4-3. Part (a) of the figure shows a wide-flange beam supported on a concrete wall and held down by anchor bolts that pass through slotted holes in the lower flange of the beam. This connection restrains the beam against vertical movement (either upward or downward) but does not prevent horizontal movement. Also, any restraint against rotation of the longitudinal axis of the beam is small and ordinarily may be disregarded. Consequently, this type of support is usually represented by a roller, as shown in part (b) of the figure.

The second example (Fig. 4-3c) is a beam-to-column connection in which the beam is attached to the column flange by bolted angles. (See the photo on the next page.) This type of support is usually assumed to restrain the beam against horizontal and vertical movement but not against rotation (restraint against rotation is slight because both the angles and the column can bend). Thus, this connection is usually represented as a pin support for the beam (Fig. 4-3d).

The last example (Fig. 4-3e/f on the next page) is a metal pole welded to a base plate that is anchored to a concrete pier embedded deep in the ground. Since the base of the pole is fully restrained against both translation and rotation, it is represented as a fixed support (Fig. 4-3f).

Beam supported on a wall: (a) actual construction, and (b) representation as a roller support. Beam-to-column connection: (c) actual construction, and (d) representation as a pin support.

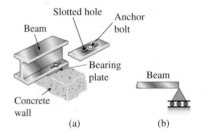

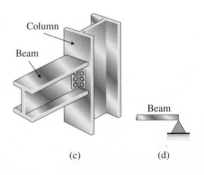

Beam-to-column connection with one beam attached to column flange and other attached to column web (Joe Gough/Shutterstock)

Fig. 4-3 (Continued)

Pole anchored to a concrete pier: (e) actual construction, and (f) representation as a fixed support

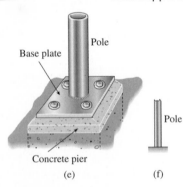

The task of representing a real structure by an **idealized model**, as illustrated by the beams shown in Fig. 4-2, is an important aspect of engineering work. The model should be simple enough to facilitate mathematical analysis and yet complex enough to represent the actual behavior of the structure with reasonable accuracy. Of course, every model is an approximation to nature. For instance, the actual supports of a beam are never perfectly rigid, and so there will always be a small amount of translation at a pin support and a small amount of rotation at a fixed support. Also, supports are never entirely free of friction, and so there will always be a small amount of restraint against translation at a roller support. In most circumstances, especially for statically determinate beams, these deviations from the idealized conditions have little effect on the action of the beam and can safely be disregarded.

## Types of Loads

Several types of loads that act on beams are illustrated in Fig. 4-2 (See page 355). When a load is applied over a very small area it may be idealized as a **concentrated load**, which is a single force. Examples are the loads $P_1$, $P_2$, $P_3$, and $P_4$ in the figure. When a load is spread along the axis of a beam, it is represented as a **distributed load**, such as the load $q$ in part (a) of the figure. Distributed loads are measured by their **intensity**, which is expressed in units of force per unit distance (for example, newtons per meter or pounds per foot). A **uniformly distributed load**, or **uniform load**, has constant intensity $q$ per unit distance (Fig. 4-2a). A varying load has an intensity that changes with distance along the axis; for instance, the **linearly varying load** of Fig. 4-2b has an intensity that varies linearly from $q_1$ to $q_2$. Another kind of load is a **couple**, illustrated by the couple of moment $M_1$ acting on the overhanging beam (Fig. 4-2c).

As mentioned in Section 4.1, we assume in this discussion that the loads act in the plane of the figure, which means that all forces must have their vectors in the plane of the figure and all couples must have their moment vectors perpendicular to the plane of the figure. Furthermore, the beam itself must be symmetric about that plane, which means that every cross section of the beam must have a vertical axis of symmetry. Under these conditions, the beam will deflect only in the *plane of bending* (the plane of the figure).

## Reactions

Finding the reactions is usually the first step in the analysis of a beam. Once the reactions are known, the shear forces and bending moments can be found, as described later in this chapter. If a beam is supported in a statically determinate manner, all reactions can be found from free-body diagrams and equations of equilibrium.

In some instances, it may be necessary to add internal releases into the beam or frame model to better represent actual conditions of construction that may have an important effect on overall structure behavior. For example, the interior span of the bridge girder shown in Fig. 4-4 is supported on roller supports at either end, which in turn rest on reinforced concrete bents (or frames), but construction details have been inserted into the girder at either end to insure that the axial force and moment at these two locations are zero. This detail also allows the bridge deck to expand or

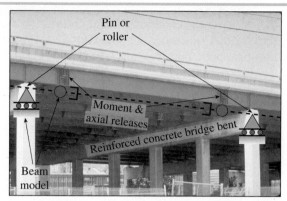

Internal releases and end supports in model of bridge beam

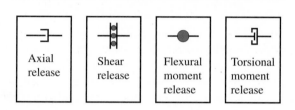

| Axial release | Shear release | Flexural moment release | Torsional moment release |

**Fig. 4-4**

Types of internal member releases for two-dimensional beam and frame members (Courtesy of the National Information Service for Earthquake Engineering EERC, University of California, Berkeley)

contract under temperature changes to avoid inducing large thermal stresses into the structure. To represent these releases in the beam model, a hinge (or internal moment release, shown as a solid circle at each end) and an axial force release (shown as a C-shaped bracket) have been included in the beam model to show that both axial force ($N$) and bending moment ($M$), but not shear ($V$), are zero at these two points along the beam. (Representations of the possible types of releases for two-dimensional beam and torsion members are shown below the photo). As examples below show, if axial, shear, or moment **releases** are present in the structure model, the structure should be broken into separate free-body diagrams (FBD) by cutting through the release; an additional equation of equilibrium is then available for use in solving for the unknown support reactions included in that FBD.

As an example, let us determine the reactions of the **simple beam** $AB$ of Fig. 4-2a. This beam is loaded by an inclined force $P_1$, a vertical force $P_2$, and a uniformly distributed load of intensity $q$. We begin by noting that the beam has three unknown reactions: a horizontal force $H_A$ at the pin support, a vertical force $R_A$ at the pin support, and a vertical force $R_B$ at the roller support. For a planar structure, such as this beam, we know from statics that we can write three independent equations of equilibrium. Thus, since there are three unknown reactions and three equations, the beam is statically determinate.

The equation of horizontal equilibrium is

$$\Sigma F_{\text{horiz}} = 0 \qquad H_A - P_1 \cos \alpha = 0$$

from which we get

$$H_A = P_1 \cos \alpha$$

**Fig. 4-2a (Repeated)**

Simple beam.

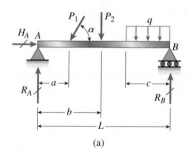

(a)

This result is so obvious from an inspection of the beam that ordinarily we would not bother to write the equation of equilibrium.

To find the vertical reactions $R_A$ and $R_B$ we write equations of moment equilibrium about points $B$ and $A$, respectively, with counter-clockwise moments being positive:

$$\Sigma M_B = 0 \quad -R_A L + (P_1 \sin \alpha)(L - a) + P_2(L - b) + qc^2/2 = 0$$

$$\Sigma M_A = 0 \quad R_B L - (P_1 \sin \alpha)(a) - P_2 b - qc(L - c/2) = 0$$

Solving for $R_A$ and $R_B$, we get

$$R_A = \frac{(P_1 \sin \alpha)(L - a)}{L} + \frac{P_2(L - b)}{L} + \frac{qc^2}{2L}$$

$$R_B = \frac{(P_1 \sin \alpha)(a)}{L} + \frac{P_2 b}{L} + \frac{qc(L - c/2)}{L}$$

As a check on these results we can write an equation of equilibrium in the vertical direction and verify that it reduces to an identity.

If the beam structure in Fig. 4-2a is modified to replace the roller support at $B$ with a pin support, it is now one degree statically indeterminate. However, if an axial force release is inserted into the model, as shown in Fig. 4-5 just to the left of the point of application of load $P_1$, the beam still can be analyzed using the laws of statics alone because the release provides one additional equilibrium equation. The beam must be cut at the release to expose the internal stress resultants $N$, $V$, and $M$; but now $N = 0$ at the release, so reactions $H_A = 0$ and $H_B = P_1 \cos \alpha$.

As a second example, consider the **cantilever beam** of Fig. 4-2b. The loads consist of an inclined force $P_3$ and a linearly varying distributed load. The latter is represented by a trapezoidal diagram of load intensity that varies from $q_1$ to $q_2$. The reactions at the fixed support are a horizontal force $H_A$, a vertical force $R_A$, and a couple $M_A$. Equilibrium of forces in the horizontal direction gives

$$H_A = \frac{5P_3}{13}$$

and equilibrium in the vertical direction gives

$$R_A = \frac{12P_3}{13} + \left(\frac{q_1 + q_2}{2}\right)b$$

In finding this reaction we used the fact that the resultant of the distributed load is equal to the area of the trapezoidal loading diagram.

The moment reaction $M_A$ at the fixed support is found from an equation of equilibrium of moments. In this example we will sum moments about point $A$ in order to eliminate both $H_A$ and $R_A$ from the moment equation. Also, for the purpose of finding the moment of the distributed load, we will divide the trapezoid into two triangles, as shown by the dashed line in Fig. 4-2b. Each load triangle can be replaced by its resultant, which is a force having its magnitude equal to the area of the triangle and having its line of action through the centroid of the

**Fig. 4-5**

Simple beam with axial release

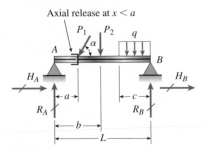

Axial release at $x < a$

**Fig. 4-2b (Repeated)**

Cantilever beam.

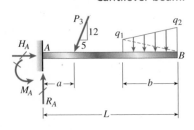

(b)

triangle. Thus, the moment about point $A$ of the lower triangular part of the load is

$$\left(\frac{q_1 b}{2}\right)\left(L - \frac{2b}{3}\right)$$

in which $q_1 b/2$ is the resultant force (equal to the area of the triangular load diagram) and $L - 2b/3$ is the moment arm (about point $A$) of the resultant.

The moment of the upper triangular portion of the load is obtained by a similar procedure, and the final equation of moment equilibrium (counterclockwise is positive) is

$$\Sigma M_A = 0 \quad M_A - \left(\frac{12 P_3}{13}\right) a - \frac{q_1 b}{2}\left(L - \frac{2b}{3}\right) - \frac{q_{2b}}{2}\left(L - \frac{b}{3}\right) = 0$$

from which

$$M_A = \frac{12 P_3 a}{13} + \frac{q_{1b}}{2}\left(L - \frac{2b}{3}\right) + \frac{q_2 b}{2}\left(L - \frac{b}{3}\right)$$

Since this equation gives a positive result, the reactive moment $M_A$ acts in the assumed direction, that is, counterclockwise. (The expressions for $R_A$ and $M_A$ can be checked by taking moments about end $B$ of the beam and verifying that the resulting equation of equilibrium reduces to an identity.)

If the cantilever beam structure in Fig. 4-2b is modified to add a roller support at $B$, it is now referred to as a one degree statically indeterminate "propped" cantilever beam. However, if a moment release is inserted into the model as shown in Fig. 4-6, just to the right of the point of application of load $P_3$, the beam can still be analyzed using the laws of statics alone because the release provides one additional equilibrium equation. The beam must be cut at the release to expose the internal stress resultants $N$, $V$, and $M$; now $M = 0$ at the release so reaction $R_B$ can be computed by summing moments in the right-hand free-body diagram. Once $R_B$ is known, reaction $R_A$ can once again be computed by summing vertical forces, and reaction moment $M_A$ can be obtained by summing moments about point $A$. Results are summarized in Fig. 4-6. Note that reaction $H_A$ is unchanged from that reported above for the original cantilever beam structure in Fig. 4-2b.

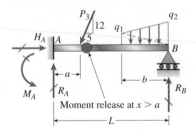

**Fig. 4-6**

Propped cantilever beam with moment release

$$R_B = \frac{\frac{1}{2} q_1 b\left(L - a - \frac{2}{3} b\right) + \frac{1}{2} q_2 b\left(L - a - \frac{b}{3}\right)}{L - a}$$

$$R_A = \frac{12}{13} P_3 + \left(\frac{q_1 + q_2}{2}\right)(b) - R_B$$

$$R_A = \frac{1}{78} \frac{-72 P_3 L + 72 P_3 a - 26\, q_1 b^2 - 13 q_2 b^2}{-L + a}$$

$$M_A = \frac{12}{13} P_3 a + q_1 \frac{b}{2}\left(L - \frac{2}{3} b\right) + q_2 \frac{b}{2}\left(L - \frac{b}{3}\right) - R_B L$$

$$M_A = \frac{1}{78} a \frac{-72 P_3 + 72 P_3 a - 26\, q_1 b^2 - 13 q_2 b^2}{-L + a}$$

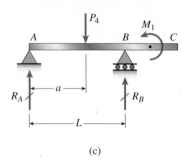

**Fig. 4-2c (Repeated)**

Beam with an overhang.

(c)

The **beam with an overhang** (Fig. 4-2c) supports a vertical force $P_4$ and a couple of moment $M_1$. Since there are no horizontal forces acting on the beam, the horizontal reaction at the pin support is nonexistent and we do not need to show it on the free-body diagram. In arriving at this conclusion, we made use of the equation of equilibrium for forces in the horizontal direction. Consequently, only two independent equations of equilibrium remain—either two moment equations or one moment equation plus the equation for vertical equilibrium.

Let us arbitrarily decide to write two moment equations, the first for moments about point $B$ and the second for moments about point $A$, as follows (counterclockwise moments are positive):

$$\Sigma M_B = 0 \quad -R_A L + P_4(L - a) + M_1 = 0$$
$$\Sigma M_A = 0 \quad -P_4 a + R_B L + M_1 = 0$$

Therefore, the reactions are

$$R_A = \frac{P_4(L - a)}{L} + \frac{M_1}{L} \qquad R_B = \frac{P_4 a}{L} - \frac{M_1}{L}$$

Again, summation of forces in the vertical direction provides a check on these results.

If the beam structure with an overhang in Fig. 4-2c is modified to add a roller support at $C$, it is now a one-degree statically indeterminate two-span beam. However, if a shear release is inserted into the model as shown in Fig. 4-7, just to the left of support $B$, the beam can be analyzed using the laws of statics alone because the release provides one additional equilibrium equation. The beam must be cut at the release to expose the internal stress resultants $N$, $V$, and $M$; now $V = 0$ at the release so reaction $R_A$ can be computed by summing forces in the left-hand free-body diagram. $R_A$ is readily seen to be equal to $P_4$. Once $R_A$ is known, reaction $R_C$ can be computed by summing moments about joint $B$, and reaction $R_B$ can be obtained by summing all vertical forces. Results are summarized below.

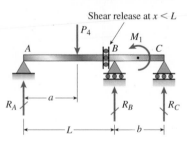

**Fig. 4-7**

Modified beam with overhang— add shear release.

Shear release at $x < L$

$$R_A = P_4$$
$$R_C = \frac{P_4 a - M_1}{b}$$
$$R_B = P_4 - R_A - R_C$$
$$R_B = \frac{M_1 - P_4 a}{b}$$

The preceding discussion illustrates how the reactions of statically determinate beams are calculated from equilibrium equations. We have intentionally used symbolic examples rather than numerical examples in order to show how the individual steps are carried out.

# 4.3 SHEAR FORCES AND BENDING MOMENTS

When a beam is loaded by forces or couples, stresses and strains are created throughout the interior of the beam. To determine these stresses and strains, we first must find the internal forces and internal couples that act on cross sections of the beam.

As an illustration of how these internal quantities are found, consider a cantilever beam $AB$ loaded by a force $P$ at its free end (Fig. 4-8a). We cut through the beam at a cross section $mn$ located at distance $x$ from the free end and isolate the left-hand part of the beam as a free body (Fig. 4-8b). The free body is held in equilibrium by the force $P$ and by the stresses that act over the cut cross section. These stresses represent the action of the right-hand part of the beam on the left-hand part. At this stage of our discussion we do not know the distribution of the stresses acting over the cross section; all we know is that the resultant of these stresses must be such as to maintain equilibrium of the free body.

From statics, we know that the resultant of the stresses acting on the cross section can be reduced to a **shear force** $V$ and a **bending moment** $M$ (Fig. 4-8b). Because the load $P$ is transverse to the axis of the beam, no axial force exists at the cross section. Both the shear force and the bending moment act in the plane of the beam, that is, the vector for the shear force lies in the plane of the figure and the vector for the moment is perpendicular to the plane of the figure.

Shear forces and bending moments, like axial forces in bars and internal torques in shafts, are the resultants of stresses distributed over the cross section. Therefore, these quantities are known collectively as **stress resultants**.

The stress resultants in statically determinate beams can be calculated from equations of equilibrium. In the case of the cantilever beam of Fig. 4-8a, we use the free-body diagram of Fig. 4-8b. Summing forces in the vertical direction and also taking moments about the cut section, we get

$$\Sigma F_{\text{vert}} = 0 \qquad P - V = 0 \quad \text{or} \quad V = P$$
$$\Sigma M = 0 \qquad M - Px = 0 \quad \text{or} \quad M = Px$$

where $x$ is the distance from the free end of the beam to the cross section where $V$ and $M$ are being determined. Thus, through the use of a free-body diagram and two equations of equilibrium, we can calculate the shear force and bending moment without difficulty.

## Sign Conventions

Let us now consider the sign conventions for shear forces and bending moments. It is customary to assume that shear forces and bending moments are positive when they act in the directions shown in Fig. 4-8b. Note that the shear force tends to rotate the material clockwise and the bending moment tends to compress the upper part of the beam and elongate the lower part. Also, in this instance, the shear force acts downward and the bending moment acts counterclockwise.

## Fig. 4-8

Shear force $V$ and bending moment $M$ in a beam

(a)

(b)

(c)

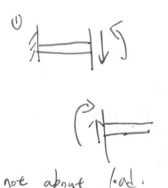

not about load.

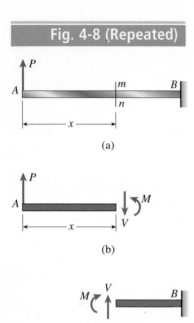

Fig. 4-8 (Repeated)

(a)

(b)

(c)

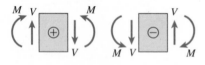

Fig. 4-9

Sign conventions for shear force
V and bending moment M

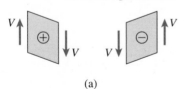

Fig. 4-10

Deformations (highly
exaggerated) of a beam element
caused by (a) shear forces, and
(b) bending moments

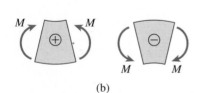

(a)

(b)

The action of these *same* stress resultants against the right-hand part of the beam is shown in Fig. 4-8c. The directions of both quantities are now reversed—the shear force acts upward and the bending moment acts clockwise. However, the shear force still tends to rotate the material clockwise and the bending moment still tends to compress the upper part of the beam and elongate the lower part.

Therefore, we must recognize that the algebraic sign of a stress resultant is determined by how it deforms the material on which it acts, rather than by its direction in space. In the case of a beam, *a positive shear force acts clockwise against the material* (Figs. 4-8b and c) *and a negative shear force acts counterclockwise against the material. Also, a positive bending moment compresses the upper part of the beam* (Figs. 4-8b and c) *and a negative bending moment compresses the lower part.*

To make these conventions clear, both positive and negative shear forces and bending moments are shown in Fig. 4-9. The forces and moments are shown acting on an element of a beam cut out between two cross sections that are a small distance apart.

The *deformations* of an element caused by both positive and negative shear forces and bending moments are sketched in Fig. 4-10. We see that a positive shear force tends to deform the element by causing the right-hand face to move downward with respect to the left-hand face, and, as already mentioned, a positive bending moment compresses the upper part of a beam and elongates the lower part.

Sign conventions for stress resultants are called **deformation sign conventions** because they are based upon how the material is deformed. For instance, we previously used a deformation sign convention in dealing with axial forces in a bar. We stated that an axial force producing elongation (or tension) in a bar is positive and an axial force producing shortening (or compression) is negative. Thus, the sign of an axial force depends upon how it deforms the material, not upon its direction in space.

By contrast, when writing equations of equilibrium we use **static sign conventions**, in which forces are positive or negative according to their directions along the coordinate axes. For instance, if we are summing forces in the $y$ direction, forces acting in the positive direction of the $y$ axis are taken as positive and forces acting in the negative direction are taken as negative.

As an example, consider Fig. 4-8b, which is a free-body diagram of part of the cantilever beam. Suppose that we are summing forces in the vertical direction and that the $y$ axis is positive upward. Then the load $P$ is given a positive sign in the equation of equilibrium because it acts upward. However, the shear force $V$ (which is a *positive* shear force) is given a negative sign because it acts downward (that is, in the negative direction of the $y$ axis). This example shows the distinction between the deformation sign convention used for the shear force and the static sign convention used in the equation of equilibrium.

The following examples illustrate the techniques for handling sign conventions and determining shear forces and bending moments in beams. The general procedure consists of constructing free-body diagrams and solving equations of equilibrium.

A simple beam *AB* supports two loads, a force *P* and a couple $M_0$, acting as shown in Fig. 4-11a.

Find the shear force *V* and bending moment *M* in the beam at cross sections located as follows: (a) a small distance to the left of the midpoint of the beam, and (b) a small distance to the right of the midpoint of the beam.

**Fig. 4-11**

Example 4-1: Shear forces and bending moment in a simple beam

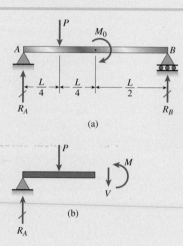

(a)

(b)

**Solution**

*Reactions.* The first step in the analysis of this beam is to find the reactions $R_A$ and $R_B$ at the supports. Taking moments about ends *B* and *A* gives two equations of equilibrium, from which we find, respectively,

$$R_A = \frac{3P}{4} - \frac{M_0}{L} \qquad R_B = \frac{P}{4} + \frac{M_0}{L} \qquad \text{(a)}$$

(a) *Shear force and bending moment to the left of the midpoint.* We cut the beam at a cross section just to the left of the midpoint and draw a free-body diagram of either half of the beam. In this example, we choose the left-hand half of the beam as the free body (Fig. 4-11b). This free body is held in equilibrium by the load *P*, the reaction $R_A$, and the two unknown stress resultants—the shear force *V* and the bending moment *M*, both of which are shown in their positive directions (see Fig. 4-9). The couple $M_0$ does not act on the free body because the beam is cut to the left of its point of application.

Summing forces in the vertical direction (upward is positive) gives

$$\Sigma F_{\text{vert}} = 0 \qquad R_A - P - V = 0$$

*Continues* ➡

**Fig. 4-11 (Continued)**

Example 4-1: Shear forces and bending moment in a simple beam (part (b) repeated)

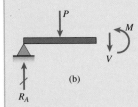

(b)

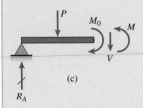

(c)

from which we get the shear force:

$$V = R_A - P = -\frac{P}{4} - \frac{M_0}{L} \qquad \text{(b)} \Longleftarrow$$

This result shows that when $P$ and $M_0$ act in the directions shown in Fig. 4-11a, the shear force (at the selected location) is negative and acts in the opposite direction to the positive direction assumed in Fig. 4-11b.

Taking moments about an axis through the cross section where the beam is cut (see Fig. 4-11b) gives

$$\Sigma M = 0 \quad -R_A\left(\frac{L}{2}\right) + P\left(\frac{L}{4}\right) + M = 0$$

in which counterclockwise moments are taken as positive. Solving for the bending moment $M$, we get

$$M = R_A\left(\frac{L}{2}\right) - P\left(\frac{L}{4}\right) = \frac{PL}{8} - \frac{M_0}{2} \qquad \text{(c)} \Longleftarrow$$

The bending moment $M$ may be either positive or negative, depending upon the magnitudes of the loads $P$ and $M_0$. If it is positive, it acts in the direction shown in the figure; if it is negative, it acts in the opposite direction.

(b) *Shear force and bending moment to the right of the midpoint.* In this case we cut the beam at a cross section just to the right of the midpoint and again draw a free-body diagram of the part of the beam to the left of the cut section (Fig. 4-11c). The difference between this diagram and the former one is that the couple $M_0$ now acts on the free body.

From two equations of equilibrium, the first for forces in the vertical direction and the second for moments about an axis through the cut section, we obtain

$$V = -\frac{P}{4} - \frac{M_0}{L} \qquad M = \frac{PL}{8} + \frac{M_0}{2} \qquad \text{(d,e)} \Longleftarrow$$

These results show that when the cut section is shifted from the left to the right of the couple $M_0$, the shear force does not change (because the vertical forces acting on the free body do not change) but the bending moment increases algebraically by an amount equal to $M_0$ (compare Eqs. c and e).

## ● ● ● Example 4-2

A beam of length $L$ is subjected to a distributed load of linearly varying intensity $q(x) = (x/L)q_0$. Three different support conditions are to be considered (see Fig. 4-12): (a) cantilever beam, (b) simply supported beam, and (c) beam with roller support at $A$ and sliding support at $B$. For each beam, find expressions for shear force $V(x)$ and bending moment $M(x)$ at a distance $x$ from point $A$ on the beam.

### Fig. 4-12

Example 4-2: Shear force and bending moment in three beams acted on by linearly varying distributed load

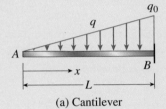

(a) Cantilever

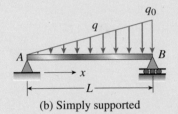

(b) Simply supported

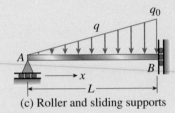

(c) Roller and sliding supports

### Solution

*Statics.* We begin by finding reactions at supports by applying *equations of statical equilibrium* (see Section 1.2) to *free-body diagrams* (FBD) of each of the three beams (Fig. 4-13). Note that we are using a *statics sign convention* for support reactions.

### Fig. 4-13

Example 4-2: Beam support reactions: (a) cantilever, (b) simply supported, and (c) roller and sliding supports

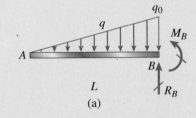

(a)

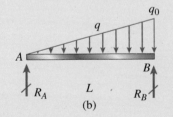

(b)

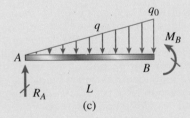

(c)

Continues ➡

*Cantilever beam.*

$$\Sigma F_y = 0 \qquad R_B = \frac{1}{2}(q_0)L \tag{a}$$

$$\Sigma M_B = 0 \qquad M_B + \frac{1}{2}(q_0 L)\left(\frac{L}{3}\right) = 0$$

$$M_B = \frac{-q_0 L^2}{6} \tag{b}$$

Use of a statics sign convention reveals that reaction moment $M_B$ is in fact clockwise, not counterclockwise, as shown in the FBD.

*Simply supported beam.*

$$\Sigma F_y = 0 \qquad R_A + R_B - \frac{1}{2}q_0 L = 0$$

$$\Sigma M_A = 0 \qquad R_B L - \frac{1}{2}(q_0 L)\left(\frac{2}{3}L\right) = 0$$

$$R_B = \frac{1}{3}q_0 L \tag{c}$$

Substituting $R_B$ in the 1st equation gives

$$R_A = -R_B + \frac{1}{2}q_0 L = \frac{1}{6}q_0 L \tag{d}$$

The reaction at $A$ carries 1/3 of the applied load and at $B$ carries 2/3 of the load. Note that the $x$-direction reaction at pin $A$ is zero by inspection, because no horizontal load or load component is applied.

*Beam with roller at A and sliding support at B.*

$$\Sigma F_y = 0 \qquad R_A = \frac{1}{2}q_0 L \tag{e}$$

$$\Sigma M_A = 0 \qquad M_B = \frac{1}{2}(q_0 L)\left(\frac{2}{3}L\right)$$

$$M_B = \frac{q_0 L^2}{3} \tag{f}$$

The reaction at $A$ carries all of the distributed load, because there is no reaction force available at joint $B$, and the moment at $B$ has twice the magnitude and an opposite sign compared to that of the cantilever beam.

*Shear force V(x) and bending moment M(x).* Now that all of the support reactions are known, we can cut the beams at some distance $x$ from joint $A$ to find expressions for the internal shear force $V$ and the bending moment $M$ at that section (Fig. 4-14). We will now apply the laws of statics to the FBDs for that portion of the beam structure to the left of the cut section, although identical results would be obtained if the FBD to the right of the cut section were used instead. It is common (although not required) to now shift to a *deformation sign convention* in which $V$ is positive if acting downward on a left-hand face of the FBD and bending moment $M$ is positive if it produces compression on top of the beam.

*Cantilever beam (Fig. 4-14a).*

$$\Sigma F_y = 0 \qquad V(x) = \frac{-1}{2}\left(\frac{x}{L}q_0\right)(x) = \frac{-q_0 x^2}{2L} \tag{g}$$

$$\Sigma M = 0 \qquad M(x) + \frac{1}{2}\left[\frac{x}{L}q_0(x)\right]\left(\frac{x}{3}\right) = 0$$

$$M(x) = \frac{-q_0 x^3}{6L} \tag{h}$$

We summed moments about the cut section to get $M(x)$ and now see that both $V$ and $M$ are zero when $x = 0$ (at joint $A$), while both $V$ and $M$ are numerically largest at $x = L$ (joint $B$). The minus signs show that both $V$ and $M$ act in opposite directions from that shown in Fig. 4-14a:

$$V_{max} = \frac{-q_0 L}{2} \qquad M_{max} = \frac{-q_0 L^2}{6} \tag{i,j}$$

---

## Fig. 4-14

Example 4-2: Beam sections:
(a) cantilever, (b) simply
supported, and (c) roller and
sliding supports

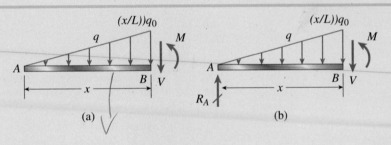

(a)

(b)

*If $M_A$, there will $N_V$ too.*

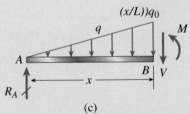

(c)

*Simply supported beam (Fig. 4-14b).*

$$\Sigma F_y = 0 \qquad V(x) = R_A - \frac{1}{2}\left(\frac{x}{L}q_0\right)(x)$$

$$V(x) = -\frac{q_0 L}{6} - \frac{1}{2}\left(\frac{x}{L}q_0\right)(x)$$

$$V(x) = \frac{q_0(L^2 - 3x^2)}{6L} \tag{k}$$

$$\Sigma M = 0 \qquad M(x) = R_A x - \frac{1}{2}\left[\frac{x}{L}q_0(x)\right]\left(\frac{x}{3}\right)$$

$$= \left(\frac{q_0 L}{6}\right)x - \frac{1}{2}\left[\frac{x}{L}q_0(x)\right]\left(\frac{x}{3}\right) = \frac{q_0 x(L^2 - x^2)}{6L} \tag{l}$$

Continues ➡

**Example 4-2 -** *Continued*

Now, when $x = 0$, $V(0) = R_A$ and $M(0) = 0$ at joint $A$. When $x = L$, $V(L) = -R_B$ and once again $M(L) = 0$, because no moment is applied at the roller support at $B$. The numerically largest shear is at $B$, where $R_B$ is twice the value of $R_A$:

$$V_{max} = \frac{-q_0 L}{3} \tag{m}$$

It is not readily apparent where along the beam the maximum moment occurs. However, if we differentiate the expression for $M(x)$, equate it to zero, and solve for $x$, we will find the point ($x_m$) at which a local maxima or minima occurs in the function $M(x)$. Solving for $x_m$ and then substituting $x_m$ into the moment expression gives

$$\frac{d}{dx}(M(x)) = \frac{d}{dx}\left[\frac{q_0 x(L^2 - x^2)}{6L}\right] = \frac{q_0(L^2 - 3x^2)}{6L} = 0$$

This results in

$$x_m = \frac{L}{\sqrt{3}}$$

and so

$$M_{max} = M(x_m) = \frac{\sqrt{3}}{27}q_0 L^2 \tag{n}$$

We note that the expression that results from $d/dx(M(x))$ is the same as that for $V(x)$ in Eq. (k). We will explore relationships between $V(x)$ and $M(x)$ in Section 4.4.

*Beam with roller at A and sliding support at B (Fig. 4-14c).*

$$\Sigma F_y = 0 \quad V(x) = R_A - \frac{1}{2}\left(\frac{x}{L}q_0\right)(x)$$

$$V(x) = \frac{q_0 L}{2} - \frac{1}{2}\left(\frac{x}{L}q_0\right)(x)$$

$$V(x) = \frac{q_0(L^2 - x^2)}{2L} \tag{o}$$

$$\Sigma M = 0 \quad M(x) = R_A x - \frac{1}{2}\left[\frac{x}{L}q_0(x)\right]\left(\frac{x}{3}\right)$$

$$= \left(\frac{q_{0L}}{2}\right)x - \frac{1}{2}\left[\frac{x}{L}q_0(x)\right]\left(\frac{x}{3}\right) = \frac{-q_0 x(x^2 - 3L^2)}{6L} \tag{p}$$

Now, when $x = 0$, $V(0) = R_A$ and $M(0) = 0$ at joint $A$. When $x = L$, $V(L) = 0$ at the sliding support, and once again, $M(L) = M_B$ at $B$. The numerically largest shear is at $A$ and is equal to the value of $R_A$:

$$V_{max} = \frac{q_0 L}{2} \tag{q}$$

The maximum moment occurs at $x = L$, so

$$M_{max} = M_B = \frac{q_0 L^2}{3} \tag{r}$$

## ● ● ● Example 4-3

A simple beam with an overhang is supported at points *A* and *B* (Fig. 4-15a). A uniform load of intensity $q$ = 200 lb/ft acts throughout the length of the beam and a concentrated load *P* = 14 k acts at a point 9 ft from the left-hand support. The span length is 24 ft and the length of the overhang is 6 ft.

Calculate the shear force *V* and bending moment *M* at cross section *D* located 15 ft from the left-hand support.

### Fig. 4-15

Example 4-3: Shear force and bending moment in a beam with an overhang

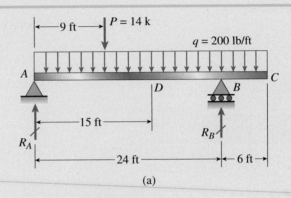

(a)

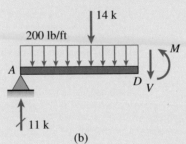

(b)

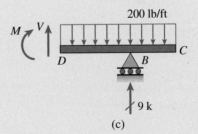

(c)

### Solution

*Reactions*. We begin by calculating the reactions $R_A$ and $R_B$ from equations of equilibrium for the entire beam considered as a free body. Thus, taking moments about the supports at *B* and *A*, respectively, we find

$$R_A = 11 \text{ k} \qquad R_B = 9 \text{ k}$$

*Continues* ⮕

**Fig. 4-15**

Example 4-3: (part (b) and (c) repeated)

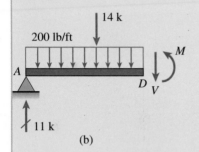

(b)

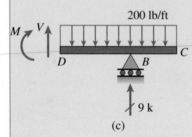

(c)

*Shear force and bending moment at section D.* Now we make a cut at section $D$ and construct a free-body diagram of the left-hand part of the beam (Fig. 4-15b). When drawing this diagram, we assume that the unknown stress resultants $V$ and $M$ are positive.

The equations of equilibrium for the free body are as follows:

$$\Sigma F_{\text{vert}} = 0 \quad 11\text{ k} - 14\text{ k} - (0.200 \text{ k/ft})(15\text{ ft}) - V = 0$$

$$\Sigma M_D = 0 - (11\text{ k})(15\text{ ft}) + (14\text{ k})(6\text{ ft}) + (0.200\text{ k/ft})(15\text{ ft})(7.5\text{ ft}) + M = 0$$

in which upward forces are taken as positive in the first equation and counterclockwise moments are taken as positive in the second equation. Solving these equations, we get

$$V = -6\text{ k} \qquad M = 58.5 \text{ k-ft}$$

The minus sign for $V$ means that the shear force is negative, that is, its direction is opposite to the direction shown in Fig. 4-15b. The positive sign for $M$ means that the bending moment acts in the direction shown in the figure.

*Alternative free-body diagram.* Another method of solution is to obtain $V$ and $M$ from a free-body diagram of the right-hand part of the beam (Fig. 4-15c). When drawing this free-body diagram, we again assume that the unknown shear force and bending moment are positive. The two equations of equilibrium are

$$\Sigma F_{\text{vert}} = 0 \quad V + 9\text{ k} - (0.200\text{ k/ft})(15\text{ ft}) = 0$$

$$\Sigma M_D = 0 \quad -M + (9\text{ k})(9\text{ ft}) - (0.200\text{ k/ft})(15\text{ ft}) = 0$$

from which

$$V = -6\text{ k} \qquad M = 585 \text{ k-ft}$$

as before. As often happens, the choice between free-body diagrams is a matter of convenience and personal preference.

# 4.4 RELATIONSHIPS BETWEEN LOADS, SHEAR FORCES, AND BENDING MOMENTS

We will now obtain some important relationships between loads, shear forces, and bending moments in beams. These relationships are quite useful when investigating the shear forces and bending moments throughout the entire length of a beam, and they are especially helpful when constructing shear-force and bending-moment diagrams (Section 4.5).

As a means of obtaining the relationships, let us consider an element of a beam cut out between two cross sections that are distance $dx$ apart (Fig. 4-16). The load acting on the top surface of the element may be a distributed load, a concentrated load, or a couple, as shown in Figs. 4-16a, b, and c, respectively. The **sign conventions** for these loads are as follows: *Distributed loads and concentrated loads are positive when they act downward on the beam and negative when they act upward. A couple acting as a load on a beam is positive when it is counterclockwise and negative when it is clockwise.* If other sign conventions are used, changes may occur in the signs of the terms appearing in the equations derived in this section.

The shear forces and bending moments acting on the sides of the element are shown in their positive directions in Fig. 4-10. In general, the shear forces and bending moments vary along the axis of the beam. Therefore, their values on the right-hand face of the element may be different from their values on the left-hand face.

In the case of a distributed load (Fig. 4-16a) the increments in $V$ and $M$ are infinitesimal, and so we denote them by $dV$ and $dM$, respectively. The corresponding stress resultants on the right-hand face are $V + dV$ and $M + dM$.

In the case of a concentrated load (Fig. 4-16b) or a couple (Fig. 4-16c) the increments may be finite, and so they are denoted $V_1$ and $M_1$. The corresponding stress resultants on the right-hand face are $V + V_1$ and $M + M_1$.

For each type of loading we can write two equations of equilibrium for the element—one equation for equilibrium of forces in the vertical direction and one for equilibrium of moments. The first of these equations gives the relationship between the load and the shear force, and the second gives the relationship between the shear force and the bending moment.

## Distributed Loads (Fig. 4-16a)

The first type of loading is a distributed load of intensity $q$, as shown in Fig. 4-16a. We will consider first its relationship to the shear force and second its relationship to the bending moment.

**Shear Force**. Equilibrium of forces in the vertical direction (upward forces are positive) gives

$$\Sigma F_{\text{vert}} = 0 \qquad V - q\,dx - (V + dV) = 0$$

or

$$\frac{dV}{dx} = -q \tag{4-1}$$

**Fig. 4-16**

Element of a beam used in deriving the relationships between loads, shear forces, and bending moments. (All loads and stress resultants are shown in their positive directions.)

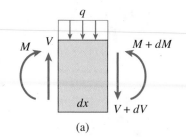

(a)

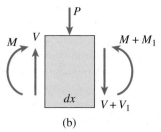

(b)

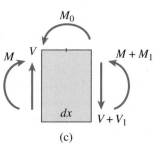

(c)

From this equation we see that the rate of change of the shear force at any point on the axis of the beam is equal to the negative of the intensity of the distributed load at that same point. (*Note:* If the sign convention for the distributed load is reversed, so that $q$ is positive upward instead of downward, then the minus sign is omitted in the preceding equation.)

Some useful relations are immediately obvious from Eq. (4-1). For instance, if there is no distributed load on a segment of the beam (that is, if $q = 0$), then $dV/dx = 0$ and the shear force is constant in that part of the beam. Also, if the distributed load is uniform along part of the beam ($q = $ constant), then $dV/dx$ is also constant and the shear force varies linearly in that part of the beam.

As a demonstration of Eq. (4-1), consider the cantilever beam with a linearly varying load that we discussed in Example 4-2 of the preceding section (see Fig. 4-12). The load on the beam (from Eq. 4-1) is

$$q = \frac{q_0 x}{L}$$

which is positive because it acts downward. Also, the shear force is

$$V = -\frac{q_0 x^2}{2L}$$

Taking the derivative $dV/dx$ gives

$$\frac{dV}{dx} = \frac{d}{dx}\left(-\frac{q_0 x^2}{2L}\right) = -\frac{q_0 x}{L} = -q$$

which agrees with Eq. (4-1).

A useful relationship pertaining to the shear forces at two different cross sections of a beam can be obtained by integrating Eq. (4-1) along the axis of the beam. To obtain this relationship, we multiply both sides of Eq. (4-1) by $dx$ and then integrate between any two points $A$ and $B$ on the axis of the beam; thus,

$$\int_A^B dV = -\int_A^B q \, dx \qquad\qquad \textbf{(4-2)}$$

where we are assuming that $x$ increases as we move from point $A$ to point $B$. The left-hand side of this equation equals the difference $(V_B - V_A)$ of the shear forces at $B$ and $A$. The integral on the right-hand side represents the area of the loading diagram between $A$ and $B$, which in turn is equal to the magnitude of the resultant of the distributed load acting between points $A$ and $B$. Thus, from Eq. (4-2), we get

$$V_B - V_A = -\int_A^B q \, dx$$

$$= -(\text{area of the loading diagram between } A \text{ and } B) \qquad \textbf{(4-3)}$$

In other words, the change in shear force between two points along the axis of the beam is equal to the negative of the total downward load between those points. The area of the loading diagram may be positive (if $q$ acts downward) or negative (if $q$ acts upward).

Because Eq. (4-1) was derived for an element of the beam subjected *only* to a distributed load (or to no load), we cannot use Eq. (4-1) at a point where a concentrated load is applied (because the *intensity* of load is not defined for a concentrated load). For the same reason, we cannot use Eq. (4-3) if a concentrated load $P$ acts on the beam between points $A$ and $B$.

**Bending Moment.** Let us now consider the moment equilibrium of the beam element shown in Fig. 4-16a. Summing moments about an axis at the left-hand side of the element (the axis is perpendicular to the plane of the figure), and taking counterclockwise moments as positive, we obtain

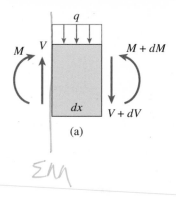

Fig. 4-16a (Repeated)

$$\Sigma M = 0 \qquad -M - q\,dx\left(\frac{dx}{2}\right) - (V + dV)dx + M + dM = 0$$

Discarding products of differentials (because they are negligible compared to the other terms), we obtain the following relationship:

$$\frac{dM}{dx} = V \tag{4-4}$$

This equation shows that the rate of change of the bending moment at any point on the axis of a beam is equal to the shear force at that same point. For instance, if the shear force is zero in a region of the beam, then the bending moment is constant in that same region.

Equation (4-4) applies only in regions where distributed loads (or no loads) act on the beam. At a point where a concentrated load acts, a sudden change (or discontinuity) in the shear force occurs and the derivative $dM/dx$ is undefined at that point.

Again using the cantilever beam of Example 4-2, we recall that the bending moment is

$$M = -\frac{q_0 x^3}{6L}$$

Therefore, the derivative $dM/dx$ is

$$\frac{dM}{dx} = \frac{d}{dx}\left(-\frac{q_0 x^3}{6L}\right) = -\frac{q_0 x^2}{2L}$$

which is equal to the shear force in the beam.

Integrating Eq. (4-4) between two points $A$ and $B$ on the beam axis gives

$$\int_A^B dM = \int_A^B V\,dx \tag{4-5}$$

The integral on the left-hand side of this equation is equal to the difference $(M_B - M_A)$ of the bending moments at points $B$ and $A$. To interpret the integral on the right-hand side, we need to consider $V$ as a function of $x$ and visualize a shear-force diagram showing the variation of $V$ with $x$. Then we see that the integral on the right-hand side represents the area below the shear-force diagram between $A$ and $B$. Therefore, we can express Eq. (4-5) in the following manner:

$$M_B - M_A = \int_A^B V \, dx$$

$$= \text{(area of the shear-force diagram between } A \text{ and } B) \quad \textbf{(4-6)}$$

This equation is valid even when concentrated loads act on the beam between points $A$ and $B$. However, it is not valid if a couple acts between $A$ and $B$. A couple produces a sudden change in the bending moment, and the left-hand side of Eq. (4-5) cannot be integrated across such a discontinuity.

## Concentrated Loads (Fig. 4-16b)

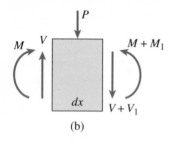

**Fig. 4-16b (Repeated)**

(b)

Now let us consider a concentrated load $P$ acting on the beam element (Fig. 4-16b). From equilibrium of forces in the vertical direction, we get

$$V - P - (V + V_1) = 0 \quad \text{or} \quad V_1 = -P \quad \textbf{(4-7)}$$

This result means that an abrupt change in the shear force occurs at any point where a concentrated load acts. As we pass from left to right through the point of load application, the shear force decreases by an amount equal to the magnitude of the downward load $P$.

From equilibrium of moments about the left-hand face of the element (Fig. 4-16b), we get

$$-M - P\left(\frac{dx}{2}\right) - (V + V_1)dx + M + M_1 = 0$$

or

$$M_1 = P\left(\frac{dx}{2}\right) + V \, dx + V_1 \, dx \quad \textbf{(4-8)}$$

Since the length $dx$ of the element is infinitesimally small, we see from this equation that the increment $M_1$ in the bending moment is also infinitesimally small. *Thus, the bending moment does not change as we pass through the point of application of a concentrated load.*

Even though the bending moment $M$ does not change at a concentrated load, its rate of change $dM/dx$ undergoes an abrupt change. At the left-hand side of the element (Fig. 4-16b), the rate of change of the bending moment (see Eq. 4-4) is $dM/dx = V$. At the right-hand side, the rate of change is $dM/dx = V + V_1 = V - P$. *Therefore, at the point of application of a concentrated load $P$, the rate of change $dM/dx$ of the bending moment decreases abruptly by an amount equal to $P$.*

## Loads in the Form of Couples (Fig. 4-16c)

The last case to be considered is a load in the form of a couple $M_0$ (Fig. 4-16c). From equilibrium of the element in the vertical direction we obtain $V_1 = 0$, which shows that *the shear force does not change at the point of application of a couple.*

Equilibrium of moments about the left-hand side of the element gives

$$-M + M_0 - (V + V_1)dx + M + M_1 = 0$$

Disregarding terms that contain differentials (because they are negligible compared to the finite terms), we obtain

$$M_1 = -M_0 \qquad \textbf{(4-9)}$$

This equation shows that the bending moment decreases by $M_0$ as we move from left to right through the point of load application. *Thus, the bending moment changes abruptly at the point of application of a couple.*

Equations (4-1) through (4-9) are useful when making a complete investigation of the shear forces and bending moments in a beam, as discussed in the next section.

**Fig. 4-16c (Repeated)**

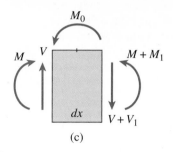

(c)

# 4.5 SHEAR-FORCE AND BENDING-MOMENT DIAGRAMS

When designing a beam, we usually need to know how the shear forces and bending moments vary throughout the length of the beam. Of special importance are the maximum and minimum values of these quantities. Information of this kind is usually provided by graphs in which the shear force and bending moment are plotted as ordinates and the distance $x$ along the axis of the beam is plotted as the abscissa. Such graphs are called **shear-force and bending-moment diagrams**.

To provide a clear understanding of these diagrams, we will explain in detail how they are constructed and interpreted for three basic loading conditions—a single concentrated load, a uniform load, and several concentrated loads. In addition, Examples 4-4 to 4-7 at the end of the section provide detailed illustration of the techniques for handling various kinds of loads, including the case of a couple acting as a load on a beam.

## Concentrated Load

Let us begin with a simple beam $AB$ supporting a concentrated load $P$ (Fig. 4-17a). The load $P$ acts at distance $a$ from the left-hand support and distance $b$ from the right-hand support. Considering the entire beam as a free body, we can readily determine the reactions of the beam from equilibrium; the results are

$$R_A = \frac{Pb}{L} \qquad R_B = \frac{Pa}{L} \qquad \textbf{(4-10a,b)}$$

**Fig. 4-17**

Shear-force and bending-moment diagrams for a simple beam with a concentrated load

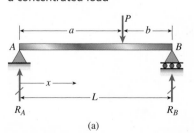

(a)

**Fig. 4-17 (Repeated)**

Shear-force and bending-moment diagrams for a simple beam with a concentrated load (part (a) repeated)

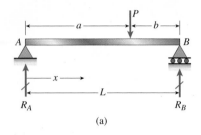

(a)

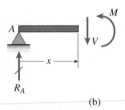

(b)

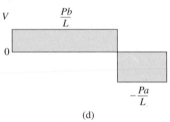

(c)

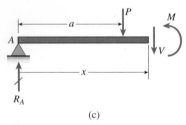

(d)

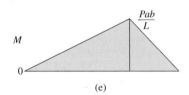

(e)

We now cut through the beam at a cross section to the left of the load $P$ and at distance $x$ from the support at $A$. Then we draw a free-body diagram of the left-hand part of the beam (Fig. 4-17b). From the equations of equilibrium for this free body, we obtain the shear force $V$ and bending moment $M$ at distance $x$ from the support:

$$V = R_A = \frac{Pb}{L} \qquad M = R_A x = \frac{Pbx}{L} \qquad (0 < x < a) \qquad \textbf{(4-11a,b)}$$

These expressions are valid only for the part of the beam to the left of the load $P$.

Next, we cut through the beam to the right of the load $P$ (that is, in the region $a < x < L$) and again draw a free-body diagram of the left-hand part of the beam (Fig. 4-17c). From the equations of equilibrium for this free body, we obtain the following expressions for the shear force and bending moment:

$$V = R_A - P = \frac{Pb}{L} - P = -\frac{Pa}{L} \qquad (a < x < L) \qquad \textbf{(4-12a)}$$

and

$$M = R_A x - P(x - a) = \frac{Pbx}{L} - P(x - a)$$
$$= \frac{Pa}{L}(L - x) \qquad (a < x < L) \qquad \textbf{(4-12b)}$$

Note that these equations are valid only for the right-hand part of the beam.

The equations for the shear forces and bending moments (Eqs. 4-11 and 4-12) are plotted below the sketches of the beam. Figure 4-17d is the *shear-force diagram* and Fig. 4-17e is the *bending-moment diagram*.

From the first diagram we see that the shear force at end $A$ of the beam ($x = 0$) is equal to the reaction $R_A$. Then it remains constant to the point of application of the load $P$. At that point, the shear force decreases abruptly by an amount equal to the load $P$. In the right-hand part of the beam, the shear force is again constant but equal numerically to the reaction at $B$.

As shown in the second diagram, the bending moment in the left-hand part of the beam increases linearly from zero at the support to $Pab/L$ at the concentrated load ($x = a$). In the right-hand part, the bending moment is again a linear function of $x$, varying from $Pab/L$ at $x = a$ to zero at the support ($x = L$). Thus, the maximum bending moment is

$$M_{max} = \frac{Pab}{L} \qquad \textbf{(4-13)}$$

and occurs under the concentrated load.

When deriving the expressions for the shear force and bending moment to the right of the load $P$ (Eqs. 4-12a and b), we considered the equilibrium of the left-hand part of the beam (Fig. 4-17c). This free body is acted upon by the forces $R_A$ and $P$ in addition to $V$ and $M$. It is slightly simpler in this particular example to consider the right-hand portion of the beam as a free body, because then only one force ($R_B$) appears in the equilibrium equations (in addition to $V$ and $M$). Of course, the final results are unchanged.

Certain characteristics of the shear-force and bending moment diagrams (Figs. 4-17d and e) may now be seen. We note first that the slope $dV/dx$ of the shear-force diagram is zero in the regions $0 < x < a$ and $a < x < L$, which is in accord with the equation $dV/dx = -q$ (Eq. 4-1). Also, in these same regions the slope $dM/dx$ of the bending moment diagram is equal to $V$ (Eq. 4-4). To the left of the load $P$, the slope of the moment diagram is positive and equal to $Pb/L$; to the right, it is negative and equal to $-Pa/L$. Thus, at the point of application of the load $P$ there is an abrupt change in the shear-force diagram (equal to the magnitude of the load $P$) and a corresponding change in the slope of the bending-moment diagram.

Now consider the *area* of the shear-force diagram. As we move from $x = 0$ to $x = a$, the area of the shear-force diagram is $(Pb/L)a$, or $Pab/L$. This quantity represents the increase in bending moment between these same two points (see Eq. 4-6). From $x = a$ to $x = L$, the area of the shear-force diagram is $-Pab/L$, which means that in this region the bending moment decreases by that amount. Consequently, the bending moment is zero at end $B$ of the beam, as expected.

If the bending moments at both ends of a beam are zero, as is usually the case with a simple beam, then the area of the shear-force diagram between the ends of the beam must be zero provided no couples act on the beam (see the discussion in Section 4.4 following Eq. 4-6).

As mentioned previously, the maximum and minimum values of the shear forces and bending moments are needed when designing beams. For a simple beam with a single concentrated load, the maximum shear force occurs at the end of the beam nearest to the concentrated load and the maximum bending moment occurs under the load itself.

## Uniform Load

A simple beam with a uniformly distributed load of constant intensity $q$ is shown in Fig. 4-18a. Because the beam and its loading are symmetric, we see immediately that each of the reactions ($R_A$ and $R_B$) is equal to $qL/2$. Therefore, the shear force and bending moment at distance $x$ from the left-hand end are

$$V = R_A - qx = \frac{qL}{2} - qx \tag{4-14a}$$

and

$$M = R_A x - qx\left(\frac{x}{2}\right) = \frac{qLx}{2} - \frac{qx^2}{2} \tag{4-14b}$$

**Fig. 4-18**

Shear-force and bending-moment diagrams for a simple beam with a uniform load

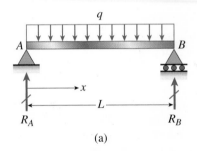

(a)

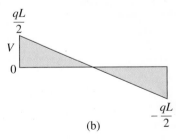

(b)

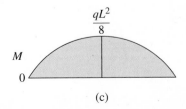

(c)

These equations, which are valid throughout the length of the beam, are plotted as shear-force and bending-moment diagrams in Figs. 4-18b and c, respectively.

The shear-force diagram consists of an inclined straight line having ordinates at $x = 0$ and $x = L$ equal numerically to the reactions. The slope of the line is $-q$, as expected from Eq. (4-1). The bending-moment diagram is a parabolic curve that is symmetric about the midpoint of the beam. At each cross section the slope of the bending-moment diagram is equal to the shear force (see Eq. 4-4):

$$\frac{dM}{dx} = \frac{d}{dx}\left(\frac{qLx}{2} - \frac{qx^2}{2}\right) = \frac{qL}{2} - qx = V$$

The maximum value of the bending moment occurs at the midpoint of the beam where both $dM/dx$ and the shear force $V$ are equal to zero. Therefore, we substitute $x = L/2$ into the expression for $M$ and obtain

$$M_{max} = \frac{qL^2}{8} \tag{4-15}$$

as shown on the bending-moment diagram.

The diagram of load intensity (Fig. 4-18a) has area $qL$, and according to Eq. (4-3) the shear force $V$ must decrease by this amount as we move along the beam from $A$ to $B$. We can see that this is indeed the case, because the shear force decreases from $qL/2$ to $-qL/2$.

The area of the shear-force diagram between $x = 0$ and $x = L/2$ is $qL^2/8$, and we see that this area represents the increase in the bending moment between those same two points (Eq. 4-6). In a similar manner, the bending moment decreases by $qL^2/8$ in the region from $x = L/2$ to $x = L$.

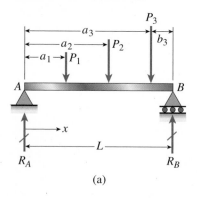

**Fig. 4-19**

Shear-force and bending-moment diagrams for a simple beam with several concentrated loads

(a)

## Several Concentrated Loads

If several concentrated loads act on a simple beam (Fig. 4-19a), expressions for the shear forces and bending moments may be determined for each segment of the beam between the points of load application. Again using free-body diagrams of the left-hand part of the beam and measuring the distance $x$ from end $A$, we obtain the following equations for the first segment of the beam:

$$V = R_A \qquad M = R_A x \qquad (0 < x < a_1) \tag{4-16a,b}$$

For the second segment, we get

$$V = R_A - P_1 \qquad M = R_A x - P_1(x - a_1) \qquad (a_1 < x < a_2) \tag{4-17a,b}$$

For the third segment of the beam, it is advantageous to consider the right-hand part of the beam rather than the left, because fewer loads act on the corresponding free body. Hence, we obtain

$$V = -R_B + P_3 \tag{4-18a}$$

$$M = R_B(L - x) - P_3(L - b_3 - x) \quad (a_2 < x < a_3) \tag{4-18b}$$

Finally, for the fourth segment of the beam, we obtain

$$V = -R_B \qquad M = R_B(L - x) \qquad (a_3 < x < L) \tag{4-19a,b}$$

Equations (4-16) through (4-19) can be used to construct the shear-force and bending-moment diagrams (Figs. 4-19b and c).

From the shear-force diagram we note that the shear force is constant in each segment of the beam and changes abruptly at every load point, with the amount of each change being equal to the load. Also, the bending moment in each segment is a linear function of $x$, and therefore the corresponding part of the bending-moment diagram is an inclined straight line. To assist in drawing these lines, we obtain the bending moments under the concentrated loads by substituting $x = a_1$, $x = a_2$, and $x = a_3$ into Eqs. (4-16b), (4-17b), and (4-18b), respectively. In this manner we obtain the following bending moments:

$$M_1 = R_A a_1 \qquad M_2 = R_A a_2 - P_1(a_2 - a_1) \qquad M_3 = R_B b_3 \tag{4-20a,b,c}$$

Knowing these values, we can readily construct the bending-moment diagram by connecting the points with straight lines.

At each discontinuity in the shear force, there is a corresponding change in the slope $dM/dx$ of the bending-moment diagram. Also, the change in bending moment between two load points equals the area of the shear-force diagram between those same two points (see Eq. 4-6). For example, the change in bending moment between loads $P_1$ and $P_2$ is $M_2 - M_1$. Substituting from Eqs. (4-20a and b), we get

$$M_2 - M_1 = (R_A - P_1)(a_2 - a_1)$$

which is the area of the rectangular shear-force diagram between $x = a_1$ and $x = a_2$.

The maximum bending moment in a beam having only concentrated loads *must* occur under one of the loads or at a reaction. To show this, recall that the slope of the bending-moment diagram is equal to the shear force. Therefore, whenever the bending moment has a maximum or minimum value, the derivative $dM/dx$ (and hence the shear force) must change sign. However, in a beam with only concentrated loads, the shear force can change sign only under a load.

Shear-force and bending-moment diagrams for a simple beam with several concentrated loads (part (a) repeated)

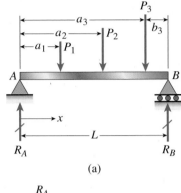

(a)

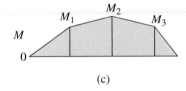

(b)

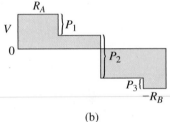

(c)

If, as we proceed along the $x$ axis, the shear force changes from positive to negative (as in Fig. 4-19b), then the slope in the bending moment diagram also changes from positive to negative. Therefore, we must have a maximum bending moment at this cross section. Conversely, a change in shear force from a negative to a positive value indicates a minimum bending moment. Theoretically, the shear-force diagram can intersect the horizontal axis at several points, although this is quite unlikely. Corresponding to each such intersection point, there is a local maximum or minimum in the bending-moment diagram. The values of all local maximums and minimums must be determined in order to find the maximum positive and negative bending moments in a beam.

## General Comments

In our discussions we frequently use the terms "maximum" and "minimum" with their common meanings of "largest" and "smallest." Consequently, we refer to "the maximum bending moment in a beam" regardless of whether the bending-moment diagram is described by a smooth, continuous function (as in Fig. 4-18c) or by a series of lines (as in Fig. 4-19c).

Furthermore, we often need to distinguish between positive and negative quantities. Therefore, we use expressions such as "maximum positive moment" and "maximum negative moment." In both of these cases, the expression refers to the numerically largest quantity; that is, the term "maximum negative moment" really means "numerically largest negative moment." Analogous comments apply to other beam quantities, such as shear forces and deflections.

The maximum positive and negative bending moments in a beam may occur at the following places: (1) a cross section where a concentrated load is applied and the shear force changes sign (see Figs. 4-17 and 4-19), (2) a cross section where the shear force equals zero (see Fig. 4-18), (3) a point of support where a vertical reaction is present, and (4) a cross section where a couple is applied. The preceding discussions and the following examples illustrate all of these possibilities.

When several loads act on a beam, the shear-force and bending-moment diagrams can be obtained by superposition (or summation) of the diagrams obtained for each of the loads acting separately. For instance, the shear-force diagram of Fig. 4-19b is actually the sum of three separate diagrams, each of the type shown in Fig. 4-17d for a single concentrated load. We can make an analogous comment for the bending-moment diagram of Fig. 4-19c. Superposition of shear-force and bending-moment diagrams is permissible because shear forces and bending moments in statically determinate beams are linear functions of the applied loads.

Computer programs are readily available for drawing shear-force and bending-moment diagrams. After you have developed an understanding of the nature of the diagrams by constructing them manually, you should feel secure in using computer programs to plot the diagrams and obtain numerical results. For convenient reference, the differential relationships used in drawing shear-force and bending-moment diagrams are summarized in the Chapter 4 Summary & Review following Example 4-7.

### ● ● ● Example 4-4

Fig. 4-20

Example 4-4: Simple beam with a uniform load over part of the span

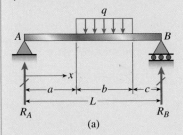

(a)

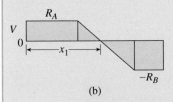

(b)

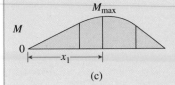

(c)

Draw the shear-force and bending-moment diagrams for a simple beam with a uniform load of intensity $q$ acting over part of the span (Fig. 4-20a).

**Solution**

*Reactions.* We begin the analysis by determining the reactions of the beam from a free-body diagram of the entire beam (Fig. 4-18a). The results are

$$R_A = \frac{qb(b + 2c)}{2L} \qquad R_B = \frac{qb(b + 2a)}{2L} \qquad \text{(4-21a,b)}$$

*Shear forces and bending moments.* To obtain the shear forces and bending moments for the entire beam, we must consider the three segments of the beam individually. For each segment we cut through the beam to expose the shear force $V$ and bending moment $M$. Then we draw a free-body diagram containing $V$ and $M$ as unknown quantities. Lastly, we sum forces in the vertical direction to obtain the shear force and take moments about the cut section to obtain the bending moment. The results for all three segments are

$$V = R_A \qquad M = R_A x \qquad (0 < x < a) \qquad \text{(4-22a,b)}$$

$$V = R_A - q(x - a) \qquad M = R_A x - \frac{q(x - a)^2}{2} \qquad (a < x < a + b) \qquad \text{(4-23a,b)}$$

$$V = -R_B \qquad M = R_B(L - x) \qquad (a + b < x < L) \qquad \text{(4-24a,b)}$$

These equations give the shear force and bending moment at every cross section of the beam and are expressed in terms of a deformation sign convention. As a partial check on these results, we can apply Eq. (4-1) to the shear forces and Eq. (4-4) to the bending moments and verify that the equations are satisfied.

We now construct the shear-force and bending-moment diagrams (Figs. 4-20b and c) from Eqs. (4-22) through (4-24). The shear-force diagram consists of horizontal straight lines in the unloaded regions of the beam and an inclined straight line with negative slope in the loaded region, as expected from the equation $dV/dx = -q$.

The bending-moment diagram consists of two inclined straight lines in the unloaded portions of the beam and a parabolic curve in the loaded portion. The inclined lines have slopes equal to $R_A$ and $-R_B$, respectively, as expected from the equation $dM/dx = V$. Also, each of these inclined lines is tangent to the parabolic curve at the point where it meets the curve. This

*Continues* ➡

**Example 4-4 -** *Continued*

### Fig. 4-20 (Repeated)

Example 4-4: Simple beam with a uniform load over part of the span

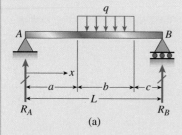

(a)

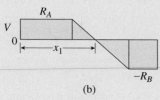

(b)

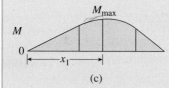

(c)

conclusion follows from the fact that there are no abrupt changes in the magnitude of the shear force at these points. Hence, from the equation $dM/dx = V$, we see that the slope of the bending-moment diagram does not change abruptly at these points. Note that, with a deformation sign convention, the bending-moment diagram is plotted on the compression side of the beam. So, the entire top surface of beam $AB$ is in compression as expected.

*Maximum bending moment.* The maximum moment occurs where the shear force equals zero. This point can be found by setting the shear force $V$ (from Eq. 4-23a) equal to zero and solving for the value of $x$, which we will denote by $x_1$. The result is

$$x_1 = a + \frac{b}{2L}(b + 2c) \tag{4-25}$$

Now we substitute $x_1$ into the expression for the bending moment (Eq. 4-23b) and solve for the maximum moment. The result is

$$M_{max} = \frac{qb}{8L^2}(b + 2c)(4aL + 2bc + b^2) \tag{4-26}$$

The maximum bending moment always occurs within the region of the uniform load, as shown by Eq. (4-25).

*Special cases.* If the uniform load is symmetrically placed on the beam ($a = c$), then we obtain the following simplified results from Eqs. (4-25) and (4-26):

$$x_1 = \frac{L}{2} \quad M_{max} = \frac{qb(2L - b)}{8} \tag{4-27a,b}$$

If the uniform load extends over the entire span, then $b = L$ and $M_{max} = qL^2/8$, which agrees with Fig. 4-18 and Eq. (4-15).

### ● ● ● Example 4-5

Draw the shear-force and bending-moment diagrams for a cantilever beam with two concentrated loads (Fig. 4-21a).

**Fig. 4-21**

Example 4-5: Cantilever beam with two concentrated loads

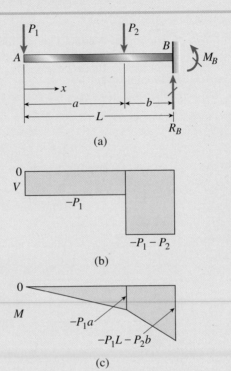

(a)

(b)

(c)

### Solution

*Reactions.* From the free-body diagram of the entire beam we find the vertical reaction $R_B$ (positive when upward) and the moment reaction $M_B$ (positive when counterclockwise in a statics sign convention):

$$R_B = P_1 + P_2 \qquad M_B = -(P_1 L + P_2 b) \qquad \text{(4-28a,b)}$$

*Shear forces and bending moments.* We obtain the shear forces and bending moments by cutting through the beam in each of the two segments, drawing the corresponding free-body diagrams, and solving the equations of equilibrium. Again measuring the distance $x$ from the left-hand end of the beam, we get (using a deformation sign convention, see Fig. 4-9)

$$V = -P_1 \qquad M = -P_1 x \qquad (0 < x < a) \qquad \text{(4-29a,b)}$$

$$V = -P_1 - P_2 \qquad M = -P_1 x - P_2(x - a) \qquad (a < x < L) \qquad \text{(4-30a,b)}$$

The corresponding shear-force and bending-moment diagrams are shown in Figs. 4-21b and c. The shear force is constant between the loads and reaches its maximum numerical value at the support, where it is equal numerically to the vertical reaction $R_B$ (Eq. 4-28a).

The bending-moment diagram consists of two inclined straight lines, each having a slope equal to the shear force in the corresponding segment of the beam. The maximum bending moment occurs at the support and is equal numerically to the moment reaction $M_B$ (Eq. 4-28b). It is also equal to the area of the entire shear-force diagram, as expected from Eq. (4-6).

## Example 4-6

In **Example 4-2**, we considered a beam of length $L$ with a distributed load of linearly varying intensity $q(x) = (x/L)q_0$. We found expressions for support reactions and equations for shear force $V(x)$ and bending moment $M(x)$ as functions of distance $x$ along the beam for three different support cases: (a) cantilever beam, (b) simply supported beam, and (c) beam with a roller support at $A$ and a sliding support at $B$.

In this example, we will plot the **shear and bending-moment diagrams** for these three beams (Fig. 4-22) using the expressions developed in Example 4-2.

### Fig. 4-22

Example 4-6: Shear and moment diagrams for three beams from Example 4-2: (a) cantilever, (b) simply supported, and (c) roller and sliding supports

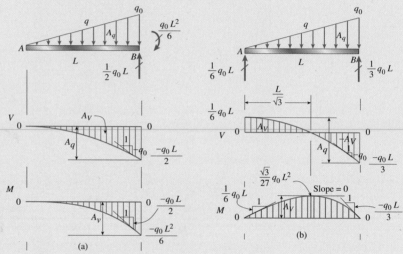

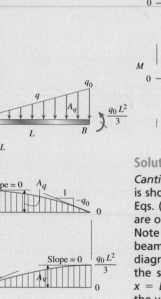

## Solution

*Cantilever Beam.* A free-body diagram of the cantilever beam of Example 4-2 is shown in Fig. 4-22a. The expressions for reactions $R_B$ and $M_B$ come from Eqs. (a, b) of Example 4-2. The shear-force and bending-moment diagrams are obtained by plotting the expressions for $V(x)$ [Eq. (g)] and $M(x)$ [Eq. (h)]. Note that the slope of the shear-force diagram at any point $x$ along the beam is equal to $-q(x)$ [see Eq. (4-1)] and the slope of the bending-moment diagram at any point $x$ is equal to $V$ [see (Eq. 4-4)]. The maximum values of the shear force and bending moment occur at the fixed support where $x = L$ [see Eqs. (i) and (j) in Example 4-2]. These values are consistent with the values for reactions $R_B$ and $M_B$.

*Alternative solution.* Instead of using free-body diagrams and equations of equilibrium, we can determine the shear forces and bending moments by integrating the differential relationships between load, shear force, and bending moment. The shear force $V$ at distance $x$ from the free end at $A$ is obtained from the load by integrating Eq. (4-3) as

$$V - V_A = V - 0 = V = -\int_0^x q(x)dx \tag{a}$$

If we integrate over the entire length of the beam, the change in shear from $A$ to $B$ is equal to the negative value of the area under the distributed load diagram ($-A_q$), as shown in Fig. 4-22a. In addition, the slope of a tangent to the shear diagram at any point $x$ is equal to the negative of the corresponding ordinate on the distributed load curve at that same point. Since the load curve is linear, the shear diagram is quadratic.

The bending moment $M$ at distance $x$ from point A is obtained from the shear force by integrating Eq. (4-6):

$$M - M_A = M - 0 = M = \int_0^x V \, dx = \int_0^x -q(x)dx \qquad \textbf{(b)}$$

If we integrate over the entire length of the beam, the change in moment from $A$ to $B$ is equal to the value of the area under the shear diagram ($A_v$), as shown in Fig. 4-22a. Also, the slope of a tangent to the moment diagram at any point $x$ is equal to the value of the corresponding ordinate on the shear-force diagram at that same point. Since the shear-force diagram is quadratic, the bending-moment diagram is cubic. For convenience, these differential relationships are summarized in the **Chapter 4 Summary & Review**.

We should note that integrating the differential relationships is quite simple in this example because the loading pattern is linear and continuous and there are no concentrated loads or couples in the regions of integration. If concentrated loads or couples were present, discontinuities in the $V$ and $M$ diagrams would exist, and we cannot integrate Eq. (4-3) through a concentrated load nor can we integrate Eq. (4-6) through a couple (see Section 4.4).

*Simply supported beam.* A free-body diagram of the simply supported beam of Example 4-2 is shown in Fig. 4-22b; the expressions for reactions $R_A$ and $R_B$ come from Eqs. (c, d) of Example 4-2. The shear-force and bending-moment diagrams are obtained by plotting the expressions from Example 4-2 for $V(x)$ [Eq. (k)] and $M(x)$ [Eq. (l)]. As with the cantilever beam discussed previously, the slope of the shear-force diagram at any point $x$ along the beam is equal to $-q(x)$ [see Eq. (4-1)] and the slope of the bending-moment diagram at any point $x$ is equal to $V$ (see Eq. 4-4). The maximum value of the shear force occurs at support $B$ where $x = L$ [see Eq. (m) of Example 4-2], and the maximum value of the moment occurs at the point at which $V = 0$. The point of maximum moment can be located by setting the expression for $V(x)$ [Eq. (k) in Example 4-2] equal to zero, then solving for $x_m = L/\sqrt{3}$. Solving for $M(x_m)$ [Eq. (l) in Example 4-2] provides the expression for $M_{max}$ shown in Fig. 4-22b.

Once again, we can use the *alternative solution approach* described above for the cantilever beam: the change in shear from $A$ to $B$ is equal to the negative value of the area under the distributed-load diagram ($-A_q$), as shown in Fig. 4-22b, and the change in moment from $A$ to $B$ is equal to the value of the area under the shear-force diagram ($A_v$).

*Beam with roller and sliding supports.* A free-body diagram of this beam from Example 4-2 is shown in Fig. 4-22c; the expressions for reactions $R_A$ and $M_B$ come from Eqs. (e, f) of Example 4-2. The shear-force and bending-moment diagrams are obtained by plotting the expressions from Example 4-2 for $V(x)$ [Eq. (o)] and M(x) [Eq. (p)]. As with the cantilever and simply supported beams discussed previously, the slope of the shear-force diagram at any point $x$ along the beam is equal to $-q(x)$ [see Eq. (4-1)] and the slope of the bending-moment diagram at any point $x$ is equal to $V$ (see Eq. 4-4). The maximum value of the shear force occurs at support $A$ where $x = 0$ [see Eq. (q) from Example 4-2], and the maximum value of the bending moment occurs at support $B$ where $V = 0$.

Once again, we can use the *alternative solution approach* described above for the cantilever and simply supported beams; the change in shear-force from A to B is equal to the negative value of the area under the distributed load diagram ($-A_q$) as shown in Fig. 4-22c, and the change in bending moment from A to B is equal to the value of the area under the shear-force diagram ($A_v$).

● ● ● **Example 4-7**

A beam $ABC$ with an overhang at the left-hand end is shown in Fig. 4-23a. The beam is subjected to a uniform load of intensity $q = 1.0$ k/ft on the overhang $AB$ and a counterclockwise couple $M_0 = 12.0$ k-ft acting midway between the supports at $B$ and $C$.

Draw the shear-force and bending-moment diagrams for this beam.

### Fig. 4-23

Example 4-7: Beam with an overhang

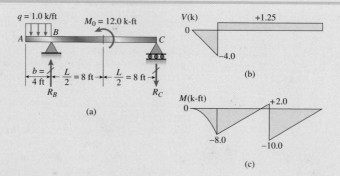

(a)

(b)

(c)

## Solution

*Reactions.* We can readily calculate the reactions $R_B$ and $R_C$ from a free-body diagram of the entire beam (Fig. 4-23a). In so doing, we find that $R_B$ is upward and $R_C$ is downward, as shown in the figure. Their numerical values are

$$R_B = 5.25 \text{ k} \qquad R_C = 1.25 \text{ k}$$

*Shear forces.* The shear force equals zero at the free end of the beam and equals $-qb$ (or $-4.0$ k) just to the left of support $B$. Since the load is uniformly distributed (that is, $q$ is constant), the slope of the shear diagram is constant and equal to $-q$ (from Eq. 4-1). Therefore, the shear diagram is an inclined straight line with negative slope in the region from $A$ to $B$ (Fig. 4-23b).

Because there are no concentrated or distributed loads between the supports, the shear-force diagram is horizontal in this region. The shear force is equal to the reaction $R_C$, or 1.25 k, as shown in the figure. (Note that the shear force does not change at the point of application of the couple $M_0$.)

The numerically largest shear force occurs just to the left of support $B$ and equals $-4.0$ k.

*Bending moments.* The bending moment is zero at the free end and decreases algebraically (but increases numerically) as we move to the right until support $B$ is reached. The slope of the moment diagram, equal to the value of the shear force (from Eq. 4-4), is zero at the free end and $-4.0$ k just to the left of support $B$. The diagram is parabolic (second degree) in this region, with the vertex at the end of the beam. The moment at point $B$ is

$$M_B = -\frac{qb^2}{2} = -\frac{1}{2}(1.0 \text{ k/ft})(4.0 \text{ ft})^2 = -8.0 \text{ k-ft}$$

which is also equal to the area of the shear-force diagram between $A$ and $B$ (see Eq. 4-6).

The slope of the bending-moment diagram from $B$ to $C$ is equal to the shear force, or 1.25 k. Therefore, the bending moment just to the left of the couple $M_0$ is

$$-8.0 \text{ k-ft} + (1.25 \text{ k})(8.0 \text{ ft}) = 2.0 \text{ k-ft}$$

as shown on the diagram. Of course, we can get this same result by cutting through the beam just to the left of the couple, drawing a free-body diagram, and solving the equation of moment equilibrium.

The bending moment changes abruptly at the point of application of the couple $M_0$, as explained earlier in connection with Eq. (4-9). Because the couple acts counterclockwise, the moment decreases by an amount equal to $M_0$. Thus, the moment just to the right of the couple $M_0$ is

$$2.0 \text{ k-ft} - 12.0 \text{ k-ft} = -10.0 \text{ k-ft}$$

From that point to support C the diagram is again a straight line with slope equal to 1.25 k. Therefore, the bending moment at the support is

$$-10.0 \text{ k-ft} + (1.25 \text{ k})(8.0 \text{ ft}) = 0$$

as expected.

Maximum and minimum values of the bending moment occur where the shear force changes sign and where the couple is applied. Comparing the various high and low points on the moment diagram, we see that the numerically largest bending moment equals $-10.0$ k-ft and occurs just to the right of the couple $M_0$. Recall that the bending-moment diagram is plotted on the compression side of the beam, so except for a small segment just to the left of midspan on BC, the entire top surface of the beam is in tension.

If a roller support is now added at joint A and a shear release is inserted just to the left of joint B (Fig. 4-23d), the support reactions must be recomputed. The beam is broken into two free-body diagrams, AB and BC, by cutting through the shear release (where $V = 0$), and reaction $R_A$ is found to be 4 kips by summing vertical forces in the left free-body diagram. Then by summing moments and forces in the entire structure, $R_B = -R_C = 0.25$ kips. Finally, shear and moment diagrams can be plotted for the modified structure. Adding the roller support at A and shear release near B results in a substantial change to both the shear-force and bending-moment diagrams compared to the original beam. For example, now the first 12 ft. of the beam has compressive rather than tensile stress on the top surface.

**Fig. 4-23 (Continued)**

Example 4-7: (d) Modified beam with overhang—add shear release.

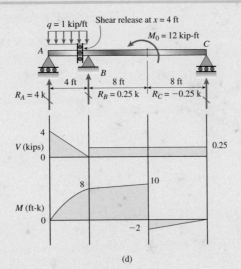

(d)

In Chapter 4, we reviewed the analysis of statically determinate beams and simple frames to find support reactions and internal stress resultants ($N$, $V$, and $M$), then plotted axial force, shear, and bending-moment diagrams to show the variation of these quantities throughout the structure. We considered clamped, sliding, pinned and roller supports, and both concentrated and distributed loadings in assembling models of a variety of structures with different support conditions. In some cases, internal releases were included in the model to represent known locations of zero values of $N$, $V$, or $M$. Some of the major concepts presented in this chapter are as follows:

1. If the structure is **statically determinate** and stable, the laws of statics alone are sufficient to solve for all values of support reaction forces and moments, as well as the magnitude of the internal axial force ($N$), shear force ($V$), and bending moment ($M$) at any location in the structure.

2. If axial, shear, or moment **releases** are present in the structure model, the structure should be broken into separate free-body diagrams (FBD) by cutting through the release; an additional equation of equilibrium is then available for use in solving for the unknown support reactions shown in that FBD.

3. Graphical displays or **diagrams** showing the variation of $N$, $V$, and $M$ over a structure are useful in design because they readily show the location of maximum values of $N$, $V$, and $M$ needed in **design** (to be considered for beams in Chapter 5). Note that, with a deformation sign convention, the moment diagram is plotted on the compression side of a structural member or portion of a member.

4. The **rules for drawing shear and bending-moment diagrams** may be summarized as follows:

   a. The ordinate on the distributed load curve ($q$) is equal to the negative of the slope on the shear diagram.

   $$\frac{dV}{dx} = -q$$

b. The difference in shear values between any two points on the shear diagram is equal to the (−) area under the distributed load curve between those same two points.

$$\int_A^B dV = -\int_A^B q \ dx$$

$$V_B - V_A = -\int_A^B q \ dx$$

$$= -(\text{area of the loading diagram between } A \text{ and } B)$$

c. The ordinate on the shear diagram ($V$) is equal to the slope on the bending-moment diagram.

$$\frac{dM}{dx} = V$$

d. The difference in values between any two points on the moment diagram is equal to the area under the shear diagram between those same two points;

$$\int_A^B dM = \int_A^B V \ dx$$

$$M_B - M_A = \int_A^B V \ dx$$

$$= (\text{area of the shear-force diagram between } A \text{ and } B)$$

e. At those points at which the shear curve crosses the reference axis (i.e., $V = 0$), the value of the moment on the moment diagram is a local maximum or minimum.

f. The ordinate on the axial force diagram ($N$) is equal to zero at an axial force release; the ordinate on the shear diagram ($V$) is zero at a shear release; and the ordinate on the moment diagram ($M$) is zero at a moment release.

# PROBLEMS CHAPTER 4

## Shear Forces and Bending Moments

**4.3-1** Calculate the shear force $V$ and bending moment $M$ at a cross section just to the left of the 1600-lb load acting on the simple beam $AB$ shown in the figure.

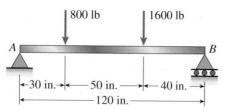

PROB. 4.3-1

**4.3-2** Determine the shear force $V$ and bending moment $M$ at the midpoint $C$ of the simple beam $AB$ shown in the figure.

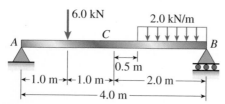

PROB. 4.3-2

**4.3-3** Determine the shear force $V$ and bending moment $M$ at the midpoint of the beam with overhangs (see figure). Note that one load acts downward and the other upward, and clockwise moments $Pb$ are applied at each support.

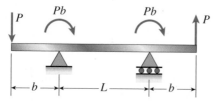

PROB. 4.3-3

**4.3-4** Calculate the shear force $V$ and bending moment $M$ at a cross section located 0.5 m from the fixed support of the cantilever beam $AB$ shown in the figure.

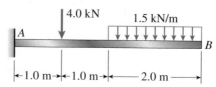

PROB. 4.3-4

**4.3-5** Consider the beam with an overhang shown in the figure.

(a) Determine the shear force $V$ and bending moment $M$ at a cross section located 18 ft from the left-hand end $A$.

(b) Find the required magnitude of load intensity $q$ acting on the right half of member $BC$ that will result in a zero shear force on the cross section 18 ft from $A$.

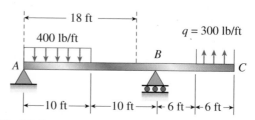

PROB. 4.3-5

**4.3-6** The beam $ABC$ shown in the figure is simply supported at $A$ and $B$ and has an overhang from $B$ to $C$. The loads consist of a horizontal force $P_1 = 4.0$ kN acting at the end of a vertical arm and a vertical force $P_2 = 8.0$ kN acting at the end of the overhang.

(a) Determine the shear force $V$ and bending moment $M$ at a cross section located 3.0 m from the left-hand support. (*Note:* Disregard the widths of the beam and vertical arm and use centerline dimensions when making calculations.)

(b) Find the value of load $P_2$ that results in $V = 0$ at a cross section located 2.0 m from the left-hand support.

(c) If $P_2 = 8$ kN, find the value of load $P_1$ that results in $M = 0$ at a cross section located 2.0 m from the left-hand support.

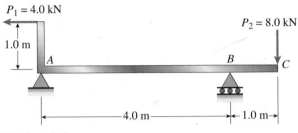

$P_1 = 4.0$ kN

1.0 m

$A$

$P_2 = 8.0$ kN

$B$

$C$

←——4.0 m——→|←1.0 m→|

PROB. 4.3-6

**4.3-7** The beam *ABCD* shown in the figure has overhangs at each end and carries a uniform load of intensity *q*.

For what ratio *b/L* will the bending moment at the midpoint of the beam be zero?

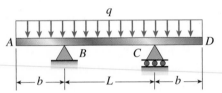

$q$

$A$

$B$

$C$

$D$

|←— *b* —→|←—— *L* ——→|←— *b* —→|

PROB. 4.3-7

**4.3-8** At full draw, an archer applies a pull of 130 N to the bowstring of the bow shown in the figure. Determine the bending moment at the midpoint of the bow.

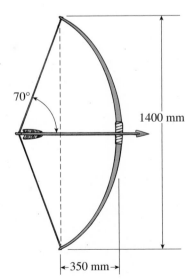

70°

1400 mm

|←350 mm→|

PROB. 4.3-8

**4.3-9** A curved bar *ABC* is subjected to loads in the form of two equal and opposite forces *P*, as shown in the figure. The axis of the bar forms a semicircle of radius *r*.

Determine the axial force *N*, shear force *V*, and bending moment *M* acting at a cross section defined by the angle θ.

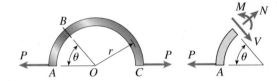

$B$

$P$

θ

$r$

$A$

$O$

$C$

$P$

$P$

$M$

$N$

$V$

θ

$A$

PROB. 4.3-9

**4.3-10** Under cruising conditions the distributed load acting on the wing of a small airplane has the idealized variation shown in the figure.

Calculate the shear force *V* and bending moment *M* at the inboard end of the wing.

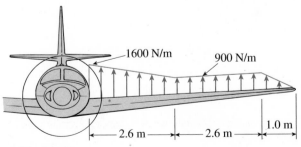

Wings of a small airplane have distributed uplift loads. (Thomasz Gulla/Shutterstock)

1600 N/m

900 N/m

|←—— 2.6 m ——→|←—— 2.6 m ——→|

1.0 m

PROB. 4.3-10

**4.3-11** A beam $ABCD$ with a vertical arm $CE$ is supported as a simple beam at $A$ and $D$ (see figure part a). A cable passes over a small pulley that is attached to the arm at $E$. One end of the cable is attached to the beam at point $B$.

(a) What is the force $P$ in the cable if the bending moment in the beam just to the left of point $C$ is equal numerically to 640 lb-ft? (*Note:* Disregard the widths of the beam and vertical arm and use centerline dimensions when making calculations.)

(b) Repeat part (a) if a roller support is added at $C$ and a shear release is inserted just left of $C$ (see figure part b).

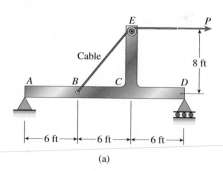

(a)

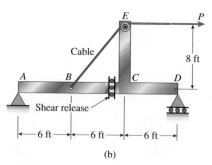

Shear release

(b)

PROB. 4.3-11

**4.3-12** A simply supported beam $AB$ supports a trapezoidally distributed load (see figure). The intensity of the load varies linearly from 50 kN/m at support $A$ to 25 kN/m at support $B$.

Calculate the shear force $V$ and bending moment $M$ at the midpoint of the beam.

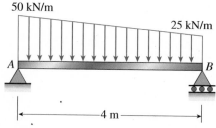

PROB. 4.3-12

**4.3-13** Beam $ABCD$ represents a reinforced-concrete foundation beam that supports a uniform load of intensity $q_1 = 3500$ lb/ft (see figure). Assume that the soil pressure on the underside of the beam is uniformly distributed with intensity $q_2$.

(a) Find the shear force $V_B$ and bending moment $M_B$ at point $B$.

(b) Find the shear force $V_m$ and bending moment $M_m$ at the midpoint of the beam.

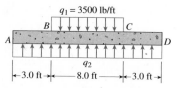

PROB. 4.3-13

**4.3-14** The simply supported beam $ABCD$ is loaded by a weight $W = 27$ kN through the arrangement shown in the figure part a. The cable passes over a small frictionless pulley at $B$ and is attached at $E$ to the end of the vertical arm.

(a) Calculate the axial force $N$, shear force $V$, and bending moment $M$ at section $C$, which is just to the left of the vertical arm. (*Note:* Disregard the widths of the beam and vertical arm and use centerline dimensions when making calculations.)

(b) Repeat part (a) if a roller support is added at $C$ and a moment release is inserted just left of $C$ (see figure part b).

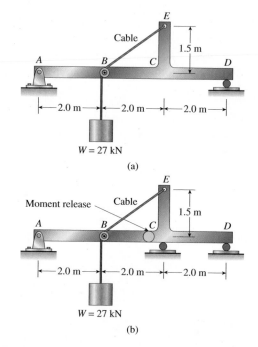

(a)

Moment release

(b)

PROB. 4.3-14

**4.3-15** The centrifuge shown in the figure rotates in a horizontal plane (the $xy$ plane) on a smooth surface about the $z$ axis (which is vertical) with an angular acceleration $\alpha$. Each of the two arms has weight $w$ per unit length and supports a weight $W = 2.0wL$ at its end.

Derive formulas for the maximum shear force and maximum bending moment in the arms, assuming $b = L/9$ and $c = L/10$.

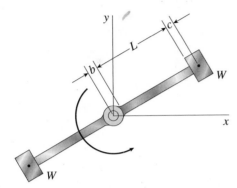

PROB. 4.3-15

# Shear-Force and Bending-Moment Diagrams

*When solving the problems for Section 4.5, draw the shear-force and bending-moment diagrams approximately to scale and label all critical ordinates, including the maximum and minimum values.*

*Probs 4.5-1 through 4.5-10 are symbolic problems and Probs. 4.5-11 through 4.5-24 are numerical problems. The remaining problems (4.5-25 through 4.5-40) involve specialized topics, such as optimization, beams with hinges, and moving loads.*

**4.5-1** Draw the shear-force and bending-moment diagrams for a simple beam $AB$ supporting two equal concentrated loads $P$ (see figure).

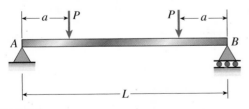

PROB. 4.5-1

**4.5-2** A simple beam $AB$ is subjected to a counterclockwise couple of moment $M_0$ acting at distance $a$ from the left-hand support (see figure).

Draw the shear-force and bending-moment diagrams for this beam.

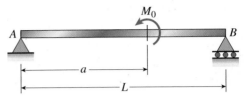

PROB. 4.5-2

**4.5-3** Draw the shear-force and bending-moment diagrams for a cantilever beam $AB$ carrying a uniform load of intensity $q$ over one-half of its length (see figure).

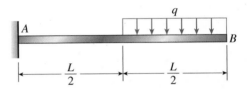

PROB. 4.5-3

**4.5-4** The cantilever beam $AB$ shown in the figure is subjected to a concentrated load $P$ at the midpoint and a counterclockwise couple of moment $M_1 = PL/4$ at the free end.

Draw the shear-force and bending-moment diagrams for this beam.

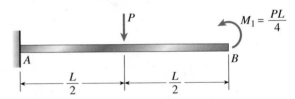

PROB. 4.5-4

**4.5-5** The simple beam $AB$ shown in the figure is subjected to a concentrated load $P$ and a clockwise couple $M_1 = PL/3$ acting at the third points.

Draw the shear-force and bending-moment diagrams for this beam.

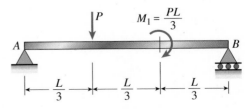

PROB. 4.5-5

**4.5-6** A simple beam $AB$ subjected to couples $M_1$ and $3M_1$ acting at the third points is shown in the figure.

Draw the shear-force and bending-moment diagrams for this beam.

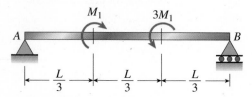

**PROB. 4.5-6**

**4.5-7** A simply supported beam $ABC$ is loaded by a vertical load $P$ acting at the end of a bracket $BDE$ (see figure).

(a) Draw the shear-force and bending-moment diagrams for beam $ABC$.

(b) Now assume that load $P$ at $E$ is directed to the right. The vertical dimension $BD$ is $L/5$. Draw axial-force, shear-force, and bending-moment diagrams for $ABC$.

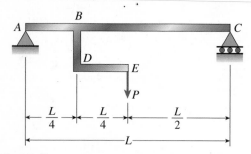

**PROB. 4.5-7**

**4.5-8** A beam $ABC$ is simply supported at $A$ and $B$ and has an overhang $BC$ (see figure). The beam is loaded by two forces $P$ and a clockwise couple of moment $Pa$ at $D$ that act through the arrangement shown.

(a) Draw the shear-force and bending-moment diagrams for beam $ABC$.

(b) If moment $Pa$ at $D$ is replaced by moment $M$, find an expression for $M$ in terms of variables $P$ and $a$ so that the reaction at $B$ goes to zero. Plot the associated shear-force and bending-moment diagrams for beam $ABC$.

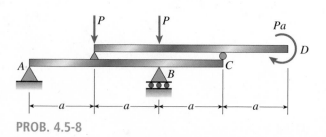

**PROB. 4.5-8**

**4.5-9** Beam $ABCD$ is simply supported at $B$ and $C$ and has overhangs at each end (see figure). The span length is $L$ and each overhang has length $L/3$. A uniform load of intensity $q$ acts along the entire length of the beam.

Draw the shear-force and bending-moment diagrams for this beam.

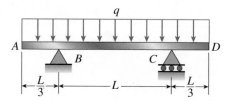

**PROB. 4.5-9**

**4.5-10** Draw the shear-force and bending-moment diagrams for a cantilever beam $AB$ acted upon by two different load cases.

(a) A distributed load with linear variation and maximum intensity $q_0$ (see figure part a).

(b) A distributed load with parabolic variation and maximum intensity $q_0$ (see figure part b).

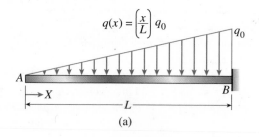

$$q(x) = \left(\frac{x}{L}\right) q_0$$

(a)

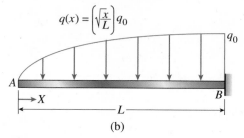

$$q(x) = \left(\sqrt{\frac{x}{L}}\right) q_0$$

(b)

**PROB. 4.5-10**

**4.5-11** The simple beam $AB$ supports a triangular load of maximum intensity $q_0 = 10$ lb/in. acting over one-half of the span and a concentrated load $P = 80$ lb acting at midspan (see figure). Draw the shear-force and bending-moment diagrams for this beam.

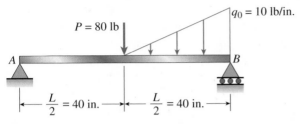

$P = 80$ lb

$q_0 = 10$ lb/in.

A            B

$\dfrac{L}{2} = 40$ in. —— $\dfrac{L}{2} = 40$ in.

PROB. 4.5-11

**4.5-12** The beam *AB* shown in the figure supports a uniform load of intensity 3000 N/m acting over half the length of the beam. The beam rests on a foundation that produces a uniformly distributed load over the entire length. Draw the shear-force and bending-moment diagrams for this beam.

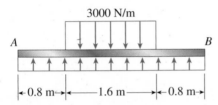

3000 N/m

A            B

←0.8 m→←——1.6 m——→←0.8 m→

PROB. 4.5-12

·**4.5-13** A cantilever beam *AB* supports a couple and a concentrated load, as shown in the figure. Draw the shear-force and bending-moment diagrams for this beam.

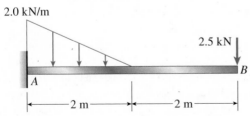

200 lb

400 lb-ft

A                    B

←——5 ft——→←——5 ft——→

PROB. 4.5-13

**4.5-14** The cantilever beam *AB* shown in the figure is subjected to a triangular load acting throughout one-half of its length and a concentrated load acting at the free end.

Draw the shear-force and bending-moment diagrams for this beam.

2.0 kN/m

2.5 kN

A                    B

←——2 m——→←——2 m——→

PROB. 4.5-14

**4.5-15** The uniformly loaded beam *ABC* has simple supports at *A* and *B* and an overhang *BC* (see figure).

Draw the shear-force and bending-moment diagrams for this beam.

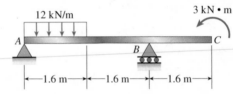

25 lb/in.

A            B            C

←———72 in.———→←———48 in.———→

PROB. 4.5-15

·**4.5-16** A beam *ABC* with an overhang at one end supports a uniform load of intensity 12 kN/m and a concentrated moment of magnitude 3 kN·m at *C* (see figure).

Draw the shear-force and bending-moment diagrams for this beam.

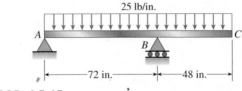

12 kN/m

3 kN•m

A            B            C

←—1.6 m—→←—1.6 m—→←—1.6 m—→

PROB. 4.5-16

**4.5-17** Consider two beams, which are loaded the same but have different support conditions. Which beam has the larger maximum moment?

First, find support reactions, then plot axial force (*N*), shear (*V*), and moment (*M*) diagrams for all three beams. *Label* all critical *N*, *V*, and *M* values and also the *distance* to points where *N*, *V*, and/or *M* is zero.

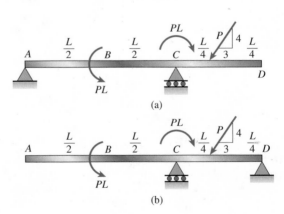

PROB. 4.5-17

**4.5-18** The three beams below are loaded the same and have the same support conditions. However, one has a *moment release* just to the left of *C*, the second has a *shear release* just to the right of *C* and the third has an axial release just to the left of *C*. Which beam has the largest maximum moment?

First, find support reactions, then plot axial force ($N$), shear ($V$), and moment ($M$) diagrams for all three beams. *Label* all critical $N$, $V$, and $M$ values and also the *distance* to points where $N$, $V$, and/or $M$ is zero.

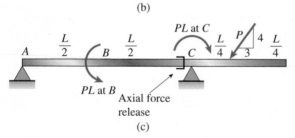

(a)

(b)

(c)

**PROB. 4.5-18**

**4.5-19** The beam *ABC* shown in the figure is simply supported at *A* and *B* and has an overhang from *B* to *C*. The loads consist of a horizontal force $P_1 = 400$ lb acting at the end of the vertical arm and a vertical force $P_2 = 900$ lb acting at the end of the overhang.

Draw the shear-force and bending-moment diagrams for this beam. (*Note:* Disregard the widths of the beam and vertical arm and use centerline dimensions when making calculations.)

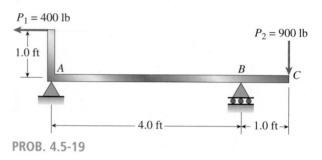

**PROB. 4.5-19**

**4.5-20** A simple beam *AB* is loaded by two segments of uniform load and two horizontal forces acting at the ends of a vertical arm (see figure).

Draw the shear-force and bending-moment diagrams for this beam.

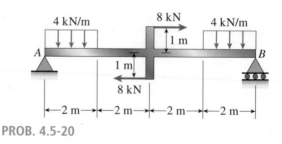

**PROB. 4.5-20**

**4.5-21** The two beams below are loaded the same and have the same support conditions. However, the location of internal *axial, shear,* and *moment releases* is different for each beam (see figures). Which beam has the larger maximum moment?

First, find support reactions, then plot axial force ($N$), shear ($V$), and moment ($M$) diagrams for both beams. *Label* all critical $N$, $V$, and $M$ values and also the *distance* to points where $N$, $V$, and/or $M$ is zero.

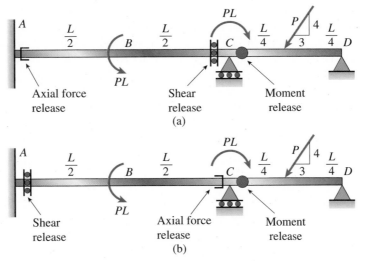

PROB. 4.5-21

**4.5-22** The beam $ABCD$ shown in the figure has overhangs that extend in both directions for a distance of 4.2 m from the supports at $B$ and $C$, which are 1.2 m apart.

Draw the shear-force and bending-moment diagrams for this overhanging beam.

(b) Repeat part (a) if a roller support is added at $C$ and a shear release is inserted just left of $C$ (see figure part b).

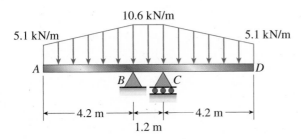

PROB. 4.5-22

**4.5-23** A beam $ABCD$ with a vertical arm $CE$ is supported as a simple beam at $A$ and $D$ (see figure). A cable passes over a small pulley that is attached to the arm at $E$. One end of the cable is attached to the beam at point $B$. The tensile force in the cable is 1800 lb.

(a) Draw the shear-force and bending-moment diagrams for beam $ABCD$. (*Note:* Disregard the widths of the beam and vertical arm and use centerline dimensions when making calculations.)

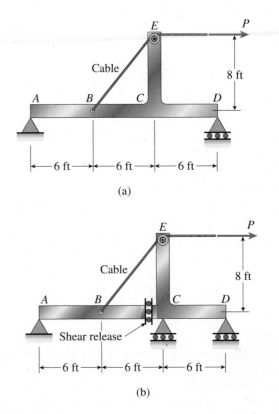

PROB. 4.5-23

**4.5-24** Beams $ABC$ and $CD$ are supported at $A$, $C$, and $D$ and are joined by a hinge (or *moment release*) just to the left of $C$. The support at $A$ is a sliding support (hence reaction $A_y = 0$ for the loading shown below). Find all support reactions, then plot shear ($V$) and moment ($M$) diagrams. **Label** all critical $V$ *and* $M$ values and also the **distance** to points where either $V$ and/or $M$ is zero.

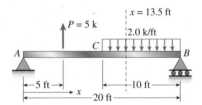

Sliding
support

Moment
release

**PROB. 4.5-24**

**4.5-25** The simple beam $AB$ shown in the figure supports a concentrated load and a segment of uniform load.

    (a) Draw the shear-force and bending-moment diagrams for this beam.

    (b) Find the value of $P$ that will result in zero shear at $x = 13.5$ ft. Draw the shear-force and bending-moment diagrams for this case.

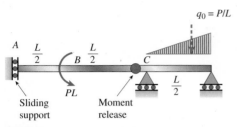

**PROB. 4.5-25**

**4.5-26** The cantilever beam shown in the figure supports a concentrated load and a segment of uniform load.

    Draw the shear-force and bending-moment diagrams for this cantilever beam.

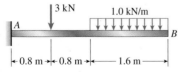

**PROB. 4.5-26**

**4.5-27** The simple beam $ACB$ shown in the figure is subjected to a triangular load of maximum intensity $q_0 = 180$ lb/ft at $a = 6.0$ ft, and a concentrated moment $M = 300$ lb-ft at $A$.

    (a) Draw the shear-force and bending-moment diagrams for this beam.

    (b) Find the value of distance $a$ that results in the maximum moment occurring at $L/2$. Draw the shear-force and bending-moment diagrams for this case.

    (c) Find the value of distance $a$ for which $M_{max}$ is the largest possible value.

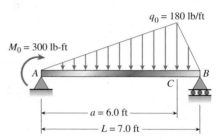

**PROB. 4.5-27**

**4.5-28** A beam with simple supports is subjected to a trapezoidally distributed load (see figure). The intensity of the load varies from 1.0 kN/m at support $A$ to 3.0 kN/m at support $B$.

    Draw the shear-force and bending-moment diagrams for this beam.

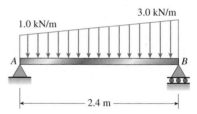

**PROB. 4.5-28**

**4.5-29** A beam of length $L$ is being designed to support a uniform load of intensity $q$ (see figure). If the supports of the beam are placed at the ends, creating a simple beam, the maximum bending moment in the beam is $qL^2/8$. However, if the supports of the beam are moved symmetrically toward the middle of the beam (as pictured), the maximum bending moment is reduced.

    Determine the distance $a$ between the supports so that the maximum bending moment in the beam has the smallest possible numerical value. Draw the shear-force and bending-moment diagrams for this condition.

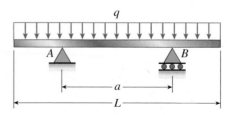

**PROB. 4.5-29**

**4.5-30** The compound beam *ABCDE* shown in the figure consists of two beams (*AD* and *DE*) joined by a hinged connection at *D*. The hinge can transmit a shear force but not a bending moment. The loads on the beam consist of a 4-kN force at the end of a bracket attached at point *B* and a 2-kN force at the midpoint of beam *DE*. Draw the shear-force and bending-moment diagrams for this compound beam.

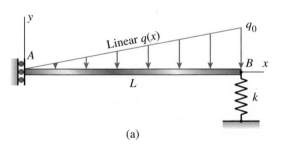

PROB. 4.5-30

**4.5-31** Draw the shear-force and bending-moment diagrams for beam *AB*, with a sliding support at *A* and an elastic support with spring constant *k* at *B* acted upon by two different load cases.

(a) A distributed load with linear variation and maximum intensity $q_0$ (see figure part a).

(b) A distributed load with parabolic variation and maximum intensity $q_0$ (see figure part b).

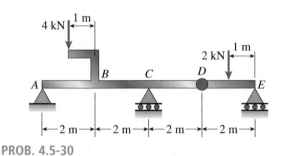

(a)

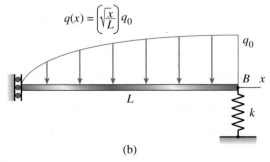

(b)

PROB. 4.5-31

**4.5-32** The shear-force diagram for a simple beam is shown in the figure.

Determine the loading on the beam and draw the bending-moment diagram, assuming that no couples act as loads on the beam.

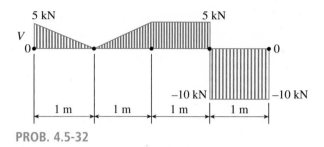

PROB. 4.5-32

**4.5-33** The shear-force diagram for a beam is shown in the figure. Assuming that no couples act as loads on the beam, determine the forces acting on the beam and draw the bending-moment diagram.

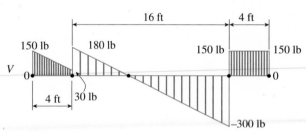

PROB. 4.5-33

**4.5-34** The compound beam below has an internal *moment release* just to the left of *B* and a *shear release* just to the right of *C*. Reactions have been computed at *A*, *C*, and *D* and are shown in the figure.

First, confirm the reaction expressions using statics, then plot shear (*V*) and moment (*M*) diagrams. *Label* all critical *V* and *M* values and also the *distance* to points where either *V* and/or *M* is zero.

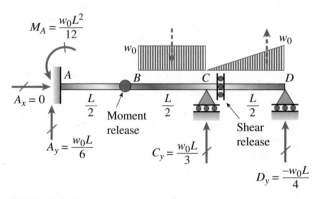

PROB. 4.5-34

**4.5-35** The compound beam below has an *shear release* just to the left of *C* and a *moment release* just to the right of *C*. A plot of the moment diagram is provided below for applied load *P* at *B* and triangular distributed loads $w(x)$ on segments *BC* and *CD*.

First, solve for reactions using statics, then plot axial force (*N*) and shear (*V*) diagrams. Confirm that the moment diagram is that shown below. *Label* all critical *N*, *V*, and *M* values and also the *distance* to points where *N*, *V*, and/or *M* is zero.

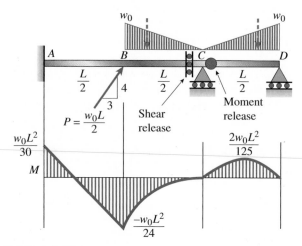

**PROB. 4.5-35**

**4.5-36** A simple beam *AB* supports two connected wheel loads *P* and *2P* that are distance *d* apart (see figure). The wheels may be placed at any distance *x* from the left-hand support of the beam.

(a) Determine the distance *x* that will produce the maximum shear force in the beam, and also determine the maximum shear force $V_{max}$.

(b) Determine the distance *x* that will produce the maximum bending moment in the beam, and also draw the corresponding bending-moment diagram. (Assume $P = 10$ kN, $d = 2.4$ m, and $L = 12$ m.)

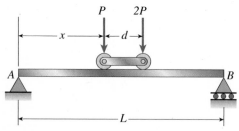

**PROB. 4.5-36**

**4.5-37** The inclined beam represents a ladder with the following applied loads: the weight (*W*) of the house painter and the distributed weight (*w*) of the ladder itself.

(a) Find support reactions at *A* and *B*, then plot axial force (*N*), shear (*V*), and moment (*M*) diagrams. *Label* all critical *N*, *V*, and *M* values and also the *distance* to points where any critical ordinates are zero. Plot *N*, *V*, and *M* diagrams normal to the inclined ladder.

(b) Repeat part (a) for the case of the ladder suspended from a pin at *B* and traveling on a roller support perpendicular to the floor at *A*.

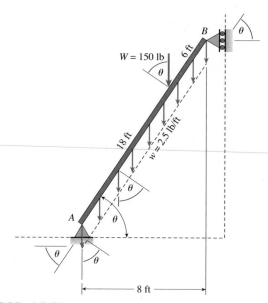

**PROB. 4.5-37**

**4.5-38** Beam *ABC* is supported by a tie rod *CD* as shown. Two configurations are possible: pin support at *A* and downward triangular load on *AB* or pin at *B* and upward load on *AB*. Which has the larger maximum moment?

First, find all support reactions, then plot axial force (*N*), shear (*V*), and moment (*M*) diagrams for *ABC* only and *label* all critical *N*, *V*, and *M* values. Label the *distance* to points where any critical ordinates are zero.

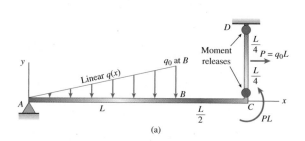

(a)

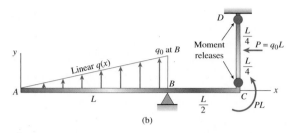

(b)

PROB. 4.5-38

**4.5-39** The plane frame below consists of column $AB$ and beam $BC$ which carries a triangular distributed load (see figure part a). Support $A$ is fixed, and there is a roller support at $C$. Column $AB$ has a moment release just below joint $B$.

(a) Find support reactions at $A$ and $C$, then plot axial force ($N$), shear-force ($V$), and bending-moment ($M$) diagrams for both members. Label all critical $N$, $V$, and $M$ values and also the distance to points where any critical ordinates are zero.

(b) Repeat part (a) if a parabolic lateral load acting to the right is now added on column $AB$ (figure part b).

**4.5-40** The plane frame shown below is part of an elevated freeway system. Supports at $A$ and $D$ are fixed but there

are *moment releases* at the base of both columns ($AB$ and $DE$), as well as in column $BC$ and at the end of beam $BE$.

Find all support reactions, then plot axial force ($N$), shear ($V$), and moment ($M$) diagrams for all beam and column members. *Label* all critical $N$, $V$, and $M$ values and also the *distance* to points where any critical ordinates are zero.

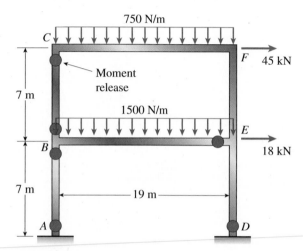

PROB. 4.5-40

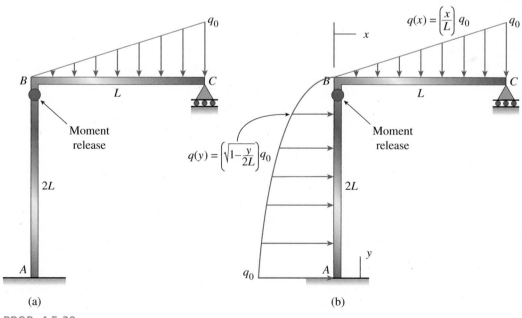

(a)                                    (b)

PROB. 4.5-39

# Stresses in Beams (Basic Topics)

Beams are essential load carrying components in modern building and bridge construction.
(© Can Stock Photo Inc./ronyzmbow)

## CHAPTER OVERVIEW

Chapter 5 is concerned with stresses and strains in beams which have loads applied in the $xy$ plane, a plane of symmetry of the cross section, resulting in beam deflection in that same plane, known as the **plane of bending**. Both **pure bending** (beam flexure under constant bending moment) and **nonuniform bending** (flexure in the presence of shear forces) are discussed (Section 5.2). We will see that strains and stresses in the beam are directly related to the **curvature** $\kappa$ of the deflection curve (Section 5.3). A **strain-curvature relation** will be developed from

consideration of longitudinal strains developed in the beam during bending; these strains vary linearly with distance from the neutral surface of the beam (Section 5.4). When Hooke's law (which applies for linearly elastic materials) is combined with the strain-curvature relation, we find that the neutral axis passes through the centroid of the cross section. As a result, $x$ and $y$ axes are seen to be principal centroidal axes. By consideration of the moment resultant of the normal stresses acting over the cross section, we next derive the **moment-curvature relation** which relates curvature ($\kappa$) to moment ($M$) and flexural rigidity ($EI$). This will lead to the differential equation of the beam elastic curve, a topic for consideration in Chapter 9 when we will discuss beam deflections in detail. Of immediate interest here, however, are beam stresses, and the moment-curvature relation is next used to develop the **flexure formula** (Section 5.5). The **flexure formula** shows that normal stresses ($\sigma_x$) vary linearly with distance ($y$) from the neutral surface and depend on bending moment ($M$) and moment of inertia ($I$) of the cross section. Next, the section modulus ($S$) of the beam cross section is defined and then used in **design** of beams in Section 5.6. In beam design, we use the maximum bending moment ($M_{max}$) (obtained from the bending moment diagram (Section 4.5)) and the allowable normal stress for the material ($\sigma_{allow}$) to compute the required section modulus, then select an appropriate beam of steel or wood from the tables in Appendices F and G. If the beam is nonprismatic (Section 5.7), the flexure formula still applies provided that changes in cross-sectional dimensions are gradual. However, we cannot assume that the maximum stresses occur at the cross section with the largest bending moment.

For beams in nonuniform bending, both normal and shear stresses are developed and must be considered in beam analysis and design. Normal stresses are computed using the **flexure formula**, as noted above, and the **shear formula** must be used to calculate shear stresses ($\tau$) which vary over the height of the beam (Sections 5.8 and 5.9). Maximum normal and shear stresses do not occur at the same location along a beam, but in most cases, maximum normal stresses control the design of the beam. Special consideration is given to shear stresses in beams with flanges (e.g., W and C shapes) (Section 5.10). **Built-up beams** fabricated of two or more pieces of material must be designed as though they were made up of one piece, and then the **connections** between the parts (e.g., nails, bolts, welds, and glue) are designed to ensure that the connections are strong enough to transmit the horizontal shear forces acting between the parts of the beam (Section 5.11). If structural members are subjected to the simultaneous action of both bending and axial loads and are not too slender so as to avoid buckling, the combined stresses can be obtained by superposition of the bending stresses and the axial stresses (Section 5.12). Finally, for beams with high localized stresses due to holes, notches, or other abrupt changes in dimensions, **stress concentrations** must be considered, especially for beams made of brittle materials or subjected to dynamic loads (Section 5.13).

## Chapter 5 is organized as follows:

**5.1** Introduction 404
**5.2** Pure Bending and Nonuniform Bending 404
**5.3** Curvature of a Beam 405
**5.4** Longitudinal Strains in Beams 407
**5.5** Normal Stress in Beams (Linearly Elastic Materials) 412
**5.6** Design of Beams for Bending Stresses 426
**5.7** Nonprismatic Beams 435
**5.8** Shear Stresses in Beams of Rectangular Cross Section 439

**5.9** Shear Stresses in Beams of Circular Cross Section 448
**5.10** Shear Stresses in the Webs of Beams with Flanges 451
*__5.11__ Built-Up Beams and Shear Flow 458
*__5.12__ Beams with Axial Loads 462
*__5.13__ Stress Concentrations in Bending 468
Chapter Summary & Review 472
Problems 476

*Advanced topics

# 5.1 INTRODUCTION

In the preceding chapter we saw how the loads acting on a beam create internal actions (or *stress resultants*) in the form of shear forces and bending moments. In this chapter we go one step further and investigate the *stresses* and *strains* associated with those shear forces and bending moments. Knowing the stresses and strains, we will be able to analyze and design beams subjected to a variety of loading conditions.

The loads acting on a beam cause the beam to bend (or *flex*), thereby deforming its axis into a curve. As an example, consider a cantilever beam *AB* subjected to a load *P* at the free end (Fig. 5-1a). The initially straight axis is bent into a curve (Fig. 5-1b), called the **deflection curve** of the beam.

For reference purposes, we construct a system of **coordinate axes** (Fig. 5-1b) with the origin located at a suitable point on the longitudinal axis of the beam. In this illustration, we place the origin at the fixed support. The positive *x* axis is directed to the right, and the positive *y* axis is directed upward. The *z* axis, not shown in the figure, is directed outward (that is, toward the viewer), so that the three axes form a right-handed coordinate system.

The beams considered in this chapter (like those discussed in Chapter 4) are assumed to be symmetric about the *xy* plane, which means that the *y* axis is an axis of symmetry of the cross section. In addition, all loads must act in the *xy* plane. As a consequence, the bending deflections occur in this same plane, known as the **plane of bending**. Thus, the deflection curve shown in Fig. 5-1b is a plane curve lying in the plane of bending.

The **deflection** of the beam at any point along its axis is the *displacement* of that point from its original position, measured in the *y* direction. We denote the deflection by the letter *v* to distinguish it from the coordinate *y* itself (see Fig. 5-1b).*

## Fig. 5-1

Bending of a cantilever beam: (a) beam with load, and (b) deflection curve

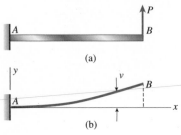

# 5.2 PURE BENDING AND NONUNIFORM BENDING

When analyzing beams, it is often necessary to distinguish between pure bending and nonuniform bending. **Pure bending** refers to flexure of a beam under a constant bending moment. Therefore, pure bending occurs only in regions of a beam where the shear force is zero (because $V = dM/dx$; see Eq. 4-6). In contrast, **nonuniform bending** refers to flexure in the presence of shear forces, which means that the bending moment changes as we move along the axis of the beam.

As an example of pure bending, consider a simple beam *AB* loaded by two couples $M_1$ having the same magnitude but acting in opposite directions (Fig. 5-2a). These loads produce a constant bending moment $M = M_1$ throughout the length of the beam, as shown by the bending moment diagram in part (b) of the figure. Note that the shear force *V* is zero at all cross sections of the beam.

## Fig. 5-2

Simple beam in pure bending ($M = M_1$)

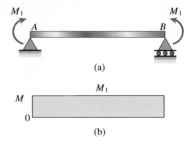

---

*In applied mechanics, the traditional symbols for displacements in the *x*, *y*, and *z* directions are *u*, *v*, and *w*, respectively.

Another illustration of pure bending is given in Fig. 5-3a, where the cantilever beam $AB$ is subjected to a clockwise couple $M_2$ at the free end. There are no shear forces in this beam, and the bending moment $M$ is constant throughout its length. The bending moment is negative ($M = -M_2$), as shown by the bending moment diagram in part (b) of Fig. 5-3.

The symmetrically loaded simple beam of Fig. 5-4a is an example of a beam that is partly in pure bending and partly in nonuniform bending, as seen from the shear-force and bending-moment diagrams (Figs. 5-4b and c). The central region of the beam is in pure bending because the shear force is zero and the bending moment is constant. The parts of the beam near the ends are in nonuniform bending because shear forces are present and the bending moments vary.

In the following two sections we will investigate the strains and stresses in beams subjected only to pure bending. Fortunately, we can often use the results obtained for pure bending even when shear forces are present, as explained later (see the last paragraph in Section 5.8).

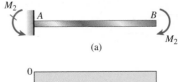

**Fig. 5-3**

Cantilever beam in pure bending ($M = -M_2$)

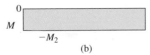

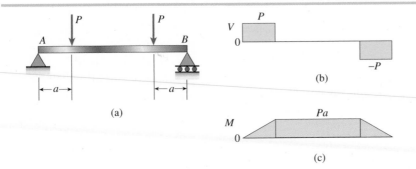

**Fig. 5-4**

Simple beam with central region in pure bending and end regions in nonuniform bending

# 5.3 CURVATURE OF A BEAM

When loads are applied to a beam, its longitudinal axis is deformed into a curve, as illustrated previously in Fig. 5-1. The resulting strains and stresses in the beam are directly related to the **curvature** of the deflection curve.

To illustrate the concept of curvature, consider again a cantilever beam subjected to a load $P$ acting at the free end (see Fig. 5-5a on the next page). The deflection curve of this beam is shown in Fig. 5-5b. For purposes of analysis, we identify two points $m_1$ and $m_2$ on the deflection curve. Point $m_1$ is selected at an arbitrary distance $x$ from the $y$ axis and point $m_2$ is located a small distance $ds$ further along the curve. At each of these points we draw a line normal to the *tangent* to the deflection curve, that is, normal to the curve itself. These normals intersect at point $O'$, which is the **center of curvature** of the deflection curve. Because most beams have very small deflections and nearly flat deflection curves, point $O'$ is usually located much farther from the beam than is indicated in the figure.

The distance $m_1O'$ from the curve to the center of curvature is called the **radius of curvature** $\rho$ (Greek letter rho), and the **curvature** $\kappa$ (Greek letter kappa) is defined as the reciprocal of the radius of curvature. Thus,

$$\kappa = \frac{1}{\rho} \qquad \qquad \textbf{(5-1)}$$

**Fig. 5-5**

Curvature of a bent beam:
(a) beam with load,
and (b) deflection curve

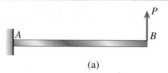

(a)

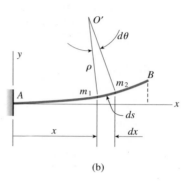

(b)

Curvature is a measure of how sharply a beam is bent. If the load on a beam is small, the beam will be nearly straight, the radius of curvature will be very large, and the curvature will be very small. If the load is increased, the amount of bending will increase—the radius of curvature will become smaller, and the curvature will become larger.

From the geometry of triangle $O'm_1m_2$ (Fig. 5-5b) we obtain

$$\rho \, d\theta = ds \qquad\qquad \textbf{(5-2)}$$

in which $d\theta$ (measured in radians) is the infinitesimal angle between the normals and $ds$ is the infinitesimal distance along the curve between points $m_1$ and $m_2$. Combining Eq. (5-2) with Eq. (5-1), we get

$$\kappa = \frac{1}{\rho} = \frac{d\theta}{ds} \qquad\qquad \textbf{(5-3)}$$

This equation for **curvature** is derived in textbooks on calculus and holds for any curve, regardless of the amount of curvature. If the curvature is *constant* throughout the length of a curve, the radius of curvature will also be constant and the curve will be an arc of a circle.

The deflections of a beam are usually very small compared to its length (consider, for instance, the deflections of the structural frame of an automobile or a beam in a building). Small deflections mean that the deflection curve is nearly flat. Consequently, the distance $ds$ along the curve may be set equal to its horizontal projection $dx$ (see Fig. 5-5b). Under these special conditions of **small deflections**, the equation for the curvature becomes

$$\kappa = \frac{1}{\rho} = \frac{d\theta}{dx} \qquad\qquad \textbf{(5-4)}$$

Both the curvature and the radius of curvature are functions of the distance $x$ measured along the $x$ axis. It follows that the position $O'$ of the center of curvature also depends upon the distance $x$.

In Section 5.5, we will see that the curvature at a particular point on the axis of a beam depends upon the bending moment at that point and upon the properties of the beam itself (shape of cross section and type of material). Therefore, if the beam is prismatic and the material is homogeneous, the curvature will vary only with the bending moment. Consequently, a beam in *pure bending* will have constant curvature and a beam in *nonuniform bending* will have varying curvature.

The **sign convention for curvature** depends upon the orientation of the coordinate axes. If the $x$ axis is positive to the right and the $y$ axis is positive upward, as shown in Fig. 5-6, then the curvature is positive when the beam is bent concave upward and the center of curvature is above the beam. Conversely, the curvature is negative when the beam is bent concave downward and the center of curvature is below the beam.

In the next section, we will see how the longitudinal strains in a bent beam are determined from its curvature, and in Chapter 9 we will see how curvature is related to the deflections of beams.

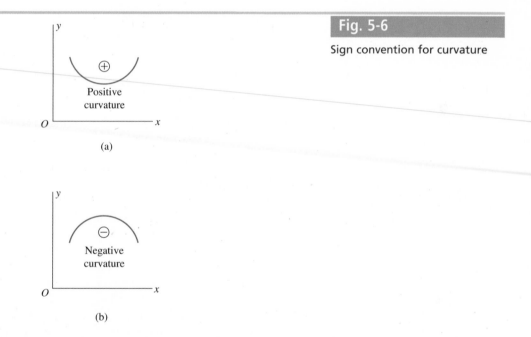

**Fig. 5-6**

Sign convention for curvature

# 5.4 LONGITUDINAL STRAINS IN BEAMS

The longitudinal strains in a beam can be found by analyzing the curvature of the beam and the associated deformations. For this purpose, let us consider a portion $AB$ of a beam in pure bending subjected to positive bending moments $M$ (Fig. 5-7a). We assume that the beam initially has a straight longitudinal axis (the $x$ axis in the figure) and that its cross section is symmetric about the $y$ axis, as shown in Fig. 5-7b.

Under the action of the bending moments, the beam deflects in the $xy$ plane (the plane of bending) and its longitudinal axis is bent into a circular curve (curve $s–s$ in Fig. 5-7c). The beam is bent concave upward, which is positive curvature (Fig. 5-6a).

**Fig. 5-7**

Deformations of a beam in pure bending: (a) side view of beam, (b) cross section of beam, and (c) deformed beam

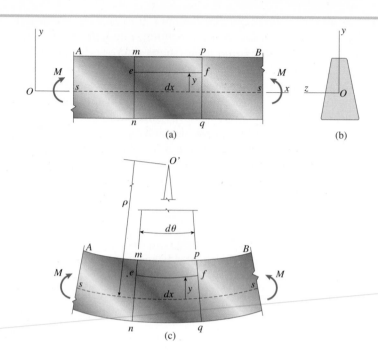

**Cross sections of the beam**, such as sections $mn$ and $pq$ in Fig. 5-7a, remain plane and normal to the longitudinal axis (Fig. 5-7c). The fact that cross sections of a beam in pure bending remain plane is so fundamental to beam theory that it is often called an assumption. However, we could also call it a theorem, because it can be proved rigorously using only rational arguments based upon symmetry (Ref. 5-1). The basic point is that the symmetry of the beam and its loading (Figs. 5-7a and b) means that all elements of the beam (such as element $mpqn$) must deform in an identical manner, which is possible only if cross sections remain plane during bending (Fig. 5-7c). This conclusion is valid for beams of any material, whether the material is elastic or inelastic, linear or nonlinear. Of course, the material properties, like the dimensions, must be symmetric about the plane of bending. (*Note:* Even though a plane cross section in pure bending remains plane, there still may be deformations in the plane itself. Such deformations are due to the effects of Poisson's ratio, as explained at the end of this discussion.)

Because of the bending deformations shown in Fig. 5-7c, cross sections $mn$ and $pq$ rotate with respect to each other about axes perpendicular to the $xy$ plane. Longitudinal lines on the lower part of the beam are elongated, whereas those on the upper part are shortened. Thus, the lower part of the beam is in tension and the upper part is in compression. Somewhere between the top and bottom of the beam is a surface in which longitudinal lines do not change in length. This surface, indicated by the dashed line $s$–$s$ in Figs. 5-7a and c, is called the **neutral surface** of the beam. Its intersection with any cross-sectional plane is called the **neutral axis** of the cross section; for instance, the $z$ axis is the neutral axis for the cross section of Fig. 5-7b.

The planes containing cross sections $mn$ and $pq$ in the deformed beam (Fig. 5-7c) intersect in a line through the center of curvature $O'$. The angle between these planes is denoted $d\theta$, and the distance from $O'$ to the neutral surface $s$–$s$ is the radius of curvature $\rho$. The initial distance $dx$ between the two planes (Fig. 5-7a) is unchanged at the neutral surface (Fig. 5-7c), hence $pd\theta = dx$. However, all other longitudinal lines between the two planes either lengthen or shorten, thereby creating **normal strains** $\varepsilon_x$.

To evaluate these normal strains, consider a typical longitudinal line $ef$ located within the beam between planes $mn$ and $pq$ (Fig. 5-7a). We identify line $ef$ by its distance $y$ from the neutral surface in the initially straight beam. Thus, we are now assuming that the $x$ axis lies along the neutral surface of the *undeformed* beam. Of course, when the beam deflects, the neutral surface moves with the beam, but the $x$ axis remains fixed in position. Nevertheless, the longitudinal line $ef$ in the deflected beam (Fig. 5-7c) is still located at the same distance $y$ from the neutral surface. Thus, the length $L_1$ of line $ef$ after bending takes place is

$$L_1 = (\rho - y)d\theta = dx - \frac{y}{\rho}dx$$

in which we have substituted $d\theta = dx/\rho$.

Since the original length of line $ef$ is $dx$, it follows that its elongation is $L_1 - dx$, or $-y\,dx/\rho$. The corresponding *longitudinal strain* is equal to the elongation divided by the initial length $dx$; therefore, the **strain-curvature relation** is

$$\varepsilon_x = -\frac{y}{\rho} = -\kappa y \qquad\qquad (5\text{-}5)$$

where $\kappa$ is the curvature [see Eq. (5-1)].

The preceding equation shows that the longitudinal strains in the beam are proportional to the curvature and vary linearly with the distance $y$ from the neutral surface. When the point under consideration is above the neutral surface, the distance $y$ is positive. If the curvature is also positive (as in Fig. 5-7c), then $\varepsilon_x$ will be a negative strain, representing a shortening. By contrast, if the point under consideration is below the neutral surface, the distance $y$ will be negative and, if the curvature is positive, the strain $\varepsilon_x$ will also be positive, representing an elongation. Note that the **sign convention** for $\varepsilon_x$ is the same as that used for normal strains in earlier chapters, namely, elongation is positive and shortening is negative.

Equation (5-5) for the normal strains in a beam was derived solely from the geometry of the deformed beam—the properties of the material did not enter into the discussion. Therefore, *the strains in a beam in pure bending vary linearly with distance from the neutral surface regardless of the shape of the stress-strain curve of the material.*

The next step in our analysis, namely, finding the stresses from the strains, requires the use of the *stress-strain curve.* This step is described in the next section for linearly elastic materials and in Section 6.10 for elasto-plastic materials.

The longitudinal strains in a beam are accompanied by *transverse strains* (that is, normal strains in the $y$ and $z$ directions) because of the effects of Poisson's ratio. However, there are no accompanying transverse stresses because beams are free to deform laterally. This stress condition is analogous to that of a prismatic bar in tension or compression, and therefore *longitudinal elements in a beam in pure bending are in a state of uniaxial stress.*

**• • • Example 5-1**

A simply supported steel beam $AB$ (Fig. 5-8a) of length $L = 8.0$ ft and height $h = 6.0$ in. is bent by couples $M_0$ into a circular arc with a downward deflection $\delta$ at the midpoint (Fig. 5-8b). The longitudinal normal strain (elongation) on the bottom surface of the beam is 0.00125, and the distance from the neutral surface to the bottom surface of the beam is 3.0 in.

Determine the radius of curvature $\rho$, the curvature $\kappa$, and the deflection $\delta$ of the beam.

*Note:* This beam has a relatively large deflection because its length is large compared to its height ($L/h = 16$) and the strain of 0.00125 is also large. (It is about the same as the yield strain for ordinary structural steel.)

---

**Fig. 5-8**

Example 5-1: Beam in pure bending: (a) beam with loads, and (b) deflection curve

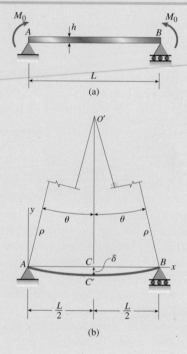

**Solution**

*Curvature.* Since we know the longitudinal strain at the bottom surface of the beam ($\varepsilon_x = 0.00125$), and since we also know the distance from the neutral surface to the bottom surface ($y = -3.0$ in.), we can use Eq. (5-5) to calculate both the radius of curvature and the curvature. Rearranging Eq. (5-5) and substituting numerical values, we get

$$\rho = -\frac{y}{\varepsilon_x} = -\frac{-3.0 \text{ in.}}{0.00125} = 2400 \text{ in.} = 200 \text{ ft} \qquad \kappa = \frac{1}{\rho} = 0.0050 \text{ ft}^{-1} \quad \Longleftarrow$$

These results show that the radius of curvature is extremely large compared to the length of the beam even when the strain in the material is large. If, as usual, the strain is less, the radius of curvature is even larger.

*Deflection.* As pointed out in Section 5.3, a constant bending moment (pure bending) produces constant curvature throughout the length of a beam. Therefore, the deflection curve is a circular arc. From Fig. 5-8b we see that the distance from the center of curvature $O'$ to the midpoint $C'$ of the deflected beam is the radius of curvature $\rho$, and the distance from $O'$ to point $C$ on the $x$ axis is $\rho \cos \theta$, where $\theta$ is angle $BO'C$. This leads to the following expression for the deflection at the midpoint of the beam:

$$\delta = \rho(1 - \cos \theta) \tag{5-6}$$

For a nearly flat curve, we can assume that the distance between supports is the same as the length of the beam itself. Therefore, from triangle $BO'C$ we get

$$\sin \theta = \frac{L/2}{\rho} \tag{5-7}$$

Substituting numerical values, we obtain

$$\sin \theta = \frac{(8.0 \text{ ft})(12 \text{ in./ft})}{2(2400 \text{ in.})} = 0.0200$$

and

$$\theta = 0.0200 \text{ rad} = 1.146°$$

Note that for practical purposes we may consider $\sin \theta$ and $\theta$ (radians) to be equal numerically because $\theta$ is a very small angle.

Now we substitute into Eq. (5-6) for the deflection and obtain

$$\delta = \rho(1 - \cos \theta) = (2400 \text{ in.})(1 - 0.999800) = 0.480 \text{ in.}$$

This deflection is very small compared to the length of the beam, as shown by the ratio of the span length to the deflection:

$$\frac{L}{\delta} = \frac{(8.0 \text{ ft})(12 \text{ in./ft})}{0.480 \text{ in.}} = 200$$

Thus, we have confirmed that the deflection curve is nearly flat in spite of the large strains. Of course, in Fig. 5-8b the deflection of the beam is highly exaggerated for clarity.

*Note:* The purpose of this example is to show the relative magnitudes of the radius of curvature, length of the beam, and deflection of the beam. However, the method used for finding the deflection has little practical value because it is limited to pure bending, which produces a circular deflected shape. More useful methods for finding beam deflections are presented later in Chapter 9.

# 5.5 NORMAL STRESSES IN BEAMS (LINEARLY ELASTIC MATERIALS)

In the preceding section we investigated the longitudinal strains $\varepsilon_x$ in a beam in pure bending (see Eq. 5-5 and Fig. 5-7). Since longitudinal elements of a beam are subjected only to tension or compression, we can use the **stress-strain curve** for the material to determine the stresses from the strains. The stresses act over the entire cross section of the beam and vary in intensity depending upon the shape of the stress-strain diagram and the dimensions of the cross section. Since the $x$ direction is longitudinal (Fig. 5-7a), we use the symbol $\sigma_x$ to denote these stresses.

The most common stress-strain relationship encountered in engineering is the equation for a **linearly elastic material**. For such materials we substitute Hooke's law for uniaxial stress ($\sigma = E\varepsilon$) into Eq. (5-5) and obtain

$$\sigma_x = E\varepsilon_x = -\frac{Ey}{\rho} = -E\kappa y \qquad \textbf{(5-8)}$$

This equation shows that the normal stresses acting on the cross section vary linearly with the distance $y$ from the neutral surface. This stress distribution is pictured in Fig. 5-9a for the case in which the bending moment $M$ is positive and the beam bends with positive curvature.

When the curvature is positive, the stresses $\sigma_x$ are negative (compression) above the neutral surface and positive (tension) below it. In the figure, compressive stresses are indicated by arrows pointing *toward* the cross section and tensile stresses are indicated by arrows pointing *away* from the cross section.

In order for Eq. (5-8) to be of practical value, we must locate the origin of coordinates so that we can determine the distance $y$. In other words, we must locate the neutral axis of the cross section. We also need to obtain a relationship between the curvature and the bending moment—so that we can substitute into Eq. (5-8) and obtain an equation relating the stresses to the bending moment. These two objectives can be accomplished by determining the resultant of the stresses $\sigma_x$ acting on the cross section.

In general, the **resultant of the normal stresses** consists of two stress resultants: (1) a force acting in the $x$ direction, and (2) a bending couple acting about the $z$ axis. However, the axial force is zero when a beam is in pure bending. Therefore, we can write the following equations of statics: (1) The resultant force in the $x$ direction is equal to zero, and (2) the resultant moment is equal to the bending moment $M$. The first equation gives the location of the neutral axis and the second gives the moment-curvature relationship.

## Location of Neutral Axis

To obtain the first equation of statics, we consider an element of area $dA$ in the cross section (Fig. 5-9b). The element is located at distance $y$ from the neutral axis, and therefore the stress $\sigma_x$ acting on the element is given by Eq. (5-8). The *force* acting on the element is equal to $\sigma_x dA$ and is

**Fig. 5-9**

Normal stresses in a beam of linearly elastic material: (a) side view of beam showing distribution of normal stresses, and (b) cross section of beam showing the $z$ axis as the neutral axis of the cross section

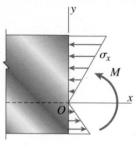

(a)

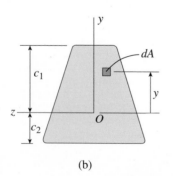

(b)

compressive when $y$ is positive. Because there is no resultant force acting on the cross section, the integral of $\sigma_x dA$ over the area $A$ of the entire cross section must vanish; thus, the *first equation of statics* is

$$\int_A \sigma_x dA = -\int_A E\kappa y\,dA = 0 \qquad \textbf{(5-9a)}$$

Because the curvature $\kappa$ and modulus of elasticity $E$ are nonzero constants at any given cross section of a bent beam, they are not involved in the integration over the cross-sectional area. Therefore, we can drop them from the equation and obtain

$$\int_A y\,dA = 0 \qquad \textbf{(5-9b)}$$

This equation states that the first moment of the area of the cross section, evaluated with respect to the $z$ axis, is zero. In other words, the $z$ axis must pass through the centroid of the cross section.*

Since the $z$ axis is also the neutral axis, we have arrived at the following important conclusion: *The neutral axis passes through the centroid of the cross-sectional area when the material follows Hooke's law and there is no axial force acting on the cross section.* This observation makes it relatively simple to determine the position of the neutral axis.

As explained in Section 5.1, our discussion is limited to beams for which the $y$ axis is an axis of symmetry. Consequently, the $y$ axis also passes through the centroid. Therefore, we have the following additional conclusion: *The origin $O$ of coordinates* (Fig. 5-9b) *is located at the centroid of the cross-sectional area.*

Because the $y$ axis is an axis of symmetry of the cross section, it follows that the $y$ axis is a *principal axis* (see Chapter 12, Section 12.9, for a discussion of principal axes). Since the $z$ axis is perpendicular to the $y$ axis, it too is a principal axis. Thus, when a beam of linearly elastic material is subjected to pure bending, *the $y$ and $z$ axes are principal centroidal axes.*

## Moment-Curvature Relationship

The *second equation of statics* expresses the fact that the moment resultant of the normal stresses $\sigma_x$ acting over the cross section is equal to the bending moment $M$ (Fig. 5-9a). The element of force $\sigma_x dA$ acting on the element of area $dA$ (Fig. 5-9b) is in the positive direction of the $x$ axis when $\sigma_x$ is positive and in the negative direction when $\sigma_x$ is negative. Since the element $dA$ is located above the neutral axis, a positive stress $\sigma_x$ acting on that element produces an element of moment equal to $\sigma_x y dA$. This element of moment acts opposite in direction to the positive bending moment $M$ shown in Fig. 5-9a. Therefore, the elemental moment is

$$dM = -\sigma_x y\ dA$$

*[handwritten: $\sigma_x \cdot A = F$]*

*[handwritten: $F \cdot y = M.$]*

---

*Centroids and first moments of areas are discussed in Chapter 12, Sections 12.2 and 12.3.

The integral of all such elemental moments over the entire cross-sectional area $A$ must equal the bending moment:

$$M = -\int_A \sigma_x y \, dA \qquad \text{(5-10a)}$$

or, upon substituting for $\sigma_x$ from Eq. (5-9),

$$M = \int_A \kappa E y^2 \, dA = \kappa E \int_A y^2 \, dA \qquad \text{(5-10b)}$$

This equation relates the curvature of the beam to the bending moment $M$.

Since the integral in the preceding equation is a property of the cross-sectional area, it is convenient to rewrite the equation as follows:

$$M = \kappa EI \qquad \text{(5-11)}$$

in which

$$I = \int_A y^2 \, dA \qquad \text{(5-12)}$$

This integral is the **moment of inertia** of the cross-sectional area with respect to the $z$ axis (that is, with respect to the neutral axis). Moments of inertia are always positive and have dimensions of length to the fourth power; for instance, typical USCS units are in.[4] and typical SI units are mm[4] when performing beam calculations.*

Equation (5-11) can now be rearranged to express the *curvature* in terms of the bending moment in the beam:

$$\kappa = \frac{1}{\rho} = \frac{M}{EI} \qquad \text{(5-13)}$$

Known as the **moment-curvature equation**, Eq. (5-13) shows that the curvature is directly proportional to the bending moment $M$ and inversely proportional to the quantity $EI$, which is called the **flexural rigidity** of the beam. Flexural rigidity is a measure of the resistance of a beam to bending, that is, the larger the flexural rigidity, the smaller the curvature for a given bending moment.

Comparing the **sign convention** for bending moments (Fig. 4-5) with that for curvature (Fig. 5-6), we see that *a positive bending moment produces positive curvature and a negative bending moment produces negative curvature* (see Fig. 5-10).

## Flexure Formula

Now that we have located the neutral axis and derived the moment-curvature relationship, we can determine the stresses in terms of the bending moment. Substituting the expression for curvature (Eq. 5-13) into the expression for the stress $\sigma_x$ (Eq. 5-8), we get

$$\sigma_x = -\frac{My}{I} \qquad \text{(5-14)}$$

---

**Fig. 5-10**

Relationships between signs of bending moments and signs of curvatures

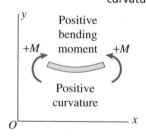

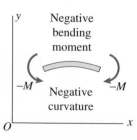

*Moments of inertia of areas are discussed in Chapter 12, Section 12.4.

This equation, called the **flexure formula**, shows that the stresses are directly proportional to the bending moment $M$ and inversely proportional to the moment of inertia $I$ of the cross section. Also, the stresses vary linearly with the distance $y$ from the neutral axis, as previously observed. Stresses calculated from the flexure formula are called **bending stresses** or **flexural stresses**.

If the bending moment in the beam is positive, the bending stresses will be positive (tension) over the part of the cross section where $y$ is negative, that is, over the lower part of the beam. The stresses in the upper part of the beam will be negative (compression). If the bending moment is negative, the stresses will be reversed. These relationships are shown in Fig. 5-11.

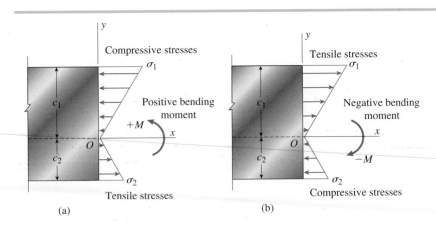

**Fig. 5-11**

Relationships between signs of bending moments and directions of normal stresses: (a) positive bending moment, and (b) negative bending moment

(a)                                                    (b)

## Maximum Stresses at a Cross Section

The maximum tensile and compressive bending stresses acting at any given cross section occur at points located farthest from the neutral axis. Let us denote by $c_1$ and $c_2$ the distances from the neutral axis to the extreme elements in the positive and negative $y$ directions, respectively (see Fig. 5-9b and Fig. 5-11). Then the corresponding **maximum normal stresses** $\sigma_1$ and $\sigma_2$ (from the flexure formula) are

$$\sigma_1 = -\frac{Mc_1}{I} = -\frac{M}{S_1} \qquad \sigma_2 = \frac{Mc_2}{I} = \frac{M}{S_2} \qquad \textbf{(5-15a,b)}$$

in which

$$S_1 = \frac{I}{c_1} \qquad S_2 = \frac{I}{c_2} \qquad \textbf{(5-16a,b)}$$

The quantities $S_1$ and $S_2$ are known as the **section moduli** of the cross-sectional area. From Eqs. (5-16a and b) we see that each section modulus has dimensions of length to the third power (for example, in.$^3$ or mm$^3$). Note that the distances $c_1$ and $c_2$ to the top and bottom of the beam are always taken as positive quantities.

The advantage of expressing the maximum stresses in terms of section moduli arises from the fact that each section modulus combines the beam's relevant cross-sectional properties into a single quantity. Then this quantity can be listed in tables and handbooks as a property of the beam, which is a convenience to designers. (Design of beams using section moduli is explained in the next section.)

## Doubly Symmetric Shapes

If the cross section of a beam is symmetric with respect to the $z$ axis as well as the $y$ axis (*doubly symmetric cross section*), then $c_1 = c_2 = c$ and the maximum tensile and compressive stresses are equal numerically:

$$\sigma_1 = -\sigma_2 = -\frac{Mc}{I} = -\frac{M}{S} \qquad \text{or} \qquad \sigma_{max} = \frac{M}{S} \qquad \textbf{(5-17a,b)}$$

in which

$$S = \frac{I}{c} \qquad \textbf{(5-18)}$$

is the only section modulus for the cross section.

For a beam of **rectangular cross section** with width $b$ and height $h$ (Fig. 5-12a), the moment of inertia and section modulus are

$$I = \frac{bh^3}{12} \qquad S = \frac{bh^3}{6} \qquad \textbf{(5-19a,b)}$$

### Fig. 5-12

Doubly symmetric
cross-sectional shapes

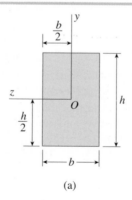

(a)

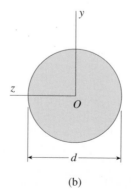

(b)

For a **circular cross section** of diameter $d$ (Fig. 5-12b), these properties are

$$I = \frac{\pi d^4}{64} \qquad S = \frac{\pi d^3}{32} \qquad \text{(5-20a,b)}$$

Properties of other doubly symmetric shapes, such as hollow tubes (either rectangular or circular) and wide-flange shapes, can be readily obtained from the preceding formulas.

## Properties of Beam Cross Sections

Moments of inertia of many plane figures are listed in Appendix E for convenient reference. Also, the dimensions and properties of standard sizes of steel and wood beams are listed in Appendixes F and G and in many engineering handbooks, as explained in more detail in the next section.

For other cross-sectional shapes, we can determine the location of the neutral axis, the moment of inertia, and the section moduli by direct calculation, using the techniques described in Chapter 12. This procedure is illustrated later in Example 5-4.

## Limitations

The analysis presented in this section is for pure bending of prismatic beams composed of homogeneous, linearly elastic materials. If a beam is subjected to nonuniform bending, the shear forces will produce warping (or out-of-plane distortion) of the cross sections. Thus, a cross section that was plane before bending is no longer plane after bending. Warping due to shear deformations greatly complicates the behavior of the beam. However, detailed investigations show that the normal stresses calculated from the flexure formula are not significantly altered by the presence of shear stresses and the associated warping (Ref. 2-1, pp. 42 and 48). Thus, we may justifiably use the theory of pure bending for calculating normal stresses in beams subjected to nonuniform bending.*

The flexure formula gives results that are accurate only in regions of the beam where the stress distribution is not disrupted by changes in the shape of the beam or by discontinuities in loading. For instance, the flexure formula is not applicable near the supports of a beam or close to a concentrated load. Such irregularities produce localized stresses, or *stress concentrations*, that are much greater than the stresses obtained from the flexure formula (see Section 5.13).

---

*Beam theory began with Galileo Galilei (1564–1642), who investigated the behavior of various types of beams. His work in mechanics of materials is described in his famous book *Two New Sciences*, first published in 1638 (Ref. 5-2). Although Galileo made many important discoveries regarding beams, he did not obtain the stress distribution that we use today. Further progress in beam theory was made by Mariotte, Jacob Bernoulli, Euler, Parent, Saint-Venant, and others (Ref. 5-3).

A high-strength steel wire of diameter $d$ is bent around a cylindrical drum of radius $R_0$ (Fig. 5-13).

Determine the bending moment $M$ and maximum bending stress $\sigma_{max}$ in the wire, assuming $d = 4$ mm and $R_0 = 0.5$ m. (The steel wire has modulus of elasticity $E = 200$ GPa GPa and proportional limit $\sigma_{p1} = 1200$ MPa.)

**Fig. 5-13**

Example 5-2: Wire bent around a drum

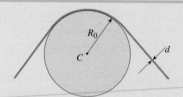

### Solution

The first step in this example is to determine the radius of curvature $\rho$ of the bent wire. Then, knowing $\rho$, we can find the bending moment and maximum stresses.

*Radius of curvature.* The radius of curvature of the bent wire is the distance from the center of the drum to the neutral axis of the cross section of the wire:

$$\rho = R_0 + \frac{d}{2} \tag{5-21}$$

*Bending moment.* The bending moment in the wire may be found from the moment-curvature relationship (Eq. 5-13):

$$M = \frac{EI}{\rho} = \frac{2EI}{2R_0 + d} \tag{5-22}$$

in which $I$ is the moment of inertia of the cross-sectional area of the wire. Substituting for $I$ in terms of the diameter $d$ of the wire (Eq. 5-20a), we get

$$M = \frac{\pi E d^4}{32(2R_0 + d)} \tag{5-23}$$

This result was obtained without regard to the *sign* of the bending moment, since the direction of bending is obvious from the figure.

*Maximum bending stresses.* The maximum tensile and compressive stresses, which are equal numerically, are obtained from the flexure formula as given by Eq. (5-17b):

$$\sigma_{max} = \frac{M}{S}$$

in which $S$ is the section modulus for a circular cross section. Substituting for $M$ from Eq. (5-23) and for $S$ from Eq. (5-20b), we get

$$\sigma_{max} = \frac{Ed}{2R_0 + d} \tag{5-24}$$

This same result can be obtained directly from Eq. (5-8) by replacing $y$ with $d/2$ and substituting for $\rho$ from Eq. (5-21).

We see by inspection of Fig. 5-13 that the stress is compressive on the lower (or inner) part of the wire and tensile on the upper (or outer) part.

*Numerical results.* We now substitute the given numerical data into Eqs. (5-23) and (5-24) and obtain the following results:

$$M = \frac{\pi E d^4}{32(2R_0 + d)} = \frac{\pi(200 \text{ GPa})(4 \text{ mm})^4}{32[2(0.5 \text{ m}) + 4 \text{ mm}]} = 5.01 \text{ N} \cdot \text{m} \quad \Longleftarrow$$

$$\sigma_{max} = \frac{Ed}{2R_0 + d} = \frac{(200 \text{ GPa})(4 \text{ mm})}{2(0.5 \text{ m}) + 4 \text{ mm}} = 797 \text{ MPa} \quad \Longleftarrow$$

Note that $\sigma_{max}$ is less than the proportional limit of the steel wire, and therefore the calculations are valid.

*Note:* Because the radius of the drum is large compared to the diameter of the wire, we can safely disregard $d$ in comparison with $2R_0$ in the denominators of the expressions for $M$ and $\sigma_{max}$. Then Eqs. (5-23) and (5-24) yield the following results:

$$M = 5.03 \text{ N} \cdot \text{m} \qquad \sigma_{max} = 800 \text{ MPa}$$

These results are on the conservative side and differ by less than 1% from the more precise values.

A simple beam *AB* of span length *L* = 22 ft (Fig. 5-14a) supports a uniform load of intensity *q* = 1.5 k/ft and a concentrated load *P* = 12 k. The uniform load includes an allowance for the weight of the beam. The concentrated load acts at a point 9.0 ft from the left-hand end of the beam. The beam is constructed of glued laminated wood and has a cross section width of *b* = 8.75 in. and a height of *h* = 27 in. (Fig. 5-14b).

(a) Determine the maximum tensile and compressive stresses in the beam due to bending.
(b) If load *q* is unchanged, find the maximum permissible value of load *P* if the allowable normal stress in tension and compression is $\sigma_a$ =1875 psi.

## Fig. 5-14

Example 5-3: Stresses in a simple beam

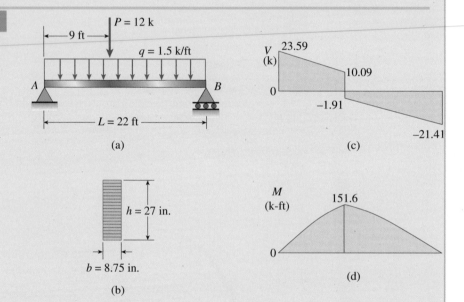

(a)

(b)

(c)

(d)

## Solution

(a) *Maximum normal stresses: reactions, shear forces, and bending moments.* We begin the analysis by calculating the reactions at supports *A* and *B*, using the techniques described in Chapter 4. The results are

$$R_A = 23.59 \text{ k} \qquad R_B = 21.41 \text{ k}$$

Knowing the reactions, we can construct the shear-force diagram, shown in Fig. 5-14c. Note that the shear force changes from positive to negative under the concentrated load P, which is at a distance of 9 ft from the left-hand support.

Next, we draw the bending-moment diagram (Fig. 5-14d) and determine the maximum bending moment, which occurs under the concentrated load where the shear force changes sign. The maximum moment is

$$M_{max} = 151.6 \text{ k-ft}$$

The maximum bending stresses in the beam occur at the cross section of maximum moment.

*Section modulus.* The section modulus of the cross-sectional area is calculated from Eq. (5-19b), as follows:

$$S = \frac{bh^2}{6} = \frac{1}{6}(8.75 \text{ in.})(27 \text{ in.})^2 = 1063 \text{ in.}^3 \qquad \textbf{(a)}$$

*Maximum stresses.* The maximum tensile and compressive stresses $\sigma_t$ and $\sigma_c$, respectively, are obtained from Eq. (5-17a):

$$\sigma_t = \sigma_2 = \frac{M_{max}}{S} = \frac{(151.6 \text{ k-ft})(12 \text{ in./ft})}{1063 \text{ in.}^3} = 1710 \text{ psi} \quad \Longleftarrow$$

$$\sigma_c = \sigma_1 = -\frac{M_{max}}{S} = -1710 \text{ psi} \qquad \Longleftarrow \textbf{(b)}$$

Because the bending moment is positive, the maximum tensile stress occurs at the bottom of the beam and the maximum compressive stress occurs at the top.

(b) *Maximum permissible load P.* Stresses $\sigma_1$ and $\sigma_2$ are only slightly below the allowable normal stress at $\sigma_a = 1875$ psi, so we do not expect $P_{max}$ to be much greater than the applied load $P = 12$ kips in part (a). Hence, the maximum moment $M_{max}$ will occur at the location of applied load $P_{max}$ (just 9 ft to the right of support A at the point at which the shear force is zero). If we let variable $a = 9$ ft, we can find the following expression for $M_{max}$ in terms of load and dimension variables:

$$M_{max} = \frac{a(L - a)(2P + Lq)}{2L} \qquad \textbf{(c)}$$

where $L = 22$ ft and $q = 1.5$ k/ft. We then equate $M_{max}$ to $\sigma_a \times S$ [see Eq. (b)], where $S = 1063$ in.$^3$ [from Eq. (a)], and then solve for $P_{max}$:

$$P_{max} = \sigma_a S \left[ \frac{L}{a(L - a)} \right] - \frac{qL}{2}$$

$$= (1.875 \text{ ksi})(1063 \text{ in.}^3) \left[ \frac{22 \text{ ft}}{9 \text{ ft}(22 \text{ ft} - 9 \text{ ft})} \right] - 1.5 \frac{\text{kip}}{\text{ft}} \left( \frac{22 \text{ ft}}{2} \right)$$

$$= 14.73 \text{ kip} \qquad \Longleftarrow \textbf{(d)}$$

## Example 5-4

The beam *ABC* shown in Fig 5-15a has simple supports at *A* and *B* and an overhang from *B* to *C*. The length of the span is $L$ = 3.0 m, and the length of the overhang is $L/2$ = 1.5 m. A uniform load of intensity $q$ = 3.2 kN/m acts throughout the entire length of the beam (4.5 m).

The beam has a cross section of channel shape with a width of $b$ = 300 mm and height of $h$ = 80 mm (Fig. 5-16a). The web thickness is $t$ = 12 mm, and the average thickness of the sloping flanges is the same. For the purpose of calculating the properties of the cross section, assume that the cross section consists of three rectangles, as shown in Fig. 5-16b.

(a) Determine the maximum tensile and compressive stresses in the beam due to the uniform load.
(b) Find the maximum permissible value of uniform load $q$ (in kN/m) if allowable stresses in tension and compression are $\sigma_{aT}$ = 110 MPa and $\sigma_{aC}$ = 92 MPa, respectively.

### Fig. 5-15

Example 5-4: Stresses in a beam with an overhang

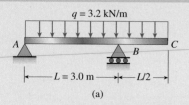

(a)

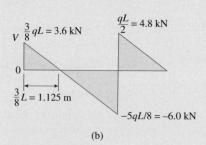

(b)

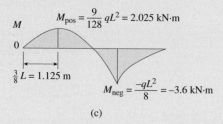

(c)

### Solution

(a) *Maximum tensile and compressive stresses. Reactions, shear forces,* and *bending moments* are computed in the analysis of this beam. First, we find the reactions at supports *A* and *B* using the techniques described in Chapter 4. The results are

$$R_A = \frac{3}{8}qL = 3.6 \text{ kN} \qquad R_B = \frac{9}{8}qL = 10.8 \text{ kN}$$

From these values, we construct the shear-force diagram (Fig. 5-15b). Note that the shear force changes sign and is equal to zero at two locations: (1) at a distance of 1.125 m from the left-hand support, and (2) at the right-hand reaction.

Next, we draw the bending-moment diagram shown in Fig. 5-15c. Both the maximum positive and maximum negative bending moments occur at the cross sections where the shear force changes sign. These maximum moments are

$$M_{pos} = \frac{9}{128} qL^2 = 2.025 \text{ kN} \cdot \text{m} \qquad M_{neg} = \frac{-qL^2}{8} = -3.6 \text{ kN} \cdot \text{m}$$

respectively.

*Neutral axis of the cross section* (Fig. 5-16b). The origin $O$ of the $yz$ coordinates is placed at the centroid of the cross-sectional area, and therefore, the $z$ axis becomes the neutral axis of the cross section. The centroid is located by using the techniques described in Chapter 12, Section 12.3, as follows.

## Fig. 5-16

Cross section of beam discussed in Example 5-4: (a) actual shape, and (b) idealized shape for use in analysis (the thickness of the beam is exaggerated for clarity.)

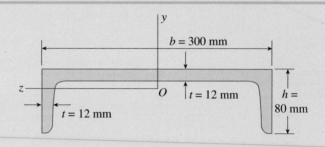

(a)

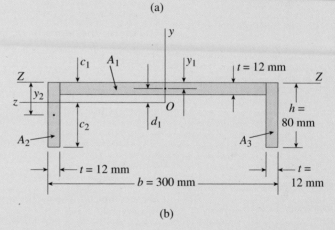

(b)

First, we divide the area into three rectangles ($A_1$, $A_2$, and $A_3$). Second, we establish a reference axis $Z$–$Z$ across the upper edge of the cross section, and we let $y_1$ and $y_2$ be the distances from the $Z$–$Z$ axis to the centroids of areas $A_1$ and $A_2$, respectively. Then the calculations for locating the centroid of the entire channel section (distances $c_1$ and $c_2$) are

Area 1:  $y_1 = t/2 = 6 \text{ mm}$
         $A_1 = (b - 2t)(t) = (276 \text{ mm})(12 \text{ mm}) = 3312 \text{ mm}^2$

Area 2:  $y_2 = h/2 = 40 \text{ mm}$
         $A_2 = ht = (80 \text{ mm})(12 \text{ mm}) = 960 \text{ mm}^2$

Area 3:  $y_3 = y_2$    $A_3 = A_2$

*Continues* ⮕

**Example 5-4** - *Continued*

$$c_1 = \frac{\Sigma y_i A_i}{\Sigma A_i} = \frac{y_1 A_1 + 2y_2 A_2}{A_1 + 2A_2}$$

$$= \frac{(6 \text{ mm})(3312 \text{ mm}^2) + 2(40 \text{ mm})(960 \text{ mm}^2)}{3312 \text{ mm}^2 + 2(960 \text{ mm}^2)} = 18.48 \text{ mm}$$

$$c_2 = h - c_1 = 80 \text{ mm} - 18.48 \text{ mm} = 61.52 \text{ mm}$$

Thus, the position of the neutral axis (the $z$ axis) is determined.

*Moment of inertia.* In order to calculate the stresses from the flexure formula, we must determine the moment of inertia of the cross-sectional area with respect to the neutral axis. These calculations require the use of the parallel axis theorem (see Chapter 12, Section 12.5).

Beginning with area $A_1$, we obtain its moment of inertia $(I_z)_1$ about the $z$ axis from the equation

$$(I_z)_1 = (I_c)_1 + A_1 d_1^2 \tag{a}$$

In this equation, $(I_c)_1$ is the moment of inertia of area $A_1$ about its own centroidal axis:

$$(I_c)_1 = \frac{1}{12}(b - 2t)(t)^3 = \frac{1}{12}(276 \text{ mm})(12 \text{ mm})^3 = 39{,}744 \text{ mm}^4$$

and $d_1$ is the distance from the centroidal axis of area $A_1$ to the $z$ axis:

$$d_1 = c_1 - t/2 = 18.48 \text{ mm} - 6 \text{ mm} = 12.48 \text{ mm}$$

Therefore, the moment of inertia of area $A_1$ about the $z$ axis [from Eq. (a)] is

$$(I_z)_1 = 39{,}744 \text{ mm}^4 + (3312 \text{ mm}^2)(12.48 \text{ mm})^2 = 555{,}600 \text{ mm}^4$$

Proceeding in the same manner for areas $A_2$ and $A_3$, we get

$$(I_z)_2 = (I_z)_3 = 956{,}600 \text{ mm}^4$$

Thus, the centroidal moment of inertia $I_z$ of the entire cross-sectional area is

$$I_z = (I_z)_1 + (I_z)_2 + (I_z)_3 = 2.469 \times 10^6 \text{ mm}^4$$

*Section moduli.* The section moduli for the top and bottom of the beam, respectively, are

$$S_1 = \frac{I_z}{c_1} = 133{,}600 \text{ mm}^3 \quad S_2 = \frac{I_z}{c_2} = 40{,}100 \text{ mm}^3$$

(see Eqs. 5-16a and b). With the cross-sectional properties determined, we can now proceed to calculate the maximum stresses from Eqs. (5-15a and b).

*Maximum stresses.* At the cross section of maximum positive bending moment, the largest tensile stress occurs at the bottom of the beam ($\sigma_2$) and the largest compressive stress occurs at the top ($\sigma_1$). Thus, from Eqs. (5-15b) and (5-15a), respectively, we get

$$\sigma_t = \sigma_2 = \frac{M_{pos}}{S_2} = \frac{2 \cdot 025 \text{ kN} \cdot \text{m}}{40{,}100 \text{ mm}^3} = 50.5 \text{ MPa}$$

$$\sigma_c = \sigma_1 = -\frac{M_{pos}}{S_1} = -\frac{2.025 \text{ kN} \cdot \text{m}}{133{,}600 \text{ mm}^3} = -15.2 \text{ MPa}$$

Similarly, the largest stresses at the section of maximum negative moment are

$$\sigma_t = \sigma_1 = -\frac{M_{neg}}{S_1} = -\frac{-3.6 \text{ kN} \cdot \text{m}}{133,600 \text{ mm}^3} = 26.9 \text{ MPa}$$

$$\sigma_c = \sigma_2 = \frac{M_{neg}}{S_2} = \frac{-3.6 \text{ kN} \cdot \text{m}}{40,100 \text{ mm}^3} = -89.8 \text{ MPa}$$

A comparison of these four stresses shows that the largest tensile stress in the beam is 50.5 MPa and occurs at the bottom of the beam at the cross section of maximum positive bending moment; thus,

$$(\sigma_t)_{max} = 50.5 \text{ MPa} \qquad \longleftarrow$$

The largest compressive stress is −89.8 MPa and occurs at the bottom of the beam at the section of maximum negative moment:

$$(\sigma_c)_{max} = -89.8 \text{ MPa} \qquad \longleftarrow$$

Thus, we have determined the maximum bending stresses due to the uniform load acting on the beam.

(b) *Maximum permissible value of uniform load q.* Next, we want to find $q_{max}$ based on the given allowable normal stresses, which are different for tension and compression. The allowable compression stress $\sigma_{aC}$ is lower than that for tension, $\sigma_{aT}$, to account for the possibility of local buckling of the flanges of the C shape (if they are in compression).

We will use the flexure formula to compute potential values of $q_{max}$ at four locations: at the top and bottom of the beam at the location of the maximum positive moment ($M_{pos}$), and at the top and bottom of the beam at the location of the maximum negative moment ($M_{neg}$). In each case, we must be sure to use the proper value of allowable stress. We assume that the C shape is used in the orientation shown in Fig. 5-16 (i.e., flanges downward), so at the location of $M_{pos}$, we note the top of the beam is in compression and the bottom is in tension, while the opposite is true at point B. Using the expressions for $M_{pos}$ and $M_{neg}$ and equating each to the appropriate product of allowable stress and section modulus, we can solve for possible values of $q_{max}$ as given here.

In beam segment AB at the *top* of beam,

$$M_{pos} = \frac{9}{128} q_1 L^2 = \sigma_{aC} S_1 \quad \text{so} \quad q_1 = \frac{128}{9L^2}(\sigma_{aC} S_1) = 19.42 \text{ kN/m}$$

In beam segment AB at the *bottom* of beam,

$$M_{pos} = \frac{9}{128} q_2 L^2 = \sigma_{aT} S_2 \quad \text{so} \quad q_2 = \frac{128}{9L^2}(\sigma_{aT} S_2) = 6.97 \text{ kN/m}$$

At joint B at the *top* of beam,

$$M_{pos} = \frac{1}{8} q_3 L^2 = \sigma_{aT} S_1 \quad \text{so} \quad q_3 = \frac{8}{L^2}(\sigma_{aT} S_1) = 13.06 \text{ kN/m}$$

At joint B at *bottom* of the beam,

$$M_{pos} = \frac{1}{8} q_4 L^2 = \sigma_{aC} S_2 \quad \text{so} \quad q_4 = \frac{8}{L^2}(\sigma_{aC} S_2) = 3.28 \text{ kN/m}$$

From these calculations, we see that the bottom of the beam near joint B (where the flange tips are in compression) does indeed control the maximum permissible value of uniform load q. Hence,

$$q_{max} = 3.28 \text{ kN/m} \qquad \longleftarrow$$

# 5.6 DESIGN OF BEAMS FOR BENDING STRESSES

The process of designing a beam requires that many factors be considered, including the type of structure (airplane, automobile, bridge, building, or whatever), the materials to be used, the loads to be supported, the environmental conditions to be encountered, and the costs to be paid. However, from the standpoint of strength, the task eventually reduces to selecting a shape and size of beam such that the actual stresses in the beam do not exceed the allowable stresses for the material. In this section, we will consider only the bending stresses (that is, the stresses obtained from the flexure formula, Eq. 5-14). Later, we will consider the effects of shear stresses (Sections 5.8, 5.9, and 5.10) and stress concentrations (Section 5.13).

When designing a beam to resist bending stresses, we usually begin by calculating the **required section modulus**. For instance, if the beam has a doubly symmetric cross section and the allowable stresses are the same for both tension and compression, we can calculate the required modulus by dividing the maximum bending moment by the allowable bending stress for the material (see Eq. 5-17):

$$S = \frac{M_{\max}}{\sigma_{\text{allow}}} \qquad\qquad \textbf{(5-25)}$$

The allowable stress is based upon the properties of the material and the desired factor of safety. To ensure that this stress is not exceeded, we must choose a beam that provides a section modulus at least as large as that obtained from Eq. (5-25).

If the cross section is not doubly symmetric, or if the allowable stresses are different for tension and compression, we usually need to determine two required section moduli—one based upon tension and the other based upon compression. Then we must provide a beam that satisfies both criteria.

To minimize weight and save material, we usually select a beam that has the least cross-sectional area while still providing the required section moduli (and also meeting any other design requirements that may be imposed).

Beams are constructed in a great variety of shapes and sizes to suit a myriad of purposes. For instance, very large steel beams are fabricated by welding (Fig. 5-17), aluminum beams are extruded as round or rectangular tubes, wood beams are cut and glued to fit special requirements, and reinforced concrete beams are cast in any desired shape by proper construction of the forms.

In addition, beams of steel, aluminum, plastic, and wood can be ordered in **standard shapes and sizes** from catalogs supplied by dealers and manufacturers. Readily available shapes include wide-flange beams, I-beams, angles, channels, rectangular beams, and tubes.

### Fig. 5-17

Welder fabricating a large wide flange steel beam (Courtesy of AISC)

# Beams of Standardized Shapes and Sizes

The dimensions and properties of many kinds of beams are listed in engineering handbooks. For instance, in the United States the shapes and sizes of structural-steel beams are standardized by the American Institute of Steel Construction (AISC), which publishes manuals giving their properties in both USCS and SI units (Ref. 5-4). The tables in these manuals list cross-sectional dimensions and properties such as weight, cross-sectional area, moment of inertia, and section modulus.

Properties of aluminum and wood beams are tabulated in a similar manner and are available in publications of the Aluminum Association (Ref. 5-5) and the American Forest and Paper Association (Ref. 5-6).

Abridged tables of steel beams and wood beams are given later in this book for use in solving problems using both USCS and SI units (see Appendixes F and G).

**Structural-steel sections** are given a designation such as W 30 $\times$ 211 in USCS units, which means that the section is of W shape (also called a wide-flange shape) with a nominal depth of 30 in. and a weight of 211 lb per ft of length (see Table F-1(a), Appendix F). The corresponding properties for each W shape are also given in SI units in Table F-1(b). For example, in SI units, the W 30 $\times$ 211 is listed as W 760 $\times$ 314 with a nominal depth of 760 millimeters and mass of 314 kilograms per meter of length.

Similar designations are used for S shapes (also called I-beams) and C shapes (also called channels), as shown in Tables F-2(a) and F-3(a) in USCS units and in Tables F-2(b) and F-3(b) in SI units. Angle sections, or L shapes, are designated by the lengths of the two legs and the thickness (see Tables F-4 and F-5). For example, L 8 $\times$ 6 $\times$ 1 [see Table F-5(a)] denotes an angle with unequal legs, one of length 8 in. and the other of length 6 in., with a thickness of 1 in. The corresponding label in SI units for this unequal leg angle is L 203 $\times$ 152 $\times$ 25.4 [see Table F-5(b)].

The standardized steel sections described above are manufactured by *rolling*, a process in which a billet of hot steel is passed back and forth between rolls until it is formed into the desired shape.

**Aluminum structural sections** are usually made by the process of *extrusion*, in which a hot billet is pushed, or extruded, through a shaped die. Since dies are relatively easy to make and the material is workable, aluminum beams can be extruded in almost any desired shape. Standard shapes of wide-flange beams, I-beams, channels, angles, tubes, and other sections are listed in the *Aluminum Design Manual* (Ref. 5-5). In addition, custom-made shapes can be ordered.

Most **wood beams** have rectangular cross sections and are designated by nominal dimensions, such as 4 $\times$ 8 inches. These dimensions represent the rough-cut size of the lumber. The net dimensions (or actual dimensions) of a wood beam are smaller than the nominal dimensions if the sides of the rough lumber have been planed, or *surfaced*, to make them smooth. Thus, a 4 $\times$ 8 wood beam has actual dimensions 3.5 $\times$ 7.25 in. after it has been surfaced. Of course, the net dimensions of surfaced lumber should be used in all engineering computations. Therefore, net dimensions and the corresponding properties (in USCS units) are given in Appendix G. Similar tables are available in SI units.

## Relative Efficiency of Various Beam Shapes

One of the objectives in designing a beam is to use the material as efficiently as possible within the constraints imposed by function, appearance, manufacturing costs, and the like. From the standpoint of strength alone, efficiency in bending depends primarily upon the shape of the cross section. In particular, the most efficient beam is one in which the material is located as far as practical from the neutral axis. The farther a given amount of material is from the neutral axis, the larger the section modulus becomes—and the larger the section modulus, the larger the bending moment that can be resisted (for a given allowable stress).

As an illustration, consider a cross section in the form of a **rectangle** of width $b$ and height $h$ (Fig. 5-18a). The section modulus (from Eq. 5-19b) is

$$S = \frac{bh^2}{6} = \frac{Ah}{6} = 0.167Ah \qquad \textbf{(5-26)}$$

where $A$ denotes the cross-sectional area. This equation shows that a rectangular cross section of given area becomes more efficient as the height $h$ is increased (and the width $b$ is decreased to keep the area constant). Of course, there is a practical limit to the increase in height, because the beam becomes laterally unstable when the ratio of height to width becomes too large. Thus, a beam of very narrow rectangular section will fail due to lateral (sideways) buckling rather than to insufficient strength of the material.

Next, let us compare a **solid circular cross section** of diameter $d$ (Fig. 5-18b) with a square cross section of the same area. The side $h$ of a square having the same area as the circle is $h = (d/2)\sqrt{\pi}$. The corresponding section moduli (from Eqs. 5-19b and 5-20b) are

$$S_{\text{square}} = \frac{h^3}{6} = \frac{\pi\sqrt{\pi}d^3}{48} = 0.1160d^3 \qquad \textbf{(5-27a)}$$

$$S_{\text{circle}} = \frac{\pi d^3}{32} = 0.0982d^3 \qquad \textbf{(5-27b)}$$

---

| **Fig. 5-18** | |
|---|---|
| Cross-sectional shapes of beams | |

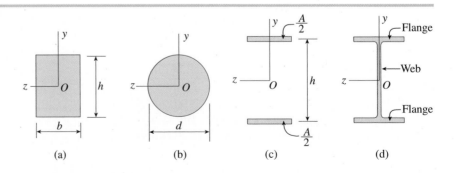

from which we get

$$\frac{S_{\text{square}}}{S_{\text{circle}}} = 1.18 \qquad\qquad \textbf{(5-28)}$$

This result shows that a beam of square cross section is more efficient in resisting bending than is a circular beam of the same area. The reason, of course, is that a circle has a relatively larger amount of material located near the neutral axis. This material is less highly stressed, and therefore it does not contribute as much to the strength of the beam.

The **ideal cross-sectional shape** for a beam of given cross-sectional area $A$ and height $h$ would be obtained by placing one-half of the area at a distance $h/2$ above the neutral axis and the other half at distance $h/2$ below the neutral axis, as shown in Fig. 5-18c. For this ideal shape, we obtain

$$I = 2\left(\frac{A}{2}\right)\left(\frac{h}{2}\right)^2 = \frac{Ah^2}{4} \qquad S = \frac{1}{h/2} = 0.5Ah \quad \textbf{(5-29a,b)}$$

These theoretical limits are approached in practice by wide-flange sections and I-sections, which have most of their material in the flanges (Fig. 5-18d). For standard wide-flange beams, the section modulus is approximately

$$S \approx 0.35Ah \qquad\qquad \textbf{(5-30)}$$

which is less than the ideal but much larger than the section modulus for a rectangular cross section of the same area and height (see Eq. 5-26).

Another desirable feature of a wide-flange beam is its greater width, and hence greater stability with respect to sideways buckling, when compared to a rectangular beam of the same height and section modulus. On the other hand, there are practical limits to how thin we can make the web of a wide-flange beam. If the web is too thin, it will be susceptible to localized buckling or it may be overstressed in shear, a topic that is discussed in Section 5.10.

The following four examples illustrate the process of selecting a beam on the basis of the allowable stresses. In these examples, only the effects of bending stresses (obtained from the flexure formula) are considered.

*Note:* When solving examples and problems that require the selection of a steel or wood beam from the tables in the appendix, we use the following rule: *If several choices are available in a table, select the lightest beam that will provide the required section modulus.*

## Example 5-5

A simply supported wood beam having a span length $L = 12$ ft carries a uniform load $q = 420$ lb/ft (Fig. 5-19). The allowable bending stress is 1800 psi, the wood weighs 35 lb/ft$^3$, and the beam is supported laterally against sideways buckling and tipping.

Select a suitable size for the beam from the table in Appendix G.

### Fig. 5-19

Example 5-5: Design of a simply supported wood beam

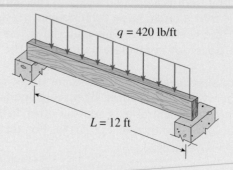

$q = 420$ lb/ft

$L = 12$ ft

### Solution

Since we do not know in advance how much the beam weighs, we will proceed by trial-and-error as follows: (1) Calculate the required section modulus based upon the given uniform load. (2) Select a trial size for the beam. (3) Add the weight of the beam to the uniform load and calculate a new required section modulus. (4) Check to see that the selected beam is still satisfactory. If it is not, select a larger beam and repeat the process.

(1) The maximum bending moment in the beam occurs at the midpoint (see Eq. 4-15):

$$M_{max} = \frac{qL^2}{8} = \frac{(420 \text{ lb/ft})(12 \text{ ft})^2(12 \text{ in./ft})}{8} = 90,720 \text{ lb-in.}$$

The required section modulus (Eq. 5-25) is

$$S = \frac{M_{max}}{\sigma_{allow}} = \frac{90,720 \text{ lb-in.}}{1800 \text{ psi}} = 50.40 \text{ in.}^3$$

(2) From the table in Appendix G we see that the lightest beam that supplies a section modulus of at least 50.40 in.$^3$ about axis 1–1 is a 3 × 12 in. beam (nominal dimensions). This beam has a section modulus equal to 52.73 in.$^3$ and weighs 6.8 lb/ft. (Note that Appendix G gives weights of beams based upon a density of 35 lb/ft$^3$.)

(3) The uniform load on the beam now becomes 426.8 lb/ft, and the corresponding required section modulus is

$$S = (50.40 \text{ in.}^3)\left(\frac{426.8 \text{ lb/ft}}{420 \text{ lb/ft}}\right) = 51.22 \text{ in.}^3$$

(4) The previously selected beam has a section modulus of 52.73 in.$^3$, which is larger than the required modulus of 51.22 in.$^3$

Therefore, a 3 × 12 in. beam is satisfactory. ⬅

*Note:* If the weight density of the wood is other than 35 lb/ft$^3$, we can obtain the weight of the beam per linear foot by multiplying the value in the last column in Appendix G by the ratio of the actual weight density to 35 lb/ft$^3$.

● ● ● **Example 5-6**

A vertical post 2.5-meters high must support a lateral load $P = 12$ kN at its upper end (Fig. 5-20). Two plans are proposed—a solid wood post and a hollow aluminum tube.

(a) What is the minimum required diameter $d_1$ of the wood post if the allowable bending stress in the wood is 15 MPa?

(b) What is the minimum required outer diameter $d_2$ of the aluminum tube if its wall thickness is to be one-eighth of the outer diameter and the allowable bending stress in the aluminum is 50 MPa?

**Fig. 5-20**

Example 5-6: (a) Solid wood post, and (b) aluminum tube

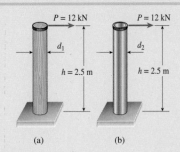

**Solution**

*Maximum bending moment.* The maximum moment occurs at the base of the post and is equal to the load $P$ times the height $h$; thus,

$$M_{max} = Ph = (12 \text{ kN})(2.5 \text{ m}) = 30 \text{ kN} \cdot \text{m}$$

(a) *Wood post.* The required section modulus $S_1$ for the wood post (see Eqs. 5-20b and 5-25) is

$$S_1 = \frac{\pi d_1^3}{32} = \frac{M_{max}}{\sigma_{allow}} = \frac{30 \text{ kN} \cdot \text{m}}{15 \text{ MPa}} = 0.0020 \text{ m}^3 = 2 \times 10^6 \text{ mm}^3$$

Solving for the diameter, we get

$$d_1 = 273 \text{ mm}$$

The diameter selected for the wood post must be equal to or larger than 273 mm if the allowable stress is not to be exceeded.

(b) *Aluminum tube.* To determine the section modulus $S_2$ for the tube, we first must find the moment of inertia $I_2$ of the cross section. The wall thickness of the tube is $d_2/8$, and therefore the inner diameter is $d_2 - d_2/4$, or $0.75d_2$. Thus, the moment of inertia (see Eq. 5-20a) is

$$I_2 = \frac{\pi}{64} [d_2^4 - (0.75d_2)^4] = 0.03356d_2^4$$

The section modulus of the tube is now obtained from Eq. (5-18) as

$$S_2 = \frac{I_2}{c} = \frac{0.03356d_2^4}{d_2/d} = 0.06712d_2^3$$

The required section modulus is obtained from Eq. (5-25):

$$S_2 = \frac{M_{max}}{\sigma_{allow}} = \frac{30 \text{ kN} \cdot \text{m}}{50 \text{ MPa}} = 0.0006 \text{ m}^3 = 600 \times 10^3 \text{ mm}^3$$

By equating the two preceding expressions for the section modulus, we can solve for the required outer diameter:

$$d_2 = \left( \frac{600 \times 10^3 \text{ mm}^3}{0.06712} \right)^{1/3} = 208 \text{ mm}$$

The corresponding inner diameter is 0.75(208 mm), or 156 mm.

## Example 5-7

A simple beam $AB$ of span length 21 ft must support a uniform load $q = 2000$ lb/ft distributed along the beam in the manner shown in Fig. 5-21a.

Considering both the uniform load and the weight of the beam, and also using an allowable bending stress of 18,000 psi, select a structural steel beam of wide-flange shape to support the loads.

## Fig. 5-21

Example 5-7: Design of a simple beam with partial uniform loads

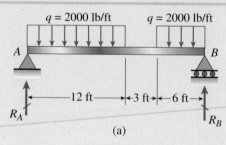

(a)

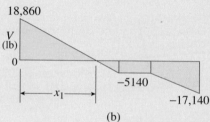

(b)

### Solution

In this example, we will proceed as follows: (1) Find the maximum bending moment in the beam due to the uniform load. (2) Knowing the maximum moment, find the required section modulus. (3) Select a trial wide-flange beam from Table F-1 in Appendix F and obtain the weight of the beam. (4) With the weight known, calculate a new value of the bending moment and a new value of the section modulus. (5) Determine whether the selected beam is still satisfactory. If it is not, select a new beam size and repeat the process until a satisfactory size of beam has been found.

*Maximum bending moment.* To assist in locating the cross section of maximum bending moment, we construct the shear-force diagram (Fig. 5-21b) using the methods described in Chapter 4. As part of that process, we determine the reactions at the supports:

$$R_A = 18,860 \text{ lb} \qquad R_B = 17,140 \text{ lb}$$

The distance $x_1$ from the left-hand support to the cross section of zero shear force is obtained from the equation

$$V = R_A - qx_1 = 0$$

which is valid in the range $0 \leq x \leq 12$ ft. Solving for $x_1$, we get

$$x_1 = \frac{R_A}{q} = \frac{18,860 \text{ lb}}{2000 \text{ lb/ft}} = 9.430 \text{ ft}$$

which is less than 12 ft, and therefore the calculation is valid.

The maximum bending moment occurs at the cross section where the shear force is zero; therefore,

$$M_{max} = R_A x_1 - \frac{q x_1^2}{2} = 88,920 \text{ lb-ft}$$

*Required section modulus.* The required section modulus (based only upon the load $q$) is obtained from Eq. (5-25):

$$S = \frac{M_{max}}{\sigma_{allow}} = \frac{(88,920 \text{ lb-ft})(12 \text{ in./ft})}{18,000 \text{ psi}} = 59.3 \text{ in.}^3$$

*Trial beam.* We now turn to Table F-1 and select the lightest wide-flange beam having a section modulus greater than 59.3 in.[3] The lightest beam that provides this section modulus is W 12 × 50 with $S = 64.7$ in.[3] This beam weighs 50 lb/ft. (Recall that the tables in Appendix F are abridged, and therefore a lighter beam may actually be available.)

We now recalculate the reactions, maximum bending moment, and required section modulus with the beam loaded by both the uniform load $q$ and its own weight. Under these combined loads the reactions are

$$R_A = 19,380 \text{ lb} \qquad R_B = 17,670 \text{ lb}$$

and the distance to the cross section of zero shear becomes

$$x_1 = \frac{19,380 \text{ lb}}{2050 \text{ lb/ft}} = 9.454 \text{ ft}$$

The maximum bending moment increases to 91,610 lb-ft, and the new required section modulus is

$$S = \frac{M_{max}}{\sigma_{allow}} = \frac{(91,610 \text{ lb-ft})(12 \text{ in./ft})}{18,000 \text{ psi}} = 61.1 \text{ in.}^3$$

Thus, we see that the W 12 × 50 beam with section modulus $S = 64.7$ in.[3] is still satisfactory.

*Note*: If the new required section modulus exceeded that of the W 12 × 50 beam, a new beam with a larger section modulus would be selected and the process repeated.

### Fig. 5-22

Example 5-8: Wood dam with horizontal planks A supported by vertical posts B

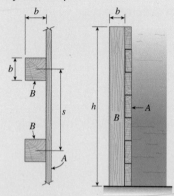

(a) Top view     (b) Side view

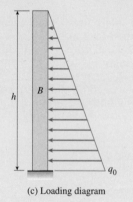

(c) Loading diagram

A temporary wood dam is constructed of horizontal planks *A* supported by vertical wood posts *B* that are sunk into the ground so that they act as cantilever beams (Fig. 5-22). The posts are of square cross section (dimensions $b \times b$) and spaced at distance $s = 0.8$ m, center to center. Assume that the water level behind the dam is at its full height $h = 2.0$ m.

Determine the minimum required dimension *b* of the posts if the allowable bending stress in the wood is $\sigma_{allow} = 8.0$ MPa.

### Solution

*Loading diagram.* Each post is subjected to a triangularly distributed load produced by the water pressure acting against the planks. Consequently, the loading diagram for each post is triangular (Fig. 5-22c). The maximum intensity $q_0$ of the load on the posts is equal to the water pressure at depth $h$ times the spacing $s$ of the posts:

$$q_0 = \gamma h s \tag{a}$$

in which $\gamma$ is the specific weight of water. Note that $q_0$ has units of force per unit distance, $\gamma$ has units of force per unit volume, and both $h$ and $s$ have units of length.

*Section modulus.* Since each post is a cantilever beam, the maximum bending moment occurs at the base and is given by the following expression:

$$M_{max} = \frac{q_0 h}{2}\left(\frac{h}{3}\right) = \frac{\gamma h^3 s}{6} \tag{b}$$

Therefore, the required section modulus (Eq. 5-25) is

$$S = \frac{M_{max}}{\sigma_{allow}} = \frac{\gamma h^3 s}{6\sigma_{allow}} \tag{c}$$

For a beam of square cross section, the section modulus is $S = b^3/6$ (see Eq. 5-19b). Substituting this expression for $S$ into Eq. (c), we get a formula for the cube of the minimum dimension *b* of the posts:

$$b^3 = \frac{\gamma h^3 s}{\sigma_{allow}} \qquad \Longleftarrow \quad (d)$$

*Numerical values.* We now substitute numerical values into Eq. (d) and obtain

$$b^3 = \frac{(9.81 \text{ kN/m}^3)(2.0 \text{ m})^3(0.8 \text{ m})}{8.0 \text{ MPa}} = 0.007848 \text{ m}^3 = 7.848 \times 10^6 \text{ mm}^3$$

from which

$$b = 199 \text{ mm} \qquad \Longleftarrow$$

Thus, the minimum required dimension *b* of the posts is 199 mm. Any larger dimension, such as 200 mm, will ensure that the actual bending stress is less than the allowable stress.

# 5.7 NONPRISMATIC BEAMS

The beam theories described in this chapter were derived for prismatic beams, that is, straight beams having the same cross sections throughout their lengths. However, nonprismatic beams are commonly used to reduce weight and improve appearance. Such beams are found in automobiles, airplanes, machinery, bridges, buildings, tools, and many other applications (Fig. 5-23). Fortunately, the flexure formula (Eq. 5-13) gives reasonably accurate values for the bending stresses in nonprismatic beams whenever the changes in cross-sectional dimensions are gradual, as in the examples shown in Fig. 5-23.

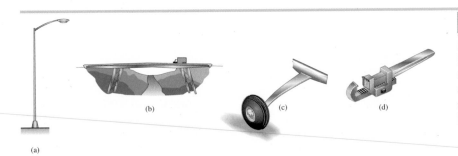

**Fig. 5-23**

Examples of nonprismatic beams: (a) street lamp, (b) bridge with tapered girders and piers, (c) wheel strut of a small airplane, and (d) wrench handle

The manner in which the bending stresses vary along the axis of a nonprismatic beam is not the same as for a prismatic beam. In a prismatic beam the section modulus $S$ is constant, and therefore the stresses vary in direct proportion to the bending moment (because $\sigma = M/S$). However, in a nonprismatic beam the section modulus also varies along the axis. Consequently, we cannot assume that the maximum stresses occur at the cross section with the largest bending moment—sometimes the maximum stresses occur elsewhere, as illustrated in Example 5-9.

## Fully Stressed Beams

To minimize the amount of material and thereby have the lightest possible beam, we can vary the dimensions of the cross sections so as to have the maximum allowable bending stress at every section. A beam in this condition is called a **fully stressed beam**, or a *beam of constant strength*.

Of course, these ideal conditions are seldom attained because of practical problems in constructing the beam and the possibility of the loads being different from those assumed in design. Nevertheless, knowing the properties of a fully stressed beam can be an important aid to the engineer when designing structures for minimum weight. Familiar examples of structures designed to maintain nearly constant maximum stress are leaf springs in automobiles, bridge girders that are tapered, and some of the structures shown in Fig. 5-23.

The determination of the shape of a fully stressed beam is illustrated in Example 5-10.

• • •    **Example 5-9**

A tapered cantilever beam *AB* of solid circular cross section supports a load *P* at the free end (Fig. 5-24). The diameter $d_B$ at the large end is twice the diameter $d_A$ at the small end:

$$\frac{d_B}{d_A} = 2$$

Determine the bending stress $\sigma_B$ at the fixed support and the maximum bending stress $\sigma_{max}$.

**Fig. 5-24**

Example 5-9: Tapered cantilever beam of circular cross section

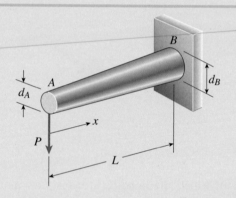

**Solution**

If the angle of taper of the beam is small, the bending stresses obtained from the flexure formula will differ only slightly from the exact values. As a guideline concerning accuracy, we note that if the angle between line *AB* (Fig. 5-24) and the longitudinal axis of the beam is about 20°, the error in calculating the normal stresses from the flexure formula is about 10%. Of course, as the angle of taper decreases, the error becomes smaller.

*Section modulus.* The section modulus at any cross section of the beam can be expressed as a function of the distance *x* measured along the axis of the beam. Since the section modulus depends upon the diameter, we first must express the diameter in terms of *x*, as

$$d_x = d_A + (d_B - d_A)\frac{x}{L} \tag{5-31}$$

in which $d_x$ is the diameter at distance *x* from the free end. Therefore, the section modulus at distance *x* from the end (Eq. 5-20b) is

$$S_x = \frac{\pi d_x^3}{32} = \frac{\pi}{32}\left[d_A + (d_B - d_A)\frac{x}{L}\right]^3 \tag{5-32}$$

*Bending stresses.* Since the bending moment equals $Px$, the maximum normal stress at any cross section is given by the equation

$$\sigma_1 = \frac{M_x}{S_x} = \frac{32Px}{\pi[d_A + (d_B - d_A)(x/L)]^3} \qquad \textbf{(5-33)}$$

We can see by inspection of the beam that the stress $\sigma_1$ is tensile at the top of the beam and compressive at the bottom.

Note that Eqs. (5-31), (5-32), and (5-33) are valid for any values of $d_A$ and $d_B$, provided the angle of taper is small. In the following discussion, we consider only the case where $d_B = 2d_A$.

*Maximum stress at the fixed support.* The maximum stress at the section of largest bending moment (end $B$ of the beam) can be found from Eq. (5-33) by substituting $x = L$ and $d_B = 2d_A$; the result is

$$\sigma_B = \frac{4PL}{\pi d_A^3} \qquad \Longleftarrow \textbf{(a)}$$

*Maximum stress in the beam.* The maximum stress at a cross section at distance $x$ from the end (Eq. 5-33) for the case where $d_B = 2d_A$ is

$$\sigma_1 = \frac{32Px}{\pi d_A^3(1 + x/L)^3} \qquad \textbf{(b)}$$

To determine the location of the cross section having the largest bending stress in the beam, we need to find the value of $x$ that makes $\sigma_1$ a maximum. Taking the derivative $d\sigma_1/dx$ and equating it to zero, we can solve for the value of $x$ that makes $\sigma_1$ a maximum; the result is

$$x = \frac{L}{2} \qquad \textbf{(c)}$$

The corresponding maximum stress, obtained by substituting $x = L/2$ into Eq. (b), is

$$\sigma_{max} = \frac{128PL}{27\pi d_A^3} = \frac{4.741PL}{\pi d_A^3} \qquad \Longleftarrow \textbf{(d)}$$

In this particular example, the maximum stress occurs at the midpoint of the beam and is 19% greater than the stress $\sigma_B$ at the built-in end.

*Note:* If the taper of the beam is reduced, the cross section of maximum normal stress moves from the midpoint toward the fixed support. For small angles of taper, the maximum stress occurs at end $B$.

## • • • Example 5-10

A cantilever beam *AB* of length *L* is being designed to support a concentrated load *P* at the free end (Fig. 5-25). The cross sections of the beam are rectangular with constant width *b* and varying height *h*. To assist them in designing this beam, the designers would like to know how the height of an idealized beam should vary in order that the maximum normal stress at every cross section will be equal to the allowable stress $\sigma_{allow}$.

Considering only the bending stresses obtained from the flexure formula, determine the height of the fully stressed beam.

### Fig. 5-25

Example 5-10: Fully stressed beam having constant maximum normal stress (theoretical shape with shear stresses disregarded)

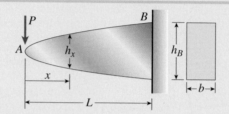

### Solution

The bending moment and section modulus at distance *x* from the free end of the beam are

$$M = Px \qquad S = \frac{bh_x^2}{6}$$

where $h_x$ is the height of the beam at distance *x*. Substituting in the flexure formula, we obtain

$$\sigma_{allow} = \frac{M}{S} = \frac{Px}{bh_x^2/6} = \frac{6Px}{bh_x^2} \tag{a}$$

Solving for the height of the beam, we get

$$h_x = \sqrt{\frac{6Px}{b\sigma_{allow}}} \qquad \longleftarrow \text{(b)}$$

At the fixed end of the beam (*x* = *L*), the height $h_B$ is

$$h_B \sqrt{\frac{6PL}{b\sigma_{allow}}} \tag{c}$$

and therefore we can express the height $h_x$ in the following form:

$$h_x = h_B \sqrt{\frac{x}{L}} \qquad \longleftarrow \text{(d)}$$

This last equation shows that the height of the fully stressed beam varies with the square root of *x*. Consequently, the idealized beam has the parabolic shape shown in Fig. 5-25.

*Note:* At the loaded end of the beam (*x* = 0) the theoretical height is zero, because there is no bending moment at that point. Of course, a beam of this shape is not practical because it is incapable of supporting the shear forces near the end of the beam. Nevertheless, the idealized shape can provide a useful starting point for a realistic design in which shear stresses and other effects are considered.

# 5.8 SHEAR STRESSES IN BEAMS OF RECTANGULAR CROSS SECTION

When a beam is in *pure bending*, the only stress resultants are the bending moments and the only stresses are the normal stresses acting on the cross sections. However, most beams are subjected to loads that produce both bending moments and shear forces (*nonuniform bending*). In these cases, both normal and shear stresses are developed in the beam. The normal stresses are calculated from the flexure formula (see Section 5.5), provided the beam is constructed of a linearly elastic material. The shear stresses are discussed in this and the following two sections.

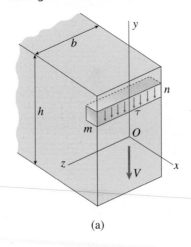

**Fig. 5-26**

Shear stresses in a beam of rectangular cross section

## Vertical and Horizontal Shear Stresses

Consider a beam of rectangular cross section (width $b$ and height $h$) subjected to a positive shear force $V$ (Fig. 5-26a). It is reasonable to assume that the shear stresses $\tau$ acting on the cross section are parallel to the shear force, that is, parallel to the vertical sides of the cross section. It is also reasonable to assume that the shear stresses are uniformly distributed across the width of the beam, although they may vary over the height. Using these two assumptions, we can determine the intensity of the shear stress at any point on the cross section.

For purposes of analysis, we isolate a small element $mn$ of the beam (Fig. 5-26a) by cutting between two adjacent cross sections and between two horizontal planes. According to our assumptions, the shear stresses $\tau$ acting on the front face of this element are vertical and uniformly distributed from one side of the beam to the other. Also, from the discussion of shear stresses in Section 1.7, we know that shear stresses acting on one side of an element are accompanied by shear stresses of equal magnitude acting on perpendicular faces of the element (see Figs. 5-26b and c). Thus, there are horizontal shear stresses acting between horizontal layers of the beam as well as vertical shear stresses acting on the cross sections. At any point in the beam, these complementary shear stresses are equal in magnitude.

The equality of the horizontal and vertical shear stresses acting on an element leads to an important conclusion regarding the shear stresses at the top and bottom of the beam. If we imagine that the element $mn$ (Fig. 5-26a) is located at either the top or the bottom, we see that the horizontal shear stresses must vanish, because there are no stresses on the outer surfaces of the beam. It follows that the vertical shear stresses must also vanish at those locations; in other words, $\tau = 0$ where $y = \pm h/2$.

The existence of horizontal shear stresses in a beam can be demonstrated by a simple experiment. Place two identical rectangular beams on simple supports and load them by a force $P$, as shown in Fig. 5-27a. If friction between the beams is small, the beams will bend independently (Fig. 5-27b). Each beam will be in compression above its own neutral axis and in tension below its neutral axis, and therefore the bottom surface of the upper beam will slide with respect to the top surface of the lower beam.

Now suppose that the two beams are glued along the contact surface, so that they become a single solid beam. When this beam is loaded, horizontal

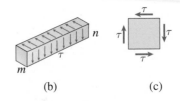

(b)          (c)

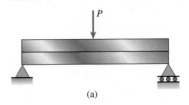

**Fig. 5-27**

Bending of two separate beams

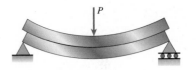

shear stresses must develop along the glued surface in order to prevent the sliding shown in Fig. 5-27b. Because of the presence of these shear stresses, the single solid beam is much stiffer and stronger than the two separate beams.

## Derivation of Shear Formula

We are now ready to derive a formula for the shear stresses $\tau$ in a rectangular beam. However, instead of evaluating the vertical shear stresses acting on a cross section, it is easier to evaluate the horizontal shear stresses acting between layers of the beam. Of course, the vertical shear stresses have the same magnitudes as the horizontal shear stresses.

With this procedure in mind, let us consider a beam in nonuniform bending (Fig. 5-28a). We take two adjacent cross sections $mn$ and $m_1n_1$, distance $dx$ apart, and consider the **element** $mm_1n_1n$. The bending moment and shear force acting on the left-hand face of this element are denoted $M$ and $V$, respectively. Since both the bending moment and shear force may change as we move along the axis of the beam, the corresponding quantities on the right-hand face (Fig. 5-28a) are denoted $M + dM$ and $V + dV$.

Because of the presence of the bending moments and shear forces, the element shown in Fig. 5-28a is subjected to normal and shear stresses on both cross-sectional faces. However, only the normal stresses are needed in the following derivation, and therefore only the normal stresses are shown in Fig. 5-28b. On cross sections $mn$ and $m_1n_1$ the normal stresses are, respectively,

$$\sigma_1 = -\frac{My}{I} \quad \text{and} \quad \sigma_2 = -\frac{(M + dM)y}{I} \qquad \textbf{(5-34a,b)}$$

as given by the flexure formula (Eq. 5-14). In these expressions, $y$ is the distance from the neutral axis and $I$ is the moment of inertia of the cross-sectional area about the neutral axis.

Next, we isolate a **subelement** $mm_1p_1p$ by passing a horizontal plane $pp_1$ through element $mm_1n_1n$ (Fig. 5-28b). The plane $pp_1$ is at distance $y_1$ from the neutral surface of the beam. The subelement is shown separately in Fig. 5-28c. We note that its top face is part of the upper surface of the beam and thus is free from stress. Its bottom face (which is parallel to the neutral surface and distance $y_1$ from it) is acted upon by the horizontal shear stresses $\tau$ existing at this level in the beam. Its cross-sectional faces $mp$ and $m_1p_1$ are acted upon by the bending stresses $\sigma_1$ and $\sigma_2$, respectively, produced by the bending moments. Vertical shear stresses also act on the cross-sectional faces; however, these stresses do not affect the equilibrium of the subelement in the horizontal direction (the $x$ direction), so they are not shown in Fig. 5-28c.

If the bending moments at cross sections $mn$ and $m_1n_1$ (Fig. 5-28b) are equal (that is, if the beam is in pure bending), the normal stresses $\sigma_1$ and $\sigma_2$ acting over the sides $mp$ and $m_1p_1$ of the subelement (Fig. 5-28c) also will be equal. Under these conditions, the subelement will be in equilibrium under the action of the normal stresses alone, and therefore the shear stresses $\tau$ acting on the bottom face $pp_1$ will vanish. This conclusion is obvious inasmuch as a beam in pure bending has no shear force and hence no shear stresses.

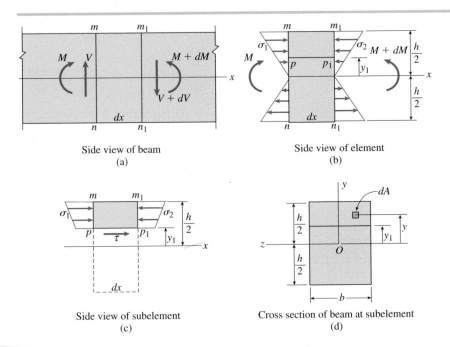

**Fig. 5-28**

Shear stresses in a beam of
rectangular cross section

Side view of beam
(a)

Side view of element
(b)

Side view of subelement
(c)

Cross section of beam at subelement
(d)

If the bending moments vary along the $x$ axis (nonuniform bending), we can determine the shear stress $\tau$ acting on the bottom face of the subelement (Fig. 5-28c) by considering the equilibrium of the subelement in the $x$ direction.

We begin by identifying an element of area $dA$ in the *cross section* at distance $y$ from the neutral axis (Fig. 5-28d). The force acting on this element is $\sigma dA$, in which $\sigma$ is the normal stress obtained from the flexure formula. If the element of area is located on the left-hand face $mp$ of the subelement (where the bending moment is $M$), the normal stress is given by Eq. (5-34a), and therefore the element of force is

$$\sigma_1 dA = \frac{My}{I} dA$$

Note that we are using only absolute values in this equation because the directions of the stresses are obvious from Fig. 5-28. Summing these elements of force over the area of face $mp$ of the subelement (Fig. 5-28c) gives the total horizontal force $F_1$ acting on that face:

$$F_1 = \int \sigma_1 dA = \int \frac{My}{I} dA \qquad \textbf{(5-35a)}$$

**Fig. 5-29**

Partial free-body diagram
of subelement showing all
horizontal forces (compare
with Fig. 5-28c.)

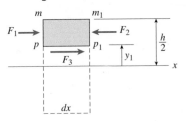

Note that this integration is performed over the area of the shaded part of the cross section shown in Fig. 5-28d, that is, over the area of the cross section from $y = y_1$ to $y = h/2$.

The force $F_1$ is shown in Fig. 5-29 on a partial free-body diagram of the subelement. (Vertical forces have been omitted).

In a similar manner, we find that the total force $F_2$ acting on the right-hand face $m_1p_1$ of the subelement (Fig. 5-29 and Fig. 5-28c) is

$$F_2 = \int \sigma_2 dA = \int \frac{(M + dM)y}{I} dA \qquad \text{(5-35b)}$$

Knowing the forces $F_1$ and $F_2$, we can now determine the horizontal force $F_3$ acting on the bottom face of the subelement.

Since the subelement is in equilibrium, we can sum forces in the $x$ direction and obtain

$$F_3 = F_2 - F_1 \qquad \text{(5-35c)}$$

or

$$F_3 = \int \frac{(M + dM)y}{I} dA - \int \frac{My}{I} dA = \int \frac{(dM)y}{I} dA$$

The quantities $dM$ and $I$ in the last term can be moved outside the integral sign because they are constants at any given cross section and are not involved in the integration. Thus, the expression for the force $F_3$ becomes

$$F_3 = \frac{dM}{I} \int y\, dA \qquad \text{(5-36)}$$

If the shear stresses $\tau$ are uniformly distributed across the width $b$ of the beam, the force $F_3$ is also equal to the following:

$$F_3 = \tau b\, dx \qquad \text{(5-37)}$$

in which $b\, dx$ is the area of the bottom face of the subelement.

Combining Eqs. (5-36) and (5-37) and solving for the shear stress $\tau$, we get

$$\tau = \frac{dM}{dx}\left(\frac{1}{Ib}\right) \int y\, dA \qquad \text{(5-38)}$$

The quantity $dM/dx$ is equal to the shear force $V$ (see Eq. 4-6), and therefore the preceding expression becomes

$$\tau = \frac{V}{Ib} \int y\, dA \qquad \text{(5-39)}$$

The integral in this equation is evaluated over the shaded part of the cross section (Fig. 5-28d), as already explained. Thus, the integral is the first moment of the shaded area with respect to the neutral axis (the $z$ axis). In other words, *the integral is the first moment of the cross-sectional area above the level at which the shear stress $\tau$ is being evaluated.* This first moment is usually denoted by the symbol $Q$:

$$Q = \int y\, dA \qquad \text{(5-40)}$$

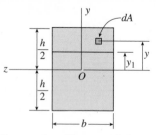

**Fig. 5-28d (Repeated)**

Cross section of beam at subelement
(d)

With this notation, the equation for the shear stress becomes

$$\tau = \frac{VQ}{Ib} \qquad \textbf{(5-41)}$$

This equation, known as the **shear formula**, can be used to determine the shear stress $\tau$ at any point in the cross section of a rectangular beam. Note that for a specific cross section, the shear force $V$, moment of inertia $I$, and width $b$ are constants. However, the first moment $Q$ (and hence the shear stress $\tau$) varies with the distance $y_1$ from the neutral axis.

## Calculation of the First Moment $Q$

If the level at which the shear stress is to be determined is above the neutral axis, as shown in Fig. 5-28d, it is natural to obtain $Q$ by calculating the first moment of the cross-sectional area *above* that level (the shaded area in the figure). However, as an alternative, we could calculate the first moment of the remaining cross-sectional area, that is, the area *below* the shaded area. Its first moment is equal to the negative of $Q$.

The explanation lies in the fact that the first moment of the entire cross-sectional area with respect to the neutral axis is equal to zero (because the neutral axis passes through the centroid). Therefore, the value of $Q$ for the area below the level $y_1$ is the negative of $Q$ for the area above that level. As a matter of convenience, we usually use the area above the level $y_1$ when the point where we are finding the shear stress is in the upper part of the beam, and we use the area below the level $y_1$ when the point is in the lower part of the beam.

Furthermore, we usually don't bother with sign conventions for $V$ and $Q$. Instead, we treat all terms in the shear formula as positive quantities and determine the direction of the shear stresses by inspection, since the stresses act in the same direction as the shear force $V$ itself. This procedure for determining shear stresses is illustrated later in Example 5-11.

## Distribution of Shear Stresses in a Rectangular Beam

We are now ready to determine the distribution of the shear stresses in a beam of rectangular cross section (Fig. 5-30a). The first moment $Q$ of the shaded part of the cross-sectional area is obtained by multiplying the area by the distance from its own centroid to the neutral axis:

$$Q = b\left(\frac{h}{2} - y_1\right)\left(y_1 + \frac{h/2 - y_1}{2}\right) = \frac{b}{2}\left(\frac{h^2}{4} - y_1^2\right) \qquad \textbf{(5-42a)}$$

Of course, this same result can be obtained by integration using Eq. (5-40):

$$Q = \int y \, dA = \int_{y_1}^{h/2} yb \, dy = \frac{b}{2}\left(\frac{h^2}{4} - y_1^2\right) \qquad \textbf{(5-42b)}$$

Substituting the expression for $Q$ into the shear formula (Eq. 5-41), we get

$$\tau = \frac{V}{2I}\left(\frac{h^2}{4} - y_1^2\right) \qquad \textbf{(5-43)}$$

**Fig. 5-30**

Distribution of shear stresses in a beam of rectangular cross section: (a) cross section of beam, and (b) diagram showing the parabolic distribution of shear stresses over the height of the beam

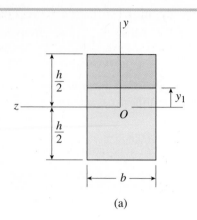

(a)

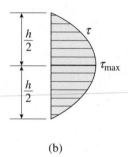

(b)

This equation shows that the shear stresses in a rectangular beam vary quadratically with the distance $y_1$ from the neutral axis. Thus, when plotted along the height of the beam, $\tau$ varies as shown in Fig. 5-30b. Note that the shear stress is zero when $y_1 = \pm h/2$.

The maximum value of the shear stress occurs at the neutral axis ($y_1 = 0$) where the first moment $Q$ has its maximum value. Substituting $y_1 = 0$ into Eq. (5-43), we get

$$\tau_{max} = \frac{Vh^2}{8I} = \frac{3V}{2A} \tag{5-44}$$

in which $A = bh$ is the cross-sectional area. Thus, the maximum shear stress in a beam of rectangular cross section is 50% larger than the average shear stress $V/A$.

Note again that the preceding equations for the shear stresses can be used to calculate either the vertical shear stresses acting on the cross sections or the horizontal shear stresses acting between horizontal layers of the beam.*

## Limitations

The formulas for shear stresses presented in this section are subject to the same restrictions as the flexure formula from which they are derived. Thus, they are valid only for beams of linearly elastic materials with small deflections.

---

*The shear-stress analysis presented in this section was developed by the Russian engineer D. J. Jourawski; see Refs. 5-7 and 5-8.

In the case of rectangular beams, the accuracy of the shear formula depends upon the height-to-width ratio of the cross section. The formula may be considered as exact for very narrow beams (height $h$ much larger than the width $b$). However, it becomes less accurate as $b$ increases relative to $h$. For instance, when the beam is square ($b = h$), the true maximum shear stress is about 13% larger than the value given by Eq. (5-44). (For a more complete discussion of the limitations of the shear formula, see Ref. 5-9.)

A common error is to apply the shear formula (Eq. 5-41) to cross-sectional shapes for which it is not applicable. For instance, it is not applicable to sections of triangular or semicircular shape. To avoid misusing the formula, we must keep in mind the following assumptions that underlie the derivation: (1) The edges of the cross section must be parallel to the $y$ axis (so that the shear stresses act parallel to the $y$ axis), and (2) the shear stresses must be uniform across the width of the cross section. These assumptions are fulfilled only in certain cases, such as those discussed in this and the next two sections.

Finally, the shear formula applies only to prismatic beams. If a beam is nonprismatic (for instance, if the beam is tapered), the shear stresses are quite different from those predicted by the formulas given here (see Refs. 5-9 and 5-10).

## Effects of Shear Strains

Because the shear stress $\tau$ varies parabolically over the height of a rectangular beam, it follows that the shear strain $\gamma = \tau/G$ also varies parabolically. As a result of these shear strains, cross sections of the beam that were originally plane surfaces become warped. This warping is shown in Fig. 5-31, where cross sections $mn$ and $pq$, originally plane, have become curved surfaces $m_1n_1$ and $p_1q_1$, with the maximum shear strain occurring at the neutral surface. At points $m_1$, $p_1$, $n_1$, and $q_1$, the shear strain is zero, and therefore the curves $m_1n_1$ and $p_1q_1$ are perpendicular to the upper and lower surfaces of the beam.

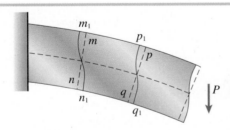

**Fig. 5-31**

Warping of the cross sections of a beam due to shear strains

If the shear force $V$ is constant along the axis of the beam, warping is the same at every cross section. Therefore, stretching and shortening of longitudinal elements due to the bending moments is unaffected by the shear strains, and the distribution of the normal stresses is the same as in pure bending. Moreover, detailed investigations using advanced methods of analysis show that warping of cross sections due to shear strains does not substantially affect the longitudinal strains even when the shear force varies continuously along the length. Thus, under most conditions it is justifiable to use the flexure formula (Eq. 5-14) for nonuniform bending, even though the formula was derived for pure bending.

### Fig. 5-32

Example 5-11: (a) Simple beam with uniform load, (b) cross section of beam, and (c) stress element showing the normal and shear stresses at point C

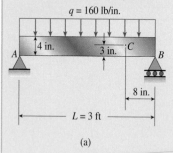

(a)

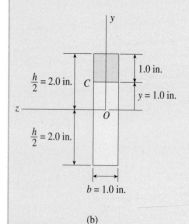

(b)

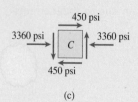

(c)

A metal beam with span $L = 3$ ft is simply supported at points $A$ and $B$ (Fig. 5-32a). The uniform load on the beam (including its own weight) is $q = 160$ lb/in. The cross section of the beam is rectangular (Fig. 5-32b) with width $b = 1$ in. and height $h = 4$ in. The beam is adequately supported against sideways buckling.

Determine the normal stress $\sigma_C$ and shear stress $\tau_C$ at point $C$, which is located 1 in. below the top of the beam and 8 in. from the right-hand support. Show these stresses on a sketch of a stress element at point $C$.

### Solution

*Shear force and bending moment.* The shear force $V_C$ and bending moment $M_C$ at the cross section through point $C$ are found by the methods described in Chapter 4. The results are

$$M_C = 17{,}920 \text{ lb-in.} \qquad V_C = -1600 \text{ lb}$$

The signs of these quantities are based upon the standard sign conventions for bending moments and shear forces (see Fig. 4-5).

*Moment of inertia.* The moment of inertia of the cross-sectional area about the neutral axis (the $z$ axis in Fig. 5-32b) is

$$I = \frac{bh^3}{12} = \frac{1}{12}(1.0 \text{ in.})(4.0 \text{ in.})^3 = 5.333 \text{ in.}^4$$

*Normal stress at point C.* The normal stress at point $C$ is found from the flexure formula (Eq. 5-14) with the distance $y$ from the neutral axis equal to 1.0 in.; thus,

$$\sigma_C = -\frac{My}{I} = -\frac{(17{,}920 \text{ lb-in.})(1.0 \text{ in.})}{5.333 \text{ in.}^4} = -3360 \text{ psi} \quad \Longleftarrow$$

The minus sign indicates that the stress is compressive, as expected.

*Shear stress at point C.* To obtain the shear stress at point $C$, we need to evaluate the first moment $Q_C$ of the cross-sectional area above point $C$ (Fig. 5-32b). This first moment is equal to the product of the area and its centroidal distance (denoted $y_C$) from the $z$ axis; thus,

$$A_C = (1.0 \text{ in.})(1.0 \text{ in.}) = 1.0 \text{ in.}^2 \qquad y_C = 1.5 \text{ in.} \qquad Q_C = A_C y_C = 1.5 \text{ in.}^3$$

Now we substitute numerical values into the shear formula (Eq. 5-41) and obtain the magnitude of the shear stress:

$$\tau_C = \frac{V_C Q_C}{Ib} = \frac{(1600 \text{ lb})(1.5 \text{ in.}^3)}{(5.333 \text{ in.}^4)(1.0 \text{ in.})} = 450 \text{ psi} \quad \Longleftarrow$$

The direction of this stress can be established by inspection, because it acts in the same direction as the shear force. In this example, the shear force acts upward on the part of the beam to the left of point $C$ and downward on the part of the beam to the right of point $C$. The best way to show the directions of both the normal and shear stresses is to draw a stress element, as follows.

*Stress element at point C.* The stress element, shown in Fig. 5-32c, is cut from the side of the beam at point $C$ (Fig. 5-32a). Compressive stresses $\sigma_C = 3360$ psi act on the cross-sectional faces of the element and shear stresses $\tau_C = 450$ psi act on the top and bottom faces as well as the cross-sectional faces.

● ● ● **Example 5-12**

## Fig. 5-33

Example 5-12: Wood beam with concentrated loads

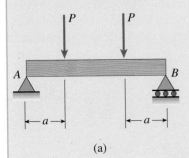

(a)

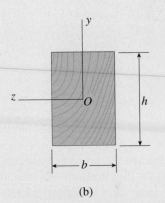

(b)

A wood beam *AB* supporting two concentrated loads *P* (Fig. 5-33a) has a rectangular cross section of width $b = 100$ mm and height $h = 150$ mm (Fig. 5-33b). The distance from each end of the beam to the nearest load is $a = 0.5$ m.

Determine the maximum permissible value $P_{max}$ of the loads if the allowable stress in bending is $\sigma_{allow} = 11$ MPa (for both tension and compression) and the allowable stress in horizontal shear is $\tau_{allow} = 1.2$ MPa. (Disregard the weight of the beam itself.)

*Note:* Wood beams are much weaker in *horizontal shear* (shear parallel to the longitudinal fibers in the wood) than in *cross-grain shear* (shear on the cross sections). Consequently, the allowable stress in horizontal shear is usually considered in design.

### Solution

The maximum shear force occurs at the supports and the maximum bending moment occurs throughout the region between the loads. Their values are

$$V_{max} = P \quad M_{max} = Pa$$

Also, the section modulus *S* and cross-sectional area *A* are

$$S = \frac{bh^2}{6} \quad A = bh$$

The maximum normal and shear stresses in the beam are obtained from the flexure and shear formulas (Eqs. 5-17 and 5-44):

$$\sigma_{max} = \frac{M_{max}}{S} = \frac{6Pa}{bh^2} \quad \tau_{max} = \frac{3V_{max}}{2A} = \frac{3P}{2bh}$$

Therefore, the maximum permissible values of the load *P* in bending and shear, respectively, are

$$P_{bending} = \frac{\sigma_{allow}bh^2}{6a} \quad P_{shear} = \frac{2\tau_{allow}bh}{3}$$

Substituting numerical values into these formulas, we get

$$P_{bending} = \frac{(11 \text{ MPa})(100 \text{ mm})(150 \text{ mm})^2}{6(0.5 \text{ m})} = 8.25 \text{ kN}$$

$$P_{shear} = \frac{2(1.2 \text{ MPa})(100 \text{ mm})(150 \text{ mm})}{3} = 12.0 \text{ kN}$$

Thus, the bending stress governs the design, and the maximum permissible load is

$$P_{max} = 8.25 \text{ kN} \qquad ⬅$$

A more complete analysis of this beam would require that the weight of the beam be taken into account, thus reducing the permissible load.

*Notes:*

(1) In this example, the maximum normal stresses and maximum shear stresses do not occur at the same locations in the beam—the normal stress is maximum in the middle region of the beam at the top and bottom of the cross section, and the shear stress is maximum near the supports at the neutral axis of the cross section.

(2) For most beams, the bending stresses (not the shear stresses) control the allowable load, as in this example.

(3) Although wood is not a homogeneous material and often departs from linearly elastic behavior, we can still obtain approximate results from the flexure and shear formulas. These approximate results are usually adequate for designing wood beams.

# 5.9 SHEAR STRESSES IN BEAMS OF CIRCULAR CROSS SECTION

When a beam has a **circular cross section** (Fig. 5-34), we can no longer assume that the shear stresses act parallel to the $y$ axis. For instance, we can easily prove that at point $m$ (on the boundary of the cross section) the shear stress $\tau$ must act *tangent* to the boundary. This observation follows from the fact that the outer surface of the beam is free of stress, and therefore the shear stress acting on the cross section can have no component in the radial direction.

---

**Fig. 5-34**

Shear stresses acting on the cross section of a circular beam

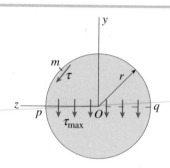

Although there is no simple way to find the shear stresses acting throughout the entire cross section, we can readily determine the shear stresses at the neutral axis (where the stresses are the largest) by making some reasonable assumptions about the stress distribution. We assume that the stresses act parallel to the $y$ axis and have constant intensity across the width of the beam (from point $p$ to point $q$ in Fig. 5-34). Since these assumptions are the same as those used in deriving the shear formula $\tau = VQ/Ib$ (Eq. 5-41), we can use the shear formula to calculate the stresses at the neutral axis.

For use in the shear formula, we need the following properties pertaining to a circular cross section having radius $r$:

$$I = \frac{\pi r^4}{4} \qquad Q = A\bar{y} = \left(\frac{\pi r^2}{2}\right)\left(\frac{4r}{3\pi}\right) = \frac{2r^3}{3} \qquad b = 2r \quad \textbf{(5-45a,b)}$$

The expression for the moment of inertia $I$ is taken from Case 9 of Appendix E, and the expression for the first moment $Q$ is based upon the formulas for a semicircle (Case 10, Appendix E). Substituting these expressions into the shear formula, we obtain

$$\tau_{max} = \frac{VQ}{Ib} = \frac{V(2r^3/3)}{(\pi r^4/4)(2r)} = \frac{4V}{3\pi r^2} = \frac{4V}{3A} \qquad \textbf{(5-46)}$$

in which $A = \pi r^2$ is the area of the cross section. This equation shows that the maximum shear stress in a circular beam is equal to 4/3 times the average vertical shear stress $V/A$.

If a beam has a **hollow circular cross section** (Fig. 5-35), we may again assume with reasonable accuracy that the shear stresses at the neutral axis are parallel to the $y$ axis and uniformly distributed across the section. Consequently, we may again use the shear formula to find the maximum stresses. The required properties for a hollow circular section are

$$I = \frac{\pi}{4}(r_2^4 - r_1^4) \qquad Q = \frac{2}{3}(r_2^3 - r_1^3) \qquad b = 2(r_2 - r_1) \textbf{(5-47a,b,c)}$$

in which $r_1$ and $r_2$ are the inner and outer radii of the cross section. Therefore, the maximum stress is

$$\tau_{max} = \frac{VQ}{Ib} = \frac{4V}{3A}\left(\frac{r_2^2 + r_2 r_1 + r_1^2}{r_2^2 + r_1^2}\right) \qquad \textbf{(5-48)}$$

in which

$$A = \pi(r_2^2 - r_1^2)$$

is the area of the cross section. Note that if $r_1 = 0$, Eq. (5-48) reduces to Eq. (5-46) for a solid circular beam.

Although the preceding theory for shear stresses in beams of circular cross section is approximate, it gives results differing by only a few percent from those obtained using the exact theory of elasticity (Ref. 5-9). Consequently, Eqs. (5-46) and (5-48) can be used to determine the maximum shear stresses in circular beams under ordinary circumstances.

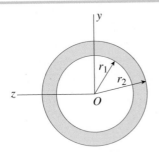

**Fig. 5-35**

Hollow circular cross section

## Example 5-13

### Fig. 5-36

Example 5-13: Shear stresses in beams of circular cross section

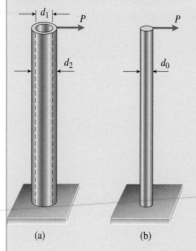

(a)                    (b)

A vertical pole consisting of a circular tube of outer diameter $d_2 = 4.0$ in. and inner diameter $d_1 = 3.2$ in. is loaded by a horizontal force $P = 1500$ lb (Fig. 5-36a).

(a) Determine the maximum shear stress in the pole.
(b) For the same load $P$ and the same maximum shear stress, what is the diameter $d_0$ of a solid circular pole (Fig. 5-36b)?

### Solution

(a) *Maximum shear stress.* For the pole having a hollow circular cross section (Fig. 5-36a), we use Eq. (5-48) with the shear force $V$ replaced by the load $P$ and the cross-sectional area $A$ replaced by the expression $\pi(r_2^2 - r_1^2)$; thus,

$$\tau_{max} = \frac{4P}{3\pi}\left(\frac{r_2^2 + r_2 r_1 + r_1^2}{r_2^4 - r_1^4}\right) \tag{a}$$

Next, we substitute numerical values, namely,

$$P = 1500 \text{ lb} \qquad r_2 = d_2/2 = 2.0 \text{ in.} \qquad r_1 = d_1/2 = 1.6 \text{ in.}$$

and obtain

$$\tau_{max} = 658 \text{ psi} \qquad \blacktriangleleft$$

which is the maximum shear stress in the pole.

(b) *Diameter of solid circular pole.* For the pole having a solid circular cross section (Fig. 5-36b), we use Eq. (5-46) with $V$ replaced by $P$ and $r$ replaced by $d_0/2$:

$$\tau_{max} = \frac{4P}{3\pi(d_0/2)^2} \tag{b}$$

Solving for $d_0$, we obtain

$$d_0^2 = \frac{16P}{3\pi\tau_{max}} = \frac{16(1500 \text{ lb})}{3\pi(658 \text{ psi})} = 3.87 \text{ in.}^2$$

from which we get

$$d_0 = 1.97 \text{ in.} \qquad \blacktriangleleft$$

In this particular example, the solid circular pole has a diameter approximately one-half that of the tubular pole.

*Note:* Shear stresses rarely govern the design of either circular or rectangular beams made of metals such as steel and aluminum. In these kinds of materials, the allowable shear stress is usually in the range 25 to 50% of the allowable tensile stress. In the case of the tubular pole in this example, the maximum shear stress is only 658 psi. In contrast, the maximum bending stress obtained from the flexure formula is 9700 psi for a relatively short pole of length 24 in. Thus, as the load increases, the allowable tensile stress will be reached long before the allowable shear stress is reached.

The situation is quite different for materials that are weak in shear, such as wood. For a typical wood beam, the allowable stress in horizontal shear is in the range of 4 to 10% of the allowable bending stress. Consequently, even though the maximum shear stress is relatively low in value, it sometimes governs the design.

# 5.10 SHEAR STRESSES IN THE WEBS OF BEAMS WITH FLANGES

When a beam of wide-flange shape (Fig. 5-37a) is subjected to shear forces as well as bending moments (nonuniform bending), both normal and shear stresses are developed on the cross sections. The distribution of the shear stresses in a wide-flange beam is more complicated than in a rectangular beam. For instance, the shear stresses in the flanges of the beam act in both vertical and horizontal directions (the $y$ and $z$ directions), as shown by the small arrows in Fig. 5-37b. The horizontal shear stresses, which are much larger than the vertical shear stresses in the flanges, are discussed later in Section 6.7.

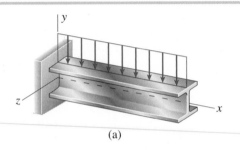

(a)

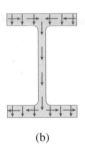

(b)

**Fig. 5-37**

(a) Beam of wide-flange shape, and (b) directions of the shear stresses acting on a cross section

The shear stresses in the web of a wide-flange beam act only in the vertical direction and are larger than the stresses in the flanges. These stresses can be found by the same techniques we used for finding shear stresses in rectangular beams.

## Shear Stresses in the Web

Let us begin the analysis by determining the shear stresses at line $ef$ in the web of a wide-flange beam (Fig. 5-38a). We will make the same assumptions as those we made for a rectangular beam; that is, we assume that the shear stresses act parallel to the $y$ axis and are uniformly distributed across the thickness of the web. Then the shear formula $\tau = VQ/Ib$ will still apply. However, the width $b$ is now the thickness $t$ of the web, and the area used in calculating the first moment $Q$ is the area between line $ef$ and the top edge of the cross section (indicated by the shaded area of Fig. 5-38a).

**Fig. 5-38**

Shear stresses in the web of a wide-flange beam: (a) Cross section of beam, and (b) distribution of vertical shear stresses in the web

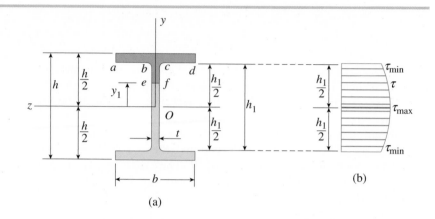

(a)

(b)

When finding the first moment $Q$ of the shaded area, we will disregard the effects of the small fillets at the juncture of the web and flange (points $b$ and $c$ in Fig. 5-38a). The error in ignoring the areas of these fillets is very small. Then we will divide the shaded area into two rectangles. The first rectangle is the upper flange itself, which has area

$$A_1 = b\left(\frac{h}{2} - \frac{h_1}{2}\right) \tag{5-49a}$$

in which $b$ is the width of the flange, $h$ is the overall height of the beam, and $h_1$ is the distance between the insides of the flanges. The second rectangle is the part of the web between $ef$ and the flange, that is, rectangle $efcb$, which has area

$$A_2 = t\left(\frac{h_1}{2} - y_1\right) \tag{5-49b}$$

in which $t$ is the thickness of the web and $y_1$ is the distance from the neutral axis to line $ef$.

The first moments of areas $A_1$ and $A_2$, evaluated about the neutral axis, are obtained by multiplying these areas by the distances from their respective centroids to the $z$-axis. Adding these first moments gives the first moment $Q$ of the combined area:

$$Q = A_1\left(\frac{h_1}{2} + \frac{h/2 - h_1/2}{2}\right) + A_2\left(y_1 + \frac{h_1/2 - y_1}{2}\right)$$

Upon substituting for $A_1$ and $A_2$ from Eqs. (5-49a) and (5-49b) and then simplifying, we get

$$Q = \frac{b}{8}(h^2 - h_1^2) + \frac{t}{8}(h_1^2 - 4y_1^2) \tag{5-50}$$

Therefore, the shear stress $\tau$ in the web of the beam at distance $y_1$ from the neutral axis is

$$\tau = \frac{VQ}{It} = \frac{V}{8It}\left[b(h^2 - h_1^2) + t(h_1^2 - 4y_1^2)\right] \tag{5-51}$$

in which the moment of inertia of the cross section is

$$I = \frac{bh^3}{12} - \frac{(b-t)h_1^3}{12} = \frac{1}{12}(bh^3 - bh_1^3 + th_1^3) \qquad \textbf{(5-52)}$$

Since all quantities in Eq. (5-51) are constants except $y_1$, we see immediately that $\tau$ varies quadratically throughout the height of the web, as shown by the graph in Fig. 5-38b. Note that the graph is drawn only for the web and does not include the flanges. The reason is simple enough—Eq. (5-51) cannot be used to determine the vertical shear stresses in the flanges of the beam (see the discussion titled "Limitations" later in this section).

## Maximum and Minimum Shear Stresses

The maximum shear stress in the web of a wide-flange beam occurs at the neutral axis, where $y_1 = 0$. The minimum shear stress occurs where the web meets the flanges ($y_1 = \pm h_1/2$). These stresses, found from Eq. (5-51), are

$$\tau_{max} = \frac{V}{8It}(bh^2 - bh_1^2 + th_1^2) \qquad \tau_{min} = \frac{Vb}{8It}(h^2 - h_1^2) \qquad \textbf{(5-53a,b)}$$

Both $\tau_{max}$ and $\tau_{min}$ are labeled on the graph of Fig. 5-38b. For typical wide-flange beams, the maximum stress in the web is from 10 to 60% greater than the minimum stress.

Although it may not be apparent from the preceding discussion, the stress $\tau_{max}$ given by Eq. (5-53a) not only is the largest shear stress in the web but also is the largest shear stress anywhere in the cross section.

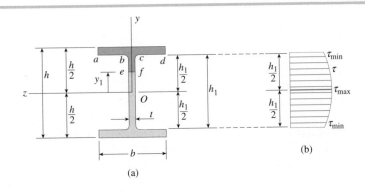

(a)

(b)

**Fig. 5-38 (Repeated)**

Shear stresses in the web of a wide-flange beam: (a) Cross section of beam, and (b) distribution of vertical shear stresses in the web

## Shear Force in the Web

The vertical shear force carried by the web alone may be determined by multiplying the area of the shear-stress diagram (Fig. 5-38b) by the thickness $t$ of the web. The shear-stress diagram consists of two parts, a rectangle of area $h_1 \tau_{min}$ and a parabolic segment of area

$$\frac{2}{3}(h_1)(\tau_{max} - \tau_{min})$$

By adding these two areas, multiplying by the thickness $t$ of the web, and then combining terms, we get the total shear force in the web:

$$V_{web} = \frac{th_1}{3}(2\tau_{max} + \tau_{min}) \qquad \textbf{(5-54)}$$

For beams of typical proportions, the shear force in the web is 90 to 98% of the total shear force $V$ acting on the cross section; the remainder is carried by shear in the flanges.

Since the web resists most of the shear force, designers often calculate an approximate value of the maximum shear stress by dividing the total shear force by the area of the web. The result is the average shear stress in the web, assuming that the web carries *all* of the shear force:

$$\tau_{aver} = \frac{V}{th_1} \qquad \textbf{(5-55)}$$

For typical wide-flange beams, the average stress calculated in this manner is within 10% (plus or minus) of the maximum shear stress calculated from Eq. (5-53a). Thus, Eq. (5-55) provides a simple way to estimate the maximum shear stress.

## Limitations

The elementary shear theory presented in this section is suitable for determining the vertical shear stresses in the web of a wide-flange beam. However, when investigating vertical shear stresses in the flanges, we can no longer assume that the shear stresses are constant across the width of the section, that is, across the width $b$ of the flanges (Fig. 5-38a). Hence, we cannot use the shear formula to determine these stresses.

To emphasize this point, consider the junction of the web and upper flange ($y_1 = h_1/2$), where the width of the section changes abruptly from $t$ to $b$. The shear stresses on the free surfaces $ab$ and $cd$ (Fig. 5-38a) must be zero, whereas the shear stress across the web at line $bc$ is $\tau_{min}$. These observations indicate that the distribution of shear stresses at the junction of the web and the flange is quite complex and cannot be investigated by elementary methods. The stress analysis is further complicated by the use of fillets at the re-entrant corners (corners $b$ and $c$). The fillets are necessary to prevent the stresses from becoming dangerously large, but they also alter the stress distribution across the web.

Thus, we conclude that the shear formula cannot be used to determine the vertical shear stresses in the flanges. However, the shear formula does give good results for the shear stresses acting *horizontally* in the flanges (Fig. 5-37b), as discussed later in Section 6.8.

The method described above for determining shear stresses in the webs of wide-flange beams can also be used for other sections having thin webs. For instance, Example 5-15 illustrates the procedure for a T-beam.

---

### Fig. 5-38 (Repeated)

Shear stresses in the web of a wide-flange beam: (a) Cross section of beam, and (b) distribution of vertical shear stresses in the web

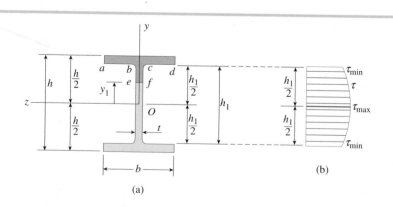

● ● ● **Example 5-14**

A beam of wide-flange shape (Fig. 5-39a) is subjected to a vertical shear force $V = 45$ kN. The cross-sectional dimensions of the beam are $b = 165$ mm, $t = 7.5$ mm, $h = 320$ mm, and $h_1 = 290$ mm.

Determine the maximum shear stress, minimum shear stress, and total shear force in the web. (Disregard the areas of the fillets when making calculations.)

**Fig. 5-39**

Example 5-14: Shear stresses in the web of a wide-flange beam

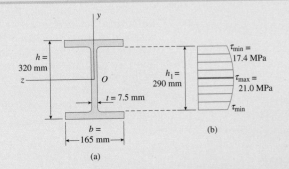

### Solution

*Maximum and minimum shear stresses.* The maximum and minimum shear stresses in the web of the beam are given by Eqs. (5-53a) and (5-53b). Before substituting into those equations, we calculate the moment of inertia of the cross-sectional area from Eq. (5-52):

$$I = \frac{1}{12} (bh^3 - bh_1^3 + th_1^3) = 130.45 \times 10^6 \text{ mm}^4$$

Now we substitute this value for $I$, as well as the numerical values for the shear force $V$ and the cross-sectional dimensions, into Eqs. (5-53a) and (5-53b):

$$\tau_{max} = \frac{V}{8It} (bh^2 - bh_1^2 + th_1^2) = 21.0 \text{ MPa} \quad \Longleftarrow$$

$$\tau_{min} = \frac{Vb}{8It} (h^2 - h_1^2) = 17.4 \text{ MPa} \quad \Longleftarrow$$

In this case, the ratio of $\tau_{max}$ to $\tau_{min}$ is 1.21, that is, the maximum stress in the web is 21% larger than the minimum stress. The variation of the shear stresses over the height $h_1$ of the web is shown in Fig. 5-39b.

*Total shear force.* The shear force in the web is calculated from Eq. (5-54) as follows:

$$V_{web} = \frac{th_1}{3} (2\tau_{max} + \tau_{min}) = 43.0 \text{ kN} \quad \Longleftarrow$$

From this result we see that the web of this particular beam resists 96% of the total shear force.

*Note:* The average shear stress in the web of the beam (from Eq. 5-55) is

$$\tau_{aver} = \frac{V}{th_1} = 20.7 \text{ MPa}$$

which is only 1% less than the maximum stress.

● ● ● **Example 5-15**

A beam having a T-shaped cross section (Fig. 5-40a) is subjected to a vertical shear force $V = 10,000$ lb. The cross-sectional dimensions are $b = 4$ in., $t = 1.0$ in., $h = 8.0$ in., and $h_1 = 7.0$ in.

Determine the shear stress $\tau_1$ at the top of the web (level $nn$) and the maximum shear stress $\tau_{max}$. (Disregard the areas of the fillets.)

**Solution**

*Location of neutral axis.* The neutral axis of the T-beam is located by calculating the distances $c_1$ and $c_2$ from the top and bottom of the beam to the centroid of the cross section (Fig. 5-40a). First, we divide the cross section into two rectangles, the flange and the web (see the dashed line in Fig. 5-40a). Then we calculate the first moment $Q_{aa}$ of these two rectangles with respect to line $aa$ at the bottom of the beam. The distance $c_2$ is equal to $Q_{aa}$ divided by the area $A$ of the entire cross section (see Chapter 12, Section 12.3, for methods for locating centroids of composite areas). The calculations are as follows:

$$A = \Sigma A_i = b(h - h_1) + th_1 = 11.0 \text{ in.}^2$$

**Fig. 5-40**

Example 5-15: Shear stresses in web of T-shaped beam

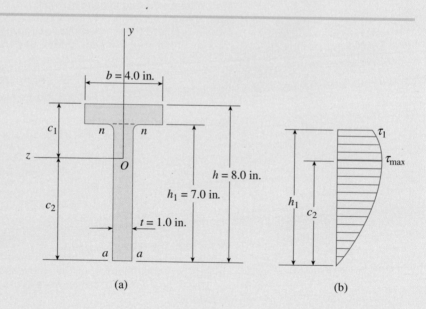

(a)

(b)

$$Q_{aa} = \Sigma y_i A_i = \left(\frac{h + h_1}{2}\right)(b)(h - h_1) + \frac{h_1}{2}(th_1) = 54.5 \text{ in.}^3$$

$$c_2 = \frac{Q_{aa}}{A} = \frac{54.5 \text{ in.}^3}{11.0 \text{ in.}^2} = 4.955 \text{ in.} \qquad c_1 = h - c_2 = 3.045 \text{ in.}$$

*Moment of inertia.* The moment of inertia $I$ of the entire cross-sectional area (with respect to the neutral axis) can be found by determining the moment of inertia $I_{aa}$ about line $aa$ at the bottom of the beam and then using the parallel-axis theorem (see Section 12.5):

$$I = I_{aa} - Ac_2^2$$

The calculations are

$$I_{aa} = \frac{bh^3}{3} - \frac{(b-t)h_1^3}{3} = 339.67 \text{ in.}^4 \quad Ac_2^2 = 270.02 \text{ in.}^4 \quad I = 69.65 \text{ in.}^4$$

*Shear stress at top of web.* To find the shear stress $\tau_1$ at the top of the web (along line $nn$) we need to calculate the first moment $Q_1$ of the area above level $nn$. This first moment is equal to the area of the flange times the distance from the neutral axis to the centroid of the flange:

$$Q_1 = b(h - h_1)\left(c_1 - \frac{h - h_1}{2}\right)$$
$$= (4 \text{ in.})(1 \text{ in.})(3.045 \text{ in.} - 0.5 \text{ in.}) = 10.18 \text{ in.}^3$$

Of course, we get the same result if we calculate the first moment of the area *below* level $nn$:

$$Q_1 = th_1\left(c_2 - \frac{h_1}{2}\right) = (1 \text{ in.})(7 \text{ in.})(4.955 \text{ in.} - 3.5 \text{ in.}) = 10.18 \text{ in.}^3$$

Substituting into the shear formula, we find

$$\tau_1 = \frac{VQ_1}{It} = \frac{(10,000 \text{ lb})(10.18 \text{ in.}^3)}{(69.65 \text{ in.}^4)(1 \text{ in.})} = 1460 \text{ psi} \quad \Leftarrow$$

This stress exists both as a vertical shear stress acting on the cross section and as a horizontal shear stress acting on the horizontal plane between the flange and the web.

*Maximum shear stress.* The maximum shear stress occurs in the web at the neutral axis. Therefore, we calculate the first moment $Q_{max}$ of the cross-sectional area below the neutral axis:

$$Q_{max} = tc_2\left(\frac{c_2}{2}\right) = (1 \text{ in.})(4.955 \text{ in.})\left(\frac{4.955 \text{ in.}}{2}\right) = 12.28 \text{ in.}^3$$

As previously indicated, we would get the same result if we calculated the first moment of the area above the neutral axis, but those calculations would be slighter longer.

Substituting into the shear formula, we obtain

$$\tau_{max} = \frac{VQ_{max}}{It} = \frac{(10,000 \text{ lb})(12.28 \text{ in.}^3)}{(69.65 \text{ in.}^4)(1 \text{ in.})} = 1760 \text{ psi} \quad \Leftarrow$$

which is the maximum shear stress in the beam.

The parabolic distribution of shear stresses in the web is shown in Fig. 5-40b.

# *5.11 BUILT-UP BEAMS AND SHEAR FLOW

**Built-up beams** are fabricated from two or more pieces of material joined together to form a single beam. Such beams can be constructed in a great variety of shapes to meet special architectural or structural needs and to provide larger cross sections than are ordinarily available.

Figure 5-41 shows some typical cross sections of built-up beams. Part (a) of the figure shows a wood **box beam** constructed of two planks, which serve as flanges, and two plywood webs. The pieces are joined together with nails, screws, or glue in such a manner that the entire beam acts as a single unit. Box beams are also constructed of other materials, including steel, plastics, and composites.

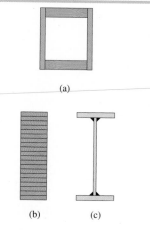

**Fig. 5-41**

Cross sections of typical built-up beams: (a) wood box beam, (b) glulam beam, and (c) plate girder

(a)

(b)          (c)

The second example is a glued laminated beam (called a **glulam beam**) made of boards glued together to form a much larger beam than could be cut from a tree as a single member. Glulam beams are widely used in the construction of small buildings.

The third example is a steel **plate girder** of the type commonly used in bridges and large buildings. These girders, consisting of three steel plates joined by welding, can be fabricated in much larger sizes than are available with ordinary wide-flange or I-beams.

Built-up beams must be designed so that the beam behaves as a single member. Consequently, the design calculations involve two phases. In the first phase, the beam is designed as though it were made of one piece, taking into account both bending and shear stresses. In the second phase, the *connections* between the parts (such as nails, bolts, welds, and glue) are designed to ensure that the beam does indeed behave as a single entity. In particular, the connections must be strong enough to transmit the horizontal shear forces acting between the parts of the beam. To obtain these forces, we make use of the concept of *shear flow*.

## Shear Flow

To obtain a formula for the horizontal shear forces acting between parts of a beam, let us return to the derivation of the shear formula (see Figs. 5-28 and 5-29 of Section 5.8). In that derivation, we cut an element $mm_1n_1n$ from a beam (Fig. 5-42a) and investigated the horizontal equilibrium of a subelement $mm_1p_1p$ (Fig. 5-42b). From the horizontal equilibrium of the subelement, we determined the force $F_3$ (Fig. 5-42c) acting on its lower surface:

$$F_3 = \frac{dM}{I} \int y\, dA \qquad \textbf{(5-56)}$$

This equation is repeated from Eq. (5-36) of Section 5.8.

Let us now define a new quantity called the **shear flow** $f$. Shear flow is the *horizontal shear force per unit distance along the longitudinal axis of the*

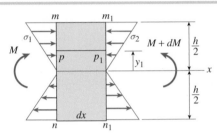

Side view of element

(a)

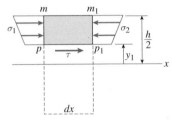

Side view of subelement

(b)

### Fig. 5-42

Horizontal shear stresses and shear forces in a beam. (*Note:* These figures are repeated from Figs. 5-28 and 5-29.)

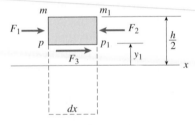

Side view of subelement

(c)

*beam.* Since the force $F_3$ acts along the distance $dx$, the shear force per unit distance is equal to $F_3$ divided by $dx$; thus,

$$f = \frac{F_3}{dx} = \frac{dM}{dx}\left(\frac{1}{I}\right)\int y\, dA$$

Replacing $dM/dx$ by the shear force $V$ and denoting the integral by $Q$, we obtain the following **shear-flow formula**:

$$f = \frac{VQ}{I} \qquad \textbf{(5-57)}$$

This equation gives the shear flow acting on the horizontal plane $pp_1$ shown in Fig. 5-42a. The terms $V$, $Q$, and $I$ have the same meanings as in the shear formula (Eq. 5-41).

If the shear stresses on plane $pp_1$ are uniformly distributed, as we assumed for rectangular beams and wide-flange beams, the shear flow $f$ equals $\tau b$. In that case, the shear-flow formula reduces to the shear formula. However, the derivation of Eq. (5-56) for the force $F_3$ does not involve any assumption about the distribution of shear stresses in the beam. Instead, the force $F_3$ is found solely from the horizontal equilibrium of the subelement (Fig. 5-42c). Therefore, we can now interpret the subelement and the force $F_3$ in more general terms than before.

The subelement may be *any* prismatic block of material between cross sections $mn$ and $m_1n_1$ (Fig. 5-42a). It does not have to be obtained by making a single horizontal cut (such as $pp_1$) through the beam. Also, since the force $F_3$ is the total horizontal shear force acting between the subelement and the rest of the beam, it may be distributed anywhere over the sides of the subelement, not just on its lower surface. These same comments apply to the shear flow $f$, since it is merely the force $F_3$ per unit distance.

Let us now return to the shear-flow formula $f = VQ/I$ (Eq. 5-57). The terms $V$ and $I$ have their usual meanings and are not affected by the choice of subelement. However, the first moment $Q$ is a property of the cross-sectional face of the subelement. To illustrate how $Q$ is determined, we will consider three specific examples of built-up beams (Fig. 5-43).

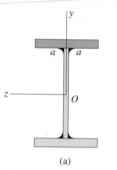

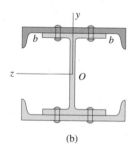

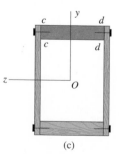

**Fig. 5-43**

Areas used when calculating the first moment $Q$

## Areas Used When Calculating the First Moment $Q$

The first example of a built-up beam is a welded steel **plate girder** (Fig. 5-43a). The welds must transmit the horizontal shear forces that act between the flanges and the web. At the upper flange, the horizontal shear force (per unit distance along the axis of the beam) is the shear flow along the contact surface $aa$. This shear flow may be calculated by taking $Q$ as the first moment of the cross-sectional area above the contact surface $aa$. In other words, $Q$ is the first moment of the flange area (shown shaded in Fig. 5-43a), calculated with respect to the neutral axis. After calculating the shear flow, we can readily determine the amount of welding needed to resist the shear force, because the strength of a weld is usually specified in terms of force per unit distance along the weld.

The second example is a **wide-flange beam** that is strengthened by riveting a channel section to each flange (Fig. 5-43b). The horizontal shear force acting between each channel and the main beam must be transmitted by the rivets. This force is calculated from the shear-flow formula using $Q$ as the first moment of the area of the entire channel (shown shaded in the figure). The resulting shear flow is the longitudinal force per unit distance acting along the contact surface $bb$, and the rivets must be of adequate size and longitudinal spacing to resist this force.

The last example is a **wood box beam** with two flanges and two webs that are connected by nails or screws (Fig. 5-43c). The total horizontal shear force between the upper flange and the webs is the shear flow acting along *both* contact surfaces $cc$ and $dd$, and therefore the first moment $Q$ is calculated for the upper flange (the shaded area). In other words, the shear flow calculated from the formula $f = VQ/I$ is the total shear flow along all contact surfaces that surround the area for which $Q$ is computed. In this case, the shear flow $f$ is resisted by the combined action of the nails on *both* sides of the beam, that is, at both $cc$ and $dd$, as illustrated in the following example.

## ● ● ● Example 5-16

### Fig. 5-44

Example 5-16: Wood box beam

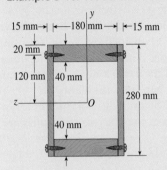

(a) Cross section

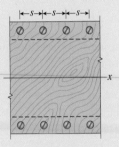

(b) Side view

A wood box beam (Fig. 5-44) is constructed of two boards, each 40 × 180 mm in cross section, that serve as flanges and two plywood webs, each 15 mm thick. The total height of the beam is 280 mm. The plywood is fastened to the flanges by wood screws having an allowable load in shear of $F = 800$ N each.

If the shear force $V$ acting on the cross section is 10.5 kN, determine the maximum permissible longitudinal spacing $s$ of the screws (Fig. 5-44b).

### Solution

*Shear flow.* The horizontal shear force transmitted between the upper flange and the two webs can be found from the shear-flow formula $f = VQ/I$, in which $Q$ is the first moment of the cross-sectional area of the flange. To find this first moment, we multiply the area $A_f$ of the flange by the distance $d_f$ from its centroid to the neutral axis:

$$A_f = 40 \text{ mm} \times 180 \text{ mm} = 7200 \text{ mm}^2 \qquad d_f = 120 \text{ mm}$$

$$Q = A_f d_f = (7200 \text{ mm}^2)(120 \text{ mm}) = 864 \times 10^3 \text{ mm}^3$$

The moment of inertia of the entire cross-sectional area about the neutral axis is equal to the moment of inertia of the outer rectangle minus the moment of inertia of the "hole" (the inner rectangle):

$$I = \frac{1}{12}(210 \text{ mm})(280 \text{ mm})^3 - \frac{1}{12}(180 \text{ mm})(200 \text{ mm})^3$$

$$= 264.2 \times 10^6 \text{ mm}^4$$

Substituting $V$, $Q$, and $I$ into the shear-flow formula (Eq. 5-52), we obtain

$$f = \frac{VQ}{I} = \frac{(10{,}500 \text{ N})(864 \times 10^3 \text{ mm}^3)}{264.2 \times 10^6 \text{ mm}^4} = 34.3 \text{ N/mm}$$

which is the horizontal shear force per millimeter of length that must be transmitted between the flange and the two webs.

*Spacing of screws.* Since the longitudinal spacing of the screws is $s$, and since there are two lines of screws (one on each side of the flange), it follows that the load capacity of the screws is $2F$ per distance $s$ along the beam. Therefore, the capacity of the screws per unit distance along the beam is $2F/s$. Equating $2F/s$ to the shear flow $f$ and solving for the spacing $s$, we get

$$s = \frac{2F}{f} = \frac{2(800 \text{ N})}{34.3 \text{ N/mm}} = 46.6 \text{ mm} \qquad \longleftarrow$$

This value of $s$ is the maximum permissible spacing of the screws, based upon the allowable load per screw. Any spacing greater than 46.6 mm would overload the screws. For convenience in fabrication, and to be on the safe side, a spacing such as $s = 45$ mm would be selected.

# *5.12 BEAMS WITH AXIAL LOADS

Structural members are often subjected to the simultaneous action of bending loads and axial loads. This happens, for instance, in aircraft frames, columns in buildings, machinery, parts of ships, and spacecraft. If the members are not too slender, the combined stresses can be obtained by superposition of the bending stresses and the axial stresses.

To see how this is accomplished, consider the cantilever beam shown in Fig. 5-45a. The only load on the beam is an inclined force $P$ acting through the centroid of the end cross section. This load can be resolved into two components, a lateral load $Q$ and an axial load $S$. These loads produce **stress resultants** in the form of bending moments $M$, shear forces $V$, and axial forces $N$ throughout the beam (Fig. 5-45b). On a typical cross section, distance $x$ from the support, these stress resultants are

$$M = Q(L - x) \qquad V = -Q \qquad N = S$$

in which $L$ is the length of the beam. The stresses associated with each of these stress resultants can be determined at any point in the cross section by means of the appropriate formula ($\sigma = -My/I$, $\tau = VQ/Ib$, and $\sigma = N/A$).

Since both the axial force $N$ and bending moment $M$ produce normal stresses, we need to combine those stresses to obtain the final stress distribution. The **axial force** (when acting alone) produces a uniform stress distribution $\sigma = N/A$ over the entire cross section, as shown by the stress diagram in Fig. 5-45c. In this particular example, the stress $\sigma$ is tensile, as indicated by the plus signs attached to the diagram.

The **bending moment** produces a linearly varying stress $\sigma = -My/I$ (Fig. 5-45d) with compression on the upper part of the beam and tension on the lower part. The distance $y$ is measured from the $z$ axis, which passes through the centroid of the cross section.

The final distribution of normal stresses is obtained by superposing the stresses produced by the axial force and the bending moment. Thus, the equation for the **combined stresses** is

$$\sigma = \frac{N}{A} - \frac{My}{I} \tag{5-58}$$

Note that $N$ is positive when it produces tension and $M$ is positive according to the bending-moment sign convention (positive bending moment produces compression in the upper part of the beam and tension in the lower part). Also, the $y$ axis is positive upward. As long as we use these sign conventions in Eq. (5-58), the normal stress $\sigma$ will be positive for tension and negative for compression.

The final stress distribution depends upon the relative algebraic values of the terms in Eq. (5-58). For our particular example, the three possibilities are shown in Figs. 5-45e, f, and g. If the bending stress at the top of the beam (Fig. 5-45d) is numerically less than the axial stress (Fig. 5-45c), the entire cross section will be in tension, as shown in Fig. 5-45e. If the bending stress at the top equals the axial stress, the distribution will be triangular (Fig. 5-45f), and if the bending stress is numerically larger than the axial stress, the cross section will be partially in compression and partially in tension (Fig. 5-45g). Of course, if the axial force is a compressive force, or if

**Fig. 5-45**

Normal stresses in a cantilever beam subjected to both bending and axial loads: (a) beam with load $P$ acting at the free end, (b) stress resultants $N$, $V$, and $M$ acting on a cross section at distance $x$ from the support, (c) tensile stresses due to the axial force $N$ acting alone, (d) tensile and compressive stresses due to the bending moment $M$ acting alone, and (e), (f), (g) possible final stress distributions due to the combined effects of $N$ and $M$

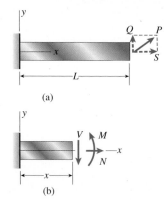

(a)

(b)

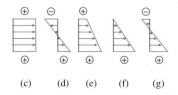

(c)  (d)  (e)  (f)  (g)

the bending moment is reversed in direction, the stress distributions will change accordingly.

Whenever bending and axial loads act simultaneously, the neutral axis (that is, the line in the cross section where the normal stress is zero) no longer passes through the centroid of the cross section. As shown in Figs. 5-45e, f, and g, respectively, the neutral axis may be outside the cross section, at the edge of the section, or within the section.

The use of Eq. (5-58) to determine the stresses in a beam with axial loads is illustrated later in Example 5-17.

## Eccentric Axial Loads

An **eccentric axial load** is an axial force that does *not* act through the centroid of the cross section. An example is shown in Fig. 5-46a, where the cantilever beam $AB$ is subjected to a tensile load $P$ acting at distance $e$ from the $x$ axis (the $x$ axis passes through the centroids of the cross sections). The distance $e$, called the *eccentricity* of the load, is positive in the positive direction of the $y$ axis.

The eccentric load $P$ is statically equivalent to an axial force $P$ acting along the $x$ axis and a bending moment $Pe$ acting about the $z$ axis (Fig. 5-46b). Note that the moment $Pe$ is a negative bending moment.

A cross-sectional view of the beam (Fig. 5-46c) shows the $y$ and $z$ axes passing through the centroid $C$ of the cross section. The eccentric load $P$ intersects the $y$ axis, which is an axis of symmetry.

Since the axial force $N$ at any cross section is equal to $P$, and since the bending moment $M$ is equal to $-Pe$, the **normal stress** at any point in the cross section (from Eq. 5-58) is

$$\sigma = \frac{P}{A} + \frac{Pey}{I} \tag{5-59}$$

in which $A$ is the area of the cross section and $I$ is the moment of inertia about the $z$ axis. The stress distribution obtained from Eq. (5-59), for the case where both $P$ and $e$ are positive, is shown in Fig. 5-46d.

Bending due to self-weight of beam and axial compression due to horizontal component of cable lifting force (Lester Lefkowitz/Getty Images)

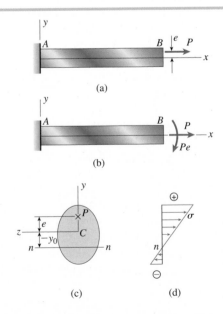

(a)

(b)

(c)          (d)

### Fig. 5-46

(a) Cantilever beam with an eccentric axial load $P$,
(b) equivalent loads $p$ and $pe$,
(c) cross section of beam, and
(d) distribution of normal stresses over the cross section

The position of the **neutral axis** *nn* (Fig. 5-46c) can be obtained from Eq. (5-59) by setting the stress $\sigma$ equal to zero and solving for the coordinate *y*, which we now denote as $y_0$. The result is

$$y_0 = -\frac{I}{Ae} \tag{5-60}$$

The coordinate $y_0$ is measured from the *z* axis (which is the neutral axis under pure bending) to the line *nn* of zero stress (the neutral axis under combined bending and axial load). Because $y_0$ is positive in the direction of the *y* axis (upward in Fig. 5-46c), it is labeled $-y_0$ when it is shown downward in the figure.

From Eq. (5-60) we see that the neutral axis lies below the *z* axis when *e* is positive and above the *z* axis when *e* is negative. If the eccentricity is reduced, the distance $y_0$ increases and the neutral axis moves away from the centroid. In the limit, as *e* approaches zero, the load acts at the centroid, the neutral axis is at an infinite distance, and the stress distribution is uniform. If the eccentricity is increased, the distance $y_0$ decreases and the neutral axis moves toward the centroid. In the limit, as *e* becomes extremely large, the load acts at an infinite distance, the neutral axis passes through the centroid, and the stress distribution is the same as in pure bending.

Eccentric axial loads are analyzed in some of the problems at the end of this chapter, beginning with Prob. 5.12-12.

## Limitations

The preceding analysis of beams with axial loads is based upon the assumption that the bending moments can be calculated without considering the deflections of the beams. In other words, when determining the bending moment *M* for use in Eq. (5-58), we must be able to use the original dimensions of the beam—that is, the dimensions *before* any deformations or deflections occur. The use of the original dimensions is valid provided the beams are relatively stiff in bending, so that the deflections are very small.

Thus, when analyzing a beam with axial loads, it is important to distinguish between a **stocky beam**, which is relatively short and therefore highly resistant to bending, and a **slender beam**, which is relatively long and therefore very flexible. In the case of a stocky beam, the lateral deflections are so small as to have no significant effect on the line of action of the axial forces. As a consequence, the bending moments will not depend upon the deflections and the stresses can be found from Eq. (5-58).

In the case of a slender beam, the lateral deflections (even though small in magnitude) are large enough to alter significantly the line of action of the axial forces. When that happens, an additional bending moment, equal to the product of the axial force and the lateral deflection, is created at every cross section. In other words, there is an interaction, or coupling, between the axial effects and the bending effects. This type of behavior is discussed in Chapter 11 on **columns**.

The distinction between a stocky beam and a slender beam is obviously not a precise one. In general, the only way to know whether interaction effects are important is to analyze the beam with and without the interaction and notice whether the results differ significantly. However, this procedure may require considerable calculating effort. Therefore, as a guideline for practical use, we usually consider a beam with a length-to-height ratio of 10 or less to be a stocky beam. Only stocky beams are considered in the problems pertaining to this section.

● ● ● ● **Example 5-17**

A tubular beam *ACB* with a length of $L = 60$ in. is pin-supported at its ends, *A* and *B*. A powered winch at *E* lifts load *W* below *C* using a cable which passes over a frictionless pulley at midlength (point *D*, Fig. 5-47a). The distance from the center of the pulley to the longitudinal axis of the tube is $d = 5.5$ in. The cross section of the tube is square (Fig. 5-47b) with an outer dimension of $b = 6.0$ in., area of $A = 20.0$ in.$^2$, and moment of inertia of $I = 86.67$ in.$^4$

(a) Determine the maximum tensile and compressive stresses in the beam due to a load $W = 3000$ lb.
(b) If the allowable normal stress in the tube is 3500 psi, find the maximum permissible load *W*. Assume that the cable, pulley, and bracket *CD* are adequate to carry load $W_{max}$.

**Fig. 5-47**

Example 5-17: Tubular beam subjected to combined bending and axial load

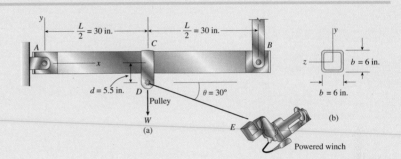

(a)

(b)

**Solution**

(a) *Maximum tensile and compressive stresses in the beam: Beam and loading.* We begin by representing the beam and its load in idealized form for the purposes of analysis (Fig. 5-48a). Since the support at end *A* resists both horizontal and vertical displacement, it is represented as a pin support. The support at *B* prevents vertical displacement but offers no resistance to horizontal displacement, so it is shown as a roller support.

We can replace the cable forces at *D* with statically equivalent forces $F_H$ and $F_V$ and moment $M_O$, all of which are applied on the axis of the beam at *C* (see Fig. 5-48a):

$$F_H = W \cos (\theta) = 2598 \text{ lb} \qquad F_V = W[1 + \sin (\theta)] = 4500 \text{ lb}$$

$$M_0 W \cos (\theta)d = 14{,}289 \text{ lb-in.}$$

*Reactions and stress resultants.* The reactions of the beam ($R_H$, $R_A$, and $R_B$) are labeled in Fig. 5-48a. Also, the diagrams of axial force *N*, shear force *V*, and bending moment *M* are shown in Figs. 5-48b, c, and d, respectively. All of these quantities are found from free-body diagrams and equations of equilibrium using the techniques described in Chapter 4. For example, using equations of statics, we find that

$$\Sigma F_H = 0: \quad R_H = -F_H = -W \cos (\theta) = -(3000 \text{ lb}) \cos (30°) = -2598 \text{ lb} \qquad \textbf{(a)}$$

Continues ➡

**Fig. 5-48**

Solution of Example 5-17:
(a) idealized beam and
loading, (b) axial-force
diagram, (c) shear-force
diagram, and (d) bending-
moment diagram

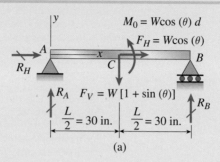

(a)

$$\Sigma M_A = 0: \quad R_B = \frac{1}{L}\left(F_V\frac{L}{2} - M_0\right) = \frac{W}{2}[1 + \sin(\theta)] - W\frac{d}{L}[\cos(\theta)] \quad \textbf{(b)}$$

$$R_B = (3000 \text{ lb})\left[\frac{1 + \sin(30°)}{2} - \left(\frac{5.5 \text{ in.}}{60 \text{ in.}}\right)\cos(30°)\right] = 2012 \text{ lb}$$

$$\Sigma F_V = 0: \quad R_A = F_V - R_B = (3000 \text{ lb})(1 + \sin(30°)) - 2012 \text{ lb} = 2488 \text{ lb} \quad \textbf{(c)}$$

**Fig. 5-48 (Continued)**

$N$

2598 lb

0
    $L/2$      $L/2$

(b)

$V$

2488 lb

0

−2012 lb

(c)

$(R_A)(L/2) = 74{,}640$ lb-in.

$M$

$(R_B)(L/2) = 60{,}360$ lb-in.

0

(d)

Next, we use the axial-force ($N$), shear-force ($V$) and bending-moment
($M$) diagrams (Figs. 5-48b, c and d, respectively) to find the combined
stresses in beam $ACB$ using Eq. (5-58).

*Stresses in the beam.* The **maximum tensile stress** in the beam occurs at
the bottom of the beam ($y = -3.0$ in.) just to the left of the midpoint $C$.
We arrive at this conclusion by noting that at this point in the beam the
tensile stress due to the axial force *adds* to the tensile stress produced by
the largest bending moment. Thus, from Eq. (5-58), we get

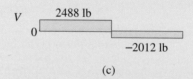

$$(\sigma_t)_{max} = \frac{N}{A} - \frac{My}{I} = \frac{2598 \text{ lb}}{20 \text{ in.}^2} - \frac{(74{,}640 \text{ lb-in.})(-3 \text{ in.})}{86.67 \text{ in.}^4}$$

$$= 130 \text{ psi} + 2583 \text{ psi} = 2713 \text{ psi}$$

The **maximum compressive stress** occurs either at the top of the beam ($y = 3.0$ in.) to the left of point $C$ or at the top of the beam to the right of point $C$. These two stresses are calculated as

$$(\sigma_c)_{\text{left}} = \frac{N}{A} - \frac{My}{I} = \frac{2598 \text{ lb}}{20 \text{ in.}^2} - \frac{(74{,}640 \text{ lb-in.})(3 \text{ in.})}{86.67 \text{ in.}^4}$$

$$= 130 \text{ psi} - 2583 \text{ psi} = -2453 \text{ psi}$$

$$(\sigma_c)_{\text{right}} = \frac{N}{A} - \frac{My}{I} = 0 - \frac{(60{,}360 \text{ lb-in.})(3 \text{ in.})}{86.67 \text{ in.}^4} = -2089 \text{ psi}$$

Thus, the maximum compressive stress is

$$(\sigma_c)_{\text{max}} = -2453 \text{ psi}$$ ⬅

and occurs at the top of the beam to the left of point $C$.

(b) *Maximum permissible load W.* From Eq. (a), we see that the *tensile stress at the bottom of the beam* just left of $C$ (equal to 2713 psi for a load $W = 3000$ lb) will reach allowable normal stress $\sigma_a = 3500$ psi first and thus will be the determining factor in finding $W_{\text{max}}$. Using expressions for reactions [Eqs. (a), (b), and (c)], we find that the axial tension force in beam segment $AC$ and the positive moment just left of $C$ are

$$N = W \cos(\theta) \qquad M = R_A \frac{L}{2} = W \left( \frac{1 + \sin(\theta)}{2} + \frac{d}{L} \cos(\theta) \right) \left( \frac{L}{2} \right)$$

From Eq. (5-58), we find

$$\sigma_a = \frac{W \cos(\theta)}{A} - \frac{W \left( \dfrac{1 + \sin(\theta)}{2} + \dfrac{d}{L} \cos(\theta) \right) \left( \dfrac{L}{2} \right) \left( \dfrac{-b}{2} \right)}{I}$$

Solving for $W = W_{\text{max}}$, we have

$$W_{\text{max}} = \frac{\sigma_a}{\dfrac{\cos(\theta)}{A} + \dfrac{bL[1 + \sin(\theta)]}{8I} + \dfrac{bd \cos(\theta)}{4I}} = 3869 \text{ lb}$$ ⬅

*Note:* This example shows how the normal stresses in a beam due to combined bending and axial load can be determined. The shear stresses acting on cross sections of the beam (due to the shear forces $V$) can be determined independently of the normal stresses, as described earlier in this chapter. Later, in Chapter 7, we will see how to determine the stresses on inclined planes when we know both the normal and shear stresses acting on cross-sectional planes.

# *5.13 STRESS CONCENTRATIONS IN BENDING

The flexure and shear formulas discussed in earlier sections of this chapter are valid for beams without holes, notches, or other abrupt changes in dimensions. Whenever such discontinuities exist, high localized stresses are produced. These **stress concentrations** can be extremely important when a member is made of brittle material or is subjected to dynamic loads. (See Chapter 2, Section 2.10, for a discussion of the conditions under which stress concentrations are important.)

For illustrative purposes, two cases of stress concentrations in beams are described in this section. The first case is a beam of rectangular cross section with a **hole at the neutral axis** (Fig. 5-49). The beam has height $h$ and thickness $b$ (perpendicular to the plane of the figure) and is in pure bending under the action of bending moments $M$.

When the diameter $d$ of the hole is small compared to the height $h$, the stress distribution on the cross section through the hole is approximately as shown by the diagram in Fig. 5-49a. At point $B$ on the edge of the hole the stress is much larger than the stress that would exist at that point if the hole were not present. (The dashed line in the figure shows the stress distribution with no hole.) However, as we go toward the outer edges of the beam (toward point $A$), the stress distribution varies linearly with distance from the neutral axis and is only slightly affected by the presence of the hole.

When the hole is relatively large, the stress pattern is approximately as shown in Fig. 5-49b. There is a large increase in stress at point $B$ and only a small change in stress at point $A$ as compared to the stress distribution in the beam without a hole (again shown by the dashed line). The stress at point $C$ is larger than the stress at $A$ but smaller than the stress at $B$.

Extensive investigations have shown that the stress at the edge of the hole (point $B$) is approximately twice the *nominal stress* at that point. The nominal stress is calculated from the flexure formula in the standard way, that is, $\sigma = My/I$, in which $y$ is the distance $d/2$ from the neutral axis to point $B$ and $I$ is the moment of inertia of the net cross section at the hole. Thus, we have the following approximate formula for the stress at point $B$:

$$\sigma_B \approx 2\,\frac{My}{I} = \frac{12Md}{b(h^3 - d^3)} \tag{5-61}$$

At the outer edge of the beam (at point $C$), the stress is approximately equal to the *nominal* stress (not the actual stress) at point $A$ (where $y = h/2$):

$$\sigma_C \approx \frac{My}{I} = \frac{6Mh}{b(h^3 - d^3)} \tag{5-62}$$

Stress distributions in a beam in pure bending with a circular hole at the neutral axis. (The beam has a rectangular cross section with height $h$ and thickness $b$.)

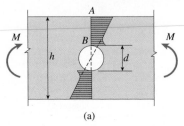

(a)

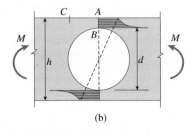

(b)

From the last two equations we see that the ratio $\sigma_B/\sigma_C$ is approximately $2d/h$. Hence we conclude that when the ratio $d/h$ of hole diameter to height of beam exceeds 1/2, the largest stress occurs at point $B$. When $d/h$ is less than 1/2, the largest stress is at point $C$.

The second case we will discuss is a **rectangular beam with notches** (Fig. 5-50). The beam shown in the figure is subjected to pure bending and has height $h$ and thickness $b$ (perpendicular to the plane of the figure). Also, the net height of the beam (that is, the distance between the bases of the notches) is $h_1$ and the radius at the base of each notch is $R$. The maximum stress in this beam occurs at the base of the notches and may be much larger than the nominal stress at that same point. The nominal stress is calculated from the flexure formula with $y = h_1/2$ and $I = bh_1^3/12$; thus,

$$\sigma_{\text{nom}} = \frac{My}{I} = \frac{6M}{bh_1^2} \qquad \textbf{(5-63)}$$

The maximum stress is equal to the stress-concentration factor $K$ times the nominal stress:

$$\sigma_{\text{max}} = K\sigma_{\text{nom}} \qquad \textbf{(5-64)}$$

The stress-concentration factor $K$ is plotted in Fig. 5-50 for a few values of the ratio $h/h_1$. Note that when the notch becomes "sharper," that is, the ratio $R/h_1$ becomes smaller, the stress-concentration factor increases. (Figure 5-50 is plotted from the formulas given in Ref. 2-9.)

The effects of the stress concentrations are confined to small regions around the holes and notches, as explained in the discussion of Saint-Venant's principle in Section 2.10. At a distance equal to $h$ or greater from the hole or notch, the stress-concentration effect is negligible and the ordinary formulas for stresses may be used.

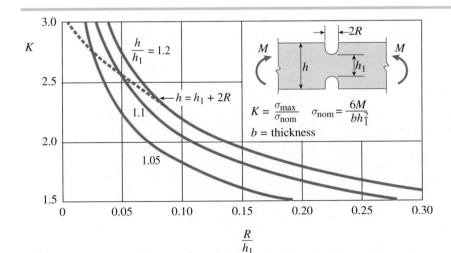

**Fig. 5-50**

Stress-concentration factor $K$ for a notched beam of rectangular cross section in pure bending ($h$ = height of beam; $b$ = thickness of beam, perpendicular to the plane of the figure). The dashed line is for semicircular notches ($h = h_1 + 2R$).

## Example 5-18

A simple beam $AB$ with rectangular cross section ($b \times h$) has a hole with a diameter of $d$ at its centerline and two notches on either side and equidistant from the beam centerline. Beam $AB$ is simply supported, and loads $P$ are applied at $L/5$ from each end of the beam. Assume that dimensions given in Fig. 5-51 are as follows: $L = 4.5$ m, $b = 50$ mm, $h = 144$ mm, $h_1 = 120$ mm, $d = 85$ mm, and $R = 10$ mm. Assume that the allowable bending stress is $\sigma_a = 150$ MPa.

(a) Find the maximum permissible value of applied load $P$.
(b) If $P = 11$ kN, find the smallest acceptable radius of the notches, $R_{min}$.
(c) If $P = 11$ kN, find the maximum acceptable diameter of hole at mid-height of beam.

### Fig. 5-51

Example 5-18: Rectangular steel beam with notches and a hole

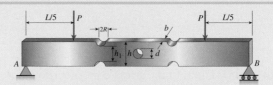

### Solution

(a) *Maximum permissible load P.* The central part of the beam between the loads $P$ ($x = L/5$ to $x = 4L/5$) is in pure bending and the maximum moment in this region is $M = PL/5$. To find $P_{max}$, we must compare the maximum bending stress (at midspan around the hole and in the notch regions) to the allowable stress value of $\sigma_a = 150$ MPa.

First, we check the *maximum stresses around the hole.* The hole diameter-to-beam depth ratio $d/h = 85$ mm/144 mm $= 0.59$ exceeds 1/2, so we know that the stress at $B$ rather than at $C$ (Fig. 5-49) will govern. Setting $\sigma_B$ equal to $\sigma_a$ and substituting $PL/5$ for $M$ in Eq. (5-61), we can develop the following expression for $P_{max}$:

$$M_{max} = \sigma_a \left[ \frac{b(h^3 - d^3)}{12d} \right] \quad \text{and} \quad P_{max1} = \frac{5}{L} \left\{ \sigma_a \left[ \frac{b(h^3 - d^3)}{12d} \right] \right\}$$

from which we can compute

$$P_{max1} = \frac{5}{4.5 \text{ m}} \left\{ 150 \text{ MPa} \left[ \frac{50 \text{ mm}[(144 \text{ mm})^3 - (85 \text{ mm})^3]}{12(85 \text{ mm})} \right] \right\} = 19.378 \text{ kN}$$

Next, we check the peak stresses at the base of the two notches to get a second value of $P_{max}$. The ratio of notch radius $R$ to height $h_1$ is equal to 0.083, and the ratio $h/h_1 = 1.2$. So from Fig. 5-50, we find that the stress concentration factor $K$ is approximately equal to 2.3 (see Fig. 5-52).

### Fig. 5-52

Stress concentration factor $K$ in notch regions of beam for part (a) of Example 5-18

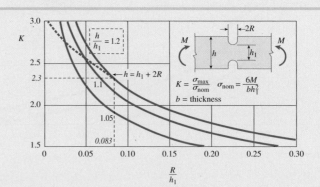

From Eqs. (5-63) and (5-64), we get the expressions:

$$\sigma_{max} = K\sigma_{nom} = K\left(\frac{6M}{bh_1^2}\right) = K\left[\frac{6}{bh_1^2}\left(\frac{PL}{5}\right)\right]$$

so

$$P_{max2} = \sigma_a\left(\frac{5bh_1^2}{6KL}\right) = 150 \text{ MPa}\left[\frac{5(50 \text{ mm})(120 \text{ mm})^2}{6(2.3) \times (4.5 \text{ m})}\right] = 8.7 \text{ kN}$$

Comparing $P_{max1}$ and $P_{max2}$, we see that the peak stress at the base of the notches controls, so

$$P_{max} = 8.7 \text{ kN} \quad \blacktriangleleft$$

(b) *Smallest acceptable radius R of the notches.* The stress concentration factor $K$ in Fig. 5-50 increases as the ratio of the notch radius $R$ to dimension $h_1$ decreases. We can compute the nominal stress using Eq. (5-63) as

$$\sigma_{nom} = \frac{6\left(\dfrac{PL}{5}\right)}{bh_1^2} = \frac{6(11 \text{ kN}) \times (4.5 \text{ m})}{5(50 \text{ mm}) \times (120 \text{ mm})^2} = 82.5 \text{ MPa}$$

Then set the maximum bending stress $\sigma_{max}$ equal to the allowable stress $\sigma_a = 150$ MPa to find the stress concentration factor $K$:

$$K = \frac{\sigma_a}{\sigma_{nom}} = \frac{150 \text{ MPa}}{82.5 \text{ MPa}} = 1.82$$

From Fig. 5-53, with $h/h_1 = 1.2$ and $K = 1.82$, we obtain

$$\frac{R}{h_1} = 0.16 \quad \text{so} \quad R_{min} = 0.16(120 \text{ mm}) = 19.2 \text{ mm} \quad \blacktriangleleft$$

(c) *Maximum acceptable diameter of the hole.* We begin by assuming that ratio $d/h > 1/2$, so we can start with Eq. (5-61) (which assumes that maximum bending stress is at $B$, as in Fig. 5-49) to find $d_{max}$. If $d/h$ turns out to be less than 1/2, we will have to use Eq. (5-62), which means that maximum bending stress is in fact at point $C$. If the peak stress is at $B$, we can write Eq. (5-61) as

$$\frac{12\left(\dfrac{PL}{5}\right)d}{b(h^3 - d^3)} = \sigma_a$$

which can be solved numerically to find that $d_{max} = 108.3$ mm. $\quad \blacktriangleleft$
    Our original assumption about the $d/h$ ratio is confirmed, since $d_{max}/h = 0.752$ exceeds 1/2, so the peak stress is indeed at $B$ and not at $C$.

## Fig. 5-53

Stress concentration factor $K$ in notch regions of beam for part (b) of Example 5-18

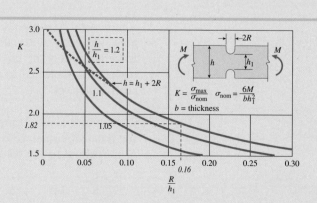

In **Chapter 5**, we investigated the behavior of beams with loads applied and bending occurring in the $xy$ plane: a plane of symmetry in the beam cross section. Both pure bending and nonuniform bending were considered. The normal stresses were seen to vary linearly from the neutral surface in accordance with the **flexure formula**, which showed that the stresses are directly proportional to the bending moment $M$ and inversely proportional to the moment of inertia I /of the cross section. Next, the relevant properties of the beam cross section were combined into a single quantity known as the **section modulus S** of the beam: a useful property in **beam design** once the maximum moment $(M_{max})$ and allowable normal stress $(\sigma_{allow})$ are known. The flexure formula was also shown to give reasonably accurate values for the bending stresses in nonprismatic beams provided the changes in cross-sectional dimensions were gradual. Next, horizontal and vertical shear stresses $(\tau)$ were computed using the **shear formula** for the case of nonuniform bending of beams with either rectangular or circular cross sections. The special cases of shear in beams with flanges and built-up beams also were considered. Finally, stocky beams with both axial and transverse loads were discussed, followed by an evaluation of localized stresses in beams with abrupt changes in cross section around notches or holes.

Some of the major concepts and findings presented in this chapter are as follows:

1. If the $xy$ plane is a plane of symmetry of a beam cross section and applied loads act in the $xy$ plane, the bending deflections occur in this same plane, known as the **plane of bending.**

2. A beam in pure bending has constant curvature $\kappa$, and a beam in nonuniform bending has varying curvature. Longitudinal strains $(\varepsilon_x)$ in a bent beam are proportional to its curvature, and the strains in a beam in pure bending vary linearly with distance from the neutral surface, regardless of the shape of the stress-strain curve of the material in accordance with Eq. (5-5):

$$\varepsilon_x = -\kappa y$$

3. The neutral axis passes through the centroid of the cross-sectional area when the material follows Hooke's law and there is no axial force acting on the cross section. When a beam of linearly elastic material is subjected to pure bending, the $y$ and $z$ axes are **principal centroidal axes**.

4. If the material of a beam is linearly elastic and follows Hooke's law, the **moment-curvature equation** shows that the curvature is directly proportional to the bending moment $M$ and inversely proportional

to the quantity $EI$, referred to as the **flexural rigidity** of the beam. The moment curvature relation was given in Eq. (5-13):

$$\kappa = \frac{M}{EI}$$

5. The **flexure formula** shows that the normal stresses $\sigma_x$ are directly proportional to the bending moment $M$ and inversely proportional to the moment of inertia $I$ of the cross section as given in Eq. (5-14):

$$\sigma_x = -\frac{My}{I}.$$

The maximum tensile and compressive bending stresses acting at any given cross section occur at points located farthest from the neutral axis

$$(y = c_1, \ y = -c_2).$$

6. The normal stresses calculated from the flexure formula are not significantly altered by the presence of shear stresses and the associated warping of the cross section for the case of nonuniform bending. However, the flexure formula is not applicable near the supports of a beam or close to a concentrated load, because such irregularities produce **stress concentrations** that are much greater than the stresses obtained from the flexure formula.

7. To **design** a beam to resist bending stresses, we calculate the required **section modulus** $S$ from the maximum moment and allowable normal stress as follows:

$$S = \frac{M_{max}}{\sigma_{allow}}$$

To minimize weight and save material, we usually select a beam from a material design manual (e.g., see sample tables in Appendices F and G for steel and wood, available online) that has the least cross-sectional area while still providing the required section modulus; wide-flange sections, and I-sections have most of their material in the flanges and the width of their flanges helps to reduce the likelihood of sideways buckling.

8. **Nonprismatic beams** (found in automobiles, airplanes, machinery, bridges, buildings, tools, and many other applications) commonly are used to reduce weight and improve appearance. The flexure formula gives reasonably accurate values for the bending stresses in nonprismatic beams, provided that the changes in cross-sectional

dimensions are gradual. However, in a nonprismatic beam, the section modulus also varies along the axis, so we cannot assume that the maximum stresses occur at the cross section with the largest bending moment.

9. Beams subjected to loads that produce both bending moments ($M$) and shear forces ($V$) (**nonuniform bending**) develop both normal and shear stresses in the beam. Normal stresses are calculated from the **flexure formula** (provided the beam is constructed of a linearly elastic material), and shear stresses are computed using the **shear formula** as follows:

$$\tau = \frac{VQ}{Ib}$$

Shear stress varies parabolically over the height of a rectangular beam, and shear strain also varies parabolically; these shear strains cause cross sections of the beam that were originally plane surfaces to become warped. The maximum values of the shear stress and strain ($\tau_{max}$, $\gamma_{max}$) occur at the neutral axis, and the shear stress and strain are zero on the top and bottom surfaces of the beam.

10. The shear formula applies only to prismatic beams and is valid only for beams of linearly elastic materials with small deflections; also, the edges of the cross section must be **parallel** to the $y$ axis. For rectangular beams, the accuracy of the shear formula depends upon the height-to-width ratio of the cross section: the formula may be considered as exact for very narrow beams but becomes less accurate as width $b$ increases relative to height $h$. Note that we can use the shear formula to calculate the shear stresses only at the neutral axis of a beam of **circular** cross section.

For rectangular cross sections,

$$\tau_{max} = \frac{3}{2} \frac{V}{A}$$

and for solid circular cross sections

$$\tau_{max} = \frac{4}{3} \frac{V}{A}$$

11. Shear stresses rarely govern the design of either circular or rectangular beams made of metals such as steel and aluminum for which the allowable shear stress is usually in the range 25 to 50% of the allowable tensile stress. However, for **materials that are weak in shear**, such as wood, the allowable stress in horizontal shear is in the range of 4 to 10% of the allowable bending stress and so may govern the design.

12. Shear stresses in the flanges of **wide-flange beams** act in both vertical and horizontal directions. The horizontal shear stresses are much larger than the vertical shear stresses in the flanges. The shear stresses in the **web of a wide-flange beam** act only in the vertical direction, are larger than the stresses in the flanges, and may be computed using the shear formula. The maximum shear stress in the web of a wide-flange beam occurs at the neutral axis, and the minimum shear stress occurs where the web meets the flanges. For beams of typical proportions, the shear force in the web is 90 to 98% of the total shear force $V$ acting on the cross section; the remainder is carried by shear in the flanges.

13. Connections between the parts in **built-up beams** (e.g., nails, bolts, welds, and glue) must be strong enough to transmit the horizontal shear forces acting between the parts of the beam. The connections are designed using the **shear flow formula**

$$f = \frac{VQ}{I}$$

to ensure that the beam behaves as a single entity. **Shear flow** $f$ is defined as horizontal shear force per unit distance along the longitudinal axis of the beam.

14. Normal stresses in **beams with axial loads** are obtained by superposing the stresses produced by the axial force $N$ and the bending moment $M$ as

$$\sigma = \frac{N}{A} - \frac{My}{I}$$

Whenever bending and axial loads act simultaneously, the neutral axis no longer passes through the centroid of the cross section and may be outside the cross section, at the edge of the section, or within the section. The discussion in Section 5.12 applies only to **stocky beams** for which the lateral deflections are so small as to have no significant effect on the line of action of the axial forces. If there is an interaction or coupling between the axial effects and the bending effects, this type of behavior is discussed in Chapter 11 on **columns**.

15. Stress distributions in beams are altered by holes, notches, or other abrupt changes in dimensions leading to high localized stresses or **stress concentrations**. These are especially important to consider when the beam material is brittle or the member is subjected to dynamic loads. The maximum stress values may be several times larger than the nominal stress.

# PROBLEMS CHAPTER 5

## Longitudinal Strains in Beams

**5.4-1** A steel wire with a diameter of $d = 1/16$ in. is bent around a cylindrical drum with a radius of $R = 36$ in. (see figure).

(a) Determine the maximum normal strain $\varepsilon_{max}$.

(b) What is the minimum acceptable radius of the drum if the maximum normal strain must remain below yield? Assume $E = 30{,}000$ ksi and $\sigma_Y = 100$ ksi.

(c) If $R = 36$ in., what is the maximum acceptable diameter of the wire if the maximum normal strain must remain below yield?

**5.4-3** A 4.75-in. outside diameter polyethylene pipe designed to carry chemical wastes is placed in a trench and bent around a quarter-circular 90° bend (see figure). The bent section of the pipe is 52 ft long.

(a) Determine the maximum compressive strain $\varepsilon_{max}$ in the pipe.

(b) If the normal strain cannot exceed $6.1 \times 10^{-3}$, what is the maximum diameter of the pipe?

(c) If $d = 4.75$ in., what is the minimum acceptable length of the bent section of the pipe?

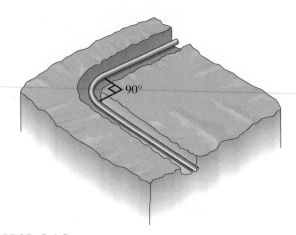

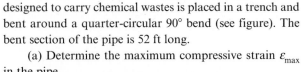

**PROB. 5.4-1**

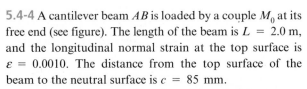

**PROB. 5.4-3**

**5.4-2** A copper wire having a diameter of $d = 4$ mm is bent into a circle and held with the ends just touching (see figure).

(a) If the maximum permissible strain in the copper is $\varepsilon_{max} = 0.0024$, what is the shortest length $L$ of wire that can be used?

(b) If $L = 5.5$ m, what is the maximum acceptable diameter of the wire if the maximum normal strain must remain below yield? Assume $E = 120$ GPa and $\sigma_Y = 300$ MPa.

**5.4-4** A cantilever beam $AB$ is loaded by a couple $M_0$ at its free end (see figure). The length of the beam is $L = 2.0$ m, and the longitudinal normal strain at the top surface is $\varepsilon = 0.0010$. The distance from the top surface of the beam to the neutral surface is $c = 85$ mm.

(a) Calculate the radius of curvature $\rho$, the curvature $\kappa$, and the vertical deflection $\delta$ at the end of the beam.

(b) If allowable strain $\varepsilon_a = 0.0008$, what is the maximum acceptable depth of the beam? (Assume that the curvature is unchanged from Part(a)).

(c) If allowable strain $\varepsilon_a = 0.0008$, $c = 85$ mm, and $L = 4$ m, what is deflection $\delta$?

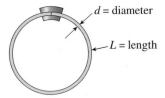

**PROB. 5.4-2**

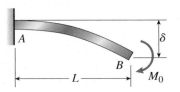

**PROB. 5.4-4**

**5.4-5** A thin strip of steel with a length of $L = 19$ in. and thickness of $t = 0.275$ in. is bent by couples $M_0$ (see figure). The deflection at the midpoint of the strip (measured from a line joining its end points) is found to be 0.30 in.

(a) Determine the longitudinal normal strain $\varepsilon$ at the top surface of the strip.

(b) If allowable strain $\varepsilon_a = 0.0008$, what is the maximum acceptable thickness of the strip?

(c) If allowable strain $\varepsilon_a = 0.0008$, $t = 0.275$ in., and $L = 32$ in., what is deflection $\delta$?

(d) If allowable strain $\varepsilon_a = 0.0008$, $t = 0.275$ in., and the deflection cannot exceed 1.0 in., what is the maximum permissible length of the strip?

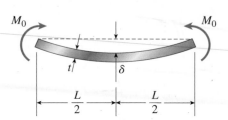

**PROB. 5.4-5**

**5.4-6** A bar of rectangular cross section is loaded and supported as shown in the figure. The distance between supports is $L = 1.75$ m, and the height of the bar is $h = 140$ mm. The deflection at the midpoint is measured as 2.5 mm.

(a) What is the maximum normal strain $\varepsilon$ at the top and bottom of the bar?

(b) If allowable strain $\varepsilon_a = 0.0006$ and the deflection cannot exceed 4.3 mm, what is the maximum permissible length of the bar?

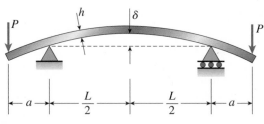

**PROB. 5.4-6**

# Normal Stresses in Beams

**5.5-1** A thin strip of hard copper ($E = 16,000$ ksi) having length $L = 90$ in. and thickness $t = 3/32$ in. is bent into a circle and held with the ends just touching (see figure).

(a) Calculate the maximum bending stress $\sigma_{max}$ in the strip.

(b) By what percent does the stress increase or decrease if the thickness of the strip is increased by 1/32 in.?

(c) Find the new length of the strip so that the stress in part (b) ($t = 1/8$ in. and $L = 90$ in.) is equal to that in part (a) ($t = 3/32$ in. and $L = 90$ in.).

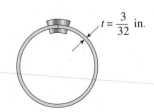

$t = \dfrac{3}{32}$ in.

**PROB. 5.5-1**

**5.5-2** A steel wire ($E = 200$ GPa) of diameter $d = 1.25$ mm is bent around a pulley of radius $R_0 = 500$ mm (see figure).

(a) What is the maximum stress $\sigma_{max}$ in the wire?

(b) By what percent does the stress increase or decrease if the radius of the pulley is increased by 25%?

(c) By what percent does the stress increase or decrease if the diameter of the wire is increased by 25% while the pulley radius remains at $R_0 = 500$ mm?

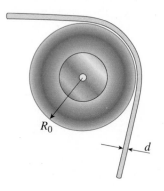

**PROB. 5.5-2**

**5.5-3** A thin, high-strength steel rule ($E = 30 \times 10^6$ psi) having thickness $t = 0.175$ in. and length $L = 48$ in. is bent by couples $M_0$ into a circular are subtending a central angle $\alpha = 40°$ (see figure).

(a) What is the maximum bending stress $\sigma_{max}$ in the rule?

(b) By what percent does the stress increase or decrease if the central angle is increased by 10%?

(c) What percent increase or decrease in rule thickness will result in the maximum stress reaching the allowable value of 42 ksi?

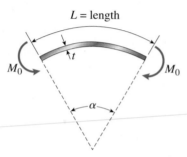

$L = $ length

PROB. 5.5-3

**5.5-4** A simply supported wood beam $AB$ with span length $L = 4$ m carries a uniform load of intensity $q = 5.8$ kN/m (see figure).

(a) Calculate the maximum bending stress $\sigma_{max}$ due to the load $q$ if the beam has a rectangular cross section with width $b = 140$ mm and height $h = 240$ mm.

(b) Repeat part (a) but use the trapezoidal distributed load shown in the figure part b.

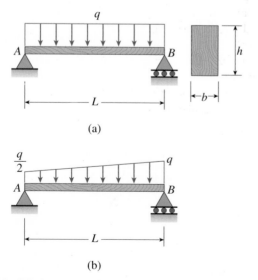

(a)

(b)

PROB. 5.5-4

**5.5-5** Each girder of the lift bridge (see figure) is 180 ft long and simply supported at the ends. The design load for each girder is a uniform load of intensity 1.6 k/ft. The girders are fabricated by welding three steel plates so as to form an I-shaped cross section (see figure) having section modulus $S = 3600$ in³.

What is the maximum bending stress $\sigma_{max}$ in a girder due to the uniform load?

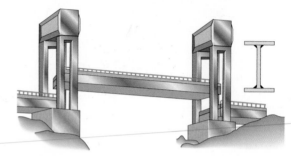

PROB. 5.5-5

**5.5-6** A freight-car axle $AB$ is loaded approximately as shown in the figure, with the forces $P$ representing the car loads (transmitted to the axle through the axle boxes) and the forces $R$ representing the rail loads (transmitted to the axle through the wheels). The diameter of the axle is $d = 82$ mm, the distance between centers of the rails is $L$, and the distance between the forces $P$ and $R$ is $b = 220$ mm.

Calculate the maximum bending stress $\sigma_{max}$ in the axle if $P = 50$ kN.

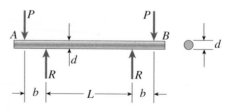

PROB. 5.5-6

**5.5-7** A seesaw weighing 3 lb/ft of length is occupied by two children, each weighing 90 lb (see figure). The center of gravity of each child is 8 ft from the fulcrum. The board is 19 ft long, 8 in. wide, and 1.5 in. thick.

What is the maximum bending stress in the board?

**PROB. 5.5-7**

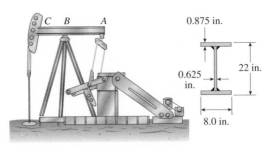

**PROB. 5.5-9**

**5.5-8** During construction of a highway bridge, the main girders are cantilevered outward from one pier toward the next (see figure). Each girder has a cantilever length of 48 m and an I-shaped cross section with dimensions shown in the figure. The load on each girder (during construction) is assumed to be 9.5 kN/m, which includes the weight of the girder.

Determine the maximum bending stress in a girder due to this load.

**5.5-10** A railroad tie (or *sleeper*) is subjected to two rail loads, each of magnitude $P = 175$ kN, acting as shown in the figure. The reaction $q$ of the ballast is assumed to be uniformly distributed over the length of the tie, which has cross-sectional dimensions $b = 300$ mm and $h = 250$ mm.

Calculate the maximum bending stress $\sigma_{max}$ in the tie due to the loads $P$, assuming the distance $L = 1500$ mm and the overhang length $a = 500$ mm.

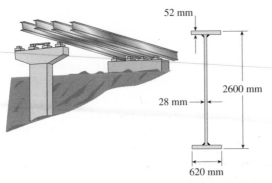

**PROB. 5.5-8**

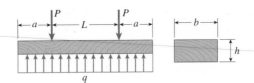

**PROB. 5.5-10**

**5.5-9** The horizontal beam *ABC* of an oil-well pump has the cross section shown in the figure. If the vertical pumping force acting at end *C* is 9 k and if the distance from the line of action of that force to point *B* is 16 ft, what is the maximum bending stress in the beam due to the pumping force?

**5.5-11** A fiberglass pipe is lifted by a sling, as shown in the figure. The outer diameter of the pipe is 6.0 in., its thickness is 0.25 in., and its weight density is 0.053 lb/in.[3] The length of the pipe is $L = 36$ ft and the distance between lifting points is $s = 11$ ft.

(a) Determine the maximum bending stress in the pipe due to its own weight.

(b) Find the spacing $s$ between lift points which will minimize the bending stress. What is the minimum bending stress?

(c) What spacing $s$ will lead to maximum bending stress? What is that stress?

Horizontal beam transfers loads as part of oil well pump
(Gabriel M. Covian/Getty Images)

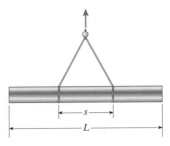

**PROB. 5.5-11**

**5.5-12** A small dam of height $h = 2.0$ m is constructed of vertical wood beams $AB$ of thickness $t = 120$ mm, as shown in the figure. Consider the beams to be simply supported at the top and bottom.

Determine the maximum bending stress $\sigma_{max}$ in the beams, assuming that the weight density of water is $\gamma = 9.81$ kN/m³.

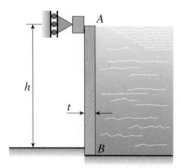

PROB. 5.5-12

**5.5-13** Determine the maximum tensile stress $\sigma_t$ (due to pure bending about a horizontal axis through C by positive bending moments $M$) for beams having cross sections as follows (see figure).

(a) A semicircle of diameter $d$

(b) An isosceles trapezoid with bases $b_1 = b$ and $b_2 = 4b/3$, and altitude $h$

(c) A circular sector with $\alpha = \pi/3$ and $r = d/2$

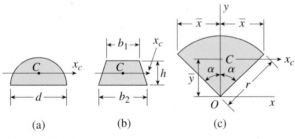

(a)      (b)      (c)

PROB. 5.5-13

**5.5-14** Determine the maximum bending stress $\sigma_{max}$ (due to pure bending by a moment $M$) for a beam having a cross section in the form of a circular core (see figure). The circle has diameter $d$ and the angle $\beta = 60°$. (*Hint:* Use the formulas given in Appendix E, Cases 9 and 15.)

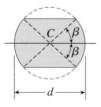

PROB. 5.5-14

**5.5-15** A simple beam $AB$ of span length $L = 24$ ft is subjected to two wheel loads acting at distance $d = 5$ ft apart (see figure). Each wheel transmits a load $P = 3.0$ k, and the carriage may occupy any position on the beam.

(a) Determine the maximum bending stress $\sigma_{max}$ due to the wheel loads if the beam is an I-beam having section modulus $S = 16.2$ in.³

(b) If $d = 5$ ft, find the required span length $L$ to reduce the maximum stress in part (a) to 18 ksi.

(c) If $L = 24$ ft, find the required wheel spacing $s$ to reduce the maximum stress in part (a) to 18 ksi.

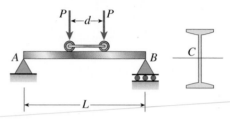

PROB. 5.5-15

**5.5-16** Determine the maximum tensile stress $\sigma_t$ and maximum compressive stress $\sigma_c$ due to the load $P$ acting on the simple beam $AB$ (see figure).

(a) Data are as follows: $P = 6.2$ kN, $L = 3.2$ m, $d = 1.25$ m, $b = 80$ mm, $t = 25$ mm, $h = 120$ mm, and $h_1 = 90$ mm.

(b) Find the value of $d$ for which tensile and compressive stresses will be largest. What are these stresses?

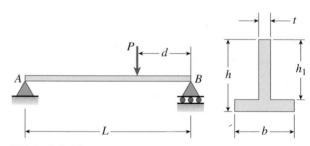

PROB. 5.5-16

**5.5-17** A cantilever beam $AB$, loaded by a uniform load and a concentrated load (see figure), is constructed of a channel section.

(a) Find the maximum tensile stress $\sigma_t$ and maximum compressive stress $\sigma_c$ if the cross section has the dimensions indicated and the moment of inertia about the $z$ axis (the neutral axis) is $I = 3.36$ in.⁴ (*Note:* The uniform load represents the weight of the beam.)

(b) Find the maximum value of the concentrated load if the maximum tensile stress cannot exceed 4 ksi and the maximum compressive stress is limited to 14.5 ksi.

(c) How far from $A$ can load $P = 250$ lb be positioned if the maximum tensile stress cannot exceed 4 ksi and the maximum compressive stress is limited to 14.5 ksi?

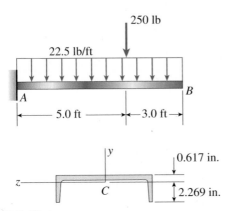

**PROB. 5.5-17**

---

**5.5-18** A cantilever beam $AB$ of isosceles trapezoidal cross section has length $L = 0.8$ m, dimensions $b_1 = 80$ mm, $b_2 = 90$ mm, and height $h = 110$ mm (see figure). The beam is made of brass weighing 85 kN/m³.

(a) Determine the maximum tensile stress $\sigma_t$ and maximum compressive stress $\sigma_c$ due to the beam's own weight.

(b) If the width $b_1$ is doubled, what happens to the stresses?

(c) If the height $h$ is doubled, what happens to the stresses?

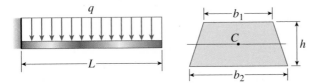

**PROB. 5.5-18**

---

**5.5-19** A beam $ABC$ with an overhang from $B$ to $C$ supports a uniform load of 200 lb/ft throughout its length (see figure). The beam is a channel section with dimensions as shown in the figure. The moment of inertia about the $z$ axis (the neutral axis) equals 8.13 in.⁴

(a) Calculate the maximum tensile stress $\sigma_t$ and maximum compressive stress $\sigma_c$ due to the uniform load.

(b) Find required span length $a$ that results in the ratio of larger to smaller compressive stress being equal to the ratio of larger to smaller tensile stress for the beam. Assume that the total length $L = a + b = 18$ ft remains unchanged.

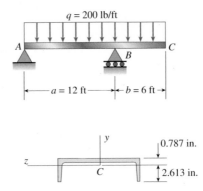

**PROB. 5.5-19**

---

**5.5-20** A frame $ABC$ travels horizontally with an acceleration $a_0$ (see figure). Obtain a formula for the maximum stress $\sigma_{max}$ in the vertical arm $AB$, which has length $L$, thickness $t$, and mass density $\rho$.

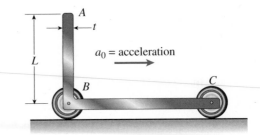

**PROB. 5.5-20**

---

**5.5-21** A beam of T-section is supported and loaded as shown in the figure. The cross section has width $b = 2$ 1/2 in., height $h = 3$ in., and thickness $t = 3/8$ in.

(a) Determine the maximum tensile and compressive stresses in the beam.

(b) If the allowable stresses in tension and compression are 18 ksi and 12 ksi, respectively, what is the required depth $h$ of the beam? Assume that thickness $t$ remains at 3/8 in. and that flange width $b = 2.5$ in.

(c) Find the new values of loads $P$ and $q$ so that allowable tension (18 ksi) and compression (12 ksi) stresses are reached simultaneously for the beam. Use the beam cross section in part (a) (see figure) and assume that $L_1$, $L_2$, and $L_3$ are unchanged.

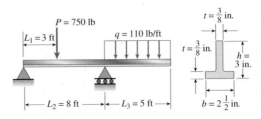

**PROB. 5.5-21**

**5.5-22** A cantilever beam $AB$ with a rectangular cross section has a longitudinal hole drilled throughout its length (see figure). The beam supports a load $P = 600$ N. The cross section is 25 mm wide and 50 mm high, and the hole has a diameter of 10 mm.

Find the bending stresses at the top of the beam, at the top of the hole, and at the bottom of the beam.

**5.5-24** Consider the compound beam with segments $AB$ and $BCD$ joined by a pin connection (moment release) just right of $B$ (see figure part a). The beam cross section is a double-T made up from three 50 mm $\times$ 150 mm wood members (actual dimensions, see figure part b).

(a) Find the centroid $C$ of the double-T cross section $(c_1, c_2)$, then compute the moment of inertia $I_z$ (mm$^4$).

(b) Find the maximum *tensile* normal stress $\sigma_t$ and maximum *compressive* normal stress $\sigma_c$ (kPa) for the loading shown. (Ignore the weight of the beam.)

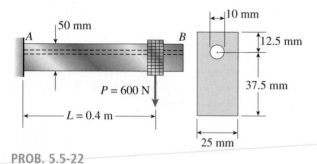

PROB. 5.5-22

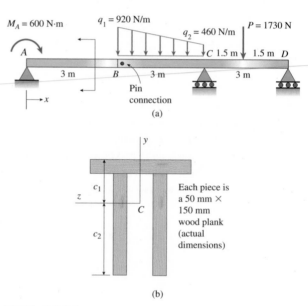

(a)

(b)

PROB. 5.5-24

**5.5-23** A small dam of height $h = 6$ ft is constructed of vertical wood beams $AB$, as shown in the figure. The wood beams, which have thickness $t = 2.5$ in., are simply supported by horizontal steel beams at $A$ and $B$.

Construct a graph showing the maximum bending stress $\sigma_{max}$ in the wood beams versus the depth $d$ of the water above the lower support at $B$. Plot the stress $\sigma_{max}$ (psi) as the ordinate and the depth $d$ (ft) as the abscissa. (*Note:* The weight density $\gamma$ of water equals 62.4 lb/ft$^3$.)

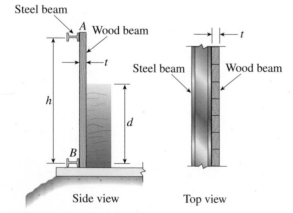

PROB. 5.5-23

**5.5-25** A steel post ($E = 30 \times 10^6$ psi) having thickness $t = 1/8$ in. and height $L = 72$ in. supports a stop sign (see figure: $s = 12.5$ in.). The height of the post $L$ is measured from the base to the centroid of the sign. The stop sign is subjected to wind pressure $p = 20$ lb/ft$^2$ normal to its surface. Assume that the post is fixed at its base.

(a) What is the resultant load on the sign? [See Appendix E, Case 25, for properties of an octagon, $n = 8$].

(b) What is the maximum bending stress $\sigma_{max}$ in the post?

(c) Repeat part (b) if the circular cut-outs are eliminated over the height of the post.

Determine the minimum value of $d$ based upon an allowable bending stress of 1125 psi in the wood tie. (Disregard the weight of the tie itself.)

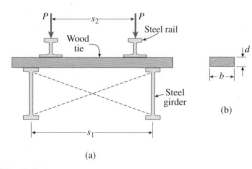

(a)

(b)

**PROB. 5.6-1**

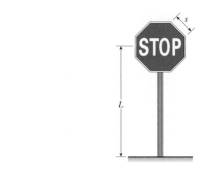

Section A–A

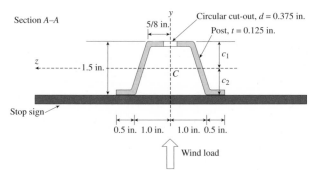

Numerical properties of post

$A = 0.578$ in.$^2$, $c_1 = 0.769$ in., $c_2 = 0.731$ in., $I_y = 0.44867$ in.$^4$, $I_z = 0.16101$ in.$^4$

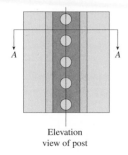

Elevation view of post

**PROB. 5.5-25**

# Design of Beams

**5.6-1** The cross section of a narrow-gage railway bridge is shown in part a of the figure. The bridge is constructed with longitudinal steel girders that support the wood cross ties. The girders are restrained against lateral buckling by diagonal bracing, as indicated by the dashed lines.

The spacing of the girders is $s_1 = 50$ in. and the spacing of the rails is $s_2 = 30$ in. The load transmitted by each rail to a single tie is $P = 1500$ lb. The cross section of a tie, shown in part b of the figure, has width $b = 5.0$ in. and depth $d$.

**5.6-2** A fiberglass bracket $ABCD$ of solid circular cross section has the shape and dimensions shown in the figure. A vertical load $P = 40$ N acts at the free end $D$.

(a) Determine the minimum permissible diameter $d_{min}$ of the bracket if the allowable bending stress in the material is 30 MPa and $b = 37$ mm. (*Note*: Disregard the weight of the bracket itself.)

(b) If $d = 10$ mm, $b = 37$ mm, and $\sigma_{allow} = 30$ MPa, what is the maximum value of load $P$ if vertical load $P$ at $D$ is replaced with horizontal loads $P$ at $B$ and $D$ (see figure part b)?

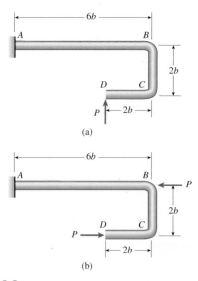

**PROB. 5.6-2**

**5.6-3** A cantilever beam $AB$ is loaded by a uniform load $q$ and a concentrated load $P$, as shown in the figure.

(a) Select the most economical steel C shape from Table F-3(a) in Appendix F; use $q = 20$ lb/ft and $P = 300$ lb (assume allowable normal stress is $\sigma_a = 18$ ksi).

(b) Select the most economical steel S shape from Table F-2(a) in Appendix F; use $q = 45$ lb/ft and $P = 2000$ lb (assume allowable normal stress is $\sigma_a = 20$ ksi).

(c) Select the most economical steel W shape from Table F-1(a) in Appendix F; use $q = 45$ lb/ft and $P = 2000$ lb (assume allowable normal stress is $\sigma_a = 20$ ksi). *However*, assume that the design requires that the W shape must be used in weak axis bending, i.e., it must bend about the 2–2 (or $y$) axis of the cross section.

*Note*: For parts (a), (b), and (c), revise your initial beam selection as needed to include the distributed weight of the beam in addition to uniform load $q$.

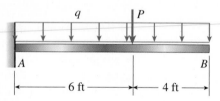

PROB. 5.6-3

**5.6-4** A simple beam of length $L = 5$ m carries a uniform load of intensity $q = 5.8$ kN/m and a concentrated load 22.5 kN (see figure).

(a) Assuming $\sigma_{allow} = 110$ MPa, calculate the required section modulus $S$. Then select the most economical wide-flange beam (W shape) from Table F-1(b) in Appendix F, and recalculate $S$, taking into account the weight of the beam. Select a new beam if necessary.

(b) Repeat part (a), but now assume that the design requires that the W shape must be used in weak axis bending (i.e., it must bend about the 2–2 (or $y$) axis of the cross section).

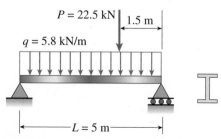

PROB. 5.6-4

**5.6-5** A simple beam $AB$ is loaded as shown in the figure.

(a) Calculate the required section modulus $S$ if $\sigma_{allow} = 18,000$ psi, $L = 32$ ft, $P = 2900$ lb, and $q = 450$ lb/ft. Then select a suitable I-beam (S shape) from Table F-2(a), Appendix F, and recalculate $S$ taking into account the weight of the beam. Select a new beam size if necessary.

(b) What is the maximum load $P$ that can be applied to your final beam selection in part (a)?

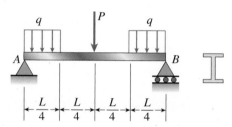

PROB. 5.6-5

**5.6-6** A pontoon bridge (see figure) is constructed of two longitudinal wood beams, known as *balks*, that span between adjacent pontoons and support the transverse floor beams, which are called *chesses*. For purposes of design, assume that a uniform floor load of 7.5 kPa acts over the chesses. (This load includes an allowance for the weights of the chesses and balks.) Also, assume that the chesses are 2.5 m long and that the balks are simply supported with a span of 3.0 m. The allowable bending stress in the wood is 15 MPa.

(a) If the balks have a square cross section, what is their minimum required width $b_{min}$?

(b) Repeat part (a) if the balk width is $1.5 b$ and the balk depth is $b$; compare the cross-sectional areas of the two designs.

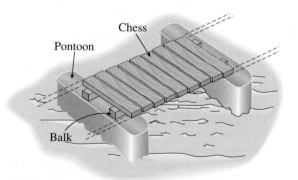

PROB. 5.6-6

**5.6-7** A floor system in a small building consists of wood planks supported by 2-in. (nominal width) joists spaced at distance $s$ and measured from center to center (see figure). The span length $L$ of each joist is 12 ft, the spacing $s$ of the joists is 16 in., and the allowable bending stress in the wood is 1250 psi. The uniform floor load is 120 lb/ft$^2$, which includes an allowance for the weight of the floor system itself.

(a) Calculate the required section modulus $S$ for the joists, and then select a suitable joist size (surfaced lumber) from Appendix G, assuming that each joist may be represented as a simple beam carrying a uniform load.

(b) What is the maximum floor load that can be applied to your final beam selection in part (a)?

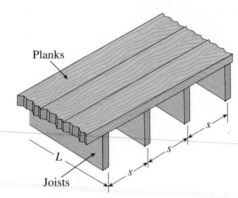

PROBs. 5.6-7 and 5.6-8

**5.6-8** The wood joists supporting a plank floor (see figure) are 38 mm × 220 mm in cross section (actual dimensions) and have a span length of $L = 4.0$ m. The floor load is 5.0 kPa, which includes the weight of the joists and the floor.

(a) Calculate the maximum permissible spacing $s$ of the joists if the allowable bending stress is 14 MPa. (Assume that each joist may be represented as a simple beam carrying a uniform load.)

(b) If spacing $s = 406$ mm, what is the required depth $h$ of the joist? Assume all other variables remain unchanged.

**5.6-9** A beam $ABC$ with an overhang from $B$ to $C$ is constructed of a C 10 × 30 channel section with flanges facing upward (see figure). The beam supports its own weight (30 lb/ft) plus a *triangular* load of maximum intensity $q_0$ acting on the overhang. The allowable stresses in tension and compression are 18 ksi and 12 ksi, respectively.

(a) Determine the allowable *triangular* load intensity $q_{0,allow}$ if the distance $L$ equals 4 ft.

(b) What is the allowable triangular load intensity $q_{0,allow}$ if the beam is rotated 180° about its longitudinal centroidal axis so that the flanges are downward.

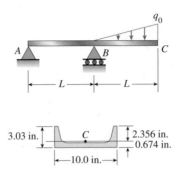

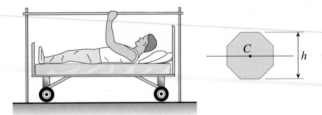

PROB. 5.6-9

**5.6-10** A so-called "trapeze bar" in a hospital room provides a means for patients to exercise while in bed (see figure). The bar is 2.1 m long and has a cross section in the shape of a regular octagon. The design load is 1.2 kN applied at the midpoint of the bar, and the allowable bending stress is 200 MPa.

Determine the minimum height $h$ of the bar. (Assume that the ends of the bar are simply supported and that the weight of the bar is negligible.)

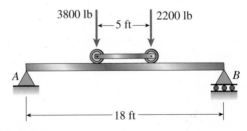

PROB. 5.6-10

**5.6-11** A two-axle carriage that is part of an overhead traveling crane in a testing laboratory moves slowly across a simple beam $AB$ (see figure). The load transmitted to the beam from the front axle is 2200 lb and from the rear axle is 3800 lb. The weight of the beam itself may be disregarded.

(a) Determine the minimum required section modulus $S$ for the beam if the allowable bending stress is 17.0 ksi, the length of the beam is 18 ft, and the wheelbase of the carriage is 5 ft.

(b) Select the most economical I-beam (S shape) from Table F-2(a), Appendix F.

PROB. 5.6-11

**5.6-12** A cantilever beam $AB$ of circular cross section and length $L$ = 750 mm supports a load $P$ = 800 N acting at the free end (see figure). The beam is made of steel with an allowable bending stress of 120 MPa.

(a) Determine the required diameter $d_{min}$ (figure part a) of the beam, considering the effect of the beam's own weight.

(b) Repeat part (a) if the beam is hollow with wall thickness $t = d/8$ (figure part b); compare the cross-sectional areas of the two designs.

density $\gamma$ = 5.5 kN/m³.) The allowable bending stress in the cantilevers is 15 MPa.

Assuming that the middle cantilever supports 50% of the load and each outer cantilever supports 25% of the load, determine the required dimensions $b$ and $h$.

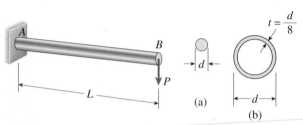

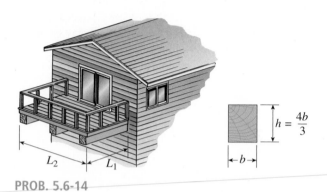

PROB. 5.6-14

PROB. 5.6-12

**5.6-13** A propped cantilever beam $ABC$ (see figure) has a shear release just right of the mid-span.

(a) Select the most economical wood beam from the table in Appendix G; assume $q$ = 55 lb/ft, $L$ = 16 ft, $\sigma_{aw}$ = 1750 psi, and $\tau_{aw}$ = 375 psi. Include the self-weight of the beam in your design.

(b) If a C 10 × 25 steel beam is now used for beam $ABC$, what is the maximum permissible value of load variable $q$? Assume $\sigma_{as}$ = 16 ksi and $L$ = 10 ft. Include the self-weight of the beam in your analysis.

**5.6-15** A beam having a cross section in the form of an unsymmetric wide-flange shape (see figure) is subjected to a negative bending moment acting about the $z$ axis.

Determine the width $b$ of the top flange in order that the stresses at the top and bottom of the beam will be in the ratio 4:3, respectively.

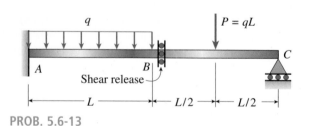

PROB. 5.6-13

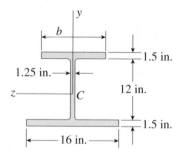

PROB. 5.6-15

**5.6-14** A small balcony constructed of wood is supported by three identical cantilever beams (see figure). Each beam has length $L_1$ = 2.1 m, width $b$, and height $h$ = $4b/3$. The dimensions of the balcony floor are $L_1 \times L_2$, with $L_2$ = 2.5 m. The design load is 5.5 kPa acting over the entire floor area. (This load accounts for all loads except the weights of the cantilever beams, which have a weight

**5.6-16** A beam having a cross section in the form of a channel (see figure) is subjected to a bending moment acting about the $z$ axis.

Calculate the thickness $t$ of the channel in order that the bending stresses at the top and bottom of the beam will be in the ratio 7:3, respectively.

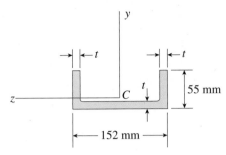

PROB. 5.6-16

**5.6-17** Determine the ratios of the weights of four beams that have the same length, are made of the same material, are subjected to the same maximum bending moment, and have the same maximum bending stress if their cross sections are (1) a rectangle with height equal to twice the width, (2) a square, (3) a circle, and (4) a pipe with outer diameter $d$ and wall thickness $t = d/8$ (see figures).

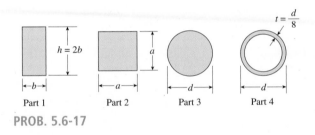

PROB. 5.6-17

**5.6-18** A horizontal shelf $AD$ of length $L = 1215$ mm, width $b = 305$ mm, and thickness $t = 22$ mm is supported by brackets at $B$ and $C$ (see part a of the figure). The brackets are adjustable and may be placed in any desired positions between the ends of the shelf. A uniform load of intensity $q$, which includes the weight of the shelf itself, acts on the shelf (see part b of the figure).

(a) Determine the maximum permissible value of the load $q$ if the allowable bending stress in the shelf is $\sigma_{allow} = 8.5$ MPa and the position of the supports is adjusted for maximum load carrying capacity.

(b) The bookshelf owner decides to reinforce the shelf with a bottom wood plate $b/2 \times t/2$ along its entire length (see figure part c). Find the new maximum permissible value of the load $q$ if the allowable bending stress in the shelf remains at $\sigma_{allow} = 8.5$ MPa.

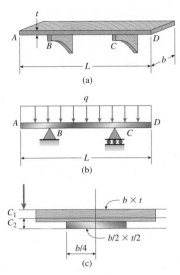

PROB. 5.6-18

**5.6-19** A steel plate (called a *cover plate*) having cross sectional dimensions 6.0 in. × 0.5 in. is welded along the full length of the bottom flange of a W 12 × 50 wide-flange beam (see figure, which shows the beam cross section).

What is the percent increase in the smaller section modulus (as compared to the wide-flange beam alone)?

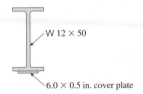

W 12 × 50

6.0 × 0.5 in. cover plate

PROB. 5.6-19

**5.6-20** A steel beam $ABC$ is simply supported at $A$ and $B$ and has an overhang $BC$ of length $L = 150$ mm (see figure). The beam supports a uniform load of intensity $q = 4.0$ kN/m over its entire span $AB$ and $1.5q$ over $BC$. The cross section of the beam is rectangular with width $b$ and height $2b$. The allowable bending stress in the steel is $\sigma_{allow} = 60$ MPa, and its weight density is $\gamma = 77.0$ kN/m$^3$.

(a) Disregarding the weight of the beam, calculate the required width $b$ of the rectangular cross section.

(b) Taking into account the weight of the beam, calculate the required width $b$.

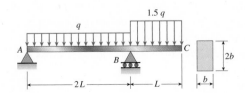

PROB. 5.6-20

**5.6-21** A retaining wall 6 ft high is constructed of horizontal wood planks 2.5 in. thick (actual dimension) that are supported by vertical wood piles of a 12-in. diameter (actual dimension), as shown in the figure. The lateral earth pressure is $p_1$ = 125 lb/ft² at the top of the wall and $p_2$ = 425 lb/ft² at the bottom.

(a) Assuming that the allowable stress in the wood is 1175 psi, calculate the maximum permissible spacing $s$ of the piles.

(b) Find the required diameter of the wood piles so that piles and planks ($t$ = 2.5 in.) reach the allowable stress at the same time.

(*Hint:* Observe that the spacing of the piles may be governed by the load-carrying capacity of either the planks or the piles. Consider the piles to act as cantilever beams subjected to a trapezoidal distribution of load, and consider the planks to act as simple beams between the piles. To be on the safe side, assume that the pressure on the bottom plank is uniform and equal to the maximum pressure.)

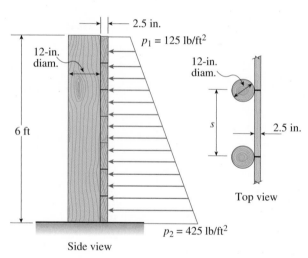

Side view

**PROB. 5.6-21**

**5.6-22** A beam of square cross section ($a$ = length of each side) is bent in the plane of a diagonal (see figure). By removing a small amount of material at the top and bottom corners, as shown by the shaded triangles in the figure, we can increase the section modulus and obtain a stronger beam, even though the area of the cross section is reduced.

(a) Determine the ratio $\beta$ defining the areas that should be removed in order to obtain the strongest cross section in bending.

(b) By what percent is the section modulus increased when the areas are removed?

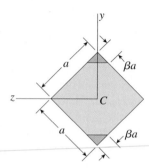

**PROB. 5.6-22**

**5.6-23** The cross section of a rectangular beam having width $b$ and height $h$ is shown in part a of the figure. For reasons unknown to the beam designer, it is planned to add structural projections of width $b/9$ and height $d$ to the top and bottom of the beam (see part b of the figure).

For what values of $d$ is the bending-moment capacity of the beam increased? For what values is it decreased?

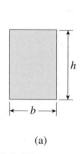
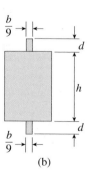

(a)                    (b)

**PROB. 5.6-23**

# Nonprismatic Beams

**5.7-1** A tapered cantilever beam $AB$ of length $L$ has square cross sections and supports a concentrated load $P$ at the free end (see figure part a). The width and height of the beam vary linearly from $h_A$ at the free end to $h_B$ at the fixed end.

Determine the distance $x$ from the free end $A$ to the cross section of maximum bending stress if $h_B = 3h_A$.

(a) What is the magnitude $\sigma_{max}$ of the maximum bending stress? What is the ratio of the maximum stress to the largest stress $B$ at the support?

(b) Repeat part (a) if load $P$ is now applied as a uniform load of intensity $q = P/L$ over the entire beam, $A$ is restrained by a roller support and $B$ is a sliding support (see figure part b).

(a) At what distance $x$ from the free end does the maximum bending stress occur? What is the magnitude $\sigma_{max}$ of the maximum bending stress? What is the ratio of the maximum stress to the largest stress $\sigma_B$ at the support?

(b) Repeat part (a) if concentrated load $P$ is applied upward at $A$ and downward uniform load $q(x) = 2P/L$ is applied over the entire beam as shown in the figure part b. What is the ratio of the maximum stress to the stress at the location of maximum moment?

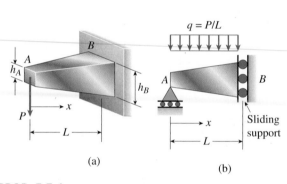

(a)

(b)

**PROB. 5.7-1**

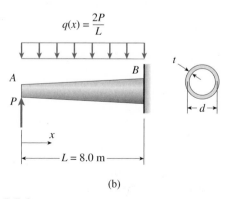

(a)

**5.7-2** A tall signboard is supported by two vertical beams consisting of thin-walled, tapered circular tubes (see figure part a). For purposes of this analysis, each beam may be represented as a cantilever $AB$ of length $L = 8.0$ m subjected to a lateral load $P = 2.4$ kN at the free end. The tubes have constant thickness $t = 10.0$ mm and average diameters $d_A = 90$ mm and $d_B = 270$ mm at ends $A$ and $B$, respectively.

Because the thickness is small compared to the diameters, the moment of inertia at any cross section may be obtained from the formula $I = \pi d^3 t/8$ (see Case 22, Appendix E), and therefore, the section modulus may be obtained from the formula $S = \pi d^2 t/4$.

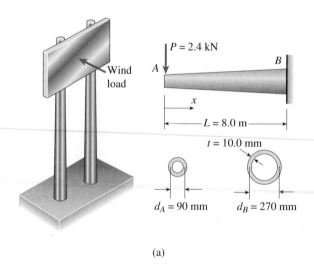

(b)

**PROB. 5.7-2**

**5.7-3** A tapered cantilever beam $AB$ having rectangular cross sections is subjected to a concentrated load $P = 50$ lb and a couple $M_0 = 800$ lb-in. acting at the free end (see figure part a). The width $b$ of the beam is constant and equal to 1.0 in., but the height varies linearly from $h_A = 2.0$ in. at the loaded end to $h_B = 3.0$ in. at the support.

(a) At what distance $x$ from the free end does the maximum bending stress $\sigma_{max}$ occur? What is the magnitude $\sigma_{max}$ of the maximum bending stress? What is the ratio of the maximum stress to the largest stress $\sigma_B$ at the support?

(b) Repeat part (a) if, in addition to $P$ and $M_0$, a triangular distributed load with peak intensity $q_0 = 3P/L$ acts upward over the entire beam as shown in the figure part b. What is the ratio of the maximum stress to the stress at the location of maximum moment?

Considering only the effects of bending due to the loads $P$ and $M_0$, determine the following quantities.

(a) The largest bending stress $\sigma_A$ at end $A$.

(b) The largest bending stress $\sigma_B$ at end $B$.

(c) The distance $x$ to the cross section of maximum bending stress.

(d) The magnitude $\sigma_{max}$ of the maximum bending stress.

(e) Repeat part (d) if uniform load $q(x) = 10P/3L$ is added to loadings $P$ and $M_0$, as shown in the figure part b.

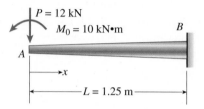

**PROB. 5.7-3**

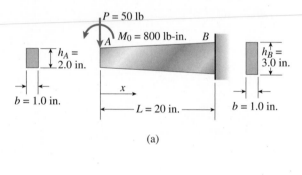

(a)

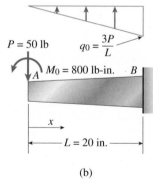

(b)

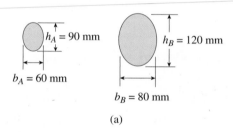

(a)

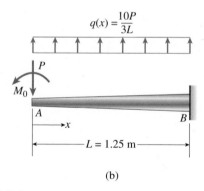

(b)

**PROB. 5.7-4**

**5.7-4** The spokes in a large flywheel are modeled as beams fixed at one end and loaded by a force $P$ and a couple $M_0$ at the other (see figure). The cross sections of the spokes are elliptical with major and minor axes (height and width, respectively) having the lengths shown in the figure part a. The cross-sectional dimensions vary linearly from end $A$ to end $B$.

**5.7-5** Refer to the tapered cantilever beam of solid circular cross section shown in Fig. 5-24 of Example 5-9.

(a) Considering only the bending stresses due to the load $P$, determine the range of values of the ratio $d_B/d_A$ for which the maximum normal stress occurs at the support.

(b) What is the maximum stress for this range of values?

## Fully Stressed Beams

*Problems 5.7-6 to 5.7-8 pertain to fully stressed beams of rectangular cross section. Consider only the bending stresses obtained from the flexure formula and disregard the weights of the beams.*

**5.7-6** A cantilever beam $AB$ having rectangular cross sections with constant width $b$ and varying height $h_x$ is subjected to a uniform load of intensity $q$ (see figure).

How should the height $h_x$ vary as a function of $x$ (measured from the free end of the beam) in order to have a fully stressed beam? (Express $h_x$ in terms of the height $h_B$ at the fixed end of the beam.)

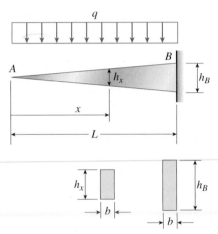

PROB. 5.7-6

**5.7-7** A simple beam $ABC$ having rectangular cross sections with constant height $h$ and varying width $b_x$ supports a concentrated load $P$ acting at the midpoint (see figure).

How should the width $b_x$ vary as a function of $x$ in order to have a fully stressed beam? (Express $b_x$ in terms of the width $b_B$ at the midpoint of the beam.)

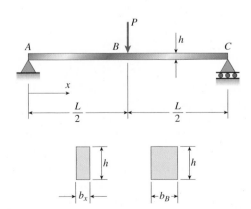

PROB. 5.7-7

**5.7-8** A cantilever beam $AB$ having rectangular cross sections with varying width $b_x$ and varying height $h_x$ is subjected to a uniform load of intensity $q$ (see figure). If the width varies linearly with $x$ according to the equation $b_x = b_B x/L$, how should the height $h_x$ vary as a function of $x$ in order to have a fully stressed beam? (Express $h_x$ in terms of the height $h_B$ at the fixed end of the beam.)

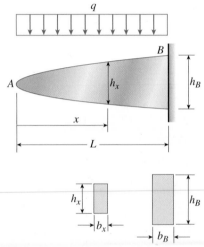

PROB. 5.7-8

## Shear Stresses in Rectangular Beams

**5.8-1** The shear stresses $\tau$ in a rectangular beam are given by Eq. (5-43):

$$\tau = \frac{V}{2I}\left(\frac{h^2}{4} - y_1^2\right)$$

in which $V$ is the shear force, $I$ is the moment of inertia of the cross-sectional area, $h$ is the height of the beam, and $y_1$ is the distance from the neutral axis to the point where the shear stress is being determined (Fig. 5-30).

By integrating over the cross-sectional area, show that the resultant of the shear stresses is equal to the shear force $V$.

**5.8-2** Calculate the maximum shear stress $\tau_{max}$ and the maximum bending stress $\sigma_{max}$ in a wood beam (see figure) carrying a uniform load of 22.5 kN/m (which includes the weight of the beam) if the length is 1.95 m and the cross section is rectangular with width 150 mm and height 300 mm, and the beam is either (a) simply supported as in the figure part a, or (b) has a sliding support at right as in the figure part b.

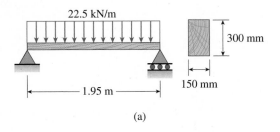

22.5 kN/m

300 mm

150 mm

1.95 m

(a)

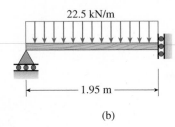

22.5 kN/m

1.95 m

(b)

PROB. 5.8-2

**5.8-3** Two wood beams, each of rectangular cross section (3.0 in. × 4.0 in., actual dimensions) are glued together to form a solid beam of dimensions 6.0 in. × 4.0 in. (see figure). The beam is simply supported with a span of 8 ft.

(a) What is the maximum moment $M_{max}$ that may be applied at the left support if the allowable shear stress in the glued joint is 200 psi? (Include the effects of the beam's own weight, assuming that the wood weighs 35 lb/ft³.)

(b) Repeat part (a) if $M_{max}$ is based on allowable bending stress of 2500 psi.

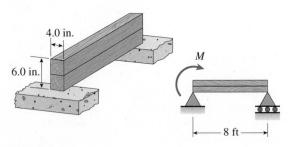

4.0 in.

6.0 in.

M

8 ft

PROB. 5.8-3

**5.8-4** A cantilever beam of length $L = 2$ m supports a load $P = 8.0$ kN (see figure). The beam is made of wood with cross-sectional dimensions 120 mm × 200 mm.

Calculate the shear stresses due to the load $P$ at points located 25 mm, 50 mm, 75 mm, and 100 mm from the top surface of the beam. From these results, plot a graph showing the distribution of shear stresses from top to bottom of the beam.

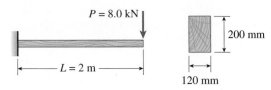

P = 8.0 kN

200 mm

L = 2 m

120 mm

PROB. 5.8-4

**5.8-5** A steel beam of length $L = 16$ in. and cross-sectional dimensions $b = 0.6$ in. and $h = 2$ in. (see figure) supports a uniform load of intensity $q = 240$ lb/in., which includes the weight of the beam.

Calculate the shear stresses in the beam (at the cross section of maximum shear force) at points located 1/4 in., 1/2 in., 3/4 in., and 1 in. from the top surface of the beam. From these calculations, plot a graph showing the distribution of shear stresses from top to bottom of the beam.

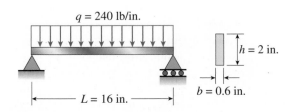

q = 240 lb/in.

h = 2 in.

L = 16 in.

b = 0.6 in.

PROB. 5.8-5

**5.8-6** A beam of rectangular cross section (width $b$ and height $h$) supports a uniformly distributed load along its entire length $L$. The allowable stresses in bending and shear are $\sigma_{allow}$ and $\tau_{allow}$, respectively.

(a) If the beam is simply supported, what is the span length $L_0$ below which the shear stress governs the allowable load and above which the bending stress governs?

(b) If the beam is supported as a cantilever, what is the length $L_0$ below which the shear stress governs the allowable load and above which the bending stress governs?

**5.8-7** A laminated wood beam on simple supports (figure part a) is built up by gluing together four 2 in. × 4 in. boards (actual dimensions) to form a solid beam 4 in. × 8 in. in cross section, as shown in the figure part b. The allowable shear stress in the glued joints is 62 psi, the allowable shear stress in the wood is 175 psi, and the allowable bending stress in the wood is 1650 psi.

(a) If the beam is 12 ft long, what is the allowable load P acting at the one-third point along the beam, as shown? (Include the effects of the beam's own weight, assuming that the wood weighs 35 lb/ft³.)

(b) Repeat part (a) if the beam is assembled by gluing together two 3 in. × 4 in. boards and a 2 in. × 4 in. board (see figure part c).

**5.8-9** A wood beam AB on simple supports with span length equal to 10 ft is subjected to a uniform load of intensity 125 lb/ft acting along the entire length of the beam, a concentrated load of magnitude 7500 lb acting at a point 3 ft from the right-hand support, and a moment at A of 18,500 ft-lb (see figure). The allowable stresses in bending and shear, respectively, are 2250 psi and 160 psi.

(a) From the table in Appendix G, select the lightest beam that will support the loads (disregard the weight of the beam).

(b) Taking into account the weight of the beam (weight density = 35 lb/ft³), verify that the selected beam is satisfactory, or if it is not, select a new beam.

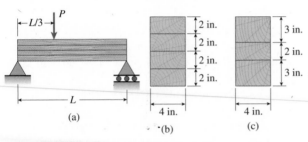

(a)

(b)    (c)

PROB. 5.8-7

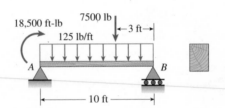

PROB. 5.8-9

**5.8-8** A laminated plastic beam of square cross section is built up by gluing together three strips, each 10 mm × 30 mm in cross section (see figure). The beam has a total weight of 3.6 N and is simply supported with span length L = 360 mm.

Considering the weight of the beam (q), calculate the maximum permissible CCW moment M that may be placed at the right support.

(a) If the allowable shear stress in the glued joints is 0.3 MPa.

(b) If the allowable bending stress in the plastic is 8 MPa.

**5.8-10** A simply supported wood beam of rectangular cross section and span length 1.2 m carries a concentrated load P at midspan in addition to its own weight (see figure). The cross section has width 140 mm and height 240 mm. The weight density of the wood is 5.4 kN/m³.

Calculate the maximum permissible value of the load P if (a) the allowable bending stress is 8.5 MPa, and (b) the allowable shear stress is 0.8 MPa.

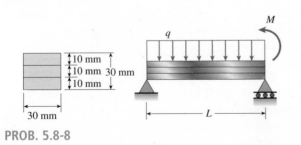

PROB. 5.8-8

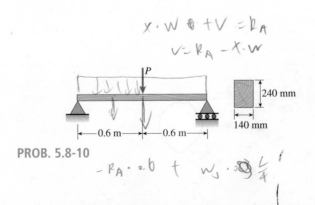

PROB. 5.8-10

**5.8-11** A square wood platform, 8 ft × 8 ft in area, rests on masonry walls (see figure). The deck of the platform is constructed of 2-in. nominal thickness tongue-and-groove planks (actual thickness 1.5 in.; see Appendix G) supported on two 8-ft long beams. The beams have 4 in. × 6 in. nominal dimensions (actual dimensions 3.5 in. × 5.5 in.).

The planks are designed to support a uniformly distributed load $w$ (lb/ft$^2$) acting over the entire top surface of the platform. The allowable bending stress for the planks is 2400 psi and the allowable shear stress is 100 psi. When analyzing the planks, disregard their weights and assume that their reactions are uniformly distributed over the top surfaces of the supporting beams.

(a) Determine the allowable platform load $w_1$ (lb/ft$^2$) based upon the bending stress in the planks.

(b) Determine the allowable platform load $w_2$ (lb/ft$^2$) based upon the shear stress in the planks.

(c) Which of the preceding values becomes the allowable load $w_{allow}$ on the platform?

(*Hints:* Use care in constructing the loading diagram for the planks, noting especially that the reactions are distributed loads instead of concentrated loads. Also, note that the maximum shear forces occur at the inside faces of the supporting beams.)

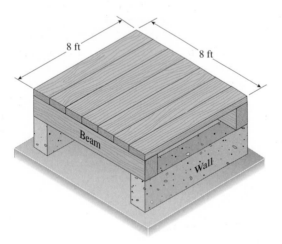

PROB. 5.8-11

**5.8-12** A wood beam $ABC$ with simple supports at $A$ and $B$ and an overhang $BC$ has height $h = 300$ mm (see figure). The length of the main span of the beam is $L = 3.6$ m and the length of the overhang is $L/3 = 1.2$ m. The beam supports a concentrated load $3P = 18$ kN at the midpoint of the main span and a moment $PL/2 = 10.8$ kN·m at the free end of the overhang. The wood has weight density $\gamma = 5.5$ kN/m$^3$.

(a) Determine the required width $b$ of the beam based upon an allowable bending stress of 8.2 MPa.

(b) Determine the required width based upon an allowable shear stress of 0.7 MPa.

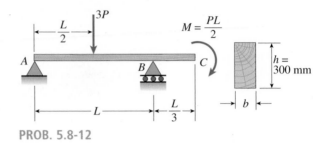

PROB. 5.8-12

## Shear Stresses in Circular Beams

**5.9-1** A wood pole of solid circular cross section ($d$ = diameter) is subjected to a triangular distributed horizontal force of peak intensity $q_0 = 20$ lb/in. (see figure). The length of the pole is $L = 6$ ft, and the allowable stresses in the wood are 1900 psi in bending and 120 psi in shear.

Determine the minimum required diameter of the pole based upon (a) the allowable bending stress, and (b) the allowable shear stress.

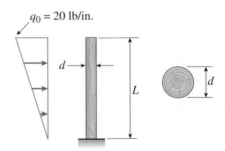

PROB. 5.9-1

**5.9-2** A simple log bridge in a remote area consists of two parallel logs with planks across them (see figure). The logs are Douglas fir with average diameter 300 mm. A truck moves slowly across the bridge, which spans 2.5 m. Assume that the weight of the truck is equally distributed between the two logs.

Because the wheelbase of the truck is greater than 2.5 m, only one set of wheels is on the bridge at a time. Thus, the wheel load on one log is equivalent to a concentrated load $W$ acting at any position along the span. In addition, the weight of one log and the planks it supports is equivalent to a uniform load of 850 N/m acting on the log.

Determine the maximum permissible wheel load $W$ based upon (a) an allowable bending stress of 7.0 MPa, and (b) an allowable shear stress of 0.75 MPa.

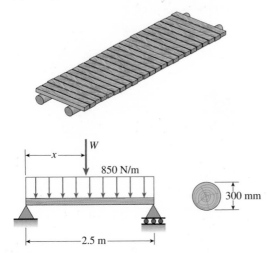

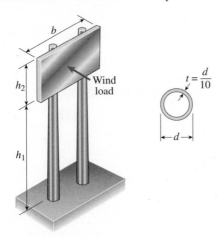

PROB. 5.9-2

**5.9-3** A sign for an automobile service station is supported by two aluminum poles of hollow circular cross section, as shown in the figure. The poles are being designed to resist a wind pressure of 75 lb/ft$^2$ against the full area of the sign. The dimensions of the poles and sign are $h_1 = 20$ ft, $h_2 = 5$ ft, and $b = 10$ ft. To prevent buckling of the walls of the poles, the thickness $t$ is specified as one-tenth the outside diameter $d$.

(a) Determine the minimum required diameter of the poles based upon an allowable bending stress of 7500 psi in the aluminum.

(b) Determine the minimum required diameter based upon an allowable shear stress of 2000 psi.

PROB. 5.9-3

**5.9-4** A steel pipe is subjected to a quadratic distributed load over its height, with the peak intensity $q_0$ at the base

(see figure). Assume the following pipe properties and dimensions: height $L$, outside diameter $d = 200$ mm, and wall thickness $t = 10$ mm. Allowable stresses for flexure and shear are $\sigma_a = 125$ MPa and $\tau_a = 30$ MPa.

(a) If $L = 2.6$ m, find $q_{0,max}$ (kN/m), assuming that allowable flexure and shear stresses in the pipe are not to be exceeded.

(b) If $q_0 = 60$ kN/m, find the maximum height $L_{max}$ (m) of the pipe if the allowable flexure and shear stresses in the pipe are not to be exceeded.

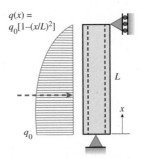

PROB. 5.9-4

# Shear Stresses in Beams with Flanges

**5.10-1 through 5.10-6** A wide-flange beam (see figure) having the cross section described below is subjected to a shear force $V$. Using the dimensions of the cross section, calculate the moment of inertia and then determine the following quantities:

(a) The maximum shear stress $\tau_{max}$ in the web.

(b) The minimum shear stress $\tau_{min}$ in the web.

(c) The average shear stress $\tau_{aver}$ (obtained by dividing the shear force by the area of the web) and the ratio $\tau_{max}/\tau_{aver}$.

(d) The shear force $V_{web}$ carried in the web and the ratio $V_{web}/V$.

(*Note:* Disregard the fillets at the junctions of the web and flanges and determine all quantities, including the moment of inertia, by considering the cross section to consist of three rectangles.)

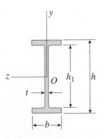

PROBS. 5.10-1 through 5.10-6

**5.10-1** Dimensions of cross section: $b = 6$ in., $t = 0.5$ in., $h = 12$ in., $h_1 = 10.5$ in., and $V = 30$ k.

**5.10-2** Dimensions of cross section: $b = 180$ mm, $t = 12$ mm, $h = 420$ mm, $h_1 = 380$ mm, and $V = 125$ kN.

**5.10-3** Wide-flange shape, W $8 \times 28$ (see Table F-1, Appendix F); $V = 10$ k.

**5.10-4** Dimensions of cross section: $b = 220$ mm, $t = 12$ mm, $h = 600$ mm, $h_1 = 570$ mm, and $V = 200$ kN.

**5.10-5** Wide-flange shape, W $18 \times 71$ (see Table F-1, Appendix F); $V = 21$ k.

**5.10-6** Dimensions of cross section: $b = 120$ mm, $t = 7$ mm, $h = 350$ mm, $h_1 = 330$ mm, and $V = 60$ kN.

**5.10-7** A cantilever beam $AB$ of length $L = 6.5$ ft supports a trapezoidal distributed load of peak intensity $q$, and minimum intensity $q/2$, that includes the weight of the beam (see figure). The beam is a steel W $12 \times 14$ wide-flange shape (see Table F-1(a), Appendix F).

Calculate the maximum permissible load $q$ based upon (a) an allowable bending stress $\sigma_{allow} = 18$ ksi, and (b) an allowable shear stress $\tau_{allow} = 7.5$ ksi. (*Note:* Obtain the moment of inertia and section modulus of the beam from Table F-1(a).)

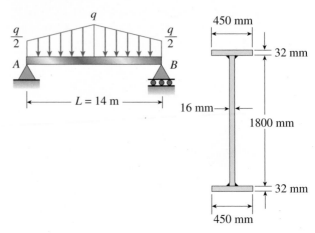

**PROB. 5.10-8**

**5.10-9** A simple beam with an overhang supports a uniform load of intensity $q = 1200$ lb/ft and a concentrated load $P = 3000$ lb at 8 ft to the right of $A$ and also at $C$ (see figure). The uniform load includes an allowance for the weight of the beam. The allowable stresses in bending and shear are 18 ksi and 11 ksi, respectively.

Select from Table F-2(a), Appendix F, the lightest I-beam (S shape) that will support the given loads.

(*Hint:* Select a beam based upon the bending stress and then calculate the maximum shear stress. If the beam is overstressed in shear, select a heavier beam and repeat.)

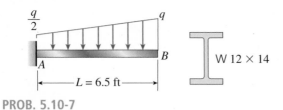

**PROB. 5.10-7**

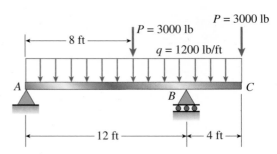

**PROB. 5.10-9**

**5.10-8** A bridge girder $AB$ on a simple span of length $L = 14$ m supports a distributed load of maximum intensity $q$ at mid-span and minimum intensity $q/2$ at supports $A$ and $B$ that includes the weight of the girder (see figure). The girder is constructed of three plates welded to form the cross section shown.

Determine the maximum permissible load $q$ based upon (a) an allowable bending stress $\sigma_{allow} = 110$ MPa, and (b) an allowable shear stress $\tau_{allow} = 50$ MPa.

**5.10-10** A hollow steel box beam has the rectangular cross section shown in the figure. Determine the maximum allowable shear force $V$ that may act on the beam if the allowable shear stress is 36 MPa.

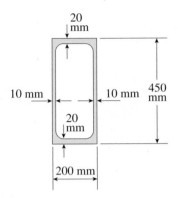

PROB. 5.10-10

**5.10-11** A hollow aluminum box beam has the square cross section shown in the figure. Calculate the maximum and minimum shear stresses $\tau_{max}$ and $\tau_{min}$ in the webs of the beam due to a shear force $V = 28$ k.

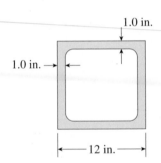

PROB. 5.10-11

**5.10-12** The T-beam shown in the figure has cross-sectional dimensions as follows: $b = 210$ mm, $t = 16$ mm, $h = 300$ mm, and $h_1 = 280$ mm. The beam is subjected to a shear force $V = 68$ kN.

Determine the maximum shear stress $\tau_{max}$ in the web of the beam.

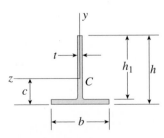

PROBS. 5.10-12 and 5.10-13

**5.10-13** Calculate the maximum shear stress $\tau_{max}$ in the web of the T-beam shown in the figure if $b = 10$ in., $t = 0.5$ in., $h = 7$ in., $h_1 = 6.2$ in., and the shear force $V = 5300$ lb.

# Built-Up Beams

**5.11-1** A prefabricated wood I-beam serving as a floor joist has the cross section shown in the figure. The allowable load in shear for the glued joints between the web and the flanges is 65 lb/in. in the longitudinal direction.

Determine the maximum allowable shear force $V_{max}$ for the beam.

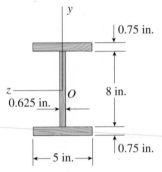

PROB. 5.11-1

**5.11-2** A welded steel girder having the cross section shown in the figure is fabricated of two 300 mm × 25 mm flange plates and a 800 mm × 16 mm web plate. The plates are joined by four fillet welds that run continuously for the length of the girder. Each weld has an allowable load in shear of 920 kN/m.

Calculate the maximum allowable shear force $V_{max}$ for the girder.

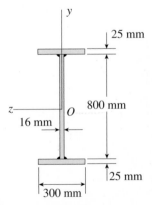

PROB. 5.11-2

**5.11-3** A welded steel girder having the cross section shown in the figure is fabricated of two 20 in. × 1 in. flange plates and a 60 in. × 5/16 in. web plate. The plates are joined by four longitudinal fillet welds that run continuously throughout the length of the girder.

If the girder is subjected to a shear force of 280 kips, what force $F$ (per inch of length of weld) must be resisted by each weld?

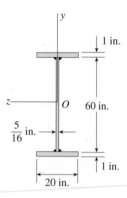

**PROB. 5.11-3**

**5.11-4** A box beam of wood is constructed of two 260 mm × 50 mm boards and two 260 mm × 25 mm boards (see figure). The boards are nailed at a longitudinal spacing $s = 100$ mm.

If each nail has an allowable shear force $F = 1200$ N, what is the maximum allowable shear force $V_{max}$?

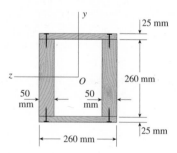

**PROB. 5.11-4**

**5.11-5** A box beam is constructed of four wood boards as shown in the figure part a. The webs are 8 in. × 1 in. and the flanges are 6 in. × 1 in. boards (actual dimensions), joined by screws for which the allowable load in shear is $F = 250$ lb per screw.

(a) Calculate the maximum permissible longitudinal spacing $S_{max}$ of the screws if the shear force $V$ is 1200 lb.

(b) Repeat part (a) if the flanges are attached to the webs using a *horizontal* arrangement of screws as shown in the figure part b.

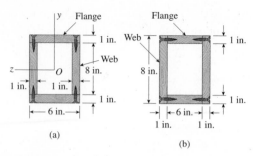

**PROB. 5.11-5**

**5.11-6** Two wood box beams (beams $A$ and $B$) have the same outside dimensions (200 mm × 360 mm) and the same thickness ($t = 20$ mm) throughout, as shown in the figure. Both beams are formed by nailing, with each nail having an allowable shear load of 250 N. The beams are designed for a shear force $V = 3.2$ kN.

(a) What is the maximum longitudinal spacing $s_A$ for the nails in beam $A$?

(b) What is the maximum longitudinal spacing $s_B$ for the nails in beam $B$?

(c) Which beam is more efficient in resisting the shear force?

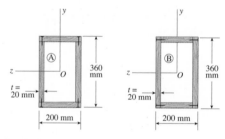

**PROB. 5.11-6**

**5.11-7** A hollow wood beam with plywood webs has the cross-sectional dimensions shown in the figure. The plywood is attached to the flanges by means of small nails. Each nail has an allowable load in shear of 30 lb.

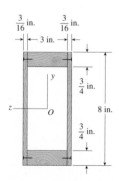

**PROB. 5.11-7**

Find the maximum allowable spacing $s$ of the nails at cross sections where the shear force $V$ is equal to (a) 200 lb, and (b) 300 lb.

**5.11-8** A beam of a T cross section is formed by nailing together two boards having the dimensions shown in the figure.

If the total shear force $V$ acting on the cross section is 1500 N and each nail may carry 760 N in shear, what is the maximum allowable nail spacing $s$?

**5.11-10** A steel beam is built up from a W 410 × 85 wide flange beam and two 180 mm × 9 mm cover plates (see figure). The allowable load in shear on each bolt is 9.8 kN. What is the required bolt spacing $s$ in the longitudinal direction if the shear force $V$ = 110 kN (*Note*: Obtain the dimensions and moment of inertia of the W shape from Table F-1(b).)

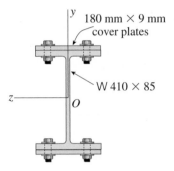

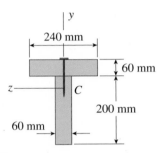

PROB. 5.11-8

PROB. 5.11-10

**5.11-9** The T-beam shown in the figure is fabricated by welding together two steel plates. If the allowable load for each weld is 1.8 k/in. in the longitudinal direction, what is the maximum allowable shear force $V$?

**5.11-11** The three beams shown have approximately the same cross-sectional area. Beam 1 is a W 14 × 82 with flange plates; Beam 2 consists of a web plate with four angles; and Beam 3 is constructed of 2 C shapes with flange plates.

(a) Which design has the largest moment capacity?

(b) Which has the largest shear capacity?

(c) Which is the most economical in bending?

(d) Which is the most economical in shear?

Assume allowable stress values are: $\sigma_a$ = 18 ksi and $\tau_a$ = 11 ksi. The most economical beam is that having the largest capacity-to-weight ratio. Neglect fabrication costs in answering parts (c) and (d) above. (*Note*: Obtain the dimensions and properties of all rolled shapes from tables in Appendix F.)

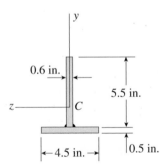

PROB. 5.11-9

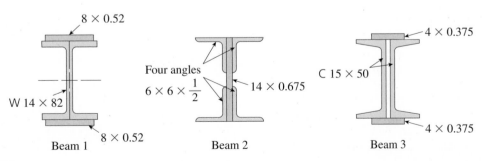

Beam 1    Beam 2    Beam 3

PROB. 5.11-11

**5.11-12** Two W 310 × 74 steel wide-flange beams are bolted together to form a built-up beam as shown in the figure. What is the maximum permissible bolt spacing $s$ if the shear force $V = 80$ kN and the allowable load in shear on each bolt is $F = 13.5$ kN [*Note*: Obtain the dimensions and properties of the W shapes from Table F-1(b).]

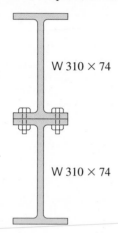

W 310 × 74

W 310 × 74

PROB. 5.11-12

## Beams with Axial Loads

*When solving the problems for Section 5.12, assume that the bending moments are not affected by the presence of lateral deflections.*

**5.12-1** While drilling a hole with a brace and bit, you exert a downward force $P = 25$ lb on the handle of the brace (see figure). The diameter of the crank arm is $d = 7/16$ in. and its lateral offset is $b = 4\text{-}7/8$ in.

Determine the maximum tensile and compressive stresses $\sigma_t$ and $\sigma_c$, respectively, in the crank.

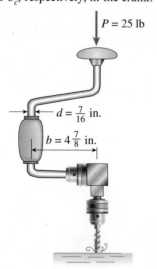

$P = 25$ lb

$d = \frac{7}{16}$ in.

$b = 4\frac{7}{8}$ in.

PROB. 5.12-1

**5.12-2** An aluminum pole for a street light weighs 4600 N and supports an arm that weighs 660 N (see figure). The center of gravity of the arm is 1.2 m from the axis of the pole. A wind force of 300 N also acts in the $(-y)$ direction at 9 m above the base. The outside diameter of the pole (at its base) is 225 mm, and its thickness is 18 mm.

Determine the maximum tensile and compressive stresses $\sigma_t$ and $\sigma_c$, respectively, in the pole (at its base) due to the weights and the wind force.

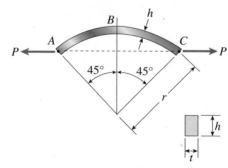

$W_2 = 660$ N

1.2 m

$P_1 = 300$ N

$W_1 = 4600$ N

9 m

18 mm

225 mm

PROB. 5.12-2

**5.12-3** A curved bar $ABC$ having a circular axis (radius $r = 12$ in.) is loaded by forces $P = 400$ lb (see figure). The cross section of the bar is rectangular with height $h$ and thickness $t$.

If the allowable tensile stress in the bar is 12,000 psi and the height $h = 1.25$ in., what is the minimum required thickness $t_{min}$?

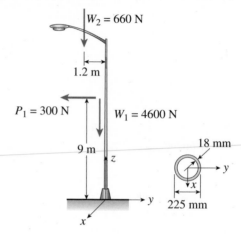

PROB. 5.12-3

**5.12-4** A rigid frame $ABC$ is formed by welding two steel pipes at $B$ (see figure). Each pipe has cross-sectional area $A = 11.31 \times 10^3$ mm², moment of inertia $I = 46.37 \times 10^6$ mm⁴, and outside diameter $d = 200$ mm.

Find the maximum tensile and compressive stresses $\sigma_t$ and $\sigma_c$, respectively, in the frame due to the load $P = 8.0$ kN if $L = H = 1.4$ m.

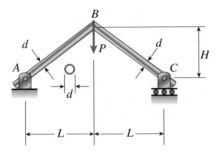

PROB. 5.12-4

**5.12-5** A palm tree weighing 1000 lb is inclined at an angle of 60° (see figure). The weight of the tree may be resolved into two resultant forces, a force $P_1 = 900$ lb acting at a point 12 ft from the base and a force $P_2 = 100$ lb acting at the top of the tree, which is 30 ft long. The diameter at the base of the tree is 14 in.

Calculate the maximum tensile and compressive stresses $\sigma_t$ and $\sigma_c$, respectively, at the base of the tree due to its weight.

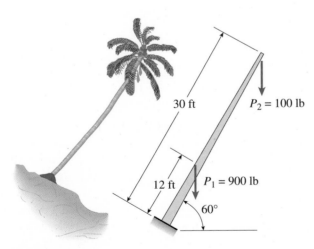

PROB. 5.12-5

**5.12-6** A vertical pole of aluminum is fixed at the base and pulled at the top by a cable having a tensile force $T$ (see figure). The cable is attached at the outer edge of a stiffened cover plate on top of the pole and makes an angle $\alpha = 20°$ at the point of attachment. The pole has length $L = 2.5$ m and a hollow circular cross section with outer

diameter $d_2 = 280$ mm and inner diameter $d_1 = 220$ mm. The circular cover plate has diameter $1.5d_2$.

Determine the allowable tensile force $T_{allow}$ in the cable if the allowable compressive stress in the aluminum pole is 90 MPa.

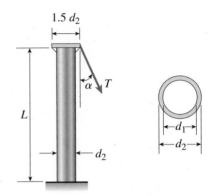

PROB. 5.12-6

**5.12-7** Because of foundation settlement, a circular tower is leaning at an angle $\alpha$ to the vertical (see figure). The structural core of the tower is a circular cylinder of height $h$, outer diameter $d_2$, and inner diameter $d_1$. For simplicity in the analysis, assume that the weight of the tower is uniformly distributed along the height.

Obtain a formula for the maximum permissible angle $\alpha$ if there is to be no tensile stress in the tower.

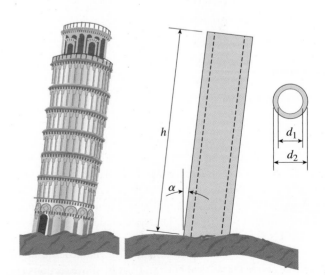

PROB. 5.12-7

**5.12-8** A steel bracket of solid circular cross section is subjected to two loads, each of which is $P = 4.5$ kN at $D$ (see figure). Let the dimension variable be $b = 240$ mm.

(a) Find the minimum permissible diameter $d_{min}$ of the bracket if the allowable normal stress is 110 MPa.

(b) Repeat part (a), including the weight of the bracket. The weight density of steel is 77.0 kN/m$^3$.

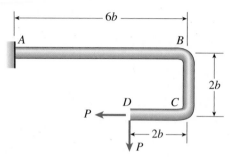

PROB. 5.12-8

**5.12-9** A cylindrical brick chimney of height $H$ weighs $w = 825$ lb/ft of height (see figure). The inner and outer diameters are $d_1 = 3$ ft and $d_2 = 4$ ft, respectively. The wind pressure against the side of the chimney is $p = 10$ lb/ft$^2$ of projected area.

Determine the maximum height $H$ if there is to be no tension in the brickwork.

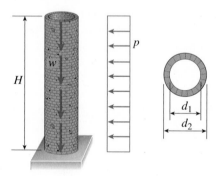

PROB. 5.12-9

**5.12-10** A flying buttress transmits a load $P = 25$ kN, acting at an angle of 60° to the horizontal, to the top of a vertical buttress $AB$ (see figure). The vertical buttress has height $h = 5.0$ m and rectangular cross section of thickness $t = 1.5$ m and width $b = 1.0$ m (perpendicular to the plane of the figure). The stone used in the construction weighs $\gamma = 26$ kN/m$^3$.

What is the required weight $W$ of the pedestal and statue above the vertical buttress (that is, above section $A$) to avoid any tensile stresses in the vertical buttress?

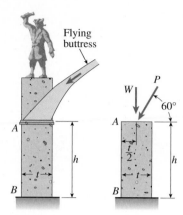

PROB. 5.12-10

**5.12-11** A plain concrete wall (i.e., a wall with no steel reinforcement) rests on a secure foundation and serves as a small dam on a creek (see figure). The height of the wall is $h = 6.0$ ft and the thickness of the wall is $t = 1.0$ ft.

(a) Determine the maximum tensile and compressive stresses $\sigma_t$ and $\sigma_c$, respectively, at the base of the wall when the water level reaches the top ($d = h$). Assume plain concrete has weight density $\gamma_c = 145$ lb/ft$^3$.

(b) Determine the maximum permissible depth $d_{max}$ of the water if there is to be no tension in the concrete.

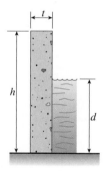

PROB. 5.12-11

## Eccentric Axial Loads

**5.12-12** A circular post, a rectangular post, and a post of cruciform cross section are each compressed by loads that produce a resultant force $P$ acting at the edge of the cross section (see figure). The diameter of the circular post and the depths of the rectangular and cruciform posts are the same.

(a) For what width $b$ of the rectangular post will the maximum tensile stresses be the same in the circular and rectangular posts?

(b) Repeat part (a) for the post with cruciform cross section.

(c) Under the conditions described in parts (a) and (b), which post has the largest compressive stress?

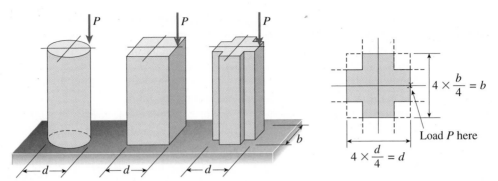

PROB. 5.12-12

**5.12-13** Two cables, each carrying a tensile force $P = 1200$ lb, are bolted to a block of steel (see figure). The block has thickness $t = 1$ in. and width $b = 3$ in.

(a) If the diameter $d$ of the cable is 0.25 in., what are the maximum tensile and compressive stresses $\sigma_t$ and $\sigma_c$, respectively, in the block?

(b) If the diameter of the cable is increased (without changing the force $P$), what happens to the maximum tensile and compressive stresses?

(a) If the end cross sections of the bar are square with sides of length $b$, what are the maximum tensile and compressive stresses $\sigma_t$ and $\sigma_c$, respectively, at cross section $mn$ within the reduced region?

(b) If the end cross sections are circular with diameter $b$, what are the maximum stresses $\sigma_t$ and $\sigma_c$?

**5.12-15** A short column constructed of a W $12 \times 35$ wide-flange shape is subjected to a resultant compressive load $P = 25$ k having its line of action at the midpoint of one flange (see figure).

(a) Determine the maximum tensile and compressive stresses $\sigma_t$ and $\sigma_c$, respectively, in the column.

(b) Locate the neutral axis under this loading condition.

(c) Recompute maximum tensile and compressive stresses if a C $10 \times 15.3$ is attached to one flange, as shown.

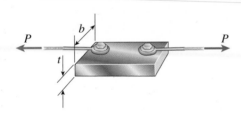

PROB. 5.12-13

**5.12-14** A bar $AB$ supports a load $P$ acting at the centroid of the end cross section (see figure). In the middle region of the bar the cross-sectional area is reduced by removing one-half of the bar.

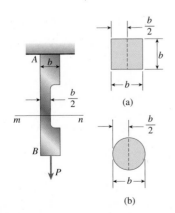

PROB. 5.12-14

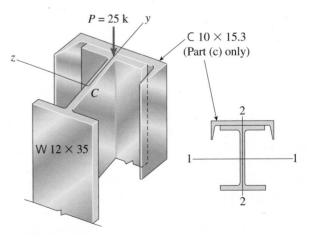

PROB. 5.12-15

**5.12-16** A short column of wide-flange shape is subjected to a compressive load that produces a resultant force $P = 55$ kN acting at the midpoint of one flange (see figure).

(a) Determine the maximum tensile and compressive stresses $\sigma_t$ and $\sigma_c$, respectively, in the column.

(b) Locate the neutral axis under this loading condition.

(c) Recompute maximum tensile and compressive stresses if a 120 mm × 10 mm cover plate is added to one flange as shown.

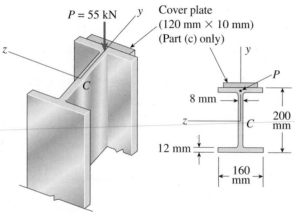

PROB. 5.12-16

**5.12-17** A tension member constructed of an L 4 × 4 × $\frac{1}{2}$ inch angle section (see Table F-4(a) in Appendix F) is subjected to a tensile load $P = 12.5$ kips that acts through the point where the mid-lines of the legs intersect (see figure part a).

(a) Determine the maximum tensile stress $\sigma_t$ in the angle section.

(b) Recompute the maximum tensile stress if two angles are used and $P$ is applied as shown in the figure part b.

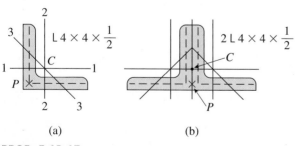

PROB. 5.12-17

**5.12-18** A short length of a C 200 × 17.1 channel is subjected to an axial compressive force $P$ that has its line of action through the midpoint of the web of the channel (see figure part a).

(a) Determine the equation of the neutral axis under this loading condition.

(b) If the allowable stresses in tension and compression are 76 MPa and 52 MPa respectively, find the maximum permissible load $P_{max}$.

(c) Repeat parts (a) and (b) if two L 76 × 76 × 6.4 angles are added to the channel as shown in the figure part b.

See Table F-3(b) in Appendix F for channel properties and Table F-4(b) for angle properties.

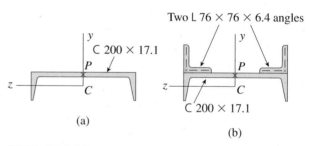

PROB. 5.12-18

## Stress Concentrations

*The problems for Section 5.13 are to be solved considering the stress-concentration factors.*

**5.13-1** The beams shown in the figure are subjected to bending moments $M = 2100$ lb-in. Each beam has a rectangular cross section with height $h = 1.5$ in. and width $b = 0.375$ in. (perpendicular to the plane of the figure).

(a) For the beam with a hole at midheight, determine the maximum stresses for hole diameters $d = 0.25, 0.50, 0.75,$ and 1.00 in.

(b) For the beam with two identical notches (inside height $h_1 = 1.25$ in.), determine the maximum stresses for notch radii $R = 0.05, 0.10, 0.15,$ and $0.20$ in.

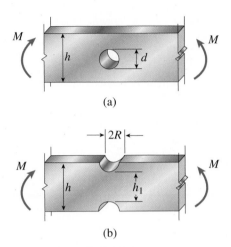

(a)

(b)

**PROBS. 5.13-1 through 5.13-4**

**5.13-2** The beams shown in the figure are subjected to bending moments $M = 250$ N·m. Each beam has a rectangular cross section with height $h = 44$ mm and width $b = 10$ mm (perpendicular to the plane of the figure).

(a) For the beam with a hole at midheight, determine the maximum stresses for hole diameters $d = 10, 16, 22,$ and 28 mm.

(b) For the beam with two identical notches (inside height $h_1 = 40$ mm), determine the maximum stresses for notch radii $R = 2, 4, 6,$ and 8 mm.

**5.13-3** A rectangular beam with semicircular notches, as shown in part b of the figure, has dimensions $h = 0.88$ in. and $h_1 = 0.80$ in. The maximum allowable bending stress in the metal beam is $\sigma_{max} = 60$ ksi, and the bending moment is $M = 600$ lb-in.

Determine the minimum permissible width $b_{min}$ of the beam.

**5.13-4** A rectangular beam with semicircular notches, as shown in part b of the figure, has dimensions $h = 120$ mm and $h_1 = 100$ mm. The maximum allowable bending stress in the plastic beam is $\sigma_{max} = 6$ MPa, and the bending moment is $M = 150$ N·m.

Determine the minimum permissible width $b_{min}$ of the beam.

**5.13-5** A rectangular beam with notches and a hole (see figure) has dimensions $h = 5.5$ in., $h_1 = 5$ in., and width $b = 1.6$ in. The beam is subjected to a bending moment $M = 130$ k-in., and the maximum allowable bending stress in the material (steel) is $\sigma_{max} = 42,000$ psi.

(a) What is the smallest radius $R_{min}$ that should be used in the notches?

(b) What is the diameter $d_{max}$ of the largest hole that should be drilled at the midheight of the beam?

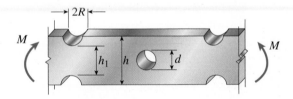

**PROB. 5.13-5**

# 6 CHAPTER

# Stresses in Beams (Advanced Topics)

A more advanced theory is required for analysis and design of composite beams and beams with unsymmetric cross sections. (© Can Stock Photo Inc./toneteam)

## CHAPTER OVERVIEW

In Chapter 6, we will consider a number of advanced topics related to shear and bending of beams of arbitrary cross section. First, stresses and strains in **composite beams**, that is beams fabricated of more than one material, is discussed in Section 6.2. First, we locate the neutral axis then find the flexure formula for a composite beam made up of two different materials. We then study the **transformed-section method** as an alternative procedure for analyzing the bending stresses in a composite beam in Section 6.3. Next, we study bending of doubly symmetric beams acted on by **inclined loads** having a line of action through the centroid of the cross section (Section 6.4). In this case, there are bending moments ($M_y$, $M_z$) about each of the principal axes of the cross section, and the neutral axis is no longer perpendicular to the longitudinal plane containing the applied loads. The final normal stresses are obtained by superposing the stresses obtained from the flexure formulas for each of the separate axes of the cross section. Next, we investigate the general case of unsymmetric beams in pure bending, removing the restriction of at least one axis of symmetry in the cross section (Section 6.5). We develop a general procedure for analyzing an unsymmetric beam subjected to any bending moment $M$ resolved into components along the principal centroidal axes

of the cross section. Of course, symmetric beams are special cases of unsymmetric beams, and therefore, the discussions also apply to symmetric beams. If the restriction of pure bending is removed and transverse loads are allowed, we note that these loads must act through the **shear center** of the cross section so that twisting of the beam about a longitudinal axis can be avoided (Sections 6.6 and 6.9). The distributions of shear stresses in the elements of the cross sections of a number of beams of thin-walled open section (such as channels, angles, and Z shapes) are calculated and then used to locate the shear center for each particular cross-sectional shape (Sections 6.7, 6.8 and 6.9). As the final topic in the chapter, the bending of elastoplastic beams is described in which the normal stresses go beyond the linear elastic range of behavior (Section 6.10).

## Chapter 6 is organized as follows:

**6.1** Introduction 508
**6.2** Composite Beams 508
**6.3** Transformed-Section Method 517
**6.4** Doubly Symmetric Beams with Inclined Loads 526
**6.5** Bending of Unsymmetric Beams 533
**6.6** The Shear-Center Concept 541
**6.7** Shear Stresses in Beams of Thin-Walled Open Cross Sections 543

**6.8** Shear Stresses in Wide-Flange Beams 546
**6.9** Shear Centers of Thin-Walled Open Sections 550
*\*6.10** Elastoplastic Bending 558
Chapter Summary & Review 566
Problems 569

*Advanced Topics

# 6.1 INTRODUCTION

In this chapter, we continue our study of the bending of beams by examining several specialized topics. These subjects are based upon the fundamental topics discussed previously in Chapter 5—topics such as curvature, normal stresses in beams (including the flexure formula), and shear stresses in beams. However, we will no longer require that beams be composed of one material only; and we will also remove the restriction that the beams have a plane of symmetry in which transverse loads must be applied. Finally, we will extend the performance into the inelastic range of behavior for beams made of elastoplastic materials.

Later, in Chapters 9 and 10, we will discuss two additional subjects of fundamental importance in beam design—deflections of beams and statically indeterminate beams.

# 6.2 COMPOSITE BEAMS

Beams that are fabricated of more than one material are called **composite beams**. Examples are bimetallic beams (such as those used in thermostats), plastic coated pipes, and wood beams with steel reinforcing plates (see Fig. 6-1).

Many other types of composite beams have been developed in recent years, primarily to save material and reduce weight. For instance, **sandwich beams** are widely used in the aviation and aerospace industries, where light weight plus high strength and rigidity are required. Such familiar objects as skis, doors, wall panels, book shelves, and cardboard boxes are also manufactured in sandwich style.

A typical sandwich beam (Fig. 6-2) consists of two thin *faces* of relatively high-strength material (such as aluminum) separated by a thick *core* of lightweight, low-strength material. Since the faces are at the greatest distance from the neutral axis (where the bending stresses are highest), they function somewhat like the flanges of an I-beam. The core serves as a filler and provides support for the faces, stabilizing them against wrinkling or buckling. Lightweight plastics and foams, as well as honeycombs and corrugations, are often used for cores.

## Strains and Stresses

The strains in composite beams are determined from the same basic axiom that we used for finding the strains in beams of one material, namely, cross sections remain plane during bending. This axiom is valid for pure bending regardless of the nature of the material (see Section 5.4). Therefore, the longitudinal strains $\varepsilon_x$ in a composite beam vary linearly from top to bottom of the beam, as expressed by Eq. (5-6), which is repeated here:

$$\varepsilon_x = -\frac{y}{\rho} = -\kappa y \tag{6-1}$$

In this equation, $y$ is the distance from the neutral axis, $\rho$ is the radius of curvature, and $\kappa$ is the curvature.

Beginning with the linear strain distribution represented by Eq. (6-1), we can determine the strains and stresses in any composite beam. To show how this is accomplished, consider the composite beam shown in Fig. 6-3.

---

### Fig. 6-1

Examples of composite beams: (a) bimetallic beam, (b) plastic-coated steel pipe, and (c) wood beam reinforced with a steel plate

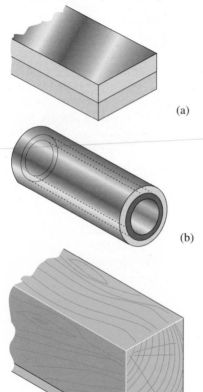

(a)

(b)

(c)

Roof structure: composite timber–steel I-beam and rafters (© Can Stock Photo Inc./tin)

This beam consists of two materials, labeled 1 and 2 in the figure, which are securely bonded so that they act as a single solid beam.

As in our previous discussions of beams (Chapter 5), we assume that the $xy$ plane is a plane of symmetry and that the $xz$ plane is the neutral plane of the beam. However, the neutral axis (the $z$ axis in Fig. 6-3b) does *not* pass through the centroid of the cross-sectional area when the beam is made of two different materials.

If the beam is bent with positive curvature, the strains $\varepsilon_x$ will vary as shown in Fig. 6-3c, where $\varepsilon_A$ is the compressive strain at the top of the beam, $\varepsilon_B$ is the tensile strain at the bottom, and $\varepsilon_C$ is the strain at the contact surface of the two materials. Of course, the strain is zero at the neutral axis (the $z$ axis).

The normal stresses acting on the cross section can be obtained from the strains by using the stress-strain relationships for the two materials. Let us assume that both materials behave in a linearly elastic manner so that Hooke's law for uniaxial stress is valid. Then the stresses in the materials are obtained by multiplying the strains by the appropriate modulus of elasticity.

Denoting the moduli of elasticity for materials 1 and 2 as $E_1$ and $E_2$, respectively, and also assuming that $E_2 > E_1$, we obtain the stress diagram shown in Fig. 6-3d. The compressive stress at the top of the beam is $\sigma_A = E_1 \varepsilon_A$ and the tensile stress at the bottom is $\sigma_B = E_2 \varepsilon_B$.

At the contact surface ($C$) the stresses in the two materials are different because their moduli are different. In material 1 the stress is $\sigma_{1C} = E_1 \varepsilon_C$ and in material 2 it is $\sigma_{2C} = E_2 \varepsilon_C$.

Using Hooke's law and Eq. (6-1), we can express the normal stresses at distance $y$ from the neutral axis in terms of the curvature:

$$\sigma_{x1} = -E_1 \kappa y \qquad \sigma_{x2} = -E_2 \kappa y \qquad \textbf{(6-2a,b)}$$

in which $\sigma_{x1}$ is the stress in material 1 and $\sigma_{x2}$ is the stress in material 2. With the aid of these equations, we can locate the neutral axis and obtain the moment-curvature relationship.

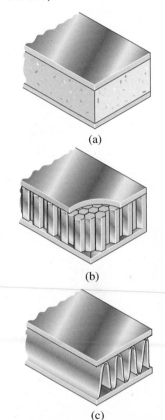

**Fig. 6-2**

Sandwich beams with: (a) plastic core, (b) honeycomb core, and (c) corrugated core (© Barry Goodno)

(a)

(b)

(c)

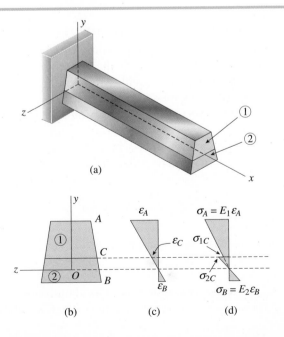

(a)

(b)                    (c)                    (d)

**Fig. 6-3**

(a) Composite beam of two materials, (b) cross section of beam, (c) distribution of strains $\varepsilon_x$ throughout the height of the beam, and (d) distribution of stresses $\sigma_x$ in the beam for the case where $E_2 > E_1$

## Neutral Axis

The position of the neutral axis (the $z$ axis) is found from the condition that the resultant axial force acting on the cross section is zero (see Section 5.5); therefore,

$$\int_1 \sigma_{x1}\,dA + \int_2 \sigma_{x2}\,dA = 0 \qquad\text{(6-3)}$$

where it is understood that the first integral is evaluated over the cross-sectional area of material 1 and the second integral is evaluated over the cross-sectional area of material 2. Replacing $\sigma_{x1}$ and $\sigma_{x2}$ in the preceding equation by their expressions from Eqs. (6-2a) and (6-2b), we get

$$-\int_1 E_1 \kappa y\,dA - \int_2 E_2 \kappa y\,dA = 0$$

Since the curvature is a constant at any given cross section, it is not involved in the integrations and can be cancelled from the equation; thus, the equation for locating the **neutral axis** becomes

$$E_1 \int_1 y\,dA + E_2 \int_2 y\,dA = 0 \qquad\text{(6-4)}$$

The integrals in this equation represent the first moments of the two parts of the cross-sectional area with respect to the neutral axis. (If there are more than two materials—a rare condition—additional terms are required in the equation.)

Equation (6-4) is a generalized form of the analogous equation for a beam of one material (Eq. 5-9). The details of the procedure for locating the neutral axis with the aid of Eq. (6-4) are illustrated later in Example 6-1.

If the cross section of a beam is **doubly symmetric**, as in the case of a wood beam with steel cover plates on the top and bottom (Fig. 6-4), the neutral axis is located at the mid-height of the cross section and Eq. (6-4) is not needed.

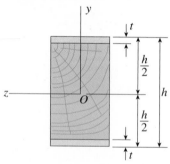

**Fig. 6-4**

Doubly symmetric cross section

## Moment-Curvature Relationship

The moment-curvature relationship for a composite beam of two materials (Fig. 6-3) may be determined from the condition that the moment resultant of the bending stresses is equal to the bending moment $M$ acting at the cross section. Following the same steps as for a beam of one material [see Eqs. (5-10) through (5-13)], and also using Eqs. (6-2a) and (6-2b), we obtain

$$M = -\int_A \sigma_x y\,dA = -\int_1 \sigma_{x1} y\,dA - \int_2 \sigma_{x2} y\,dA$$

$$= \kappa E_1 \int_1 y^2\,dA + \kappa E_2 \int_2 y^2\,dA \qquad\text{(6-5a)}$$

This equation can be written in the simpler form

$$M = \kappa(E_1 I_1 + E_2 I_2) \qquad\text{(6-5b)}$$

in which $I_1$ and $I_2$ are the moments of inertia about the neutral axis (the $z$ axis) of the cross-sectional areas of materials 1 and 2, respectively. Note that $I = I_1 + I_2$, where $I$ is the moment of inertia of the *entire* cross-sectional area about the neutral axis.

Equation (6-5b) can now be solved for the curvature in terms of the bending moment:

$$\kappa = \frac{1}{\rho} = \frac{M}{E_1 I_1 + E_2 I_2} \qquad \textbf{(6-6)}$$

This equation is the **moment-curvature relationship** for a beam of two materials (compare with Eq. 5-13 for a beam of one material). The denominator on the right-hand side is the **flexural rigidity** of the composite beam.

## Normal Stresses (Flexure Formulas)

The normal stresses (or bending stresses) in the beam are obtained by substituting the expression for curvature (Eq. 6-6) into the expressions for $\sigma_{x1}$ and $\sigma_{x2}$ [Eqs. (6-2a) and (6-2b)]; thus,

$$\sigma_{x1} = -\frac{M y E_1}{E_1 I_1 + E_2 I_2} \qquad \sigma_{x2} = -\frac{M y E_2}{E_1 I_2 + E_2 I_2} \qquad \textbf{(6-7a,b)}$$

These expressions, known as the **flexure formulas for a composite beam**, give the normal stresses in materials 1 and 2, respectively. If the two materials have the same modulus of elasticity ($E_1 = E_2 = E$), then both equations reduce to the flexure formula for a beam of one material (Eq. 5-14).

The analysis of composite beams, using Eqs. (6-4) through (6-7), is illustrated in Examples 6-1 and 6-2 at the end of this section.

## Approximate Theory for Bending of Sandwich Beams

Sandwich beams having doubly symmetric cross sections and composed of two linearly elastic materials (Fig. 6-5) can be analyzed for bending using Eqs. (6-6) and (6-7), as described previously. However, we can also develop an approximate theory for bending of sandwich beams by introducing some simplifying assumptions.

If the material of the faces (material 1) has a much larger modulus of elasticity than does the material of the core (material 2), it is reasonable to disregard the normal stresses in the core and assume that the faces resist all of the longitudinal bending stresses. This assumption is equivalent to saying that the modulus of elasticity $E_2$ of the core is zero. Under these conditions the flexure formula for material 2 (Eq. 6-7b) gives $\sigma_{x2} = 0$ (as expected), and the flexure formula for material 1 (Eq. 6-7a) gives

$$\sigma_{x1} = -\frac{M y}{I_1} \qquad \textbf{(6-8)}$$

which is similar to the ordinary flexure formula (Eq. 5-14). The quantity $I_1$ is the moment of inertia of the two faces evaluated with respect to the neutral axis; thus,

$$I_1 = \frac{b}{12}\left(h^3 - h_c^3\right) \qquad \textbf{(6-9)}$$

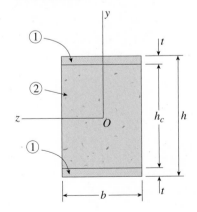

**Fig. 6-5**

Cross section of a sandwich beam having two axes of symmetry (doubly symmetric cross section) in which $b$ is the width of the beam, $h$ is the overall height of the beam, and $h_c$ is the height of the core. Note that $h_c = h - 2t$ where $t$ is the thickness of the faces.

The maximum normal stresses in the sandwich beam occur at the top and bottom of the cross section where $y = h/2$ and $-h/2$, respectively. Thus, from Eq. (6-8), we obtain

$$\sigma_{top} = -\frac{Mh}{2I_1} \qquad \sigma_{bottom} = \frac{Mh}{2I_1} \qquad \textbf{(6-10a,b)}$$

If the bending moment $M$ is positive, the upper face is in compression and the lower face is in tension. (These equations are conservative because they give stresses in the faces that are higher than those obtained from Eqs. 6-7a and 6-7b.)

If the faces are thin compared to the thickness of the core (that is, if $t$ is small compared to $h_c$), we can disregard the shear stresses in the faces and assume that the core carries all of the shear stresses. Under these conditions the average shear stress and average shear strain in the core are, respectively,

$$\tau_{aver} = \frac{V}{bh_c} \qquad \gamma_{aver} = \frac{V}{bh_c G_c} \qquad \textbf{(6-11a,b)}$$

in which $V$ is the shear force acting on the cross section and $G_c$ is the shear modulus of elasticity for the core material. (Although the maximum shear stress and maximum shear strain are larger than the average values, the average values are often used for design purposes.)

## Limitations

**Fig. 6-6**

Reinforced concrete beam with longitudinal reinforcing bars and vertical stirrups

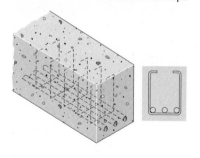

Throughout the preceding discussion of composite beams, we assumed that both materials followed Hooke's law and that the two parts of the beam were adequately bonded so that they acted as a single unit. Thus, our analysis is highly idealized and represents only a first step in understanding the behavior of composite beams and composite materials. Methods for dealing with nonhomogeneous and nonlinear materials, bond stresses between the parts, shear stresses on the cross sections, buckling of the faces, and other such matters are treated in reference books dealing specifically with composite construction.

**Reinforced concrete beams** are one of the most complex types of composite construction (Fig. 6-6), and their behavior differs significantly from that of the composite beams discussed in this section. Concrete is strong in compression but extremely weak in tension. Consequently, its tensile strength is usually disregarded entirely. Under those conditions, *the formulas given in this section do not apply*. Working stress design in which the portion of the beam in tension is removed from the composite beam cross section is used in evaluating deflections of reinforced concrete beams, so an allowable stress approach will be presented in Example 6-4 to show the general analysis procedure. Example 6-4 uses a "cracked section analysis" to demonstrate this reinforced-concrete analysis procedure.

Note that most reinforced concrete beams are not designed on the basis of linearly elastic behavior—instead, more realistic design methods (based upon load-carrying capacity instead of allowable stresses) are used. The design of reinforced concrete members is a highly specialized subject that is presented in courses and textbooks devoted solely to that subject.

• • • **Example 6-1**

### Fig. 6-7

Example 6-1: Cross section of a composite beam of wood and steel

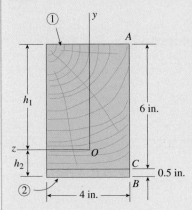

A composite beam (Fig. 6-7) is constructed from a wood beam (4.0 in. × 6.0 in. actual dimensions) and a steel reinforcing plate (4.0 in. wide and 0.5 in. thick). The wood and steel are securely fastened to act as a single beam. The beam is subjected to a positive bending moment $M = 60$ k-in.

Calculate the largest tensile and compressive stresses in the wood (material 1) and the maximum and minimum tensile stresses in the steel (material 2) if $E_1 = 1500$ ksi and $E_2 = 30{,}000$ ksi.

#### Solution

*Neutral axis.* The first step in the analysis is to locate the neutral axis of the cross section. For that purpose, let us denote the distances from the neutral axis to the top and bottom of the beam as $h_1$ and $h_2$, respectively. To obtain these distances, we use Eq. (6-4). The integrals in that equation are evaluated by taking the first moments of areas 1 and 2 about the $z$ axis, as follows:

$$\int_1 y\,dA = \bar{y}_1 A_1 = (h_1 - 3 \text{ in.})(4 \text{ in.} \times 6 \text{ in.}) = (h_1 - 3 \text{ in.})(24 \text{ in.}^2)$$

$$\int_1 y\,dA = \bar{y}_2 A_2 = -(6.25 \text{ in.} - h_1)(4 \text{ in.} \times 0.5 \text{ in.}) = (h_1 - 6.25 \text{ in.})(2 \text{ in.}^2)$$

in which $A_1$ and $A_2$ are the areas of parts 1 and 2 of the cross section, $\bar{y}_1$ and $\bar{y}_2$ are the $y$ coordinates of the centroids of the respective areas, and $h_1$ has units of inches.

Substituting the preceding expressions into Eq. (6-4) gives the equation for locating the neutral axis, as follows:

$$E_1 \int_1 y\,dA + E_2 \int_2 y\,dA = 0$$

or

$$(1500 \text{ ksi})(h_1 - 3 \text{ in.})(24 \text{ in.}^2) + (30{,}000 \text{ ksi})(h_1 - 6.25 \text{ in.})(2 \text{ in.}^2) = 0$$

Solving this equation, we obtain the distance $h_1$ from the neutral axis to the top of the beam:

$$h_1 = 5.031 \text{ in.}$$

Also, the distance $h_2$ from the neutral axis to the bottom of the beam is

$$h_2 = 6.5 \text{ in.} - h_1 = 1.469 \text{ in.}$$

Thus, the position of the neutral axis is established.

*Moments of inertia.* The moments of inertia $I_1$ and $I_2$ of areas $A_1$ and $A_2$ with respect to the neutral axis can be found by using the parallel-axis theorem (see Section 12.5 of Chapter 12). Beginning with area 1 (Fig. 6-7), we get

$$I_1 = \frac{1}{12}(4 \text{ in.})(6 \text{ in.})^3 + (4 \text{ in.})(6 \text{ in.})(h_1 - 3 \text{ in.})^2 = 171.0 \text{ in.}^4$$

Similarly, for area 2 we get

$$I_2 = \frac{1}{12}(4 \text{ in.})(0.5 \text{ in.})^3 + (4 \text{ in.})(0.5 \text{ in.})(h_2 - 0.25 \text{ in.})^2 = 3.01 \text{ in.}^4$$

*Continues* ➡

To check these calculations, we can determine the moment of inertia $I$ of the entire cross-sectional area about the $z$ axis as

$$I = \frac{1}{3}(4 \text{ in.})h_1^3 + \frac{1}{3}(4 \text{ in.})h_2^3 = 169.8 + 4.2 = 174.0 \text{ in.}^4$$

which agrees with the sum of $I_1$ and $I_2$.

*Normal stresses.* The stresses in materials 1 and 2 are calculated from the flexure formulas for composite beams (Eqs. 6-7a and b). The largest compressive stress in material 1 occurs at the top of the beam (A) where $y = h_1 = 5.031$ in. Denoting this stress by $\sigma_{1A}$ and using Eq. (6-7a), we get

$$\sigma_{1A} = -\frac{Mh_1 E_1}{E_1 I_1 + E_2 I_2}$$

$$= -\frac{(60 \text{ k-in.})(5.031 \text{ in.})(1500 \text{ ksi})}{(1500 \text{ ksi})(171.0 \text{ in.}^4) + (30,000 \text{ ksi})(3.01 \text{ in.}^4)} = -1310 \text{ psi} \quad \Longleftarrow$$

The largest tensile stress in material 1 occurs at the contact plane between the two materials (C) where $y = -(h_2 - 0.5 \text{ in.}) = -0.969$ in. Proceeding as in the previous calculation, we get

$$\sigma_{1C} = -\frac{(60 \text{ k-in.})(-0.969 \text{ in.})(1500 \text{ ksi})}{(1500 \text{ ksi})(171.0 \text{ in.}^4) + (30,000 \text{ ksi})(3.01 \text{ in.}^4)} = 251 \text{ psi} \quad \Longleftarrow$$

Thus, we have found the largest compressive and tensile stresses in the wood.

The steel plate (material 2) is located below the neutral axis, and therefore it is entirely in tension. The maximum tensile stress occurs at the bottom of the beam (B) where $y = -h_2 = -1.469$ in. Hence, from Eq. (6-7b) we get

$$\sigma_{2B} = -\frac{M(-h_2)E_2}{E_1 I_1 + E_2 I_2}$$

$$= -\frac{(60 \text{ k-in.})(-1.469 \text{ in.})(30,000 \text{ ksi})}{(1500 \text{ ksi})(171.0 \text{ in.}^4) + (30,000 \text{ ksi})(3.01 \text{ in.}^4)} = 7620 \text{ psi} \quad \Longleftarrow$$

The minimum tensile stress in material 2 occurs at the contact plane (C) where $y = -0.969$ in. Thus,

$$\sigma_{2C} = -\frac{(60 \text{ k-in.})(-0.969 \text{ in.})(30,000 \text{ ksi})}{(1500 \text{ ksi})(171.0 \text{ in.}^4) + (30,000 \text{ ksi})(3.01 \text{ in.}^4)} = 5030 \text{ psi} \quad \Longleftarrow$$

These stresses are the maximum and minimum tensile stresses in the steel. The stress distribution over the cross section of the composite wood–steel beam is shown in Fig. 6-8.

*Note:* At the contact plane the ratio of the stress in the steel to the stress in the wood is

$$\sigma_{2C}/\sigma_{1C} = 5030 \text{ psi}/251 \text{ psi} = 20$$

which is equal to the ratio $E_2/E_1$ of the moduli of elasticity (as expected). Although the strains in the steel and wood are equal at the contact plane, the stresses are different because of the different moduli.

**Fig. 6.8**

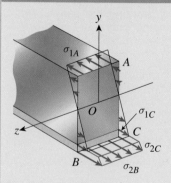

A sandwich beam having aluminum-alloy faces enclosing a plastic core (Fig. 6-9) is subjected to a bending moment $M = 3.0$ kN·m. The thickness of the faces is $t = 5$ mm and their modulus of elasticity is $E_1 = 72$ GPa. The height of the plastic core is $h_c = 150$ mm and its modulus of elasticity is $E_2 = 800$ MPa. The overall dimensions of the beam are $h = 160$ mm and $b = 200$ mm.

Determine the maximum tensile and compressive stresses in the faces and the core using: (a) the general theory for composite beams, and (b) the approximate theory for sandwich beams.

**Fig. 6-9**

Example 6-2: Cross section of sandwich beam having aluminum-alloy faces and a plastic core

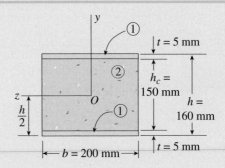

**Solution**

*Neutral axis.* Because the cross section is doubly symmetric, the neutral axis (the $z$ axis in Fig. 6-9) is located at mid-height.

*Moments of inertia.* The moment of inertia $I_1$ of the cross-sectional areas of the faces (about the $z$ axis) is

$$I_1 = \frac{b}{12}(h^3 - h_c^3) = \frac{200 \text{ mm}}{12}\left[(160 \text{ mm})^3 - (150 \text{ mm})^3\right]$$

$$= 12.017 \times 10^6 \text{ mm}^4$$

and the moment of inertia $I_2$ of the plastic core is

$$I_2 = \frac{b}{12}(h_c^3) = \frac{200 \text{ mm}}{12}(150 \text{ mm})^3 = 56.250 \times 10^6 \text{ mm}^4$$

As a check on these results, note that the moment of inertia of the entire cross-sectional area about the $z$ axis ($I = bh^3/12$) is equal to the sum of $I_1$ and $I_2$.

(a) *Normal stresses calculated from the general theory for composite beams.* To calculate these stresses, we use Eqs. (6-7a) and (6-7b).

*Continues* ➡

As a preliminary matter, we will evaluate the term in the denominator of those equations (that is, the flexural rigidity of the composite beam):

$$E_1 I_1 + E_2 I_2 = (72 \text{ GPa})(12.017 \times 10^6 \text{ mm}^4) + (800 \text{ MPa})(56.250 \times 10^6 \text{ mm}^4)$$

$$= 910{,}200 \text{ N} \cdot \text{m}^2$$

The maximum tensile and compressive stresses in the aluminum faces are found from Eq. (6-7a):

$$(\sigma_1)_{max} = \pm \frac{M(h/2)(E_1)}{E_1 I_1 + E_2 I_2}$$

$$= \pm \frac{(3.0 \text{ kN} \cdot \text{m})(80 \text{ mm})(72 \text{ GPa})}{910{,}200 \text{ N} \cdot \text{m}^2} = \pm 19.0 \text{ MPa} \quad \Leftarrow$$

The corresponding quantities for the plastic core (from Eq. 6-7b) are

$$(\sigma_2)_{max} = \pm \frac{M(h_c/2)(E_2)}{E_1 I_1 + E_2 I_2}$$

$$= \pm \frac{(3.0 \text{ kN} \cdot \text{m})(75 \text{ mm})(800 \text{ MPa})}{910{,}200 \text{ N} \cdot \text{m}^2} = \pm 0.198 \text{ MPa} \quad \Leftarrow$$

The maximum stresses in the faces are 96 times greater than the maximum stresses in the core, primarily because the modulus of elasticity of the aluminum is 90 times greater than that of the plastic.

(b) *Normal stresses calculated from the approximate theory for sandwich beams.* In the approximate theory we disregard the normal stresses in the core and assume that the faces transmit the entire bending moment. Then the maximum tensile and compressive stresses in the faces can be found from Eqs. (6-10a) and (6-10b), as follows:

$$(\sigma_1)_{max} = \pm \frac{Mh}{2I_1} = \pm \frac{(3.0 \text{ kN} \cdot \text{m})(80 \text{ mm})}{12.017 \times 10^6 \text{ mm}^4} = \pm 20.0 \text{ MPa} \quad \Leftarrow$$

As expected, the approximate theory gives slightly higher stresses in the faces than does the general theory for composite beams.

# 6.3 TRANSFORMED-SECTION METHOD

The transformed-section method is an alternative procedure for analyzing the bending stresses in a composite beam. The method is based upon the theories and equations developed in the preceding section, and therefore it is subject to the same limitations (for instance, it is valid only for linearly elastic materials) and gives the same results. Although the transformed-section method does not reduce the calculating effort, many designers find that it provides a convenient way to visualize and organize the calculations.

The method consists of transforming the cross section of a composite beam into an equivalent cross section of an imaginary beam that is composed of only one material. This new cross section is called the **transformed section**. Then the imaginary beam with the transformed section is analyzed in the customary manner for a beam of one material. As a final step, the stresses in the transformed beam are converted to those in the original beam.

## Neutral Axis and Transformed Section

If the transformed beam is to be equivalent to the original beam, *its neutral axis must be located in the same place and its moment-resisting capacity must be the same*. To show how these two requirements are met, consider again a composite beam of two materials (Fig. 6-10a). The **neutral axis** of the cross section is obtained from Eq. (6-4), which is repeated here:

$$E_1 \int_1 y \, dA + E_2 \int_2 y \, dA = 0 \qquad \textbf{(6-12)}$$

In this equation, the integrals represent the first moments of the two parts of the cross section with respect to the neutral axis.

Let us now introduce the notation

$$n = \frac{E_2}{E_1} \qquad \textbf{(6-13)}$$

where $n$ is the **modular ratio**. With this notation, we can rewrite Eq. (6-12) in the form

$$\int_1 y \, dA + \int_2 yn \, dA = 0 \qquad \textbf{(6-14)}$$

Since Eqs. (6-12) and (6-14) are equivalent, the preceding equation shows that the neutral axis is unchanged if each element of area $dA$ in material 2 is multiplied by the factor $n$, provided that the $y$ coordinate for each such element of area is not changed.

Therefore, we can create a new cross section consisting of two parts: (1) area 1 with its dimensions unchanged, and (2) area 2 with its *width* (that is, its dimension parallel to the neutral axis) multiplied by $n$. This new cross section (the transformed section) is shown in Fig. 6-10b for the case where $E_2 > E_1$ (and therefore $n > 1$). Its neutral axis is in the same position as the neutral axis of the original beam. (Note that all dimensions perpendicular to the neutral axis remain the same.)

Composite beam of two materials: (a) actual cross section, and (b) transformed section consisting only of material 1

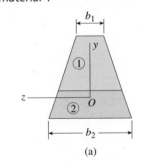

(a)

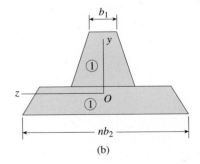

(b)

Since the stress in the material (for a given strain) is proportional to the modulus of elasticity ($\sigma = E\varepsilon$), we see that multiplying the width of material 2 by $n = E_2/E_1$ is equivalent to transforming it to material 1. For instance, suppose that $n = 10$. Then the area of part 2 of the cross section is now 10 times wider than before. If we imagine that this part of the beam is now material 1, we see that it will carry the same force as before because its modulus is *reduced* by a factor of 10 (from $E_2$ to $E_1$) at the same time that its area is *increased* by a factor of 10. Thus, the new section (the transformed section) consists only of material 1.

## Moment-Curvature Relationship

The *moment-curvature relationship* for the transformed beam must be the same as for the original beam. To show that this is indeed the case, we note that the stresses in the transformed beam (since it consists only of material 1) are given by Eq. (5-8) of Section 5.5:

$$\sigma_x = -E_1 \kappa y$$

Using this equation, and also following the same procedure as for a beam of one material (see Section 5.5), we can obtain the moment-curvature relation for the transformed beam:

$$M = -\int_A \sigma_x y \, dA = -\int_1 \sigma_x y \, dA - \int_2 \sigma_x y \, dA$$

$$= E_1 \kappa \int_1 y^2 \, dA + E_1 \kappa \int_2 y^2 \, dA = \kappa(E_1 I_1 + E_1 n I_2)$$

or

$$M = \kappa(E_1 I_1 + E_2 I_2) \tag{6-15}$$

This equation is the same as Eq. (6-5), thereby demonstrating that the moment-curvature relationship for the transformed beam is the same as for the original beam.

## Normal Stresses

Since the transformed beam consists of only one material, the *normal stresses* (or *bending stresses*) can be found from the standard flexure formula (Eq. 5-14). Thus, the normal stresses in the beam transformed to material 1 (Fig. 6-10b) are

$$\sigma_{x1} = -\frac{My}{I_T} \tag{6-16}$$

where $I_T$ is the moment of inertia of the transformed section with respect to the neutral axis. By substituting into this equation, we can calculate the stresses at any point in the *transformed* beam. (As explained later, the stresses in the transformed beam match those in the original beam in the part of the original beam consisting of material 1; however, in the part of the original beam consisting of material 2, the stresses are different from those in the transformed beam.)

We can easily verify Eq. (6-16) by noting that the moment of inertia of the transformed section (Fig. 6-10b) is related to the moment of inertia of the original section (Fig. 6-10a) by the following relation:

$$I_T = I_1 + nI_2 = I_1 + \frac{E_2}{E_1} I_2 \qquad \text{(6-17)}$$

Substituting this expression for $I_T$ into Eq. (6-16) gives

$$\sigma_{x1} = -\frac{MyE_1}{E_1I_1 + E_2I_2} \qquad \text{(6-18a)}$$

which is the same as Eq. (6-7a), thus demonstrating that the stresses in material 1 in the original beam are the same as the stresses in the corresponding part of the transformed beam.

As mentioned previously, the stresses in material 2 in the original beam are *not* the same as the stresses in the corresponding part of the transformed beam. Instead, the stresses in the transformed beam (Eq. 6-16) must be multiplied by the modular ratio $n$ to obtain the stresses in material 2 of the original beam:

$$\sigma_{x2} = -\frac{My}{I_T} n \qquad \text{(6-18b)}$$

We can verify this formula by noting that when Eq. (6-17) for $I_T$ is substituted into Eq. (6-18b), we get

$$\sigma_{x2} = -\frac{MynE_1}{E_1I_1 + E_2I_2} = -\frac{MyE_2}{E_1I_1 + E_2I_2} \qquad \text{(6-18c)}$$

which is the same as Eq. (6-7b).

## General Comments

In this discussion of the transformed-section method, we chose to transform the original beam to a beam consisting entirely of material 1. It is also possible to transform the beam to material 2. In that case the stresses in the original beam in material 2 will be the same as the stresses in the corresponding part of the transformed beam. However, the stresses in material 1 in the original beam must be obtained by multiplying the stresses in the corresponding part of the transformed beam by the modular ratio $n$, which in this case is defined as $n = E_1/E_2$.

It is also possible to transform the original beam into a material having any arbitrary modulus of elasticity $E$, in which case all parts of the beam must be transformed to the fictitious material. Of course, the calculations are simpler if we transform to one of the original materials. Finally, with a little ingenuity it is possible to extend the transformed-section method to composite beams of more than two materials.

The composite beam shown in Fig. 6-11a is formed of a wood beam (4.0 in. × 6.0 in. actual dimensions) and a steel reinforcing plate (4.0 in. wide and 0.5 in. thick). The beam is subjected to a positive bending moment $M = 60$ k-in.

Using the transformed-section method, calculate the largest tensile and compressive stresses in the wood (material 1) and the maximum and minimum tensile stresses in the steel (material 2) if $E_1 = 1500$ ksi and $E_2 = 30{,}000$ ksi.

*Note:* This same beam was analyzed previously in Example 6-1 of Section 6.2.

## Fig. 6-11

Example 6-3: Composite beam of Example 6-1 analyzed by the transformed-section method: (a) cross section of original beam, and (b) transformed section (material 1)

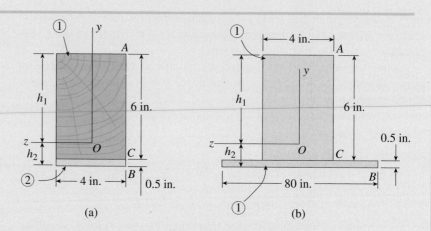

(a)

(b)

### Solution

*Transformed section.* We will transform the original beam into a beam of material 1, which means that the modular ratio is defined as

$$n = \frac{E_2}{E_1} = \frac{30{,}000 \text{ ksi}}{1500 \text{ ksi}} = 20$$

The part of the beam made of wood (material 1) is not altered but the part made of steel (material 2) has its width multiplied by the modular ratio. Thus, the width of this part of the beam becomes

$$n(4 \text{ in.}) = 20(4 \text{ in.}) = 80 \text{ in.}$$

in the transformed section (Fig. 6-11b).

*Neutral axis.* Because the transformed beam consists of only one material, the neutral axis passes through the centroid of the cross-sectional area. Therefore, with the top edge of the cross section serving as a reference line, and with the distance $y_i$ measured positive downward, we can calculate the distance $h_1$ to the centroid as

$$h_1 = \frac{\Sigma y_i A_i}{\Sigma A_i} = \frac{(3\ \text{in.})(4\ \text{in.})(6\ \text{in.}) + (6.25\ \text{in.})(80\ \text{in.})(0.5\ \text{in.})}{(4\ \text{in.})(6\ \text{in.}) + (80\ \text{in.})(0.5\ \text{in.})}$$

$$= \frac{322.0\ \text{in.}^3}{64.0\ \text{in.}^2} = 5.031\ \text{in.}$$

Also, the distance $h_2$ from the lower edge of the section to the centroid is

$$h_2 = 6.5\ \text{in.} - h_1 = 1.469\ \text{in.}$$

Thus, the location of the neutral axis is determined.

*Moment of inertia of the transformed section.* Using the parallel-axis theorem (see Section 12.5 of Chapter 12), we can calculate the moment of inertia $I_T$ of the entire cross-sectional area with respect to the neutral axis as

$$I_T = \frac{1}{12}\ (4\ \text{in.})(6\ \text{in.})^3 + (4\ \text{in.})(6\ \text{in.})(h_1 - 3\ \text{in.})^2 +$$

$$\frac{1}{12}\ (80\ \text{in.})(0.5\ \text{in.})^3 + (80\ \text{in.})(0.5\ \text{in.})(h_2 - 0.25\ \text{in.})^2$$

$$= 171.0\ \text{in.}^4 + 60.3\ \text{in.}^4 = 231.3\ \text{in.}^4$$

*Normal stresses in the wood (material 1).* The stresses in the transformed beam (Fig. 6-11b) at the top of the cross section (A) and at the contact plane between the two parts (C) are the same as in the original beam (Fig. 6-11a). These stresses can be found from the flexure formula in Eq. (6-16) as

$$\sigma_{1A} = -\frac{My}{I_T} = -\frac{(60\ \text{k-in.})(5.031\ \text{in.})}{231.3\ \text{in.}^4} = -1310\ \text{psi} \quad \Longleftarrow$$

$$\sigma_{1C} = -\frac{My}{I_T} = -\frac{(60\ \text{k-in.})(-0.969\ \text{in.})}{231.3\ \text{in.}^4} = 251\ \text{psi} \quad \Longleftarrow$$

These are the largest tensile and compressive stresses in the wood (material 1) in the original beam. The stress $\sigma_{1A}$ is compressive and the stress $\sigma_{1C}$ is tensile.

*Normal stresses in the steel (material 2).* The maximum and minimum stresses in the steel plate are found by multiplying the corresponding stresses in the transformed beam by the modular ratio $n$ in Eq. (6-18b). The

*Continues* ➡

**Example 6-3 -** *Continued*

maximum stress occurs at the lower edge of the cross section (*B*) and the minimum stress occurs at the contact plane (*C*):

$$\sigma_{2B} = -\frac{My}{I_T}n = -\frac{(60 \text{ k-in.})(-1.469 \text{ in.})}{231.3 \text{ in.}^4}(20) = 7620 \text{ psi} \quad \Longleftarrow$$

$$\sigma_{2C} = -\frac{My}{I_T}n = -\frac{(60 \text{ k-in.})(-0.969 \text{ in.})}{231.3 \text{ in.}^4}(20) = 5030 \text{ psi} \quad \Longleftarrow$$

Both of these stresses are tensile.

Note that the stresses calculated by the transformed-section method agree with those found in Example 6-1 by direct application of the formulas for a composite beam.

*Balanced design.* As a final evaluation of the wood–steel composite beam considered here and in Example 6-1, we note that neither wood nor steel has reached typical allowable stress levels. Perhaps some redesigning of this beam would be of interest; we will consider only the steel plate here (but could re-size the wood beam too if we wish).

A *balanced design* is one in which wood and steel reach their allowable stress values at the same time under the design moment; this could be regarded as a more efficient design of this beam. *First*, holding the steel plate thickness at $t_s = 0.5$ in., we will find the required width $b_s$ of the steel plate, so the wood and steel reach allowable stress values simultaneously under design moment $M_D$. Then setting $b_s = 4$ in., we will repeat the previous process but also find required plate thickness $t_s$ to achieve the same objective. Assume that the allowable stress values for wood and steel are $\sigma_{aw} = 1850$ psi and $\sigma_{as} = 14$ ksi, respectively. Also assume that the wood beam dimensions are unchanged.

Using the transformed-section approach, we can write the expressions for the stresses at the top of the wood and bottom of the steel. We can equate each to its allowable value as

$$\sigma_{aw} = \frac{-M_D h_1}{I_T} \quad \text{and} \quad \sigma_{as} = \frac{-M_D h_2 n}{I_T} \tag{a,b}$$

Next, we solve each of Eqs. (a) and (b) for ratio $M_D/I_T$, then equate the two expressions to find the $h_1/h_2$ ratio for which allowable stress levels are reached in both materials:

$$\frac{h_1}{h_2} = n\frac{\sigma_{aw}}{\sigma_{as}} \tag{c}$$

Expressions for $h_1$ and $h_2$ can be obtained in terms of the transformed-section dimensions $b$, $h$, $b_s$, and $t_s$ (Figs. 6-11c and d) by taking first moments about the *z* axis to get

$$h_1 = \frac{h}{2} + \frac{(b_s n t_s^2) + (b_s h n t_s)}{(2bh) + (2b_s n t_s)} \quad \text{and} \quad h_2 = \frac{n b_s t_s\left(\dfrac{t_s}{2}\right) + bh\left(t_s + \dfrac{h}{2}\right)}{(n b_s t_s) + (bh)} \tag{d}$$

## Fig. 6-11

Example 6-3: Balanced design of composite beam: (c) original beam, and (d) transformed beam

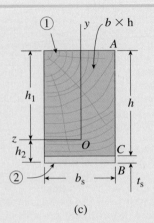

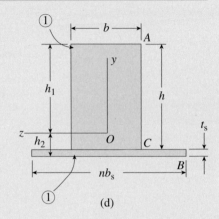

(c)

(d)

With some effort (and perhaps with computer assistance), we can re-write Eq. (c) as

$$\frac{bh^2 + (2b_snht_s) + (b_snt_s^2)}{bh^2 + (2bht_s) + (b_snt_s^2)} = n\frac{\sigma_{aw}}{\sigma_{as}} \qquad \text{(e)}$$

Collecting terms and solving for the required width of $b_s$ for the steel plate (with thickness $t_s$ unchanged), then substituting numerical values, we get for $b_s$ (instead of the original width of 4 in.):

$$b = 4 \text{ in.} \quad h = 6 \text{ in.} \quad t_s = 0.5 \text{ in.} \quad n = 20 \quad \sigma_{aw} = 1850 \text{ psi} \quad \sigma_{as} = 14 \text{ ksi}$$

$$b_s = \frac{\left(n\dfrac{\sigma_{aw}}{\sigma_{as}}\right)\left(bh^2 + 2bht_s\right) - bh^2}{(2nht_s) + nt_s^2\left(1 - n\dfrac{\sigma_{aw}}{\sigma_{as}}\right)} = 2.68 \text{ in.} \qquad \text{(f)}$$

So for a 4 in. $\times$ 6 in. wood beam reinforced by a 2.68 in. $\times$ 0.5 in. steel plate (Fig. 6-11a) under any applied moment $M$ that is less than or equal to $M_D$, the stress ratio $\sigma_{1A}/\sigma_{2B}$ will be equal to $\sigma_{aw}/\sigma_{as}$. If $M = M_D$, then $\sigma_{1A} = \sigma_{aw}$ and $\sigma_{2B} = \sigma_{as}$.

Alternatively, we could reformulate Eq.(e) to get a quadratic equation for the steel plate thickness $t_s$ (with the original width $b_s = 4$ in.) to obtain

$$t_s^2\left[nb_s\left(1 - n\frac{\sigma_{aw}}{\sigma_{as}}\right)\right] + t_s\left\{2h\left[nb_s - b\left(n\frac{\sigma_{aw}}{\sigma_{as}}\right)\right]\right\} + bh^2\left(1 - n\frac{\sigma_{aw}}{\sigma_{as}}\right) = 0 \qquad \text{(g)}$$

The solution of Eq. (g) results in a revised steel plate thickness leading to a *balanced design* of the wood–steel composite beam as $t_s = 0.298$ in. Once again, for $b_s = 4$ in. and $t_s = 0.298$ in., the stress ratio $\sigma_{1A}/\sigma_{2B} = \sigma_{aw}/\sigma_{as}$ for applied moments $M$ which are less than or equal to $M_D$.

### Fig. 6-12

Example 6-4: Cross section of a singly reinforced concrete inverted T-beam

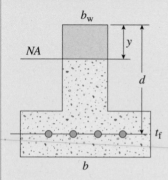

An inverted precast concrete T-beam is used to support precast double-tee floor beams in a parking deck (see Fig. 6-12 and photo). Beam dimensions are $b = 500$ mm, $b_w = 300$ mm, $d = 600$ mm and $t_f = 100$ mm. Steel reinforcement consists of four bars each with a 25-mm diameter. The modulus of elasticity for the concrete is $E_c = 25$ GPa, while that of the steel is $E_s = 200$ GPa. Allowable stresses for concrete and steel are $\sigma_{ac} = 9.3$ MPa and $\sigma_{as} = 137$ MPa, respectively.

(a) Use the transformed section in Fig. 6-13 (in which the concrete in tension is neglected and the steel reinforcing bars are converted to the equivalent concrete) to find the maximum permissible moment that can be applied to this beam.

(b) Repeat part (a) if the beam is rotated 180°, as shown in Fig. 6-14, and if the steel reinforcement remains in the bottom tension zone.

### Fig. 6-13

Transformed section for singly reinforced concrete inverted T-beam

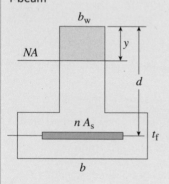

(Photo used by permission of Dr. Lawrence F. Kahn, Georgia Tech)

### Solution

(a) *Inverted T-beam.* We start by finding the *neutral axis* (at some distance $y$ down from the top of the beam) for the transformed section shown in Fig. 6-13. Equating the first moments of areas of concrete in compression ($b_w \times y$) and transformed area of steel in tension ($n \times A_s$) leads to a quadratic equation. The solution for $y$ gives the position of the neutral axis as

$$b_w y \frac{y}{2} - nA_s(d - y) = 0 \quad \text{where} \quad n = \frac{E_s}{E_c} = 8 \quad \textbf{(a)}$$

$$y = \sqrt{\left(\frac{nA_s}{b_w}\right)^2 + 2d\left(\frac{nA_s}{b_w}\right)} - \left(\frac{nA_s}{b_w}\right) = 0.204 \text{ m} \quad \textbf{(b)}$$

Now using Eq. (6-17), we compute the moment of inertia of the transformed section:

$$I_T = \frac{b_w y^3}{3} + nA_s[(d-y)^2] = 3.312 \times 10^{-3} \text{ m}^4 \tag{c}$$

Finally, we obtain the moment capacity of the beam by solving Eqs. (6-16) (allowable stress in concrete controls) and Eq. (6-18b) (allowable stress in steel controls) for $M$, where the lower value based on allowable stress in the steel governs:

$$M_c = \frac{\sigma_{ac}}{y} I_T = \left(\frac{9.3 \text{ MPa}}{0.204 \text{ m}}\right)\left(3.312 \times 10^{-3} \text{ m}^4\right) = 151 \text{ kN} \cdot \text{m} \tag{d}$$

$$M_s = \frac{\sigma_{as}}{n(d-y)} I_T = \frac{137 \text{ MPa}}{8(0.6 \text{ m} - 0.204 \text{ m})} (3.312 \times 10^{-3} \text{ m}^4) \tag{e}$$

$$= 143.2 \text{ kN} \cdot \text{m}$$

(b) *T-beam.* Now the flange of the T-beam with a thickness of $t_f$ is on top, so we start by assuming that neutral axis location distance $y$ is greater than $t_f$. We can divide the compression of the concrete for the transformed section (Fig. 6-14) into three rectangles, then equate the first moments of areas of the concrete in compression and transformed area of the steel ($n \times A_s$) to get a quadratic equation for distance $y$. The solution for $y$ gives the position of the neutral axis as

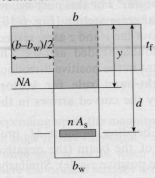

### Fig. 6-14

Transformed section for singly reinforced concrete T-beam

$$(b - b_w)t_f\left(y - \frac{t_f}{2}\right) + b_w y \frac{y}{2} - nA_s(d-y) = 0 \tag{f}$$

Solving Eq. (f) for $y$, we get

$$y = 0.1702 \text{ m}$$

The moment of inertia of the transformed section is now

$$I_T = \frac{b_w y^3}{3} + \frac{(b - b_w)t_f^3}{12} + (b - b_w)t_f\left(y - \frac{t_f}{2}\right)^2 + nA_s(d-y)^2$$

$$= 3.7 \times 10^{-3} \text{ m}^4$$

Finally, we repeat the solutions for maximum permissible moment $M$ in Eqs. (d) and (e) as

$$M_c = \frac{\sigma_{ac}}{y} I_T = \frac{(9.3 \text{ MPa})}{0.1702 \text{ m}} (3.7 \times 10^{-3} \text{ m}^4) = 202 \text{ kN} \cdot \text{m} \tag{g}$$

$$M_s = \frac{\sigma_{as}}{n(d-y)} I_T = \frac{137 \text{ MPa}}{8(0.6 \text{ m} - 0.1702 \text{ m})} (3.7 \times 10^{-3} \text{ m}^4) \tag{h}$$

$$= 147.4 \text{ kN} \cdot \text{m}$$

and find that once again, the lower value of moment $M$ based on allowable stress in the steel governs. Since the allowable stress in the reinforcing steel bars controls both beams, their moment capacities [Eqs.(e) and (h)] are approximately the same.

● ● ● **Example 6-6** - *Continued*

**Fig. 6-22**

Solution to Example 6-6:
(a) Components of the uniform
load, (b) bending moments
acting on a cross section, and
(c) normal stress distribution

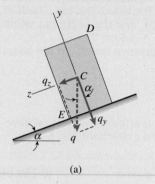

(a)

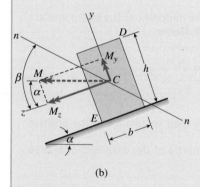

(b)

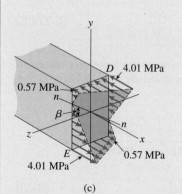

(c)

From the orientation of the cross section and the directions of the loads and bending moments (Fig. 6-22), it is apparent that the maximum compressive stress occurs at point $D$ (where $y = h/2$ and $z = -b/2$) and the maximum tensile stress occurs at point $E$ (where $y = -h/2$ and $z = b/2$). Substituting these coordinates into Eq. (6-28) and then simplifying, we obtain expressions for the maximum and minimum stresses in the beam:

$$\sigma_E = -\sigma_D = \frac{3qL^2}{4bh}\left(\frac{\sin\alpha}{b} + \frac{\cos\alpha}{h}\right) \quad \Longleftarrow \quad (6\text{-}29)$$

*Numerical values.* The maximum tensile and compressive stresses can be calculated from the preceding equation by substituting the given data:

$$q = 3.0 \text{ kN/m} \quad L = 1.6 \text{ m} \quad b = 100 \text{ mm} \quad h = 150 \text{ mm} \quad \alpha = 26.57°$$

The results are

$$\sigma_E = -\sigma_D = 4.01 \text{ MPa} \quad \Longleftarrow$$

*Neutral axis.* In addition to finding the stresses in the beam, it is often useful to locate the neutral axis. The equation of this line is obtained by setting the stress (Eq. 6-28) equal to zero:

$$\frac{\sin\alpha}{b^2}z - \frac{\cos\alpha}{h^2}y = 0 \quad (6\text{-}30)$$

The neutral axis is shown in Fig. 6-22b as line $nn$. The angle $\beta$ from the $z$ axis to the neutral axis is obtained from Eq. (6-30) as

$$\tan\beta = \frac{y}{z} = \frac{h^2}{b^2}\tan\alpha \quad (6\text{-}31)$$

Substituting numerical values, we get

$$\tan\beta = \frac{h^2}{b^2}\tan\alpha = \frac{(150 \text{ mm})^2}{(100 \text{ mm})^2}(\tan 26.57°) = 1.125 \quad \beta = 48.4° \quad \Longleftarrow$$

Since the angle $\beta$ is not equal to the angle $\alpha$, the neutral axis is inclined to the plane of loading (which is vertical).

From the orientation of the neutral axis (Fig. 6-22b), we see that points $D$ and $E$ are the farthest from the neutral axis, thus confirming our assumption that the maximum stresses occur at those points. The part of the beam above and to the right of the neutral axis is in compression, and the part to the left and below the neutral axis is in tension.

# 6.5 BENDING OF UNSYMMETRIC BEAMS

In our previous discussions of bending, we assumed that the beams had cross sections with at least one axis of symmetry. Now we will abandon that restriction and consider beams having unsymmetric cross sections. We begin by investigating beams in pure bending, and then in later sections (Sections 6.6 through 6.9) we will consider the effects of lateral loads. As in earlier discussions, it is assumed that the beams are made of linearly elastic materials.

Suppose that a beam having an unsymmetric cross section is subjected to a bending moment $M$ acting at the end cross section (Fig. 6-23a). We would like to know the stresses in the beam and the position of the neutral axis. Unfortunately, at this stage of the analysis there is no direct way of determining these quantities. Therefore, we will use an indirect approach—instead of starting with a bending moment and trying to find the neutral axis, we will start with an assumed neutral axis and find the associated bending moment.

## Neutral Axis

We begin by constructing two perpendicular axes (the $y$ and $z$ axes) at an arbitrarily selected point in the plane of the cross section (Fig. 6-23b). The axes may have any orientation, but for convenience we will orient them horizontally and vertically. Next, we *assume* that the beam is bent in such a manner that the $z$ axis is the neutral axis of the cross section. Consequently, the beam deflects in the $xy$ plane, which becomes the plane of bending. Under these conditions, the normal stress acting on an element of area $dA$ located at distance $y$ from the neutral axis (see Fig. 6-23b and Eq. 5-8 of Chapter 5) is

$$\sigma_x = -E\kappa_y y \qquad \textbf{(6-32)}$$

The minus sign is needed because the part of the beam above the $z$ axis (the neutral axis) is in compression when the curvature is positive. (The sign convention for curvature when the beam is bent in the $xy$ plane is shown in Fig. 6-24a.)

The force acting on the element of area $dA$ is $\sigma_x dA$, and the resultant force acting on the entire cross section is the integral of this elemental force over the cross-sectional area $A$. Since the beam is in pure bending, the resultant force must be zero; hence,

$$\int_A \sigma_x dA = -\int_A E\kappa_x y\, dA = 0$$

The modulus of elasticity and the curvature are constants at any given cross section, and therefore

$$\int_A y\, dA = 0 \qquad \textbf{(6-33)}$$

**Fig. 6-23**

Unsymmetric beam subjected to a bending moment $M$

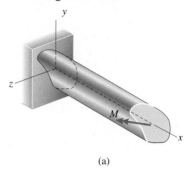

(a)

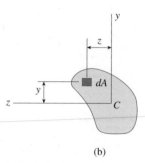

(b)

Unsymmetric composite beam made up from channel section and old wood beam
(Franz Pflueg/Shutterstock)

### Fig. 6-24

Sign conventions for curvatures $k_y$ and $k_z$ in the $xy$ and $xz$ planes, respectively

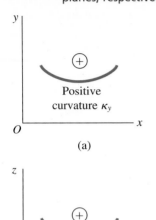

(a)

(b)

### Fig. 6-25

Bending moments $M_y$ and $M_z$ acting about the $y$ and $z$ axes, respectively

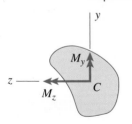

This equation shows that the $z$ axis (the neutral axis) passes through the centroid $C$ of the cross section.

Now assume that the beam is bent in such a manner that the $y$ axis is the neutral axis and the $xz$ plane is the plane of bending. Then the normal stress acting on the element of area $dA$ (Fig. 6-23b) is

$$\sigma_x = -E\kappa_z z \tag{6-34}$$

The sign convention for the curvature $\kappa_z$ in the $xz$ plane is shown in Fig. 6-24b. The minus sign is needed in Eq. (6-34) because positive curvature in the $xz$ plane produces compression on the element $dA$. The resultant force for this case is

$$\int_A \sigma_x \, dA = -\int_A E\kappa_z z \, dA = 0$$

from which we get

$$\int_A z \, dA = 0 \tag{6-35}$$

and again we see that the neutral axis must pass through the centroid. Thus, we have established that *the origin of the $y$ and $z$ axes for an unsymmetric beam must be placed at the centroid $C$.*

Now let us consider the moment resultant of the stresses $\sigma_x$. Once again we assume that bending takes place with the $z$ axis as the neutral axis, in which case the stresses $\sigma_x$ are given by Eq. (6-32). The corresponding bending moments $M_z$ and $M_y$ about the $z$ and $y$ axes, respectively (Fig. 6-25), are

$$M_z = -\int_A \sigma_x y \, dA = \kappa_y E \int_A y^2 dA = \kappa_y E I_z \tag{6-36a}$$

$$M_y = \int_A \sigma_x z \, dA = -\kappa_y E \int_A yz \, dA = -\kappa_y E I_{yz} \tag{6-36b}$$

In these equations, $I_z$ is the moment of inertia of the cross-sectional area with respect to the $z$ axis and $I_{yz}$ is the *product of inertia* with respect to the $y$ and $z$ axes.*

From Eqs. (6-36a) and (6-36b) we can draw the following conclusions: (1) If the $z$ axis is selected in an arbitrary direction through the centroid, it will be the neutral axis *only* if moments $M_y$ and $M_z$ act about the $y$ and $z$ axes and *only* if these moments are in the ratio established by Eqs. (6-36a) and (6-36b). (2) If the $z$ axis is selected as a *principal axis*, then the product of inertia $I_{yz}$ equals zero and the only bending moment is $M_z$. In that case, the $z$ axis is the neutral axis, bending takes place in the $xy$ plane, and the moment $M_z$ acts in that same plane. Thus, bending occurs in a manner analogous to that of a symmetric beam.

In summary, an unsymmetric beam bends in the same general manner as a symmetric beam provided the $z$ axis is a *principal centroidal axis*

---

*Products of inertia are discussed in Section 12.7 of Chapter 12.

and the only bending moment is the moment $M_z$ acting about that same axis.

If we now assume that the $y$ axis is the neutral axis, we will arrive at similar conclusions. The stresses $\sigma_x$ are given by Eq. (6-34) and the bending moments are

$$M_y = \int_A \sigma_x z \, dA = -\kappa_z E \int_A z^2 \, dA = -\kappa_z E I_y \qquad \textbf{(6-37a)}$$

$$M_z = -\int_A \sigma_x y \, dA = \kappa_z E \int_A yz \, dA = \kappa_z E I_{yz} \qquad \textbf{(6-37b)}$$

in which $I_y$ is the moment of inertia with respect to the $y$ axis. Again we observe that if the neutral axis (the $y$ axis in this case) is oriented arbitrarily, moments $M_y$ and $M_z$ must exist. However, if the $y$ axis is a principal axis, the only moment is $M_y$ and we have ordinary bending in the $xz$ plane. Therefore, we can state that an unsymmetric beam bends in the same general manner as a symmetric beam when the $y$ axis is a *principal centroidal axis* and the only bending moment is the moment $M_y$ acting about that same axis.

One further observation—since the $y$ and $z$ axes are orthogonal, we know that if *either* axis is a principal axis, then the other axis is automatically a principal axis.

We have now arrived at the following important conclusion: *When an unsymmetric beam is in pure bending, the plane in which the bending moment acts is perpendicular to the neutral surface only if the $y$ and $z$ axes are principal centroidal axes of the cross section and the bending moment acts in one of the two principal planes (the $xy$ plane or the $xz$ plane).* In such a case, the principal plane in which the bending moment acts becomes the plane of bending and the usual bending theory (including the flexure formula) is valid.

Having arrived at this conclusion, we now have a direct method for finding the stresses in an unsymmetric beam subjected to a bending moment acting in an arbitrary direction.

## Procedure for Analyzing an Unsymmetric Beam

We will now describe a general procedure for analyzing an unsymmetric beam subjected to any bending moment $M$ (Fig. 6-26). We begin by locating the centroid $C$ of the cross section and constructing a set of principal axes at that point (the $y$ and $z$ axes in the figure).* Next, the bending moment $M$ is resolved into components $M_y$ and $M_z$, positive in the directions shown in the figure. These components are

$$M_y = M \sin \theta \qquad M_z = M \cos \theta \qquad \textbf{(6-38a,b)}$$

in which $\theta$ is the angle between the moment vector $M$ and the $z$ axis (Fig. 6-26). Since each component acts in a principal plane, it produces

*Principal axes are discussed in Sections 12.8 and 12.9 of Chapter 12.

### Fig. 6-26

Unsymmetric cross section with the bending moment $M$ resolved into components $M_y$ and $M_z$ acting about the principal centroidal axes

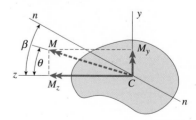

pure bending in that same plane. Thus, the usual formulas for pure bending apply, and we can readily find the stresses due to the moments $M_y$ and $M_z$ acting separately. The bending stresses obtained from the moments acting separately are then superposed to obtain the stresses produced by the original bending moment $M$. (Note that this general procedure is similar to that described in the preceding section for analyzing doubly symmetric beams with inclined loads.)

The superposition of the bending stresses in order to obtain the resultant stress at any point in the cross section is given by Eq. (6-19):

$$\sigma_x = \frac{M_y z}{I_y} - \frac{M_z y}{I_z} = \frac{(M \sin \theta)z}{I_y} - \frac{(M \cos \theta)y}{I_z} \qquad \textbf{(6-39)}$$

in which $y$ and $z$ are the coordinates of the point under consideration.

Also, the equation of the neutral axis $nn$ (Fig. 6-26) is obtained by setting $\sigma_x$ equal to zero and simplifying:

$$\frac{\sin \theta}{I_y} z - \frac{\cos \theta}{I_z} y = 0 \qquad \textbf{(6-40)}$$

The angle $\beta$ between the neutral axis and the $z$ axis can be obtained from the preceding equation, as

$$\tan \beta = \frac{y}{z} = \frac{I_z}{I_y} \tan \theta \qquad \textbf{(6-41)}$$

This equation shows that in general the angles $\beta$ and $\theta$ are not equal, hence the neutral axis is generally not perpendicular to the plane in which the applied couple $M$ acts. The only exceptions are the three special cases described in the preceding section in the paragraph following Eq. (6-24).

In this section we have focused our attention on unsymmetric beams. Of course, symmetric beams are special cases of unsymmetric beams, and therefore the discussions of this section also apply to symmetric beams. If a beam is singly symmetric, the axis of symmetry is one of the centroidal principal axes of the cross section; the other principal axis is perpendicular to the axis of symmetry at the centroid. If a beam is doubly symmetric, the two axes of symmetry are centroidal principal axes.

In a strict sense the discussions of this section apply only to pure bending, which means that no shear forces act on the cross sections. When shear forces do exist, the possibility arises that the beam will twist about the longitudinal axis. However, twisting is avoided when the shear forces act through the *shear center*, which is described in the next section.

The following examples illustrate the analysis of a beam having one axis of symmetry. (The calculations for an unsymmetric beam having no axes of symmetry proceed in the same general manner, except that the determination of the various cross-sectional properties is much more complex.)

## • • • Example 6-7

Example 6-7: Channel section subjected to a bending moment $M$ acting at an angle $\theta$ to the $z$ axis

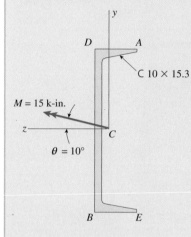

A channel section (C 10 × 15.3) is subjected to a bending moment $M = 15$ k-in. oriented at an angle $\theta = 10°$ to the $z$ axis (Fig. 6-27).

    Calculate the bending stresses $\sigma_A$ and $\sigma_B$ at points $A$ and $B$, respectively, and determine the position of the neutral axis.

### Solution

*Properties of the cross section.* The centroid $C$ is located on the axis of symmetry (the $z$ axis) at a distance

$$c = 0.634 \text{ in.}$$

from the back of the channel (Fig. 6-28).* The $y$ and $z$ axes are principal centroidal axes with moments of inertia

$$I_y = 2.28 \text{ in.}^4 \qquad I_z = 67.4 \text{ in.}^4$$

Also, the coordinates of points $A$, $B$, $D$, and $E$ are

$$y_A = 5.00 \text{ in.} \quad z_A = -2.600 \text{ in.} + 0.634 \text{ in.} = -1.966 \text{ in.}$$
$$y_B = -5.00 \text{ in.} \quad z_B = 0.634 \text{ in.}$$
$$y_D = y_A, z_D = z_B$$
$$y_E = y_B, z_E = z_A$$

*Bending moments.* The bending moments about the $y$ and $z$ axes (Fig. 6-28) are

$$M_y = M \sin \theta = (15 \text{ k-in.})(\sin 10°) = 2.605 \text{ k-in.}$$
$$M_z = M \cos \theta = (15 \text{ k-in.})(\cos 10°) = 14.77 \text{ k-in.}$$

*Bending stresses.* We now calculate the stress at point $A$ from Eq. (6-39):

$$\sigma_A = \frac{M_y z_A}{I_y} - \frac{M_z y_A}{I_z}$$

$$= \frac{(2.605 \text{ k-in.})(-1.966 \text{ in.})}{2.28 \text{ in.}^4} - \frac{(14.77 \text{ k-in.})(5.00 \text{ in.})}{67.4 \text{ in.}^4}$$

$$= -2246 \text{ psi} - 1096 \text{ psi} = -3340 \text{ psi} \qquad \Longleftarrow$$

By a similar calculation, we obtain the stress at point $B$:

$$\sigma_B = \frac{M_y z_B}{I_y} - \frac{M_z y_B}{I_z}$$

$$= \frac{(2.605 \text{ k-in.})(0.634 \text{ in.})}{2.28 \text{ in.}^4} - \frac{(14.77 \text{ k-in.})(-5.00 \text{ in.})}{67.4 \text{ in.}^4}$$

$$= 724 \text{ psi} + 1096 \text{ psi} = 1820 \text{ psi} \qquad \Longleftarrow$$

These stresses are the maximum compressive and tensile stresses in the beam.

(a) Solution to Example 6-7

(a)

---

*See Table F-3, Appendix F, for dimensions and properties of channel sections.

*Continues*

**Fig. 6-28 (Continued)**

(b) Normal stress distribution in channel section

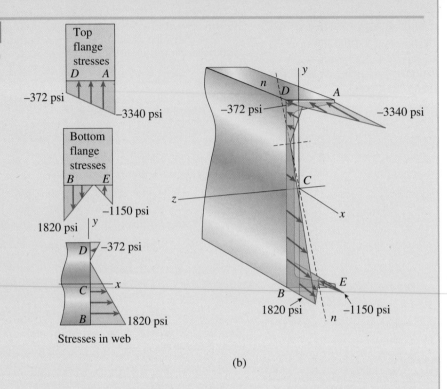

(b)

The normal stresses at points D and E also can be computed using the procedure shown.

$$\sigma_D = -372 \text{ psi}, \qquad \sigma_E = -1150 \text{ psi}$$

The normal stresses acting on the cross section are shown in Fig. 6-28(b).

*Neutral axis.* The angle $\beta$ that locates the neutral axis (Eq. 6-41) is found as

$$\tan \beta = \frac{I_z}{I_y} \tan \theta = \frac{67.4 \text{ in.}^4}{2.28 \text{ in.}^4} \tan 10° = 5.212 \quad \beta = 79.1° \quad \Longleftarrow$$

The neutral axis $nn$ is shown in Fig. 6-28, and we see that points A and B are located at the farthest distances from the neutral axis, thus confirming that $\sigma_A$ and $\sigma_B$ are the largest stresses in the beam.

   In this example, the angle $\beta$ between the z axis and the neutral axis is much larger than the angle $\theta$ (Fig. 6-28) because the ratio $I_z/I_y$ is large. The angle $\beta$ varies from 0 to 79.1° as the angle $\theta$ varies from 0 to 10°. As discussed previously in Example 6-5 of Section 6.4, beams with large $I_z/I_y$ ratios are very sensitive to the direction of loading. Thus, beams of this kind should be provided with lateral support to prevent excessive lateral deflections.

## Example 6-8

A Z-section is subjected to bending moment $M = 3$ kN·m at an angle $\theta = -20$ degrees to the $z$ axis, as shown. Find the normal stresses at $A$, $B$, $D$, and $E$ ($\sigma_A$, $\sigma_B$, $\sigma_D$, and $\sigma_E$, respectively) and also find the position of the neutral axis. Use the following numerical data: $h = 200$ mm, $b = 90$ mm, thickness $t = 15$ mm.

### Solution

*Properties of the cross section.* Use the results of Example 12-7.

$$I_z = 32.6(10^6) \text{ mm}^4 \quad I_Y = 2.4(10^6) \text{ mm}^4$$

$$\theta_{p1} = 19.2° \quad \theta_{p1} = (19.2)\frac{\pi}{180} \text{ radians}$$

**Coordinates** $(y, z)$ of points $A$, $B$, $D$, $D'$, $E$, and $E'$:

$$\theta = -20\left(\frac{\pi}{180}\right) \text{ radians}$$

$$y_A = \frac{h}{2}\cos(\theta_{p1}) + \left(b - \frac{t}{2}\right)\sin(\theta_{p1}) \qquad y_A = 121.569 \text{ mm}$$

$$y_B = -y_A \qquad y_B = -121.569 \text{ mm}$$

$$y_D = \frac{h}{2}\cos(\theta_{p1}) - \frac{t}{2}\sin(\theta_{p1}) \qquad y_D = 91.971 \text{ mm}$$

$$y_{D'} = \frac{h}{2}\cos(\theta_{p1}) \qquad y_{D'} = 94.438 \text{ mm}$$

$$y_{E'} = -y_{D'} \qquad y_{E'} = -94.438 \text{ mm}$$

$$y_E = -y_D \qquad y_E = -91.971 \text{ mm}$$

$$z_A = \left(b - \frac{t}{2}\right)\cos(\theta_{p1}) - \frac{h}{2}\sin(\theta_{p1}) \qquad z_A = 45.024 \text{ mm}$$

$$z_B = -z_A \qquad z_B = -45.024 \text{ mm}$$

$$z_D = \frac{-h}{2}\sin(\theta_{p1}) - \frac{t}{2}\cos(\theta_{p1}) \qquad z_D = -39.969 \text{ mm}$$

$$z_{D'} = \frac{-h}{2}\sin(\theta_{p1}) \qquad z_{D'} = -32.887 \text{ mm}$$

$$z_{E'} = -z_{D'} \qquad z_{E'} = 32.887 \text{ mm}$$

$$z_E = -z_D \qquad z_E = 39.969 \text{ mm}$$

*Bending moments* (kN·m) $M = 3$ kN·m:

$$M_y = M\sin(\theta) \qquad M_y = -1.026 \text{ kN·m}$$

$$M_z = M\cos(\theta) \qquad M_z = 2.819 \text{ kN·m}$$

*Bending stresses at $A$, $B$, $D$, and $E$* (see plots of normal stresses in Fig. 6-29b):

$$\sigma_A = \frac{M_y z_A}{I_y} - \frac{M_z y_A}{I_z} = -19.249 - 10.513 = -29.8 \text{ MPa}$$

$$\sigma_B = \frac{M_y z_B}{I_y} - \frac{M_z y_B}{I_z} = 19.249 + 10.513 = 29.8 \text{ MPa}$$

$$\sigma_D = \frac{M_y z_D}{I_y} - \frac{M_z y_D}{I_z} = 17.088 - 7.953 = 9.14 \text{ MPa}$$

*Continues* ➡

**Fig. 6-29**

(a) Z section subjected to bending moment *M* at angle *θ* to Z axis (b) normal stress distribution in Z section

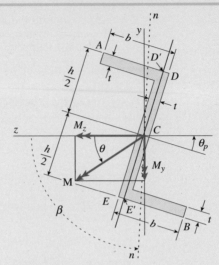

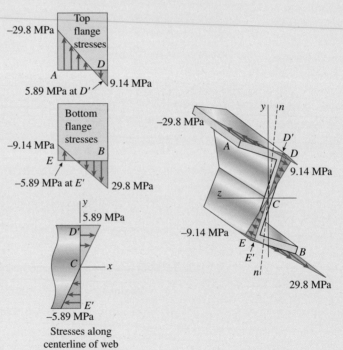

$$\sigma_{D'} = \frac{M_y z_{D'}}{I_y} - \frac{M_z y_{D'}}{I_z} = 14.06 - 8.167 = 5.89 \text{ MPa} = -\sigma_{E'}$$

$$\sigma_{E} = \frac{M_y z_{E}}{I_y} - \frac{M_z y_{E}}{I_z} = -17.088 + 7.953 = -9.14 \text{ MPa}$$

*Location of neutral axis:*

$$\tan(\beta) = \frac{I_z}{I_z} \tan(\theta)$$

$$\beta = -89.1°$$

# 6.6 THE SHEAR-CENTER CONCEPT

In the preceding sections of this chapter, we were concerned with determining the bending stresses in beams under a variety of special conditions. For instance, in Section 6.4 we considered symmetrical beams with inclined loads, and in Section 6.5 we considered unsymmetrical beams. However, lateral loads acting on a beam produce shear forces as well as bending moments, and therefore in this and the next three sections we will examine the effects of shear.

In Chapter 5, we saw how to determine the shear stresses in beams when the loads act in a plane of symmetry, and we derived the shear formula for calculating those stresses for certain shapes of beams. Now we will examine the shear stresses in beams when the lateral loads act in a plane that is *not* a plane of symmetry. We will find that the loads must be applied at a particular point in the cross section, called the *shear center*, if the beam is to bend without twisting.

Consider a cantilever beam of singly symmetric cross section supporting a load $P$ at the free end (see Fig. 6-30a). A beam having the cross section shown in Fig. 6-30b is called an *unbalanced I-beam*. Beams of I-shape, whether balanced or unbalanced, are usually loaded in the plane of symmetry (the $xz$ plane), but in this case the line of action of the force $P$ is perpendicular to that plane. Since the origin of coordinates is taken at the centroid $C$ of the cross section, and since the $z$ axis is an axis of symmetry of the cross section, both the $y$ and $z$ axes are principal centroidal axes.

Let us assume that under the action of the load $P$ the beam bends with the $xz$ plane as the neutral plane, which means that the $xy$ plane is the plane of bending. Under these conditions, two stress resultants exist at intermediate cross sections of the beam (Fig. 6-30b): a bending moment $M_0$ acting about the $z$ axis and having its moment vector in the negative direction of the $z$ axis, and a shear force of magnitude $P$ acting in the negative $y$ direction. For a given beam and loading, both $M_0$ and $P$ are known quantities.

The normal stresses acting on the cross section have a resultant that is the bending moment $M_0$, and the shear stresses have a resultant that is the shear force (equal to $P$). If the material follows Hooke's law, the normal stresses vary linearly with the distance from the neutral axis (the $z$ axis) and can be calculated from the flexure formula. Since the shear stresses acting on a cross section are determined from the normal stresses solely from equilibrium considerations (see the derivation of the shear formula in Section 5.8), it follows that the distribution of shear stresses over the cross section is also determined. The resultant of these shear stresses is a vertical force equal in magnitude to the force $P$ and having its line of action through some point $S$ lying on the $z$ axis (Fig. 6-30b). This point is known as the **shear center** (also called the *center of flexure*) of the cross section.

In summary, by assuming that the $z$ axis is the neutral axis, we can determine not only the distribution of the normal stresses but also the distribution of the shear stresses and the position of the resultant shear force. Therefore, we now recognize that a load $P$ applied at the end of the beam (Fig. 6-30a) must act through a particular point (the shear center) if bending is to occur with the $z$ axis as the neutral axis.

If the load is applied at some other point on the $z$ axis (say, at point $A$ in Fig. 6-31), it can be replaced by a statically equivalent system consisting

## Fig. 6-30

Cantilever beam with singly symmetric cross section:
(a) beam with load, and
(b) intermediate cross section of beam showing stress resultants $P$ and $M_0$, centroid $C$, and shear center $S$

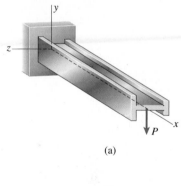

(a)

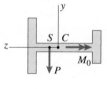

(b)

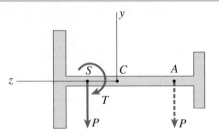

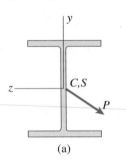

(a)

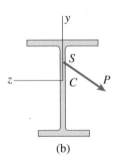

(b)

of a force $P$ acting at the shear center and a torque $T$. The force acting at the shear center produces bending about the $z$ axis and the torque produces torsion. Therefore, we now recognize that *a lateral load acting on a beam will produce bending without twisting only if it acts through the shear center.*

The shear center (like the centroid) lies on any axis of symmetry, and therefore the shear center $S$ and the centroid $C$ coincide for a **doubly symmetric cross section** (Fig. 6-32a). A load $P$ acting through the centroid produces bending about the $y$ and $z$ axes without torsion, and the corresponding bending stresses can be found by the method described in Section 6.4 for doubly symmetric beams.

If a beam has a **singly symmetric cross section** (Fig. 6-32b), both the centroid and the shear center lie on the axis of symmetry. A load $P$ acting through the shear center can be resolved into components in the $y$ and $z$ directions. The component in the $y$ direction will produce bending in the $xy$ plane with the $z$ axis as the neutral axis, and the component in the $z$ direction will produce bending (without torsion) in the $xz$ plane with the $y$ axis as the neutral axis. The bending stresses produced by these components can be superposed to obtain the stresses caused by the original load.

Finally, if a beam has an **unsymmetric cross section** (Fig. 6-33), the bending analysis proceeds as follows (provided the load acts through the shear center). First, locate the centroid $C$ of the cross section and determine the orientation of the principal centroidal axes $y$ and $z$. Then resolve the load into components (acting at the shear center) in the $y$ and $z$ directions and determine the bending moments $M_y$ and $M_z$ about the principal axes. Lastly, calculate the bending stresses using the method described in Section 6.5 for unsymmetric beams.

Now that we have explained the *significance* of the shear center and its use in beam analysis, it is natural to ask, "How do we *locate* the shear center?" For doubly symmetric shapes the answer, of course, is simple—it is at the centroid. For singly symmetric shapes the shear center lies on the axis of symmetry, but the precise location on that axis may not be easy to determine. Locating the shear center is even more difficult if the cross section is unsymmetric (Fig. 6-33). In such cases, the task requires more advanced methods than are appropriate for this book. (A few engineering handbooks give formulas for locating shear centers; e.g., see Ref. 2-9.)

Beams of **thin-walled open cross sections**, such as wide-flange beams, channels, angles, T-beams, and Z-sections, are a special case. Not only are they in common use for structural purposes, they also are very weak in torsion. Consequently, it is especially important to locate their shear centers. Cross sections of this type are considered in the following three sections— in Sections 6.7 and 6.8 we discuss how to find the shear stresses in such beams, and in Section 6.9 we show how to locate their shear centers.

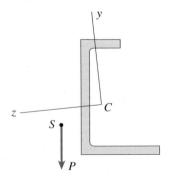

# 6.7 SHEAR STRESSES IN BEAMS OF THIN-WALLED OPEN CROSS SECTIONS

The distribution of shear stresses in rectangular beams, circular beams, and in the webs of beams with flanges was described previously in Sections 5.8, 5.9, and 5.10, and we derived the shear formula (Eq. 5-41) for calculating the stresses:

$$\tau = \frac{VQ}{Ib} \qquad (6\text{-}42)$$

In this formula, $V$ represents the shear force acting on the cross section, $I$ is the moment of inertia of the cross-sectional area (with respect to the neutral axis), $b$ is the width of the beam at the location where the shear stress is to be determined, and $Q$ is the first moment of the cross-sectional area outside of the location where the stress is being found.

Now we will consider the shear stresses in a special class of beams known as beams of **thin-walled open cross section**. Beams of this type are distinguished by two features: (1) The wall thickness is small compared to the height and width of the cross section, and (2) the cross section is open, as in the case of an I-beam or channel beam, rather than closed, as in the case of a hollow box beam. Examples are shown in Fig. 6-34. Beams of this type are also called **structural sections** or **profile sections**.

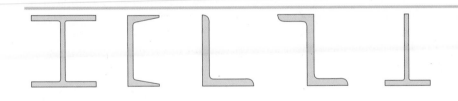

**Fig. 6-34**

Typical beams of thin-walled open cross section (wide-flange beam or I-beam, channel beam, angle section, Z-section, and T-beam)

We can determine the shear stresses in thin-walled beams of open cross section by using the same techniques we used when deriving the shear formula (Eq. 6-42). To keep the derivation as general as possible, we will consider a beam having its cross-sectional centerline $mm$ of arbitrary shape (Fig. 6-35a). The $y$ and $z$ axes are principal centroidal axes of the cross section, and the load $P$ acts parallel to the $y$ axis through the shear center $S$ (Fig. 6-35b). Therefore, bending will occur in the $xy$ plane with the $z$ axis as the neutral axis.

Under these conditions, we can obtain the normal stress at any point in the beam from the flexure formula:

$$\sigma_x = -\frac{M_z y}{I_z} \qquad (6\text{-}43)$$

where $M_z$ is the bending moment about the $z$ axis (positive as defined in Fig. 6-17) and $y$ is a coordinate of the point under consideration.

Now consider a volume element $abcd$ cut out between two cross sections distance $dx$ apart (Fig. 6-35a). Note that the element begins at the edge of the cross section and has length $s$ measured along the centerline $mm$ (Fig. 6-35b). To determine the shear stresses, we isolate the element

**Fig. 6-35**

Shear stresses in a beam of
thin-walled open cross section.
(The *y* and *z* axes are principal
centroidal axes.)

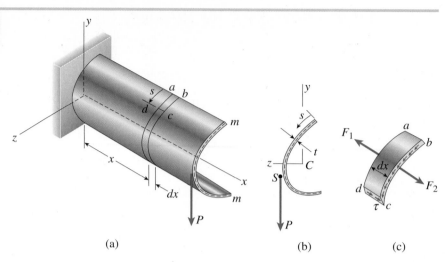

(a)                    (b)                    (c)

as shown in Fig. 6-35c. The resultant of the normal stresses acting on face
*ad* is the force $F_1$ and the resultant on face *bc* is the force $F_2$. Since the normal stresses acting on face *ad* are larger than those acting on face *bc*
(because the bending moment is larger), the force $F_1$ will be larger than
$F_2$. Therefore, shear stresses $\tau$ must act along face *cd* in order for the element to be in equilibrium. These shear stresses act parallel to the top and
bottom surfaces of the element and must be accompanied by complementary shear stresses acting on the cross-sectional faces *ad* and *bc*, as shown
in the figure.

To evaluate these shear stresses, we sum forces in the *x* direction for
element *abcd* (Fig. 6-35c); thus,

$$\tau t\, dx + F_2 - F_1 = 0 \quad \text{or} \quad \tau t\, dx = F_1 - F_2 \qquad \textbf{(6-44)}$$

where *t* is the thickness of the cross section at face *cd* of the element. In
other words, *t* is the thickness of the cross section at distance *s* from the
free edge (Fig. 6-35b). Next, we obtain an expression for the force $F_1$ by
using Eq. (6-43):

$$F_1 = \int_0^s \sigma_x\, dA = -\frac{M_{z1}}{I_z} \int_0^s y\, dA \qquad \textbf{(6-45a)}$$

where *dA* is an element of area on side *ad* of the volume element *abcd*, *y*
is a coordinate to the element *dA*, and $M_{z1}$ is the bending moment at the
cross section. An analogous expression is obtained for the force $F_2$:

$$F_2 = \int_0^s \sigma_x\, dA = -\frac{M_{z2}}{I_z} \int_0^s y\, dA \qquad \textbf{(6-45b)}$$

Substituting these expressions for $F_1$ and $F_2$ into Eq. (a), we get

$$\tau = \left(\frac{M_{z2} - M_{z1}}{dx}\right)\frac{1}{I_z t} \int_0^s y\, dA \qquad \textbf{(6-46)}$$

The quantity $(M_{z2} - M_{z1})/dx$ is the rate of change $dM/dx$ of the bending moment and is equal to the shear force acting on the cross section (see Eq. 4-6):

$$\frac{dM}{dx} = \frac{M_{z2} - M_{z1}}{dx} = V_y \qquad \textbf{(6-47)}$$

The shear force $V_y$ is parallel to the $y$ axis and positive in the negative direction of the $y$ axis, that is, positive in the direction of the force $P$ (Fig. 6-35). This convention is consistent with the sign convention previously adopted in Chapter 4 (see Fig. 4-5 for the sign convention for shear forces).

Substituting from Eq. (6-47) into Eq. (6-46), we get the following equation for the shear stress $\tau$:

$$\tau = \frac{V_y}{I_z t} \int_0^s y \, dA \qquad \textbf{(6-48)}$$

This equation gives the shear stresses at any point in the cross section at distance $s$ from the free edge. The integral on the right-hand side represents the first moment with respect to the $z$ axis (the neutral axis) of the area of the cross section from $s = 0$ to $s = s$. Denoting this first moment by $Q_z$, we can write the equation for the **shear stresses** $\tau$ in the simpler form

$$\tau = \frac{V_y Q_z}{I_z t} \qquad \textbf{(6-49)}$$

which is analogous to the standard shear formula (Eq. 6-42).

The shear stresses are directed along the centerline of the cross section and act parallel to the edges of the section. Furthermore, we tacitly assumed that these stresses have constant intensity across the thickness $t$ of the wall, which is a valid assumption when the thickness is small. (Note that the wall thickness need not be constant but may vary as a function of the distance $s$.)

The **shear flow** at any point in the cross section, equal to the product of the shear stress and the thickness at that point, is

$$f = \tau t = \frac{V_y Q_z}{I_z} \qquad \textbf{(6-50)}$$

Because $V_y$ and $I_z$ are constants, the shear flow is directly proportional to $Q_z$. At the top and bottom edges of the cross section, $Q_z$ is zero and hence the shear flow is also zero. The shear flow varies continuously between these end points and reaches its maximum value where $Q_z$ is maximum, which is at the neutral axis.

Now suppose that the beam shown in Fig. 6-35 is bent by loads that act parallel to the $z$ axis and through the shear center. Then the beam will bend in the $xz$ plane and the $y$ axis will be the neutral axis. In this case we can repeat the same type of analysis and arrive at the following equations for the shear stresses and shear flow (compare with Eqs. 6-49 and 6-50):

$$\tau = \frac{V_z Q_y}{I_y t} \qquad f = \tau t = \frac{V_z Q_y}{I_y} \qquad \textbf{(6-51a,b)}$$

In these equations, $V_z$ is the shear force parallel to the $z$ axis and $Q_y$ is the first moment with respect to the $y$ axis.

In summary, we have derived expressions for the shear stresses in beams of thin-walled open cross sections with the stipulations that the shear force must act through the shear center and must be parallel to one of the principal centroidal axes. If the shear force is inclined to the $y$ and $z$ axes (but still acts through the shear center), it can be resolved into components parallel to the principal axes. Then two separate analyses can be made, and the results can be superimposed.

To illustrate the use of the shear-stress equations, we will consider the shear stresses in a wide-flange beam in the next section. Later, in Section 6.9, we will use the shear-stress equations to locate the shear centers of several thin-walled beams with open cross sections.

## 6.8 SHEAR STRESSES IN WIDE-FLANGE BEAMS

We will now use the concepts and equations discussed in the preceding section to investigate the shear stresses in wide-flange beams. For discussion purposes, consider the wide-flange beam of Fig. 6-36a on the next page. This beam is loaded by a force $P$ acting in the plane of the web, that is, through the shear center, which coincides with the centroid of the cross

**Fig. 6-36**

Shear stresses in a wide-flange beam

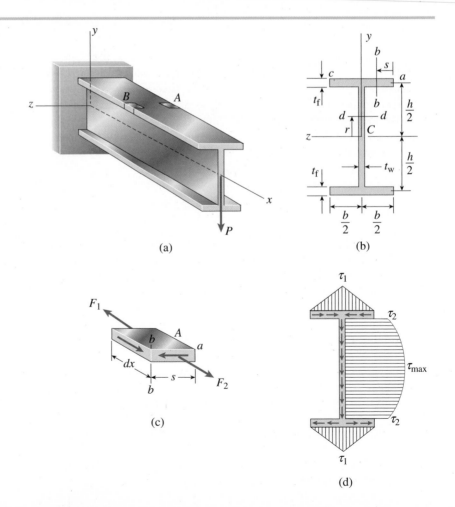

section. The cross-sectional dimensions are shown in Fig. 6-36b, where we note that $b$ is the flange width, $h$ is the height between *centerlines* of the flanges, $t_f$ is the flange thickness, and $t_w$ is the web thickness.

## Shear Stresses in the Upper Flange

We begin by considering the shear stresses at section $bb$ in the right-hand part of the upper flange (Fig. 6-36b). Since the distance $s$ has its origin at the edge of the section (point $a$), the cross-sectional area between point $a$ and section $bb$ is $st_f$. Also, the distance from the centroid of this area to the neutral axis is $h/2$, and therefore its first moment $Q_z$ is equal to $st_f h/2$. Thus, the shear stress $\tau_f$ in the flange at section $bb$ (from Eq. 6-49) is

$$\tau_f = \frac{V_y Q_z}{I_z t} = \frac{P(st_f h/2)}{I_z t_f} = \frac{shP}{2I_z} \qquad \textbf{(6-52)}$$

The direction of this stress can be determined by examining the forces acting on element $A$, which is cut out of the flange between point $a$ and section $bb$ (see Figs. 6-36a and b).

The element is drawn to a larger scale in Fig. 6-36c in order to show clearly the forces and stresses acting on it. We recognize immediately that the tensile force $F_1$ is larger than the force $F_2$, because the bending moment is larger on the rear face of the element than it is on the front face. It follows that the shear stress on the left-hand face of element $A$ must act toward the reader if the element is to be in equilibrium. From this observation it follows that the shear stresses on the front face of element $A$ must act toward the left.

Returning now to Fig. 6-36b, we see that we have completely determined the magnitude and direction of the shear stress at section $bb$, which may be located anywhere between point $a$ and the junction of the top flange and the web. Thus, the shear stresses throughout the right-hand part of the flange are horizontal, act to the left, and have a magnitude given by Eq. (6-52). As seen from that equation, the shear stresses increase linearly with the distance $s$.

The variation of the stresses in the upper flange is shown graphically in Fig. 6-36d, and we see that the stresses vary from zero at point $a$ (where $s = 0$) to a maximum value $\tau_1$ at $s = b/2$:

$$\tau_1 = \frac{bhP}{4I_z} \qquad \textbf{(6-53)}$$

The corresponding shear flow is

$$f_1 = \tau_1 t_f = \frac{bht_f P}{4I_z} \qquad \textbf{(6-54)}$$

Note that we have calculated the shear stress and shear flow at the junction of the *centerlines* of the flange and web, using only centerline dimensions of the cross section in the calculations. This approximate procedure simplifies the calculations and is satisfactory for thin-walled cross sections.

By beginning at point $c$ on the left-hand part of the top flange (Fig. 6-36b) and measuring $s$ toward the right, we can repeat the same type of analysis. We will find that the magnitude of the shear stresses is again given by Eqs. (6-52) and (6-53). However, by cutting out an element $B$ (Fig. 6-36a) and considering its equilibrium, we find that the shear stresses on the cross section now act toward the right, as shown in Fig. 6-36d.

## Shear Stresses in the Web

The next step is to determine the shear stresses acting in the web. Considering a horizontal cut at the top of the web (at the junction of the flange and web), we find the first moment about the neutral axis to be $Q_z = bt_f h/2$, so that the corresponding shear stress is

$$\tau_2 = \frac{bht_f P}{2I_z t_w} \tag{6-55}$$

The associated shear flow is

$$f_2 = \tau_2 t_w = \frac{bht_f P}{2I_z} \tag{6-56}$$

Note that the shear flow $f_2$ is equal to twice the shear flow $f_1$, which is expected since the shear flows in the two halves of the upper flange combine to produce the shear flow at the top of the web.

The shear stresses in the web act downward and increase in magnitude until the neutral axis is reached. At section $dd$, located at distance $r$ from the neutral axis (Fig. 6-36b), the shear stress $\tau_w$ in the web is calculated as follows:

$$Q_z = \frac{bt_f h}{2} + \left(\frac{h}{2} - r\right)(t_w)\left(\frac{h/2 + r}{2}\right) = \frac{bt_f h}{2} + \frac{t_w}{2}\left(\frac{h^2}{4} - r^2\right)$$

$$\tau_w = \left(\frac{bt_f h}{t_w} + \frac{h^2}{4} - r^2\right)\frac{P}{2I_z} \tag{6-57}$$

When $r = h/2$, this equation reduces to Eq. (6-55), and when $r = 0$, it gives the maximum shear stress:

$$\tau_{max} = \left(\frac{bt_f}{t_w} + \frac{h}{4}\right)\frac{Ph}{2I_z} \tag{6-58}$$

Again it should be noted that we have made all calculations on the basis of the centerline dimensions of the cross section. For this reason, the shear stresses in the web of a wide-flange beam calculated from Eq. (6-57) may be slightly different from those obtained by the more exact analysis made in Chapter 5 (see Eq. 5-51 of Section 5.10).

The shear stresses in the web vary parabolically, as shown in Fig. 6-36d, although the variation is not large. The ratio of $\tau_{max}$ to $\tau_2$ is

$$\frac{\tau_{max}}{\tau_2} = 1 + \frac{ht_w}{4bt_f} \tag{6-59}$$

For instance, if we assume $h = 2b$ and $t_f = 2t_w$, the ratio is $\tau_{max}/\tau_2 = 1.25$.

## Shear Stresses in the Lower Flange

As the final step in the analysis, we can investigate the shear stresses in the lower flange using the same methods we used for the top flange. We will find that the magnitudes of the stresses are the same as in the top flange, but the directions are as shown in Fig. 6-36d.

## General Comments

From Fig. 6-36d we see that the shear stresses on the cross section "flow" inward from the outermost edges of the top flange, then down through the web, and finally outward to the edges of the bottom flange. Because this flow is always continuous in any structural section, it serves as a convenient method for determining the directions of the stresses. For instance, if the shear force acts downward on the beam of Fig. 6-36a, we know immediately that the shear flow in the web must also be downward. Knowing the direction of the shear flow in the web, we also know the directions of the shear flows in the flanges because of the required continuity in the flow. Using this simple technique to get the directions of the shear stresses is easier than visualizing the directions of the forces acting on elements such as $A$ (Fig. 6-36c) cut out from the beam.

The resultant of all the shear stresses acting on the cross section is clearly a vertical force, because the horizontal stresses in the flanges produce no resultant. The shear stresses in the web have a resultant $R$, which can be found by integrating the shear stresses over the height of the web, as follows:

$$R = \int \tau \, dA = 2 \int_0^{h/2} \tau t_w \, dr$$

Substituting from Eq. (6-57), we get

$$R = 2t_w \int_0^{h/2} \left( \frac{bt_f h}{t_w} + \frac{h^2}{4} - r^2 \right) \left( \frac{P}{2I_z} \right) dr = \left( \frac{bt_f}{t_w} + \frac{h}{6} \right) \frac{h^2 t_w P}{2I_z} \quad \textbf{(6-60)}$$

The moment of inertia $I_z$ can be calculated as follows (using centerline dimensions):

$$I_z = \frac{t_w h^3}{12} + \frac{bt_f h^2}{2} \quad \textbf{(6-61)}$$

in which the first term is the moment of inertia of the web and the second term is the moment of inertia of the flanges. When this expression for $I_z$ is substituted into Eq. (6-60), we get $R = P$, which demonstrates that the resultant of the shear stresses acting on the cross section is equal to the load. Furthermore, the line of action of the resultant is in the plane of the web, and therefore the resultant passes through the shear center.

The preceding analysis provides a more complete picture of the shear stresses in a wide-flange or I-beam because it includes the flanges (recall that in Chapter 5 we investigated only the shear stresses in the web). Furthermore, this analysis illustrates the general techniques for finding shear stresses in beams of thin-walled open cross section. Other illustrations can be found in the next section, where the shear stresses in a channel section and an angle section are determined as part of the process of locating their shear centers.

# 6.9 SHEAR CENTERS OF THIN-WALLED OPEN SECTIONS

Shear center $S$ of a channel section

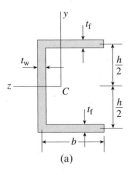

(a)

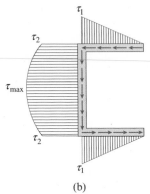

(b)

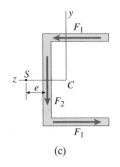

(c)

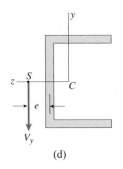

(d)

In Sections 6.7 and 6.8 we developed methods for finding the shear stresses in beams of thin-walled open cross section. Now we will use those methods to locate the shear centers of several shapes of beams. Only beams with singly symmetric or unsymmetric cross sections will be considered, because we already know that the shear center of a doubly symmetric cross section is located at the centroid.

The procedure for locating the shear center consists of two principal steps: first, evaluating the shear stresses acting on the cross section when bending occurs about one of the principal axes, and second, determining the resultant of those stresses. The shear center is located on the line of action of the resultant. By considering bending about *both* principal axes, we can determine the position of the shear center.

As in Sections 6.7 and 6.8, we will use only centerline dimensions when deriving formulas and making calculations. This procedure is satisfactory if the beam is thin-walled, that is, if the thickness of the beam is small compared to the other dimensions of the cross section.

## Channel Section

The first beam to be analyzed is a singly symmetric channel section (Fig. 6-37a). From the general discussion in Section 6.6 we know immediately that the shear center is located on the axis of symmetry (the $z$ axis). To find the position of the shear center on the $z$ axis, we assume that the beam is bent about the $z$ axis as the neutral axis, and then we determine the line of action of the resultant shear force $V_y$ acting parallel to the $y$ axis. The shear center is located where the line of action of $V_y$ intersects the $z$ axis. (Note that the origin of axes is at the centroid $C$, so that both the $y$ and $z$ axes are principal centroidal axes.)

Based upon the discussions in Section 6.8, we conclude that the shear stresses in a channel vary linearly in the flanges and parabolically in the web (Fig. 6-37b). We can find the resultant of those stresses if we know the maximum stress $\tau_1$ in the flange, the stress $\tau_2$ at the top of the web, and the maximum stress $\tau_{max}$ in the web.

To find the stress $\tau_1$ in the flange, we use Eq. (6-49) with $Q_z$ equal to the first moment of the flange area about the $z$ axis:

$$Q_z = \frac{bt_f h}{2} \tag{6-62}$$

in which $b$ is the flange width, $t_f$ is the flange thickness, and $h$ is the height of the beam. (Note again that the dimensions $b$ and $h$ are measured along the centerline of the section.) Thus, the stress $\tau_1$ in the flange is

$$\tau_1 = \frac{V_y Q_z}{I_z t_f} = \frac{bh V_y}{2 I_z} \tag{6-63}$$

where $I_z$ is the moment of inertia about the $z$ axis.

The stress $\tau_2$ at the top of the web is obtained in a similar manner but with the thickness equal to the web thickness instead of the flange thickness:

$$\tau_2 = \frac{V_y Q_z}{I_z t_w} = \frac{b t_f h V_y}{2 t_w I_z} \tag{6-64}$$

Also, at the neutral axis the first moment of area is

$$Q_z = \frac{b t_f h}{2} + \frac{h t_w}{2}\left(\frac{h}{4}\right) = \left(b t_f + \frac{h t_w}{4}\right)\frac{h}{2} \tag{6-65}$$

Therefore, the maximum stress is

$$\tau_{max} = \frac{V_y Q_z}{I_z t_w} = \left(\frac{b t_f}{t_w} + \frac{h}{4}\right)\frac{h V_y}{2 I_z} \tag{6-66}$$

The stresses $\tau_1$ and $\tau_2$ in the lower half of the beam are equal to the corresponding stresses in the upper half (Fig. 6-37b).

The horizontal shear force $F_1$ in either flange (Fig. 6-37c) can be found from the triangular stress diagrams. Each force is equal to the area of the stress triangle multiplied by the thickness of the flange:

$$F_1 = \left(\frac{\tau_1 b}{2}\right)(t_f) = \frac{h b^2 t_f V_y}{4 I_z} \tag{6-67}$$

The vertical force $F_2$ in the web must be equal to the shear force $V_y$, since the forces in the flanges have no vertical components. As a check, we can verify that $F_2 = V_y$ by considering the parabolic stress diagram of Fig. 6-37b. The diagram is made up of two parts—a rectangle of area $\tau_2 h$ and a parabolic segment of area

$$\frac{2}{3}(\tau_{max} - \tau_2)h$$

Thus, the shear force $F_2$, equal to the area of the stress diagram times the web thickness $t_w$, is

$$F_2 = \tau_2 h t_w + \frac{2}{3}(\tau_{max} - \tau_2)h t_w$$

Substituting the expressions for $\tau_2$ and $\tau_{max}$ (Eqs. 6-64 and 6-66) into the preceding equation, we obtain

$$F_2 = \left(\frac{t_w h^3}{12} + \frac{b h^2 t_f}{2}\right)\frac{V_y}{I_z} \tag{6-68}$$

Finally, we note that the expression for the moment of inertia is

$$I_z = \frac{t_w h^3}{12} + \frac{bh^2 t_f}{2}$$

(6-69)

in which we again base the calculations upon centerline dimensions. Substituting this expression for $I_z$ into Eq. (6-68) for $F_2$, we get

$$F_2 = V_y$$

(6-70)

as expected.

The three forces acting on the cross section (Fig. 6-37c) have a resultant $V_y$ that intersects the $z$ axis at the shear center $S$ (Fig. 6-37d). Hence, the moment of the three forces about any point in the cross section must be equal to the moment of the force $V_y$ about that same point. This moment relationship provides an equation from which the position of the shear center may be found.

As an illustration, let us select the shear center itself as the center of moments. In that case, the moment of the three forces (Fig. 6-37c) is $F_1 h - F_2 e$, where $e$ is the distance from the centerline of the web to the shear center, and the moment of the resultant force $V_y$ is zero (Fig. 6-37d). Equating these moments gives

$$F_1 h - F_2 e = 0$$

(6-71)

Substituting for $F_1$ from Eq. (6-67) and for $F_2$ from Eq. (6-70), and then solving for $e$, we get

$$e = \frac{b^2 h^2 t_f}{4 I_z}$$

(6-72)

When the expression for $I_z$ (Eq. 6-69) is substituted, Eq. (6-72) becomes

$$e = \frac{3 b^2 t_f}{h t_w + 6 b t_f}$$

(6-73)

Thus, we have determined the position of the shear center of a channel section.

As explained in Section 6.6, a channel beam will undergo bending without twisting whenever it is loaded by forces acting through the shear center. If the loads act parallel to the $y$ axis but through some point other than the shear center (for example, if the loads act in the plane of the web), they can be replaced by a statically equivalent force system consisting of loads through the shear center and twisting couples. We then have a combination of bending and torsion of the beam. If the loads act along the $z$ axis, we have simple bending about the $y$ axis. If the loads act in skew directions through the shear center, they can be replaced by statically equivalent loads acting parallel to the $y$ and $z$ axes.

## Angle Section

The next shape to be considered is an equal-leg angle section (Fig. 6-38a), in which each leg of the angle has length $b$ and thickness $t$. The $z$ axis is an axis of symmetry and the origin of coordinates is at the centroid $C$; therefore, both the $y$ and $z$ axes are principal centroidal axes.

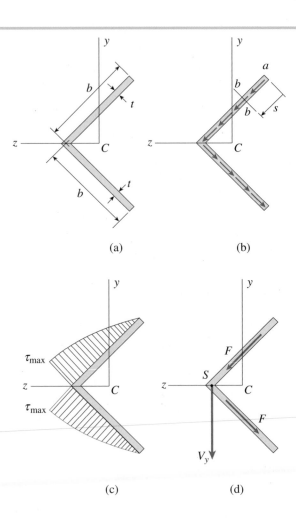

**Fig. 6-38**

Shear center of an equal-leg angle section

To locate the shear center, we will follow the same general procedure as that described for a channel section, because we wish to determine the distribution of the shear stresses as part of the analysis. However, as we will see later, the shear center of an angle section can be determined by inspection.

We begin by assuming that the section is subjected to a shear force $V_y$ acting parallel to the $y$ axis. Then we use Eq. (6-49) to find the corresponding shear stresses in the legs of the angle. For this purpose we need the first moment of the cross-sectional area between point $a$ at the outer edge of the beam (Fig. 6-38b) and section $bb$ located at distance $s$ from point $a$. The area is equal to $st$ and its centroidal distance from the neutral axis is

$$\frac{b - s/2}{\sqrt{2}}$$

Thus, the first moment of the area is

$$Q_z = st\left(\frac{b - s/2}{\sqrt{2}}\right) \tag{6-74}$$

**Fig. 6-38 (Repeated)**

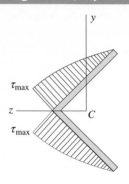

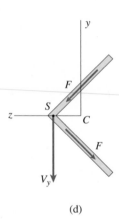

(c)

(d)

Substituting into Eq. (6-49), we get the following expression for the shear stress at distance $s$ from the edge of the cross section:

$$\tau = \frac{V_y Q_z}{I_z t} = \frac{V_y s}{I_z \sqrt{2}} \left( b - \frac{s}{2} \right) \tag{6-75}$$

The moment of inertia $I_z$ can be obtained from Case 24 of Appendix E with $\beta = 45°$:

$$I_z = 2I_{BB} = 2\left( \frac{t b^3}{6} \right) = \frac{t b^3}{3} \tag{6-76}$$

Substituting this expression for $I_z$ into Eq. (6-75), we get

$$\tau = \frac{3 V_y s}{b^3 t \sqrt{2}} \left( b - \frac{s}{2} \right) \tag{6-77}$$

This equation gives the shear stress at any point along the leg of the angle. The stress varies quadratically with $s$, as shown in Fig. 6-38c.

The maximum value of the shear stress occurs at the intersection of the legs of the angle and is obtained from Eq. (6-77) by substituting $s = b$:

$$\tau_{max} = \frac{3 V_y}{2 b t \sqrt{2}} \tag{6-78}$$

The shear force $F$ in each leg (Fig. 6-38d) is equal to the area of the parabolic stress diagram (Fig. 6-38c) times the thickness $t$ of the legs:

$$F = \frac{2}{3} (\tau_{max} b)(t) = \frac{V_y}{\sqrt{2}} \tag{6-79}$$

Since the horizontal components of the forces $F$ cancel each other, only the vertical components remain. Each vertical component is equal to $F/\sqrt{2}$, or $V_y/2$, and therefore the resultant vertical force is equal to the shear force $V_y$, as expected.

Since the resultant force passes through the intersection point of the lines of action of the two forces $F$ (Fig. 6-38d), we see that the shear center $S$ is located at the junction of the two legs of the angle.

## Sections Consisting of Two Intersecting Narrow Rectangles

In the preceding discussion of an angle section we evaluated the shear stresses and the forces in the legs in order to illustrate the general methodology for analyzing thin-walled open sections. However, if our sole objective had been to locate the shear center, it would not have been necessary to evaluate the stresses and forces.

Since the shear stresses are parallel to the centerlines of the legs (Fig. 6-38b), we would have known immediately that their resultants are two forces $F$ (Fig. 6-38d). The resultant of those two forces is a single force that passes through their point of intersection. Consequently, this point is the shear center. Thus, we can determine the location of the shear center of an equal-leg angle section by a simple line of reasoning (without making *any* calculations).

The same line of reasoning is valid for all cross sections consisting of two thin, intersecting rectangles (Fig. 6-39). In each case the resultants of the shear stresses are forces that intersect at the junction of the rectangles. Therefore, the shear center $S$ is located at that point.

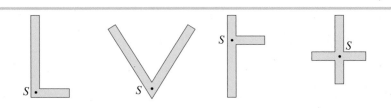

### Fig. 6-39

Shear centers of sections consisting of two intersecting narrow rectangles

## Z-Section

Let us now determine the location of the shear center of a Z-section having thin walls (Fig. 6-40a). The section has no axes of symmetry but is symmetric about the centroid $C$ (see Section 12.2 of Chapter 12 for a discussion of *symmetry about a point*). The $y$ and $z$ axes are principal axes through the centroid.

We begin by assuming that a shear force $V_y$ acts parallel to the $y$ axis and causes bending about the $z$ axis as the neutral axis. Then the shear stresses in the flanges and web will be directed as shown in Fig. 6-40a. From symmetry considerations we conclude that the forces $F_1$ in the two flanges must be equal to each other (Fig. 6-40b). The resultant of the three forces acting on the cross section ($F_1$ in the flanges and $F_2$ in the web) must be equal to the shear force $V_y$. The forces $F_1$ have a resultant $2F_1$ acting through the centroid and parallel to the flanges. This force intersects the force $F_2$ at the centroid $C$, and therefore we conclude that the line of action of the shear force $V_y$ must be through the centroid.

If the beam is subjected to a shear force $V_z$ parallel to the $z$ axis, we arrive at a similar conclusion, namely, that the shear force acts through the centroid. Since the shear center is located at the intersection of the lines of action of the two shear forces, we conclude that the shear center of the Z section coincides with the centroid.

This conclusion applies to any Z section that is symmetric about the centroid, that is, any Z section having identical flanges (same width and same thickness). Note, however, that the thickness of the web does not have to be the same as the thickness of the flanges.

The locations of the shear centers of many other structural shapes are given in the problems at the end of this chapter.*

### Fig. 6-40

Shear center of a thin-walled Z-section

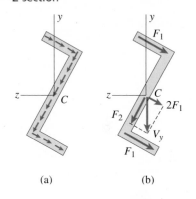

(a)          (b)

*The first determination of a shear center was made by S. P. Timoshenko in 1913 (Ref. 6-1).

● ● ● **Example 6-9**

A thin-walled semicircular cross section of radius $r$ and thickness $t$ is shown in Fig. 6-41a. Determine the distance $e$ from the center $O$ of the semicircle to the shear center $S$.

**Fig. 6-41**

Example 6-9: Shear center of a thin-walled semicircular section

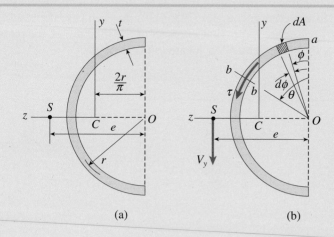

(a)                                    (b)

### Solution

We know immediately that the shear center is located somewhere on the axis of symmetry (the $z$ axis). To determine the exact position, we assume that the beam is bent by a shear force $V_y$ acting parallel to the $y$ axis and producing bending about the $z$ axis as the neutral axis (Fig. 6-41b).

*Shear stresses.* The first step is to determine the shear stresses $\tau$ acting on the cross section (Fig. 6-41b). We consider a section $bb$ defined by the distance $s$ measured along the centerline of the cross section from point $a$. The central angle subtended between point $a$ and section $bb$ is denoted $\theta$. Therefore, the distance $s$ equals $r\theta$, where $r$ is the radius of the centerline and $\theta$ is measured in radians.

To evaluate the first moment of the cross-sectional area between point $a$ and section $bb$, we identify an element of area $dA$ (shown shaded in the figure) and integrate as follows:

$$Q_z = \int y\, dA = \int_0^\theta (r\cos\phi) = r^2 t \sin\theta \qquad \text{(a)}$$

in which $\phi$ is the angle to the element of area and $t$ is the thickness of the section. Thus, the shear stress $\tau$ at section $bb$ is

$$\tau = \frac{V_y Q_z}{I_z t} = \frac{V_y r^2 \sin\theta}{I_z} \qquad \text{(b)}$$

Substituting $I_z = \pi r^3 t / 2$ (see Case 22 or Case 23 of Appendix E), we get

$$\tau = \frac{2V_y \sin\theta}{\pi rt} \qquad \text{(6-80)}$$

When $\theta = 0$ or $\theta = \pi$, this expression gives $\tau = 0$, as expected. When $\theta = \pi/2$, it gives the maximum shear stress.

*Location of shear center.* The resultant of the shear stresses must be the vertical shear force $V_y$. Therefore, the moment $M_0$ of the shear stresses about the center $O$ must equal the moment of the force $V_y$ about that same point:

$$M_0 = V_y e \tag{c}$$

To evaluate $M_0$, we begin by noting that the shear stress $\tau$ acting on the element of area $dA$ (Fig. 6-41b) is

$$\tau = \frac{2V_y \sin \phi}{\pi r t}$$

as found from Eq. (6-80). The corresponding force is $\tau\, dA$, and the moment of this force is

$$dM_0 = r(\tau\, dA) = \frac{2V_y \sin \phi\, dA}{\pi t}$$

Since $dA = tr\, d\phi$, this expression becomes

$$dM_0 = \frac{2rV_y \sin \phi\, d\phi}{\pi}$$

Therefore, the moment produced by the shear stresses is

$$M_0 = \int dM_0 = \int_0^\pi \frac{2rV_y \sin \phi\, d\phi}{\pi} = \frac{4rV_y}{\pi} \tag{d}$$

It follows from Eq. (c) that the distance $e$ to the shear center is

$$e = \frac{M_0}{V_y} = \frac{4r}{\pi} \approx 1.27r \qquad \text{⟵} \tag{6-81}$$

This result shows that the shear center $S$ is located outside of the semicircular section.

*Note:* The distance from the center $O$ of the semicircle to the centroid $C$ of the cross section (Fig. 6-41a) is $2r/\pi$ (from Case 23 of Appendix E), which is one-half of the distance $e$. Thus, the centroid is located midway between the shear center and the center of the semicircle.

The location of the shear center in a more general case of a thin-walled circular section is determined in Prob. 6.9-13.

## *6.10 ELASTOPLASTIC BENDING

**Fig. 6-42**

Idealized stress-strain diagram
for an elastoplastic material

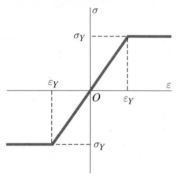

In our previous discussions of bending we assumed that the beams were made of materials that followed Hooke's law (linearly elastic materials). Now we will consider the bending of elastoplastic beams when the material is strained beyond the linear region. When that happens, the distribution of the stresses is no longer linear but varies according to the shape of the stress-strain curve.

**Elastoplastic materials** were discussed earlier when we analyzed axially loaded bars in Section 2.12. As explained in that section, elastoplastic materials follow Hooke's law up to the yield stress $\sigma_Y$ and then yield plastically under constant stress (see the stress-strain diagram of Fig. 6-42). From the figure, we see that an elastoplastic material has a region of linear elasticity between regions of perfect plasticity. Throughout this section, we will assume that the material has the same yield stress $\sigma_Y$ and same yield strain $\varepsilon_Y$ in both tension and compression.

Structural steels are excellent examples of elastoplastic materials because they have sharply defined yield points and undergo large strains during yielding. Eventually the steels begin to strain harden, and then the assumption of perfect plasticity is no longer valid. However, strain hardening provides an increase in strength, and therefore the assumption of perfect plasticity is on the side of safety.

### Yield Moment

**Fig. 6-43**

Beam of elastoplastic material
subjected to a positive bending
moment $M$

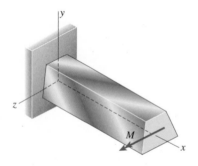

Let us consider a beam of elastoplastic material subjected to a bending moment $M$ that causes bending in the $xy$ plane (Fig. 6-43). When the bending moment is small, the maximum stress in the beam is less than the yield stress $\sigma_Y$, and therefore the beam is in the same condition as a beam in ordinary elastic bending with a linear stress distribution, as shown in Fig. 6-44b. Under these conditions, the neutral axis passes through the centroid of the cross section and the normal stresses are obtained from the flexure formula ($\sigma = -M_y/I$). Since the bending moment is positive, the stresses are compressive above the $z$ axis and tensile below it.

The preceding conditions exist until the stress in the beam at the point farthest from the neutral axis reaches the yield stress $\sigma_Y$, either in tension or in compression (Fig. 6-44c). The bending moment in the beam when the maximum stress just reaches the yield stress, called the **yield moment** $M_Y$, can be obtained from the flexure formula:

$$M_Y = \frac{\sigma_Y I}{c} = \sigma_Y S \tag{6-82}$$

in which $c$ is the distance to the point farthest from the neutral axis and $S$ is the corresponding section modulus.

### Plastic Moment and Neutral Axis

If we now increase the bending moment above the yield moment $M_Y$, the strains in the beam will continue to increase and the maximum strain will exceed the yield strain $\varepsilon_Y$. However, because of perfectly plastic yielding, the maximum stress will remain constant and equal to $\sigma_Y$, as pictured in Fig. 6-44d. Note that the outer regions of the beam have become fully plastic while a central core (called the **elastic core**) remains linearly elastic.

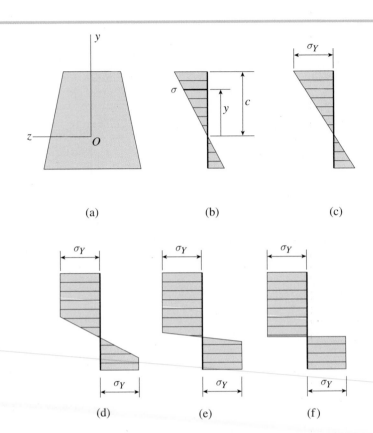

**Fig. 6-44**

Stress distributions in a beam of elastoplastic material

If the $z$ axis is not an axis of symmetry (singly symmetric cross section), the neutral axis moves away from the centroid when the yield moment is exceeded. This shift in the location of the neutral axis is not large, and in the case of the trapezoidal cross section of Fig. 6-44, it is too small to be seen in the figure. If the cross section is doubly symmetric, the neutral axis passes through the centroid even when the yield moment is exceeded.

As the bending moment increases still further, the plastic region enlarges and moves inward toward the neutral axis until the condition shown in Fig. 6-44e is reached. At this stage the maximum strain in the beam (at the farthest distance from the neutral axis) is perhaps 10 or 15 times the yield strain $\varepsilon_Y$ and the elastic core has almost disappeared. Thus, for practical purposes the beam has reached its ultimate moment-resisting capacity, and we can idealize the ultimate stress distribution as consisting of two rectangular parts (Fig. 6-44f). The bending moment corresponding to this idealized stress distribution, called the **plastic moment** $M_P$, represents the maximum moment that can be sustained by a beam of elastoplastic material.

To find the plastic moment $M_P$, we begin by locating the **neutral axis** of the cross section under fully plastic conditions. For this purpose, consider the cross section shown in Fig. 6-45a on the next page and let the $z$ axis be the neutral axis. Every point in the cross section above the neutral axis is subjected to a compressive stress $\sigma_Y$ (Fig. 6-45b), and every point below the neutral axis is subjected to a tensile stress $\sigma_Y$. The resultant compressive force $C$ is equal to $\sigma_Y$ times the cross-sectional area $A_1$ above the

**Fig. 6-45**

Location of the neutral axis and determination of the plastic moment $M_P$ under fully plastic conditions

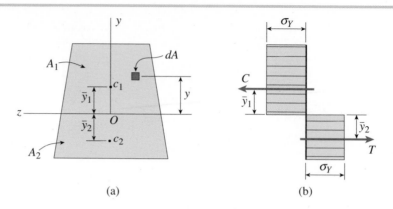

(a)          (b)

neutral axis (Fig. 6-45a), and the resultant tensile force $T$ equals $\sigma_Y$ times the area $A_2$ below the neutral axis. Since the resultant force acting on the cross section is zero, it follows that

$$T = C \quad \text{or} \quad A_1 = A_2 \tag{6-83a,b}$$

Because the total area $A$ of the cross section is equal to $A_1 + A_2$, we see that

$$A_1 = A_2 = \frac{A}{2} \tag{6-84}$$

Therefore, under fully plastic conditions, **the neutral axis divides the cross section into two equal areas**.

As a result, the location of the neutral axis for the plastic moment $M_P$ may be different from its location for linearly elastic bending. For instance, in the case of a trapezoidal cross section that is narrower at the top than at the bottom (Fig. 6-45a), the neutral axis for fully plastic bending is slightly below the neutral axis for linearly elastic bending.

Since the plastic moment $M_P$ is the moment resultant of the stresses acting on the cross section, it can be found by integrating over the cross-sectional area $A$ (Fig. 6-45a):

$$M_P = -\int_A \sigma y \, dA = -\int_{A1} (-\sigma_Y) y \, dA - \int_{A2} \sigma_Y y \, dA \tag{6-85}$$

$$= \sigma_Y(\bar{y}_1 A_1) - \sigma_Y(-\bar{y}_2 A_2) = \frac{\sigma_Y A(\bar{y}_1 + \bar{y}_2)}{2}$$

in which $y$ is the coordinate (positive upward) of the element of area $dA$ and $\bar{y}_1$ and $\bar{y}_2$ are the distances from the neutral axis to the centroids $c_1$ and $c_2$ of areas $A_1$ and $A_2$, respectively.

An easier way to obtain the plastic moment is to evaluate the moments about the neutral axis of the forces $C$ and $T$ (Fig. 6-45b):

$$M_P = C\bar{y}_1 + T\bar{y}_2 \tag{6-86}$$

Replacing $T$ and $C$ by $\sigma_Y A/2$, we get

$$M_P = \frac{\sigma_Y A(\bar{y}_1 + \bar{y}_2)}{2} \tag{6-87}$$

which is the same as Eq. (6-85).

The procedure for obtaining the **plastic moment** is to divide the cross section of the beam into two equal areas, locate the centroid of each half, and then use Eq. (6-87) to calculate $M_P$.

## Plastic Modulus and Shape Factor

The expression for the plastic moment can be written in a form similar to that for the yield moment (Eq. 6-82), as

$$M_P = \sigma_Y Z \qquad \textbf{(6-88)}$$

in which

$$Z = \frac{A(\bar{y}_1 + \bar{y}_2)}{2} \qquad \textbf{(6-89)}$$

is the **plastic modulus** (or the *plastic section modulus*) for the cross section. The plastic modulus may be interpreted geometrically as the first moment (evaluated with respect to the neutral axis) of the area of the cross section above the neutral axis plus the first moment of the area below the neutral axis.

The ratio of the plastic moment to the yield moment is solely a function of the shape of the cross section and is called the **shape factor** $f$:

$$f = \frac{M_P}{M_Y} = \frac{Z}{S} \qquad \textbf{(6-90)}$$

This factor is a measure of the reserve strength of the beam after yielding first begins. It is highest when most of the material is located near the neutral axis (for instance, a beam having a solid circular section), and lowest when most of the material is away from the neutral axis (for instance, a beam having a wide-flange section). Values of $f$ for cross sections of rectangular, wide-flange, and circular shapes are given in the remainder of this section. Other shapes are considered in the problems at the end of the chapter.

## Beams of Rectangular Cross Section

Now let us determine the properties of a beam of rectangular cross section (Fig. 6-46) when the material is elastoplastic. The section modulus is $S = bh^2/6$, and therefore the **yield moment** (Eq. 6-82) is

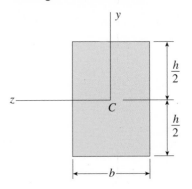

**Fig. 6-46**

Rectangular cross section

$$M_Y = \frac{\sigma_Y bh^2}{6} \qquad \textbf{(6-91)}$$

in which $b$ is the width and $h$ is the height of the cross section.

Because the cross section is doubly symmetric, the neutral axis passes through the centroid even when the beam is loaded into the plastic range. Consequently, the distances to the centroids of the areas above and below the neutral axis are

$$\bar{y}_1 = \bar{y}_2 = \frac{h}{4} \qquad \textbf{(6-92)}$$

Therefore, the **plastic modulus** (Eq. 6-89) is

$$Z = \frac{A(\bar{y}_1 + \bar{y}_2)}{2} = \frac{bh}{2}\left(\frac{h}{4} + \frac{h}{4}\right) = \frac{bh^2}{4} \qquad \textbf{(6-93)}$$

and the **plastic moment** (Eq. 6-88) is

$$M_P = \frac{\sigma_Y bh^2}{4}$$

(6-94)

Finally, the **shape factor** for a rectangular cross section is

$$f = \frac{M_P}{M_Y} = \frac{Z}{S} = \frac{3}{2}$$

(6-95)

which means that the plastic moment for a rectangular beam is 50% greater than the yield moment.

Next, we consider the stresses in a rectangular beam when the bending moment $M$ is greater than the yield moment but has not yet reached the plastic moment. The outer parts of the beam will be at the yield stress $\sigma_Y$ and the inner part (the **elastic core**) will have a linearly varying stress distribution (Figs. 6-47a and b). The fully plastic zones are shaded in Fig. 6-47a, and the distances from the neutral axis to the inner edges of the plastic zones (or the outer edges of the elastic core) are denoted by $e$.

The stresses acting on the cross section have the force resultants $C_1$, $C_2$, $T_1$, and $T_2$, as shown in Fig. 6-47c. The forces $C_1$ and $T_1$ in the plastic zones are each equal to the yield stress times the cross-sectional area of the zone:

$$C_1 = T_1 = \sigma_Y b\left(\frac{h}{2} - e\right)$$

(6-96)

The forces $C_2$ and $T_2$ in the elastic core are each equal to the area of the stress diagram times the width $b$ of the beam:

$$C_2 = T_2 = \frac{\sigma_Y e}{2} b$$

(6-97)

---

| **Fig. 6-47** |
| --- |

Stress distribution in a beam of rectangular cross section with an elastic core ($M_Y \le M \le M_P$)

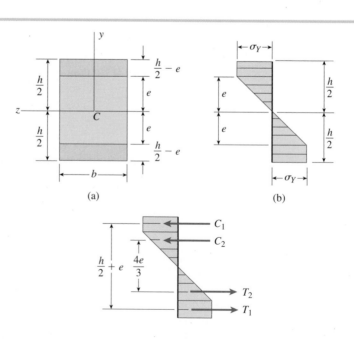

Thus, the **bending moment** (see Fig. 6-47c) is

$$M = C_1\left(\frac{h}{2} + e\right) + C_2\left(\frac{4e}{3}\right) = \sigma_Y b\left(\frac{h}{2} - e\right)\left(\frac{h}{2} + e\right) + \frac{\sigma_Y be}{2}\left(\frac{4e}{3}\right)$$

$$= \frac{\sigma_Y bh^2}{6}\left(\frac{3}{2} - \frac{2e^2}{h^2}\right) = M_Y\left(\frac{3}{2} - \frac{2e^2}{h^2}\right) \quad M_Y \leq M \leq M_P \quad \textbf{(6-98)}$$

Note that when $e = h/2$, the equation gives $M = M_Y$, and when $e = 0$, it gives $M = 3M_Y/2$, which is the plastic moment $M_P$.

Equation (6-98) can be used to determine the bending moment when the dimensions of the elastic core are known. However, a more common requirement is to determine the **size of the elastic core** when the bending moment is known. Therefore, we solve Eq. (6-98) for $e$ in terms of the bending moment:

$$e = h\sqrt{\frac{1}{2}\left(\frac{3}{2} - \frac{M}{M_Y}\right)} \quad M_Y \leq M \leq M_P \quad \textbf{(6-99)}$$

Again we note the limiting conditions: When $M = M_Y$, the equation gives $e = h/2$, and when $M = M_P = 3M_Y/2$, it gives $e = 0$, which is the fully plastic condition.

## Beams of Wide-Flange Shape

For a doubly symmetric wide-flange beam (Fig. 6-48), the plastic modulus $Z$ (Eq. 6-89) is calculated by taking the first moment about the neutral axis of the area of one flange plus the upper half of the web and then multiplying by 2. The result is

$$Z = 2\left[(bt_f)\left(\frac{h}{2} - \frac{t_f}{2}\right) + (t_w)\left(\frac{h}{2} - t_f\right)\left(\frac{1}{2}\right)\left(\frac{h}{2} - t_f\right)\right]$$

$$= bt_f(h - t_f) + t_w\left(\frac{h}{2} - t_f\right)^2 \quad \textbf{(6-100)}$$

With a little rearranging, we can express $Z$ in a more convenient form:

$$Z = \frac{1}{4}\left[bh^2 - (b - t_w)(h - 2t_f)^2\right] \quad \textbf{(6-101)}$$

After calculating the plastic modulus from Eq. (6-101), we can obtain the plastic moment $M_P$ from Eq. (6-88).

Values of $Z$ for commercially available shapes of wide-flange beams are listed in the AISC manual (Ref. 5-4). The shape factor $f$ for wide-flange beams is typically in the range from 1.1 to 1.2, depending upon the proportions of the cross section.

Other shapes of elastoplastic beams can be analyzed in a manner similar to that described for rectangular and wide-flange beams (see the following examples and the problems at the end of the chapter).

**Fig. 6-48**

Cross section of a wide-flange beam

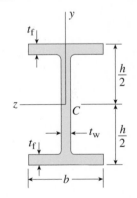

### Fig. 6-49

Example 6-10: Cross section of a circular beam (elastoplastic material)

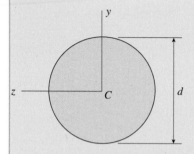

Determine the yield moment, plastic modulus, plastic moment, and shape factor for a beam of circular cross section with diameter $d$ (Fig. 6-49).

### Solution

As a preliminary matter, we note that since the cross section is doubly symmetric, the neutral axis passes through the center of the circle for both linearly elastic and elastoplastic behavior.

The yield moment $M_Y$ is found from the flexure formula (Eq. 6-82) as

$$M_Y = \frac{\sigma_Y I}{c} = \frac{\sigma_Y(\pi d^4/64)}{d/2} = \sigma_Y\left(\frac{\pi d^3}{32}\right) \qquad \text{(6-102)}$$

The plastic modulus $Z$ is found from Eq. (6-89) in which $A$ is the area of the circle and $\bar{y}$ and $\bar{y}_2$ are the distances to the centroids $c_1$ and $c_2$ of the two halves of the circle (Fig. 6-50). Thus, from Cases 9 and 10 of Appendix E, we get

### Fig. 6-50

Solution to Example 6-10

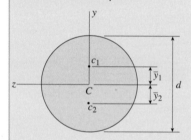

$$A = \frac{\pi d^2}{4} \qquad \bar{y}_1 = \bar{y}_2 = \frac{2d}{3\pi}$$

Now substituting into Eq. (6-89) for the plastic modulus, we find

$$Z = \frac{A(\bar{y}_1 + \bar{y}_2)}{2} = \frac{d^3}{6} \qquad \text{(6-103)}$$

Therefore, the plastic moment $M_P$ (Eq. 6-88) is

$$M_P = \sigma_Y Z = \frac{\sigma_Y d^3}{6} \qquad \text{(6-104)}$$

and the shape factor $f$ (Eq. 6-90) is

$$f = \frac{M_P}{M_Y} = \frac{16}{3\pi} \approx 1.70 \qquad \text{(6-105)}$$

This result shows that the maximum bending moment for a circular beam of elastoplastic material is about 70% larger than the bending moment when the beam first begins to yield.

**Fig. 6-51**

Example 6-11: Cross section of a hollow box beam (elastoplastic material)

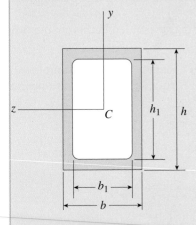

**Fig. 6-52**

Solution to Example 6-11

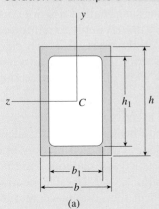

(a)

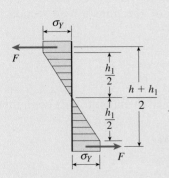

(b)

A doubly symmetric hollow box beam (Fig. 6-51) of elastoplastic material ($\sigma_Y$ = 33 ksi) is subjected to a bending moment $M$ of such magnitude that the flanges yield but the webs remain linearly elastic.

Determine the magnitude of the moment $M$ if the dimensions of the cross section are $b$ = 5.0 in., $b_1$ = 4.0 in., $h$ = 9.0 in., and $h_1$ = 7.5 in.

### Solution

The cross section of the beam and the distribution of the normal stresses are shown in Figs. 6-52a and b, respectively. From the figure, we see that the stresses in the webs increase linearly with distance from the neutral axis and the stresses in the flanges equal the yield stress $\sigma_Y$. Therefore, the bending moment $M$ acting on the cross section consists of two parts:

(1) a moment $M_1$ corresponding to the elastic core, and
(2) a moment $M_2$ produced by the yield stresses $\sigma_Y$ in the flanges.

The bending moment supplied by the core is found from the flexure formula (Eq. 6-82) with the section modulus calculated for the webs alone; thus,

$$S_1 = \frac{(b - b_1)h_1^2}{6} \tag{6-106}$$

and

$$M_1 = \sigma_Y S_1 = \frac{\sigma_Y(b - b_1)h_1^2}{6} \tag{6-107}$$

To find the moment supplied by the flanges, we note that the resultant force $F$ in each flange (Fig. 6-52b) is equal to the yield stress multiplied by the area of the flange:

$$F = \sigma_Y b\left(\frac{h - h_1}{2}\right) \tag{a}$$

The force in the top flange is compressive and the force in the bottom flange is tensile if the bending moment $M$ is positive. Together, the two forces create the bending moment $M_2$:

$$M_2 = F\left(\frac{h + h_1}{2}\right) = \frac{\sigma_Y b(h^2 - h_1^2)}{4} \tag{6-108}$$

Therefore, the total moment acting on the cross section, after some rearranging, is

$$M = M_1 + M_2 = \frac{\sigma_Y}{12}\left[3bh^2 - (b + 2b_1)h_1^2\right] \quad \Longleftarrow \tag{6-109}$$

Substituting the given numerical values, we obtain

$$M = 1330 \text{ k-in.} \quad \Longleftarrow$$

*Note:* The yield moment $M_Y$ and the plastic moment $M_p$ for the beam in this example have the following values (determined in Prob. 6.10-13):

$$M_Y = 1196 \text{ k-in.} \quad M_p = 1485 \text{ k-in.}$$

The bending moment $M$ is between these values, as expected.

In Chapter 6, we considered a number of specialized topics related to the bending of beams, including the analysis of **composite beams** (that is, beams of more than one material), beams with **inclined loads**, **unsymmetric beams**, **shear stresses** in thin-walled beams, **shear centers**, and **elastoplastic bending**.

Some of the major concepts and findings presented in this chapter are as follows:

1. In the introduction to **composite beams**, specialized moment-curvature relationship and flexure formulas for composite beams of two materials were developed:

$$\kappa = \frac{1}{\rho} = \frac{M}{E_1 I_1 + E_2 I_2}$$

$$\sigma_{x1} = -\frac{M y E_1}{E_1 I_1 + E_2 I_2} \qquad \sigma_{x2} = -\frac{M y E_2}{E_1 I_1 + E_2 I_2}$$

We assumed that both materials follow Hooke's law and that the two parts of the beam are adequately bonded so that they act as a single unit. Advanced topics such as nonhomogeneous and nonlinear materials, bond stresses between the parts, shear stresses on the cross sections, buckling of the faces, and other such matters are not considered. In particular, the formulas above do not apply to reinforced concrete beams which are not designed on the basis of linearly elastic behavior. However, a transformed section approach (see below and see Example 6-4) can be used as part of a cracked section analysis of reinforced concrete beams.

2. The **transformed-section method** offers a convenient way of transforming the cross section of a composite beam into an equivalent cross section of an imaginary beam that is composed of only one material. The ratio of the modulus of elasticity of material 2 to that of material 1 is known as the **modular ratio**, $n = E_2/E_1$. The neutral axis of the transformed beam is located in the same place, and its moment-resisting capacity is the same as that of the original composite beam. The moment of inertia of the transformed section is defined as

$$I_T = I_1 + n I_2 = I_1 + \frac{E_2}{E_1} I_2$$

Normal stresses in the beam transformed to material 1 are computed using the simplified flexure formula:

$$\sigma_{x1} = -\frac{M y}{I_T}$$

while those in material 2 are computed as

$$\sigma_{x2} = -\frac{My}{l_T}n$$

3. If **inclined loads** act through the centroid of the cross section of beams with two axes of symmetry in the cross section, there will be no twisting of the beam about the longitudinal axis. For these beams, we determined the bending stresses by resolving the inclined load into two components, one acting in each plane of symmetry. The bending stresses were obtained from the flexure formula for each load component acting separately, and the final stresses obtained by superposing the separate stresses. Also, in general, the angle of inclination of the neutral axis ($\beta$) is not equal to the angle of the inclined loads ($\theta$). As a result, except in special cases, the neutral axis is not perpendicular to the longitudinal plane containing the load. In this case, the stresses in the beam are very sensitive to slight changes in the direction of the load and to irregularities in the alignment of the beam itself.

4. When the restriction of symmetry about at least one axis of the cross section was removed, we found that for pure bending the plane in which the bending moment acts is perpendicular to the neutral surface only if the $y$ and $z$ axes are principal centroidal axes of the cross section and the bending moment acts in one of the two principal planes (the $xy$ plane or the $xz$ plane). We then established a **general procedure for computing normal stresses in *unsymmetric beams*** acted on by any moment $M$. First, the centroid is found, and then, normal stresses are obtained by superposing results of the flexure formula about the two principal centroidal axes.

5. A lateral load acting on a beam will produce bending without twisting only if it acts through the **shear center**. The shear center (like the centroid) lies on any axis of symmetry; the shear center $S$ and the centroid $C$ coincide for a doubly symmetric cross section.

6. Beams of **thin-walled open cross sections** (such as wide-flange beams, channels, angles, T-beams, and Z-sections) are in common use for structural purposes, but are **very weak in torsion**.

7. We derived expressions for the **shear stresses** in beams of thin-walled open cross sections for the case of the shear force acting through the shear center and parallel to one of the principal centroidal axes. We used these expressions to find the shear-stress distributions in the flanges and webs of wide-flange beams, channels and angles. We saw that the shear stresses on the cross section *flow* inward from the outermost edges, then down through the web, and finally outward to the edges of the bottom flange.

8. The procedure for **locating the shear center** was illustrated for several thin-walled open sections. First, the shear stresses acting on the cross section when bending occurs about one of the principal axes were computed, and then, the resultant force associated with those stresses was determined. The shear center was seen to lie on the line of action of the resultant.

9. Any Z-section that is symmetric about the centroid (i.e., any Z-section having identical flanges—same width and same thickness) has its shear center at the centroid of the cross section. The locations of the shear centers of many other structural shapes are given in both the examples and the problems at the end of the chapter.

10. Finally, we considered **elastoplastic materials**, which follow Hooke's law up to the yield stress $\sigma_Y$ and then yield plastically under constant stress. Structural steels are excellent examples of elastoplastic materials, because they have sharply defined yield points and undergo large strains during yielding. First, we found the *yield moment*

$$M_Y = \sigma_Y \times S$$

using the flexure formula. Then, we continued on to the *plastic moment*

$$M_P = \sigma_Y \times Z$$

where $S$ and $Z$ are the section modulus and plastic section modulus of the cross section, respectively. $M_Y$ is the bending moment in the beam when the maximum stress just reaches the yield stress, and $M_P$ is the maximum moment that can be sustained by a beam of elastoplastic material. We defined the shape factor

$$f = M_P/M_Y = Z/S$$

as a measure of the reserve strength of the beam after yielding first begins.

# PROBLEMS CHAPTER 6

## Composite Beams

*When solving the problems for Section 6.2, assume that the component parts of the beams are securely bonded by adhesives or connected by fasteners. Also, be sure to use the general theory for composite beams described in Section 6.2.*

**6.2-1** A composite beam consisting of fiberglass faces and a core of particle board has the cross section shown in the figure. The width of the beam is 2.0 in., the thickness of the faces is 0.10 in., and the thickness of the core is 0.50 in. The beam is subjected to a bending moment of 250 lb-in. acting about the *z* axis.

Find the maximum bending stresses $\sigma_{face}$ and $\sigma_{core}$ in the faces and the core, respectively, if their respective moduli of elasticity are $4 \times 10^6$ psi and $1.5 \times 10^6$ psi.

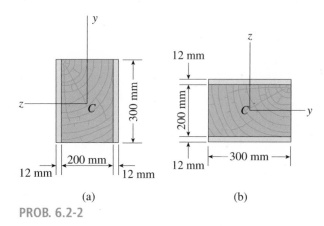

PROB. 6.2-2

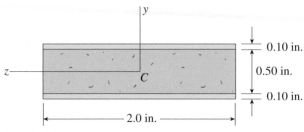

PROB. 6.2-1

**6.2-2** A wood beam with cross-sectional dimensions 200 mm × 300 mm is reinforced on its sides by steel plates 12 mm thick (see figure). The moduli of elasticity for the steel and wood are $E_s = 190$ GPa and $E_w = 11$ GPa, respectively. Also, the corresponding allowable stresses are $\sigma_s = 110$ MPa and $\sigma_w = 7.5$ MPa.

(a) Calculate the maximum permissible bending moment $M_{max}$ when the beam is bent about the *z* axis.

(b) Repeat part (a) if the beam is now bent about its *y* axis.

(c) Find the required thickness of the steel plates on the beam bent about the *y* axis so that $M_{max}$ is the same for both beam orientations.

**6.2-3** A hollow box beam is constructed with webs of Douglas-fir plywood and flanges of pine, as shown in the figure in a cross-sectional view. The plywood is 1 in. thick and 12 in. wide; the flanges are 2 in. × 4 in. (nominal size). The modulus of elasticity for the plywood is 1,800,000 psi and for the pine is 1,400,000 psi.

(a) If the allowable stresses are 2000 psi for the plywood and 1750 psi for the pine, find the allowable bending moment $M_{max}$ when the beam is bent about the *z* axis.

(b) Repeat part (a) if the beam is now bent about its *y* axis.

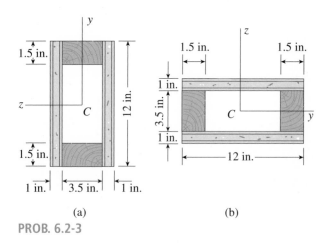

PROB. 6.2-3

**6.2-4** A round steel tube of outside diameter $d_2$ and a brass core of diameter $d_1$ are bonded to form a composite beam, as shown in the figure.

(a) Derive formulas for the allowable bending moment $M$ that can be carried by the beam based upon an allowable stress $\sigma_S$ in the steel and an allowable stress $\sigma_B$ in the brass. (Assume that the moduli of elasticity for the steel and brass are $E_S$ and $E_B$, respectively.)

(b) If $d_2 = 50$ mm, $d_1 = 40$ mm, $E_s = 210$ GPa, $E_B = 110$ GPa, $\sigma_s = 150$ MPa, and $\sigma_B = 100$ MPa, what is the maximum bending moment $M$?

(c) What new value of brass diameter $d_1$ will result in a balanced design? (i.e., a balanced design is that in which steel and brass reach allowable stress values at the same time.)

(a) Calculate the maximum bending stresses $\sigma_s$ in the steel plates and $\sigma_w$ in the wood member due to the applied loads.

(b) If the allowable bending stress in the steel plates is $\sigma_{as} = 14,000$ psi and that in the wood is $\sigma_{aw} = 900$ psi, find $q_{max}$. (Assume that the moment at $B$, $M_0$, remains at 300 ft-lb.)

(c) If $q = 660$ lb/ft and allowable stress values in part (b) apply, what is $M_{0,max}$ at $B$?

**6.2-6** A plastic-lined steel pipe has the cross-sectional shape shown in the figure. The steel pipe has outer diameter $d_3 = 100$ mm and inner diameter $d_2 = 94$ mm. The plastic liner has inner diameter $d_1 = 82$ mm. The modulus of elasticity of the steel is 75 times the modulus of the plastic.

(a) Determine the allowable bending moment $M_{allow}$ if the allowable stress in the steel is 35 MPa and in the plastic is 600 kPa.

(b) If pipe and liner diameters remain unchanged, what new value of allowable stress for the steel pipe will result in the steel pipe and plastic liner reaching their allowable stress values under the same maximum moment (i.e., a balanced design)? What is the new maximum moment?

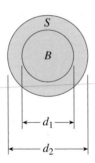

**PROB. 6.2-4**

**6.2-5** A beam with a guided support and 10-ft span supports a distributed load of intensity $q = 660$ lb/ft over its first half (see figure part a) and a moment $M_0 = 300$ ft-lb at joint $B$. The beam consists of a wood member (nominal dimensions 6 in. × 12 in., actual dimensions 5.5 in. × 11.5 in. in cross section, as shown in the figure part b) that is reinforced by 0.25-in.-thick steel plates on top and bottom. The moduli of elasticity for the steel and wood are $E_s = 30 \times 10^6$ psi and $E_w = 1.5 \times 10^6$ psi, respectively.

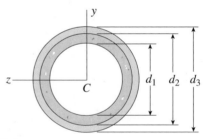

**PROB. 6.2-6**

**6.2-7** The cross section of a sandwich beam consisting of aluminum alloy faces and a foam core is shown in the figure. The width $b$ of the beam is 8.0 in., the thickness $t$ of the faces is 0.25 in., and the height $h_c$ of the core is 5.5 in. (total height $h = 6.0$ in.). The moduli of elasticity are

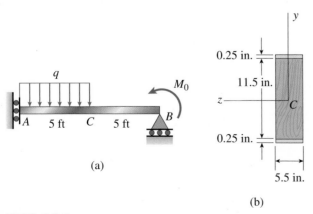

**PROB. 6.2-5**

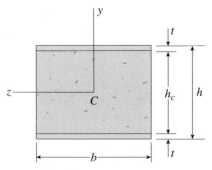

**PROBS. 6.2-7 and 6.2-8**

$10.5 \times 10^6$ psi for the aluminum faces and 12,000 psi for the foam core. A bending moment $M = 40$ k-in. acts about the $z$ axis.

Determine the maximum stresses in the faces and the core using (a) the general theory for composite beams, and (b) the approximate theory for sandwich beams.

**6.2-8** The cross section of a sandwich beam consisting of fiberglass faces and a lightweight plastic core is shown in the figure. The width $b$ of the beam is 50 mm, the thickness $t$ of the faces is 4 mm, and the height $h_c$ of the core is 92 mm (total height $h = 100$ mm). The moduli of elasticity are 75 GPa for the fiberglass and 1.2 GPa for the plastic. A bending moment $M = 275$ N·m acts about the $z$ axis.

Determine the maximum stresses in the faces and the core using (a) the general theory for composite beams, and (b) the approximate theory for sandwich beams.

**6.2-9** A bimetallic beam used in a temperature-control switch consists of strips of aluminum and copper bonded together as shown in the figure, which is a cross-sectional view. The width of the beam is 1.0 in., and each strip has a thickness of 1/16 in.

Under the action of a bending moment $M = 12$ lb-in. acting about the $z$ axis, what are the maximum stresses $\sigma_a$ and $\sigma_c$ in the aluminum and copper, respectively? (Assume $E_a = 10.5 \times 10^6$ psi and $E_c = 16.8 \times 10^6$ psi.)

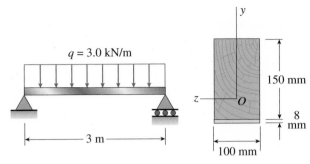

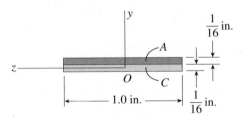

**PROB. 6.2-9**

**6.2-10** A simply supported composite beam 3 m long carries a uniformly distributed load of intensity $q = 30$ kN/m (see figure). The beam is constructed of a wood member, 100 mm wide by 150 mm deep, reinforced on its lower side by a steel plate 8 mm thick and 100 mm wide.

(a) Find the maximum bending stresses $\sigma_w$ and $\sigma_s$ in the wood and steel, respectively, due to the uniform load if the moduli of elasticity are $E_w = 10$ GPa for the wood and $E_s = 210$ GPa for the steel.

(b) Find the required thickness of the steel plate so that the steel plate and wood reach their allowable stress values, $\sigma_{as} = 100$ MPa and $\sigma_{aw} = 8.5$ MPa, simultaneously under the maximum moment.

**PROB. 6.2-10**

**6.2-11** A simply supported wooden I-beam with a 12-ft span supports a distributed load of intensity $q = 90$ lb/ft over its length (see figure part a). The beam is constructed with a web of Douglas-fir plywood and flanges of pine glued to the web, as shown in the figure part b. The plywood is 3/8 in. thick; the flanges are 2 in. × 2 in. (actual size). The modulus of elasticity for the plywood is 1,600,000 psi and for the pine is 1,200,000 psi.

(a) Calculate the maximum bending stresses in the pine flanges and in the plywood web.

(b) What is $q_{max}$ if allowable stresses are 1600 psi in the flanges and 1200 psi in the web?

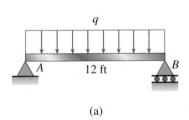

(a)

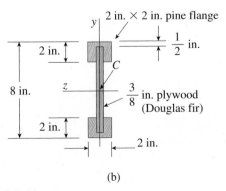

(b)

**PROB. 6.2-11**

**6.2-12** A simply supported composite beam with a 3.6 m span supports a triangularly distributed load of peak intensity $q_0$ at midspan (see figure part a). The beam is constructed of two wood joists, each 50 mm × 280 mm, fastened to two steel plates, one of dimensions 6 mm × 80 mm and the lower plate of dimensions 6 mm × 120 mm (see figure part b). The modulus of elasticity for the wood is 11 GPa and for the steel is 210 GPa.

If the allowable stresses are 7 MPa for the wood and 120 MPa for the steel, find the allowable peak load intensity $q_{0,max}$ when the beam is bent about the $z$ axis. Neglect the weight of the beam.

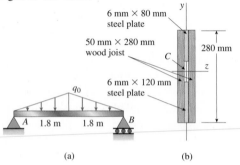

(a)                                      (b)

**PROB. 6.2-12**

## Transformed-Section Method

*When solving the problems for Section 6.3, assume that the component parts of the beams are securely bonded by adhesives or connected by fasteners. Also, be sure to use the transformed-section method in the solutions.*

**6.3-1** A wood beam 8 in. wide and 12 in. deep (nominal dimensions) is reinforced on top and bottom by 0.25-in.-thick steel plates (see figure part a).

(a) Find the allowable bending moment $M_{max}$ about the $z$ axis if the allowable stress in the wood is 1100 psi and in the steel is 15,000 psi. (Assume that the ratio of the moduli of elasticity of steel and wood is 20.)

(b) Compare the moment capacity of the beam in part a with that shown in the figure part b which has two 4 in. × 12 in. joists (nominal dimensions) attached to a 1/4 in. × 11.0 in. steel plate.

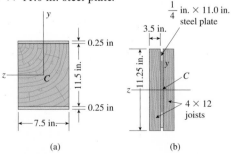

(a)                                      (b)

**PROB. 6.3-1**

**6.3-2** A simple beam of span length 3.2 m carries a uniform load of intensity 48 kN/m. The cross section of the beam is a hollow box with wood flanges and steel side plates, as shown in the figure. The wood flanges are 75 mm by 100 mm in cross section, and the steel plates are 300 mm deep.

What is the required thickness $t$ of the steel plates if the allowable stresses are 120 MPa for the steel and 6.5 MPa for the wood? (Assume that the moduli of elasticity for the steel and wood are 210 GPa and 10 GPa, respectively, and disregard the weight of the beam.)

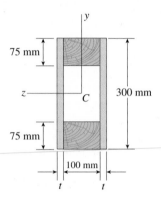

**PROB. 6.3-2**

**6.3-3** A simple beam that is 18 ft long supports a uniform load of intensity $q$. The beam is constructed of two C 8 × 11.5 sections (channel sections or C shapes) on either side of a 4 × 8 (actual dimensions) wood beam (see the cross section shown in the figure part a). The modulus of elasticity of the steel ($E_s$ = 30,000 ksi) is 20 times that of the wood ($E_w$).

(a) If the allowable stresses in the steel and wood are 12,000 psi and 900 psi, respectively, what is the allowable load $q_{allow}$? (*Note*: Disregard the weight of the beam, and see Table F-3a of Appendix F for the dimensions and properties of the C-shape beam.)

(b) If the beam is rotated 90° to bend about its $y$ axis (see figure part b), and uniform load q = 250 lb/ft is applied, find the maximum stresses $\sigma_s$ and $\sigma_w$ in the steel and wood, respectively. Include the weight of the beam. (Assume weight densities of 35 lb/ft$^3$ and 490 lb/ft$^3$ for the wood and steel, respectively.)

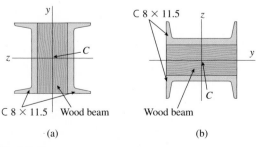

(a)                                      (b)

**PROB. 6.3-3**

**6.3-4** The composite beam shown in the figure is simply supported and carries a total uniform load of 40 kN/m on a span length of 4.0 m. The beam is built of a southern pine wood member having cross-sectional dimensions of 150 mm × 250 mm and two brass plates of cross-sectional dimensions 30 mm × 150 mm.

(a) Determine the maximum stresses $\sigma_B$ and $\sigma_w$ in the brass and wood, respectively, if the moduli of elasticity are $E_B$ = 96 GPa and $E_w$ = 14 GPa. (Disregard the weight of the beam.)

(b) Find the required thickness of the brass plates so that the plate and wood reach their allowable stress values of $\sigma_{aB}$ = 70 MPa and $\sigma_{aw}$ = 8.5 MPa simultaneously under the maximum moment. What is the maximum moment?

**6.3-6** Consider the preceding problem if the beam has width $b$ = 75 mm, the aluminum strips have thickness $t$ = 3 mm, the plastic segments have heights $d$ = 40 mm and $3d$ = 120 mm, and the total height of the beam is $h$ = 212 mm. Also, the moduli of elasticity are $E_a$ = 75 GPa and $E_P$ = 3 GPa, respectively.

Determine the maximum stresses $\sigma_a$ and $\sigma_p$ in the aluminum and plastic, respectively, due to a bending moment of 1.0 kN·m.

**6.3-7** A simple beam that is 18 ft long supports a uniform load of intensity $q$. The beam is constructed of two angle sections, each L 6 × 4 × 1/2, on either side of a 2 in. × 8 in. (actual dimensions) wood beam (see the cross section shown in the figure part a). The modulus of elasticity of the steel is 20 times that of the wood.

(a) If the allowable stresses in the steel and wood are 12,000 psi and 900 psi, respectively, what is the allowable load $q_{allow}$? (*Note:* Disregard the weight of the beam, and see Table F-5a of Appendix F for the dimensions and properties of the angles.)

(b) Repeat part (a) if a 1 in. × 10 in. wood flange (actual dimensions) is added (see figure part b).

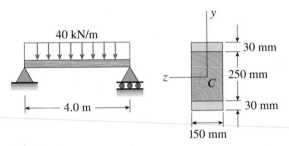

PROB. 6.3-4

**6.3-5** The cross section of a beam made of thin strips of aluminum separated by a lightweight plastic is shown in the figure. The beam has width $b$ = 3.0 in., the aluminum strips have thickness $t$ = 0.1 in., and the plastic segments have heights $d$ = 1.2 in. and $3d$ = 3.6 in. The total height of the beam is $h$ = 6.4 in.

The moduli of elasticity for the aluminum and plastic are $E_a$ = 11 × 10⁶ psi and $E_P$ = 440 × 10³ psi, respectively.

Determine the maximum stresses $\sigma_a$ and $\sigma_p$ in the aluminum and plastic, respectively, due to a bending moment of 6.0 k-in.

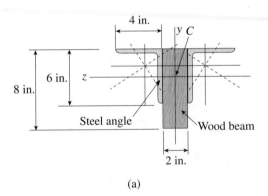

(a)

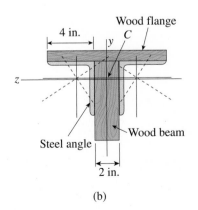

(b)

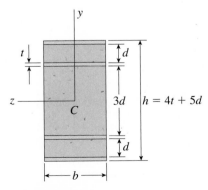

PROBS. 6.3-5 and 6.3-6

PROB. 6.3-7

**6.3-8** The cross section of a composite beam made of aluminum and steel is shown in the figure. The moduli of elasticity are $E_a = 75$ GPa and $E_s = 200$ GPa.

(a) Under the action of a bending moment that produces a maximum stress of 50 MPa in the aluminum, what is the maximum stress $\sigma_s$ in the steel?

(b) If the height of the beam remains at 120 mm and allowable stresses in steel and aluminum are defined as 94 MPa and 40 MPa, respectively, what heights $h_a$ and $h_s$ are required for aluminum and steel, respectively, so that both steel and aluminum reach their allowable stress values under the maximum moment?

**6.3-10** The cross section of a bimetallic strip is shown in the figure. Assuming that the moduli of elasticity for metals $A$ and $B$ are $E_A = 168$ GPa and $E_B = 90$ GPa, respectively, determine the smaller of the two section moduli for the beam. (Recall that section modulus is equal to bending moment divided by maximum bending stress.) In which material does the maximum stress occur?

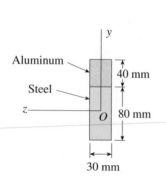

PROB. 6.3-8

PROB. 6.3-10

**6.3-9** A beam is constructed of two angle sections, each L 5 × 3 × 1/2, which reinforce a 2 × 8 (actual dimensions) wood plank (see the cross section shown in the figure). The modulus of elasticity for the wood is $E_w = 1.2 \times 10^6$ psi and for the steel is $E_s = 30 \times 10^6$ psi.

Find the allowable bending moment $M_{allow}$ for the beam if the allowable stress in the wood is $\sigma_w = 1100$ psi and in the steel is $\sigma_s = 12{,}000$ psi. (*Note:* Disregard the weight of the beam, and see Table F-5a of Appendix F for the dimensions and properties of the angles.)

**6.3-11** A W 12 × 50 steel wide-flange beam and a segment of a 4-inch thick concrete slab (see figure) jointly resist a positive bending moment of 95 k-ft. The beam and slab are joined by shear connectors that are welded to the steel beam. (These connectors resist the horizontal shear at the contact surface.) The moduli of elasticity of the steel and the concrete are in the ratio 12 to 1.

Determine the maximum stresses $\sigma_s$ and $\sigma_c$ in the steel and concrete, respectively. (*Note:* See Table F-1a of Appendix F for the dimensions and properties of the steel beam.)

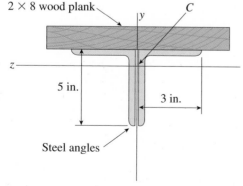

PROB. 6.3-9

PROB. 6.3-11

**6.3-12** A reinforced concrete beam (see figure) is acted on by a positive bending moment of $M = 160$ kN·m. Steel reinforcement consists of 4 bars of 28 mm diameter. The modulus of elasticity for the concrete is $E_c = 25$ GPa while that of the steel is $E_s = 200$ GPa.

(a) Find the maximum stresses in steel and concrete.

(b) If *allowable* stresses for concrete and steel are $\sigma_{ac} = 9.2$ MPa and $\sigma_{as} = 135$ MPa, respectively, what is the maximum permissible positive bending moment?

(c) What is the required area of steel reinforcement, $A_s$, if a balanced condition must be achieved? What is the allowable positive bending moment? (Recall that in a balanced design, both steel and concrete reach allowable stress values simultaneously under the design moment.)

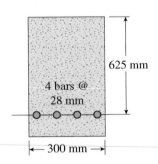

PROB. 6.3-12

**6.3-13** A reinforced concrete T-beam (see figure) is acted on by a positive bending moment of $M = 175$ ft-kips. Steel reinforcement consists of four bars of 1.41-inch diameter. The modulus of elasticity for the concrete is $E_c = 3000$ ksi while that of the steel is $E_s = 29,000$ ksi. Let $b = 48$ in., $t_f = 4$ in., $b_w = 15$ in., and $d = 24$ in.

(a) Find the maximum stresses in steel and concrete.

(b) If *allowable* stresses for concrete and steel are $\sigma_{ac} = 1400$ psi and $\sigma_{as} = 18$ ksi, respectively, what is the maximum permissible positive bending moment?

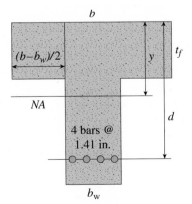

PROB. 6.3-13

**6.3-14** A reinforced concrete slab (see figure) is reinforced with 13-mm bars spaced 160 mm apart at $d = 105$ mm from the top of the slab. The modulus of elasticity for the concrete is $E_c = 25$ GPa, while that of the steel is $E_s = 200$ GPa. Assume that *allowable* stresses for concrete and steel are $\sigma_{ac} = 9.2$ MPa and $\sigma_{as} = 135$ MPa.

(a) Find the maximum permissible positive bending moment for a 1-m-wide strip of the slab.

(b) What is the required area of steel reinforcement, $A_s$, if a *balanced condition* must be achieved? What is the allowable positive bending moment? (Recall that in a balanced design, both steel and concrete reach allowable stress values simultaneously under the design moment.)

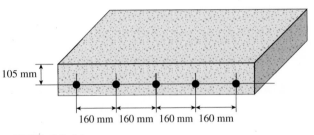

PROB. 6.3-14

**6.3-15** A wood beam reinforced by an aluminum channel section is shown in the figure. The beam has a cross section of dimensions 6 in. by 10 in., and the channel has a uniform thickness of 0.25 in. If the allowable stresses in the wood and aluminum are 1160 psi and 5500 psi, respectively, and if their moduli of elasticity are in the ratio 1 to 6, what is the maximum allowable bending moment for the beam?

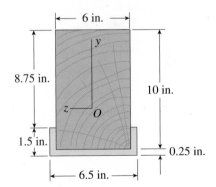

PROB. 6.3-15

# Beams with Inclined Loads

*When solving the problems for Section 6.4, be sure to draw a sketch of the cross section showing the orientation of the neutral axis and the locations of the points where the stresses are being found.*

**6.4-1** A beam of rectangular cross section supports an inclined load $P$ having its line of action along a diagonal of the cross section (see figure). Show that the neutral axis lies along the other diagonal.

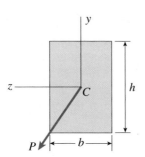

PROB. 6.4-1

**6.4-2** A wood beam of rectangular cross section (see figure) is simply supported on a span of length $L$. The longitudinal axis of the beam is horizontal, and the cross section is tilted at an angle $\alpha$. The load on the beam is a vertical uniform load of intensity $q$ acting through the centroid $C$.

Determine the orientation of the neutral axis and calculate the maximum tensile stress $\sigma_{max}$ if $b = 80$ mm, $h = 140$ mm, $L = 1.75$ m, $\alpha = 22.5°$, and $q = 7.5$ kN/m.

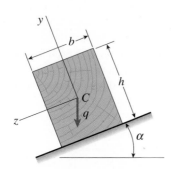

PROBS. 6.4-2 and 6.4-3

**6.4-3** Solve the preceding problem for the following data: $b = 6$ in., $h = 10$ in., $L = 12.0$ ft, $\tan \alpha = 1/3$, and $q = 325$ lb/ft.

**6.4-4** A simply supported wide-flange beam of span length $L$ carries a vertical concentrated load $P$ acting through the centroid $C$ at the midpoint of the span (see figure). The beam is attached to supports inclined at an angle $\alpha$ to the horizontal.

Determine the orientation of the neutral axis and calculate the maximum stresses at the outside corners of the cross section (points $A$, $B$, $D$, and $E$) due to the load $P$. Data for

the beam are as follows: W 250 $\times$ 44.8 section, $L = 3.5$ m, $P = 18$ kN, and $\alpha = 26.57°$. (*Note:* See Table F-1(b) of Appendix F for the dimensions and properties of the beam.)

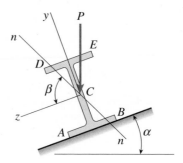

PROBS. 6.4-4 and 6.4-5

**6.4-5** Solve the preceding problem using the following data: W 8 $\times$ 21 section, $L = 84$ in., $P = 4.5$ k, and $\alpha = 22.5°$.

**6.4-6** A wood cantilever beam of rectangular cross section and length $L$ supports an inclined load $P$ at its free end (see figure).

Determine the orientation of the neutral axis and calculate the maximum tensile stress $\sigma_{max}$ due to the load $P$. Data for the beam are as follows: $b = 80$ mm, $h = 140$ mm, $L = 2.0$ m, $P = 575$ N, and $\alpha = 30°$.

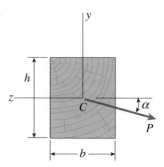

PROBS. 6.4-6 and 6.4-7

**6.4-7** Solve the preceding problem for a cantilever beam with data as follows: $b = 4$ in., $h = 9$ in., $L = 10.0$ ft, $P = 325$ lb, and $\alpha = 45°$.

**6.4-8** A steel beam of I-section (see figure) is simply supported at the ends. Two equal and oppositely directed bending moments $M_0$ act at the ends of the beam, so that the beam is in pure bending. The moments act in plane $mm$, which is oriented at an angle $\alpha$ to the $xy$ plane.

Determine the orientation of the neutral axis and calculate the maximum tensile stress $\sigma_{max}$ due to the moments $M_0$. Data for the beam are as follows: S 200 × 27.4 section, $M_0 = 4$ kN·m., and $\alpha = 24°$. (*Note:* See Table F-2(b) of Appendix F for the dimensions and properties of the beam.)

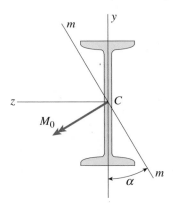

PROB. 6.4-8

**6.4-9** A cantilever beam of wide-flange cross section and length $L$ supports an inclined load $P$ at its free end (see figure).

Determine the orientation of the neutral axis and calculate the maximum tensile stress $\sigma_{max}$ due to the load $P$.

Data for the beam are as follows: W 10 × 45 section, $L = 8.0$ ft, $P = 1.5$ k, and $\alpha = 55°$. (*Note:* See Table F-1 of Appendix F for the dimensions and properties of the beam.)

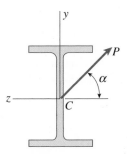

PROBS. 6.4-9 and 6.4-10

**6.4-10** Solve the preceding problem using the following data: W 310 × 129 section, $L = 1.8$ m, $P = 9.5$ kN, and $\alpha = 60°$. (*Note:* See Table F-1(b) of Appendix F for the dimensions and properties of the beam.)

**6.4-11** A cantilever beam of W 12 × 14 section and length $L = 9$ ft supports a slightly inclined load $P = 500$ lb at the free end (see figure).

(a) Plot a graph of the stress $\sigma_A$ at point $A$ as a function of the angle of inclination $\alpha$.

(b) Plot a graph of the angle $\beta$, which locates the neutral axis $nn$, as a function of the angle $\alpha$. (When plotting the graphs, let $\alpha$ vary from 0 to 10°.) (*Note:* See Table F-1 of Appendix F for the dimensions and properties of the beam.)

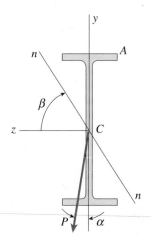

PROB. 6.4-11

**6.4-12** A cantilever beam built up from two channel shapes, each C 200 × 17.1 and of length $L$, supports an inclined load $P$ at its free end (see figure).

Determine the orientation of the neutral axis and calculate the maximum tensile stress $\sigma_{max}$ due to the load $P$. Data for the beam are as follows: $L = 4.5$ m, $P = 500$ N, and $\alpha = 30°$.

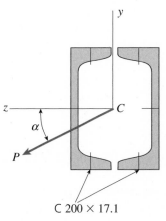

C 200 × 17.1

PROB. 6.4-12

**6.4-13** A built-up steel beam of I-section with channels attached to the flanges (see figure part a) is simply supported at the ends. Two equal and oppositely directed bending moments $M_0$ act at the ends of the beam, so that the beam is in pure bending. The moments act in plane $mm$, which is oriented at an angle $\alpha$ to the $xy$ plane.

(a) Determine the orientation of the neutral axis and calculate the maximum tensile stress $\sigma_{max}$ due to the moments $M_0$.

(b) Repeat part (a) if the channels are now with their flanges pointing away from the beam flange, as shown in the figure part b. Data for the beam are as follows: S 6 × 12.5 section with C 4 × 5.4 sections attached to the flanges, $M_0 = 45$ k-in., and $\alpha = 40°$. (*Note:* See Tables F-2(a) and F-3(a) of Appendix F for the dimensions and properties of the S and C shapes.)

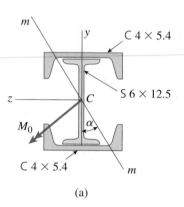

(a)

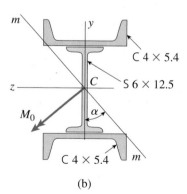

(b)

PROB. 6.4-13

## Bending of Unsymmetric Beams

*When solving the problems for Section 6.5, be sure to draw a sketch of the cross section showing the orientation of the neutral axis and the locations of the points where the stresses are being found.*

**6.5-1** A beam of channel section is subjected to a bending moment $M$ having its vector at an angle $\theta$ to the $z$ axis (see figure).

Determine the orientation of the neutral axis and calculate the maximum tensile stress $\sigma_t$ and maximum compressive stress $\sigma_c$ in the beam.

Use the following data: C 8 × 11.5 section, $M = 20$ k-in., $\tan \theta = 1/3$. (*Note:* See Table F-3 of Appendix F for the dimensions and properties of the channel section.)

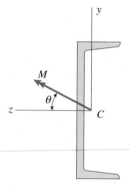

PROBS. 6.5-1 and 6.5-2

**6.5-2** A beam of channel section is subjected to a bending moment $M$ having its vector at an angle $\theta$ to the $z$ axis (see figure).

Determine the orientation of the neutral axis and calculate the maximum tensile stress $\sigma_t$ and maximum compressive stress $\sigma_c$ in the beam. Use a C 200 × 20.5 channel section with $M = 0.75$ kN·m and $\theta = 20°$.

**6.5-3** An angle section with equal legs is subjected to a bending moment $M$ having its vector directed along the 1–1 axis, as shown in the figure.

Determine the orientation of the neutral axis and calculate the maximum tensile stress $\sigma_t$ and maximum compressive stress $\sigma_c$ if the angle is an L 6 × 6 × 3/4 section and $M = 20$ k-in. (*Note:* See Table F-4 of Appendix F for the dimensions and properties of the angle section.)

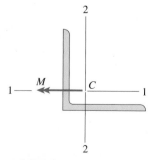

PROBS. 6.5-3 and 6.5-4

**6.5-4** An angle section with equal legs is subjected to a bending moment $M$ having its vector directed along the 1–1 axis, as shown in the figure.

Determine the orientation of the neutral axis and calculate the maximum tensile stress $\sigma_t$ and maximum compressive stress $\sigma_c$ if the section is an L $152 \times 152 \times 12.7$ section and $M = 2.5$ kN · m. (*Note*: See Table F-4(b) of Appendix F for the dimensions and properties of the angle section.)

**6.5-5** A beam made up of two unequal leg angles is subjected to a bending moment $M$ having its vector at an angle $\theta$ to the $z$ axis (see figure part a).

(a) For the position shown in the figure, determine the orientation of the neutral axis and calculate the maximum tensile stress $\sigma_t$ and maximum compressive stress $\sigma_c$ in the beam. Assume that $\theta = 30°$ and $M = 30$ k-in.

(b) The two angles are now inverted and attached back-to-back to form a lintel beam which supports two courses of brick façade (see figure part b). Find the new orientation of the neutral axis and calculate the maximum tensile stress $\sigma_t$ and maximum compressive stress $\sigma_c$ in the beam using $\theta = 30°$ and $M = 30$ k-in.

**6.5-6** The Z-section of Example 12-7 is subjected to $M = 5$ kN · m, as shown.

Determine the orientation of the neutral axis and calculate the maximum tensile stress $\sigma_t$ and maximum compressive stress $\sigma_c$ in the beam. Use the following numerical data: height $h = 200$ mm, width $b = 90$ mm, constant thickness $t = 15$ mm, and $\theta_p = 19.2°$. Use $I_1 = 32.6 \times 10^6$ mm⁴ and $I_2 = 2.4 \times 10^6$ mm⁴ from Example 12-7.

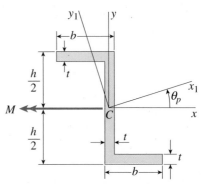

**PROB. 6.5-6**

**6.5-7** The cross section of a steel beam is constructed of a W $18 \times 71$ wide-flange section with a 6 in. $\times$ 1/2 in. cover plate welded to the top flange and a C $10 \times 30$ channel section welded to the bottom flange. This beam is subjected to a bending moment $M$ having its vector at an angle $\theta$ to the $z$ axis (see figure).

Determine the orientation of the neutral axis and calculate the maximum tensile stress $\sigma_t$ and maximum compressive stress $\sigma_c$ in the beam. Assume that $\theta = 30°$ and $M = 75$ k-in. (*Note*: The cross-sectional properties of this beam were computed in Examples 12-2 and 12-5.)

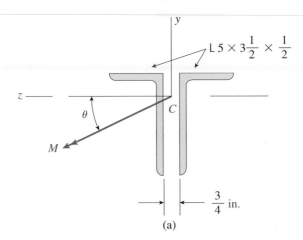

(a)

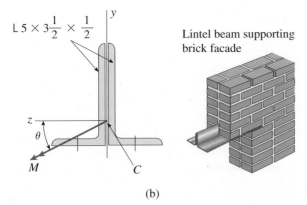

(b)

**PROB. 6.5-5**

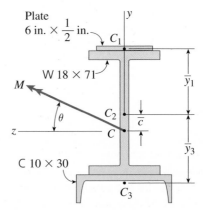

**PROB. 6.5-7**

**6.5-8** The cross section of a steel beam is shown in the figure. This beam is subjected to a bending moment $M$ having its vector at an angle $\theta$ to the $z$ axis.

Determine the orientation of the neutral axis and calculate the maximum tensile stress $\sigma_t$ and maximum compressive stress $\sigma_c$ in the beam. Assume that $\theta = 22.5°$ and $M = 4.5 \text{ kN} \cdot \text{m}$. Use cross sectional properties $I_{x_1} = 93.14 \times 10^6 \text{ mm}^4$, $I_{y_1} = 152.7 \times 10^6 \text{ mm}^4$, and $\theta_P = 27.3°$.

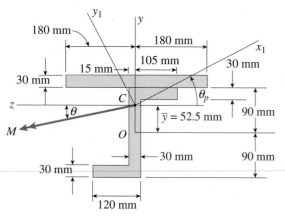

PROB. 6.5-8

**6.5-9** A beam of semicircular cross section of radius $r$ is subjected to a bending moment $M$ having its vector at an angle $\theta$ to the $z$ axis (see figure).

Derive formulas for the maximum tensile stress $\sigma_t$ and the maximum compressive stress $\sigma_c$ in the beam for $\theta = 0$, $45°$, and $90°$. (*Note*: Express the results in the form $\alpha\, M/r^3$, where $\alpha$ is a numerical value.)

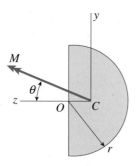

PROB. 6.5-9

**6.5-10** A built-up beam supporting a condominium balcony is made up of a structural T (one half of a W 200 × 31.3) for the top flange and web and two angles (2 L 102 × 76 × 6.4, long legs back-to-back) for the bottom flange and web, as shown. The beam is subjected to a

bending moment $M$ having its vector at an angle $\theta$ to the $z$ axis (see figure).

Determine the orientation of the neutral axis and calculate the maximum tensile stress $\sigma_t$ and maximum compressive stress $\sigma_c$ in the beam. Assume that $\theta = 30°$ and $M = 15 \text{ kN} \cdot \text{m}$.

Use the following numerical properties: $c_1 = 4.111$ mm, $c_2 = 4.169$ mm, $b_f = 134$ mm, $L_s = 76$ mm, $A = 4144 \text{ mm}^2$, $I_y = 3.88 \times 10^6 \text{ mm}^4$, and $I_z = 34.18 \times 10^6 \text{ mm}^4$.

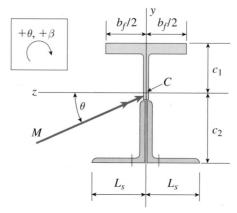

Built-up beam
(© Barry Goodno)

PROB. 6.5-10

**6.5-11** A steel post ($E = 30 \times 10^6$ psi) having thickness $t = 1/8$ in. and height $L = 72$ in. supports a stop sign (see figure). The stop sign post is subjected to a bending moment $M$ having its vector at an angle $\theta$ to the $z$ axis.

Determine the orientation of the neutral axis and calculate the maximum tensile stress $\sigma_t$ and maximum compressive stress $\sigma_c$ in the beam. Assume that $\theta = 30°$ and $M = 5.0$ k-in.

Use the following numerical properties for the post: $A = 0.578$ in.$^2$, $c_1 = 0.769$ in., $c_2 = 0.731$ in., $I_y = 0.44867$ in.$^4$, and $I_z = 0.16101$ in.$^4$.

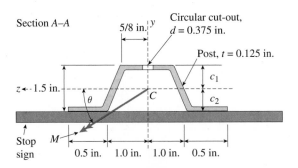

Section A–A

5/8 in.

Circular cut-out, $d = 0.375$ in.

Post, $t = 0.125$ in.

$c_1$

$z$ ←1.5 in.

$\theta$

$C$

$c_2$

Stop sign    $M$

0.5 in.    1.0 in.    1.0 in.    0.5 in.

STOP

$L$

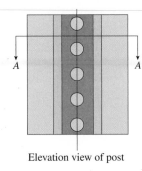

Elevation view of post

Steel post
(© Barry Goodno)

PROB. 6.5-11

**6.5-12** A C 200 × 17.1 channel section has an angle with equal legs attached as shown; the angle serves as a lintel beam. The combined steel section is subjected to a bending moment $M$ having its vector directed along the $z$ axis, as shown in the figure. The centroid $C$ of the combined section is located at distances $x_c$ and $y_c$ from the centroid ($C_1$) of the channel alone. Principal axes $x_1$ and $y_1$ are also shown in the figure and properties $I_{x_1}$, $I_{y_1}$, and $\theta_P$ are given below.

Find the orientation of the neutral axis and calculate the maximum tensile stress $\sigma_t$ and maximum compressive stress $\sigma_c$ if the angle is an L 76 × 76 × 6.4 section and

$M = 3.5$ kN·m. Use the following properties for principal axes for the combined section: $I_{x_1} = 18.49 \times 10^6$ mm$^4$, $I_{y_1} = 1.602 \times 10^6$ mm$^4$, $\theta_P = 7.448°$ (CW), $x_c = 10.70$ mm, $y_c = 24.07$ mm.

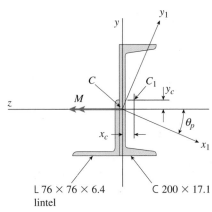

$y_1$

$y$

$C$

$C_1$    $y_c$

$M$

$z$

$\theta_P$

$x_c$

$x_1$

L 76 × 76 × 6.4
lintel

C 200 × 17.1

PROB. 6.5.12

## Shear Stresses in Wide-Flange Beams

*When solving the problems for Section 6.8, assume that the cross sections are thin-walled. Use centerline dimensions for all calculations and derivations, unless otherwise specified.*

**6.8-1** A simple beam of W 10 × 30 wide-flange cross section supports a uniform load of intensity $q = 3.0$ k/ft on a span of length $L = 12$ ft (see figure). The dimensions of the cross section are $h = 10.5$ in., $b = 5.81$ in., $t_f = 0.510$ in., and $t_w = 0.300$ in.

(a) Calculate the maximum shear stress $\tau_{max}$ on cross section $A$–$A$ located at distance $d = 2.5$ ft from the end of the beam.

(b) Calculate the shear stress $\tau$ at point $B$ on the cross section. Point $B$ is located at a distance $a = 1.5$ in. from the edge of the lower flange.

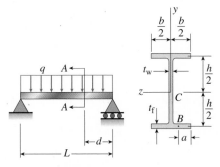

$q$    $A$

$A$

$L$    $d$

$\frac{b}{2}$    $\frac{b}{2}$    $y$

$t_w$

$z$    $C$    $\frac{h}{2}$

$t_f$    $\frac{h}{2}$

$B$

$a$

PROBS. 6.8-1 and 6.8-2

**6.8-2** Solve the preceding problem for a W $250 \times 44.8$ wide-flange shape with the following data: $L = 3.5$ m, $q = 45$ kN/m, $h = 267$ mm, $b = 148$ mm, $t_f = 13$ mm, $t_w = 7.62$ mm, $d = 0.5$ m, and $a = 50$ mm.

**6.8-3** A beam of wide-flange shape, W $8 \times 28$, has the cross section shown in the figure. The dimensions are $b = 6.54$ in., $h = 8.06$ in., $t_w = 0.285$ in., and $t_f = 0.465$ in. The loads on the beam produce a shear force $V = 7.5$ k at the cross section under consideration.

(a) Using centerline dimensions, calculate the maximum shear stress $\tau_{max}$ in the web of the beam.

(b) Using the more exact analysis of Section 5.10 in Chapter 5, calculate the maximum shear stress in the web of the beam and compare it with the stress obtained in part (a).

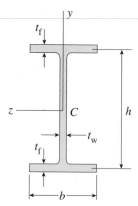

PROBS. 6.8-3 and 6.8-4

**6.8-4** Solve the preceding problem for a W $200 \times 41.7$ shape with the following data: $b = 166$ mm, $h = 205$ mm, $t_w = 7.24$ mm, $t_f = 11.8$ mm, and $V = 38$ kN.

# Shear Centers of Thin-Walled Open Sections

*When locating the shear centers in the problems for Section 6.9, assume that the cross sections are thin-walled and use centerline dimensions for all calculations and derivations.*

**6.9-1** Calculate the distance $e$ from the centerline of the web of a C $15 \times 40$ channel section to the shear center $S$ (see figure). (*Note*: For purposes of analysis, consider the flanges to be rectangles with thickness $t_f$ equal to the average flange thickness given in Table F-3(a) in Appendix F.)

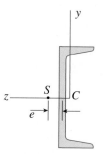

PROBS. 6.9-1 and 6.9-2

**6.9-2** Calculate the distance $e$ from the centerline of the web of a C $310 \times 45$ channel section to the shear center $S$ (see figure). (*Note*: For purposes of analysis, consider the flanges to be rectangles with thickness $t_f$ equal to the average flange thickness given in Table F-3(b) in Appendix F.)

**6.9-3** The cross section of an unbalanced wide-flange beam is shown in the figure. Derive the following formula for the distance $h_1$ from the centerline of one flange to the shear center $S$:

$$h_1 = \frac{t_2 b_2^3 h}{t_1 b_1^3 + t_2 b_2^3}$$

Also, check the formula for the special cases of a T-beam ($b_2 = t_2 = 0$) and a balanced wide-flange beam ($t_2 = t_1$ and $b_2 = b_1$).

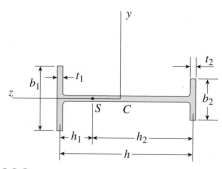

PROB. 6.9-3

**6.9-4** The cross section of an unbalanced wide-flange beam is shown in the figure. Derive the following formula for the distance $e$ from the centerline of the web to the shear center $S$:

$$e = \frac{3t_f(b_2^2 - b_1^2)}{ht_w + 6t_t(b_1 + b_2)}$$

Also, check the formula for the special cases of a channel section ($b_1 = 0$ and $b_2 = b$) and a doubly symmetric beam ($b_1 = b_2 = b/2$).

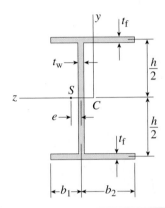

PROB. 6.9-4

**6.9-5** The cross section of a channel beam with double flanges and constant thickness throughout the section is shown in the figure.

Derive the following formula for the distance $e$ from the centerline of the web to the shear center $S$:

$$e = \frac{3b^2(h_1^2 + h_2^2)}{h_2^3 + 6b(h_1^2 + h_2^2)}$$

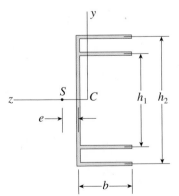

PROB. 6.9-5

**6.9-6** The cross section of a slit circular tube of constant thickness is shown in the figure.

(a) Show that the distance $e$ from the center of the circle to the shear center $S$ is equal to $2r$ in the figure part a.

(b) Find an expression for $e$ if flanges with the same thickness as that of the tube are added, as shown in the figure part b.

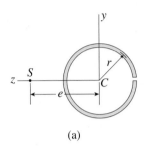

(a)

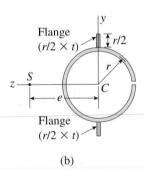

(b)

PROB. 6.9-6

**6.9-7** The cross section of a slit square tube of constant thickness is shown in the figure. Derive the following formula for the distance $e$ from the corner of the cross section to the shear center $S$:

$$e = \frac{b}{2\sqrt{2}}$$

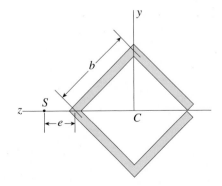

PROB. 6.9-7

**6.9-8** The cross section of a slit rectangular tube of constant thickness is shown in the figures.

(a) Derive the following formula for the distance $e$ from the centerline of the wall of the tube in the figure part a to the shear center $S$:

$$e = \frac{b(2h + 3b)}{2(h + 3b)}$$

(b) Find an expression for $e$ if flanges with the same thickness as that of the tube are added as shown in figure part b.

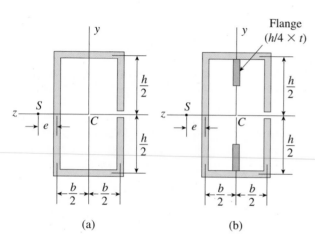

(a)                (b)

PROB. 6.9-8

**6.9-9** A U-shaped cross section of constant thickness is shown in the figure. Derive the following formula for the distance $e$ from the center of the semicircle to the shear center $S$:

$$e = \frac{2(2r^2 + b^2 + \pi br)}{4b + \pi r}$$

Also, plot a graph showing how the distance $e$ (expressed as the nondimensional ratio $e/r$) varies as a function of the ratio $b/r$. (Let $b/r$ range from 0 to 2.)

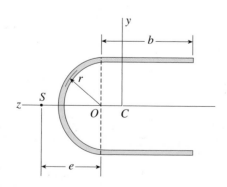

PROB. 6.9-9

**6.9-10** Derive the following formula for the distance $e$ from the centerline of the wall to the shear center $S$ for the C-section of constant thickness shown in the figure:

$$e = \frac{3bh^2(b + 2a) - 8ba^3}{h^2(h + 6b + 6a) + 4a^2(2a - 3h)}$$

Also, check the formula for the special cases of a channel section ($a = 0$) and a slit rectangular tube ($a = h/2$).

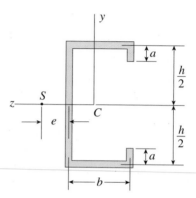

PROB. 6.9-10

**6.9-11** Derive the following formula for the distance $e$ from the centerline of the wall to the shear center $S$ for the hat section of constant thickness shown in the figure:

$$e = \frac{3bh^2(b + 2a) - 8ba^3}{h^2(h + 6b + 6a) + 4a^2(2a + 3h)}$$

Also, check the formula for the special case of a channel section ($a = 0$).

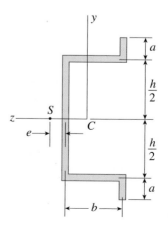

PROB. 6.9-11

**6.9-12** The cross section of a sign post of constant thickness is shown in the figure.

Derive the formula for the distance $e$ from the centerline of the wall of the post to the shear center $S$:

$$e = \frac{1}{3} t \, ba \, (4a^2 + 3ab \, \sin{(\beta)} + 3ab$$

$$+ \, 2\sin{(\beta)}b^2) \frac{\cos{(\beta)}}{I_z}$$

where $I_z =$ moment of inertia about the $z$ axis

Also, compare this formula with that given in Prob. 6.9-11 for the special case of $\beta = 0$ here and $a = h/2$ in both formulas.

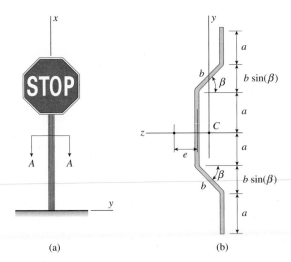

(a)     (b)

PROB. 6.9-12

**6.9-13** A cross section in the shape of a circular arc of constant thickness is shown in the figure. Derive the following formula for the distance $e$ from the center of the arc to the shear center $S$:

$$e = \frac{2r(\sin{\beta} - \beta\cos{\beta})}{\beta - \sin{\beta}\cos{\beta}}$$

in which $\beta$ is in radians. Also, plot a graph showing how the distance $e$ varies as $\beta$ varies from 0 to $\pi$.

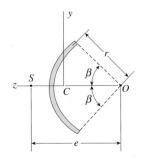

PROB. 6.9-13

# Elastoplastic Bending

*The problems for Section 6.10 are to be solved using the assumption that the material is elastoplastic with yield stress $\sigma_Y$.*

**6.10-1** Determine the shape factor $f$ for a cross section in the shape of a double trapezoid having the dimensions shown in the figure.

Also, check your result for the special cases of a rhombus ($b_1 = 0$) and a rectangle ($b_1 = b_2$).

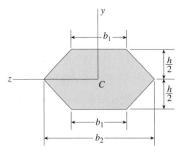

PROB. 6.10-1

**6.10-2** (a) Determine the shape factor $f$ for a hollow circular cross section having inner radius $r_1$ and outer radius $r_2$ (see figure). (b) If the section is very thin, what is the shape factor?

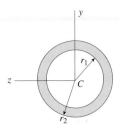

PROB. 6.10-2

**6.10-3** A propped cantilever beam of length $L = 54$ in. with a sliding support supports a uniform load of intensity $q$ (see figure). The beam is made of steel ($\sigma_Y = 36$ ksi) and has a rectangular cross section of width $b = 4.5$ in. and height $h = 6.0$ in.

What load intensity $q$ will produce a fully plastic condition in the beam?

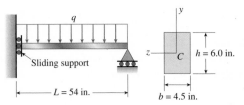

PROB. 6.10-3

**6.10-4** A steel beam of rectangular cross section is 40 mm wide and 80 mm high (see figure). The yield stress of the steel is 210 MPa.

(a) What percent of the cross-sectional area is occupied by the elastic core if the beam is subjected to a bending moment of 12.0 kN·m acting about the $z$ axis?

(b) What is the magnitude of the bending moment that will cause 50% of the cross section to yield?

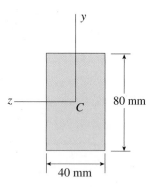

PROB. 6.10-4

**6.10-5** Calculate the shape factor $f$ for the wide-flange beam shown in the figure if $h = 12.2$ in., $b = 8.08$ in., $t_f = 0.64$ in., and $t_w = 0.37$ in.

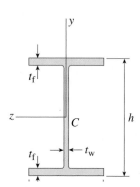

PROBS. 6.10-5 and 6.10-6

**6.10-6** Solve the preceding problem for a wide-flange beam with $h = 404$ mm, $b = 140$ mm, $t_f = 11.2$ mm, and $t_w = 6.99$ mm.

**6.10-7** Determine the plastic modulus $Z$ and shape factor $f$ for a W $12 \times 14$ wide-flange beam. (*Note*: Obtain the cross-sectional dimensions and section modulus of the beam from Table F-1(a) in Appendix F.)

**6.10-8** Solve the preceding problem for a W $250 \times 89$ wide-flange beam. (*Note*: Obtain the cross-sectional dimensions and section modulus of the beam from Table F-1(b) in Appendix F.)

**6.10-9** Determine the yield moment $M_Y$, plastic moment $M_P$, and shape factor $f$ for a W $16 \times 100$ wide-flange beam if $\sigma_Y = 36$ ksi. (*Note*: Obtain the cross-sectional dimensions and section modulus of the beam from Table F-1(a) in Appendix F.)

**6.10-10** Solve the preceding problem for a W $410 \times 85$ wide-flange beam. Assume that $\sigma_Y = 250$ MPa. (*Note*: Obtain the cross-sectional dimensions and section modulus of the beam from Table F-1(b) in Appendix F.)

**6.10-11** A hollow box beam with height $h = 16$ in., width $b = 8$ in., and constant wall thickness $t = 0.75$ in. is shown in the figure. The beam is constructed of steel with yield stress $\sigma_Y = 32$ ksi.

Determine the yield moment $M_Y$, plastic moment $M_P$, and shape factor $f$.

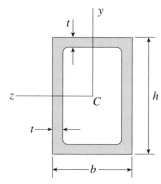

PROBS. 6.10-11 and 6.10-12

**6.10-12** Solve the preceding problem for a box beam with dimensions $h = 0.5$ m, $b = 0.18$ m, and $t = 22$ mm. The yield stress of the steel is 210 MPa.

**6.10-13** A hollow box beam with height $h = 9.5$ in., inside height $h_1 = 8.0$ in., width $b = 5.25$ in., and inside width $b_1 = 4.5$ in. is shown in the figure.

Assuming that the beam is constructed of steel with yield stress $\sigma_Y = 42$ ksi, calculate the yield moment $M_Y$, plastic moment $M_P$, and shape factor $f$.

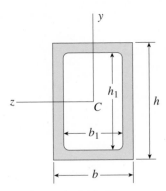

PROBS. 6.10-13 through 6.10-16

**6.10-14** Solve the preceding problem for a box beam with dimensions $h = 200$ mm, $h_1 = 160$ mm, $b = 150$ mm, and $b_1 = 130$ mm. Assume that the beam is constructed of steel with yield stress $\sigma_Y = 220$ MPa.

**6.10-15** The hollow box beam shown in the figure is subjected to a bending moment $M$ of such magnitude that the flanges yield but the webs remain linearly elastic.

(a) Calculate the magnitude of the moment $M$ if the dimensions of the cross section are $h = 15$ in., $h_1 = 12.75$ in., $b = 9$ in., and $b_1 = 7.5$ in. Also, the yield stress is $\sigma_Y = 33$ ksi.

(b) What percent of the moment $M$ is produced by the elastic core?

**6.10-16** Solve the preceding problem for a box beam with dimensions $h = 400$ mm, $h_1 = 360$ mm, $b = 200$ mm, and $b_1 = 160$ mm, and with yield stress $\sigma_Y = 220$ MPa.

**6.10-17** A W 10 × 60 wide-flange beam is subjected to a bending moment $M$ of such magnitude that the flanges yield but the web remains linearly elastic.

(a) Calculate the magnitude of the moment $M$ if the yield stress is $\sigma_Y = 36$ ksi.

(b) What percent of the moment $M$ is produced by the elastic core?

**6.10-18** A singly symmetric beam of T-section (see figure) has cross-sectional dimensions $b = 140$ mm, $a = 190.8$ mm, $t_w = 6.99$ mm, and $t_f = 11.2$ mm.

Calculate the plastic modulus $Z$ and the shape factor $f$.

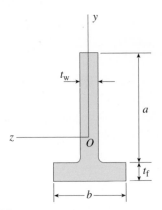

PROB. 6.10-18

**6.10-19** A wide-flange beam of unbalanced cross section has the dimensions shown in the figure.

Determine the plastic moment $M_P$ if $\sigma_Y = 36$ psi.

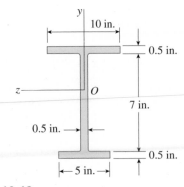

PROB. 6.10-19

**6.10-20** Determine the plastic moment $M_P$ for a beam having the cross section shown in the figure if $\sigma_Y = 210$ MPa.

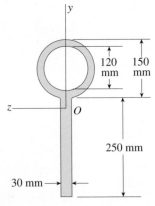

PROB. 6.10-20

# Analysis of Stress and Strain

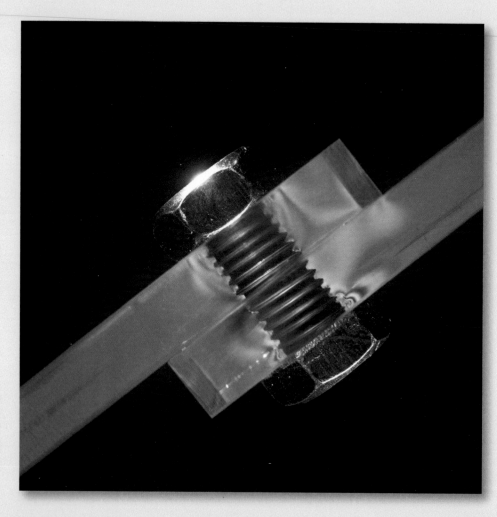

Photoelasticity is an experimental method that can be used to find the complex state of stress near a bolt connecting two plates. (Alfred Pasieka/Peter Arnold, Inc.)

# | CHAPTER OVERVIEW

Chapter 7 is concerned with finding normal and shear stresses acting on inclined sections cut through a member, because these stresses may be larger than those on a stress element aligned with the cross section. In two dimensions, a stress element displays the **state of plane stress** at a point (normal stresses $\sigma_x$, $\sigma_y$, and shear stress $\tau_{xy}$) (Section 7.2), and transformation equations (Section 7.3) are needed to find the stresses acting on an element rotated by some angle $\theta$ from that position. The resulting expressions for normal and shear stresses can be reduced to those examined in Section 2.6 for uniaxial stress ($\sigma_x \neq 0$, $\sigma_y = 0$, $\tau_{xy} = 0$) and in Section 3.5 for pure shear ($\sigma_x = 0$, $\sigma_y = 0$, $\tau_{xy} \neq 0$). Maximum values of stress are needed for design, and the transformation equations can be used to find these **principal stresses** and the planes on which they act (Section 7.3). There are no shear stresses acting on the principal planes, but a separate analysis can be made to find the maximum shear stress ($\tau_{max}$) and the inclined plane on which it acts. **Maximum shear stress** is shown to be equal to one-half of the difference between the principal normal stresses ($\sigma_1$, $\sigma_2$). A graphical representation of the transformation equations for plane stress, known as **Mohr's Circle**, provides a convenient way of calculating stresses on

any inclined plane of interest and those on principal planes, in particular (Section 7.4). Mohr's Circle also can be used to represent strains (Section 7.7) and moments of inertia. In Section 7.5, normal and shear strains ($\varepsilon_x$, $\varepsilon_y$, $\gamma_{xy}$) are studied, and **Hooke's law for plane stress** is derived, which relates elastic moduli $E$ and $G$ and Poisson's ratio $v$ for homogeneous and isotropic materials. The general expressions for Hooke's law can be simplified to the stress-strain relationships for biaxial stress, uniaxial stress, and pure shear. Further examination of strains leads to an expression for unit volume change (or **dilatation** $e$) as well as the strain-energy density in plane stress (Section 7.5). Next, **triaxial stress** is discussed (Section 7.6). Special cases of triaxial stress, known as **spherical stress** and **hydrostatic stress** are then explained: for spherical stress, the three normal stresses are equal and tensile, while for hydrostatic stress, they are equal and compressive. Finally, the transformation equations for **plane strain** (Section 7.7) are derived, relating strains on inclined sections to those in the reference axes directions, and then compared to **plane stress**. The **plane-strain transformation equations** are needed for evaluation of strain measurements obtained using strain gages in field or laboratory experiments of actual structures.

## The discussions in Chapter 7 are organized as follows:

**7.1** Introduction 590

**7.2** Plane Stress 590

**7.3** Principal Stresses and Maximum Shear Stresses 598

**7.4** Mohr's Circle for Plane Stress 607

**7.5** Hooke's Law for Plane Stress 623

**7.6** Triaxial Stress 629

**7.7** Plane Strain 633

Chapter Summary & Review 648

Problems 652

# 7.1 INTRODUCTION

Normal and shear stresses in beams, shafts, and bars can be calculated from the basic formulas discussed in the preceding chapters. For instance, the stresses in a beam are given by the flexure and shear formulas ($\sigma = My/I$ and $\tau = VQ/Ib$), and the stresses in a shaft are given by the torsion formula ($\tau = T\rho/I_p$). The stresses calculated from these formulas act on cross sections of the members, but larger stresses may occur on **inclined sections**. Therefore, we will begin our analysis of stresses and strains by discussing methods for finding the normal and shear stresses acting on inclined sections cut through a member.

We have already derived expressions for the normal and shear stresses acting on inclined sections in both *uniaxial stress* and *pure shear* (see Sections 2.6 and 3.5, respectively). In the case of uniaxial stress, we found that the maximum shear stresses occur on planes inclined at 45° to the axis, whereas the maximum normal stresses occur on the cross sections. In the case of pure shear, we found that the maximum tensile and compressive stresses occur on 45° planes. In an analogous manner, the stresses on inclined sections cut through a beam may be larger than the stresses acting on a cross section. To calculate such stresses, we need to determine the stresses acting on inclined planes under a more general stress state known as **plane stress** (Section 7.2).

In our discussions of plane stress we will use **stress elements** to represent the state of stress at a point in a body. Stress elements were discussed previously in a specialized context (see Sections 2.6 and 3.5), but now we will use them in a more formalized manner. We will begin our analysis by considering an element on which the stresses are known, and then we will derive the **transformation equations** that give the stresses acting on the sides of an element oriented in a different direction.

When working with stress elements, we must always keep in mind that only one intrinsic **state of stress** exists at a point in a stressed body, regardless of the orientation of the element being used to portray that state of stress. When we have two elements with different orientations at the same point in a body, the stresses acting on the faces of the two elements are different, but they still represent the same state of stress, namely, the stress at the point under consideration. This situation is analogous to the representation of a force vector by its components—although the components are different when the coordinate axes are rotated to a new position, the force itself is the same.

Furthermore, we must always keep in mind that stresses are *not* vectors. This fact can sometimes be confusing, because we customarily represent stresses by arrows just as we represent force vectors by arrows. *Although the arrows used to represent stresses have magnitude and direction, they are not vectors because they do not combine according to the parallelogram law of addition.* Instead, stresses are much more complex quantities than are vectors, and in mathematics they are called **tensors**. Other tensor quantities in mechanics are strains and moments of inertia.

# 7.2 PLANE STRESS

The stress conditions that we encountered in earlier chapters when analyzing bars in tension and compression, shafts in torsion, and beams in bending are examples of a state of stress called **plane stress**. To explain plane stress, we will consider the stress element shown in Fig. 7-1a. This element is infinitesimal in size and can be sketched either as a cube or as a rectangular parallelepiped. The *xyz* axes are parallel to the edges of the element, and the faces of the element are designated by the directions of their outward normals, as explained previously in Section 1.6. For instance, the right-hand face of the element is referred to as the positive *x* face, and the left-hand face (hidden from the viewer) is referred to as the negative *x* face. Similarly, the top face is the positive *y* face, and the front face is the positive *z* face.

When the material is in plane stress in the *xy* plane, only the *x* and *y* faces of the element are subjected to stresses, and all stresses act parallel to the *x* and *y* axes, as shown in Fig. 7-1a. This stress condition is very common because it exists at the surface of any stressed body, except at points where external loads act on the surface. When the element shown in Fig. 7-1a is located at the free surface of a body, the *z* axis is normal to the surface and the *z* face is in the plane of the surface.

The symbols for the stresses shown in Fig. 7-1a have the following meanings. A **normal stress** $\sigma$ has a subscript that identifies the face on which the stress acts; for instance, the stress $\sigma_x$ acts on the *x* face of the element and the stress $\sigma_y$ acts on the *y* face of the element. Since the element is infinitesimal in size, equal normal stresses act on the opposite faces. The **sign convention for normal stresses** is the familiar one, namely, tension is positive and compression is negative.

A **shear stress** $\tau$ has two subscripts—the first subscript denotes the face on which the stress acts, and the second gives the direction on that face. Thus, the stress $\tau_{xy}$ acts on the *x* face in the direction of the *y* axis (Fig. 7-1a), and the stress $\tau_{yx}$ acts on the *y* face in the direction of the *x* axis.

The **sign convention for shear stresses** is as follows. A shear stress is positive when it acts on a positive face of an element in the positive direction of an axis, and it is negative when it acts on a positive face of an element in the negative direction of an axis. Therefore, the stresses $\tau_{xy}$ and $\tau_{yx}$ shown on the positive *x* and *y* faces in Fig. 7-1a are positive shear stresses. Similarly, on a negative face of the element, a shear stress is positive when it acts in the negative direction of an axis. Hence, the stresses $\tau_{xy}$ and $\tau_{yx}$ shown on the negative *x* and *y* faces of the element are also positive.

This sign convention for shear stresses is easy to remember if we state it as follows:

*A shear stress is positive when the directions associated with its subscripts are plus-plus or minus-minus; the stress is negative when the directions are plus-minus or minus-plus.*

The preceding sign convention for shear stresses is consistent with the equilibrium of the element, because we know that shear stresses on opposite faces of an infinitesimal element must be equal in magnitude and opposite in direction. Hence, according to our sign convention, a positive stress $\tau_{xy}$ acts upward on the positive face (Fig. 7-1a) and downward on the negative face. In a similar manner, the stresses $\tau_{yx}$ acting on the top and bottom faces of the element are positive although they have opposite directions.

We also know that shear stresses on perpendicular planes are equal in magnitude and have directions such that both stresses point toward, or both point away from, the line of intersection of the faces. Inasmuch

**Fig. 7-1**

Elements in plane stress: (a) three-dimensional view of an element oriented to the *xyz* axes, (b) two-dimensional view of the same element, and (c) two-dimensional view of an element oriented to the $x_1y_1z_1$ axes

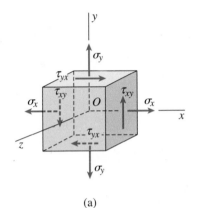

(a)

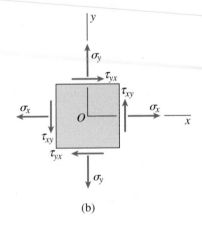

(b)

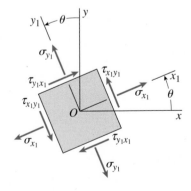

(c)

**Fig. 7-1 (Repeated)**

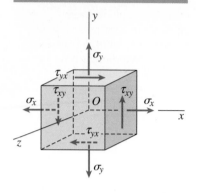

(a)

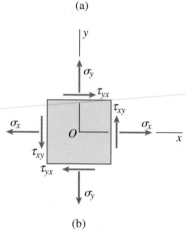

(b)

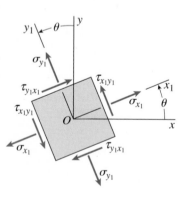

(c)

as $\tau_{xy}$ and $\tau_{yx}$ are positive in the directions shown in the figure, they are consistent with this observation. Therefore, we note that

$$\tau_{xy} = \tau_{yx} \qquad (7\text{-}1)$$

This relationship was derived previously from equilibrium of the element (see Section 1.7).

For convenience in sketching plane-stress elements, we usually draw only a two-dimensional view of the element, as shown in Fig. 7-1b. Although a figure of this kind is adequate for showing all stresses acting on the element, we must still keep in mind that the element is a solid body with a thickness perpendicular to the plane of the figure.

## Stresses on Inclined Sections

We are now ready to consider the stresses acting on inclined sections, assuming that the stresses $\sigma_x$, $\sigma_y$, and $\tau_{xy}$ (Figs. 7-1a and b) are known. To portray the stresses acting on an inclined section, we consider a new stress element (Fig. 7-1c) that is located at the same point in the material as the original element (Fig. 7-1b). However, the new element has faces that are parallel and perpendicular to the inclined direction. Associated with this new element are axes $x_1$, $y_1$, and $z_1$, such that the $z_1$ axis coincides with the $z$ axis and the $x_1 y_1$ axes are rotated counterclockwise through an angle $\theta$ with respect to the $xy$ axes.

The normal and shear stresses acting on this new element are denoted $\sigma_{x_1}$, $\sigma_{y_1}$, $\tau_{x_1y_1}$, and $\tau_{y_1x_1}$, using the same subscript designations and sign conventions described previously for the stresses acting on the $xy$ element. The previous conclusions regarding the shear stresses still apply, so that

$$\tau_{x_1y_1} = \tau_{y_1x_1} \qquad (7\text{-}2)$$

From this equation and the equilibrium of the element, we see that *the shear stresses acting on all four side faces of an element in plane stress are known if we determine the shear stress acting on any one of those faces.*

The stresses acting on the inclined $x_1 y_1$ element (Fig. 7-1c) can be expressed in terms of the stresses on the $xy$ element (Fig. 7-1b) by using equations of equilibrium. For this purpose, we choose a **wedge-shaped stress element** (Fig. 7-2a) having an inclined face that is the same as the $x_1$ face of the inclined element shown in Fig. 7-1c. The other two side faces of the wedge are parallel to the $x$ and $y$ axes.

In order to write equations of equilibrium for the wedge, we need to construct a free-body diagram showing the forces acting on the faces. Let us denote the area of the left-hand side face (that is, the negative $x$ face) as $A_0$. Then the normal and shear forces acting on that face are $\sigma_x A_0$ and $\tau_{xy} A_0$, as shown in the free-body diagram of Fig. 7-2b. The area of the bottom face (or negative $y$ face) is $A_0 \tan\theta$, and the area of the inclined face (or positive $x_1$ face) is $A_0 \sec\theta$. Thus, the normal and shear forces acting on these faces have the magnitudes and directions shown in Fig. 7-2b.

The forces acting on the left-hand and bottom faces can be resolved into orthogonal components acting in the $x_1$ and $y_1$ directions. Then we can obtain two equations of equilibrium by summing forces in those directions. The first equation, obtained by summing forces in the $x_1$ direction, is

$$\sigma_{x_1} A_0 \sec \theta - \sigma_x A_0 \cos \theta - \tau_{xy} A_0 \sin \theta$$
$$- \sigma_y A_0 \tan \theta \sin \theta - \tau_{yx} A_0 \tan \theta \cos \theta = 0$$

In the same manner, summation of forces in the $y_1$ direction gives

$$\tau_{x_1 y_1} A_0 \sec \theta + \sigma_x A_0 \sin \theta - \tau_{xy} A_0 \cos \theta$$
$$- \sigma_y A_0 \tan \theta \cos \theta + \tau_{yx} A_0 \tan \theta \sin \theta = 0$$

Using the relationship $\tau_{xy} = \tau_{yx}$, and also simplifying and rearranging, we obtain the following two equations:

$$\sigma_{x_1} = \sigma_x \cos^2 \theta + \sigma_y \sin^2 \theta + 2\tau_{xy} \sin \theta \cos \theta \qquad \textbf{(7-3a)}$$

$$\tau_{x_1 y_1} = -(\sigma_x - \sigma_y) \sin \theta \cos \theta + \tau_{xy}(\cos^2 \theta - \sin^2 \theta) \qquad \textbf{(7-3b)}$$

Equations (7-3a) and (7-3b) give the normal and shear stresses acting on the $x_1$ plane in terms of the angle $\theta$ and the stresses $\sigma_x$, $\sigma_y$, and $\tau_{xy}$ acting on the $x$ and $y$ planes.

For the special case when $\theta = 0$, we note that Eqs. (7-3a) and (7-3b) give $\sigma_{x_1} = \sigma_x$ and $\tau_{x_1 y_1} = \tau_{xy}$, as expected. Also, when $\theta = 90°$, the equations give $\sigma_{x_1} = \sigma_y$ and $\tau_{x_1 y_1} = -\tau_{xy} = -\tau_{yx}$. In the latter case, since the $x_1$ axis is vertical when $\theta = 90°$, the stress $\tau_{x_1 y_1}$ will be positive when it acts to the left. However, the stress $\tau_{yx}$ acts to the right, and therefore $\tau_{x_1 y_1} = -\tau_{yx}$.

## Transformation Equations for Plane Stress

Equations (7-3a) and (7-3b) for the stresses on an inclined section can be expressed in a more convenient form by introducing the following trigonometric identities (see Appendix D):

$$\cos^2 \theta = \frac{1}{2}(1 + \cos 2\theta) \qquad \sin^2 \theta = \frac{1}{2}(1 - \cos 2\theta)$$

$$\sin \theta \cos \theta = \frac{1}{2} \sin 2\theta$$

When these substitutions are made, the equations become

$$\sigma_{x_1} = \frac{\sigma_x + \sigma_y}{2} + \frac{\sigma_x - \sigma_y}{2} \cos 2\theta + \tau_{xy} \sin 2\theta \qquad \textbf{(7-4a)}$$

$$\tau_{x_1 y_1} = -\frac{\sigma_x - \sigma_y}{2} \sin 2\theta + \tau_{xy} \cos 2\theta \qquad \textbf{(7-4b)}$$

These equations are usually called the **transformation equations for plane stress** because they transform the stress components from one set of axes to another. However, as explained previously, the intrinsic state of stress at the point under consideration is the same whether represented by stresses acting on the $xy$ element (Fig. 7-1b) or by stresses acting on the inclined $x_1 y_1$ element (Fig. 7-1c).

**Fig. 7-2**

Wedge-shaped stress element in plane stress: (a) stresses acting on the element, and (b) forces acting on the element (free-body diagram)

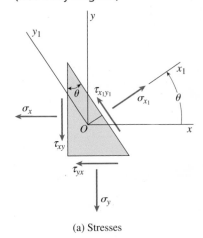

(a) Stresses

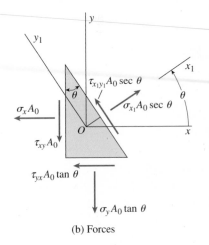

(b) Forces

Since the transformation equations were derived solely from equilibrium of an element, they are applicable to stresses in any kind of material, whether linear or nonlinear, elastic or inelastic.

An important observation concerning the normal stresses can be obtained from the transformation equations. As a preliminary matter, we note that the normal stress $\sigma_{y_1}$ acting on the $y_1$ face of the inclined element (Fig. 7-1c) can be obtained from Eq. (7-4a) by substituting $\theta + 90°$ for $\theta$. The result is the following equation for $\sigma_{y_1}$:

$$\sigma_{y_1} = \frac{\sigma_x + \sigma_y}{2} - \frac{\sigma_x - \sigma_y}{2} \cos 2\theta - \tau_{xy} \sin 2\theta \qquad \textbf{(7-5)}$$

Summing the expressions for $\sigma_{x_1}$ and $\sigma_{y_1}$ (Eqs. 7-4a and 7-5), we obtain the following equation for plane stress:

$$\sigma_{x_1} + \sigma_{y_1} = \sigma_x + \sigma_y \qquad \textbf{(7-6)}$$

This equation shows that the sum of the normal stresses acting on perpendicular faces of plane-stress elements (at a given point in a stressed body) is constant and independent of the angle $\theta$.

The manner in which the normal and shear stresses vary is shown in Fig. 7-3, which is a graph of $\sigma_{x_1}$ and $\tau_{x_1y_1}$ versus the angle $\theta$ (from Eqs. 7-4a and 7-4b). The graph is plotted for the particular case of $\sigma_y = 0.2\sigma_x$ and $\tau_{xy} = 0.8\sigma_x$. We see from the plot that the stresses vary continuously as the orientation of the element is changed. At certain angles, the normal stress reaches a maximum or minimum value; at other angles, it becomes zero. Similarly, the shear stress has maximum, minimum, and zero values at certain angles. A detailed investigation of these maximum and minimum values is made in Section 7.3.

**Fig. 7-3**

Graph of normal stress $\sigma_{x_1}$ and shear stress $\tau_{x_1y_1}$ versus the angle $\theta$ (for $\sigma_y = 0.2\sigma_x$ and $\tau_{xy} = 0.8\sigma_x$)

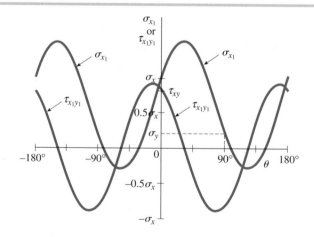

## Special Cases of Plane Stress

The general case of plane stress reduces to simpler states of stress under special conditions. For instance, if all stresses acting on the $xy$ element (Fig. 7-1b) are zero except for the normal stress $\sigma_x$, then the element is in **uniaxial stress** (Fig. 7-4). The corresponding transformation equations, obtained by setting $\sigma_y$ and $\tau_{xy}$ equal to zero in Eqs. (7-4a) and (7-4b), are

$$\sigma_{x_1} = \frac{\sigma_x}{2}(1 + \cos 2\theta) \qquad \tau_{x_1y_1} = -\frac{\sigma_x}{2}(\sin 2\theta) \qquad \textbf{(7-7a,b)}$$

These equations agree with the equations derived previously in Section 2.6 (see Eqs. 2-29a and 2-29b), except that now we are using a more generalized notation for the stresses acting on an inclined plane.

Another special case is **pure shear** (Fig. 7-5), for which the transformation equations are obtained by substituting $\sigma_x = 0$ and $\sigma_y = 0$ into Eqs. (7-4a) and (7-4b):

$$\sigma_{x_1} = \tau_{xy} \sin 2\theta \qquad \tau_{x_1y_1} = \tau_{xy} \cos 2\theta \qquad \textbf{(7-8a,b)}$$

Again, these equations correspond to those derived earlier (see Eqs. 3-30a and 3-30b in Section 3.5).

Finally, we note the special case of **biaxial stress**, in which the $xy$ element is subjected to normal stresses in both the $x$ and $y$ directions but without any shear stresses (Fig. 7-6). The equations for biaxial stress are obtained from Eqs. (7-4a) and (7-4b) simply by dropping the terms containing $\tau_{xy}$, as follows:

$$\sigma_{x_1} = \frac{\sigma_x + \sigma_y}{2} + \frac{\sigma_x - \sigma_y}{2} \cos 2\theta \qquad \textbf{(7-9a)}$$

$$\tau_{x_1y_1} = -\frac{\sigma_x - \sigma_y}{2} \sin 2\theta \qquad \textbf{(7-9b)}$$

Biaxial stress occurs in many kinds of structures, including thin-walled pressure vessels (see Sections 8.2 and 8.3).

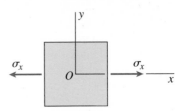

**Fig. 7-4**

Element in uniaxial stress

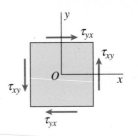

**Fig. 7-5**

Element in pure shear

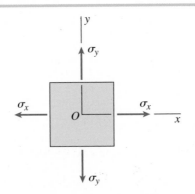

**Fig. 7-6**

Element in biaxial stress

**Fig. 7-7**

Example 7-1: (a) Cylindrical pressure vessel with stress element at C, (b) element C in plane stress, and (c) element C inclined at an angle $\theta = 45°$

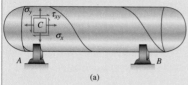

(a)

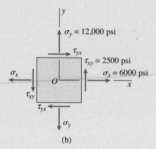

(b)

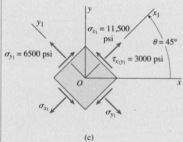

(c)

Fuel storage tanks (© Barry Goodno)

A cylindrical pressure vessel rests on simple supports at A and B (see Fig. 7-7). The vessel is under internal pressure resulting in longitudinal stress $\sigma_x = 6{,}000$ psi and circumferential stress $\sigma_y = 12{,}000$ psi on a stress element at point C on the wall of the vessel. In addition, differential settlement after an earthquake has caused the support at B to rotate, which applies a torsional moment to the vessel leading to shear stress $\tau_{xy} = 2500$ psi. Find the stresses acting on the element at C when rotated through angle $\theta = 45°$.

## Solution

*Transformation equations.* To determine the stresses acting on an inclined element, we will use the transformation equations given in Eqs. (7-4a) and (7-4b). From the given numerical data, we obtain the following values for substitution into those equations:

$$\frac{\sigma_x + \sigma_y}{2} = 9000 \text{ psi} \qquad \frac{\sigma_x - \sigma_y}{2} = -3000 \text{ psi} \qquad \tau_{xy} = 2500 \text{ psi}$$

$$\sin 2\theta = \sin 90° = 1 \qquad \cos 2\theta = \cos 90° = 0$$

Substituting these values into Eqs. (7-4a) and (7-4b), we get

$$\sigma_{x_1} = \frac{\sigma_x + \sigma_y}{2} + \frac{\sigma_x - \sigma_y}{2} \cos 2\theta + \tau_{xy} \sin 2\theta$$

$$= 9000 \text{ psi} + (-3000 \text{ psi})(0) + (2500 \text{ psi})(1) = 11{,}500 \text{ psi} \quad \Longleftarrow$$

$$\tau_{x_1 y_1} = -\frac{\sigma_x - \sigma_y}{2} \sin 2\theta + \tau_{xy} \cos 2\theta$$

$$= -(-3000 \text{ psi})(1) + (2500 \text{ psi})(0) = 3000 \text{ psi} \quad \Longleftarrow$$

In addition, the stress $\sigma_{y_1}$ may be obtained from Eq. (7-5):

$$\sigma_{y_1} = \frac{\sigma_x + \sigma_y}{2} - \frac{\sigma_x - \sigma_y}{2} \cos 2\theta - \tau_{xy} \sin 2\theta$$

$$= 9000 \text{ psi} - (-3000 \text{ psi})(0) - (2500 \text{ psi})(1) = 6500 \text{ psi} \quad \Longleftarrow$$

*Stress elements.* From these results, we can readily obtain the stresses acting on all sides of an element oriented at $\theta = 45°$, as shown in Fig. 7-7c. The arrows show the true directions in which the stresses act. Note especially the directions of the shear stresses, all of which have the same magnitude. Also, observe that the sum of the normal stresses remains constant and equal to 18,000 psi [see Eq. (7-6)].

*Note:* The stresses shown in Fig. 7-7b represent the same intrinsic state of stress as do the stresses shown in Fig. 7-7a. However, the stresses have different values, because the elements on which they act have different orientations.

## ● ● ● Example 7-2

Fig. 7-8

Example 7-2: (a) Cylindrical pressure vessel with stress element at *D*, (b) element *D* in plane stress, and (c) element *D* inclined at an angle $\theta = -35°$

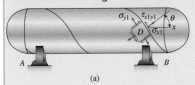

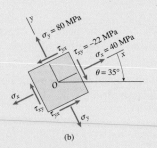

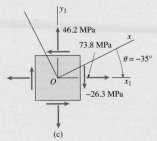

A cylindrical pressure vessel rests on simple supports at *A* and *B* (see Fig. 7-8). The vessel has a helical weld joint oriented at $\theta = 35°$ to the longitudinal axis. The vessel is under internal pressure and also has some torsional shear stress due to differential settlement of the support at *B*. The state of stress on the element at *D* along and perpendicular to the weld seam is known and is given in Fig. 7-8b. Find the equivalent stress state for the element at *D* when rotated through angle $\theta = -35°$ so that the element is aligned with the longitudinal axis of the vessel.

### Solution

The stresses acting on the original element (Fig 7-8b) have the following values:

$$\sigma_x = 40 \text{ MPa} \quad \sigma_y = 80 \text{ MPa} \quad \tau_{xy} = -22 \text{ MPa}$$

An element oriented at a clockwise angle of $-35°$ is shown in Fig. 7-8c, where the $x_1$ axis is at an angle $\theta = -35°$ with respect to the $x$ axis.

*Stress transformation equations.* We can readily calculate the stresses on the $x_1$ face of the element oriented at $\theta = -35°$ by using the transformation equations given in Eqs. (7-4a) and (7-4b). The calculations proceed as follows:

$$\frac{\sigma_x + \sigma_y}{2} = 60 \text{ MPa} \qquad \frac{\sigma_x - \sigma_y}{2} = -20 \text{ MPa}$$

$$\sin 2\theta = \sin (-70°) = -0.94 \qquad \cos 2\theta = \cos (-70°) = 0.342$$

Substituting into the transformation equations, we get

$$\sigma_{x_1} = \frac{\sigma_x + \sigma_y}{2} + \frac{\sigma_x - \sigma_y}{2} \cos (2\theta) + \tau_{xy} \sin (2\theta)$$

$$= 60 \text{ MPa} + (-20 \text{ MPa})(0.342) + (-22 \text{ MPa})(-0.94) = 73.8 \text{ MPa}$$

$$\tau_{x_1 y_1} = -\left(\frac{\sigma_x - \sigma_y}{2}\right) \sin (2\theta) + \tau_{xy} \cos (2\theta)$$

$$= -(-20 \text{ MPa})(-0.94) + (-22 \text{ MPa})(0.342) = -26.3 \text{ MPa}$$

The normal stress acting on the $y_1$ face [see Eq. (7-5)] is

$$\sigma_{y_1} = \frac{\sigma_x + \sigma_y}{2} - \frac{\sigma_x - \sigma_y}{2} \cos (2\theta) - \tau_{xy} \sin (2\theta)$$

$$= 60 \text{ MPa} - (-20 \text{ MPa})(0.342) - (-22 \text{ MPa})(-0.94) = 46.2 \text{ MPa}$$

As a check on the results, we note that $\sigma_{x_1} + \sigma_{y_1} = \sigma_x + \sigma_y$.

The stresses acting on the inclined element are shown in Fig. 7-8c, where the arrows indicate the true directions of the stresses. Again, we note that both stress elements shown in Fig. 7-8 represent the same state of stress.

Fuel storage tank supported on pedestals. (© Barry Goodno)

(a) Photo of a crane-hook

(b) Photoelastic fringe pattern

# 7.3 PRINCIPAL STRESSES AND MAXIMUM SHEAR STRESSES

The transformation equations for plane stress show that the normal stresses $\sigma_{x_1}$ and the shear stresses $\tau_{x_1y_1}$ vary continuously as the axes are rotated through the angle $\theta$. This variation is pictured in Fig. 7-3 for a particular combination of stresses. From the figure, we see that both the normal and shear stresses reach maximum and minimum values at 90° intervals. Not surprisingly, these maximum and minimum values are usually needed for design purposes. For instance, fatigue failures of structures such as machines and aircraft are often associated with the maximum stresses, and hence their magnitudes and orientations should be determined as part of the design process (see Fig. 7-9).

## Principal Stresses

The maximum and minimum normal stresses, called the **principal stresses**, can be found from the transformation equation for the normal stress $\sigma_{x_1}$ (Eq. 7-4a). By taking the derivative of $\sigma_{x_1}$ with respect to $\theta$ and setting it equal to zero, we obtain an equation from which we can find the values of $\theta$ at which $\sigma_{x_1}$ is a maximum or a minimum. The equation for the derivative is

$$\frac{d\sigma_{x_1}}{d\theta} = -(\sigma_x - \sigma_y)\sin 2\theta + 2\tau_{xy}\cos 2\theta = 0 \qquad \textbf{(7-10)}$$

from which we get

$$\tan 2\theta_p = \frac{2\tau_{xy}}{\sigma_x - \sigma_y} \qquad \textbf{(7-11)}$$

The subscript $p$ indicates that the angle $\theta_p$ defines the orientation of the **principal planes**, that is, the planes on which the principal stresses act.

Two values of the angle $2\theta_p$ in the range from 0 to 360° can be obtained from Eq. (7-11). These values differ by 180°, with one value between 0 and 180° and the other between 180° and 360°. Therefore, the angle $\theta_p$ has two values that differ by 90°, one value between 0 and 90° and the other between 90° and 180°. The two values of $\theta_p$ are known as the **principal angles**. For one of these angles, the normal stress $\sigma_{x_1}$ is a *maximum* principal stress; for the other, it is a *minimum* principal stress. Because the principal angles differ by 90°, we see that *the principal stresses occur on mutually perpendicular planes.*

The principal stresses can be calculated by substituting each of the two values of $\theta_p$ into the first stress-transformation equation (Eq. 7-4a) and solving for $\sigma_{x_1}$. By determining the principal stresses in this manner, we not only obtain the values of the principal stresses but we also learn which principal stress is associated with which principal angle.

We can also obtain general formulas for the principal stresses. To do so, refer to the right triangle in Fig. 7-10, which is constructed from

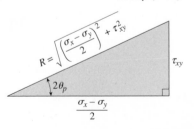

$$R = \sqrt{\left(\frac{\sigma_x - \sigma_y}{2}\right)^2 + \tau_{xy}^2}$$

$$2\theta_p$$

$$\tau_{xy}$$

$$\frac{\sigma_x - \sigma_y}{2}$$

Eq. (7-11). Note that the hypotenuse of the triangle, obtained from the Pythagorean theorem, is

$$R = \sqrt{\left(\frac{\sigma_x - \sigma_y}{2}\right)^2 + \tau_{xy}^2} \qquad \textbf{(7-12)}$$

The quantity $R$ is always a positive number and, like the other two sides of the triangle, has units of stress. From the triangle we obtain two additional relations:

$$\cos 2\theta_p = \frac{\sigma_x - \sigma_y}{2R} \qquad \sin 2\theta_p = \frac{\tau_{xy}}{R} \qquad \textbf{(7-13a,b)}$$

Now we substitute these expressions for $\cos 2\theta_p$ and $\sin 2\theta_p$ into Eq. (7-4a) and obtain the algebraically larger of the two principal stresses, denoted by $\sigma_1$:

$$\sigma_1 = \sigma_{x_1} = \frac{\sigma_x + \sigma_y}{2} + \frac{\sigma_x - \sigma_y}{2} \cos 2\theta_p + \tau_{xy} \sin 2\theta_p$$

$$= \frac{\sigma_x + \sigma_y}{2} + \frac{\sigma_x - \sigma_y}{2}\left(\frac{\sigma_x - \sigma_y}{2R}\right) + \tau_{xy}\left(\frac{\tau_{xy}}{R}\right)$$

After substituting for $R$ from Eq. (7-12) and performing some algebraic manipulations, we obtain

$$\sigma_1 = \frac{\sigma_x + \sigma_y}{2} + \sqrt{\left(\frac{\sigma_x - \sigma_y}{2}\right)^2 + \tau_{xy}^2} \qquad \textbf{(7-14)}$$

The smaller of the principal stresses, denoted by $\sigma_2$, may be found from the condition that the sum of the normal stresses on perpendicular planes is constant (see Eq. 7-6):

$$\sigma_1 + \sigma_2 = \sigma_x + \sigma_y \qquad \textbf{(7-15)}$$

Substituting the expression for $\sigma_1$ into Eq. (7-15) and solving for $\sigma_2$, we get

$$\sigma_2 = \sigma_x + \sigma_y - \sigma_1$$

$$= \frac{\sigma_x + \sigma_y}{2} - \sqrt{\left(\frac{\sigma_x - \sigma_y}{2}\right)^2 + \tau_{xy}^2} \qquad \textbf{(7-16)}$$

This equation has the same form as the equation for $\sigma_1$ but differs by the presence of the minus sign before the square root.

The preceding formulas for $\sigma_1$ and $\sigma_2$ can be combined into a single formula for the **principal stresses**:

$$\sigma_{1,2} = \frac{\sigma_x + \sigma_y}{2} \pm \sqrt{\left(\frac{\sigma_x - \sigma_y}{2}\right)^2 + \tau_{xy}^2} \qquad \textbf{(7-17)}$$

The plus sign gives the algebraically larger principal stress and the minus sign gives the algebraically smaller principal stress.

## Principal Angles

Let us now denote the two angles defining the principal planes as $\theta_{p_1}$ and $\theta_{p_2}$, corresponding to the principal stresses $\sigma_1$ and $\sigma_2$, respectively. Both angles can be determined from the equation for $\tan 2\theta_p$ (Eq. 7-11). However, we cannot tell from that equation which angle is $\theta_{p_1}$ and which is $\theta_{p_2}$. A simple procedure for making this determination is to take one of the values and substitute it into the equation for $\sigma_{x_1}$ (Eq. 7-4a). The resulting value of $\sigma_{x_1}$ will be recognized as either $\sigma_1$ or $\sigma_2$ (assuming we have already found $\sigma_1$ and $\sigma_2$ from Eq. 7-17), thus correlating the two principal angles with the two principal stresses.

Another method for correlating the principal angles and principal stresses is to use Eqs. (7-13a) and (7-13b) to find $\theta_p$, since the only angle that satisfies *both* of those equations is $\theta_{p_1}$. Thus, we can rewrite those equations as follows:

$$\cos 2\theta_{p_1} = \frac{\sigma_x - \sigma_y}{2R} \qquad \sin 2\theta_{p_1} = \frac{\tau_{xy}}{R} \qquad \textbf{(7-18a,b)}$$

Only one angle exists between 0 and 360° that satisfies both of these equations. Thus, the value of $\theta_{p_1}$ can be determined uniquely from Eqs. (7-18a) and (7-18b). The angle $\theta_{p_2}$, corresponding to $\sigma_2$, defines a plane that is perpendicular to the plane defined by $\theta_{p_1}$. Therefore, $\theta_{p_2}$ can be taken as 90° larger or 90° smaller than $\theta_{p_1}$.

## Shear Stresses on the Principal Planes

An important characteristic of the principal planes can be obtained from the transformation equation for the shear stresses (Eq. 7-4b). If we set the shear stress $\tau_{x_1y_1}$ equal to zero, we get an equation that is the same as Eq. (7-10). Therefore, if we solve that equation for the angle $2\theta$, we get the same expression for $\tan 2\theta$ as before (Eq. 7-11). In other words, the angles to the planes of zero shear stress are the same as the angles to the principal planes.

Thus, we can make the following important observation: *The shear stresses are zero on the principal planes.*

## Special Cases

The principal planes for elements in **uniaxial stress** and **biaxial stress** are the $x$ and $y$ planes themselves (Fig. 7-11), because $\tan 2\theta_p = 0$ (see Eq. 7-11)

| **Fig. 7-11** |
|---|

Elements in uniaxial and biaxial stress

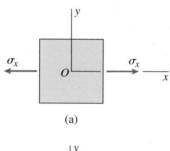

(a)

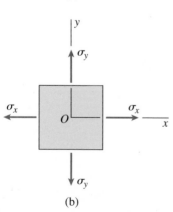

(b)

and the two values of $\theta_p$ are 0 and 90°. We also know that the $x$ and $y$ planes are the principal planes from the fact that the shear stresses are zero on those planes.

For an element in **pure shear** (Fig. 7-12a), the principal planes are oriented at 45° to the $x$ axis (Fig. 7-12b), because $\tan 2\theta_p$ is infinite and the two values of $\theta_p$ are 45° and 135°. If $\tau_{xy}$ is positive, the principal stresses are $\sigma_1 = \tau_{xy}$ and $\sigma_2 = -\tau_{xy}$ (see Section 3.5 for a discussion of pure shear).

## The Third Principal Stress

The preceding discussion of principal stresses refers only to rotation of axes in the $xy$ plane, that is, rotation about the $z$ axis (Fig. 7-13a). Therefore, the two principal stresses determined from Eq. (7-17) are called the **in-plane principal stresses**. However, we must not overlook the fact that the stress element is actually three-dimensional and has three (not two) principal stresses acting on three mutually perpendicular planes.

By making a more complete three-dimensional analysis, it can be shown that the three principal planes for a plane-stress element are the two principal planes already described plus the $z$ face of the element. These principal planes are shown in Fig. 7-13b, where a stress element has been oriented at the principal angle $\theta_{p_1}$, which corresponds to the principal stress $\sigma_1$. The principal stresses $\sigma_1$ and $\sigma_2$ are given by Eq. (7-17), and the third principal stress ($\sigma_3$) equals zero.

By definition, $\sigma_1$ is algebraically larger than $\sigma_2$, but $\sigma_3$ may be algebraically larger than, between, or smaller than $\sigma_1$ and $\sigma_2$. Of course, it is also possible for some or all of the principal stresses to be equal. Note again that there are no shear stresses on any of the principal planes.*

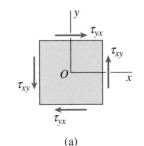

**Fig. 7-12**

(a) Element in pure shear, and (b) principal stresses.

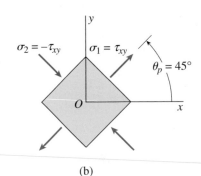

(a)

(b)

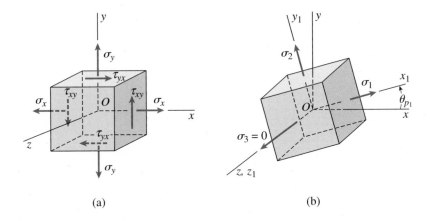

(a)                                    (b)

**Fig. 7-13**

Elements in plane stress: (a) original element, and (b) element oriented to the three principal planes and three principal stresses

## Maximum Shear Stresses

Having found the principal stresses and their directions for an element in plane stress, we now consider the determination of the maximum shear stresses and the planes on which they act. The shear stresses $\tau_{x_1 y_1}$ acting on

---

*The determination of principal stresses is an example of a type of mathematical analysis known as *eigenvalue analysis*, which is described in books on matrix algebra. The stress-transformation equations and the concept of principal stresses are due to the French mathematicians A. L. Cauchy (1789–1857) and Barré de Saint-Venant (1797–1886) and to the Scottish scientist and engineer W. J. M. Rankine (1820–1872); see Refs. 7-1, 7-2, and 7-3, respectively.

inclined planes are given by the second transformation equation [Eq. (7-4b)]. Taking the derivative of $\tau_{x_1 y_1}$ with respect to $\theta$ and setting it equal to zero, we obtain

$$\frac{d\tau_{x_1 y_1}}{d\theta} = -(\sigma_x - \sigma_y)\cos 2\theta - 2\tau_{xy}\sin 2\theta = 0 \qquad \textbf{(7-19)}$$

from which

$$\tan 2\theta_s = -\frac{\sigma_x - \sigma_y}{2\tau_{xy}} \qquad \textbf{(7-20)}$$

The subscript $s$ indicates that the angle $\theta_s$ defines the orientation of the planes of maximum positive and negative shear stresses.

Equation (7-20) yields one value of $\theta_s$ between 0 and 90° and another between 90° and 180°. Furthermore, these two values differ by 90°, and therefore the maximum shear stresses occur on perpendicular planes. Because shear stresses on perpendicular planes are equal in absolute value, the maximum positive and negative shear stresses differ only in sign.

Comparing Eq. (7-20) for $\theta_s$ with Eq. (7-11) for $\theta_p$ shows that

$$\tan 2\theta_s = -\frac{1}{\tan 2\theta_p} = -\cot 2\theta_p \qquad \textbf{(7-21)}$$

From this equation we can obtain a relationship between the angles $\theta_s$ and $\theta_p$. First, we rewrite the preceding equation in the form

$$\frac{\sin 2\theta_s}{\cos 2\theta_s} + \frac{\cos 2\theta_p}{\sin 2\theta_p} = 0$$

Multiplying by the terms in the denominator, we get

$$\sin 2\theta_s \sin 2\theta_p + \cos 2\theta_s \cos 2\theta_p = 0$$

which is equivalent to the following expression (see Appendix D):

$$\cos (2\theta_s - 2\theta_p) = 0$$

Therefore,

$$2\theta_s - 2\theta_p = \pm 90°$$

and

$$\theta_s = \theta_p \pm 45° \qquad \textbf{(7-22)}$$

This equation shows that *the planes of maximum shear stress occur at 45°
to the principal planes.*

The plane of the maximum positive shear stress $\tau_{max}$ is defined by the
angle $\theta_{s_1}$, for which the following equations apply:

$$\cos 2\theta_{s_1} = \frac{\tau_{xy}}{R} \qquad \sin 2\theta_{s_1} = -\frac{\sigma_x - \sigma_y}{2R} \qquad \textbf{(7-23a,b)}$$

in which $R$ is given by Eq. (7-12). Also, the angle $\theta_{s_1}$ is related to the
angle $\theta_{p_1}$ [see Eqs. (7-18a) and (7-18b)] as

$$\theta_{s_1} = \theta_{p_1} - 45° \qquad \textbf{(7-24)}$$

The corresponding maximum shear stress is obtained by substituting the
expressions for $\cos 2\theta_{s_1}$ and $\sin 2\theta_{s_1}$ into the second transformation equation [Eq. (7-4b)], yielding

$$\tau_{max} = \sqrt{\left(\frac{\sigma_x - \sigma_y}{2}\right)^2 + \tau_{xy}^2} \qquad \textbf{(7-25)}$$

The maximum negative shear stress $\tau_{min}$ has the same magnitude but
opposite sign.

Another expression for the maximum shear stress can be obtained
from the principal stresses $\sigma_1$ and $\sigma_2$, both of which are given by Eq. (7-17).
Subtracting the expression for $\sigma_2$ from that for $\sigma_1$, and then comparing
with Eq. (7-25), we see that

$$\tau_{max} = \frac{\sigma_1 - \sigma_2}{2} \qquad \textbf{(7-26)}$$

Thus, *the maximum shear stress is equal to one-half the difference of the
principal stresses.*

The planes of maximum shear stress also contain normal stresses. The
**normal stress** acting on the planes of maximum positive shear stress can
be determined by substituting the expressions for the angle $\theta_{s_1}$ [Eqs. (7-23a)
and (7-23b)] into the equation for $\sigma_{x_1}$ [Eq. (7-4a)]. The resulting stress is
equal to the average of the normal stresses on the $x$ and $y$ planes:

$$\sigma_{aver} = \frac{\sigma_x + \sigma_y}{2} \qquad \textbf{(7-27)}$$

This same normal stress acts on the planes of maximum negative shear
stress.

In the particular cases of **uniaxial stress** and **biaxial stress** (Fig. 7-11), the planes of maximum shear stress occur at 45° to the $x$ and $y$ axes. In the case of **pure shear** (Fig. 7-12), the maximum shear stresses occur on the $x$ and $y$ planes.

## In-Plane and Out-of-Plane Shear Stresses

The preceding analysis of shear stresses has dealt only with **in-plane shear stresses**, that is, stresses acting in the $xy$ plane. To obtain the maximum in-plane shear stresses [Eqs. (7-25) and (7-26)], we considered elements that were obtained by rotating the $xyz$ axes about the $z$ axis, which is a principal axis (Fig. 7-13a). We found that the maximum shear stresses occur on planes at 45° to the principal planes. The principal planes for the element of Fig. 7-13a are shown in Fig. 7-13b, where $\sigma_1$ and $\sigma_2$ are the principal stresses. Therefore, the maximum in-plane shear stresses are found on an element obtained by rotating the $x_1 y_1 z_1$ axes (Fig. 7-13b) about the $z_1$ axis through an angle of 45°. These stresses are given by Eq. (7-25) or Eq. (7-26).

We can also obtain maximum shear stresses by 45° rotations about the other two principal axes (the $x_1$ and $y_1$ axes in Fig. 7-13b). As a result, we obtain three sets of **maximum positive and maximum negative shear stresses** [compare with Eq. (7-26)]:

$$(\tau_{max})_{x_1} = \pm\frac{\sigma_2}{2} \qquad (\tau_{max})_{y_1} = \pm\frac{\sigma_1}{2} \qquad \textbf{(7-28a,b,c)}$$

$$(\tau_{max})_{z_1} = \pm\frac{\sigma_1 - \sigma_2}{2}$$

in which the subscripts indicate the principal axes about which the 45° rotations take place. The stresses obtained by rotations about the $x_1$ and $y_1$ axes are called **out-of-plane shear stresses**.

The algebraic values of $\sigma_1$ and $\sigma_2$ determine which of the preceding expressions gives the numerically largest shear stress. If $\sigma_1$ and $\sigma_2$ have the same sign, then one of the first two expressions is numerically largest; if they have opposite signs, the last expression is largest.

**Fig. 7-13 (Repeated)**

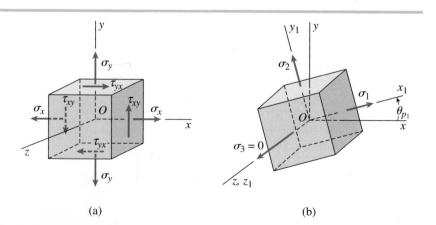

(a)                    (b)

### Fig. 7-14

Example 7-3: (a) Beam structure, (b) element at C in plane stress, (c) principal stresses, and (d) maximum shear stresses

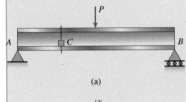

(a)

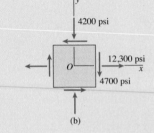

(b)

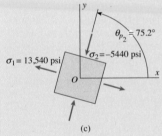

(c)

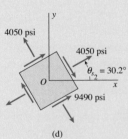

(d)

A simply supported wide-flange beam has a concentrated load $P$ applied at mid-span (Fig. 7-14a). The state of stress in the beam web at element $C$ is known (Fig. 7-14b) to be $\sigma_x = 12,300$ psi, $\sigma_y = -4200$ psi, and $\tau_{xy} = -4700$ psi.

(a) Determine the principal stresses and show them on a sketch of a properly oriented element.
(b) Determine the maximum shear stresses and show them on a sketch of a properly oriented element. (Consider only the in-plane stresses.)

### Solution

(a) *Principal stresses.* The principal angles $\theta_p$ that locate the principal planes can be obtained from Eq. (7-11):

$$\tan 2\theta_p = \frac{2\tau_{xy}}{\sigma_x - \sigma_y} = \frac{2(-4700 \text{ psi})}{12,300 \text{ psi} - (-4200 \text{ psi})} = -0.5697$$

Solving for the angles, we get the following two sets of values:

$$2\theta_p = 150.3° \text{ and } \theta_p = 75.2°$$
$$2\theta_p = 330.3° \text{ and } \theta_p = 165.2°$$

The principal stresses may be obtained by substituting the two values of $2\theta_p$ into the transformation equation for $\sigma_{x_1}$ from Eq. (7-4a). As a preliminary calculation, we determine the following quantities:

$$\frac{\sigma_x + \sigma_y}{2} = \frac{12,300 \text{ psi} - 4200 \text{ psi}}{2} = 4050 \text{ psi}$$

$$\frac{\sigma_x - \sigma_y}{2} = \frac{12,300 \text{ psi} + 4200 \text{ psi}}{2} = 8250 \text{ psi}$$

Now we substitute the first value of $2\theta_p$ into Eq. (7-4a) and obtain

$$\sigma_{x_1} = \frac{\sigma_x + \sigma_y}{2} + \frac{\sigma_x - \sigma_y}{2} \cos 2\theta + \tau_{xy} \sin 2\theta$$

$$= 4050 \text{ psi} + (8250 \text{ psi})(\cos 150.3°) - (4700 \text{ psi})(\sin 150.3°)$$

$$= -5440 \text{ psi}$$

In a similar manner, we substitute the second value of $2\theta_p$ and obtain $\sigma_{x_1} = 13,540$ psi. Thus, the principal stresses and their corresponding principal angles are

$$\sigma_1 = 13,540 \text{ psi} \quad \text{and} \quad \theta_{p_1} = 165.2° \quad \Leftarrow$$
$$\sigma_2 = -5440 \text{ psi} \quad \text{and} \quad \theta_{p_2} = 75.2° \quad \Leftarrow$$

Note that $\theta_{p_1}$ and $\theta_{p_2}$ differ by 90° and that $\sigma_1 + \sigma_2 = \sigma_x + \sigma_y$.

The principal stresses are shown on a properly oriented element in Fig. 7-14c. Of course, no shear stresses act on the principal planes.

*Continues* ➡

**Example 7-3 -** *Continued*

**Fig. 7-14 (Repeated)**

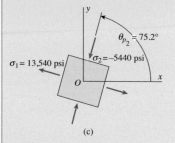

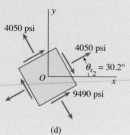

(c)

(d)

*Alternative solution for the principal stresses.* The principal stresses may also be calculated directly from Eq. (7-17):

$$\sigma_{1,2} = \frac{\sigma_x + \sigma_y}{2} \pm \sqrt{\left(\frac{\sigma_x - \sigma_y}{2}\right)^2 + \tau_{xy}^2}$$

$$= 4050 \text{ psi} \pm \sqrt{(8250 \text{ psi})^2 + (-4700 \text{ psi})^2}$$

$$\sigma_{1,2} = 4050 \text{ psi} \pm 9490 \text{ psi}$$

Therefore,

$$\sigma_1 = 13{,}540 \text{ psi} \qquad \sigma_2 = -5440 \text{ psi}$$

The angle $\theta_{p_1}$ to the plane on which $\sigma_1$ acts is obtained from Eqs. (7-18a) and (7-18b):

$$\cos 2\theta_{p_1} = \frac{\sigma_x - \sigma_y}{2R} = \frac{8250 \text{ psi}}{9490 \text{ psi}} = 0.869$$

$$\sin 2\theta_{p_1} = \frac{\tau_{xy}}{R} = \frac{-4700 \text{ psi}}{9490 \text{ psi}} = -0.495$$

in which $R$ is given by Eq. (7-12) and is equal to the square-root term in the preceding calculation for the principal stresses $\sigma_1$ and $\sigma_2$.

The only angle between 0 and 360° having the specified sine and cosine is $2\theta_{p_1} = 330.3°$; hence, $\theta_{p_1} = 165.2°$. This angle is associated with the algebraically larger principal stress $\sigma_1 = 13{,}540$ psi. The other angle is 90° larger or smaller than $\theta_{p_1}$; hence, $\theta_{p_2} = 75.2°$. This angle corresponds to the smaller principal stress $\sigma_2 = -5440$ psi. Note that these results for the principal stresses and principal angles agree with those found previously.

(b) *Maximum shear stresses.* The maximum in-plane shear stresses are given by Eq. (7-25):

$$\tau_{max} = \sqrt{\left(\frac{\sigma_x - \sigma_y}{2}\right)^2 + \tau_{xy}^2}$$

$$= \sqrt{(8250 \text{ psi})^2 + (-4700 \text{ psi})^2} = 9490 \text{ psi} \quad \Leftarrow$$

The angle $\theta_{s_1}$ to the plane having the maximum positive shear stress is calculated from Eq. (7-24):

$$\theta_{s_1} = \theta_{p_1} - 45° = 165.2° - 45° = 120.2° \quad \Leftarrow$$

It follows that the maximum negative shear stress acts on the plane for which $\theta_{s_2} = 120.2° - 90° = 30.2°$.

The normal stresses acting on the planes of maximum shear stresses are calculated from Eq. (7-27):

$$\sigma_{aver} = \frac{\sigma_x + \sigma_y}{2} = 4050 \text{ psi} \quad \Leftarrow$$

Finally, the maximum shear stresses and associated normal stresses are shown on the stress element of Fig. 7-14d.

As an alternative approach to finding the maximum shear stresses, we can use Eq. (7-20) to determine the two values of the angles $\theta_s$, and then we can use the second transformation equation (Eq. 7-4b) to obtain the corresponding shear stresses.

# 7.4 MOHR'S CIRCLE FOR PLANE STRESS

The transformation equations for plane stress can be represented in graphical form by a plot known as **Mohr's circle**. This graphical representation is extremely useful because it enables you to visualize the relationships between the normal and shear stresses acting on various inclined planes at a point in a stressed body. It also provides a means for calculating principal stresses, maximum shear stresses, and stresses on inclined planes. Furthermore, Mohr's circle is valid not only for stresses but also for other quantities of a similar mathematical nature, including strains and moments of inertia.*

## Equations of Mohr's Circle

The equations of Mohr's circle can be derived from the transformation equations for plane stress in Eqs. (7-4a) and (7-4b). The two equations are repeated here, but with a slight rearrangement of the first equation:

$$\sigma_{x_1} - \frac{\sigma_x + \sigma_y}{2} = \frac{\sigma_x - \sigma_y}{2} \cos 2\theta + \tau_{xy} \sin 2\theta \quad \textbf{(7-29a)}$$

$$\tau_{x_1 y_1} = -\frac{\sigma_x - \sigma_y}{2} \sin 2\theta + \tau_{xy} \cos 2\theta \quad \textbf{(7-29b)}$$

From analytic geometry, we might recognize that these two equations are the equations of a circle in parametric form. The angle $2\theta$ is the parameter and the stresses $\sigma_{x_1}$ and $\tau_{x_1 y_1}$ are the coordinates. However, it is not necessary to recognize the nature of the equations at this stage—if we eliminate the parameter, the significance of the equations will become apparent.

To eliminate the parameter $2\theta$, we square both sides of each equation and then add the two equations. The equation that results is

$$\left(\sigma_{x_1} - \frac{\sigma_x + \sigma_y}{2}\right)^2 + \tau_{x_1 y_1}^2 = \left(\frac{\sigma_x - \sigma_y}{2}\right)^2 + \tau_{xy}^2 \quad \textbf{(7-30)}$$

This equation can be written in simpler form by using the following notation from Section 7.3 [see Eqs. (7-27) and (7-12), respectively]:

$$\sigma_{\text{aver}} = \frac{\sigma_x + \sigma_y}{2} \qquad R = \sqrt{\left(\frac{\sigma_x - \sigma_y}{2}\right)^2 + \tau_{xy}^2} \quad \textbf{(7-31a,b)}$$

Equation (7-30) now becomes

$$(\sigma_{x_1} - \sigma_{\text{aver}})^2 + \tau_{x_1 y_1}^2 = R^2 \quad \textbf{(7-32)}$$

---

*Mohr's circle is named after the famous German civil engineer Otto Christian Mohr (1835–1918), who developed the circle in 1882 (Ref. 7-4).

which is the equation of a circle in standard algebraic form. The coordinates are $\sigma_{x_1}$ and $\tau_{x_1 y_1}$, the radius is $R$, and the center of the circle has coordinates $\sigma_{x_1} = \sigma_{aver}$ and $\tau_{x_1 y_1} = 0$.

## Two Forms of Mohr's Circle

Mohr's circle can be plotted from Eqs. (7-29) and (7-32) in either of two forms. In the first form of Mohr's circle, we plot the normal stress $\sigma_{x_1}$ positive to the right and the shear stress $\tau_{x_1 y_1}$ positive downward, as shown in Fig. 7-15a. The advantage of plotting shear stresses positive downward is that the angle $2\theta$ on Mohr's circle will be positive when counterclockwise, which agrees with the positive direction of $2\theta$ in the derivation of the transformation equations (see Figs. 7-1 and 7-2).

In the second form of Mohr's circle, $\tau_{x_1 y_1}$ is plotted positive upward but the angle $2\theta$ is now positive clockwise (Fig. 7-15b), which is opposite to its usual positive direction.

Both forms of Mohr's circle are mathematically correct, and either one can be used. However, it is easier to visualize the orientation of the stress element if the positive direction of the angle $2\theta$ is the same in Mohr's circle as it is for the element itself. Furthermore, a counterclockwise rotation agrees with the customary right-hand rule for rotation.

Therefore, we will choose the first form of Mohr's circle (Fig. 7-15a) in which *positive shear stress is plotted downward and a positive angle $2\theta$ is plotted counterclockwise*.

## Fig. 7-15

Two forms of Mohr's circle: (a) $\tau_{x_1 y_1}$ is positive downward and the angle $2\theta$ is positive counterclockwise, and (b) $\tau_{x_1 y_1}$ is positive upward and the angle $2\theta$ is positive clockwise. (*Note:* The first form is used in this book.)

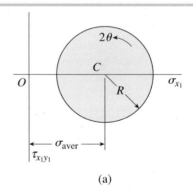

(a)

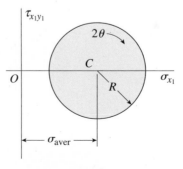

(b)

## Construction of Mohr's Circle

Mohr's circle can be constructed in a variety of ways, depending upon which stresses are known and which are to be found. For our immediate purpose, which is to show the basic properties of the circle, let us assume that we know the stresses $\sigma_x$, $\sigma_y$, and $\tau_{xy}$ acting on the $x$ and $y$ planes of an element in plane stress (Fig. 7-16a). As we will see, this information is sufficient to construct the circle. Then, with the circle drawn, we can determine the stresses $\sigma_{x_1}$, $\sigma_{y_1}$, and $\tau_{x_1y_1}$ acting on an inclined element (Fig. 7-16b). We can also obtain the principal stresses and maximum shear stresses from the circle.

With $\sigma_x$, $\sigma_y$, and $\tau_{xy}$ known, the **procedure for constructing Mohr's circle** is as follows (see Fig. 7-16c):

1.  Draw a set of coordinate axes with $\sigma_{x_1}$ as abscissa (positive to the right) and $\tau_{x_1y_1}$ as ordinate (positive downward).

2.  Locate the center $C$ of the circle at the point having coordinates $\sigma_{x_1} = \sigma_{aver}$ and $\tau_{x_1y_1} = 0$ [see Eqs. (7-31a) and (7-32)].

3.  Locate point $A$, representing the stress conditions on the $x$ face of the element shown in Fig. 7-16a, by plotting its coordinates $\sigma_{x_1} = \sigma_x$ and $\tau_{x_1y_1} = \tau_{xy}$. Note that point $A$ on the circle corresponds to $\theta = 0$. Also, note that the $x$ face of the element (Fig. 7-16a) is labeled "$A$" to show its correspondence with point $A$ on the circle.

4.  Locate point $B$, representing the stress conditions on the $y$ face of the element shown in Fig. 7-16a, by plotting its coordinates $\sigma_{x_1} = \sigma_y$ and $\tau_{x_1y_1} = -\tau_{xy}$. Note that point $B$ on the circle corresponds to $\theta = 90°$.

**Fig. 7-16**

Construction of Mohr's circle for plane stress

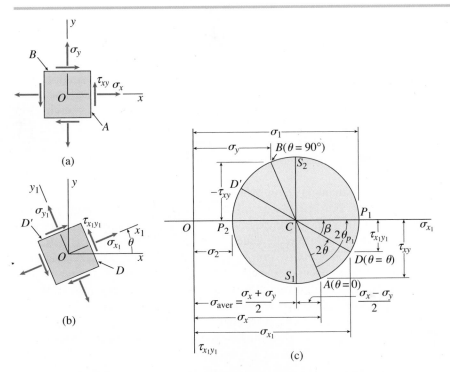

(a)

(b)

(c)

In addition, the $y$ face of the element (Fig. 7-16a) is labeled "$B$" to show its correspondence with point $B$ on the circle.

5. Draw a line from point $A$ to point $B$. This line is a diameter of the circle and passes through the center $C$. Points $A$ and $B$, representing the stresses on planes at 90° to each other (Fig. 7-16a), are at opposite ends of the diameter (and therefore are 180° apart on the circle).

6. Using point $C$ as the center, draw Mohr's circle through points $A$ and $B$. The circle drawn in this manner has radius $R$ (Eq. 7-31b), as shown in the next paragraph.

Now that we have drawn the circle, we can verify by geometry that lines $CA$ and $CB$ are radii and have lengths equal to $R$. We note that the abscissas of points $C$ and $A$ are $(\sigma_x + \sigma_y)/2$ and $\sigma_x$, respectively. The difference in these abscissas is $(\sigma_x - \sigma_y)/2$, as dimensioned in the figure. Also, the ordinate to point $A$ is $\tau_{xy}$. Therefore, line $CA$ is the hypotenuse of a right triangle having one side of length $(\sigma_x - \sigma_y)/2$ and the other side of length $\tau_{xy}$. Taking the square root of the sum of the squares of these two sides gives the radius $R$:

$$R = \sqrt{\left(\frac{\sigma_x - \sigma_y}{2}\right)^2 + \tau_{xy}^2}$$

which is the same as Eq. (7-31b). By a similar procedure, we can show that the length of line $CB$ is also equal to the radius $R$ of the circle.

## Stresses on an Inclined Element

Now we will consider the stresses $\sigma_{x_1}$, $\sigma_{y_1}$, and $\tau_{x_1y_1}$ acting on the faces of a plane-stress element oriented at an angle $\theta$ from the $x$ axis (Fig. 7-16b). If the angle $\theta$ is known, these stresses can be determined from Mohr's circle. The procedure is as follows.

On the circle (Fig. 7-16c), we measure an angle $2\theta$ counterclockwise from radius $CA$, because point $A$ corresponds to $\theta = 0$ and is the reference point from which we measure angles. The angle $2\theta$ locates point $D$ on the circle, which (as shown in the next paragraph) has coordinates $\sigma_{x_1}$ and $\tau_{x_1y_1}$. Therefore, point $D$ represents the stresses on the $x_1$ face of the element of Fig. 7-16b. Consequently, this face of the element is labeled "$D$" in Fig. 7-16b.

Note that an angle $2\theta$ on Mohr's circle corresponds to an angle $\theta$ on a stress element. For instance, point $D$ on the circle is at an angle $2\theta$ from point $A$, but the $x_1$ face of the element shown in Fig. 7-16b (the face labeled "$D$") is at an angle $\theta$ from the $x$ face of the element shown in Fig. 7-16a (the face labeled "$A$"). Similarly, points $A$ and $B$ are 180° apart on the circle, but the corresponding faces of the element (Fig. 7-16a) are 90° apart.

To show that the coordinates $\sigma_{x_1}$ and $\tau_{x_1y_1}$ of point $D$ on the circle are indeed given by the stress-transformation equations given in Eqs. (7-4a) and (7-4b), we again use the geometry of the circle. Let $\beta$ be the angle between the radial line $CD$ and the $\sigma_{x_1}$ axis. Then, from the geometry of the figure, we obtain the following expressions for the coordinates of point $D$:

Fig. 7-16 (Repeated)

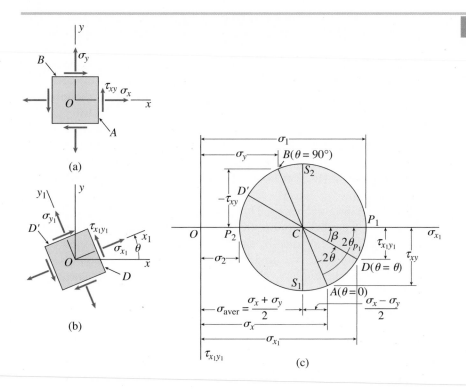

(a)

(b)

(c)

$$\sigma_{x_1} = \frac{\sigma_x + \sigma_y}{2} + R \cos \beta \qquad \tau_{x_1 y_1} = R \sin \beta \quad \textbf{(7-33a,b)}$$

Noting that the angle between the radius $CA$ and the horizontal axis is $2\theta + \beta$, we get

$$\cos(2\theta + \beta) = \frac{\sigma_x - \sigma_y}{2R} \qquad \sin(2\theta + \beta) = \frac{\tau_{xy}}{R}$$

Expanding the cosine and sine expressions (see Appendix D) gives

$$\cos 2\theta \cos \beta - \sin 2\theta \sin \beta = \frac{\sigma_x - \sigma_y}{2R} \qquad \textbf{(7-34a)}$$

$$\sin 2\theta \cos \beta + \cos 2\theta \sin \beta = \frac{\tau_{xy}}{R} \qquad \textbf{(7-34b)}$$

Multiplying the first of these equations by $\cos 2\theta$ and the second by $\sin 2\theta$ and then adding, we obtain

$$\cos \beta = \frac{1}{R} \left( \frac{\sigma_x - \sigma_y}{2} \cos 2\theta + \tau_{xy} \sin 2\theta \right) \qquad \textbf{(7-34c)}$$

Also, multiplying Eq. (7-34a) by $\sin 2\theta$ and Eq. (7-34b) by $\cos 2\theta$ and then subtracting, we get

$$\sin \beta = \frac{1}{R} \left( -\frac{\sigma_x - \sigma_y}{2} \sin 2\theta + \tau_{xy} \cos 2\theta \right) \qquad \textbf{(7-34d)}$$

When these expressions for $\cos \beta$ and $\sin \beta$ are substituted into Eqs. (7-33a) and (7-33b), we obtain the stress-transformation equations for $\sigma_{x_1}$ and $\tau_{x_1 y_1}$ (Eqs. 7-4a and 7-4b). Thus, we have shown that point $D$ on Mohr's circle, defined by the angle $2\theta$, represents the stress conditions on the $x_1$ face of the stress element defined by the angle $\theta$ (Fig. 7-16b).

Point $D'$, which is diametrically opposite point $D$ on the circle, is located by an angle $2\theta$ (measured from line $CA$) that is 180° greater than the angle $2\theta$ to point $D$. Therefore, point $D'$ on the circle represents the stresses on a face of the stress element (Fig. 7-16b) at 90° from the face represented by point $D$. Thus, point $D'$ on the circle gives the stresses $\sigma_{y_1}$ and $-\tau_{x_1 y_1}$ on the $y_1$ face of the stress element (the face labeled "$D'$" in Fig. 7-16b).

From this discussion we see how the stresses represented by points on Mohr's circle are related to the stresses acting on an element. The stresses on an inclined plane defined by the angle $\theta$ (Fig. 7-16b) are found on the circle at the point where the angle from the reference point (point $A$) is $2\theta$. Thus, as we rotate the $x_1 y_1$ axes counterclockwise through an angle $\theta$ (Fig. 7-16b), the point on Mohr's circle corresponding to the $x_1$ face moves counterclockwise through an angle $2\theta$. Similarly, if we rotate the axes clockwise through an angle, the point on the circle moves clockwise through an angle twice as large.

## Principal Stresses

The determination of principal stresses is probably the most important application of Mohr's circle. Note that as we move around Mohr's circle (Fig. 7-16c), we encounter point $P_1$ where the normal stress reaches its algebraically largest value and the shear stress is zero. Hence, point $P_1$ represents a **principal stress** and a **principal plane**. The abscissa $\sigma_1$ of point $P_1$ gives the algebraically larger principal stress and its angle $2\theta_{p_1}$ from the reference point $A$ (where $\theta = 0$) gives the orientation of the principal plane. The other principal plane, associated with the algebraically smallest normal stress, is represented by point $P_2$, diametrically opposite point $P_1$.

From the geometry of the circle, we see that the algebraically larger principal stress is

$$\sigma_1 = OC + \overline{CP_1} = \frac{\sigma_x + \sigma_y}{2} + R$$

which, upon substitution of the expression for $R$ (Eq. 7-31b), agrees with the earlier equation for this stress (Eq. 7-14). In a similar manner, we can verify the expression for the algebraically smaller principal stress $\sigma_2$.

The principal angle $\theta_{p_1}$ between the $x$ axis (Fig. 7-16a) and the plane of the algebraically larger principal stress is one-half the angle $2\theta_{p_1}$, which is the angle on Mohr's circle between radii $CA$ and $CP_1$. The cosine and sine of the angle $2\theta_{p_1}$ can be obtained by inspection from the circle:

$$\cos 2\theta_{p_1} = \frac{\sigma_x - \sigma_y}{2R} \qquad \sin 2\theta_{p_1} = \frac{\tau_{xy}}{R}$$

These equations agree with Eqs. (7-18a) and (7-18b), and so once again we see that the geometry of the circle matches the equations derived earlier. On the circle, the angle $2\theta_{p_2}$ to the other principal point (point $P_2$) is 180° larger than $2\theta_{p_1}$; hence, $\theta_{p_2} = \theta_{p_1} + 90°$, as expected.

## Maximum Shear Stresses

Points $S_1$ and $S_2$, representing the planes of maximum positive and maximum negative shear stresses, respectively, are located at the bottom and top of Mohr's circle (Fig. 7-16c). These points are at angles $2\theta = 90°$ from points $P_1$ and $P_2$, which agrees with the fact that the planes of maximum shear stress are oriented at 45° to the principal planes.

The maximum shear stresses are numerically equal to the radius $R$ of the circle [compare Eq. (7-31b) for $R$ with Eq. (7-25) for $\tau_{max}$]. Also, the normal stresses on the planes of maximum shear stress are equal to the abscissa of point $C$, which is the average normal stress $\sigma_{aver}$ [see Eq. (7-31a)].

## Alternative Sign Convention for Shear Stresses

An alternative sign convention for shear stresses is sometimes used when constructing Mohr's circle. In this convention, the direction of a shear stress acting on an element of the material is indicated by the sense of the rotation that it tends to produce (Figs. 7-17a and b). If the shear stress $\tau$ tends to rotate the stress element clockwise, it is called a *clockwise shear stress*, and if it tends to rotate it counterclockwise, it is called a *counterclockwise stress*. Then, when constructing Mohr's circle, clockwise shear stresses are plotted upward and counterclockwise shear stresses are plotted downward (Fig. 7-17c).

It is important to realize that *the alternative sign convention produces a circle that is identical to the circle already described* (Fig. 7-16c). The reason is that a positive shear stress $\tau_{x_1y_1}$ is also a counterclockwise shear stress, and both are plotted downward. Also, a negative shear stress $\tau_{x_1y_1}$ is a clockwise shear stress, and both are plotted upward.

Thus, the alternative sign convention merely provides a different point of view. Instead of thinking of the vertical axis as having negative shear stresses plotted upward and positive shear stresses plotted

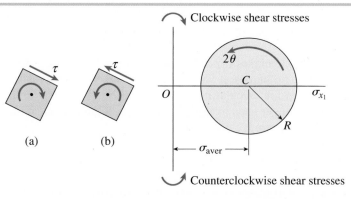

(a)  (b)  (c)

### Fig. 7-17

Alternative sign convention for shear stresses: (a) clockwise shear stress, (b) counterclockwise shear stress, and (c) axes for Mohr's circle. (Note that clockwise shear stresses are plotted upward and counterclockwise shear stresses are plotted downward.)

downward (which is a bit awkward), we can think of the vertical axis as having clockwise shear stresses plotted upward and counterclockwise shear stresses plotted downward (Fig. 7-17c).

## General Comments about the Circle

From the preceding discussions in this section, it is apparent that we can find the stresses acting on any inclined plane, as well as the principal stresses and maximum shear stresses, from Mohr's circle. However, only rotations of axes in the $xy$ plane (that is, rotations about the $z$ axis) are considered, and therefore *all stresses on Mohr's circle are in-plane stresses*.

For convenience, the circle of Fig. 7-16 was drawn with $\sigma_x$, $\sigma_y$, and $\tau_{xy}$ as positive stresses, but the same procedures may be followed if one or more of the stresses is negative. If one of the normal stresses is negative, part or all of the circle will be located to the left of the origin, as illustrated in Example 7-6 that follows.

Point $A$ in Fig. 7-16c, representing the stresses on the plane $\theta = 0$, may be situated anywhere around the circle. However, the angle $2\theta$ is always measured counterclockwise from the radius $CA$, regardless of where point $A$ is located.

In the special cases of *uniaxial stress*, *biaxial stress*, and *pure shear*, the construction of Mohr's circle is simpler than in the general case of plane stress. These special cases are illustrated in Example 7-4 and in Probs. 7.4-1 through 7.4-9.

Besides using Mohr's circle to obtain the stresses on inclined planes when the stresses on the $x$ and $y$ planes are known, we can also use the circle in the opposite manner. If we know the stresses $\sigma_{x_1}$, $\sigma_{y_1}$, and $\tau_{x_1y_1}$ acting on an inclined element oriented at a known angle $\theta$, we can easily construct the circle and determine the stresses $\sigma_x$, $\sigma_y$, and $\tau_{xy}$ for the angle $\theta = 0$. The procedure is to locate points $D$ and $D'$ from the known stresses and then draw the circle using line $DD'$ as a diameter. By measuring the angle $2\theta$ in a negative sense from radius $CD$, we can locate point $A$, corresponding to the $x$ face of the element. Then we can locate point $B$ by constructing a diameter from $A$. Finally, we can determine the coordinates of points $A$ and $B$, and thereby obtain the stresses acting on the element for which $\theta = 0$.

If desired, we can construct Mohr's circle to scale and measure values of stress from the drawing. However, it is usually preferable to obtain the stresses by numerical calculations, either directly from the various equations or by using trigonometry and the geometry of the circle.

Mohr's circle makes it possible to visualize the relationships between stresses acting on planes at various angles, and it also serves as a simple memory device for calculating stresses. Although many graphical techniques are no longer used in engineering work, Mohr's circle remains valuable because it provides a simple and clear picture of an otherwise complicated analysis.

Mohr's circle is also applicable to the transformations for plane? strain and moments of inertia of plane areas, because these quantities follow the same transformation laws as do stresses (see Sections 7.7, 12.8, and 12.9).

## • • • Example 7-4

**Fig. 7-18a**

Example 7-4: (a) Hydraulic
cylinder on construction
equipment (© Can Stock Photo
Inc./zoomzoom)

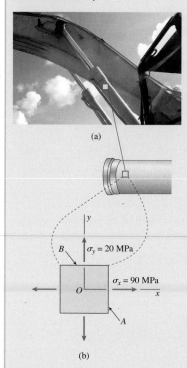

(a)

At a point on the surface of a hydraulic ram on a piece of construction
equipment (Fig. 7-18a), the material is subjected to biaxial stresses
$\sigma_x = 90$ MPa and $\sigma_y = 20$ MPa, as shown on the stress element of Fig. 7-18b.
Using Mohr's circle, determine the stresses acting on an element inclined at
an angle $\theta = 30°$. (Consider only the in-plane stresses, and show the results
on a sketch of a properly oriented element.)

### Solution

*Construction of Mohr's circle.* We begin by setting up the axes for the normal
and shear stresses, with $\sigma_{x_1}$ positive to the right and $\tau_{x_1 y_1}$ positive downward,
as shown in Fig. 7-18c. Then we place the center $C$ of the circle on the $\sigma_{x_1}$ axis
at the point where the stress equals the average normal stress (Eq. 7-31a):

$$\sigma_{aver} = \frac{\sigma_x + \sigma_y}{2} = \frac{90 \text{ MPa} + 20 \text{ MPa}}{2} = 55 \text{ MPa}$$

Point $A$, representing the stresses on the $x$ face of the element ($\theta = 0$), has
coordinates

$$\sigma_{x_1} = 90 \text{ MPa} \qquad \tau_{x_1 y_1} = 0$$

Similarly, the coordinates of point $B$, representing the stresses on the $y$ face
($\theta = 90°$), are

$$\sigma_{x_1} = 20 \text{ MPa} \qquad \tau_{x_1 y_1} = 0$$

**Fig. 7-18b,c**

Example 7-4: (b) Element on
hydraulic ram in plane stress,
and (c) the corresponding
Mohr's circle (*Note:* All stresses
on the circle have units of
MPa.)

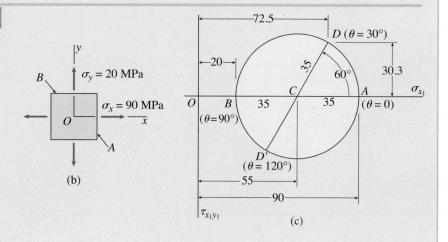

(b)

(c)

*Continues* ➡

**Example 7-4 -** *Continued*

Now we draw the circle through points $A$ and $B$ with center at $C$ and radius $R$ [see Eq. (7-31b)] equal to

$$R = \sqrt{\left(\frac{\sigma_x - \sigma_y}{2}\right)^2 + \tau_{xy}^2} = \sqrt{\left(\frac{90 \text{ MPa} - 20 \text{ MPa}}{2}\right)^2 + 0} = 35 \text{ MPa}$$

*Stresses on an element inclined at $\theta = 30°$.* The stresses acting on a plane oriented at an angle $\theta = 30°$ are given by the coordinates of point $D$, which is at an angle $2\theta = 60°$ from point $A$ (Fig. 7-18c). By inspection of the circle, we see that the coordinates of point $D$ are

(Point $D$)  $\sigma_{x_1} = \sigma_{\text{aver}} + R \cos 60°$

$= 55 \text{ MPa} + (35 \text{ MPa})(\cos 60°) = 72.5 \text{ MPa}$  ⬅

$\tau_{x_1 y_1} = -R \sin 60° = -(35 \text{ MPa})(\sin 60°) = -30.3 \text{ MPa}$  ⬅

In a similar manner, we can find the stresses represented by point $D'$, which corresponds to an angle $\theta = 120°$ (or $2\theta = 240°$):

(Point $D'$)  $\sigma_{x_1} = \sigma_{\text{aver}} - R \cos 60°$

$= 55 \text{ MPa} - (35 \text{ MPa})(\cos 60°) = 37.5 \text{ MPa}$  ⬅

$\tau_{x_1 y_1} = R \sin 60° = (35 \text{ MPa})(\sin 60°) = 30.3 \text{ MPa}$  ⬅

These results are shown in Fig. 7-19 on a sketch of an element oriented at an angle $\theta = 30°$, with all stresses shown in their true directions. Note that the sum of the normal stresses on the inclined element is equal to $\sigma_x + \sigma_y$, or 110 MPa.

**Fig. 7-19**

Example 7-4 (continued):
Stresses acting on an element
oriented at an angle $\theta = 30°$

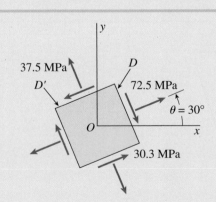

## • • • Example 7-5

### Fig. 7-20a

Example 7-5: (a) Oil drilling pumps (Can Stock Photo Inc. ssvaphoto)

(a)

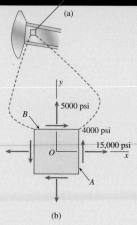

(b)

An element in plane stress on the surface of an oil-drilling pump arm (Fig. 7-20a) is subjected to stresses $\sigma_x = 15{,}000$ psi, $\sigma_y = 5000$ psi, and $\tau_{xy} = 4000$ psi, as shown in Fig. 7-20b.

Using Mohr's circle, determine the following quantities: (a) the stresses acting on an element inclined at an angle $\theta = 40°$, (b) the principal stresses, and (c) the maximum shear stresses. (Consider only the in-plane stresses, and show all results on sketches of properly oriented elements.)

### Solution

*Construction of Mohr's circle.* The first step in the solution is to set up the axes for Mohr's circle, with $\sigma_{x_1}$ positive to the right and $\tau_{x_1y_1}$ positive downward (Fig. 7-20c). The center $C$ of the circle is located on the $\sigma_{x_1}$ axis at the point where $\sigma_{x_1}$ equals the average normal stress (Eq. 7-31a):

$$\sigma_{aver} = \frac{\sigma_x + \sigma_y}{2} = \frac{15{,}000 \text{ psi} + 5000 \text{ psi}}{2} = 10{,}000 \text{ psi}$$

Point $A$, representing the stresses on the $x$ face of the element ($\theta = 0$), has coordinates

$$\sigma_{x_1} = 15{,}000 \text{ psi} \qquad \tau_{x_1y_1} = 4000 \text{ psi}$$

Similarly, the coordinates of point $B$, representing the stresses on the $y$ face ($\theta = 90°$) are

$$\sigma_{x_1} = 5000 \text{ psi} \qquad \tau_{x_1y_1} = -4000 \text{ psi}$$

### Fig. 7-20b,c

Example 7-5: (b) Element in plane stress, and (c) the corresponding Mohr's circle (*Note:* All stresses on the circle have units of psi.)

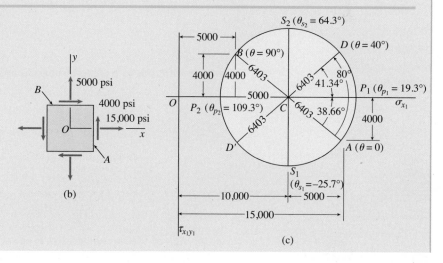

(b)

(c)

Continues ➡

• • • **Example 7-5 -** *Continued*

**Fig. 7-21**

Example 7-5 (continued):
(a) Stresses acting on an element oriented at $\theta = 40°$,
(b) principal stresses, and
(c) maximum shear stresses

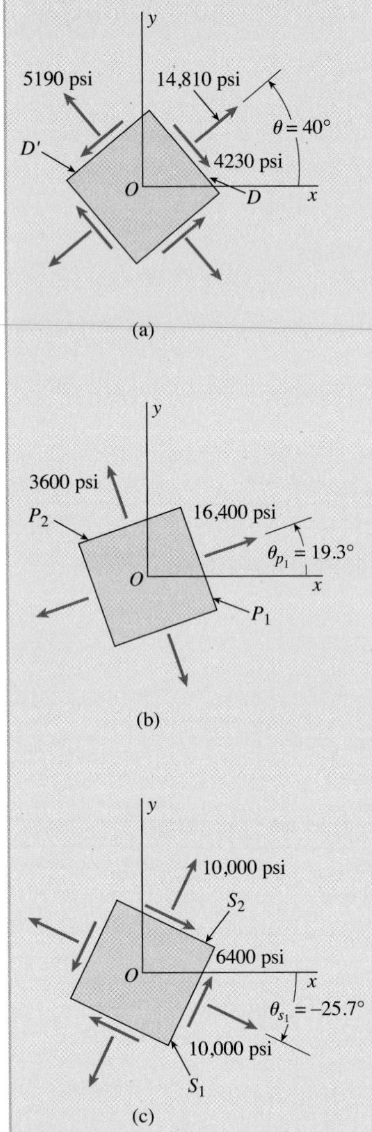

(a)

(b)

(c)

The circle is now drawn through points $A$ and $B$ with center at $C$. The radius of the circle, from Eq. (7-31b), is

$$R = \sqrt{\left(\frac{\sigma_x - \sigma_y}{2}\right)^2 + \tau_{xy}^2}$$

$$= \sqrt{\left(\frac{15{,}000 \text{ psi} - 5000 \text{ psi}}{2}\right)^2 + (4000 \text{ psi})^2} = 6403 \text{ psi}$$

(a) *Stresses on an element inclined at $\theta = 40°$.* The stresses acting on a plane oriented at an angle $\theta = 40°$ are given by the coordinates of point $D$, which is at an angle $2\theta = 80°$ from point $A$ (Fig. 7-20c). To evaluate these coordinates, we need to know the angle between line $CD$ and the $\sigma_{x_1}$ axis (that is, angle $DCP_1$), which in turn requires that we know the angle between line $CA$ and the $\sigma_{x_1}$ axis (angle $ACP_1$). These angles are found from the geometry of the circle, as follows:

$$\tan \overline{ACP_1} = \frac{4000 \text{ psi}}{5000 \text{ psi}} = 0.8 \qquad \overline{ACP_1} = 38.66°$$

$$\overline{DCP_1} = 80° - \overline{ACP_1} = 80° - 38.66° = 41.34°$$

Knowing these angles, we can determine the coordinates of point $D$ directly from the Fig. 7-21a:

(Point $D$) $\quad \sigma_{x_1} = 10{,}000 \text{ psi} + (6403 \text{ psi})(\cos 41.34°) = 14{,}810 \text{ psi}$ ◀

$\qquad\qquad \tau_{x_1 y_1} = -(6403 \text{ psi})(\sin 41.34°) = -4230 \text{ psi}$ ◀

In an analogous manner, we can find the stresses represented by point $D'$, which corresponds to a plane inclined at an angle $\theta = 130°$ (or $2\theta = 260°$):

(Point $D'$) $\quad \sigma_{x_1} = 10{,}000 \text{ psi} - (6403 \text{ psi})(\cos 41.34°) = 5190 \text{ psi}$ ◀

$\qquad\qquad \tau_{x_1 y_1} = (6403 \text{ psi})(\sin 41.34°) = 4230 \text{ psi}$ ◀

These stresses are shown in Fig. 7-21a on a sketch of an element oriented at an angle $\theta = 40°$ (all stresses are shown in their true directions). Also, note that the sum of the normal stresses is equal to $\sigma_x + \sigma_y$, or 20,000 psi.

(b) *Principal stresses*. The principal stresses are represented by points $P_1$ and $P_2$ on Mohr's circle (Fig. 7-20c). The algebraically larger principal stress (point $P_1$) is

$$\sigma_1 = 10{,}000 \text{ psi} + 6400 \text{ psi} = 16{,}400 \text{ psi}$$

as seen by inspection of the circle. The angle $2\theta_{p_1}$ to point $P_1$ from point $A$ is the angle $ACP_1$ on the circle, that is,

$$\overline{ACP_1} = 2\theta_{p_1} = 38.66° \qquad \theta_{p_1} = 19.3°$$

Thus, the plane of the algebraically larger principal stress is oriented at an angle $\theta_{p_1} = 19.3°$, as shown in Fig. 7-21b.

The algebraically smaller principal stress (represented by point $P_2$) is obtained from the circle in a similar manner:

$$\sigma_2 = 10{,}000 \text{ psi} - 6400 \text{ psi} = 3600 \text{ psi}$$

The angle $2\theta_{p_2}$ to point $P_2$ on the circle is $38.66° + 180° = 218.66°$; thus, the second principal plane is defined by the angle $\theta_{p_2} = 109.3°$. The principal stresses and principal planes are shown in Fig. 7-21b, and again we note that the sum of the normal stresses is equal to 20,000 psi.

(c) *Maximum shear stresses*. The maximum shear stresses are represented by points $S_1$ and $S_2$ on Mohr's circle; therefore, the maximum in-plane shear stress (equal to the radius of the circle) is

$$\tau_{max} = 6400 \text{ psi}$$

The angle $ACS_1$ from point $A$ to point $S_1$ is $90° - 38.66° = 51.34°$, and therefore the angle $2\theta_{s_1}$ for point $S_1$ is

$$2\theta_{s_1} = -51.34°$$

This angle is negative because it is measured clockwise on the circle. The corresponding angle $\theta_{s_1}$ to the plane of the maximum positive shear stress is one-half that value, or $\theta_{s_1} = -25.7°$, as shown in Figs. 7-20c and 7-21c. The maximum negative shear stress (point $S_2$ on the circle) has the same numerical value as the maximum positive stress (6,400 psi).

The normal stresses acting on the planes of maximum shear stress are equal to $\sigma_{aver}$, which is the abscissa of the center $C$ of the circle (10,000 psi). These stresses are also shown in Fig. 7-21c. Note that the planes of maximum shear stress are oriented at 45° to the principal planes.

At a point on the surface of a metal-working lathe the stresses are $\sigma_x = -50$ MPa, $\sigma_y = 10$ MPa, and $\tau_{xy} = -40$ MPa, as shown in Fig. 7-22a.

Using Mohr's circle, determine the following quantities: (a) the stresses acting on an element inclined at an angle $\theta = 45°$, (b) the principal stresses, and (c) the maximum shear stresses. (Consider only the in-plane stresses, and show all results on sketches of properly oriented elements.)

**Solution**

*Construction of Mohr's circle.* The axes for the normal and shear stresses are shown in Fig. 7-22b, with $\sigma_{x_1}$ positive to the right and $\tau_{x_1y_1}$ positive downward. The center $C$ of the circle is located on the $\sigma_{x_1}$ axis at the point where the stress equals the average normal stress (Eq. 7-31a):

$$\sigma_{aver} = \frac{\sigma_x + \sigma_y}{2} = \frac{-50 \text{ MPa} + 10 \text{ MPa}}{2} = -20 \text{ MPa}$$

Point $A$, representing the stresses on the $x$ face of the element ($\theta = 0$), has coordinates

$$\sigma_{x_1} = -50 \text{ MPa} \qquad \tau_{x_1y_1} = -40 \text{ MPa}$$

Similarly, the coordinates of point $B$, representing the stresses on the $y$ face ($\theta = 90°$), are

$$\sigma_{x_1} = 10 \text{ MPa} \qquad \tau_{x_1y_1} = 40 \text{ MPa}$$

The circle is now drawn through points $A$ and $B$ with center at $C$ and radius $R$ (from Eq. 7-31b) equal to

$$R = \sqrt{\left(\frac{\sigma_x - \sigma_y}{2}\right)^2 + \tau_{xy}^2}$$

$$= \sqrt{\left(\frac{-50 \text{ MPa} - 10 \text{ MPa}}{2}\right)^2 + (-40 \text{ MPa})^2} = 50 \text{ MPa}$$

**Fig. 7-22**

Example 7-6: (a) Element in plane stress, and (b) the corresponding Mohr's circle. (*Note:* All stresses on the circle have units of MPa.)

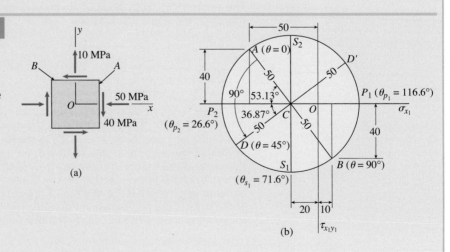

(a)

(b)

(a) *Stresses on an element inclined at $\theta = 45°$.* The stresses acting on a plane oriented at an angle $\theta = 45°$ are given by the coordinates of point $D$, which is at an angle $2\theta = 90°$ from point $A$ (Fig. 7-22b). To evaluate these coordinates, we need to know the angle between line $CD$ and the negative $\sigma_{x_1}$ axis (that is, angle $DCP_2$), which in turn requires that we know the angle between line $CA$ and the negative $\sigma_{x_1}$ axis (angle $ACP_2$). These angles are found from the geometry of the circle as follows:

$$\tan \overline{ACP_2} = \frac{40 \text{ MPa}}{30 \text{ MPa}} = \frac{4}{3} \qquad \overline{ACP_2} = 53.13°$$

$$\overline{DCP_2} = 90° - \overline{ACP_2} = 90° - 53.13° = 36.87°$$

Knowing these angles, we can obtain the coordinates of point $D$ directly from Fig. 7-23a:

(Point $D$)    $\sigma_{x_1} = -20 \text{ MPa} - (50 \text{ MPa})(\cos 36.87°) = -60 \text{ MPa}$   ⬅

$\tau_{x_1 y_1} = (50 \text{ MPa})(\sin 36.87°) = 30 \text{ MPa}$   ⬅

In an analogous manner, we can find the stresses represented by point $D'$, which corresponds to a plane inclined at an angle $\theta = 135°$ (or $2\theta = 270°$):

(Point $D'$)    $\sigma_{x_1} = -20 \text{ MPa} + (50 \text{ MPa})(\cos 36.87°) = 20 \text{ MPa}$   ⬅

$\tau_{x_1 y_1} = (-50 \text{ MPa})(\sin 36.87°) = -30 \text{ MPa}$   ⬅

These stresses are shown in Fig. 7-23a on a sketch of an element oriented at an angle $\theta = 45°$ (all stresses are shown in their true directions). Also, note that the sum of the normal stresses is equal to $\sigma_x + \sigma_y$, or $-40$ MPa.

**Fig. 7-23**

Example 7-6 (continued):
(a) Stresses acting on an element oriented at $\theta = 45°$,
(b) principal stresses, and
(c) maximum shear stresses

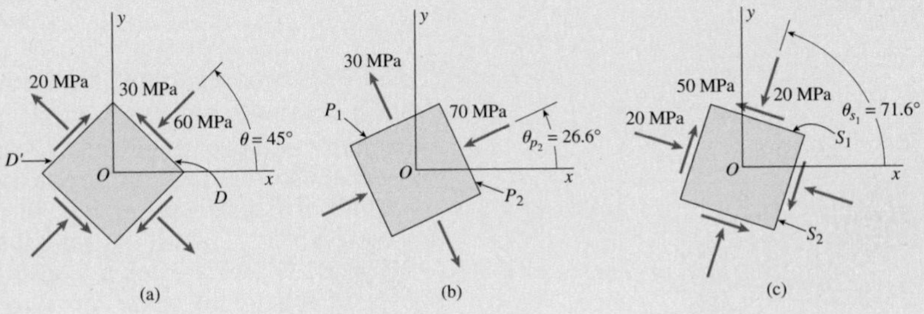

(a)
(b)
(c)

*Continues* ⮕

**Example 7-6 -** *Continued*

(b) *Principal stresses.* The principal stresses are represented by points $P_1$ and $P_2$ on Mohr's circle. The algebraically larger principal stress (represented by point $P_1$) is

$$\sigma_1 = -20 \text{ MPa} + 50 \text{ MPa} = 30 \text{ MPa}$$

as seen by inspection of the circle. The angle $2\theta_{p_1}$ to point $P_1$ from point $A$ is the angle $ACP_1$ measured counterclockwise on the circle, that is,

$$\overline{ACP_1} = 2\theta_{p_1} = 53.13° + 180° = 233.13° \qquad \theta_{p_1} = 116.6°$$

Thus, the plane of the algebraically larger principal stress is oriented at an angle $\theta_{p_1} = 116.6°$.

The algebraically smaller principal stress (point $P_2$) is obtained from the circle in a similar manner:

$$\sigma_2 = -20 \text{ MPa} - 50 \text{ MPa} = -70 \text{ MPa}$$

The angle $2\theta_{p_2}$ to point $P_2$ on the circle is 53.13°; thus, the second principal plane is defined by the angle $\theta_{p_2} = 26.6°$.

The principal stresses and principal planes are shown in Fig. 7-23b, and again we note that the sum of the normal stresses is equal to $\sigma_x + \sigma_y$, or $-40$ MPa.

(c) *Maximum shear stresses.* The maximum positive and negative shear stresses are represented by points $S_1$ and $S_2$ on Mohr's circle (Fig. 7-22b). Their magnitudes, equal to the radius of the circle, are

$$\tau_{max} = 50 \text{ MPa}$$

The angle $ACS_1$ from point $A$ to point $S_1$ is $90° + 53.13° = 143.13°$, and therefore the angle $2\theta_{s_1}$ for point $S_1$ is

$$2\theta_{s_1} = 143.13°$$

The corresponding angle $\theta_{s_1}$ to the plane of the maximum positive shear stress is one-half that value, or $\theta_{s_1} = 71.6°$, as shown in Fig. 7-23c. The maximum negative shear stress (point $S_2$ on the circle) has the same numerical value as the positive stress (50 MPa).

The normal stresses acting on the planes of maximum shear stress are equal to $\sigma_{aver}$, which is the coordinate of the center $C$ of the circle ($-20$ MPa). These stresses are also shown in Fig. 7-23c. Note that the planes of maximum shear stress are oriented at 45° to the principal planes.

# 7.5 HOOKE'S LAW FOR PLANE STRESS

The stresses acting on inclined planes when the material is subjected to plane stress (Fig. 7-24) were discussed in Sections 7.2, 7.3, and 7.4. The stress-transformation equations derived in those discussions were obtained solely from equilibrium, and therefore the properties of the materials were not needed. Now, in this section, we will investigate the *strains* in the material, which means that the material properties must be considered. However, we will limit our discussion to materials that meet two important conditions: first, *the material is uniform throughout the body and has the same properties in all directions* (homogeneous and isotropic material), and second, *the material follows Hooke's law* (linearly elastic material). Under these conditions, we can readily obtain the relationships between the stresses and strains in the body.

Let us begin by considering the **normal strains** $\varepsilon_x$, $\varepsilon_y$, and $\varepsilon_z$ in plane stress. The effects of these strains are pictured in Fig. 7-25, which shows the changes in dimensions of a small element having edges of lengths $a$, $b$, and $c$. All three strains are shown positive (elongation) in the figure. The strains can be expressed in terms of the stresses (Fig. 7-24) by superimposing the effects of the individual stresses.

For instance, the strain $\varepsilon_x$ in the $x$ direction due to the stress $\sigma_x$ is equal to $\sigma_x/E$, where $E$ is the modulus of elasticity. Also, the strain $\varepsilon_x$ due to the stress $\sigma_y$ is equal to $-v\sigma_y/E$, where $v$ is Poisson's ratio (see Section 1.6). Of course, the shear stress $\tau_{xy}$ produces no normal strains in the $x$, $y$, or $z$ directions. Thus, the resultant strain in the $x$ direction is

$$\varepsilon_x = \frac{1}{E}(\sigma_x - v\sigma_y) \qquad \textbf{(7-35a)}$$

In a similar manner, we obtain the strains in the $y$ and $z$ directions:

$$\varepsilon_y = \frac{1}{E}(\sigma_y - v\sigma_x) \qquad \varepsilon_z = -\frac{v}{E}(\sigma_x + \sigma_y) \qquad \textbf{(7-35b,c)}$$

These equations may be used to find the normal strains (in plane stress) when the stresses are known.

The shear stress $\tau_{xy}$ (Fig. 7-24) causes a distortion of the element such that each $z$ face becomes a rhombus (Fig. 7-26). The **shear strain** $\gamma_{xy}$ is the decrease in angle between the $x$ and $y$ faces of the element and is related to the shear stress by Hooke's law in shear, as follows:

$$\gamma_{xy} = \frac{\tau_{xy}}{G} \qquad \textbf{(7-36)}$$

where $G$ is the shear modulus of elasticity. Note that the normal stresses $\sigma_x$ and $\sigma_y$ have no effect on the shear strain $\gamma_{xy}$. Consequently, Eqs. (7-35) and (7-36) give the strains (in plane stress) when all stresses ($\sigma_x$, $\sigma_y$, and $\tau_{xy}$) act simultaneously.

The first two equations (Eqs. 7-35a and 7-35b) give the strains $\varepsilon_x$ and $\varepsilon_y$ in terms of the stresses. These equations can be solved simultaneously for the stresses in terms of the strains:

$$\sigma_x = \frac{E}{1 - v^2}(\varepsilon_x + v\varepsilon_y) \qquad \sigma_y = \frac{E}{1 - v^2}(\varepsilon_y + v\varepsilon_x) \qquad \textbf{(7-37a,b)}$$

**Fig. 7-24**

Element of material in plane stress ($\sigma_z = 0$)

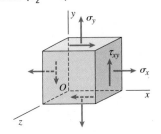

**Fig. 7-25**

Element of material subjected to normal strains $\varepsilon_x$, $\varepsilon_y$, and $\varepsilon_z$

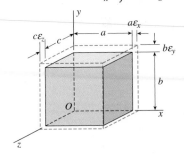

**Fig. 7-26**

Shear strain $\gamma_{xy}$

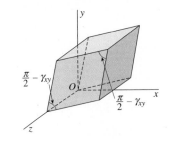

In addition, we have the following equation for the shear stress in terms of the shear strain:

$$\tau_{xy} = G\gamma_{xy} \tag{7-38}$$

Equations (7-37) and (7-38) may be used to find the stresses (in plane stress) when the strains are known. Of course, the normal stress $\sigma_z$ in the $z$ direction is equal to zero.

Equations (7-35) through (7-38) are known collectively as **Hooke's law for plane stress**. They contain three material constants ($E$, $G$, and $v$), but only two are independent because of the relationship

$$G = \frac{E}{2(1 + v)} \tag{7-39}$$

which was derived previously in Section 3.6.

## Special Cases of Hooke's Law

In the special case of **biaxial stress** (Fig. 7-11b), we have $\tau_{xy} = 0$, and therefore Hooke's law for plane stress simplifies to

$$\varepsilon_x = \frac{1}{E}(\sigma_x - v\sigma_y) \qquad \varepsilon_y = \frac{1}{E}(\sigma_y - v\sigma_x)$$

$$\varepsilon_z = -\frac{v}{E}(\sigma_x + \sigma_y) \tag{7-40a,b,c}$$

$$\sigma_x = \frac{E}{1 - v^2}(\varepsilon_x + v\varepsilon_y) \qquad \sigma_y = \frac{E}{1 - v^2}(\varepsilon_y + v\varepsilon_x) \tag{7-41a,b}$$

These equations are the same as Eqs. (7-35) and (7-37) because the effects of normal and shear stresses are independent of each other.

For **uniaxial stress**, with $\sigma_y = 0$ (Fig. 7-11a), the equations of Hooke's law simplify even further:

$$\varepsilon_x = \frac{\sigma_x}{E} \qquad \varepsilon_y = \varepsilon_z = -\frac{v\sigma_x}{E} \qquad \sigma_x = E\varepsilon_x \tag{7-42a,b,c}$$

Finally, we consider **pure shear** (Fig. 7-12a), which means that $\sigma_x = \sigma_y = 0$. Then we obtain

$$\varepsilon_x = \varepsilon_y = \varepsilon_z = 0 \qquad \gamma_{xy} = \frac{\tau_{xy}}{G} \tag{7-43a,b}$$

In all three of these special cases, the normal stress $\sigma_z$ is equal to zero.

## Volume Change

When a solid object undergoes strains, both its dimensions and its volume will change. The change in volume can be determined if the normal strains in three perpendicular directions are known. To show how this is accomplished, let us again consider the small element of material shown

**Fig. 7-11 (Repeated)**

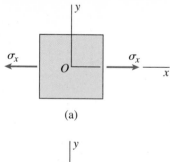

(a)

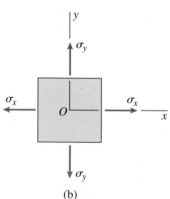

(b)

in Fig. 7-25. The original element is a rectangular parallelepiped having sides of lengths $a$, $b$, and $c$ in the $x$, $y$, and $z$ directions, respectively. The strains $\varepsilon_x$, $\varepsilon_y$, and $\varepsilon_z$ produce the changes in dimensions shown by the dashed lines. Thus, the increases in the lengths of the sides are $a\varepsilon_x$, $b\varepsilon_y$, and $c\varepsilon_z$.

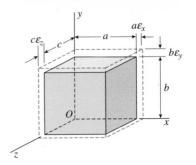

**Fig. 7-25 (Repeated)**

The original volume of the element is

$$V_0 = abc \qquad \textbf{(7-44a)}$$

and its final volume is

$$
\begin{aligned}
V_1 &= (a + a\varepsilon_x)(b + b\varepsilon_y)(c + c\varepsilon_z) \\
&= abc(1 + \varepsilon_x)(1 + \varepsilon_y)(1 + \varepsilon_z) \qquad \textbf{(7-44b)}
\end{aligned}
$$

By referring to Eq. (7-44a), we can express the final volume of the element (Eq. 7-44b) in the form

$$V_1 = V_0(1 + \varepsilon_x)(1 + \varepsilon_y)(1 + \varepsilon_z) \qquad \textbf{(7-45a)}$$

Upon expanding the terms on the right-hand side, we obtain the following equivalent expression:

$$V_1 = V_0(1 + \varepsilon_x + \varepsilon_y + \varepsilon_z + \varepsilon_x\varepsilon_y + \varepsilon_x\varepsilon_z + \varepsilon_y\varepsilon_z + \varepsilon_x\varepsilon_y\varepsilon_z) \qquad \textbf{(7-45b)}$$

The preceding equations for $V_1$ are valid for both large and small strains.

If we now limit our discussion to structures having only very small strains (as is usually the case), we can disregard the terms in Eq. (7-45b) that consist of products of small strains. Such products are themselves small in comparison to the individual strains $\varepsilon_x$, $\varepsilon_y$, and $\varepsilon_z$. Then the expression for the final volume simplifies to

$$V_1 = V_0(1 + \varepsilon_x + \varepsilon_y + \varepsilon_z) \qquad \textbf{(7-46)}$$

and the **volume change** is

$$\Delta V = V_1 - V_0 = V_0(\varepsilon_x + \varepsilon_y + \varepsilon_z) \qquad \textbf{(7-47)}$$

This expression can be used for any volume of material *provided the strains are small and remain constant throughout the volume*. Note also that the material does not have to follow Hooke's law. Furthermore, the expression is not limited to plane stress, but is valid for any stress conditions. (As a final note, we should mention that shear strains produce no change in volume.)

The **unit volume change** $e$, also known as the **dilatation**, is defined as the change in volume divided by the original volume; thus,

$$e = \frac{\Delta V}{V_0} = \varepsilon_x + \varepsilon_y + \varepsilon_z \qquad \textbf{(7-48)}$$

By applying this equation to a differential element of volume and then integrating, we can obtain the change in volume of a body even when the normal strains vary throughout the body.

The preceding equations for volume changes apply to both tensile and compressive strains, inasmuch as the strains $\varepsilon_x$, $\varepsilon_y$, and $\varepsilon_z$ are algebraic quantities (positive for elongation and negative for shortening). With this sign convention, positive values for $\Delta V$ and $e$ represent increases in volume, and negative values represent decreases.

Let us now return to materials that follow **Hooke's law** and are subjected only to **plane stress** (Fig. 7-24). In this case the strains $\varepsilon_x$, $\varepsilon_y$, and $\varepsilon_z$ are given by Eqs. (7-35a, b, and c). Substituting those relationships into Eq. (7-48), we obtain the following expression for the unit volume change in terms of stresses:

$$e = \frac{\Delta V}{V_0} = \frac{1 - 2v}{E}(\sigma_x + \sigma_y) \qquad \textbf{(7-49)}$$

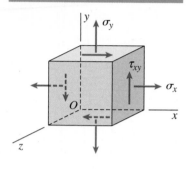

**Fig. 7-24 (Repeated)**

Note that this equation also applies to **biaxial stress**.

In the case of a prismatic bar in tension, that is, **uniaxial stress**, Eq. (7-49) simplifies to

$$e = \frac{\Delta V}{V_0} = \frac{\sigma_x}{E}(1 - 2v) \qquad \textbf{(7-50)}$$

From this equation we see that the maximum possible value of Poisson's ratio for common materials is 0.5, because a larger value means that the volume decreases when the material is in tension, which is contrary to ordinary physical behavior.

## Strain-Energy Density in Plane Stress

The strain-energy density $u$ is the strain energy stored in a unit volume of the material (see the discussions in Sections 2.7 and 3.9). For an element in plane stress, we can obtain the strain-energy density by referring to the elements pictured in Figs. 7-25 and 7-26. Because the normal and shear strains occur independently, we can add the strain energies from these two elements to obtain the total energy.

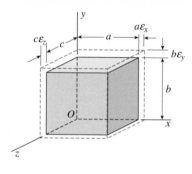

**Fig. 7-25 (Repeated)**

Let us begin by finding the strain energy associated with the normal strains (Fig. 7-25). Since the stress acting on the $x$ face of the element is $\sigma_x$ (see Fig. 7-24), we find that the force acting on the $x$ face of the element (Fig. 7-25) is equal to $\sigma_x bc$. Of course, as the loads are applied to the structure, this force increases gradually from zero to its maximum value. At the same time, the $x$ face of the element moves through the distance $a\varepsilon_x$. Therefore, the work done by this force is

$$\frac{1}{2}(\sigma_x bc)(a\varepsilon_x)$$

provided Hooke's law holds for the material. Similarly, the force $\sigma_y ac$ acting on the $y$ face does work equal to

$$\frac{1}{2}(\sigma_y ac)(b\varepsilon_y)$$

The sum of these two terms gives the strain energy stored in the element:

$$\frac{abc}{2}(\sigma_x\varepsilon_x + \sigma_y\varepsilon_y)$$

Thus, the strain-energy density (strain energy per unit volume) due to the normal stresses and strains is

$$u_1 = \frac{1}{2}(\sigma_x \varepsilon_x + \sigma_y \varepsilon_y) \qquad \textbf{(7-51a)}$$

The strain-energy density associated with the shear strains (Fig. 7-26) was evaluated previously in Section 3.9 (see Eq. (d) of that section):

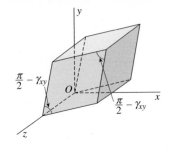

**Fig. 7-26 (Repeated)**

$$u_2 = \frac{\tau_{xy}\gamma_{xy}}{2} \qquad \textbf{(7-51b)}$$

By combining the strain-energy densities for the normal and shear strains, we obtain the following formula for the **strain-energy density in plane stress**:

$$u = \frac{1}{2}(\sigma_x \varepsilon_x + \sigma_y \varepsilon_y + \tau_{xy}\gamma_{xy}) \qquad \textbf{(7-52)}$$

Substituting for the strains from Eqs. (7-35) and (7-36), we obtain the strain-energy density in terms of stresses alone:

$$u = \frac{1}{2E}(\sigma_x^2 + \sigma_y^2 - 2\nu\sigma_x\sigma_y) + \frac{\tau_{xy}^2}{2G} \qquad \textbf{(7-53)}$$

In a similar manner, we can substitute for the stresses from Eqs. (7-37) and (7-38) and obtain the strain-energy density in terms of strains alone:

$$u = \frac{E}{2(1 - \nu^2)}(\varepsilon_x^2 + \varepsilon_y^2 + 2\nu\varepsilon_x\varepsilon_y) + \frac{G\gamma_{xy}^2}{2} \qquad \textbf{(7-54)}$$

To obtain the strain-energy density in the special case of **biaxial stress**, we simply drop the shear terms in Eqs. (7-52), (7-53), and (7-54).

For the special case of **uniaxial stress**, we substitute the following values

$$\sigma_y = 0 \qquad \tau_{xy} = 0 \qquad \varepsilon_y = -\nu\varepsilon_x \qquad \gamma_{xy} = 0$$

into Eqs. (7-53) and (7-54) and obtain, respectively,

$$u = \frac{\sigma_x^2}{2E} \qquad u = \frac{E\varepsilon_x^2}{2} \qquad \textbf{(7-55a,b)}$$

These equations agree with Eqs. (2-44a) and (2-44b) of Section 2.7.

Also, for **pure shear** we substitute

$$\sigma_x = \sigma_y = 0 \qquad \varepsilon_x = \varepsilon_y = 0$$

into Eqs. (7-53) and (7-54) and obtain

$$u = \frac{\tau_{xy}^2}{2G} \qquad u = \frac{G\gamma_{xy}^2}{2} \qquad \textbf{(7-56a,b)}$$

These equations agree with Eqs. (3-55a) and (3-55b) of Section 3.9.

## • • • Example 7-7

Strain gages $A$ and $B$ (oriented in the $x$ and $y$ directions, respectively) are attached to a rectangular aluminum plate with a thickness of $t = 0.275$ in. The plate is subjected to uniform normal stresses $\sigma_x$ and $\sigma_y$, as shown in Fig. 7-27, and the gage readings for normal strains are $\varepsilon_x = -0.00075$ (shortening, gage $A$) and $\varepsilon_y = 0.00125$ (elongation, gage $B$). The modulus of elasticity is $E = 10{,}600$ ksi, and Poisson's ratio is $v = 0.33$. Find the stresses $\sigma_x$ and $\sigma_y$ and the change $\Delta t$ in the thickness of the plate. Also, find the unit volume change (or dilatation) $e$ and the strain-energy density $u$ for the plate.

### Fig. 7-27

Example 7-7: Rectangular aluminum plate with strain gages $A$ and $B$

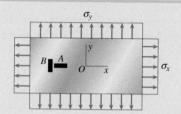

### Solution

For a plate in *biaxial stress*, we can use Eqs. (7-41a) and (7-41b) to find the normal stresses $\sigma_x$ and $\sigma_y$ in the $x$ and $y$ directions, respectively, based upon the measured normal strains $\varepsilon_x$ and $\varepsilon_y$:

$$\sigma_x = \frac{E}{1 - v^2}(\varepsilon_x + v\varepsilon_y) = \frac{10{,}600 \text{ ksi}}{1 - 0.33^2}[-0.00075 + (0.33)(0.00125)]$$

$$= -4.01 \text{ ksi} \qquad \Leftarrow$$

$$\sigma_y = \frac{E}{1 - v^2}(\varepsilon_y + v\varepsilon_x) = \frac{10{,}600 \text{ ksi}}{1 - 0.33^2}[0.00125 + (0.33)(-0.00075)]$$

$$= 11.93 \text{ ksi} \qquad \Leftarrow$$

The normal strain in the $z$ direction is then computed from Eq. (7-40c) as

$$\varepsilon_z = \frac{-v}{E}(\sigma_x + \sigma_y) = \frac{-(0.33)}{10{,}600 \text{ ksi}}(-4.01 \text{ ksi} + 11.93 \text{ ksi})$$

$$= -2.466 \times 10^{-4}$$

The change (i.e., here a decrease) in the thickness of the plate is then

$$\Delta t = \varepsilon_z t = [-2.466(10^{-4})](0.275 \text{ in.}) = -6.78 \times 10^{-5} \text{ in.} \qquad \Leftarrow$$

We use Eq. (7-49) to find the dilatation or unit volume change $e$ of the plate as

$$e = \frac{1 - 2v}{E}(\sigma_x + \sigma_y) = 2.537 \times 10^{-4} \qquad \Leftarrow$$

The positive sign for $e$ means that the plate increases in volume (although the increase is very small). Finally, we compute the strain-energy density of the plate using Eq. 7-53 (deleting the shear term):

$$u = \frac{1}{2E}(\sigma_x^2 + \sigma_y^2 - 2v\sigma_x\sigma_y)$$

$$= \frac{1}{2(10{,}600 \text{ ksi})}[(-4.01 \text{ ksi})^2 + (11.93 \text{ ksi})^2 - 2(0.33)(-4.01 \text{ ksi})(11.93 \text{ ksi})]$$

$$= 8.96 \text{ psi}$$

# 7.6 TRIAXIAL STRESS

**Fig. 7-28**

Element in triaxial stress

An element of material subjected to normal stresses $\sigma_x$, $\sigma_y$, and $\sigma_z$ acting in three mutually perpendicular directions is said to be in a state of **triaxial stress** (Fig. 7-28a). Since there are no shear stresses on the $x$, $y$, and $z$ faces, the stresses $\sigma_x$, $\sigma_y$, and $\sigma_z$ are the *principal stresses* in the material.

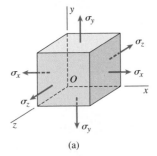

(a)

If an inclined plane parallel to the $z$ axis is cut through the element (Fig. 7-28b), the only stresses on the inclined face are the normal stress $\sigma$ and shear stress $\tau$, both of which act parallel to the $xy$ plane. These stresses are analogous to the stresses $\sigma_{x_1}$ and $\tau_{x_1y_1}$ encountered in our earlier discussions of plane stress (see, for instance, Fig. 7-2a). Because the stresses $\sigma$ and $\tau$ (Fig. 7-28b) are found from equations of force equilibrium in the $xy$ plane, they are independent of the normal stress $\sigma_z$. Therefore, we can use the transformation equations of plane stress, as well as Mohr's circle for plane stress, when determining the stresses $\sigma$ and $\tau$ in triaxial stress. The same general conclusion holds for the normal and shear stresses acting on inclined planes cut through the element parallel to the $x$ and $y$ axes.

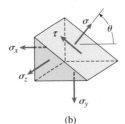

(b)

## Maximum Shear Stresses

From our previous discussions of plane stress, we know that the maximum shear stresses occur on planes oriented at 45° to the principal planes. Therefore, for a material in triaxial stress (Fig. 7-28a), the maximum shear stresses occur on elements oriented at angles of 45° to the $x$, $y$, and $z$ axes. For example, consider an element obtained by a 45° rotation about the $z$ axis. The maximum positive and negative shear stresses acting on this element are

$$(\tau_{max})_z = \pm \frac{\sigma_x - \sigma_y}{2} \qquad \textbf{(7-57)}$$

Similarly, by rotating about the $x$ and $y$ axes through angles of 45°, we obtain the following maximum shear stresses:

$$(\tau_{max})_x = \pm \frac{\sigma_y - \sigma_z}{2} \qquad (\tau_{max})_y = \pm \frac{\sigma_x - \sigma_z}{2} \quad \textbf{(7-58a,b)}$$

**Fig. 7-29**

Mohr's circles for an element in triaxial stress

The absolute maximum shear stress is the numerically largest of the stresses determined from Eqs. (7-57, 7-58a and b). It is equal to one-half the difference between the algebraically largest and algebraically smallest of the three principal stresses.

The stresses acting on elements oriented at various angles to the $x$, $y$, and $z$ axes can be visualized with the aid of **Mohr's circles**. For elements oriented by rotations about the $z$ axis, the corresponding circle is labeled $A$ in Fig. 7-29. Note that this circle is drawn for the case in which $\sigma_x > \sigma_y$ and both $\sigma_x$ and $\sigma_y$ are tensile stresses.

In a similar manner, we can construct circles $B$ and $C$ for elements oriented by rotations about the $x$ and $y$ axes, respectively. The radii of the circles represent the maximum shear stresses given by Eqs. (7-57, 7-58a and b),

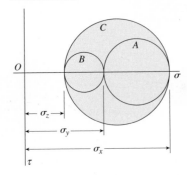

and the absolute maximum shear stress is equal to the radius of the largest circle. The normal stresses acting on the planes of maximum shear stresses have magnitudes given by the abscissas of the centers of the respective circles.

In the preceding discussion of triaxial stress we only considered stresses acting on planes obtained by rotating about the $x$, $y$, and $z$ axes. Thus, every plane we considered is parallel to one of the axes. For instance, the inclined plane of Fig. 7-28b is parallel to the $z$ axis, and its normal is parallel to the $xy$ plane. Of course, we can also cut through the element in **skew directions**, so that the resulting inclined planes are skew to all three coordinate axes. The normal and shear stresses acting on such planes can be obtained by a more complicated three-dimensional analysis. However, the normal stresses acting on skew planes are intermediate in value between the algebraically maximum and minimum principal stresses, and the shear stresses on those planes are smaller (in absolute value) than the absolute maximum shear stress obtained from Eqs. (7-58a, b, and c).

## Hooke's Law for Triaxial Stress

If the material follows Hooke's law, we can obtain the relationships between the normal stresses and normal strains by using the same procedure as for plane stress (see Section 7.5). The strains produced by the stresses $\sigma_x$, $\sigma_y$, and $\sigma_z$ acting independently are superimposed to obtain the resultant strains. Thus, we readily arrive at the following equations for the **strains in triaxial stress**:

$$\varepsilon_x = \frac{\sigma_x}{E} - \frac{v}{E}(\sigma_y + \sigma_z) \tag{7-59a}$$

$$\varepsilon_y = \frac{\sigma_y}{E} - \frac{v}{E}(\sigma_z + \sigma_x) \tag{7-59b}$$

$$\varepsilon_z = \frac{\sigma_z}{E} - \frac{v}{E}(\sigma_x + \sigma_y) \tag{7-59c}$$

In these equations, the standard sign conventions are used; that is, tensile stress $\sigma$ and extensional strain $\varepsilon$ are positive.

The preceding equations can be solved simultaneously for the **stresses in terms of the strains**:

$$\sigma_x = \frac{E}{(1 + v)(1 - 2v)}[(1 - v)\varepsilon_x + v(\varepsilon_y + \varepsilon_z)] \tag{7-60a}$$

$$\sigma_y = \frac{E}{(1 + v)(1 - 2v)}[(1 - v)\varepsilon_y + v(\varepsilon_z + \varepsilon_x)] \tag{7-60b}$$

$$\sigma_z = \frac{E}{(1 + v)(1 - 2v)}[(1 - v)\varepsilon_z + v(\varepsilon_x + \varepsilon_y)] \tag{7-60c}$$

Equations (7-59) and (7-60) represent **Hooke's law for triaxial stress**.

In the special case of **biaxial stress** (Fig. 7-11b), we can obtain the equations of Hooke's law by substituting $\sigma_z = 0$ into the preceding equations. The resulting equations reduce to Eqs. (7-40) and (7-41) of Section 7.5.

## Unit Volume Change

The unit volume change (or *dilatation*) for an element in triaxial stress is obtained in the same manner as for plane stress (see Section 7.5). If the element is subjected to strains $\varepsilon_x$, $\varepsilon_y$, and $\varepsilon_z$, we may use Eq. (7-48) for the unit volume change:

$$e = \varepsilon_x + \varepsilon_y + \varepsilon_z \tag{7-61}$$

This equation is valid for any material provided the strains are small.

If Hooke's law holds for the material, we can substitute for the strains $\varepsilon_x$, $\varepsilon_y$, and $\varepsilon_z$ from Eqs. (7-59a, b, and c) and obtain

$$e = \frac{1 - 2v}{E}(\sigma_x + \sigma_y + \sigma_z) \tag{7-62}$$

Equations (7-61) and (7-62) give the unit volume change in triaxial stress in terms of the strains and stresses, respectively.

## Strain-Energy Density

The strain-energy density for an element in triaxial stress is obtained by the same method used for plane stress. When stresses $\sigma_x$ and $\sigma_y$ act alone (biaxial stress), the strain-energy density (from Eq. 7-52 with the shear term discarded) is

$$u = \frac{1}{2}(\sigma_x \varepsilon_x + \sigma_y \varepsilon_y)$$

When the element is in triaxial stress and subjected to stresses $\sigma_x$, $\sigma_y$, and $\sigma_z$, the expression for strain-energy density becomes

$$u = \frac{1}{2}(\sigma_x \varepsilon_x + \sigma_y \varepsilon_y + \sigma_z \varepsilon_z) \tag{7-63a}$$

Substituting for the strains from Eqs. (7-59a, b, and c), we obtain the strain-energy density in terms of the stresses:

$$u = \frac{1}{2E}(\sigma_x^2 + \sigma_y^2 + \sigma_z^2) - \frac{v}{E}(\sigma_x \sigma_y + \sigma_x \sigma_z + \sigma_y \sigma_z) \tag{7-63b}$$

In a similar manner, but using Eqs. (7-60a, b, and c), we can express the strain-energy density in terms of the strains:

$$u = \frac{E}{2(1 + v)(1 - 2v)}[(1 - v)(\varepsilon_x^2 + \varepsilon_y^2 + \varepsilon_z^2) +$$
$$2v(\varepsilon_x \varepsilon_y + \varepsilon_x \varepsilon_z + \varepsilon_y \varepsilon_z)] \tag{7-63c}$$

When calculating from these expressions, we must be sure to substitute the stresses and strains with their proper algebraic signs.

Fig. 7-30

Element in spherical stress

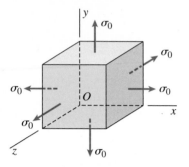

## Spherical Stress

A special type of triaxial stress, called **spherical stress**, occurs whenever all three normal stresses are equal (Fig. 7-30):

$$\sigma_x = \sigma_y = \sigma_z = \sigma_0 \tag{7-64}$$

Under these stress conditions, *any* plane cut through the element will be subjected to the same normal stress $\sigma_0$ and will be free of shear stress. Thus, we have equal normal stresses in every direction and no shear stresses anywhere in the material. Every plane is a principal plane, and the three Mohr's circles shown in Fig. 7-29 reduce to a single point.

The normal strains in spherical stress are also the same in all directions, provided the material is homogeneous and isotropic. If Hooke's law applies, the normal strains are

$$\varepsilon_0 = \frac{\sigma_0}{E}(1 - 2\nu) \tag{7-65}$$

as obtained from Eqs. (7-59a, b, and c).

Since there are no shear strains, an element in the shape of a cube changes in size but remains a cube. In general, any body subjected to spherical stress will maintain its relative proportions but will expand or contract in volume depending upon whether $\sigma_0$ is tensile or compressive.

The expression for the unit volume change can be obtained from Eq. (7-61) by substituting for the strains from Eq. (7-65). The result is

$$e = 3\varepsilon_0 = \frac{3\sigma_0(1 - 2\nu)}{E} \tag{7-66}$$

Equation (7-66) is usually expressed in more compact form by introducing a new quantity $K$ called the **volume modulus of elasticity**, or **bulk modulus of elasticity**, which is defined as follows:

$$K = \frac{E}{3(1 - 2\nu)} \tag{7-67}$$

With this notation, the expression for the unit volume change becomes

$$e = \frac{\sigma_0}{K} \tag{7-68}$$

and the volume modulus is

$$K = \frac{\sigma_0}{e} \tag{7-69}$$

Thus, the volume modulus can be defined as the ratio of the spherical stress to the volumetric strain, which is analogous to the definition of the modulus $E$ in uniaxial stress. Note that the preceding formulas for $e$ and $K$ are based upon the assumptions that *the strains are small and Hooke's law holds for the material.*

From Eq. (7-61) for $K$, we see that if Poisson's ratio $\nu$ equals 1/3, the moduli $K$ and $E$ are numerically equal. If $\nu = 0$, then $K$ has the value $E/3$, and if $\nu = 0.5$, $K$ becomes infinite, which corresponds to a rigid material having no change in volume (that is, the material is incompressible).

The preceding formulas for spherical stress were derived for an element subjected to uniform tension in all directions, but of course the formulas also apply to an element in uniform compression. In the case of uniform compression, the stresses and strains have negative signs. Uniform compression occurs when the material is subjected to uniform pressure in all directions; for example, an object submerged in water or rock deep within the earth. This state of stress is often called **hydrostatic stress**.

Although uniform compression is relatively common, a state of uniform tension is difficult to achieve. It can be realized by suddenly and uniformly heating the outer surface of a solid metal sphere, so that the outer layers are at a higher temperature than the interior. The tendency of the outer layers to expand produces uniform tension in all directions at the center of the sphere.

# 7.7 PLANE STRAIN

The strains at a point in a loaded structure vary according to the orientation of the axes, in a manner similar to that for stresses. In this section we will derive the transformation equations that relate the strains in inclined directions to the strains in the reference directions. These transformation equations are widely used in laboratory and field investigations involving measurements of strains.

Strains are customarily measured by *strain gages*; for example, gages are placed in aircraft to measure structural behavior during flight, and gages are placed in buildings to measure the effects of earthquakes. Since each gage measures the strain in one particular direction, it is usually necessary to calculate the strains in other directions by means of the transformation equations.

## Plane Strain Versus Plane Stress

Let us begin by explaining what is meant by plane strain and how it relates to plane stress. Consider a small element of material having sides of lengths $a$, $b$, and $c$ in the $x$, $y$, and $z$ directions, respectively (Fig. 7-31a). If the only deformations are those in the $xy$ plane, then three strain components may exist—the normal strain $\varepsilon_x$ in the $x$ direction (Fig. 7-31b), the normal strain $\varepsilon_y$ in the $y$ direction (Fig. 7-31c), and the shear strain $\gamma_{xy}$ (Fig. 7-31d). An element of material subjected to these strains (and *only* these strains) is said to be in a state of **plane strain**.

It follows that an element in plane strain has no normal strain $\varepsilon_z$ in the $z$ direction and no shear strains $\gamma_{xz}$ and $\gamma_{yz}$ in the $xz$ and $yz$ planes, respectively. Thus, plane strain is defined by the following conditions:

$$\varepsilon_z = 0 \qquad \gamma_{xz} = 0 \qquad \gamma_{yz} = 0 \qquad \textbf{(7-70a,b,c)}$$

The remaining strains ($\varepsilon_x$, $\varepsilon_y$, and $\gamma_{xy}$) may have nonzero values.

From the preceding definition, we see that plane strain occurs when the front and rear faces of an element of material (Fig. 7-31a) are fully restrained against displacement in the $z$ direction—an idealized condition that is seldom reached in actual structures. However, this does not mean that the transformation equations of plane strain are not useful. It turns out that they are extremely useful because they also apply to the strains in plane stress, as explained in the following paragraphs.

## Fig. 7-31

Strain components $\varepsilon_x$, $\varepsilon_y$, and $\gamma_{xy}$ in the $xy$ plane (plane strain)

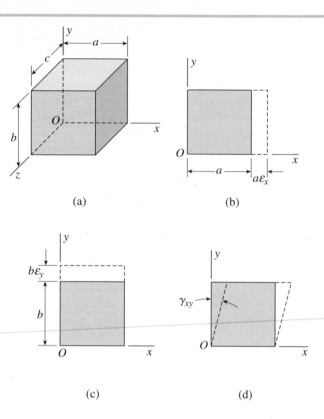

(a)

(b)

(c)

(d)

The definition of plane strain (Eqs. 7-70a, b, and c) is analogous to that for plane stress. In plane stress, the following stresses must be zero:

$$\sigma_z = 0 \quad \tau_{xz} = 0 \quad \tau_{yz} = 0 \qquad \textbf{(7-71a,b,c)}$$

whereas the remaining stresses ($\sigma_x$, $\sigma_y$, and $\tau_{xy}$) may have nonzero values. A comparison of the stresses and strains in plane stress and plane strain is given in Fig. 7-32.

It should not be inferred from the similarities in the definitions of plane stress and plane strain that both occur simultaneously. In general, an element in plane stress will undergo a strain in the $z$ direction (Fig. 7-32); hence, it is *not* in plane strain. Also, an element in plane strain usually will have stresses $\sigma_z$ acting on it because of the requirement that $\varepsilon_z = 0$; therefore, it is *not* in plane stress. Thus, under ordinary conditions plane stress and plane strain do not occur simultaneously.

An exception occurs when an element in plane stress is subjected to equal and opposite normal stresses (that is, when $\sigma_x = -\sigma_y$) and Hooke's law holds for the material. In this special case, there is no normal strain in the $z$ direction, as shown by Eq. (7-35c), and therefore the element is in a state of plane strain as well as plane stress. Another special case, albeit a hypothetical one, is when a material has Poisson's ratio equal to zero ($v = 0$); then every plane stress element is also in plane strain because $\varepsilon_z = 0$ (Eq. 7-35c).*

---

*In the discussions of this chapter we are omitting the effects of temperature changes and prestrains, both of which produce additional deformations that may alter some of our conclusions.

**Fig. 7-32**

Comparison of plane stress and plane strain

| | Plane stress | Plane strain |
|---|---|---|
| | $y$ $\sigma_y$ $\tau_{xy}$ $\sigma_x$ $O$ $x$ $z$ | $y$ $\varepsilon_y$ $\gamma_{xy}$ $\varepsilon_x$ $O$ $x$ $z$ |
| **Stresses** | $\sigma_z = 0$ $\quad$ $\tau_{xz} = 0$ $\quad$ $\tau_{yz} = 0$<br><br>$\sigma_x$, $\sigma_y$, and $\tau_{xy}$ may have<br>nonzero values | $\tau_{xz} = 0$ $\quad$ $\tau_{yz} = 0$<br><br>$\sigma_x$, $\sigma_y$, $\sigma_z$, and $\tau_{xy}$ may have<br>nonzero values |
| **Strains** | $\gamma_{xz} = 0$ $\quad$ $\gamma_{yz} = 0$<br><br>$\varepsilon_x$, $\varepsilon_y$, $\varepsilon_z$, and $\gamma_{xy}$ may have<br>nonzero values | $\varepsilon_z = 0$ $\quad$ $\gamma_{xz} = 0$ $\quad$ $\gamma_{yz} = 0$<br><br>$\varepsilon_x$, $\varepsilon_y$, and $\gamma_{xy}$ may have<br>nonzero values |

## Application of the Transformation Equations

The stress-transformation equations derived for plane stress in the $xy$ plane (Eqs. 7-4a and 7-4b) are valid even when a normal stress $\sigma_z$ is present. The explanation lies in the fact that the stress $\sigma_z$ does not enter the equations of equilibrium used in deriving Eqs. (7-4a) and (7-4b). Therefore, *the transformation equations for plane stress can also be used for the stresses in plane strain.*

An analogous situation exists for plane strain. Although we will derive the strain-transformation equations for the case of plane strain in the $xy$ plane, the equations are valid even when a strain $\varepsilon_z$ exists. The reason is simple enough—the strain $\varepsilon_z$ does not affect the geometric relationships used in the derivations. Therefore, *the transformation equations for plane strain can also be used for the strains in plane stress.*

Finally, we should recall that the transformation equations for plane stress were derived solely from equilibrium and therefore are valid for any material, whether linearly elastic or not. The same conclusion applies to the transformation equations for plane strain—since they are derived solely from geometry, *they are independent of the material properties.*

## Transformation Equations for Plane Strain

In the derivation of the transformation equations for plane strain, we will use the coordinate axes shown in Fig. 7-33. We will assume that the normal strains $\varepsilon_x$ and $\varepsilon_y$ and the shear strain $\gamma_{xy}$ associated with the $xy$ axes are known (Fig. 7-31). The objectives of our analysis are to determine the normal strain $\varepsilon_{x_1}$ and the shear strain $\gamma_{x_1y_1}$ associated with the $x_1y_1$ axes, which are rotated counterclockwise through an angle $\theta$ from the $xy$ axes. (It is not necessary to derive a separate equation for the normal strain $\varepsilon_{y_1}$ because it can be obtained from the equation for $\varepsilon_{x_1}$ by substituting $\theta + 90°$ for $\theta$.)

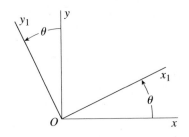

**Normal strain $\varepsilon_{x_1}$.** To determine the normal strain $\varepsilon_{x_1}$ in the $x_1$ direction, we consider a small element of material selected so that the $x_1$ axis is along a diagonal of the $z$ face of the element and the $x$ and $y$ axes are along the sides of the element (Fig. 7-34a). The figure shows a two-dimensional view of the element, with the $z$ axis toward the viewer. Of course, the element is actually three dimensional, as in Fig. 7-31a, with a dimension in the $z$ direction.

Consider first the strain $\varepsilon_x$ in the $x$ direction (Fig. 7-34a). This strain produces an elongation in the $x$ direction equal to $\varepsilon_x dx$, where $dx$ is the length of the corresponding side of the element. As a result of this elongation, the diagonal of the element increases in length by an amount

$$\varepsilon_x dx \cos\theta \tag{7-72a}$$

as shown in Fig. 7-34a.

Next, consider the strain $\varepsilon_y$ in the $y$ direction (Fig. 7-34b). This strain produces an elongation in the $y$ direction equal to $\varepsilon_y dy$, where $dy$ is the length of the side of the element parallel to the $y$ axis. As a result of this elongation, the diagonal of the element increases in length by an amount

$$\varepsilon_y dy \sin\theta \tag{7-72b}$$

which is shown in Fig. 7-34b.

Finally, consider the shear strain $\gamma_{xy}$ in the $xy$ plane (Fig. 7-34c). This strain produces a distortion of the element such that the angle at the lower left corner of the element decreases by an amount equal to the shear strain. Consequently, the upper face of the element moves to the right (with respect to the lower face) by an amount $\gamma_{xy} dy$. This deformation results in an increase in the length of the diagonal equal to

$$\gamma_{xy} dy \cos\theta \tag{7-72c}$$

as shown in Fig. 7-34c.

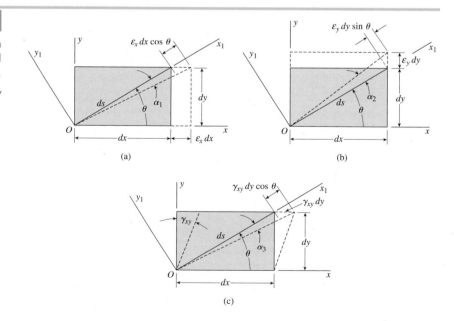

### Fig. 7-34

Deformations of an element in plane strain due to (a) normal strain $\varepsilon_x$, (b) normal strain $\varepsilon_y$, and (c) shear strain $\gamma_{xy}$

The total increase $\Delta d$ in the length of the diagonal is the sum of the preceding three expressions; thus,

$$\Delta d = \varepsilon_x \, dx \cos \theta + \varepsilon_y \, dy \sin \theta + \gamma_{xy} \, dy \cos \theta \qquad \textbf{(7-73)}$$

The normal strain $\varepsilon_{x_1}$ in the $x_1$ direction is equal to this increase in length divided by the initial length $ds$ of the diagonal:

$$\varepsilon_{x_1} = \frac{\Delta d}{ds} = \varepsilon_x \frac{dx}{ds} \cos \theta + \varepsilon_y \frac{dy}{ds} \sin \theta + \gamma_{xy} \frac{dy}{ds} \cos \theta \qquad \textbf{(7-74)}$$

Observing that $dx/ds = \cos \theta$ and $dy/ds = \sin \theta$, we obtain the following equation for the **normal strain**:

$$\varepsilon_{x_1} = \varepsilon_x \cos^2 \theta + \varepsilon_y \sin^2 \theta + \gamma_{xy} \sin \theta \cos \theta \qquad \textbf{(7-75)}$$

Thus, we have obtained an expression for the normal strain in the $x_1$ direction in terms of the strains $\varepsilon_x$, $\varepsilon_y$, and $\gamma_{xy}$ associated with the $xy$ axes.

As mentioned previously, the normal strain $\varepsilon_{y_1}$ in the $y_1$ direction is obtained from the preceding equation by substituting $\theta + 90°$ for $\theta$.

**Shear strain $\gamma_{x_1y_1}$.** Now we turn to the shear strain $\gamma_{x_1y_1}$ associated with the $x_1y_1$ axes. This strain is equal to the decrease in angle between lines in the material that were initially along the $x_1$ and $y_1$ axes. To clarify this idea, consider Fig. 7-35, which shows both the $xy$ and $x_1y_1$ axes, with the angle $\theta$ between them. Let line $Oa$ represent a line in the material that *initially* was along the $x_1$ axis (that is, along the diagonal of the element in Fig. 7-34). The deformations caused by the strains $\varepsilon_x$, $\varepsilon_y$, and $\gamma_{xy}$ (Fig. 7-34) cause line $Oa$ to rotate through a counterclockwise angle $\alpha$ from the $x_1$ axis to the position shown in Fig. 7-35. Similarly, line $Ob$ was originally along the $y_1$ axis, but because of the deformations it rotates through a clockwise angle $\beta$. The shear strain $\gamma_{x_1y_1}$ is the decrease in angle between the two lines that originally were at right angles; therefore,

$$\gamma_{x_1y_1} = \alpha + \beta \qquad \textbf{(7-76)}$$

Thus, in order to find the shear strain $\gamma_{x_1y_1}$, we must determine the angles $\alpha$ and $\beta$.

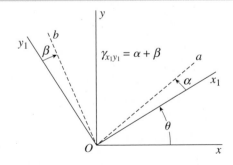

**Fig. 7-35**

Shear strain $\gamma_{x_1y_1}$ associated with the $x_1y_1$ axes

The angle $\alpha$ can be found from the deformations pictured in Fig. 7-34 as follows. The strain $\varepsilon_x$ (Fig. 7-34a) produces a clockwise rotation of the diagonal of the element. Let us denote this angle of rotation as $\alpha_1$. The angle $\alpha_1$ is equal to the distance $\varepsilon_x \, dx \sin \theta$ divided by the length $ds$ of the diagonal:

$$\alpha_1 = \varepsilon_x \frac{dx}{ds} \sin \theta \qquad \textbf{(7-77a)}$$

Similarly, the strain $\varepsilon_y$ produces a counterclockwise rotation of the diagonal through an angle $\alpha_2$ (Fig. 7-34b). This angle is equal to the distance $\varepsilon_y \, dy \cos \theta$ divided by $ds$:

$$\alpha_2 = \varepsilon_y \frac{dy}{ds} \cos \theta \qquad \textbf{(7-77b)}$$

Finally, the strain $\gamma_{xy}$ produces a clockwise rotation through an angle $\alpha_3$ (Fig. 7-34c) equal to the distance $\gamma_{xy} \, dy \sin \theta$ divided by $ds$:

$$\alpha_3 = \gamma_{xy} \frac{dy}{ds} \sin \theta \qquad \textbf{(7-77c)}$$

Therefore, the resultant counterclockwise rotation of the diagonal (Fig. 7-34), equal to the angle $\alpha$ shown in Fig. 7-35, is

$$\alpha = -\alpha_1 + \alpha_2 - \alpha_3$$

$$= -\varepsilon_x \frac{dx}{ds} \sin \theta + \varepsilon_y \frac{dy}{ds} \cos \theta - \gamma_{xy} \frac{dy}{ds} \sin \theta \qquad \textbf{(7-78)}$$

Again observing that $dx/ds = \cos \theta$ and $dy/ds = \sin \theta$, we obtain

$$\alpha = -(\varepsilon_x - \varepsilon_y) \sin \theta \cos \theta - \gamma_{xy} \sin^2 \theta \qquad \textbf{(7-79)}$$

The rotation of line $Ob$ (Fig. 7-35), which initially was at 90° to line $Oa$, can be found by substituting $\theta + 90°$ for $\theta$ in the expression for $\alpha$. The resulting expression is counterclockwise when positive (because $\alpha$ is counterclockwise when positive), hence it is equal to the negative of the angle $\beta$ (because $\beta$ is positive when clockwise). Thus,

$$\beta = (\varepsilon_x - \varepsilon_y) \sin (\theta + 90°) \cos (\theta + 90°) + \gamma_{xy} \sin^2 (\theta + 90°)$$

$$= -(\varepsilon_x - \varepsilon_y) \sin \theta \cos \theta + \gamma_{xy} \cos^2 \theta \qquad \textbf{(7-80)}$$

Adding $\alpha$ and $\beta$ gives the shear strain $\gamma_{x_1 y_1}$ (see Eq. 7-67):

$$\gamma_{x_1 y_1} = -2(\varepsilon_x - \varepsilon_y) \sin \theta \cos \theta + \gamma_{xy}(\cos^2 \theta - \sin^2 \theta) \qquad \textbf{(7-81)}$$

To put the equation in a more useful form, we divide each term by 2:

$$\frac{\gamma_{x_1 y_1}}{2} = -(\varepsilon_x - \varepsilon_y) \sin \theta \cos \theta + \frac{\gamma_{xy}}{2} (\cos^2 \theta - \sin^2 \theta) \qquad \textbf{(7-82)}$$

We have now obtained an expression for the **shear strain** $\gamma_{x_1y_1}$ associated with the $x_1y_1$ axes in terms of the strains $\varepsilon_x$, $\varepsilon_y$, and $\gamma_{xy}$ associated with the $xy$ axes.

**Transformation equations for plane strain.** The equations for plane strain [Eqs. (7-75) and (7-82)] can be expressed in terms of the angle $2\theta$ by using the following trigonometric identities:

$$\cos^2\theta = \frac{1}{2}(1 + \cos 2\theta) \quad \sin^2\theta = \frac{1}{2}(1 - \cos 2\theta)$$

$$\sin\theta\cos\theta = \frac{1}{2}\sin 2\theta$$

Thus, the transformation equations for plane strain become

$$\varepsilon_{x_1} = \frac{\varepsilon_x + \varepsilon_y}{2} + \frac{\varepsilon_x - \varepsilon_y}{2}\cos 2\theta + \frac{\gamma_{xy}}{2}\sin 2\theta \qquad \textbf{(7-83a)}$$

and

$$\frac{\gamma_{x_1y_1}}{2} = -\frac{\varepsilon_x - \varepsilon_y}{2}\sin 2\theta + \frac{\gamma_{xy}}{2}\cos 2\theta \qquad \textbf{(7-83b)}$$

These equations are the counterparts of Eqs. (7-4a) and (7-4b) for plane stress.

When comparing the two sets of equations, note that $\varepsilon_{x_1}$ corresponds to $\sigma_{x_1}$, $\gamma_{x_1y_1}/2$ corresponds to $\tau_{x_1y_1}$, $\varepsilon_x$ corresponds to $\sigma_x$, $\varepsilon_y$ corresponds to $\sigma_y$, and $\gamma_{xy}/2$ corresponds to $\tau_{xy}$. The corresponding variables in the two sets of transformation equations are listed in Table 7-1.

The analogy between the transformation equations for plane stress and those for plane strain shows that all of the observations made in Sections 7.2, 7.3, and 7.4 concerning plane stress, principal stresses, maximum shear stresses, and Mohr's circle have their counterparts in plane strain. For instance, the sum of the normal strains in perpendicular directions is a constant [compare with Eq. (7-6)]:

$$\varepsilon_{x_1} + \varepsilon_{y_1} = \varepsilon_x + \varepsilon_y \qquad \textbf{(7-84)}$$

This equality can be verified easily by substituting the expressions for $\varepsilon_{x_1}$ [from Eq. (7-83a)] and $\varepsilon_{y_1}$ [from Eq. (7-83a) with $\theta$ replaced by $\theta + 90°$].

## Principal Strains

Principal strains exist on perpendicular planes with the principal angles $\theta_p$ calculated from the following equation [compare with Eq. (7-11)]:

$$\tan 2\theta_p = \frac{\gamma_{xy}}{\varepsilon_x - \varepsilon_y} \qquad \textbf{(7-85)}$$

The principal strains can be calculated from the equation

$$\varepsilon_{1,2} = \frac{\varepsilon_x + \varepsilon_y}{2} \pm \sqrt{\left(\frac{\varepsilon_x - \varepsilon_y}{2}\right)^2 + \left(\frac{\gamma_{xy}}{2}\right)^2} \qquad \textbf{(7-86)}$$

### Table 7-1

Corresponding Variables in the Transformation Equations for Plane Stress (Eqs. 7-4a and b) and Plane Strain (Eqs. 7-83a and b)

| Stresses | Strains |
|---|---|
| $\sigma_x$ | $\varepsilon_x$ |
| $\sigma_y$ | $\varepsilon_y$ |
| $\tau_{xy}$ | $\gamma_{xy}/2$ |
| $\sigma_{x_1}$ | $\varepsilon_{x_1}$ |
| $\tau_{x_1y_1}$ | $\gamma_{x_1y_1}/2$ |

which corresponds to Eq. (7-17) for the principal stresses. The two principal strains (in the $xy$ plane) can be correlated with the two principal directions using the technique described in Section 7.3 for the principal stresses. (This technique is illustrated later in Example 7-8.) Finally, note that in plane strain the third principal strain is $\varepsilon_z = 0$. Also, the shear strains are zero on the principal planes.

## Maximum Shear Strains

The maximum shear strains in the $xy$ plane are associated with axes at 45° to the directions of the principal strains. The algebraically maximum shear strain (in the $xy$ plane) is given by the following equation (compare with Eq. 7-25):

$$\frac{\gamma_{max}}{2} = \sqrt{\left(\frac{\varepsilon_x - \varepsilon_y}{2}\right)^2 + \left(\frac{\gamma_{xy}}{2}\right)^2} \qquad \textbf{(7-87)}$$

The minimum shear strain has the same magnitude but is negative. In the directions of maximum shear strain, the normal strains are

$$\varepsilon_{aver} = \frac{\varepsilon_x + \varepsilon_y}{2} \qquad \textbf{(7-88)}$$

which is analogous to Eq. (7-27) for stresses. The maximum out-of-plane shear strains, that is, the shear strains in the $xz$ and $yz$ planes, can be obtained from equations analogous to Eq. (7-87).

An element in plane stress that is oriented to the principal directions of stress (see Fig. 7-13b) has no shear stresses acting on its faces. Therefore, the shear strain $\gamma_{x_1 y_1}$ for this element is zero. It follows that the normal strains in this element are the principal strains. Thus, at a given point in a stressed body, *the principal strains and principal stresses occur in the same directions*.

## Mohr's Circle for Plane Strain

Mohr's circle for plane strain is constructed in the same manner as the circle for plane stress, as illustrated in Fig. 7-36. Normal strain $\varepsilon_{x_1}$ is plotted as the abscissa (positive to the right) and one-half the shear strain ($\gamma_{x_1 y_1}/2$) is plotted as the ordinate (positive downward). The center $C$ of the circle has an abscissa equal to $\varepsilon_{aver}$ (Eq. 7-88).

Point $A$, representing the strains associated with the $x$ direction ($\theta = 0$), has coordinates $\varepsilon_x$ and $\gamma_{xy}/2$. Point $B$, at the opposite end of a diameter from $A$, has coordinates $\varepsilon_y$ and $-\gamma_{xy}/2$, representing the strains associated with a pair of axes rotated through an angle $\theta = 90°$.

The strains associated with axes rotated through an angle $\theta$ are given by point $D$, which is located on the circle by measuring an angle $2\theta$ counterclockwise from radius $CA$. The principal strains are represented by points $P_1$ and $P_2$, and the maximum shear strains by points $S_1$ and $S_2$. All of these strains can be determined from the geometry of the circle or from the transformation equations.

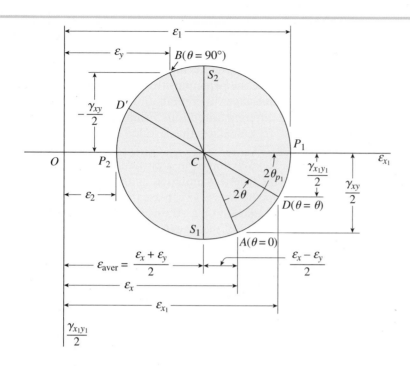

**Fig. 7-36**

Mohr's circle for plane strain

## Strain Measurements

An electrical-resistance **strain gage** is a device for measuring normal strains on the surface of a stressed object. These gages are quite small, with lengths typically in the range from one-eighth to one-half of an inch. The gages are bonded securely to the surface of the object so that they change in length in proportion to the strains in the object itself.

Each gage consists of a fine metal grid that is stretched or shortened when the object is strained at the point where the gage is attached. The grid is equivalent to a continuous wire that goes back and forth from one end of the grid to the other, thereby effectively increasing its length (Fig. 7-37). The electrical resistance of the wire is altered when it

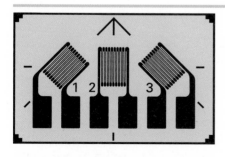

(a) 45° strain gages three-element rosette

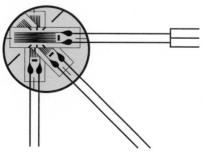

(b) Three-element strain-gage rosettes prewired

**Fig. 7-37**

Three electrical-resistance strain gages arranged as a 45° strain rosette (magnified view). (Courtesy of Micro-Measurements Division of Vishay Precision Group, Raleigh, NC, USA)

stretches or shortens—then this change in resistance is converted into a measurement of strain. The gages are extremely sensitive and can measure strains as small as $1 \times 10^{-6}$.

Since each gage measures the normal strain in only one direction, and since the directions of the principal stresses are usually unknown, it is necessary to use three gages in combination, with each gage measuring the strain in a different direction. From three such measurements, it is possible to calculate the strains in any direction, as illustrated in Example 7-9.

A group of three gages arranged in a particular pattern is called a **strain rosette**. Because the rosette is mounted on the surface of the body, where the material is in plane stress, we can use the transformation equations for plane strain to calculate the strains in various directions. (As explained earlier in this section, the transformation equations for plane strain can also be used for the strains in plane stress.)

## Calculation of Stresses from the Strains

The strain equations presented in this section are derived solely from geometry, as already pointed out. Therefore, the equations apply to any material, whether linear or nonlinear, elastic or inelastic. However, if it is desired to determine the stresses from the strains, the material properties must be taken into account.

If the material follows Hooke's law, we can find the stresses using the appropriate stress-strain equations from either Section 7.5 (for plane stress) or Section 7.6 (for triaxial stress).

As a first example, suppose that the material is in plane stress and that we know the strains $\varepsilon_x$, $\varepsilon_y$, and $\gamma_{xy}$, perhaps from strain-gage measurements. Then we can use the stress-strain equations for plane stress (Eqs. 7-37 and 7-38) to obtain the stresses in the material.

Now consider a second example. Suppose we have determined the three principal strains $\varepsilon_1$, $\varepsilon_2$, and $\varepsilon_3$ for an element of material (if the element is in plane strain, then $\varepsilon_3 = 0$). Knowing these strains, we can find the principal stresses using Hooke's law for triaxial stress (see Eqs. 7-60a, b, and c). Once the principal stresses are known, we can find the stresses on inclined planes using the transformation equations for plane stress (see the discussion at the beginning of Section 7.6).

### • • • Example 7-8

An element of material in plane strain undergoes the following strains:

$$\varepsilon_x = 340 \times 10^{-6} \quad \varepsilon_y = 110 \times 10^{-6} \quad \gamma_{xy} = 180 \times 10^{-6}$$

These strains are shown highly exaggerated in Fig. 7-38a, which shows the deformations of an element of unit dimensions. Since the edges of the element have unit lengths, the changes in linear dimensions have the same magnitudes as the normal strains $\varepsilon_x$ and $\varepsilon_y$. The shear strain $\gamma_{xy}$ is the decrease in angle at the lower-left corner of the element.

Determine the following quantities: (a) the strains for an element oriented at an angle $\theta = 30°$, (b) the principal strains, and (c) the maximum shear strains. (Consider only the in-plane strains, and show all results on sketches of properly oriented elements.)

### Fig. 7-38

Example 7-8: Element of material in plane strain: (a) element oriented to the x and y axes, (b) element oriented at an angle $\theta = 30°$, (c) principal strains, and (d) maximum shear strains (*Note:* The edges of the elements have unit lengths.)

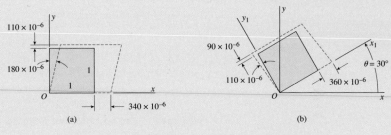

(a)     (b)

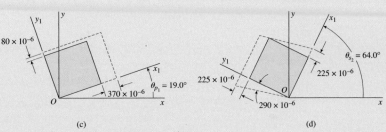

(c)     (d)

### Solution

(a) *Element oriented at an angle $\theta = 30°$.* The strains for an element oriented at an angle $\theta$ to the x axis can be found from the transformation equations of Eqs. (7-83a) and (7-83b). As a preliminary matter, we make the following calculations:

$$\frac{\varepsilon_x + \varepsilon_y}{2} = \frac{(340 + 110)10^{-6}}{2} = 225 \times 10^{-6}$$

$$\frac{\varepsilon_x - \varepsilon_y}{2} = \frac{(340 - 110)10^{-6}}{2} = 115 \times 10^{-6}$$

$$\frac{\gamma_{xy}}{2} = 90 \times 10^{-6}$$

*Continues* ➡

Now substituting into Eqs. (7-83a) and (7-83b), we get

$$\varepsilon_{x_1} = \frac{\varepsilon_x + \varepsilon_y}{2} + \frac{\varepsilon_x - \varepsilon_y}{2} \cos 2\theta + \frac{\gamma_{xy}}{2} \sin 2\theta$$

$$= (225 \times 10^{-6}) + (115 \times 10^{-6})(\cos 60°) + (90 \times 10^{-6})(\sin 60°)$$

$$= 360 \times 10^{-6}$$

$$\frac{\gamma_{x_1y_1}}{2} = -\frac{\varepsilon_x - \varepsilon_y}{2} \sin 2\theta + \frac{\gamma_{xy}}{2} \cos 2\theta$$

$$= -(115 \times 10^{-6})(\sin 60°) + (90 \times 10^{-6})(\cos 60°)$$

$$= -55 \times 10^{-6}$$

Therefore, the shear strain is

$$\gamma_{x_1y_1} = -110 \times 10^{-6}$$

The strain $\varepsilon_{y_1}$ can be obtained from Eq. (7-84), as

$$\varepsilon_{y_1} = \varepsilon_x + \varepsilon_y - \varepsilon_{x_1} = (340 + 110 - 360)10^{-6} = 90 \times 10^{-6}$$

The strains $\varepsilon_{x_1}$, $\varepsilon_{y_1}$, and $\gamma_{x_1y_1}$ are shown in Fig. 7-38b for an element oriented at $\theta = 30°$. Note that the angle at the lower-left corner of the element increases because $\gamma_{x_1y_1}$ is negative.

(b) *Principal strains.* The principal strains are readily determined from Eq. (7-86), as

$$\varepsilon_{1,2} = \frac{\varepsilon_x + \varepsilon_y}{2} \pm \sqrt{\left(\frac{\varepsilon_x - \varepsilon_y}{2}\right)^2 + \left(\frac{\gamma_{xy}}{2}\right)^2}$$

$$= 225 \times 10^{-6} \pm \sqrt{(115 \times 10^{-6})^2 + (90 \times 10^{-6})^2}$$

$$= 225 \times 10^{-6} \pm 146 \times 10^{-6}$$

Thus, the principal strains are

$$\varepsilon_1 = 370 \times 10^{-6} \qquad \varepsilon_2 = 80 \times 10^{-6}$$

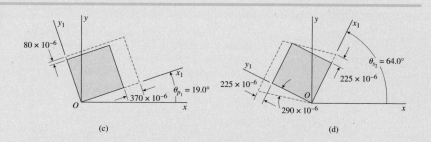

Fig. 7-38c,d (Repeated)

(c)     (d)

in which $\varepsilon_1$ denotes the algebraically larger principal strain and $\varepsilon_2$ denotes the algebraically smaller principal strain. (Recall that we are considering only in-plane strains in this example.)

The angles to the principal directions can be obtained from Eq. (7-85):

$$\tan 2\theta_p = \frac{\gamma_{xy}}{\varepsilon_x - \varepsilon_y} = \frac{180}{340 - 110} = 0.7826$$

The values of $2\theta_p$ between 0 and 360° are 38.0° and 218.0°, and therefore the angles to the principal directions are

$$\theta_p = 19.0° \text{ and } 109.0°$$

To determine the value of $\theta_p$ associated with each principal strain, we substitute $\theta_p = 19.0°$ into the first transformation equation (Eq. 7-83a) and solve for the strain:

$$\varepsilon_{x_1} = \frac{\varepsilon_x + \varepsilon_y}{2} + \frac{\varepsilon_x + \varepsilon_y}{2} \cos 2\theta + \frac{\gamma_{xy}}{2} \sin 2\theta$$

$$= (225 \times 10^{-6}) + (115 \times 10^{-6})(\cos 38.0°) + (90 \times 10^{-6})(\sin 38.0°)$$

$$= 370 \times 10^{-6}$$

This result shows that the larger principal strain $\varepsilon_1$ is at the angle $\theta_{p_1} = 19.0°$. The smaller strain $\varepsilon_2$ acts at 90° from that direction ($\theta_{p_2} = 109.0°$). Thus,

$$\varepsilon_1 = 370 \times 10^{-6} \quad \text{and} \quad \theta_{p_1} = 19.0° \quad \Longleftarrow$$

$$\varepsilon_2 = 80 \times 10^{-6} \quad \text{and} \quad \theta_{p_2} = 109.0° \quad \Longleftarrow$$

Note that $\varepsilon_1 + \varepsilon_2 = \varepsilon_x + \varepsilon_y$.

*Continues* ➡

**Example 7-8 -** *Continued*

The principal strains are portrayed in Fig. 7-38c. There are, of course, no shear strains on the principal planes.

(c) *Maximum shear strain*. The maximum shear strain is calculated from Eq. (7-87):

$$\frac{\gamma_{max}}{2} = \sqrt{\left(\frac{\varepsilon_x - \varepsilon_y}{2}\right)^2 + \left(\frac{\gamma_{xy}}{2}\right)^2} = 146 \times 10^{-6} \qquad \gamma_{max} = 290 \times 10^{-6} \quad \Longleftarrow$$

The element having the maximum shear strains is oriented at 45° to the principal directions; therefore, $\theta_s = 19.0° + 45° = 64.0°$ and $2\theta_s = 128.0°$. By substituting this value of $2\theta_s$ into the second transformation equation (Eq. 7-83b), we can determine the sign of the shear strain associated with this direction. The calculations are

$$\frac{\gamma_{x_1 y_1}}{2} = -\frac{\varepsilon_x - \varepsilon_y}{2} \sin 2\theta + \frac{\gamma_{xy}}{2} \cos 2\theta$$

$$= -(115 \times 10^{-6})(\sin 128.0°) + (90 \times 10^{-6})(\cos 128.0°)$$

$$= -146 \times 10^{-6}$$

This result shows that an element oriented at an angle $\theta_{s_2} = 64.0°$ has the maximum negative shear strain.

We can arrive at the same result by observing that the angle $\theta_{s_1}$ to the direction of maximum positive shear strain is always 45° less than $\theta_{p_1}$. Hence,

$$\theta_{s_1} = \theta_{p_1} - 45° = 19.0° - 45° = -26.0° \quad \Longleftarrow$$

$$\theta_{s_2} = \theta_{s_1} + 90° = 64.0° \quad \Longleftarrow$$

The shear strains corresponding to $\theta_{s_1}$ and $\theta_{s_2}$ are $\gamma_{max} = 290 \times 10^{-6}$ and $\gamma_{min} = -290 \times 10^{-6}$, respectively.

The normal strains on the element having the maximum and minimum shear strains are

$$\varepsilon_{aver} = \frac{\varepsilon_x + \varepsilon_y}{2} = 225 \times 10^{-6} \quad \Longleftarrow$$

A sketch of the element having the maximum in-plane shear strains is shown in Fig. 7-38d.

In this example, we solved for the strains by using the transformation equations. However, all of the results can be obtained just as easily from Mohr's circle.

### ● ● ● Example 7-9

A 45° strain rosette (also called a *rectangular rosette*) consists of three electrical-resistance strain gages arranged to measure strains in two perpendicular directions and also at a 45° angle between them, as shown in Fig. 7-39a. The rosette is bonded to the surface of the structure before it is loaded. Gages A, B, and C measure the normal strains $\varepsilon_a$, $\varepsilon_b$, and $\varepsilon_c$ in the directions of lines Oa, Ob, and Oc, respectively.

Explain how to obtain the strains $\varepsilon_{x_1}$, $\varepsilon_{y_1}$, and $\gamma_{x_1 y_1}$ associated with an element oriented at an angle $\theta$ to the xy axes (Fig. 7-39b).

### Solution

At the surface of the stressed object, the material is in plane stress. Since the strain-transformation equations (Eqs. 7-83a and 7-83b) apply to plane stress as well as to plane strain, we can use those equations to determine the strains in any desired direction.

*Strains associated with the xy axes.* We begin by determining the strains associated with the xy axes. Because gages A and C are aligned with the x and y axes, respectively, they give the strains $\varepsilon_x$ and $\varepsilon_y$ directly:

$$\varepsilon_x = \varepsilon_a \qquad \varepsilon_y = \varepsilon_c \qquad \text{(7-89a,b)}$$

To obtain the shear strain $\gamma_{xy}$, we use the transformation equation for normal strains (Eq. 7-83a):

$$\varepsilon_{x_1} = \frac{\varepsilon_x + \varepsilon_y}{2} + \frac{\varepsilon_x - \varepsilon_y}{2} \cos 2\theta + \frac{\gamma_{xy}}{2} \sin 2\theta$$

For an angle $\theta = 45°$, we know that $\varepsilon_{x_1} = \varepsilon_b$ (Fig. 7-39a); therefore, the preceding equation gives

$$\varepsilon_b = \frac{\varepsilon_a + \varepsilon_c}{2} + \frac{\varepsilon_a - \varepsilon_c}{2} (\cos 90°) + \frac{\gamma_{xy}}{2} (\sin 90°)$$

Solving for $\gamma_{xy}$, we get

$$\gamma_{xy} = 2\varepsilon_b - \varepsilon_a - \varepsilon_c \qquad \text{(7-90)}$$

Thus, the strains $\varepsilon_x$, $\varepsilon_y$, and $\gamma_{xy}$ are easily determined from the given strain-gage readings.

*Strains associated with the $x_1 y_1$ axes.* Knowing the strains $\varepsilon_x$, $\varepsilon_y$, and $\gamma_{xy}$, we can calculate the strains for an element oriented at any angle $\theta$ (Fig. 7-39b) from the strain-transformation equations (Eqs. 7-83a and 7-83b) or from Mohr's circle. We can also calculate the principal strains and the maximum shear strains from Eqs. (7-86) and (7-87), respectively.

### FIG. 7-39

Example 7-9: (a) 45° strain rosette, and (b) element oriented at an angle $\theta$ to the xy axes

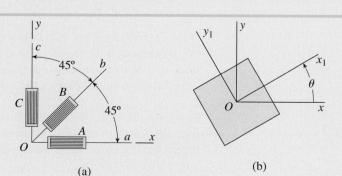

(a)                          (b)

In Chapter 7, we investigated the **state of stress** at a point on a stressed body and then displayed it on a stress element. In two dimensions, **plane stress** was discussed and we derived transformation equations that gave different, but equivalent, expressions of the state of normal and shear stresses at that point. **Principal normal stresses** and **maximum shear stress**, and their orientations, were seen to be the most important information for design. A graphical representation of the transformation equations, **Mohr's circle**, was found to be a convenient way of exploring various representations of the state of stress at a point, including those orientations of the stress element at which principal stresses and maximum shear stress occur. Later, strains were introduced and **Hooke's law for plane stress** was derived (for homogeneous and isotropic materials) and then specialized to obtain stress-strain relationships for **biaxial stress**, uniaxial stress, and pure shear. The stress state in three dimensions, referred to as triaxial stress, was then introduced along with Hooke's law for triaxial stress. **Spherical stress** and **hydrostatic stress** were defined as special cases of triaxial stress. Finally, plane strain was defined for use in experimental stress analysis and compared to plane stress. The major concepts presented in this chapter may be summarized as follows:

1.  The **stresses on inclined sections** cut through a body, such as a beam, may be larger than the stresses acting on a stress element aligned with the cross section.

2.  Stresses are tensors, not vectors, so we used equilibrium of a wedge element to transform the stress components from one set of axes to another. Since the transformation equations were derived solely from equilibrium of an element, they are applicable to stresses in any kind of material, whether linear, nonlinear, elastic, or inelastic. The **transformation equations for plane stress** are

$$\sigma_{x_1} = \frac{\sigma_x + \sigma_y}{2} + \frac{\sigma_x - \sigma_y}{2} \cos 2\theta + \tau_{xy} \sin 2\theta$$

$$\tau_{x_1y_1} = -\frac{\sigma_x - \sigma_y}{2} \sin 2\theta + \tau_{xy} \cos 2\theta$$

$$\sigma_{y_1} = \frac{\sigma_x + \sigma_y}{2} - \frac{\sigma_x - \sigma_y}{2} \cos 2\theta - \tau_{xy} \sin 2\theta$$

3.  If we use two elements with different orientations to display the **state of plane stress** at the same point in a body, the stresses acting on the faces of the two elements are different, but they still represent the same intrinsic state of stress at that point.

4.  From equilibrium, we showed that the shear stresses acting on all four side faces of a stress element in plane stress are known if we determine the shear stress acting on any one of those faces.

5. The sum of the normal stresses acting on perpendicular faces of plane-stress elements (at a given point in a stressed body) is constant and independent of the angle $\theta$:

$$\sigma_{x_1} + \sigma_{y_1} = \sigma_x + \sigma_y$$

6. The maximum and minimum normal stresses (called the **principal stresses** $\sigma_1$, $\sigma_2$) can be found from the transformation equation for normal stress as

$$\sigma_{1,2} = \frac{\sigma_x + \sigma_y}{2} \pm \sqrt{\left(\frac{\sigma_x - \sigma_y}{2}\right)^2 + \tau_{xy}^2}$$

We also can find the principal planes, at orientation $\theta_p$, on which they act. The shear stresses are zero on the principal planes, the planes of maximum shear stress occur at 45° to the principal planes, and the maximum shear stress is equal to one-half the difference of the principal stresses. Maximum shear stress can be computed from the normal and shear stresses on the original element, or from the principal stresses as

$$\tau_{max} = \sqrt{\left(\frac{\sigma_x - \sigma_y}{2}\right)^2 + \tau_{xy}^2}$$

$$\tau_{max} = \frac{\sigma_1 - \sigma_2}{2}$$

7. The transformation equations for plane stress can be represented in graphical form by a plot known as **Mohr's circle** which displays the relationship between normal and shear stresses acting on various inclined planes at a point in a stressed body. It also is used for calculating principal stresses, maximum shear stresses, and the orientations of the elements on which they act.

8. **Hooke's law for plane stress** provides the relationships between normal strains and stresses for homogeneous and isotropic materials which follow Hooke's law. These relationships contain three material constants $(E, G, \text{ and } \mu)$. When the normal stresses in plane stress are known, the normal strains in the $x$, $y$ and $z$ directions are

$$\varepsilon_x = \frac{1}{E}(\sigma_x - v\sigma_y)$$

$$\varepsilon_y = \frac{1}{E}(\sigma_y - v\sigma_x)$$

$$\varepsilon_z = -\frac{v}{E}(\sigma_x + \sigma_y)$$

These equations can be solved simultaneously to give the $x$ and $y$ normal stresses in terms of the strains:

$$\sigma_x = \frac{E}{1 - v^2}(\varepsilon_x + v\varepsilon_y)$$

$$\sigma_y = \frac{E}{1 - v^2}(\varepsilon_y + v\varepsilon_x)$$

9. The **unit volume change** $e$, or the **dilatation** of a solid body, is defined as the change in volume divided by the original volume and is equal to the sum of the normal strains in three perpendicular directions:

$$e = \frac{\Delta V}{V_0} = \varepsilon_x + \varepsilon_y + \varepsilon_z$$

10. The **strain-energy density** for plane stress, or the strain energy stored in a unit volume of the material, is computed as one-half of the sum of the products of stress times corresponding strain, provided Hooke's law holds for the material.

$$u = \frac{1}{2}(\sigma_x\varepsilon_x + \sigma_y\varepsilon_y + \sigma_z\varepsilon_z)$$

11. A state of **triaxial stress** exists in an element if it is subjected to normal stresses in three mutually perpendicular directions and there are no shear stresses on the faces of the element; the stresses are seen to be the principal stresses in the material. A special type of triaxial stress (called **spherical stress**) occurs when all three normal stresses are equal and tensile. If all three stresses are equal and compressive, the triaxial stress state is referred to as **hydrostatic stress**.

12. Finally, **transformation equations for plane strain** may be derived for use in interpretation of experimental measurements made with strain gages. Plane strains at any orientation can be represented in graphical form using Mohr's circle for plane strain. Plane stress and plane strain are compared in Fig. 7-32, and under ordinary conditions do not occur simultaneously. The transformation equations for plane strain were derived solely from geometry and are independent of the material properties. At a given point in a stressed body, the **principal strains and principal stresses** occur in the same directions. Lastly, the transformation equations for plane stress also can be used

for the stresses in plane strain, and the transformation equations for plane strain also can be used for the strains in plane stress. The transformation equations for plane strain are

$$\varepsilon_{x_1} = \frac{\varepsilon_x + \varepsilon_y}{2} + \frac{\varepsilon_x - \varepsilon_y}{2} \cos 2\theta + \frac{\gamma_{xy}}{2} \sin 2\theta$$

$$\frac{\gamma_{x_1 y_1}}{2} = -\frac{\varepsilon_x - \varepsilon_y}{2} \sin 2\theta + \frac{\gamma_{xy}}{2} \cos 2\theta$$

# PROBLEMS CHAPTER 7

## Plane Stress

**7.2-1** The stresses on the bottom surface of a fuel tanker (figure part a) are known to be $\sigma_x = 7750$ psi, $\sigma_y = 1175$ psi, and $\tau_{xy} = 940$ psi (figure part b).

Determine the stresses acting on an element oriented at an angle $\theta = 55°$ from the $x$ axis, where the angle $\theta$ is positive when counterclockwise. Show these stresses on a sketch of an element oriented at the angle $\theta$.

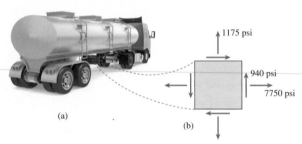

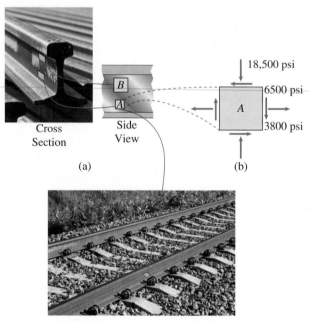

(a)

(b)

**PROB. 7.2-1** ((a) Can Stock Photo Inc/Johan H)

**7.2-2** Solve the preceding problem for an element in *plane stress* on the bottom surface of a fuel tanker (figure part a); stresses are $\sigma_x = 105$ MPa, $\sigma_y = 75$ MPa, and $\tau_{xy} = 25$ MPa.

Determine the stresses acting on an element oriented at an angle $\theta = 40°$ from the $x$ axis, where the angle $\theta$ is positive when counterclockwise. Show these stresses on a sketch of an element oriented at the angle $\theta$.

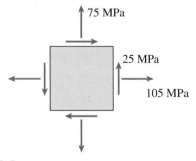

**PROB. 7.2-2**

**7.2-3** The stresses acting on element $A$ on the web of a train rail (see figure part a) are found to be 6500 psi tension in the horizontal direction and 18,500 psi compression in

the vertical direction (see figure part b). Also, shear stresses with a magnitude of 3800 psi act in the directions shown.

Determine the stresses acting on an element oriented at a counterclockwise angle of $30°$ from the horizontal. Show these stresses on a sketch of an element oriented at this angle.

Cross
Section

Side
View

(a)

(b)

**PROB. 7.2-3** ((a) Can Stock Photo Inc./corepics; (b) Can Stock Photo Inc./scanrail)

**7.2-4** Solve the preceding problem if the stresses acting on element $A$ on the web of a train rail (see figure part a of Prob. 7.2-3) are found to be 40 MPa in tension in the horizontal direction and 160 MPa in compression in the vertical direction. Also, shear stresses of magnitude 54 MPa act in the directions shown in the figure.

Determine the stresses acting on an element oriented at a counterclockwise angle of $52°$ from the horizontal. Show these stresses on a sketch of an element oriented at this angle.

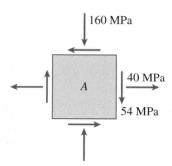

160 MPa

40 MPa

54 MPa

*A*

**PROB. 7.2-4**

**7.2-5** The stresses acting on element *B* on the web of a train rail (see figure part a of Prob. 7.2-3) are found to be 5700 psi in compression in the horizontal direction and 2300 psi in compression in the vertical direction (see figure). Also, shear stresses of magnitude 2500 psi act in the directions shown.

Determine the stresses acting on an element oriented at a counterclockwise angle of 50° from the horizontal. Show these stresses on a sketch of an element oriented at this angle.

2300 psi

2500 psi

*B*

5700 psi

**PROB. 7.2-5**

**7.2-6** An element in *plane stress* on the fuselage of an airplane (figure part a) is subjected to compressive stresses with a magnitude of 42 MPa in the horizontal direction and tensile stresses with a magnitude of 9.5 MPa in the vertical direction (see figure part b). Also, shear stresses with a magnitude of 15.5 MPa act in the directions shown.

Determine the stresses acting on an element oriented at a clockwise angle of 40° from the horizontal. Show these stresses on a sketch of an element oriented at this angle.

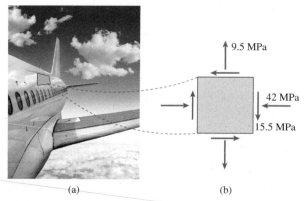

9.5 MPa

42 MPa

15.5 MPa

(a)                                          (b)

**PROB. 7.2-6 ((a) Daboost/Shutterstock)**

**7.2-7** The stresses acting on element *B* (see figure part a) on the web of a wide-flange beam are found to be 14,500 psi in compression in the horizontal direction and 2530 psi in compression in the vertical direction (see figure part b). Also, shear stresses with a magnitude of 3500 psi act in the directions shown.

Determine the stresses acting on an element oriented at a counterclockwise angle of 38° from the horizontal. Show these stresses on a sketch of an element oriented at this angle.

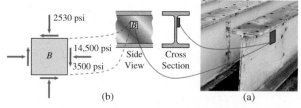

2530 psi

14,500 psi

3500 psi

*B*

Side View

Cross Section

(b)                                          (a)

**PROB. 7.2-7 ((a) Can Stock Photo Inc./rekemp)**

**7.2-8** Solve the preceding problem if the normal and shear stresses acting on element $B$ are 56 MPa, 17 MPa, and 27 MPa (in the directions shown in the figure) and the angle is 40° (clockwise).

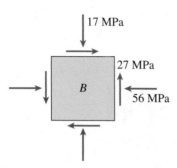

**PROB. 7.2-8**

**7.2-9** The polyethylene liner of a settling pond is subjected to stresses $\sigma_x = 350$ psi, $\sigma_y = 112$ psi, and $\tau_{xy} = -120$ psi, as shown by the plane-stress element in the figure part a.

Determine the normal and shear stresses acting on a seam oriented at an angle of 30° to the element, as shown in the figure part b. Show these stresses on a sketch of an element having its sides parallel and perpendicular to the seam.

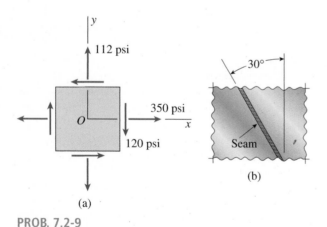

(a)

**PROB. 7.2-9**

**7.2-10** Solve the preceding problem if the normal and shear stresses acting on the element are $\sigma_x = 2100$ kPa, $\sigma_y = 300$ kPa, and $\tau_{xy} = -560$ kPa, and the seam is oriented at an angle of 22.5° to the element.

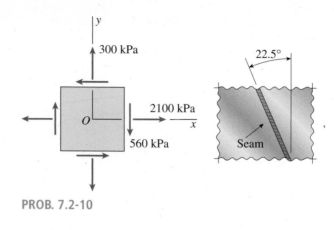

**PROB. 7.2-10**

**7.2-11** A rectangular plate of dimensions 3.0 in. × 5.0 in. is formed by welding two triangular plates (see figure). The plate is subjected to a tensile stress of 500 psi in the long direction and a compressive stress of 350 psi in the short direction.

Determine the normal stress $\sigma_w$ acting perpendicular to the line of the weld and the shear stress $\tau_w$ acting parallel to the weld. (Assume that the normal stress $\sigma_w$ is positive when it acts in tension against the weld and the shear stress $\tau_w$ is positive when it acts counterclockwise against the weld.)

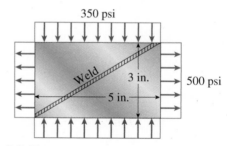

**PROB. 7.2-11**

**7.2-12** Solve the preceding problem for a plate of dimensions 100 mm × 250 mm subjected to a compressive stress of 2.5 MPa in the long direction and a tensile stress of 12.0 MPa in the short direction (see figure).

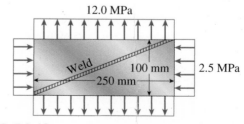

**PROB. 7.2-12**

**7.2-13** At a point on the surface of an elliptical exercise machine the material is in *biaxial stress* with $\sigma_x = 1400$ psi and $\sigma_y = -900$ psi, as shown in the figure part a. The figure part b shows an inclined plane *aa* cut through the same point in the material but oriented at an angle $\theta$.

Determine the value of the angle $\theta$ between zero and 90° such that no normal stress acts on plane *aa*. Sketch a stress element having plane *aa* as one of its sides and show all stresses acting on the element.

**7.2-15** An element in *plane stress* from the frame of a racing car is oriented at a known angle $\theta$ (see figure). On this inclined element, the normal and shear stresses have the magnitudes and directions shown in the figure.

Determine the normal and shear stresses acting on an element whose sides are parallel to the *xy* axes, that is, determine $\sigma_x$, $\sigma_y$, and $\tau_{xy}$. Show the results on a sketch of an element oriented at $\theta = 0°$.

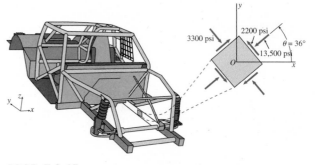

**PROB. 7.2-15**

**7.2-16** Solve the preceding problem for the element shown in the figure.

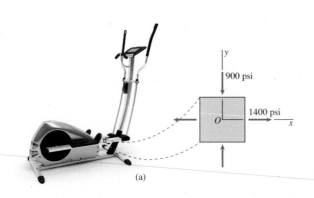

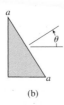

(a)

(b)

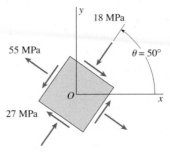

**PROB. 7.2-16**

**PROB. 7.2-13** ((a) Ali Ender Birer/Shutterstock)

**7.2-14** Solve the preceding problem for $\sigma_x = 11$ MPa and $\sigma_y = -20$ MPa (see figure).

**7.2-17** A gusset plate on a truss bridge is in *plane stress* with normal stresses $\sigma_x$ and $\sigma_y$ and *shear stress* $\tau_{xy}$, as shown in the figure. At counterclockwise angles $\theta = 32°$ and $\theta = 78°$ from the *x* axis, the normal stress is 4200 psi in tension.

If the stress $\sigma_x$ equals 2650 psi in tension, what are the stresses $\sigma_y$ and $\tau_{xy}$?

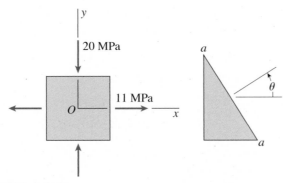

**PROB. 7.2-14**

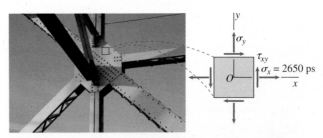

**PROB. 7.2-17** (Photo Courtesy fo John A. Weeks III)

**7.2-18** The surface of an airplane wing is subjected to *plane stress* with normal stresses $\sigma_x$ and $\sigma_y$ and shear stress $\tau_{xy}$, as shown in the figure. At a counterclockwise angle $\theta = 32°$ from the $x$ axis, the normal stress is 29 MPa in tension, and at an angle $\theta = 46°$, it is 17 MPa in compression.

If the stress $\sigma_x$ equals 105 MPa in tension, what are the stresses $\sigma_y$ and $\tau_{xy}$?

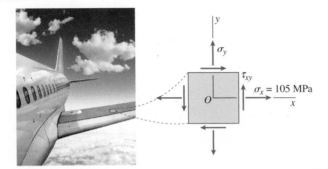

PROB. 7.2-18  (Daboost/Shutterstock)

**7.2-19** At a point on the web of a girder on an overhead bridge crane in a manufacturing facility, the stresses are known to be $\sigma_x = -4300$ psi, $\sigma_y = 1700$ psi, and $\tau_{xy} = 3100$ psi (the sign convention for these stresses is shown in Fig. 7-1). A stress element located at the same point in the structure (but oriented at a counterclockwise angle $\theta_1$ with respect to the $x$ axis) is subjected to the stresses shown in the figure ($\sigma_b$, $\tau_b$, and 2100 psi).

Assuming that the angle $\theta_1$ is between zero and 90°, calculate the normal stress $\sigma_b$, the shear stress $\tau_b$, and the angle $\theta_1$.

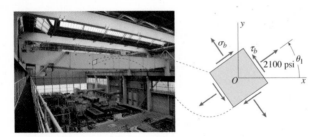

PROB. 7.2-19  (© Paul Rapson/Alamy)

# Principal Stresses and Maximum Shear Stresses

*When solving the problems for Section 7.3, consider only the in-plane stresses (the stresses in the xy plane).*

**7.3-1** An element in *plane stress* is subjected to stresses $\sigma_x = 5750$ psi, $\sigma_y = 1100$ psi, and $\tau_{xy} = 750$ psi (see the figure for Prob. 7.2-1).

Determine the principal stresses and show them on a sketch of a properly oriented element.

**7.3-2** An element in *plane stress* is subjected to stresses $\sigma_x = 105$ MPa, $\sigma_y = 75$ MPa, and $\tau_{xy} = 25$ MPa (see the figure for Prob. 7.2-2).

Determine the principal stresses and show them on a sketch of a properly oriented element.

**7.3-3** An element in *plane stress* is subjected to stresses $\sigma_x = -5500$ psi, $\sigma_y = -2000$ psi, and $\tau_{xy} = 1900$ psi (see the figure for Prob. 7.2-3).

Determine the principal stresses and show them on a sketch of a properly oriented element.

**7.3-4** The stresses acting on element $A$ in the web of a train rail are found to be 40 MPa tension in the horizontal direction and 160 MPa compression in the vertical direction (see figure). Also, shear stresses of magnitude 54 MPa act in the directions shown (see the figure for Prob. 7.2-4).

Determine the principal stresses and show them on a sketch of a properly oriented element.

**7.3-5** The normal and shear stresses acting on element $A$ are 6500 psi, 17,300 psi, and 2900 psi (see the figure for Prob. 7.2-4).

Determine the maximum shear stresses and associated normal stresses and show them on a sketch of a properly oriented element.

**7.3-6** An element in *plane stress* from the fuselage of an airplane is subjected to compressive stresses of magnitude 35 MPa in the horizontal direction and tensile stresses of magnitude 6.5 MPa in the vertical direction. Also, shear stresses of magnitude 12.5 MPa act in the directions shown (see the figure for Prob. 7.2-6).

Determine the maximum shear stresses and associated normal stresses and show them on a sketch of a properly oriented element.

**7.3-7** The stresses acting on element $B$ in the web of a wide-flange beam are found to be 14,000 psi compression in the horizontal direction and 2600 psi compression in the vertical direction. Also, shear stresses of magnitude 3800 psi act in the directions shown (see the figure for Prob. 7.2-7).

Determine the maximum shear stresses and associated normal stresses and show them on a sketch of a properly oriented element.

**7.3-8** The normal and shear stresses acting on element $B$ are $\sigma_x = -46$ MPa, $\sigma_y = -13$ MPa, and $\tau_{xy} = 21$ MPa (see figure for Prob. 7.2-8).

Determine the maximum shear stresses and associated normal stresses and show them on a sketch of a properly oriented element.

**7.3-9** A shear wall in a reinforced concrete building is subjected to a vertical uniform load of intensity $q$ and a horizontal force $H$, as shown in the first part of the figure. (The force $H$ represents the effects of wind and earthquake loads.) As a consequence of these loads, the stresses at point $A$ on the surface of the wall have the values shown in the second part of the figure (compressive stress equal to 1100 psi and shear stress equal to 480 psi).

(a) Determine the principal stresses and show them on a sketch of a properly oriented element.

(b) Determine the maximum shear stresses and associated normal stresses and show them on a sketch of a properly oriented element.

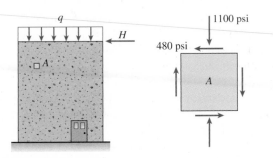

**PROB. 7.3-9**

**7.3-10** A propeller shaft subjected to combined torsion and axial thrust is designed to resist a shear stress of 57 MPa and a compressive stress of 105 MPa (see figure).

(a) Determine the principal stresses and show them on a sketch of a properly oriented element.

(b) Determine the maximum shear stresses and associated normal stresses and show them on a sketch of a properly oriented element.

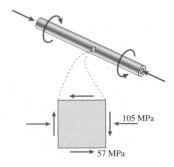

**PROB. 7.3-10**

**7.3-11** The stresses at a point along a beam supporting a sign (see figure) are $\sigma_x = 2250$ psi, $\sigma_y = 1175$ psi, and $\tau_{xy} = -820$ psi.

(a) Find the principal stresses. Show them on a sketch of a properly oriented element.

(b) Find the maximum shear stresses and associated normal stresses. Show them on a sketch of a properly oriented element.

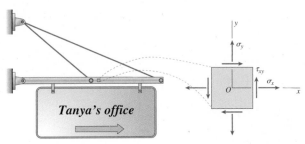

**PROB. 7.3-11**

**7.3-12 through 7.3-16** An element in *plane stress* (see figure) is subjected to stresses $\sigma_x$, $\sigma_y$, and $\tau_{xy}$.

(a) Determine the principal stresses and show them on a sketch of a properly oriented element.

(b) Determine the maximum shear stresses and associated normal stresses and show them on a sketch of a properly oriented element.

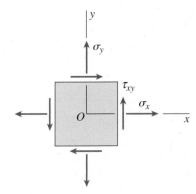

**PROBS. 7.3-12 through 7.3-16**

**7.3-12** $\sigma_x = 2150$ kPa, $\sigma_y = 375$ kPa, $\tau_{xy} = -460$ kPa

**7.3-13** $\sigma_x = 14{,}500$ psi, $\sigma_y = 1070$ psi, $\tau_{xy} = 1900$ psi

**7.3-14** $\sigma_x = 16.5$ MPa, $\sigma_y = -91$ MPa, $\tau_{xy} = -39$ MPa

**7.3-15** $\sigma_x = -3300$ psi, $\sigma_y = -11{,}000$ psi, $\tau_{xy} = 4500$ psi

**7.3-16** $\sigma_x = -108$ MPa, $\sigma_y = 58$ MPa, $\tau_{xy} = -58$ MPa

**7.3-17** At a point on the web of a girder on a gantry crane, the stresses acting on the $x$ face of a stress element are $\sigma_x = 6250$ psi and $\tau_{xy} = 1425$ psi (see figure).

What is the allowable range of values for the stress $\sigma_y$ if the maximum shear stress is limited to $\tau_0 = 2150$ psi?

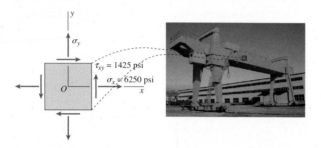

**PROB. 7.3-17** (ZCW/Shutterstock)

**7.3-18** The stresses acting on a stress element on the arm of a power excavator (see figure) are $\sigma_x = 52$ MPa and $\tau_{xy} = 33$ MPa (see figure).

What is the allowable range of values for the stress $\sigma_y$ if the maximum shear stress is limited to $\tau_0 = 37$ MPa?

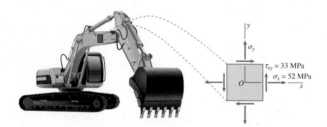

**PROB. 7.3-18** (Can Stock Photo Inc./busja)

**7.3-19** The stresses at a point on the down tube of a bicycle frame are $\sigma_x = 4800$ psi and $\tau_{xy} = -1950$ psi (see figure). It is known that one of the principal stresses equals 6375 psi in tension.

(a) Determine the stress $\sigma_y$.

(b) Determine the other principal stress and the orientation of the principal planes, then show the principal stresses on a sketch of a properly oriented element.

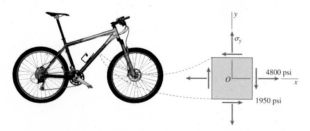

**PROB. 7.3-19** (Can Stock Photo Inc./Aviafan)

**7.3-20** An element in *plane stress* on the surface of an automobile drive shaft (see figure) is subjected to stresses of $\sigma_x = -45$ MPa and $\tau_{xy} = 39$ MPa (see figure). It is known that one of the principal stresses equals 41 MPa in tension.

(a) Determine the stress $\sigma_y$.

(b) Determine the other principal stress and the orientation of the principal planes, then show the principal stresses on a sketch of a properly oriented element.

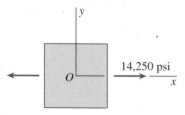

**PROB. 7.3-20** (Courtesy of www.rietzusa.com)

# Mohr's Circle

*The problems for Section 7.4 are to be solved using Mohr's circle. Consider only the in-plane stresses (the stresses in the xy plane).*

**7.4-1** An element in *uniaxial stress* is subjected to tensile stresses $\sigma_x = 14{,}250$ psi, as shown in the figure. Using Mohr's circle, determine the following.

(a) The stresses acting on an element oriented at a counterclockwise angle $\theta = 29°$ from the $x$ axis.

(b) The maximum shear stresses and associated normal stresses.

Show all results on sketches of properly oriented elements.

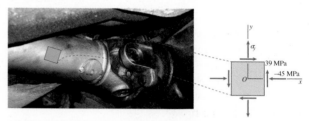

**PROB. 7.4-1**

**7.4-2** An element in *uniaxial stress* is subjected to tensile stresses $\sigma_x = 57$ MPa, as shown in the figure. Using Mohr's circle, determine the following.

(a) The stresses acting on an element oriented at an angle $\theta = -33°$ from the $x$ axis (minus means clockwise).

(b) The maximum shear stresses and associated normal stresses.

Show all results on sketches of properly oriented elements.

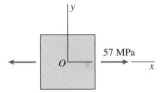

**PROB. 7.4-2**

**7.4-3** An element on the gusset plate in Prob. 7.2-17 in *uniaxial stress* is subjected to compressive stresses of magnitude 6750 psi, as shown in the figure. Using Mohr's circle, determine the following.

(a) The stresses acting on an element oriented at a slope of 1 on 2 (see figure).

(b) The maximum shear stresses and associated normal stresses.

Show all results on sketches of properly oriented elements.

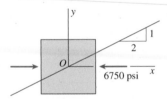

**PROB. 7.4-3**

**7.4-4** An element on the top surface of the fuel tanker in Prob. 7.2-1 is in *biaxial stress* and is subjected to stresses $\sigma_x = -48$ MPa and $\sigma_y = 19$ MPa, as shown in the figure. Using Mohr's circle, determine the following.

(a) The stresses acting on an element oriented at a counterclockwise angle $\theta = 25°$ from the $x$ axis.

(b) The maximum shear stresses and associated normal stresses.

Show all results on sketches of properly oriented elements.

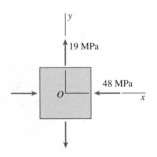

**PROB. 7.4-4**

**7.4-5** An element on the top surface of the fuel tanker in Prob. 7.2-1 is in *biaxial stress* and is subjected to stresses $\sigma_x = 6250$ psi and $\sigma_y = -1750$ psi, as shown in the figure. Using Mohr's circle, determine the following.

(a) The stresses acting on an element oriented at a counterclockwise angle $\theta = 55°$ from the $x$ axis.

(b) The maximum shear stresses and associated normal stresses.

Show all results on sketches of properly oriented elements.

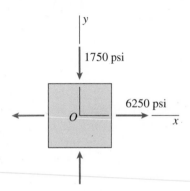

**PROB. 7.4-5**

**7.4-6** An element in *biaxial stress* is subjected to stresses $\sigma_x = -29$ MPa and $\sigma_y = 57$ MPa, as shown in the figure. Using Mohr's circle, determine the following.

(a) The stresses acting on an element oriented at a slope of 1 on 2.5 (see figure).

(b) The maximum shear stresses and associated normal stresses.

Show all results on sketches of properly oriented elements.

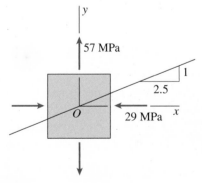

**PROB. 7.4-6**

**7.4-7** An element on the surface of a drive shaft is in *pure shear* and is subjected to stresses $\tau_{xy}$ = 2700 psi, as shown in the figure. Using Mohr's circle, determine the following.

(a) The stresses acting on an element oriented at a counterclockwise angle $\theta$ = 52° from the $x$ axis.

(b) The principal stresses.

Show all results on sketches of properly oriented elements.

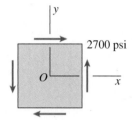

**PROB. 7.4-7**

**7.4-8** The rotor shaft of a helicopter (see figure part a) drives the rotor blades that provide the lifting force and is subjected to a combination of torsion and axial loading (see figure part b).

It is known that normal stress $\sigma_y$ = 68 MPa and shear stress $\tau_{xy}$ = −100 MPa. Using Mohr's circle, determine the following.

(a) The stresses acting on an element oriented at a counterclockwise angle $\theta$ = 22.5° from the $x$ axis

(b) Find the maximum tensile stress, maximum compressive stress, and maximum shear stress in the shaft.

Show all results on sketches of properly oriented elements.

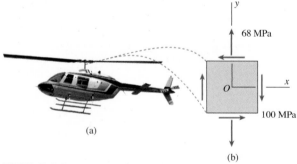

**PROB. 7.4-8** (iker canikligil/Shutterstock)

**7.4-9** An element in *pure shear* is subjected to stresses $\tau_{xy}$ = 3750 psi, as shown in the figure. Using Mohr's circle, determine the following.

(a) The stresses acting on an element oriented at a slope of 3 on 4 (see figure).

(b) The principal stresses.

Show all results on sketches of properly oriented elements.

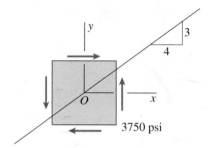

**PROB. 7.4-9**

**7.4-10 through 7.4-15** An element in *plane stress* is subjected to stresses $\sigma_x$, $\sigma_y$, and $\tau_{xy}$ (see figure).

Using Mohr's circle, determine the stresses acting on an element oriented at an angle $\theta$ from the $x$ axis. Show these stresses on a sketch of an element oriented at the angle $\theta$. (*Note:* The angle $\theta$ is positive when counterclockwise and negative when clockwise.)

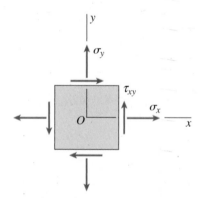

**PROBS. 7.4-10 through 7.4-15**

**7.4-10**  $\sigma_x$ = 27 MPa,  $\sigma_y$ = 14 MPa,  $\tau_{xy}$ = 6 MPa, $\theta$ = 40°

**7.4-11**  $\sigma_x$ = 3500 psi, $\sigma_y$ = 12,200 psi, $\tau_{xy}$ = −3300 psi, $\theta$ = −51°

**7.4-12**  $\sigma_x$ = −47 MPa, $\sigma_y$ = −186 MPa, $\tau_{xy}$ = −29 MPa, $\theta$ = −33°

**7.4-13**  $\sigma_x$ = −1720 psi,  $\sigma_y$ = −680 psi,  $\tau_{xy}$ = 320 psi, $\theta$ = 14°

**7.4-14**  $\sigma_x$ = 33 MPa,  $\sigma_y$ = −9 MPa,  $\tau_{xy}$ = 29 MPa, $\theta$ = 35°

**7.4-15**  $\sigma_x$ = −5700 psi,  $\sigma_y$ = 950 psi,  $\tau_{xy}$ = −2100 psi, $\theta$ = 65°

**7.4-16 through 7.4-23** An element in *plane stress* is subjected to stresses $\sigma_x$, $\sigma_y$, and $\tau_{xy}$ (see figure).

Using Mohr's circle, determine (a) the principal stresses, and (b) the maximum shear stresses and associated normal stresses. Show all results on sketches of properly oriented elements.

shown in the figure. Strain gages $A$ and $B$, oriented in the $x$ and $y$ directions, respectively, are attached to the plate. The gage readings give normal strains $\varepsilon_x = 0.00065$ (elongation) and $\varepsilon_y = 0.00040$ (elongation).

Knowing that $E = 30 \times 10^6$ psi and $v = 0.3$, determine the stresses $\sigma_x$ and $\sigma_y$ and the change $\Delta t$ in the thickness of the plate.

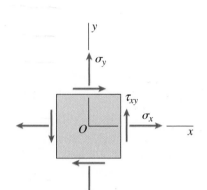

PROBS. 7.4-16 through 7.4-23

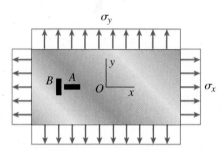

PROBS. 7.5-1 and 7.5-2

**7.4-16** $\sigma_x = 2900$ kPa, $\sigma_y = 9100$ kPa, $\tau_{xy} = -3750$ kPa

**7.4-17** $\sigma_x = 800$ psi, $\sigma_y = -2200$ psi, $\tau_{xy} = 2900$ psi

**7.4-18** $\sigma_x = -3.3$ MPa, $\sigma_y = 8.9$ MPa, $\tau_{xy} = -14.1$ MPa

**7.4-19** $\sigma_x = -11,500$ psi, $\sigma_y = -18,250$ psi, $\tau_{xy} = -7200$ psi

**7.4-20** $\sigma_x = -29.5$ MPa, $\sigma_y = 29.5$ MPa, $\tau_{xy} = 27$ MPa

**7.4-21** $\sigma_x = 2050$ psi, $\sigma_y = 6100$ psi, $\tau_{xy} = 2750$ psi

**7.4-22** $\sigma_x = 0$ MPa, $\sigma_y = -23.4$ MPa, $\tau_{xy} = -9.6$ MPa

**7.4-23** $\sigma_x = 7300$ psi, $\sigma_y = 0$ psi, $\tau_{xy} = 1300$ psi

**7.5-2** Solve the preceding problem if the thickness of the steel plate is $t = 12$ mm, the gage readings are $\varepsilon_x = 530 \times 10^{-6}$ (elongation) and $\varepsilon_y = -210 \times 10^{-6}$ (shortening), the modulus is $E = 200$ GPa, and Poisson's ratio is $v = 0.30$.

**7.5-3** Assume that the normal strains $\varepsilon_x$ and $\varepsilon_y$ for an element in *plane stress* (see figure) are measured with strain gages.

(a) Obtain a formula for the normal strain $\varepsilon_z$ in the $z$ direction in terms of $\varepsilon_x$, $\varepsilon_y$, and Poisson's ratio $v$.

(b) Obtain a formula for the dilatation $e$ in terms of $\varepsilon_x$, $\varepsilon_y$, and Poisson's ratio $v$.

# Hooke's Law for Plane Stress

*When solving the problems for Section 7.5, assume that the material is linearly elastic with modulus of elasticity E and Poisson's ratio v.*

**7.5-1** A rectangular steel plate with thickness $t = 5/8$ in. is subjected to uniform normal stresses $\sigma_x$ and $\sigma_y$, as

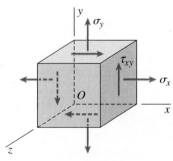

PROB. 7.5-3

**7.5-4** A cast-iron plate in *biaxial stress* is subjected to tensile stresses $\sigma_x = 31$ MPa and $\sigma_y = 17$ MPa (see figure). The corresponding strains in the plate are $\varepsilon_x = 240 \times 10^{-6}$ and $\varepsilon_y = 85 \times 10^{-6}$.

Determine Poisson's ratio $v$ and the modulus of elasticity $E$ for the material.

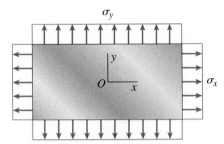

PROBS. 7.5-4 through 7.5-7

**7.5-5** Solve the preceding problem for a steel plate with $\sigma_x = 11,600$ psi (tension), $\sigma_y = -5700$ psi (compression), $\varepsilon_x = 450 \times 10^{-6}$ (elongation), and $\varepsilon_y = -310 \times 10^{-6}$ (shortening).

**7.5-6** A rectangular plate in *biaxial stress* (see figure) is subjected to normal stresses $\sigma_x = 67$ MPa (tension) and $\sigma_y = -23$ MPa (compression). The plate has dimensions $400 \times 550 \times 20$ mm and is made of steel with $E = 200$ GPa and $v = 0.30$.

(a) Determine the maximum in-plane shear strain $\gamma_{max}$ in the plate.

(b) Determine the change $\Delta t$ in the thickness of the plate.

(c) Determine the change $\Delta V$ in the volume of the plate.

**7.5-7** Solve the preceding problem for an aluminum plate with $\sigma_x = 12,000$ psi (tension), $\sigma_y = -3000$ psi (compression), dimensions $20 \times 30 \times 0.5$ in., $E = 10.5 \times 10^6$ psi, and $v = 0.33$.

**7.5-8** A brass cube of 48 mm on each edge is compressed in two perpendicular directions by forces $P = 160$ kN (see figure).

(a) Calculate the change $\Delta V$ in the volume of the cube and the strain energy $U$ stored in the cube, assuming $E = 100$ GPa and $v = 0.34$.

(b) Repeat part (a) if the cube is made of an aluminum alloy with $E = 73$ GPa and $v = 0.33$.

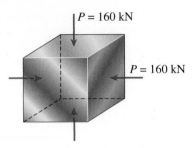

PROB. 7.5-8

**7.5-9** A 4.0-inch cube of concrete ($E = 4.5 \times 10^6$ psi, $v = 0.2$) is compressed in *biaxial stress* by means of a framework that is loaded as shown in the figure.

Assuming that each load $F$ equals 25 k, determine the change $\Delta V$ in the volume of the cube and the strain energy $U$ stored in the cube.

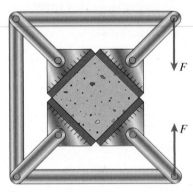

PROB. 7.5-9

**7.5-10** A square plate of width $b$ and thickness $t$ is loaded by normal forces $P_x$ and $P_y$ and by shear forces $V$, as shown in the figure. These forces produce uniformly distributed stresses acting on the side faces of the plate.

(a) Calculate the change $\Delta V$ in the volume of the plate and the strain energy $U$ stored in the plate if the dimensions are $b = 600$ mm and $t = 40$ mm; the plate is made of magnesium with $E = 41$ GPa and $v = 0.35$; and the forces are $P_x = 420$ kN, $P_y = 210$ kN, and $V = 96$ kN.

(b) Find the maximum permissible thickness of the plate when the strain energy $U$ must be at least 62 J. (Assume that all other numerical values in part (a) are unchanged.)

(c) Find the minimum with $b$ of the square plate of thickness $t = 40$ mm when the change in volume of the plate cannot exceed 0.018% of the original volume.

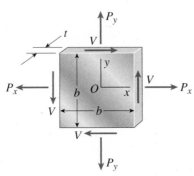

**PROBS. 7.5-10 and 7.5-11**

**7.5-11** Solve the preceding problem for an aluminum plate with $b = 10$ in., $t = 0.75$ in., $E = 10,600$ ksi, $v = 0.33$, $P_x = 96$ k, $P_y = 24$ k, and $V = 18$ k.

For part (b), assume that the required strain energy stored is 640 in.-lb. In part (c), the change in volume cannot exceed 0.05%.

**7.5-12** A circle of diameter $d = 200$ mm is etched on a brass plate (see figure). The plate has dimensions of $400 \times 400 \times 20$ mm. Forces are applied to the plate, producing uniformly distributed normal stresses $\sigma_x = 59$ MPa and $\sigma_y = -17$ MPa. Calculate the following quantities: (a) the change in length $\Delta ac$ of diameter $ac$; (b) the change in length $\Delta bd$ of diameter $bd$; (c) the change $\Delta t$ in the thickness of the plate; (d) the change $\Delta V$ in the volume of the plate, and (e) the strain energy $U$ stored in the plate; (f) the maximum permissible thickness of the plate when strain energy $U$ must be at least 78.4 J; (g) the maximum permissible value of normal stress $\sigma_x$ when the change in volume of the plate cannot exceed 0.015% of the original volume. (Assume $E = 100$ GPa and $v = 0.34$.)

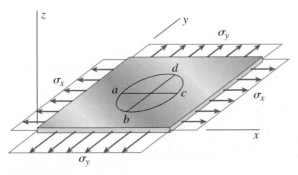

**PROB. 7.5-12**

# Triaxial Stress

*When solving the problems for Section 7.6, assume that the material is linearly elastic with modulus of elasticity E and Poisson's ratio v.*

**7.6-1** An element of aluminum in the form of a rectangular parallelepiped (see figure) of dimensions $a = 5.5$ in., $b = 4.5$ in., and $c = 3.5$ in. is subjected to *triaxial stresses* $\sigma_x = 12,500$ psi, $\sigma_y = -5000$ psi, and $\sigma_z = -1400$ psi acting on the $x$, $y$, and $z$ faces, respectively.

Determine the following quantities: (a) the maximum shear stress $\tau_{max}$ in the material; (b) the changes $\Delta a$, $\Delta b$, and $\Delta c$ in the dimensions of the element; (c) the change $\Delta V$ in the volume; (d) the strain energy $U$ stored in the element; (e) the maximum value of $\sigma_x$ when the change in volume must be limited to 0.021%; and (f) the required value of $\sigma_x$ when the strain energy must be 900 in.-lb. (Assume $E = 10,400$ ksi and $v = 0.33$.)

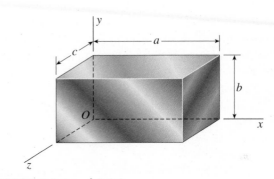

**PROBS. 7.6-1 and 7.6-2**

**7.6-2** Solve the preceding problem if the element is steel ($E = 200$ GPa, $v = 0.30$) with dimensions $a = 300$ mm, $b = 150$ mm, and $c = 150$ mm and the stresses $\sigma_x = -62$ MPa, $\sigma_y = -45$ MPa, and $\sigma_z = -45$ MPa.

For part (e), find the maximum value of $\sigma_x$ if the change in volume must be limited to $-0.028\%$. For part (f), find the required value of $\sigma_x$ if the strain energy must be 60 J.

**7.6-3** A cube of cast iron with sides of length $a = 4.0$ in. (see figure) is tested in a laboratory under *triaxial stress*. Gages mounted on the testing machine show that the compressive strains in the material are $\varepsilon_x = -225 \times 10^{-6}$ and $\varepsilon_y = \varepsilon_z = -37.5 \times 10^{-6}$.

Determine the following quantities: (a) the normal stresses $\sigma_x$, $\sigma_y$, and $\sigma_z$ acting on the $x$, $y$, and $z$ faces of the cube; (b) the maximum shear stress $\tau_{max}$ in the material; (c) the change $\Delta V$ in the volume of the cube; (d) the strain energy $U$ stored in the cube; (e) the maximum value of $\sigma_x$ when the change in volume must be limited to 0.028%; and (f) the required value of $\varepsilon_x$ when the strain energy must be 38 in.-lb. (Assume $E = 14{,}000$ ksi and $v = 0.25$.)

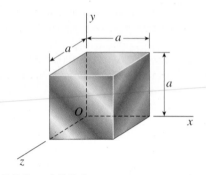

**PROBS. 7.6-3 and 7.6-4**

**7.6-4** Solve the preceding problem if the cube is granite ($E = 80$ GPa, $v = 0.25$) with dimensions $a = 89$ mm and compressive strains $\varepsilon_x = 690 \times 10^{-6}$ and $\varepsilon_y = \varepsilon_z = 255 \times 10^{-6}$. For part (e), find the maximum value of $\sigma_x$ when the change in volume must be limited to 0.11%. For part (f), find the required value of $\varepsilon_x$ when the strain energy must be 33 J.

**7.6-5** An element of aluminum in subjected to *triaxial stress* (see figure).

(a) Find the bulk modulus $K$ for the aluminum if the following stress and strain data is known: normal stresses are $\sigma_x = 5200$ psi (tension), $\sigma_y = -4750$ psi (compression), and $\sigma_z = -3090$ psi (compression) and normal strains in the $x$ and $y$ directions are $\varepsilon_x = 713.8 \times 10^{-6}$ (elongation) and $\varepsilon_y = -502.3 \times 10^{-6}$ (shortening).

(b) If the element is replaced by one of magnesium, find the modulus of elasticity $E$ and Poisson's ratio $v$ if the following data is given: bulk modulus $K = 6.8 \times 10^6$ psi; normal stresses are $\sigma_x = 4550$ psi (tension), $\sigma_y = -1700$ psi (compression), and $\sigma_z = -1090$ psi (compression); and normal strain in the $x$ direction is $\varepsilon_x = 900 \times 10^{-6}$ (elongation).

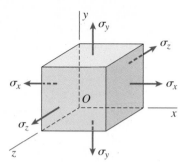

**PROBS. 7.6-5 and 7.6-6**

**7.6-6** Solve the preceding problem if the material is nylon.

(a) Find the bulk modulus $K$ for the nylon if the following stress and strain data is known: normal stresses are $\sigma_x = -3.9$ MPa, $\sigma_y = -3.2$ MPa, and $\sigma_z = -1.8$ MPa, and normal strains in the $x$ and $y$ directions are $\varepsilon_x = -640 \times 10^{-6}$ (shortening) and $\varepsilon_y = -310 \times 10^{-6}$ (shortening).

(b) If the element is replaced by one of polyethylene, find the modulus of elasticity $E$ and Poisson's ratio $v$ if the following data is given: bulk modulus $K = 2162$ MPa; normal stresses are $\sigma_x = -3.6$ MPa (compression) $\sigma_y = -2.1$ MPa (compression), and $\sigma_z = -2.1$ MPa (compression); and normal strain in the $x$ direction is $\varepsilon_x = -1480 \times 10^{-6}$ (shortening).

**7.6-7** A rubber cylinder $R$ of length $L$ and cross-sectional area $A$ is compressed inside a steel cylinder $S$ by a force $F$ that applies a uniformly distributed pressure to the rubber (see figure).

(a) Derive a formula for the lateral pressure $p$ between the rubber and the steel. (Disregard friction between the rubber and the steel, and assume that the steel cylinder is rigid when compared to the rubber.)

(b) Derive a formula for the shortening $\delta$ of the rubber cylinder.

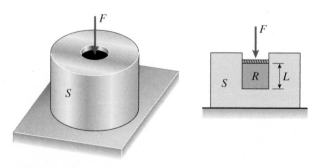

**PROB. 7.6-7**

**7.6-8** A block $R$ of rubber is confined between plane parallel walls of a steel block $S$ (see figure). A uniformly distributed pressure $p_0$ is applied to the top of the rubber block by a force $F$.

(a) Derive a formula for the lateral pressure $p$ between the rubber and the steel. (Disregard friction between the rubber and the steel, and assume that the steel block is rigid when compared to the rubber.)

(b) Derive a formula for the dilatation $e$ of the rubber.

(c) Derive a formula for the strain-energy density $u$ of the rubber.

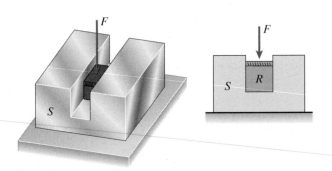

**PROB. 7.6-8**

**7.6-9** A solid spherical ball of magnesium alloy ($E = 6.5 \times 10^6$ psi, $v = 0.35$) is lowered into the ocean to a depth of 8000 ft. The diameter of the ball is 9.0 in.

(a) Determine the decrease $\Delta d$ in diameter, the decrease $\Delta V$ in volume, and the strain energy $U$ of the ball.

(b) At what depth will the volume change be equal to 0.0324% of the original volume?

**7.6-10** A solid steel sphere ($E = 210$ GPa, $v = 0.3$) is subjected to hydrostatic pressure $p$ such that its volume is reduced by 0.4%.

(a) Calculate the pressure $p$.

(b) Calculate the volume modulus of elasticity $K$ for the steel.

(c) Calculate the strain energy $U$ stored in the sphere if its diameter is $d = 150$ mm.

**7.6-11** A solid bronze sphere (volume modulus of elasticity $K = 14.5 \times 10^6$ psi) is suddenly heated around its outer surface. The tendency of the heated part of the sphere to expand produces uniform tension in all directions at the center of the sphere.

If the stress at the center is 12,000 psi, what is the strain? Also, calculate the unit volume change $e$ and the strain-energy density $u$ at the center.

## Plane Strain

*When solving the problems for Section 7.7, consider only the in-plane strains (the strains in the xy plane) unless stated otherwise. Use the transformation equations of plane strain except when Mohr's circle is specified (Prob. 7.7-23 through 7.7-28).*

**7.7-1** A thin rectangular plate in *biaxial stress* is subjected to stresses $\sigma_x$ and $\sigma_y$, as shown in part a of the figure. The width and height of the plate are $b = 7.5$ in. and $h = 2.5$ in., respectively. Measurements show that the normal strains in the $x$ and $y$ directions are $\varepsilon_x = 285 \times 10^{-6}$ and $\varepsilon_y = -190 \times 10^{-6}$, respectively.

With reference to part b of the figure, which shows a two-dimensional view of the plate, determine the following quantities.

(a) The increase $\Delta d$ in the length of diagonal $Od$.

(b) The change $\Delta\phi$ in the angle $\phi$ between diagonal $Od$ and the $x$ axis.

(c) The change $\Delta\psi$ in the angle $\psi$ between diagonal $Od$ and the $y$ axis.

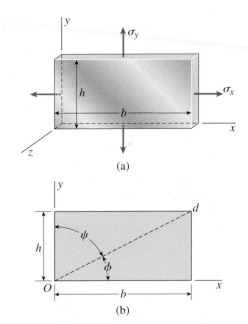

**PROBS. 7.7-1 and 7.7-2**

**7.7-2** Solve the preceding problem if $b = 180$ mm, and $h = 70$ mm, respectively. Measurements show that the normal strains in the $x$ and $y$ directions are $\varepsilon_x = 390 \times 10^{-6}$, and $\varepsilon_y = -240 \times 10^{-6}$, respectively.

**7.7-3** A thin square plate in *biaxial stress* is subjected to stresses $\sigma_x$ and $\sigma_y$, as shown in part a of the figure. The width of the plate is $b = 12.0$ in. Measurements show that the normal strains in the $x$ and $y$ directions are $\varepsilon_x = 427 \times 10^{-6}$ and $\varepsilon_y = 113 \times 10^{-6}$, respectively.

With reference to part b of the figure, which shows a two-dimensional view of the plate, determine the following quantities.

(a) The increase $\Delta d$ in the length of diagonal $Od$.

(b) The change $\Delta\phi$ in the angle $\phi$ between diagonal $Od$ and the $x$ axis.

(c) The shear strain $\gamma$ associated with diagonals $Od$ and $cf$ (that is, find the decrease in angle $ced$).

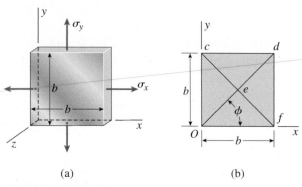

(a)                                (b)

**PROBS. 7.7-3 and 7.7-4**

**7.7-4** Solve the preceding problem if $b = 225$ mm, $\varepsilon_x = 845 \times 10^{-6}$, and $\varepsilon_y = 211 \times 10^{-6}$.

**7.7-5** An element of material subjected to *plane strain* (see figure) has strains of $\varepsilon_x = 280 \times 10^{-6}$, $\varepsilon_y = 420 \times 10^{-6}$, and $\gamma_{xy} = 150 \times 10^{-6}$.

Calculate the strains for an element oriented at an angle $\theta = 35°$. Show these strains on a sketch of a properly oriented element.

**7.7-6** Solve the preceding problem for the following data: $\varepsilon_x = 190 \times 10^{-6}$, $\varepsilon_y = -230 \times 10^{-6}$, $\gamma_{xy} = 160 \times 10^{-6}$, and $\theta = 40°$.

**7.7-7** The strains for an element of material in *plane strain* (see figure) are as follows: $\varepsilon_x = 480 \times 10^{-6}$, $\varepsilon_y = 140 \times 10^{-6}$, and $\gamma_{xy} = -350 \times 10^{-6}$.

Determine the principal strains and maximum shear strains, and show these strains on sketches of properly oriented elements.

**7.7-8** Solve the preceding problem for the following strains: $\varepsilon_x = 120 \times 10^{-6}$, $\varepsilon_y = -450 \times 10^{-6}$, and $\gamma_{xy} = -360 \times 10^{-6}$.

**7.7-9** An element of material in *plane strain* (see figure) is subjected to strains $\varepsilon_x = 480 \times 10^{-6}$, $\varepsilon_y = 70 \times 10^{-6}$, and $\gamma_{xy} = 420 \times 10^{-6}$.

Determine the following quantities: (a) the strains for an element oriented at an angle $\theta = 75°$, (b) the principal strains, and (c) the maximum shear strains. Show the results on sketches of properly oriented elements.

**7.7-10** Solve the preceding problem for the following data: $\varepsilon_x = -1120 \times 10^{-6}$, $\varepsilon_y = -430 \times 10^{-6}$, $\gamma_{xy} = 780 \times 10^{-6}$, and $\theta = 45°$.

**7.7-11** A brass plate with a modulus of elasticity $E = 16 \times 10^6$ psi and Poisson's ratio $v = 0.34$ is loaded in *biaxial stress* by normal stresses $\sigma_x$ and $\sigma_y$ (see figure). A strain gage is bonded to the plate at an angle $\phi = 35°$.

If the stress $\sigma_x$ is 10,700 psi and the strain measured by the gage is $\varepsilon = 390 \times 10^{-6}$, what is the maximum in-plane shear stress $(\tau_{max})_{xy}$ and shear strain $(\gamma_{max})_{xy}$? What is the maximum shear strain $(\gamma_{max})_{xz}$ in the $xz$ plane? What is the maximum shear strain $(\gamma_{max})_{yz}$ in the $yz$ plane?

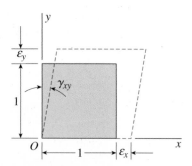

**PROBS. 7.7-5 through 7.7-10**

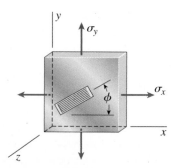

**PROBS. 7.7-11 and 7.7-12**

**7.7-12** Solve the preceding problem if the plate is made of aluminum with $E = 72$ GPa and Poisson's ratio $v = 0.33$. The plate is loaded in *biaxial stress* with normal stress $\sigma_x = 79$ MPa, angle $\phi = 18°$ and the strain measured by the gage is $\varepsilon = 925 \times 10^{-6}$.

**7.7-13** An element in *plane stress* is subjected to stresses $\sigma_x = -8400$ psi, $\sigma_y = 1100$ psi, and $\tau_{xy} = -1700$ psi (see figure). The material is aluminum with modulus of elasticity $E = 10,000$ ksi and Poisson's ratio $v = 0.33$.

Determine the following quantities: (a) the strains for an element oriented at an angle $\theta = 30°$, (b) the principal strains, and (c) the maximum shear strains. Show the results on sketches of properly oriented elements.

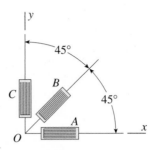

PROBS. 7.7-15 and 7.7-16

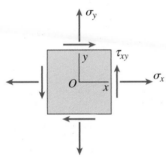

PROBS. 7.7-13 and 7.7-14

**7.7-16** A 45° strain rosette (see figure) mounted on the surface of an automobile frame gives the following readings: gage $A$, $310 \times 10^{-6}$; gage $B$, $180 \times 10^{-6}$; and gage $C$, $-160 \times 10^{-6}$.

Determine the principal strains and maximum shear strains, and show them on sketches of properly oriented elements.

**7.7-17** A solid circular bar with a diameter of $d = 1.25$ in. is subjected to an axial force $P$ and a torque $T$ (see figure). Strain gages $A$ and $B$ mounted on the surface of the bar give readings $\varepsilon_A = 140 \times 10^{-6}$ and $\varepsilon_B = -60 \times 10^{-6}$. The bar is made of steel having $E = 30 \times 10^6$ psi and $v = 0.29$.

(a) Determine the axial force $P$ and the torque $T$.

(b) Determine the maximum shear strain $\gamma_{max}$ and the maximum shear stress $\tau_{max}$ in the bar.

**7.7-14** Solve the preceding problem for the following data: $\sigma_x = -150$ MPa, $\sigma_y = -210$ MPa, $\tau_{xy} = -16$ MPa, and $\theta = 50°$. The material is brass with $E = 100$ GPa and $v = 0.34$.

**7.7-15** During a test of an airplane wing, the strain gage readings from a 45° rosette (see figure) are as follows: gage $A$, $520 \times 10^{-6}$; gage $B$, $360 \times 10^{-6}$; and gage $C$, $-80 \times 10^{-6}$.

Determine the principal strains and maximum shear strains, and show them on sketches of properly oriented elements.

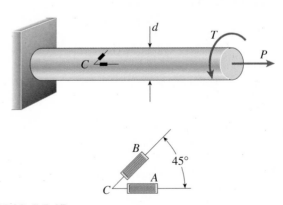

PROB. 7.7-17

**7.7-18** A cantilever beam of rectangular cross section (width $b = 20$ mm, height $h = 175$ mm) is loaded by a force $P$ that acts at the mid-height of the beam and is inclined at an angle $\alpha$ to the vertical (see figure). Two strain gages are placed at point $C$, which also is at the mid-height of the beam. Gage $A$ measures the strain in the horizontal direction, and gage $B$ measures the strain at an angle $\beta = 60°$ to the horizontal. The measured strains are $\varepsilon_A = 145 \times 10^{-6}$ and $\varepsilon_B = -165 \times 10^{-6}$.

Determine the force $P$ and the angle $\alpha$, assuming the material is steel with $E = 200$ GPa and $v = 1/3$.

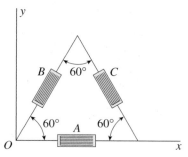

**PROB. 7.7-20**

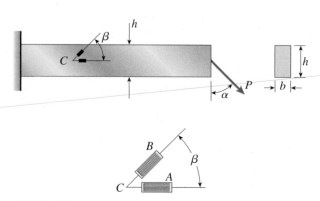

**PROBS. 7.7-18 and 7.7-19**

**7.7-19** Solve the preceding problem if the cross-sectional dimensions are $b = 1.5$ in. and $h = 5.0$ in., the gage angle is $\beta = 75°$, the measured strains are $\varepsilon_A = 209 \times 10^{-6}$ and $\varepsilon_B = -110 \times 10^{-6}$, and the material is a magnesium alloy with modulus $E = 6.0 \times 10^6$ psi and Poisson's ratio $v = 0.35$.

**7.7-20** A 60° strain rosette, or *delta rosette*, consists of three electrical-resistance strain gages arranged as shown in the figure. Gage $A$ measures the normal strain $\varepsilon_a$ in the direction of the $x$ axis. Gages $B$ and $C$ measure the strains $\varepsilon_b$ and $\varepsilon_c$ in the inclined directions shown.

Obtain the equations for the strains $\varepsilon_x$, $\varepsilon_y$, and $\gamma_{xy}$ associated with the $xy$ axes.

**7.7-21** On the surface of a structural component in a space vehicle, the strains are monitored by means of three strain gages arranged as shown in the figure. During a certain maneuver, the following strains were recorded: $\varepsilon_a = 1100 \times 10^{-6}$, $\varepsilon_b = 200 \times 10^{-6}$, and $\varepsilon_c = 200 \times 10^{-6}$.

Determine the principal strains and principal stresses in the material, which is a magnesium alloy for which $E = 6000$ ksi and $v = 0.35$. (Show the principal strains and principal stresses on sketches of properly oriented elements.)

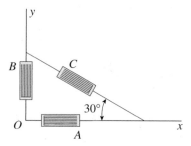

**PROB. 7.7-21**

**7.7-22** The strains on the surface of an experimental device made of pure aluminum ($E = 70$ GPa, $v = 0.33$) and tested in a space shuttle were measured by means of strain gages. The gages were oriented as shown in the figure, and the measured strains were $\varepsilon_a = 1100 \times 10^{-6}$, $\varepsilon_b = 1496 \times 10^{-6}$, and $\varepsilon_c = -39.44 \times 10^{-6}$.

What is the stress $\sigma_x$ in the $x$ direction?

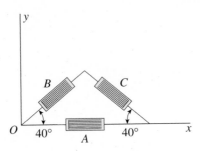

**PROB. 7.7-22**

**7.7-23** Solve Prob. 7.7-5 by using Mohr's circle for plane strain.

**7.7-24** Solve Prob. 7.7-6 by using Mohr's circle for plane strain.

**7.7-25** Solve Prob. 7.7-7 by using Mohr's circle for plane strain.

**7.7-26** Solve Prob. 7.7-8 by using Mohr's circle for plane strain.

**7.7-27** Solve Prob. 7.7-9 by using Mohr's circle for plane strain.

**7.7-28** Solve Prob. 7.7-10 by using Mohr's circle for plane strain.

# Applications of Plane Stress
## (Pressure Vessels, Beams, and Combined Loadings)

Airships such as this blimp rely on internal pressure to maintain their shape using a gas lighter than air for buoyant lift. (Courtesy of Christian Michel, www.modernairships.info)

# CHAPTER OVERVIEW

Chapter 8 deals with a number of applications of plane stress, a topic discussed in detail in Sections 7.2 through 7.5 of the previous chapter. **Plane stress** is a common stress condition that exists in all ordinary structures, including buildings, machines, vehicles, and aircraft. First, thin-wall shell theory is presented describing the behavior of **spherical** (Section 8.2) and **cylindrical** (Section 8.3) **pressure vessels** under internal pressure and having walls whose thickness $t$ is small compared with radius $r$ of the cross section (i.e., $r/t > 10$). We will determine the stresses and strains in the walls of these structures due to the internal pressures from the compressed gases or liquids. Only positive internal pressure (not the effects of external loads, reactions, the weight of the contents, and the weight of the structure) is considered. Linear-elastic behavior is assumed, and the formulas for **membrane stresses** in spherical tanks and **hoop** and **axial stresses** in cylindrical tanks are only valid in regions of the tank away from stress concentra-

tions caused by openings and support brackets or legs. Next, the variation in **principal stresses and maximum shear stresses in beams** is investigated (Section 8.4), building upon the discussions of stresses in beams in Chapter 5. The variation in these stress quantities across the beam can be displayed using either **stress trajectories** or **stress contours**. Stress trajectories give the directions of the principal stresses, while stress contours connect points of equal principal stress at points throughout the beam. Finally, stresses at points of interest in structures under **combined loadings** (axial, shear, torsion, bending, and possibly internal pressure) are assessed (Section 8.5). Our objective is to determine the maximum normal and shear stresses at various points in these structures. Linear-elastic behavior is assumed so that superposition can be used to combine normal and shear stresses due to various loadings, all of which contribute to the state of plane stress at that point.

## Chapter 8 is organized as follows:

**8.1** Introduction 672
**8.2** Spherical Pressure Vessels 672
**8.3** Cylindrical Pressure Vessels 678
**8.4** Maximum Stresses in Beams 685

**8.5** Combined Loadings 694
Chapter Summary & Review 712
Problems 714

# 8.1 INTRODUCTION

We will now investigate some practical examples of structures and components in states of plane stress or strain, building upon the concepts presented in Chapter 7. First, stresses and strains in the walls of thin pressure vessels are examined. Next, the variations in stresses at various points of interest in beams will be considered. Finally, structures acted upon by combined loadings will be evaluated to find the maximum normal and shear stresses which govern their design.

# 8.2 SPHERICAL PRESSURE VESSELS

**Pressure vessels** are closed structures containing liquids or gases under pressure. Familiar examples include tanks, pipes, and pressurized cabins in aircraft and space vehicles. When pressure vessels have walls that are thin in comparison to their overall dimensions, they are included within a more general category known as **shell structures**. Other examples of shell structures are roof domes, airplane wings, and submarine hulls.

In this section we consider thin-walled pressure vessels of spherical shape, like the compressed-air tank shown in Fig. 8-1. The term **thin-walled** is not precise, but as a general rule, pressure vessels are considered to be thin-walled when the ratio of radius $r$ to wall thickness $t$ (Fig. 8-2) is greater than 10. When this condition is met, we can determine the stresses in the walls with reasonable accuracy using statics alone.

We assume in the following discussions that the internal pressure $p$ (Fig. 8-2) exceeds the pressure acting on the outside of the shell. Otherwise, the vessel may collapse inward due to buckling.

A sphere is the theoretically ideal shape for a vessel that resists internal pressure. We only need to contemplate the familiar soap bubble to recognize that a sphere is the "natural" shape for this purpose. To determine the stresses in a spherical vessel, let us cut through the sphere on a vertical diametral plane (Fig. 8-3a) and isolate half of the shell *and its fluid contents* as a single free body (Fig. 8-3b). Acting on this free body are the tensile stresses $\sigma$ in the wall of the vessel and the fluid pressure $p$. This pressure acts horizontally against the plane circular area of fluid remaining inside the hemisphere. Since the pressure is uniform, the resultant pressure force $P$ (Fig. 8-3b) is

$$P = p(\pi r^2) \tag{8-1}$$

where $r$ is the inner radius of the sphere.

Note that the pressure $p$ is not the absolute pressure inside the vessel but is the net internal pressure, or the **gage pressure**. Gage pressure is the internal pressure *above* the pressure acting on the outside of the vessel. If the internal and external pressures are the same, no stresses are developed in the wall of the vessel—only the excess of internal pressure over external pressure has any effect on these stresses.

Because of the symmetry of the vessel and its loading (Fig. 8-3b), the tensile stress $\sigma$ is uniform around the circumference. Furthermore, since

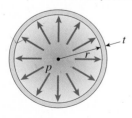

Thin-walled spherical pressure vessel used for storage of propane in this oil refinery (Wayne Eastep/Getty Images)

**Fig. 8-1**

Spherical pressure vessel

Welded seam

**Fig. 8-2**

Cross section of spherical pressure vessel showing inner radius $r$, wall thickness $t$, and internal pressure $p$

the wall is thin, we can assume with good accuracy that the stress is uniformly distributed across the thickness $t$. The accuracy of this approximation increases as the shell becomes thinner and decreases as it becomes thicker.

The resultant of the tensile stresses $\sigma$ in the wall is a horizontal force equal to the stress $\sigma$ times the area over which it acts, or

$$\sigma(2\pi r_m t)$$

where $t$ is the thickness of the wall and $r_m$ is its mean radius:

$$r_m = r + \frac{t}{2} \tag{8-2}$$

Thus, equilibrium of forces in the horizontal direction (Fig. 8-3b) gives

$$\Sigma F_{\text{horiz}} = 0: \quad \sigma(2\pi r_m t) - p(\pi r^2) = 0 \tag{8-3}$$

from which we obtain the *tensile stresses* in the wall of the vessel:

$$\sigma = \frac{pr^2}{2r_m t} \tag{8-4}$$

Since our analysis is valid only for thin shells, we can disregard the small difference between the two radii appearing in Eq. (8-4) and replace $r$ by $r_m$ or replace $r_m$ by $r$. While either choice is satisfactory for this approximate analysis, it turns out that the stresses are closer to the theoretically exact stresses if we use the inner radius $r$ instead of the mean radius $r_m$. Therefore, we will adopt the following formula for calculating the **tensile stresses in the wall of a spherical shell**:

$$\sigma = \frac{pr}{2t} \tag{8-5}$$

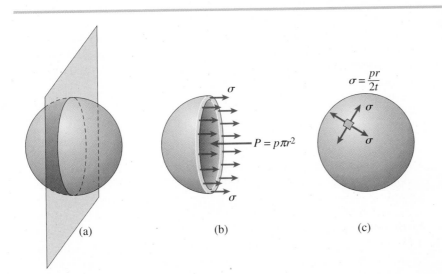

### Fig. 8-3

Tensile stresses $\sigma$ in the wall of a spherical pressure vessel

As is evident from the symmetry of a spherical shell, we obtain the same equation for the tensile stresses when we cut a plane through the center of the sphere in any direction whatsoever. Thus, we reach the following conclusion: *The wall of a pressurized spherical vessel is subjected to uniform tensile stresses $\sigma$ in all directions.* This stress condition is represented in Fig. 8-3c by the small stress element with stresses $\sigma$ acting in mutually perpendicular directions.

Stresses that act tangentially to the curved surface of a shell, such as the stresses $\sigma$ shown in Fig. 8-3c, are known as **membrane stresses**. The name arises from the fact that these are the only stresses that exist in true membranes, such as soap films.

## Stresses at the Outer Surface

The outer surface of a spherical pressure vessel is usually free of any loads. Therefore, the element shown in Fig. 8-3c is in *biaxial stress*. To aid in analyzing the stresses acting on this element, we show it again in Fig. 8-4a, where a set of coordinate axes is oriented parallel to the sides of the element. The $x$ and $y$ axes are tangential to the surface of the sphere, and the $z$ axis is perpendicular to the surface. Thus, the normal stresses $\sigma_x$ and $\sigma_y$ are the same as the membrane stresses $\sigma$, and the normal stress $\sigma_z$ is zero. No shear stresses act on the sides of this element.

If we analyze the element of Fig. 8-4a by using the transformation equations for plane stress (see Fig. 7-1 and Eqs. 7-4a and 7-4b of Section 7.2), we find

$$\sigma_{x_1} = \sigma \quad \text{and} \quad \tau_{x_1 y_1} = 0$$

as expected. In other words, when we consider elements obtained by rotating the axes about the $z$ axis, the normal stresses remain constant and there are no shear stresses. *Every plane is a principal plane and every direction is a principal direction.* Thus, the **principal stresses** for the element are

$$\sigma_1 = \sigma_2 = \frac{pr}{2t} \quad \sigma_3 = 0 \tag{8-6a,b}$$

The stresses $\sigma_1$ and $\sigma_2$ lie in the $xy$ plane and the stress $\sigma_3$ acts in the $z$ direction.

To obtain the **maximum shear stresses**, we must consider out-of-plane rotations, that is, rotations about the $x$ and $y$ axes (because all in-plane shear stresses are zero). Elements oriented by making 45° rotations about the $x$ and $y$ axes have maximum shear stresses equal to $\sigma/2$ and normal stresses equal to $\sigma/2$. Therefore,

$$\tau_{max} = \frac{\sigma}{2} = \frac{pr}{4t} \tag{8-7}$$

These stresses are the largest shear stresses in the element.

## Stresses at the Inner Surface

At the inner surface of the wall of a spherical vessel, a stress element (Fig. 8-4b) has the same membrane stresses $\sigma_x$ and $\sigma_y$ as does an element

### Fig. 8-4

Stresses in a spherical pressure vessel at (a) the outer surface, and (b) the inner surface

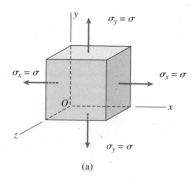

(a)

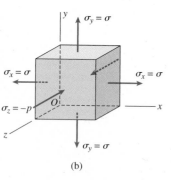

(b)

at the outer surface (Fig. 8-4a). In addition, a compressive stress $\sigma_z$ equal to the pressure $p$ acts in the $z$ direction (Fig. 8-4b). This compressive stress decreases from $p$ at the inner surface of the sphere to zero at the outer surface.

The element shown in Fig. 8-4b is in triaxial stress with principal stresses

$$\sigma_1 = \sigma_2 = \frac{pr}{2t} \qquad \sigma_3 = -p \qquad \textbf{(8-8a,b)}$$

The in-plane shear stresses are zero, but the maximum out-of-plane shear stress (obtained by a 45° rotation about either the $x$ or $y$ axis) is

$$\tau_{max} = \frac{\sigma + p}{2} = \frac{pr}{4t} + \frac{p}{2} = \frac{p}{2}\left(\frac{r}{2t} + 1\right) \qquad \textbf{(8-9)}$$

When the vessel is thin-walled and the ratio $r/t$ is large, we can disregard the number 1 in comparison with the term $r/2t$. In other words, the principal stress $\sigma_3$ in the $z$ direction is small when compared with the principal stresses $\sigma_1$ and $\sigma_2$. Consequently, we can consider the stress state at the inner surface to be the same as at the outer surface (biaxial stress). This approximation is consistent with the approximate nature of thin-shell theory, and therefore we will use Eqs. (8-5), (8-6a,b), and (8-7) to obtain the stresses in the wall of a spherical pressure vessel.

## General Comments

Pressure vessels usually have openings in their walls (to serve as inlets and outlets for the fluid contents) as well as fittings and supports that exert forces on the shell (Fig. 8-1). These features result in nonuniformities in the stress distribution, or *stress concentrations*, that cannot be analyzed by the elementary formulas given here. Instead, more advanced methods of analysis are needed. Other factors that affect the design of pressure vessels include corrosion, accidental impacts, and temperature changes.

Some of the limitations of thin-shell theory as applied to pressure vessels are listed here:

1. The wall thickness must be small in comparison to the other dimensions (the ratio $r/t$ should be 10 or more).

2. The internal pressure must exceed the external pressure (to avoid inward buckling).

3. The analysis presented in this section is based only on the effects of internal pressure (the effects of external loads, reactions, the weight of the contents, and the weight of the structure are not considered).

4. The formulas derived in this section are valid throughout the wall of the vessel *except* near points of stress concentrations.

The following example illustrates how the principal stresses and maximum shear stresses are used in the analysis of a spherical shell.

**Fig. 8-5**

Example 8-1: Spherical pressure vessel. (Attachments and supports are shown in photo.)

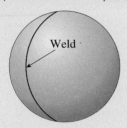

Spherical tanks at oil refinery
(© Kevin Burke/Corbis)

A compressed-air tank having an inner diameter of 18 feet and a wall thickness of 1.75 inches is formed by welding two steel hemispheres (Fig. 8-5).

(a) If the allowable tensile stress in the steel is 13.5 ksi, what is the maximum permissible air pressure $p_a$ in the tank?

(b) If the allowable shear stress in the steel is 6200 psi, what is the maximum permissible pressure $p_b$?

(c) If the normal strain at the outer surface of the tank is not to exceed 0.0003, what is the maximum permissible pressure $p_c$? (Assume that Hooke's law is valid and that the modulus of elasticity for the steel is $29 \times 10^6$ psi and Poisson's ratio is 0.28.)

(d) Tests on the welded seam show that failure occurs when the tensile load on the welds exceeds 42 kips per inch of weld. If the required factor of safety against failure of the weld is 2.5, what is the maximum permissible pressure $p_d$?

(e) Considering the four preceding factors, what is the allowable pressure $p_{allow}$ in the tank?

**Solution**

(a) *Allowable pressure based upon the tensile stress in the steel.* The maximum tensile stress in the wall of the tank is given by the formula $\sigma = pr/2t$ (see Eq. 8-5). Solving this equation for the pressure in terms of the allowable stress, we get

$$p_a = \frac{2t\sigma_{allow}}{r} = \frac{2(1.75 \text{ in.})(13{,}500 \text{ psi})}{108 \text{ in.}} = 437.5 \text{ psi}$$ ◄

Thus, the maximum allowable pressure based upon tension in the wall of the tank is $p_a = 437$ psi. (Note that in a calculation of this kind, we round downward, not upward.)

(b) *Allowable pressure based upon the shear stress in the steel.* The maximum shear stress in the wall of the tank is given by Eq. (8-7), from which we get the following equation for the pressure:

$$p_b = \frac{4t\tau_{allow}}{r} = \frac{4(1.75 \text{ in.})(6200 \text{ psi})}{108 \text{ in.}} = 401.9 \text{ psi}$$ ◄

Therefore, the allowable pressure based upon shear is $p_b = 401$ psi.

(c) *Allowable pressure based upon the normal strain in the steel.* The normal strain is obtained from Hooke's law for biaxial stress (Eq. 7-40a):

$$\varepsilon_x = \frac{1}{E}(\sigma_x - \nu\sigma_y) \tag{a}$$

Substituting $\sigma_x = \sigma_y = \sigma = pr/2t$ (see Fig. 8-4a), we obtain

$$\varepsilon_x = \frac{\sigma}{E}(1 - \nu) = \frac{pr}{2tE}(1 - \nu) \tag{8-10}$$

This equation can be solved for the pressure $p_c$:

$$p_c = \frac{2tE\varepsilon_{allow}}{r(1-\nu)} = \frac{2(1.75 \text{ in.})(29{,}000 \text{ ksi})(0.0003)}{108 \text{ in.}(1-0.28)} = 391.6 \text{ psi} \quad \Longleftarrow$$

Thus, the allowable pressure based upon the normal strain in the wall is $p_c = 391$ psi.

(d) *Allowable pressure based upon the tension in the welded seam.* The allowable tensile load on the welded seam is equal to the failure load divided by the factor of safety:

$$T_{allow} = \frac{T_{failure}}{n} = \frac{42 \text{ kip/in.}}{2.5} = 16{,}800 \text{ lb/in.}$$

The corresponding allowable tensile stress is equal to the allowable load on a one-inch length of weld divided by the cross-sectional area of a one-inch length of weld:

$$\sigma_{allow} = \frac{T_{allow}(1.0 \text{ in.})}{(1.0 \text{ in.})(t)} = \frac{16{,}800 \text{ lb/in.}(1.0 \text{ in.})}{(1.0 \text{ in.})(1.75 \text{ in.})} = 9600 \text{ psi}$$

Finally, we solve for the internal pressure by using Eq. (8-5):

$$p_d = \frac{2t\sigma_{allow}}{r} = \frac{2(1.75 \text{ in.})(9600 \text{ psi})}{108 \text{ in.}} = 311.1 \text{ psi} \quad \Longleftarrow$$

This result gives the allowable pressure based upon tension in the welded seam.

(e) *Allowable pressure.* Comparing the preceding results for $p_a$, $p_b$, $p_c$, and $p_d$, we see that tension in the welded seam governs and the allowable pressure in the tank is

$$p_{allow} = 311 \text{ psi} \quad \Longleftarrow$$

This example illustrates how various stresses and strains enter into the design of a spherical pressure vessel.

*Note:* When the internal pressure is at its maximum allowable value (311 psi), the tensile stresses in the shell are

$$\sigma = \frac{pr}{2t} = \frac{311 \text{ psi}(108 \text{ in.})}{2(1.75 \text{ in.})} = 9597 \text{ psi}$$

Thus, at the inner surface of the shell (Fig. 8-4b), the ratio of the principal stress in the z direction (311 psi) to the in-plane principal stresses (9597 psi) is only 0.032. Therefore, our earlier assumption that we can disregard the principal stress $\sigma_3$ in the z direction and consider the entire shell to be in biaxial stress is justified.

**Fig. 8-6**

Cylindrical pressure vessels with circular cross sections

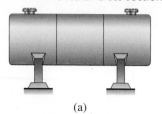

(a)

(b)

Cylindrical storage tanks in a petrochemical plant (Perov Stanislav/Shutterstock)

# 8.3 CYLINDRICAL PRESSURE VESSELS

Cylindrical pressure vessels with a circular cross section (Fig. 8-6) are found in industrial settings (compressed air tanks and rocket motors), in homes (fire extinguishers and spray cans), and in the countryside (propane tanks and grain silos). Pressurized pipes, such as water-supply pipes and penstocks, are also classified as cylindrical pressure vessels.

We begin our analysis of cylindrical vessels by determining the normal stresses in a *thin-walled circular tank AB* subjected to internal pressure (Fig. 8-7a). A *stress element* with its faces parallel and perpendicular to the axis of the tank is shown on the wall of the tank. The normal stresses $\sigma_1$ and $\sigma_2$ acting on the side faces of this element are the membrane stresses in the wall. No shear stresses act on these faces because of the symmetry of the vessel and its loading. Therefore, the stresses $\sigma_1$ and $\sigma_2$ are principal stresses.

Because of their directions, the stress $\sigma_1$ is called the **circumferential stress** or the **hoop stress**, and the stress $\sigma_2$ is called the **longitudinal stress** or the **axial stress**. Each of these stresses can be calculated from equilibrium by using appropriate free-body diagrams.

## Circumferential Stress

To determine the circumferential stress $\sigma_1$, we make two cuts (*mn* and *pq*) perpendicular to the longitudinal axis and distance *b* apart (Fig. 8-7a). Then we make a third cut in a vertical plane through the longitudinal axis of the tank, resulting in the free body shown in Fig. 8-7b. This free body consists not only of the half-circular piece of the tank but also of the fluid contained within the cuts. Acting on the longitudinal cut (plane *mpqn*) are the circumferential stresses $\sigma_1$ and the internal pressure *p*.

Stresses and pressures also act on the left-hand and right-hand faces of the free body. However, these stresses and pressures are not shown in the figure because they do not enter the equation of equilibrium that we will use. As in our analysis of a spherical vessel, we will disregard the weight of the tank and its contents.

**Fig. 8-7**

Stresses in a circular cylindrical pressure vessel

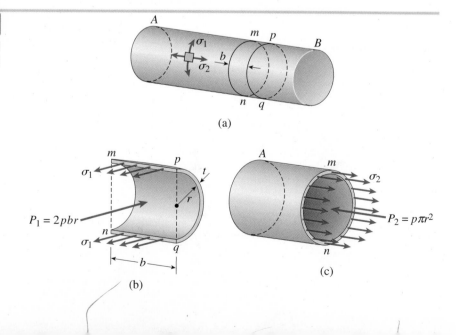

(a)

(b)

(c)

The circumferential stresses $\sigma_1$ acting in the wall of the vessel have a resultant equal to $\sigma_1(2bt)$, where $t$ is the thickness of the wall. Also, the resultant force $P_1$ of the internal pressure is equal to $2pbr$, where $r$ is the inner radius of the cylinder. Hence, we have the following equation of equilibrium:

$$\sigma_1(2bt) - 2pbr = 0$$

From this equation we obtain the following formula for the *circumferential stress in a pressurized cylinder*:

$$\sigma_1 = \frac{pr}{t} \qquad\qquad \textbf{(8-11)}$$

This stress is uniformly distributed over the thickness of the wall, provided the thickness is small compared to the radius.

## Longitudinal Stress

The longitudinal stress $\sigma_2$ is obtained from the equilibrium of a free body of the part of the vessel to the left of cross section $mn$ (Fig. 8-7c). Again, the free body includes not only part of the tank but also its contents. The stresses $\sigma_2$ act longitudinally and have a resultant force equal to $\sigma_2(2\pi rt)$. Note that we are using the inner radius of the shell in place of the mean radius, as explained in Section 8.2.

The resultant force $P_2$ of the internal pressure is a force equal to $p\pi r^2$. Thus, the equation of equilibrium for the free body is

$$\sigma_2(2\pi rt) - p\pi r^2 = 0$$

Solving this equation for $\sigma_2$, we obtain the following formula for the *longitudinal stress* in a cylindrical pressure vessel:

$$\sigma_2 = \frac{pr}{2t} \qquad\qquad \textbf{(8-12)}$$

This stress is equal to the membrane stress in a spherical vessel (Eq. 8-5).

Comparing Eqs. (8-11) and (8-12), we see that the circumferential stress in a cylindrical vessel is equal to twice the longitudinal stress:

$$\sigma_1 = 2\sigma_2 \qquad\qquad \textbf{(8-13)}$$

From this result we note that a longitudinal welded seam in a pressurized tank must be twice as strong as a circumferential seam.

## Stresses at the Outer Surface

The principal stresses $\sigma_1$ and $\sigma_2$ at the outer surface of a cylindrical vessel are shown on the stress element of Fig. 8-8a. Since the third principal stress (acting in the $z$ direction) is zero, the element is in *biaxial stress*.

The maximum *in-plane shear stresses* occur on planes that are rotated $45°$ about the $z$ axis; these stresses are

$$(\tau_{max})_z = \frac{\sigma_1 - \sigma_2}{2} = \frac{\sigma_1}{4} = \frac{pr}{4t} \qquad\qquad \textbf{(8-14)}$$

The maximum *out-of-plane shear stresses* are obtained by $45°$ rotations about the $x$ and $y$ axes, respectively; thus,

$$(\tau_{max})_x = \frac{\sigma_1}{2} = \frac{pr}{2t} \qquad (\tau_{max})_y = \frac{\sigma_2}{2} = \frac{pr}{4t} \qquad \textbf{(8-15a,b)}$$

**Fig. 8-8**

Stresses in a circular cylindrical pressure vessel at (a) the outer surface, and (b) the inner surface

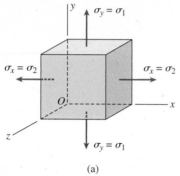

(a)

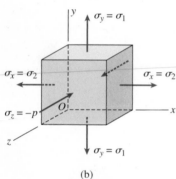

(b)

Comparing the preceding results, we see that the *absolute maximum shear stress* is

$$\tau_{max} = \frac{\sigma_1}{2} = \frac{pr}{2t} \qquad \text{(8-16)}$$

This stress occurs on a plane that has been rotated 45° about the $x$ axis.

## Stresses at the Inner Surface

The stress conditions at the inner surface of the wall of the vessel are shown in Fig. 8-8b. The principal stresses are

$$\sigma_1 = \frac{pr}{t} \qquad \sigma_2 = \frac{pr}{2t} \qquad \sigma_3 = -p \qquad \text{(8-17a,b,c)}$$

The three maximum shear stresses, obtained by 45° rotations about the $x$, $y$, and $z$ axes, are

$$(\tau_{max})_x = \frac{\sigma_1 - \sigma_3}{2} = \frac{pr}{2t} + \frac{p}{2} \qquad \text{(8-18a)}$$

$$(\tau_{max})_y = \frac{\sigma_2 - \sigma_3}{2} = \frac{pr}{4t} + \frac{p}{2} \qquad \text{(8-18b)}$$

$$(\tau_{max})_z = \frac{\sigma_1 - \sigma_2}{2} = \frac{pr}{4t} \qquad \text{(8-18c)}$$

The first of these three stresses is the largest. However, as explained in the discussion of shear stresses in a spherical shell, we may disregard the additional term $p/2$ in Eqs. (8-18a and b) when the shell is thin-walled. Equations (8-18a, b, and c) then become the same as Eqs. (8-15) and (8-14), respectively.

Therefore, in all of our examples and problems pertaining to cylindrical pressure vessels, *we will disregard the presence of the compressive stress in the z direction.* (This compressive stress varies from $p$ at the inner surface to zero at the outer surface.) With this approximation, the stresses at the inner surface become the same as the stresses at the outer surface (biaxial stress). As explained in the discussion of spherical pressure vessels, this procedure is satisfactory when we consider the numerous other approximations in this theory.

## General Comments

The preceding formulas for stresses in a circular cylinder are valid in parts of the cylinder away from any discontinuities that cause stress concentrations, as discussed previously for spherical shells. An obvious discontinuity exists at the ends of the cylinder where the heads are attached, because the geometry of the structure changes abruptly. Other stress concentrations occur at openings, at points of support, and wherever objects or fittings are attached to the cylinder. The stresses at such points cannot be determined solely from equilibrium equations; instead, more advanced methods of analysis (such as shell theory and finite-element analysis) must be used.

Some of the limitations of the elementary theory for thin-walled shells are listed in Section 8.2.

### ● ● ● Example 8-2

**Fig. 8-9**

Example 8-2: Cylindrical pressure
vessel with a helical weld

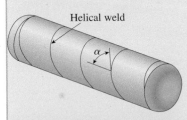

Helical weld

Cylindrical pressure vessel on
simple supports (Perov Stanislav/
Shutterstock)

A cylindrical pressure vessel is constructed from a long, narrow steel plate by
wrapping the plate around a mandrel and then welding along the edges of
the plate to make a helical joint (Fig. 8-9). The helical weld makes an angle
$\alpha = 55°$ with the longitudinal axis. The vessel has inner radius $r = 1.8$ m and
wall thickness $t = 20$ mm. The material is steel with modulus $E = 200$ GPa
and Poisson's ratio $v = 0.30$. The internal pressure $p$ is 800 kPa.

   Calculate the following quantities for the cylindrical part of the vessel:
(a) the circumferential and longitudinal stresses $\sigma_1$ and $\sigma_2$, respectively;
(b) the maximum in-plane and out-of-plane shear stresses; (c) the circumferential and longitudinal strains $\varepsilon_1$ and $\varepsilon_2$, respectively; and (d) the normal
stress $\sigma_w$ and shear stress $\tau_w$ acting perpendicular and parallel, respectively,
to the welded seam.

### Solution

(a) *Circumferential and longitudinal stresses.* The circumferential and longitudinal stresses $\sigma_1$ and $\sigma_2$, respectively, are pictured in Fig. 8-10a, where
   they are shown acting on a stress element at point $A$ on the wall of the
   vessel. The magnitudes of the stresses can be calculated from Eqs. (8-11)
   and (8-12):

$$\sigma_1 = \frac{pr}{t} = \frac{(800 \text{ kPa})(1.8 \text{ m})}{20 \text{ mm}} = 72 \text{ MPa} \qquad \sigma_2 = \frac{pr}{2t} = \frac{\sigma_1}{2} = 36 \text{ MPa} \Longleftarrow$$

The stress element at point $A$ is shown again in Fig. 8-10b, where the
$x$ axis is in the longitudinal direction of the cylinder and the $y$ axis is in

**Fig. 8-10**

Solution to Example 8-2

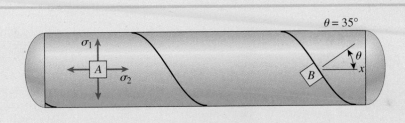

$\theta = 35°$

(a)

$\sigma_y = \sigma_1 = 72$ MPa

$\sigma_x = \sigma_2 = 36$ MPa

(b)

60.2 MPa

47.8 MPa

$\theta = 35°$

16.9 MPa

(c)

*Continues* ➡

the circumferential direction. Since there is no stress in the *z* direction ($\sigma_3 = 0$), the element is in biaxial stress.

Note that the ratio of the internal pressure (800 kPa) to the smaller in-plane principal stress (36 MPa) is 0.022. Therefore, our assumption that we may disregard any stresses in the *z* direction and consider all elements in the cylindrical shell, even those at the inner surface, to be in biaxial stress is justified.

(b) *Maximum shear stresses.* The largest in-plane shear stress is obtained from Eq. (8-14):

$$(\tau_{max})_z = \frac{\sigma_1 - \sigma_2}{2} = \frac{\sigma_1}{4} = \frac{pr}{4t} = 18 \text{ MPa}$$

Because we are disregarding the normal stress in the *z* direction, the largest out-of-plane shear stress is obtained from Eq. (8-15a):

$$\tau_{max} = \frac{\sigma_1}{2} = \frac{pr}{2t} = 36 \text{ MPa}$$

This last stress is the absolute maximum shear stress in the wall of the vessel.

(c) *Circumferential and longitudinal strains.* Since the largest stresses are well below the yield stress of steel (see Table I-3, Appendix I), we may assume that Hooke's law applies to the wall of the vessel. Then we can obtain the strains in the *x* and *y* directions (Fig. 8-10b) from Eqs. (7-40a) and (7-40b) for biaxial stress:

$$\varepsilon_x = \frac{1}{E}(\sigma_x - v\sigma_y) \qquad \varepsilon_y = \frac{1}{E}(\sigma_y - v\sigma_x) \qquad \text{(a,b)}$$

We note that the strain $\varepsilon_x$ is the same as the principal strain $\varepsilon_2$ in the longitudinal direction and that the strain $\varepsilon_y$ is the same as the principal strain $\varepsilon_1$ in the circumferential direction. Also, the stress $\sigma_x$ is the same as the stress $\sigma_2$, and the stress $\sigma_y$ is the same as the stress $\sigma_1$. Therefore, the preceding two equations can be written in the following forms:

$$\varepsilon_2 = \frac{\sigma_2}{E}(1 - 2v) = \frac{pr}{2tE}(1 - 2v) \qquad \text{(8-19a)}$$

$$\varepsilon_1 = \frac{\sigma_1}{2E}(2 - v) = \frac{pr}{2tE}(2 - v) \qquad \text{(8-19b)}$$

Substituting numerical values, we find

$$\varepsilon_2 = \frac{\sigma_2}{E}(1 - 2v) = \frac{(36 \text{ MPa})[1 - 2(0.30)]}{200 \text{ GPa}} = 72 \times 10^{-6}$$

$$\varepsilon_1 = \frac{\sigma_1}{2E}(2 - v) = \frac{(72 \text{ MPa})(2 - 0.30)}{2(200 \text{ GPa})} = 306 \times 10^{-6}$$

These are the longitudinal and circumferential strains in the cylinder.

**Fig. 8-10c (Repeated)**

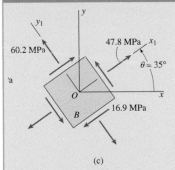

(c)

(d) *Normal and shear stresses acting on the welded seam*. The stress element at point $B$ in the wall of the cylinder (Fig. 8-10a) is oriented so that its sides are parallel and perpendicular to the weld. The angle $\theta$ for the element is

$$\theta = 90° - \alpha = 35°$$

as shown in Fig. 8-10c. Either the stress-transformation equations or Mohr's circle may be used to obtain the normal and shear stresses acting on the side faces of this element.

*Stress-transformation equations*. The normal stress $\sigma_{x_1}$ and the shear stress $\tau_{x_1y_1}$ acting on the $x_1$ face of the element (Fig. 8-10c) are obtained from Eqs. (7-4a) and (7-4b), which are repeated here:

$$\sigma_{x_1} = \frac{\sigma_x + \sigma_y}{2} + \frac{\sigma_x - \sigma_y}{2} \cos 2\theta + \tau_{xy} \sin 2\theta \qquad \text{(8-20a)}$$

$$\tau_{x_1y_1} = -\frac{\sigma_x - \sigma_y}{2} \sin 2\theta + \tau_{xy} \cos 2\theta \qquad \text{(8-20b)}$$

Substituting $\sigma_x = \sigma_2 = pr/2t$, $\sigma_y = \sigma_1 = pr/t$, and $\tau_{xy} = 0$, we obtain

$$\sigma_{x_1} = \frac{pr}{4t}(3 - \cos 2\theta) \qquad \tau_{x_1y_1} = \frac{pr}{4t} \sin 2\theta \qquad \text{(8-21a,b)}$$

These equations give the normal and shear stresses acting on an inclined plane oriented at an angle $\theta$ with the longitudinal axis of the cylinder.

Substituting $pr/4t = 18$ MPa and $\theta = 35°$ into Eqs. (8-21a and b), we obtain

$$\sigma_{x_1} = 47.8 \text{ MPa} \qquad \tau_{x_1y_1} = 16.9 \text{ MPa}$$

These stresses are shown on the stress element of Fig. 8-10c.

To complete the stress element, we can calculate the normal stress $\sigma_{y_1}$ acting on the $y_1$ face of the element from the sum of the normal stresses on perpendicular faces (Eq. 7-6):

$$\sigma_1 + \sigma_2 = \sigma_{x_1} + \sigma_{y_1} \qquad \text{(8-22)}$$

Substituting numerical values, we get

$$\sigma_{y_1} = \sigma_1 + \sigma_2 - \sigma_{x_1} = 72 \text{ MPa} + 36 \text{ MPa} - 47.8 \text{ MPa} = 60.2 \text{ MPa}$$

as shown in Fig. 8-10c.

From the figure, we see that the normal and shear stresses acting perpendicular and parallel, respectively, to the welded seam are

$$\sigma_w = 47.8 \text{ MPa} \qquad \tau_w = 16.9 \text{ MPa}$$

*Continues* ⮕

**Example 8-2 -** *Continued*

### Fig. 8-11

Mohr's circle for the biaxial stress element of Fig. 8-10b. (*Note:* All stresses on the circle have units of MPa.)

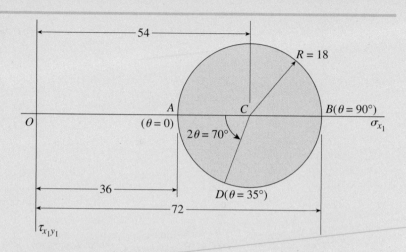

*Mohr's circle.* The Mohr's circle construction for the biaxial stress element of Fig. 8-10b is shown in Fig. 8-11. Point A represents the stress $\sigma_2 = 36$ MPa on the x face ($\theta = 0$) of the element, and point B represents the stress $\sigma_1 = 72$ MPa on the y face ($\theta = 90°$). The center C of the circle is at a stress of 54 MPa, and the radius of the circle is

$$R = \frac{72 \text{ MPa} - 36 \text{ MPa}}{2} = 18 \text{ MPa}$$

A counterclockwise angle $2\theta = 70°$ (measured on the circle from point A) locates point D, which corresponds to the stresses on the $x_1$ face ($\theta = 35°$) of the element. The coordinates of point D (from the geometry of the circle) are

$$\sigma_{x_1} = 54 \text{ MPa} - R \cos 70° = 54 \text{ MPa} - (18 \text{ MPa})(\cos 70°) = 47.8 \text{ MPa}$$

$$\tau_{x_1y_1} = R \sin 70° = (18 \text{ MPa})(\sin 70°) = 16.9 \text{ MPa}$$

These results are the same as those found earlier from the stress-transformation equations.

*Note:* When seen in a side view, a **helix** follows the shape of a sine curve (Fig. 8-12). The pitch of the helix is

### Fig. 8-12

Side view of a helix

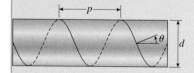

$$p = \pi d \tan \theta \qquad (8\text{-}23)$$

where $d$ is the diameter of the circular cylinder and $\theta$ is the angle between a normal to the helix and a longitudinal line. The width of the flat plate that wraps into the cylindrical shape is

$$w = \pi d \sin \theta \qquad (8\text{-}24)$$

Thus, if the diameter of the cylinder and the angle $\theta$ are given, both the pitch and the plate width are established. For practical reasons, the angle $\theta$ is usually in the range from 20° to 35°.

# 8.4 MAXIMUM STRESSES IN BEAMS

The stress analysis of a beam usually begins by finding the normal and shear stresses acting on cross sections. For instance, when Hooke's law holds, we can obtain the normal and shear stresses from the **flexure and shear formulas** (Eqs. 5-14 and 5-41, respectively, of Chapter 5):

$$\sigma = -\frac{My}{I} \qquad \tau = \frac{VQ}{Ib} \qquad \textbf{(8-25a,b)}$$

In the flexure formula, $\sigma$ is the normal stress acting on the cross section, $M$ is the bending moment, $y$ is the distance from the neutral axis, and $I$ is the moment of inertia of the cross-sectional area with respect to the neutral axis. (The sign conventions for $M$ and $y$ in the flexure formula are shown in Figs. 5-9 and 5-10 of Chapter 5.)

In the case of the shear formula, $\tau$ is the shear stress at any point in the cross section, $V$ is the shear force, $Q$ is the first moment of the cross-sectional area outside of the point in the cross section where the stress is being found, and $b$ is the width of the cross section. (The shear formula is usually written without regard to signs because the directions of the shear stresses are apparent from the directions of the loads.)

The normal stresses obtained from the flexure formula have their maximum values at the farthest distances from the neutral axis, whereas the shear stresses obtained from the shear formula usually have their highest values at the neutral axis. The normal stresses are calculated at the cross section of maximum bending moment, and the shear stresses are calculated at the cross section of maximum shear force. In most circumstances, these are the only stresses that are needed for design purposes.

However, to obtain a more complete picture of the stresses in a beam, we need to determine the principal stresses and maximum shear stresses at various points in the beam. We will begin by discussing the stresses in a rectangular beam.

## Beams of Rectangular Cross Section

We can gain an understanding of how the stresses in a beam vary by considering the simple beam of rectangular cross section shown in Fig. 8-13a. For the purposes of this discussion, we choose a cross section to the left of the load and then select five points ($A$, $B$, $C$, $D$, and $E$) on the side of the beam. Points $A$ and $E$ are at the top and bottom of the beam, respectively, point $C$ is at the midheight of the beam, and points $B$ and $D$ are in between.

If Hooke's law applies, the normal and shear stresses at each of these five points can be readily calculated from the flexure and shear formulas. Since these stresses act on the cross section, we can picture them on stress elements having vertical and horizontal faces, as shown in Fig. 8-13b. Note that all elements are in plane stress, because there are no stresses acting perpendicular to the plane of the figure.

At point $A$ the normal stress is compressive and there are no shear stresses. Similarly, at point $E$ the normal stress is tensile and again there are no shear stresses. Thus, the elements at these locations are in uniaxial stress. At the neutral axis (point $C$) the element is in pure shear. At the

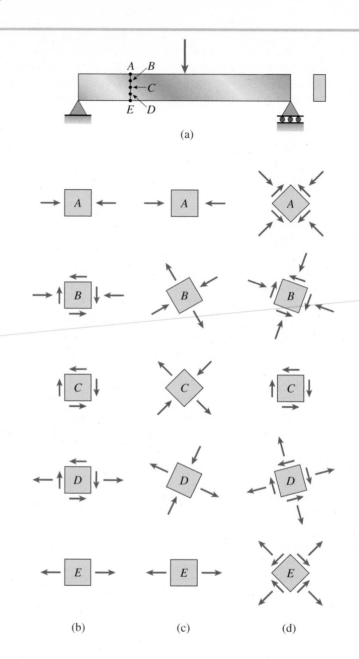

other two locations (points *B* and *D*), both normal and shear stresses act
on the stress elements.

To find the principal stresses and maximum shear stresses at each point,
we may use either the transformation equations of plane stress or Mohr's
circle. The directions of the principal stresses are shown in Fig. 8-13c, and
the directions of the maximum shear stresses are shown in Fig. 8-13d. (Note
that we are considering only the in-plane stresses.)

Now let us examine the **principal stresses** in more detail. From the
sketches in Fig. 8-13c, we can observe how the principal stresses change as we
go from top to bottom of the beam. Let us begin with the compressive
principal stress. At point *A* the compressive stress acts in the horizontal direction and the other principal stress is zero. As we move toward the neutral axis

the compressive principal stress becomes inclined, and at the neutral axis (point $C$) it acts at 45° to the horizontal. At point $D$ the compressive principal stress is further inclined from the horizontal, and at the bottom of the beam its direction becomes vertical (except that its magnitude is now zero).

Thus, the direction and magnitude of the compressive principal stress vary continuously from top to bottom of the beam. If the chosen cross section is located in a region of large bending moment, the largest compressive principal stress occurs at the top of the beam (point $A$), and the smallest compressive principal stress (zero) occurs at the bottom of the beam (point $E$). If the cross section is located in a region of small bending moment and large shear force, then the largest compressive principal stress is at the neutral axis.

Analogous comments apply to the tensile principal stress, which also varies in both magnitude and direction as we move from point $A$ to point $E$. At point $A$ the tensile stress is zero and at point $E$ it has its maximum value. (Graphs showing how the principal stresses vary in magnitude for a particular beam and particular cross section are given later in Fig. 8-19 of Example 8-3.)

The **maximum shear stresses** (Fig. 8-13d) at the top and bottom of the beam occur on 45° planes (because the elements are in uniaxial stress). At the neutral axis, the maximum shear stresses occur on horizontal and vertical planes (because the element is in pure shear). At all points, the maximum shear stresses occur on planes oriented at 45° to the principal planes. In regions of high bending moment, the largest shear stresses occur at the top and bottom of the beam; in regions of low bending moment and high shear force, the largest shear stresses occur at the neutral axis.

By investigating the stresses at many cross sections of the beam, we can determine how the principal stresses vary throughout the beam. Then we can construct two systems of orthogonal curves, called **stress trajectories**, that give the directions of the principal stresses. Examples of stress trajectories for rectangular beams are shown in Fig. 8-14. Part (a) of the figure shows a cantilever beam with a load acting at the free end, and part (b)

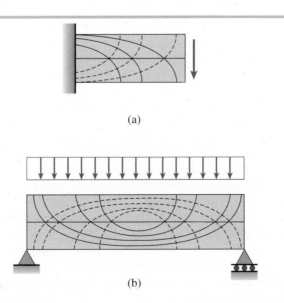

(a)

(b)

### Fig. 8-14

Principal-stress trajectories for beams of rectangular cross section: (a) cantilever beam, and (b) simple beam. (Solid lines represent tensile principal stresses and dashed lines represent compressive principal stresses.)

shows a simple beam with a uniform load. Solid lines are used for tensile principal stresses and dashed lines for compressive principal stresses. The curves for tensile and compressive principal stresses always intersect at right angles, and every trajectory crosses the longitudinal axis at 45°. At the top and bottom surfaces of the beam, where the shear stress is zero, the trajectories are either horizontal or vertical.*

Another type of curve that may be plotted from the principal stresses is a **stress contour**, which is a curve connecting points of equal principal stress. Stress contours for a cantilever beam of rectangular cross section are shown in Fig. 8-15 (for tensile principal stresses only). The contour of largest stress is at the upper left part of the figure. As we move downward in the figure, the tensile stresses represented by the contours become smaller and smaller. The contour line of zero tensile stress is at the lower edge of the beam. Thus, the largest tensile stress occurs at the support, where the bending moment has its largest value.

Note that stress trajectories (Fig. 8-14) give the directions of the principal stresses but give no information about the magnitudes of the stresses. In general, the magnitudes of the principal stresses vary as we move along a trajectory. In contrast, the magnitudes of the principal stresses are constant as we move along a stress contour (Fig. 8-15), but the contours give no information about the directions of the stresses. In particular, the principal stresses are neither parallel nor perpendicular to a stress contour.

The stress trajectories and contours of Figs. 8-14 and 8-15 were plotted from the flexure and shear formulas (Eqs. 8-25a and b). Stress concentrations near the supports and near the concentrated loads, as well as the direct compressive stresses caused by the uniform load bearing on the top of the beam (Fig. 8-14b), were disregarded in plotting these figures.

## Wide-Flange Beams

Beams having other cross-sectional shapes, such as wide-flange beams, can be analyzed for the principal stresses in a manner similar to that described previously for rectangular beams. For instance, consider the simply supported wide-flange beam shown in Fig. 8-16a. Proceeding as for a rectangular beam, we identify points $A$, $B$, $C$, $D$, and $E$ from top to bottom of the beam (Fig. 8-16b). Points $B$ and $D$ are in the web where it meets the flange, and point $C$ is at the neutral axis. We can think of these points as being located either on the side of the beam (Figs. 8-16b and c) or inside the beam along a vertical axis of symmetry (Fig. 8-16d). The stresses determined from the flexure and shear formulas are the same at both sets of points.

*Stress trajectories were originated by the German engineer Karl Culmann (1821–1881); see Ref. 8-1.

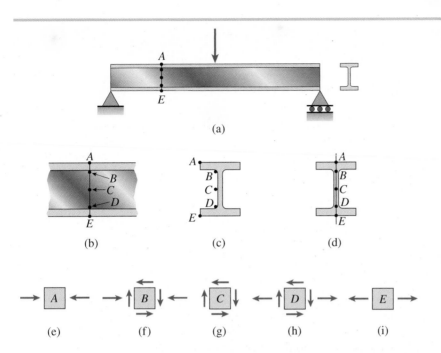

**Fig. 8-16**

Stresses in a wide-flange beam

Stress elements at points $A$, $B$, $C$, $D$, and $E$ (as seen in a side view of the beam) are shown in parts (e) through (i) of Fig. 8-16. These elements have the same general appearance as those for a rectangular beam (Fig. 8-13b).

The largest **principal stresses** usually occur at the top and bottom of the beam (points $A$ and $E$) where the stresses obtained from the flexure formula have their largest values. However, depending upon the relative magnitudes of the bending moment and shear force, the largest stresses sometimes occur in the web where it meets the flange (points $B$ and $D$). The explanation lies in the fact that the normal stresses at points $B$ and $D$ are only slightly smaller than those at points $A$ and $E$, whereas the shear stresses (which are zero at points $A$ and $E$) may be significant at points $B$ and $D$ because of the thin web. (*Note:* Fig. 5-38 in Chapter 5 shows how the shear stresses vary in the web of a wide-flange beam.)

The **maximum shear stresses** acting on a cross section of a wide-flange beam always occur at the neutral axis, as shown by the shear formula of Eq. (8-25b). However, the maximum shear stresses acting on inclined planes usually occur either at the top and bottom of the beam (points $A$ and $E$) or in the web where it meets the flange (points $B$ and $D$) because of the presence of normal stresses.

When analyzing a wide-flange beam for the maximum stresses, remember that high stresses may exist near supports, points of loading, fillets, and holes. Such *stress concentrations* are confined to the region very close to the discontinuity and cannot be calculated by elementary beam formulas.

The following example illustrates the procedure for determining the principal stresses and maximum shear stresses at a selected cross section in a rectangular beam. The procedures for a wide-flange beam are similar.

**Fig. 8-13b (Repeated)**

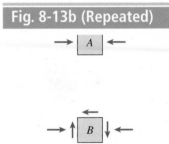

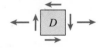

(b)

**Example 8-3**

A simple beam $AB$ with span length $L = 6$ ft supports a concentrated load $P = 10,800$ lb acting at distance $c = 2$ ft from the right-hand support (Fig. 8-17). The beam is made of steel and has a rectangular cross section of width $b = 2$ in. and height $h = 6$ in.

Investigate the principal stresses and maximum shear stresses at cross section $mn$, located at distance $x = 9$ in. from end $A$ of the beam. (Consider only the in-plane stresses.)

**Fig. 8-17**

Example 8-3: Beam of rectangular cross section

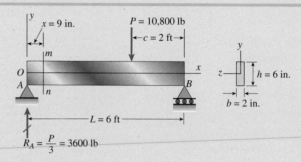

**Solution**

We begin by using the flexure and shear formulas to calculate the stresses acting on cross section $mn$. Once those stresses are known, we can determine the principal stresses and maximum shear stresses from the equations of plane stress. Finally, we can plot graphs of these stresses to show how they vary over the height of the beam.

As a preliminary matter, we note that the reaction of the beam at support $A$ is $R_A = P/3 = 3600$ lb, and therefore the bending moment and shear force at section $mn$ are

$$M = R_A x = (3600 \text{ lb})(9 \text{ in.}) = 32,400 \text{ lb-in.} \quad V = R_A = 3600 \text{ lb}$$

*Normal stresses on cross section mn.* These stresses are found from the flexure formula (Eq. 8-25a), as follows:

$$\sigma_x = -\frac{My}{I} = -\frac{12My}{bh^3} = -\frac{12(32,400 \text{ lb-in.})y}{(2 \text{ in.})(6 \text{ in.})^3} = -900y \quad \textbf{(a)}$$

in which $y$ has units of inches (in.) and $\sigma_x$ has units of pounds per square inch (psi). The stresses calculated from Eq. (a) are positive when in tension and negative when in compression. For instance, note that a positive value of $y$ (upper half of the beam) gives a negative stress, as expected.

A stress element cut from the side of the beam at cross section $mn$ (Fig. 8-25) is shown in Fig. 8-18. For reference purposes, a set of $xy$ axes is associated with the element. The normal stress $\sigma_x$ and the shear stress $\tau_{xy}$ are shown acting on the element in their positive directions. (Note that in this example there is no normal stress $\sigma_y$ acting on the element.)

*Shear stresses on cross section mn.* The shear stresses are given by the shear formula (Eq. 8-25b) in which the first moment $Q$ for a rectangular cross section is

$$Q = b\left(\frac{h}{2} - y\right)\left(y + \frac{h/2 - y}{2}\right) = \frac{b}{2}\left(\frac{h^2}{4} - y^2\right) \quad \textbf{(8-26)}$$

**Fig. 8-18**

Example 8-3: Plane-stress element at cross section $mn$ of the beam of Fig. 8-17

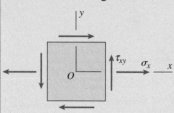

Thus, the shear formula becomes

$$\tau = \frac{VQ}{Ib} = \frac{12V}{(bh^3)(b)}\left(\frac{b}{2}\right)\left(\frac{h^2}{4} - y^2\right) = \frac{6V}{bh^3}\left(\frac{h^2}{4} - y^2\right) \quad \text{(8-27)}$$

The shear stresses $\tau_{xy}$ acting on the $x$ face of the stress element (Fig. 8-18) are positive upward, whereas the actual shear stresses $\tau$ (Eq. 8-27) act downward. Therefore, the shear stresses $\tau_{xy}$ are given by

$$\tau_{xy} = -\frac{6V}{bh^3}\left(\frac{h^2}{4} - y^2\right) \quad \text{(8-28)}$$

Substituting numerical values into this equation gives

$$\tau_{xy} = -\frac{6(3600 \text{ lb})}{(2 \text{ in.})(6 \text{ in.})^3}\left(\frac{(6 \text{ in.})^2}{4} - y^2\right) = -50(9 - y^2) \quad \text{(b)}$$

in which $y$ has units of inches (in.) and $\tau_{xy}$ has units of pounds per square inch (psi).

*Calculation of stresses.* For the purpose of calculating the stresses at cross section *mn*, let us divide the height of the beam into six equal intervals and label the corresponding points from A to G, as shown in the side view of the beam (Fig. 8-19a). The $y$ coordinates of these points are listed in column 2 of Table 8-1 and the corresponding stresses $\sigma_x$ and $\tau_{xy}$ (calculated from Eqs. a and b, respectively) are listed in columns 3 and 4. These stresses are plotted in Figs. 8-19b and c. The normal stresses vary linearly from a compressive stress of $-2700$ psi at the top of the beam (point A) to a tensile stress of 2700 psi at the bottom of the beam (point G). The shear stresses have a parabolic distribution with the maximum stress at the neutral axis (point D).

*Principal stresses and maximum shear stresses.* The principal stresses at each of the seven points A through G may be determined from Eq. (7-17):

$$\sigma_{1,2} = \frac{\sigma_x + \sigma_y}{2} \pm \sqrt{\left(\frac{\sigma_x - \sigma_y}{2}\right)^2 + \tau_{xy}^2} \quad \text{(8-29)}$$

Since there is no normal stress in the $y$ direction (Fig. 8-18), this equation simplifies to

$$\sigma_{1,2} = \frac{\sigma_x}{2} \pm \sqrt{\left(\frac{\sigma_x}{2}\right)^2 + \tau_{xy}^2} \quad \text{(8-30)}$$

Also, the maximum shear stresses (from Eq. 7-25) are

$$\tau_{\max} = \sqrt{\left(\frac{\sigma_x - \sigma_y}{2}\right)^2 + \tau_{xy}^2} \quad \text{(8-31)}$$

*Continues* ➡

**Example 8-3** - *Continued*

### Fig. 8-19

Example 8-3: Stresses in the beam of Fig. 8-17 (a) Points $A$, $B$, $C$, $D$, $E$, $F$, and $G$ at cross section $mn$; (b) normal stresses $\sigma_x$ acting on cross section $mn$; (c) shear stresses $\tau_{xy}$ acting on cross section $mn$; (d) principal tensile stresses $\sigma_1$; (e) principal compressive stresses $\sigma_2$; and (f) maximum shear stresses $\tau_{max}$. (*Note:* All stresses have units of psi.)

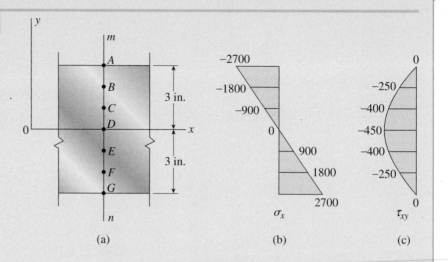

(a)    (b)    (c)

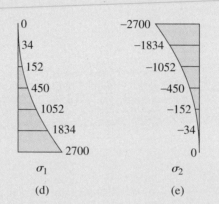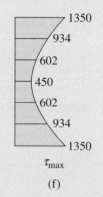

(d)    (e)    (f)

### Table 8-1

Stresses at Cross Section $mn$ in the Beam of Fig. 8-17

| (1) Point | (2) $y$ (in.) | (3) $\sigma_x$ (psi) | (4) $\tau_{xy}$ (psi) | (5) $\sigma_1$ (psi) | (6) $\sigma_2$ (psi) | (7) $\tau_{max}$ (psi) |
|---|---|---|---|---|---|---|
| $A$ | 3 | −2700 | 0 | 0 | −2700 | 1350 |
| $B$ | 2 | −1800 | −250 | 34 | −1834 | 934 |
| $C$ | 1 | −900 | −400 | 152 | −1052 | 602 |
| $D$ | 0 | 0 | −450 | 450 | −450 | 450 |
| $E$ | −1 | 900 | −400 | 1052 | −152 | 602 |
| $F$ | −2 | 1800 | −250 | 1834 | −34 | 934 |
| $G$ | −3 | 2700 | 0 | 2700 | 0 | 1350 |

which simplifies to

$$\tau_{max} = \sqrt{\left(\frac{\sigma_x}{2}\right)^2 + \tau_{xy}^2} \qquad \text{(8-32)}$$

Thus, by substituting the values of $\sigma_x$ and $\tau_{xy}$ (from Table 8-1) into Eqs. (8-30) and (8-32), we can calculate the principal stresses $\sigma_1$ and $\sigma_2$ and the maximum shear stress $\tau_{max}$. These quantities are listed in the last three columns of Table 8-1 and are plotted in Figs. 8-19d, e, and f.

The tensile principal stresses $\sigma_1$ increase from zero at the top of the beam to a maximum of 2700 psi at the bottom (Fig. 8-19d). The directions of the stresses also change, varying from vertical at the top to horizontal at the bottom. At midheight, the stress $\sigma_1$ acts on a 45° plane. Similar comments apply to the compressive principal stress $\sigma_2$, except in reverse. For instance, the stress is largest at the top of the beam and zero at the bottom (Fig. 8-19e).

The maximum shear stresses at cross section $mn$ occur on 45° planes at the top and bottom of the beam. These stresses are equal to one-half of the normal stresses $\sigma_x$ at the same points. At the neutral axis, where the normal stress $\sigma_x$ is zero, the maximum shear stresses occur on the horizontal and vertical planes.

*Note 1:* If we consider other cross sections of the beam, the maximum normal and shear stresses will be different from those shown in Fig. 8-19. For instance, at a cross section between section $mn$ and the concentrated load (Fig. 8-17), the normal stresses $\sigma_x$ are larger than shown in Fig. 8-19b because the bending moment is larger. However, the shear stresses $\tau_{xy}$ are the same as those shown in Fig. 8-19c because the shear force doesn't change in that region of the beam. Consequently, the principal stresses $\sigma_1$ and $\sigma_2$ and maximum shear stresses $\tau_{max}$ will vary in the same general manner as shown in Figs. 8-19d, e, and f but with different numerical values.

The largest tensile stress anywhere in the beam is the normal stress at the bottom of the beam at the cross section of maximum bending moment. This stress is

$$(\sigma_{tens})_{max} = 14{,}400 \text{ psi}$$

The largest compressive stress has the same numerical value and occurs at the top of the beam at the same cross section.

The largest shear stress $\tau_{xy}$ acting on a cross section of the beam occurs to the right of the load $P$ (Fig. 8-17) because the shear force is larger in that region of the beam ($V = R_B = 7200$ lb). Therefore, the largest value of $\tau_{xy}$, which occurs at the neutral axis, is

$$(\tau_{xy})_{max} = 900 \text{ psi}$$

The largest shear stress anywhere in the beam occurs on 45° planes at either the top or bottom of the beam at the cross section of maximum bending moment:

$$\tau_{max} = \frac{14{,}400 \text{ psi}}{2} = 7200 \text{ psi}$$

*Note 2:* In the practical design of ordinary beams, the principal stresses and maximum shear stresses are rarely calculated. Instead, the tensile and compressive stresses to be used in design are calculated from the flexure formula at the cross section of maximum bending moment, and the shear stress to be used in design is calculated from the shear formula at the cross section of maximum shear force.

**Fig. 8-20**

Examples of structures subjected to combined loadings: (a) wide-flange beam supported by a cable (combined bending and axial load), (b) cylindrical pressure vessel supported as a beam, and (c) shaft in combined torsion and bending

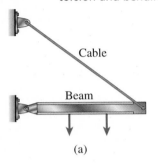

Cable

Beam

(a)

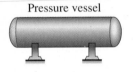

Pressure vessel

(b)

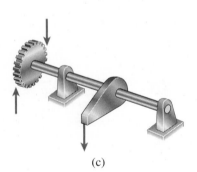

(c)

# 8.5 COMBINED LOADINGS

In previous chapters we analyzed structural members subjected to a single type of loading. For instance, we analyzed axially loaded bars in Chapters 1 and 2, shafts in torsion in Chapter 3, and beams in bending in Chapters 4, 5, and 6. We also analyzed pressure vessels earlier in this chapter. For each type of loading, we developed methods for finding stresses, strains, and deformations.

However, in many structures the members are required to resist more than one kind of loading. For example, a beam may be subjected to the simultaneous action of bending moments and axial forces (Fig. 8-20a), a pressure vessel may be supported so that it also functions as a beam (Fig. 8-20b), or a shaft in torsion may carry a bending load (Fig. 8-20c). Known as **combined loadings**, situations similar to those shown in Fig. 8-20 occur in a great variety of machines, buildings, vehicles, tools, equipment, and many other kinds of structures.

A structural member subjected to combined loadings can often be analyzed by superimposing the stresses and strains caused by each load acting separately. However, superposition of both stresses and strains is permissible only under certain conditions, as explained in earlier chapters. One requirement is that the stresses and strains must be linear functions of the applied loads, which in turn requires that the material follow Hooke's law and the displacements remain small.

A second requirement is that there must be no interaction between the various loads, that is, the stresses and strains due to one load must not be affected by the presence of the other loads. Most ordinary structures satisfy these two conditions, and therefore the use of superposition is very common in engineering work.

## Method of Analysis

While there are many ways to analyze a structure subjected to more than one type of load, the procedure usually includes the following steps:

1. Select a point in the structure where the stresses and strains are to be determined. (The point is usually selected at a cross section where the stresses are large, such as at a cross section where the bending moment has its maximum value.)

2. For each load on the structure, determine the stress resultants at the cross section containing the selected point. (The possible stress resultants are an axial force, a twisting moment, a bending moment, and a shear force.)

3. Calculate the normal and shear stresses at the selected point due to each of the stress resultants. Also, if the structure is a pressure vessel, determine the stresses due to the internal pressure. (The stresses are found from the stress formulas derived previously; for instance, $\sigma = P/A$, $\tau = T\rho/I_P$, $\sigma = My/I$, $\tau = VQ/Ib$, and $\sigma = pr/t$.)

4. Combine the individual stresses to obtain the resultant stresses at the selected point. In other words, obtain the stresses $\sigma_x$, $\sigma_y$, and $\tau_{xy}$ acting on a stress element at the point. (Note that in this chapter we are dealing only with elements in plane stress.)

5. Determine the principal stresses and maximum shear stresses at the selected point, using either the stress-transformation equations or Mohr's circle. If required, determine the stresses acting on other inclined planes.

6. Determine the strains at the point with the aid of Hooke's law for plane stress.

7. Select additional points and repeat the process. Continue until enough stress and strain information is available to satisfy the purposes of the analysis.

## Illustration of the Method

To illustrate the procedure for analyzing a member subjected to combined loadings, we will discuss in general terms the stresses in the cantilever bar of circular cross section shown in Fig. 8-21a. This bar is subjected to two types of load—a torque $T$ and a vertical load $P$, both acting at the free end of the bar.

Let us begin by arbitrarily selecting two points $A$ and $B$ for investigation (Fig. 8-21a). Point $A$ is located at the top of the bar and point $B$ is located on the side. Both points are located at the same cross section.

The stress resultants acting at the cross section (Fig. 8-21b) are a twisting moment equal to the torque $T$, a bending moment $M$ equal to the load $P$ times the distance $b$ from the free end of the bar to the cross section, and a shear force $V$ equal to the load $P$.

The stresses acting at points $A$ and $B$ are shown in Fig. 8-21c. The twisting moment $T$ produces torsional shear stresses

$$\tau_1 = \frac{Tr}{I_P} = \frac{2T}{\pi r^3} \qquad \textbf{(8-33)}$$

in which $r$ is the radius of the bar and $I_P = \pi r^4/2$ is the polar moment of inertia of the cross-sectional area. The stress $\tau_1$ acts horizontally to the left at point $A$ and vertically downward at point $B$, as shown in the figure.

The bending moment $M$ produces a tensile stress at point $A$:

$$\sigma_A = \frac{Mr}{I} = \frac{4M}{\pi r^3} \qquad \textbf{(8-34)}$$

in which $I = \pi r^4/4$ is the moment of inertia about the neutral axis. However, the bending moment produces no stress at point $B$, because $B$ is located on the neutral axis.

The shear force $V$ produces no shear stress at the top of the bar (point $A$), but at point $B$ the shear stress is as follows (see Eq. 5-46 in Chapter 5):

$$\tau_2 = \frac{4V}{3A} = \frac{4V}{3\pi r^2} \qquad \textbf{(8-35)}$$

in which $A = \pi r^2$ is the cross-sectional area.

The stresses $\sigma_A$ and $\tau_1$ acting at point $A$ (Fig. 8-21c) are shown acting on a stress element in Fig. 8-22a. This element is cut from the top of the bar at point $A$. A two-dimensional view of the element, obtained by looking vertically downward on the element, is shown in Fig. 8-22b. For the purpose of determining the principal stresses and maximum shear stresses, we construct

### Fig. 8-21

Cantilever bar subjected to combined torsion and bending: (a) loads acting on the bar, (b) stress resultants at a cross section, and (c) stresses at points $A$ and $B$

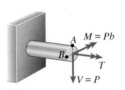

(a)

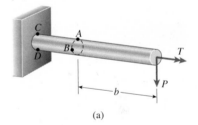

(b)

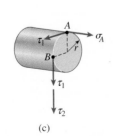

(c)

The circumferential stress $\sigma_x$ is due to the internal pressure of oil and gas and is computed using Eq. 8-11 as

$$\sigma_x = \frac{pr}{t} = \frac{[15 \text{ MPa} \times (100 \text{ mm})]}{18 \text{ mm}} = 83.3 \text{ MPa}$$

The longitudinal stress $\sigma_y$ is caused by the axial compressive force $P$ (due to self-weight) and is divided by casing cross sectional area $A$. The longitudinal tensile stress $\sigma_L$ is due to internal pressure [see Eq. (8-12) to find $\sigma_L$, which is non-zero when the well is capped and not operational]. Here we assume that oil and gas are flowing, so $\sigma_L$ is zero and $\sigma_y$ is computed as

$$\sigma_y = \frac{-P}{A} = \frac{-(175 \text{ kN})}{\pi\left[r^2 - (r - t)^2\right]} = -17 \text{ MPa}$$

The shear stress $\tau_{xy}$ is obtained from the torsion formula [see Eq. (3-13) of Section 3.3]:

$$\tau_{xy} = \frac{Tr}{I_p} = \frac{(14 \text{ kN} \cdot \text{m}) \times (100 \text{ mm})}{8.606(10^{-5}) \text{ m}^4} = 16.3 \text{ MPa}$$

The shear stress is positive in accordance with the sign convention established in Section 1.7.

Knowing the stresses $\sigma_x$, $\sigma_y$, and $\tau_{xy}$, we now can obtain the principal stresses and maximum shear stresses by the methods described in Section 7.3. The principal stresses are obtained from Eq. (7-17):

$$\sigma_{1,2} = \frac{\sigma_x + \sigma_y}{2} \pm \sqrt{\left(\frac{\sigma_x - \sigma_y}{2}\right)^2 + \tau_{xy}^2}$$

Substituting $\sigma_x = 83.3$ MPa, $\sigma_y = -17$ MPa, and $\tau_{xy} = 16.3$ MPa, we get

$$\sigma_{1,2} = 33.2 \text{ MPa} \pm 52.7 \text{ MPa} \quad \text{or} \quad \sigma_1 = 85.9 \text{ MPa} \quad \sigma_2 = -19.5 \text{ MPa} \quad \Longleftarrow$$

These are the maximum tensile and compressive stresses in the drill casing. The maximum in-plane shear stresses from Eq. (7-25) are

$$\tau_{max} = \sqrt{\left(\frac{\sigma_x - \sigma_y}{2}\right)^2 + \tau_{xy}^2} = 52.7 \text{ MPa} \quad \Longleftarrow$$

Because the principal stresses $\sigma_1$ and $\sigma_2$ have opposite signs, the maximum in-plane shear stresses are larger than the maximum out-of-plane shear stresses [see Eqs. (7-28a, b, and c) and the accompanying discussion]. Therefore, the maximum shear stress in the drill casing is 52.7 MPa.

but the *maximum out-of-plane shear s...*

$$\tau_{max} = \frac{\sigma_1}{2}$$

Because the principal stresses have
that one of the out-of-plane shear stre
(see the discussion following Eqs. (7-28

*At point B,* the stress element is l...
vessel and (as we look up at it from th...
shown in Fig. 8-26c. The *x* axis is parall...
sure vessel, and the *y* axis is circumfere...
ing on element *B* due to load *q*, beca...
surface, but normal tensile stress is m...
are computed as

$$\sigma_x = \sigma_l$$

where $I_z$ for the vessel is

$$I_z = \frac{\pi}{4}\left[(r + t)^4 - r^4 \right]$$

so

$$\sigma_x = \frac{pr}{2t} + \frac{\left(\dfrac{qL^2}{40}\right)(r +}{I_z}$$

$$= \frac{105 \text{ psi}(48 \text{ in.})}{2(0.75 \text{ in.})}$$

$$\sigma_x = 3360 \text{ psi} + 230$$

$$\sigma_y = \sigma_r = \frac{pr}{t} = \frac{105}{}$$

Because there are no shear stresses
are the principal normal stresses, (i.
in-plane and out-of-plane shear str
and c).
The *maximum in-plane shear str*

$$\tau_{max} = \frac{\sigma_1 -}{}$$

but the *maximum out-of-plane she*

$$\tau_{max} =$$

The cylindrical pressure vessel from Example 8-2 (see photo) is now placed
on simple supports and is acted on by uniformly distributed load
$q = 10{,}500$ lb/ft, which includes the weight of the tank and its contents.
The 20-ft-long tank has an inner radius of $r = 4$ ft and a wall thickness of
$t = 0.75$ in. The material is steel with a modulus of $E = 29{,}000$ ksi and the
internal pressure $p = 105$ psi.

In Example 8-2, we investigated the longitudinal and circumferential
stresses and strains, as well as the maximum in-plane and out-of-plane shear
stresses. Now, we will investigate the effect of distributed load *q* to find *states
of stress at element locations A and B* (see Fig. 8-25) due to the combined
effects of internal pressure and transverse shear and bending moment
(shear-force and bending-moment diagrams are given in Figs. 8-25c and d).
Element *A* is on the outer surface of the vessel, just to the right of the left-hand
support; element *B* is located on the bottom surface of the tank at the midspan.

### Fig. 8-25

Example 8-5: Cylindrical
pressure vessel subjected to
combined internal pressure *p*
and transverse load *q*

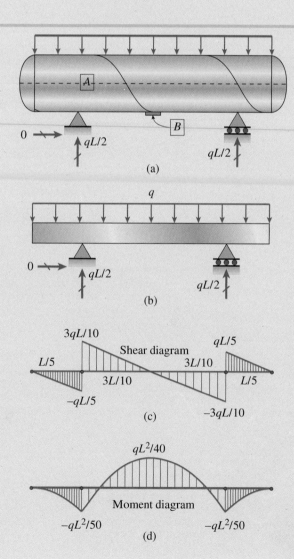

*Continues* ➡

## ● ● ● Example 8-5 - *Continued*

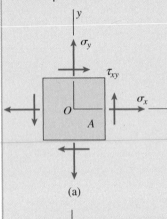

### Fig. 8-26

Stresses in a cylindrical pressure vessel for solution to Example 8-5

(a)

(b)

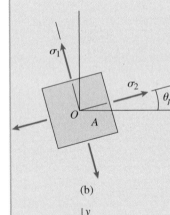

(c)

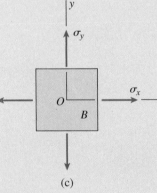

### Solution

The stresses in the wall of the pressure v
*action* of internal pressure and transverse

At point A, we isolate a stress el
Fig. 8-26a. The *x* axis is parallel to the
vessel, and the *y* axis is circumferential. 
element *A* due to load *q* (we assume we 
support so that any stress concentrat
stresses are computed as

$$\sigma_x = \sigma_L = \frac{pr}{2t} = \frac{105 \text{ psi}}{2(0.}$$

$$\sigma_y = \sigma_r = \frac{pr}{t} = \frac{105 \text{ psi}}{0.7}$$

where $\sigma_L$ is the longitudinal stress and 
stress due to internal pressure *p*. There a
ing moment because the longitudinal a
plane for bending. Next we compute she
from the shear diagram, *V* = 3*qL*/10. W

$$\tau_{xy} = \frac{-4}{3}\frac{V}{A}\left(\frac{r_1^2 + r_1 r_2 + r_2^2}{r_1^2 + r_2^2}\right)$$

$$\tau_{xy} = \frac{-4}{3}\frac{\left[\frac{3}{10}(10{,}500 \text{ lb/ft})(20 \text{ ft})\right]}{\pi[(48.75 \text{ in.})^2 - (48 \text{ in.})^2]}(48$$

$$= -552.7 \text{ psi}$$

Shear stress $\tau_{xy}$ is negative (downward o
in accordance with the sign convention
*Principal stresses and maximum she*
stresses are obtained from Eq. (7-17), w

$$\sigma_{1,2} = \frac{\sigma_x + \sigma_y}{2} \pm \sqrt{(}$$

so

$$\sigma_1 = 5040 \text{ psi} + 17$$

$$\sigma_2 = 5040 \text{ psi} - 17$$

The principal stresses are shown on an 
in Fig. 8-26b.
The *maximum in-plane shear stres*

$$\tau_{max} = \frac{\sigma_1 - \sigma}{2}$$

---

## ● ● ● Example

### Fig. 8-27

Example 8-6
against a sig
bending, tor
of the pole)

0.5 m

### Fig. 8-28

Solution to Exa

---

# CHAPTER SUMMARY & REVIEW

In Chapter 8, we investigated some practical examples of structures in states of plane stress, building upon the material presented in Sections 7.2 through 7.5 in the previous chapter. First, we considered the **stresses in thin-walled spherical and cylindrical vessels**, such as storage tanks containing compressed gases or liquids. Then we investigated the **distribution of principal stresses and maximum shear stresses in beams** and plotted either **stress trajectories** or **stress contours** to display the variation of these stresses over the length of the beam. Finally, we evaluated the maximum normal and shear stresses at various points in structures or components acted upon by **combined loadings**. The major concepts and findings presented in this chapter are as follows:

1. **Plane stress** is a common stress condition that exists in all ordinary structures, such as in the walls of pressure vessels, in the webs and/or flanges of beams of various shapes, and in a wide variety of structures subject to the combined effects of axial, shear, and bending loads, as well as internal pressure.

2. The wall of a pressurized **thin-walled spherical vessel** is in a state of plane stress—specifically, biaxial stress—with uniform tensile stresses known as membrane stresses $\sigma$ acting in all directions. The tensile stresses $\sigma$ in the wall of a spherical shell may be calculated as

$$\sigma = \frac{pr}{2t}$$

Only the excess of internal pressure over external pressure, or gage pressure, has any effect on these stresses. Additional important considerations for more detailed analysis or design of spherical vessels include: stress concentrations around openings, effects of external loads and self weight (including contents), and influence of corrosion, impacts, and temperature changes.

3. The walls of **thin-walled cylindrical pressure vessels** with circular cross sections are also in a state of biaxial stress. The circumferential stress $\sigma_1$ is referred to as the hoop stress, and the stress parallel to the axis of the tank is called the longitudinal stress or the axial stress $\sigma_2$. The circumferential stress is equal to twice the longitudinal stress. Both are principal stresses. The formulas for $\sigma_1$ and $\sigma_2$ are

$$\sigma_1 = \frac{pr}{t} \quad \sigma_2 = \frac{pr}{2t}$$

The formulas were derived using elementary theory for thin-walled shells and are only valid in parts of the cylinder away from any discontinuities that cause stress concentrations.

4. If Hooke's law applies, the flexure and shear formulas (Chapter 5) are used to find normal and shear stresses at points of interest along a beam. By investigating the stresses at many cross sections of the beam for a given loading, we can find the variation in principal stresses throughout the beam and then construct two systems of orthogonal curves (called **stress trajectories**) that give the directions of the principal stresses. We can also construct curves connecting points of equal principal stress, known as **stress contours**.

5. **Stress trajectories** give the directions of the principal stresses but give no information about the magnitudes of the stresses. In contrast, the magnitudes of the principal stresses are constant along a **stress contour**, but the contours give no information about the directions of the stresses.

6. In the practical design of ordinary beams, the principal stresses and maximum shear stresses are rarely calculated. Instead, the tensile and compressive stresses to be used in design are calculated from the **flexure formula** at the cross section of maximum bending moment

$$\sigma = -\frac{My}{I}$$

and the shear stress to be used in design is calculated from the **shear formula** at the cross section of maximum shear force

$$\tau = \frac{VQ}{Ib}$$

7. A structural member subjected to **combined loadings** often can be analyzed by superimposing the stresses and strains caused by each load acting separately. However, the stresses and strains must be linear functions of the applied loads, which in turn requires that the material follow Hooke's law and the displacements remain small. There must be no interaction between the various loads, that is, the stresses and strains due to one load must not be affected by the presence of the other loads.

8. A **detailed approach for analysis** of critical points in a structure or component subjected to more than one type of load is presented in Section 8.5.

# PROBLEMS CHAPTER 8

## Spherical Pressure Vessels

*When solving the problems for Section 8.2, assume that the given radius or diameter is an inside dimension and that all internal pressures are gage pressures.*

**8.2-1** A large spherical tank (see figure) contains gas at a pressure of 420 psi. The tank is 45 ft in diameter and is constructed of high-strength steel having a yield stress in tension of 80 ksi.

(a) Determine the required thickness (to the nearest 1/4 inch) of the wall of the tank if a factor of safety of 3.5 with respect to yielding is required.

(b) If the tank wall thickness is 2.25 in., what is the maximum permissible internal pressure?

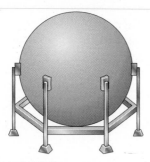

**PROBS. 8.2-1 and 8.2-2**

**8.2-2** Solve the preceding problem if the internal pressure is 3.85 MPa, the diameter is 20 m, the yield stress is 590 MPa, and the factor of safety is 3.0.

(a) Determine the required thickness to the nearest millimeter.

(b) If the tank wall thickness is 85 mm, what is the maximum permissible internal pressure?

**8.2-3** A hemispherical window (or *viewport*) in a decompression chamber (see figure) is subjected to an internal air pressure of 85 psi. The port is attached to the wall of the chamber by 14 bolts.

(a) Find the tensile force $F$ in each bolt and the tensile stress $\sigma$ in the viewport if the radius of the hemisphere is 7.5 in. and its thickness is 1.25 in.

(b) If the yield stress for each of the 14 bolts is 50 ksi and the factor of safety is 3.0, find the required bolt diameter.

(c) If the stress in the viewport is limited to 250 psi, find the required radius of the hemisphere.

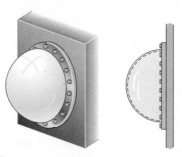

**PROB. 8.2-3**

**8.2-4** A rubber ball (see figure) is inflated to a pressure of 65 kPa. At that pressure the diameter of the ball is 240 mm and the wall thickness is 1.25 mm. The rubber has modulus of elasticity $E = 3.7$ MPa and Poisson's ratio $v = 0.48$.

(a) Determine the maximum stress and strain in the ball.

(b) If the strain must be limited to 0.425, find the minimum required wall thickness of the ball.

**PROB. 8.2-4**

**8.2-5** (a) Solve the preceding problem if the pressure is 8.5 psi, the diameter is 10 in., the wall thickness is 0.05 in., the modulus of elasticity is 200 psi, and Poisson's ratio is 0.48.

(b) If the strain must be limited to 1.01, find the maximum acceptable inflation pressure.

**PROB. 8.2-5**

**8.2-6** A spherical steel pressure vessel (diameter 500 mm, thickness 10 mm) is coated with brittle lacquer that cracks when the strain reaches $150 \times 10^{-6}$ (see figure).

(a) What internal pressure $p$ will cause the lacquer to develop cracks? (Assume $E = 205$ GPa and $v = 0.30$.)

(b) If the strain is measured at $125 \times 10^{-6}$, what is the internal pressure at that point?

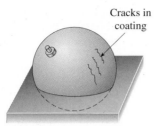

Cracks in coating

PROB. 8.2-6

**8.2-7** A spherical tank of diameter 48 in. and wall thickness 1.75 in. contains compressed air at a pressure of 2200 psi. The tank is constructed of two hemispheres joined by a welded seam (see figure).

(a) What is the tensile load $f$ (lb per in. of length of weld) carried by the weld?

(b) What is the maximum shear stress $\tau_{max}$ in the wall of the tank?

(c) What is the maximum normal strain $\varepsilon$ in the wall? (For steel, assume $E = 30 \times 10^6$ psi and $v = 0.29$.)

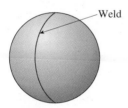

Weld

PROBS. 8.2-7 and 8.2-8

**8.2-8** Solve the preceding problem for the following data: diameter 1.0 m, thickness 48 mm, pressure 22 MPa, modulus 210 GPa, and Poisson's ratio 0.29.

**8.2-9** A spherical stainless-steel tank having a diameter of 26 in. is used to store propane gas at a pressure of 2075 psi. The properties of the steel are as follows: yield stress in tension, 140,000 psi; yield stress in shear, 65,000 psi; modulus of elasticity, $30 \times 10^6$ psi; and Poisson's ratio, 0.28. The desired factor of safety with respect to yielding is 2.8. Also, the normal strain must not exceed $1250 \times 10^{-6}$.

(a) Determine the minimum permissible thickness $t_{min}$ of the tank.

(b) If the tank thickness is 0.30 in. and normal strain is measured at $990 \times 10^{-6}$, what is the internal pressure in the tank at that point?

**8.2-10** Solve the preceding problem if the diameter is 480 mm, the pressure is 20 MPa, the yield stress in tension is 975 MPa, the yield stress in shear is 460 MPa, the factor of safety is 2.75,

the modulus of elasticity is 210 GPa, Poisson's ratio is 0.28, and the normal strain must not exceed $1190 \times 10^{-6}$. For part (b) assume that the tank thickness is 8 mm and the measured normal strain is $990 \times 10^{-6}$.

**8.2-11** A hollow pressurized sphere having radius $r = 4.8$ in. and wall thickness $t = 0.4$ in. is lowered into a lake (see figure). The compressed air in the tank is at a pressure of 24 psi (gage pressure when the tank is out of the water).

At what depth $D_0$ will the wall of the tank be subjected to a compressive stress of 90 psi?

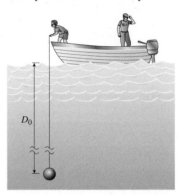

$D_0$

PROB. 8.2-11

# Cylindrical Pressure Vessels

*When solving the problems for Section 8.3, assume that the given radius or diameter is an inside dimension and that all internal pressures are gage pressures.*

**8.3-1** A scuba tank (see figure) is being designed for an internal pressure of 2640 psi with a factor of safety of 2.0 with respect to yielding. The yield stress of the steel is 65,000 psi in tension and 32,000 psi in shear.

(a) If the diameter of the tank is 7.0 in., what is the minimum required wall thickness?

(b) If the wall thickness is 0.25 in., what is the maximum acceptable internal pressure?

PROB. 8.3-1

**8.3-2** A tall standpipe with an open top (see figure) has diameter $d = 2.2$ m and wall thickness $t = 20$ mm.

(a) What height $h$ of water will produce a circumferential stress of 12 MPa in the wall of the standpipe?

(b) What is the axial stress in the wall of the tank due to the water pressure?

**PROB. 8.3-2**

**8.3-3** An inflatable structure used by a traveling circus has the shape of a half-circular cylinder with closed ends (see figure). The fabric and plastic structure is inflated by a small blower and has a radius of 40 ft when fully inflated. A longitudinal seam runs the entire length of the "ridge" of the structure.

If the longitudinal seam along the ridge tears open when it is subjected to a tensile load of 540 pounds per inch of seam, what is the factor of safety $n$ against tearing when the internal pressure is 0.5 psi and the structure is fully inflated?

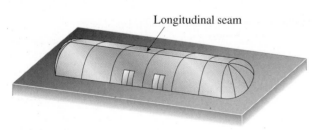

**PROB. 8.3-3**

**8.3-4** A thin-walled cylindrical pressure vessel of radius $r$ is subjected simultaneously to internal gas pressure $p$ and a compressive force $F$ acting at the ends (see figure).

(a) What should be the magnitude of the force $F$ in order to produce pure shear in the wall of the cylinder?

(b) If force $F = 190$ kN, internal pressure $p = 12$ MPa, inner diameter $= 200$ mm, and allowable

normal and shear stresses are 110 MPa and 60 MPa, respectively, what is the required thickness of the vessel?

**PROB. 8.3-4**

**8.3-5** A strain gage is installed in the longitudinal direction on the surface of an aluminum beverage can (see figure). The radius-to-thickness ratio of the can is 200. When the lid of the can is popped open, the strain changes by $\varepsilon_0 = 187 \times 10^{-6}$.

(a) What was the internal pressure $p$ in the can? (Assume $E = 10 \times 10^6$ psi and $\nu = 0.33$.)

(b) What is the change in strain in the radial direction when the lid is opened?

**PROB. 8.3-5**

**8.3-6** A circular cylindrical steel tank (see figure) contains a volatile fuel under pressure. A strain gage at point $A$ records the longitudinal strain in the tank and transmits this information to a control room. The ultimate shear stress in the wall of the tank is 98 MPa, and a factor of safety of 2.8 is required.

(a) At what value of the strain should the operators take action to reduce the pressure in the tank? (Data for the steel are as follows: modulus of elasticity $E = 210$ GPa and Poisson's ratio $\nu = 0.30$.)

(b) What is the associated strain in the radial direction?

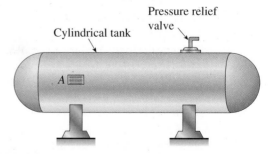

**PROB. 8.3-6**

**8.3-7** A cylinder filled with oil is under pressure from a piston, as shown in the figure. The diameter $d$ of the piston is 1.80 in. and the compressive force $F$ is 3500 lb. The maximum allowable shear stress $\tau_{allow}$ in the wall of the cylinder is 5500 psi.

What is the minimum permissible thickness $t_{min}$ of the cylinder wall? (See figure.)

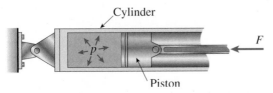

Cylinder

$F$

Piston

**PROBS. 8.3-7 and 8.3-8**

**8.3-8** Solve the preceding problem if $d = 90$ mm, $F = 42$ kN, and $\tau_{allow} = 40$ MPa.

**8.3-9** A standpipe in a water-supply system (see figure) is 12 ft in diameter and 6 in. thick. Two horizontal pipes carry water out of the standpipe; each is 2 ft in diameter and 1 in. thick. When the system is shut down and water fills the pipes but is not moving, the hoop stress at the bottom of the standpipe is 130 psi.

(a) What is the height $h$ of the water in the standpipe?

(b) If the bottoms of the pipes are at the same elevation as the bottom of the standpipe, what is the hoop stress in the pipes?

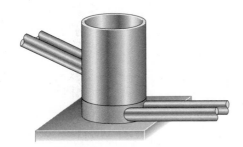

**PROB. 8.3-9**

**8.3-10** A cylindrical tank with hemispherical heads is constructed of steel sections that are welded circumferentially (see figure). The tank diameter is 1.25 m, the wall thickness is 22 mm, and the internal pressure is 1750 kPa.

(a) Determine the maximum tensile stress $\sigma_h$ in the heads of the tank.

(b) Determine the maximum tensile stress $\sigma_c$ in the cylindrical part of the tank.

(c) Determine the tensile stress $\sigma_w$ acting perpendicular to the welded joints.

(d) Determine the maximum shear stress $\tau_h$ in the heads of the tank.

(e) Determine the maximum shear stress $\tau_c$ in the cylindrical part of the tank.

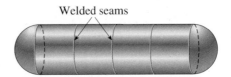

Welded seams

**PROBS. 8.3-10 and 8.3-11**

**8.3-11** A cylindrical tank with diameter $d = 18$ in. is subjected to internal gas pressure $p = 450$ psi. The tank is constructed of steel sections that are welded circumferentially (see figure). The heads of the tank are hemispherical. The allowable tensile and shear stresses are 8200 psi and 3000 psi, respectively. Also, the allowable tensile stress perpendicular to a weld is 6250 psi.

Determine the minimum required thickness $t_{min}$ of (a) the cylindrical part of the tank, and (b) the hemispherical heads.

**8.3-12** A pressurized steel tank is constructed with a helical weld that makes an angle $\alpha = 55°$ with the longitudinal axis (see figure). The tank has radius $r = 0.6$ m, wall thickness $t = 18$ mm, and internal pressure $p = 2.8$ MPa. Also, the steel has modulus of elasticity $E = 200$ GPa and Poisson's ratio $v = 0.30$.

Determine the following quantities for the cylindrical part of the tank.

(a) The circumferential and longitudinal stresses.

(b) The maximum in-plane and out-of-plane shear stresses.

(c) The circumferential and longitudinal strains.

(d) The normal and shear stresses acting on planes parallel and perpendicular to the weld (show these stresses on a properly oriented stress element).

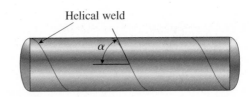

Helical weld

$\alpha$

**PROBS. 8.3-12 and 8.3-13**

**8.3-13** Solve the preceding problem for a welded tank with $\alpha = 62°$, $r = 19$ in., $t = 0.65$ in., $p = 240$ psi, $E = 30 \times 10^6$ psi, and $v = 0.30$.

# Maximum Stresses in Beams

*When solving the problems for Section 8.4, consider only the in-plane stresses and disregard the weights of the beams.*

**8.4-1** A cantilever beam ($L = 6$ ft) with a rectangular cross section ($b = 3.5$ in., $h = 12$ in.) supports an upward load $P = 35$ kips at its free end.

(a) Find the state of stress ($\sigma_x$, $\sigma_y$, $\tau_{xy}$ in MPa) on a plane-stress element at $L/2$ which is $d = 8$ in. up from the bottom of the beam. Find the principal normal stresses and maximum shear stress. Show these stresses on sketches of properly oriented elements.

(b) Repeat part (a) if axial compressive centroidal load $N = 40$ kips is added at $B$.

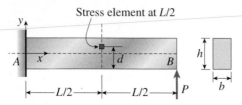

PROBS. 8.4-1 and 8.4-2

**8.4-2** Solve the preceding problem for the following data: $P = 160$ kN, $N = 200$ kN, $L = 2$ m, $b = 95$ mm, $h = 300$ mm, and $d = 200$ mm.

**8.4-3** A simple beam with a rectangular cross section (width, 3.5 in.; height, 12 in.) carries a trapezoidally distributed load of 1400 lb/ft at $A$ and 1000 lb/ft at $B$ on a span of 14 ft (see figure).

Find the principal stresses $\sigma_1$ and $\sigma_2$ and the maximum shear stress $\tau_{max}$ at a cross section 2 ft from the left-hand support at each of the following locations: (a) the neutral axis, (b) 2 in. above the neutral axis, and (c) the top of the beam. (Disregard the direct compressive stresses produced by the uniform load bearing against the top of the beam.)

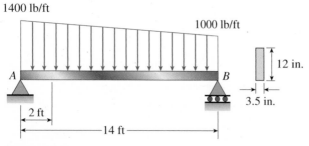

PROB. 8.4-3

**8.4-4** An overhanging beam $ABC$ has a guided support at $A$, a rectangular cross section, and supports an upward uniform load $q = P/L$ over $AB$ and a downward concentrated load $P$ at the free end $C$ (see figure). The span length from $A$ to $B$ is $L$, and the length of the overhang is $L/2$. The cross section has a width of $b$ and a height $h$. Point $D$ is located midway between the supports at a distance $d$ from the top face of the beam.

Knowing that the maximum tensile stress (principal stress) at point $D$ is $\sigma_1 = 38$ MPa, determine the magnitude of the load $P$. Data for the beam are as follows: $L = 1.75$ m, $b = 50$ mm, $h = 220$ mm, and $d = 55$ mm.

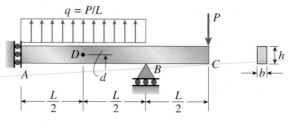

PROBS. 8.4-4 and 8.4-5

**8.4-5** Solve the preceding problem if the stress and dimensions are as follows: $\sigma_1 = 2450$ psi, $L = 80$ in., $b = 2.5$ in., $h = 10$ in., and $d = 2.5$ in.

**8.4-6** A beam of wide-flange cross section (see figure) has the following dimensions: $b = 120$ mm, $t = 10$ mm, $h = 300$ mm, and $h_1 = 260$ mm. The beam is simply supported with span length $L = 3.0$ m. A concentrated load $P = 120$ kN acts at the midpoint of the span.

At a cross section located 1.0 m from the left-hand support, determine the principal stresses $\sigma_1$ and $\sigma_2$ and the maximum shear stress $\tau_{max}$ at each of the following locations: (a) the top of the beam, (b) the top of the web, and (c) the neutral axis.

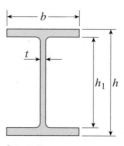

PROBS. 8.4-6 and 8.4-7

**8.4-7** A beam of wide-flange cross section (see figure) has the following dimensions: $b = 5$ in., $t = 0.5$ in., $h = 12$ in., and $h_1 = 10.5$ in. The beam is simply supported with span length $L = 10$ ft and supports a uniform load $q = 6$ k/ft.

Calculate the principal stresses $\sigma_1$ and $\sigma_2$ and the maximum shear stress $\tau_{max}$ at a cross section located 3 ft from the left-hand support at each of the following locations: (a) the bottom of the beam, (b) the bottom of the web, and (c) the neutral axis.

**8.4-8** A W $200 \times 41.7$ wide-flange beam (see Table F-1(b), Appendix F) is simply supported with a span length of 2.5 m (see figure). The beam supports a concentrated load of 100 kN at 0.9 m from support $B$.

At a cross section located 0.7 m from the left-hand support, determine the principal stresses $\sigma_1$ and $\sigma_2$ and the maximum shear stress $\tau_{max}$ at each of the following locations: (a) the top of the beam, (b) the top of the web, and (c) the neutral axis.

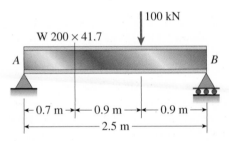

PROB. 8.4-8

**8.4-9** A W $12 \times 14$ wide-flange beam (see Table F-1(a), Appendix F) is simply supported with a span length of 120 in. (see figure). The beam supports two anti-symmetrically placed concentrated loads of 7.5 k each.

At a cross section located 20 in. from the right-hand support, determine the principal stresses $\sigma_1$ and $\sigma_2$ and the maximum shear stress $\tau_{max}$ at each of the following locations: (a) the top of the beam, (b) the top of the web, and (c) the neutral axis.

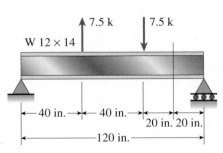

PROB. 8.4-9

**8.4-10** A cantilever beam of T-section is loaded by an inclined force of magnitude 6.5 kN (see figure). The line of action of the force is inclined at an angle of 60° to the horizontal and intersects the top of the beam at the end cross section. The beam is 2.5 m long and the cross section has the dimensions shown.

Determine the principal stresses $\sigma_1$ and $\sigma_2$ and the maximum shear stress $\tau_{max}$ at points $A$ and $B$ in the web of the beam near the support.

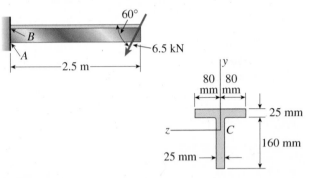

PROB. 8.4-10

**8.4-11** Beam $ABCD$ has a sliding support at $A$, roller supports at $C$ and $D$, and a pin connection at $B$ (see figure). Assume that the beam has a rectangular cross section ($b = 4$ in., $h = 12$ in.). Uniform load $q$ acts on $ABC$ and a concentrated moment is applied at $D$. Let load variable $q = 1750$ lb/ft, and assume that dimension variable $L = 4$ ft. First, use statics to confirm the reaction moment at $A$ and the reaction forces at $C$ and $D$, as given in the figure. Then find the *ratio* of the magnitudes of the principal stresses $(\sigma_1/\sigma_2)$ just left of support $C$ at a distance $d = 8$ in. up from the bottom.

**8.4-12** Solve the preceding problem using the following numerical data: $b = 90$ mm, $h = 280$ mm, $d = 210$ mm, $q = 14$ kN/m and $L = 1.2$ m.

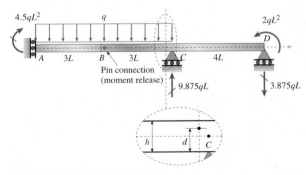

PROBS. 8.4-11 and 8.4-12

# Combined Loadings

*The problems for Section 8.5 are to be solved assuming that the structures behave linearly elastically and that the stresses caused by two or more loads may be superimposed to obtain the resultant stresses acting at a point. Consider both in-plane and out-of-plane shear stresses unless otherwise specified.*

**8.5-1** A cylindrical tank having diameter $d = 2.5$ in. is subjected to internal gas pressure $p = 600$ psi and an external tensile load $T = 1000$ lb (see figure).

Determine the minimum thickness $t$ of the wall of the tank based upon an allowable shear stress of 3000 psi.

PROB. 8.5-1

**8.5-2** A cylindrical tank subjected to internal pressure $p$ is simultaneously compressed by an axial force $F = 72$ kN (see figure). The cylinder has diameter $d = 100$ mm and wall thickness $t = 4$ mm.

Calculate the maximum allowable internal pressure $p_{max}$ based upon an allowable shear stress in the wall of the tank of 60 MPa.

PROB. 8.5-2

**8.5-3** A cylindrical pressure vessel having radius $r = 14$ in. and wall thickness $t = 0.5$ in. is subjected to internal pressure $p = 375$ psi. In addition, a torque $T = 90$ k-ft acts at each end of the cylinder (see figure).

(a) Determine the maximum tensile stress $\sigma_{max}$ and the maximum in-plane shear stress $\tau_{max}$ in the wall of the cylinder.

(b) If the allowable in-plane shear stress is 4.5 ksi, what is the maximum allowable torque $T$?

(c) If $T = 150$ k-ft and allowable in-plane shear and allowable normal stresses are 4.5 ksi and 11.5 ksi, respectively, what is the minimum required wall thickness?

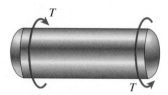

PROB. 8.5-3

**8.5-4** A pressurized cylindrical tank with flat ends is loaded by torques $T$ and tensile forces $P$ (see figure). The tank has radius of $r = 125$ mm and wall thickness $t = 6.5$ mm. The internal pressure $p = 7.25$ MPa and the torque $T = 850$ N·m.

(a) What is the maximum permissible value of the forces $P$ if the allowable tensile stress in the wall of the cylinder is 80 MPa?

(b) If forces $P = 114$ kN, what is the maximum acceptable internal pressure in the tank?

PROB. 8.5-4

**8.5-5** A cylindrical pressure vessel with flat ends is subjected to a torque $T$ and a bending moment $M$ (see figure). The outer radius is 12.0 in. and the wall thickness is 1.0 in. The loads are as follows: $T = 800$ k-in., $M = 1000$ k-in., and the internal pressure $p = 900$ psi.

Determine the maximum tensile stress $\sigma_t$, maximum compressive stress $\sigma_c$, and maximum shear stress $\tau_{max}$ in the wall of the cylinder.

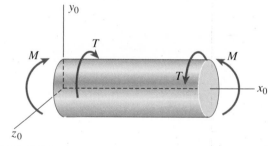

PROB. 8.5-5

**8.5-6** The torsional pendulum shown in the figure consists of a horizontal circular disk of mass $M = 60$ kg suspended by a vertical steel wire ($G = 80$ GPa) of length $L = 2$ m and diameter $d = 4$ mm.

Calculate the maximum permissible angle of rotation $\phi_{max}$ of the disk (that is, the maximum amplitude of torsional vibrations) so that the stresses in the wire do not exceed 100 MPa in tension or 50 MPa in shear.

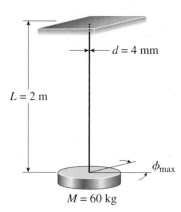

**PROB. 8.5-6**

**8.5-7** The hollow drill pipe for an oil well (see figure) is 6.2 in. in outer diameter and 0.75 in. in thickness. Just above the bit, the compressive force in the pipe (due to the weight of the pipe) is 62 k and the torque (due to drilling) is 185 k-in.

Determine the maximum tensile, compressive, and shear stresses in the drill pipe.

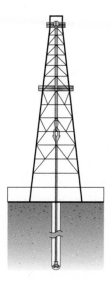

**PROB. 8.5-7**

**8.5-8** A segment of a generator shaft is subjected to a torque $T$ and an axial force $P$, as shown in the figure. The shaft is hollow (outer diameter $d_2 = 300$ mm and inner diameter $d_1 = 250$ mm) and delivers 1800 kW at 4.0 Hz.

If the compressive force $P = 540$ kN, what are the maximum tensile, compressive, and shear stresses in the shaft?

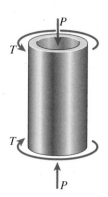

**PROBS. 8.5-8 and 8.5-9**

**8.5-9** A segment of a generator shaft of hollow circular cross section is subjected to a torque $T = 240$ k-in. (see figure). The outer and inner diameters of the shaft are 8.0 in. and 6.25 in., respectively.

What is the maximum permissible compressive load $P$ that can be applied to the shaft if the allowable in-plane shear stress is $\tau_{allow} = 6250$ psi?

**8.5-10** A post having a hollow, circular cross section supports a load $P = 3.2$ kN acting at the end of an arm that is $b = 1.5$ m long (see figure). The height of the post is $L = 9$ m, and its section modulus is $S = 2.65 \times 10^5$ mm$^3$. Assume that the outer radius of the post is $r_2 = 123$ mm, and the inner radius is $r_1 = 117$ mm.

(a) Calculate the maximum tensile stress $\sigma_{max}$ and maximum in-plane shear stress $\tau_{max}$ at point $A$ on the outer surface of the post along the $x$ axis due to the load $P$. Load $P$ acts at $B$ along line $BC$.

(b) If the maximum tensile stress and maximum inplane shear stress at point $A$ are limited to 90 MPa and 38 MPa, respectively, what is the largest permissible value of the load $P$?

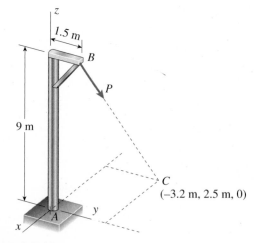

**PROB. 8.5-10**

**8.5-11** A sign is supported by a pole of hollow circular cross section, as shown in the figure. The outer and inner diameters of the pole are 10.5 in. and 8.5 in., respectively. The pole is 42 ft high and weighs 8.8 lb. The sign has dimensions 8 ft × 3 ft and weighs 500 lb. Note that its center of gravity is 53.25 in. from the axis of the pole. The wind pressure against the sign is 35 lb/ft².

(a) Determine the stresses acting on a stress element at point $A$, which is on the outer surface of the pole at the "front" of the pole, that is, the part of the pole nearest to the viewer.

(b) Determine the maximum tensile, compressive, and shear stresses at point $A$.

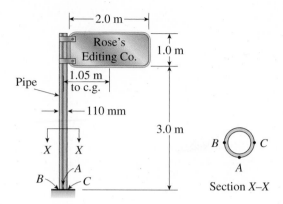

PROB. 8.5-12

**8.5-13** A bracket $ABCD$ having a hollow circular cross section consists of a vertical arm $AB$ ($L = 6$ ft), a horizontal arm $BC$ parallel to the $x_0$ axis, and a horizontal arm $CD$ parallel to the $z_0$ axis (see figure). The arms $BC$ and $CD$ have lengths $b_1 = 3.6$ ft and $b_2 = 2.2$ ft, respectively. The outer and inner diameters of the bracket are $d_2 = 7.5$ in. and $d_1 = 6.8$ in. An inclined load $P = 2200$ lb acts at point $D$ along line $DH$. Determine the maximum tensile, compressive, and shear stresses in the vertical arm.

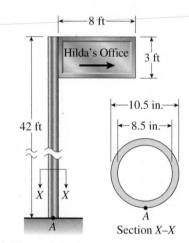

PROB. 8.5-11

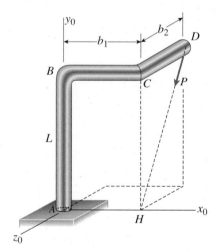

PROB. 8.5-13

**8.5-12** A sign is supported by a pipe (see figure) having outer diameter 110 mm and inner diameter 90 mm. The dimensions of the sign are 2.0 m × 1.0 m, and its lower edge is 3.0 m above the base. Note that the center of gravity of the sign is 1.05 m from the axis of the pipe. The wind pressure against the sign is 1.5 kPa.

Determine the maximum in-plane shear stresses due to the wind pressure on the sign at points $A$, $B$, and $C$, located on the outer surface at the base of the pipe.

**8.5-14** A gondola on a ski lift is supported by two bent arms, as shown in the figure. Each arm is offset by the distance $b = 180$ mm from the line of action of the weight force $W$. The allowable stresses in the arms are 100 MPa in tension and 50 MPa in shear.

If the loaded gondola weighs 12 kN, what is the minimum diameter $d$ of the arms?

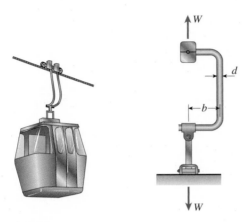

**PROB. 8.5-14**

8.5-15 Determine the maximum tensile, compressive, and shear stresses at points $A$ and $B$ on the bicycle pedal crank shown in the figure.

The pedal and crank are in a horizontal plane and points $A$ and $B$ are located on the top of the crank. The load $P = 160$ lb acts in the vertical direction and the distances (in the horizontal plane) between the line of action of the load and points $A$ and $B$ are $b_1 = 5.0$ in., $b_2 = 2.5$ in., and $b_3 = 1.0$ in. Assume that the crank has a solid circular cross section with diameter $d = 0.6$ in.

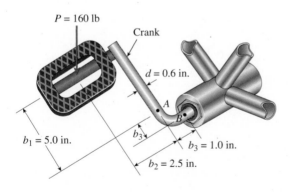

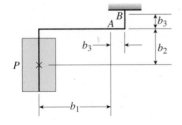

Top view

**PROB. 8.5-15**

8.5-16 A semicircular bar $AB$ lying in a horizontal plane is supported at $B$ (see figure part a). The bar has centerline radius $R$ and weight $q$ per unit of length (total weight of the bar equals $\pi qR$). The cross section of the bar is circular with diameter $d$.

(a) Obtain formulas for the maximum tensile stress $\sigma_t$, maximum compressive stress $\sigma_c$, and maximum in-plane shear stress $\tau_{max}$ at the top of the bar at the support due to the weight of the bar.

(b) Repeat part (a) if the bar is a quarter-circular segment (see figure part b) but has the same total weight as the semicircular bar.

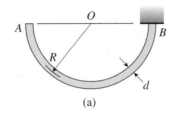 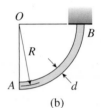

(a)                                    (b)

**PROB. 8.5-16**

8.5-17 An L-shaped bracket lying in a horizontal plane supports a load $P = 150$ lb (see figure). The bracket has a hollow rectangular cross section with thickness $t = 0.125$ in. and outer dimensions $b = 2.0$ in. and $h = 3.5$ in. The centerline lengths of the arms are $b_1 = 20$ in. and $b_2 = 30$ in.

Considering only the load $P$, calculate the maximum tensile stress $\sigma_t$, maximum compressive stress $\sigma_c$, and maximum shear stress $\tau_{max}$ at point $A$, which is located on the top of the bracket at the support.

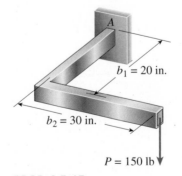 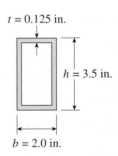

**PROB. 8.5-17**

**8.5-18** A horizontal bracket $ABC$ consists of two perpendicular arms $AB$ of length 0.75 m, and $BC$ of length of 0.5 m. The bracket has a solid circular cross section with diameter equal to 65 mm. The bracket is inserted in a frictionless sleeve at $A$ (which is slightly larger in diameter) so is free to rotate about the $z_0$ axis at $A$, and is supported by a pin at $C$. Moments are applied at point $C$ as follows: $M_1 = 1.5$ kN·m in the $x$ direction and $M_2 = 1.0$ kN·m acts in the $(-z)$ direction.

Considering only the moments $M_1$ and $M_2$, calculate the maximum tensile stress $\sigma_t$, the maximum compressive stress $\sigma_c$, and the maximum in-plane shear stress $\tau_{max}$ at point $p$, which is located at support $A$ on the side of the bracket at midheight.

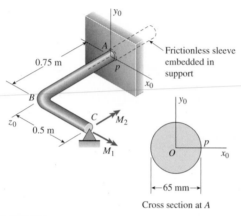

PROB. 8.5-18

**8.5-19** An arm $ABC$ lying in a horizontal plane and supported at $A$ (see figure) is made of two identical solid steel bars $AB$ and $BC$ welded together at a right angle. Each bar is 22 in. long.

(a) Knowing that the maximum tensile stress (principal stress) at the top of the bar at support $A$ due solely to the weights of the bars is 1025 psi, determine the diameter $d$ of the bars.

(b) If the allowable tensile stress is 1475 psi and each bar has diameter $d = 2.0$ in., what is the maximum downward load $P$ that can be applied at $C$ (in addition to self-weight)?

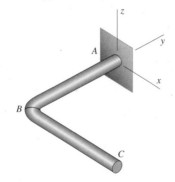

PROB. 8.5-19

**8.5-20** A crank arm consists of a solid segment of length $b_1$ and diameter $d$, a segment of length $b_2$, and a segment of length $b_3$, as shown in the figure. Two loads $P$ act as shown: one parallel to $-x$ and another parallel to $-y$. Each load $P$ equals 1.2 kN. The crankshaft dimensions are $b_1 = 75$ mm, $b_2 = 125$ mm, and $b_3 = 35$ mm. The diameter of the upper shaft is $d = 22$ mm.

(a) Determine the maximum tensile, compressive, and shear stresses at point $A$, which is located on the surface of the shaft at the $z$ axis.

(b) Determine the maximum tensile, compressive, and shear stresses at point $B$, which is located on the surface of the shaft at the $y$ axis.

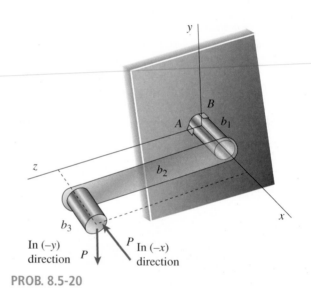

PROB. 8.5-20

**8.5-21** A moveable steel stand supports an automobile engine weighing $W = 750$ lb as shown in the figure part a. The stand is constructed of 2.5 in. × 2.5 in. × 1/8 in.-thick steel tubing. Once in position, the stand is restrained by pin supports at $B$ and $C$. Of interest are stresses at point $A$ at the base of the vertical post; point $A$ has coordinates $(x = 1.25, y = 0, z = 1.25)$ in inches. Neglect the weight of the stand.

(a) Initially, the engine weight acts in the $-z$ direction through point $Q$, which has coordinates (24, 0, 1.25). Find the maximum tensile, compressive, and shear stresses at point $A$.

(b) Repeat part (a) assuming now that, during repair, the engine is rotated about its own longitudinal axis (which is parallel to the $x$ axis) so that $W$ acts through $Q'$ (with coordinates (24, 6, 1.25)) and force $F_y = 200$ lb is applied parallel to the $y$ axis at distance $d = 30$ in.

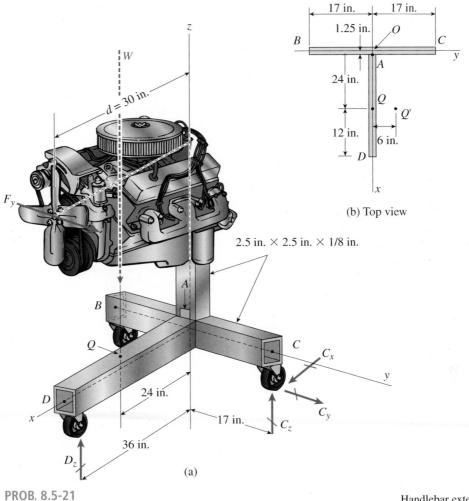

17 in.    17 in.

1.25 in.    $O$

$B$    $C$    $y$

$A$

24 in.

$Q$
$Q'$

12 in.
6 in.

$D$

$x$

(b) Top view

$W$

$d = 30$ in.

$F_y$

2.5 in. × 2.5 in. × 1/8 in.

$A$

$B$

$Q$    $C$    $C_x$

$y$

$D$    24 in.    $C_y$

$x$    17 in.    $C_z$

36 in.

$D_z$

(a)

**PROB. 8.5-21**

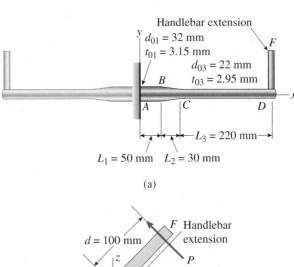

Handlebar extension
$y$    $d_{01} = 32$ mm
$t_{01} = 3.15$ mm
$d_{03} = 22$ mm
$B$    $t_{03} = 2.95$ mm
$A$    $C$    $D$    $x$

$L_3 = 220$ mm

$L_1 = 50$ mm    $L_2 = 30$ mm

(a)

$F$ Handlebar
extension
$d = 100$ mm    $P$
$z$    45°
$d_{03}$    $y$
$D$
Handlebar

(b) Section $D$–$F$

**8.5-22** A mountain bike rider going uphill applies force $P = 65$ N to each end of the handlebars $ABCD$, made of aluminum alloy 7075-T6, by pulling on the handlebar extenders ($DF$ on right handlebar segment). Consider the right half of the handlebar assembly only (assume the bars are fixed at the fork at $A$). Segments $AB$ and $CD$ are prismatic with lengths $L_1$ and $L_3$ and with outer diameters and thicknesses $d_{01}$, $t_{01}$ and $d_{03}$, $t_{03}$, respectively, as shown. Segment $BC$ of length $L_2$, however, is tapered, and outer diameter and thickness vary linearly between dimensions at $B$ and $C$. Consider shear, torsion, and bending effects only for segment $AD$; assume $DF$ is rigid.

Find maximum tensile, compressive, and shear stresses adjacent to support $A$. Show where each maximum stress value occurs.

**PROB. 8.5-22**

**8.5-23** Determine the maximum tensile, compressive, and shear stresses acting on the cross section of the tube at point $A$ of the hitch bicycle rack shown in the figure.

The rack is made up of 2 in. × 2 in. steel tubing which is 1/8 in. thick. Assume that the weight of each of four bicycles is distributed evenly between the two support arms so that the rack can be represented as a cantilever beam ($ABCDEF$) in the $xy$ plane. The overall weight of the rack alone is $W = 60$ lb directed through $C$, and the weight of each bicycle is $B = 30$ lb.

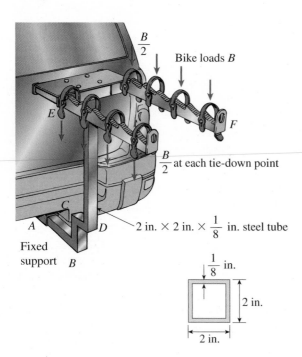

2 in. × 2 in. × $\frac{1}{8}$ in. steel tube

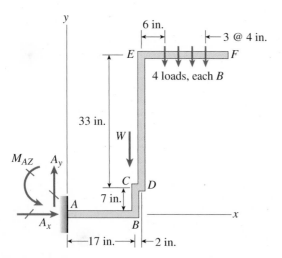

PROB. 8.5-23

**8.5-24** The mountain bike from Prob. 1.2-26 is shown in the figure. To account for impact, crashes, and other loading uncertainties, a *design load* $P = 5000$ N is used to design the seat post. The length of the seat post is $L = 254$ mm.

(a) Find the required diameter of the seat post if it is to be constructed using an *aluminum alloy* with the ultimate stress $\sigma_U = 550$ MPa and a factor of safety of 2.8. Consider only axial and flexural normal stresses in your design.

(b) Repeat part (a) if a *titanium alloy* is used instead. Assume the ultimate stress $\sigma_U = 900$ MPa and a factor of safety of 2.5.

(hamurishi/Shutterstock)
PROB. 8.5-24

**8.5-25** A plumber's valve wrench is used to replace valves in plumbing fixtures. A simplified model of the wrench (see figure part a) consists of pipe $AB$ (length $L$, outer diameter $d_2$, inner diameter $d_1$) which is fixed at $A$ and has holes of diameter $d_b$ on either side of the pipe at $B$. A solid, cylindrical bar $CBD$ (length $a$, diameter $d_b$) is inserted into the holes at $B$ and *only one force* $F = 55$ lb is applied in the $-z$ direction at $C$ to loosen the fixture valve at $A$ (see figure part c). Let $G = 11,800$ ksi, $v = 0.30$, $L = 4$ in. $a = 4.5$ in., $d_2 = 1.25$ in., $d_1 = 1$ in., $d_b = 0.25$ in.

*Find the state of plane stress* on the top of the pipe near $A$ (at coordinates $x = 0$, $y = 0$, $z = d_2/2$), and show all stresses on a plane stress element (see figure part b). Compute the principal stresses and maximum shear stress, and show on properly rotated stress elements.

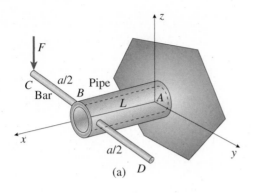

(a)

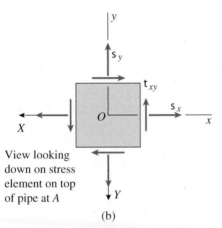

View looking
down on stress
element on top
of pipe at A

(b)

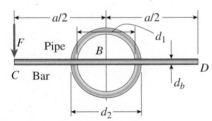

View of end cross section at $x = L$

(c)

**PROB. 8.5-25**

(a) Calculate the maximum normal stresses and maximum in-plane shear stress on the bottom surface of the beam at support $A$.

(b) Repeat part (a) for a plane stress element located at midheight of the beam at $A$.

(c) If the maximum tensile stress and maximum in-plane shear stress at point $A$ are limited to 90 MPa and 42 MPa, respectively, what is the largest permissible value of the cable force $P$?

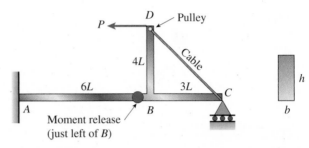

**PROB. 8.5-26**

**8.5-27** A steel hanger bracket $ABCD$ has a solid, circular cross section with a diameter of $d = 2$ in. The dimension variable is $b = 6$ in. (see figure). Load $P = 1200$ lb is applied at $D$ along a line $DH$; the coordinates of point $H$ are $(8b, -5b, 3b)$. Find normal and shear stresses on a plane stress element on the surface of the bracket at $A$. Then find the principal stresses and maximum shear stress. Show each stress state on properly rotated elements.

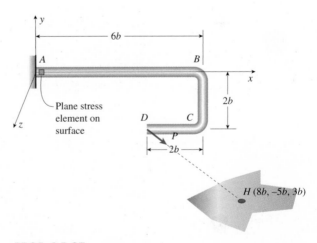

**8.5-26** A compound beam $ABCD$ has a cable with force $P$ anchored at $C$. The cable passes over a pulley at $D$ and force $P$ acts in the $-x$ direction. There is a moment release just left of $B$. *Neglect the self-weight of the beam and cable.* Cable force $P = 450$ N and dimension variable $L = 0.25$ m. The beam has a rectangular cross section ($b = 20$ mm, $h = 50$ mm).

**PROB. 8.5-27**

# 9

CHAPTER

# Deflections of Beams

Deflection of beams is an important consideration in their initial design; deflections also must be monitored during construction. (Glen Jones/Shutterstock)

# CHAPTER OVERVIEW

In Chapter 9, methods for calculation of beam deflections are presented. Beam deflections, in addition to beam stresses and strains discussed in Chapters 5 and 6, are an essential consideration in their analysis and design. A beam may be strong enough to carry a range of static or dynamic loadings (see the discussion in Sections 1.8 and 5.6), but if it deflects too much or vibrates under applied loadings, it fails to meet the "serviceability" requirements which are an important element of its overall design. Chapter 9 covers a range of methods that can be used to compute either deflections (both translations and rotations) **at specific points** along the beam or **the deflected shape of the entire beam**. The beam may be prismatic or nonprismatic (Section 9.7), acted on by concentrated or distributed loads (or both), or the "loading" may be a difference in temperature between the top and bottom of the beam (Section 9.11). In general, the beam is assumed to behave in a linearly elastic manner and is restricted to small displacements (i.e., small compared to its own length). First, methods based on **integration of the differential equation of the elastic curve** are discussed (Sections 9.2 through 9.4). Beam deflection results for a wide range of loadings acting on either cantilever or simple beams are summarized in Appendix H and are available for use in the **method of superposition** (Section 9.5). Next, a method based on the **area of the bending moment diagram** is described (Section 9.6). The concepts of work and **strain energy** are presented (Section 9.8) followed by an application of these principles to computation of beam deflections known as **Castigliano's theorem**. Finally, the specialized topic of beam deflections due to **impact** is discussed (Section 9.10).

## Chapter 9 is organized as follows:

**9.1** Introduction 730
**9.2** Differential Equations of the Deflection Curve 730
**9.3** Deflections by Integration of the Bending-Moment Equation 735
**9.4** Deflections by Integration of the Shear-Force and Load Equations 746
**9.5** Method of Superposition 752
**9.6** Moment-Area Method 760
**9.7** Nonprismatic Beams 769
**9.8** Strain Energy of Bending 774
* **9.9** Castigliano's Theorem 779
***9.10** Deflections Produced by Impact 791
***9.11** Temperature Effects 793
Chapter Summary & Review 798
Problems 800

*Advanced topics

## 9.1 INTRODUCTION

When a beam with a straight longitudinal axis is loaded by lateral forces, the axis is deformed into a curve, called the **deflection curve** of the beam. In Chapter 5, we used the curvature of the bent beam to determine the normal strains and stresses in the beam. However, we did not develop a method for finding the deflection curve itself. In this chapter, we will determine the equation of the deflection curve and also find deflections at specific points along the axis of the beam.

The calculation of deflections is an important part of structural analysis and design. For example, finding deflections is an essential ingredient in the analysis of statically indeterminate structures (Chapter 10). Deflections are also important in dynamic analyses, as when investigating the vibrations of aircraft or the response of buildings to earthquakes.

Deflections are sometimes calculated in order to verify that they are within tolerable limits. For instance, specifications for the design of buildings usually place upper limits on the deflections. Large deflections in buildings are unsightly (and even unnerving) and can cause cracks in ceilings and walls. In the design of machines and aircraft, specifications may limit deflections in order to prevent undesirable vibrations.

## 9.2 DIFFERENTIAL EQUATIONS OF THE DEFLECTION CURVE

Most procedures for finding beam deflections are based on the differential equations of the deflection curve and their associated relationships. Consequently, we will begin by deriving the basic equation for the deflection curve of a beam.

For discussion purposes, consider a cantilever beam with a concentrated load acting upward at the free end (Fig. 9-1a). Under the action of this load, the axis of the beam deforms into a curve, as shown in Fig. 9-1b. The reference axes have their origin at the fixed end of the beam, with the $x$ axis directed to the right and the $y$ axis directed upward. The $z$ axis is directed outward from the figure (toward the viewer).

As in our previous discussions of beam bending in Chapter 5, we assume that the $xy$ plane is a plane of symmetry of the beam, and we assume that all loads act in this plane (the *plane of bending*).

The **deflection** $v$ is the displacement in the $y$ direction of any point on the axis of the beam (Fig. 9-1b). Because the $y$ axis is positive upward, the deflections are also positive when upward.*

To obtain the equation of the deflection curve, we must express the deflection $v$ as a function of the coordinate $x$. Therefore, let us now consider the deflection curve in more detail. The deflection $v$ at any point $m_1$ on the deflection curve is shown in Fig. 9-2a. Point $m_1$ is located at distance $x$ from the origin (measured along the $x$ axis). A second point $m_2$, located at distance $x + dx$ from the origin, is also shown. The deflection

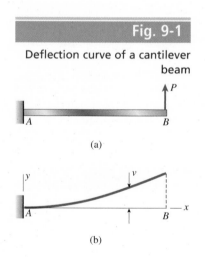

**Fig. 9-1**

Deflection curve of a cantilever beam

(a)

(b)

---

*As mentioned in Section 5.1, the traditional symbols for displacements in the *x, y,* and *z* directions are *u, v,* and *w,* respectively. The advantage of this notation is that it emphasizes the distinction between a *coordinate* and a *displacement*.

at this second point is $v + dv$, where $dv$ is the increment in deflection as we move along the curve from $m_1$ to $m_2$.

When the beam is bent, there is not only a deflection at each point along the axis but also a rotation. The **angle of rotation** $\theta$ of the axis of the beam is the angle between the $x$ axis and the tangent to the deflection curve, as shown for point $m_1$ in the enlarged view of Fig. 9-2b. For our choice of axes ($x$ positive to the right and $y$ positive upward), the angle of rotation is positive when counterclockwise. (Other names for the angle of rotation are *angle of inclination* and *angle of slope*.)

The angle of rotation at point $m_2$ is $\theta + d\theta$, where $d\theta$ is the increase in angle as we move from point $m_1$ to point $m_2$. It follows that if we construct lines normal to the tangents (Figs. 9-2a and b), the angle between these normals is $d\theta$. Also, as discussed earlier in Section 5.3, the point of intersection of these normals is the **center of curvature** $O'$ (Fig. 9-2a) and the distance from $O'$ to the curve is the **radius of curvature** $\rho$. From Fig. 9-2a we see that

$$\rho \, d\theta = ds \qquad \textbf{(9-1)}$$

in which $d\theta$ is in radians and $ds$ is the distance along the deflection curve between points $m_1$ and $m_2$. Therefore, the **curvature** $\kappa$ (equal to the reciprocal of the radius of curvature) is given by the equation

$$\kappa = \frac{1}{\rho} = \frac{d\theta}{ds} \qquad \textbf{(9-2)}$$

The **sign convention** for curvature is pictured in Fig. 9-3, which is repeated from Fig. 5-6 of Section 5.3. Note that curvature is positive when the angle of rotation increases as we move along the beam in the positive $x$ direction.

The **slope of the deflection curve** is the first derivative $dv/dx$ of the expression for the deflection $v$. In geometric terms, the slope is the increment $dv$ in the deflection (as we go from point $m_1$ to point $m_2$ in Fig. 9-2) divided by the increment $dx$ in the distance along the $x$ axis. Since $dv$ and

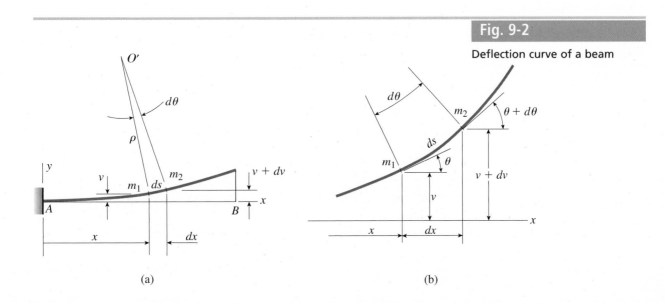

**Fig. 9-2**

Deflection curve of a beam

(a)                    (b)

**Fig. 9-3**

Sign convention for curvature

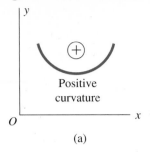

Positive
curvature

(a)

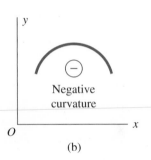

Negative
curvature

(b)

$dx$ are infinitesimally small, the slope $dv/dx$ is equal to the tangent of the angle of rotation $\theta$ (Fig. 9-2b). Thus,

$$\frac{dv}{dx} = \tan\theta \qquad \theta = \arctan\frac{dv}{dx} \qquad \textbf{(9-3a,b)}$$

In a similar manner, we also obtain the following relationships:

$$\cos\theta = \frac{dx}{ds} \qquad \sin\theta = \frac{dv}{ds} \qquad \textbf{(9-4a,b)}$$

Note that when the $x$ and $y$ axes have the directions shown in Fig. 9-2a, the slope $dv/dx$ is positive when the tangent to the curve slopes upward to the right.

Equations (9-2) through (9-4) are based only upon geometric considerations, and therefore they are valid for beams of any material. Furthermore, there are no restrictions on the magnitudes of the slopes and deflections.

## Beams with Small Angles of Rotation

The structures encountered in everyday life, such as buildings, automobiles, aircraft, and ships, undergo relatively small changes in shape while in service. The changes are so small as to be unnoticed by a casual observer. Consequently, the deflection curves of most beams and columns have very small angles of rotation, very small deflections, and very small curvatures. Under these conditions we can make some mathematical approximations that greatly simplify beam analysis.

Consider, for instance, the deflection curve shown in Fig. 9-2. If the angle of rotation $\theta$ is a very small quantity (and hence the deflection curve is nearly horizontal), we see immediately that the distance $ds$ along the deflection curve is practically the same as the increment $dx$ along the $x$ axis. This same conclusion can be obtained directly from Eq. (9-4a). Since $\cos\approx 1$ when the angle $\theta$ is small, Eq. (9-4a) gives

$$ds \approx dx \qquad \textbf{(9-5)}$$

With this approximation, the curvature becomes (see Eq. 9-2)

$$\kappa = \frac{1}{\rho} = \frac{d\theta}{dx} \qquad \textbf{(9-6)}$$

Also, since $\tan\theta \approx \theta$ when $\theta$ is small, we can make the following approximation to Eq. (9-3a):

$$\theta \approx \tan\theta = \frac{dv}{dx} \qquad \textbf{(9-7)}$$

Thus, if the rotations of a beam are small, we can assume that the angle of rotation $\theta$ and the slope $dv/dx$ are equal. (Note that the angle of rotation must be measured in radians.)

Taking the derivative of $\theta$ with respect to $x$ in Eq. (9-7), we get

$$\frac{d\theta}{dx} = \frac{d^2v}{dx^2} \qquad \textbf{(9-8)}$$

Combining this equation with Eq. (9-6), we obtain a relation between the **curvature** of a beam and its deflection:

$$\kappa = \frac{1}{\rho} = \frac{d^2v}{dx^2} \qquad \textbf{(9-9)}$$

This equation is valid for a beam of any material, provided the rotations are small quantities.

If the material of a beam is **linearly elastic** and follows Hooke's law, the curvature [from Eq. 5-13, Chapter 5] is

$$\kappa = \frac{1}{\rho} = \frac{M}{EI} \qquad \textbf{(9-10)}$$

in which $M$ is the bending moment and $EI$ is the flexural rigidity of the beam. Equation (9-10) shows that a positive bending moment produces positive curvature and a negative bending moment produces negative curvature, as shown earlier in Fig. 5-10.

Combining Eq. (9-9) with Eq. (9-10) yields the basic **differential equation of the deflection curve** of a beam:

$$\frac{d^2v}{dx^2} = \frac{M}{EI} \qquad \textbf{(9-11)}$$

This equation can be integrated in each particular case to find the deflection $v$, provided the bending moment $M$ and flexural rigidity $EI$ are known as functions of $x$.

As a reminder, the **sign conventions** to be used with the preceding equations are repeated here: (1) The $x$ and $y$ axes are positive to the right and upward, respectively; (2) the deflection $v$ is positive upward; (3) the slope $dv/dx$ and angle of rotation $\theta$ are positive when counterclockwise with respect to the positive $x$ axis; (4) the curvature $\kappa$ is positive when the beam is bent concave upward; and (5) the bending moment $M$ is positive when it produces compression in the upper part of the beam.

Additional equations can be obtained from the relations between bending moment $M$, shear force $V$, and intensity $q$ of distributed load. In Chapter 4 we derived the following equations between $M$, $V$, and $q$ (see Eqs. 4-4 and 4-6):

$$\frac{dV}{dx} = -q \qquad \frac{dM}{dx} = V \qquad \textbf{(9-12a,b)}$$

The sign conventions for these quantities are shown in Fig. 9-4. By differentiating Eq. (9-11) with respect to $x$ and then substituting the preceding equations for shear force and load, we can obtain the additional equations. In so doing, we will consider two cases, nonprismatic beams and prismatic beams.

## Nonprismatic Beams

In the case of a nonprismatic beam, the flexural rigidity $EI$ is variable, and therefore we write Eq. (9-11) in the form

$$EI_x \frac{d^2v}{dx^2} = M \qquad \textbf{(9-13a)}$$

### Fig. 9-4

Sign conventions for bending moment $M$, shear force $V$, and intensity $q$ of distributed load

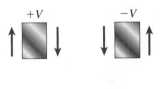

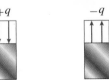

where the subscript $x$ is inserted as a reminder that the flexural rigidity may vary with $x$. Differentiating both sides of this equation and using Eqs. (9-12a) and (9-12b), we obtain

$$\frac{d}{dx}\left(EI_x\frac{d^2v}{dx^2}\right) = \frac{dM}{dx} = V \qquad \textbf{(9-13b)}$$

$$\frac{d^2}{dx^2}\left(EI_x\frac{d^2v}{dx^2}\right) = \frac{dV}{dx} = -q \qquad \textbf{(9-13c)}$$

The deflection of a nonprismatic beam can be found by solving (either analytically or numerically) any one of the three preceding differential equations. The choice usually depends upon which equation provides the most efficient solution.

## Prismatic Beams

In the case of a prismatic beam (constant $EI$), the differential equations become

$$EI\frac{d^2v}{dx^2} = M \qquad EI\frac{d^3v}{dx^3} = V \qquad EI\frac{d^4v}{dx^4} = -q \qquad \textbf{(9-14a,b,c)}$$

To simplify the writing of these and other equations, **primes** are often used to denote differentiation:

$$v' \equiv \frac{dv}{dx} \qquad v'' \equiv \frac{d^2v}{dx^2} \qquad v''' \equiv \frac{d^3v}{dx^3} \qquad v'''' \equiv \frac{d^4v}{dx^4} \qquad \textbf{(9-15)}$$

Using this notation, we can express the differential equations for a prismatic beam in the following forms:

$$EIv'' = M \qquad EIv''' = V \qquad EIv'''' = -q \qquad \textbf{(9-16a,b,c)}$$

We will refer to these equations as the **bending-moment equation**, the **shear-force equation**, and the **load equation**, respectively.

In the next two sections we will use the preceding equations to find deflections of beams. The general procedure consists of integrating the equations and then evaluating the constants of integration from boundary and other conditions pertaining to the beam.

When deriving the differential equations [Eqs. (9-13), (9-14), and (9-16)], we assumed that the material followed Hooke's law and that the slopes of the deflection curve were very small. We also assumed that any shear deformations were negligible; consequently, we considered only the deformations due to pure bending. All of these assumptions are satisfied by most beams in common use.

## Exact Expression for Curvature

If the deflection curve of a beam has large slopes, we cannot use the approximations given by Eqs. (9-5) and (9-7). Instead, we must resort to the exact expressions for curvature and angle of rotation [see Eqs. (9-2) and (9-3b)]. Combining those expressions, we get

$$\kappa = \frac{1}{\rho} = \frac{d\theta}{ds} = \frac{d(\arctan v')}{dx}\frac{dx}{ds} \qquad \textbf{(9-17)}$$

From Fig. 9-2 we see that

$$ds^2 = dx^2 + dv^2 \quad \text{or} \quad ds = [dx^2 + dv^2]^{1/2} \tag{9-18a,b}$$

Dividing both sides of Eq. (9-18b) by $dx$ gives

$$\frac{ds}{dx} = \left[1 + \left(\frac{dv}{dx}\right)^2\right]^{1/2} = [1 + (v')^2]^{1/2} \quad \text{or} \quad \frac{dx}{ds} = \frac{1}{[1 + (v')^2]^{1/2}} \tag{9-18c,d}$$

Also, differentiation of the arctangent function (see Appendix D) gives

$$\frac{d}{dx}(\arctan v') = \frac{v''}{1 + (v')^2} \tag{9-18e}$$

Substitution of expressions Eqs. (9-18d and e) into the equation for curvature from Eq. (9-17) yields

$$\kappa = \frac{1}{\rho} = \frac{v''}{[1 + (v')^2]^{3/2}} \tag{9-19}$$

Comparing this equation with Eq. (9-9), we see that the assumption of small rotations is equivalent to disregarding $(v')^2$ in comparison to one. Equation (9-19) should be used for the curvature whenever the slopes are large.*

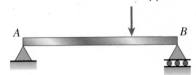

Fig. 9-5

Boundary conditions at simple supports

$v_A = 0$ $\qquad v_B = 0$

# 9.3 DEFLECTIONS BY INTEGRATION OF THE BENDING-MOMENT EQUATION

We are now ready to solve the differential equations of the deflection curve and obtain deflections of beams. The first equation we will use is the bending-moment equation (Eq. 9-16a). Since this equation is of second order, two integrations are required. The first integration produces the slope $v' = dv/dx$, and the second produces the deflection $v$.

We begin the analysis by writing the equation (or equations) for the bending moments in the beam. Since only statically determinate beams are considered in this chapter, we can obtain the bending moments from free-body diagrams and equations of equilibrium, using the procedures described in Chapter 4. In some cases a single bending-moment expression holds for the entire length of the beam, as illustrated in Examples 9-1 and 9-2. In other cases the bending moment changes abruptly at one or more points along the axis of the beam. Then we must write separate bending-moment expressions for each region of the beam between points where changes occur, as illustrated in Example 9-3.

---

*The basic relationship stating that the curvature of a beam is proportional to the bending moment (Eq. 9-10) was first obtained by Jacob Bernoulli, although he obtained an incorrect value for the constant of proportionality. The relationship was used later by Euler, who solved the differential equation of the deflection curve for both large deflections (using Eq. 9-19) and small deflections (using Eq. 9-11). For the history of deflection curves, see Ref. 9-1.

Regardless of the number of bending-moment expressions, the general procedure for solving the differential equations is as follows. For each region of the beam, we substitute the expression for $M$ into the differential equation and integrate to obtain the slope $v'$. Each such integration produces one constant of integration. Next, we integrate each slope equation to obtain the corresponding deflection $v$. Again, each integration produces a new constant. Thus, there are two constants of integration for each region of the beam. These constants are evaluated from known conditions pertaining to the slopes and deflections. The conditions fall into three categories: (1) boundary conditions, (2) continuity conditions, and (3) symmetry conditions.

**Boundary conditions** pertain to the deflections and slopes at the supports of a beam. For example, at a simple support (either a pin or a roller) the deflection is zero (Fig. 9-5), and at a fixed support both the deflection and the slope are zero (Fig. 9-6). Each such boundary condition supplies one equation that can be used to evaluate the constants of integration.

**Continuity conditions** occur at points where the regions of integration meet, such as at point $C$ in the beam of Fig. 9-7. The deflection curve of this beam is physically continuous at point $C$, and therefore the deflection at point $C$ as determined for the left-hand part of the beam must be equal to the deflection at point $C$ as determined for the right-hand part. Similarly, the slopes found for each part of the beam must be equal at point $C$. Each of these continuity conditions supplies an equation for evaluating the constants of integration.

**Symmetry conditions** may also be available. For instance, if a simple beam supports a uniform load throughout its length, we know in advance that the slope of the deflection curve at the midpoint must be zero. This condition supplies an additional equation, as illustrated in Example 9-1.

Each boundary, continuity, and symmetry condition leads to an equation containing one or more of the constants of integration. Since the number of *independent* conditions always matches the number of constants of integration, we can always solve these equations for the constants. (The boundary and continuity conditions alone are always sufficient to determine the constants. Any symmetry conditions provide additional equations, but they are not independent of the other equations. The choice of which conditions to use is a matter of convenience.)

Once the constants are evaluated, they can be substituted back into the expressions for slopes and deflections, thus yielding the final equations of the deflection curve. These equations can then be used to obtain the deflections and angles of rotation at particular points along the axis of the beam.

The preceding method for finding deflections is sometimes called the **method of successive integrations.** The following examples illustrate the method in detail.

**Note:** When sketching deflection curves, such as those shown in the following examples and in Figs. 9-5, 9-6, and 9-7, we greatly exaggerate the deflections for clarity. However, it should always be kept in mind that the actual deflections are very small quantities.

**Fig. 9-6**

Boundary conditions at a fixed support

$v_A = 0$

$v'_A = 0$

**Fig. 9-7**

Continuity conditions at point $C$

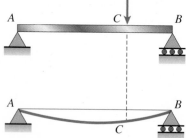

At point $C$:  $(v)_{AC} = (v)_{CB}$

$(v')_{AC} = (v')_{CB}$

Determine the equation of the deflection curve for a simple beam *AB* supporting a uniform load of intensity *q* acting throughout the span of the beam (Fig. 9-8a).

Also, determine the maximum deflection $\delta_{max}$ at the midpoint of the beam and the angles of rotation $\theta_A$ and $\theta_B$ at the supports (Fig. 9-8b). (*Note:* The beam has length *L* and constant flexural rigidity *EI*.)

### Fig. 9-8

Example 9-1: Deflections of a simple beam with a uniform load

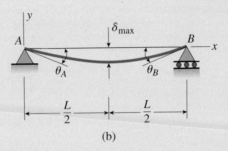

(a)

(b)

### Solution

*Bending moment in the beam.* The bending moment at a cross section distance *x* from the left-hand support is obtained from the free-body diagram of Fig. 9-9. Since the reaction at the support is *qL/2*, the equation for the bending moment is

### Fig. 9-9

Example 9-1: Free-body diagram used in determining the bending moment *M*

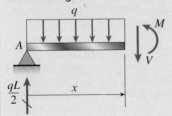

$$M = \frac{qL}{2}(x) - qx\left(\frac{x}{2}\right) = \frac{qLx}{2} - \frac{qx^2}{2} \qquad \textbf{(9-20)}$$

*Differential equation of the deflection curve.* By substituting the expression for the bending moment (Eq. 9-20) into the differential equation (Eq. 9-16a), we obtain

$$EIv'' = \frac{qLx}{2} - \frac{qx^2}{2} \qquad \textbf{(9-21)}$$

This equation can now be integrated to obtain the slope and deflection of the beam.

*Slope of the beam.* Multiplying both sides of the differential equation by *dx*, we get the following equation:

$$EIv''\,dx = \frac{qLx}{2}\,dx - \frac{qx^2}{2}\,dx$$

*Continues* ➡

Integrating each term, we obtain

$$EI \int v'' \, dx = \int \frac{qLx}{2} \, dx - \int \frac{qx^2}{2} \, dx$$

or

$$EIv' = \frac{qLx^2}{4} - \frac{qx^3}{6} + C_1 \qquad \text{(a)}$$

in which $C_1$ is a constant of integration.

To evaluate the constant $C_1$, we observe from the symmetry of the beam and its load that the slope of the deflection curve at midspan is equal to zero. Thus, we have the following symmetry condition:

$$v' = 0 \quad \text{when} \quad x = \frac{L}{2}$$

This condition may be expressed more succinctly as

$$v'\left(\frac{L}{2}\right) = 0$$

Applying this condition to Eq. (a) gives

$$0 = \frac{qL}{4}\left(\frac{L}{2}\right)^2 - \frac{q}{6}\left(\frac{L}{2}\right)^3 + C_1 \quad \text{or} \quad C_1 = -\frac{qL^3}{24}$$

The equation for the slope of the beam (Eq. a) then becomes

$$EIv' = \frac{qLx^2}{4} - \frac{qx^3}{6} - \frac{qL^3}{24} \qquad \text{(b)}$$

or

$$v' = -\frac{q}{24EI}(L^3 - 6Lx^2 + 4x^3) \qquad \Longleftarrow \text{(9-22)}$$

As expected, the slope is negative (i.e., clockwise) at the left-hand end of the beam ($x = 0$), positive at the right-hand end ($x = L$), and equal to zero at the midpoint ($x = L/2$).

*Deflection of the beam.* The deflection is obtained by integrating the equation for the slope. Thus, upon multiplying both sides of Eq. (b) by $dx$ and integrating, we obtain

$$EIv = \frac{qLx^3}{12} - \frac{qx^3}{24} - \frac{qL^3x}{24} + C_2 \qquad \text{(c)}$$

The constant of integration $C_2$ may be evaluated from the condition that the deflection of the beam at the left-hand support is equal to zero; that is, $v = 0$ when $x = 0$, or

$$v(0) = 0$$

Applying this condition to Eq. (c) yields $C_2 = 0$; hence, the equation for the deflection curve is

$$EIv = \frac{qLx^3}{12} - \frac{qx^4}{24} - \frac{qL^3x}{24} \qquad \text{(d)}$$

or

$$v = -\frac{qx}{24EI}(L^3 - 2Lx^2 + x^3) \qquad \Longleftarrow \text{(9-23)}$$

This equation gives the deflection at any point along the axis of the beam. Note that the deflection is zero at both ends of the beam ($x = 0$ and $x = L$) and negative elsewhere (recall that downward deflections are negative).

*Maximum deflection.* From symmetry we know that the maximum deflection occurs at the midpoint of the span (Fig. 9-8b). Thus, setting $x$ equal to $L/2$ in Eq. (9-23), we obtain

$$v\left(\frac{L}{2}\right) = -\frac{5qL^4}{384EI}$$

in which the negative sign means that the deflection is downward (as expected). Since $\delta_{max}$ represents the magnitude of this deflection, we obtain

$$\delta_{max} = \left|v\left(\frac{L}{2}\right)\right| = \frac{5qL^4}{384EI} \qquad \Longleftarrow \text{(9-24)}$$

*Angles of rotation.* The maximum angles of rotation occur at the supports of the beam. At the left-hand end of the beam, the angle $\theta_A$, which is a clockwise angle (Fig. 9-8b), is equal to the negative of the slope $v'$. Thus, by substituting $x = 0$ into Eq. (9-22), we find

$$\theta_A = -v'(0) = \frac{qL^3}{24EI} \qquad \Longleftarrow \text{(9-25)}$$

In a similar manner, we can obtain the angle of rotation $\theta_B$ at the right-hand end of the beam. Since $\theta_B$ is a counterclockwise angle, it is equal to the slope at the end:

$$\theta_B = v'(L) = \frac{qL^3}{24EI} \qquad \Longleftarrow \text{(9-26)}$$

Because the beam and loading are symmetric about the midpoint, the angles of rotation at the ends are equal.

This example illustrates the process of setting up and solving the differential equation of the deflection curve. It also illustrates the process of finding slopes and deflections at selected points along the axis of a beam.

*Note:* Now that we have derived formulas for the maximum deflection and maximum angles of rotation (see Eqs. 9-24, 9-25, and 9-26), we can evaluate those quantities numerically and observe that the deflections and angles are indeed small, as the theory requires.

Consider a steel beam on simple supports with a span length $L = 6$ ft. The cross section is rectangular with width $b = 3$ in. and height $h = 6$ in. The intensity of uniform load is $q = 8000$ lb/ft, which is relatively large because it produces a stress in the beam of 24,000 psi. (Thus, the deflections and slopes are larger than would normally be expected.)

Substituting into Eq. (9-24), and using $E = 30 \times 10^6$ psi, we find that the maximum deflection is $\delta_{max} = 0.144$ in., which is only 1/500 of the span length. Also, from Eq. (9-25), we find that the maximum angle of rotation is $\theta_A = 0.0064$ radians, or 0.37°, which is a very small angle.

Thus, our assumption that the slopes and deflections are small is validated.

Determine the equation of the deflection curve for a cantilever beam $AB$ subjected to a uniform load of intensity $q$ (Fig. 9-10a).

Also, determine the angle of rotation $\theta_B$ and the deflection $\delta_B$ at the free end (Fig. 9-10b). (*Note:* The beam has length $L$ and constant flexural rigidity $EI$.)

## Fig. 9-10

Example 9-2: Deflections of a cantilever beam with a uniform load

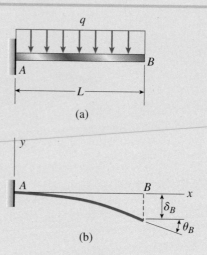

(a)

(b)

## Solution

*Bending moment in the beam.* The bending moment at distance $x$ from the fixed support is obtained from the free-body diagram of Fig. 9-11. Note that the vertical reaction at the support is equal to $qL$ and the moment reaction is equal to $qL^2/2$. Consequently, the expression for the bending moment $M$ is

## Fig. 9-11

Example 9-2: Free-body diagram used in determining the bending moment $M$

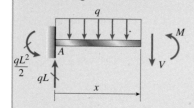

$$M = -\frac{qL^2}{2} + qLx - \frac{qx^2}{2} \qquad (9\text{-}27)$$

*Differential equation of the deflection curve.* When the preceding expression for the bending moment is substituted into the differential equation (Eq. 9-16a), we obtain

$$EIv'' = -\frac{qL^2}{2} + qLx - \frac{qx^2}{2} \qquad (9\text{-}28)$$

We now integrate both sides of this equation to obtain the slopes and deflections.

*Slope of the beam.* The first integration of Eq. (9-28) gives the following equation for the slope:

$$EIv' = -\frac{qL^2x}{2} + \frac{qLx^2}{2} - \frac{qx^3}{6} + C_1 \qquad \text{(a)}$$

The constant of integration $C_1$ can be found from the boundary condition that the slope of the beam is zero at the support; thus, we have the following condition:

$$v'(0) = 0$$

When this condition is applied to Eq. (a) we get $C_1 = 0$. Therefore, Eq. (a) becomes

$$EIv' = -\frac{qL^2x}{2} + \frac{qLx^2}{2} - \frac{qx^3}{6} \qquad \textbf{(b)}$$

and the slope is

$$v' = -\frac{qx}{6EI}(3L^2 - 3Lx + x^2) \qquad \Longleftarrow \textbf{(9-29)}$$

As expected, the slope obtained from this equation is zero at the support ($x = 0$) and negative (i.e., clockwise) throughout the length of the beam.
    *Deflection of the beam.* Integration of the slope equation (Eq. b) yields

$$EIv = -\frac{qL^2x^2}{4} + \frac{qLx^3}{6} - \frac{qx^4}{24} + C_2 \qquad \textbf{(c)}$$

The constant $C_2$ is found from the boundary condition that the deflection of the beam is zero at the support:

$$v(0) = 0$$

When this condition is applied to Eq. (c), we see immediately that $C_2 = 0$. Therefore, the equation for the deflection $v$ is

$$v = -\frac{qx^2}{24EI}(6L^2 - 4Lx + x^2) \qquad \Longleftarrow \textbf{(9-30)}$$

As expected, the deflection obtained from this equation is zero at the support ($x = 0$) and negative (that is, downward) elsewhere.
    *Angle of rotation at the free end of the beam.* The clockwise angle of rotation $\theta_B$ at end $B$ of the beam (Fig. 9-10b) is equal to the negative of the slope at that point. Thus, using Eq. (9-29), we get

$$\theta_B = -v'(L) = \frac{qL^3}{6EI} \qquad \Longleftarrow \textbf{(9-31)}$$

This angle is the maximum angle of rotation for the beam.
    *Deflection at the free end of the beam.* Since the deflection $\delta_B$ is downward (Fig. 9-10b), it is equal to the negative of the deflection obtained from Eq. (9-30):

$$\delta_B = -v(L) = \frac{qL^4}{8EI} \qquad \Longleftarrow \textbf{(9-32)}$$

This deflection is the maximum deflection of the beam.

A simple beam *AB* supports a concentrated load *P* acting at distances *a* and *b* from the left-hand and right-hand supports, respectively (Fig. 9-12a).

Determine the equations of the deflection curve, the angles of rotation $\theta_A$ and $\theta_B$ at the supports, the maximum deflection $\delta_{max}$, and the deflection $\delta_C$ at the midpoint *C* of the beam (Fig. 9-12b). (*Note:* The beam has length *L* and constant flexural rigidity *EI*.)

### Fig. 9-12

Example 9-3: Deflections of a simple beam with a concentrated load

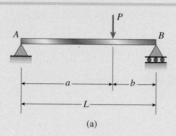

(a)

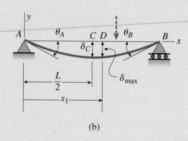

(b)

### Solution

*Bending moments in the beam.* In this example the bending moments are expressed by two equations, one for each part of the beam. Using the free-body diagrams of Fig. 9-13, we arrive at the following equations:

### Fig. 9-13

Example 9-3: Free-body diagrams used in determining the bending moments

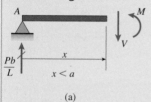

(a)

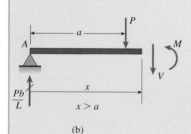

(b)

$$M = \frac{Pbx}{L} \quad (0 \le x \le a) \tag{9-33a}$$

$$M = \frac{Pbx}{L} - P(x - a) \quad (a \le x \le L) \tag{9-33b}$$

*Differential equations of the deflection curve.* The differential equations for the two parts of the beam are obtained by substituting the bending-moment expressions (Eqs. 9-33a and b) into Eq. (9-16a). The results are

$$EIv'' = \frac{Pbx}{L} \quad (0 \le x \le a) \tag{9-34a}$$

$$EIv'' = \frac{Pbx}{L} - P(x - a) \quad (a \le x \le L) \tag{9-34b}$$

*Slopes and deflections of the beam.* The first integrations of the two differential equations yield the following expressions for the slopes:

$$EIv' = \frac{Pbx^2}{2L} + C_1 \quad (0 \le x \le a) \tag{a}$$

$$EIv' = \frac{Pbx^2}{2L} - \frac{P(x - a)^2}{2} + C_2 \quad (a \le x \le L) \tag{b}$$

in which $C_1$ and $C_2$ are constants of integration. A second pair of integrations gives the deflections:

$$EIv = \frac{Pbx^3}{6L} + C_1x + C_3 \qquad (0 \le x \le a) \qquad \textbf{(c)}$$

$$EIv = \frac{Pbx^3}{6L} - \frac{P(x-a)^3}{6} + C_2x + C_4 \qquad (a \le x \le L) \qquad \textbf{(d)}$$

These equations contain two additional constants of integration, making a total of four constants to be evaluated.

*Constants of integration.* The four constants of integration can be found from the following four conditions:

1. At $x = a$, the slopes $v'$ for the two parts of the beam are the same.
2. At $x = a$, the deflections $v$ for the two parts of the beam are the same.
3. At $x = 0$, the deflection $v$ is zero.
4. At $x = L$, the deflection $v$ is zero.

The first two conditions are continuity conditions based upon the fact that the axis of the beam is a continuous curve. Conditions (3) and (4) are boundary conditions that must be satisfied at the supports.

Condition (1) means that the slopes determined from Eqs. (a) and (b) must be equal when $x = a$; therefore,

$$\frac{Pba^2}{2L} + C_1 = \frac{Pba^2}{2L} + C_2 \quad \text{or} \quad C_1 = C_2$$

Condition (2) means that the deflections found from Eqs. (c) and (d) must be equal when $x = a$; therefore,

$$\frac{Pba^3}{6L} + C_1a + C_3 = \frac{Pba^3}{6L} + C_2a + C_4$$

Inasmuch as $C_1 = C_2$, this equation gives $C_3 = C_4$.

Next, we apply condition (3) to Eq. (c) and obtain $C_3 = 0$; therefore,

$$C_3 = C_4 = 0 \qquad \textbf{(e)}$$

Finally, we apply condition (4) to Eq. (d) and obtain

$$\frac{PbL^2}{6} - \frac{Pb^3}{6} + C_2L = 0$$

Therefore,

$$C_1 = C_2 = -\frac{Pb(L^2 - b^2)}{6L} \qquad \textbf{(f)}$$

*Equations of the deflection curve.* We now substitute the constants of integration (Eqs. e and f) into the equations for the deflections (Eqs. c and d) and obtain the deflection equations for the two parts of the beam. The resulting equations, after a slight rearrangement, are

$$v = -\frac{Pbx}{6LEI}(L^2 - b^2 - x^2) \qquad (0 \le x \le a) \qquad \Longleftarrow \textbf{(9-35a)}$$

$$v = -\frac{Pbx}{6LEI}(L^2 - b^2 - x^2) - \frac{P(x-a)^3}{6EI} \qquad (a \le x \le L) \qquad \Longleftarrow \textbf{(9-35b)}$$

*Continues* ➡

The first of these equations gives the deflection curve for the part of the beam to the left of the load $P$, and the second gives the deflection curve for the part of the beam to the right of the load.

The slopes for the two parts of the beam can be found either by substituting the values of $C_1$ and $C_2$ into Eqs. (a) and (b) or by taking the first derivatives of the deflection equations (Eqs. 9-35a and b). The resulting equations are

$$v' = -\frac{Pb}{6LEI}(L^2 - b^2 - 3x^2) \qquad (0 \leq x \leq a) \qquad \text{(9-36a)}$$

$$v' = -\frac{Pb}{6LEI}(L^2 - b^2 - 3x^2) - \frac{P(x - a)^2}{2EI} \qquad (a \leq x \leq L) \qquad \text{(9-36b)}$$

The deflection and slope at any point along the axis of the beam can be calculated from Eqs. (9-35) and (9-36).

*Angles of rotation at the supports.* To obtain the angles of rotation $\theta_A$ and $\theta_B$ at the ends of the beam (Fig. 9-12b), we substitute $x = 0$ into Eq. (9-36a) and $x = L$ into Eq. (9-36b):

$$\theta_A = -v'(0) = \frac{Pb(L^2 - b^2)}{6LEI} = \frac{Pab(L + b)}{6LEI} \qquad \text{(9-37a)}$$

$$\theta_B = v'(L) = \frac{Pb(2L^2 - 3bL + b^2)}{6LEI} = \frac{Pab(L + a)}{6LEI} \qquad \text{(9-37b)}$$

Note that the angle $\theta_A$ is clockwise and the angle $\theta_B$ is counterclockwise, as shown in Fig. 9-12b.

The angles of rotation are functions of the position of the load and reach their largest values when the load is located near the midpoint of the beam. In the case of the angle of rotation $\theta_A$, the maximum value of the angle is

$$(\theta_A)_{max} = \frac{PL^2\sqrt{3}}{27EI} \qquad \text{(9-38)}$$

and occurs when $b = L/\sqrt{3} = 0.577L$ (or $a = 0.423L$). This value of $b$ is obtained by taking the derivative of $\theta_A$ with respect to $b$ (using the first of the two expressions for $\theta_A$ in Eq. 9-37a) and then setting it equal to zero.

*Maximum deflection of the beam.* The maximum deflection $\delta_{max}$ occurs at point $D$ (Fig. 9-12b) where the deflection curve has a horizontal tangent. If the load is to the right of the midpoint, that is, if $a > b$, point $D$ is in the part of the beam to the left of the load. We can locate this point by equating the slope $v'$ from Eq. (9-36a) to zero and solving for the distance $x$, which we now denote as $x_1$. In this manner we obtain the following formula for $x_1$:

$$x_1 = \sqrt{\frac{L^2 - b^2}{3}} \qquad (a \geq b) \qquad \text{(9-39)}$$

From this equation we see that as the load $P$ moves from the middle of the beam ($b = L/2$) to the right-hand end ($b = 0$), the distance $x_1$ varies from $L/2$ to $L/\sqrt{3} = 0.577L$. Thus, the maximum deflection occurs at a point very

**Fig. 9-12b (Repeated)**

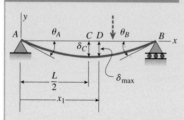

close to the midpoint of the beam, and this point is always between the midpoint of the beam and the load.

The maximum deflection $\delta_{max}$ is found by substituting $x_1$ (from Eq. 9-39) into the deflection equation (Eq. 9-35a) and then inserting a minus sign:

$$\delta_{max} = -(v)_{x=x_1} = \frac{Pb(L^2 - b^2)^{3/2}}{9\sqrt{3}\ LEI} \qquad (a \geq b) \quad \Longleftarrow \quad (9\text{-}40)$$

The minus sign is needed because the maximum deflection is downward (Fig. 9-12b) whereas the deflection $v$ is positive upward.

The maximum deflection of the beam depends on the position of the load $P$, that is, on the distance $b$. The maximum value of the maximum deflection (the "max-max" deflection) occurs when $b = L/2$ and the load is at the midpoint of the beam. This maximum deflection is equal to $PL^3/48EI$.

*Deflection at the midpoint of the beam.* The deflection $\delta_C$ at the midpoint C when the load is acting to the right of the midpoint (Fig. 9-12b) is obtained by substituting $x = L/2$ into Eq. (9-35a), as

$$\delta_C = -v\left(\frac{L}{2}\right) = \frac{Pb(3L^2 - 4b^2)}{48EI} \qquad (a \geq b) \quad \Longleftarrow \quad (9\text{-}41)$$

Because the maximum deflection always occurs near the midpoint of the beam, Eq. (9-41) yields a close approximation to the maximum deflection. In the most unfavorable case (when $b$ approaches zero), the difference between the maximum deflection and the deflection at the midpoint is less than 3% of the maximum deflection, as demonstrated in Prob. 9.3-7.

*Special case (load at the midpoint of the beam).* An important special case occurs when the load $P$ acts at the midpoint of the beam ($a = b = L/2$). Then we obtain the following results from Eqs. (9-36a), (9-35a), (9-37), and (9-40), respectively:

$$v' = -\frac{P}{16EI}(L^2 - 4x^2) \qquad \left(0 \leq x \leq \frac{L}{2}\right) \qquad (9\text{-}42)$$

$$v = -\frac{Px}{48EI}(3L^2 - 4x^2) \qquad \left(0 \leq x \leq \frac{L}{2}\right) \qquad (9\text{-}43)$$

$$\theta_A = \theta_B = \frac{PL^2}{16EI} \qquad (9\text{-}44)$$

$$\delta_{max} = \delta_C = \frac{PL^3}{48EI} \qquad (9\text{-}45)$$

Since the deflection curve is symmetric about the midpoint of the beam, the equations for $v'$ and $v$ are given only for the left-hand half of the beam in Eqs. (9-42) and (9-43). If needed, the equations for the right-hand half can be obtained from Eqs. (9-36b) and (9-35b) by substituting $a = b = L/2$.

# 9.4 DEFLECTIONS BY INTEGRATION OF THE SHEAR-FORCE AND LOAD EQUATIONS

The equations of the deflection curve in terms of the shear force $V$ and the load $q$ (Eqs. 9-16b and c, respectively) may also be integrated to obtain slopes and deflections. Since the loads are usually known quantities, whereas the bending moments must be determined from free-body diagrams and equations of equilibrium, many analysts prefer to start with the load equation. For this same reason, most computer programs for finding deflections begin with the load equation and then perform numerical integrations to obtain the shear forces, bending moments, slopes, and deflections.

The procedure for solving either the load equation or the shear-force equation is similar to that for solving the bending-moment equation, except that more integrations are required. For instance, if we begin with the load equation, four integrations are needed in order to arrive at the deflections. Thus, four constants of integration are introduced for each load equation that is integrated. As before, these constants are found from boundary, continuity, and symmetry conditions. However, these conditions now include conditions on the shear forces and bending moments as well as conditions on the slopes and deflections.

Conditions on the shear forces are equivalent to conditions on the third derivative (because $EIv''' = V$). In a similar manner, conditions on the bending moments are equivalent to conditions on the second derivative (because $EIv'' = M$). When the shear-force and bending-moment conditions are added to those for the slopes and deflections, we always have enough independent conditions to solve for the constants of integration.

The following examples illustrate the techniques of analysis in detail. The first example begins with the load equation and the second begins with the shear-force equation.

### ● ● ● Example 9-4

Determine the equation of the deflection curve for a cantilever beam $AB$ supporting a triangularly distributed load of maximum intensity $q_0$ (Fig. 9-14a).

Also, determine the deflection $\delta_B$ and angle of rotation $\theta_B$ at the free end (Fig. 9-14b). Use the fourth-order differential equation of the deflection curve (the load equation). (*Note:* The beam has length $L$ and constant flexural rigidity $EI$.)

### Fig. 9-14

Example 9-4: Deflections of a cantilever beam with a triangular load

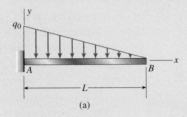

(a)

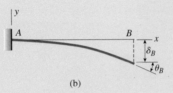

(b)

### Solution

*Differential equation of the deflection curve.* The intensity of the distributed load is given by the following equation (see Fig. 9-14a):

$$q = \frac{q_0(L - x)}{L} \tag{9-46}$$

Consequently, the fourth-order differential equation (Eq. 9-16c) becomes

$$EIv'''' = -q = -\frac{q_0(L - x)}{L} \tag{a}$$

*Shear force in the beam.* The first integration of Eq. (a) gives

$$EIv''' = \frac{q_0}{2L}(L - x)^2 + C_1 \tag{b}$$

The right-hand side of this equation represents the shear force $V$ (see Eq. 9-16b). Because the shear force is zero at $x = L$, we have the following boundary condition:

$$v'''(L) = 0$$

Using this condition with Eq. (b), we get $C_1 = 0$. Therefore, Eq. (b) simplifies to

$$EIv''' = \frac{q_0}{2L}(L - x)^2 \tag{c}$$

Continues ➡

and the shear force in the beam is

$$V = EIv''' = \frac{q_0}{2L}(L - x)^2 \qquad \text{(9-47)}$$

*Bending moment in the beam.* Integrating a second time, we obtain the following equation from Eq. (c):

$$EIv'' = -\frac{q_0}{6L}(L - x)^3 + C_2 \qquad \text{(d)}$$

This equation is equal to the bending moment $M$ (see Eq. 9-16a). Since the bending moment is zero at the free end of the beam, we have the following boundary condition:

$$v''(L) = 0$$

Applying this condition to Eq. (d), we obtain $C_2 = 0$, and therefore the bending moment is

$$M = EIv'' = -\frac{q_0}{6L}(L - x)^3 \qquad \text{(9-48)}$$

*Slope and deflection of the beam.* The third and fourth integrations yield

$$EIv' = \frac{q_0}{24L}(L - x)^4 + C_3 \qquad \text{(e)}$$

$$EIv = -\frac{q_0}{120L}(L - x)^5 + C_3 x + C_4 \qquad \text{(f)}$$

The boundary conditions at the fixed support, where both the slope and deflection equal to zero, are

$$v'(0) = 0 \qquad v(0) = 0$$

Applying these conditions to Eqs. (e) and (f), respectively, we find

$$C_3 = -\frac{q_0 L^3}{24} \qquad C_4 = \frac{q_0 L^4}{120}$$

Substituting these expressions for the constants into Eqs. (e) and (f), we obtain the following equations for the slope and deflection of the beam:

$$v' = -\frac{q_0 x}{24LEI}(4L^3 - 6L^2 x + 4Lx^2 - x^3) \qquad \Longleftarrow \text{(9-49)}$$

$$v = -\frac{q_0 x^2}{120LEI}(10L^3 - 10L^2 x + 5Lx^2 - x^3) \qquad \Longleftarrow \text{(9-50)}$$

*Angle of rotation and deflection at the free end of the beam.* The angle of rotation $\theta_B$ and deflection $\delta_B$ at the free end of the beam (Fig. 9-14b) are obtained from Eqs. (9-49) and (9-50), respectively, by substituting $x = L$. The results are

$$\theta_B = -v'(L) = \frac{q_0 L^3}{24EI} \qquad \delta_B = -v(L) = \frac{q_0 L^4}{30EI} \qquad \Longleftarrow \text{(9-51a,b)}$$

Thus, we have determined the required slopes and deflections of the beam by solving the fourth-order differential equation of the deflection curve.

Cantilever portion of roof structure (Courtesy of the National Information Service, for Earthquake Engineering EERC, University of California, Berkeley.)

● ● ● **Example 9-5**

A simple beam $AB$ with an overhang $BC$ supports a concentrated load $P$ at the end of the overhang (Fig. 9-15a). The main span of the beam has length $L$ and the overhang has length $L/2$.

Determine the equations of the deflection curve and the deflection $\delta_C$ at the end of the overhang (Fig. 9-15b). Use the third-order differential equation of the deflection curve (the shear-force equation). (*Note:* The beam has constant flexural rigidity $EI$.)

**Fig. 9-15**

Example 9-5: Deflections of a beam with an overhang

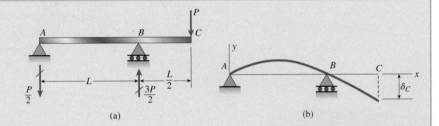

(a)                              (b)

**Solution**

*Differential equations of the deflection curve.* Because reactive forces act at supports $A$ and $B$, we must write separate differential equations for parts $AB$ and $BC$ of the beam. Therefore, we begin by finding the shear forces in each part of the beam.

The downward reaction at support $A$ is equal to $P/2$, and the upward reaction at support $B$ is equal to $3P/2$ (see Fig. 9-15a). It follows that the shear forces in parts $AB$ and $BC$ are

$$V = -\frac{P}{2} \quad (0 < x < L) \tag{9-52a}$$

$$V = P \quad \left(L < x < \frac{3L}{2}\right) \tag{9-52b}$$

in which $x$ is measured from end $A$ of the beam (Fig. 9-16b).

The third-order differential equations for the beam now become [see Eq. (9-16b):

$$EIv''' = -\frac{P}{2} \quad (0 < x < L) \tag{a}$$

$$EIv''' = P \quad \left(L < x < \frac{3L}{2}\right) \tag{b}$$

*Bending moments in the beam.* Integration of the preceding two equations yields the bending-moment equations:

$$M = EIv'' = -\frac{Px}{2} + C_1 \quad (0 \le x \le L) \tag{c}$$

$$M = EIv'' = Px + C_2 \quad \left(L \le x \le \frac{3L}{2}\right) \tag{d}$$

*Continues* ➡

Bridge girder with overhang during transport to the construction site (Tom Brakefield/Getty Images)

The bending moments at points $A$ and $C$ are zero; hence, we have the following boundary conditions:

$$v''(0) = 0 \qquad v''\left(\frac{3L}{2}\right) = 0$$

Using these conditions with Eqs. (c) and (d), we get

$$C_1 = 0 \qquad C_2 = -\frac{3PL}{2}$$

Therefore, the bending moments are

$$M = EIv'' = -\frac{Px}{2} \qquad (0 \le x \le L) \tag{9-53a}$$

$$M = EIv'' = -\frac{P(3L - 2x)}{2} \qquad \left(L \le x \le \frac{3L}{2}\right) \tag{9-53b}$$

These equations can be verified by determining the bending moments from free-body diagrams and equations of equilibrium.

*Slopes and deflections of the beam.* The next integrations yield the slopes:

$$EIv' = -\frac{Px^2}{4} + C_3 \qquad (0 \le x \le L)$$

$$EIv' = -\frac{Px(3L - x)}{2} + C_4 \qquad \left(L \le x \le \frac{3L}{2}\right)$$

The only condition on the slopes is the continuity condition at support $B$. According to this condition, the slope at point $B$ as found for part $AB$ of the beam is equal to the slope at the same point as found for part $BC$ of the beam. Therefore, we substitute $x = L$ into each of the two preceding equations for the slopes and obtain

$$-\frac{PL^2}{4} + C_3 = -PL^2 + C_4$$

This equation eliminates one constant of integration because we can express $C_4$ in terms of $C_3$:

$$C_4 = C_3 + \frac{3PL^2}{4} \tag{e}$$

The third and last integrations give

$$EIv = -\frac{Px^3}{12} + C_3 x + C_5 \qquad (0 \le x \le L) \tag{f}$$

$$EIv = -\frac{Px^2(9L - 2x)}{12} + C_4 x + C_6 \qquad \left(L \le x \le \frac{3L}{2}\right) \tag{g}$$

For part $AB$ of the beam (Fig. 9-15a), we have two boundary conditions on the deflections, namely, the deflection is zero at points $A$ and $B$:

$$v(0) = 0 \quad \text{and} \quad v(L) = 0$$

Applying these conditions to Eq. (f), we obtain

$$C_5 = 0 \qquad C_3 = \frac{PL^2}{12} \qquad \text{(h,i)}$$

Substituting the preceding expression for $C_3$ in Eq. (e), we get

$$C_4 = \frac{5PL^2}{6} \qquad \text{(j)}$$

For part $BC$ of the beam, the deflection is zero at point $B$. Therefore, the boundary condition is

$$v(L) = 0$$

Applying this condition to Eq. (g), and also substituting Eq. (j) for $C_4$, we get

$$C_6 = -\frac{PL^3}{4} \qquad \text{(k)}$$

All constants of integration have now been evaluated.

The deflection equations are obtained by substituting the constants of integration (Eqs. h, i, j, and k) into Eqs. (f) and (g). The results are

$$v = \frac{Px}{12EI}(L^2 - x^2) \qquad (0 \leq x \leq L) \qquad \text{(9-54a)}$$

$$v = -\frac{P}{12EI}(3L^3 - 10L^2x + 9Lx^2 - 2x^3) \qquad \left(L \leq x \leq \frac{3L}{2}\right) \qquad \text{(9-54b)}$$

Note that the deflection is always positive (upward) in part $AB$ of the beam (Eq. 9-54a) and always negative (downward) in the overhang $BC$ (Eq. 9-54b).

*Deflection at the end of the overhang.* We can find the deflection $\delta_C$ at the end of the overhang (Fig. 9-15b) by substituting $x = 3L/2$ in Eq. (9-54b):

$$\delta_C = -v\left(\frac{3L}{2}\right) = \frac{PL^3}{8EI} \qquad \text{(9-55)}$$

Thus, we have determined the required deflections of the overhanging beam (Eqs. 9-54 and 9-55) by solving the third-order differential equation of the deflection curve.

**Fig. 9-15 (Repeated)**

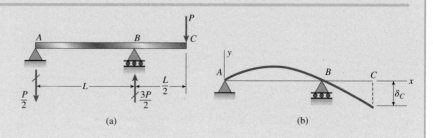

(a)                                        (b)

# 9.5 METHOD OF SUPERPOSITION

The **method of superposition** is a practical and commonly used technique for obtaining deflections and angles of rotation of beams. The underlying concept is quite simple and may be stated as follows:

*Under suitable conditions, the deflection of a beam produced by several different loads acting simultaneously can be found by superposing the deflections produced by the same loads acting separately.*

For instance, if $v_1$ represents the deflection at a particular point on the axis of a beam due to a load $q_1$, and if $v_2$ represents the deflection at that same point due to a different load $q_2$, then the deflection at that point due to loads $q_1$ and $q_2$ acting simultaneously is $v_1 + v_2$. (The loads $q_1$ and $q_2$ are independent loads and each may act anywhere along the axis of the beam.)

The justification for superposing deflections lies in the nature of the differential equations of the deflection curve [Eqs. (9-16a, b, and c)]. These equations are *linear* differential equations, because all terms containing the deflection $v$ and its derivatives are raised to the first power. Therefore, the solutions of these equations for several loading conditions may be added algebraically, or *superposed*. (The conditions for superposition to be valid are described later in the subsection "Principle of Superposition.")

As an **illustration** of the superposition method, consider the simple beam $ACB$ shown in Fig. 9-16a. This beam supports two loads: (1) a uniform load of intensity $q$ acting throughout the span, and (2) a concentrated load $P$ acting at the midpoint. Suppose we wish to find the deflection $\delta_C$ at the midpoint and the angles of rotation $\theta_A$ and $\theta_B$ at the ends (Fig. 9-16b). Using the method of superposition, we obtain the effects of each load acting separately and then combine the results.

For the uniform load acting alone, the deflection at the midpoint and the angles of rotation are obtained from the formulas of Example 9-1 [see Eqs. (9-24), (9-25), and (9-26)]:

$$(\delta_C)_1 = \frac{5qL^4}{384EI} \qquad (\theta_A)_1 = (\theta_B)_1 = \frac{qL^3}{24EI}$$

in which $EI$ is the flexural rigidity of the beam and $L$ is its length.

For the load $P$ acting alone, the corresponding quantities are obtained from the formulas of Example 9-3 [see Eqs. (9-44) and (9-45)]:

$$(\delta_C)_2 = \frac{PL^3}{48EI} \qquad (\theta_A)_2 = (\theta_B)_2 = \frac{PL^2}{16EI}$$

The deflection and angles of rotation due to the combined loading (Fig. 9-16a) are obtained by summation:

$$\delta_C = (\delta_C)_1 + (\delta_C)_2 = \frac{5qL^4}{384EI} + \frac{PL^3}{48EI} \qquad \textbf{(9-56a)}$$

$$\theta_A = \theta_B = (\theta_A)_1 + (\theta_A)_2 = \frac{qL^3}{24EI} + \frac{PL^2}{16EI} \qquad \textbf{(9-56b)}$$

**Fig. 9-16**

Simple beam with two loads

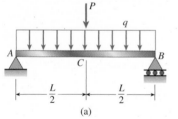

(a)

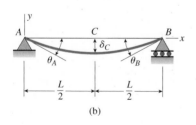

(b)

The deflections and angles of rotation at other points on the beam axis can be found by this same procedure. However, the method of superposition is not limited to finding deflections and angles of rotation at single points. The method may also be used to obtain general equations for the slopes and deflections of beams subjected to more than one load.

## Tables of Beam Deflections

The method of superposition is useful only when formulas for deflections and slopes are readily available. To provide convenient access to such formulas, tables for both cantilever and simple beams are given in Appendix H at the back of the book. Similar tables can be found in engineering handbooks. Using these tables and the method of superposition, we can find deflections and angles of rotation for many different loading conditions, as illustrated in the examples at the end of this section.

## Distributed Loads

Sometimes we encounter a distributed load that is not included in a table of beam deflections. In such cases, superposition may still be useful. We can consider an element of the distributed load as though it were a concentrated load, and then we can find the required deflection by integrating throughout the region of the beam where the load is applied.

To illustrate this process of integration, consider a simple beam $ACB$ with a triangular load acting on the left-hand half (Fig. 9-17a). We wish to obtain the deflection $\delta_C$ at the midpoint $C$ and the angle of rotation $\theta_A$ at the left-hand support (Fig. 9-17c).

We begin by visualizing an element $q\,dx$ of the distributed load as a concentrated load (Fig. 9-17b). Note that the load acts to the left of the midpoint of the beam. The deflection at the midpoint due to this concentrated load is obtained from Case 5 of Table H-2, Appendix H. The formula given there for the midpoint deflection (for the case in which $a \leq b$) is

$$\frac{Pa}{48EI}(3L^2 - 4a^2)$$

In our example (Fig. 9-17b), we substitute $q\,dx$ for $P$ and $x$ for $a$:

$$\frac{(q\,dx)(x)}{48EI}(3L^2 - 4x^2) \tag{9-57}$$

This expression gives the deflection at point $C$ due to the element $q\,dx$ of the load.

Next, we note that the intensity of the uniform load (Figs. 9-17a and b) is

$$q = \frac{2q_0 x}{L} \tag{9-58}$$

where $q_0$ is the maximum intensity of the load. With this substitution for $q$, the formula for the deflection (Eq. 9-57) becomes

$$\frac{q_0 x^2}{24LEI}(3L^2 - 4x^2)dx$$

### Fig. 9-17

Simple beam with a triangular load

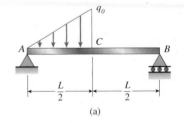

(a)

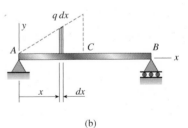

(b)

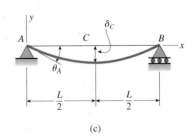

(c)

Finally, we integrate throughout the region of the load to obtain the deflection $\delta_C$ at the midpoint of the beam due to the entire triangular load:

$$\delta_C = \int_0^{L/2} \frac{q_0 x^2}{24LEI}(3L^2 - 4x^2)dx$$

$$= \frac{q_0}{24LEI} \int_0^{L/2} (3L^2 - 4x^2)x^2\, dx = \frac{q_0 L^4}{240EI} \qquad \textbf{(9-59)}$$

By a similar procedure, we can calculate the angle of rotation $\theta_A$ at the left-hand end of the beam (Fig. 9-17c). The expression for this angle due to a concentrated load $P$ (see Case 5 of Table H-2) is

$$\frac{Pab(L + b)}{6LEI}$$

Replacing $P$ with $2q_0 x\, dx/L$, $a$ with $x$, and $b$ with $L - x$, we obtain

$$\frac{2q_0 x^2(L - x)(L + L - x)}{6L^2 EI}dx \quad \text{or} \quad \frac{q_0}{3L^2 EI}(L - x)(2L - x)x^2 dx$$

Finally, we integrate throughout the region of the load:

$$\theta_A = \int_0^{L/2} \frac{q_0}{3L^2 EI}(L - x)(2L - x)x^2 dx = \frac{41q_0 L^3}{2880EI} \qquad \textbf{(9-60)}$$

This is the angle of rotation produced by the triangular load.

This example illustrates how we can use superposition and integration to find deflections and angles of rotation produced by distributed loads of almost any kind. If the integration cannot be performed easily by analytical means, numerical methods can be used.

## Principle of Superposition

The method of superposition for finding beam deflections is an example of a more general concept known in mechanics as the **principle of superposition**. This principle is valid whenever the quantity to be determined is a linear function of the applied loads. When that is the case, the desired quantity may be found due to each load acting separately, and then these results may be superposed to obtain the desired quantity due to all loads acting simultaneously. In ordinary structures, the principle is usually valid for stresses, strains, bending moments, and many other quantities besides deflections.

In the particular case of **beam deflections**, the principle of super position is valid under the following conditions: (1) Hooke's law holds for the material, (2) the deflections and rotations are small, and (3) the presence of the deflections does not alter the actions of the applied loads. These requirements ensure that the differential equations of the deflection curve are linear.

The following examples provide additional illustrations in which the principle of superposition is used to calculate deflections and angles of rotation of beams.

A cantilever beam $AB$ supports a uniform load of intensity $q$ acting over part of the span and a concentrated load $P$ acting at the free end (Fig. 9-18a).

Determine the deflection $\delta_B$ and angle of rotation $\theta_B$ at end $B$ of the beam (Fig. 9-18b). (*Note:* The beam has length $L$ and constant flexural rigidity $EI$.)

### Fig. 9-18

Example 9-6: Cantilever beam with a uniform load and a concentrated load

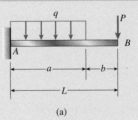

(a)

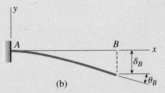

(b)

### Solution

We can obtain the deflection and angle of rotation at end $B$ of the beam by combining the effects of the loads acting separately. If the uniform load acts alone, the deflection and angle of rotation (obtained from Case 2 of Table H-1, Appendix H) are

$$(\delta_B)_1 = \frac{qa^3}{24EI}(4L - a) \qquad (\theta_B)_1 = \frac{qa^3}{6EI}$$

If the load $P$ acts alone, the corresponding quantities (from Case 4, Table H-1) are

$$(\delta_B)_2 = \frac{PL^3}{3EI} \qquad (\theta_B)_2 = \frac{PL^2}{2EI}$$

Therefore, the deflection and angle of rotation due to the combined loading (Fig. 9-18a) are

$$\delta_B = (\delta_B)_1 + (\delta_B)_2 = \frac{qa^3}{24EI}(4L - a) + \frac{PL^3}{3EI} \qquad \Longleftarrow \quad \textbf{(9-61)}$$

$$\theta_B = (\theta_B)_1 + (\theta_B)_2 = \frac{qa^3}{6EI} + \frac{PL^2}{2EI} \qquad \Longleftarrow \quad \textbf{(9-62)}$$

Thus, we have found the required quantities by using tabulated formulas and the method of superposition.

A cantilever beam *AB* with a uniform load of intensity *q* acting on the right-hand half of the beam is shown in Fig. 9-19a.

Obtain formulas for the deflection $\delta_B$ and angle of rotation $\theta_B$ at the free end (Fig. 9-19c). (*Note:* The beam has length *L* and constant flexural rigidity *EI*.)

**Fig. 9-19**

Example 9-7: Cantilever beam with a uniform load acting on the right-hand half of the beam

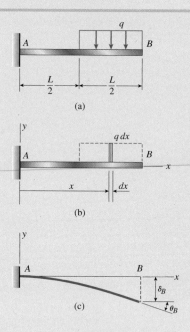

**Solution**

In this example we will determine the deflection and angle of rotation by treating an element of the uniform load as a concentrated load and then integrating (see Fig. 9-19b). The element of load has magnitude *q dx* and is located at distance *x* from the support. The resulting differential deflection $d\delta_B$ and differential angle of rotation $d\theta_B$ at the free end are found from the corresponding formulas in Case 5 of Table H-1, Appendix H, by replacing *P* with *q dx* and *a* with *x*; thus,

$$d\delta_B = \frac{(q\,dx)(x^2)(3L - x)}{6EI} \qquad d\theta_B = \frac{(q\,dx)(x^2)}{2EI}$$

By integrating over the loaded region, we get

$$\delta_B = \int d\delta_B = \frac{q}{6EI} \int_{L/2}^{L} x^2(3L - x)\,dx = \frac{41qL^4}{384EI} \qquad \text{(9-63)}$$

$$\theta_B = \int d\theta_B = \frac{q}{2EI} \int_{L/2}^{L} x^2\,dx = \frac{7qL^3}{48EI} \qquad \text{(9-64)}$$

*Note:* These same results can be obtained by using the formulas in Case 3 of Table H-1 and substituting *a* = *b* = *L*/2.

● ● ● ● **Example 9-8**

## Fig. 9-20

Example 9-8: Compound beam with a hinge

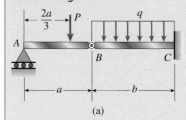

(a)

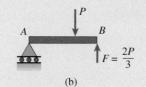

(b)

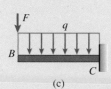

(c)

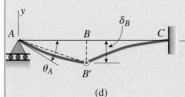

(d)

A compound beam $ABC$ has a roller support at $A$, an internal hinge (i.e., moment release) at $B$, and a fixed support at $C$ (Fig. 9-20a). Segment $AB$ has length $a$ and segment $BC$ has length $b$. A concentrated load $P$ acts at distance $2a/3$ from support $A$ and a uniform load of intensity $q$ acts between points $B$ and $C$.

Determine the deflection $\delta_B$ at the hinge and the angle of rotation $\theta_A$ at support $A$ (Fig. 9-20d). (*Note:* The beam has constant flexural rigidity $EI$.)

### Solution

For purposes of analysis, we will consider the compound beam to consist of two individual beams: (1) a simple beam $AB$ of length $a$, and (2) a cantilever beam $BC$ of length $b$. The two beams are linked together by a pin connection at $B$.

If we separate beam $AB$ from the rest of the structure (Fig. 9-20b), we see that there is a vertical force $F$ at end $B$ equal to $2P/3$. This same force acts downward at end $B$ of the cantilever (Fig. 9-20c). Consequently, the cantilever beam $BC$ is subjected to two loads: a uniform load and a concentrated load. The deflection at the end of this cantilever (which is the same as the deflection $\delta_B$ of the hinge) is readily found from Cases 1 and 4 of Table H-1, Appendix H:

$$\delta_B = \frac{qb^4}{8EI} + \frac{Fb^3}{3EI}$$

or, since $F = 2P/3$,

$$\delta_B = \frac{qb^4}{8EI} + \frac{2Pb^3}{9EI} \qquad \Longleftarrow \quad \textbf{(9-65)}$$

The angle of rotation $\theta_A$ at support $A$ (Fig. 9-20d) consists of two parts: (1) an angle $BAB'$ produced by the downward displacement of the hinge, and (2) an additional angle of rotation produced by the bending of beam $AB$ (or beam $AB'$) as a simple beam. The angle $BAB'$ is

$$(\theta_A)_1 = \frac{\delta_B}{a} = \frac{qb^4}{8aEI} + \frac{2Pb^3}{9aEI}$$

The angle of rotation at the end of a simple beam with a concentrated load is obtained from Case 5 of Table H-2. The formula given there is

$$\frac{Pab(L + b)}{6LEI}$$

in which $L$ is the length of the simple beam, $a$ is the distance from the left-hand support to the load, and $b$ is the distance from the right-hand support to the load. Thus, in the notation of our example (Fig. 9-20a), the angle of rotation is

$$(\theta_A)_2 = \frac{P\left(\dfrac{2a}{3}\right)\left(\dfrac{a}{3}\right)\left(a + \dfrac{a}{3}\right)}{6aEI} = \frac{4Pa^2}{81EI}$$

Combining the two angles, we obtain the total angle of rotation at support $A$:

$$\theta_A = (\theta_A)_1 + (\theta_A)_2 = \frac{qb^4}{8aEI} + \frac{2Pb^3}{9aEI} + \frac{4Pa^2}{81EI} \qquad \Longleftarrow \quad \textbf{(9-66)}$$

This example illustrates how the method of superposition can be adapted to handle a seemingly complex situation in a relatively simple manner.

A simple beam $AB$ of span length $L$ has an overhang $BC$ of length $a$ (Fig. 9-21a). The beam supports a uniform load of intensity $q$ throughout its length.

Obtain a formula for the deflection $\delta_C$ at the end of the overhang (Fig. 9-21c). (*Note:* The beam has constant flexural rigidity $EI$.)

**Fig. 9-21**

Example 9-9: Simple beam with an overhang

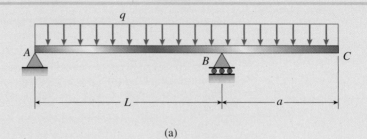

(a)

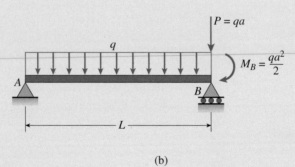

(b)

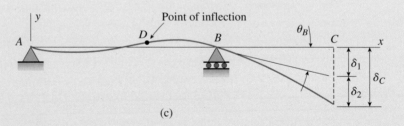

(c)

Beam with overhang loaded by gravity uniform load (Courtesy of the National Information Service for Earthquake Engineering EERC, University of California, Berkeley.)

### Solution

We can find the deflection of point $C$ by imagining the overhang $BC$ (Fig. 9-21a) to be a cantilever beam subjected to two actions. The first action is the rotation of the support of the cantilever through an angle $\theta_B$, which is the angle of rotation of beam $ABC$ at support $B$ (Fig. 9-21c). (We assume that a clockwise angle $\theta_B$ is positive.) This angle of rotation causes a rigid-body rotation of the overhang $BC$, resulting in a downward displacement $\delta_1$ of point $C$.

The second action is the bending of $BC$ as a cantilever beam supporting a uniform load. This bending produces an additional downward displacement $\delta_2$ (Fig. 9-21c). The superposition of these two dis placements gives the total displacement $\delta_C$ at point $C$.

*Deflection $\delta_1$.* Let us begin by finding the deflection $\delta_1$ caused by the angle of rotation $\theta_B$ at point $B$. To find this angle, we observe that part $AB$ of the beam is in the same condition as a simple beam (Fig. 9-21b) subjected to the following loads: (1) a uniform load of intensity $q$, (2) a couple $M_B$

(equal to $qa^2/2$), and (3) a vertical load $P$ (equal to $qa$). Only the loads $q$ and $M_B$ produce angles of rotation at end $B$ of this simple beam. These angles are found from Cases 1 and 7 of Table H-2, Appendix H. Thus, the angle $\theta_B$ is

$$\theta_B = -\frac{qL^3}{24EI} + \frac{M_B L}{3EI} = -\frac{qL^3}{24EI} + \frac{qa^2 L}{6EI} = \frac{qL(4a^2 - L^2)}{24EI} \qquad \textbf{(9-67)}$$

in which a clockwise angle is positive, as shown in Fig. 9-21c.

The downward deflection $\delta_1$ of point $C$, due solely to the angle of rotation $\theta_B$, is equal to the length of the overhang times the angle (Fig. 9-21c):

$$\delta_1 = a\theta_B = \frac{qaL(4a^2 - L^2)}{24EI} \qquad \textbf{(a)}$$

*Deflection $\delta_2$.* Bending of the overhang $BC$ produces an additional downward deflection $\delta_2$ at point $C$. This deflection is equal to the deflection of a cantilever beam of length $a$ subjected to a uniform load of intensity $q$ (see Case 1 of Table H-1):

$$\delta_2 = \frac{qa^4}{8EI} \qquad \textbf{(b)}$$

*Deflection $\delta_C$.* The total downward deflection of point $C$ is the algebraic sum of $\delta_1$ and $\delta_2$:

$$\delta_C = \delta_1 + \delta_2 = \frac{qaL(4a^2 - L^2)}{24EI} + \frac{qa^4}{8EI} = \frac{qa}{24EI}[L(4a^2 - L^2) + 3a^3]$$

or

$$\delta_C = \frac{qa}{24EI}(a + L)(3a^3 + aL - L^2) \qquad \Longleftarrow \textbf{(9-68)}$$

From the preceding equation we see that the deflection $\delta_C$ may be upward or downward, depending upon the relative magnitudes of the lengths $L$ and $a$. If $a$ is relatively large, the last term in the equation (the three-term expression in parentheses) is positive and the deflection $\delta_C$ is downward. If $a$ is relatively small, the last term is negative and the deflection is upward. The deflection is zero when the last term is equal to zero:

$$3a^2 + aL - L^2 = 0$$

or

$$a = \frac{L(\sqrt{13} - 1)}{6} = 0.4343L \qquad \textbf{(c)}$$

From this result, we see that if $a$ is greater than $0.4343L$, the deflection of point $C$ is downward; if $a$ is less than $0.4343L$, the deflection is upward.

*Deflection curve.* The shape of the deflection curve for the beam in this example is shown in Fig. 9-21c for the case where $a$ is large enough $(a > 0.4343L)$ to produce a downward deflection at $C$ and small enough $(a < L)$ to ensure that the reaction at $A$ is upward. Under these conditions the beam has a positive bending moment between support $A$ and a point such as $D$. The deflection curve in region $AD$ is concave upward (positive curvature). From $D$ to $C$, the bending moment is negative, and therefore the deflection curve is concave downward (negative curvature).

*Point of inflection.* At point $D$ the curvature of the deflection curve is zero because the bending moment is zero. A point such as $D$ where the curvature and bending moment *change signs* is called a **point of inflection** (or *point of contraflexure*). The bending moment $M$ and the second derivative $d^2v/dx^2$ always vanish at an inflection point.

However, a point where $M$ and $d^2v/dx^2$ equal zero is not necessarily an inflection point because it is possible for those quantities to be zero without changing signs at that point; for example, they could have maximum or minimum values.

# 9.6 MOMENT-AREA METHOD

In this section we will describe another method for finding deflections and angles of rotation of beams. Because the method is based upon two theorems related to the *area of the bending-moment diagram*, it is called the **moment-area method**.

The assumptions used in deriving the two theorems are the same as those used in deriving the differential equations of the deflection curve. Therefore, the moment-area method is valid only for linearly elastic beams with small slopes.

## First Moment-Area Theorem

To derive the first theorem, consider a segment $AB$ of the deflection curve of a beam in a region where the curvature is positive (Fig. 9-22). Of course, the deflections and slopes shown in the figure are highly exaggerated for clarity. At point $A$ the tangent $AA'$ to the deflection curve is at an angle $\theta_A$ to the $x$ axis, and at point $B$ the tangent $BB'$ is at an angle $\theta_B$. These two tangents meet at point $C$.

The **angle between the tangents**, denoted $\theta_{B/A}$, is equal to the difference between $\theta_B$ and $\theta_A$:

$$\theta_{B/A} = \theta_B - \theta_A \tag{9-69}$$

**Fig. 9-22**

Derivation of the first moment-area theorem

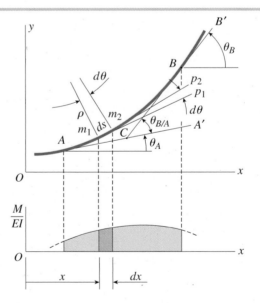

Thus, the angle $\theta_{B/A}$ may be described as the angle to the tangent at $B$ measured relative to, or with respect to, the tangent at $A$. Note that the angles $\theta_A$ and $\theta_B$, which are the angles of rotation of the beam axis at points $A$ and $B$, respectively, are also equal to the slopes at those points, because in reality the slopes and angles are very small quantities.

Next, consider two points $m_1$ and $m_2$ on the deflected axis of the beam (Fig. 9-22). These points are a small distance $ds$ apart. The tangents to the deflection curve at these points are shown in the figure as lines $m_1 p_1$ and $m_2 p_2$. The normals to these tangents intersect at the center of curvature (not shown in the figure).

The angle $d\theta$ between the normals (Fig. 9-22) is given by the following equation:

$$d\theta = \frac{ds}{\rho} \qquad \textbf{(9-70a)}$$

in which $\rho$ is the radius of curvature and $d\theta$ is measured in radians [see Eq. (9-2)]. Because the normals and the tangents ($m_1 p_1$ and $m_2 p_2$) are perpendicular, it follows that the angle between the tangents is also equal to $d\theta$.

For a beam with small angles of rotation, we can replace $ds$ with $dx$, as explained in Section 9.2. Thus,

$$d\theta = \frac{dx}{\rho} \qquad \textbf{(9-70b)}$$

Also, from Eq. (9-10) we know that

$$\frac{1}{\rho} = \frac{M}{EI} \qquad \textbf{(9-71)}$$

and therefore

$$d\theta = \frac{M\,dx}{EI} \qquad \textbf{(9-72)}$$

in which $M$ is the bending moment and $EI$ is the flexural rigidity of the beam.

The quantity $M\,dx/EI$ has a simple geometric interpretation. To see this, refer to Fig. 9-22 where we have drawn the $M/EI$ diagram directly below the beam. At any point along the $x$ axis, the height of this diagram is equal to the bending moment $M$ at that point divided by the flexural rigidity $EI$ at that point. Thus, the $M/EI$ diagram has the same shape as the bending-moment diagram whenever $EI$ is constant. The term $M\,dx/EI$ is the area of the shaded strip of width $dx$ within the $M/EI$ diagram. (Note that since the curvature of the deflection curve in Fig. 9-22 is positive, the bending moment $M$ and the area of the $M/EI$ diagram are also positive.)

Let us now integrate $d\theta$ from Eq. (9-72) between points $A$ and $B$ of the deflection curve:

$$\int_A^B d\theta = \int_A^B \frac{M\,dx}{EI} \qquad \textbf{(9-73)}$$

When evaluated, the integral on the left-hand side becomes $\theta_B - \theta_A$, which is equal to the angle $\theta_{B/A}$ between the tangents at $B$ and $A$ from Eq. (9-69).

The integral on the right-hand side of Eq. (9-73) is equal to the area of the $M/EI$ diagram between points $A$ and $B$. (Note that the area of the $M/EI$ diagram is an algebraic quantity and may be positive or negative, depending upon whether the bending moment is positive or negative.)

Now we can write Eq. (9-73) as

$$
\theta_{B/A} = \int_A^B \frac{M\,dx}{EI}
$$

$$
= \text{Area of the } M/EI \text{ diagram between points } A \text{ and B} \qquad \textbf{(9-74)}
$$

This equation may be stated as a theorem:

**First moment-area theorem:** The angle $\theta_{B/A}$ between the tangents to the deflection curve at two points $A$ and $B$ is equal to the area of the $M/EI$ diagram between those points.

The sign conventions used in deriving the preceding theorem are as follows:

1.  The angles $\theta_A$ and $\theta_B$ are positive when counterclockwise.

2.  The angle $\theta_{B/A}$ between the tangents is positive when the angle $\theta_B$ is algebraically larger than the angle $\theta_A$. Also, note that point $B$ must be to the right of point $A$; that is, it must be further along the axis of the beam as we move in the $x$ direction.

3.  The bending moment $M$ is positive according to our usual sign convention; that is, $M$ is positive when it produces compression in the upper part of the beam.

4.  The area of the $M/EI$ diagram is given a positive or negative sign according to whether the bending moment is positive or negative. If part of the bending-moment diagram is positive and part is negative, then the corresponding parts of the $M/EI$ diagram are given those same signs.

The preceding sign conventions for $\theta_A$, $\theta_B$, and $\theta_{B/A}$ are often ignored in practice because (as explained later) the directions of the angles of rotation are usually obvious from an inspection of the beam and its loading. When this is the case, we can simplify the calculations by ignoring signs and using only absolute values when applying the first moment-area theorem.

## Second Moment-Area Theorem

Now we turn to the second theorem, which is related primarily to deflections rather than to angles of rotation. Consider again the deflection curve between points $A$ and $B$ (Fig. 9-23). We draw the tangent at point $A$ and

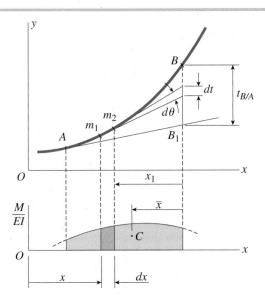

Fig. 9-23

Derivation of the second moment-area theorem

note that its intersection with a vertical line through point $B$ is at point $B_1$. The vertical distance between points $B$ and $B_1$ is denoted $t_{B/A}$ in the figure. This distance is referred to as the **tangential deviation** of $B$ with respect to $A$. More precisely, the distance $t_{B/A}$ is the vertical deviation of point $B$ on the deflection curve from the tangent at point $A$. The tangential deviation is positive when point $B$ is above the tangent at $A$.

To determine the tangential deviation, we again select two points $m_1$ and $m_2$ a small distance apart on the deflection curve (Fig. 9-23). The angle between the tangents at these two points is $d\theta$, and the segment on line $BB_1$ between these tangents is $dt$. Since the angles between the tangents and the $x$ axis are actually very small, we see that the vertical distance $dt$ is equal to $x_1 \, d\theta$, where $x_1$ is the horizontal distance from point $B$ to the small element $m_1m_2$. Since $d\theta = M \, dx/EI$ (Eq. 9-72), we obtain

$$dt = x_1 d\theta = x_1 \frac{M \, dx}{EI} \qquad \textbf{(9-75)}$$

The distance $dt$ represents the contribution made by the bending of element $m_1m_2$ to the tangential deviation $t_{B/A}$. The expression $x_1 M \, dx/EI$ may be interpreted geometrically as the first moment of the area of the shaded strip of width $dx$ within the $M/EI$ diagram. This first moment is evaluated with respect to a vertical line through point $B$.

Integrating Eq. (9-75) between points $A$ and $B$, we get

$$\int_A^B dt = \int_A^B x_1 \frac{M \, dx}{EI} \qquad \textbf{(9-76)}$$

The integral on the left-hand side is equal to $t_{B/A}$, that is, it is equal to the deviation of point $B$ from the tangent at $A$. The integral on the right-hand side represents the first moment with respect to point $B$ of the area of the $M/EI$ diagram between $A$ and $B$. Therefore, we can write Eq. (9-76) as

$$
t_{B/A} = \int_A^B x_1 \frac{M\,dx}{EI}
$$

$$
= \text{First moment of the area of the } M/EI \text{ diagram} \\
\text{between points } A \text{ and } B, \text{ evaluated with respect to } B \quad \textbf{(9-77)}
$$

This equation represents the second theorem:

> **Second moment-area theorem:** The tangential deviation $t_{B/A}$ of point $B$ from the tangent at point $A$ is equal to the first moment of the area of the $M/EI$ diagram between $A$ and $B$, evaluated with respect to $B$.

If the bending moment is positive, then the first moment of the $M/EI$ diagram is also positive, provided point $B$ is to the right of point $A$. Under these conditions the tangential deviation $t_{B/A}$ is positive and point $B$ is above the tangent at $A$ (as shown in Fig. 9-23). If, as we move from $A$ to $B$ in the $x$ direction, the area of the $M/EI$ diagram is negative, then the first moment is also negative and the tangential deviation is negative, which means that point $B$ is below the tangent at $A$.

The first moment of the area of the $M/EI$ diagram can be obtained by taking the product of the area of the diagram and the distance $\bar{x}$ from point $B$ to the centroid $C$ of the area (Fig. 9-23). This procedure is usually more convenient than integrating, because the $M/EI$ diagram usually consists of familiar geometric figures such as rectangles, triangles, and parabolic segments. The areas and centroidal distances of such figures are tabulated in Appendix E.

As a method of analysis, the moment-area method is feasible only for relatively simple kinds of beams. Therefore, it is usually obvious whether the beam deflects upward or downward and whether an angle of rotation is clockwise or counterclockwise. Consequently, it is seldom necessary to follow the formal (and somewhat awkward) sign conventions described previously for the tangential deviation. Instead, we can determine the directions by inspection and use only absolute values when applying the moment-area theorems.

By integratin...
we can expre...
ing forms:

Note that $M$...
tion of $x$. W...
and the sec...
known. (Exa...

In the de...
of the bendi...
energy will b...
relatively sm...
beams in w...
$L/d > 8$). T...
be disregard...

## Deflecti...

If a beam su...
$M_0$, the cor...
be determin...

In the d...
*ding deflect*...
load is app...
of the load...

In the d...
*angle of ro*...
where the d...

Since t...
load, and s...
the followi...

The first e...
ond equat...
from Eqs.

As ex...
angles of...
deflection...
(or angle)...
ple). How...
Example...

## 9.7 NONF...

The methods pr...
of prismatic bea...
varying momen...
shown in Fig. 9...
tia, and the se...
moments of ine...
increasing the n...
is largest.

Although n...
matic beam is n...
moment of inert...
in the examples...

In the first...
of inertia), the d...
of the deflectio...
ing two differe...
is used.

These two e...
relatively simple...
as tapered beam...
usually required...
beam deflection:

---

• • • Example 9-10

### Fig. 9-24

Example 9-10: Cantilever beam with a concentrated load

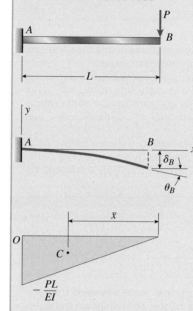

Determine the angle of rotation $\theta_B$ and deflection $\delta_B$ at the free end $B$ of a cantilever beam $AB$ supporting a concentrated load $P$ (Fig. 9-24). (*Note:* The beam has length $L$ and constant flexural rigidity $EI$.)

### Solution

By inspection of the beam and its loading, we know that the angle of rotation $\theta_B$ is clockwise and the deflection $\delta_B$ is downward (Fig. 9-24). Therefore, we can use absolute values when applying the moment-area theorems.

*M/EI diagram.* The bending-moment diagram is triangular in shape with the moment at the support equal to $-PL$. Since the flexural rigidity $EI$ is constant, the $M/EI$ diagram has the same shape as the bending-moment diagram, as shown in the last part of Fig. 9-24.

*Angle of rotation.* From the first moment-area theorem, we know that the angle $\theta_{B/A}$ between the tangents at points $B$ and $A$ is equal to the area of the $M/EI$ diagram between those points. This area, which we will denote as $A_1$, is determined as follows:

$$A_1 = \frac{1}{2}(L)\left(\frac{PL}{EI}\right) = \frac{PL^2}{2EI}$$

Note that we are using only the absolute value of the area.

The relative angle of rotation between points $A$ and $B$ (from the first theorem) is

$$\theta_{B/A} = \theta_B - \theta_A = A_1 = \frac{PL^2}{2EI}$$

Since the tangent to the deflection curve at support $A$ is horizontal ($\theta_A = 0$), we obtain

$$\theta_B = \frac{PL^2}{2EI} \qquad \longleftarrow \text{(9-78)}$$

This result agrees with the formula for $\theta_B$ given in Case 4 of Table H-1, Appendix H.

*Deflection.* The deflection $\delta_B$ at the free end can be obtained from the second moment-area theorem. In this case, the tangential deviation $t_{B/A}$ of point $B$ from the tangent at $A$ is equal to the deflection $\delta_B$ itself (see Fig. 9-24). The first moment of the area of the $M/EI$ diagram, evaluated with respect to point $B$, is

$$Q_1 = A_1\bar{x} = \left(\frac{PL^2}{2EI}\right)\left(\frac{2L}{3}\right) = \frac{PL^3}{3EI}$$

Note again that we are disregarding signs and using only absolute values.

From the second moment-area theorem, we know that the deflection $\delta_B$ is equal to the first moment $Q_1$. Therefore,

$$\delta_B = \frac{PL^3}{3EI} \qquad \longleftarrow \text{(9-79)}$$

This result also appears in Case 4 of Table H-1.

Example 9

Beam in pure
pl

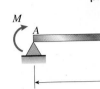

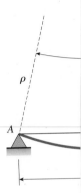

$M$

$A$

$\rho$

$A$

Diagram sh
tionship
moments

$M$

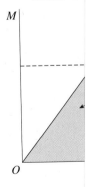

$O$

## Example 9-15

### Fig. 9-33

Example 9-15: Strain energy of
a beam

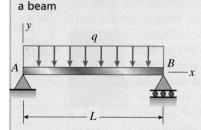

A simple beam $AB$ of length $L$ supports a uniform load of intensity $q$ (Fig. 9-33).
(a) Evaluate the strain energy of the beam from the bending moment in the
beam. (b) Evaluate the strain energy of the beam from the equation of the
deflection curve. (*Note:* The beam has constant flexural rigidity $EI$.)

### Solution

(a) *Strain energy from the bending moment.* The reaction of the beam at
support $A$ is $qL/2$, and therefore the expression for the bending moment
in the beam is

$$M = \frac{qLx}{2} - \frac{qx^2}{2} = \frac{q}{2}(Lx - x^2) \tag{a}$$

The strain energy of the beam (from Eq. 9-95a) is

$$U = \int_0^L \frac{M^2 \, dx}{2EI} = \frac{1}{2EI} \int_0^L \left[\frac{q}{2}(Lx - x^2)\right]^2 dx = \frac{q^2}{8EI} \int_0^L (L^2x^2 - 2Lx^2 + x^4)dx \tag{b}$$

from which we get

$$U = \frac{q^2L^5}{240EI} \tag{9-98}$$

Note that the load $q$ appears to the second power, which is consistent
with the fact that strain energy is always positive. Furthermore, Eq. (9-98)
shows that strain energy is *not* a linear function of the loads, even
though the beam itself behaves in a linearly elastic manner.

(b) *Strain energy from the deflection curve.* The equation of the deflection
curve for a simple beam with a uniform load is given in Case 1 of
Table H-2, Appendix H, as

$$v = -\frac{qx}{24EI}(L^3 - 2Lx^2 + x^3) \tag{c}$$

Taking two derivatives of this equation, we get

$$\frac{dv}{dx} = -\frac{q}{24EI}(L^3 - 6Lx^2 + 4x^3) \qquad \frac{d^2v}{dx^2} = \frac{q}{2EI}(Lx - x^2)$$

Substituting the latter expression into the equation for strain energy
(Eq. 9-95b), we obtain

$$U = \int_0^L \frac{EI}{2}\left(\frac{d^2v}{dx^2}\right)^2 dx = \frac{EI}{2} \int_0^L \left[\frac{q}{2EI}(Lx - x^2)\right]^2 dx$$

$$= \frac{q^2}{8EI} \int_0^L (L^2x^2 - 2Lx^3 + x^4)dx \tag{d}$$

Since the final integral in this equation is the same as the final integral
in Eq. (b), we get the same result as before (Eq. 9-98).

### ● ● ● Example 9-16

**Fig. 9-34**

Example 9-16: Strain energy of a beam

(a)

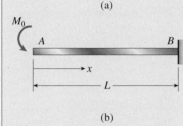

(b)

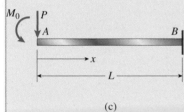

(c)

A cantilever beam $AB$ (Fig. 9-34) is subjected to three different loading conditions: (a) a concentrated load $P$ at its free end, (b) a couple $M_0$ at its free end, and (c) both loads acting simultaneously.

For each loading condition, determine the strain energy of the beam. Also, determine the vertical deflection $\delta_A$ at end $A$ of the beam due to the load $P$ acting alone (Fig. 9-34a), and determine the angle of rotation $\theta_A$ at end $A$ due to the moment $M_0$ acting alone (Fig. 9-34b). (*Note:* The beam has constant flexural rigidity $EI$.)

### Solution

(a) *Beam with concentrated load P* (Fig. 9-34a). The bending moment in the beam at distance $x$ from the free end is $M = -Px$. Substituting this expression for $M$ into Eq. (9-95a), we get the following expression for the strain energy of the beam:

$$U = \int_0^L \frac{M^2 dx}{2EI} = \int_0^L \frac{(-Px)^2 dx}{2EI} = \frac{P^2 L^3}{6EI} \quad \Longleftarrow \quad \textbf{(9-99)}$$

To obtain the vertical deflection $\delta_A$ under the load $P$, we equate the work done by the load to the strain energy:

$$W = U \quad \text{or} \quad \frac{P\delta_A}{2} = \frac{P^2 L^3}{6EI}$$

from which

$$\delta_A = \frac{PL^3}{3EI} \quad \Longleftarrow$$

The deflection $\delta_A$ is the only deflection we can find by this procedure, because it is the only deflection that corresponds to the load $P$.

(b) *Beam with moment* $M_0$ (Fig. 9-34b). In this case the bending moment is constant and equal to $-M_0$. Therefore, the strain energy (from Eq. 9-95a) is

$$U = \int_0^L \frac{M^2 dx}{2EI} = \int_0^L \frac{(-M_0)^2 dx}{2EI} = \frac{M_0^2 L}{2EI} \quad \Longleftarrow \quad \textbf{(9-100)}$$

Continues ➡

The work $W$ done by the couple $M_0$ during loading of the beam is $M_0 \theta_A / 2$, where $\theta_A$ is the angle of rotation at end $A$. Therefore,

$$W = U \quad \text{or} \quad \frac{M_0 \theta_A}{2} = \frac{M_0^2 L}{2EI}$$

and

$$\theta_A = \frac{M_0 L}{EI}$$

The angle of rotation has the same sense as the moment (counterclockwise in this example).

(c) *Beam with both loads acting simultaneously* (Fig. 9-34c). When both loads act on the beam, the bending moment in the beam is

$$M = -Px - M_0$$

Therefore, the strain energy is

$$U = \int_0^L \frac{M^2 \, dx}{2EI} = \frac{1}{2EI} \int_0^L (-Px - M_0)^2 \, dx$$

$$= \frac{P_2 L^3}{6EI} + \frac{PM_0 L^2}{2EI} + \frac{M_0^2 L}{2EI} \qquad \text{(9-101)}$$

The first term in this result gives the strain energy due to $P$ acting alone (Eq. 9-99), and the last term gives the strain energy due to $M_0$ alone (Eq. 9-100). However, when both loads act simultaneously, an additional term appears in the expression for strain energy.

Therefore, we conclude that *the strain energy in a structure due to two or more loads acting simultaneously cannot be obtained by adding the strain energies due to the loads acting separately.* The reason is that strain energy is a quadratic function of the loads, not a linear function. Therefore, *the principle of superposition does not apply to strain energy.*

We also observe that we cannot calculate a deflection for a beam with two or more loads by equating the work done by the loads to the strain energy. For instance, if we equate work and energy for the beam of Fig. 9-34c, we get

$$W = U \quad \text{or} \quad \frac{P\delta_{A2}}{2} + \frac{M_0 \theta_{A2}}{2} = \frac{P^2 L^3}{6EI} + \frac{PM_0 L^2}{2EI} + \frac{M_0^2 L}{2EI} \qquad \text{(a)}$$

in which $\delta_{A2}$ and $\theta_{A2}$ represent the deflection and angle of rotation at end $A$ of the beam with two loads acting simultaneously (Fig. 9-34c). Although the work done by the two loads is indeed equal to the strain energy, and Eq. (a) is quite correct, we cannot solve for either $\delta_{A2}$ or $\theta_{A2}$ because there are two unknowns and only one equation.

# *9.9 CASTIGLIANO'S THEOREM

**Castigliano's theorem** provides a means for finding the deflections of a structure from the strain energy of the structure. To illustrate what we mean by that statement, consider a cantilever beam with a concentrated load $P$ acting at the free end (Fig. 9-35a). The strain energy of this beam is obtained from Eq. (9-99) of Example 9-16:

$$ U = \frac{P^2 L^3}{6EI} \tag{9-102a} $$

Now take the derivative of this expression with respect to the load $P$:

$$ \frac{dU}{dP} = \frac{d}{dP}\left( \frac{P^2 L^3}{6EI} \right) = \frac{PL^3}{3EI} \tag{9-102b} $$

We immediately recognize this result as the deflection $\delta_A$ at the free end $A$ of the beam (see Fig. 9-35b). Note especially that the deflection $\delta_A$ *corresponds* to the load $P$ itself. (Recall that a deflection corresponding to a concentrated load is the deflection at the point where the concentrated load is applied. Furthermore, the deflection is in the direction of the load.) Thus, Eq. (9-102b) shows that *the derivative of the strain energy with respect to the load is equal to the deflection corresponding to the load.* Castigliano's theorem is a generalized statement of this observation, and we will now derive it in more general terms.

## Derivation of Castigliano's Theorem

Let us consider a beam subjected to any number of loads, say $n$ loads $P_1, P_2, \ldots, P_i, \ldots, P_n$ (Fig. 9-36a). The deflections of the beam corresponding to the various loads are denoted $\delta_1, \delta_2, \ldots, \delta_i, \ldots, \delta_n$, as shown in Fig. 9-36b. As in our earlier discussions of deflections and strain energy, we assume that the principle of superposition is applicable to the beam and its loads.

Now we will determine the strain energy of this beam. When the loads are applied to the beam, they gradually increase in magnitude from zero to their maximum values. At the same time, each load moves through its corresponding displacement and does work. The total work $W$ done by the loads is equal to the strain energy $U$ stored in the beam:

$$ W = U \tag{9-103} $$

Note that $W$ (and hence $U$) is a function of the loads $P_1, P_2, \ldots, P_n$ acting on the beam.

Next, let us suppose that one of the loads, say the $i$th load, is increased slightly by the amount $dP_i$ while the other loads are held constant. This increase in load will cause a small increase $dU$ in the strain energy of the beam. This increase in strain energy may be expressed as the rate of change of $U$ with respect to $P_i$ times the small increase in $P_i$. Thus, the increase in strain energy is

$$ dU = \frac{\partial U}{\partial P_i}\, dP_i \tag{9-104} $$

**Fig. 9-35**

Beam supporting a single load $P$

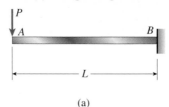

(a)

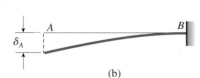

(b)

**Fig. 9-36**

Beam supporting $n$ loads

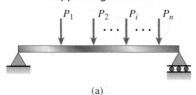

(a)

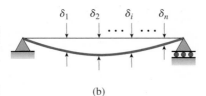

(b)

where $\partial U/\partial P_i$ is the rate of change of $U$ with respect to $P_i$. (Since $U$ is a function of *all* the loads, the derivative with respect to any one of the loads is a partial derivative.) The final strain energy of the beam is

$$U + dU = U + \frac{\partial U}{\partial P_i} dP_i \qquad \textbf{(9-105)}$$

in which $U$ is the strain energy referred to in Eq. (9-103).

Because the principle of superposition holds for this beam, the total strain energy is independent of the order in which the loads are applied. That is, the final displacements of the beam (and the work done by the loads in reaching those displacements) are the same regardless of the order in which the loads are applied. In arriving at the strain energy given by Eq. (9-105), we first applied the $n$ loads $P_1, P_2, \ldots, P_n$, and then we applied the load $dP_i$. However, we can reverse the order of application and apply the load $dP_i$ first, followed by the loads $P_1, P_2, \ldots, P_n$. The final amount of strain energy is the same in either case.

When the load $dP_i$ is applied first, it produces strain energy equal to one-half the product of the load $dP_i$ and its corresponding displacement $d\delta_i$. Thus, the amount of strain energy due to the load $dP_i$ is

$$\frac{dP_i\, d\delta_i}{2} \qquad \textbf{(9-106a)}$$

When the loads $P_1, P_2, \ldots, P_n$ are applied, they produce the same displacements as before ($\delta_1, \delta_2, \ldots, \delta_n$) and do the same amount of work as before (Eq. 9-103). However, during the application of these loads, the force $dP_i$ automatically moves through the displacement $\delta_i$. In so doing, it produces additional work equal to the product of the force and the distance through which it moves. (Note that the work does not have a factor 1/2, because the force $dP_i$ acts at full value throughout this displacement.) Thus, the additional work, equal to the additional strain energy, is

$$dP_i\, \delta_i \qquad \textbf{(9-106b)}$$

Therefore, the final strain energy for the second loading sequence is

$$\frac{dP_i\, d\delta_i}{2} + U + dP_i\, \delta_i \qquad \textbf{(9-106c)}$$

Equating this expression for the final strain energy to the earlier expression (Eq. 9-105), which was obtained for the first loading sequence, we get

$$\frac{dP_i\, d\delta_i}{2} + U + dP_i\, \delta_i = U + \frac{\partial U}{\partial P_i} dP_i \qquad \textbf{(9-106d)}$$

We can discard the first term because it contains the product of two differentials and is infinitesimally small compared to the other terms. We then obtain the following relationship:

$$\delta_i = \frac{\partial U}{\partial P_i} \qquad \textbf{(9-107)}$$

This equation is known as **Castigliano's theorem**.*

Although we derived Castigliano's theorem by using a beam as an illustration, we could have used any other type of structure (for example, a truss) and any other kinds of loads (for example, loads in the form of couples). The important requirements are that the structure be linearly elastic and that the principle of superposition be applicable. Also, note that the strain energy must be expressed as a function of the loads (and not as a function of the displacements), a condition which is implied in the theorem itself, since the partial derivative is taken with respect to a load. With these limitations in mind, we can state Castigliano's theorem in general terms as follows:

> *The partial derivative of the strain energy of a structure with respect to any load is equal to the displacement corresponding to that load.*

The strain energy of a linearly elastic structure is a *quadratic* function of the loads [for instance, see Eq. (9-102a)], and therefore, the partial derivatives and the displacements (Eq. 9-107) are *linear* functions of the loads (as expected).

When using the terms *load* and *corresponding displacement* in connection with Castigliano's theorem, it is understood that these terms are used in a generalized sense. The load $P_i$ and corresponding displacement $\delta_i$ may be a force and a corresponding translation, or a couple and a corresponding rotation, or some other set of corresponding quantities.

## Application of Castigliano's Theorem

As an application of Castigliano's theorem, let us consider a cantilever beam $AB$ carrying a concentrated load $P$ and a couple of moment $M_0$ acting at the free end (Fig. 9-37a). We wish to determine the vertical deflection $\delta_A$ and angle of rotation $\theta_A$ at the end of the beam (Fig. 9-37b). Note that $\delta_A$ is the deflection corresponding to the load $P$, and $\theta_A$ is the angle of rotation corresponding to the moment $M_0$.

The first step in the analysis is to determine the strain energy of the beam. For that purpose, we write the equation for the bending moment as

$$M = -Px - M_0 \tag{9-108}$$

in which $x$ is the distance from the free end (Fig. 9-37a). The strain energy is found by substituting this expression for $M$ into Eq. (9-95a):

$$U = \int_0^L \frac{M^2 dx}{2EI} = \frac{1}{2EI} \int_0^L (-Px - M_0)^2 dx$$

$$= \frac{P^2 L^3}{6EI} + \frac{PM_0 L^2}{2EI} + \frac{M_0^2 L}{2EI} \tag{9-109}$$

in which $L$ is the length of the beam and $EI$ is its flexural rigidity. Note that the strain energy is a quadratic function of the loads $P$ and $M_0$.

---

**Fig. 9-37**

Application of Castigliano's theorem to a beam

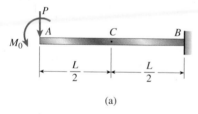

(a)

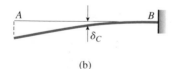

(b)

To obtain the vertical deflection $\delta_A$ at the end of the beam, we use Castigliano's theorem (Eq. 9-107) and take the partial derivative of the strain energy with respect to $P$:

$$\delta_A = \frac{\partial U}{\partial P} = \frac{PL^3}{3EI} + \frac{M_0 L^2}{2EI} \qquad \textbf{(9-110)}$$

This expression for the deflection can be verified by comparing it with the formulas of Cases 4 and 6 of Table H-1, Appendix H.

In a similar manner, we can find the angle of rotation $\theta_A$ at the end of the beam by taking the partial derivative with respect to $M_0$:

$$\theta_A = \frac{\partial U}{\partial M_0} = \frac{PL^2}{2EI} + \frac{M_0 L}{EI} \qquad \textbf{(9-111)}$$

This equation can also be verified by comparing with the formulas of Cases 4 and 6 of Table H-1.

## Use of a Fictitious Load

The only displacements that can be found from Castigliano's theorem are those that correspond to loads acting on the structure. If we wish to calculate a displacement at a point on a structure where there is no load, then a fictitious load *corresponding to the desired displacement* must be applied to the structure. We can then determine the displacement by evaluating the strain energy and taking the partial derivative with respect to the fictitious load. The result is the displacement produced by the actual loads and the fictitious load acting simultaneously. By setting the fictitious load equal to zero, we obtain the displacement produced only by the actual loads.

To illustrate this concept, suppose we wish to find the vertical deflection $\delta_C$ at the midpoint $C$ of the cantilever beam shown in Fig. 9-38a. Since the deflection $\delta_C$ is downward (Fig. 9-38b), the load corresponding to that deflection is a downward vertical force acting at the same point.

**Fig. 9-38**

Beam supporting loads $P$ and $M_0$

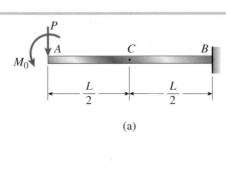

(a)

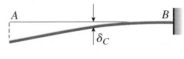

(b)

Therefore, we must supply a fictitious load $Q$ acting at point $C$ in the downward direction (Fig. 9-39a). Then we can use Castigliano's theorem to determine the deflection $(\delta_C)_0$ at the midpoint of this beam (Fig. 9-39b). From that deflection, we can obtain the deflection $\delta_C$ in the beam of Fig. 9-38 by setting $Q$ equal to zero.

We begin by finding the bending moments in the beam of Fig. 9-39a:

$$M = -Px - M_0 \quad \left(0 \le x \le \frac{L}{2}\right) \tag{9-112a}$$

$$M = -Px - M_0 - Q\left(x - \frac{L}{2}\right) \quad \left(\frac{L}{2} \le x \le L\right) \tag{9-112b}$$

Next, we determine the strain energy of the beam by applying Eq. (9-95a) to each half of the beam. For the left-hand half of the beam (from point $A$ to point $C$), the strain energy is

$$U_{AC} = \int_0^{L/2} \frac{M^2\,dx}{2EI} = \frac{1}{2EI} \int_0^{L/2} (-Px - M_0)^2\,dx$$

$$= \frac{P^2 L^3}{48EI} + \frac{PM_0 L^2}{8EI} + \frac{M_0^2 L}{4EI} \tag{9-113a}$$

For the right-hand half, the strain energy is

$$U_{CB} = \int_{L/2}^L \frac{M^2\,dx}{2EI} = \frac{1}{2EI} \int_{L/2}^L \left[-Px - M_0 - Q\left(x - \frac{L}{2}\right)\right]^2 dx$$

$$= \frac{7P^2 L^3}{48EI} + \frac{3PM_0 L^2}{8EI} + \frac{5PQL^3}{48EI} + \frac{M_0^2 L}{4EI} + \frac{M_0 Q^2}{8EI} + \frac{Q^2 L^3}{48EI} \tag{9-113b}$$

which requires a very lengthy process of integration. Adding the strain energies for the two parts of the beam, we obtain the strain energy for the entire beam (Fig. 9-39a):

$$U = U_{AC} + U_{CB}$$

$$= \frac{P^2 L^3}{6EI} + \frac{PM_0 L^2}{2EI} + \frac{5PQL^3}{48EI} + \frac{M_0^2 L}{2EI} + \frac{M_0 QL^2}{8EI} + \frac{Q^2 L^3}{48EI} \tag{9-114}$$

The deflection at the midpoint of the beam shown in Fig. 9-39a can now be obtained from Castigliano's theorem:

$$(\delta_C)_0 = \frac{\partial U}{\partial Q} = \frac{5PL^3}{48EI} + \frac{M_0 L^2}{8EI} + \frac{QL^3}{24EI} \tag{9-115}$$

This equation gives the deflection at point $C$ produced by all three loads acting on the beam. To obtain the deflection produced by the loads $P$ and $M_0$ only, we set the load $Q$ equal to zero in the preceding equation. The result is the deflection at the midpoint $C$ for the beam with two loads (Fig. 9-38a):

$$\delta_C = \frac{5PL^3}{48EI} + \frac{M_0 L^2}{8EI} \tag{9-116}$$

Thus, the deflection in the original beam has been obtained.

**Fig. 9-39**

Beam with a fictitious load $Q$

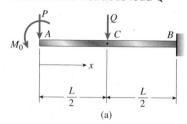

(a)

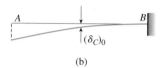

(b)

This method is sometimes called the *dummy-load method*, because of the introduction of a fictitious, or dummy, load.

## Differentiation Under the Integral Sign

As we saw in the preceding example, the use of Castigliano's theorem for determining beam deflections may lead to lengthy integrations, especially when more than two loads act on the beam. The reason is clear—finding the strain energy requires the integration of the *square* of the bending moment (Eq. 9-95a). For instance, if the bending moment expression has three terms, its square may have as many as six terms, each of which must be integrated.

After the integrations are completed and the strain energy has been determined, we differentiate the strain energy to obtain the deflections. However, we can bypass the step of finding the strain energy by differentiating *before* integrating. This procedure does not eliminate the integrations, but it does make them much simpler.

To derive this method, we begin with the equation for the strain energy (Eq. 9-95a) and apply Castigliano's theorem (Eq. 9-107):

$$\delta_i = \frac{\partial U}{\partial P_i} = \frac{\partial}{\partial P_i} \int \frac{M^2 dx}{2EI} \qquad \textbf{(9-117)}$$

Following the rules of calculus, we can differentiate the integral by differentiating under the integral sign:

$$\delta_i = \frac{\partial}{\partial P_i} \int \frac{M^2 dx}{2EI} = \int \left(\frac{M}{EI}\right)\left(\frac{\partial M}{\partial P_i}\right) dx \qquad \textbf{(9-118)}$$

We will refer to this equation as the **modified Castigliano's theorem**.

When using the modified theorem, we integrate the product of the bending moment and its derivative. By contrast, when using the standard Castigliano's theorem [see Eq. (9-117)], we integrate the square of the bending moment. Since the derivative is a shorter expression than the moment itself, this new procedure is much simpler. To show this, we will now solve the preceding examples using the modified theorem (Eq. 9-118).

Let us begin with the beam shown in Fig. 9-37 and recall that we wish to find the deflection and angle of rotation at the free end. The bending moment and its derivatives [see Eq. (9-108)] are

$$M = -Px - M_0$$
$$\frac{\partial M}{\partial P} = -x \qquad \frac{\partial M}{\partial M_0} = -1$$

From Eq. (9-118) we obtain the deflection $\delta_A$ and angle of rotation $\theta_A$:

$$\delta_A = \frac{1}{EI} \int_0^L (-Px - M_0)(-x)dx = \frac{PL^3}{3EI} + \frac{M_0 L^2}{2EI} \qquad \textbf{(9-119a)}$$

$$\theta_A = \frac{1}{EI} \int_0^L (-Px - M_0)(-1)dx = \frac{PL^2}{2EI} + \frac{M_0 L}{EI} \qquad \textbf{(9-119b)}$$

These equations agree with the earlier results in Eqs. (9-110) and (9-111). However, the calculations are shorter than those performed earlier, because we did not have to integrate the square of the bending moment [see Eq. (9-109)].

The advantages of differentiating under the integral sign are even more apparent when there are more than two loads acting on the structure, as in the example of Fig. 9-38. In that example, we wished to determine the deflection $\delta_C$ at the midpoint $C$ of the beam due to the loads $P$ and $M_0$. To do so, we added a fictitious load $Q$ at the midpoint (Fig. 9-39). We then proceeded to find the deflection $(\delta_C)_0$ at the midpoint of the beam when all three loads ($P$, $M_0$, and $Q$) were acting. Finally, we set $Q = 0$ to obtain the deflection $\delta_C$ due to $P$ and $M_0$ alone. The solution was time-consuming, because the integrations were extremely long. However, if we use the modified theorem and differentiate first, the calculations are much shorter.

With all three loads acting (Fig. 9-39), the bending moments and their derivatives are as follows [see Eqs. (9-112) and (9-113)]:

$$M = -Px - M_0 \qquad \frac{\partial M}{\partial Q} = 0 \qquad \left(0 \leq x \leq \frac{L}{2}\right)$$

$$M = -Px - M_0 - Q\left(x - \frac{L}{2}\right) \qquad \frac{\partial M}{\partial Q} = -\left(x - \frac{L}{2}\right) \qquad \left(\frac{L}{2} \leq x \leq L\right)$$

Therefore, the deflection $(\delta_C)_0$, from Eq. (9-118), is

$$(\delta_C)_0 = \frac{1}{EI} \int_0^{L/2} (-Px - M_0)(0)dx$$

$$+ \frac{1}{EI} \int_{L/2}^{L} \left[-Px - M_0 - Q\left(x - \frac{L}{2}\right)\right]\left[-\left(x - \frac{L}{2}\right)\right]dx$$

Since $Q$ is a fictitious load, and since we have already taken the partial derivatives, we can set $Q$ equal to zero before integrating and obtain the deflection $\delta_C$ due to the two loads $P$ and $M_0$ as

$$\delta_C = \frac{1}{EI} \int_{L/2}^{L} [-Px - M_0]\left[-\left(x - \frac{L}{2}\right)\right]dx = \frac{5PL^3}{48EI} + \frac{M_0L^2}{8EI}$$

which agrees with the earlier result in Eq. (9-116). Again, the integrations are greatly simplified by differentiating under the integral sign and using the modified theorem.

The partial derivative that appears under the integral sign in Eq. (9-118) has a simple physical interpretation. It represents the rate of change of the bending moment $M$ with respect to the load $P_i$, that is, it is equal to the bending moment $M$ produced by a load $P_i$ of unit value. This observation leads to a method of finding deflections known as the *unit-load method*. Castigliano's theorem also leads to a method of structural analysis known as the *flexibility method*. Both the unit-load method and the flexibility method are widely used in structural analysis and are described in textbooks on that subject.

The following examples provide additional illustrations of the use of Castigliano's theorem for finding deflections of beams. However, it should be remembered that the theorem is not limited to finding beam deflections—it applies to any kind of linearly elastic structure for which the principle of superposition is valid.

A simple beam $AB$ supports a uniform load of intensity $q = 1.5$ k/ft and a concentrated load $P = 5$ k (Fig. 9-40). The load $P$ acts at the midpoint $C$ of the beam. The beam has length $L = 8.0$ ft, modulus of elasticity $E = 30 \times 10^6$ psi, and moment of inertia $I = 75.0$ in.$^4$

Determine the downward deflection $\delta_C$ at the midpoint of the beam by the following methods: (1) Obtain the strain energy of the beam and use Castigliano's theorem, and (2) use the modified form of Castigliano's theorem (differentiation under the integral sign).

### Solution

*Method (1).* Because the beam and its loading are symmetrical about the midpoint, the strain energy for the entire beam is equal to twice the strain energy for the left-hand half of the beam. Therefore, we need to analyze only the left-hand half of the beam.

The reaction at the left-hand support $A$ (Figs. 9-40 and 9-41) is

$$R_A = \frac{P}{2} + \frac{qL}{2}$$

and therefore, the bending moment $M$ is

$$M = R_A x - \frac{qx^2}{2} = \frac{Px}{2} + \frac{qLx}{2} - \frac{qx^2}{2} \tag{a}$$

in which $x$ is measured from support $A$.

The strain energy of the entire beam (from Eq. 9-95a) is

$$U = \int \frac{M^2 dx}{2EI} = 2 \int_0^{L/2} \frac{1}{2EI} \left( \frac{Px}{2} + \frac{qLx}{2} - \frac{qx^2}{2} \right)^2 dx$$

### Fig. 9-40

Example 9-17: Simple beam with two loads

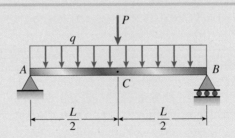

### Fig. 9-41

Example 9-17: Free-body diagram for determining the bending moment $M$ in the left-hand half of the beam

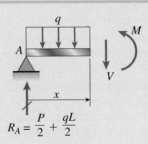

After squaring the term in parentheses and performing a lengthy integration, we find

$$U = \frac{P^2L^3}{96EI} + \frac{5PqL^4}{384EI} + \frac{q^2L^5}{240EI}$$

Since the deflection at the midpoint C (Fig. 9-40) corresponds to the load P, we can find the deflection by using Castigliano's theorem (Eq. 9-107):

$$\delta_C = \frac{\partial U}{\partial P} = \frac{\partial}{\partial P}\left(\frac{P^2L^3}{96EI} + \frac{5PqL^4}{384EI} + \frac{q^2L^5}{240EI}\right) = \frac{PL^3}{48EI} + \frac{5qL^4}{384EI} \quad \Longleftarrow \text{ (b)}$$

*Method (2).* By using the modified form of Castigliano's theorem (Eq. 9-118), we avoid the lengthy integration for finding the strain energy. The bending moment in the left-hand half of the beam has already been determined [see Eq. (a)], and its partial derivative with respect to the load P is

$$\frac{\partial M}{\partial P} = \frac{x}{2}$$

Therefore, the modified Castigliano's theorem becomes

$$\delta_C = \int \left(\frac{M}{EI}\right)\left(\frac{\partial M}{\partial P}\right)dx$$

$$= 2\int_0^{L/2} \frac{1}{EI}\left(\frac{Px}{2} + \frac{qLx}{2} - \frac{qx^2}{2}\right)\left(\frac{x}{2}\right)dx = \frac{PL^3}{48EI} + \frac{5qL^4}{384EI} \quad \Longleftarrow \text{ (c)}$$

which agrees with the earlier result (Eq. b), but requires a much simpler integration.

*Numerical solution.* Now that we have an expression for the deflection at point C, we can substitute numerical values, as follows:

$$\delta_C = \frac{PL^3}{48EI} + \frac{5qL^4}{383EI}$$

$$= \frac{(5 \text{ k})(96 \text{ in.})^3}{48(30 \times 10^6 \text{ psi})(75.0 \text{ in.}^4)} + \frac{5(1.5 \text{ k/ft})(1/12 \text{ ft/in.})(96 \text{ in.})^4}{383(30 \times 10^6 \text{ psi})(75.0 \text{ in.}^4)}$$

$$= 0.0410 \text{ in.} + 0.0614 \text{ in.} = 0.1024 \text{ in.} \quad \Longleftarrow$$

*Note:* Numerical values cannot be substituted until *after* the partial derivative is obtained. If numerical values are substituted prematurely, either in the expression for the bending moment or the expression for the strain energy, it may be impossible to take the derivative.

● ● ●    **Example 9-18**

A simple beam with an overhang supports a uniform load of intensity $q$ on span $AB$ and a concentrated load $P$ at end $C$ of the overhang (Fig. 9-42).

Determine the deflection $\delta_C$ and angle of rotation $\theta_C$ at point $C$. (Use the modified form of Castigliano's theorem.)

## Fig. 9-42

Example 9-18: Beam with an overhang

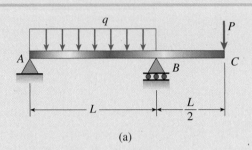

(a)

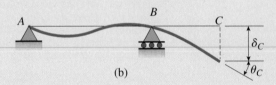

(b)

## Solution

*Deflection $\delta_C$ at the end of the overhang (Fig. 9-42b).* Since the load $P$ corresponds to this deflection, we do not need to supply a fictitious load. Instead, we can begin immediately to find the bending moments throughout the length of the beam. The reaction at support $A$ is

$$R_A = \frac{qL}{2} - \frac{P}{2}$$

## Fig. 9-43

Reaction at support $A$ and coordinates $x_1$ and $x_2$ for the beam of Example 9-18

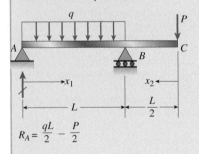

$$R_A = \frac{qL}{2} - \frac{P}{2}$$

as shown in Fig. 9-43. Therefore, the bending moment in span $AB$ is

$$M_{AB} = R_A x_1 - \frac{qx_1^2}{2} = \frac{qLx_1}{2} - \frac{Px_1}{2} - \frac{qx_1^2}{2} \qquad (0 \le x_1 \le L)$$

where $x_1$ is measured from support $A$ (Fig. 9-43). The bending moment in the overhang is

$$M_{BC} = -Px_2 \qquad \left(0 \le x_2 \le \frac{L}{2}\right)$$

where $x_2$ is measured from point $C$ (Fig. 9-43).

Next, we determine the partial derivatives with respect to the load $P$:

$$\frac{\partial M_{AB}}{\partial P} = -\frac{x_1}{2} \qquad (0 \le x_1 \le L)$$

$$\frac{\partial M_{BC}}{\partial P} = -x_2 \qquad \left(0 \le x_2 \le \frac{L}{2}\right)$$

Now we are ready to use the modified form of Castigliano's theorem (Eq. 9-118) to obtain the deflection at point C:

$$\delta_C = \int \left(\frac{M}{EI}\right)\left(\frac{\partial M}{\partial P}\right)dx$$

$$= \frac{1}{EI}\int_0^L M_{AB}\left(\frac{\partial M_{AB}}{\partial P}\right)dx + \frac{1}{EI}\int_0^{L/2} M_{BC}\left(\frac{\partial M_{BC}}{\partial P}\right)dx$$

Substituting the expressions for the bending moments and partial derivatives, we get

$$\delta_C = \frac{1}{EI}\int_0^L \left(\frac{qLx_1}{2} - \frac{Px_1}{2} - \frac{qx_1^2}{2}\right)\left(-\frac{x_1}{2}\right)dx_1 + \frac{1}{EI}\int_0^{L/2}(-Px_2)(-x_2)dx_2$$

By performing the integrations and combining terms, we obtain the deflection:

$$\delta_C = \frac{PL^3}{8EI} - \frac{qL^4}{48EI} \qquad \textbf{(9-119c)}$$

Since the load P acts downward, the deflection $\delta_C$ is also positive downward. In other words, if the preceding equation produces a positive result, the deflection is downward. If the result is negative, the deflection is upward.

Comparing the two terms in Eq. (9-119c), we see that the deflection at the end of the overhang is downward when $P > qL/6$ and upward when $P < qL/6$.

*Angle of rotation $\theta_C$ at the end of the overhang* (Fig. 9-42b). Since there is no load on the original beam (Fig. 9-42a) corresponding to this angle of rotation, we must supply a fictitious load. Therefore, we place a couple of moment $M_C$ at point C (Fig. 9-44). Note that the couple $M_C$ acts at the point on the beam where the angle of rotation is to be determined. Furthermore, it has the same clockwise direction as the angle of rotation (Fig. 9-42).

We now follow the same steps as when determining the deflection at C. First, we note that the reaction at support A (Fig. 9-44) is

$$R_A = \frac{qL}{2} - \frac{P}{2} - \frac{M_C}{L}$$

Consequently, the bending moment in span AB becomes

$$M_{AB} = R_A x_1 - \frac{qx_1^2}{2} = \frac{qLx_1}{2} - \frac{Px_1}{2} - \frac{M_C x_1}{L} - \frac{qx_1^2}{2} \qquad (0 \le x_1 \le L)$$

**Fig. 9-44**

Fictitious moment $M_C$ acting on the beam of Example 9-18

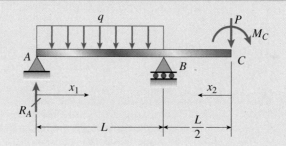

*Continues* ➡

Also, the bending moment in the overhang becomes

$$M_{BC} = -Px_2 - M_C \qquad \left(0 \leq x_2 \leq \frac{L}{2}\right)$$

The partial derivatives are taken with respect to the moment $M_C$, which is the load corresponding to the angle of rotation. Therefore,

$$\frac{\partial M_{AB}}{\partial M_C} = -\frac{x_1}{L} \qquad (0 \leq x_1 \leq L)$$

$$\frac{\partial M_{BC}}{\partial M_C} = -1 \qquad \left(0 \leq x_2 \leq \frac{L}{2}\right)$$

Now we use the modified form of Castigliano's theorem (Eq. 9-118) to obtain the angle of rotation at point C:

$$\theta_C = \int \left(\frac{M}{EI}\right)\left(\frac{\partial M}{\partial M_C}\right)dx$$

$$= \frac{1}{EI}\int_0^L M_{AB}\left(\frac{\partial M_{AB}}{\partial M_C}\right)dx + \frac{1}{EI}\int_0^{L/2} M_{BC}\left(\frac{\partial M_{BC}}{\partial M_C}\right)dx$$

Substituting the expressions for the bending moments and partial derivatives, we obtain

$$\theta_C = \frac{1}{EI}\int_0^L \left(\frac{qLx_1}{2} - \frac{Px_1}{2} - \frac{M_C x_1}{L} - \frac{qx_1^2}{2}\right)\left(-\frac{x_1}{L}\right)dx_1$$

$$+ \frac{1}{EI}\int_0^{L/2}(-Px_2 - M_C)(-1)dx_2$$

Since $M_C$ is a fictitious load, and since we have already taken the partial derivatives, we can set $M_C$ equal to zero at this stage of the calculations and simplify the integrations:

$$\theta_C = \frac{1}{EI}\int_0^L \left(\frac{qLx_1}{2} - \frac{Px_1}{2} - \frac{qx_1^2}{2}\right)\left(-\frac{x_1}{L}\right)dx_1 + \frac{1}{EI}\int_0^{L/2}(-Px_2)(-1)\,dx_2$$

After carrying out the integrations and combining terms, we obtain

$$\theta_C = \frac{7PL^2}{24EI} - \frac{qL^3}{24EI} \qquad \text{(9-120)}$$

If this equation produces a positive result, the angle of rotation is clockwise. If the result is negative, the angle is counterclockwise.

Comparing the two terms in Eq. (9-120), we see that the angle of rotation is clockwise when $P > qL/7$ and counterclockwise when $P < qL/7$.

If numerical data are available, it is now a routine matter to substitute numerical values into Eqs. (9-119c) and (9-120) and calculate the deflection and angle of rotation at the end of the overhang.

# *9.10 DEFLECTIONS PRODUCED BY IMPACT

In this section we will discuss the impact of an object falling onto a beam (Fig. 9-45a). We will determine the dynamic deflection of the beam by equating the potential energy lost by the falling mass to the strain energy acquired by the beam. This approximate method was described in detail in Section 2.8 for a mass striking an axially loaded bar; consequently, Section 2.8 should be fully understood before proceeding.

Most of the assumptions described in Section 2.8 apply to beams as well as to axially loaded bars. Some of these assumptions are as follows: (1) The falling weight sticks to the beam and moves with it, (2) no energy losses occur, (3) the beam behaves in a linearly elastic manner, (4) the deflected shape of the beam is the same under a dynamic load as under a static load, and (5) the potential energy of the beam due to its change in position is relatively small and may be disregarded. In general, these assumptions are reasonable if the mass of the falling object is very large compared to the mass of the beam. Otherwise, this approximate analysis is not valid and a more advanced analysis is required.

As an example, we will consider the simple beam $AB$ shown in Fig. 9-45. The beam is struck at its midpoint by a falling body of mass $M$ and weight $W$. Based upon the preceding idealizations, we may assume that all of the potential energy lost by the body during its fall is transformed into elastic strain energy that is stored in the beam. Since the distance through which the body falls is $h + \delta_{max}$, where $h$ is the initial height above the beam (Fig. 9-45a) and $\delta_{max}$ is the maximum dynamic deflection of the beam (Fig. 9-45b), the potential energy lost is

$$\text{Potential energy} = W(h + \delta_{max}) \qquad \textbf{(9-121)}$$

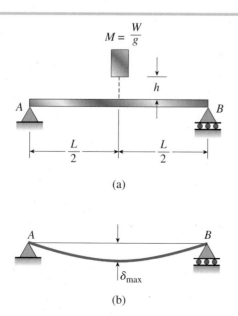

**Fig. 9-45**

Deflection of a beam struck by a falling body

The strain energy acquired by the beam can be determined from the deflection curve by using Eq. (9-95b), which is repeated here:

$$U = \int \frac{EI}{2}\left(\frac{d^2v}{dx^2}\right)^2 dx \qquad \textbf{(9-122)}$$

The deflection curve for a simple beam subjected to a concentrated load acting at the midpoint (see Case 4 of Table H-2, Appendix H) is

$$v = -\frac{Px}{48EI}(3L^2 - 4x^2) \qquad \left(0 \le x \le \frac{L}{2}\right) \qquad \textbf{(9-123)}$$

Also, the maximum deflection of the beam is

$$\delta_{max} = \frac{PL^3}{48EI} \qquad \textbf{(9-124)}$$

Eliminating the load $P$ between Eqs. (9-123) and (9-124), we get the equation of the deflection curve in terms of the maximum deflection:

$$v = -\frac{\delta_{max}\, x}{L^3}(3L^2 - 4x^2) \qquad \left(0 \le x \le \frac{L}{2}\right) \qquad \textbf{(9-125)}$$

Taking two derivatives, we find

$$\frac{d^2v}{dx^2} = \frac{24\delta_{max}\, x}{L^3} \qquad \textbf{(9-126)}$$

Finally, we substitute the second derivative into Eq. (9-122) and obtain the following expression for the strain energy of the beam in terms of the maximum deflection:

$$U = 2\int_0^{L/2} \frac{EI}{2}\left(\frac{d^2v}{dx^2}\right)^2 dx = EI\int_0^{L/2}\left(\frac{24\delta_{max}\, x}{L^3}\right)^2 dx = \frac{24EI\delta_{max}^2}{L^3}$$

$$\textbf{(9-127)}$$

Equating the potential energy lost by the falling mass in Eq. (9-121) to the strain energy acquired by the beam in Eq. (9-127), we get

$$W(h + \delta_{max}) = \frac{24EI\delta_{max}^2}{L^3} \qquad \textbf{(9-128)}$$

This equation is quadratic in $\delta_{max}$ and can be solved for its positive root:

$$\delta_{max} = \frac{WL^3}{48EI} + \left[\left(\frac{WL^3}{48EI}\right)^2 + 2h\left(\frac{WL^3}{48EI}\right)\right]^{1/2} \qquad \textbf{(9-129)}$$

We see that the maximum dynamic deflection increases if either the weight of the falling object or the height of fall is increased, and it decreases if the stiffness $EI/L^3$ of the beam is increased.

To simplify the preceding equation, we will denote the *static deflection* of the beam due to the weight $W$ as $\delta_{st}$:

$$\delta_{st} = \frac{WL^3}{48EI} \qquad \textbf{(9-130)}$$

Then Eq. (9-129) for the maximum dynamic deflection becomes

$$\delta_{max} = \delta_{st} + (\delta_{st}^2 + 2h\delta_{st})^{1/2} \qquad \textbf{(9-131)}$$

This equation shows that the dynamic deflection is always larger than the static deflection.

If the height $h$ equals zero, which means that the load is applied suddenly but without any free fall, the dynamic deflection is twice the static deflection. If $h$ is very large compared to the deflection, then the term containing $h$ in Eq. (9-131) predominates, and the equation can be simplified to

$$\delta_{max} = \sqrt{2h\delta_{st}} \qquad \textbf{(9-132)}$$

These observations are analogous to those discussed previously in Section 2.8 for impact on a bar in tension or compression.

The deflection $\delta_{max}$ calculated from Eq. (9-131) generally represents an upper limit, because we assumed there were no energy losses during impact. Several other factors also tend to reduce the deflection, including localized deformation of the contact surfaces, the tendency of the falling mass to bounce upward, and inertia effects of the mass of the beam. Thus, we see that the phenomenon of impact is quite complex, and if a more accurate analysis is needed, books and articles devoted specifically to that subject must be consulted.

# *9.11 TEMPERATURE EFFECTS

In the preceding sections of this chapter we considered the deflections of beams due to lateral loads. In this section, we will consider the deflections caused by **nonuniform temperature changes**. As a preliminary matter, recall that the effects of *uniform* temperature changes have already been described in Section 2.5, where it was shown that a uniform temperature increase causes an unconstrained bar or beam to have its length increased by the amount

$$\delta_T = \alpha(\Delta T)L \tag{9-133}$$

In this equation, $\alpha$ is the coefficient of thermal expansion, $\Delta T$ is the uniform increase in temperature, and $L$ is the length of the bar [see Fig. 2-20 and Eq. (2-16) in Chapter 2].

If a beam is supported in such a manner that longitudinal expansion is free to occur, as is the case for all of the statically determinate beams considered in this chapter, then a uniform temperature change will not produce any stresses in the beam. Also, there will be no lateral deflections of such a beam, because there is no tendency for the beam to bend.

The behavior of a beam is quite different if the temperature is not constant across its height. For example, assume that a simple beam, initially straight and at a uniform temperature $T_0$, has its temperature changed to $T_1$ on its upper surface and $T_2$ on its lower surface, as pictured in Fig. 9-46a. If we assume that the variation in temperature is linear between the top and bottom of the beam, then the *average temperature* of the beam is

$$T_{\text{aver}} = \frac{T_1 + T_2}{2} \tag{9-134}$$

and occurs at midheight. Any difference between this average temperature and the initial temperature $T_0$ results in a change in length of the beam, given by Eq. (9-96), as follows:

$$\delta_T = \alpha(T_{\text{aver}} - T_0)L = \alpha\left(\frac{T_1 + T_2}{2} - T_0\right)L \tag{9-135}$$

In addition, the temperature differential $T_2 - T_1$ between the bottom and top of the beam produces a *curvature* of the axis of the beam, with the accompanying lateral deflections (Fig. 9-46b).

To investigate the deflections due to a temperature differential, consider an element of length $dx$ cut out from the beam (Figs. 9-46a and c). The changes in length of the element at the bottom and top are $\alpha(T_2 - T_0)dx$ and $\alpha(T_1 - T_0)dx$, respectively. If $T_2$ is greater than $T_1$, the sides of the element will rotate with respect to each other through an angle $d\theta$, as shown in Fig. 9-46c. The angle $d\theta$ is related to the changes in dimension by the following equation, obtained from the geometry of the figure:

$$h\,d\theta = \alpha(T_2 - T_0)dx - \alpha(T_1 - T_0)dx$$

from which we get

$$\frac{d\theta}{dx} = \frac{\alpha(T_2 - T_1)}{h} \tag{9-136}$$

in which $h$ is the height of the beam.

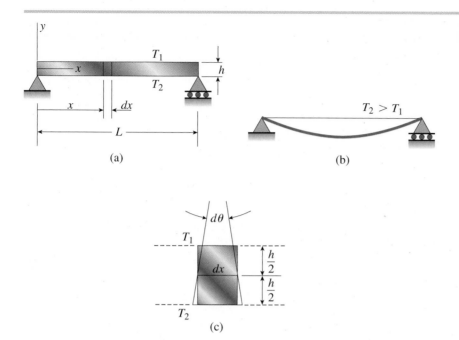

Fig. 9-46

Temperature effects in a beam

We have already seen that the quantity $d\theta/dx$ represents the curvature of the deflection curve of the beam (see Eq. 9-6). Since the curvature is equal to $d^2v/dx^2$ (Eq. 9-9), we may write the following **differential equation of the deflection curve**:

$$\frac{d^2v}{dx^2} = \frac{\alpha(T_2 - T_1)}{h} \qquad \textbf{(9-137)}$$

Note that when $T_2$ is greater than $T_1$, the curvature is positive and the beam is bent concave upward, as shown in Fig. 9-46b. The quantity $\alpha(T_2 - T_1)/h$ in Eq. (9-100) is the counterpart of the quantity $M/EI$, which appears in the basic differential equation (Eq. 9-7).

We can solve Eq. (9-100) by the same integration techniques described earlier for the effects of bending moments (see Section 9.3). We can integrate the differential equation to obtain $dv/dx$ and $v$, and we can use boundary or other conditions to evaluate the constants of integration. In this manner we can obtain the equations for the slopes and deflections of the beam, as illustrated by Probs. 9.11-1 through 9.11-5 at the end of this chapter.

If the beam is able to change in length and deflect freely, there will be no stresses associated with the temperature changes described in this section. However, if the beam is restrained against longitudinal expansion or lateral deflection, or if the temperature changes do not vary linearly from top to bottom of the beam, internal temperature stresses will develop. The determination of such stresses requires the use of more advanced methods of analysis. Statically indeterminate beams subject to temperature effects will be studied in Section 10.5.

## Example 9-19

An overhanging beam *ABC* of height *h* has a pin support at *A* and a roller support at *B*. The beam is heated to a temperature $T_1$ on the top and $T_2$ on the bottom (see Fig. 9-47). Determine the equation of the deflection curve of the beam, the angle of rotation $\theta_C$ at end *C*, and the deflection $\delta_C$ at end *C*.

### Fig. 9-47

Example 9-19: Simple beam with overhang and temperature change

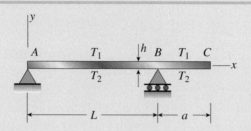

### Solution

We investigated the displacement of this beam at selected points due to a concentrated load at *C* in Example 9-5, under a uniform load *q* in Example 9-9, and with uniform load *q* on *AB* and load *P* at *C* in Example 9-18. Now we will consider the effect of a temperature differential $(T_2 - T_1)$ on the deflection $v(x)$ of the beam using Eq. (9-136).

$$\frac{d^2}{dx^2} v(x) = \frac{\alpha}{h}(T_2 - T_1) \qquad \textbf{(9-137, repeated)}$$

Integrating, we obtain two constants of integration, $C_1$ and $C_2$, which must be determined using two independent bounday conditions:

$$\frac{d}{dx} v(x) = \frac{\alpha}{h}(T_2 - T_1)x + C_1 \qquad \textbf{(a)}$$

$$v(x) = \frac{\alpha}{h}(T_2 - T_1)\frac{x^2}{2} + C_1 x + C_2 \qquad \textbf{(b)}$$

The boundary conditions are $v(0) = 0$ and $v(L) = 0$. So $v(0) = 0$, which gives $C_2 = 0$.

And $v(L) = 0$, which leads to                                             **(c)**

$$C_1 = \frac{1}{L}\left[\frac{-\alpha L^2}{2h}(T_2 - T_1)\right] = -\left[\frac{L\alpha(T_2 - T_1)}{2h}\right] \qquad \textbf{(d)}$$

Substituting $C_1$ and $C_2$ into Eq. (b) results in the equation of the elastic curve of the beam due to temperature differential $(T_2 - T_1)$ as

*Continues* ➡

$$v(x) = \frac{\alpha x(T_2 - T_1)(x - L)}{2h} \tag{e}$$

If $x = L + a$ in Eq. (e), we get an expression for the deflection of the beam at $C$:

$$\delta_C = v(L + a) = \frac{\alpha(L + a)(T_2 - T_1)(L + a - L)}{2h} = \frac{\alpha(T_2 - T_1)a(L + a)}{2h} \tag{f}$$

We have assumed linear elastic behavior here and in earlier examples, so (if desired) the *principle of superposition* can be used to find the total deflection at $C$ due to simultaneous application of all loads considered in Examples 9-5, 9-9, and 9-18 and for the temperature differential studied here.

*Numerical example.* If beam $ABC$ is a steel wide flange W 30 × 211 [see Table F-1(a)], with a length of $L = 30$ ft and with an overhang $a = L/2$, we can compare the deflection at $C$ due to self-weight (see Example 9-9; let $q = 211$ lb/ft) to the deflection at $C$ due to temperature differential $(T_2 - T_1) = 5°$ Fahrenheit. From Table I-4, the coefficient of thermal expansion for structural steel is $\alpha = 6.5 \times 10^{-6}/°$F. The modulus for steel is 30,000 ksi.

From Eq. (9-68), the deflection at $C$ due to self weight is

$$\delta_{C_q} = \frac{qa}{24EI_z}(a + L)(3a^2 + aL - L^2) \tag{g}$$

$$= \frac{\left(211\frac{\text{lb}}{\text{ft}} \cdot \frac{\text{ft}}{12\text{in.}}\right)(180 \text{ in.})}{24\left(30 \times 10^6\frac{\text{lb}}{\text{in.}^2}\right)(10,300 \text{ in.}^4)}(180 \text{ in.} + 360 \text{ in.})[3(180 \text{ in.})^2 + 180 \text{ in.}(360 \text{ in.}) - (360 \text{ in.})^2]$$

$$= 7.467 \times 10^{-3} \text{ in.}$$

where $a = 15$ ft $= 180$ in. and $L = 30$ ft $= 360$ in.

The deflection at $C$ due to a temperature differential of only $5°$ Fahrenheit is from Eq. (f):

$$\delta_{CT} = \frac{\alpha(T_2 - T_1)a(L + a)}{2h}$$

$$= \frac{(6.5 \times 10^{-6})(5)(180 \text{ in.})(360 \text{ in.} + 180 \text{ in.})}{2(30 \text{ in.})} = 0.053 \text{ in.} \tag{h}$$

The deflection at $C$ due to a temperature differential is seven times that due to self-weight.

In Chapter 9, we investigated the linear elastic, small displacement behavior of beams of different types, with different support conditions, acted upon by a wide variety of loadings including impact and temperature effects. We studied methods based on integration of the second-, third- or fourth-order differential equations of the deflection curve. We computed displacements (both translations and rotations) at specific points along the beam and also found the equation describing the deflected shape of the entire beam. Using solutions for a number of standard cases (tabulated in Appendix H), we used the powerful principle of superposition to solve more complicated beams and loadings by combining the simpler standard solutions. We also considered a method for calculating displacements of beams based on the area of the moment diagram. Finally, we studied an energy-based method for computing beam displacements. The major concepts presented in this chapter may be summarized as follows:

1. By combining expressions for linear curvature ($\kappa = d^2v/dx^2$) and the moment curvature relation ($\kappa = M/EI$), we obtained the **ordinary differential equation of the deflection curve** for a beam, which is valid only for linear elastic behavior.

$$EI\frac{d^2v}{dx^2} = M$$

2. The differential equation of the deflection curve may be differentiated once to obtain a third-order equation relating shear force $V$ and first derivative of moment, $dM/dx$, or twice to obtain a fourth-order equation relating intensity of distributed load $q$ and first derivative of shear, $dV/dx$.

$$EI\frac{d^3v}{dx^3} = V$$

$$EI\frac{d^4v}{dx^4} = -q$$

The choice of second-, third-, or fourth-order differential equations depends on which is most efficient for a particular beam support case and applied loading.

3. We must write expressions for either moment ($M$), shear ($V$), or load intensity ($q$) for each separate region of the beam (e.g., whenever $q$,

$V$, $M$, or $EI$ vary) and then apply **boundary**, **continuity**, or **symmetry conditions**, as appropriate, to solve for unknown constants of integration which arise as we apply the method of successive integrations; the beam deflection equation, $v(x)$, may be evaluated at a particular value of $x$ to find the translational displacement at that point; evaluation of $dv/dx$ at that same point provides the slope of the deflection equation.

4. The **method of superposition** may be used to solve for displacements and rotations for more complicated beams and loadings; the actual beam first must be broken down into the sum of a number of simpler cases whose solutions already are known (see Appendix H); superposition is only applicable to beams undergoing small displacements and behaving in a linear elastic manner.

5. The **moment-area method** is an alternative approach for finding beam displacements; it is based on two theorems which are related to the area of the bending moment diagram.

6. Equating the strain energy of bending ($U$) to the work ($W$) of a concentrated load or moment, and then taking a partial derivative with respect to a particular load ($P$, $M$), provides still another method for computing beam deflections and rotations; this method is known as **Castigliano's Theorem**; however, the method has limited application because loads may not be applied at locations where deflections and rotations are of interest; in this case, a fictitious load must be applied at the point where displacements are to be computed.

7. By equating the potential energy of a falling mass to strain energy acquired by the beam, **deflections due to impact** may be approximated.

8. Finally, if a beam experiences a temperature change which is not constant across its height (i.e., a **temperature differential**, $T_2 - T_1$, over height $h$), it produces a curvature of the axis of the beam:

$$\kappa = d\theta/dx = d^2v/dx^2 = \alpha(T_2 - T_1)/h$$

This equation can be integrated to obtain the equation of the deflection curve using successive integration as described above.

# PROBLEMS CHAPTER 9

## Differential Equations of the Deflection Curve

*The beams described in the problems for Section 9.2 have constant flexural rigidity EI.*

**9.2-1** The deflection curve for a simple beam $AB$ (see figure) is given by the following equation:

$$v = -\frac{q_0 x}{360 LEI}(7L^4 - 10L^2x^2 + 3x^4)$$

Describe the load acting on the beam.

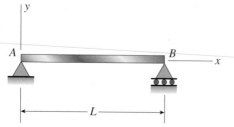

PROBS. 9.2-1 and 9.2-2

**9.2-2** The deflection curve for a simple beam $AB$ (see figure) is given by the following equation:

$$v = -\frac{q_0 L^4}{\pi^4 EI}\sin\frac{\pi x}{L}$$

(a) Describe the load acting on the beam.
(b) Determine the reactions $R_A$ and $R_B$ at the supports.
(c) Determine the maximum bending moment $M_{max}$.

**9.2-3** The deflection curve for a cantilever beam $AB$ (see figure) is given by the following equation:

$$v = -\frac{q_0 x^2}{120 LEI}(10L^3 - 10L^2x + 5Lx^2 - x^3)$$

Describe the load acting on the beam.

PROBS. 9.2-3 and 9.2-4

**9.2-4** The deflection curve for a cantilever beam $AB$ (see figure) is given by the following equation:

$$v = -\frac{q_0 x^2}{360 L^2 EI}(45L^4 - 40L^3x + 15L^2x^2 - x^4)$$

(a) Describe the load acting on the beam.
(b) Determine the reactions $R_A$ and $M_A$ at the support.

## Deflection Formulas

*Problems 9.3-1 through 9.3-7 require the calculation of deflections using the formulas derived in Examples 9-1, 9-2, and 9-3. All beams have constant flexural rigidity EI.*

**9.3-1** A wide-flange beam (W 12 × 35) supports a uniform load on a simple span of length $L = 14$ ft (see figure).

Calculate the maximum deflection $\delta_{max}$ at the midpoint and the angles of rotation $\theta$ at the supports if $q = 1.8$ k/ft and $E = 30 \times 10^6$ psi. Use the formulas of Example 9-1.

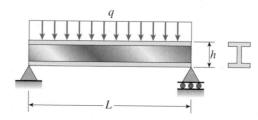

PROBS. 9.3-1, 9.3-2, and 9.3-3

**9.3-2** A uniformly loaded steel wide-flange beam with simple supports (see figure) has a downward deflection of 10 mm at the midpoint and angles of rotation equal to 0.01 radians at the ends.

Calculate the height $h$ of the beam if the maximum bending stress is 90 MPa and the modulus of elasticity is 200 GPa. (*Hint:* Use the formulas of Example 9-1.)

**9.3-3** What is the span length $L$ of a uniformly loaded simple beam of wide-flange cross section (see figure) if the maximum bending stress is 12,000 psi, the maximum deflection is 0.1 in., the height of the beam is 12 in., and the modulus of elasticity is $30 \times 10^6$ psi? (Use the formulas of Example 9-1.)

**9.3-4** Calculate the maximum deflection $\delta_{max}$ of a uniformly loaded simple beam (see figure on the next page) if the span length $L = 2.0$ m, the intensity of the uniform load $q = 2.0$ kN/m, and the maximum bending stress $\sigma = 60$ MPa.

The cross section of the beam is square, and the material is aluminum having modulus of elasticity $E = 70$ GPa. (Use the formulas of Example 9-1.)

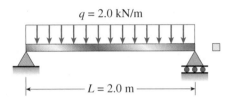

$q = 2.0$ kN/m

$L = 2.0$ m

**PROB. 9.3-4**

**9.3-5** A cantilever beam with a uniform load (see figure) has a height $h$ equal to 1/8 of the length $L$. The beam is a steel wide-flange section with $E = 28 \times 10^6$ psi and an allowable bending stress of 17,500 psi in both tension and compression. Calculate the ratio $\delta/L$ of the deflection at the free end to the length, assuming that the beam carries the maximum allowable load. (Use the formulas of Example 9-2.)

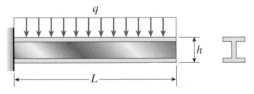

$q$

$L$

$h$

**PROB. 9.3-5**

**9.3-6** A gold-alloy microbeam attached to a silicon wafer behaves like a cantilever beam subjected to a uniform load (see figure). The beam has length $L = 27.5$ $\mu$m and rectangular cross section of width $b = 4.0$ $\mu$m and thickness $t = 0.88$ $\mu$m. The total load on the beam is 17.2 $\mu$N. If the deflection at the end of the beam is 2.46 $\mu$m, what is the modulus of elasticity $E_g$ of the gold alloy? (Use the formulas of Example 9-2.)

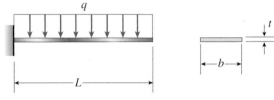

$q$

$L$

$t$

$b$

**PROB. 9.3-6**

**9.3-7** Obtain a formula for the ratio $\delta_C/\delta_{max}$ of the deflection at the midpoint to the maximum deflection for a simple beam supporting a concentrated load $P$ (see figure).

From the formula, plot a graph of $\delta_C/\delta_{max}$ versus the ratio $a/L$ that defines the position of the load $(0.5 < a/L < 1)$. What conclusion do you draw from the graph? (Use the formulas of Example 9-3.)

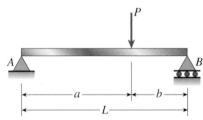

$P$

$A$

$B$

$a$

$b$

$L$

**PROB. 9.3-7**

# Deflections by Integration of the Bending-Moment Equation

*Problems 9.3-8 through 9.3-17 are to be solved by integrating the second-order differential equation of the deflection curve (the bending-moment equation). The origin of coordinates is at the left-hand end of each beam, and all beams have constant flexural rigidity EI.*

**9.3-8** Derive the equation of the deflection curve for a cantilever beam $AB$ supporting a load $P$ at the free end (see figure). Also, determine the deflection $\delta_B$ and angle of rotation $\theta_B$ at the free end. (*Note:* Use the second-order differential equation of the deflection curve.)

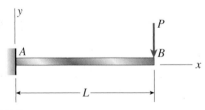

$y$

$P$

$A$

$B$

$x$

$L$

**PROB. 9.3-8**

**9.3-9** Derive the equation of the deflection curve for a simple beam $AB$ loaded by a couple $M_0$ at the left-hand support (see figure). Also, determine the maximum deflection $\delta_{max}$. (*Note:* Use the second-order differential equation of the deflection curve.)

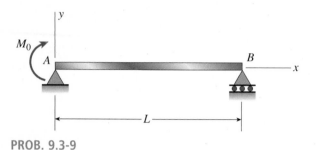

**PROB. 9.3-9**

**9.3-10** A cantilever beam $AB$ supporting a triangularly distributed load of maximum intensity $q_0$ is shown in the figure.

Derive the equation of the deflection curve and then obtain formulas for the deflection $\delta_B$ and angle of rotation $\theta_B$ at the free end. (*Note:* Use the second-order differential equation of the deflection curve.)

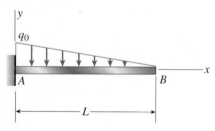

**PROB. 9.3-10**

**9.3-11** A cantilever beam $AB$ is acted upon by a uniformly distributed moment (bending moment, not torque) of intensity $m$ per unit distance along the axis of the beam (see figure).

Derive the equation of the deflection curve and then obtain formulas for the deflection $\delta_B$ and angle of rotation $\theta_B$ at the free end. (*Note:* Use the second-order differential equation of the deflection curve.)

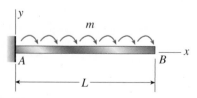

**PROB. 9.3-11**

**9.3-12** The beam shown in the figure has a guided support at $A$ and a spring support at $B$. The guided support permits vertical movement but no rotation. Derive the equation of the deflection curve and determine the deflection $\delta_B$ at end $B$ due to the uniform load of intensity $q$. (*Note:* Use the second-order differential equation of the deflection curve.)

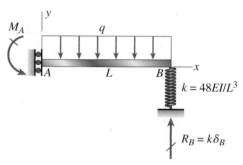

**PROB. 9.3-12**

**9.3-13** Derive the equations of the deflection curve for a simple beam $AB$ loaded by a couple $M_0$ acting at distance $a$ from the left-hand support (see figure). Also, determine the deflection $\delta_0$ at the point where the load is applied. (*Note:* Use the second-order differential equation of the deflection curve.)

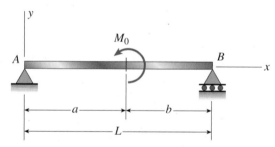

**PROB. 9.3-13**

**9.3-14** Derive the equations of the deflection curve for a cantilever beam $AB$ carrying a uniform load of intensity $q$ over part of the span (see figure). Also, determine the deflection $\delta_B$ at the end of the beam. (*Note:* Use the second-order differential equation of the deflection curve.)

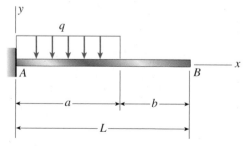

**PROB. 9.3-14**

**9.3-15** Derive the equations of the deflection curve for a cantilever beam *AB* supporting a distributed load of peak intensity $q_0$ acting over one-half of the length (see figure). Also, obtain formulas for the deflections $\delta_B$ and $\delta_C$ at points *B* and *C*, respectively. (*Note*: Use the second-order differential equation of the deflection curve.)

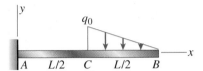

PROB. 9.3-15

**9.3-16** Derive the equations of the deflection curve for a simple beam *AB* with a distributed load of peak intensity $q_0$ acting over the left-hand half of the span (see figure). Also, determine the deflection $\delta_C$ at the midpoint of the beam. (*Note*: Use the second-order differential equation of the deflection curve.)

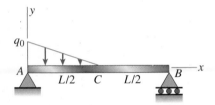

PROB. 9.3-16

**9.3-17** The beam shown in the figure has a guided support at *A* and a roller support at *B*. The guided support permits vertical movement but no rotation. Derive the equation of the deflection curve and determine the deflection $\delta_A$ at end *A* and also $\delta_C$ at point *C* due to the uniform load of intensity $q = P/L$ applied over segment *CB* and load *P* at $x = L/3$. (*Note*: Use the second-order differential equation of the deflection curve.)

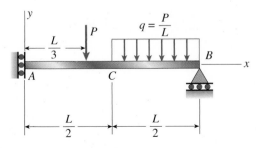

PROB. 9.3-17

# Deflections by Integration of the Shear-Force and Load Equations

*The beams described in the problems for Section 9.4 have constant flexural rigidity EI. Also, the origin of coordinates is at the left-hand end of each beam.*

**9.4-1** Derive the equation of the deflection curve for a cantilever beam *AB* when a couple $M_0$ acts counterclockwise at the free end (see figure). Also, determine the deflection $\delta_B$ and slope $\theta_B$ at the free end. Use the third-order differential equation of the deflection curve (the shear-force equation).

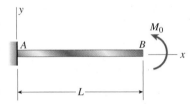

PROB. 9.4-1

**9.4-2** A simple beam *AB* is subjected to a distributed load of intensity $q = q_0 \sin \pi x/L$, where $q_0$ is the maximum intensity of the load (see figure).

Derive the equation of the deflection curve, and then determine the deflection $\delta_{max}$ at the midpoint of the beam. Use the fourth-order differential equation of the deflection curve (the load equation).

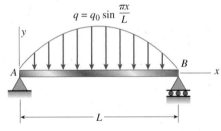

PROB. 9.4-2

**9.4-3** The simple beam *AB* shown in the figure has moments $2M_0$ and $M_0$ acting at the ends.

Derive the equation of the deflection curve, and then determine the maximum deflection $\delta_{max}$. Use the third-order differential equation of the deflection curve (the shear-force equation).

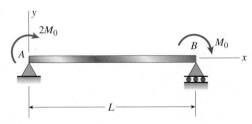

PROB. 9.4-3

**9.4-4** A beam with a uniform load has a guided support at one end and spring support at the other. The spring has stiffness $k = 48EI/L^3$. Derive the equation of the deflection curve by starting with the third-order differential equation (the shear-force equation). Also, determine the angle of rotation $\theta_B$ at support $B$.

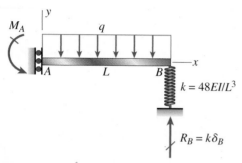

PROB. 9.4-4

**9.4-5** The distributed load acting on a cantilever beam $AB$ has an intensity $q$ given by the expression $q_0 \cos \pi x/2L$, where $q_0$ is the maximum intensity of the load (see figure).

Derive the equation of the deflection curve, and then determine the deflection $\delta_B$ at the free end. Use the fourth-order differential equation of the deflection curve (the load equation).

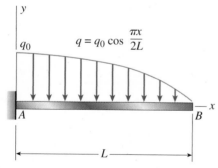

PROB. 9.4-5

**9.4-6** A cantilever beam $AB$ is subjected to a parabolically varying load of intensity $q = q_0(L^2 - x^2)/L^2$, where $q_0$ is the maximum intensity of the load (see figure).

Derive the equation of the deflection curve, and then determine the deflection $\delta_B$ and angle of rotation $\theta_B$ at the free end. Use the fourth-order differential equation of the deflection curve (the load equation).

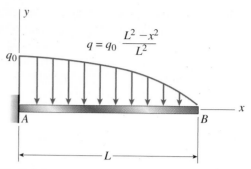

PROB. 9.4-6

**9.4-7** A beam on simple supports is subjected to a parabolically distributed load of intensity $q = 4q_0x(L - x)/L^2$, where $q_0$ is the maximum intensity of the load (see figure).

Derive the equation of the deflection curve, and then determine the maximum deflection $\delta_{max}$. Use the fourth-order differential equation of the deflection curve (the load equation).

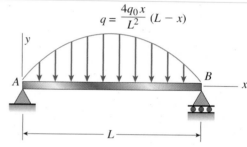

PROB. 9.4-7

**9.4-8** Derive the equation of the deflection curve for beam $AB$, with guided support at $A$ and roller at $B$, carrying a triangularly distributed load of maximum intensity $q_0$ (see figure). Also, determine the maximum deflection $\delta_{max}$ of the beam. Use the fourth-order differential equation of the deflection curve (the load equation).

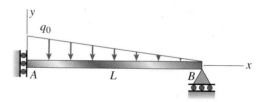

PROB. 9.4-8

**9.4-9** Derive the equations of the deflection curve for beam $ABC$, with guided support at $A$ and roller support at $B$, supporting a uniform load of intensity $q$ acting on the overhang portion of the beam (see figure). Also, determine deflection

$\delta_C$ and angle of rotation $\theta_C$. Use the fourth-order differential equation of the deflection curve (the load equation).

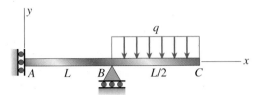

PROB. 9.4-9

**9.4-10** Derive the equations of the deflection curve for beam $AB$, with guided support at $A$ and roller support at $B$, supporting a distributed load of maximum intensity $q_0$ acting on the right-hand half of the beam (see figure). Also, determine deflection $\delta_A$, angle of rotation $\theta_B$, and deflection $\delta_C$ at the midpoint. Use the fourth-order differential equation of the deflection curve (the load equation).

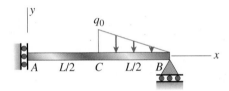

PROB. 9.4-10

# Method of Superposition

*The problems for Section 9.5 are to be solved by the method of superposition. All beams have constant flexural rigidity EI.*

**9.5-1** A cantilever beam $AB$ carries three equally spaced concentrated loads, as shown in the figure. Obtain formulas for the angle of rotation $\theta_B$ and deflection $\delta_B$ at the free end of the beam.

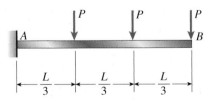

PROB. 9.5-1

**9.5-2** A simple beam $AB$ supports five equally spaced loads $P$ (see figure).

(a) Determine the deflection $\delta_1$ at the midpoint of the beam.

(b) If the same total load ($5P$) is distributed as a uniform load on the beam, what is the deflection $\delta_2$ at the midpoint?

(c) Calculate the ratio of $\delta_1$ to $\delta_2$.

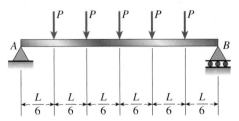

PROB. 9.5-2

**9.5-3** The cantilever beam $AB$ shown in the figure has an extension $BCD$ attached to its free end. A force $P$ acts at the end of the extension.

(a) Find the ratio $a/L$ so that the vertical deflection of point $B$ will be zero.

(b) Find the ratio $a/L$ so that the angle of rotation at point $B$ will be zero.

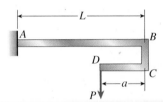

PROB. 9.5-3

**9.5-4** Beam $ACB$ hangs from two springs, as shown in the figure. The springs have stiffnesses $k_1$ and $k_2$ and the beam has flexural rigidity $EI$.

(a) What is the downward displacement of point $C$, which is at the midpoint of the beam, when the moment $M_0$ is applied? Data for the structure are as follows: $M_0 = 10.0 \text{ kN} \cdot \text{m}$, $L = 1.8 \text{ m}$, $EI = 216 \text{ kN} \cdot \text{m}^2$, $k_1 = 250 \text{ kN/m}$, and $k_2 = 160 \text{ kN/m}$.

(b) Repeat part (a), but remove $M_0$ and apply uniform load $q = 3.5 \text{ kN/m}$ to the entire beam.

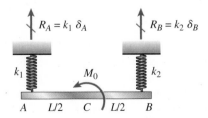

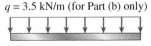

PROB. 9.5-4

**9.5-5** What must be the equation $y = f(x)$ of the axis of the slightly curved beam $AB$ (see figure) *before* the load is applied in order that the load $P$, moving along the bar, always stays at the same level?

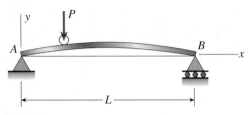

**PROB. 9.5-5**

**9.5-6** Determine the angle of rotation $\theta_B$ and deflection $\delta_B$ at the free end of a cantilever beam $AB$ having a uniform load of intensity $q$ acting over the middle third of its length (see figure).

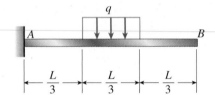

**PROB. 9.5-6**

**9.5-7** The cantilever beam $ACB$ shown in the figure has flexural rigidity $EI = 2.1 \times 10^6$ k-in.$^2$ Calculate the downward deflections $\delta_C$ and $\delta_B$ at points $C$ and $B$, respectively, due to the simultaneous action of the moment of 35 k-in. applied at point $C$ and the concentrated load of 2.5 k applied at the free end $B$.

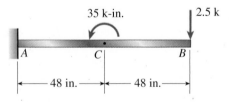

**PROB. 9.5-7**

**9.5-8** A cantilever beam is subjected to load $P$ at midspan and Counterclockwise moment $M$ at $B$ (see figure).

(a) Find an *expression* for moment $M$ in terms of the load $P$ so that the reaction moment at $A$, $M_A$, is equal to zero.

(b) Find an *expression* for moment $M$ in terms of the load $P$ so that the deflection is $\delta_B = 0$; *also*, what is rotation $\theta_B$?

(c) Find an *expression* for moment $M$ in terms of the load $P$ so that the rotation $\theta_B = 0$; *also*, what is deflection $\delta_B$?

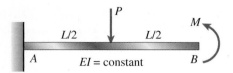

**PROB. 9.5-8**

**9.5-9** A cantilever beam is subjected to a quadratic distributed load $q(x)$ over the length of the beam (see figure). Find an *expression* for moment $M$ in terms of the peak distributed load intensity $q_0$ so that the deflection is $\delta_B = 0$.

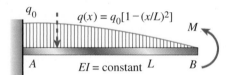

**PROB. 9.5-9**

**9.5-10** A beam $ABCD$ consisting of a simple span $BD$ and an overhang $AB$ is loaded by a force $P$ acting at the end of the bracket $CEF$ (see figure).

(a) Determine the deflection $\delta_A$ at the end of the overhang.

(b) Under what conditions is this deflection upward? Under what conditions is it downward?

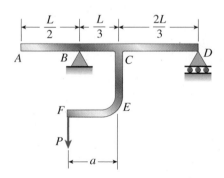

**PROB. 9.5-10**

**9.5-11** A horizontal load $P$ acts at end $C$ of the bracket $ABC$ shown in the figure.

(a) Determine the deflection $\delta_C$ of point $C$.

(b) Determine the maximum upward deflection $\delta_{max}$ of member $AB$.

*Note:* Assume that the flexural rigidity $EI$ is constant throughout the frame. Also, disregard the effects of axial deformations and consider only the effects of bending due to the load $P$.

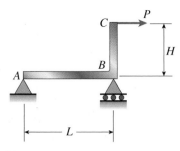

PROB. 9.5-11

**9.5-12** A beam *ABC* having flexural rigidity $EI = 75 \text{ kN} \cdot \text{m}^2$ is loaded by a force $P = 800$ N at end *C* and tied down at end *A* by a wire having axial rigidity $EA = 900$ kN (see figure).

What is the deflection at point *C* when the load *P* is applied?

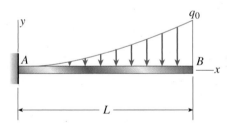

PROB. 9.5-12

**9.5-13** Determine the angle of rotation $\theta_B$ and deflection $\delta_B$ at the free end of a cantilever beam *AB* supporting a parabolic load defined by the equation $q(x) = q_0 x^2/L^2$ (see figure).

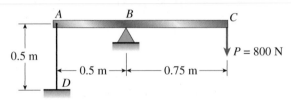

PROB. 9.5-13

**9.5-14** A simple beam *AB* supports a uniform load of intensity *q* acting over the middle region of the span (see figure).

Determine the angle of rotation $\theta_A$ at the left-hand support and the deflection $\delta_{max}$ at the midpoint.

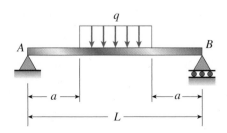

PROB. 9.5-14

**9.5-15** The overhanging beam *ABCD* supports two concentrated loads *P* and *Q* (see figure).

(a) For what ratio $P/Q$ will the deflection at point *B* be zero?

(b) For what ratio will the deflection at point *D* be zero?

(c) If *Q* is replaced by uniform load with intensity *q* (on the overhang), repeat parts (a) and (b), but find ratio $P/(qa)$.

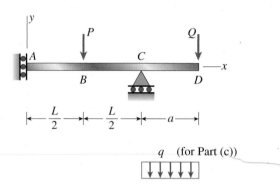

PROB. 9.5-15

**9.5-16** A thin metal strip of total weight *W* and length *L* is placed across the top of a flat table of width $L/3$ as shown in the figure.

What is the clearance $\delta$ between the strip and the middle of the table? (The strip of metal has flexural rigidity *EI*.)

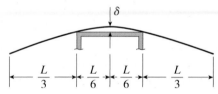

PROB. 9.5-16

**9.5-17** An overhanging beam *ABC* with flexural rigidity $EI = 15$ k-in.$^2$ is supported by a guided support at *A* and by a spring of stiffness *k* at point *B* (see figure). Span *AB* has length $L = 30$ in. and carries a uniform load. The overhang *BC* has length $b = 15$ in. For what stiffness *k* of the spring will the uniform load produce no deflection at the free end *C*?

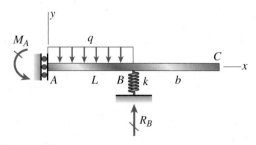

PROB. 9.5-17

**9.5-18** A beam $ABCD$ rests on simple supports at $B$ and $C$ (see figure). The beam has a slight initial curvature so that end $A$ is 18 mm above the elevation of the supports and end $D$ is 12 mm above. What moments $M_1$ and $M_2$, acting at points $A$ and $D$, respectively, will move points $A$ and $D$ downward to the level of the supports? (The flexural rigidity $EI$ of the beam is $2.5 \times 10^6$ N·m$^2$ and $L = 2.5$ m).

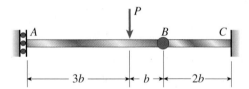

PROB. 9.5-18

**9.5-19** The compound beam $ABC$ shown in the figure has a guided support at $A$ and a fixed support at $C$. The beam consists of two members joined by a pin connection (i.e., moment release) at $B$. Find the deflection $\delta$ under the load $P$.

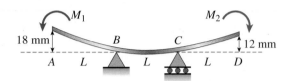

PROB. 9.5-19

**9.5-20** A compound beam $ABCDE$ (see figure) consists of two parts ($ABC$ and $CDE$) connected by a hinge (i.e., moment release) at $C$. The elastic support at $B$ has stiffness $k = EI/b^3$. Determine the deflection $\delta_E$ at the free end $E$ due to the load $P$ acting at that point.

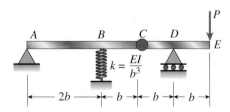

PROB. 9.5-20

**9.5-21** A steel beam $ABC$ is simply supported at $A$ and held by a high-strength steel wire at $B$ (see figure). A load $P = 240$ lb acts at the free end $C$. The wire has axial rigidity $EA = 1500 \times 10^3$ lb, and the beam has flexural rigidity $EI = 36 \times 10^6$ lb-in.$^2$

What is the deflection $\delta_C$ of point $C$ due to the load $P$?

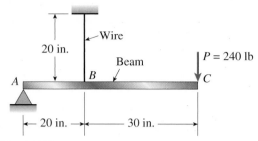

PROB. 9.5-21

PROB. 9.5-21

**9.5-22** The compound beam shown in the figure consists of a cantilever beam $AB$ (length $L$) that is pin-connected to a simple beam $BD$ (length $2L$). After the beam is constructed, a clearance $c$ exists between the beam and a support at $C$, midway between points $B$ and $D$. Subsequently, a uniform load is placed along the entire length of the beam.

What intensity $q$ of the load is needed to close the gap at $C$ and bring the beam into contact with the support?

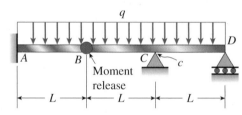

PROB. 9.5-22

**9.5-23** Find the horizontal deflection $\delta_h$ and vertical deflection $\delta_v$ at the free end $C$ of the frame $ABC$ shown in the figure. (The flexural rigidity $EI$ is constant throughout the frame.)

*Note:* Disregard the effects of axial deformations and consider only the effects of bending due to the load $P$.

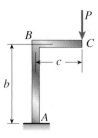

PROB. 9.5-23

**9.5-24** The frame *ABCD* shown in the figure is squeezed by two collinear forces *P* acting at points *A* and *D*. What is the decrease $\delta$ in the distance between points *A* and *D* when the loads *P* are applied? (The flexural rigidity *EI* is constant throughout the frame.)

*Note:* Disregard the effects of axial deformations and consider only the effects of bending due to the loads *P*.

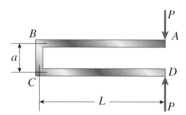

PROB. 9.5-24

**9.5-25** A framework *ABCD* is acted on by counterclockwise moment *M* at *A* (see figure). Assume that *EI* is constant.

(a) Find expressions for reactions at supports *B* and *C*.

(b) Find expressions for angles of rotation at *A*, *B*, *C*, and *D*.

(c) Find expressions for horizontal deflections $\delta_A$ and $\delta_D$.

(d) If length $L_{AB} = L/2$, find length $L_{CD}$ in terms of *L* for the absolute value of the ratio $|\delta_A/\delta_D| = 1$.

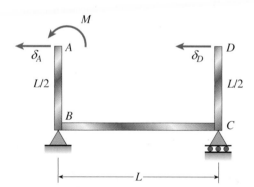

PROB. 9.5-25

**9.5-26** A framework *ABCD* is acted on by force *P* at 2*L*/3 from *B* (see figure). Assume that *EI* is constant.

(a) Find expressions for reactions at supports *B* and *C*.

(b) Find expressions for angles of rotation at *A*, *B*, *C*, and *D*.

(c) Find expressions for horizontal deflections $\delta_A$ and $\delta_D$.

(d) If length $L_{AB} = L/2$, find length $L_{CD}$ in terms of *L* for the absolute value of the ratio $|\delta_A/\delta_D| = 1$.

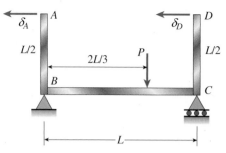

PROB. 9.5-26

**9.5-27** A beam *ABCDE* has simple supports at *B* and *D* and symmetrical overhangs at each end (see figure). The center span has length *L* and each overhang has length *b*. A uniform load of intensity *q* acts on the beam.

(a) Determine the ratio *b/L* so that the deflection $\delta_C$ at the midpoint of the beam is equal to the deflections $\delta_A$ and $\delta_E$ at the ends.

(b) For this value of *b/L*, what is the deflection $\delta_C$ at the midpoint?

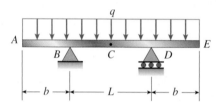

PROB. 9.5-27

**9.5-28** A frame *ABC* is loaded at point *C* by a force *P* acting at an angle $\alpha$ to the horizontal (see figure). Both members of the frame have the same length and the same flexural rigidity.

Determine the angle $\alpha$ so that the deflection of point *C* is in the same direction as the load. (Disregard the effects of axial deformations and consider only the effects of bending due to the load *P*.)

*Note:* A direction of loading such that the resulting deflection is in the same direction as the load is called a *principal direction*. For a given load on a planar structure, there are two principal directions, perpendicular to each other.

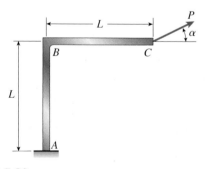

PROB. 9.5-28

# Moment-Area Method

*The problems for Section 9.6 are to be solved by the moment-area method. All beams have constant flexural rigidity EI.*

**9.6-1** A cantilever beam $AB$ is subjected to a uniform load of intensity $q$ acting throughout its length (see figure). Determine the angle of rotation $\theta_B$ and the deflection $\delta_B$ at the free end.

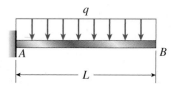

PROB. 9.6-1

**9.6-2** The load on a cantilever beam $AB$ has a triangular distribution with maximum intensity $q_0$ (see figure). Determine the angle of rotation $\theta_B$ and the deflection $\delta_B$ at the free end.

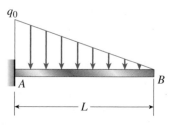

PROB. 9.6-2

**9.6-3** A cantilever beam $AB$ is subjected to a concentrated load $P$ and a couple $M_0$ acting at the free end (see figure).

Obtain formulas for the angle of rotation $\theta_B$ and the deflection $\delta_B$ at end $B$.

PROB. 9.6-3

**9.6-4** Determine the angle of rotation $\theta_B$ and the deflection $\delta_B$ at the free end of a cantilever beam $AB$ with a uniform load of intensity $q$ acting over the middle third of the length (see figure).

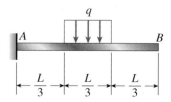

PROB. 9.6-4

**9.6-5** Calculate the deflections $\delta_B$ and $\delta_C$ at points $B$ and $C$, respectively, of the cantilever beam $ACB$ shown in the figure. Assume $M_0 = 36$ k-in., $P = 3.8$ k, $L = 8$ ft, and $EI = 2.25 \times 10^9$ lb-in.$^2$

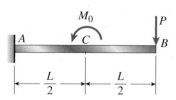

PROB. 9.6-5

**9.6-6** A cantilever beam $ACB$ supports two concentrated loads $P_1$ and $P_2$ as shown in the figure.

Calculate the deflections $\delta_B$ and $\delta_C$ at points $B$ and $C$, respectively. Assume $P_1 = 10$ kN, $P_2 = 5$ kN, $L = 2.6$ m, $E = 200$ GPa, and $I = 20.1 \times 10^6$ mm$^4$.

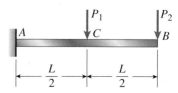

PROB. 9.6-6

**9.6-7** Obtain formulas for the angle of rotation $\theta_A$ at support $A$ and the deflection $\delta_{max}$ at the midpoint for a simple beam $AB$ with a uniform load of intensity $a$ (see figure).

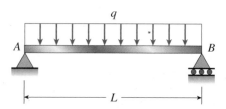

PROB. 9.6-7

**9.6-8** A simple beam $AB$ supports two concentrated loads $P$ at the positions shown in the figure. A support $C$ at the midpoint of the beam is positioned at distance $d$ below the beam before the loads are applied.

Assuming that $d = 10$ mm, $L = 6$ m, $E = 200$ GPa, and $I = 198 \times 10^6$ mm$^4$, calculate the magnitude of the loads $P$ so that the beam just touches the support at $C$.

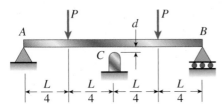

PROB. 9.6-8

**9.6-9** A simple beam $AB$ is subjected to a load in the form of a couple $M_0$ acting at end $B$ (see figure).

Determine the angles of rotation $\theta_A$ and $\theta_B$ at the supports and the deflection $\delta$ at the midpoint.

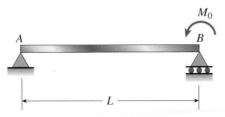

PROB. 9.6-9

**9.6-10** The simple beam $AB$ shown in the figure supports two equal concentrated loads $P$, one acting downward and the other upward.

Determine the angle of rotation $\theta_A$ at the left-hand end, the deflection $\delta_1$ under the downward load, and the deflection $\delta_2$ at the midpoint of the beam.

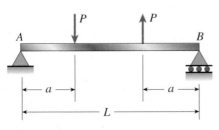

PROB. 9.6-10

**9.6-11** A simple beam $AB$ is subjected to couples $M_0$ and $2M_0$ acting as shown in the figure. Determine the angles of rotation $\theta_A$ and $\theta_B$ at the ends of the beam and the deflection $\delta$ at point $D$ where the load $M_0$ is applied.

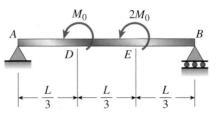

PROB. 9.6-11

# Nonprismatic Beams

**9.7-1** The cantilever beam $ACB$ shown in the figure has moments of inertia $I_2$ and $I_1$ in parts $AC$ and $CB$, respectively.

(a) Using the method of superposition, determine the deflection $\delta_B$ at the free end due to the load $P$.

(b) Determine the ratio $r$ of the deflection $\delta_B$ to the deflection $\delta_1$ at the free end of a prismatic cantilever with moment of inertia $I_1$ carrying the same load.

(c) Plot a graph of the deflection ratio $r$ versus the ratio $I_2/I_1$ of the moments of inertia. (Let $I_2/I_1$ vary from 1 to 5.)

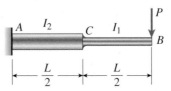

PROB. 9.7-1

**9.7-2** The cantilever beam $ACB$ shown in the figure supports a uniform load of intensity $q$ throughout its length. The beam has moments of inertia $I_2$ and $I_1$ in parts $AC$ and $CB$, respectively.

(a) Using the method of superposition, determine the deflection $\delta_B$ at the free end due to the uniform load.

(b) Determine the ratio $r$ of the deflection $\delta_B$ to the deflection $\delta_1$ at the free end of a prismatic cantilever with moment of inertia $I_1$ carrying the same load.

(c) Plot a graph of the deflection ratio $r$ versus the ratio $I_2/I_1$ of the moments of inertia. (Let $I_2/I_1$ vary from 1 to 5.)

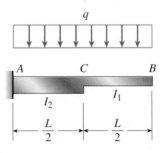

PROB. 9.7-2

**9.7-3** Beam $ACB$ hangs from two springs, as shown in the figure. The springs have stiffnesses $k_1$ and $k_2$ and the beam has flexural rigidity $EI$.

(a) What is the downward displacement of point $C$, which is at the midpoint of the beam, when the moment $M_0$ is applied? Data for the structure are as follows: $M_0 = 7.5$ k-ft, $L = 6$ ft, $EI = 520$ k-ft$^2$, $k_1 = 17$ k/ft, and $k_2 = 11$ k/ft.

(b) Repeat part (a), but remove $M_0$ and instead, apply uniform load $q$ over the entire beam.

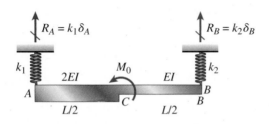

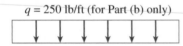

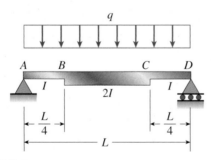

**PROB. 9.7-3**

**9.7-4** A simple beam $ABCD$ has moment of inertia $I$ near the supports and moment of inertia $2I$ in the middle region, as shown in the figure. A uniform load of intensity $q$ acts over the entire length of the beam.

Determine the equations of the deflection curve for the left-hand half of the beam. Also, find the angle of rotation $\theta_A$ at the left-hand support and the deflection $\delta_{max}$ at the midpoint.

**PROB. 9.7-4**

**9.7-5** A beam $ABC$ has a rigid segment from $A$ to $B$ and a flexible segment with moment of inertia $I$ from $B$ to $C$ (see figure). A concentrated load $P$ acts at point $B$.

Determine the angle of rotation $\theta_A$ of the rigid segment, the deflection $\delta_B$ at point $B$, and the maximum deflection $\delta_{max}$.

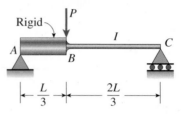

**PROB. 9.7-5**

**9.7-6** A simple beam $ABC$ has moment of inertia $1.5I$ from $A$ to $B$ and $I$ from $B$ to $C$ (see figure). A concentrated load $P$ acts at point $B$.

Obtain the equations of the deflection curves for both parts of the beam. From the equations, determine the angles of rotation $\theta_A$ and $\theta_C$ at the supports and the deflection $\delta_B$ at point $B$.

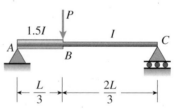

**PROB. 9.7-6**

**9.7-7** The tapered cantilever beam $AB$ shown in the figure has thin-walled, hollow circular cross sections of constant thickness $t$. The diameters at the ends $A$ and $B$ are $d_A$ and $d_B = 2d_A$, respectively. Thus, the diameter $d$ and moment of inertia $I$ at distance $x$ from the free end are, respectively,

$$d = \frac{d_A}{L}(L + x)$$

$$I = \frac{\pi t d^3}{8} = \frac{\pi t d_A^3}{8L^3}(L + x)^3 = \frac{I_A}{L^3}(L + x)^3$$

in which $I_A$ is the moment of inertia at end $A$ of the beam.

Determine the equation of the deflection curve and the deflection $\delta_A$ at the free end of the beam due to the load $P$.

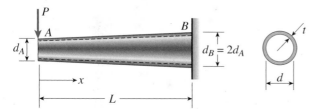

PROB. 9.7-7

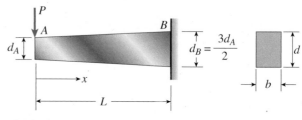

PROB. 9.7-9

**9.7-8** The tapered cantilever beam $AB$ shown in the figure has a solid circular cross section. The diameters at the ends $A$ and $B$ are $d_A$ and $d_B = 2d_A$, respectively. Thus, the diameter $d$ and moment of inertia $I$ at distance $x$ from the free end are, respectively,

$$d = \frac{d_A}{L}(L + x)$$

$$I = \frac{\pi d^4}{64} = \frac{\pi d_A^4}{64L^4}(L + x)^4 = \frac{I_A}{L^4}(L + x)^4$$

in which $I_A$ is the moment of inertia at end $A$ of the beam.

Determine the equation of the deflection curve and the deflection $\delta_A$ at the free end of the beam due to the load $P$.

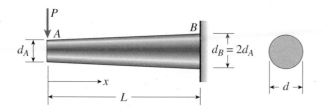

PROB. 9.7-8

**9.7-9** A tapered cantilever beam $AB$ supports a concentrated load $P$ at the free end (see figure). The cross sections of the beam are rectangular with constant width $b$, depth $d_A$ at support $A$, and depth $d_B = 3d_A/2$ at the support. Thus, the depth $d$ and moment of inertia $I$ at distance $x$ from the free end are, respectively,

$$d = \frac{d_A}{2L}(2L + x)$$

$$I = \frac{bd^3}{12} = \frac{bd_A^3}{96L^3}(2L + x)^3 = \frac{I_A}{8L^3}(2L + x)^3$$

in which $I_A$ is the moment of inertia at end $A$ of the beam.

Determine the equation of the deflection curve and the deflection $\delta_A$ at the free end of the beam due to the load $P$.

**9.7-10** A tapered cantilever beam $AB$ supports a concentrated load $P$ at the free end (see figure). The cross sections of the beam are rectangular tubes with constant width $b$ and *outer tube* depth $d_A$ at $A$, and *outer tube* depth $d_B = 3d_A/2$ at support $B$. The tube thickness is constant, $t = d_A/20$. $I_A$ is the moment of inertia of the *outer tube* at end $A$ of the beam.

If the moment of inertia of the tube is approximated as $I_a(x)$ as defined, find the *equation* of the deflection curve and the deflection $\delta_A$ at the free end of the beam due to the load $P$.

$$I_a(x) = I_A \times \left(\frac{3}{4} + \frac{10 \times x}{27L}\right)^3 \quad I_A = \frac{b \times d^3}{12}$$

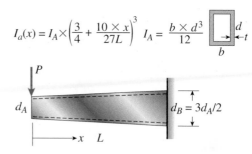

**9.7-11** Repeat Prob. 9.7-10, but now use the tapered propped cantilever tube $AB$, with guided support at $B$, shown in the figure which supports a concentrated load $P$ at the guided end.

Find the equation of the deflection curve and the deflection $\delta_B$ at the guided end of the beam due to the load $P$.

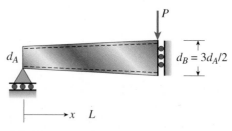

PROB. 9.7-11

**9.7-12** A simple beam $ACB$ is constructed with square cross sections and a double taper (see figure). The depth of the beam at the supports is $d_A$ and at the midpoint is $d_C = 2d_A$. Each half of the beam has length $L$. Thus, the depth $d$ and moment of inertia $I$ at distance $x$ from the left-hand end are, respectively,

$$d = \frac{d_A}{L}(L + x)$$

$$I = \frac{d^4}{12} = \frac{d_A^4}{12L^4}(L + x)^4 = \frac{I_A}{L^4}(L + x)^4$$

in which $I_A$ is the moment of inertia at end $A$ of the beam. (These equations are valid for $x$ between 0 and $L$, that is, for the left-hand half of the beam.)

(a) Obtain equations for the slope and deflection of the left-hand half of the beam due to the uniform load.

(b) From those equations obtain formulas for the angle of rotation $\theta_A$ at support $A$ and the deflection $\delta_C$ at the midpoint.

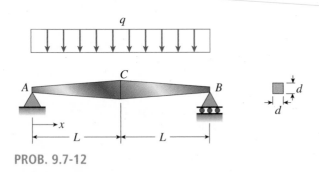

PROB. 9.7-12

## Strain Energy

*The beams described in the problems for Section 9.8 have constant flexural rigidity EI.*

**9.8-1** A uniformly loaded simple beam $AB$ (see figure) of span length $L$ and rectangular cross section ($b$ = width, $h$ = height) has a maximum bending stress $\sigma_{max}$ due to the uniform load.

Determine the strain energy $U$ stored in the beam.

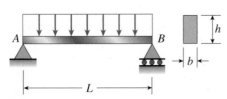

PROB. 9.8-1

**9.8-2** A simple beam $AB$ of length $L$ supports a concentrated load $P$ at the midpoint (see figure).

(a) Evaluate the strain energy of the beam from the bending moment in the beam.

(b) Evaluate the strain energy of the beam from the equation of the deflection curve.

(c) From the strain energy, determine the deflection $\delta$ under the load $P$.

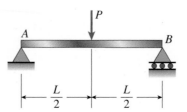

PROB. 9.8-2

**9.8-3** A propped cantilever beam $AB$ of length $L$, and with guided support at $A$, supports a uniform load of intensity $q$ (see figure).

(a) Evaluate the strain energy of the beam from the bending moment in the beam.

(b) Evaluate the strain energy of the beam from the equation of the deflection curve.

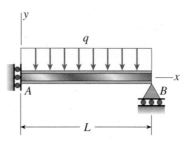

PROB. 9.8-3

**9.8-4** A simple beam $AB$ of length $L$ is subjected to loads that produce a symmetric deflection curve with maximum deflection $\delta$ at the midpoint of the span (see figure).

How much strain energy $U$ is stored in the beam if the deflection curve is (a) a parabola, and (b) a half wave of a sine curve?

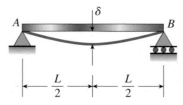

PROB. 9.8-4

**9.8-5** A beam $ABC$ with simple supports at $A$ and $B$ and an overhang $BC$ supports a concentrated load $P$ at the free end $C$ (see figure).

(a) Determine the strain energy $U$ stored in the beam due to the load $P$.

(b) From the strain energy, find the deflection $\delta_C$ under the load $P$.

(c) Calculate the numerical values of $U$ and $\delta_C$ if the length $L$ is 8 ft, the overhang length $a$ is 3 ft, the beam is a W 10 × 12 steel wide-flange section, and the load $P$ produces a maximum stress of 12,000 psi in the beam. (Use $E = 29 \times 10^6$ psi.)

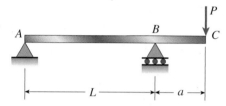

**PROB. 9.8-5**

**9.8-6** A simple beam $ACB$ supporting a uniform load $q$ over the first half of the beam and a couple of moment $M_0$ at end $B$ is shown in the figure.

Determine the strain energy $U$ stored in the beam due to the load $q$ and the couple $M_0$ acting simultaneously.

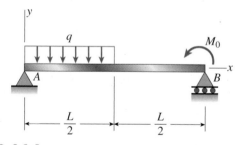

**PROB. 9.8-6**

**9.8-7** The frame shown in the figure consists of a beam $ACB$ supported by a strut $CD$. The beam has length $2L$ and is continuous through joint $C$. A concentrated load $P$ acts at the free end $B$.

Determine the vertical deflection $\delta_B$ at point $B$ due to the load $P$.

*Note*: Let $EI$ denote the flexural rigidity of the beam, and let $EA$ denote the axial rigidity of the strut. Disregard axial and shearing effects in the beam, and disregard any bending effects in the strut.

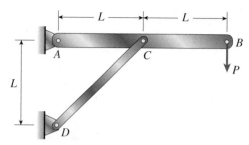

**PROB. 9.8-7**

# Castigliano's Theorem

*The beams described in the problems for Section 9.9 have constant flexural rigidity EI.*

**9.9-1** A simple beam $AB$ of length $L$ is loaded at the left-hand end by a couple of moment $M_0$ (see figure).

Determine the angle of rotation $\theta_A$ at support $A$. (Obtain the solution by determining the strain energy of the beam and then using Castigliano's theorem.)

**PROB. 9.9-1**

**9.9-2** The simple beam shown in the figure supports a concentrated load $P$ acting at distance $a$ from the left-hand support and distance $b$ from the right-hand support.

Determine the deflection $\delta_D$ at point $D$ where the load is applied. (Obtain the solution by determining the strain energy of the beam and then using Castigliano's theorem.)

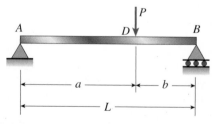

**PROB. 9.9-2**

**9.9-3** An overhanging beam *ABC* supports a concentrated load *P* at the end of the overhang (see figure). Span *AB* has length *L* and the overhang has length *a*.

Determine the deflection $\delta_C$ at the end of the overhang. (Obtain the solution by determining the strain energy of the beam and then using Castigliano's theorem.)

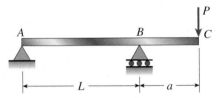

PROB. 9.9-3

**9.9-4** The cantilever beam shown in the figure supports a triangularly distributed load of maximum intensity $q_0$.

Determine the deflection $\delta_B$ at the free end *B*. (Obtain the solution by determining the strain energy of the beam and then using Castigliano's theorem.)

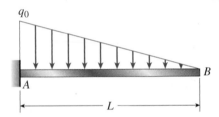

PROB. 9.9-4

**9.9-5** A simple beam *ACB* supports a uniform load of intensity *q* on the left-hand half of the span (see figure).

Determine the angle of rotation $\theta_B$ at support *B*. (Obtain the solution by using the modified form of Castigliano's theorem.)

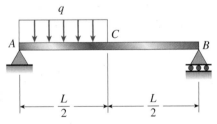

PROB. 9.9-5

**9.9-6** A cantilever beam *ACB* supports two concentrated loads $P_1$ and $P_2$, as shown in the figure. Determine the deflections $\delta_C$ and $\delta_B$ at points *C* and *B*, respectively. (Obtain the solution by using the modified form of Castigliano's theorem.)

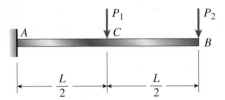

PROB. 9.9-6

**9.9-7** The cantilever beam *ACB* shown in the figure is subjected to a uniform load of intensity *q* acting between points *A* and *C*.

Determine the angle of rotation $\theta_A$ at the free end *A*. (Obtain the solution by using the modified form of Castigliano's theorem.)

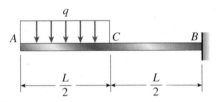

PROB. 9.9-7

**9.9-8** The frame *ABC* supports a concentrated load *P* at point *C* (see figure). Members *AB* and *BC* have lengths *h* and *b*, respectively.

Determine the vertical deflection $\delta_C$ and angle of rotation $\theta_C$ at end *C* of the frame. (Obtain the solution by using the modified form of Castigliano's theorem.)

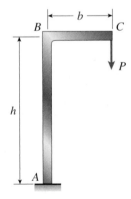

PROB. 9.9-8

**9.9-9** A simple beam *ABCDE* supports a uniform load of intensity *q* (see figure). The moment of inertia in the central part of the beam (*BCD*) is twice the moment of inertia in the end parts (*AB* and *DE*).

Find the deflection $\delta_C$ at the midpoint *C* of the beam. (Obtain the solution by using the modified form of Castigliano's theorem.)

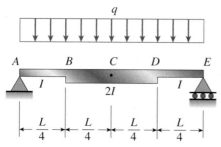

PROB. 9.9-9

**9.9-10** An overhanging beam *ABC* is subjected to a couple $M_A$ at the free end (see figure). The lengths of the overhang and the main span are *a* and *L*, respectively.

Determine the angle of rotation $\theta_A$ and deflection $\delta_A$ at end *A*. (Obtain the solution by using the modified form of Castigliano's theorem.)

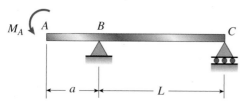

PROB. 9.9-10

**9.9-11** An overhanging beam *ABC* rests on a simple support at *A* and a spring support at *B* (see figure). A concentrated load *P* acts at the end of the overhang. Span *AB* has length *L*, the overhang has length *a*, and the spring has stiffness *k*.

Determine the downward displacement $\delta_C$ of the end of the overhang. (Obtain the solution by using the modified form of Castigliano's theorem.)

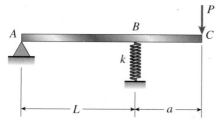

PROB. 9.9-11

**9.9-12** A symmetric beam *ABCD* with overhangs at both ends supports a uniform load of intensity *q* (see figure).

Determine the deflection $\delta_D$ at the end of the overhang. (Obtain the solution by using the modified form of Castigliano's theorem.)

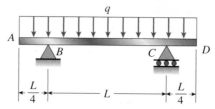

PROB. 9.9-12

# Deflections Produced by Impact

*The beams described in the problems for Section 9.10 have constant flexural rigidity EI. Disregard the weights of the beams themselves, and consider only the effects of the given loads.*

**9.10-1** A heavy object of weight *W* is dropped onto the midpoint of a simple beam *AB* from a height *h* (see figure).

Obtain a formula for the maximum bending stress $\sigma_{max}$ due to the falling weight in terms of *h*, $\sigma_{st}$, and $\delta_{st}$, where $\sigma_{st}$ is the maximum bending stress and $\delta_{st}$ is the deflection at the midpoint when the weight *W* acts on the beam as a statically applied load.

Plot a graph of the ratio $\sigma_{max}/\sigma_{st}$ (that is, the ratio of the dynamic stress to the static stress) versus the ratio $h/\delta_{st}$. (Let $h/\delta_{st}$ vary from 0 to 10.)

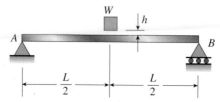

PROB. 9.10-1

**9.10-2** An object of weight *W* is dropped onto the midpoint of a simple beam *AB* from a height *h* (see figure). The beam has a rectangular cross section of area *A*.

Assuming that *h* is very large compared to the deflection of the beam when the weight *W* is applied statically, obtain a formula for the maximum bending stress $\sigma_{max}$ in the beam due to the falling weight.

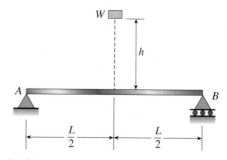

PROB. 9.10-2

**9.10-3** A cantilever beam $AB$ of length $L = 6$ ft is constructed of a W $8 \times 21$ wide-flange section (see figure). A weight $W = 1500$ lb falls through a height $h = 0.25$ in. onto the end of the beam.

Calculate the maximum deflection $\delta_{max}$ of the end of the beam and the maximum bending stress $\sigma_{max}$ due to the falling weight. (Assume $E = 30 \times 10^6$ psi.)

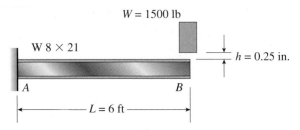

PROB. 9.10-3

**9.10-4** A weight $W = 20$ kN falls through a height $h = 1.0$ mm onto the midpoint of a simple beam of length $L = 3$ m (see figure). The beam is made of wood with square cross section (dimension $d$ on each side) and $E = 12$ GPa.

If the allowable bending stress in the wood is $\sigma_{allow} = 10$ MPa, what is the minimum required dimension $d$?

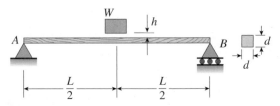

PROB. 9.10-4

**9.10-5** A weight $W = 4000$ lb falls through a height $h = 0.5$ in. onto the midpoint of a simple beam of length $L = 10$ ft (see figure).

Assuming that the allowable bending stress in the beam is $\sigma_{allow} = 18,000$ psi and $E = 30 \times 10^6$ psi, select the lightest wide-flange beam listed in Table F-1 in Appendix F that will be satisfactory.

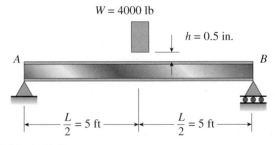

PROB. 9.10-5

**9.10-6** An overhanging beam $ABC$ of rectangular cross section has the dimensions shown in the figure. A weight $W = 750$ N drops onto end $C$ of the beam.

If the allowable normal stress in bending is 45 MPa, what is the maximum height $h$ from which the weight may be dropped? (Assume $E = 12$ GPa.)

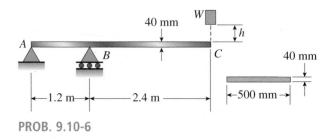

PROB. 9.10-6

**9.10-7** A heavy flywheel rotates at an angular speed $\omega$ (radians per second) around an axle (see figure). The axle is rigidly attached to the end of a simply supported beam of flexural rigidity $EI$ and length $L$ (see figure). The flywheel has mass moment of inertia $I_m$ about its axis of rotation.

If the flywheel suddenly freezes to the axle, what will be the reaction $R$ at support $A$ of the beam?

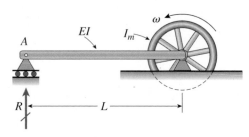

PROB. 9.10-7

# Temperature Effects

*The beams described in the problems for Section 9.11 have constant flexural rigidity EI. In every problem, the temperature varies linearly between the top and bottom of the beam.*

**9.11-1** A simple beam $AB$ of length $L$ and height $h$ undergoes a temperature change such that the bottom of the beam is at temperature $T_2$ and the top of the beam is at temperature $T_1$ (see figure).

Determine the equation of the deflection curve of the beam, the angle of rotation $\theta_A$ at the left-hand support, and the deflection $\delta_{max}$ at the midpoint.

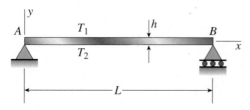

**PROB. 9.11-1**

**9.11-2** A cantilever beam $AB$ of length $L$ and height $h$ (see figure) is subjected to a temperature change such that the temperature at the top is $T_1$ and at the bottom is $T_2$.

Determine the equation of the deflection curve of the beam, the angle of rotation $\theta_B$ at end $B$, and the deflection $\delta_B$ at end $B$.

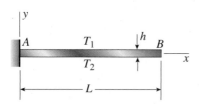

**PROB. 9.11-2**

**9.11-3** An overhanging beam $ABC$ of height $h$ has a guided support at $A$ and a roller at $B$. The beam is heated to a temperature $T_1$ on the top and $T_2$ on the bottom (see figure).

Determine the equation of the deflection curve of the beam, the angle of rotation $\theta_C$ at end $C$, and the deflection $\delta_C$ at end $C$.

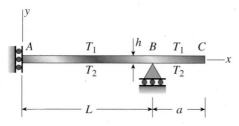

**PROB. 9.11-3**

**9.11-4** A simple beam $AB$ of length $L$ and height $h$ (see figure) is heated in such a manner that the temperature difference $T_2 - T_1$ between the bottom and top of the beam is proportional to the distance from support $A$; that is,

assume the temperature difference varies *linearly* along the beam:

$$T_2 - T_1 = T_0 x$$

in which $T_0$ is a constant having units of temperature (degrees) per unit distance.

(a) Determine the maximum deflection $\delta_{max}$ of the beam.

(b) Repeat for *quadratic* temperature variation along the beam, $T_2 - T_1 = T_0 x^2$.

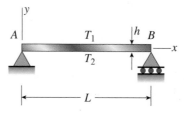

**PROB. 9.11-4**

**9.11-5** Beam $AB$, with elastic support $k_R$ at $A$ and pin support at $B$, of length $L$ and height $h$ (see figure), is heated in such a manner that the temperature difference $T_2 - T_1$ between the bottom and top of the beam is proportional to the distance from support $A$; that is, assume the temperature difference varies *linearly* along the beam:

$$T_2 - T_1 = T_0 x$$

in which $T_0$ is a constant having units of temperature (degrees) per unit distance. Assume the spring at $A$ is unaffected by the temperature change.

(a) Determine the maximum deflection $\delta_{max}$ of the beam.

(b) Repeat for *quadratic* temperature variation along the beam, $T_2 - T_1 = T_0 x^2$.

(c) What is $\delta_{max}$ for parts (a) and (b) if $k_R$ goes to infinity?

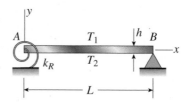

**PROB. 9.11-5**

# 10

# Statically Indeterminate Beams

Large and complex structures, such as this crane which is loading a container ship, are usually statically indeterminate and require a computer to find forces in members and displacements of joints. (© david sanger photography/Alamy)

# CHAPTER OVERVIEW

In Chapter 10, statically indeterminate beams are considered. Here, the beam structure has more unknown reaction forces than available equations of statical equilibrium, so it is said to be **statically indeterminate**. The number of excess unknown reactions defines the **degree of indeterminacy**. Solution of statically indeterminate beams requires that additional equations be developed based on the deformations of the structure, in addition to the equations from statics. First, a number of different types of statically indeterminate beams are defined (Section 10.2), along with some common terminology (e.g., **primary structure, released structure,** and **redundant**) used in representing the solution. Then a solution approach based on integration of the equation of the elastic curve and application of boundary conditions to find unknown constants is presented (Section 10.3). This procedure can only be applied in relatively simple cases, so a more general approach based on superposition is described (Section 10.4) and is applicable to beams undergoing small displacements and behaving in a linearly elastic manner. Here, the **equations of equilibrium** are augmented by the **equations of compatibility**; applied loads and resulting beam deflections are related by the **force-displacement equations** for beams derived in Chapter 9. The general superposition solution approach follows that introduced in Section 2.4 for axially loaded members and in Section 3.8 for circular shafts acted upon by torsional moments. Finally, a number of specialized and advanced topics are introduced at the end of the chapter. In Section 10.5, the effect of **differential temperature loading** is discussed, and in Section 10.6, the effect of **curvature shortening** due to bending alone is presented.

## Chapter 10 is organized as follows:

10.1 Introduction 822
10.2 Types of Statically Indeterminate Beams 822
10.3 Analysis by the Differential Equations of the Deflection Curve 825
10.4 Method of Superposition 832
*10.5 Temperature Effects 845
*10.6 Longitudinal Displacements at the Ends of a Beam 853
Chapter Summary & Review 856
Problems 858
*Advanced topics

## 10.1 INTRODUCTION

In this chapter we will analyze beams in which the number of reactions exceeds the number of independent equations of equilibrium. Since the reactions of such beams cannot be determined by statics alone, the beams are said to be **statically indeterminate**.

The analysis of statically indeterminate beams is quite different from that of statically determinate beams. When a beam is statically determinate, we can obtain all reactions, shear forces, and bending moments from free-body diagrams and equations of equilibrium. Then, knowing the shear forces and bending moments, we can obtain the stresses and deflections.

However, when a beam is statically indeterminate, the equilibrium equations are not sufficient and additional equations are needed. The most fundamental method for analyzing a statically indeterminate beam is to solve the differential equations of the deflection curve, as described later in Section 10.3. Although this method serves as a good starting point in our analysis, it is practical for only the simplest types of statically indeterminate beams.

Therefore, we also discuss the method of superposition (Section 10.4), a method that is applicable to a wide variety of structures. In the method of superposition, we supplement the equilibrium equations with compatibility equations and force-displacement equations. (This same method was described earlier in Section 2.4, where we analyzed statically indeterminate bars subjected to tension and compression.)

In the last part of this chapter we discuss two specialized topics pertaining to statically indeterminate beams, namely, beams with temperature changes (Section 10.5), and longitudinal displacements at the ends of beams (Section 10.6). Throughout this chapter, we assume that the beams are made of **linearly elastic materials**.

Although only statically indeterminate beams are discussed in this chapter, the fundamental ideas have much wider application. Most of the structures we encounter in everyday life, including automobile frames, buildings, and aircraft, are statically indeterminate. However, they are much more complex than beams and must be designed by very sophisticated analytical techniques. Many of these techniques rely on the concepts described in this chapter, and therefore, this chapter may be viewed as an introduction to the analysis of statically indeterminate structures of all kinds.

## 10.2 TYPES OF STATICALLY INDETERMINATE BEAMS

Statically indeterminate beams are usually identified by the arrangement of their supports. For instance, a beam that is fixed at one end and simply supported at the other (Fig. 10-1a) is called a **propped cantilever beam**. The reactions of the beam shown in the figure consist of horizontal and vertical forces at support $A$, a moment at support $A$, and a vertical force at support $B$. Because there are only three independent equations of equilibrium for this beam, it is not possible to calculate all

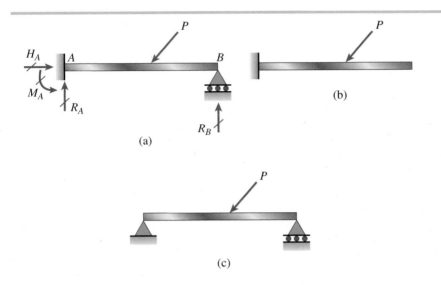

**Fig. 10-1**

Propped cantilever beam:
(a) beam with load and
reactions; (b) released structure
when the reaction at end *B* is
selected as the redundant; and
(c) released structure when the
moment reaction at end *A* is
selected as the redundant

four of the reactions from equilibrium alone. The number of reactions in *excess* of the number of equilibrium equations is called the **degree of static indeterminacy**. Thus, a propped cantilever beam is statically indeterminate to the first degree.

The excess reactions are called **static redundants** and must be selected in each particular case. For example, the reaction $R_B$ of the propped cantilever beam shown in Fig. 10-1a may be selected as the redundant reaction. Since this reaction is in excess of those needed to maintain equilibrium, it can be released from the structure by removing the support at *B*. When support *B* is removed, we are left with a cantilever beam (Fig. 10-1b). The structure that remains when the redundants are released is called the **released structure** or the **primary structure**. The released structure must be stable (so that it is capable of carrying loads), and it must be statically determinate (so that all force quantities can be determined by equilibrium alone).

Another possibility for the analysis of the propped cantilever beam of Fig. 10-1a is to select the reactive moment $M_A$ as the redundant. Then, when the moment restraint at support *A* is removed, the released structure is a simple beam with a pin support at one end and a roller support at the other (Fig. 10-1c).

A special case arises if all loads acting on the beam are vertical (Fig. 10-2). Then the horizontal reaction at support *A* vanishes, and three

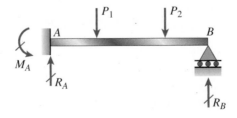

**Fig. 10-2**

Propped cantilever beam
with vertical loads only

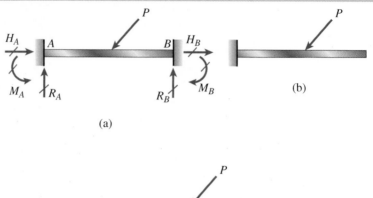

(a)

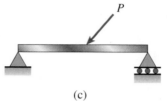

(c)

reactions remain. However, only two independent equations of equilibrium are now available, and therefore the beam is still statically indeterminate to the first degree. If the reaction $R_B$ is chosen as the redundant, the released structure is a cantilever beam; if the moment $M_A$ is chosen, the released structure is a simple beam.

Another type of statically indeterminate beam, known as a **fixed-end beam**, is shown in Fig. 10-3a. This beam has fixed supports at both ends, resulting in a total of six unknown reactions (two forces and a moment at each support). Because there are only three equations of equilibrium, the beam is statically indeterminate to the third degree. (Other names for this type of beam are *clamped beam* and *built-in beam*.)

If we select the three reactions at end *B* of the beam as the redundants, and if we remove the corresponding restraints, we are left with a cantilever beam as the released structure (Fig. 10-3b). If we release the two fixed-end moments and one horizontal reaction, the released structure is a simple beam (Fig. 10-3c).

Again considering the special case of vertical loads only (Fig. 10-4), we find that the fixed-end beam now has only four nonzero reactions (one force and one moment at each support). The number of available equilibrium equations is two, and therefore the beam is statically indeterminate

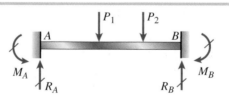

to the second degree. If the two reactions at end *B* are selected as the redundants, the released structure is a cantilever beam; if the two moment reactions are selected, the released structure is a simple beam.

The beam shown in Fig. 10-5a is an example of a **continuous beam**, so called because it has more than one span and is continuous over an interior support. This particular beam is statically indeterminate to the first degree because there are four reactive forces and only three equations of equilibrium.

If the reaction $R_B$ at the interior support is selected as the redundant, and if we remove the corresponding support from the beam, then there remains a released structure in the form of a statically determinate simple beam (Fig. 10-5b). If the reaction $R_C$ is selected as the redundant, the released structure is a simple beam with an overhang (Fig. 10-5c).

In the following sections, we will discuss two methods for analyzing statically indeterminate beams. The objective in each case is to determine the redundant reactions. Once they are known, all remaining reactions (plus the shear forces and bending moments) can be found from equations of equilibrium. In effect, the structure has become statically determinate. Therefore, as the final step in the analysis, the stresses and deflections can be found by the methods described in preceding chapters.

Long-span bridges are often constructed using continuous beams (Lopatinsky Vladislav/ Shutterstock)

## 10.3 ANALYSIS BY THE DIFFERENTIAL EQUATIONS OF THE DEFLECTION CURVE

Statically indeterminate beams may be analyzed by solving any one of the three differential equations of the deflection curve: (1) the second-order equation in terms of the bending moment (Eq. 9-16a), (2) the third-order equation in terms of the shear force (Eq. 9-16b), or (3) the fourth-order equation in terms of the intensity of distributed load (Eq. 9-16c).

The procedure is essentially the same as that for a statically determinate beam (see Sections 9.2, 9.3, and 9.4) and consists of writing the differential equation, integrating to obtain its general solution, and then applying boundary and other conditions to evaluate the unknown quantities. The unknowns consist of the redundant reactions as well as the constants of integration.

The differential equation for a beam may be solved in symbolic terms only when the beam and its loading are relatively simple and uncomplicated. The resulting solutions are in the form of general purpose formulas. However, in more complex situations the differential equations must be solved numerically, using computer programs intended for that purpose. In such cases the results apply only to specific numerical problems.

The following examples illustrate the analysis of statically indeterminate beams by solving the differential equations in symbolic terms.

### Fig. 10-5

Example of a continuous beam: (a) beam with loads and reactions; (b) released structure when the reaction at support *B* is selected as the redundant; and (c) released structure when the reaction at end *C* is selected as the redundant

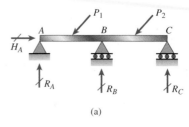

(a)

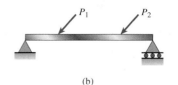

(b)

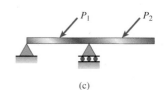

(c)

### Fig. 10-6

Example 10-1: Propped cantilever beam with a uniform load

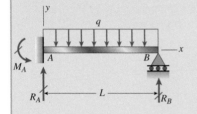

A propped cantilever beam *AB* of length *L* supports a uniform load of intensity *q* (Fig. 10-6). Analyze this beam by solving the second-order differential equation of the deflection curve (the bending-moment equation). Determine the reactions, shear forces, bending moments, slopes, and deflections of the beam.

### Solution

Because the load on this beam acts in the vertical direction (Fig. 10-6), we conclude that there is no horizontal reaction at the fixed support. Therefore, the beam has three unknown reactions ($M_A$, $R_A$, and $R_B$). Only two equations of equilibrium are available for determining these reactions, and therefore, the beam is statically indeterminate to the first degree.

Since we will be analyzing this beam by solving the bending-moment equation, we must begin with a general expression for the moment. This expression will be in terms of both the load and the selected redundant.

*Redundant reaction.* Let us choose the reaction $R_B$ at the simple support as the redundant. Then, by considering the equilibrium of the entire beam, we can express the other two reactions in terms of $R_B$:

$$R_A = qL - R_B \qquad M_A = \frac{qL^2}{2} - R_B L \qquad \text{(a,b)}$$

*Bending moment.* The bending moment $M$ at distance $x$ from the fixed support can be expressed in terms of the reactions as

$$M = R_A x - M_A - \frac{qx^2}{2} \qquad \text{(c)}$$

This equation can be obtained by the customary technique of constructing a free-body diagram of part of the beam and solving an equation of equilibrium.

Substituting into Eq. (c) from Eqs. (a) and (b), we obtain the bending moment in terms of the load and the redundant reaction:

$$M = qLx - R_B x - \frac{qL^2}{2} + R_B L - \frac{qx^2}{2} \qquad \text{(d)}$$

*Differential equation.* The second-order differential equation of the deflection curve (Eq. 9-16a) now becomes

$$EIv'' = M = qLx - R_B x - \frac{qL^2}{2} + R_B L - \frac{qx^2}{2} \qquad \text{(e)}$$

After two successive integrations, we obtain the following equations for the slopes and deflections of the beam:

$$EIv' = \frac{qLx^2}{2} - \frac{R_B x^2}{2} - \frac{qL^2 x}{2} + R_B L_x - \frac{qx^3}{6} + C_1 \qquad \text{(f)}$$

$$EIv = \frac{qLx^3}{6} - \frac{R_B x^3}{6} - \frac{qL^2 x^2}{4} + \frac{R_B Lx^2}{2} - \frac{qx^4}{24} + C_1 x + C_2 \qquad \text{(g)}$$

These equations contain three unknown quantities ($C_1$, $C_2$, and $R_B$).

*Boundary conditions.* Three boundary conditions pertaining to the deflections and slopes of the beam are apparent from an inspection of Fig. 10-6. These conditions are as follows: (1) the deflection at the fixed support is zero, (2) the slope at the fixed support is zero, and (3) the deflection at the simple support is zero. Thus,

$$v(0) = 0 \quad v'(0) = 0 \quad v(L) = 0$$

Applying these conditions to the equations for slopes and deflections given in Eqs. (f) and (g), we find $C_1 = 0$, $C_2 = 0$, and

$$R_B = \frac{3qL}{8} \qquad \text{(10-1)}$$

Thus, the redundant reaction $R_B$ is now known.

*Reactions.* With the value of the redundant established, we can find the remaining reactions from Eqs. (a) and (b). The results are

$$R_A = \frac{5qL}{8} \qquad M_A = \frac{qL^2}{8} \qquad \text{(10-2a,b)}$$

Knowing these reactions, we can find the shear forces and bending moments in the beam.

*Shear forces and bending moments.* These quantities can be obtained by the usual techniques involving free-body diagrams and equations of equilibrium. The results are

$$V = R_A - qx = \frac{5qL}{8} - qx \qquad \text{(10-3)}$$

$$M = R_A x - M_A - \frac{qx^2}{2} = \frac{5qLx}{8} - \frac{qL^2}{8} - \frac{qx^2}{2} \qquad \text{(10-4)}$$

Shear-force and bending-moment diagrams for the beam can be drawn with the aid of these equations (see Fig. 10-7).

From the diagrams, we see that the maximum shear force occurs at the fixed support and is equal to

$$V_{max} = \frac{5qL}{8} \qquad \text{(10-5)}$$

Also, the maximum positive and negative bending moments are

$$M_{pos} = \frac{9qL^2}{128} \qquad M_{neg} = -\frac{qL^2}{8} \qquad \text{(10-6a,b)}$$

Finally, we note that the bending moment is equal to zero at distance $x = L/4$ from the fixed support.

*Slopes and deflections of the beam.* Returning to Eqs. (f) and (g) for the slopes and deflections, we now substitute the values of the constants of

## Fig. 10-7

Shear-force and bending-moment diagrams for the propped cantilever beam of Fig. 10-6

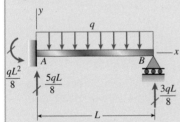

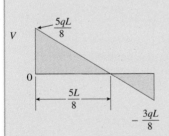

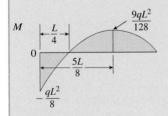

Continues ➡

integration ($C_1 = 0$ and $C_2 = 0$) as well as the expression for the redundant $R_B$ (Eq. 10-1) and obtain

$$v' = \frac{qx}{48EI}(-6L^2 + 15Lx - 8x^2) \qquad \blacktriangleleft \quad (10\text{-}7)$$

$$v = -\frac{qx^2}{48EI}(3L^2 - 5Lx + 2x^2) \qquad \blacktriangleleft \quad (10\text{-}8)$$

The deflected shape of the beam as obtained from Eq. (10-8) is shown in Fig. 10-8.

To determine the maximum deflection of the beam, we set the slope (Eq. 10-7) equal to zero and solve for the distance $x_1$ to the point where this deflection occurs:

$$v' = 0 \quad \text{or} \quad -6L^2 + 15Lx - 8x^2 = 0$$

from which

$$x_1 = \frac{15 - \sqrt{33}}{16}L = 0.5785L \qquad (10\text{-}9)$$

Substituting this value of $x$ into the equation for the deflection (Eq. 10-8) and also changing the sign, we get the maximum deflection:

$$\delta_{max} = -(v)_{x=x_1} = \frac{qL^4}{65,536\,EI}(39 + 55\sqrt{33})$$

$$= \frac{qL^4}{184.6\,EI} = 0.005416\frac{qL^4}{EI} \qquad (10\text{-}10)$$

The point of inflection is located where the bending moment is equal to zero, that is, where $x = L/4$. The corresponding deflection $\delta_0$ of the beam [from Eq. (10-8)] is

$$\delta_0 = -(v)_{x=L/4} = \frac{5qL^4}{2048\,EI} = 0.002441\frac{qL^4}{EI} \qquad (10\text{-}11)$$

Note that when $x < L/4$, both the curvature and the bending moment are negative, and when $x > L/4$, the curvature and bending moment are positive.

To determine the angle of rotation $\theta_B$ at the simply supported end of the beam, we use Eq. (10-7), as

$$\theta_B = (v')_{x=L} = \frac{qL^3}{48EI} \qquad (10\text{-}12)$$

Slopes and deflections at other points along the axis of the beam can be obtained by similar procedures.

*Note:* In this example, we analyzed the beam by taking the reaction $R_B$ (Fig. 10-6) as the redundant reaction. An alternative approach is to take the reactive moment $M_A$ as the redundant. Then we can express the bending moment $M$ in terms of $M_A$, substitute the resulting expression into the second-order differential equation, and solve as before. Still another approach is to begin with the fourth-order differential equation, as illustrated in the next example.

**Fig. 10-8**

Deflection curve for the propped cantilever beam of Fig. 10-6

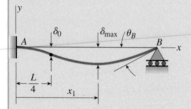

● ● ● **Example 10-2**

The fixed-end beam *ACB* shown in Fig. 10-9 supports a concentrated load *P* at the midpoint. Analyze this beam by solving the fourth-order differential equation of the deflection curve (the load equation). Determine the reactions, shear forces, bending moments, slopes, and deflections of the beam.

### Fig. 10-9

Example 10-2: Fixed-end beam with a concentrated load at the midpoint

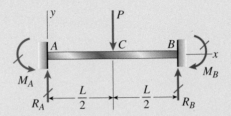

### Solution

Because the load on this beam acts only in the vertical direction, we know that there are no horizontal reactions at the supports. Therefore, the beam has four unknown reactions, two at each support. Since only two equations of equilibrium are available, the beam is statically indeterminate to the second degree.

However, we can simplify the analysis by observing from the symmetry of the beam and its loading that the forces and moments at supports *A* and *B* are equal, that is,

$$R_A = R_B \quad \text{and} \quad M_A = M_B$$

Since the vertical reactions at the supports are equal, we know from equilibrium of forces in the vertical direction that each force is equal to *P*/2:

$$R_A = R_B = \frac{P}{2} \quad \text{(10-13)}$$

Thus, the only unknown quantities that remain are the moment reactions $M_A$ and $M_B$. For convenience, we will select the moment $M_A$ as the redundant quantity.

*Differential equation.* Because there is no load acting on the beam between points *A* and *C*, the fourth-order differential equation (Eq. 9-16c) for the left-hand half of the beam is

$$EIv'''' = -q = 0 \quad (0 < x < L/2) \quad \text{(a)}$$

Successive integrations of this equation yield the following equations, which are valid for the left-hand half of the beam:

$$EIv''' = C_1 \quad \text{(b)}$$

$$EIv'' = C_1x + C_2 \quad \text{(c)}$$

$$EIv' = \frac{C_1x^2}{2} + C_2x + C_3 \quad \text{(d)}$$

$$EIv = \frac{C_1x^3}{6} + \frac{C_2x^2}{2} + C_3x + C_4 \quad \text{(e)}$$

*Continues* ➡

**Example 10-2 -** *Continued*

These equations contain four unknown constants of integration. Since we now have five unknowns ($C_1$, $C_2$, $C_3$, $C_4$, and $M_A$), we need five boundary conditions.

*Boundary conditions.* The boundary conditions applicable to the left-hand half of the beam are as follows:

(1) The shear force in the left-hand segment of the beam is equal to $R_A$, or $P/2$. Therefore, from Eq. (9-16b) we find

$$EIv''' = V = \frac{P}{2}$$

Combining this equation with Eq. (b), we obtain $C_1 = P/2$.

(2) The bending moment at the left-hand support is equal to $-M_A$. Therefore, from Eq. (9-16a) we get

$$EIv'' = M = -M_A \quad \text{at} \quad x = 0$$

Combining this equation with Eq. (c), we obtain $C_2 = -M_A$.

(3) The slope of the beam at the left-hand support ($x = 0$) is equal to zero. Therefore, Eq. (d) yields $C_3 = 0$.

(4) The slope of the beam at the midpoint ($x = L/2$) is also equal to zero (from symmetry). Therefore, from Eq. (d) we find

$$M_A = M_B = \frac{PL}{8} \quad \text{(10-14)}$$

Thus, the reactive moments at the ends of the beam have been determined.

(5) The deflection of the beam at the left-hand support ($x = 0$) is equal to zero. Therefore, from Eq. (e) we find $C_4 = 0$.

In summary, the four constants of integration are

$$C_1 = \frac{P}{2} \quad C_2 = -M_A = -\frac{PL}{8} \quad C_3 = 0 \quad C_4 = 0 \quad \text{(f,g,h,i)}$$

*Shear forces and bending moments.* The shear forces and bending moments can be found by substituting the appropriate constants of integration into Eqs. (b) and (c). The results are

$$EIv''' = V = \frac{P}{2} \quad (0 < x < L/2) \quad \text{(10-15)}$$

$$EIv'' = M = \frac{Px}{2} - \frac{PL}{8} \quad (0 \leq x \leq L/2) \quad \text{(10-16)}$$

Since we know the reactions of the beam, we can also obtain these expressions directly from free-body diagrams and equations of equilibrium.

The shear-force and bending moment diagrams are shown in Fig. 10-10.

### Fig. 10-10

Shear-force and bending-moment diagrams for the fixed-end beam of Fig. 10-9

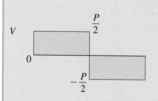

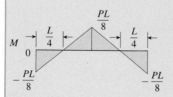

*Slopes and deflections.* The slopes and deflections in the left-hand half of the beam can be found from Eqs. (d) and (e) by substituting the expressions for the constants of integration. In this manner, we find

$$v' = -\frac{Px}{8EI}(L - 2x) \quad (0 \le x \le L/2) \qquad \blacktriangleleft \quad \textbf{(10-17)}$$

$$v = -\frac{Px^2}{48EI}(3L - 4x) \quad (0 \le x \le L/2) \qquad \blacktriangleleft \quad \textbf{(10-18)}$$

The deflection curve of the beam is shown in Fig. 10-11.

To find the maximum deflection $\delta_{max}$ we set $x$ equal to $L/2$ in Eq. (10-18) and change the sign; thus,

$$\delta_{max} = -(v)_{x=L/2} = \frac{PL^3}{192EI} \qquad \textbf{(10-19)}$$

The point of inflection in the left-hand half of the beam occurs where the bending moment $M$ is equal to zero, that is, where $x = L/4$ (see Eq. 10-16). The corresponding deflection $\delta_0$ (from Eq. 10-18) is

$$\delta_0 = -(v)_{x=L/4} = \frac{PL^3}{384EI} \qquad \textbf{(10-20)}$$

which is equal numerically to one-half of the maximum deflection. A second point of inflection occurs in the right-hand half of the beam at distance $L/4$ from end $B$.

*Notes:* As we observed in this example, the number of boundary and other conditions is always sufficient to evaluate not only the constants of integration but also the redundant reactions.

Sometimes it is necessary to set up differential equations for more than one region of the beam and use conditions of continuity between regions, as illustrated in Examples 9-3 and 9-5 of Chapter 9 for statically determinate beams. Such analyses are likely to be long and tedious because of the large number of conditions that must be satisfied. However, if deflections and angles of rotation are needed at only one or two specific points, the method of superposition may be useful (see the next section).

**Fig. 10-11**

Deflection curve for the fixed-end beam of Fig. 10-9

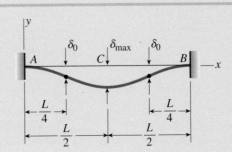

# 10.4 METHOD OF SUPERPOSITION

The method of superposition is of fundamental importance in the analysis of statically indeterminate bars, trusses, beams, frames, and many other kinds of structures. We have already used the superposition method to analyze statically indeterminate structures composed of bars in tension and compression (Section 2.4) and shafts in torsion (Section 3.8). In this section, we will apply the method to beams.

We begin the analysis by noting the degree of static indeterminacy and selecting the redundant reactions. Then, having identified the redundants, we can write **equations of equilibrium** that relate the other unknown reactions to the redundants and the loads.

Next, we assume that both the original loads and the redundants act upon the released structure. Then we find the deflections in the released structure by superposing the separate deflections due to the loads and the redundants. The sum of these deflections must match the deflections in the original beam. However, the deflections in the original beam (at the points where restraints were removed) are either zero or have known values. Therefore, we can write **equations of compatibility** (or *equations of superposition*) expressing the fact that the deflections of the released structure (at the points where restraints were removed) are the same as the deflections in the original beam (at those same points).

Since the released structure is statically determinate, we can easily determine its deflections by using the techniques described in Chapter 9. The relationships between the loads and the deflections of the released structure are called **force-displacement relations**. When these relations are substituted into the equations of compatibility, we obtain equations in which the redundants are the unknown quantities. Therefore, we can solve those equations for the redundant reactions. Then, with the redundants known, we can determine all other reactions from the equations of equilibrium. Furthermore, we can also determine the shear forces and bending moments from equilibrium.

The steps described in general terms in the preceding paragraphs can be made clearer by considering a particular case, namely, a propped cantilever beam supporting a uniform load (Fig. 10-12a). We will make two analyses, the first with the force reaction $R_B$ selected as the redundant and the second with the moment reaction $M_A$ as the redundant. (This same beam was analyzed in Example 10-1 of Section 10.3 by solving the differential equation of the deflection curve.)

## Analysis with $R_B$ as Redundant

In this first illustration we select the reaction $R_B$ at the simple support (Fig. 10-12a) as the redundant. Then the *equations of equilibrium* that express the other unknown reactions in terms of the redundant are as follows.

$$R_A = qL - R_B \qquad M_B = \frac{qL^2}{2} - R_B L \qquad \textbf{(10-21a,b)}$$

These equations are obtained from equations of equilibrium that apply to the entire beam taken as a free body (Fig. 10-12a).

**Fig. 10-12**

Analysis of a propped cantilever beam by the method of superposition with the reaction $R_B$ selected as the redundant

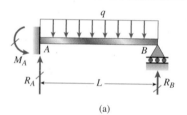

(a)

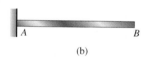

(b)

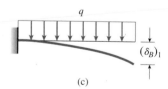

(c)

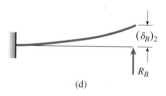

(d)

The next step is to remove the restraint corresponding to the redundant (in this case, we remove the support at end *B*). The *released structure* that remains is a cantilever beam (Fig. 10-12b). The uniform load *q* and the redundant force $R_B$ are now applied as loads on the released structure (Figs. 10-12c and d).

The deflection at end *B* of the released structure due solely to the uniform load is denoted $(\delta_B)_1$, and the deflection at the same point due solely to the redundant is denoted $(\delta_B)_2$. The deflection $\delta_B$ at point *B* in the original structure is obtained by superposing these two deflections. Since the deflection in the original beam is equal to zero, we obtain the following *equation of compatibility*:

$$\delta_B = (\delta_B)_1 - (\delta_B)_2 = 0 \qquad \textbf{(10-22)}$$

The minus sign appears in this equation because $(\delta_B)_1$ is positive downward whereas $(\delta_B)_2$ is positive upward.

The *force-displacement relations* that give the deflections $(\delta_B)_1$ and $(\delta_B)_2$ in terms of the uniform load *q* and the redundant $R_B$, respectively, are found with the aid of Table H-1 in Appendix H (see Cases 1 and 4). Using the formulas given there, we obtain

$$(\delta_B)_1 = \frac{qL^4}{8EI} \qquad (\delta_B)_2 = \frac{R_B L^3}{3EI} \qquad \textbf{(10-23a,b)}$$

Substituting these force-displacement relations into the equation of compatibility yields

$$\delta_B = \frac{qL^4}{8EI} - \frac{R_B L^3}{3EI} = 0 \qquad \textbf{(10-23c)}$$

which can be solved for the *redundant reaction*:

$$R_B = \frac{3qL}{8} \qquad \textbf{(10-24)}$$

Note that this equation gives the redundant in terms of the loads acting on the original beam.

The remaining reactions ($R_A$ and $M_A$) can be found from the equilibrium equations [Eqs. (10-21a and b)]; the results are

$$R_A = \frac{5qL}{8} \qquad M_A = \frac{qL^2}{8} \qquad \textbf{(10-25a,b)}$$

Knowing all reactions, we can now obtain the shear forces and bending moments throughout the beam and plot the corresponding diagrams (see Fig. 10-7 for these diagrams).

We can also determine the *deflections and slopes* of the original beam by means of the principle of superposition. The procedure consists of superposing the deflections of the released structure when acted upon by the loads shown in Figs. 10-12c and d. For instance, the equations of the deflection curves for those two loading systems are obtained from Cases 1 and 4, respectively, of Table H-1, Appendix H:

$$v_1 = -\frac{qx^2}{24EI}(6L^2 - 4Lx + x^2)$$

$$v_2 = \frac{R_B x^2}{6EI}(3L - x)$$

Substituting for $R_B$ from Eq. (10-24) and then adding the deflections $v_1$ and $v_2$, we obtain the following equation for the deflection curve of the original statically indeterminate beam (Fig. 10-12a):

$$v = v_1 + v_2 = -\frac{qx^2}{48EI}(3L^2 - 5Lx + 2x^2)$$

This equation agrees with Eq. (10-8) of Example 10-1. Other deflection quantitites can be found in an analogous manner.

## Analysis with $M_A$ as Redundant

We will now analyze the same propped cantilever beam by selecting the moment reaction $M_A$ as the redundant (Fig. 10-13). In this case, the released structure is a simple beam (Fig. 10-13b). The equations of equilibrium for the reactions $R_A$ and $R_B$ in the original beam are

$$R_A = \frac{qL}{2} + \frac{M_A}{L} \qquad R_B = \frac{qL}{2} - \frac{M_A}{L} \qquad \textbf{(10-26a,b)}$$

**Fig. 10-13**

Analysis of a propped cantilever beam by the method of superposition with the moment reaction $M_A$ selected as the redundant

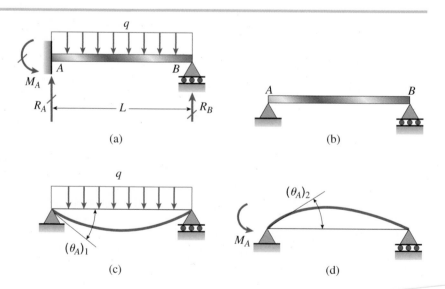

The equation of compatibility expresses the fact that the angle of rotation $\theta_A$ at the fixed end of the original beam is equal to zero. Since this angle is obtained by superposing the angles of rotation $(\theta_A)_1$ and $(\theta_A)_2$ in the released structure (Figs. 10-13c and d), the *compatibility equation* becomes

$$\theta_A = (\theta_A)_1 - (\theta_A)_2 = 0 \qquad \textbf{(10-27a)}$$

In this equation, the angle $(\theta_A)_1$ is assumed to be positive when clockwise and the angle $(\theta_A)_2$ is assumed to be positive when counterclockwise.

The angles of rotation in the released structure are obtained from the formulas given in Table H-2 of Appendix H (see Cases 1 and 7). Thus, the *force-displacement relations* are

$$(\theta_A)_1 = \frac{qL^3}{24EI} \qquad (\theta_A)_2 = \frac{M_A L}{3EI}$$

Substituting into the compatibility equation (Eq. 10-27a), we get

$$\theta_A = \frac{qL^3}{24EI} - \frac{M_A L}{3EI} = 0 \qquad \textbf{(10-27b)}$$

Solving this equation for the redundant, we get $M_A = qL^2/8$, which agrees with the previous result (Eq. 10-25b). Also, the equations of equilibrium [Eqs. (10-26a and b)] yield the same results as before for the reactions $R_A$ and $R_B$ [see Eqs. (10-25a) and (10-24), respectively].

Now that all reactions have been found, we can determine the shear forces, bending moments, slopes, and deflections by the techniques already described.

## General Comments

The method of superposition described in this section is also called the *flexibility method* or the *force method*. The latter name arises from the use of force quantities (forces and moments) as the redundants; the former name is used because the coefficients of the unknown quantities in the compatibility equation [terms such as $L^3/3EI$ in Eq. (10-27a) and $L/3EI$ in Eq. (10-27b)] are *flexibilities* (that is, deflections or angles produced by a unit load).

Since the method of superposition involves the superposition of deflections, it is applicable only to linearly elastic structures. (Recall that this same limitation applies to all topics discussed in this chapter.)

In the following examples, and also in the problems at the end of the chapter, we are concerned primarily with finding the reactions, since this is the key step in the solutions.

A two-span continuous beam *ABC* supports a uniform load of intensity *q*, as shown in Fig. 10-14a. Each span of the beam has length *L*. Using the method of superposition, determine all reactions for this beam.

## Fig. 10-14

Example 10-3: Two-span continuous beam with a uniform load

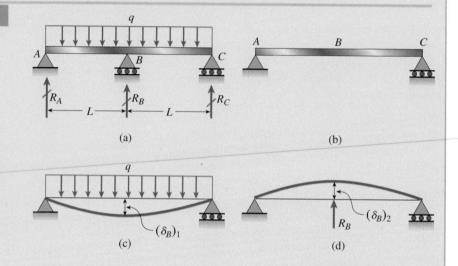

### Solution

This beam has three unknown reactions ($R_A$, $R_B$, and $R_C$). Since there are two equations of equilibrium for the beam as a whole, it is statically indeterminate to the first degree. For convenience, let us select the reaction $R_B$ at the middle support as the redundant.

*Equations of equilibrium.* We can express the reactions $R_A$ and $R_C$ in terms of the redundant $R_B$ by means of two equations of equilibrium. The first equation, which is for equilibrium of moments about point *B*, shows that $R_A$ and $R_C$ are equal. The second equation, which is for equilibrium in the vertical direction, yields the following result:

$$R_A = R_C = qL - \frac{R_B}{2} \tag{a}$$

*Equation of compatibility.* Because the reaction $R_B$ is selected as the redundant, the released structure is a simple beam with supports at *A* and *C* (Fig. 10-14b). The deflections at point *B* in the released structure due to the uniform load *q* and the redundant $R_B$ are shown in Figs. 10-14c and d, respectively. Note that the deflections are denoted $(\delta_B)_1$ and $(\delta_B)_2$. The superposition of these deflections must produce the deflection $\delta_B$ in the original beam at point *B*. Since the latter deflection is equal to zero, the equation of compatibility is

$$\delta_B = (\delta_B)_1 - (\delta_B)_2 = 0 \tag{b}$$

in which the deflection $(\delta_B)_1$ is positive downward and the deflection $(\delta_B)_2$ is positive upward.

*Force-displacement relations.* The deflection $(\delta_B)_1$ caused by the uniform load acting on the released structure (Fig. 10-14c) is obtained from Table H-2, Case 1 as

$$(\delta_B)_1 = \frac{5q(2L)^4}{384EI} = \frac{5qL^4}{24EI}$$

where $2L$ is the length of the released structure. The deflection $(\delta_B)_2$ produced by the redundant (Fig. 10-14d) is

$$(\delta_B)_2 = \frac{R_B(2L)^3}{48EI} = \frac{R_B L^3}{6EI}$$

as obtained from Table H-2, Case 4.

*Reactions.* The equation of compatibility pertaining to the vertical deflection at point $B$ (Eq. b) now becomes

$$\delta_B = \frac{5qL^4}{24EI} - \frac{R_B L^3}{6EI} = 0 \qquad \text{(c)}$$

from which we find the reaction at the middle support:

$$R_B = \frac{5qL}{4} \qquad \Longleftarrow \quad \text{(10-28)}$$

The other reactions are obtained from Eq. (a):

$$R_A = R_C = \frac{3qL}{8} \qquad \Longleftarrow \quad \text{(10-29)}$$

With the reactions known, we can find the shear forces, bending moments, stresses, and deflections without difficulty.

*Note:* The purpose of this example is to provide an illustration of the method of superposition, and therefore we have described all steps in the analysis. However, this particular beam (Fig. 10-14a) can be analyzed by inspection because of the symmetry of the beam and its loading.

From symmetry we know that the slope of the beam at the middle support must be zero, and therefore, each half of the beam is in the same condition as a propped cantilever beam with a uniform load (see, for instance, Fig. 10-6). Consequently, all of our previous results for a propped cantilever beam with a uniform load (Eqs. 10-1 to 10-12) can be adapted immediately to the continuous beam of Fig. 10-14a.

### Fig. 10-15

Example 10-4: Fixed-end beam with a concentrated load

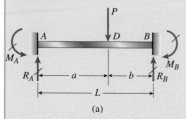

(a)

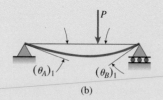

(b)

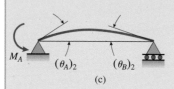

(c)

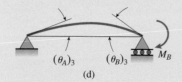

(d)

A fixed-end beam $AB$ (Fig. 10-15a) is loaded by a force $P$ acting at an intermediate point $D$. Find the reactive forces and moments at the ends of the beam using the method of superposition. Also, determine the deflection at point $D$ where the load is applied.

### Solution

This beam has four unknown reactions (a force and a moment at each support), but only two independent equations of equilibrium are available. Therefore, the beam is statically indeterminate to the second degree. In this example, we will select the reactive moments $M_A$ and $M_B$ as the redundants.

*Equations of equilibrium.* The two unknown force reactions ($R_A$ and $R_B$) can be expressed in terms of the redundants ($M_A$ and $M_B$) with the aid of two equations of equilibrium. The first equation is for moments about point $B$, and the second is for moments about point $A$. The resulting expressions are

$$R_A = \frac{Pb}{L} + \frac{M_A}{L} - \frac{M_B}{L} \qquad R_B = \frac{Pa}{L} - \frac{M_A}{L} + \frac{M_B}{L} \qquad \textbf{(a,b)}$$

*Equations of compatibility.* When both redundants are released by removing the rotational restraints at the ends of the beam, we are left with a simple beam as the released structure (Figs. 10-15b, c, and d). The angles of rotation at the ends of the released structure due to the concentrated load $P$ are denoted $(\theta_A)_1$ and $(\theta_B)_1$, as shown in Fig. 10-15b. In a similar manner, the angles at the ends due to the redundant $M_A$ are denoted $(\theta_A)_2$ and $(\theta_B)_2$, and the angles due to the redundant $M_B$ are denoted $(\theta_A)_3$ and $(\theta_B)_3$.

Since the angles of rotation at the supports of the original beam are equal to zero, the two equations of compatibility are

$$\theta_A = (\theta_A)_1 - (\theta_A)_2 - (\theta_A)_3 = 0 \qquad \textbf{(c)}$$

$$\theta_B = (\theta_B)_1 - (\theta_B)_2 - (\theta_B)_3 = 0 \qquad \textbf{(d)}$$

in which the signs of the various terms are determined by inspection from the figures.

*Force-displacement relations.* The angles at the ends of the beam due to the load $P$ (Fig. 10-15b) are obtained from Case 5 of Table H-2:

$$(\theta_A)_1 = \frac{Pab(L + b)}{6LEI} \qquad (\theta_B)_1 = \frac{Pab(L + a)}{6LEI}$$

in which $a$ and $b$ are the distances from the supports to point $D$ where the load is applied.

Also, the angles at the ends due to the redundant moment $M_A$ are (see Case 7 of Table H-2):

$$(\theta_A)_2 = \frac{M_A L}{3EI} \qquad (\theta_B)_2 = \frac{M_A L}{6EI}$$

Similarly, the angles due to the moment $M_B$ are

$$(\theta_A)_3 = \frac{M_B L}{6EI} \qquad (\theta_B)_3 = \frac{M_B L}{3EI}$$

*Reactions.* When the preceding expressions for the angles are substituted into the equations of compatibility (Eqs. c and d), we arrive at two simultaneous equations containing $M_A$ and $M_B$ as unknowns:

$$\frac{M_A L}{3EI} + \frac{M_B L}{6EI} = \frac{Pab(L + b)}{6LEI} \qquad \textbf{(e)}$$

$$\frac{M_A L}{6EI} + \frac{M_B L}{3EI} = \frac{Pab(L + a)}{6LEI} \qquad \textbf{(f)}$$

Solving these equations for the redundants, we obtain

$$M_A = \frac{Pab^2}{L^2} \qquad M_B = \frac{Pa^2 b}{L^2} \qquad \Longleftarrow \textbf{(10-30a,b)}$$

Substituting these expressions for $M_A$ and $M_B$ into the equations of equilibrium (Eqs. a and b), we obtain the vertical reactions:

$$R_A = \frac{Pb^2}{L^3}(L + 2a) \qquad R_B = \frac{Pa^2}{L^3}(L + 2b) \qquad \Longleftarrow \textbf{(10-31a,b)}$$

Thus, all reactions for the fixed-end beam have been determined.

The reactions at the supports of a beam with fixed ends are commonly referred to as **fixed-end moments** and **fixed-end forces**. They are widely used in structural analysis, and formulas for these quantities are listed in engineering handbooks.

*Deflection at point D.* To obtain the deflection at point $D$ in the original fixed-end beam (Fig. 10-15a), we again use the principle of superposition. The deflection at point $D$ is equal to the sum of three deflections: (1) the downward deflection $(\delta_D)_1$ at point $D$ in the released structure due to the load $P$ (Fig. 10-15b); (2) the upward deflection $(\delta_D)_2$ at the same point in the released structure due to the redundant $M_A$ (Fig. 10-15c); and (3) the upward deflection $(\delta_D)_3$ at the same point in the released structure due to the redundant $M_B$ (Fig. 10-15d). This superposition of deflections is expressed by the following equation:

$$\delta_D = (\delta_D)_1 - (\delta_D)_2 - (\delta_D)_3 \qquad \textbf{(g)}$$

in which $\delta_D$ is the downward deflection in the original beam.

*Continues*

● ● ● **Example 10-4** - *Continued*

The deflections appearing in Eq. (g) can be obtained from the formulas given in Table H-2 of Appendix H (see Cases 5 and 7) by making the appropriate substitutions and algebraic simplifications. The results of these manipulations are as follows:

$$(\delta_D)_1 = \frac{Pa^2b^2}{3LEI} \qquad (\delta_D)_2 = \frac{M_Aab}{6LEI}(L + b) \qquad (\delta_D)_3 = \frac{M_Bab}{6LEI}(L + a)$$

Substituting the expressions for $M_A$ and $M_B$ from Eqs. (10-30a and b) into the last two expressions, we get

$$(\delta_D)_2 = \frac{Pa^2b^3}{6L^3EI}(L + b) \qquad (\delta_D)_3 = \frac{Pa^3b^2}{6L^3EI}(L + a)$$

Therefore, the deflection at point $D$ in the original beam, obtained by substituting $(\delta_D)_1$, $(\delta_D)_2$, and $(\delta_D)_3$ into Eq. (g) and simplifying, is

$$\delta_D = \frac{Pa^3b^3}{3L^3EI} \qquad \qquad \blacktriangleleft \textbf{(10-32)}$$

The method described in this example for finding the deflection $\delta_D$ can be used not only to find deflections at individual points, but also to find the equations of the deflection curve.

*Concentrated load acting at the midpoint of the beam.* When the load $P$ acts at the midpoint $C$ (Fig. 10-16), the reactions of the beam [from Eqs. (10-30) and (10-31) with $a = b = L/2$] are

$$M_A = M_B = \frac{PL}{8} \qquad R_A = R_B = \frac{P}{2} \qquad \textbf{(10-33a,b)}$$

Also, the deflection at the midpoint [from Eq. (10-32)] is

$$\delta_C = \frac{PL^3}{192EI} \qquad \qquad \textbf{(10-34)}$$

This deflection is only one-fourth of the deflection at the midpoint of a simple beam with the same load, which shows the stiffening effect of clamping the ends of the beam.

The preceding results for the reactions at the ends and the deflection at the middle (Eqs. 10-32 and 10-33) agree with those found in Example 10-2 by solving the differential equation of the deflection curve [see Eqs. (10-13), (10-14), and (10-19)].

**Fig. 10-16**

Fixed-end beam with a concentrated load acting at the midpoint

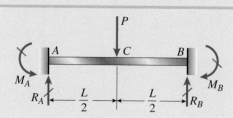

## • • • Example 10-5

A fixed-end beam $AB$ supports a uniform load of intensity $q$ acting over part of the span (Fig. 10-17a). Determine the reactions of this beam (that is, find the fixed-end moments and fixed-end forces).

### Fig. 10-17

Example 10-5: (a) Fixed-end beam with a uniform load over part of the span, and (b) reactions produced by an element $q\,dx$ of the uniform load

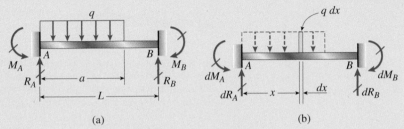

(a)                                    (b)

### Solution

*Procedure.* We can find the reactions of this beam by using the principle of superposition together with the results obtained in the preceding example (Example 10-4). In that example we found the reactions of a fixed-end beam subjected to a concentrated load $P$ acting at distance $a$ from the left-hand end [see Fig. 10-15a and Eqs. (10-30) and (10-31)].

In order to apply those results to the uniform load of Fig. 10-17a, we will treat an element of the uniform load as a concentrated load of magnitude $q\,dx$ acting at distance $x$ from the left-hand end (Fig. 10-17b). Then, using the formulas derived in Example 10-4, we can obtain the reactions caused by this element of load. Finally, by integrating over the length $a$ of the uniform load, we can obtain the reactions due to the entire uniform load.

*Fixed-end moments.* Let us begin with the moment reactions, for which we use Eqs. (10-30a and b) of Example 10-4. To obtain the moments caused by the element $q\,dx$ of the uniform load (compare Fig. 10-17b with Fig. 10-15a), we replace $P$ with $q\,dx$, $a$ with $x$, and $b$ with $L - x$. Thus, the fixed-end moments due to the element of load (Fig. 10-17b) are

$$dM_A = \frac{qx(L-x)^2\,dx}{L^2} \qquad dM_B = \frac{qx^2(L-x)\,dx}{L^2}$$

Integrating over the loaded part of the beam, we get the fixed-end moments due to the entire uniform load:

$$M_A = \int dM_A = \frac{q}{L^2}\int_0^a x(L-x)^2\,dx = \frac{qa^2}{12L^2}(6L^2 - 8aL + 3a^2) \quad \Longleftarrow \quad \textbf{(10-35a)}$$

$$M_B = \int dM_B = \frac{q}{L^2}\int_0^a x^2(L-x)\,dx = \frac{qa^3}{12L^2}(4L - 3a) \quad \Longleftarrow \quad \textbf{(10-35b)}$$

*Fixed-end forces.* Proceeding in a similar manner as for the fixed-end moments, but using Eqs. (10-31a and b), we obtain the following expressions for the fixed-end forces due to the element $q\,dx$ of load:

$$dR_A = \frac{q(L-x)^2(L+2x)\,dx}{L^3} \qquad dR_B = \frac{qx^2(3L-2x)\,dx}{L^3}$$

*Continues* ➡

Integration gives

$$R_A = \int dR_A = \frac{q}{L^3} \int_0^a (L-x)^2(L+2x)dx = \frac{qa}{2L^3}(2L^3 - 2a^2 + a^3) \quad \text{(10-36a)}$$

$$R_B = \int dR_B = \frac{q}{L^3} \int_0^a x^2(3L-2x)dx = \frac{qa^3}{2L^3}(2L-a) \quad \text{(10-36b)}$$

Thus, all reactions (fixed-end moments and fixed-end forces) have been found.

*Uniform load acting over the entire length of the beam.* When the load acts over the entire span (Fig. 10-18), we can obtain the reactions by substituting $a = L$ into the preceding equations, yielding

$$M_A = M_B = \frac{qL^2}{12} \qquad R_A = R_B = \frac{qL}{2} \qquad \text{(10-37a,b)}$$

| **Fig. 10-18** | |
|---|---|
| Fixed-end beam with a uniform load | 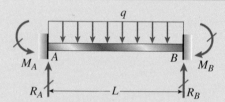 |

The deflection at the midpoint of a uniformly loaded beam is also of interest. The simplest procedure for obtaining this deflection is to use the method of superposition. The first step is to remove the moment restraints at the supports and obtain a released structure in the form of a simple beam. The downward deflection at the midpoint of a simple beam due to a uniform load (from Case 1, Table H-2) is

$$(\delta_C)_1 = \frac{5qL^4}{384EI} \qquad \text{(a)}$$

and the upward deflection at the midpoint due to the end moments (from Case 10, Table H-2) is

$$(\delta_C)_2 = \frac{M_AL^2}{8EI} = \frac{(qL^2/12)L^2}{8EI} = \frac{qL^4}{96EI} \qquad \text{(b)}$$

Thus, the final downward deflection of the original fixed-end beam (Fig. 10-18) is

$$\delta_C = (\delta_C)_1 - (\delta_C)_2$$

Substituting for the deflections from Eqs. (a) and (b), we get

$$\delta_C = \frac{qL^4}{384EI} \qquad \text{(10-38)}$$

This deflection is one-fifth of the deflection at the midpoint of a simple beam with a uniform load (Eq. a), again illustrating the stiffening effect of fixity at the ends of the beam.

A beam *ABC* (Fig. 10-19a) rests on simple supports at points *A* and *B* and is supported by a cable at point *C*. The beam has total length 2*L* and supports a uniform load of intensity *q*. Prior to the application of the uniform load, there is no force in the cable nor is there any slack in the cable.

When the uniform load is applied, the beam deflects downward at point *C* and a tensile force *T* develops in the cable. Find the magnitude of this force.

**Fig. 10-19**

Example 10-6: Beam *ABC* with one end supported by a cable

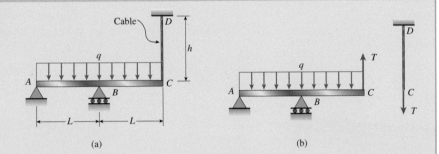

(a)          (b)

**Solution**

*Redundant force.* The structure *ABCD*, consisting of the beam and cable, has three vertical reactions (at points *A*, *B*, and *D*). However, only two equations of equilibrium are available from a free-body diagram of the entire structure. Therefore, the structure is statically indeterminate to the first degree, and we must select one redundant quantity for purposes of analysis.

The tensile force *T* in the cable is a suitable choice for the redundant. We can release this force by removing the connection at point *C*, thereby cutting the structure into two parts (Fig. 10-19b). The released structure consists of the beam *ABC* and the cable *CD* as separate elements, with the redundant force *T* acting upward on the beam and downward on the cable.

*Equation of compatibility.* The deflection at point *C* of beam *ABC* (Fig. 10-19b) consists of two parts, a downward deflection $(\delta_C)_1$ due to the uniform load and an upward deflection $(\delta_C)_2$ due to the force *T*. At the same time, the lower end *C* of cable *CD* displaces downward by an amount $(\delta_C)_3$, equal to the elongation of the cable due to the force *T*. Therefore, the *equation of compatibility*, which expresses the fact that the downward deflection of end *C* of the beam is equal to the elongation of the cable, is

$$(\delta_C)_1 - (\delta_C)_2 = (\delta_C)_3 \tag{a}$$

Having formulated this equation, we now turn to the task of evaluating all three displacements.

*Force-displacement relations.* The deflection $(\delta_C)_1$ at the end of the overhang (point *C* in beam *ABC*) due to the uniform load can be found from the results given in Example 9-9 of Section 9.5 (see Fig. 9-21). Using Eq. (9-68) of that example, and substituting *a* = *L*, we get

$$(\delta_C)_1 = \frac{qL^4}{4E_b I_b} \tag{b}$$

*Continues*

**Example 10-6** - *Continued*

where $E_b I_b$ is the flexural rigidity of the beam.

The deflection of the beam at point C due to the force T can be taken from the answer to Prob. 9.8-5 or Prob. 9.9-3. Those answers give the deflection $(\delta_C)_2$ at the end of the overhang when the length of the overhang is a:

$$(\delta_C)_2 = \frac{Ta^2(L + a)}{3E_b I_b}$$

Now substituting $a = L$, we obtain the desired deflection:

$$(\delta_C)_2 = \frac{2TL^3}{3E_b I_b} \tag{c}$$

Finally, the elongation of the cable is

$$(\delta_C)_3 = \frac{Th}{E_c A_c} \tag{d}$$

where h is the length of the cable and $E_c A_c$ is its axial rigidity.

*Force in the cable.* By substituting the three displacements (Eqs. b, c, and d) into the equation of compatibility (Eq. a), we get

$$\frac{qL^4}{4E_b I_b} - \frac{2TL^3}{3E_b I_b} = \frac{Th}{E_c A_c}$$

Solving for the force T, we find

$$T = \frac{3qL^4 E_c A_c}{8L^3 E_c A_c + 12hE_b I_b} \quad \longleftarrow \tag{10-39}$$

With the force T known, we can find all reactions, shear forces, and bending moments by means of free-body diagrams and equations of equilibrium.

This example illustrates how an internal force quantity (instead of an external reaction) can be used as the redundant.

**Fig. 10-19 (Repeated)**

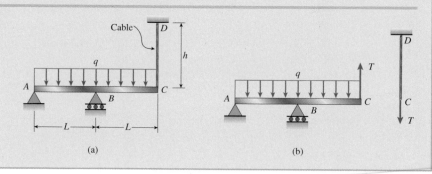

(a)     (b)

# *10.5 TEMPERATURE EFFECTS

Temperature changes may produce changes in lengths of bars and lateral deflections of beams as discussed previously in Sections 2.5 and 9.13. If these length changes and lateral deflections are restrained, thermal stresses will be produced in the material. In Section 2.5 we saw how to find these stresses in statically indeterminate bars, and now we will consider some of the effects of temperature changes in statically indeterminate beams.

The stresses and deflections produced by temperature changes in a statically indeterminate beam can be analyzed by methods that are similar to those already described for the effects of loads. To begin the discussion, consider the propped cantilever beam $AB$ shown in Fig. 10-20. We assume that the beam was originally at a uniform temperature $T_0$, but later its temperature is increased to $T_1$ on the upper surface and $T_2$ on the lower surface. The variation of temperature over the height $h$ of the beam is assumed to be linear.

Because the temperature varies linearly, the *average* temperature of the beam is

$$T_{\text{aver}} = \frac{T_1 + T_2}{2} \tag{10-40}$$

## Fig. 10-20

Propped cantilever beam with a temperature differential

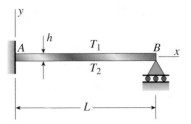

and occurs at midheight of the beam. The difference between this average temperature and the initial temperature $T_0$ results in a tendency for the beam to change in length. If the beam is free to expand longitudinally, its length will increase by an amount $\delta_T$ given by Eq. (9-134), which is repeated here:

$$\delta_T = \alpha(T_{\text{aver}} - T_0)L = \alpha\left(\frac{T_1 + T_2}{2} - T_0\right)L \tag{10-41}$$

In this equation, $\alpha$ is the coefficient of thermal expansion of the material and $L$ is the length of the beam. If longitudinal expansion is free to occur, no axial stresses will be produced by the temperature changes. However, if longitudinal expansion is restrained, axial stresses will develop, as described in Section 2.5.

Now consider the effects of the temperature differential $T_2 - T_1$, which tends to produce a *curvature* of the beam but no change in length. Curvature due to temperature changes is described in Section 9.11, where the following differential equation of the deflection curve is derived [see Eq. (9-136)]:

$$\frac{d^2v}{dx^2} = \frac{\alpha(T_2 - T_1)}{h} \tag{10-42}$$

This equation applies to a beam that is unrestrained by supports and therefore is free to deflect and rotate. Note that when $T_2$ is greater than $T_1$, the curvature is positive and the beam tends to bend concave upward. Deflections and rotations of simple beams and cantilever beams due to a temperature differential can be determined with the aid of Eq. (10-42), as discussed in Section 9.11. We can now use those results when analyzing statically indeterminate beams by the method of superposition.

## Method of Superposition

To illustrate the use of superposition, let us determine the reactions of the fixed-end beam of Fig. 10-21a due to the temperature differential. As usual, we begin the analysis by selecting the redundant reactions. Although other choices result in more efficient calculations, we will select the reactive force $R_B$ and reactive moment $M_B$ as the redundants in order to illustrate the general methodology.

**Fig. 10-21**

(a) Fixed-end beam with a temperature differential, (b) released structure, and (c) deflection curve for the released structure

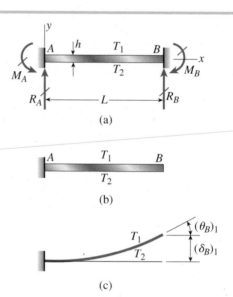

When the supports corresponding to the redundants are removed, we obtain the released structure shown in Fig. 10-21b (a cantilever beam). The deflection and angle of rotation at end $B$ of this cantilever (due to the temperature differential) are as follows (see Fig. 10-21c):

$$(\delta_B)_1 = \frac{\alpha(T_2 - T_1)L^2}{2h} \qquad (\theta_B)_1 = \frac{\alpha(T_2 - T_1)L}{h}$$

These equations are obtained from the solution to Prob. 9.11-2 in the preceding chapter. Note that when $T_2$ is greater than $T_1$, the deflection $(\delta_B)_1$ is upward and the angle of rotation $(\theta_B)_1$ is counterclockwise.

Next, we need to find the deflections and angles of rotation in the released structure (Fig. 10-21b) due to the redundants $R_B$ and $M_B$. These quantities are obtained from Cases 4 and 6, respectively, of Table H-1:

$$(\delta_B)_2 = \frac{R_B L^3}{3EI} \qquad (\theta_B)_2 = \frac{R_B L^2}{2EI}$$

$$(\delta_B)_3 = -\frac{M_B L^2}{2EI} \qquad (\theta_B)_3 = -\frac{M_B L}{EI}$$

In these expressions, upward deflection and counterclockwise rotation are positive (as in Fig. 10-21c).

We can now write the equations of compatibility for the deflection and angle of rotation at support $B$ as follows:

$$\delta_B = (\delta_B)_1 + (\delta_B)_2 + (\delta_B)_3 = 0 \qquad \textbf{(10-43a)}$$

$$\theta_B = (\theta_B)_1 + (\theta_B)_2 + (\theta_B)_3 = 0 \qquad \textbf{(10-43b)}$$

or, upon substituting the appropriate expressions,

$$\frac{\alpha(T_2 - T_1)L^2}{2h} + \frac{R_B L^3}{3EI} - \frac{M_B L^2}{2EI} = 0 \qquad \textbf{(10-43c)}$$

$$\frac{\alpha(T_2 - T_1)L}{h} + \frac{R_B L^2}{2EI} - \frac{M_B L}{EI} = 0 \qquad \textbf{(10-43d)}$$

These equations can be solved simultaneously for the two redundants:

$$R_B = 0 \qquad M_B = \frac{\alpha EI(T_2 - T_1)}{h}$$

The fact that $R_B$ is zero could have been anticipated initially from the symmetry of the fixed-end beam. If we had utilized this fact from the outset, the preceding solution would have been simplified because only one equation of compatibility would have been required.

We also know from symmetry (or from equations of equilibrium) that the reaction $R_B$ is equal to the reaction $R_A$ and that the moment $M_A$ is equal to the moment $M_B$. Therefore, the reactions for the fixed-end beam shown in Fig. 10-21a are as follows:

$$R_A = R_B = 0 \qquad M_A = M_B = \frac{\alpha EI(T_2 - T_1)}{h} \qquad \textbf{(10-44a,b)}$$

From these results we see that the beam is subjected to a constant bending moment due to the temperature changes.

## Differential Equation of the Deflection Curve

We can also analyze the fixed-end beam of Fig. 10-21a by solving the differential equation of the deflection curve. When a beam is subjected to both a bending moment $M$ and a temperature differential $T_2 - T_1$, the differential equation becomes [see Eqs. (9-11) and (10-42)]:

$$\frac{d^2v}{dx^2} = \frac{M}{EI} + \frac{\alpha(T_2 - T_1)}{h} \qquad \textbf{(10-45a)}$$

or

$$EIv'' = M + \frac{\alpha EI(T_2 - T_1)}{h} \qquad \textbf{(10-45b)}$$

For the fixed-end beam of Fig. 10-21a, the expression for the bending moment in the beam is

$$M = R_A x - M_A \qquad \textbf{(10-46)}$$

where $x$ is measured from support $A$. Substituting into the differential equation and integrating, we obtain the following equation for the slope of the beam:

$$EIv' = \frac{R_A x^2}{2} - M_A x + \frac{\alpha EI(T_2 - T_1)x}{h} + C_1 \qquad \textbf{(10-47)}$$

The two boundary conditions on the slope ($v' = 0$ when $x = 0$ and $x = L$) give $C_1 = 0$ and

$$\frac{R_A L}{2} - M_A = -\frac{\alpha EI(T_2 - T_1)}{h} \qquad \textbf{(10-48)}$$

A second integration gives the deflection of the beam:

$$EIv = \frac{R_A x^3}{6} - \frac{M_A x^2}{2} + \frac{\alpha EI(T_2 - T_1)x^2}{2h} + C_2 \qquad \textbf{(10-49)}$$

The boundary conditions on the deflection ($v = 0$ when $x = 0$ and $x = L$) give $C_2 = 0$ and

$$\frac{R_A L}{3} - M_A = -\frac{\alpha EI(T_2 - T_1)}{h} \qquad \textbf{(10-50)}$$

Solving simultaneously Eqs. (g) and (i), we find

$$R_A = 0 \qquad M_A = \frac{\alpha EI(T_2 - T_1)}{h}$$

From the equilibrium of the beam, we obtain $R_B = 0$ and $M_B = M_A$. Thus, these results agree with those found by the method of superposition [see Eqs. (10-44a and b)].

Note that we carried out the preceding solution without taking advantage of symmetry because we wished to illustrate the general approach of the integration method.

Knowing the reactions of the beam, we can now find the shear forces, bending moments, slopes, and deflections. The simplicity of the results may surprise you.

The two-span beam *ABC* in Fig. 10-22 below has a pin support at *A*, a roller support at *B*, and *either* a roller or elastic spring support (spring constant *k*) at *C*. The beam has a height of *h* and is subjected to a temperature differential with temperature $T_1$ on its upper surface and $T_2$ on its lower surface (see Fig. 10-22a and b). Assume that the elastic spring is unaffected by the temperature change.

(a) If support *C* is a roller support, find all support reactions using the *method of superposition*.
(b) Find all support reactions if the roller at *C* is replaced by the *elastic spring support*; also find the displacement at *C*.

### Fig. 10-22

Example 10-7: Two-span beam subject to differential temperature change

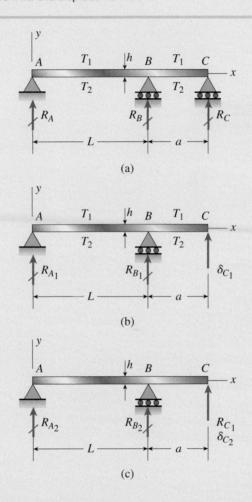

(a)

(b)

(c)

### Solution

(a) *Roller support at C.* This beam (Fig. 10-22a) is statically indeterminate to the first degree (see discussion in Example 10-3 solution). We select reaction $R_C$ as the redundant which allows us to use the analyses of the released structure (with the support at *C* removed) presented in Examples 9-5 (concentrated load applied at *C*) and 9-19 (subject to temperature differential). We will use the method of superposition, also known as the *force* or *flexibility* method, to find the solution.

*Continues* ➡

• • • **Example 10-7** - *Continued*

*Superposition.* The superposition process is shown in the Figs. 10-22b and 10-22c in which redundant $R_C$ is removed to produce a *released* (or statically determinate) structure. We first apply the "actual loads" [here, temperature differential $(T_2 - T_1)$], and then *we apply the redundant as a load* to the second released structure.

*Equilibrium.* Summing forces in the y direction in Fig. 10-22a (using a statics sign convention in which forces in the positive y direction are positive), we find that

$$R_A + R_B = -R_C \tag{a}$$

Summing moments about *B* (again using a statics sign convention in which counterclockwise is positive), we find

$$-R_A L + R_C a = 0$$

so

$$R_A = \left(\frac{a}{L}\right) R_C \tag{b}$$

which can be substituted back into Eq. (a) to give

$$R_B = -R_C - \left(\frac{a}{L}\right) R_C = -R_C \left(1 + \frac{a}{L}\right) \tag{c}$$

(Note that we could also find reactions $R_A$ and $R_B$ using superposition of the reactions shown in Figs. 10-22b and c, as follows: $R_A = R_{A1} + R_{A2}$ and $R_B = R_{B1} + R_{B2}$, where $R_{A1}$ and $R_{B1}$ are known to be zero.)

*Compatibility.* Displacement $\delta_C = 0$ in the actual structure (Fig. 10-22a), so compatibility of displacements requires that

$$\delta_{C1} + \delta_{C2} = \delta_C = 0 \tag{d}$$

where $\delta_{C1}$ and $\delta_{C2}$ are shown in Figs. 10-22b and 10-22c for the released structures subject to temperature differential and applied redundant force $R_C$, respectively. Initially, $\delta_{C1}$ and $\delta_{C2}$ are assumed positive (upward) when using a statics sign convention, and a negative result indicates that the reverse is true.

*Force-displacement and temperature-displacement relations.* We can now use the results of Example 9-5 and 9-19 to find displacements $\delta_{C1}$ and $\delta_{C2}$. First, from Eq. (f) in Example 9-19, we see that

$$\delta_{C1} = \frac{\alpha(T_2 - T_1)a(L + a)}{2h} \tag{e}$$

and from Eq. (9-55) (modified to include variable *a* as the length of member *BC* and replacing load *P* with redundant force $R_C$:

$$\delta_{C2} = \frac{R_C a^2(L + a)}{3EI} \tag{f}$$

*Reactions.* We can now substitute Eqs. (e) and (f) into Eq. (d) and then solve for redundant $R_C$:

$$\frac{\alpha(T_2 - T_1)}{2h}(a)(L + a) + \frac{R_C a^2(L + a)}{3EI} = 0$$

so

$$R_C = \frac{-3EI\alpha(T_2 - T_1)}{2ah} \quad \Longleftarrow \textbf{(g)}$$

noting that the negative result means that reaction force $R_C$ is downward [for positive temperature differential $(T_2 - T_1)$]. The result for $R_C$ is now substituted into Eqs. (b) and (c) to find reactions $R_A$ and $R_B$ as

$$R_A = \left(\frac{a}{L}\right)R_C = \left(\frac{a}{L}\right)\left[\frac{-3EI\alpha(T_2 - T_1)}{2ah}\right] = \frac{-3EI\alpha(T_2 - T_1)}{2Lh} \quad \Longleftarrow \textbf{(h)}$$

$$R_B = -R_C\left(1 + \frac{a}{L}\right) = \frac{3EI\alpha(T_2 - T_1)}{2ah}\left(1 + \frac{a}{L}\right)$$

$$= \frac{3EI\alpha(T_2 - T_1)(L + a)}{2Lah} \quad \Longleftarrow \textbf{(i)}$$

where $R_A$ acts downward and $R_B$ acts upward.

*Numerical example.* In Example 9-19, we computed the upward displacement at joint $C$ [see Eq. (h), Example 9-19] assuming that beam $ABC$ is a steel wide flange, W 30 × 211 (see Table F-1a), with a length $L = 30$ ft, an overhang $a = L/2$, and is subject to temperature differential $(T_2 - T_1) = 5°$ Fahrenheit. From Table I-4, the coefficient of thermal expansion for structural steel is $\alpha = 6.5 \times 10^{-6}/°F$. The modulus for steel is 30,000 ksi. Now we can find numerical values of reactions $R_A$, $R_B$, and $R_C$ using Eqs. (g), (h), and (i):

$$R_A = \frac{-3EI\alpha(T_2 - T_1)}{2Lh} = \frac{-3(30,000 \text{ ksi})(10,300 \text{ in.}^4)(6.5 \times 10^{-6})(5)}{2(360 \text{ in.})(30 \text{ in.})}$$

$$= -1.395 \text{ kip (downward)}$$

$$R_B = \frac{3EI\alpha(T_2 - T_1)(L + a)}{2Lah}$$

$$= \frac{3(30,000 \text{ ksi})(10,300 \text{ in.}^4)(6.5 \times 10^{-6})(5)(360 \text{ in.} + 180 \text{ in.})}{2(360 \text{ in.})(180 \text{ in.})(30 \text{ in.})}$$

$$= 4.18 \text{ kip (upward)}$$

$$R_C = \frac{-3EI\alpha(T_2 - T_1)}{2ah} = \frac{-3(30,000 \text{ ksi})(10,300 \text{ in.}^4)(6.5 \times 10^{-6})(5)}{2(180 \text{ in.})30 \text{ in.}}$$

$$= -2.79 \text{ kip (downward)}$$

We note that the reactions sum to zero as required for equilibrium.

(b) *Spring support at C.* Once again, we select reaction $R_C$ as the redundant. However, $R_C$ is now at the base of the elastic spring support. When redundant reaction $R_C$ is applied to the second released structure, it will first compress the spring and then be applied to the beam at $C$, causing upward deflection.

*Superposition.* The superposition solution approach (i.e., force or flexibility method) follows that used previously and is shown in Fig. 10-23.

### Fig. 10-23

Example 10-7: Two-span beam with elastic support and subject to differential temperature change

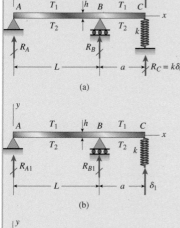

(a)

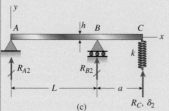

(b)

(c)

Continues ➡

**Example 10-7 -** *Continued*

*Equilibrium.* The addition of the spring support at C does not alter the expressions of static equilibrium in Eqs. (a), (b), and (c).

*Compatibility.* The compatibility equation is now written for the *base of the spring* (not the top of the spring), where it is attached to the beam at C. From Fig. 10-23, we see that compatibility of displacements requires:

$$\delta_1 + \delta_2 = \delta = 0 \tag{j}$$

*Force-displacement and temperature-displacement relations.* The spring is assumed to be unaffected by the differential change, so we conclude that the top and base of the spring displace the same in Fig. 10-23b, which means that Eq. (e) is still valid and $\delta_1 = \delta_{C1}$. However, we must include the compression of the spring in the expression for $\delta_2$, so we get

$$\delta_2 = \frac{R_C}{k} + \delta_{C2} = \frac{R_C}{k} + \frac{R_C a^2(L + a)}{3EI} \tag{k}$$

where the expression for $\delta_{C2}$ comes from Eq. (f).

*Reactions.* We can now substitute Eqs. (e) and (k) into compatibility with Eq. (j) and then solve for redundant $R_C$:

$$\frac{\alpha(T_2 - T_1)}{2h}(a)(L + a) + \frac{R_C a^2(L + a)}{3EI} + \frac{R_C}{k} = 0$$

so

$$R_C = \frac{-a\alpha(T_2 - T_1)(L + a)}{2h\left[\dfrac{1}{k} + \dfrac{a^2(L + a)}{3EI}\right]} \tag{l}$$

From statics [Eqs. (b), (c)], we find:

$$R_A = \left(\frac{a}{L}\right)R_C = \frac{-a\alpha(T_2 - T_1)a(L + a)}{2Lh\left[\dfrac{1}{k} + \dfrac{a^2(L + a)}{3EI}\right]} \tag{m}$$

$$R_B = -R_C\left(1 + \frac{a}{L}\right) = \frac{a\alpha(T_2 - T_1)(L + a)^2}{2Lh\left[\dfrac{1}{k} + \dfrac{a^2(L + a)}{3EI}\right]} \tag{n}$$

Once again, the minus signs for $R_A$ and $R_C$ indicate that they are downward [for positive $(T_2 - T_1)$], while $R_B$ is upward. Finally, if spring constant $k$ goes to infinity, the support at C is once again a roller support, as in Fig. 10-22, and Eqs. (l), (m), and (n) reduce to Eqs. (g), (h), and (i).

# *10.6 LONGITUDINAL DISPLACEMENTS AT THE ENDS OF A BEAM

When a beam is bent by lateral loads, the ends of the beam move closer together. It is common practice to disregard these longitudinal displacements because usually they have no noticeable effect on the behavior of the beam. In this section, we will show how to evaluate these displacements and determine whether or not they are important.

Consider a simple beam $AB$ that is pin-supported at one end and free to displace longitudinally at the other (Fig. 10-24a). When this beam is bent by lateral loads, the deflection curve has the shape shown in part (b) of the figure. In addition to the lateral deflections, there is a longitudinal displacement at end $B$ of the beam. End $B$ moves horizontally from point $B$ to point $B'$ through a small distance $\lambda$, called the **curvature shortening** of the beam.

As the name implies, curvature shortening is due to bending of the axis of the beam and is *not* due to axial strains produced by tensile or compressive forces. As we see from Fig. 10-24b, the curvature shortening is equal to the difference between the initial length $L$ of the straight beam and the length of the chord $AB'$ of the bent beam. Of course, both the lateral deflections and the curvature shortening are highly exaggerated in the figure.

## Curvature Shortening

To determine the curvature shortening, we begin by considering an element of length $ds$ measured along the curved axis of the beam (Fig. 10-24b). The projection of this element on the horizontal axis has length $dx$. The relationship between the length of the element and the length of its horizontal projection is obtained from the Pythagorean theorem:

$$(ds)^2 = (dx)^2 + (dv)^2$$

where $dv$ is the increment in the deflection $v$ of the beam as we move through the distance $dx$. Thus,

$$ds = \sqrt{(dx)^2 + (dv)^2} = dx\sqrt{1 + \left(\frac{dv}{dx}\right)^2} \qquad \textbf{(10-51a)}$$

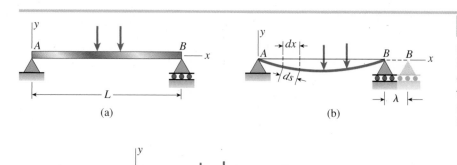

**Fig. 10-24**

(a) Simple beam with lateral loads, (b) horizontal displacement $\lambda$ at the end of the beam, and (c) horizontal reactions $H$ for a beam with immovable supports

The difference between the length of the element and the length of its horizontal projection is

$$ds - dx = dx\sqrt{1 + \left(\frac{dv}{dx}\right)^2} - dx = dx\left[\sqrt{1 + \left(\frac{dv}{dx}\right)^2} - 1\right] \quad \textbf{(10-51b)}$$

Now let us introduce the following binomial series (see Appendix D):

$$\sqrt{1 + t} = 1 + \frac{t}{2} - \frac{t^2}{8} + \frac{t^2}{16} - \cdots \quad \textbf{(10-52)}$$

which converges when $t$ is numerically less than 1. If $t$ is very small compared to 1, we can disregard the terms involving $t^2$, $t^3$, and so on, in comparison with the first two terms. Then we obtain

$$\sqrt{1 + t} \approx 1 + \frac{t}{2} \quad \textbf{(10-53)}$$

The term $(dv/dx)^2$ in Eq. (10-51b) is ordinarily very small compared to 1. Therefore, we can use Eq. (10-53) with $t = (dv/dx)^2$ and rewrite Eq. (10-51b) as

$$ds - dx = dx\left[1 + \frac{1}{2}\left(\frac{dv}{dx}\right)^2 - 1\right] = \frac{1}{2}\left(\frac{dv}{dx}\right)^2 dx \quad \textbf{(10-54)}$$

If the left- and right-hand sides of this expression are integrated over the length of the beam, we obtain an expression for the difference between the length of the beam and the length of the chord $AB'$ (Fig. 10-24b):

$$L - \overline{AB'} = \int_0^L \frac{1}{2}\left(\frac{dv}{dx}\right)^2 dx$$

Thus, the curvature shortening is

$$\lambda = \frac{1}{2}\int_0^L \left(\frac{dv}{dx}\right)^2 dx \quad \textbf{(10-55)}$$

This equation is valid provided the deflections and slopes are small.

Note that when the equation of the deflection curve is known, we can substitute into Eq. (10-55) and determine the shortening $\lambda$.

## Horizontal Reactions

Now suppose that the ends of the beam are prevented from translating longitudinally by immovable supports (Fig. 10-24c). Because the ends cannot move toward each other, a horizontal reaction $H$ will develop at each end. This force will cause the axis of the beam to elongate as bending occurs.

In addition, the force $H$ itself will have an effect upon the bending moments in the beam, because an additional bending moment (equal to $H$ times the deflection) will exist at every cross section. Thus, the deflection curve of the beam depends not only upon the lateral loads but also upon the reaction $H$, which in turn depends upon the shape of the deflection curve, as shown by Eq. (10-55).

Rather than attempt an exact analysis of this complicated problem, let us obtain an approximate expression for the force $H$ in order to ascertain its importance. For that purpose, we can use any reasonable approximation to the deflection curve. In the case of a pin-ended beam with downward loads (Fig. 10-24c), a good approximation is a parabola having the equation

$$v = -\frac{4\delta x(L - x)}{L^2} \qquad \textbf{(10-56)}$$

where $\delta$ is the downward deflection at the midpoint of the beam. The curvature shortening $\lambda$ corresponding to this assumed deflected shape can be found by substituting the expression for the deflection $v$ into Eq. (10-55) and integrating; the result is

$$\lambda = \frac{8\delta^2}{3L} \qquad \textbf{(10-57)}$$

The horizontal force $H$ required to elongate the beam by this amount is

$$H = \frac{EA\lambda}{L} = \frac{8EA\delta^2}{3L^2} \qquad \textbf{(10-58)}$$

in which $EA$ is the axial rigidity of the beam. The corresponding axial tensile stress in the beam is

$$\sigma_t = \frac{H}{A} = \frac{8E\delta^2}{3L^2} \qquad \textbf{(10-59)}$$

This equation gives a close estimate of the tensile stress produced by the immovable supports of a simple beam.

## General Comments

Now let us substitute some numerical values so that we can assess the significance of the curvature shortening. The deflection $\delta$ at the midpoint of the beam is usually very small compared to the length; for example, the ratio $\delta/L$ might be 1/500 or smaller. Using this value, and also assuming that the material is steel with $E = 30 \times 10^6$ psi, we find from Eq. (10-59) that the tensile stress is only 320 psi. Since the allowable tensile stress in the steel is typically 15,000 psi or larger, it becomes clear that the axial stress due to the horizontal force $H$ may be disregarded when compared to the ordinary working stresses in the beam.

Furthermore, in the derivation of Eq. (10-55) we assumed that the ends of the beam were held rigidly against horizontal displacements, which is not physically possible. In reality, small longitudinal displacements always occur, thereby reducing the axial stress calculated from Eq. (10-55).*

From the preceding discussions, we conclude that the customary practice of disregarding the effects of any longitudinal restraints and assuming that one end of the beam is on a roller support (regardless of the actual construction) is justified. The stiffening effect of longitudinal restraints is significant only when the beam is very long and slender and supports large loads. This behavior is sometimes referred to as "string action," because it is analogous to the action of a cable, or string, supporting a load.

---

*For a more complete analysis of beams with immovable supports, see Ref. 10-1.

In Chapter 10, we investigated the behavior of statically indeterminate beams acted on by concentrated and also distributed loads, such as self-weight; thermal effects and longitudinal displacement due to curvature shortening also were considered as specialized topics at the end of the chapter. We developed two analysis approaches: (1) **integration** of the equation of the elastic curve using available boundary conditions to solve for unknown constants of integration and redundant reactions, and (2) the more general approach (used earlier in Chapters 2 and 3 for axial and torsional structures, respectively) based on **superposition**. In the superposition procedure, we augmented the **equilibrium** equations from statics with **compatibility** equations to generate a sufficient number of equations to solve for all unknown forces. The **force-displacement relations** were used with the compatibility equations to generate the additional equations needed to solve the problem. The number of additional equations required was seen to be dependent on the **degree of statical indeterminacy** of the beam structure. The superposition approach is limited to beam structures made of linearly elastic materials. The major concepts presented in this chapter are as follows:

1. Several types of statically indeterminate beam structures, such as propped cantilever, fixed-end, and continuous beams were discussed. The **degree of statical indeterminacy** was determined for each beam type, and a **released** structure was defined for each case by removing different **redundant** reaction forces.

2. The released structure must be statically determinate and **stable** under the action of the applied loadings. Note that it is also possible to insert **internal releases** on axial force, shear, and moment (see discussion in Chapter 4) to produce the released structure, as will be discussed in later courses on structural analysis.

3. For simple statically indeterminate beam structures, the **differential equation of the elastic curve** can be written as a second-, third-, or fourth-order equation in terms of moment, shear force, and distributed load, respectively. By applying boundary and other conditions, one can solve for the constants of integration and the redundant reactions.

4. A more general solution approach for more complex beam and other types of structures is the **method of superposition** (also known as the **force** or **flexibility** method). Here, additional equations which describe the **compatibility** of displacements and incorporate the appropriate **force-displacement relations** for beams are used to supplement the **equilibrium** equations. The number of compatibility equations required for solution is equal to the degree of statical indeterminacy of the beam structure.

5. In most cases, there are multiple paths to the same solution depending upon the choice of the redundant reaction.

6. Differential **temperature changes** and **longitudinal displacements** induce reaction forces only in statically indeterminate beams; if the beam is statically determinate, joint displacements will occur, but no internal forces will result from these effects.

# PROBLEMS CHAPTER 10

## Differential Equations of the Deflection Curve

*The problems for Section 10.3 are to be solved by integrating the differential equations of the deflection curve. All beams have constant flexural rigidity EI. When drawing shear-force and bending-moment diagrams, be sure to label all critical ordinates, including maximum and minimum values.*

**10.3-1** A propped cantilever beam $AB$ of length $L$ is loaded by a counterclockwise moment $M_0$ acting at support $B$ (see figure).

Beginning with the second-order differential equation of the deflection curve (the bending-moment equation), obtain the reactions, shear forces, bending moments, slopes, and deflections of the beam. Construct the shear-force and bending-moment diagrams, labeling all critical ordinates.

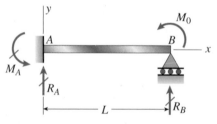

PROB. 10.3-1

**10.3-2** A fixed-end beam $AB$ of length $L$ supports a uniform load of intensity $q$ (see figure).

Beginning with the second-order differential equation of the deflection curve (the bending-moment equation), obtain the reactions, shear forces, bending moments, slopes, and deflections of the beam. Construct the shear-force and bending-moment diagrams, labeling all critical ordinates.

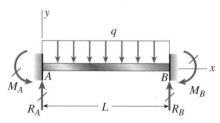

PROB. 10.3-2

**10.3-3** A cantilever beam $AB$ of length $L$ has a fixed support at $A$ and a roller support at $B$ (see figure). The support at $B$ is moved downward through a distance $\delta_B$.

Using the fourth-order differential equation of the deflection curve (the load equation), determine the reactions of the beam and the equation of the deflection curve. (*Note:* Express all results in terms of the imposed displacement $\delta_B$.)

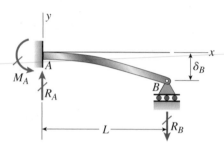

PROB. 10.3-3

**10.3-4** A cantilever beam of length $L$ and loaded by uniform load of intensity $q$ has a fixed support at $A$ and spring support at $B$ with rotational stiffness $k_R$. A rotation at $B$, $\theta_B$, results in a reaction moment $M_B = k_R \times \theta_B$.

Find rotation $\theta_B$ and displacement $\delta_B$ at end $B$. Use the second-order differential equation of the deflection curve to solve for displacements at end $B$.

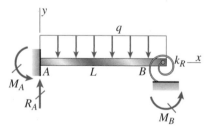

PROB. 10.3-4

**10.3-5** A cantilever beam of length $L$ and loaded by a triangularly distributed load of maximum intensity $q_0$ at $B$.

Use the fourth-order differential equation of the deflection curve to solve for reactions at $A$ and $B$ and also the equation of the deflection curve.

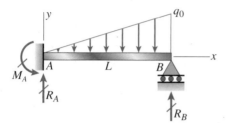

PROB. 10.3-5

**10.3-6** A propped cantilever beam of length $L$ is loaded by a parabolically distributed load with maximum intensity $q_0$ at $B$.

(a) Use the fourth-order differential equation of the deflection curve to solve for reactions at $A$ and $B$ and also the equation of the deflection curve.

(b) Repeat part (a) if the parabolic load is replaced by $q_0 \sin(\pi x/2L)$.

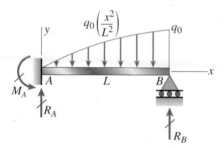

PROB. 10.3-6

**10.3-7** A propped cantilever beam of length $L$ is loaded by a parabolically distributed load with maximum intensity $q_0$ at $A$.

(a) Use the fourth-order differential equation of the deflection curve to solve for reactions at $A$ and $B$ and also the equation of the deflection curve.

(b) Repeat part (a) if the parabolic load is replaced by $q_0 \cos(\pi x/2L)$.

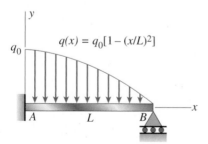

$$q(x) = q_0[1 - (x/L)^2]$$

PROB. 10.3-7

**10.3-8** A fixed-end beam of length $L$ is loaded by a distributed load in the form of a cosine curve with maximum intensity $q_0$ at $A$.

(a) Use the fourth-order differential equation of the deflection curve to solve for reactions at $A$ and $B$ and also the equation of the deflection curve.

(b) Repeat part (a) using the distributed load $q_0 \sin(\pi x/L)$.

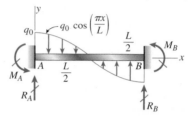

$$q_0 \cos\left(\frac{\pi x}{L}\right)$$

PROB. 10.3-8

**10.3-9** A fixed-end beam of length $L$ is loaded by a distributed load in the form of a cosine curve with maximum intensity $q_0$ at $A$.

(a) Use the fourth-order differential equation of the deflection curve to solve for reactions at $A$ and $B$ and also the equation of the deflection curve.

(b) Repeat part (a) if the distributed load is now $q_0(1 - x^2/L^2)$.

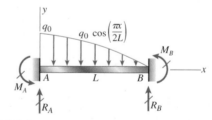

$$q_0 \cos\left(\frac{\pi x}{2L}\right)$$

PROB. 10.3-9

**10.3-10** A fixed-end beam of length $L$ is loaded by triangularly distributed load of maximum intensity $q_0$ at $B$.

Use the fourth-order differential equation of the deflection curve to solve for reactions at $A$ and $B$ and also the equation of the deflection curve.

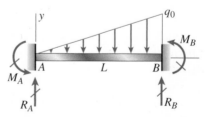

PROB. 10.3-10

**10.3-11** A counterclockwise moment $M_0$ acts at the midpoint of a fixed-end beam $ACB$ of length $L$ (see figure).

Beginning with the second-order differential equation of the deflection curve (the bending-moment equation), determine all reactions of the beam and obtain the equation of the deflection curve for the left-hand half of the beam.

Then construct the shear-force and bending-moment diagrams for the entire beam, labeling all critical ordinates. Also, draw the deflection curve for the entire beam.

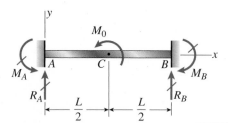

PROB. 10.3-11

**10.3-12** A propped cantilever beam of length $L$ is loaded by a concentrated moment $M_0$ at midpoint $C$. Use the second-order differential equation of the deflection curve to solve for reactions at $A$ and $B$. Draw shear-force and bending-moment diagrams for the entire beam. Also find the equations of the deflection curves for both halves of the beam, and draw the deflection curve for the entire beam.

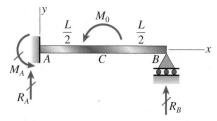

PROB. 10.3-12

# Method of Superposition

*The problems for Section 10.4 are to be solved by the method of superposition. All beams have constant flexural rigidity EI unless otherwise stated. When drawing shear-force and bending-moment diagrams, be sure to label all critical ordinates, including maximum and minimum values.*

**10.4-1** A propped cantilever beam $AB$ of length $L$ carries a concentrated load $P$ acting at the position shown in the figure.

Determine the reactions $R_A$, $R_B$, and $M_A$ for this beam. Also, draw the shear-force and bending-moment diagrams, labeling all critical ordinates.

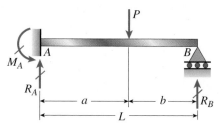

PROB. 10.4-1

**10.4-2** A beam with a guided support at $B$ is loaded by a uniformly distributed load with intensity $q$. Use the method of superposition to solve for all reactions. Also draw shear-force and bending-moment diagrams, labeling all critical ordinates.

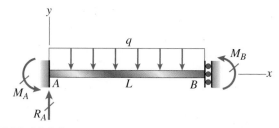

PROB. 10.4-2

**10.4-3** A propped cantilever beam of length $2L$ with support at $B$ is loaded by a uniformly distributed load with intensity $q$. Use the method of superposition to solve for all reactions. Also draw shear-force and bending-moment diagrams, labeling all critical ordinates.

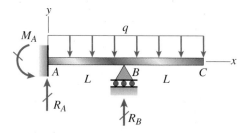

PROB. 10.4-3

**10.4-4** The continuous frame $ABC$ has a pin support at $A$, roller supports at $B$ and $C$, and a rigid corner connection at $B$ (see figure). Members $AB$ and $BC$ each have flexural rigidity *EI*. A moment $M_0$ acts counterclockwise at *B*. (*Note:* Disregard axial deformations in member *AB* and consider only the effects of bending.)

(a) Find all reactions of the frame.

(b) Find joint rotations $\theta$ at *A*, *B*, and *C*.

(c) Find the required new length of member *BC* in terms of *L*, so that $\theta_B$ in part (b) is doubled in size.

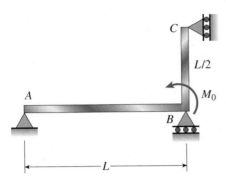

PROB. 10.4-4

**10.4-5** The continuous frame *ABC* has a pin support at *A*, roller supports at *B* and *C*, and a rigid corner connection at *B* (see figure). Members *AB* and *BC* each have flexural rigidity *EI*. A moment $M_0$ acts counterclockwise at *A*. (*Note*: Disregard axial deformations in member *AB* and consider only the effects of bending.)

(a) Find all reactions of the frame.

(b) Find joint rotations $\theta$ at *A*, *B*, and *C*.

(c) Find the required new length of member *AB* in terms of *L*, so that $\theta_A$ in part (b) is doubled in size.

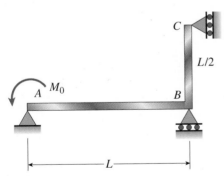

PROB. 10.4-5

**10.4-6** Beam *AB* has a pin support at *A* and a roller support at *B*. Joint *B* is also restrained by a linearly elastic rotational spring with stiffness $k_R$, which provides a resisting moment $M_B$ due to rotation at *B*. Member *AB* has flexural rigidity *EI*. A moment $M_0$ acts counterclockwise at *B*.

(a) Use the method of superposition to solve for all reactions.

(b) Find an expression for joint rotation $\theta_A$ in terms of spring stiffness $k_R$. What is $\theta_A$ when $k_R \to 0$? What is $\theta_A$ when $k_R \to \infty$? What is $\theta_A$ when $k_R = 6EI/L$?

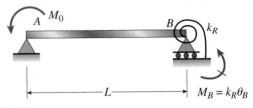

PROB. 10.4-6

**10.4-7** The continuous frame *ABCD* has a pin support at *B*, roller supports at *A*, *C*, and *D*, and rigid corner connections at *B* and *C* (see figure). Members *AB*, *BC*, and *CD* each have flexural rigidity *EI*. Moment $M_0$ acts counterclockwise at *B* and clockwise at *C*. (*Note*: Disregard axial deformations in member *AB* and consider only the effects of bending.)

(a) Find all reactions of the frame.

(b) Find joint rotations $\theta$ at *A*, *B*, *C*, and *D*.

(c) Repeat parts (a) and (b) if both moments $M_0$ are counterclockwise.

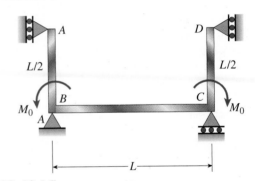

PROB. 10.4-7

**10.4-8** Two flat beams *AB* and *CD*, lying in horizontal planes, cross at right angles and jointly support a vertical load *P* at their midpoints (see figure). Before the load *P* is applied, the beams just touch each other. Both beams are made of the same material and have the same widths. Also, the ends of both beams are simply supported. The lengths of beams *AB* and *CD* are $L_{AB}$ and $L_{CD}$, respectively.

What should be the ratio $t_{AB}/t_{CD}$ of the thicknesses of the beams if all four reactions are to be the same?

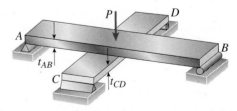

PROB. 10.4-8

**10.4-9** A propped cantilever beam of length $2L$ is loaded by a uniformly distributed load with intensity $q$. The beam is supported at $B$ by a linearly elastic spring with stiffness $k$. Use the method of superposition to solve for all reactions. Also draw shear-force and bending-moment diagrams, labeling all critical ordinates. Let $k = 6EI/L^3$.

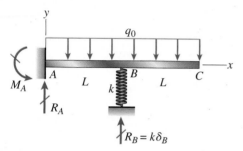

PROB. 10.4-9

**10.4-10** A propped cantilever beam of length $2L$ is loaded by a uniformly distributed load with intensity $q$. The beam is supported at $B$ by a linearly elastic rotational spring with stiffness $k_R$, which provides a resisting moment $M_B$ due to rotation $\theta_B$. Use the method of superposition to solve for all reactions. Also draw shear-force and bending-moment diagrams, labeling all critical ordinates. Let $k_R = EI/L$.

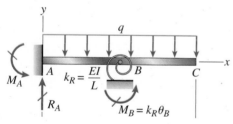

PROB. 10.4-10

**10.4-11** Determine the fixed-end moments ($M_A$ and $M_B$) and fixed-end forces ($R_A$ and $R_B$) for a beam of length $L$ supporting a triangular load of maximum intensity $q_0$ (see figure). Then draw the shear-force and bending-moment diagrams, labeling all critical ordinates.

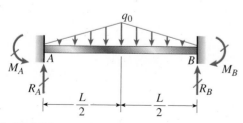

PROB. 10.4-11

**10.4-12** A continuous beam $ABC$ with two unequal spans, one of length $L$ and one of length $2L$, supports a uniform load of intensity $q$ (see figure).

Determine the reactions $R_A$, $R_B$, and $R_C$ for this beam. Also, draw the shear-force and bending-moment diagrams, labeling all critical ordinates.

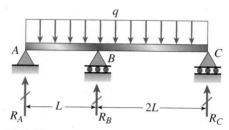

PROB. 10.4-12

**10.4-13** Beam $ABC$ is fixed at support $A$ and rests (at point $B$) upon the midpoint of beam $DE$ (see the first part of the figure). Thus, beam $ABC$ may be represented as a propped cantilever beam with an overhang $BC$ and a linearly elastic support of stiffness $k$ at point $B$ (see the second part of the figure).

The distance from $A$ to $B$ is $L = 10$ ft, the distance from $B$ to $C$ is $L/2 = 5$ ft, and the length of beam $DE$ is $L = 10$ ft. Both beams have the same flexural rigidity $EI$. A concentrated load $P = 1700$ lb acts at the free end of beam $ABC$.

Determine the reactions $R_A$, $R_B$, and $M_A$ for beam $ABC$. Also, draw the shear-force and bending-moment diagrams for beam $ABC$, labeling all critical ordinates.

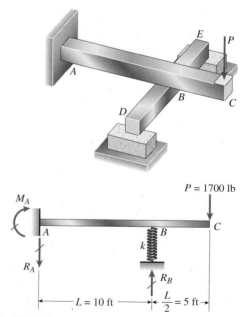

PROB. 10.4-13

**10.4-14** A propped cantilever beam has flexural rigidity $EI = 4.5$ MN·m². When the loads shown are applied to the beam, it settles at joint $B$ by 5 mm. Find the reaction at joint $B$.

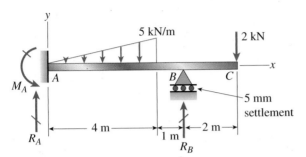

PROB. 10.4-14

**10.4-15** A cantilever beam is supported by a tie rod at $B$ as shown. Both the tie rod and the beam are steel with $E = 30 \times 10^6$ psi. The tie rod is just taut before the distributed load $q = 200$ lb/ft is applied.

(a) Find the tension force in the tie rod.

(b) Draw shear-force and bending-moment diagrams for the beam, labeling all critical ordinates.

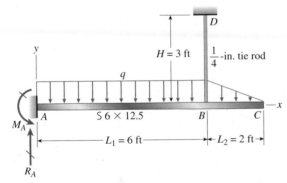

PROB. 10.4-15

**10.4-16** The figure shows a nonprismatic, propped cantilever beam $AB$ with flexural rigidity $2EI$ from $A$ to $C$ and $EI$ from $C$ to $B$.

Determine all reactions of the beam due to the uniform load of intensity $q$. (*Hint:* Use the results of Probs. 9.7-1 and 9.7-2.)

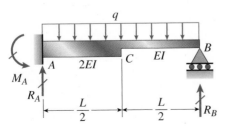

PROB. 10.4-16

**10.4-17** A beam $ABC$ is fixed at end $A$ and supported by beam $DE$ at point $B$ (see figure). Both beams have the same cross section and are made of the same material.

(a) Determine all reactions due to the load $P$.

(b) What is the numerically largest bending moment in either beam?

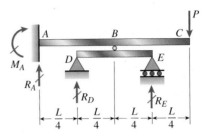

PROB. 10.4-17

**10.4-18** A three-span continuous beam $ABCD$ with three equal spans supports a uniform load of intensity $q$ (see figure).

Determine all reactions of this beam and draw the shear-force and bending-moment diagrams, labeling all critical ordinates.

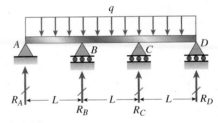

PROB. 10.4-18

**10.4-19** A beam rests on supports at $A$ and $B$ and is loaded by a distributed load with intensity $q$ as shown. A small gap $\Delta$ exists between the unloaded beam and the support at $C$. Assume that span length $L = 40$ in. and flexural rigidity of the beam $EI = 0.4 \times 10^9$ lb-in². Plot a graph of the bending moment at $B$ as a function of the load intensity $q$.

(*Hint:* See Example 9-9 for guidance on computing the deflection at $C$.)

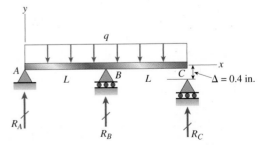

PROB. 10.4-19

**10.4-20** A fixed-end beam $AB$ of length $L$ is subjected to a moment $M_0$ acting at the position shown in the figure.

(a) Determine all reactions for this beam.

(b) Draw shear-force and bending-moment diagrams for the special case in which $a = b = L/2$.

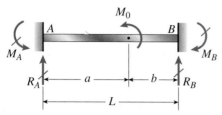

PROB. 10.4-20

**10.4-21** A temporary wood flume serving as a channel for irrigation water is shown in the figure. The vertical boards forming the sides of the flume are sunk in the ground, which provides a fixed support. The top of the flume is held by tie rods that are tightened so that there is no deflection of the boards at that point. Thus, the vertical boards may be modeled as a beam $AB$, supported and loaded as shown in the last part of the figure.

Assuming that the thickness $t$ of the boards is 1.5 in., the depth $d$ of the water is 40 in., and the height $h$ to the tie rods is 50 in., what is the maximum bending stress $\sigma$ in the boards? (*Hint:* The numerically largest bending moment occurs at the fixed support.)

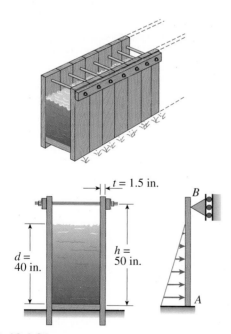

PROB. 10.4-21

**10.4-22** Two identical, simply supported beams $AB$ and $CD$ are placed so that they cross each other at their midpoints (see figure). Before the uniform load is applied, the beams just touch each other at the crossing point.

Determine the maximum bending moments $(M_{AB})_{max}$ and $(M_{CD})_{max}$ in beams $AB$ and $CD$, respectively, due to the uniform load if the intensity of the load is $q = 6.4$ kN/m and the length of each beam is $L = 4$ m.

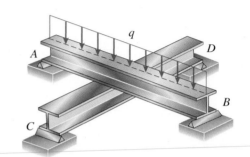

PROB. 10.4-22

**10.4-23** The cantilever beam $AB$ shown in the figure is an S 6 × 12.5 steel I-beam with $E = 30 × 10^6$ psi. The simple beam $DE$ is a wood beam 4 in. × 12 in. (nominal dimensions) in cross section with $E = 1.5 × 10^6$ psi. A steel rod $AC$ of diameter 0.25 in., length 10 ft, and $E = 30 × 10^6$ psi serves as a hanger joining the two beams. The hanger fits snugly between the beams before the uniform load is applied to beam $DE$.

Determine the tensile force $F$ in the hanger and the maximum bending moments $M_{AB}$ and $M_{DE}$ in the two beams due to the uniform load, which has intensity $q = 400$ lb/ft. (*Hint:* To aid in obtaining the maximum bending moment in beam $DE$, draw the shear-force and bending-moment diagrams.)

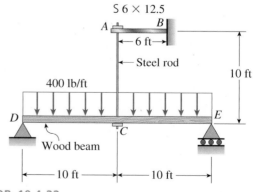

PROB. 10.4-23

**10.4-24** The beam $AB$ shown in the figure is simply supported at $A$ and $B$ and supported on a spring of stiffness $k$ at its midpoint $C$. The beam has flexural rigidity $EI$ and length $2L$.

What should be the stiffness $k$ of the spring in order that the maximum bending moment in the beam (due to the uniform load) will have the smallest possible value?

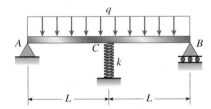

PROB. 10.4-24

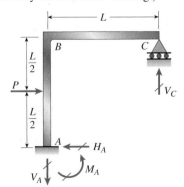

Ship container handling cranes made up of two plane frames (Courtesy of the National Information Service for Earthquake Engineering EERC, University of California Berkeley.)

**10.4-25** The continuous frame $ABC$ has a fixed support at $A$, a roller support at $C$, and a rigid corner connection at $B$ (see figure). Members $AB$ and $BC$ each have length $L$ and flexural rigidity $EI$. A horizontal force $P$ acts at midheight of member $AB$.

(a) Find all reactions of the frame.

(b) What is the largest bending moment $M_{max}$ in the frame? (*Note:* Disregard axial deformations in member $AB$ and consider only the effects of bending.)

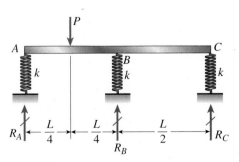

PROB. 10.4-25

**10.4-26** The continuous frame $ABC$ has a pinned support at $A$, a guided support at $C$, and a rigid corner connection at $B$ (see figure). Members $AB$ and $BC$ each have length $L$ and flexural rigidity $EI$. A horizontal force $P$ acts at midheight of member $AB$.

(a) Find all reactions of the frame.

(b) What is the largest bending moment $M_{max}$ in the frame? (*Note:* Disregard axial deformations in members $AB$ and $BC$ and consider only the effects of bending.)

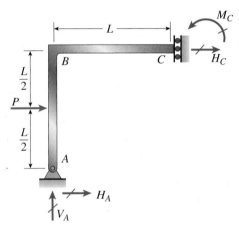

PROB. 10.4-26

**10.4-27** A wide-flange beam $ABC$ rests on three identical spring supports at points $A$, $B$, and $C$ (see figure). The flexural rigidity of the beam is $EI = 6912 \times 10^6$ lb - in.², and each spring has stiffness $k = 62{,}500$ lb/in. The length of the beam is $L = 16$ ft.

If the load $P$ is 6000 lb, what are the reactions $R_A$, $R_B$, and $R_C$? Also, draw the shear-force and bending-moment diagrams for the beam, labeling all critical ordinates.

PROB. 10.4-27

**10.4-28** A fixed-end beam $AB$ of length $L$ is subjected to a uniform load of intensity $q$ acting over the middle region of the beam (see figure).

(a) Obtain a formula for the fixed-end moments $M_A$ and $M_B$ in terms of the load $q$, the length $L$, and the length $b$ of the loaded part of the beam.

(b) Plot a graph of the fixed-end moment $M_A$ versus the length $b$ of the loaded part of the beam. For convenience, plot the graph in the following nondimensional form:

$$\frac{M_A}{qL^2/12} \quad \text{versus} \quad \frac{b}{L}$$

with the ratio $b/L$ varying between its extreme values of 0 and 1.

(c) For the special case in which $a = b = L/3$, draw the shear-force and bending-moment diagrams for the beam, labeling all critical ordinates.

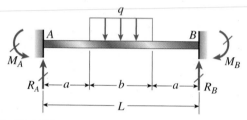

PROB. 10.4-28

**10.4-29** A beam supporting a uniform load of intensity $q$ throughout its length rests on pistons at points $A$, $C$, and $B$ (see figure). The cylinders are filled with oil and are connected by a tube so that the oil pressure on each piston is the same. The pistons at $A$ and $B$ have diameter $d_1$, and the piston at $C$ has diameter $d_2$.

(a) Determine the ratio of $d_2$ to $d_1$ so that the largest bending moment in the beam is as small as possible.

(b) Under these optimum conditions, what is the largest bending moment $M_{max}$ in the beam?

(c) What is the difference in elevation between point $C$ and the end supports?

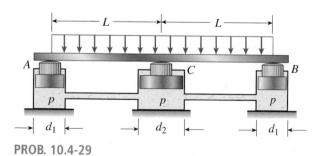

PROB. 10.4-29

**10.4-30** A thin steel beam $AB$ used in conjunction with an electromagnet in a high-energy physics experiment is securely bolted to rigid supports (see figure). A magnetic field produced by coils $C$ results in a force acting on the beam. The force is trapezoidally distributed with maximum intensity $q_0 = 18$ kN/m. The length of the beam between supports is $L = 200$ mm and the dimension $c$ of the trapezoidal load is 50 mm. The beam has a rectangular cross section with width $b = 60$ mm and height $h = 20$ mm.

Determine the maximum bending stress $\sigma_{max}$ and the maximum deflection $\delta_{max}$ for the beam. (Disregard any effects of axial deformations and consider only the effects of bending. Use $E = 200$ GPa.)

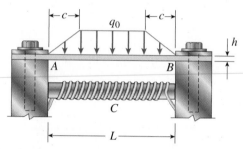

PROB. 10.4-30

# Temperature Effects

*The beams described in the problems for Section 10.5 have constant flexural rigidity EI.*

**10.5-1** A cable $CD$ of length $H$ is attached to the third point of a simple beam $AB$ of length $L$ (see figure). The moment of inertia of the beam is $I$, and the effective cross-sectional area of the cable is $A$. The cable is initially taut but without any initial tension.

(a) Obtain a formula for the tensile force $S$ in the cable when the temperature drops uniformly by $\Delta T$ degrees, assuming that the beam and cable are made of the same material (modulus of elasticity $E$ and coefficient of thermal expansion $\alpha$). Use the method of superposition in the solution.

(b) Repeat part (a) assuming a wood beam and steel cable.

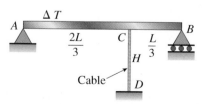

PROB. 10.5-1

**10.5-2** A propped cantilever beam, fixed at the left-hand end $A$ and simply supported at the right-hand end $B$, is subjected to a temperature differential with temperature $T_1$ on its upper surface and $T_2$ on its lower surface (see figure).

(a) Find all reactions for this beam. Use the method of superposition in the solution. Assume the spring support is unaffected by temperature.

(b) What are the reactions when $k \rightarrow \infty$?

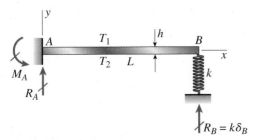

**PROBS. 10.5-2 and 10.5-3**

**10.5-3** Solve the preceding problem by integrating the differential equation of the deflection curve.

**10.5-4** A two-span beam with spans of lengths $L$ and $L/3$ is subjected to a temperature differential with temperature $T_1$ on its upper surface and $T_2$ on its lower surface (see figure).

(a) Determine all reactions for this beam. Use the method of superposition in the solution. Assume the spring support is unaffected by temperature.

(b) What are the reactions when $k \rightarrow \infty$?

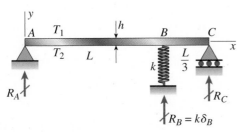

**PROBS. 10.5-4 and 10.5-5**

**10.5-5** Solve the preceding problem by integrating the differential equation of the deflection curve

## Longitudinal Displacements at the Ends of Beams

**10.6-1** Assume that the deflected shape of a beam $AB$ with *immovable* pinned supports (see figure) is given by the equation $v = -\delta \sin \pi x/L$, where $\delta$ is the deflection at the midpoint of the beam and $L$ is the length. Also, assume that the beam has constant axial rigidity $EA$.

(a) Obtain formulas for the longitudinal force $H$ at the ends of the beam and the corresponding axial tensile stress $\sigma_t$.

(b) For an aluminum-alloy beam with $E = 10 \times 10^6$ psi, calculate the tensile stress $\sigma_t$ when the ratio of the deflection $\delta$ to the length $L$ equals 1/200, 1/400, and 1/600.

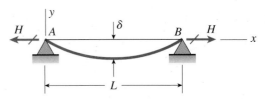

**PROB. 10.6-1**

**10.6-2** (a) A simple beam $AB$ with length $L$ and height $h$ supports a uniform load of intensity $q$ (see the *first part* of the figure). Obtain a formula for the curvature shortening $\lambda$ of this beam. Also, obtain a formula for the maximum bending stress $\sigma_b$ in the beam due to the load $q$.

(b) Now assume that the ends of the beam are pinned so that curvature shortening is prevented and a horizontal force $H$ develops at the supports (see the *second part* of the figure). Obtain a formula for the corresponding axial tensile stress $\sigma_t$.

(c) Using the formulas obtained in parts (a) and (b), calculate the curvature shortening $\lambda$, the maximum bending stress $\sigma_b$, and the tensile stress $\sigma_t$ for the following steel beam: length $L = 3$ m, height $h = 300$ mm, modulus of elasticity $E = 200$ GPa, and moment of inertia $I = 36 \times 10^6$ mm⁴. Also, the load on the beam has intensity $q = 25$ kN/m.

Compare the tensile stress $\sigma_t$ produced by the axial forces with the maximum bending stress $\sigma_b$ produced by the uniform load.

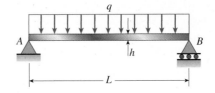

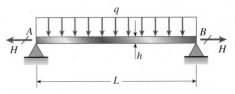

**PROB. 10.6-2**

# Columns

Critical load carrying elements in structures such as columns and other slender compression members are susceptible to buckling failure. (Jose AS Reyes/Shutterstock)

# CHAPTER OVERVIEW

Chapter 11 is primarily concerned with the **buckling of slender columns** which support compressive loads in structures. First, the critical axial load which indicates the onset of buckling is defined and computed for a number of simple models composed of rigid bars and elastic springs (Section 11.2). **Stable**, **neutral**, and **unstable equilibrium conditions** are described for these idealized rigid structures. Then **linear elastic buckling** of slender columns with pinned-end conditions is considered (Section 11.3). The differential equation of the deflection curve is derived and solved to obtain expressions for the **Euler buckling load** ($P_{cr}$) and associated **buckled shape** for the fundamental mode. Critical stress ($\sigma_{cr}$) and **slenderness ratio** ($L/r$) are defined, and the effects of large deflections, column imperfections, inelastic behavior, and optimum shapes of columns are explained. Critical loads and

buckled mode shapes are then computed for **three additional column support cases** (fixed-free, fixed-fixed, and fixed-pinned) (Section 11.4), and the concept of **effective length** ($L_e$) is introduced. If the axial compressive load is not applied at the centroid of the cross section of the column, the eccentricity of the load must be considered in the differential equation of the deflection curve (Section 11.5), and the behavior of the column is changed as shown in the load-deflection diagram. Maximum stresses in columns with **eccentric loads** may be computed using the **secant formula** (Section 11.6). If the material is stressed beyond the proportional limit, **inelastic buckling** must be considered (Sections 11.7 and 11.8) using one of three available theories. Finally, **formulas for design of columns** of various materials are explained (Section 11.9) and illustrated by examples.

## Chapter 11 is organized as follows:

**11.1** Introduction 870
**11.2** Buckling and Stability 870
**11.3** Columns with Pinned Ends 878
**11.4** Columns with Other Support Conditions 889
**11.5** Columns with Eccentric Axial Loads 899
**11.6** The Secant Formula for Columns 904

**11.7** Elastic and Inelastic Column Behavior 909
**11.8** Inelastic Buckling 911
**11.9** Design Formulas for Columns 916
  Chapter Summary & Review 934
  Problems 936

Fig. 11-1

Buckling of a slender column
due to an axial compressive
load *P*

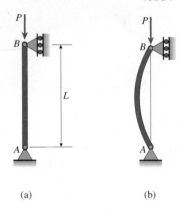

(a)               (b)

# 11.1 INTRODUCTION

Load-carrying structures may fail in a variety of ways, depending upon the type of structure, the conditions of support, the kinds of loads, and the materials used. For instance, an axle in a vehicle may fracture suddenly from repeated cycles of loading, or a beam may deflect excessively, so that the structure is unable to perform its intended functions. These kinds of failures are prevented by designing structures so that the maximum stresses and maximum displacements remain within tolerable limits. Thus, **strength** and **stiffness** are important factors in design, as discussed throughout the preceding chapters.

Another type of failure is **buckling**, which is the subject matter of this chapter. We will consider specifically the buckling of **columns**, which are long, slender structural members loaded axially in compression (Fig. 11-1a). If a compression member is relatively slender, it may deflect laterally and fail by bending (Fig. 11-1b) rather than failing by direct compression of the material. You can demonstrate this behavior by compressing a plastic ruler or other slender object. When lateral bending occurs, we say that the column has *buckled*. Under an increasing axial load, the lateral deflections will increase too, and eventually the column will collapse completely.

The phenomenon of buckling is not limited to columns. Buckling can occur in many kinds of structures and can take many forms. When you step on the top of an empty aluminum can, the thin cylindrical walls buckle under your weight and the can collapses. When a large bridge collapsed a few years ago, investigators found that failure was caused by the buckling of a thin steel plate that wrinkled under compressive stresses. Buckling is one of the major causes of failures in structures, and therefore the possibility of buckling should always be considered in design.

# 11.2 BUCKLING AND STABILITY

To illustrate the fundamental concepts of buckling and stability, we will analyze the **idealized structure**, or **buckling model**, shown in Fig. 11-2a. This hypothetical structure consists of two rigid bars *AB* and *BC*, each of length *L*/2. They are joined at *B* by a pin connection and held in a vertical position by a rotational spring having stiffness $\beta_R$.*

This idealized structure is analogous to the column of Fig. 11-1a, because both structures have simple supports at the ends and are compressed by an axial load *P*. However, the elasticity of the idealized structure is "concentrated" in the rotational spring, whereas a real column can bend throughout its length (Fig. 11-1b).

In the idealized structure, the two bars are perfectly aligned and the axial load *P* has its line of action along the longitudinal axis (Fig. 11-2a). Consequently, the spring is initially unstressed and the bars are in direct compression.

---

*The general relationship for a rotational spring is $M = \beta_R\theta$, where *M* is the moment acting on the spring, $\beta_R$ is the rotational stiffness of the spring, and $\theta$ is the angle through which the spring rotates. Thus, rotational stiffness has units of moment divided by angle, such as lb-in./rad or N·m/rad. The analogous relationship for a translational spring is $F = \beta\delta$, where *F* is the force acting on the spring, $\beta$ is the translational stiffness of the spring (or spring constant), and $\delta$ is the change in length of the spring. Thus, translational stiffness has units of force divided by length, such as lb/in. or N/m.

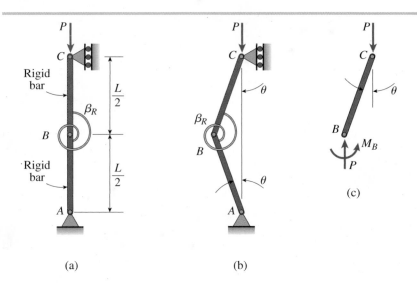

**Fig. 11-2**

Buckling of an idealized structure consisting of two rigid bars and a rotational spring

(a)    (b)

Now suppose that the structure is disturbed by some external force that causes point $B$ to move a small distance laterally (Fig. 11-2b). The rigid bars rotate through small angles $\theta$ and a moment develops in the spring. The direction of this moment is such that it tends to return the structure to its original straight position, and therefore it is called a **restoring moment**. At the same time, however, the tendency of the axial compressive force is to increase the lateral displacement. Thus, these two actions have opposite effects—the restoring moment tends to *decrease* the displacement and the axial force tends to *increase* it.

Now consider what happens when the disturbing force is removed. If the axial force $P$ is relatively small, the action of the restoring moment will predominate over the action of the axial force and the structure will return to its initial straight position. Under these conditions, the structure is said to be **stable**. However, if the axial force $P$ is large, the lateral displacement of point $B$ will increase and the bars will rotate through larger and larger angles until the structure collapses. Under these conditions, the structure is **unstable** and fails by lateral buckling.

## Critical Load

The transition between the stable and unstable conditions occurs at a special value of the axial force known as the **critical load** (denoted by the symbol $P_{cr}$). We can determine the critical load of our buckling model by considering the structure in the disturbed position (Fig. 11-2b) and investigating its equilibrium.

First, we consider the entire structure as a free body and sum moments about support $A$. This step leads to the conclusion that there is no horizontal reaction at support $C$. Second, we consider bar $BC$ as a free body (Fig. 11-2c) and note that it is subjected to the action of the axial forces $P$ and the moment $M_B$ in the spring. The moment $M_B$ is equal to the rotational stiffness $\beta_R$ times the angle of rotation $2\theta$ of the spring; thus,

$$M_B = 2\beta_R\theta \tag{11-1a}$$

Since the angle $\theta$ is a small quantity, the lateral displacement of point $B$ is $\theta L/2$. Therefore, we obtain the following equation of equilibrium by summing moments about point $B$ for bar $BC$ (Fig. 11-2c):

$$M_B - P\left(\frac{\theta L}{2}\right) = 0 \qquad \textbf{(11-1b)}$$

or, upon substituting from Eq. (11-1a),

$$\left(2\beta_R - \frac{PL}{2}\right)\theta = 0 \qquad \textbf{(11-2)}$$

One solution of this equation is $\theta = 0$, which is a trivial solution and merely means that the structure is in equilibrium when it is perfectly straight, regardless of the magnitude of the force $P$.

A second solution is obtained by setting the term in parentheses equal to zero and solving for the load $P$, which is the *critical load*:

$$P_{cr} = \frac{4\beta_R}{L} \qquad \textbf{(11-3)}$$

At the critical value of the load the structure is in equilibrium regardless of the magnitude of the angle $\theta$ [provided the angle remains small, because we made that assumption when deriving Eq. (11-1b)].

From the preceding analysis we see that the critical load is the *only* load for which the structure will be in equilibrium in the disturbed position. At this value of the load, the restoring effect of the moment in the spring just matches the buckling effect of the axial load. Therefore, the critical load represents the boundary between the stable and unstable conditions.

If the axial load is less than $P_{cr}$, the effect of the moment in the spring predominates and the structure returns to the vertical position after a slight disturbance; if the axial load is larger than $P_{cr}$, the effect of the axial force predominates and the structure buckles:

If $P < P_{cr}$, the structure is *stable*.
If $P > P_{cr}$, the structure is *unstable*.

From Eq. (11-3) we see that the stability of the structure is increased either by *increasing its stiffness* or by *decreasing its length*. Later in this chapter, when we determine critical loads for various types of columns, we will see that these same observations apply.

## Summary

Let us now summarize the behavior of the idealized structure (Fig. 11-2a) as the axial load $P$ increases from zero to a large value.

When the axial load is less than the critical load ($0 < P < P_{cr}$), the structure is in equilibrium when it is perfectly straight. Because the equilibrium is **stable**, the structure returns to its initial position after being disturbed. Thus, the structure is in equilibrium *only* when it is perfectly straight ($\theta = 0$).

When the axial load is greater than the critical load ($P > P_{cr}$), the structure is still in equilibrium when $\theta = 0$ (because it is in direct compression

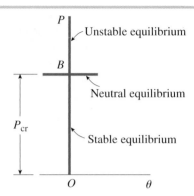

**Fig. 11-3**

Equilibrium diagram for buckling of an idealized structure

and there is no moment in the spring), but the equilibrium is **unstable** and cannot be maintained. The slightest disturbance will cause the structure to buckle.

At the critical load ($P = P_{cr}$), the structure is in equilibrium even when point $B$ is displaced laterally by a small amount. In other words, the structure is in equilibrium for *any* small angle $\theta$, including $\theta = 0$. However, the structure is neither stable nor unstable—it is at the boundary between stability and instability. This condition is referred to as **neutral equilibrium**.

The three equilibrium conditions for the idealized structure are shown in the graph of axial load $P$ versus angle of rotation $\theta$ (Fig. 11-3). The two heavy lines, one vertical and one horizontal, represent the equilibrium conditions. Point $B$, where the equilibrium diagram branches, is called a *bifurcation point*.

The horizontal line for neutral equilibrium extends to the left and right of the vertical axis because the angle $\theta$ may be clockwise or counterclockwise. The line extends only a short distance, however, because our analysis is based upon the assumption that $\theta$ is a small angle. (This assumption is quite valid, because $\theta$ is indeed small when the structure first departs from its vertical position. If buckling continues and $\theta$ becomes large, the line labeled "Neutral equilibrium" curves upward, as shown later in Fig. 11-12.)

The three equilibrium conditions represented by the diagram of Fig. 11-3 are analogous to those of a ball placed upon a smooth surface (Fig. 11-4). If the surface is concave upward, like the inside of a dish, the equilibrium is stable and the ball always returns to the low point when disturbed. If the surface is convex upward, like a dome, the ball can theoretically be in equilibrium on top of the surface, but the equilibrium is unstable and in reality the ball rolls away. If the surface is perfectly flat, the ball is in neutral equilibrium and remains wherever it is placed.

As we will see in the next section, the behavior of an ideal elastic column is analogous to that of the buckling model shown in Fig. 11-2. Furthermore, many other kinds of structural and mechanical systems fit this model.

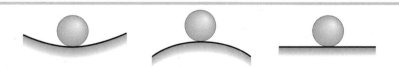

**Fig. 11-4**

Ball in stable, unstable, and neutral equilibrium

**Example 11-1**

Two idealized columns are shown in Fig. 11-5. Both columns are initially straight and vertical. The first column (Structure 1, Fig. 11-5a) consists of a single rigid bar *ABCD* which is pinned at *D* and laterally supported at *B* by a spring with translational stiffness $\beta$. The second column (*Structure 2*, Fig. 11-5b) is comprised of rigid bars *ABC* and *CD* that are joined at *C* by an elastic connection with rotational stiffness $\beta_R = (2/5)\beta L^2$. Structure 2 is pinned at *D* and has a roller support at *B*. Find an expression for critical load $P_{cr}$ for each column.

### Fig. 11-5

Example 11-1: Buckled positions of two idealized structures, (a) one supported laterally by a translational spring, and (b) the other supported by a rotational elastic connection

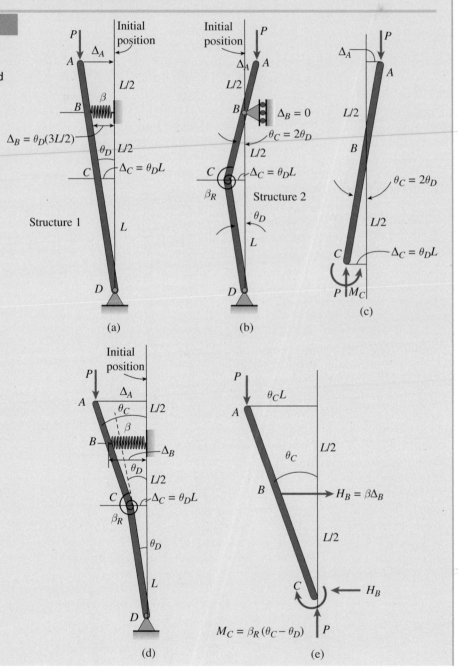

## Solution

*Structure 1.* We begin by considering the equilibrium of Structure 1 in a disturbed position caused by some external load and defined by small rotation angle $\theta_D$ (Fig. 11-5a). Summing moments about $D$, we get the following equilibrium equation:

$$\Sigma M_D = 0 \qquad P\Delta_A = H_B\left(\frac{3L}{2}\right) \tag{a}$$

where
$$\Delta_A = \theta_D\left(L + 2\frac{L}{2}\right) = \theta_D(2L) \tag{b}$$

and
$$H_B = \beta\Delta_B = \beta\left[\theta_D\left(\frac{3L}{2}\right)\right] \tag{c}$$

Since the angle $\theta_D$ is small, lateral displacement $\Delta_A$ is obtained using Eq. (b). The force $H_B$ in the translational spring at $B$ is the product of spring constant $\beta$ and small horizontal displacement $\Delta_B$. Substituting the expression for $\Delta_A$ from Eq. (b) and the expression for $H_B$ from Eq (c) into Eq. (a), and solving for P, we find that the critical load $P_{cr}$ for Structure 1 is

$$P_{cr} = \frac{H_B}{\Delta_A}\left(\frac{3L}{2}\right) = \frac{\beta\theta_D\left(\dfrac{3L}{2}\right)}{\theta_D(2L)}\left(\frac{3L}{2}\right) = \frac{9}{8}\beta L \qquad \text{(d)} \Longleftarrow$$

The buckled *mode shape* for Structure 1 is the disturbed position shown in Fig. 11-5a.

*Structure 2.* The translational spring at $B$ is now replaced by a roller support, and the structure is assembled using two rigid bars (*ABC* and *CD*) joined by a rotational spring having stiffness $\beta_R$. If we sum moments about $D$ for the *undisturbed* structure, we conclude that horizontal reaction $H_B$ is zero. Next, we consider the equilibrium of Structure 2 in a *disturbed* position, once again defined by small rotation angle $\theta_D$ (Fig. 11-5b). Using a free-body diagram of the upper bar *ABC* (Fig. 11-5c) and noting that the moment $M_c$ is equal to rotational stiffness $\beta_R$ times the total relative rotation of the spring, we have

$$M_C = \beta_R(\theta_C + \theta_D) = \beta_R(2\theta_D + \theta_D) = \beta_R(3\theta_D) \tag{e}$$

We see that equilibrium of bar *ABC* requires that

$$\Sigma M_C = 0 \qquad M_C - P(\Delta_A + \Delta_C) = 0 \tag{f}$$

Substituting expressions for $M_c$, $\Delta_A$, and $\Delta_c$ into Eq. (f), we obtain

$$P_{cr} = \frac{M_C}{\Delta_A + \Delta_C} = \frac{\beta_R(3\theta_D)}{\theta_C\left(\dfrac{L}{2}\right) + \theta_D(L)} = \frac{\beta_R(3\theta_D)}{\theta_D(2L)}$$

So the critical load $P_{cr}$ for Structure 2 is

$$P_{cr} = \frac{3\beta_R}{2L} \quad \text{or} \quad P_{cr} = \frac{3}{2L}\left(\frac{2}{5}\beta L^2\right) = \frac{3}{5}\beta L \qquad \Longleftarrow \text{(g)}$$

The buckled *mode shape* for Structure 2 is the disturbed position shown in Fig. 11-5b.

*Combined Model and Analysis.* We can create a more advanced or complex structure model by combining the features of Structure 1 and Structure 2 into a single structure, as shown in Fig. 11-5d. This idealized

*Continues* ➡

structure is shown in its disturbed position and now has both translational spring $\beta$ at $B$ and rotational elastic connection $\beta_R$ at joint $C$ where rigid bars $ABC$ and $CD$ are joined. Note that two rotation angles, $\theta_C$ and $\theta_D$, are now required to uniquely describe any arbitrary position of the disturbed structure (alternatively, we could use translations $\Delta_B$ and $\Delta_C$, for example, instead of $\theta_C$ and $\theta_D$). We will refer to position angles $\theta_C$ and $\theta_D$ as *degrees of freedom*. Hence, the *combined* structure has *two degrees of freedom* and, therefore, has two possible buckled mode shapes and two different critical loads, each of which causes the associated buckling mode. In contrast, we see now that Structures 1 and 2 are *single degree of freedom* structures, because only $\theta_D$ is needed (or alternatively, $\Delta_C$) to define the buckled shape of each structure depicted in Figs. 11-5a and b.

We can now observe that if rotational spring $\beta_R$ becomes infinitely stiff in the combined structure (Fig. 11-5d) (but $\beta$ remains finite), the two degree of freedom (2DOF) combined model reduces to the single degree of freedom (SDOF) model of Fig. 11-5a. Similarly, if translational spring $\beta$ becomes infinitely stiff in Fig. 11-5d (while $\beta_R$ remains finite), the elastic support at $B$ becomes a roller support. *We conclude that the solutions for $P_{cr}$ for Structures 1 and 2 in Eqs. (d) and (g) are simply two special-case solutions of the general combined model in Fig. 11-5d.*

Our goal now is to find a general solution for the 2DOF model in Fig. 11-5d and then show that solutions for $P_{cr}$ for Structures 1 and 2 can be obtained from this general solution.

First, we consider the equilibrium of the entire 2DOF model in the disturbed position shown in Fig. 11-5d. Summing moments about $D$, we get

$$\Sigma M_D = 0 \qquad P\Delta_A - H_B\left(\frac{3L}{2}\right) = 0$$

where

$$\Delta_A = (\theta_C + \theta_D)L$$

and

$$H_B = \beta\Delta_B = \beta\left(\theta_C\frac{L}{2} + \theta_D L\right)$$

Combining these expressions, we obtain the following equation in terms of the two unknown position angles ($\theta_C$ and $\theta_D$) as

$$\theta_C\left(P - \frac{3}{4}\beta L\right) + \theta_D\left(P - \frac{3}{2}\beta L\right) = 0 \tag{h}$$

We can obtain a second equation which describes the equilibrium of the disturbed structure from the free-body diagram of bar $ABC$ alone (Fig. 11-5e). The moment at $C$ is equal to rotational spring stiffness $\beta_R$ times the relative rotation at $C$, and the spring force $H_B$ is equal to the spring constant $\beta$ times the total translational displacement at $B$:

$$M_C = \beta_R(\theta_C - \theta_D) \tag{i}$$

and

$$H_B = \beta\Delta_B = \beta\left(\theta_C\frac{L}{2} + \theta_D L\right) \tag{j}$$

Summing moments about $C$ in Fig. 11-5e, we get the second equilibrium equation for the combined model as

$$\Sigma M_C = 0 \qquad P(\theta_C L) - M_C - H_B \frac{L}{2} = 0 \qquad \textbf{(k)}$$

Inserting expressions for $M_C$ using Eq. (i) and $H_B$ using Eq. (j) into Eq. (k) and simplifying gives

$$\theta_C\left(P - \frac{1}{4}\beta L - \frac{\beta_R}{L}\right) + \theta_D\left(\frac{\beta_R}{L} - \frac{1}{2}\beta L\right) = 0 \qquad \textbf{(l)}$$

We now have two algebraic equations in Eqs. (h) and (l) and two unknowns $(\theta_C, \theta_D)$. These equations can have nonzero (i.e., nontrivial) solutions only if the determinant of the coefficients of $\theta_C$ and $\theta_D$ is equal to zero. Substituting the assumed expression for $\beta_R (2/5\beta L^2)$ and then evaluating the determinant produces the following *characteristic equation* for the system:

$$P^2 - \left(\frac{41}{20}\beta L\right)P + \frac{9}{10}(\beta L)^2 = 0 \qquad \textbf{(m)}$$

Solving Eq. (m) using the quadratic formula results in two possible values of the critical load:

$$P_{cr1} = \beta L\left(\frac{41 - \sqrt{241}}{40}\right) = 0.637\beta L$$

$$P_{cr2} = \beta L\left(\frac{41 + \sqrt{241}}{40}\right) = 1.413\beta L$$

These are the *eigenvalues* of the combined 2DOF system. Usually, the lower value of the critical load is of more interest, because the structure will buckle first at this lower load value. If we substitute $P_{cr1}$ and $P_{cr2}$ back into Eqs. (h) and (l), we can find the buckled mode shape (i.e., *eigenvector*) associated with each critical load.

*Application of combined model to Structures 1 and 2.* If the rotational spring stiffness $\beta_R$ goes to infinity while the translational spring stiffness $\beta$ remains finite, the combined model (Fig. 11-5d) reduces to Structure 1 because the rotation angles $\theta_C$ and $\theta_D$ are equal, as shown in Fig. 11-5a. Equating $\theta_C$ and $\theta_D$ in Eq. (h) and solving for $P$ results in $P_{cr} = (9/8)\beta L$, which is the critical load for Structure 1 [see Eq. (d)].

If the rotational spring stiffness $\beta_R$ remains finite while the translational spring stiffness $\beta$ goes to infinity, the combined model (Fig. 11-5d) reduces to Structure 2. The translational spring becomes a roller support, so $\Delta_B = 0$ (i.e., $H_B = 0$) while rotation angle $\theta_C = -2\theta_D$ (i.e., $\theta_C$ is clockwise, so negative, as shown in Fig. 11-5b). Inserting $\beta = 0$ and $\theta_C = -2\theta_D$ into Eq. (l) gives the critical load for Structure 2 [see Eq. (g)].

## 11.3 COLUMNS WITH PINNED ENDS

We begin our investigation of the stability behavior of columns by analyzing a slender column with pinned ends (Fig. 11-6a). The column is loaded by a vertical force $P$ that is applied through the centroid of the end cross section. The column itself is perfectly straight and is made of a linearly elastic material that follows Hooke's law. Since the column is assumed to have no imperfections, it is referred to as an **ideal column**.

For purposes of analysis, we construct a coordinate system with its origin at support $A$ and with the $x$ axis along the longitudinal axis of the column. The $y$ axis is directed to the left in the figure, and the $z$ axis (not shown) comes out of the plane of the figure toward the viewer. We assume that the $xy$ plane is a plane of symmetry of the column and that any bending takes place in that plane (Fig. 11-6b). The coordinate system is identical to the one used in our previous discussions of beams, as can be seen by rotating the column clockwise through an angle of 90°.

When the axial load $P$ has a small value, the column remains perfectly straight and undergoes direct axial compression. The only stresses are the uniform compressive stresses obtained from the equation $\sigma = P/A$. The column is in **stable equilibrium**, which means that it returns to the straight position after a disturbance. For instance, if we apply a small lateral load and cause the column to bend, the deflection will disappear and the column will return to its original position when the lateral load is removed.

As the axial load $P$ is gradually increased, we reach a condition of **neutral equilibrium** in which the column may have a bent shape. The corresponding value of the load is the **critical load** $P_{cr}$. At this load the column may undergo small lateral deflections with no change in the axial force. For instance, a small lateral load will produce a bent shape that does not disappear when the lateral load is removed. Thus, the critical load can maintain the column in equilibrium *either* in the straight position or in a slightly bent position.

At higher values of the load, the column is **unstable** and may collapse by buckling, that is, by excessive bending. For the ideal case that we are discussing, the column will be in equilibrium in the straight position even

---

**Fig. 11-6**

Column with pinned ends: (a) ideal column, (b) buckled shape, and (c) axial force $P$ and bending moment $M$ acting at a cross section

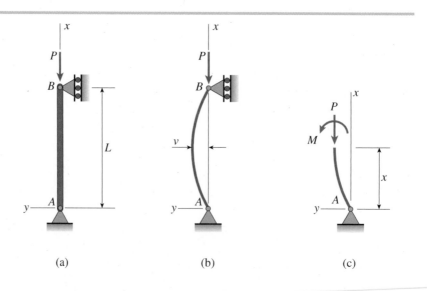

(a)                    (b)                    (c)

when the axial force $P$ is greater than the critical load. However, since the equilibrium is unstable, the smallest imaginable disturbance will cause the column to deflect sideways. Once that happens, the deflections will immediately increase and the column will fail by buckling. The behavior is similar to that described in the preceding section for the idealized buckling model (Fig. 11-2).

The behavior of an ideal column compressed by an axial load $P$ (Figs. 11-6a and b) may be summarized as follows:

If $P < P_{cr}$, the column is in stable equilibrium in the straight position.

If $P = P_{cr}$, the column is in neutral equilibrium in either the straight or a slightly bent position.

If $P > P_{cr}$, the column is in unstable equilibrium in the straight position and will buckle under the slightest disturbance.

Of course, a real column does not behave in this idealized manner because imperfections are always present. For instance, the column is not *perfectly* straight, and the load is not *exactly* at the centroid. Nevertheless, we begin by studying ideal columns because they provide insight into the behavior of real columns.

## Differential Equation for Column Buckling

To determine the critical loads and corresponding deflected shapes for an ideal pin-ended column (Fig. 11-6a), we use one of the differential equations of the deflection curve of a beam (see Eqs. 9-16a, b, and c in Section 9.2). These equations are applicable to a buckled column because the column bends as though it were a beam (Fig. 11-6b).

Although both the fourth-order differential equation (the load equation) and the third-order differential equation (the shear-force equation) are suitable for analyzing columns, we will elect to use the second-order equation (the bending-moment equation) because its general solution is usually the simplest. The **bending-moment equation** (Eq. 9-16a) is

$$EIv'' = M \qquad \textbf{(11-4)}$$

in which $M$ is the bending moment at any cross section, $v$ is the lateral deflection in the $y$ direction, and $EI$ is the flexural rigidity for bending in the $xy$ plane.

The bending moment $M$ at distance $x$ from end $A$ of the buckled column is shown acting in its positive direction in Fig. 11-6c. Note that the bending moment sign convention is the same as that used in earlier chapters, namely, positive bending moment produces positive curvature (see Figs. 9-3 and 9-4).

The axial force $P$ acting at the cross section is also shown in Fig. 11-6c. Since there are no horizontal forces acting at the supports, there are no shear forces in the column. Therefore, from equilibrium of moments about point $A$, we obtain

$$M + Pv = 0 \quad \text{or} \quad M = -Pv \qquad \textbf{(11-5)}$$

where $v$ is the deflection at the cross section.

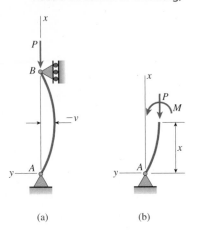

**Fig. 11-7**

Column with pinned ends (alternative direction of buckling)

(a)         (b)

This same expression for the bending moment is obtained if we assume that the column buckles to the right instead of to the left (Fig. 11-7a). When the column deflects to the right, the deflection itself is $-v$ but the moment of the axial force about point $A$ also changes sign. Thus, the equilibrium equation for moments about point $A$ (see Fig. 11-7b) is

$$M - P(-v) = 0$$

which gives the same expression for the bending moment $M$ as before.

The **differential equation of the deflection curve** (Eq. 11-4) now becomes

$$EIv'' + Pv = 0 \qquad \textbf{(11-6)}$$

By solving this equation, which is a *homogeneous, linear, differential equation of second order with constant coefficients*, we can determine the magnitude of the critical load and the deflected shape of the buckled column. Note that we are analyzing the buckling of columns by solving the same basic differential equation as the one we solved in Chapters 9 and 10 when finding beam deflections. However, there is a fundamental difference in the two types of analysis. In the case of beam deflections, the bending moment $M$ appearing in Eq. (11-4) is a function of the loads only—it does not depend upon the deflections of the beam. In the case of buckling, the bending moment is a function of the deflections themselves (Eq. 11-5).

Thus, we now encounter a new aspect of bending analysis. In our previous work, the deflected shape of the structure was not considered, and the equations of equilibrium were based upon the geometry of the *undeformed* structure. Now, however, the geometry of the *deformed* structure is taken into account when writing equations of equilibrium.

## Solution of the Differential Equation

For convenience in writing the solution of the differential equation of Eq. (11-6), we introduce the notation

$$k^2 = \frac{P}{EI} \quad \text{or} \quad k = \sqrt{\frac{P}{EI}} \qquad \textbf{(11-7a,b)}$$

in which $k$ is always taken as a positive quantity. Note that $k$ has units of the reciprocal of length, and therefore quantities such as $kx$ and $kL$ are nondimensional.

Using this notation, we can rewrite Eq. (11-6) in the form

$$v'' + k^2 v = 0 \qquad \textbf{(11-8)}$$

From mathematics we know that the general solution of this equation is

$$v = C_1 \sin kx + C_2 \cos kx \qquad \textbf{(11-9)}$$

in which $C_1$ and $C_2$ are constants of integration (to be evaluated from the boundary conditions, or end conditions, of the column). Note that the number of arbitrary constants in the solution (two in this case) agrees with the order of the differential equation. Also, note that we can verify the solution by substituting the expression for $v$ (Eq. 11-9) into the differential equation (Eq. 11-8) and reducing it to an identity.

To evaluate the **constants of integration** appearing in the solution (Eq. 11-9), we use the boundary conditions at the ends of the column; namely, the deflection is zero when $x = 0$ and $x = L$ (see Fig. 11-5b):

$$v(0) = 0 \quad \text{and} \quad v(L) = 0 \qquad \textbf{(11-10a,b)}$$

The first condition gives $C_2 = 0$, and therefore,

$$v = C_1 \sin kx \qquad \textbf{(11-10c)}$$

The second condition gives

$$C_1 \sin kL = 0 \qquad \textbf{(11-10d)}$$

From this equation we conclude that either $C_1 = 0$ or $\sin kL = 0$. We will consider both of these possibilities.

**Case 1.** If the constant $C_1$ equals zero, the deflection $v$ is also zero [see Eq. (11-10c)], and therefore the column remains straight. In addition, we note that when $C_1$ equals zero, Eq. (11-10d) is satisfied for *any* value of the quantity $kL$. Consequently, the axial load $P$ may also have any value [see Eq. (11-7b)]. This solution of the differential equation (known in mathematics as the *trivial solution*) is represented by the vertical axis of the load-deflection diagram (Fig. 11-8). It gives the behavior of an ideal column that is in equilibrium (either stable or unstable) in the straight position (no deflection) under the action of the compressive load $P$.

**Case 2.** The second possibility for satisfying Eq. (11-10d) is given by the following equation, known as the **buckling equation**:

$$\sin kL = 0 \qquad \textbf{(11-11)}$$

This equation is satisfied when $kL = 0, \pi, 2\pi, \ldots$. However, since $kL = 0$ means that $P = 0$, this solution is not of interest. Therefore, the solutions we will consider are

$$kL = n\pi \qquad n = 1, 2, 3, \ldots \qquad \textbf{(11-12)}$$

or [see Eq. (11-6a)]:

$$P = \frac{n^2 \pi^2 EI}{L^2} \qquad n = 1, 2, 3, \ldots \qquad \textbf{(11-13)}$$

This formula gives the values of $P$ that satisfy the buckling equation and provide solutions (other than the trivial solution) to the differential equation.

The equation of the **deflection curve** from Eqs. (11-10c) and (11-12) is

$$v = C_1 \sin kx = C_1 \sin \frac{n\pi x}{L} \qquad n = 1, 2, 3, \ldots \qquad \textbf{(11-14)}$$

Only when $P$ has one of the values given by Eq. (11-13) is it theoretically possible for the column to have a bent shape [given by Eq. (11-14)]. For

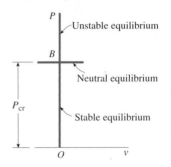

**Fig. 11-8**

Load-deflection diagram for an ideal, linearly elastic column

all other values of $P$, the column is in equilibrium only if it remains straight. Therefore, the values of $P$ given by Eq. (11-13) are the **critical loads** for this column.

## Critical Loads

The lowest critical load for a column with pinned ends (Fig. 11-9a) is obtained when $n = 1$:

$$P_{cr} = \frac{\pi^2 EI}{L^2} \tag{11-15}$$

The corresponding buckled shape (sometimes called a *mode shape*) is

$$v = C_1 \sin \frac{\pi x}{L} \tag{11-16}$$

as shown in Fig. 11-9b. The constant $C_1$ represents the deflection at the midpoint of the column and may have any small value, either positive or negative. Therefore, the part of the load-deflection diagram corresponding to $P_{cr}$ is a horizontal straight line (Fig. 11-8). Thus, the deflection at the critical load is *undefined*, although it must remain small for our equations to be valid. Above the bifurcation point $B$ the equilibrium is unstable, and below point $B$ it is stable.

Buckling of a pinned-end column in the first mode is called the **fundamental case** of column buckling.

The type of buckling described in this section is called **Euler buckling**, and the critical load for an ideal elastic column is often called the **Euler load**. The famous mathematician Leonhard Euler (1707–1783), generally recognized as the greatest mathematician of all time, was the first person to investigate the buckling of a slender column and determine its critical load (Euler published his results in 1744); see Ref. 11-1.

By taking higher values of the index $n$ in Eqs. (11-13) and (11-14), we obtain an infinite number of critical loads and corresponding mode shapes. The mode shape for $n = 2$ has two half-waves, as pictured in Fig. 11-9c. The corresponding critical load is four times larger than the critical load for the fundamental case. The magnitudes of the critical loads

---

**Fig. 11-9**

Buckled shapes for an ideal column with pinned ends: (a) initially straight column, (b) buckled shape for $n = 1$, and (c) buckled shape for $n = 2$

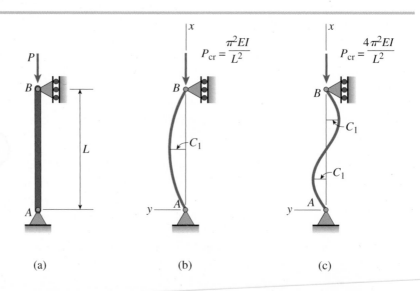

are proportional to the square of *n*, and the number of half-waves in the buckled shape is equal to *n*.

Buckled shapes for the **higher modes** are often of no practical interest because the column buckles when the axial load *P* reaches its lowest critical value. The only way to obtain modes of buckling higher than the first is to provide lateral support of the column at intermediate points, such as at the midpoint of the column shown in Fig. 11-9 (see Example 11-2 at the end of this section).

## General Comments

From Eq. (11-15), we see that the critical load of a column is proportional to the flexural rigidity *EI* and inversely proportional to the square of the length. Of particular interest is the fact that the *strength* of the material itself, as represented by a quantity such as the proportional limit or the yield stress, does not appear in the equation for the critical load. Therefore, increasing a strength property does not raise the critical load of a slender column. It can only be raised by increasing the flexural rigidity, reducing the length, or providing additional lateral support.

The *flexural rigidity* can be increased by using a "stiffer" material (that is, a material with larger modulus of elasticity *E*) or by distributing the material in such a way as to increase the moment of inertia *I* of the cross section, just as a beam can be made stiffer by increasing the moment of inertia. The moment of inertia is increased by distributing the material farther from the centroid of the cross section. Hence, a hollow tubular member is generally more economical for use as a column than a solid member having the same cross-sectional area.

Reducing the *wall thickness* of a tubular member and increasing its lateral dimensions (while keeping the cross-sectional area constant) also increases the critical load because the moment of inertia is increased. This process has a practical limit, however, because eventually the wall itself will become unstable. When that happens, localized buckling occurs in the form of small corrugations or wrinkles in the walls of the column. Thus, we must distinguish between *overall buckling* of a column, which is discussed in this chapter, and *local buckling* of its parts. The latter requires more detailed investigations and is beyond the scope of this book.

In the preceding analysis (see Fig. 11-9), we assumed that the *xy* plane was a plane of symmetry of the column and that buckling took place in that plane. The latter assumption will be met if the column has lateral supports perpendicular to the plane of the figure, so that the column is constrained to buckle in the *xy* plane. If the column is supported only at its ends and is free to buckle in *any* direction, then bending will occur about the principal centroidal axis having the smaller moment of inertia.

For instance, consider the rectangular and wide-flange cross sections shown in Fig. 11-10. In each case, the moment of inertia $I_1$ is greater than the moment of inertia $I_2$; hence the column will buckle in the 1–1 plane, and the smaller moment of inertia $I_2$ should be used in the formula for the critical load. If the cross section is square or circular, all centroidal axes have the same moment of inertia and buckling may occur in any longitudinal plane.

### Fig. 11-10

Cross sections of columns showing principal centroidal axes with $I_1 > I_2$

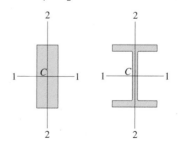

## Critical Stress

After finding the critical load for a column, we can calculate the corresponding **critical stress** by dividing the load by the cross-sectional area. For the fundamental case of buckling (Fig. 11-9b), the critical stress is

$$\sigma_{cr} = \frac{P_{cr}}{A} = \frac{\pi^2 EI}{AL^2} \tag{11-17}$$

in which $I$ is the moment of inertia for the principal axis about which buckling occurs. This equation can be written in a more useful form by introducing the notation

$$r = \sqrt{\frac{I}{A}} \tag{11-18}$$

in which $r$ is the **radius of gyration** of the cross section in the plane of bending.* Then the equation for the critical stress becomes

$$\sigma_{cr} = \frac{\pi^2 E}{(L/r)^2} \tag{11-19}$$

in which $L/r$ is a nondimensional ratio called the **slenderness ratio**:

$$\text{Slenderness ratio} = \frac{L}{r} \tag{11-20}$$

Note that the slenderness ratio depends only on the dimensions of the column. A column that is long and slender will have a high slenderness ratio and therefore a low critical stress. A column that is short and stubby will have a low slenderness ratio and will buckle at a high stress. Typical values of the slenderness ratio for actual columns are between 30 and 150.

The critical stress is the average compressive stress on the cross section at the instant the load reaches its critical value. We can plot a graph of this stress as a function of the slenderness ratio and obtain a curve known as **Euler's curve** (Fig. 11-11). The curve shown in the figure is plotted for a structural steel with $E = 30 \times 10^3$ ksi. The curve is valid only when the critical stress is less than the proportional limit of the steel, because the equations were derived using Hooke's law. Therefore, we draw a horizontal line on the graph at the proportional limit of the steel (assumed to be 36 ksi) and terminate Euler's curve at that level of stress.**

### Fig. 11-11

Graph of Euler's curve from Eq. (11-19) for structural steel with $E = 30 \times 10^3$ ksi and $\sigma_{pl} = 36$ ksi

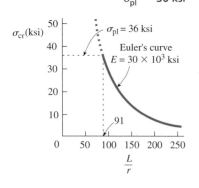

---

*Radius of gyration is described in Section 12.4.

**Euler's curve is not a common geometric shape. It is sometimes mistakenly called a hyperbola, but hyperbolas are plots of polynomial equations of the second degree in two variables, whereas Euler's curve is a plot of an equation of the third degree in two variables.

# Effects of Large Deflections, Imperfections, and Inelastic Behavior

The equations for critical loads were derived for ideal columns, that is, columns for which the loads are precisely applied, the construction is perfect, and the material follows Hooke's law. As a consequence, we found that the magnitudes of the small deflections at buckling were undefined.[*] Thus, when $P = P_{cr}$, the column may have any small deflection, a condition represented by the horizontal line labeled $A$ in the load-deflection diagram of Fig. 11-12. (In this figure, we show only the right-hand half of the diagram, but the two halves are symmetric about the vertical axis.)

The theory for ideal columns is limited to small deflections because we used the second derivative $v''$ for the curvature. A more exact analysis, based upon the exact expression for curvature [Eq. (9-19) in Section 9.2], shows that there is no indefiniteness in the magnitudes of the deflections at buckling. Instead, for an ideal, linearly elastic column, the load-deflection diagram goes upward in accord with curve $B$ of Fig. 11-12. Thus, after a linearly elastic column begins to buckle, an increasing load is required to cause an increase in the deflections.

Now suppose that the column is not constructed perfectly; for instance, the column might have an imperfection in the form of a small initial curvature, so that the unloaded column is not perfectly straight. Such imperfections produce deflections from the onset of loading, as shown by curve $C$ in Fig. 11-12. For small deflections, curve $C$ approaches line $A$ as an asymptote. However, as the deflections become large, it approaches curve $B$. The larger the imperfections, the further curve $C$ moves to the right, away from the vertical line. Conversely, if the column is constructed with considerable accuracy, curve $C$ approaches the vertical axis and the horizontal line labeled $A$. By comparing lines $A$, $B$, and $C$, we see that for practical purposes the critical load represents the maximum load-carrying capacity of an elastic column, because large deflections are not acceptable in most applications.

Finally, consider what happens when the stresses exceed the proportional limit and the material no longer follows Hooke's law. Of course, the load-deflection diagram is unchanged up to the level of load at which the proportional limit is reached. Then the curve for inelastic behavior (curve $D$) departs from the elastic curve, continues upward, reaches a maximum, and turns downward.

The precise shapes of the curves in Fig. 11-12 depend upon the material properties and column dimensions, but the general nature of the behavior is typified by the curves shown.

Only extremely slender columns remain elastic up to the critical load. Stockier columns behave inelastically and follow a curve such as $D$. Thus, the maximum load that can be supported by an inelastic column may be considerably less than the Euler load for that same column. Furthermore, the descending part of curve $D$ represents sudden and catastrophic collapse, because it takes smaller and smaller loads to maintain larger and larger deflections. By contrast, the curves for elastic

**Fig. 11-12**

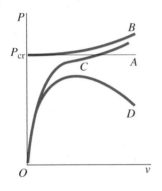

Load-deflection diagram for columns: Line $A$, ideal elastic column with small deflections; Curve $B$, ideal elastic column with large deflections; Curve $C$, elastic column with imperfections; and Curve $D$, inelastic column with imperfections

---

[*]In mathematical terminology, we solved a *linear eigenvalue problem*. The critical load is an *eigenvalue* and the corresponding buckled mode shape is an *eigenfunction*.

columns are quite stable, because they continue upward as the deflections increase, and therefore it takes larger and larger loads to cause an increase in deflection. (Inelastic buckling is described in more detail in Sections 11.7 and 11.8.)

## Optimum Shapes of Columns

Compression members usually have the same cross sections throughout their lengths, and therefore only prismatic columns are analyzed in this chapter. However, prismatic columns are not the optimum shape if minimum weight is desired. The critical load of a column consisting of a given amount of material may be increased by varying the shape so that the column has larger cross sections in those regions where the bending moments are larger. Consider, for instance, a column of solid circular cross section with pinned ends. A column shaped as shown in Fig. 11-13a will have a larger critical load than a prismatic column made from the same volume of material. As a means of approximating this optimum shape, prismatic columns are sometimes reinforced over part of their lengths (Fig. 11-13b).

Now consider a prismatic column with pinned ends that is free to buckle in *any* lateral direction (Fig. 11-14a). Also, assume that the column has a solid cross section, such as a circle, square, triangle, rectangle, or hexagon (Fig. 11-14b). An interesting question arises: For a given cross-sectional area, which of these shapes makes the most efficient column? Or, in more precise terms, which cross section gives the largest critical load? Of course, we are assuming that the critical load is calculated from the Euler formula $P_{cr} = \pi^2 EI/L^2$ using the smallest moment of inertia for the cross section.

While a common answer to this question is "the circular shape," you can readily demonstrate that a cross section in the shape of an equilateral triangle gives a 21% higher critical load than does a circular cross section of the same area (see Prob. 11.3-11). The critical load for an equilateral triangle is also higher than the loads obtained for the other shapes; hence, an equilateral triangle is the optimum cross section (based only upon theoretical considerations). For a mathematical analysis of optimum column shapes, including columns with varying cross sections, see Ref. 11-4.

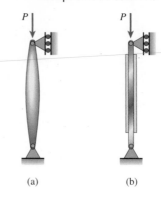

**Fig. 11-13**

Nonprismatic columns

(a)      (b)

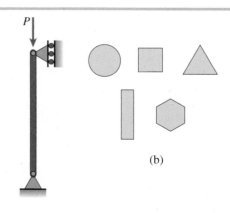

**Fig. 11-14**

Which cross-sectional shape is the optimum shape for a prismatic column?

(b)

(a)

### ● ● ● Example 11-2

A long, slender column *ABC* is pin-supported at the ends and compressed by an axial load *P* (Fig. 11-15). Lateral support is provided at the midpoint *B* in the plane of the figure. However, lateral support perpendicular to the plane of the figure is provided only at the ends.

The column is constructed of a steel wide-flange section W 200 × 34 having modulus of elasticity $E = 200$ GPa and proportional limit $\sigma_{pl} = 300$ MPa. The total length of the column is $L = 8$ m.

Determine the allowable load $P_{allow}$ using a factor of safety $n = 2.5$ with respect to Euler buckling of the column.

### Fig. 11-15

Example 11-2: Euler buckling of a slender column

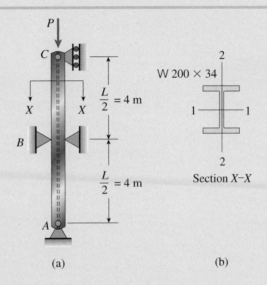

(a)                                        (b)

Slender steel column with lateral support near midheight (Lester Lefkowitz/Getty Images)

### Solution

Because of the manner in which it is supported, this column may buckle in either of the two principal planes of bending. As one possibility, it may buckle in the plane of the figure, in which case the distance between lateral supports is $L/2 = 4$ m and bending occurs about axis 2–2 (see Fig. 11-9c for the mode shape of buckling).

As a second possibility, the column may buckle perpendicular to the plane of the figure with bending about axis 1–1. Because the only lateral support in this direction is at the ends, the distance between lateral supports is $L = 8$ m (see Fig. 11-9b for the mode shape of buckling).

*Column properties.* From Table F-2, Appendix F, we obtain the following moments of inertia and cross-sectional area for a W 200 × 34 column:

$$I_1 = 2690 \text{ cm}^4 \qquad I_2 = 178 \text{ cm}^4 \qquad A = 43.6 \text{ cm}^2$$

*Continues* ➡

*Critical loads.* If the column buckles in the plane of the figure, the critical load is

$$P_{cr} = \frac{\pi^2 EI_2}{(L/2)^2} = \frac{4\pi^2 EI_2}{L^2}$$

Substituting numerical values, we obtain

$$P_{cr} = \frac{4\pi^2 EI_2}{L^2} = \frac{4\pi^2 (200 \text{ GPa})(178 \text{ cm}^4)}{(8 \text{ m})^2} = 220 \text{ kN}$$

If the column buckles perpendicular to the plane of the figure, the critical load is

$$P_{cr} = \frac{\pi^2 EI_1}{L^2} = \frac{\pi^2 (200 \text{ GPa})(2690 \text{ cm}^4)}{(8 \text{ m})^2} = 830 \text{ kN}$$

Therefore, the critical load for the column (the smaller of the two preceding values) is

$$P_{cr} = 220 \text{ kN}$$

and buckling occurs in the plane of the figure.

*Critical stresses.* Since the calculations for the critical loads are valid only if the material follows Hooke's law, we need to verify that the critical stresses do not exceed the proportional limit of the material. In the case of the larger critical load, we obtain the following critical stress:

$$\sigma_{cr} = \frac{P_{cr}}{A} = \frac{830 \text{ kN}}{43.6 \text{ cm}^2} = 190.4 \text{ MPa}$$

Since this stress is less than the proportional limit ($\sigma_{pl} = 300$ MPa), both critical-load calculations are satisfactory.

*Allowable load.* The allowable axial load for the column, based on Euler buckling, is

$$P_{allow} = \frac{P_{cr}}{n} = \frac{220 \text{ kN}}{2.5} = 88 \text{ kN}$$

in which $n = 2.5$ is the desired factor of safety.

# 11.4 COLUMNS WITH OTHER SUPPORT CONDITIONS

Buckling of a column with pinned ends (described in the preceding section) is usually considered as the most basic case of buckling. However, in practice we encounter many other end conditions, such as fixed ends, free ends, and elastic supports. The critical loads for columns with various kinds of support conditions can be determined from the differential equation of the deflection curve by following the same procedure that we used when analyzing a pinned-end column.

The **procedure** is as follows. First, with the column assumed to be in the buckled state, we obtain an expression for the bending moment in the column. Second, we set up the differential equation of the deflection curve, using the bending-moment equation ($EIv'' = M$). Third, we solve the equation and obtain its general solution, which contains two constants of integration plus any other unknown quantities. Fourth, we apply boundary conditions pertaining to the deflection $v$ and the slope $v'$ and obtain a set of simultaneous equations. Finally, we solve those equations to obtain the critical load and the deflected shape of the buckled column.

This straightforward mathematical procedure is illustrated in the following discussion of three types of columns.

Slender concrete columns fixed at the base and free at the top during construction (Digital Vision/Getty Images)

## Column Fixed at the Base and Free at the Top

The first case we will consider is an ideal column that is fixed at the base, free at the top, and subjected to an axial load $P$ (Fig. 11-16a).* The deflected shape of the buckled column is shown in Fig. 11-16b. From this figure we see that the bending moment at distance $x$ from the base is

$$M = P(\delta - v) \qquad \text{(11-21)}$$

where $\delta$ is the deflection at the free end of the column. The **differential equation** of the deflection curve then becomes

$$EIv'' = M = P(\delta - v) \qquad \text{(11-22)}$$

in which $I$ is the moment of inertia for buckling in the $xy$ plane.

Using the notation $k^2 = P/EI$ from Eq. (11-7a), we can rearrange Eq. (11-22) into the form

$$v'' + k^2v = k^2\delta \qquad \text{(11-23)}$$

which is a linear differential equation of second order with constant coefficients. However, it is a more complicated equation than the equation for a column with pinned ends [see Eq. (11-8)] because it has a nonzero term on the right-hand side.

The **general solution** of Eq. (11-23) consists of two parts: (1) the *homogeneous solution*, which is the solution of the homogeneous equation obtained by replacing the right-hand side with zero, and (2) the *particular*

---

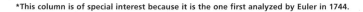

*This column is of special interest because it is the one first analyzed by Euler in 1744.

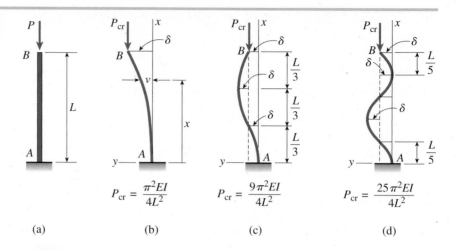

Fig. 11-16

Ideal column fixed at the base and free at the top: (a) initially straight column, (b) buckled shape for $n = 1$, (c) buckled shape for $n = 3$, and (d) buckled shape for $n = 5$

$$P_{cr} = \frac{\pi^2 EI}{4L^2}$$

$$P_{cr} = \frac{9\pi^2 EI}{4L^2}$$

$$P_{cr} = \frac{25\pi^2 EI}{4L^2}$$

(a)        (b)        (c)        (d)

*solution*, which is the solution of Eq. (11-23) that produces the term on the right-hand side.

The homogeneous solution (also called the *complementary solution*) is the same as the solution of Eq. (11-8); hence,

$$v_H = C_1 \sin kx + C_2 \cos kx \qquad \textbf{(11-24a)}$$

where $C_1$ and $C_2$ are constants of integration. Note that when $v_H$ is substituted into the left-hand side of the differential equation of Eq. (11-23), it produces zero.

The particular solution of the differential equation is

$$v_P = \delta \qquad \textbf{(11-24b)}$$

When $v_P$ is substituted into the left-hand side of the differential equation, it produces the right-hand side, that is, it produces the term $k^2\delta$. Consequently, the *general solution* of the equation, equal to the sum of $v_H$ and $v_P$, is

$$v = C_1 \sin kx + C_2 \cos kx + \delta \qquad \textbf{(11-25)}$$

This equation contains three unknown quantities ($C_1$, $C_2$, and $\delta$), and therefore three **boundary conditions** are needed to complete the solution.

At the base of the column, the deflection and slope are each equal to zero. Therefore, we obtain the following boundary conditions:

$$v(0) = 0 \qquad v'(0) = 0$$

Applying the first condition to Eq. (11-25), we find

$$C_2 = -\delta \qquad \textbf{(11-26)}$$

To apply the second condition, we first differentiate Eq. (11-25) to obtain the slope:

$$v' = C_1 k \cos kx - C_2 k \sin kx \qquad \textbf{(11-27)}$$

Applying the second condition to this equation, we find $C_1 = 0$.

Now we can substitute the expressions for $C_1$ and $C_2$ into the general solution of Eq. (11-25) and obtain the **equation of the deflection curve** for the buckled column:

$$v = \delta(1 - \cos kx) \qquad \textbf{(11-28)}$$

Note that this equation gives only the *shape* of the deflection curve—the amplitude $\delta$ remains undefined. Thus, when the column buckles, the deflection given by Eq. (11-28) may have any arbitrary magnitude, except that it must remain small (because the differential equation is based upon small deflections).

The third boundary condition applies to the upper end of the column, where the deflection $v$ is equal to $\delta$:

$$v(L) = \delta$$

Using this condition with Eq. (11-28), we get

$$\delta \cos kL = 0 \qquad \textbf{(11-29)}$$

From this equation we conclude that either $\delta = 0$ or $\cos kL = 0$. If $\delta = 0$, there is no deflection of the bar (see Eq. 11-28) and we have the *trivial solution*—the column remains straight and buckling does not occur. In that case, Eq. (11-29) will be satisfied for any value of the quantity $kL$, that is, for any value of the load $P$. This conclusion is represented by the vertical line in the load-deflection diagram of Fig. 11-8.

The other possibility for solving Eq. (11-29) is

$$\cos kL = 0 \qquad \textbf{(11-30)}$$

which is the **buckling equation**. In this case, Eq. (11-29) is satisfied regardless of the value of the deflection $\delta$. Thus, as already observed, $\delta$ is undefined and may have any small value.

The equation $\cos kL = 0$ is satisfied when

$$kL = \frac{n\pi}{2} \qquad n = 1, 3, 5, \ldots \qquad \textbf{(11-31)}$$

Using the expression $k^2 = P/EI$, we obtain the following formula for the **critical loads**:

$$P_{cr} = \frac{n^2 \pi^2 EI}{4L^2} \qquad n = 1, 3, 5, \ldots \qquad \textbf{(11-32)}$$

Also, the **buckled mode shapes** are obtained from Eq. (11-28):

$$v = \delta\left(1 - \cos \frac{n\pi x}{2L}\right) \qquad n = 1, 3, 5, \ldots \qquad \textbf{(11-33)}$$

The lowest critical load is obtained by substituting $n = 1$ in Eq. (11-32):

$$P_{cr} = \frac{\pi^2 EI}{4L^2} \tag{11-34}$$

The corresponding buckled shape (from Eq. 11-33) is

$$v = \delta\left(1 - \cos\frac{\pi x}{2L}\right) \tag{11-35}$$

and is shown in Fig. 11-16b.

By taking higher values of the index $n$, we can theoretically obtain an infinite number of critical loads from Eq. (11-32). The corresponding buckled mode shapes have additional waves in them. For instance, when $n = 3$ the buckled column has the shape shown in Fig. 11-16c and $P_{cr}$ is nine times larger than for $n = 1$. Similarly, the buckled shape for $n = 5$ has even more waves (Fig. 11-16d) and the critical load is twenty-five times larger.

## Effective Lengths of Columns

The critical loads for columns with various support conditions can be related to the critical load of a pinned-end column through the concept of an **effective length**. To demonstrate this idea, consider the deflected shape of a column fixed at the base and free at the top (Fig. 11-17a). This column buckles in a curve that is one-quarter of a complete sine wave. If we extend the deflection curve (Fig. 11-17b), it becomes one-half of a complete sine wave, which is the deflection curve for a pinned-end column.

The effective length $L_e$ for any column is the length of the equivalent pinned-end column, that is, it is the length of a pinned-end column having a deflection curve that exactly matches all or part of the deflection curve of the original column.

Another way of expressing this idea is to say that the effective length of a column is the distance between points of inflection (that is, points of zero moment) in its deflection curve, assuming that the curve is extended (if necessary) until points of inflection are reached. Thus, for a fixed-free column (Fig. 11-17), the effective length is

$$L_e = 2L \tag{11-36}$$

Because the effective length is the length of an equivalent pinned-end column, we can write a general formula for critical loads as

$$P_{cr} = \frac{\pi^2 EI}{L_e^2} \tag{11-37}$$

If we know the effective length of a column (no matter how complex the end conditions may be), we can substitute into the preceding equation and determine the critical load. For instance, in the case of a fixed-free column, we can substitute $L_e = 2L$ and obtain Eq. (11-34).

### Fig. 11-17

Deflection curves showing the effective length $L_e$ for a column fixed at the base and free at the top

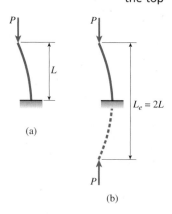

(a)

(b)

The effective length is often expressed in terms of an **effective-length factor** $K$:

$$L_e = KL \qquad \textbf{(11-38)}$$

where $L$ is the actual length of the column. Thus, the critical load is

$$P_{cr} = \frac{\pi^2 EI}{(KL)^2} \qquad \textbf{(11-39)}$$

The factor $K$ equals 2 for a column fixed at the base and free at the top and equals 1 for a pinned-end column. The effective-length factor is often included in design formulas for columns, as illustrated later in Section 11.9.

## Column with Both Ends Fixed Against Rotation

Next, let us consider a column with both ends fixed against rotation (Fig. 11-18a). Note that in this figure we use the standard symbol for the fixed support at the base of the column. However, since the column is free to shorten under an axial load, we must introduce a new symbol at the top of the column. This new symbol shows a rigid block that is constrained in such a manner that rotation and horizontal displacement are prevented but vertical movement can occur. (As a convenience when drawing sketches, we often replace this more accurate symbol with the standard symbol for a fixed support—see Fig. 11-18b—with the understanding that the column is free to shorten.)

The buckled shape of the column in the first mode is shown in Fig. 11-18c. Note that the deflection curve is symmetrical (with zero slope at the midpoint) and has zero slope at the ends. Because rotation at the ends is prevented, reactive moments $M_0$ develop at the supports. These moments, as well as the reactive force at the base, are shown in the figure.

From our previous solutions of the differential equation, we know that the equation of the deflection curve involves sine and cosine functions. Also, we know that the curve is symmetric about the midpoint. Therefore, we see immediately that the curve must have inflection points at distances $L/4$ from the ends. It follows that the middle portion of the

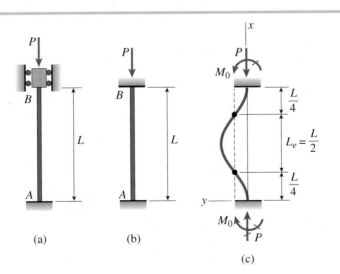

(a)          (b)

(c)

**Fig. 11-18**

Buckling of a column with both ends fixed against rotation

deflection curve has the same shape as the deflection curve for a pinned-end column. Thus, the effective length of a column with fixed ends, equal to the distance between inflection points, is

$$L_e = \frac{L}{2} \qquad \textbf{(11-40)}$$

Substituting into Eq. (11-37) gives the critical load:

$$P_{cr} = \frac{4\pi^2 EI}{L^2} \qquad \textbf{(11-41)}$$

This formula shows that the critical load for a column with fixed ends is four times that for a column with pinned ends. As a check, this result may be verified by solving the differential equation of the deflection curve (see Prob. 11.4-9).

## Column Fixed at the Base and Pinned at the Top

The critical load and buckled mode shape for a column that is fixed at the base and pinned at the top (Fig. 11-19a) can be determined by solving the differential equation of the deflection curve. When the column buckles (Fig. 11-19b), a reactive moment $M_0$ develops at the base because there can be no rotation at that point. Then from the equilibrium of the entire column, we know that there must be horizontal reactions $R$ at each end such that

$$M_0 = RL \qquad \textbf{(11-42)}$$

The bending moment in the buckled column, at distance $x$ from the base, is

$$M = M_0 - Pv - Rx = -Pv + R(L - x) \qquad \textbf{(11-43)}$$

and therefore, the **differential equation** is

$$EIv'' = M = -Pv + R(L - x) \qquad \textbf{(11-44)}$$

Again substituting $k^2 = P/EI$ and rearranging, we get

$$v'' + k^2 v = \frac{R}{EI}(L - x) \qquad \textbf{(11-45)}$$

The **general solution** of this equation is

$$v = C_1 \sin kx + C_2 \cos kx + \frac{R}{P}(L - x) \qquad \textbf{(11-46)}$$

in which the first two terms on the right-hand side constitute the homogeneous solution and the last term is the particular solution. This solution can be verified by substitution into the differential equation of Eq. (11-44).

Since the solution contains three unknown quantities ($C_1$, $C_2$, and $R$), we need three **boundary conditions**. They are

$$v(0) = 0 \qquad v'(0) = 0 \qquad v(L) = 0$$

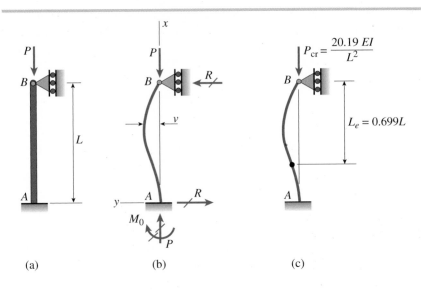

**Fig. 11-19**

Column fixed at the base and pinned at the top

(a)                    (b)                    (c)

Applying these conditions to Eq. (11-46) yields

$$C_2 + \frac{RL}{P} = 0 \qquad C_1 k - \frac{R}{P} = 0$$

$$C_1 \tan kL + C_2 = 0 \qquad \textbf{(11-47a,b,c)}$$

All three equations are satisfied if $C_1 = C_2 = R = 0$, in which case we have the trivial solution and the deflection is zero.

To obtain the solution for buckling, we must solve Eqs. (11-47a, b, and c) in a more general manner. One method of solution is to eliminate $R$ from the first two equations, which yields

$$C_1 kL + C_2 = 0 \quad \text{or} \quad C_2 = -C_1 kL \qquad \textbf{(11-47d)}$$

Next, we substitute this expression for $C_2$ into Eq. (11-47c) and obtain the **buckling equation**:

$$kL = \tan kL \qquad \textbf{(11-48)}$$

The solution of this equation gives the critical load.

Since the buckling equation is a transcendental equation, it cannot be solved explicitly.[*] Nevertheless, the values of $kL$ that satisfy the equation can be determined numerically by using a computer program for finding roots of equations. The smallest nonzero value of $kL$ that satisfies Eq. (11-48) is

$$kL = 4.4934 \qquad \textbf{(11-49)}$$

The corresponding **critical load** is

$$P_{cr} = \frac{20.19EI}{L^2} = \frac{2.046\pi^2 EI}{L^2} \qquad \textbf{(11-50)}$$

---

[*]In a transcendental equation, the variables are contained within transcendental functions. A transcendental function cannot be expressed by a finite number of algebraic operations; hence, trigonometric, logarithmic, exponential, and other such functions are transcendental.

which (as expected) is higher than the critical load for a column with pinned ends and lower than the critical load for a column with fixed ends [see Eqs. (11-15) and (11-41)].

The **effective length** of the column may be obtained by comparing Eqs. (11-50) and (11-37); thus,

$$L_e = 0.699L \approx 0.7L \tag{11-51}$$

This length is the distance from the pinned end of the column to the point of inflection in the buckled shape (Fig. 11-19c).

The equation of the **buckled mode shape** is obtained by substituting $C_2 = -C_1 kL$ [Eq. (11-47d)] and $R/P = C_1 k$ [Eq. (11-47b)] into the general solution [Eq. (11-46)]:

$$v = C_1[\sin kx - kL \cos kx + k(L - x)] \tag{11-52}$$

in which $k = 4.4934/L$. The term in brackets gives the mode shape for the deflection of the buckled column. However, the amplitude of the deflection curve is undefined because $C_1$ may have any value (within the usual limitation that the deflections must remain small).

## Limitations

In addition to the requirement of small deflections, the Euler buckling theory used in this section is valid only if the column is perfectly straight before the load is applied, the column and its supports have no imperfections, and the column is made of a linearly elastic material that follows Hooke's law. These limitations were explained previously in Section 11.3.

## Summary of Results

The lowest critical loads and corresponding effective lengths for the four columns we have analyzed are summarized in Fig. 11-20.

**Fig. 11-20**

Critical loads, effective lengths, and effective-length factors for ideal columns

| (a) Pinned-pinned column | (b) Fixed-free column | (c) Fixed-fixed column | (d) Fixed-pinned column |
|---|---|---|---|
| $P_{cr} = \dfrac{\pi^2 EI}{L^2}$ | $P_{cr} = \dfrac{\pi^2 EI}{4L^2}$ | $P_{cr} = \dfrac{4\pi^2 EI}{L^2}$ | $P_{cr} = \dfrac{2.046\,\pi^2 EI}{L^2}$ |
| $L_e = L$ | $L_e = 2L$ | $L_e = 0.5L$ | $L_e = 0.699L$ |
| $K = 1$ | $K = 2$ | $K = 0.5$ | $K = 0.699$ |

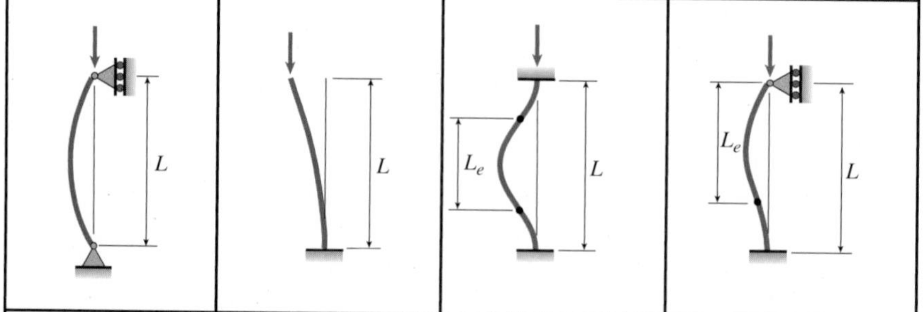

● ● ● **Example 11-3**

A viewing platform in a wild-animal park (Fig. 11-21a) is supported by a row of aluminum pipe columns having length $L$ = 10 ft 8 in. and outer diameter $d$ = 4 in. The bases of the columns are set in concrete footings and the tops of the columns are supported laterally by the platform. The columns are being designed to support compressive loads $P$ = 22.5 kips.

Determine the minimum required thickness $t$ of the columns (Fig. 11-21b) if a factor of safety $n$ = 3 is required with respect to Euler buckling. (For the aluminum, use 10,400 ksi for the modulus of elasticity and use 70 ksi for the proportional limit.)

**Fig. 11-21**

Example 11-3: Aluminum pipe column

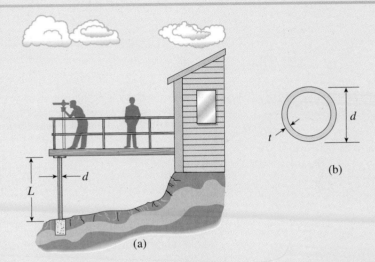

(a)

(b)

**Solution**

*Critical load.* Because of the manner in which the columns are constructed, we will model each column as a fixed-pinned column (see Fig. 11-20d). Therefore, the critical load is

$$P_{cr} = \frac{2.046\pi^2 EI}{L^2} \qquad \text{(a)}$$

in which $I$ is the moment of inertia of the tubular cross section:

$$I = \frac{\pi}{64}[d^4 - (d - 2t)^4] \qquad \text{(b)}$$

Substituting $d$ = 4 in., we get

$$I = \frac{\pi}{64}[(4 \text{ in.})^4 - (4 \text{ in.} - 2t)^4] \qquad \text{(c)}$$

in which $t$ is expressed in inches.

*Continues* ➡

*Required thickness of the columns.* Since the load per column is 22.5 kips and the factor of safety is 3, each column must be designed for the following critical load:

$$P_{cr} = nP = 3(22.5 \text{ kip}) = 67.5 \text{ kip}$$

Substituting this value for $P_{cr}$ in Eq. (a), and also replacing $I$ with its expression from Eq. (c), we obtain

$$67.5 \text{ kip} = \frac{2.046 \, \pi^2 (10,400 \text{ ksi})}{(128 \text{ in.})^2} \left[ \frac{\pi}{64} \left[ (4 \text{ in.})^4 - (4 \text{ in.} - 2t)^4 \right] \right]$$

Note that all terms in this equation are expressed in units of kips and inches. Solving the previous equation, we find that the minimum required thickness of the column to meet the specified conditions is

$$t_{min} = 0.254 \text{ in.} \qquad \Longleftarrow$$

*Supplementary calculations.* Knowing the diameter and thickness of the column, we can now calculate its moment of inertia, cross-sectional area, and radius of gyration. Using the minimum thickness of 0.254 in. we obtain

$$I = \frac{\pi}{64} [d^4 - (d - 2t)^4] = 5.267 \text{ in.}^4$$

$$A = \frac{\pi}{4} [d^2 - (d - 2t)^2] = 2.989 \text{ in.}^2 \qquad r = \sqrt{\frac{I}{A}} = 1.327 \text{ in.}$$

The slenderness ratio $L/r$ of the column is approximately 96.5, which is in the customary range for slender columns, and the diameter-to-thickness ratio $d/t$ is approximately 16, which should be adequate to prevent local buckling of the walls of the column.

The critical stress in the column must be less than the proportional limit of the aluminum if the formula for the critical load of Eq. (a) is to be valid. The critical stress is

$$\sigma_{cr} = \frac{P_{cr}}{A} = \frac{67.5 \text{ kip}}{2.989 \text{ in.}^2} = 22.6 \text{ ksi}$$

which is less than the proportional limit (70 ksi). Therefore, our calculation for the critical load using Euler buckling theory is satisfactory.

# 11.5 COLUMNS WITH ECCENTRIC AXIAL LOADS

**Fig. 11-22**

Column with eccentric axial loads

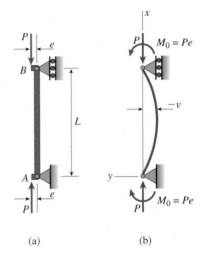

In Sections 11.3 and 11.4, we analyzed ideal columns in which the axial loads acted through the centroids of the cross sections. Under these conditions, the columns remain straight until the critical loads are reached, after which bending may occur.

Now we will assume that a column is compressed by loads $P$ that are applied with a small eccentricity $e$ measured from the axis of the column (Fig. 11-22a). Each eccentric axial load is equivalent to a centric load $P$ and a couple of moment $M_0 = Pe$ (Fig. 11-22b). This moment exists from the instant the load is first applied, and therefore the column begins to deflect at the onset of loading. The deflection then becomes steadily larger as the load increases.

To analyze the pin-ended column shown in Fig. 11-22, we make the same assumptions as in previous sections; namely, the column is initially perfectly straight, the material is linearly elastic, and the $xy$ plane is a plane of symmetry. The bending moment in the column at distance $x$ from the lower end (Fig. 11-22b) is

$$M = M_0 + P(-v) = Pe - Pv \tag{11-53}$$

where $v$ is the deflection of the column (positive when in the positive direction of the $y$ axis). Note that the deflections of the column are negative when the eccentricity of the load is positive.

The **differential equation** of the deflection curve is

$$EIv'' = M = Pe - Pv \tag{11-54}$$

or

$$v'' + k^2v = k^2e \tag{11-55}$$

in which $k^2 = P/EI$, as before. The general solution of this equation is

$$v = C_1 \sin kx + C_2 \cos kx + e \tag{11-56}$$

in which $C_1$ and $C_2$ are constants of integration in the homogeneous solution and $e$ is the particular solution. As usual, we can verify the solution by substituting it into the differential equation.

The **boundary conditions** for determining the constants $C_1$ and $C_2$ are obtained from the deflections at the ends of the column (Fig. 11-22b):

$$v(0) = 0 \qquad v(L) = 0$$

These conditions yield

$$C_2 = -e \qquad C_1 = -\frac{e(1 - \cos kL)}{\sin kL} = -e \tan \frac{kL}{2}$$

Therefore, the **equation of the deflection curve** is

$$v = -e\left(\tan \frac{kL}{2} \sin kx + \cos kx - 1\right) \tag{11-57}$$

For a column with known loads $P$ and known eccentricity $e$, we can use this equation to calculate the deflection at any point along the $x$ axis.

The behavior of a column with an eccentric load is quite different from that of a centrally loaded column, as can be seen by comparing Eq. (11-57) with Eqs. (11-16), (11-33), and (11-52). Equation (11-57) shows that each value of the eccentric load $P$ produces a definite value of the deflection, just as each value of the load on a beam produces a definite deflection. In contrast, the deflection equations for centrally loaded columns give the buckled mode shape (when $P = P_{cr}$) but with the amplitude undefined.

Because the column shown in Fig. 11-22 has pinned ends, its critical load (when centrally loaded) is

$$P_{cr} = \frac{\pi^2 EI}{L^2} \tag{11-58}$$

We will use this formula as a reference quantity in some of the equations that follow.

## Maximum Deflection

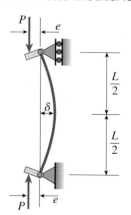

**Fig. 11-23**

Maximum deflection $\delta$ of a column with eccentric axial loads

The maximum deflection $\delta$ produced by the eccentric loads occurs at the midpoint of the column (Fig. 11-23) and is obtained by setting $x$ equal to $L/2$ in Eq. (11-57):

$$\delta = -v\left(\frac{L}{2}\right) = e\left( \tan\frac{kL}{2} \ \sin\frac{kL}{2} \ + \ \cos\frac{kL}{2} - 1 \right)$$

or, after simplifying,

$$\delta = e\left( \sec\frac{kL}{2} - 1 \right) \tag{11-59}$$

This equation can be written in a slightly different form by replacing the quantity $k$ with its equivalent in terms of the critical load [see Eq. (11-58)]:

$$k = \sqrt{\frac{P}{EI}} = \sqrt{\frac{P\pi^2}{P_{cr}L^2}} = \frac{\pi}{L}\sqrt{\frac{P}{P_{cr}}} \tag{11-60}$$

Thus, the nondimensional term $kL$ becomes

$$kL = \pi\sqrt{\frac{P}{P_{cr}}} \tag{11-61}$$

and Eq. (11-59) for the **maximum deflection** becomes

$$\delta = e\left[ \sec\left( \frac{\pi}{2}\sqrt{\frac{P}{P_{cr}}} \right) - 1 \right] \tag{11-62}$$

As special cases, we note the following: (1) The deflection $\delta$ is zero when the eccentricity $e$ is zero and $P$ is not equal to $P_{cr}$, (2) the deflection is zero when the axial load $P$ is zero, and (3) the deflection becomes infinitely large as $P$ approaches $P_{cr}$. These characteristics are shown in the **load-deflection diagram** of Fig. 11-24.

To plot the load-deflection diagram, we select a particular value $e_1$ of the eccentricity and then calculate $\delta$ for various values of the load $P$. The resulting curve is labeled $e = e_1$ in Fig. 11-24. We note immediately that the deflection $\delta$ increases as $P$ increases, but the relationship is nonlinear. Therefore, we *cannot* use the principle of superposition for calculating deflections due to more than one load, even though the material of the column is linearly elastic. As an example, the deflection due to an axial load $2P$ is *not* equal to twice the deflection caused by an axial load $P$.

Additional curves, such as the curve labeled $e = e_2$, are plotted in a similar manner. Since the deflection $\delta$ is linear with $e$ in Eq. (11-62), the curve for $e = e_2$ has the same *shape* as the curve for $e = e_1$, but the abscissas are larger by the ratio $e_2/e_1$.

As the load $P$ approaches the critical load, the deflection $\delta$ increases without limit and the horizontal line corresponding to $P = P_{cr}$ becomes an asymptote for the curves. In the limit, as $e$ approaches zero, the curves on the diagram approach two straight lines, one vertical and one horizontal (compare with Fig. 11-8). Thus, as expected, an ideal column with a centrally applied load ($e = 0$) is the limiting case of a column with an eccentric load ($e > 0$).

Although the curves plotted in Fig. 11-24 are mathematically correct, we must keep in mind that the differential equation is valid only for small deflections. Therefore, when the deflections become large, the curves are no longer physically valid and must be modified to take into account the presence of large deflections and (if the proportional limit of the material is exceeded) inelastic bending effects (see Fig. 11-12).

The reason for the nonlinear relationship between loads and deflections, even when the deflections are small and Hooke's law holds, can be understood if we observe once again that the axial loads $P$ are equivalent to centrally applied loads $P$ plus couples $Pe$ acting at the ends of the column (Fig. 11-22b). The couples $Pe$, if acting alone, would produce bending deflections of the column in the same manner as for a beam. In a beam, the presence of the deflections does not change the action of the loads, and the bending moments are the same whether the deflections exist or not. However, when an axial load is applied to the member, the existence of deflections increases the bending moments (the increases are equal to the product of the axial load and the deflections). When the bending moments increase, the deflections are further increased—hence, the moments increase even more, and so on. Thus, the bending moments in a column depend upon the deflections, which in turn depend upon the bending moments. This type of behavior results in a nonlinear relationship between the axial loads and the deflections.

In general, a straight structural member subjected to both bending loads and axial compressive loads is called a **beam-column**. In the case of a column with eccentric loads (Fig. 11-22), the bending loads are the moments $M_0 = Pe$ and the axial loads are the forces $P$.

**Fig. 11-24**

Load-deflection diagram for a column with eccentric axial loads [see Fig. 11-23 and Eq. (11-62)]

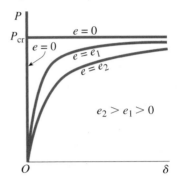

## Maximum Bending Moment

The maximum bending moment in an eccentrically loaded column occurs at the midpoint where the deflection is a maximum (Fig. 11-23):

$$M_{max} = P(e + \delta) \qquad \text{(11-63)}$$

Substituting for $\delta$ from Eqs. (11-59) and (11-62), we obtain

$$M_{max} = Pe\sec\frac{kL}{2} = Pe\sec\left(\frac{\pi}{2}\sqrt{\frac{P}{P_{cr}}}\right) \qquad \text{(11-64)}$$

The manner in which $M_{max}$ varies as a function of the axial load $P$ is shown in Fig. 11-25.

When $P$ is small, the maximum moment is equal to $Pe$, which means that the effect of the deflections is negligible. As $P$ increases, the bending moment grows nonlinearly and theoretically becomes infinitely large as $P$ approaches the critical load. However, as explained before, our equations are valid only when the deflections are small, and they cannot be used when the axial load approaches the critical load. Nevertheless, the preceding equations and accompanying graphs indicate the general behavior of beam-columns.

## Other End Conditions

The equations given in this section were derived for a pinned-end column, as shown in Figs. 11-22 and 11-23. If a column is fixed at the base and free at the top (Fig. 11-20b), we can use Eqs. 11-59 and 11-64 by replacing the actual length $L$ with the equivalent length $2L$ (see Prob. 11.5-9). However, the equations do not apply to a column that is fixed at the base and pinned at the top (Fig. 11-20d). The use of an equivalent length equal to $0.699L$ gives erroneous results; instead, we must return to the differential equation and derive a new set of equations.

In the case of a column with both ends fixed against rotation (Fig. 11-20c), the concept of an eccentric axial load acting at the end of the column has no meaning. Any moment applied at the end of the column is resisted directly by the supports and produces no bending of the column itself.

---

**Fig. 11-25**

Maximum bending moment in a column with eccentric axial loads [see Fig. 11-23 and Eq. (11-64)]

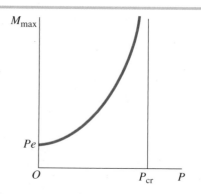

## ● ● ● Example 11-4

Fig. 11-26

Example 11-4: Brass bar with an eccentric axial load

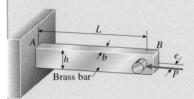

Brass bar

A brass bar $AB$ projecting from the side of a large machine is loaded at end $B$ by a force $P = 7$ kN acting with an eccentricity $e = 11$ mm (Fig. 11-26). The bar has a rectangular cross section with height $h = 30$ mm and width $b = 15$ mm.

What is the longest permissible length $L_{max}$ of the bar if the deflection at the end is limited to 3 mm? (For the brass, use $E = 110$ GPa.)

### Solution

*Critical load.* We will model this bar as a slender column that is fixed at end $A$ and free at end $B$. Therefore, the critical load (see Fig. 11-20b) is

$$P_{cr} = \frac{\pi^2 EI}{4L^2} \qquad \text{(a)}$$

The moment of inertia for the axis about which bending occurs is

$$I = \frac{hb^3}{12} = \frac{(30 \text{ mm})(15 \text{ mm})^3}{12} = 0.844 \text{ cm}^4$$

Therefore, the expression for the critical load becomes

$$P_{cr} = \frac{\pi^2 (110 \text{ GPa})(0.844 \text{ cm}^4)}{4L^2} = \frac{2.29 \text{ kN} \cdot \text{m}^2}{L^2} \qquad \text{(b)}$$

in which $P_{cr}$ has units of KN and $L$ has units of meters.

*Deflection.* The deflection at the end of the bar is given by Eq. (11-62), which applies to a fixed-free column as well as a pinned-end column:

$$\delta = e \left[ \sec \left( \frac{\pi}{2} \sqrt{\frac{P}{P_{cr}}} \right) - 1 \right] \qquad \text{(c)}$$

In this equation, $P_{cr}$ is given by Eq. (a).

*Length.* To find the maximum permissible length of the bar, we substitute for $\delta$ its limiting value of 3 mm. Also, we substitute $e = 11$ mm and $P = 7$ kN, and we substitute for $P_{cr}$ from Eq. (b). Thus,

$$3 \text{ mm} = (11 \text{ mm}) \left[ \sec \left( \frac{\pi}{2} \sqrt{\frac{7 \text{ kN}}{2.29/L^2}} \right) - 1 \right]$$

The only unknown in this equation is the length $L$ (meters). To solve for $L$, we perform the various arithmetic operations in the equation and then rearrange the terms. The result is

$$0.2727 = \sec (2.746 L) - 1$$

Using radians and solving this equation, we get $L = 0.243$ m. Thus, the maximum permissible length of the bar is

$$L_{max} = 0.243 \text{ m} \qquad \longleftarrow$$

If a longer bar is used, the deflection will exceed the allowable value of 3 mm.

# 11.6 THE SECANT FORMULA FOR COLUMNS

In the preceding section we determined the maximum deflection and maximum bending moment in a pin-ended column subjected to eccentric axial loads. In this section, we will investigate the maximum stresses in the column and obtain a special formula for calculating them.

The maximum stresses in a column with eccentric axial loads occur at the cross section where the deflection and bending moment have their largest values, that is, at the midpoint (Fig. 11-27a). Acting at this cross section are the compressive force $P$ and the bending moment $M_{max}$ (Fig. 11-27b). The stresses due to the force $P$ are equal to $P/A$, where $A$ is the cross-sectional area of the column, and the stresses due to the bending moment $M_{max}$ are obtained from the flexure formula.

Thus, the maximum compressive stress, which occurs on the concave side of the column, is

$$\sigma_{max} = \frac{P}{A} + \frac{M_{max}c}{I} \qquad \textbf{(11-65)}$$

in which $I$ is the moment of inertia in the plane of bending and $c$ is the distance from the centroidal axis to the extreme point on the concave side of the column. Note that in this equation we consider compressive stresses to be positive, since these are the important stresses in a column.

The bending moment $M_{max}$ is obtained from Eq. (11-64), which is repeated here:

$$M_{max} = Pe \sec\left(\frac{\pi}{2}\sqrt{\frac{P}{P_{cr}}}\right)$$

Since $P_{cr} = \pi^2 EI/L^2$ for a pinned-end column, and since $I = Ar^2$, where $r$ is the radius of gyration in the plane of bending, the preceding equation becomes

$$M_{max} = Pe \sec\left(\frac{L}{2r}\sqrt{\frac{P}{EA}}\right) \qquad \textbf{(11-66)}$$

Substituting into Eq. (11-65), we obtain the following formula for the **maximum compressive stress**:

$$\sigma_{max} = \frac{P}{A} + \frac{Pec}{I}\sec\left(\frac{L}{2r}\sqrt{\frac{P}{EA}}\right)$$

or

$$\sigma_{max} = \frac{P}{A}\left[1 + \frac{ec}{r^2}\sec\left(\frac{L}{2r}\sqrt{\frac{P}{EA}}\right)\right] \qquad \textbf{(11-67)}$$

This equation is commonly known as the **secant formula** for an eccentrically loaded column with pinned ends.

The secant formula gives the maximum compressive stress in the column as a function of the average compressive stress $P/A$, the modulus of elasticity $E$, and two nondimensional ratios—the slenderness ratio $L/r$ from Eq. (11-20) and the **eccentricity ratio**:

$$\text{Eccentricity ratio} = \frac{ec}{r^2} \qquad \textbf{(11-68)}$$

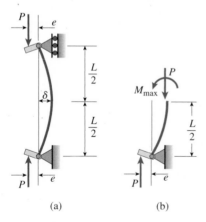

**Fig. 11-27**

Column with eccentric axial loads

As the name implies, the eccentricity ratio is a measure of the eccentricity of the load as compared to the dimensions of the cross section. Its numerical value depends upon the position of the load, but typical values are in the range from 0 to 3, with the most common values being less than 1.

When analyzing a column, we can use the secant formula to calculate the maximum compressive stress whenever the axial load $P$ and its eccentricity $e$ are known. Then the maximum stress can be compared with the allowable stress to determine if the column is adequate to support the load.

We can also use the secant formula in the reverse manner, that is, if we know the allowable stress, we can calculate the corresponding value of the load $P$. However, because the secant formula is transcendental, it is not practical to derive a formula for the load $P$. Instead, we can solve Eq. (11-67) numerically in each individual case.

A **graph of the secant formula** is shown in Fig. 11-28. The abscissa is the slenderness ratio $L/r$, and the ordinate is the average compressive stress $P/A$. The graph is plotted for a steel column with modulus of elasticity $E = 30 \times 10^3$ ksi and maximum stress $\sigma_{max} = 36$ ksi. Curves are plotted for several values of the eccentricity ratio $ec/r^2$. These curves are valid only when the maximum stress is less than the proportional limit of the material, because the secant formula was derived using Hooke's law.

A special case arises when the eccentricity of the load disappears ($e = 0$), because then we have an ideal column with a centrally applied load. Under these conditions the maximum load is the critical load ($P_{cr} = \pi^2 EI/L^2$) and the corresponding maximum stress is the critical stress [see Eqs. (11-17) and (11-19)]:

$$\sigma_{cr} = \frac{P_{cr}}{A} = \frac{\pi^2 EI}{AL^2} = \frac{\pi^2 E}{(L/r)^2} \qquad \textbf{(11-69)}$$

Since this equation gives the stress $P/A$ in terms of the slenderness ratio $L/r$, we can plot it on the graph of the secant formula (Fig. 11-28) as **Euler's curve**.

Let us now assume that the proportional limit of the material is the same as the selected maximum stress, that is, 36 ksi. Then we construct a

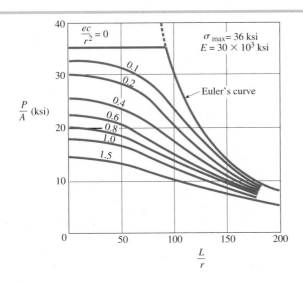

**Fig. 11-28**

Graph of the secant formula (Eq. 11-67) for $\sigma_{max} = 36$ ksi and $E = 30 \times 10^3$ ksi

horizontal line on the graph at a value of 36 ksi, and we terminate Euler's curve at that stress. The horizontal line and Euler's curve represent the limits of the secant-formula curves as the eccentricity $e$ approaches zero.

## Discussion of the Secant Formula

The graph of the secant formula shows that the load-carrying capacity of a column decreases significantly as the slenderness ratio $L/r$ increases, especially in the intermediate region of $L/r$ values. Thus, long slender columns are much less stable than short, stocky columns. The graph also shows that the load-carrying capacity decreases with increasing eccentricity $e$; furthermore, this effect is relatively greater for short columns than for long ones.

The secant formula was derived for a column with pinned ends, but it can also be used for a column that is fixed at the base and free at the top. All that is required is to replace the length $L$ in the secant formula with the equivalent length $2L$. However, because it is based upon Eq. (11-64), the secant formula is not valid for the other end conditions that we discussed.

Now let us consider an actual column, which inevitably differs from an ideal column because of imperfections, such as initial curvature of the longitudinal axis, imperfect support conditions, and nonhomogeneity of the material. Furthermore, even when the load is supposed to be centrally applied, there will be unavoidable eccentricities in its direction and point of application. The extent of these imperfections varies from one column to another, and therefore there is considerable scatter in the results of laboratory tests performed with actual columns.

All imperfections have the effect of producing bending in addition to direct compression. Therefore, it is reasonable to assume that the behavior of an imperfect, centrally loaded column is similar to that of an ideal, eccentrically loaded column. In such cases, the secant formula can be used by choosing an approximate value of the eccentricity ratio $ec/r^2$ to account for the combined effects of the various imperfections. For instance, a commonly used value of the eccentricity ratio for pinned-end columns in structural-steel design is $ec/r^2 = 0.25$. The use of the secant formula in this manner for columns with centrally applied loads provides a rational means of accounting for the effects of imperfections, rather than accounting for them simply by increasing the factor of safety. (For further discussions of the secant formula and the effects of imperfections, see Ref. 11-5 and textbooks on buckling and stability.)

The procedure for analyzing a centrally loaded column by means of the secant formula depends upon the particular conditions. For instance, if the objective is to determine the allowable load, the procedure is as follows. Assume a value of the eccentricity ratio $ec/r^2$ based upon test results, code values, or practical experience. Substitute this value into the secant formula, along with the values of $L/r$, $A$, and $E$ for the actual column. Assign a value to $\sigma_{max}$, such as the yield stress $\sigma_Y$ or the proportional limit $\sigma_{pl}$ of the material. Then solve the secant formula for the load $P_{max}$ that produces the maximum stress. (This load will always be less than the critical load $P_{cr}$ for the column.) The allowable load on the column equals the load $P_{max}$ divided by the factor of safety $n$.

The following example illustrates how the secant formula may be used to determine the maximum stress in a column when the load is known, and also how to determine the load when the maximum stress is given.

● ● ● **Example 11-5**

**Fig. 11-29**

Example 11-5: Column with an eccentrically applied axial load

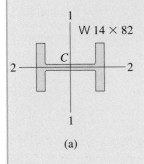

(a)

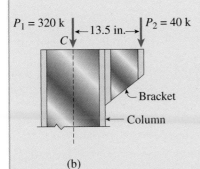

(b)

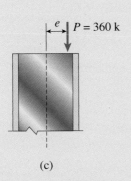

(c)

A steel wide-flange column of W 14 × 82 shape (Fig. 11-29a) is pin-supported at the ends and has a length of 25 ft. The column supports a centrally applied load $P_1 = 320$ k and an eccentrically applied load $P_2 = 40$ k (Fig. 11-29b). Bending takes place about axis 1–1 of the cross section, and the eccentric load acts on axis 2–2 at a distance of 13.5 in. from the centroid C.

(a) Using the secant formula, and assuming $E = 30,000$ ksi, calculate the maximum compressive stress in the column.
(b) If the yield stress for the steel is $\sigma_Y = 42$ ksi, what is the factor of safety with respect to yielding?

**Solution**

(a) *Maximum compressive stress.* The two loads $P_1$ and $P_2$ acting as shown in Fig. 11-29b are statically equivalent to a single load $P = 360$ k acting with an eccentricity $e = 1.5$ in. (Fig. 11-29c). Since the column is now loaded by a single force $P$ having an eccentricity $e$, we can use the secant formula to find the maximum stress.

The required properties of the W 14 × 82 wide-flange shape are obtained from Table F-1 in Appendix F:

$$A = 24.1 \text{ in.}^2 \qquad r = 6.05 \text{ in.} \qquad c = \frac{14.31 \text{ in.}}{2} = 7.155 \text{ in.}$$

The required terms in the secant formula of Eq. (11-67) are calculated as

$$\frac{P}{A} = \frac{360 \text{ k}}{24.1 \text{ in.}^2} = 14.94 \text{ ksi}$$

$$\frac{ec}{r^2} = \frac{(1.5 \text{ in.})(7.155 \text{ in.})}{(6.05 \text{ in.})^2} = 0.2932$$

$$\frac{L}{r} = \frac{(25 \text{ ft})(12 \text{ in./ft})}{6.05 \text{ in.}} = 49.59$$

$$\frac{P}{EA} = \frac{360 \text{ k}}{(30,000 \text{ ksi})(24.1 \text{ in.}^2)} = 497.9 \times 10^{-6}$$

Substituting these values into the secant formula, we get

$$\sigma_{max} = \frac{P}{A}\left[1 + \frac{ec}{r^2} \sec\left(\frac{L}{2r}\sqrt{\frac{P}{EA}}\right)\right]$$

$$= (14.94 \text{ ksi})(1 + 0.345) = 20.1 \text{ ksi} \qquad \Longleftarrow$$

This compressive stress occurs at midheight of the column on the concave side (the right-hand side in Fig. 11-29b).

*Continues* ➡

• • • **Example 11-5** - *Continued*

(b) *Factor of safety with respect to yielding.* To find the factor of safety, we need to determine the value of the load $P$, acting at the eccentricity $e$, that will produce a maximum stress equal to the yield stress $\sigma_Y = 42$ ksi. Since this value of the load is just sufficient to produce initial yielding of the material, we will denote it as $P_Y$.

Note that we cannot determine $P_Y$ by multiplying the load $P$ (equal to 360 k) by the ratio $\sigma_Y/\sigma_{max}$. The reason is that we are dealing with a non-linear relationship between load and stress. Instead, we must substitute $\sigma_{max} = \sigma_Y = 42$ ksi in the secant formula and then solve for the corresponding load $P$, which becomes $P_Y$. In other words, we must find the value of $P_Y$ that satisfies the following equation:

$$\sigma_Y = \frac{P_Y}{A}\left[1 + \frac{ec}{r^2}\sec\left(\frac{L}{2r}\sqrt{\frac{P_Y}{EA}}\right)\right] \tag{11-70}$$

Substituting numerical values, we obtain

$$42 \text{ ksi} = \frac{P_Y}{24.1 \text{ in.}^2}\left[1 + 0.2939\sec\left(\frac{49.59}{2}\sqrt{\frac{P_Y}{(30{,}000 \text{ ksi})(24.1 \text{ in.}^2)}}\right)\right]$$

or

$$1012 \text{ k} = P_Y\left[1 + 0.2939\sec\left(0.02916\sqrt{P_Y}\right)\right]$$

in which $P_Y$ has units of kips. Solving this equation numerically, we get

$$P_Y = 716 \text{ k}$$

This load will produce yielding of the material (in compression) at the cross section of maximum bending moment.

Since the actual load is $P = 360$ k, the factor of safety against yielding is

$$n = \frac{P_Y}{P} = \frac{716 \text{ k}}{360 \text{ k}} = 1.99$$

This example illustrates two of the many ways in which the secant formula may be used. Other types of analysis are illustrated in the problems at the end of the chapter.

# 11.7 ELASTIC AND INELASTIC COLUMN BEHAVIOR

In the preceding sections we described the behavior of columns when the material is stressed below the proportional limit. We began by considering an ideal column subjected to a centrally applied load (Euler buckling), and we arrived at the concept of a critical load $P_{cr}$. Then we considered columns with eccentric axial loads and derived the secant formula. We portrayed the results of these analyses on a diagram of average compressive stress $P/A$ versus the slenderness ratio $L/r$ (see Fig. 11-28). The behavior of an ideal column is represented in Fig. 11-28, by Euler's curve, and the behavior of columns with eccentric loads is represented by the family of curves having various values of the eccentricity ratio $ec/r^2$.

We will now extend our discussion to include **inelastic buckling**, that is, the buckling of columns when the proportional limit is exceeded. We will portray the behavior on the same kind of diagram as before, namely, a diagram of average compressive stress $P/A$ versus slenderness ratio $L/r$ (see Fig. 11-30). Note that Euler's curve is shown on this diagram as curve $ECD$. This curve is valid only in the region $CD$ where the stress is below the proportional limit $\sigma_{pl}$ of the material. Therefore, the part of Euler's curve above the proportional limit is shown by a dashed line.

The value of slenderness ratio above which Euler's curve is valid is obtained by setting the critical stress from Eq. (11-69) equal to the proportional limit $\sigma_{pl}$ and solving for the slenderness ratio. Thus, letting $(L/r)_c$ represent the **critical slenderness ratio** (Fig. 11-30), we get

$$\left(\frac{L}{r}\right)_c = \sqrt{\frac{\pi^2 E}{\sigma_{pl}}} \qquad \textbf{(11-71)}$$

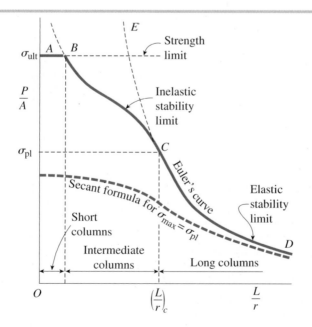

### Fig. 11-30

Diagram of average compressive stress $P/A$ versus slenderness ratio $L/r$

As an example, consider structural steel with $\sigma_{pl}$ = 36 ksi and $E$ = 30,000 ksi. Then the critical slenderness ratio $(L/r)_c$ is equal to 90.7. Above this value, an ideal column buckles elastically and the Euler load is valid. Below this value, the stress in the column exceeds the proportional limit and the column buckles inelastically.

If we take into account the effects of eccentricities in loading or imperfections in construction, but still assume that the material follows Hooke's law, we obtain a curve such as the one labeled "Secant formula" in Fig. 11-30. This curve is plotted for a maximum stress $\sigma_{max}$ equal to the proportional limit $\sigma_{pl}$.

When comparing the secant-formula curve with Euler's curve, we must keep in mind an important distinction. In the case of Euler's curve, the stress $P/A$ not only is proportional to the applied load $P$, but also is the actual maximum stress in the column when buckling occurs. Consequently, as we move from $C$ to $D$ along Euler's curve, both the maximum stress $P/A$ (equal to the critical stress) and the axial load $P$ decrease. However, in the case of the secant-formula curve, the *average* stress $P/A$ decreases as we move from left to right along the curve (and therefore the axial load $P$ also decreases) but the maximum stress (equal to the proportional limit) remains constant.

From Euler's curve, we see that **long columns** with large slenderness ratios buckle at low values of the average compressive stress $P/A$. This condition cannot be improved by using a higher-strength material, because collapse results from instability of the column as a whole and not from failure of the material itself. The stress can only be raised by reducing the slenderness ratio $L/r$ or by using a material with higher modulus of elasticity $E$.

When a compression member is very **short**, it fails by yielding and crushing of the material, and no buckling or stability considerations are involved. In such a case, we can define an ultimate compressive stress $\sigma_{ult}$ as the failure stress for the material. This stress establishes a **strength limit** for the column, represented by the horizontal line $AB$ in Fig. 11-30. The strength limit is much higher than the proportional limit, since it represents the ultimate stress in compression.

Between the regions of short and long columns, there is a range of **intermediate slenderness ratios** too small for elastic stability to govern and too large for strength considerations alone to govern. Such an intermediate-length column fails by inelastic buckling, which means that the maximum stresses are above the proportional limit when buckling occurs. Because the proportional limit is exceeded, the slope of the stress-strain curve for the material is less than the modulus of elasticity; hence, the critical load for inelastic buckling is always less than the Euler load (see Section 11.8).

The dividing lines between short, intermediate, and long columns are not precise. Nevertheless, it is useful to make these distinctions because the maximum load-carrying capacity of columns in each category is based upon quite different types of behavior. The maximum load-carrying capacity of a particular column (as a function of its length) is represented by curve $ABCD$ in Fig. 11-30. If the length is very small (region $AB$), the column fails by direct compression; if the column is longer (region $BC$), it fails by inelastic buckling; and if it is even longer (region $CD$), it fails by elastic buckling (that is, Euler buckling). Curve $ABCD$ applies to columns with various support conditions if the length $L$ in the slenderness ratio is replaced by the effective length $L_e$.

The results of **load tests** on columns are in reasonably good agreement with curve *ABCD*. When test results are plotted on the diagram, they generally form a band that lies just below this curve. Considerable scatter of test results is to be expected, because column performance is sensitive to such matters as the accuracy of construction, the alignment of loads, and the details of support conditions. To account for these variables, we usually obtain the allowable stress for a column by dividing the maximum stress (from curve *ABCD*) by a suitable **factor of safety**, which often has a value of about 2. Because imperfections are apt to increase with increase in length, a variable factor of safety (increasing as $L/r$ increases) is sometimes used. In Section 11.9 we will give some typical formulas for allowable stresses.

# 11.8 INELASTIC BUCKLING

The critical load for elastic buckling is valid only for relatively long columns, as explained previously (see curve *CD* in Fig. 11-30). If a column is of intermediate length, the stress in the column will reach the proportional limit before buckling begins (curve *BC* in Fig. 11-30). To calculate critical loads in this intermediate range, we need a theory of **inelastic buckling**. Three such theories are described in this section: the tangent-modulus theory, the reduced-modulus theory, and the Shanley theory.

## Tangent-Modulus Theory

Let us again consider an ideal, pinned-end column subjected to an axial force $P$ (Fig. 11-31a). The column is assumed to have a slenderness ratio $L/r$ that is less than the critical slenderness ratio (Eq. 11-71), and therefore the axial stress $P/A$ reaches the proportional limit before the critical load is reached.

The compressive **stress-strain diagram** for the material of the column is shown in Fig. 11-32. The proportional limit of the material is indicated as $\sigma_{pl}$, and the actual stress $\sigma_A$ in the column (equal to $P/A$) is represented by point $A$ (which is above the proportional limit). If the load is increased, so that a small increase in stress occurs, the relationship between the increment of stress and the corresponding increment of strain is given by the *slope* of the stress-strain diagram at point $A$. This slope, equal to the slope of the tangent line at $A$, is called the **tangent modulus** and is denoted by $E_t$; thus,

$$E_t = \frac{d\sigma}{d\varepsilon} \qquad \textbf{(11-72)}$$

Note that the tangent modulus *decreases* as the stress increases beyond the proportional limit. When the stress is below the proportional limit, the tangent modulus is the same as the ordinary elastic modulus $E$.

According to the **tangent-modulus theory** of inelastic buckling, the column shown in Fig. 11-31a remains straight until the inelastic critical load is reached. At that value of load, the column may undergo a small lateral deflection (Fig. 11-31b). The resulting bending stresses are superimposed

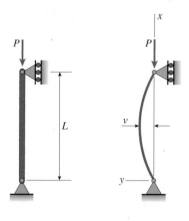

**Fig. 11-31**

Ideal column of intermediate length that buckles inelastically

(a)                    (b)

Fig. 11-32

Compression stress-strain diagram for the material of the column shown in Fig. 11-31

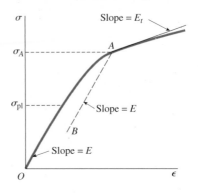

upon the axial compressive stresses $\sigma_A$. Since the column starts bending from a straight position, the initial bending stresses represent only a small increment of stress. Therefore, the relationship between the bending stresses and the resulting strains is given by the tangent modulus. Since the strains vary linearly across the cross section of the column, the initial bending stresses also vary linearly, and therefore the expressions for curvature are the same as those for linearly elastic bending except that $E_t$ replaces $E$:

$$\kappa = \frac{1}{\rho} = \frac{d^2v}{dx^2} = \frac{M}{E_t I} \tag{11-73}$$

(compare with Eqs. 9-5 and 9-7).

Because the bending moment $M = -Pv$ (see Fig. 11-31b), the differential equation of the deflection curve is

$$E_t I v'' + Pv = 0 \tag{11-74}$$

This equation has the same form as the equation for elastic buckling (Eq. 11-6) except that $E_t$ appears in place of $E$. Therefore, we can solve the equation in the same manner as before and obtain the following equation for the **tangent-modulus load**:

$$P_t = \frac{\pi^2 E_t I}{L^2} \tag{11-75}$$

This load represents the critical load for the column according to the tangent-modulus theory. The corresponding critical stress is

$$\sigma_t = \frac{P_t}{A} = \frac{\pi^2 E_t}{(L/r)^2} \tag{11-76}$$

which is similar in form to Eq. (11-69) for the Euler critical stress.

Since the tangent modulus $E_t$ varies with the compressive stress $\sigma = P/A$ (Fig. 11-32), we usually obtain the tangent-modulus load by an iterative procedure. We begin by estimating the value of $P_t$. This trial value, call it $P_1$, should be slightly larger than $\sigma_{pl} A$, which is the axial load when the stress just reaches the proportional limit. Knowing $P_1$, we can calculate the corresponding axial stress $\sigma_1 = P_1/A$ and determine the tangent modulus $E_t$ from the stress-strain diagram. Next, we use Eq. (11-75) to obtain a second estimate of $P_t$. Let us call this value $P_2$. If $P_2$ is very close to $P_1$, we may accept the load $P_2$ as the tangent-modulus load. However, it is more likely that additional cycles of iteration will be required until we reach a load that is in close agreement with the preceding trial load. This value is the tangent-modulus load.

A diagram showing how the critical stress $\sigma_t$ varies with the slenderness ratio $L/r$ is given in Fig. 11-33 for a typical metal column with pinned ends. Note that the curve is above the proportional limit and below Euler's curve.

The tangent-modulus formulas may be used for columns with various support conditions by using the effective length $L_e$ in place of the actual length $L$.

## Reduced-Modulus Theory

The tangent-modulus theory is distinguished by its simplicity and ease of use. However, it is conceptually deficient because it does not account for the complete behavior of the column. To explain the difficulty, we will consider again the column shown in Fig. 11-31a. When this column first departs from the straight position (Fig. 11-31b), bending stresses are added to the existing compressive stresses $P/A$. These additional stresses are compressive on the concave side of the column and tensile on the convex side. Therefore, the compressive stresses in the column become larger on the concave side and smaller on the other side.

Now imagine that the axial stress $P/A$ is represented by point $A$ on the stress-strain curve (Fig. 11-32). On the concave side of the column (where the compressive stress is increased), the material follows the tangent modulus $E_t$. However, on the convex side (where the compressive stress is decreased), the material follows the unloading line $AB$ on the stress-strain diagram. This line is parallel to the initial linear part of the diagram, and therefore its slope is equal to the elastic modulus $E$. Thus, at the onset of bending, the column behaves as if it were made of two different materials, a material of modulus $E_t$ on the concave side and a material of modulus $E$ on the convex side.

A bending analysis of such a column can be made using the bending theories for a beam of two materials (Sections 6.2 and 6.3). The results of such analyses show that the column bends as though the material had a modulus of elasticity between the values of $E$ and $E_t$. This "effective modulus" is known as the **reduced modulus** $E_r$, and its value depends not only upon the magnitude of the stress (because $E_t$ depends upon the magnitude of the stress), but also upon the shape of the cross section of the column. Thus, the reduced modulus $E_r$ is more difficult to determine than is the tangent modulus $E_t$. In the case of a column having a *rectangular cross section*, the equation for the reduced modulus is

$$E_r = \frac{4EE_t}{(\sqrt{E} + \sqrt{E_t})^2} \qquad \textbf{(11-77)}$$

For a *wide-flange beam* with the area of the web disregarded, the reduced modulus for bending about the strong axis is

$$E_r = \frac{2EE_t}{E + E_t} \qquad \textbf{(11-78)}$$

The reduced modulus $E_r$ is also called the *double modulus*.

Since the reduced modulus represents an effective modulus that governs the bending of the column when it first departs from the straight position, we can formulate a reduced-modulus theory of inelastic buckling. Proceeding in the same manner as for the tangent-modulus theory, we begin with an equation for the curvature and then we write the differential equation of the deflection curve. These equations are the same as

**Fig. 11-33**

Diagram of critical stress versus slenderness ratio

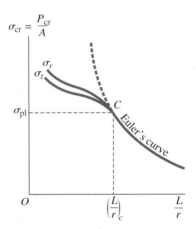

Eqs. (11-73) and (11-74), except that $E_r$ appears instead of $E_t$. Thus, we arrive at the following equation for the **reduced-modulus load**:

$$P_r = \frac{\pi^2 E_r I}{L^2} \qquad \textbf{(11-79)}$$

The corresponding equation for the critical stress is

$$\sigma_r = \frac{\pi^2 E_r}{(L/r)^2} \qquad \textbf{(11-80)}$$

To find the reduced-modulus load $P_r$, we again must use an iterative procedure, because $E_r$ depends upon $E_t$. The critical stress according to the reduced-modulus theory is shown in Fig. 11-33. Note that the curve for $\sigma_r$ is above that for $\sigma_t$, because $E_r$ is always greater than $E_t$.

The reduced-modulus theory is difficult to use in practice because $E_r$ depends upon the shape of the cross section as well as the stress-strain curve and must be evaluated for each particular column. Moreover, this theory also has a conceptual defect. In order for the reduced modulus $E_r$ to apply, the material on the convex side of the column must be undergoing a reduction in stress. However, such a reduction in stress cannot occur until bending actually takes place. Therefore, the axial load $P$, applied to an ideal straight column, can never actually reach the reduced-modulus load $P_r$. To reach that load would require that bending already exist, which is a contradiction.

## Shanley Theory

From the preceding discussions we see that neither the tangent-modulus theory nor the reduced-modulus theory is entirely rational in explaining the phenomenon of inelastic buckling. Nevertheless, an understanding of both theories is necessary in order to develop a more complete and logically consistent theory. Such a theory was developed by F. R. Shanley in 1946 (see the historical note that follows), and today it is called the *Shanley theory of inelastic buckling.*

The Shanley theory overcomes the difficulties with both the tangent-modulus and reduced-modulus theories by recognizing that it is not possible for a column to buckle inelastically in a manner that is analogous to Euler buckling. In Euler buckling, a critical load is reached at which the column is in neutral equilibrium, represented by a horizontal line on the load-deflection diagram (Fig. 11-34). As already explained, neither the tangent-modulus load $P_t$ nor the reduced-modulus load $P_r$ can represent this type of behavior. In both cases, we are led to a contradiction if we try to associate the load with a condition of neutral equilibrium.

Instead of neutral equilibrium, wherein a deflected shape suddenly becomes possible with no change in load, we must think of a column that has an ever-increasing axial load. When the load reaches the tangent-modulus load (which is less than the reduced-modulus load), bending can begin only if the load continues to increase. Under these conditions, bending occurs simultaneously with an increase in load, resulting in a decrease in strain on the convex side of the column. Thus, the effective modulus of the material throughout the cross section becomes greater than $E_t$, and

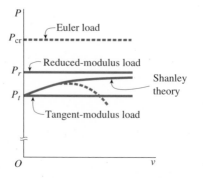

**Fig. 11-34**

Load-deflection diagram for elastic and inelastic buckling

therefore, an increase in load is possible. However, the effective modulus is not as great as $E_r$, because $E_r$ is based upon full strain reversal on the convex side of the column. In other words, $E_r$ is based upon the amount of strain reversal that exists if the column bends without a change in the axial force, whereas the presence of an increasing axial force means that the reduction in strain is not as great.

Thus, instead of neutral equilibrium, where the relationship between load and deflection is undefined, we now have a definite relationship between each value of the load and the corresponding deflection. This behavior is shown by the curve labeled "Shanley theory" in Fig. 11-34. Note that buckling begins at the tangent-modulus load; then the load increases, but does not reach the reduced-modulus load until the deflection becomes infinitely large (theoretically). However, other effects become important as the deflection increases, and in reality the curve eventually goes downward, as shown by the dashed line.

The Shanley concept of inelastic buckling has been verified by numerous investigators and by many tests. However, the maximum load attained by real columns (see the dashed curve trending downward in Fig. 11-34) is only slightly above the tangent-modulus load $P_t$. In addition, the tangent-modulus load is very simple to calculate. Therefore, for many practical purposes it is reasonable to adopt the **tangent-modulus load** as the critical load for inelastic buckling of columns.

The preceding discussions of elastic and inelastic buckling are based upon idealized conditions. Although theoretical concepts are important in understanding column behavior, the actual design of columns must take into account additional factors not considered in the theory. For instance, steel columns always contain residual stresses produced by the rolling process. These stresses vary greatly in different parts of the cross section, and therefore the stress level required to produce yielding varies throughout the cross section. For such reasons, a variety of empirical design formulas have been developed for use in designing columns. Some of the commonly used formulas are given in the next section.

**Historical Note**    Over 200 years had elapsed between the first calculation of a buckling load by Euler (in 1744) and the final development of the theory by Shanley (in 1946). Several famous investigators in the field of mechanics contributed to this development, and their work is described in this note.

After Euler's pioneering studies (Ref. 11-1), little progress was made until 1845, when the French engineer A. H. E. Lamarle pointed out that Euler's formula should be used only for slenderness ratios beyond a certain limit and that experimental data should be relied upon for columns with smaller ratios (Ref. 11-6). Then in 1889, another French engineer, A. G. Considère, published the results of the first comprehensive tests on columns (Ref. 11-7). He pointed out that the stresses on the concave side of the column increased with $E_t$ and the stresses on the convex side decreased with $E$. Thus, he showed why the Euler formula was not applicable to inelastic buckling, and he stated that the effective modulus was between $E$ and $E_t$. Although he made no attempt to evaluate the effective modulus, Considère was responsible for beginning the reduced-modulus theory.

In the same year, and quite independently, the German engineer F. Engesser suggested the tangent-modulus theory (Ref. 11-8). He denoted the tangent modulus by the symbol $T$ (equal to $d\sigma/d\varepsilon$) and proposed that $T$ be substituted for $E$ in Euler's formula for the critical load. Later, in March 1895, Engesser again presented the tangent-modulus theory (Ref. 11-9), obviously without knowledge of Considère's work. Today, the tangent-modulus theory is often called the *Engesser theory*.

Three months later, Polish-born F. S. Jasinsky, then a professor in St. Petersburg, pointed out that Engesser's tangent-modulus theory was incorrect, called attention to Considère's work, and presented the reduced-modulus theory (Ref. 11-10). He also stated that the reduced modulus could not be calculated theoretically. In response, and only one month later, Engesser acknowledged the error in the tangent-modulus approach and showed how to calculate the reduced modulus for any cross section (Ref. 11-11). Thus, the reduced-modulus theory is also known as the *Considère-Engesser theory*.

The reduced-modulus theory was also presented by the famous scientist Theodore von Kármán in 1908 and 1910 (Refs. 11-12, 11-13, and 11-14), apparently independently of the earlier investigations. In Ref. 11-13 he derived the formulas for $E_r$ for both rectangular and idealized wide-flange sections (that is, wide-flange sections without a web). He extended the theory to include the effects of eccentricities of the buckling load, and he showed that the maximum load decreases rapidly as the eccentricity increases.

The reduced-modulus theory was the accepted theory of inelastic buckling until 1946, when the American aeronautical-engineering professor F. R. Shanley pointed out the logical paradoxes in both the tangent-modulus and reduced-modulus theories. In a remarkable one-page paper (Ref. 11-15), Shanley not only explained what was wrong with the generally accepted theories, but also proposed his own theory that resolved the paradoxes. In a second paper, five months later, he gave further analyses to support his earlier theory and gave results from tests on columns (Ref. 11-16). Since that time, many other investigators have confirmed and expanded Shanley's concept.

For excellent discussions of the column-buckling problem, see the comprehensive papers by Hoff (Refs. 11-17 and 11-18), and for a historical account, see the paper by Johnston (Ref. 11-19).

## 11.9 DESIGN FORMULAS FOR COLUMNS

In the preceding sections of this chapter we discussed the theoretical load-carrying capacity of columns for both elastic and inelastic buckling. With that background in mind, we are now ready to examine some practical formulas that are used in the design of columns. These design formulas are based not only upon the theoretical analyses, but also upon the behavior of real columns as observed in laboratory tests.

The theoretical results are represented by the column curves shown in Figs. 11-30 and 11-33. A common design approach is to approximate these curves in the inelastic buckling range (low values of slenderness ratio) by empirical formulas and to use Euler's formula in the elastic range (high values of slenderness ratio). Of course, a factor of safety must be applied to obtain the allowable loads from the maximum loads (or to obtain the allowable stresses from the maximum stresses).

The following examples of column design formulas are based on an Allowable Stress Design (ASD) approach and are applicable to centrally loaded columns of structural steel, aluminum, and wood. The formulas give the allowable stresses in terms of the column properties, such as length, cross-sectional dimensions, and conditions of support. Thus, for a given column, the allowable stress can be readily obtained.*

Once the allowable stress is known, we can determine the **allowable load** by multiplying by the cross-sectional area:

$$P_{\text{allow}} = \sigma_{\text{allow}}A \qquad \textbf{(11-81)}$$

The allowable load must be larger than the actual load if the allowable stress is not to be exceeded.

The selection of a column often requires an iterative or trial-and-error procedure. Such a procedure is necessary whenever we do not know in advance which design formula to use. Since each formula is valid only for a certain range of slenderness ratios, and since the slenderness ratio is unknown until after the column is selected, we usually do not know which formula is applicable until we have made at least one trial.

A common **trial-and-error procedure** for selecting a column to support a given axial load is the following:

1. Estimate the allowable stress $\sigma_{\text{allow}}$. (Note that an upper limit for $\sigma_{\text{allow}}$ is the allowable stress for a column of zero length. This stress is readily obtained from the design formulas, and the estimated stress should be equal to or less than this upper limit.)

2. Calculate an approximate value of the cross-sectional area $A$ by dividing the given axial load $P$ by the estimated allowable stress.

3. Determine a column size and/or shape that supplies the required area, either by calculating a required dimension or by selecting a column from a table of available shapes.

4. Knowing the dimensions of a trial column from step (3), determine the allowable stress $\sigma_{\text{allow}}$ in the column from the appropriate design formula.

5. Using Eq. (11-81), calculate the allowable load $P_{\text{allow}}$ and compare it with the actual load $P$.

6. If the column is not adequate to support the given load, select a larger column and repeat the process. If the column appears to be overdesigned (because the allowable load is much larger than the given load), select a smaller column and repeat the process. A suitable column can usually be obtained with only two or three trials.

---

*The design formulas given in this section are samples of the many formulas in use around the world. They are intended for use in solving the problems at the end of the chapter and should not be used in actual design, which requires many additional considerations. See the subsection titled "Limitations" at the end of this section.

Many variations of this procedure are possible, depending upon the type of column and what quantities are known in advance. Sometimes a direct design procedure, bypassing the trial-and-error steps, can be devised.

## Structural Steel

Let us begin with design formulas for centrally loaded, structural-steel columns based on **Allowable Stress Design** (ASD). The following formulas were adopted by the American Institute of Steel Construction (AISC), which is a technical organization that prepares specifications for structural-steel designers (Ref. 5-4) and provides many other services to engineers. The AISC formulas for the allowable stress in a column are obtained by dividing the maximum stress by an appropriate factor of safety. The term 'maximum stress' means the stress that is obtained by taking the maximum load (or ultimate load) the column can carry and dividing it by the cross-sectional area.

We begin by defining the *elastic buckling stress* $\sigma_e$ (in ksi or MPa, referred to as $F_e$ in AISC) as

$$\sigma_e = \frac{\pi^2 E}{\left(\dfrac{KL}{r}\right)^2} \tag{11-82}$$

where the effective length $KL$ is used so that the formula may be applied to a variety of support conditions.

Equation (11-82) is valid only when the stresses in the column are less than the proportional limit $\sigma_{pl}$. Under ordinary conditions, we assume that the proportional limit of steel is equal to the yield stress $\sigma_Y$. However, rolled steel sections (such as wide-flange sections) contain significant residual stresses—stresses that may be as large as one-half of the yield stress. To determine the smallest slenderness ratio for which Eq. (11-82) is applicable, we set $\sigma_e$ equal to $0.445\sigma_Y$ and solve for the corresponding value of $KL/r$, which is known as the *critical slenderness ratio*:

$$\left(\frac{KL}{r}\right)_c = 4.71\sqrt{\frac{E}{\sigma_Y}} \tag{11-83}$$

Next, we find the *critical stress* $\sigma_{cr}$ for the region of *inelastic buckling* as

$$\sigma_{cr} = \left(0.658^{\frac{\sigma_Y}{\sigma_e}}\right)\sigma_Y$$

when

$$\frac{KL}{r} \le 4.71\sqrt{\frac{E}{\sigma_Y}} \tag{11-84}$$

and for *elastic buckling*

$$\sigma_{cr} = 0.877\sigma_e$$

when

$$\frac{KL}{r} > 4.71\sqrt{\frac{E}{\sigma_Y}} \qquad \textbf{(11-85)}$$

The critical-stress to yield-stress ratio, $\sigma_{cr}/\sigma_Y$, from Eqs. (11-84) and (11-85) is plotted in Fig. 11-35. At the critical slenderness ratio of Eq. (11-83), the two curves merge smoothly at ratio $\sigma_{cr}/\sigma_Y = 0.445$. The validity of the formulas for use in design has been verified by numerous tests.

Finally, we introduce the factor of safety of 1.67 specified by the AISC to get the allowable stress $\sigma_{allow}$:

$$\sigma_{allow} = \frac{\sigma_{cr}}{1.67} \qquad \textbf{(11-86)}$$

At the critical slenderness ratio, the normalized curves merge at $\sigma_{allow}/\sigma_Y = 0.267$ (Fig. 11-35).

The AISC specifications place an upper limit of 200 on the slenderness ratio $KL/r$ and specify the modulus of elasticity $E$ as 29,000 ksi. Also, the symbols used in the AISC specifications differ slightly from those in the preceding formulas. For instance, the elastic buckling stress is denoted $F_e$, the yield stress is denoted $F_y$, and the critical stress is $F_{cr}$.

All of the preceding design formulas for structural steel may be used with either USCS or SI units, as illustrated in Examples 11-6 and 11-7. The formulas are applicable to wide-flange and other rolled shapes, as well as to columns with rectangular and circular cross sections.

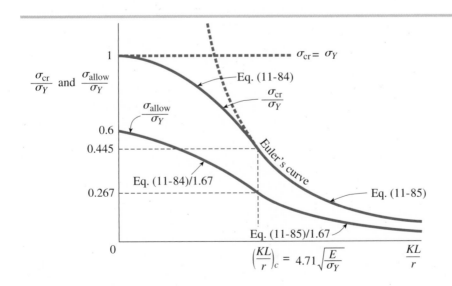

**Fig. 11-35**

Design formulas for structural-steel columns

Fig. 11-36

Design formulas for
aluminum columns

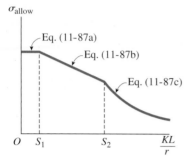

## Aluminum

The design formulas presented here for aluminum columns are taken from the Specification for Aluminum Structures of the Aluminum Association (Ref. 5-5). Like the formulas for steel design, the formulas for aluminum are based upon the theoretical curves given in Figs. 11-30 and 11-33.

The general shape of the design curves for aluminum is shown in Fig. 11-36, where the ordinate is the allowable stress and the abscissa is the effective slenderness ratio $KL/r$. $S_1$ is the slenderness ratio at the intersection of the equations for yielding and inelastic buckling (i.e., transition from short to intermediate columns); and $S_2$ is the slenderness ratio at the intersection of the equations for inelastic buckling and elastic buckling (i.e., transition from intermediate to long columns, as shown in Fig. 11-30). The allowable stress in the short-column region is based upon the yield strength of the material; in the intermediate-column region it is based upon the tangent-modulus formula; and in the long-column region it is based upon Euler's formula. Note that the design curves (Fig. 11-36) meet with distinctly different slopes at the slenderness ratio $S_2$.

For aluminum columns in direct compression, the general design formulas are expressed as

$$\sigma_{\text{allow}} = \frac{\sigma_Y}{\Omega_c} \qquad 0 \le \frac{KL}{r} \le S_1 \tag{11-87a}$$

$$\sigma_{\text{allow}} = \frac{0.85}{\Omega_c}\left( B_c - D_c \frac{KL}{r} \right) \qquad S_1 \le \frac{KL}{r} \le S_2 \tag{11-87b}$$

$$\sigma_{\text{allow}} = \frac{0.85\,\pi^2 E}{\Omega_c (KL/r)^2} \qquad \frac{KL}{r} \ge S_2 \tag{11-87c}$$

In these equations, $KL/r$ is the effective slenderness ratio, the stress $\sigma_Y$ is the compressive yield stress (0.2% offset), the factor of safety $\Omega_c = 1.65$ in ASD for building-type structures, and $B_c$ and $D_c$ are constants. The 0.85 factor accounts for the effects of load eccentricity and initial crookedness, both of which reduce buckling strength.

The values of the various quantities appearing in Eqs. (11-87a, b, and c) depend upon the particular aluminum alloy, the temper of the finished product, and the use to which it will be put. Numerous alloys and tempers are available, so the Aluminum Association gives tables of values based upon the material and usage.

As examples, the following formulas apply to two alloys used in buildings. In these particular cases, the short-column region is nonexistent and is combined with the intermediate-column region; thus, for these materials, the slenderness ratio $S_1$ is taken as zero.

**Case 1.** Alloy 2014-T6 extrusions: $S_1 = 0$, $S_2 = 55$
The allowable stresses in ksi are

$$\sigma_{\text{allow}} = 30.9 - 0.229\left(\frac{KL}{r}\right) \text{ksi} \qquad 0 \le \frac{KL}{r} \le 55 \tag{11-88a}$$

$$\sigma_{\text{allow}} = \frac{55{,}400 \text{ ksi}}{(KL/r)^2} \qquad \frac{KL}{r} \ge 55 \tag{11-88b}$$

If Eqs. (11-88a, b) are expressed in SI units, the allowable stresses are

$$\sigma_{\text{allow}} = \left[ 213 - 1.577\left(\frac{KL}{r}\right) \right] \text{MPa} \qquad 0 \le \frac{KL}{r} \le 55 \quad \textbf{(11-88c)}$$

$$\sigma_{\text{allow}} = \frac{3.81 \times 10^5 \text{ MPa}}{\left(\dfrac{KL}{r}\right)^2} \qquad \frac{KL}{r} \ge 55 \qquad \textbf{(11-88d)}$$

**Case 2.** Alloy 6061-T6 extrusions: $S_1 = 0$, $S_2 = 66$
  The allowable stresses in ksi are

$$\sigma_{\text{allow}} = 20.3 - 0.127\left(\frac{KL}{r}\right) \text{ksi} \qquad 0 \le \frac{KL}{r} \le 66 \quad \textbf{(11-89a)}$$

$$\sigma_{\text{allow}} = \frac{51{,}400 \text{ ksi}}{(KL/r)^2} \qquad \frac{KL}{r} \ge 66 \qquad \textbf{(11-89b)}$$

If Eqs. (11-89a, b) are expressed in SI units, the allowable stresses are

$$\sigma_{\text{allow}} = \left[ 140 - 0.874\left(\frac{KL}{r}\right) \right] \text{MPa} \qquad 0 \le \frac{KL}{r} \le 66 \quad \textbf{(11-89c)}$$

$$\sigma_{\text{allow}} = \frac{3.53 \times 10^5 \text{ MPa}}{\left(\dfrac{KL}{r}\right)^2} \qquad \frac{KL}{r} \ge 66 \qquad \textbf{(11-89d)}$$

Note that these formulas give the allowable stresses, hence, they already incorporate the safety factor $\Omega_c$, which is 1.65 for building structures.

## Wood

Wood structural members are readily available in the form of sawn lumber, glued-laminated timbers, and round poles and piles. Their strength depends upon many factors, the most important being the species (such as Douglas fir or southern pine) and the grade (such as Select Structural or Construction). Among the other factors affecting strength are moisture content and duration of loading (wood will support greater loads for short durations than for long durations).

The design of wood structural members, like those of steel and aluminum, is governed by codes and specifications. In the United States, the most widely used design codes for wood are those of the American Forest and Paper Association (Ref. 5-6), which publishes the *National Design Specifications for Wood Construction* and related manuals. The formulas and requirements described in this section are taken from those specifications. We will limit our discussion to columns of rectangular cross section constructed of either sawn lumber or glued-laminated timber.

The allowable stress in compression, parallel to the grain of the wood, on the cross section of a column is denoted in the specifications as $F'_c$, which is the same as $\sigma_{allow}$ in the notation of this book. Therefore, the **allowable axial load** on a centrally loaded column is

$$P_{allow} = \sigma_{allow}A = F'_c A \tag{11-90}$$

in which $A$ is the cross-sectional area of the column.

The **allowable stress** $F'_c$ for use in the preceding equation is given in the specifications as

$$F'_c = F_c C^* C_P = F^*_c C_P \tag{11-91}$$

in which $F_c$ is the compressive design stress for the particular species and grade of wood, $C^*$ is an adjustment factor for various service conditions, $C_P$ is the column stability factor, and $F^*_c$ is the adjusted compressive design stress (equal to the product of $F_c$ and the adjustment factor $C^*$). Each of these terms will now be described.

The **design stress** $F_c$ is based upon laboratory tests of wood specimens and is listed in tables in the specifications. For instance, typical values of $F_c$ for structural grades of Douglas fir and southern pine are in the range of 700 to 2000 psi (5 to 14 MPa).

The **adjustment factor** $C^*$ takes into account the service conditions, that is, the actual conditions of use, including duration of loading, wet conditions, and high temperatures. When solving problems in this book, we will assume $C^* = 1.0$, which is not unreasonable for ordinary indoor conditions.

The **column stability factor** $C_P$ is based upon buckling considerations analogous to those described in connection with Figs. 11-30 and 11-33. For wood columns, a single buckling formula has been devised that covers the entire region of column behavior, including short, intermediate, and long columns. The formula, which follows as Eq. (11-93), gives the stability factor $C_P$ in terms of several variables, one of which is the **wood slenderness ratio**:

$$\text{Wood slenderness ratio} = \frac{L_e}{d} \tag{11-92}$$

in which $L_e$ is the effective length for buckling and $d$ is the depth of the cross section in the plane of buckling.

The **effective length** $L_e$ appearing in the wood slenderness ratio is the same as the effective length $KL$ in our earlier discussions (see Fig. 11-20). However, note carefully that the slenderness ratio $L_e/d$ is **not** the same as the slenderness ratio $L/r$ used previously [see Eq. (11-20)]. The dimension $d$ is the *depth* of the cross section in the plane of buckling, whereas $r$ is the *radius of gyration* of the cross section in the plane of buckling. Also, note that the maximum permissible value of the wood slenderness ratio $L_e/d$ is 50.

The **column stability factor** $C_P$ is calculated from the following formula:

$$C_P = \frac{1 + (F_{cE}/F^*_c)}{2c} - \sqrt{\left[\frac{1 + (F_{cE}/F^*_c)}{2c}\right]^2 - \frac{F_{cE}/F^*_c}{c}} \tag{11-93}$$

in which $F_{cE}$ is the Euler buckling coefficient (Eq. 11-94), $F_c^*$ is the adjusted compressive design stress [see Eq. (11-91)], and $c$ is a constant depending upon the type of column (for instance, $c = 0.8$ for sawn lumber and 0.9 for glued-laminated timber).

The **Euler buckling coefficient** is defined as

$$F_{cE} = \frac{0.822 E'_{min}}{(L_e/d)^2} \qquad \textbf{(11-94)}$$

in which $E'_{min}$ is an adjusted modulus of elasticity, and $L_e/d$ is the wood slenderness ratio.

The adjusted modulus $E'_{min}$ is equal to the modulus of elasticity $E$ multiplied by an adjustment factor for service conditions. When solving problems in this book, we will assume that these adjustment factors equal 1.0, and therefore, $E'_{min} = E$. Typical values of the modulus $E$ for structural lumber are in the range of 1,200,000 to 2,000,000 psi (8 to 14 GPa).

In summary, Eqs. (11-90) through (11-94) are the **general equations** for the buckling of wood columns. However, when solving problems in this book, we assume the following **specific conditions**:

1. The columns have rectangular cross sections and are constructed of either sawn lumber or glued-laminated timber.

2. The adjustment factor $C^* = 1.0$, and therefore, the following three relations may be used:

$$F_c' = \sigma_{allow} = F_c C_P \qquad F_c^* = F_c \qquad \textbf{(11-95a,b)}$$

$$P_{allow} = F_c' A = F_c C_P A \qquad \textbf{(11-96)}$$

3. The constant $c = 0.8$ or 0.9 (for sawn lumber and glued-laminated timber, respectively).

4. The modulus $E'_{min} = E$.

With these conditions, the equation for the Euler buckling coefficient (Eq. 11-94) becomes

$$F_{cE} = \frac{0.822 E}{(L_e/d)^2} \qquad \textbf{(11-97)}$$

and the nondimensional ratio $F_{cE}/F_c^*$, which we will denote by the Greek letter $\phi$ (phi), becomes

$$\phi = \frac{F_{cE}}{F_c^*} = \frac{0.822 E}{F_c(L_e/d)^2} \qquad \textbf{(11-98)}$$

With this simplified notation, the equation for the column stability factor becomes

$$C_P = \frac{1 + \phi}{2c} - \sqrt{\left[\frac{1 + \phi}{2c}\right]^2 - \frac{\phi}{c}} \qquad \textbf{(11-99)}$$

Note that the slenderness ratio $L_e/d$ enters the calculation of $C_P$ through the ratio $\phi$.

A **graph of the stability factor** is shown in Fig. 11-37. The curves for $C_P$ are plotted for two values of the ratio $E/F_c$. Note that both curves have zero slope for $L_e/d$ equal to zero, and both curves terminate at $L_e/d = 50$, which is the upper limit permitted by the specifications. Although these curves are plotted for specific values of the various parameters, they show in general how the stability factor varies with the slenderness ratio $L_e/d$.

## Limitations

The preceding formulas for the design of steel, aluminum, and wood columns are intended solely for use in solving problems in this book. They should not be used for the design of actual columns, because they represent only a small part of the complete design process. Many factors besides those discussed here enter into the design of columns, and therefore, textbooks or other references on structural design should be consulted before designing a column for a specific application.

Furthermore, all design formulas presented in specifications and codes, such as the formulas given in this section, require informed judgment in their use. There are many cases of structures that "met the code," but nevertheless collapsed or failed to perform adequately. Meeting the code requirements is not enough for a safe design—practical design experience is also essential.

**Fig. 11-37**

Typical curves for the column stability factor $C_P$ (rectangular wood columns)

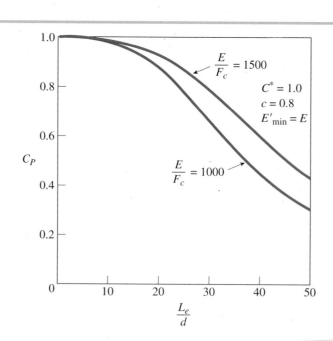

## ● ● ● Example 11-6

### Fig. 11-38

Example 11-6: Steel
wide-flange column

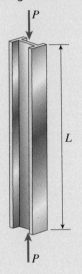

A steel column is constructed from a W 250 × 89 wide-flange section
(Fig. 11-38). Assume that the column has pin supports and may buckle in
any direction. Also, assume that the steel has a modulus of elasticity of
$E = 200$ GPa and a yield stress of $\sigma_Y = 250$ MPa.

(a) If the length of the column is $L = 6.5$ m, what is the allowable axial
load?
(b) If the column is subjected to an axial load $P = 890$ kN, what is the maximum permissible length?

### Solution

We will use the AISC formulas of Eqs. (11-82) through (11-86) when analyzing
this column. Since the column has pin supports, the effective-length factor
$K = 1$. Also, since the column will buckle about the weak axis of bending,
we will use the smaller radius of gyration $r = 65.3$ mm, as obtained from
Table F-1(b), Appendix F. The critical slenderness ratio (Eq. 11-83) is

$$\left(\frac{KL}{r}\right)_c = 4.71\sqrt{\frac{E}{\sigma_Y}} = 4.71\sqrt{\frac{200 \text{ GPa}}{250 \text{ MPa}}} = 133.2 \qquad \text{(a)}$$

(a) *Allowable axial load.* If the length $L = 6.5$ m, the slenderness ratio of
the column is

$$\frac{L}{r} = \frac{6.5 \text{ m}}{65.3 \text{ mm}} = 99.5$$

which is less than the critical ratio of Eq. (a), so we obtain the critical stress
using Eq. (11-84) as

$$\sigma_{cr} = \left(0.658^{\frac{\sigma_Y}{\sigma_e}}\right)\sigma_Y = \left(0.658^{\frac{250 \text{ MPa}}{199.4 \text{ MPa}}}\right) 250 \text{ MPa} = 147.9 \text{ MPa}$$

where

$$\sigma_e = \frac{\pi^2 E}{\left(\dfrac{KL}{r}\right)^2} = \frac{\pi^2 (200 \text{ GPa})}{99.5^2} = 199.4 \text{ MPa}$$

The allowable stress is

$$\sigma_{allow} = \frac{\sigma_{cr}}{1.67} = \frac{147.9 \text{ MPa}}{1.67} = 88.6 \text{ MPa}$$

Since the cross-sectional area of the column is $A = 11,400$ mm² [from
Table F-1(b)], the allowable axial load is

$$P_{allow} = \sigma_{allow} A = 88.6 \text{ MPa} (11,400 \text{ mm}^2) = 1010 \text{ kN} \quad ⬅$$

*Continues* ➡

**Example 11-6 -** *Continued*

(b) *Maximum permissible length.* To determine the maximum length when the axial load $P = 890$ kN, we begin with an estimated value of the length and then use a *trial-and-error procedure.* Note that when the load $P = 890$ kN, the maximum length is greater than 6.5 m (because a length of 6.5 m corresponds to an axial load of 1010 kN). Therefore, as a trial value, we will assume $L = 7$ m. The corresponding slenderness ratio is

$$\frac{L}{r} = \frac{7000 \text{ mm}}{65.3 \text{ mm}} = 107.2$$

which is less than the critical ratio in Eq. (a). First, we use Eq. (11-82) to find the elastic buckling stress as

$$\sigma_e = \frac{\pi^2 E}{\left(\dfrac{KL}{r}\right)^2} = \frac{\pi^2 (200 \text{ GPa})}{107.2^2} = 171.8 \text{ MPa}$$

Then we use Eqs. (11-84) and (11-86) to find the allowable stress:

$$\sigma_{allow} = \frac{\sigma_{cr}}{1.67} = \frac{\left(0.658^{\frac{250 \text{ MPa}}{171.8 \text{ MPa}}}\right)250 \text{ MPa}}{1.67} = 81.4 \text{ MPa}$$

The allowable load is

$$P_{allow} = \sigma_{allow} A = 81.4 \text{ MPa} (11{,}400 \text{ mm}^2) = 928 \text{ kN}$$

which is greater than the given load of 890 kN, so the permissible length is greater than 7 m. With further trials, we find that

$$L = 7.2 \text{ m} \qquad P_{allow} = 896 \text{ kN}$$

$$L = 7.3 \text{ m} \qquad P_{allow} = 880 \text{ kN}$$

Interpolating between these results, we find that the maximum permissible length is approximately 7.24 m, as confirmed by the following calculations:

$$\frac{L}{r} = \frac{7240 \text{ mm}}{65.3 \text{ mm}} = 110.9 \qquad \text{so} \qquad \sigma_e = \frac{\pi^2 (200 \text{ GPa})}{110.9^2} = 160.5 \text{ MPa}$$

and

$$\sigma_{allow} = \frac{\left(0.658^{\frac{250 \text{ MPa}}{160.5 \text{ MPa}}}\right)250 \text{ MPa}}{1.67} = 78 \text{ MPa}$$

resulting in

$$P_{allow} = (78 \text{ MPa})(11{,}400 \text{ mm}^2) = 889.2 \text{ kN}$$

Hence, the maximum permissible length of the column is 7.24 m.

## ● ● ● Example 11-7

Find the minimum required thickness $t_{min}$ for a steel pipe column with a length of $L = 12$ ft and outer diameter of $d = 6.5$ in. supporting an axial load of $P = 54$ kips (Fig. 11-39). The column is fixed at the base and free at the top. (Use $E = 29{,}000$ ksi and $\sigma_Y = 36$ ksi.)

### Fig. 11-39

Example 11-7: Steel pipe column

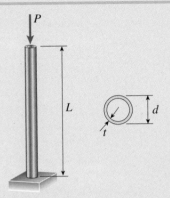

### Solution

We will use the AISC formulas of Eqs. (11-82) through (11-86) when analyzing this column. Since the column has fixed and free end conditions, the effective length is

$$L_e = KL = 2(12\text{ ft}) = 24\text{ ft}$$

Also, the critical slenderness ratio of Eq. (11-83) is

$$\left(\frac{KL}{r}\right)_c = 4.71\sqrt{\frac{E}{\sigma_Y}} = 4.71\sqrt{\frac{29{,}000\text{ ksi}}{36\text{ ksi}}} = 133.7 \qquad \textbf{(a)}$$

*First trial.* To determine the required thickness of the column, we will use a trial-and-error method. Let us start by assuming a trial value $t = 0.5$ in. Then the moment of inertia of the cross-sectional area is

$$I = \frac{\pi}{64}[d^4 - (d - 2t)^4] = \frac{\pi}{64}[(6.5\text{ in.})^4 - (5.5\text{ in.})^4] = 42.706\text{ in.}^4$$

Also, the cross-sectional area and radius of gyration are

$$A = \frac{\pi}{4}[d^2 - (d - 2t)^2] = \frac{\pi}{4}[(6.5\text{ in.})^2 - (5.5\text{ in.})^2] = 9.425\text{ in.}^2$$

*Continues* ➡

and

$$r = \sqrt{\frac{I}{A}} = \sqrt{\frac{42.706 \text{ in.}^4}{9.425 \text{ in.}^2}} = 2.129 \text{ in.}$$

Therefore, the slenderness ratio of the column is

$$\frac{KL}{r} = \frac{2(144 \text{ in.})}{2.129 \text{ in.}} = 135.3$$

Since this ratio is larger than the critical slenderness ratio of Eq. (a), we obtain the allowable stress from Eqs. (11-82), (11-85), and (11-86):

$$\sigma_{allow} = \frac{0.877\sigma_e}{1.67} = \frac{0.877\left[\dfrac{\pi^2 E}{\left(\dfrac{KL}{r}\right)^2}\right]}{1.67} = \frac{0.877\left[\dfrac{\pi^2(29{,}000 \text{ ksi})}{135.3^2}\right]}{1.67} = 8.211 \text{ ksi}$$

Thus, the allowable axial load is

$$P_{allow} = \sigma_{allow} A = 8.211 \text{ ksi }(9.425 \text{ in.}^2) = 77.4 \text{ kip}$$

Since this load is greater than the required load of 54 kips, we must try a smaller value of the thickness *t*.

*Additional trials.* Performing similar calculations for $t = 0.25$ in. and $t = 0.375$ in., we get the following results:

$$t = 0.25 \text{ in.} \qquad P_{allow} = 43.5 \text{ kip}$$

$$t = 0.375 \text{ in.} \qquad P_{allow} = 61.5 \text{ kip}$$

$$t = 0.5 \text{ in.} \qquad P_{allow} = 77.4 \text{ kip}$$

By interpolation, we see that $t = 0.32$ in. corresponds to a load of 54 kips. Therefore, the required thickness of the pipe column is

$$t_{min} = 0.32 \text{ in.}$$

### ● ● ● Example 11-8

An aluminum tube (alloy 2014-T6) with an effective length $L = 405$ mm is compressed by an axial force $P = 22$ kN (Fig. 11-40).

Determine the minimum required outer diameter $d$ if the thickness $t$ equals one-tenth the outer diameter.

### Fig. 11-40

Example 11-8: Aluminum tube in compression

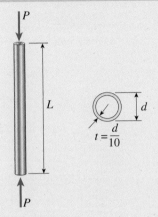

#### Solution

We will use the Aluminum Association formulas for alloy 2014-T6 given as Eqs. (11-88c and d) for analyzing this column. However, we must make an initial guess as to which formula is applicable, because each formula applies to a different range of slenderness ratios. Let us assume that the slenderness ratio of the tube is less than 55, in which case we use Eq. (11-88c) with $K = 1$:

$$\sigma_{\text{allow}} = 213 - 1.577\left(\frac{L}{r}\right) \text{ MPa} \qquad \text{(a)}$$

In this equation, we can replace the allowable stress by the actual stress $P/A$, that is, by the axial load divided by the cross-sectional area. The cross-sectional area is

$$A = \frac{\pi}{4}[d^2 - (d - 2t)^2] = \frac{\pi}{4}[d^2 - (0.8d)^2] = 0.2827d^2 \qquad \text{(b)}$$

Therefore, the stress $P/A$ is

$$\frac{P}{A} = \frac{22{,}000 \text{ N}}{0.2827d^2} = \frac{77{,}821.0}{d^2}$$

*Continues* ➡

**Example 11-8 -** *Continued*

in which $P/A$ has units of Newtons per square millimeter (MPa) and $d$ has units of mm. Substituting into Eq. (a), we get

$$\frac{77{,}821}{d^2} = 213 - 1.577\left(\frac{L}{r}\right) \text{ MPa} \qquad \text{(c)}$$

The slenderness ratio $L/r$ can also be expressed in terms of the diameter $d$. First, we find the moment of inertia and radius of gyration of the cross section:

$$I = \frac{\pi}{64}[d^4 - (d - 2t)^4] = \frac{\pi}{64}[d^4 - (0.8d)^4] = 0.02898d^4$$

$$r = \sqrt{\frac{I}{A}} = \sqrt{\frac{0.02898d^4}{0.2827d^2}} = 0.3202d$$

Therefore, the slenderness ratio is

$$\frac{L}{r} = \frac{405 \text{ mm}}{0.3202d} = \frac{1265.8}{d} \qquad \text{(d)}$$

where (as before) the diameter $d$ has units of millimeters.

Substituting into Eq. (c), we obtain the following equation, in which $d$ is the only unknown quantity:

$$\frac{77{,}821}{d^2} = 213 - 1.577\left(\frac{1265.8}{d}\right)$$

With a little rearranging, this equation becomes

$$213.0d^2 - 1996.1666d - 77{,}821.0 = 0$$

from which we find

$$d = 24.4 \text{ mm}$$

This result is satisfactory provided the slenderness ratio is less than 55, as required for Eq. (a) to be valid. To verify that this is the case, we calculate the slenderness ratio from Eq. (d):

$$\frac{L}{r} = \frac{1265.8 \text{ mm}}{d} = \frac{1265.8 \text{ mm}}{24.4 \text{ mm}} = 51.9$$

Therefore, the solution is valid, and the minimum required diameter is

$$d_{min} = 24.4 \text{ mm}$$

A wood post of rectangular cross section (Fig. 11-41) is constructed of Douglas fir lumber having a compressive design stress $F_c = 1600$ psi and modulus of elasticity $E = 1900$ ksi. The length of the post is $L$ and the cross-sectional dimensions are $b$ and $h$. The supports at the ends of the post provide pinned-end conditions, so the length $L$ becomes the effective length $L_e$. Also, buckling is free to occur about either principal axis of the cross section. (*Note:* Since the post is made of sawn lumber, the constant $c$ equals 0.8.)

(a) Determine the allowable axial load $P_{\text{allow}}$ if $L = 7$ ft, $b = 4.75$ in., and $h = 6.25$ in.
(b) Determine the maximum allowable length $L_{\text{max}}$ if the axial load $P = 40$ kips, $b = 4.75$ in., and $h = 6.25$ in.
(c) Determine the minimum width $b_{\text{min}}$ of the cross section if the column is square, $P = 38$ kips, and $L = 8.5$ ft.

**Fig. 11-41**

Example 11-9: Wood post in compression

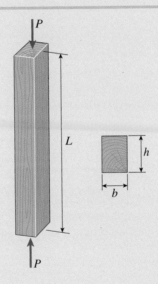

**Solution**

(a) *Allowable axial load.* The allowable load from Eq. (11-96) is

$$P_{\text{allow}} = F_c' A = F_c C_p A$$

in which $F_c = 1600$ psi and

$$A = bh = (4.75 \text{ in.})(6.25 \text{ in.}) = 29.6875 \text{ in.}^2$$

*Continues* ➡

To find the stability factor $C_p$, we first calculate the slenderness ratio, as follows:

$$\frac{L_e}{d} = \frac{7 \text{ ft}}{4.75 \text{ in.}} = 17.684$$

in which $d$ is the smaller dimension of the cross section. Next, we obtain the ratio $\phi$ from Eq. (11-98):

$$\phi = \frac{F_{cE}}{F_c^*} = \frac{0.822E}{F_c(L_e/d)^2} = \frac{0.822(1900 \text{ ksi})}{(1600 \text{ psi})(17.684)^2} = 3.121$$

Then we substitute $\phi$ into Eq. (11-99) for $C_p$, while also using $c = 0.8$, and we obtain

$$C_p = \frac{1 + 3.121}{1.6} - \sqrt{\left(\frac{1 + 3.121}{1.6}\right)^2 - \frac{3.121}{0.8}} = 0.923$$

Finally, the allowable axial load is

$$P_{allow} = F_c C_p A = 1600 \text{ psi}(0.923)(29.6875 \text{ in.}^2) = 43.8 \text{ kip} \quad \blacktriangleleft$$

(b) *Maximum allowable length.* We begin by determining the required value of $C_p$. Rearranging Eq. (11-96) and replacing $P_{allow}$ by the load $P$, we obtain the formula for $C_p$ shown below. Then we substitute numerical values and obtain the following result:

$$C_p = \frac{P}{F_c A} = \frac{40 \text{ kip}}{1600 \text{ psi}(29.6875 \text{ in.}^2)} = 0.84211$$

Substituting this value of $C_p$ into Eq. (11-99), and also setting $c$ equal to 0.8, we get the following equation in which $\phi$ is the only unknown quantity:

$$C_p = 0.84211 = \frac{1 + \phi}{1.6} - \sqrt{\left[\frac{1 + \phi}{1.6}\right]^2 - \frac{\phi}{0.8}}$$

Solving numerically by trial and error, we find

$$\phi = 1.74039$$

Finally, from Eq. (11-98), we get

$$\frac{L}{d} = \sqrt{\frac{0.822E}{\phi F_c}} = \sqrt{\frac{0.822(1900 \text{ ksi})}{1.74039(1600 \text{ psi})}} = 23.683$$

and

$$L_{max} = 23.683\,d = 23.683\,(4.75 \text{ in.}) = 9.37 \text{ ft} \quad \longleftarrow$$

Any larger value of the length $L$ will produce a smaller value of $C_p$ and, hence, a load $P$ that is less than the actual load of 40 kips.

(c) *Minimum width of square cross section*. The minimum width $b_{min}$ can be found by trial and error, using the procedure described in part (a). The steps are:

1. Select a trial value of $b$ (feet)
2. Calculate the slenderness ratio $L/d = 8.5/b$ (nondimensional)
3. Calculate the ratio $\phi$ from Eq. (11-98):

$$\phi = \frac{0.822E}{F_c(L_e/d)^2} = \frac{0.822(1900 \text{ ksi})}{1.6 \text{ ksi}\left(\dfrac{8.5}{b}\right)^2} = 13.51b^2 \quad \text{(nondimensional)}$$

4. Substitute $\phi$ into Eq. (11-99) and calculate $C_p$ (nondimensional)
5. Calculate the load $P$ from Eq. (11-96):

$$P = F_c C_p A = (1600 \text{ psi})\left(\frac{144 \text{ in.}^2}{\text{ft}^2}\right)(C_p)\,b^2 = 230.4\,C_p b^2 \quad \text{(kips)}$$

6. Compare the calculated value of $P$ with the given load of 38 kips. If $P$ is less than 38 kips, select a larger trial value for $b$ and repeat steps (2) through (5). If $P$ is larger than 38 kips by a significant amount, select a smaller value for $b$ and repeat the steps. Continue until $P$ reaches a satisfactory value.

Let us take a trial value of $b$ equal to 5 in., or 0.417 ft. Then steps (2) through (5) produce the following results:

$$\frac{L}{d} = \frac{8.5}{b} = \frac{8.5}{0.417} = 20.384 \qquad \phi = 13.51b^2 = 13.51(0.417^2) = 2.349$$

$$C_p = 0.891 \qquad P = 230.4\,C_p b^2 = 230.4(0.891)(0.417^2) = 35.7 \text{ kips}$$

Since the given load is 38 kips, we select a larger value of $b$, say 5.2 in., for the next trial. Proceeding in this manner with successive trials, we obtain the following results:

$$b = 5.2 \text{ in.}, \quad P = 38.97 \text{ kips}$$

$$b = 5.15 \text{ in.}, \quad P = 38.13 \text{ kips}$$

Therefore, the minimum width of the square cross section is

$$b_{min} = 5.15 \text{ in.} \quad \longleftarrow$$

In Chapter 11, we investigated the elastic and inelastic behavior of axially loaded members known as columns. First, the concepts of **buckling and stability** of these slender compression elements were discussed using equilibrium of simple column models made up of rigid bars and elastic springs. Then elastic columns with pinned ends, acted on by centroidal compressive loads, were considered and the differential equation of the deflection curve was solved to obtain the **buckling load** ($P_{cr}$) and **buckled mode shape**; linear elastic behavior was assumed. Three additional support cases were investigated, and the buckling load for each case was expressed in terms of the **column's effective length**, that is, the length of an equivalent pinned-end column. Behavior of pinned-end columns with **eccentric axial loads** was discussed and the **secant formula** was derived that defines the maximum stress in these columns. Three theories for inelastic buckling of columns were presented. Finally, formulas for **design of columns** made of steel, aluminum, and wood were presented and discussed.

The major concepts presented in this chapter are as follows:

1. Buckling instability of slender columns is an important mode of failure that must be considered in their design (in addition to strength and stiffness).

2. A slender column with pinned ends and length $L$, acted on by a compressive load at the centroid of the cross section, and restricted to linear elastic behavior, will buckle at the **Euler buckling load**

$$P_{cr} = \pi^2 EI/L^2$$

in the fundamental mode; hence, the buckling load depends on the flexural rigidity ($EI$) and length ($L$), but not the strength of the material.

3. Changing the support conditions, or providing additional lateral supports, changes the critical buckling load. However, $P_{cr}$ for these **other support cases** may be obtained by replacing the actual column length ($L$) by the **effective length** ($L_e$) in the formula for $P_{cr}$ above. Three additional support cases are shown in Fig. 11-20. We can express the effective length $L_e$ in terms of an effective-length factor $K$ as

$$L_e = KL$$

where $K = 1$ for a pinned-end column and $K = 2$ for a column fixed at its base. The critical load $P_{cr}$ then is expressed as

$$P_{cr} = \frac{\pi^2 EI}{(KL)^2}$$

Effective-length factor $K$ is often used in column design formulas (Section 11.9).

4. Columns with **eccentric axial loads** behave quite differently from those with centroidal loads. The maximum compressive stress in pinned-end columns with load $P$ applied at eccentricity $e$ is defined by the **secant formula**; a graph of this formula (Fig. 11-28) shows that column load-carrying capacity decreases with increasing eccentricity. The secant formula gives the maximum compressive stress or $\sigma_{max}$ in an eccentrically loaded, pinned-end column in terms of average compressive stress $P/A$, modulus of elasticity $E$, slenderness ratio $L/r$, and eccentricity ratio $ec/r^2$ as

$$\sigma_{max} = \frac{P}{A}\left[1 + \frac{ec}{r^2} \sec\left(\frac{L}{2r}\sqrt{\frac{P}{EA}}\right)\right]$$

5. **Long columns** (i.e., large slenderness ratios $L/r$) buckle at low values of compressive stress; **short columns** (i.e., low $L/r$) fail by yielding and crushing of the material; and **intermediate columns** (with values of $L/r$ which lie between those for long and short columns) fail by inelastic buckling. The critical buckling load for inelastic buckling is always less than the Euler buckling load; the dividing lines between short, intermediate, and long columns are not precisely defined.

6. Three **theories for inelastic buckling of intermediate columns** are: the tangent-modulus theory, the reduced-modulus theory, and the Shanley theory. However, empirical formulas are actually used for the design of columns because the theoretical formulas do not account for such things as **residual stresses** in steel columns and other factors.

7. **Design formulas** for actual columns of various materials (e.g., steel, aluminum, wood) are based on both theory and observed behavior in laboratory tests.

# PROBLEMS CHAPTER 11

## Idealized Buckling Models

**11.2-1** The figure shows an idealized structure consisting of one or more **rigid bars** with pinned connections and linearly elastic springs. Rotational stiffness is denoted $\beta_R$, and translational stiffness is denoted $\beta$.

Determine the critical load $P_{cr}$ for the structure.

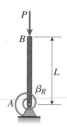

**PROB. 11.2-1**

**11.2-2** The figure shows an idealized structure consisting of one or more rigid bars with pinned connections and linearly elastic springs. Rotational stiffness is denoted $\beta_R$, and translational stiffness is denoted $\beta$.

(a) Determine the critical load $P_{cr}$ for the structure from the figure part a.

(b) Find $P_{cr}$ if another rotational spring is added at $B$ from the figure part b.

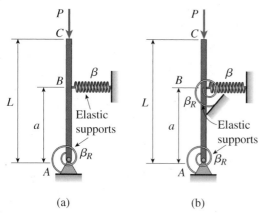

(a)                    (b)

**PROB. 11.2-2**

**11.2-3** The figure shows an idealized structure consisting of one or more **rigid bars** with pinned connections and linearly elastic springs. Rotational stiffness is denoted $\beta_R$, and translational stiffness is denoted $\beta$.

Determine the critical load $P_{cr}$ for the structure.

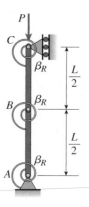

**PROB. 11.2-3**

**11.2-4** The figure shows an idealized structure consisting of bars $AB$ and $BC$ which are connected using a hinge at $B$ and linearly elastic springs at $A$ and $B$. Rotational stiffness is denoted $\beta_R$ and translational stiffness is denoted $\beta$.

(a) Determine the critical load $P_{cr}$ for the structure from the figure part a.

(b) Find $P_{cr}$ if an elastic connection is now used to connect bar segments $AB$ and $BC$ from the figure part b.

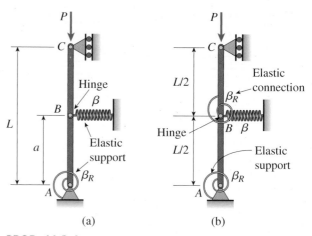

(a)                    (b)

**PROB. 11.2-4**

**11.2-5** The figure shows an idealized structure consisting of two rigid bars joined by an elastic connection with rotational stiffness $\beta_R$. Determine the critical load $P_{cr}$ for the structure.

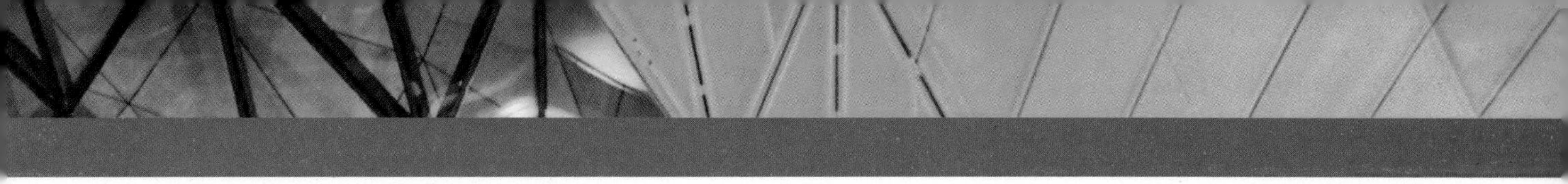

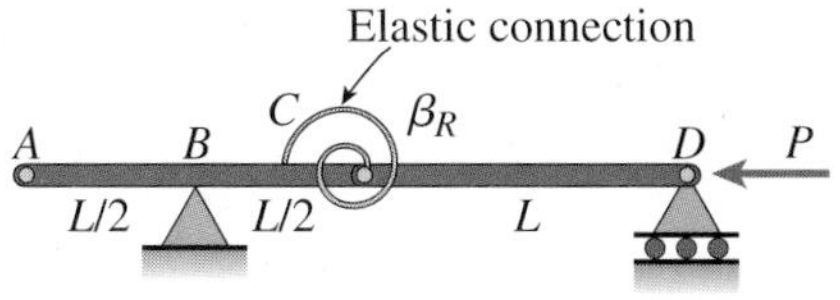


PROB. 11.2-5

**11.2-6** The figure shows an idealized structure consisting of rigid bars $ABC$ and $DEF$ joined by linearly elastic spring $\beta$ between $C$ and $D$. The structure is also supported by translational elastic support $\beta$ at $B$ and rotational elastic support $\beta_R$ at $E$.

Determine the critical load $P_{cr}$ for the structure.

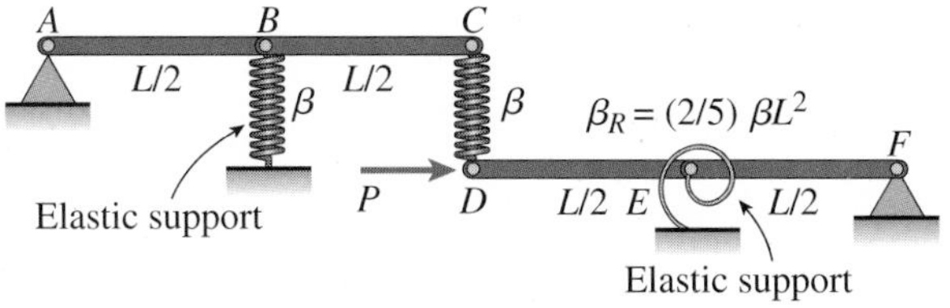


PROB. 11.2-6

**11.2-7** The figure shows an idealized structure consisting of an L-shaped rigid bar structure supported by linearly elastic springs at $A$ and $C$. Rotational stiffness is denoted $\beta_R$ and translational stiffness is denoted $\beta$.

Determine the critical load $P_{cr}$ for the structure.

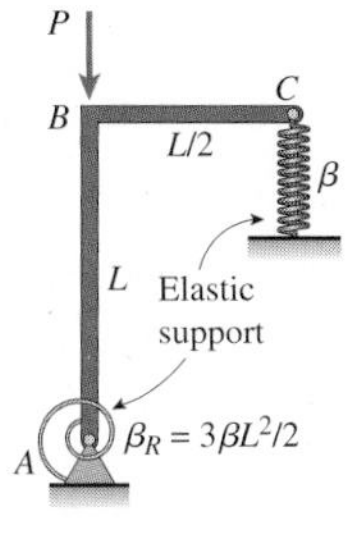


PROB. 11.2-7

## Critical Loads of Columns with Pinned Supports

*The problems for Section 11.3 are to be solved using the assumptions of ideal, slender, prismatic, linearly elastic columns (Euler buckling). Buckling occurs in the plane of the figure unless stated otherwise.*

**11.3-1** Calculate the critical load $P_{cr}$ for a W $8 \times 35$ steel column (see figure) having length $L = 24$ ft and $E = 30 \times 10^6$ psi under the following conditions:

(a) The column buckles by bending about its strong axis (axis 1–1), and (b) the column buckles by bending about its weak axis (axis 2–2). In both cases, assume that the column has pinned ends.

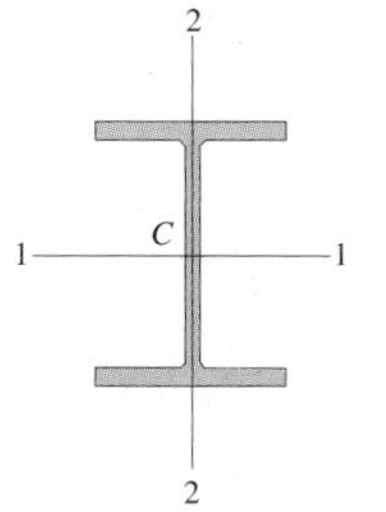


PROBS. 11.3-1 through 11.3-3

**11.3-2** Solve the preceding problem for a W $250 \times 89$ steel column having length $L = 10$ m. Let $E = 200$ GPa.

**11.3-3** Solve Prob. 11.3-1 for a W $10 \times 45$ steel column having length $L = 28$ ft.

**11.3-4** A horizontal beam $AB$ is pin-supported at end $A$ and carries a clockwise moment $M$ at joint $B$, as shown in the figure. The beam is also supported at $C$ by a pinned-end column of length $L$; the column is restrained laterally at $0.6L$ from the base at $D$. Assume the column can only buckle in the plane of the frame. The column is a solid steel bar ($E = 200$ GPa) of square cross section having length $L = 2.4$ m and side dimensions $b = 70$ mm. Let dimensions $d = L/2$. Based upon the critical load of the column, determine the allowable moment $M$ if the factor of safety with respect to buckling is $n = 2.0$.

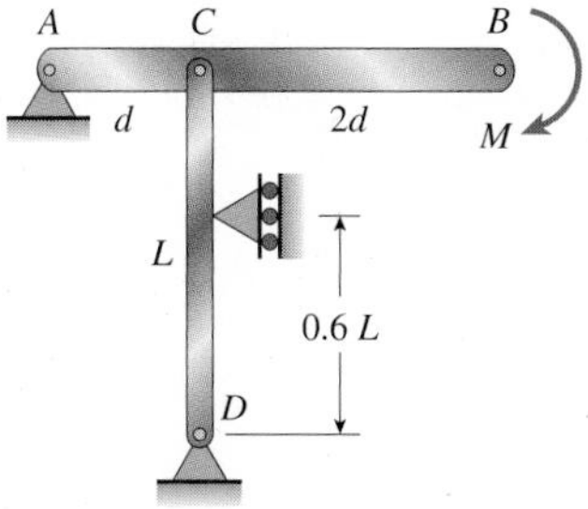


PROB. 11.3-4

**11.3-5** A horizontal beam $AB$ is pin-supported at end $A$ and carries a load $Q$ at joint $B$, as shown in the figure. The beam is also supported at $C$ by a pinned-end column of length $L$; the column is restrained laterally at $0.6L$ from the base at $D$. Assume the column can only buckle in the plane of the frame. The column is a solid aluminum bar ($E = 10 \times 10^6$ psi) of square cross section having length $L = 30$ in. and side dimensions $b = 1.5$ in. Let dimension $d = L/2$. Based upon the critical load of the column, determine the allowable force $Q$ if the factor of safety with respect to buckling is $n = 1.8$.

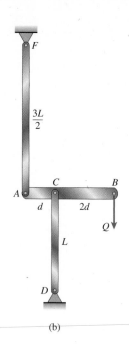

(b)

PROB. 11.3-6

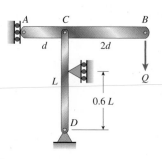

PROB. 11.3-5

**11.3-6** A horizontal beam $AB$ is supported at end A and carries a load $Q$ at joint $B$, as shown in the figure part a. The beam is also supported at $C$ by a pinned-end column of length $L$. The column has flexural rigidity $EI$.

(a) For the case of a guided support at $A$ (figure part a), what is the critical load $Q_{cr}$? (In other words, at what load $Q_{cr}$ does the system collapse because of Euler buckling of the column $DC$?)

(b) Repeat part (a) if the guided support at $A$ is replaced by column $AF$ with length $3L/2$ and flexural rigidity $EI$ (see figure part b).

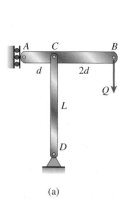

(a)

**11.3-7** A horizontal beam $AB$ has a guided support at end $A$ and carries a load $Q$ at end $B$, as shown in the figure part a. The beam is supported at $C$ and $D$ by two identical pinned-end columns of length $L$. Each column has flexural rigidity $EI$.

(a) Find an expression for the critical load $Q_{cr}$. (In other words, at what load $Q_{cr}$ does the system collapse because of Euler buckling of the columns?

(b) Repeat part (a), but assume a pin support at $A$. Find an expression for the critical moment $M_{cr}$ (i.e., find the moment $M$ at $B$ at which the system collapses because of Euler buckling of the columns).

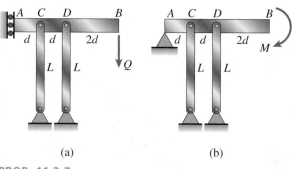

(a)                    (b)

PROB. 11.3-7

**11.3-8** A slender bar $AB$ with pinned ends and length $L$ is held between immovable supports (see figure).

What increase $\Delta T$ in the temperature of the bar will produce buckling at the Euler load?

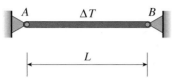

**PROB. 11.3-8**

**11.3-9** A rectangular column with cross-sectional dimensions $b$ and $h$ is pin-supported at ends $A$ and $C$ (see figure). At midheight, the column is restrained in the plane of the figure, but is free to deflect perpendicularly to the plane of the figure.

Determine the ratio $h/b$ such that the critical load is the same for buckling in the two principal planes of the column.

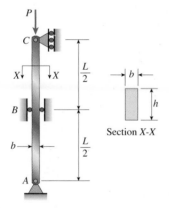

**PROB. 11.3-9**

**11.3-10** Three identical, solid circular rods, each of radius $r$ and length $L$, are placed together to form a compression member (see the cross section shown in the figure).

Assuming pinned-end conditions, determine the critical load $P_{cr}$ as follows: (a) The rods act independently as individual columns, and (b) the rods are bonded by epoxy throughout their lengths so that they function as a single member.

What is the effect on the critical load when the rods act as a single member?

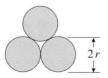

**PROB. 11.3-10**

**11.3-11** Three pinned-end columns of the same material have the same length and the same cross-sectional area (see figure). The columns are free to buckle in any direction. The columns have cross sections as follows: (1) a circle, (2) a square, and (3) an equilateral triangle.

Determine the ratios $P_1 : P_2 : P_3$ of the critical loads for these columns.

**PROB. 11.3-11**

**11.3-12** A long slender column $ABC$ is pinned at ends $A$ and $C$ and compressed by an axial force $P$ (see figure). At the midpoint $B$, lateral support is provided to prevent deflection in the plane of the figure. The column is a steel wide-flange section (W 250 × 67) with $E = 200$ GPa. The distance between lateral supports is $L = 5.5$ m.

Calculate the allowable load $P$ using a factor of safety $n = 2.4$, taking into account the possibility of Euler buckling about either principal centroidal axis (i.e., axis 1–1 or axis 2–2).

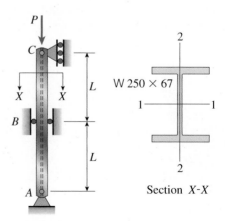

**PROB. 11.3-12**

**11.3-13** The roof over a concourse at an airport is supported by the use of pretensioned cables. At a typical joint in the roof structure, a strut $AB$ is compressed by the action of tensile forces $F$ in a cable that makes an angle $\alpha = 75°$ with the strut (see figure and photo). The strut is a circular tube of steel ($E = 30,000$ ksi) with outer diameter $d_2 = 2.5$ in. and inner diameter $d_1 = 2.0$ in. The strut is 5.75 ft long and is assumed to be pin-connected at both ends.

Using a factor of safety $n = 2.5$ with respect to the critical load, determine the allowable force $F$ in the cable.

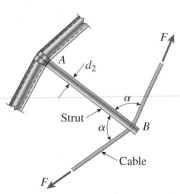

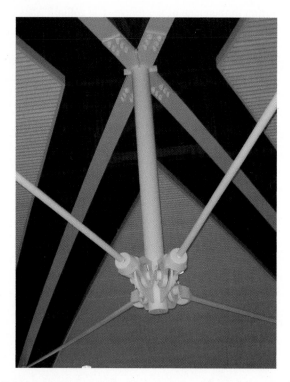

PROB. 11.3-13

Cable and strut at typical joint of airport concourse roof (© Barry Goodno)

**11.3-14** The hoisting arrangement for lifting a large pipe is shown in the figure. The spreader is a steel tubular section with outer diameter 70 mm and inner diameter 57 mm. Its length is 2.6 m and its modulus of elasticity is 200 GPa.

Based upon a factor of safety of 2.25 with respect to Euler buckling of the spreader, what is the maximum weight of pipe that can be lifted? (Assume pinned conditions at the ends of the spreader.)

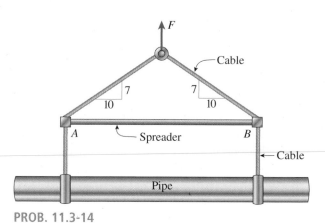

PROB. 11.3-14

**11.3-15** A pinned-end strut of aluminum ($E = 10,400$ ksi) with length $L = 6$ ft is constructed of circular tubing with outside diameter $d = 2$ in. (see figure). The strut must resist an axial load $P = 4$ kips with a factor of safety $n = 2.0$ with respect to the critical load.

Determine the required thickness $t$ of the tube.

$d = 2$ in.

PROB. 11.3-15

**11.3-16** The cross section of a column built up of two steel I-beams (S 150 × 25.7 sections) is shown in the figure. The beams are connected by spacer bars, or *lacing*, to ensure that they act together as a single column. (The lacing is represented by dashed lines in the figure.)

The column is assumed to have pinned ends and may buckle in any direction. Assuming $E = 200$ GPa and $L = 8.5$ m, calculate the critical load $P_{cr}$ for the column.

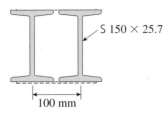

**PROB. 11.3-16**

**11.3-17** The truss $ABC$ shown in the figure supports a vertical load $W$ at joint $B$. Each member is a slender circular steel pipe ($E = 30,000$ ksi) with outside diameter 4 in. and wall thickness 0.25 in. The distance between supports is 23 ft. Joint $B$ is restrained against displacement perpendicular to the plane of the truss.

Determine the critical value $W_{cr}$ of the load.

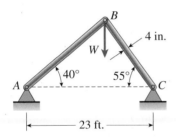

**PROB. 11.3-17**

**11.3-18** A truss $ABC$ supports a load $W$ at joint $B$, as shown in the figure. The length $L_1$ of member $AB$ is fixed, but the length of strut $BC$ varies as the angle $\theta$ is changed. Strut $BC$ has a solid circular cross section. Joint $B$ is restrained against displacement perpendicular to the plane of the truss.

Assuming that collapse occurs by Euler buckling of the strut, determine the angle $\theta$ for minimum weight of the strut.

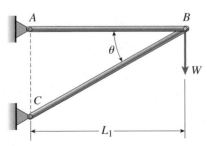

**PROB. 11.3-18**

**11.3-19** An S 6 × 12.5 steel cantilever beam $AB$ is supported by a steel tie rod at $B$ as shown. The tie rod is just taut when a roller support is added at $C$ at a distance $s$ to the left of $B$, then the distributed load $q$ is applied to beam segment $AC$. Assume $E = 30 \times 10^6$ psi and neglect the self weight of the beam and tie rod. See Table F-2(a) in Appendix F for the properties of the S-shape beam.

(a) What value of uniform load $q$ will, if exceeded, result in buckling of the tie rod if $L_1 = 6$ ft, $s = 2$ ft, $H = 3$ ft, and $d = 0.25$ in.?

(b) What minimum beam moment of inertia $I_b$ is required to prevent buckling of the tie rod if $q = 200$ lb/ft, $L_1 = 6$ ft, $H = 3$ ft, $d = 0.25$ in., and $s = 2$ ft?

(c) For what distance $S$ will the tie rod be just on the verge of buckling if $q = 200$ lb/ft, $L_1 = 6$ ft, $H = 3$ ft, and $d = 0.25$ in.?

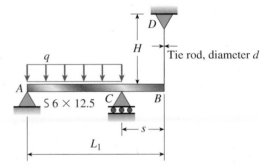

**PROB. 11.3-19**

**11.3-20** The plane truss shown in the figure supports vertical loads $F$ at joint $D$, $2F$ at joint $C$, and $3F$ at joint $B$. Each member is a slender circular pipe ($E = 70$ GPa) with an outside diameter of 60 mm and wall thickness of 5 mm. Joint $B$ is restrained against displacement perpendicular to the plane of the truss. Determine the critical value of load variable $F$ (kN) at which member $BF$ fails by Euler buckling.

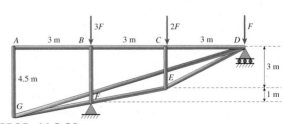

**PROB. 11.3-20**

**11.3-21** A space truss is restrained at joints $O$, $A$, $B$, and $C$, as shown in the figure. Load $P$ is applied at joint $A$ and load $2P$ acts downward at joint $C$. Each member is a slender, circular pipe ($E$ = 10,600 ksi) with an outside diameter of 3.5 in. and wall thickness of 0.25 in. Length variable $L$ = 11 ft. Determine the critical value of load variable $P$ (kips) at which member $OB$ fails by Euler buckling.

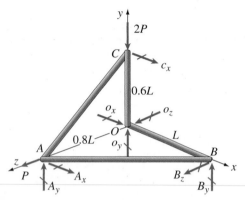

**PROB. 11.3-21**

## Columns with Other Support Conditions

*The problems for Section 11.4 are to be solved using the assumptions of ideal, slender, prismatic, linearly elastic columns (Euler buckling). Buckling occurs in the plane of the figure unless stated otherwise.*

**11.4-1** An aluminum pipe column ($E$ = 10,400 ksi) with length $L$ = 10.0 ft has inside and outside diameters $d_1$ = 5.0 in. and $d_2$ = 6.0 in., respectively (see figure). The column is supported only at the ends and may buckle in any direction.

Calculate the critical load $P_{cr}$ for the following end conditions: (1) pinned-pinned, (2) fixed-free, (3) fixed-pinned, and (4) fixed-fixed.

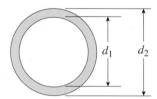

**PROBS. 11.4-1 and 11.4-2**

**11.4-2** Solve the preceding problem for a steel pipe column ($E$ = 210 GPa) with length $L$ = 1.2 m, inner diameter $d_1$ = 36 mm, and outer diameter $d_2$ = 40 mm.

**11.4-3** A wide-flange steel column ($E$ = 30 × 10⁶ psi) of W 12 × 87 shape (see figure) has length $L$ = 28 ft. It is supported only at the ends and may buckle in any direction.

Calculate the allowable load $P_{allow}$ based upon the critical load with a factor of safety $n$ = 2.5. Consider the following end conditions: (1) pinned-pinned, (2) fixed-free, (3) fixed-pinned, and (4) fixed-fixed.

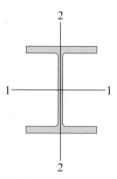

**PROBS. 11.4-3 and 11.4-4**

**11.4-4** Solve the preceding problem for a W 250 × 89 shape with length $L$ = 7.5 m and $E$ = 200 GPa.

**11.4-5** The upper end of a W 8 × 21 wide-flange steel column ($E$ = 30 × 10³ ksi) is supported laterally between two pipes (see figure). The pipes are not attached to the column, and friction between the pipes and the column is unreliable. The base of the column provides a fixed support, and the column is 13 ft long.

Determine the critical load for the column, considering Euler buckling in the plane of the web and also perpendicular to the plane of the web.

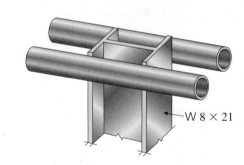

W 8 × 21

**PROB. 11.4-5**

**11.4-6** A vertical post $AB$ is embedded in a concrete foundation and held at the top by two cables (see figure). The post is a hollow steel tube with modulus of elasticity 200 GPa, outer diameter of 40 mm, and thickness of 5 mm. The cables are tightened equally by turnbuckles.

If a factor of safety of 3.0 against Euler buckling in the plane of the figure is desired, what is the maximum allowable tensile force $T_{allow}$ in the cables?

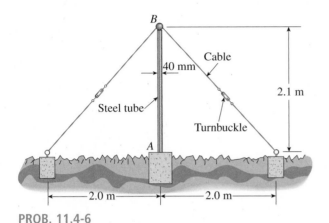

**PROB. 11.4-6**

**11.4-7** The horizontal beam $ABC$ shown in the figure is supported by columns $BD$ and $CE$. The beam is prevented from moving horizontally by the pin support at end $A$. Each column is pinned at its upper end to the beam, but at the lower ends, support $D$ is a guided support and support $E$ is pinned. Both columns are solid steel bars ($E = 30 \times 10^6$ psi) of square cross section with width equal to 0.625 in. A load $Q$ acts at distance a from column $BD$.

(a) If the distance $a = 12$ in., what is the critical value $Q_{cr}$ of the load?

(b) If the distance $a$ can be varied between 0 and 40 in., what is the maximum possible value of $Q_{cr}$? What is the corresponding value of the distance $a$?

**11.4-8** The roof beams of a warehouse are supported by pipe columns (see figure) having outer diameter $d_2 = 100$ mm and inner diameter $d_1 = 90$ mm. The columns have length $L = 4.0$ m, modulus $E = 210$ GPa, and fixed supports at the base.

Calculate the critical load $P_{cr}$ of one of the columns using the following assumptions: (1) the upper end is pinned and the beam prevents horizontal displacement; (2) the upper end is fixed against rotation and the beam prevents horizontal displacement; (3) the upper end is pinned, but the beam is free to move horizontally; and (4) the upper end is fixed against rotation, but the beam is free to move horizontally.

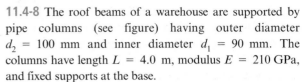

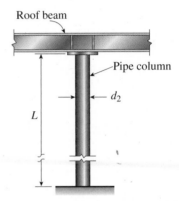

**PROB. 11.4-8**

**11.4-9** Determine the critical load $P_{cr}$ and the equation of the buckled shape for an ideal column with ends fixed against rotation (see figure) by solving the differential equation of the deflection curve. (See also Fig. 11-18.)

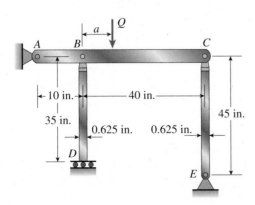

**PROB. 11.4-7**

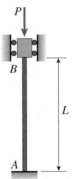

**PROB. 11.4-9**

**11.4-10** An aluminum tube $AB$ of circular cross section has a guided support at the base and is pinned at the top to a horizontal beam supporting a load $Q = 200$ kN (see figure).

Determine the required thickness $t$ of the tube if its outside diameter $d$ is 200 mm and the desired factor of safety with respect to Euler buckling is $n = 3.0$. (Assume $E = 72$ GPa.)

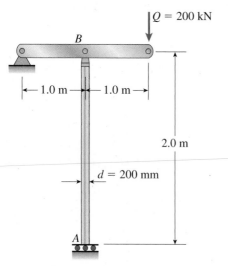

**PROB. 11.4-10**

**11.4-11** The frame $ABC$ consists of two members $AB$ and $BC$ that are rigidly connected at joint $B$, as shown in part a of the figure. The frame has pin supports at $A$ and $C$. A concentrated load $P$ acts at joint $B$, thereby placing member $AB$ in direct compression.

To assist in determining the buckling load for member $AB$, we represent it as a pinned-end column, as shown in part b of the figure. At the top of the column, a rotational spring of stiffness $\beta_R$ represents the restraining action of the horizontal beam $BC$ on the column (note that the horizontal beam provides resistance to rotation of joint $B$ when the column buckles). Also, consider only bending effects in the analysis (i.e., disregard the effects of axial deformations).

(a) By solving the differential equation of the deflection curve, derive the following buckling equation for this column:

$$\frac{\beta_R L}{EI}(kL \cot kL - 1) - k^2 L^2 = 0$$

in which $L$ is the length of the column and $EI$ is its flexural rigidity.

(b) For the particular case when member $BC$ is identical to member $AB$, the rotational stiffness $\beta_R$ equals $3EI/L$ (see Case 7, Table H-2, Appendix H). For this special case, determine the critical load $P_{cr}$.

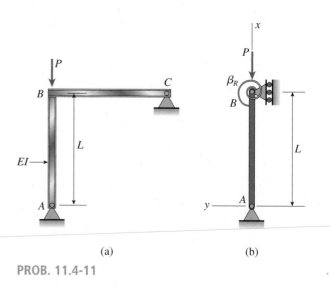

(a)                    (b)

**PROB. 11.4-11**

# Columns with Eccentric Axial Loads

*When solving the problems for Section 11.5, assume that bending occurs in the principal plane containing the eccentric axial load.*

**11.5-1** An aluminum bar having a rectangular cross section (2.0 in. × 1.0 in.) and length $L = 30$ in. is compressed by axial loads that have a resultant $P = 2800$ lb acting at the midpoint of the long side of the cross section (see figure).

Assuming that the modulus of elasticity $E$ is equal to $10 \times 10^6$ psi and that the ends of the bar are pinned, calculate the maximum deflection $\delta$ and the maximum bending moment $M_{max}$.

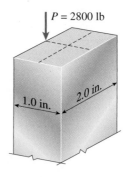

**PROB. 11.5-1**

**11.5-2** A steel bar having a square cross section (50 mm × 50 mm) and length $L = 2.0$ m is compressed by axial loads that have a resultant $P = 60$ kN acting at the midpoint of one side of the cross section (see figure).

Assuming that the modulus of elasticity $E$ is equal to 210 GPa and that the ends of the bar are pinned, calculate the maximum deflection $\delta$ and the maximum bending moment $M_{max}$.

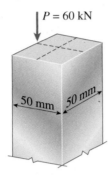

$P = 60$ kN

50 mm    50 mm

**PROB. 11.5-2**

**11.5-3** Determine the bending moment $M$ in the pinned-end column with eccentric axial loads shown in the figure. Then plot the bending-moment diagram for an axial load $P = 0.3P_{cr}$.

*Note:* Express the moment as a function of the distance $x$ from the end of the column, and plot the diagram in nondimensional form with $M/Pe$ as ordinate and $x/L$ as abscissa.

**11.5-4** Plot the load-deflection diagram for a pinned-end column with eccentric axial loads (see figure) if the eccentricity $e$ of the load is 5 mm and the column has length $L = 3.6$ m, moment of inertia $I = 9.0 \times 10^6$ mm⁴, and modulus of elasticity $E = 210$ GPa.

*Note:* Plot the axial load as ordinate and the deflection at the midpoint as abscissa.

**11.5-5** Solve the preceding problem for a column with $e = 0.20$ in., $L = 12$ ft, $I = 21.7$ in.⁴, and $E = 30 \times 10^6$ psi.

**11.5-6** A wide-flange member (W 200 × 22.5) is compressed by axial loads that have a resultant $P$ acting at the point shown in the figure. The member has modulus of elasticity $E = 200$ GPa and pinned conditions at the ends. Lateral supports prevent any bending about the weak axis of the cross section.

If the length of the member is 6.2 m and the deflection is limited to 6.5 mm, what is the maximum allowable load $P_{allow}$?

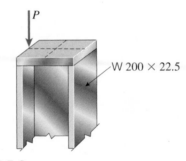

$P$

W 200 × 22.5

**PROB. 11.5-6**

**11.5-7** A wide-flange member (W 10 × 30) is compressed by axial loads that have a resultant $P = 20$ k acting at the point shown in the figure. The material is steel with modulus of elasticity $E = 29,000$ ksi. Assuming pinned-end conditions, determine the maximum permissible length $L_{max}$ if the deflection is not to exceed 1/400th of the length.

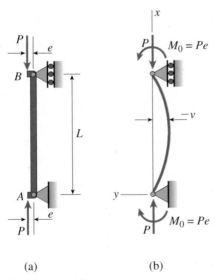

(a)                           (b)

**PROBS. 11.5-3 through 11.5-5**

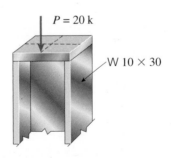

$P = 20$ k

W 10 × 30

**PROBS. 11.5-7 and 11.5-8**

**11.5-8** Solve the preceding problem (W 250 × 44.8) if the resultant force $P$ equals 110 kN and $E = 200$ GPa.

**11.5-9** The column shown in the figure is fixed at the base and free at the upper end. A compressive load $P$ acts at the top of the column with an eccentricity $e$ from the axis of the column.

Beginning with the differential equation of the deflection curve, derive formulas for the maximum deflection $\delta$ of the column and the maximum bending moment $M_{max}$ in the column.

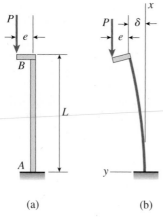

(a)                    (b)

**PROB. 11.5-9**

**11.5-10** An aluminum box column of square cross section is fixed at the base and free at the top (see figure). The outside dimension $b$ of each side is 100 mm and the thickness $t$ of the wall is 8 mm. The resultant of the compressive loads acting on the top of the column is a force $P = 50$ kN acting at the outer edge of the column at the midpoint of one side.

What is the longest permissible length $L_{max}$ of the column if the deflection at the top is not to exceed 30 mm? (Assume $E = 73$ GPa.)

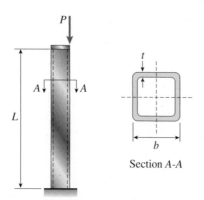

Section A-A

**PROBS. 11.5-10 and 11.5-11**

**11.5-11** Solve the preceding problem for an aluminum column with $b = 6.0$ in., $t = 0.5$ in., $P = 30$ k, and $E = 10.6 \times 10^3$ ksi. The deflection at the top is limited to 2.0 in.

**11.5-12** A steel post $AB$ of hollow circular cross section is fixed at the base and free at the top (see figure). The inner and outer diameters are $d_1 = 96$ mm and $d_2 = 110$ mm, respectively, and the length $L = 4.0$ m.

A cable $CBD$ passes through a fitting that is welded to the side of the post. The distance between the plane of the cable (plane $CBD$) and the axis of the post is $e = 100$ mm, and the angles between the cable and the ground are $\alpha = 53.13°$. The cable is pretensioned by tightening the turnbuckles.

If the deflection at the top of the post is limited to $\delta = 20$ mm, what is the maximum allowable tensile force $T$ in the cable? (Assume $E = 205$ GPa.)

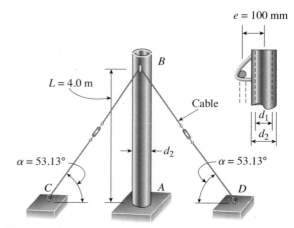

**PROB. 11.5-12**

**11.5-13** A frame $ABCD$ is constructed of steel wide-flange members (W 8 × 21; $E = 30 \times 10^6$ psi) and subjected to triangularly distributed loads of maximum intensity $q_0$ acting along the vertical members (see figure). The distance between supports is $L = 20$ ft and the height of the frame is $h = 4$ ft. The members are rigidly connected at $B$ and $C$.

(a) Calculate the intensity of load $q_0$ required to produce a maximum bending moment of 80 k-in. in the horizontal member $BC$.

(b) If the load $q_0$ is reduced to one-half of the value calculated in part (a), what is the maximum bending moment in member $BC$? What is the ratio of this moment to the moment of 80 k-in. in part (a)?

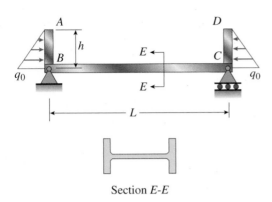

Section *E-E*

**PROB. 11.5-13**

# The Secant Formula

*When solving the problems for Section 11.6, assume that bending occurs in the principal plane containing the eccentric axial load.*

**11.6-1** A steel bar has a square cross section of width $b = 2.0$ in. (see figure). The bar has pinned supports at the ends and is 3.0 ft long. The axial forces acting at the end of the bar have a resultant $P = 20$ k located at distance $e = 0.75$ in. from the center of the cross section. Also, the modulus of elasticity of the steel is 29,000 ksi.

(a) Determine the maximum compressive stress $\sigma_{max}$ in the bar.

(b) If the allowable stress in the steel is 18,000 psi, what is the maximum permissible length $L_{max}$ of the bar?

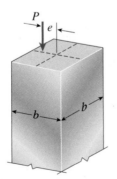

**PROBS. 11.6-1 through 11.6-3**

**11.6-2** A brass bar ($E = 100$ GPa) with a square cross section is subjected to axial forces having a resultant $P$ acting at distance $e$ from the center (see figure). The bar is pin supported at the ends and is 0.6 m in length. The side dimension $b$ of the bar is 30 mm and the eccentricity $e$ of the load is 10 mm.

If the allowable stress in the brass is 150 MPa, what is the allowable axial force $P_{allow}$?

**11.6-3** A square aluminum bar with pinned ends carries a load $P = 25$ k acting at distance $e = 2.0$ in. from the center (see figure). The bar has length $L = 54$ in. and modulus of elasticity $E = 10,600$ ksi.

If the stress in the bar is not to exceed 6 ksi, what is the minimum permissible width $b_{min}$ of the bar?

**11.6-4** A pinned-end column of length $L = 2.1$ m is constructed of steel pipe ($E = 210$ GPa) having inside diameter $d_1 = 60$ mm and outside diameter $d_2 = 68$ mm (see figure). A compressive load $P = 10$ kN acts with eccentricity $e = 30$ mm.

(a) What is the maximum compressive stress $\sigma_{max}$ in the column?

(b) If the allowable stress in the steel is 50 MPa, what is the maximum permissible length $L_{max}$ of the column?

**PROBS. 11.6-4 through 11.6-6**

**11.6-5** A pinned-end strut of length $L = 5.2$ ft is constructed of steel pipe ($E = 30 \times 10^3$ ksi) having inside diameter $d_1 = 2.0$ in. and outside diameter $d_2 = 2.2$ in. (see figure). A compressive load $P = 2.0$ k is applied with eccentricity $e = 1.0$ in.

(a) What is the maximum compressive stress $\sigma_{max}$ in the strut?

(b) What is the allowable load $P_{allow}$ if a factor of safety $n = 2$ with respect to yielding is required? (Assume that the yield stress $\sigma_Y$ of the steel is 42 ksi.)

**11.6-6** A circular aluminum tube with pinned ends supports a load $P = 18$ kN acting at distance $e = 50$ mm from the center (see figure). The length of the tube is 3.5 m and its modulus of elasticity is 73 GPa.

If the maximum permissible stress in the tube is 20 MPa, what is the required outer diameter $d_2$ if the ratio of diameters is to be $d_1/d_2 = 0.9$?

**11.6-7** A steel column ($E = 30 \times 10^3$ ksi) with pinned ends is constructed of a W 10 × 60 wide-flange shape (see figure). The column is 24 ft long. The resultant of the axial loads acting on the column is a force $P$ acting with an eccentricity $e = 2.0$ in.

(a) If $P = 120$ k, determine the maximum compressive stress $\sigma_{max}$ in the column.

(b) Determine the allowable load $P_{allow}$ if the yield stress is $\sigma_Y = 42$ ksi and the factor of safety with respect to yielding of the material is $n = 2.5$.

wide-flange member (see figure). The column is 9.0 ft long. The force $P$ acting at the top of the column has an eccentricity $e = 1.25$ in.

(a) If $P = 40$ k, what is the maximum compressive stress in the column?

(b) If the yield stress is 36 ksi and the required factor of safety with respect to yielding is 2.1, what is the allowable load $P_{allow}$?

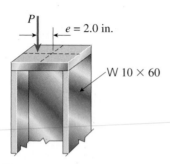

PROB. 11.6-7

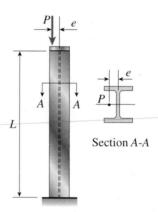

Section A-A

PROBS. 11.6-9 and 11.6-10

**11.6-8** A W 410 × 85 steel column is compressed by a force $P = 340$ kN acting with an eccentricity $e = 38$ mm., as shown in the figure. The column has pinned ends and length $L$. Also, the steel has modulus of elasticity $E = 200$ GPa and yield stress $\sigma_Y = 250$ MPa.

(a) If the length $L = 3$ m, what is the maximum compressive stress $\sigma_{max}$ in the column?

(b) If a factor of safety $n = 2.0$ is required with respect to yielding, what is the longest permissible length $L_{max}$ of the column?

**11.6-10** A W 310 × 74 wide-flange steel column with length $L = 3.8$ m is fixed at the base and free at the top (see figure). The load $P$ acting on the column is intended to be centrally applied, but because of unavoidable discrepancies in construction, an eccentricity ratio of 0.25 is specified. Also, the following data are supplied: $E = 200$ GPa, $\sigma_Y = 290$ MPa, and $P = 310$ kN.

(a) What is the maximum compressive stress $\sigma_{max}$ in the column?

(b) What is the factor of safety $n$ with respect to yielding of the steel?

**11.6-11** A pinned-end column with length $L = 18$ ft is constructed from a W 12 × 87 wide-flange shape (see figure). The column is subjected to a centrally applied load $P_1 = 180$ k and an eccentrically applied load $P_2 = 75$ k. The load $P_2$ acts at distance $s = 5.0$ in. from the centroid of the cross section. The properties of the steel are $E = 29,000$ ksi and $\sigma_Y = 36$ ksi.

(a) Calculate the maximum compressive stress in the column.

(b) Determine the factor of safety with respect to yielding.

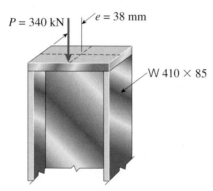

PROB. 11.6-8

**11.6-9** A steel column ($E = 30 \times 10^3$ ksi) that is fixed at the base and free at the top is constructed of a W 8 × 35

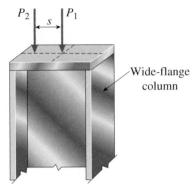

PROBS. 11.6-11 and 11.6-12

**11.6-12** The wide-flange pinned-end column shown in the figure carries two loads, a force $P_1 = 450$ kN acting at the centroid and a force $P_2 = 270$ kN acting at distance $s = 100$ mm, from the centroid. The column is a W 250 × 67 shape with $L = 4.2$ m, $E = 200$ GPa, and $\sigma_Y = 290$ MPa.

(a) What is the maximum compressive stress in the column?

(b) If the load $P_1$ remains at 450 kN, what is the largest permissible value of the load $P_2$ in order to maintain a factor of safety of 2.0 with respect to yielding?

**11.6-13** A W 14 × 53 wide-flange column of length $L = 15$ ft is fixed at the base and free at the top (see figure). The column supports a centrally applied load $P_1 = 120$ k and a load $P_2 = 40$ k supported on a bracket. The distance from the centroid of the column to the load $P_2$ is $s = 12$ in. Also, the modulus of elasticity is $E = 29,000$ ksi and the yield stress is $\sigma_Y = 36$ ksi.

(a) Calculate the maximum compressive stress in the column.

(b) Determine the factor of safety with respect to yielding.

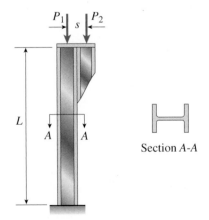

Section A-A

PROBS. 11.6-13 and 11.6-14

**11.6-14** A wide-flange column with a bracket is fixed at the base and free at the top (see figure). The column supports a load $P_1 = 340$ kN acting at the centroid and a load $P_2 = 110$ kN acting on the bracket at distance $s = 250$ mm, from the load $P_1$. The column is a W 310 × 52 shape with $L = 5$ m, $E = 200$ GPa, and $\sigma_Y = 290$ MPa.

(a) What is the maximum compressive stress in the column?

(b) If the load $P_1$ remains at 340 kN, what is the largest permissible value of the load $P_2$ in order to maintain a factor of safety of 1.8 with respect to yielding?

## Design Formulas for Columns

*The problems for Section 11.9 are to be solved assuming that the axial loads are centrally applied at the ends of the columns. Unless otherwise stated, the columns may buckle in any direction.*

## Steel Columns

**11.9-1** Determine the allowable axial load $P_{\text{allow}}$ for a W 10 × 45 steel wide-flange column with pinned ends (see figure) for each of the following lengths: $L = 8$ ft, 16 ft, 24 ft, and 32 ft. (Assume $E = 29,000$ ksi and $\sigma_Y = 36$ ksi.)

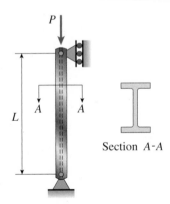

Section A-A

PROBS. 11.9-1 through 11.9-6

**11.9-2** Determine the allowable axial load $P_{\text{allow}}$ for a W 310 × 129 steel wide-flange column with pinned ends (see figure) for each of the following lengths: $L = 3$ m, 6 m, 9 m, and 12 m. (Assume $E = 200$ GPa and $\sigma_Y = 340$ MPa.)

**11.9-3** Determine the allowable axial load $P_{\text{allow}}$ for a W 10 × 60 steel wide-flange column with pinned ends (see figure) for each of the following lengths: $L = 10$ ft, 20 ft, 30 ft, and 40 ft. (Assume $E = 29,000$ ksi and $\sigma_Y = 36$ ksi.)

**11.9-4** Select a steel wide-flange column of nominal depth 250 mm. (W 250 shape) to support an axial load $P = 800$ kN (see figure). The column has pinned ends and length $L = 4.25$ m. Assume $E = 200$ GPa and $\sigma_Y = 250$ MPa. (*Note:* The selection of columns is limited to those listed in Table F-1(b), Appendix F.)

**11.9-5** Select a steel wide-flange column of nominal depth 12 in. (W 12 shape) to support an axial load $P = 175$ k (see figure). The column has pinned ends and length $L = 35$ ft. Assume $E = 29,000$ ksi and $\sigma_Y = 36$ ksi. (*Note:* The selection of columns is limited to those listed in Table F-1, Appendix F.)

**11.9-6** Select a steel wide-flange column of nominal depth 360 mm (W 360 shape) to support an axial load $P = 1100$ kN (see figure). The column has pinned ends and length $L = 6$ m. Assume $E = 200$ GPa and $\sigma_Y = 340$ MPa. (*Note:* The selection of columns is limited to those listed in Table F-1(b), Appendix F.)

**11.9-7** Determine the allowable axial load $P_{allow}$ for a steel *pipe column with pinned ends* having an outside diameter of 4.5 in. and wall thickness of 0.237 in. for each of the following lengths: $L = 6$ ft, 12 ft, 18 ft, and 24 ft. (Assume $E = 29,000$ ksi and $\sigma_Y = 36$ ksi.)

**11.9-8** Determine the allowable axial load $P_{allow}$ for a steel *pipe column with pinned ends* having an outside diameter of 220 mm and wall thickness of 12 mm for each of the following lengths: $L = 2.5$ m, 5 m, 7.5 m, and 10 m. (Assume $E = 200$ GPa and $\sigma_Y = 250$ MPa.)

**11.9-9** Determine the allowable axial load $P_{allow}$ for a steel pipe column that is fixed at the base and free at the top (see figure) for each of the following lengths: $L = 6$ ft, 9 ft, 12 ft, and 15 ft. The column has outside diameter $d = 6.625$ in. and wall thickness $t = 0.280$ in. (Assume $E = 29,000$ ksi and $\sigma_Y = 36$ ksi.)

**11.9-10** Determine the allowable axial load $P_{allow}$ for a steel pipe column that is fixed at the base and free at the top (see figure) for each of the following lengths: $L = 2.6$ m, 2.8 m, 3.0 m, and 3.2 m. The column has outside diameter $d = 140$ mm and wall thickness $t = 7$ mm. (Assume $E = 200$ GPa and $\sigma_Y = 250$ MPa.)

**11.9-11** Determine the maximum permissible length $L_{max}$ for a steel pipe column that is fixed at the base and free at the top and must support an axial load $P = 40$ k (see figure). The column has outside diameter $d = 4.0$ in., wall thickness $t = 0.226$ in., $E = 29,000$ ksi, and $\sigma_Y = 42$ ksi.

**11.9-12** Determine the maximum permissible length $L_{max}$ for a steel pipe column that is fixed at the base and free at the top and must support an axial load $P = 500$ kN (see figure). The column has outside diameter $d = 200$ mm, wall thickness $t = 10$ mm, $E = 200$ GPa, and $\sigma_Y = 250$ MPa.

**11.9-13** A steel pipe column with *pinned ends* supports an axial load $P = 21$ k. The pipe has outside and inside diameters of 3.5 in. and 2.9 in., respectively. What is the maximum permissible length $L_{max}$ of the column if $E = 29,000$ ksi and $\sigma_Y = 36$ ksi?

**11.9-14** The steel columns used in a college recreation center are 16.75 m long and are formed by welding three wide-flange sections (see figure). The columns are pin-supported at the ends and may buckle in any direction.

Calculate the allowable load $P_{allow}$ for one column, assuming $E = 200$ GPa and $\sigma_Y = 250$ MPa.

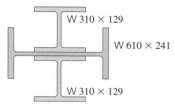

W 310 × 129
W 610 × 241
W 310 × 129

**PROB. 11.9-14**

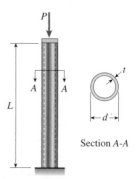

Section A-A

**PROBS. 11.9-9 through 11.9-12**

**11.9-15** A W 8 × 28 steel wide-flange column with pinned ends carries an axial load $P$. What is the maximum permissible length $L_{max}$ of the column if (a) $P = 50$ k, and (b) $P = 100$ k? (Assume $E = 29,000$ ksi and $\sigma_Y = 36$ ksi.)

**PROBS. 11.9-15 and 11.9-16**

**11.9-16** A W 250 × 67 steel wide-flange column with pinned ends carries an axial load $P$. What is the maximum permissible length $L_{max}$ of the column if (a) $P = 560$ kN, and (b) $P = 890$ kN? (Assume $E = 200$ GPa and $\sigma_Y = 290$ MPa.)

**11.9-17** Find the required outside diameter $d$ for a steel pipe column (see figure) of length $L = 20$ ft that is pinned at both ends and must support an axial load $P = 25$ k. Assume that the wall thickness $t$ is equal to $d/20$. (Use $E = 29,000$ ksi and $\sigma_Y = 36$ ksi.)

**PROBS. 11.9-17 through 11.9-20**

**11.9-18** Find the required outside diameter $d$ for a steel pipe column (see figure) of length $L = 3.5$ m that is pinned at both ends and must support an axial load $P = 130$ kN. Assume that the wall thickness $t$ is equal to $d/20$. (Use $E = 200$ GPa and $\sigma_Y = 275$ MPa).

**11.9-19** Find the required outside diameter $d$ for a steel pipe column (see figure) of length $L = 11.5$ ft that is pinned at both ends and must support an axial load $P = 80$ k. Assume that the wall thickness $t$ is 0.30 in. (Use $E = 29,000$ ksi and $\sigma_Y = 42$ ksi.)

**11.9-20** Find the required outside diameter $d$ for a steel pipe column (see figure) of length $L = 3.0$ m that is pinned at both ends and must support an axial load $P = 800$ kN. Assume that the wall thickness $t$ is 9 mm. (Use $E = 200$ GPa and $\sigma_Y = 300$ MPa.)

## Aluminum Columns

**11.9-21** An aluminum pipe column (alloy 2014-T6) with pinned ends has outside diameter $d_2 = 5.60$ in. and inside diameter $d_1 = 4.80$ in. (see figure).

Determine the allowable axial load $P_{allow}$ for each of the following lengths: $L = 6$ ft, 8 ft, 10 ft, and 12 ft.

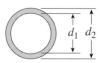

**PROBS. 11.9-21 through 11.9-24**

**11.9-22** An aluminum pipe column (alloy 2014-T6) with pinned ends has outside diameter $d_2 = 120$ mm and inside diameter $d_1 = 110$ mm (see figure).

Determine the allowable axial load $P_{allow}$ for each of the following lengths: $L = 1.0$ m, 2.0 m, 3.0 m, and 4.0 m.

**11.9-23** An aluminum pipe column (alloy 6061-T6) that is fixed at the base and free at the top has outside diameter $d_2 = 3.25$ in. and inside diameter $d_1 = 3.00$ in. (see figure).

Determine the allowable axial load $P_{allow}$ for each of the following lengths: $L = 2$ ft, 3 ft, 4 ft, and 5 ft.

**11.9-24** An aluminum pipe column (alloy 6061-T6) that is fixed at the base and free at the top has outside diameter $d_2 = 80$ mm and inside diameter $d_1 = 72$ mm (see figure).

Determine the allowable axial load $P_{allow}$ for each of the following lengths: $L = 0.6$ m, 0.8 m, 1.0 m, and 1.2 m.

**11.9-25** A solid round bar of aluminum having diameter $d$ (see figure) is compressed by an axial force $P = 60$ k. The bar has pinned supports and is made of alloy 2014-T6.

(a) If the diameter $d = 2.0$ in., what is the maximum allowable length $L_{max}$ of the bar?

(b) If the length $L = 30$ in., what is the minimum required diameter $d_{min}$?

**PROBS. 11.9-25 through 11.9-28**

**11.9-26** A solid round bar of aluminum having diameter $d$ (see figure) is compressed by an axial force $P = 175$ kN. The bar has pinned supports and is made of alloy 2014-T6.

(a) If the diameter $d = 40$ mm, what is the maximum allowable length $L_{max}$ of the bar?

(b) If the length $L = 0.6$ m, what is the minimum required diameter $d_{min}$?

**11.9-27** A solid round bar of aluminum having diameter $d$ (see figure) is compressed by an axial force $P = 10$ k. The bar has pinned supports and is made of alloy 6061-T6.

(a) If the diameter $d = 1.0$ in., what is the maximum allowable length $L_{max}$ of the bar?

(b) If the length $L = 20$ in., what is the minimum required diameter $d_{min}$?

**11.9-28** A solid round bar of aluminum having diameter $d$ (see figure) is compressed by an axial force $P = 60$ kN. The bar has pinned supports and is made of alloy 6061-T6.

(a) If the diameter $a = 30$ mm, what is the maximum allowable length $L_{max}$ of the bar?

(b) If the length $L = 0.6$ m, what is the minimum required diameter $d_{min}$?

## Wood Columns

*When solving the problems for wood columns, assume that the columns are constructed of sawn lumber ($c = 0.8$) and have pinned-end conditions. Also, buckling may occur about either principal axis of the cross section.*

**11.9-29** A wood post of rectangular cross section (see figure) is constructed of 4 in. $\times$ 6 in. structural grade, Douglas fir lumber ($F_c = 2000$ psi, $E = 1,800,000$ psi). The net cross-sectional dimensions of the post are $b = 3.5$ in. and $h = 5.5$ in. (see Appendix G).

Determine the allowable axial load $P_{allow}$ for each of the following lengths: $L = 5.0$ ft, 7.5 ft, and 10.0 ft.

PROBS. 11.9-29 through 11.9-32

**11.9-30** A wood post of rectangular cross section (see figure) is constructed of structural grade, southern pine lumber ($F_c = 14$ MPa, $E = 12$ GPa). The cross-sectional dimensions of the post (actual dimensions) are $b = 100$ mm and $h = 150$ mm.

Determine the allowable axial load $P_{allow}$ for each of the following lengths: $L = 1.5$ m, 2.0 m, and 2.5 m.

**11.9-31** A wood column of rectangular cross section (see figure) is constructed of 4 in. $\times$ 8 in. construction grade, western hemlock lumber ($F_c = 1000$ psi, $E = 1,300,000$ psi). The net cross-sectional dimensions of the column are $b = 3.5$ in. and $h = 7.25$ in. (see Appendix G).

Determine the allowable axial load $P_{allow}$ for each of the following lengths: $L = 6$ ft, 8 ft, and 10 ft.

**11.9-32** A wood column of rectangular cross section (see figure) is constructed of structural grade, Douglas fir lumber ($F_c = 12$ MPa, $E = 10$ GPa). The cross-sectional dimensions of the column (actual dimensions) are $b = 140$ mm and $h = 210$ mm.

Determine the allowable axial load $P_{allow}$ for each of the following lengths: $L = 2.5$ m, 3.5 m, and 4.5 m.

**11.9-33** A square wood column with side dimensions $b$ (see figure) is constructed of a structural grade of Douglas fir for which $F_c = 1700$ psi and $E = 1,400,000$ psi. An axial force $P = 40$ k acts on the column.

(a) If the dimension $b = 5.5$ in., what is the maximum allowable length $L_{max}$ of the column?

(b) If the length $L = 11$ ft, what is the minimum required dimension $b_{min}$?

PROBS. 11.9-33 through 11.9-36

**11.9-34** A square wood column with side dimensions $b$ (see figure) is constructed of a structural grade of southern pine for which $F_c = 10.5$ MPa and $E = 12$ GPa. An axial force $P = 200$ kN acts on the column.

(a) If the dimension $b = 150$ mm, what is the maximum allowable length $L_{max}$ of the column?

(b) If the length $L = 4.0$ m, what is the minimum required dimension $b_{min}$?

**11.9-35** A square wood column with side dimensions $b$ (see figure) is constructed of a structural grade of spruce for which $F_c = 900$ psi and $E = 1,500,000$ psi. An axial force $P = 8.0$ k acts on the column.

(a) If the dimension $b = 3.5$ in., what is the maximum allowable length $L_{max}$ of the column?

(b) If the length $L = 10$ ft, what is the minimum required dimension $b_{min}$?

**11.9-36** A square wood column with side dimensions $b$ (see figure) is constructed of a structural grade of eastern white pine for which $F_c = 8.0$ MPa and $E = 8.5$ GPa. An axial force $P = 100$ kN acts on the column.

(a) If the dimension $b = 120$ mm, what is the maximum allowable length $L_{max}$ of the column?

(b) If the length $L = 4.0$ m, what is the minimum required dimension $b_{min}$?

# 12

# Review of Centroids and Moments of Inertia

Steel members come in a wide variety of shapes; the properties of the cross section are needed for analysis and design. (© Fresh Picked/Alamy)

# | CHAPTER OVERVIEW

Topics covered in Chapter 12 include centroids and how to locate them (Sections 12.2 and 12.3), moments of inertia (Section 12.4), parallel-axis theorems (Section 12.5), polar moments of inertia (Section 12.6), products of inertia (Section 12.7), rotation of axes (Section 12.8), and principal axes (Section 12.9).

Only plane areas are considered. There are numerous examples within the chapter and problems at the end of the chapter available for review.

A table of centroids and moments of inertia for a variety of common geometric shapes is given in Appendix E for convenient reference.

## Chapter 12 is organized as follows:

**12.1** Introduction 956
**12.2** Centroids of Plane Areas 956
**12.3** Centroids of Composite Areas 959
**12.4** Moments of Inertia of Plane Areas 962
**12.5** Parallel-Axis Theorem for Moments of Inertia 965

**12.6** Polar Moments of Inertia 969
**12.7** Products of Inertia 971
**12.8** Rotation of Axes 974
**12.9** Principal Axes and Principal Moments of Inertia 976
Problems 980

## 12.1 INTRODUCTION

This chapter is a review of the definitions and formulas pertaining to centroids and moments of inertia of plane areas. The word "review" is appropriate because these topics are usually covered in earlier courses, such as mathematics and engineering statics, and therefore most readers will already have been exposed to the material. However, since centroids and moments of inertia are used repeatedly throughout the preceding chapters, they must be clearly understood by the reader and the essential definitions and formulas must be readily accessible.

The terminology used in this and earlier chapters may appear puzzling to some readers. For instance, the term "moment of inertia" is clearly a misnomer when referring to properties of an area, since no mass is involved. Even the word "area" is used inappropriately. When we say "plane area," we really mean "plane surface." Strictly speaking, area is a measure of the *size* of a surface and is not the same thing as the surface itself. In spite of its deficiencies, the terminology used in this book is so entrenched in the engineering literature that it rarely causes confusion.

## 12.2 CENTROIDS OF PLANE AREAS

The position of the centroid of a plane area is an important geometric property. To obtain formulas for locating centroids, we will refer to Fig. 12-1, which shows a plane area of irregular shape with its centroid at point $C$. The $xy$ coordinate system is oriented arbitrarily with its origin at any point $O$. The **area** of the geometric figure is defined by the following integral:

$$A = \int dA \qquad \textbf{(12-1)}$$

in which $dA$ is a differential element of area having coordinates $x$ and $y$ (Fig. 12-1) and $A$ is the total area of the figure.

The **first moments** of the area with respect to the $x$ and $y$ axes are defined, respectively, as follows:

$$Q_x = \int y \, dA \qquad Q_y = \int x \, dA \qquad \textbf{(12-2a,b)}$$

Thus, the first moments represent the sums of the products of the differential areas and their coordinates. First moments may be positive or negative, depending upon the position of the $xy$ axes. Also, first moments have units of length raised to the third power; for instance, in.$^3$ or mm$^3$.

The coordinates $\bar{x}$ and $\bar{y}$ of the **centroid** $C$ (Fig. 12-1) are equal to the first moments divided by the area:

$$\bar{x} = \frac{Q_y}{A} = \frac{\int x \, dA}{\int dA} \qquad \bar{y} = \frac{Q_x}{A} = \frac{\int y \, dA}{\int dA} \qquad \textbf{(12-3a,b)}$$

**Fig. 12-1**

Plane area of arbitrary shape with centroid $C$

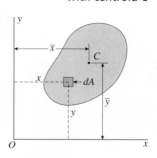

If the boundaries of the area are defined by simple mathematical expressions, we can evaluate the integrals appearing in Eqs. (12-3a) and (12-3b) in closed form and thereby obtain formulas for $\bar{x}$ and $\bar{y}$. The formulas listed in Appendix E were obtained in this manner. In general, the coordinates $\bar{x}$ and $\bar{y}$ may be positive or negative, depending upon the position of the centroid with respect to the reference axes.

If an area is **symmetric about an axis**, the centroid must lie on that axis because the first moment about an axis of symmetry equals zero. For example, the centroid of the singly symmetric area shown in Fig. 12-2 must lie on the x axis, which is the axis of symmetry. Therefore, only one coordinate must be calculated in order to locate the centroid C.

If an area has **two axes of symmetry**, as illustrated in Fig. 12-3, the position of the centroid can be determined by inspection because it lies at the intersection of the axes of symmetry.

An area of the type shown in Fig. 12-4 is **symmetric about a point**. It has no axes of symmetry, but there is a point (called the **center of symmetry**) such that every line drawn through that point contacts the area in a symmetrical manner. The centroid of such an area coincides with the center of symmetry, and therefore, the centroid can be located by inspection.

If an area has **irregular boundaries** not defined by simple mathematical expressions, we can locate the centroid by numerically evaluating the integrals in Eqs. (12-3a) and (12-3b). The simplest procedure is to divide the geometric figure into small finite elements and replace the integrations with summations. If we denote the area of the ith element by $\Delta A_i$, then the expressions for the summations are

$$A = \sum_{i=1}^{n} \Delta A_i \qquad Q_x = \sum_{i=1}^{n} \bar{y}_i \Delta A_i \qquad Q_y = \sum_{i=1}^{n} \bar{x}_i \Delta A_i \qquad \textbf{(12-4a,b,c)}$$

in which n is the total number of elements, $\bar{y}_i$ is the y coordinate of the centroid of the ith element, and $\bar{x}_i$ is the x coordinate of the centroid of the ith element. Replacing the integrals in Eqs. (12-3a) and (12-3b) by the corresponding summations, we obtain the following formulas for the coordinates of the centroid:

$$\bar{x} = \frac{Q_y}{A} = \frac{\sum_{i=1}^{n} \bar{x}_i \Delta A_i}{\sum_{i=1}^{n} \Delta A_i} \qquad \bar{y} = \frac{Q_x}{A} = \frac{\sum_{i=1}^{n} \bar{y}_i \Delta A_i}{\sum_{i=1}^{n} \Delta A_i} \qquad \textbf{(12-5a,b)}$$

The accuracy of the calculations for $\bar{x}$ and $\bar{y}$ depends upon how closely the selected elements fit the actual area. If they fit exactly, the results are exact. Many computer programs for locating centroids use a numerical scheme similar to the one expressed by Eqs. (12-5a) and (12-5b).

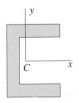

**Fig. 12-2**

Area with one axis of symmetry

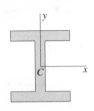

**Fig. 12-3**

Area with two axes of symmetry

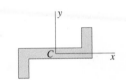

**Fig. 12-4**

Area that is symmetric about a point

The centroid of wide-flange steel sections lies at the intersection of the axes of symmetry (Photo courtesy of Louis Geschwinder)

## Example 12-1

### Fig. 12-5

Example 12-1: Centroid of a parabolic semisegment

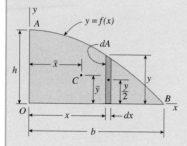

A parabolic semisegment $OAB$ is bounded by the $x$ axis, the $y$ axis, and a parabolic curve having its vertex at $A$ (Fig. 12-5). The equation of the curve is

$$y = f(x) = h\left(1 - \frac{x^2}{b^2}\right) \tag{a}$$

in which $b$ is the base and $h$ is the height of the semisegment.

Locate the centroid $C$ of the semisegment.

### Solution

To determine the coordinates $\bar{x}$ and $\bar{y}$ of the centroid $C$ (Fig. 12-5), we will use Eqs. (12-3a) and (12-3b). We begin by selecting an element of area $dA$ in the form of a thin vertical strip of width $dx$ and height $y$. The area of this differential element is

$$dA = y\, dx = h\left(1 - \frac{x^2}{b^2}\right)dx \tag{b}$$

Therefore, the area of the parabolic semisegment is

$$A = \int dA = \int_0^b h\left(1 - \frac{x^2}{b^2}\right)dx = \frac{2bh}{3} \tag{c}$$

Note that this area is 2/3 of the area of the surrounding rectangle.

The first moment of an element of area $dA$ with respect to an axis is obtained by multiplying the area of the element by the distance from its centroid to the axis. Since the $x$ and $y$ coordinates of the centroid of the element shown in Fig. 12-5 are $x$ and $y/2$, respectively, the first moments of the element with respect to the $x$ and $y$ axes are

$$Q_x = \int \frac{y}{2}\, dA = \int_0^b \frac{h^2}{2}\left(1 - \frac{x^2}{b^2}\right)^2 dx = \frac{4bh^2}{15} \tag{d}$$

$$Q_y = \int x\, dA = \int_0^b hx\left(1 - \frac{x^2}{b^2}\right)dx = \frac{b^2 h}{4} \tag{e}$$

in which we have substituted for $dA$ from Eq. (b).

We can now determine the coordinates of the centroid $C$:

$$\bar{x} = \frac{Q_y}{A} = \frac{3b}{8} \qquad \bar{y} = \frac{Q_x}{A} = \frac{2h}{5} \tag{f,g}$$

These results agree with the formulas listed in Appendix E, Case 17.

*Notes:* The centroid $C$ of the parabolic semisegment may also be located by taking the element of area $dA$ as a horizontal strip of height $dy$ and width

$$x = b\sqrt{1 - \frac{y}{h}} \tag{h}$$

This expression is obtained by solving Eq. (a) for $x$ in terms of $y$.

# 12.3 CENTROIDS OF COMPOSITE AREAS

Fig. 12-6

Centroid of a composite area consisting of two parts

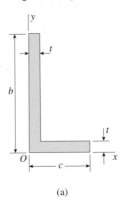

In engineering work we rarely need to locate centroids by integration, because the centroids of common geometric figures are already known and tabulated. However, we frequently need to locate the centroids of areas composed of several parts, each part having a familiar geometric shape, such as a rectangle or a circle. Examples of such **composite areas** are the cross sections of beams and columns, which usually consist of rectangular elements (for instance, see Figs. 12-2, 12-3, and 12-4).

The **areas and first moments** of composite areas may be calculated by summing the corresponding properties of the component parts. Let us assume that a composite area is divided into a total of $n$ parts, and let us denote the area of the $i$th part as $A_i$. Then we can obtain the area and first moments by the following summations:

$$A = \sum_{i=1}^{n} A_i \qquad Q_x = \sum_{i=1}^{n} \bar{y}_i A_i \qquad Q_y = \sum_{i=1}^{n} \bar{x}_i A_i \qquad \textbf{(12-6a,b,c)}$$

(a)

in which $\bar{x}_i$ and $\bar{y}_i$ are the coordinates of the centroid of the $i$th part.

The **coordinates of the centroid** of the composite area are

$$\bar{x} = \frac{Q_y}{A} = \frac{\sum_{i=1}^{n} \bar{x}_i A_i}{\sum_{i=1}^{n} A_i} \qquad \bar{y} = \frac{Q_x}{A} = \frac{\sum_{i=1}^{n} \bar{y}_i A_i}{\sum_{i=1}^{n} A_i} \qquad \textbf{(12-7a,b)}$$

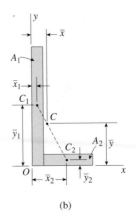

Since the composite area is represented exactly by the $n$ parts, the preceding equations give exact results for the coordinates of the centroid.

To illustrate the use of Eqs. (12-7a) and (12-7b), consider the L-shaped area (or angle section) shown in Fig. 12-6a. This area has side dimensions $b$ and $c$ and thickness $t$. The area can be divided into two rectangles of areas $A_1$ and $A_2$ with centroids $C_1$ and $C_2$, respectively (Fig. 12-6b). The areas and centroidal coordinates of these two parts are

(b)

$$A_1 = +bt \qquad \bar{x}_1 = \frac{t}{2} \qquad \bar{y}_1 = \frac{b}{2}$$

$$A_2 = (c - t)t \qquad \bar{x}_2 = \frac{c + t}{2} \qquad \bar{y}_2 = \frac{t}{2}$$

Therefore, the area and first moments of the composite area (from Eqs. 12-6a, b, and c) are

$$A = A_1 + A_2 = t(b + c - t)$$

$$Q_x = \bar{y}_1 A_1 + \bar{y}_2 A_2 = \frac{t}{2}(b^2 + ct - t^2)$$

$$Q_y = \bar{x}_1 A_1 + \bar{x}_2 A_2 = \frac{t}{2}(bt + c^2 - t^2)$$

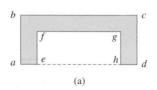

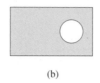

Composite areas with a cutout and a hole

(a)

(b)

Finally, we can obtain the coordinates $\bar{x}$ and $\bar{y}$ of the centroid $C$ of the composite area (Fig. 12-6b) from Eqs. (12-7a) and (12-7b):

$$\bar{x} = \frac{Q_y}{A} = \frac{bt + c^2 - t^2}{2(b + c - t)} \qquad \bar{y} = \frac{Q_x}{A} = \frac{b^2 + ct - t^2}{2(b + c - t)} \quad \textbf{(12-8a,b)}$$

A similar procedure can be used for more complex areas, as illustrated in Example 12-2.

**Note 1:** When a composite area is divided into only two parts, the centroid $C$ of the entire area lies on the line joining the centroids $C_1$ and $C_2$ of the two parts (as shown in Fig. 12-6b for the L-shaped area).

**Note 2:** When using the formulas for composite areas [see Eqs. (12-6) and (12-7)], we can handle the *absence* of an area by subtraction. This procedure is useful when there are cutouts or holes in the figure.

For instance, consider the area shown in Fig. 12-7a. We can analyze this figure as a composite area by subtracting the properties of the inner rectangle *efgh* from the corresponding properties of the outer rectangle *abcd*. (From another viewpoint, we can think of the outer rectangle as a "positive area" and the inner rectangle as a "negative area.")

Similarly, if an area has a hole (Fig. 12-7b), we can subtract the properties of the area of the hole from those of the outer rectangle. (Again, the same effect is achieved if we treat the outer rectangle as a "positive area" and the hole as a "negative area.")

Cutouts in beams must be considered in centroid and moment of inertia calculations (Don Farrall/Getty Images)

**Example 12-2**

## Fig. 12-8

Example 12-2: Centroid of a composite area

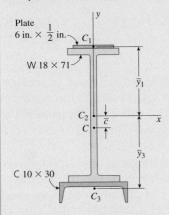

The cross section of a steel beam is constructed of a W 18 × 71 wide-flange section with a 6 in. × 1/2 in. cover plate welded to the top flange and a C 10 × 30 channel section welded to the bottom flange (Fig. 12-8).
Locate the centroid C of the cross-sectional area.

### Solution

Let us denote the areas of the cover plate, the wide-flange section, and the channel section as areas $A_1$, $A_2$, and $A_3$, respectively. The centroids of these three areas are labeled $C_1$, $C_2$, and $C_3$, respectively, in Fig. 12-8. Note that the composite area has an axis of symmetry, and therefore all centroids lie on that axis. The three partial areas are

$$A_1 = (6 \text{ in.})(0.5 \text{ in.}) = 3.0 \text{ in.}^2 \quad A_2 = 20.8 \text{ in.}^2 \quad A_3 = 8.82 \text{ in.}^2$$

in which the areas $A_2$ and $A_3$ are obtained from Tables F-1 and F-3 of Appendix F.

Let us place the origin of the x and y axes at the centroid $C_2$ of the wide-flange section. Then the distances from the x axis to the centroids of the three areas are as follows:

$$\bar{y}_1 = \frac{18.47 \text{ in.}}{2} + \frac{0.5 \text{ in.}}{2} = 9.485 \text{ in.}$$

$$\bar{y}_2 = 0 \quad \bar{y}_3 = \frac{18.47 \text{ in.}}{2} + 0.649 \text{ in.} = 9.884 \text{ in.}$$

in which the pertinent dimensions of the wide-flange and channel sections are obtained from Tables F-1 and F-3.

The area A and first moment $Q_x$ of the entire cross section are obtained from Eqs. (12-6a) and (12-6b) as follows:

$$A = \sum_{i=1}^{n} A_i = A_1 + A_2 + A_3$$

$$= 3.0 \text{ in.}^2 + 20.8 \text{ in.}^2 + 8.82 \text{ in.}^2 = 32.62 \text{ in.}^2$$

$$Q_x = \sum_{i=1}^{n} \bar{y}_i A_i = \bar{y}_1 A_1 + \bar{y}_2 A_2 + \bar{y}_3 A_3$$

$$= (9.485 \text{ in.})(3.0 \text{ in.}^2) + 0 - (9.884 \text{ in.})(8.82 \text{ in.}^2) = -58.72 \text{ in.}^3$$

Now we can obtain the coordinate $\bar{y}$ to the centroid C of the composite area from Eq. (12-7b):

$$\bar{y} = \frac{Q_x}{A} = \frac{-58.72 \text{ in.}^3}{32.62 \text{ in.}^2} = -1.80 \text{ in.}$$

Since $\bar{y}$ is positive in the positive direction of the y axis, the minus sign means that the centroid C of the composite area is located below the x axis, as shown in Fig. 12-8. Thus, the distance $\bar{c}$ between the x axis and the centroid C is

$$\bar{c} = 1.80 \text{ in.}$$

Note that the position of the reference axis (the x axis) is arbitrary; however, in this example we placed it through the centroid of the wide-flange section because it slightly simplifies the calculations.

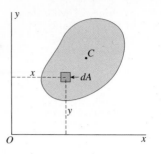

**Fig. 12-9**

Plane area of arbitrary shape

# 12.4 MOMENTS OF INERTIA OF PLANE AREAS

The **moments of inertia** of a plane area (Fig. 12-9) with respect to the $x$ and $y$ axes, respectively, are defined by the integrals

$$I_x = \int y^2 \, dA \qquad I_y = \int x^2 \, dA \qquad \textbf{(12-9a,b)}$$

in which $x$ and $y$ are the coordinates of the differential element of area $dA$. Because the element $dA$ is multiplied by the square of the distance from the reference axis, moments of inertia are also called **second moments of area**. Also, we see that moments of inertia of areas (unlike first moments) are always positive quantities.

To illustrate how moments of inertia are obtained by integration, we will consider a rectangle having width $b$ and height $h$ (Fig. 12-10). The $x$ and $y$ axes have their origin at the centroid $C$. For convenience, we use a differential element of area $dA$ in the form of a thin horizontal strip of width $b$ and height $dy$ (therefore, $dA = b \, dy$). Since all parts of the elemental strip are the same distance from the $x$ axis, we can express the moment of inertia $I_x$ with respect to the $x$ axis as follows:

$$I_x = \int y^2 \, dA = \int_{-h/2}^{h/2} y^2 b \, dy = \frac{bh^3}{12} \qquad \textbf{(12-10)}$$

In a similar manner, we can use an element of area in the form of a vertical strip with area $dA = h \, dx$ and obtain the moment of inertia with respect to the $y$ axis:

$$I_y = \int x^2 \, dA = \int_{-b/2}^{b/2} x^2 h \, dx = \frac{hb^3}{12} \qquad \textbf{(12-11)}$$

**Fig. 12-10**

Moments of inertia of a rectangle

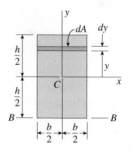

If a different set of axes is selected, the moments of inertia will have different values. For instance, consider axis $BB$ at the base of the rectangle (Fig. 12-10). If this axis is selected as the reference, we must define $y$ as the coordinate distance from that axis to the element of area $dA$. Then the calculations for the moment of inertia become

$$I_{BB} = \int y^2 \, dA = \int_0^h y^2 b \, dy = \frac{bh^3}{3} \qquad \textbf{(12-12)}$$

Note that the moment of inertia with respect to axis $BB$ is larger than the moment of inertia with respect to the centroidal $x$ axis. In general, the moment of inertia increases as the reference axis is moved parallel to itself farther from the centroid.

The moment of inertia of a **composite area** with respect to any particular axis is the sum of the moments of inertia of its parts with respect to that same axis. An example is the hollow box section shown in Fig. 12-11a, where the $x$ and $y$ axes are axes of symmetry through the centroid $C$. The moment of inertia $I_x$ with respect to the $x$ axis is equal to the algebraic sum of the moments of inertia of the outer and inner rectangles. (As explained earlier, we can think of the inner rectangle as a "negative area" and the outer rectangle as a "positive area.") Therefore,

$$I_x = \frac{bh^3}{12} - \frac{b_1 h_1^3}{12} \qquad \textbf{(12-13)}$$

This same formula applies to the channel section shown in Fig. 12-11b, where we may consider the cutout as a "negative area."

For the hollow box section, we can use a similar technique to obtain the moment of inertia $I_y$ with respect to the vertical axis. However, in the case of the channel section, the determination of the moment of inertia $I_y$ requires the use of the parallel-axis theorem, which is described in the next section (Section 12.5).

**Formulas for moments of inertia** are listed in Appendix E. For shapes not shown, the moments of inertia can usually be obtained by using the listed formulas in conjunction with the parallel-axis theorem. If an area is of such irregular shape that its moments of inertia cannot be obtained in this manner, then we can use numerical methods. The procedure is to divide the area into small elements of area $\Delta A_i$, multiply each such area by the square of its distance from the reference axis, and then sum the products.

## Radius of Gyration

A distance known as the **radius of gyration** is occasionally encountered in mechanics. Radius of gyration of a plane area is defined as the square root of the moment of inertia of the area divided by the area itself; thus,

$$r_x = \sqrt{\frac{I_x}{A}} \qquad r_y = \sqrt{\frac{I_y}{A}} \qquad \textbf{(12-14a,b)}$$

in which $r_x$ and $r_y$ denote the radii of gyration with respect to the $x$ and $y$ axes, respectively. Since moment of inertia has units of length to the fourth power and area has units of length to the second power, radius of gyration has units of length.

Although the radius of gyration of an area does not have an obvious physical meaning, we may consider it to be the distance (from the reference axis) at which the entire area could be concentrated and still have the same moment of inertia as the original area.

**Fig. 12-11**

Composite areas

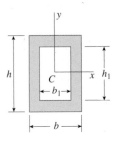

(a)

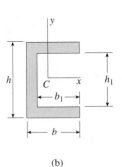

(b)

Example 12-3

### Fig. 12-12

Example 12-3: Moments of inertia of a parabolic semisegment

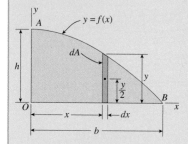

Determine the moments of inertia $I_x$ and $I_y$ for the parabolic semisegment OAB shown in Fig. 12-12. The equation of the parabolic boundary is

$$y = f(x) = h\left(1 - \frac{x^2}{b^2}\right) \tag{a}$$

(This same area was considered previously in Example 12-1.)

### Solution

To determine the moments of inertia by integration, we will use Eqs. (12-9a) and (12-9b). The differential element of area $dA$ is selected as a vertical strip of width $dx$ and height $y$, as shown in Fig. 12-12. The area of this element is

$$dA = y\,dx = h\left(1 - \frac{x^2}{b^2}\right)dx \tag{b}$$

Since every point in this element is at the same distance from the $y$ axis, the moment of inertia of the element with respect to the $y$ axis is $x^2\,dA$. Therefore, the moment of inertia of the entire area with respect to the $y$ axis is obtained as follows:

$$I_y = \int x^2\,dA = \int_0^b x^2 h\left(1 - \frac{x^2}{b^2}\right)dx = \frac{2hb^3}{15} \tag{c}$$

To obtain the moment of inertia with respect to the $x$ axis, we note that the differential element of area $dA$ has a moment of inertia $dI_x$ with respect to the $x$ axis equal to

$$dI_x = \frac{1}{3}(dx)y^3 = \frac{y^3}{3}\,dx$$

as obtained from Eq. (12-12). Hence, the moment of inertia of the entire area with respect to the $x$ axis is

$$I_x = \int_0^b \frac{y^3}{3}\,dx = \int_0^b \frac{h^3}{3}\left(1 - \frac{x^2}{b^2}\right)^3 dx = \frac{16bh^3}{105} \tag{d}$$

These same results for $I_x$ and $I_y$ can be obtained by using an element in the form of a horizontal strip of area $dA = x\,dy$ or by using a rectangular element of area $dA = dx\,dy$ and performing a double integration. Also, note that the preceding formulas for $I_x$ and $I_y$ agree with those given in Case 17 of Appendix E.

# 12.5 PARALLEL-AXIS THEOREM FOR MOMENTS OF INERTIA

In this section we will derive a very useful theorem pertaining to moments of inertia of plane areas. Known as the **parallel-axis theorem**, it gives the relationship between the moment of inertia with respect to a centroidal axis and the moment of inertia with respect to any parallel axis.

To derive the theorem, we consider an area of arbitrary shape with centroid $C$ (Fig. 12-13). We also consider two sets of coordinate axes: (1) the $x_c y_c$ axes with origin at the centroid, and (2) a set of parallel $xy$ axes with origin at any point $O$. The distances between the two sets of parallel axes are denoted $d_1$ and $d_2$. Also, we identify an element of area $dA$ having coordinates $x$ and $y$ with respect to the centroidal axes.

From the definition of moment of inertia, we can write the following equation for the moment of inertia $I_x$ with respect to the $x$ axis:

$$I_x = \int (y + d_1)^2 \, dA = \int y^2 \, dA + 2d_1 \int y \, dA + d_1^2 \int dA \quad \textbf{(12-15)}$$

The first integral on the right-hand side is the moment of inertia $I_{x_c}$ with respect to the $x_c$ axis. The second integral is the first moment of the area with respect to the $x_c$ axis (this integral equals zero because the $x_c$ axis passes through the centroid). The third integral is the area $A$ itself. Therefore, the preceding equation reduces to

$$I_x = I_{x_c} + Ad_1^2 \quad \textbf{(12-16a)}$$

Proceeding in the same manner for the moment of inertia with respect to the $y$ axis, we obtain

$$I_y = I_{y_c} + Ad_2^2 \quad \textbf{(12-16b)}$$

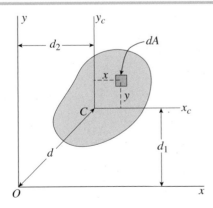

Derivation of parallel-axis theorem

**Fig. 12-10 (Repeated)**

Moments of inertia of a
rectangle

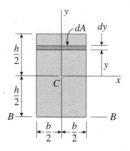

Equations (12-16a) and (12-16b) represent the **parallel-axis theorem for moments of inertia**:

*The moment of inertia of an area with respect to any axis in its plane is equal to the moment of inertia with respect to a parallel centroidal axis plus the product of the area and the square of the distance between the two axes.*

To illustrate the use of the theorem, consider again the rectangle shown in Fig. 12-10. Knowing that the moment of inertia about the $x$ axis, which is through the centroid, is equal to $bh^3/12$ [see Eq. (12-10) of Section 12.4), we can determine the moment of inertia $I_{BB}$ about the base of the rectangle by using the parallel-axis theorem:

$$I_{BB} = I_x + Ad^2 = \frac{bh^3}{12} + bh\left(\frac{h}{2}\right)^2 = \frac{bh^3}{3}$$

This result agrees with the moment of inertia obtained previously by integration [see Eq. (12-12) of Section 12.4].

From the parallel-axis theorem, we see that the moment of inertia increases as the axis is moved parallel to itself farther from the centroid. Therefore, the moment of inertia about a centroidal axis is the least moment of inertia of an area (for a given direction of the axis).

When using the parallel-axis theorem, it is essential to remember that one of the two parallel axes *must* be a centroidal axis. If it is necessary to find the moment of inertia $I_2$ about a noncentroidal axis 2-2 (Fig. 12-14) when the moment of inertia $I_1$ about another noncentroidal (and parallel) axis 1-1 is known, we must apply the parallel-axis theorem twice. First, we find the centroidal moment of inertia $I_{x_c}$ from the known moment of inertia $I_1$:

$$I_{x_c} = I_1 - Ad_1^2 \qquad \textbf{(12-17)}$$

Then we find the moment of inertia $I_2$ from the centroidal moment of inertia:

$$I_2 = I_{x_c} + Ad_2^2 = I_1 + A(d_2^2 - d_1^2) \qquad \textbf{(12-18)}$$

This equation shows again that the moment of inertia increases with increasing distance from the centroid of the area.

**Fig. 12-14**

Plane area with two parallel
noncentroidal axes
(axes 1-1 and 2-2)

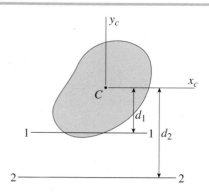

● ● ● **Example 12-4**

The parabolic semisegment *OAB* shown in Fig. 12-15 has base *b* and height *h*. Using the parallel-axis theorem, determine the moments of inertia $I_{x_c}$ and $I_{y_c}$ with respect to the centroidal axes $x_c$ and $y_c$.

### Fig. 12-15

Example 12-4: Parallel-axis theorem

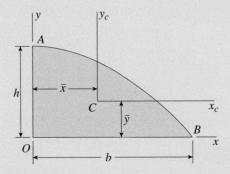

### Solution

We can use the parallel-axis theorem (rather than integration) to find the centroidal moments of inertia because we already know the area *A*, the centroidal coordinates $\bar{x}$ and $\bar{y}$, and the moments of inertia $I_x$ and $I_y$ with respect to the *x* and *y* axes. These quantities were obtained earlier in Examples 12-1 and 12-3. They also are listed in Case 17 of Appendix E and are repeated here:

$$A = \frac{2bh}{3} \qquad \bar{x} = \frac{3b}{8} \qquad \bar{y} = \frac{2h}{5} \qquad I_x = \frac{16bh^3}{105} \qquad I_y = \frac{2hb^3}{15}$$

To obtain the moment of inertia with respect to the $x_c$ axis, we use Eq. (12-17) and write the parallel-axis theorem as follows:

$$I_{x_c} = I_x - A\bar{y}^2 = \frac{16bh^3}{105} - \frac{2bh}{3}\left(\frac{2h}{5}\right)^2 = \frac{8bh^3}{175} \qquad \text{(12-19a)} \; \Longleftarrow$$

In a similar manner, we obtain the moment of inertia with respect to the $y_c$ axis:

$$I_{y_c} = I_c - A\bar{x}^2 = \frac{2hb^3}{15} - \frac{2bh}{3}\left(\frac{3b}{8}\right)^2 = \frac{19hb^3}{480} \qquad \text{(12-19b)} \; \Longleftarrow$$

Thus, we have found the centroidal moments of inertia of the semisegment.

**Example 12-5**

### Fig. 12-16

Example 12-5: Moment of inertia of a composite area

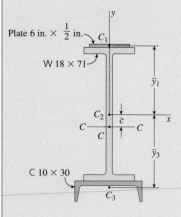

Plate 6 in. $\times \frac{1}{2}$ in.

W 18 × 71

C 10 × 30

Determine the moment of inertia $I_c$ with respect to the horizontal axis C–C through the centroid C of the beam cross section shown in Fig. 12-16. (The position of the centroid C was determined previously in Example 12-2 of Section 12.3.)

*Note:* From beam theory (Chapter 5), we know that axis C–C is the neutral axis for bending of this beam, and therefore the moment of inertia $I_c$ must be determined in order to calculate the stresses and deflections of this beam.

### Solution

We will determine the moment of inertia $I_c$ with respect to axis C–C by applying the parallel-axis theorem to each individual part of the composite area. The area divides naturally into three parts: (1) the cover plate, (2) the wide-flange section, and (3) the channel section. The following areas and centroidal distances were obtained previously in Example 12-2:

$$A_1 = 3.0 \text{ in.}^2 \qquad A_2 = 20.8 \text{ in.}^2 \qquad A_3 = 8.82 \text{ in.}^2$$

$$\bar{y}_1 = 9.485 \text{ in.} \qquad \bar{y}_2 = 0 \qquad \bar{y}_3 = 9.884 \text{ in.} \qquad \bar{c} = 1.80 \text{ in.}$$

The moments of inertia of the three parts with respect to horizontal axes through their own centroids $C_1$, $C_2$, and $C_3$ are

$$I_1 = \frac{bh^3}{12} = \frac{1}{12}(6.0 \text{ in.})(0.5 \text{ in.})^3 = 0.063 \text{ in.}^4$$

$$I_2 = 1170 \text{ in.}^4 \qquad I_3 = 3.94 \text{ in.}^4$$

The moments of inertia $I_2$ and $I_3$ are obtained from Tables F-1 and F-3, respectively, of Appendix F.

Now we can use the parallel-axis theorem to calculate the moments of inertia about axis C–C for each of the three parts of the composite area:

$$(I_c)_1 = I_1 + A_1(\bar{y}_1 + \bar{c})^2 = 0.063 \text{ in.}^4 + (3.0 \text{ in.}^2)(11.28 \text{ in.})^2 = 382 \text{ in.}^4$$

$$(I_c)_2 = I_2 + A_2\bar{c}^2 = 1170 \text{ in.}^4 + (20.8 \text{ in.}^2)(1.80 \text{ in.})^2 = 1240 \text{ in.}^4$$

$$(I_c)_3 = I_3 + A_3(\bar{y}_3 - \bar{c})^2 = 3.94 \text{ in.}^4 + (8.82 \text{ in.}^2)(8.084 \text{ in.})^2 = 580 \text{ in.}^4$$

The sum of these individual moments of inertia gives the moment of inertia of the entire cross-sectional area about its centroidal axis C–C:

$$I_c = (I_c)_1 + (I_c)_2 + (I_c)_3 = 2200 \text{ in.}^4$$

This example shows how to calculate moments of inertia of composite areas by using the parallel-axis theorem.

# 12.6 POLAR MOMENTS OF INERTIA

The moments of inertia discussed in the preceding sections are defined with respect to axes lying in the plane of the area itself, such as the $x$ and $y$ axes in Fig. 12-17. Now we will consider an axis *perpendicular* to the plane of the area and intersecting the plane at the origin $O$. The moment of inertia with respect to this perpendicular axis is called the **polar moment of inertia** and is denoted by the symbol $I_P$.

The polar moment of inertia with respect to an axis through $O$ perpendicular to the plane of the figure is defined by the integral

$$I_P = \int \rho^2 \, dA \qquad \textbf{(12-20)}$$

in which $\rho$ is the distance from point $O$ to the differential element of area $dA$ (Fig. 12-17). This integral is similar in form to those for moments of inertia $I_x$ and $I_y$ [see Eqs. (12-9a) and (12-9b)].

Inasmuch as $\rho^2 = x^2 + y^2$, where $x$ and $y$ are the rectangular coordinates of the element $dA$, we obtain the following expression for $I_P$:

$$I_P = \int \rho^2 \, dA = \int (x^2 + y^2) \, dA = \int x^2 \, dA + \int y^2 \, dA$$

Thus, we obtain the important relationship

$$I_P = I_x + I_y \qquad \textbf{(12-21)}$$

This equation shows that the polar moment of inertia with respect to an axis perpendicular to the plane of the figure at any point $O$ is equal to the sum of the moments of inertia with respect to *any* two perpendicular axes $x$ and $y$ passing through that same point and lying in the plane of the figure.

For convenience, we usually refer to $I_P$ simply as the polar moment of inertia with respect to point $O$, without mentioning that the axis is perpendicular to the plane of the figure. Also, to distinguish them from **polar** moments of inertia, we sometimes refer to $I_x$ and $I_y$ as **rectangular** moments of inertia.

Polar moments of inertia with respect to various points in the plane of an area are related by the **parallel-axis theorem for polar moments of inertia**. We can derive this theorem by referring again to Fig. 12-13. Let us denote the polar moments of inertia with respect to the origin $O$ and the centroid $C$ by $(I_P)_O$ and $(I_P)_C$, respectively. Then, using Eq. (12-21), we can write the following equations:

$$(I_P)_O = I_x + I_y \qquad (I_P)_C = I_{x_c} + I_{y_c} \qquad \textbf{(12-22)}$$

Now refer to the parallel-axis theorems derived in Section 12.5 for rectangular moments of inertia [see Eqs. (12-16a) and (12-16b)]. Adding those two equations, we get

$$I_x + I_y = I_{x_c} + I_{y_c} + A(d_1^2 + d_2^2)$$

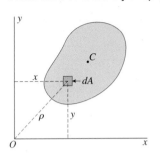

**Fig. 12-17**

Plane area of arbitrary shape

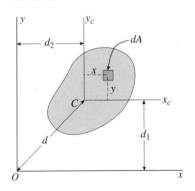

**Fig. 12-13 (Repeated)**

Derivation of parallel-axis theorem

Substituting from Eqs. (12-22), and also noting that $d^2 = d_1^2 + d_2^2$ (Fig. 12-13), we obtain

$$(I_P)_O = (I_P)_C + Ad^2 \qquad \textbf{(12-23)}$$

This equation represents the **parallel-axis theorem** for polar moments of inertia:

> *The polar moment of inertia of an area with respect to any point O in its plane is equal to the polar moment of inertia with respect to the centroid C plus the product of the area and the square of the distance between points O and C.*

To illustrate the determination of polar moments of inertia and the use of the parallel-axis theorem, consider a circle of radius $r$ (Fig. 12-18). Let us take a differential element of area $dA$ in the form of a thin ring of radius $\rho$ and thickness $d\rho$ (thus, $dA = 2\pi\rho\, d\rho$). Since every point in the element is at the same distance $\rho$ from the center of the circle, the polar moment of inertia of the entire circle with respect to the center is

$$(I_P)_C = \int \rho^2 dA = \int_0^r 2\pi\rho^3 d\rho = \frac{\pi r^4}{2} \qquad \textbf{(12-24)}$$

This result is listed in Case 9 of Appendix E.

The polar moment of inertia of the circle with respect to any point $B$ on its circumference (Fig. 12-18) can be obtained from the parallel-axis theorem:

$$(I_P)_B = (I_P)_C + Ad^2 = \frac{\pi r^4}{2} + \pi r^2(r^2) = \frac{3\pi r^4}{2} \qquad \textbf{(12-25)}$$

As an incidental matter, note that the polar moment of inertia has its smallest value when the reference point is the centroid of the area.

A circle is a special case in which the polar moment of inertia can be determined by integration. However, most of the shapes encountered in engineering work do not lend themselves to this technique. Instead, polar moments of inertia are usually obtained by summing the rectangular moments of inertia for two perpendicular axes (Eq. 12-21).

**Fig. 12-18**

Polar moment of inertia
of a circle

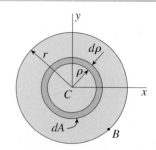

# 12.7 PRODUCTS OF INERTIA

The product of inertia of a plane area is defined with respect to a set of perpendicular axes lying in the plane of the area. Thus, referring to the area shown in Fig. 12-19, we define the **product of inertia** with respect to the $x$ and $y$ axes as follows:

$$I_{xy} = \int xy\, dA \qquad \textbf{(12-26)}$$

**Fig. 12-19**

Plane area of arbitrary shape

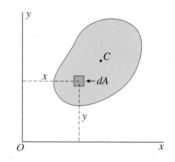

From this definition we see that each differential element of area $dA$ is multiplied by the product of its coordinates. As a consequence, products of inertia may be positive, negative, or zero, depending upon the position of the $xy$ axes with respect to the area.

   If the area lies entirely in the first quadrant of the axes (as in Fig. 12-19), then the product of inertia is positive because every element $dA$ has positive coordinates $x$ and $y$. If the area lies entirely in the second quadrant, the product of inertia is negative because every element has a positive $y$ coordinate and a negative $x$ coordinate. Similarly, areas entirely within the third and fourth quadrants have positive and negative products of inertia, respectively. When the area is located in more than one quadrant, the sign of the product of inertia depends upon the distribution of the area within the quadrants.

   A special case arises when one of the axes is an **axis of symmetry** of the area. For instance, consider the area shown in Fig. 12-20, which is symmetric about the $y$ axis. For every element $dA$ having coordinates $x$ and $y$, there exists an equal and symmetrically located element $dA$ having the same $y$ coordinate but an $x$ coordinate of opposite sign. Therefore, the products $xy\, dA$ cancel each other and the integral in Eq. (12-26) vanishes. Thus, *the product of inertia of an area is zero with respect to any pair of axes in which at least one axis is an axis of symmetry of the area.*

   As examples of the preceding rule, the product of inertia $I_{xy}$ equals zero for the areas shown in Figs. 12-10, 12-11, 12-16, and 12-18. In contrast, the product of inertia $I_{xy}$ has a positive nonzero value for the area shown in Fig. 12-15. (These observations are valid for products of inertia with respect to the particular $xy$ axes shown in the figures. If the axes are shifted to another position, the product of inertia may change.)

   Products of inertia of an area with respect to parallel sets of axes are related by a **parallel-axis theorem** that is analogous to the corresponding theorems for rectangular moments of inertia and polar moments of inertia. To obtain this theorem, consider the area shown in Fig. 12-21, which has centroid $C$ and centroidal $x_c y_c$ axes. The product of inertia $I_{xy}$ with respect to any other set of axes, parallel to the $x_c y_c$ axes, is

**Fig. 12-20**

The product of inertia equals zero when one axis is an axis of symmetry

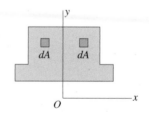

$$I_{xy} = \int (x + d_2)(y + d_1)dA$$

$$= \int xy\, dA + d_1 \int x\, dA + d_2 \int y\, dA + d_1 d_2 \int dA$$

in which $d_1$ and $d_2$ are the coordinates of the centroid $C$ with respect to the $xy$ axes (thus, $d_1$ and $d_2$ may have positive or negative values).

**Fig. 12-21**

Plane area of arbitrary shape

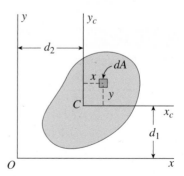

The first integral in the last expression is the product of inertia $I_{x_c y_c}$ with respect to the centroidal axes; the second and third integrals equal zero because they are the first moments of the area with respect to the centroidal axes; and the last integral is the area $A$. Therefore, the preceding equation reduces to

$$I_{xy} = I_{x_c y_c} + A d_1 d_2 \qquad \textbf{(12-27)}$$

This equation represents the **parallel-axis theorem for products of inertia**:

> *The product of inertia of an area with respect to any pair of axes in its plane is equal to the product of inertia with respect to parallel centroidal axes plus the product of the area and the coordinates of the centroid with respect to the pair of axes.*

To demonstrate the use of this parallel-axis theorem, let us determine the product of inertia of a rectangle with respect to $xy$ axes having their origin at point $O$ at the lower left-hand corner of the rectangle (Fig. 12-22). The product of inertia with respect to the centroidal $x_c y_c$ axes is zero because of symmetry. Also, the coordinates of the centroid with respect to the $xy$ axes are

$$d_1 = \frac{h}{2} \qquad d_2 = \frac{b}{2}$$

Substituting into Eq. (12-20), we obtain

$$I_{xy} = I_{x_c y_c} + A d_1 d_2 = 0 + bh\left(\frac{h}{2}\right)\left(\frac{b}{2}\right) = \frac{b^2 h^2}{4} \qquad \textbf{(12-28)}$$

This product of inertia is positive because the entire area lies in the first quadrant. If the $xy$ axes are translated horizontally so that the origin moves to point $B$ at the lower right-hand corner of the rectangle (Fig. 12-22), the entire area lies in the second quadrant and the product of inertia becomes $-b^2 h^2/4$.

The following example also illustrates the use of the parallel-axis theorem for products of inertia.

**Fig. 12-22**

Parallel-axis theorem for products of inertia

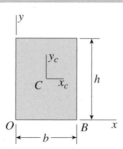

● ● ● **Example 12-6**

Determine the product of inertia $I_{xy}$ of the Z-section shown in Fig. 12-23. The section has width $b$, height $h$, and constant thickness $t$.

**Fig. 12-23**

Example 12-6: Product of inertia of a Z-section

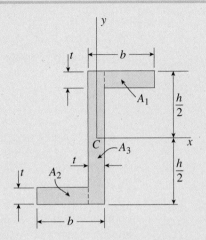

**Solution**

To obtain the product of inertia with respect to the $xy$ axes through the centroid, we divide the area into three parts and use the parallel-axis theorem. The parts are as follows: (1) a rectangle of width $b - t$ and thickness $t$ in the upper flange, (2) a similar rectangle in the lower flange, and (3) a web rectangle with height $h$ and thickness $t$.

The product of inertia of the web rectangle with respect to the $xy$ axes is zero (from symmetry). The product of inertia $(I_{xy})_1$ of the upper flange rectangle (with respect to the $xy$ axes) is determined by using the parallel-axis theorem:

$$(I_{xy})_1 = I_{x_c y_c} + A d_1 d_2 \tag{a}$$

in which $I_{x_c y_c}$ is the product of inertia of the rectangle with respect to its own centroid, $A$ is the area of the rectangle, $d_1$ is the $y$ coordinate of the centroid of the rectangle, and $d_2$ is the $x$ coordinate of the centroid of the rectangle. Thus,

$$I_{x_c y_c} = 0 \qquad A = (b - t)(t) \qquad d_1 = \frac{h}{2} - \frac{t}{2} \qquad d_2 = \frac{b}{2}$$

Substituting into Eq. (a), we obtain the product of inertia of the rectangle in the upper flange:

$$(I_{xy})_1 = I_{x_c y_c} + A d_1 d_2 = 0 + (b - t)(t)\left(\frac{h}{2} - \frac{t}{2}\right)\left(\frac{b}{2}\right) = \frac{bt}{4}(h - t)(b - t)$$

The product of inertia of the rectangle in the lower flange is the same. Therefore, the product of inertia of the entire Z-section is twice $(I_{xy})_1$, or

$$I_{xy} = \frac{bt}{2}(h - t)(b - t) \tag{12-29}$$

Note that this product of inertia is positive because the flanges lie in the first and third quadrants.

# 12.8 ROTATION OF AXES

The moments of inertia of a plane area depend upon the position of the origin and the orientation of the reference axes. For a given origin, the moments and product of inertia vary as the axes are rotated about that origin. The manner in which they vary, and the magnitudes of the maximum and minimum values, are discussed in this and the following section.

Let us consider the plane area shown in Fig. 12-24, and let us assume that the $xy$ axes are a pair of arbitrarily located reference axes. The moments and products of inertia with respect to those axes are

$$I_x = \int y^2 dA \quad I_y = \int x^2 dA \quad I_{xy} = \int xy \, dA \quad \textbf{(12-30a,b,c)}$$

in which $x$ and $y$ are the coordinates of a differential element of area $dA$.

The $x_1 y_1$ axes have the same origin as the $xy$ axes but are rotated through a counterclockwise angle $\theta$ with respect to those axes. The moments and product of inertia with respect to the $x_1 y_1$ axes are denoted $I_{x_1}$, $I_{y_1}$, and $I_{x_1 y_1}$, respectively. To obtain these quantities, we need the coordinates of the element of area $dA$ with respect to the $x_1 y_1$ axes. These coordinates may be expressed in terms of the $xy$ coordinates and the angle $\theta$ by geometry, as follows:

$$x_1 = x \cos \theta + y \sin \theta \quad y_1 = y \cos \theta - x \sin \theta \quad \textbf{(12-31a,b)}$$

Then the moment of inertia with respect to the $x_1$ axis is

$$I_{x_1} = \int y_1^2 \, dA = \int (y \cos \theta - x \sin \theta)^2 \, dA$$

$$= \cos^2 \theta \int y^2 \, dA + \sin^2 \theta \int x^2 \, dA - 2 \sin \theta \cos \theta \int xy \, dA$$

or, by using Eqs. (12-30a, b, and c),

$$I_{x_1} = I_x \cos^2 \theta + I_y \sin^2 \theta - 2I_{xy} \sin \theta \cos \theta \quad \textbf{(12-32)}$$

Now we introduce the following trigonometric identities:

$$\cos^2 \theta = \frac{1}{2}(1 + \cos 2\theta) \quad \sin^2 \theta = \frac{1}{2}(1 - \cos 2\theta)$$

$$2 \sin \theta \cos \theta = \sin 2\theta$$

**Fig. 12-24**

Rotation of axes

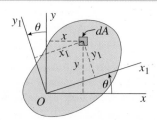

Then Eq. (12-32) becomes

$$I_{x_1} = \frac{I_x + I_y}{2} + \frac{I_x - I_y}{2} \cos 2\theta - I_{xy} \sin 2\theta \qquad \textbf{(12-33)}$$

In a similar manner, we can obtain the product of inertia with respect to the $x_1 y_1$ axes:

$$I_{x_1 y_1} = \int x_1 y_1 \, dA = \int (x \cos \theta + y \sin \theta)(y \cos \theta - x \sin \theta) dA$$

$$= (I_x - I_y) \sin \theta \cos \theta + I_{xy}(\cos^2 \theta - \sin^2 \theta) \qquad \textbf{(12-34)}$$

Again using the trigonometric identities, we obtain

$$I_{x_1 y_1} = \frac{I_x - I_y}{2} \sin 2\theta + I_{xy} \cos 2\theta \qquad \textbf{(12-35)}$$

Equations (12-33) and (12-35) give the moment of inertia $I_{x_1}$ and the product of inertia $I_{x_1 y_1}$ with respect to the rotated axes in terms of the moments and product of inertia for the original axes. These equations are called the **transformation equations for moments and products of inertia**.

Note that these transformation equations have the same form as thetransformation equations for plane stress [Eqs. (7-4a) and (7-4b) of Section 7.2]. Upon comparing the two sets of equations, we see that $I_{x_1}$ corresponds to $\sigma_{x_1}$, $I_{x_1 y_1}$ corresponds to $-\tau_{x_1 y_1}$, $I_x$ corresponds to $\sigma_x$, $I_y$ corresponds to $\sigma_y$, and $I_{xy}$ corresponds to $-\tau_{xy}$. Therefore, we can also analyze moments and products of inertia by the use of **Mohr's circle** (see Section 7.4).

The moment of inertia $I_{y_1}$ may be obtained by the same procedure that we used for finding $I_{x_1}$ and $I_{x_1 y_1}$. However, a simpler procedure is to replace $\theta$ with $\theta + 90°$ in Eq. (12-33). The result is

$$I_{y_1} = \frac{I_x + I_y}{2} - \frac{I_x - I_y}{2} \cos 2\theta + I_{xy} \sin 2\theta \qquad \textbf{(12-36)}$$

This equation shows how the moment of inertia $I_{y_1}$ varies as the axes are rotated about the origin.

A useful equation related to moments of inertia is obtained by taking the sum of $I_{x_1}$ and $I_{y_1}$ (Eqs. 12-33 and 12-36). The result is

$$I_{x_1} + I_{y_1} = I_x + I_y \qquad \textbf{(12-37)}$$

This equation shows that the sum of the moments of inertia with respect to a pair of axes remains constant as the axes are rotated about the origin. This sum is the polar moment of inertia of the area with respect to the origin. Note that Eq. (12-37) is analogous to Eq. (7-6) for stresses and Eq. (7-72) for strains.

# 12.9 PRINCIPAL AXES AND PRINCIPAL MOMENTS OF INERTIA

The transformation equations for moments and products of inertia (Eqs. 12-33, 12-35, and 12-36) show how the moments and products of inertia vary as the angle of rotation $\theta$ varies. Of special interest are the maximum and minimum values of the moment of inertia. These values are known as the **principal moments of inertia**, and the corresponding axes are known as **principal axes**.

## Principal Axes

To find the values of the angle $\theta$ that make the moment of inertia $I_{x_1}$ a maximum or a minimum, we take the derivative with respect to $\theta$ of the expression on the right-hand side of Eq. (12-33) and set it equal to zero:

$$(I_x - I_y) \sin 2\theta + 2I_{xy} \cos 2\theta = 0 \qquad \textbf{(12-38)}$$

Solving for $\theta$ from this equation, we get

$$\tan 2\theta_p = -\frac{2I_{xy}}{I_x - I_y} \qquad \textbf{(12-39)}$$

in which $\theta_p$ denotes the angle defining a principal axis. This same result is obtained if we take the derivative of $I_{y_1}$ (Eq. 12-36).

Equation (12-39) yields two values of the angle $2\theta_p$ in the range from 0 to 360°; these values differ by 180°. The corresponding values of $\theta_p$ differ by 90° and define the two perpendicular principal axes. One of these axes corresponds to the maximum moment of inertia and the other corresponds to the minimum moment of inertia.

Now let us examine the variation in the product of inertia $I_{x_1y_1}$ as $\theta$ changes (see Eq. 12-35). If $\theta = 0$, we get $I_{x_1y_1} = I_{xy}$, as expected. If $\theta = 90°$, we obtain $I_{x_1y_1} = -I_{xy}$. Thus, during a 90° rotation the product of inertia changes sign, which means that for an intermediate orientation of the axes, the product of inertia must equal zero. To determine this orientation, we set $I_{x_1y_1}$ (Eq. 12-35) equal to zero:

$$(I_x - I_y) \sin 2\theta + 2I_{xy} \cos 2\theta = 0$$

This equation is the same as Eq. (12-38), which defines the angle $\theta_p$ to the principal axes. Therefore, we conclude that *the product of inertia is zero for the principal axes.*

In Section 12.7 we showed that the product of inertia of an area with respect to a pair of axes equals zero if at least one of the axes is an axis of symmetry. It follows that if an area has an axis of symmetry, that axis and any axis perpendicular to it constitute a set of principal axes.

The preceding observations may be summarized as follows: (1) principal axes through an origin $O$ are a pair of orthogonal axes for which the moments of inertia are a maximum and a minimum; (2) the orientation of the principal axes is given by the angle $\theta_p$ obtained from Eq. (12-39); (3) the product of inertia is zero for principal axes; and (4) an axis of symmetry is always a principal axis.

# Principal Points

Now consider a pair of principal axes with origin at a given point $O$. If there exists a *different* pair of principal axes through that same point, then *every* pair of axes through that point is a set of principal axes. Furthermore, the moment of inertia must be constant as the angle $\theta$ is varied.

The preceding conclusions follow from the nature of the transformation equation for $I_{x_1}$ (Eq. 12-33). Because this equation contains trigonometric functions of the angle $2\theta$, there is one maximum value and one minimum value of $I_{x_1}$ as $2\theta$ varies through a range of 360° (or as $\theta$ varies through a range of 180°). If a second maximum exists, then the only possibility is that $I_{x_1}$ remains constant, which means that every pair of axes is a set of principal axes and all moments of inertia are the same.

A point so located that every axis through the point is a principal axis, and hence the moments of inertia are the same for all axes through the point, is called a **principal point**.

An illustration of this situation is the rectangle of width $2b$ and height $b$ shown in Fig. 12-25. The $xy$ axes, with origin at point $O$, are principal axes of the rectangle because the $y$ axis is an axis of symmetry. The $x'y'$ axes, with the same origin, are also principal axes because the product of inertia $I_{x'y'}$ equals zero (because the triangles are symmetrically located with respect to the $x'$ and $y'$ axes). It follows that every pair of axes through $O$ is a set of principal axes and every moment of inertia is the same (and equal to $2b^4/3$). Therefore, point $O$ is a principal point for the rectangle. (A second principal point is located where the $y$ axis intersects the upper side of the rectangle.)

A useful corollary of the concepts described in the preceding four paragraphs applies to axes through the centroid of an area. Consider an area having *two different pairs* of centroidal axes such that at least one axis in each pair is an axis of symmetry. In other words, there exist two different axes of symmetry that are not perpendicular to each other. Then it follows that the centroid is a principal point.

Two examples, a square and an equilateral triangle, are shown in Fig. 12-26. In each case the $xy$ axes are principal centroidal axes because their origin is at the centroid $C$ and at least one of the two axes is an axis of symmetry. In addition, a second pair of centroidal axes (the $x'y'$ axes) has at least one axis of symmetry. It follows that both the $xy$ and $x'y'$ axes are principal axes. Therefore, every axis through the centroid $C$ is a principal axis, and every such axis has the same moment of inertia.

If an area has *three different axes of symmetry*, even if two of them are perpendicular, the conditions described in the preceding paragraph are automatically fulfilled. Therefore, if an area has three or more axes of symmetry, the centroid is a principal point and every axis through the centroid is a principal axis and has the same moment of inertia. These conditions are fulfilled for a circle, for all regular polygons (equilateral triangle, square, regular pentagon, regular hexagon, and so on), and for many other symmetric shapes.

In general, every plane area has two principal points. These points lie equidistant from the centroid on the principal centroidal axis having the larger principal moment of inertia. A special case occurs when the two principal centroidal moments of inertia are equal; then the two principal points merge at the centroid, which becomes the sole principal point.

**Fig. 12-25**

Rectangle for which every axis (in the plane of the area) through point $O$ is a principal axis

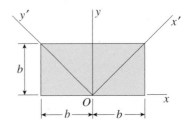

**Fig. 12-26**

Examples of areas for which every centroidal axis is a principal axis and the centroid $C$ is a principal point

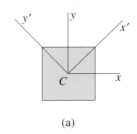

(a)

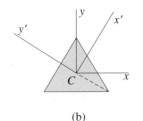

(b)

## Principal Moments of Inertia

Let us now determine the principal moments of inertia, assuming that $I_x$, $I_y$, and $I_{xy}$ are known. One method is to determine the two values of $\theta_p$ (differing by 90°) from Eq. (12-39) and then substitute these values into Eq. (12-33) for $I_{x_1}$. The resulting two values are the principal moments of inertia, denoted by $I_1$ and $I_2$. The advantage of this method is that we know which of the two principal angles $\theta_p$ corresponds to each principal moment of inertia.

It is also possible to obtain general formulas for the principal moments of inertia. We note from Eq. (12-39) and Fig. 12-27 [which is a geometric representation of Eq. (12-39)] that

$$\cos 2\theta_p = \frac{I_x - I_y}{2R} \qquad \sin 2\theta_p = \frac{-I_{xy}}{R} \qquad \textbf{(12-40a,b)}$$

in which

$$R = \sqrt{\left(\frac{I_x - I_y}{2}\right)^2 + I_{xy}^2} \qquad \textbf{(12-41)}$$

**Fig. 12-27**

Geometric representation of Eq. (12-39)

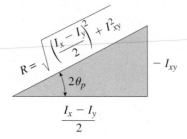

is the hypotenuse of the triangle. When evaluating $R$, we always take the positive square root.

Now we substitute the expressions for $\cos 2\theta_p$ and $\sin 2\theta_p$ [from Eqs. (12-40a and b)] into Eq. (12-33) for $I_{x_1}$ and obtain the algebraically larger of the two principal moments of inertia, denoted by the symbol $I_1$:

$$I_1 = \frac{I_x + I_y}{2} + \sqrt{\left(\frac{I_x - I_y}{2}\right)^2 + I_{xy}^2} \qquad \textbf{(12-42a)}$$

The smaller principal moment of inertia, denoted as $I_2$, may be obtained from the equation

$$I_1 + I_2 = I_x + I_y$$

[see Eq. (12-37)]. Substituting the expression for $I_1$ into this equation and solving for $I_2$, we get

$$I_2 = \frac{I_x + I_y}{2} - \sqrt{\left(\frac{I_x - I_y}{2}\right)^2 + I_{xy}^2} \qquad \textbf{(12-42b)}$$

Eq. (12-42a and b) provide a convenient way to calculate the principal moments of inertia.

The following example illustrates the method for locating the principal axes and determining the principal moments of inertia.

Determine the orientations of the principal centroidal axes and the magnitudes of the principal centroidal moments of inertia for the cross-sectional area of the Z-section shown in Fig. 12-28. Use the following numerical data: height $h = 200$ mm, width $b = 90$ mm, and constant thickness $t = 15$ mm.

### Fig. 12-28

Example 12-7: Principal axes and principal moments of inertia for a Z-section

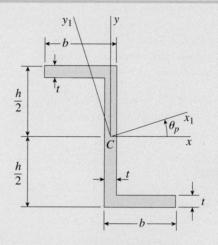

### Solution

Let us use the $xy$ axes (Fig. 12-28) as the reference axes through the centroid C. The moments and product of inertia with respect to these axes can be obtained by dividing the area into three rectangles and using the parallel-axis theorems. The results of such calculations are as follows:

$$I_x = 29.29 \times 10^6 \text{ mm}^4 \quad I_y = 5.667 \times 10^6 \text{ mm}^4$$

$$I_{xy} = -9.366 \times 10^6 \text{ mm}^4$$

Substituting these values into the equation for the angle $\theta_p$ (Eq. 12-39), we get

$$\tan 2\theta_p = -\frac{2I_{xy}}{I_x - I_y} = 0.7930 \quad 2\theta_p = 38.4° \text{ and } 218.4°$$

Thus, the two values of $\theta_p$ are

$$\theta_p = 19.2° \text{ and } 109.2°$$

Using these values of $\theta_p$ in the transformation equation for $I_{x_1}$ (Eq. 12-33), we find $I_{x_1} = 32.6 \times 10^6$ mm$^4$ and $2.4 \times 10^6$ mm$^4$, respectively. These same values are obtained if we substitute into Eqs. (12-42a and b). Thus, the principal moments of inertia and the angles to the corresponding principal axes are

$$I_1 = 32.6 \times 10^6 \text{ mm}^4 \quad \theta_{p_1} = 19.2°$$

$$I_2 = 2.4 \times 10^6 \text{ mm}^4 \quad \theta_{p_2} = 109.2°$$

The principal axes are shown in Fig. 12-28 as the $x_1y_1$ axes.

# PROBLEMS CHAPTER 12

## Centroids of Areas

*The problems for Section 12.2 are to be solved by integration.*

**12.2-1** Determine the distances $\bar{x}$ and $\bar{y}$ to the centroid $C$ of a right triangle having base $b$ and altitude $h$ (see Case 6, Appendix E).

**12.2-2** Determine the distance $\bar{y}$ to the centroid $C$ of a trapezoid having bases $a$ and $b$ and altitude $h$ (see Case 8, Appendix E).

**12.2-3** Determine the distance $\bar{y}$ to the centroid $C$ of a semicircle of radius $r$ (see Case 10, Appendix E).

**12.2-4** Determine the distances $\bar{x}$ and $\bar{y}$ to the centroid $C$ of a parabolic spandrel of base $b$ and height $h$ (see Case 18, Appendix E).

**12.2-5** Determine the distances $\bar{x}$ and $\bar{y}$ to the centroid $C$ of a semisegment of $n$th degree having base $b$ and height $h$ (see Case 19, Appendix E).

## Centroids of Composite Areas

*The problems for Section 12.3 are to be solved by using the formulas for composite areas.*

**12.3-1** Determine the distance $\bar{y}$ to the centroid $C$ of a trapezoid having bases $a$ and $b$ and altitude $h$ (see Case 8, Appendix E) by dividing the trapezoid into two triangles.

**12.3-2** One quarter of a square of side $a$ is removed (see figure). What are the coordinates $\bar{x}$ and $\bar{y}$ of the centroid $C$ of the remaining area?

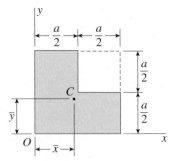

**PROBS. 12.3-2 and 12.5-2**

**12.3-3** Calculate the distance $\bar{y}$ to the centroid $C$ of the channel section shown in the figure if $a = 6$ in., $b = 1$ in., and $c = 2$ in.

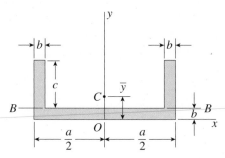

**PROBS. 12.3-3, 12.3-4, and 12.5-3**

**12.3-4** What must be the relationship between the dimensions $a$, $b$, and $c$ of the channel section shown in the figure in order that the centroid $C$ will lie on line $BB$?

**12.3-5** The cross section of a beam constructed of a W 24 × 162 wide-flange section with an 8 in. × 3/4 in. cover plate welded to the top flange is shown in the figure.

Determine the distance $\bar{y}$ from the base of the beam to the centroid $C$ of the cross-sectional area.

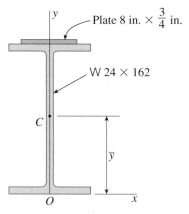

**PROBS. 12.3-5 and 12.5-5**

**12.3-6** Determine the distance $\bar{y}$ to the centroid $C$ of the composite area shown in the figure.

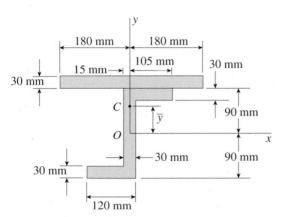

PROBS. 12.3-6, 12.5-6, and 12.7-6

**12.3-7** Determine the coordinates $\bar{x}$ and $\bar{y}$ of the centroid $C$ of the L-shaped area shown in the figure.

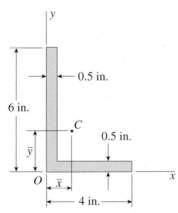

PROBS. 12.3-7, 12.4-7, 12.5-7, and 12.7-7

**12.3-8** Determine the coordinates $\bar{x}$ and $\bar{y}$ of the centroid $C$ of the area shown in the figure.

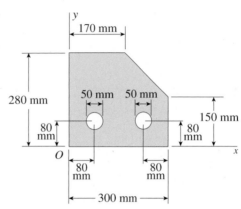

PROB. 12.3-8

## Moments of Inertia

*Problems 12.4-1 through 12.4-4 are to be solved by integration.*

**12.4-1** Determine the moment of inertia $I_x$ of a triangle of base $b$ and altitude $h$ with respect to its base (see Case 4, Appendix E).

**12.4-2** Determine the moment of inertia $I_{BB}$ of a trapezoid having bases $a$ and $b$ and altitude $h$ with respect to its base (see Case 8, Appendix E).

**12.4-3** Determine the moment of inertia $I_x$ of a parabolic spandrel of base $b$ and height $h$ with respect to its base (see Case 18, Appendix E).

**12.4-4** Determine the moment of inertia $I_x$ of a circle of radius $r$ with respect to a diameter (see Case 9, Appendix E).

*Problems 12.4-5 through 12.4-9 are to be solved by considering the area to be a composite area.*

**12.4-5** Determine the moment of inertia $I_{BB}$ of a rectangle having sides of lengths $b$ and $h$ with respect to a diagonal of the rectangle (see Case 2, Appendix E).

**12.4-6** Calculate the moment of inertia $I_x$ for the composite circular area shown in the figure. The origin of the axes is at the center of the concentric circles, and the three diameters are 20, 40, and 60 mm.

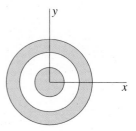

PROB. 12.4-6

**12.4-7** Calculate the moments of inertia $I_x$ and $I_y$ with respect to the $x$ and $y$ axes for the L-shaped area shown in the figure for Prob. 12.3-7.

**12.4-8** A semicircular area of radius 150 mm has a rectangular cutout of dimensions 50 mm $\times$ 100 mm (see figure).

Calculate the moments of inertia $I_x$ and $I_y$ with respect to the $x$ and $y$ axes. Also, calculate the corresponding radii of gyration $r_x$ and $r_y$.

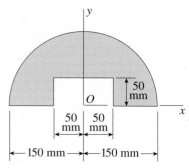

PROB. 12.4-8

**12.4-9** Calculate the moments of inertia $I_1$ and $I_2$ of a W 16 $\times$ 100 wide-flange section using the cross-sectional dimensions given in Table F-l, Appendix F. (Disregard the cross-sectional areas of the fillets.) Also, calculate the corresponding radii of gyration $r_1$ and $r_2$, respectively.

# Parallel-Axis Theorem

**12.5-1** Calculate the moment of inertia $I_b$ of a W 12 $\times$ 50 wide-flange section with respect to its base. (Use data from Table F-l, Appendix F.)

**12.5-2** Determine the moment of inertia $I_c$ with respect to an axis through the centroid $C$ and parallel to the $x$ axis for the geometric figure described in Prob. 12.3-2.

**12.5-3** For the channel section described in Prob. 12.3-3, calculate the moment of inertia $I_{x_c}$ with respect to an axis through the centroid $C$ and parallel to the $x$ axis.

**12.5-4** The moment of inertia with respect to axis 1–1 of the scalene triangle shown in the figure is 90 $\times$ 10$^3$ mm$^4$. Calculate its moment of inertia $I_2$ with respect to axis 2–2.

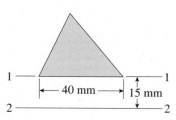

PROB. 12.5-4

**12.5-5** For the beam cross section described in Prob. 12.3-5, calculate the centroidal moments of inertia $I_{x_c}$ and $I_{y_c}$ with respect to axes through the centroid $C$ such that the $x_c$ axis is parallel to the $x$ axis and the $y_c$ axis coincides with the $y$ axis.

**12.5-6** Calculate the moment of inertia $I_{x_c}$ with respect to an axis through the centroid $C$ and parallel to the $x$ axis for the composite area shown in the figure for Prob. 12.3-6.

**12.5-7** Calculate the centroidal moments of inertia $I_{x_c}$ and $I_{y_c}$ with respect to axes through the centroid $C$ and parallel to the $x$ and $y$ axes, respectively, for the L-shaped area shown in the figure for Prob. 12.3-7.

**12.5-8** The wide-flange beam section shown in the figure has a total height of 250 mm and a constant thickness of 15 mm.

Determine the flange width $b$ if it is required that the centroidal moments of inertia $I_x$ and $I_y$ be in the ratio 3 to 1, respectively.

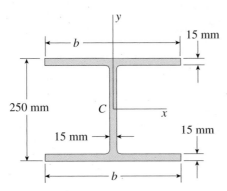

**PROB. 12.5-8**

## Polar Moments of Inertia

**12.6-1** Determine the polar moment of inertia $I_P$ of an isosceles triangle of base $b$ and altitude $h$ with respect to its apex (see Case 5, Appendix E).

**12.6-2** Determine the polar moment of inertia $(I_P)_C$ with respect to the centroid $C$ for a circular sector (see Case 13, Appendix E).

**12.6-3** Determine the polar moment of inertia $I_P$ for a W 8 × 21 wide-flange section with respect to one of its outermost corners.

**12.6-4** Obtain a formula for the polar moment of inertia $I_P$ with respect to the midpoint of the hypotenuse for a right triangle of base $b$ and height $h$ (see Case 6, Appendix E).

**12.6-5** Determine the polar moment of inertia $(I_P)_C$ with respect to the centroid $C$ for a quarter-circular spandrel (see Case 12, Appendix E).

## Products of Inertia

**12.7-1** Using integration, determine the product of inertia $I_{xy}$ for the parabolic semisegment shown in Fig. 12-5 (see also Case 17 in Appendix E).

**12.7-2** Using integration, determine the product of inertia $I_{xy}$ for the quarter-circular spandrel shown in Case 12, Appendix E.

**12.7-3** Find the relationship between the radius $r$ and the distance $b$ for the composite area shown in the figure in order that the product of inertia $I_{xy}$ will be zero.

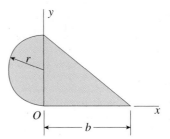

**PROB. 12.7-3**

**12.7-4** Obtain a formula for the product of inertia $I_{xy}$ of the symmetrical L-shaped area shown in the figure.

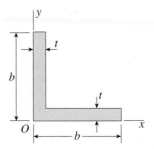

**PROB. 12.7-4**

**12.7-5** Calculate the product of inertia $I_{12}$ with respect to the centroidal axes 1–1 and 2–2 for an L 6 in. × 6 in. × 1 in. angle section (see Table F-4, Appendix F). (Disregard the cross-sectional areas of the fillet and rounded corners.)

**12.7-6** Calculate the product of inertia $I_{xy}$ for the composite area shown in Prob. 12.3-6.

**12.7-7** Determine the product of inertia $I_{x_c y_c}$ with respect to centroidal axes $x_c$ and $y_c$ parallel to the $x$ and $y$ axes, respectively, for the L-shaped area shown in Prob. 12.3-7.

# Rotation of Axes

*The problems for Section 12.8 are to be solved by using the transformation equations for moments and products of inertia.*

**12.8-1** Determine the moments of inertia $I_{x_1}$ and $I_{y_1}$ and the product of inertia $I_{x_1y_1}$ for a square with sides $b$, as shown in the figure. (Note that the $x_1y_1$ axes are centroidal axes rotated through an angle $\theta$ with respect to the $xy$ axes.)

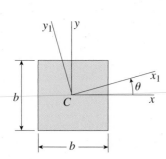

PROB. 12.8-1

**12.8-2** Determine the moments and product of inertia with respect to the $x_1y_1$ axes for the rectangle shown in the figure. (Note that the $x_1$ axis is a diagonal of the rectangle.)

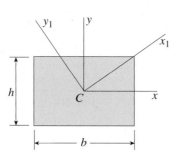

PROB. 12.8-2

**12.8-3** Calculate the moment of inertia $I_d$ for a W 12 × 50 wide-flange section with respect to a diagonal passing through the centroid and two outside corners of

the flanges. (Use the dimensions and properties given in Table F-1.)

**12.8-4** Calculate the moments of inertia $I_{x_1}$ and $I_{y_1}$ and the product of inertia $I_{x_1y_1}$ with respect to the $x_1y_1$ axes for the L-shaped area shown in the figure if $a = 150$ mm, $b = 100$ mm, $t = 15$ mm, and $\theta = 30°$.

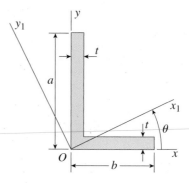

PROBS. 12.8-4 and 12.9-4

**12.8-5** Calculate the moments of inertia $I_{x_1}$ and $I_{y_1}$ and the product of inertia $I_{x_1y_1}$ with respect to the $x_1y_1$ axes for the Z-section shown in the figure if $b = 3$ in., $h = 4$ in., $t = 0.5$ in., and $\theta = 60°$.

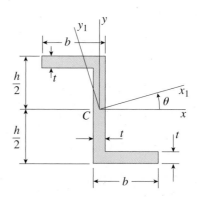

PROBS. 12.8-5, 12.8-6, 12.9-5, and 12.9-6

**12.8-6** Solve the preceding problem if $b = 80$ mm, $h = 120$ mm, $t = 12$ mm, and $\theta = 30°$.

# Principal Axes, Principal Points, and Principal Moments of Inertia

**12.9-1** An ellipse with major axis of length $2a$ and minor axis of length $2b$ is shown in the figure.

(a) Determine the distance $c$ from the centroid $C$ of the ellipse to the principal points $P$ on the minor axis ($y$ axis).

(b) For what ratio $a/b$ do the principal points lie on the circumference of the ellipse?

(c) For what ratios do they lie inside the ellipse?

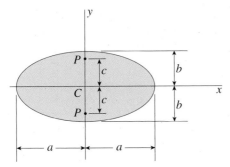

**PROB. 12.9-1**

**12.9-2** Demonstrate that the two points $P_1$ and $P_2$, located as shown in the figure, are the principal points of the isosceles right triangle.

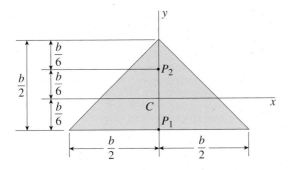

**PROB. 12.9-2**

**12.9-3** Determine the angles $\theta_{p_1}$ and $\theta_{p_2}$ defining the orientations of the principal axes through the origin $O$ for the right triangle shown in the figure if $b = 6$ in. and $h = 8$ in. Also, calculate the corresponding principal moments of inertia $I_1$ and $I_2$.

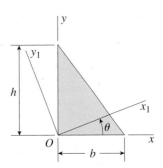

**PROB. 12.9-3**

**12.9-4** Determine the angles $\theta_{p_1}$ and $\theta_{p_2}$ defining the orientations of the principal axes through the origin $O$ and the corresponding principal moments of inertia $I_1$ and $I_2$ for the L-shaped area described in Prob. 12.8-4 ($a = 150$ mm, $b = 100$ mm, and $t = 15$ mm).

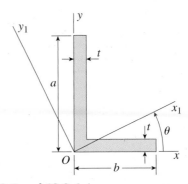

**PROBS. 12.8-4 and 12.9-4**

**12.9-5** Determine the angles $\theta_{p_1}$ and $\theta_{p_2}$ defining the orientations of the principal axes through the centroid $C$ and the corresponding principal centroidal moments of inertia $I_1$ and $I_2$ for the Z-section described in Prob. 12.8-5 ($b = 3$ in., $h = 4$ in., and $t = 0.5$ in.).

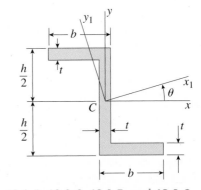

**PROBS. 12.8-5, 12.8-6, 12.9-5, and 12.9-6**

**12.9-6** Solve the preceding problem for the Z-section described in Prob. 12.8-6 ($b$ = 80 mm, $h$ = 120 mm, and $t$ = 12 mm).

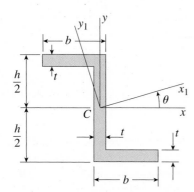

PROBS. 12.8-5, 12.8-6, 12.9-5, and 12.9-6

**12.9-7** Determine the angles $\theta_{p_1}$ and $\theta_{p_2}$ defining the orientations of the principal axes through the centroid $C$ for the right triangle shown in the figure if $h$ = 2$b$. Also, determine the corresponding principal centroidal moments of inertia $I_1$ and $I_2$.

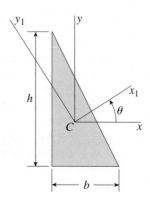

PROB. 12.9-7

**12.9-8** Determine the angles $\theta_{p_1}$ and $\theta_{p_2}$ defining the orientations of the principal centroidal axes and the corresponding principal moments of inertia $I_1$ and $I_2$ for the L-shaped area shown in the figure if $a$ = 80 mm, $b$ = 150 mm, and $t$ = 16 mm.

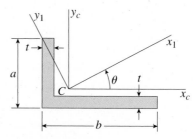

PROBS. 12.9-8 and 12.9-9

**12.9-9** Solve the preceding problem if $a$ = 3 in., $b$ = 6 in., and $t$ = 5/8 in.

# References and Historical Notes

**1-1** Timoshenko, S. P., *History of Strength of Materials*, Dover Publications, Inc., New York, 1983 (originally published by McGraw-Hill Book Co., Inc., New York, 1953).

S. P. Timoshenko
(1878–1972)

*Courtesy of Stanford University Archives*

*Note:* Stephen P. Timoshenko (1878–1972) was a famous scientist, engineer, and teacher. Born in Russia, he came to the United States in 1922. He was a researcher with the Westinghouse Research Laboratory, a professor at the University of Michigan, and later a professor at Stanford University, where he retired in 1944.

Timoshenko made many original contributions, both theoretical and experimental, to the field of applied mechanics, and he wrote twelve pioneering textbooks that revolutionized the teaching of mechanics in the United States. These books, which were published in as many as five editions and translated into as many as 35 languages, covered the subjects of statics, dynamics, mechanics of materials, vibrations, structural theory, stability, elasticity, plates, and shells.

**1-2** Todhunter, I., and Pearson, K., *A History of the Theory of Elasticity and of the Strength of Materials*, Vols. I and II, Dover Publications, Inc., New York, 1960 (originally published by the Cambridge University Press in 1886 and 1893). *Note:* Isaac Todhunter (1820–1884) and Karl Pearson (1857–1936) were English mathematicians and educators. Pearson was especially noteworthy for his original contributions to statistics.

**1-3** Love, A. E. H., *A Treatise on the Mathematical Theory of Elasticity*, 4th Ed., Dover Publications, Inc., New York, 1944 (originally published by the Cambridge University Press in 1927); see "Historical Introduction," pp. 1–31.

*Note:* Augustus Edward Hough Love (1863–1940) was an outstanding English elastician who taught at Oxford University. His many important investigations included the analysis of seismic surface waves, now called *Love waves* by geophysicists.

**1-4** Jacob Bernoulli (1654–1705), also known by the names James, Jacques, and Jakob, was a member of the famous family of mathematicians and scientists of Basel, Switzerland (see Ref. 9-1). He did important work in connection with elastic curves of beams. Bernoulli also developed polar coordinates and became famous for his work in theory of probability, analytic geometry, and other fields.

Jean Victor Poncelet (1788–1867) was a Frenchman who fought in Napoleon's campaign against Russia and was given up for dead on the battlefield. He survived, was taken prisoner, and later returned to France to continue his work in mathematics. His major contributions to mathematics are in geometry; in mechanics he is best known for his work on properties of materials and dynamics. (For the work of Bernoulli and Poncelet in connection with stress-strain diagrams, see Ref. 1-1, p. 88, and Ref. 1-2, Vol. I, pp. 10, 533, and 873.)

**1-5** James and James, *Mathematics Dictionary*, Van Nostrand Reinhold, New York (latest edition).

**1-6** Robert Hooke (1635–1703) was an English scientist who performed experiments with elastic bodies and developed improvements in timepieces. He also formulated the laws of gravitation independently of Newton, of whom he was a contemporary. Upon the founding of the Royal Society of London in 1662, Hooke was appointed its first curator. (For the origins of Hooke's law, see Ref. 1-1, pp. 17–20, and Ref. 1-2, Vol. I, p. 5.)

**1-7** Thomas Young (1773–1829) was an outstanding English scientist who did pioneering work in optics, sound, impact, and other subjects. (For information about his work with materials, see Ref. 1-1, pp. 90–98, and Ref. 1-2, Vol. I, pp. 80–86.)

Thomas Young
(1773–1829)

**1-8** Siméon Denis Poisson (1781–1840) was a great French mathematician. He made many contributions to both mathematics and mechanics, and his name is preserved in numerous ways besides Poisson's ratio. For instance, we have Poisson's equation in partial differential equations and the Poisson distribution in theory of probability. (For information about Poisson's theories of material behavior, see Ref. 1-1, pp. 111–114; Ref. 1-2, Vol. I, pp. 208–318; and Ref. 1-3, p. 13.)

S. D. Poisson
(1781–1840)

**2-1** Timoshenko, S. P., and Goodier, J. N., *Theory of Elasticity*, 3rd Ed., McGraw-Hill Book Co., Inc., New York, 1970 (see p. 110). *Note:* James Norman Goodier (1905–1969) was well known for his research contributions to theory of elasticity, stability, wave propagation in solids, and other branches of applied mechanics. Born in England, he studied at Cambridge University and later at the University of Michigan. He was a professor at Cornell University and subsequently at Stanford University, where he headed the program in applied mechanics.

**2-2** Leonhard Euler (1707–1783) was a famous Swiss mathematician, perhaps the greatest mathematician of all time. Ref. 11-1 gives information about his life and works. (For his work on statically indeterminate structures, see Ref. 1-1, p. 36, and Ref. 2-3, p. 650.)

**2-3** Oravas, G. A., and McLean, L., "Historical development of energetical principles in elastomechanics," *Applied Mechanics Reviews,* Part I, Vol. 19, No. 8, August 1966, pp. 647–658, and Part II, Vol. 19, No. 11, November 1966, pp. 919–933.

**2-4** Louis Marie Henri Navier (1785–1836), a famous French mathematician and engineer, was one of the founders of the mathematical theory of elasticity. He contributed to beam, plate, and shell theory, to theory of vibrations, and to the theory of viscous fluids. (See Ref. 1-1, p. 75; Ref. 1-2, Vol. I, p. 146; and Ref. 2-3, p. 652, for his analysis of statically indeterminate structures.)

**2-5** Piobert, G., Morin, A.-J., and Didion, I., "Commission des Principes du Tir," *Mémorial de l'Artillerie*, Vol. 5, 1842, pp. 501–552.

*Note:* This paper describes experiments made by firing artillery projectiles against iron plating. On page 505 appears the description of the markings that are the slip bands. The description is quite brief, and there is no indication that the authors attributed the markings to inherent material characteristics. Guillaume Piobert (1793– 1871) was a French general and mathematician who made many studies of ballistics; when this paper was written, he was a captain in the artillery.

**2-6** Lüders, W., "Ueber die Äusserung der elasticität an stahlartigen Eisenstäben und Stahlstäben, und über eine beim Biegen solcher Stäbe beobachtete Molecularbewegung," *Dingler's Polytechnisches Journal*, Vol. 155, 1860, pp. 18–22.

*Note:* This paper clearly describes and illustrates the bands that appear on the surface of a polished steel specimen during yielding. Of course, these bands are only the surface manifestation of three-dimensional zones of deformation; hence, the zones should probably be characterized as "wedges" rather than bands.

**2-7** Benoit Paul Emile Clapeyron (1799–1864) was a famous French structural engineer and bridge designer; he taught engineering at the École des Ponts et Chaussées in

Paris. It appears that Clapeyron's theorem, which states that the work of the external loads acting on a linearly elastic body is equal to the strain energy, was first published in 1833. (See Ref. 1-1, pp. 118 and 288; Ref. 1-2, Vol. I, p. 578; and Ref. 1-2, Vol. II, p. 418.)

**2-8** Poncelet investigated longitudinal vibrations of a bar due to impact loads (see Ref. 1-1, p. 88). See Ref. 1-4 for additional information about his life and works.

**2-9** Budynas, R., and Young, W. C., *Roark's Formulas for Stress and Strain*, McGraw-Hill Book Co., Inc., New York, 2002.

**2-10** Barré de Saint-Venant (1797–1886) is generally recognized as the most outstanding elastician of all time. Born near Paris, he studied briefly at the École Polytechnique and later graduated from the École des Ponts et Chaussées. His later professional career suffered greatly from his refusal, as a matter of conscience and politics, to join his schoolmates in preparing for the defense of Paris in March 1814, just prior to Napoleon's abdication. As a consequence, his achievements received greater recognition in other countries than they did in France.

Some of his most famous contributions are the formulation of the fundamental equations of elasticity and the development of the exact theories of bending and torsion. He also developed theories for plastic deformations and vibrations. His full name was Adéhmar Jean Claude Barré, Count de Saint-Venant. (See Ref. 1-1, pp. 229–242; Ref. 1-2, Vol. I, pp. 833–872, Vol. II, Part I, pp. 1–286, Vol. II, Part II, pp. 1–51; and Ref. 2-1, pp. 39–40.)

**2-11** Zaslavsky, A., "A note on Saint-Venant's principle," *Israel Journal of Technology*, Vol. 20, 1982, pp. 143–144.

**2-12** Ramberg, W. A., and Osgood, W. R., "Description of stress-strain curves by three parameters," *National Advisory Committee for Aeronautics*, Technical Note No. 902, July 1943.

**3-1** The relationship between torque and angle of twist in a circular bar was correctly established in 1784 by Charles Augustin de Coulomb (1736–1806), a famous French scientist (see Ref. 1-1, pp. 51–53, 82, and 92, and Ref. 1-2, Vol. I, p. 69). Coulomb made contributions in electricity and magnetism, viscosity of fluids, friction, beam bending, retaining walls and arches, torsion and torsional vibrations, and other subjects (see Ref. 1-1, pp. 47–54).

Thomas Young (Ref. 1-7) observed that the applied torque is balanced by the shear stresses on the cross section and that the shear stresses are proportional to the distance from the axis. The French engineer Alphonse J. C. B. Duleau (1789–1832) performed tests on bars in torsion and also developed a theory for circular bars (see Ref. 1-1, p. 82).

C. A. de Coulomb
(1736–1806)

**3-2** Bredt, R., "Kritische Bemerkungen zur Drehungselastizität," *Zeitschrift des Vereines Deutscher Ingenieure*, Vol. 40, 1896, pp. 785–790, and 813–817.

*Note:* Rudolph Bredt (1842–1900) was a German engineer who studied in Karlsruhe and ZÜrich. Then he worked for a while in Crewe, England, at a train factory, where he learned about the design and construction of cranes. This experience formed the basis for his later work as a crane manufacturer in Germany. His theory of torsion was developed in connection with the design of box-girder cranes.

**5-1** A proof of the theorem that cross sections of a beam in pure bending remain plane can be found in the paper by Fazekas, G. A., "A note on the bending of Euler beams," *Journal of Engineering Education*, Vol. 57, No. 5, January 1967. The validity of the theorem has long been recognized, and it was used by early investigators such as Jacob Bernoulli (Ref. 1-4) and L. M. H. Navier (Ref. 2-4). For a discussion of the work done by Bernoulli and Navier in connection with bending of beams, see Ref. 1-1, pp. 25–27 and 70–75.

**5-2** Galilei, Galileo, *Dialogues Concerning Two New Sciences*, translated from the Italian and Latin into English by Henry Crew and Alfonso De Salvio, The Macmillan Company, New York, 1933 (translation first published in 1914.)

*Note:* This book was published in 1638 by Louis Elzevir in Leida, now Leiden, Netherlands. *Two New Sciences* represents the culmination of Galileo's work on dynamics and mechanics of materials. It can truly be said that these two subjects, as we know them today, began with Galileo and the publication of this famous book.

Galileo Galilei was born in Pisa in 1564. He made many famous experiments and discoveries, including those on falling bodies and pendulums that initiated the science of dynamics. Galileo was an eloquent lecturer and attracted students from many countries. He pioneered in astronomy and developed a telescope with which he made many astronomical discoveries, including the mountainous character of the moon, Jupiter's satellites, the phases of Venus, and sunspots. Because his scientific views of the solar system were contrary to theology, he was condemned by the church in Rome and spent the last years of his life in seclusion in Florence; during this period he wrote *Two New Sciences*. Galileo died in 1642 and was buried in Florence.

Galileo Galilei
(1564–1642)

**5-3** The history of beam theory is described in Ref. 1-1, pp. 11–47 and 135–141, and in Ref. 1-2. Edme Mariotte (1620–1684) was a French physicist who made developments in dynamics, hydrostatics, optics, and mechanics. He made tests on beams and developed a theory for calculating load-carrying capacity; his theory was an improvement on Galileo's work, but still not correct. Jacob Bernoulli (1654–1705), who is described in Ref. 1-4, first determined that the curvature is proportional to the bending moment. However, his constant of proportionality was incorrect.

Leonhard Euler (1707–1783) obtained the differential equation of the deflection curve of a beam and used it to solve many problems of both large and small deflections (Euler's life and work are described in Ref. 11-1). The first person to obtain the distribution of stresses in a beam and correctly relate the stresses to the bending moment probably was Antoine Parent (1666–1716), a French physicist and mathematician. Later, a rigorous investigation of strains and stresses in beams was made by Saint-Venant (1797–1886);

see Ref. 2-10. Important contributions were also made by Coulomb (Ref. 3-1) and Navier (Ref. 2-4).

**5-4** Manual of Steel Construction (ASD/LRFD), published by the American Institute of Steel Construction, Inc., One East Wacker Drive, (Suite 3100), Chicago, Illinois 60601. (For other publications and additional information, go to their website: www.aisc.org.)

**5-5** *Aluminum Design Manual,* published by the Aluminum Association, Inc., 900 19th Street NW, Washington, D.C. 20006. (For other publications and additional information, go to their website: www.aluminum.org.)

**5-6** National Design Specification for Wood Construction (ASD/LRFD), published by the American Wood Council, a division of the American Forest and Paper Association, 1111 19th Street NW, (Suite 800), Washington, D.C. 20036. (For other publications and additional information, go to their websites: www.awc.org and www.afandpa.org.)

**5-7** D. J. Jourawski (1821–1891) was a Russian bridge and railway engineer who developed the now widely used approximate theory for shear stresses in beams (see Ref. 1-1, pp. 141–144, and Ref. 1-2, Vol. II, Part I, pp. 641– 642). In 1844, only two years after graduating from the Institute of Engineers of Ways of Communication in St. Petersburg, he was assigned the task of designing and constructing a major bridge on the first railway line from Moscow to St. Petersburg. He noticed that some of the large timber beams split longitudinally in the centers of the cross sections, where he knew the bending stresses were zero. Jourawski drew free-body diagrams and quickly discovered the existence of horizontal shear stresses in the beams. He derived the shear formula and applied his theory to various shapes of beams. Jourawski's paper on shear in beams is cited in Ref. 5-8. His name is sometimes transliterated as Dimitrii Ivanovich Zhuravskii.

**5-8** Jourawski, D. J., "Sur la résistance d'un corps prismatique . . . ," *Annales des Ponts et Chaussés*, Mémoires et Documents, 3rd Series, Vol. 12, Part 2, 1856, pp. 328–351.

**5-9** Zaslavsky, A., "On the limitations of the shearing stress formula," *International Journal of Mechanical Engineering Education,* Vol. 8, No. 1, 1980, pp. 13–19. (See also Ref. 2-1, pp. 358–359.)

**5-10** Maki, A. C., and Kuenzi, E. W., "Deflection and stresses of tapered wood beams," Research Paper FPL 34, U. S. Forest Service, Forest Products Laboratory, Madison, Wisconsin, September 1965, 54 pages.

**6-1** Timoshenko, S. P., "Use of stress functions to study flexure and torsion of prismatic bars," (in Russian), St. Petersburg, 1913 (reprinted in Vol. 82 of the *Memoirs of the Institute of Ways of Communication*, pp. 1–21).

*Note:* In this paper, the point in the cross section of a beam through which a concentrated force should act in order to eliminate rotation was found. Thus, this work contains the first determination of a shear center. The particular beam under investigation had a solid semicircular cross section (see Ref. 2-1, pp. 371–373).

**7-1** Augustin Louis Cauchy (1789–1857) was one of the greatest mathematicians. Born in Paris, he entered the École Polytechnique at the age of 16, where he studied under Lagrange, Laplace, Fourier, and Poisson. He was quickly recognized for his mathematical prowess, and at age 27 he became a professor at the École and a member of the Academy of Sciences. His major works in pure mathematics were in group theory, number theory, series, integration, differential equations, and analytical functions.

In applied mathematics, Cauchy introduced the concept of stress as we know it today, developed the equations of theory of elasticity, and introduced the notion of principal stresses and principal strains (see Ref. 1-1, pp. 107–111). An entire chapter is devoted to his work on theory of elasticity in Ref. 1-2 (see Vol. I, pp. 319–376).

**7-2** See Ref. 1-1, pp. 229–242. *Note:* Saint-Venant was a pioneer in many aspects of theory of elasticity, and Todhunter and Pearson dedicated their book, *A History of the Theory of Elasticity* (Ref. 1-2), to him. For further information about Saint-Venant, see Ref. 2-10.

**7-3** William John Macquorn Rankine (1820–1872) was born in Edinburgh, Scotland, and taught engineering at Glasgow University. He derived the stress transformation equations in 1852 and made many other contributions to theory of elasticity and applied mechanics (see Ref. 1-1, pp. 197–202, and Ref. 1-2, Vol. II, Part I, pp. 86 and 287–322). His engineering subjects included arches, retaining walls, and structural theory.

Rankine also achieved scientific fame for his work with fluids, light, sound, and behavior of crystals, and he is especially well known for his contributions to molecular physics and thermodynamics. His name is preserved by the Rankine cycle in thermodynamics and the Rankine absolute temperature scale.

**7-4** The famous German civil engineer Otto Christian Mohr (1835–1918) was both a theoretician and a practical designer. He was a professor at the Stuttgart Polytechnikum and later at the Dresden Polytechnikum.

He developed the circle of stress in 1882 (Ref. 7-5 and Ref. 1-1, pp. 283–288).

Mohr made numerous contributions to the theory of structures, including the Williot-Mohr diagram for truss displacements, the moment-area method for beam deflections, and the Maxwell-Mohr method for analyzing statically indeterminate structures. (*Note:* Joseph Victor Williot, 1843–1907, was a French engineer, and James Clerk Maxwell, 1831–1879, was a famous British scientist.)

**7-5** Mohr, O., "Über die Darstellung des Spannungszustandes und des Deformationszustandes eines Körperelementes," *Zivilingenieur*, 1882, p. 113.

**8-1** Karl Culmann (1821–1881) was a famous German bridge and railway engineer. In 1849–1850 he spent two years traveling in England and the United States to study bridges, which he later wrote about in Germany. He designed numerous bridge structures in Europe, and in 1855 he became professor of structures at the newly organized Zürich Polytechnicum. Culmann made many developments in graphical methods and wrote the first book on graphic statics, published in Zürich in 1866. Stress trajectories are one of the original topics presented in this book (see Ref. 1-1, pp. 190–197).

**9-1** The work of Jacob Bernoulli, Euler, and many others with respect to elastic curves is described in Ref. 1-1, pp. 27 and 30–36, and Ref. 1-2. Another member of the Bernoulli family, Daniel Bernoulli (1700–1782), proposed to Euler that he obtain the differential equation of the deflection curve by minimizing the strain energy, which Euler did. Daniel Bernoulli, a nephew of Jacob Bernoulli, is renowned for his work in hydrodynamics, kinetic theory of gases, beam vibrations, and other subjects. His father, John Bernoulli (1667–1748), a younger brother of Jacob, was an

Jacob Bernoulli
(1654–1705)

equally famous mathematician and scientist who first formulated the principle of virtual displacements, and solved the problem of the brachystochrone.

John Bernoulli established the rule for obtaining the limiting value of a fraction when both the numerator and denominator tend to zero. He communicated this last rule to G. F. A. de l'Hôpital (1661–1704), a French nobleman who wrote the first book on calculus (1696) and included this theorem, which consequently became known as *L'Hôpital's rule.*

Daniel's nephew, Jacob Bernoulli (1759–1789), also known as James or Jacques, was a pioneer in the theory of plate bending and plate vibrations.

Much interesting information about the many prominent members of the Bernoulli family, as well as other pioneers in mechanics and mathematics, can be found in books on the history of mathematics.

**9-2** Castigliano, A., *Théorie de l'équilibre des systèmes élastiques et ses applications,* A. F. Negro, Turin, 1879, 480 pages.

*Note:* In this book Castigliano presented in very complete form many fundamental concepts and principles of structural analysis. Although Castigliano was Italian, he wrote this book in French in order to gain a wider audience for his work. It was translated into both German and English (Refs. 9-3 and 9-4). The English translation was republished in 1966 by Dover Publications and is especially valuable because of the introductory material by Gunhard A. Oravas (Refs. 9-5 and 9-6).

Castigliano's first and second theorems appear on pp. 15–16 of the 1966 edition of his book. He identified them as Part 1 and Part 2 of the "Theorem of the Differential Coefficients of the Internal Work." In mathematical form, they appear in his book as

$$F_p = \frac{dW_i}{dr_p} \quad \text{and} \quad r_p = \frac{dW_i}{dF_p}$$

where $W_i$ is the internal work (or strain energy), $F_p$ represents any one of the external forces, and $r_p$ is the displacement of the point of application of $F_p$.

Castigliano did not claim complete originality for the first theorem, although he stated in the Preface to his book that his presentation and proof were more general than anything published previously. The second theorem was original with him and was part of his thesis for the civil engineering degree at the Polytechnic Institute of Turin in 1873.

Carlo Alberto Pio Castigliano was born of a poor family in Asti in 1847 and died of pneumonia in 1884, while at the height of his productivity. The story of his life is told by Oravas in the introduction to the 1966 edition,

and a bibliography of Castigliano's works and a list of his honors and awards are also given there. His contributions are also documented in Refs. 2-3 and 1-1. He used the name Alberto Castigliano when signing his writings.

**9-3** Hauff, E., *Theorie des Gleichgewichtes elastischer Systeme und deren Anwendung,* Carl Gerold's Sohn, Vienna, 1886. (A translation of Castigliano's book, Ref. 9-2.)

**9-4** Andrews, E. S., *Elastic Stresses in Structures,* Scott, Greenwood and Son, London, 1919. (A translation of Castigliano's book, Ref. 9-2.)

**9-5** Castigliano, C. A. P., *The Theory of Equilibrium of Elastic Systems and Its Applications,* translated by E. S. Andrews, with a new introduction and biographical portrait by G. A. Oravas, Dover Publications, Inc., New York, 1966. (A republication of Ref. 9-4 but with the addition of historical material by Oravas.)

**9-6** Oravas, G. A., "Historical Review of Extremum Principles in Elastomechanics," an introductory section (pp. xx–xlvi) of the book, *The Theory of Equilibrium of Elastic Systems and Its Applications,* by C. A. P. Castigliano, translated by E. S. Andrews, Dover Publications, Inc., New York, 1966 (Ref. 9-5).

**9-7** Macaulay, W. H., "Note on the deflection of beams," *The Messenger of Mathematics,* vol. XLVIII, May 1918–April 1919, Cambridge, 1919, pp. 129–130.

*Note*: William Herrick Macaulay, 1853–1936, was an English mathematician and Fellow of King's College, Cambridge. In this paper he defined "by $\{f(x)\}_a$ a function of $x$ which is zero when $x$ is less than $a$ and equal to $f(x)$ when $x$ is equal to or greater than $a$." Then he showed how to use this function when finding beam deflections. Unfortunately, he did not give any references to the earlier work of Clebsch and Föppl; see Refs. 9-8 through 9-10.

**9-8** Clebsch, A., *Theorie der Elasticität fester Körper,* B. G. Teubner, Leipzig, 1862, 424 pages. (Translated into French and annotated by Saint-Venant, *Théorie de l'Élasticité des Corps Solides,* Paris, 1883. Saint-Venant's notes increased Clebsch's book threefold in size.)

*Note*: The method of finding beam deflections by integrating across points of discontinuity was presented first in this book; see Ref. 1-1, pp. 258–259 and Ref. 9-10. Rudolf Friedrich Alfred Clebsch, 1933–1872, was a German mathematician and scientist. He was a professor of engineering at the Karlsruhe Polytechnicum and later a professor of mathematics at Göttingen University.

**9-9** Föppl, A., *Vorlesungen über technische Mechanik*, Vol. III: Festigkeitslehre, B. G. Teubner, Leipzig, 1897.

*Note*: In this book, Föppl extended Clebsch's method for finding beam deflections. August Föppl, 1854–1924, was a German mathematician and engineer. He was a professor at the University of Leipzig and later at the Polytechnic Institute of Munich.

**9-10** Pilkey, W. D., "Clebsch's method for beam deflections," *Journal of Engineering Education*, vol. 54, no. 5, January 1964, pp. 170–174. This paper describes Clebsch's method and gives a very complete historical account, with many references.

**10-1** Zaslavsky, A., "Beams on immovable supports," *Publications of the International Association for Bridge and Structural Engineering*, Vol. 25, 1965, pp. 353–362.

**11-1** Euler, L., "Methodus inveniendi lineas curvas maximi minimive proprietate gaudentes . . . ," Appendix I, "De curvis elasticis," Bousquet, Lausanne and Geneva, 1744. (English translation: Oldfather, W. A., Ellis, C. A., and Brown, D. M., *Isis*, Vol. 20, 1933, pp. 72–160. Also, republished in *Leonhardi Euleri Opera Omnia*, series 1, Vol. 24, 1952.)

*Note:* Leonhard Euler (1707–1783) made many remarkable contributions to mathematics and mechanics, and he is considered by most mathematicians to be the most productive mathematician of all time. His name, pronounced "oiler," appears repeatedly in present-day textbooks; for instance, in mechanics we have Euler's equations of motion of a rigid body, Euler's angles, Euler's equations of fluid flow, the Euler load in column buckling, and much more; and in mathematics we encounter the famous Euler constant, as well as Euler's numbers, the Euler identity ($e^{i\theta} = \cos \theta + i \sin \theta$), Euler's formula ($e^{i\pi} + 1 = 0$), Euler's differential equation, Euler's

equation of a variational problem, Euler's quadrature formula, the Euler summation formula, Euler's theorem on homogeneous functions, Euler's integrals, and even Euler squares (square arrays of numbers possessing special properties).

In applied mechanics, Euler was the first to derive the formula for the critical buckling load of an ideal, slender column and the first to solve the problem of the elastica. This work was published in 1744, as cited previously. He dealt with a column that is fixed at the base and free at the top. Later, he extended his work on columns (Ref. 11-2). Euler's numerous books include treatises on celestial mechanics, dynamics, and hydromechanics, and his papers include subjects such as vibrations of beams and plates and statically indeterminate structures.

In the field of mathematics, Euler made outstanding contributions to trigonometry, algebra, number theory, differential and integral calculus, infinite series, analytic geometry, differential equations, calculus of variations, and many other subjects. He was the first to conceive of trigonometric values as the ratios of numbers and the first to present the famous equation $e^{i\theta} = \cos \theta + i \sin \theta$. Within his books on mathematics, all of which were classical references for many generations, we find the first development of the calculus of variations as well as such intriguing items as the proof of Fermat's "last theorem" for $n = 3$ and $n = 4$. Euler also solved the famous problem of the seven bridges of Königsberg, a problem of topology, another field in which he pioneered.

Euler was born near Basel, Switzerland, and attended the University of Basel, where he studied under John Bernoulli (1667–1748). From 1727 to 1741 he lived and worked in St. Petersburg, where he established a great reputation as a mathematician. In 1741 he moved to Berlin upon the invitation of Frederick the Great, King of Prussia. He continued his mathematical research in Berlin until the year 1766, when he returned to St. Petersburg at the request of Catherine II, Empress of Russia.

Euler continued to be prolific until his death in St. Petersburg at the age of 76; during this final period of his life he wrote more than 400 papers. In his entire lifetime, the number of books and papers written by Euler totaled 886; he left many manuscripts at his death and they continued to be published by the Russian Academy of Sciences in St. Petersburg for 47 years afterward. All this in spite of the fact that one of his eyes went blind in 1735 and the other in 1766. The story of Euler's life is told in Ref. 1-1, pp. 28–30, and some of his contributions to mechanics are described in Ref. 1-1, pp. 30–36 (see also Refs. 1-2, 1-3, 2-2, and 5-3).

**11-2** Euler, L., "Sur la force des colonnes," *Histoire de L'Académie Royale des Sciences et Belles Lettres*, 1757,

Leonhard Euler
(1707–1783)

published in *Memoires* of the Academie, Vol. 13, Berlin, 1759, pp. 252–282. (See Ref. 11-3 for a translation and discussion of this paper.)

**11-3** Van den Broek, J. A., "Euler's classic paper 'On the strength of columns,'" *American Journal of Physics*, Vol. 15, No. 4, July–August 1947, pp. 309–318.

**11-4** Keller, J. B., "The shape of the strongest column," *Archive for Rational Mechanics and Analysis*, Vol. 5, No. 4, 1960, pp. 275–285.

**11-5** Young, D. H., "Rational design of steel columns," *Transactions of the American Society of Civil Engineers*, Vol. 101, 1936, pp. 422–451. *Note:* Donovan Harold Young (1904–1980) was a well-known engineering educator. He was a professor at the University of Michigan and later at Stanford University. His five textbooks in the field of applied mechanics, written with S. P. Timoshenko, were translated into many languages and used throughout the world.

**11-6** Lamarle, A. H. E., "Mémoire sur la flexion du bois," *Annales des Travaux Publiques de Belgique*, Part 1, Vol. 3, 1845, pp. 1–64, and Part 2, Vol. 4, 1846, pp. 1–36. *Note:* Anatole Henri Ernest Lamarle (1806–1875) was an engineer and professor. He was born in Calais, studied in Paris, and became a professor at the University of Ghent, Belgium. For his work on columns, see Ref. 1-1, p. 208.

**11-7** Considère, A., "Résistance des pièces comprimées," *Congrès International des Procédés de Construction*, Paris, September 9–14, 1889, proceedings published by Librairie Polytechnique, Paris, Vol. 3, 1891, p. 371. *Note:* Armand Gabriel Considère (1841–1914) was a French engineer.

**11-8** Engesser, F., "Ueber die Knickfestigkeit gerader Stäbe," *Zeitschrift für Architektur und Ingenieurwesen*, Vol. 35, No. 4, 1889, pp. 455–462. *Note:* Friedrich Engesser (1848–1931) was a German railway and bridge engineer. Later, he became a professor at the Karlsruhe Polytechnic Institute, where he made important advances in the theory of structures, especially in buckling and energy methods. For his work on columns, see Ref. 1-1, pp. 292 and 297–299.

**11-9** Engesser, F., "Knickfragen," *Schweizerische Bauzeitung*, Vol. 25, No. 13, March 30, 1895, pp. 88–90.

**11-10** Jasinski, F., "Noch ein Wort zu den 'Knickfragen,'" *Schweizerische Bauzeitung*, Vol. 25, No. 25, June 22, 1895, pp. 172–175. *Note:* Félix S. Jasinski (1856–1899) was born in Warsaw and studied in Russia. He became a professor at the Institute of Engineers of Ways of Communication in St. Petersburg.

**11-11** Engesser, F., "Ueber Knickfragen," *Schweizerische Bauzeitung*, Vol. 26, No. 4, July 27, 1895, pp. 24–26.

**11-12** von Kármán, T., "Die Knickfestigkeit gerader Stäbe," *Physikalische Zeitschrift*, Vol. 9, No. 4, 1908, pp. 136–140 (this paper also appears in Vol. I of Ref. 11-14).

*Note:* Theodore von Kármán (1881–1963) was born in Hungary and later worked at the University of Göttingen in the field of aerodynamics. After coming to the United States in 1929, he founded the Jet Propulsion Laboratory and pioneered in aircraft and rocket problems. His research also included inelastic buckling of columns and stability of shells.

**11-13** von Kármán, T., "Untersuchungen über Knickfestigkeit," *Mitteilungen über Forschungsarbeiten auf dem Gebiete des Ingenieurwesens, Verein Deutscher Ingenieure*, Berlin, Heft 81, 1910 (this paper also appears in Ref. 11-14).

**11-14** *Collected Works of Theodore von Kármán*, Vols. I–IV, Butterworths Scientific Publications, London, 1956.

**11-15** Shanley, F. R., "The column paradox," *Journal of the Aeronautical Sciences*, Vol. 13, No. 12, December 1946, p. 678. *Note:* Francis Reynolds Shanley (1904–1968) was a professor of aeronautical engineering at the University of California, Los Angeles.

**11-16** Shanley, F. R., "Inelastic column theory," *ibid.*, Vol. 14, No. 5, May 1947, pp. 261–267.

**11-17** Hoff, N. J., "Buckling and Stability," The Forty-First Wilbur Wright Memorial Lecture, *Journal of the Royal Aeronautical Society*, Vol. 58, January 1954, pp. 3–52.

**11-18** Hoff, N. J., "The idealized column," *Ingenieur-Archiv*, Vol. 28, 1959 (Festschrift Richard Grammel), pp. 89–98.

**11-19** Johnston, B. G., "Column buckling theory: Historical highlights," *Journal of Structural Engineering, Structural Division, American Society of Civil Engineers*, Vol. 109, No. 9, September 1983, pp. 2086–2096.

# FE Exam Review Problems

The Fundamentals of Engineering (FE) examination [see *http://www.ncees .org/Exams.php*] is the first step on the path to registration as a Professional Engineer (P.E.). In its current form, the FE exam is an 8-hour exam consisting of 120 multiple choice questions in the 4-hour morning session, followed by 60 multiple choice questions in the 4-hour afternoon session. The exam is usually taken by recent graduates of accredited college engineering programs and covers a broad range of topics presented in their undergraduate courses. The afternoon portion is usually focused on questions related to the student's specific engineering subdiscipline (chemical, civil, electrical, environmental, industrial, mechanical, and "other").

In the past, approximately 10 to 15% of the questions have been based on principles presented in undergraduate courses in engineering mechanics. This appendix presents 106 FE-type review problems in *Mechanics of Materials*, many of which are based upon modifications of problems presented at the end of each chapter throughout this text. The problems cover all of the major topics presented in the text and are thought to be representative of those likely to appear on an FE exam. Most of these problems are in SI units, which is the system of units used on the FE Exam itself, and require use of an engineering calculator to carry out the solution. Each of the 106 problems is presented in the FE Exam format. The student must select from four available answers (A, B, C, or D), only one of which is the correct answer. The correct answer choices are listed in the Answers section at the back of this text, and the detailed solution for each problem is available for download on the student website. It is expected that careful review of these problems will serve as a useful guide to the student in preparing for this important examination.

**A-1.1:** A plane truss has downward applied load $P$ at joint 2 and another load $P$ applied leftward at joint 5. The force in member 3–5 is:

(A) 0

(B) $-P/2$

(C) $-P$

(D) $+1.5 P$

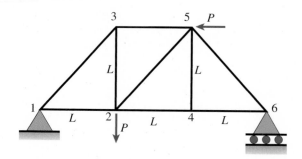

**A-1.2:** The force in member $FE$ of the plane truss below is approximately:

(A) $-1.5$ kN

(B) $-2.2$ kN

(C) $3.9$ kN

(D) $4.7$ kN

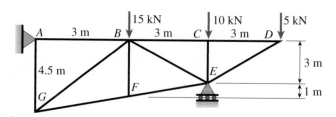

**A-1.3:** The moment reaction at $A$ in the plane frame below is approximately:

(A) $+1400$ N·m

(B) $-2280$ N·m

(C) $-3600$ N·m

(D) $+6400$ N·m

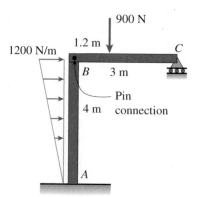

**A-1.4:** A hollow circular post $ABC$ (see figure) supports a load $P_1 = 16$ kN acting at the top. A second load $P_2$ is uniformly distributed around the cap plate at $B$. The diameters and thicknesses of the upper and lower parts of the post are $d_{AB} = 30$ mm, $t_{AB} = 12$ mm, $d_{BC} = 60$ mm, and $t_{BC} = 9$ mm, respectively. The lower part of the post must have the same compressive stress as the upper part. The required magnitude of the load $P_2$ is approximately:

(A)  18 kN
(B)  22 kN
(C)  28 kN
(D)  46 kN

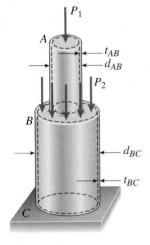

**A-1.5:** A circular aluminum tube of length $L = 650$ mm is loaded in compression by forces $P$. The outside and inside diameters are 80 mm and 68 mm, respectively. A strain gage on the outside of the bar records a normal strain in the longitudinal direction of $400 \times 10^{-6}$. The shortening of the bar is approximately:

(A)  0.12 mm
(B)  0.26 mm
(C)  0.36 mm
(D)  0.52 mm

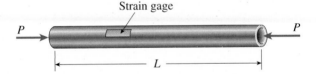

**A-1.6:** A steel plate weighing 27 kN is hoisted by a cable sling that has a clevis at each end. The pins through the clevises are 22 mm in diameter. Each half of the cable is at an angle of $35°$ to the vertical. The average shear stress in each pin is approximately:

(A)  22 MPa
(B)  28 MPa
(C)  40 MPa
(D)  48 MPa

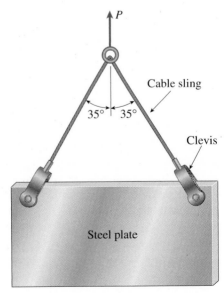

**A-1.7:** A steel wire hangs from a high-altitude balloon. The steel has unit weight 77 kN/m³ and yield stress of 280 MPa. The required factor of safety against yield is 2.0. The maximum permissible length of the wire is approximately:

(A) 1800 m
(B) 2200 m
(C) 2600 m
(D) 3000 m

**A-1.8:** An aluminum bar ($E = 72$ GPa, $v = 0.33$) of diameter 50 mm cannot exceed a diameter of 50.1 mm when compressed by axial force P. The maximum acceptable compressive load $P$ is approximately:

(A) 190 kN
(B) 200 kN
(C) 470 kN
(D) 860 kN

**A-1.9:** An aluminum bar ($E = 70$ GPa, $v = 0.33$) of diameter 20 mm is stretched by axial forces $P$, causing its diameter to decrease by 0.022 mm. The load $P$ is approximately:

(A) 73 kN
(B) 100 kN
(C) 140 kN
(D) 339 kN

**A-1.10:** An polyethylene bar ($E = 1.4$ GPa, $v = 0.4$) of diameter 80 mm is inserted in a steel tube of inside diameter 80.2 mm and then compressed by axial force $P$. The gap between steel tube and polyethylene bar will close when compressive load $P$ is approximately:

(A) 18 kN
(B) 25 kN
(C) 44 kN
(D) 60 kN

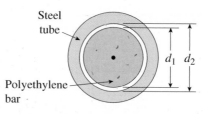

**A-1.11:** A pipe ($E = 110$ GPa) carries a load $P_1 = 120$ kN at $A$ and a uniformly distributed load $P_2 = 100$ kN on the cap plate at $B$. Initial pipe diameters and thicknesses are $d_{AB} = 38$ mm, $t_{AB} = 12$ mm, $d_{BC} = 70$ mm, and $t_{BC} = 10$ mm. Under loads $P_1$ and $P_2$, wall thickness $t_{BC}$ increases by 0.0036 mm. Poisson's ratio $v$ for the pipe material is approximately:

(A) 0.27
(B) 0.30
(C) 0.31
(D) 0.34

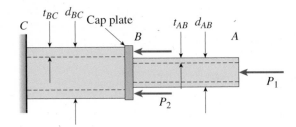

**A-1.12:** A titanium bar ($E$ = 100 GPa, $v$ = 0.33) with square cross section ($b$ = 75 mm) and length $L$ = 3.0 m is subjected to tensile load $P$ = 900 kN. The increase in volume of the bar is approximately:

(A) 1400 mm$^3$
(B) 3500 mm$^3$
(C) 4800 mm$^3$
(D) 9200 mm$^3$

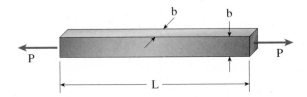

**A-1.13:** An elastomeric bearing pad is subjected to a shear force $V$ during a static loading test. The pad has dimensions $a$ = 150 mm and $b$ = 225 mm, and thickness $t$ = 55 mm. The lateral displacement of the top plate with respect to the bottom plate is 14 mm under a load $P$ = 16 kN. The shear modulus of elasticity $G$ of the elastomer is approximately:

(A) 1.0 MPa
(B) 1.5 MPa
(C) 1.7 MPa
(D) 1.9 MPa

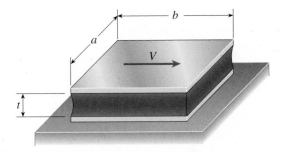

**A-1.14:** A bar of diameter $d$ = 18 mm and length $L$ = 0.75 m is loaded in tension by forces $P$. The bar has modulus $E$ = 45 GPa and allowable normal stress of 180 MPa. The elongation of the bar must not exceed 2.7 mm. The allowable value of forces $P$ is approximately:

(A) 41 kN
(B) 46 kN
(C) 56 kN
(D) 63 kN

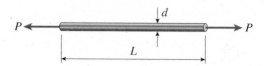

**A-1.15:** Two flanged shafts are connected by eight 18-mm bolts. The diameter of the bolt circle is 240 mm. The allowable shear stress in the bolts is 90 MPa. Ignore friction between the flange plates. The maximum value of torque $T_0$ is approximately:

(A)  19 kN·m
(B)  22 kN·m
(C)  29 kN·m
(D)  37 kN·m

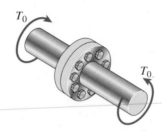

**A-1.16:** A copper tube with wall thickness of 8 mm must carry an axial tensile force of 175 kN. The allowable tensile stress is 90 MPa. The minimum required outer diameter is approximately:

(A)  60 mm
(B)  72 mm
(C)  85 mm
(D)  93 mm

**A-2.1:** Two wires, one copper and the other steel, of equal length stretch the same amount under an applied load $P$. The moduli of elasticity for each is $E_s = 210$ GPa and $E_c = 120$ GPa. The ratio of the diameter of the copper wire to that of the steel wire is approximately:

(A)  1.00
(B)  1.08
(C)  1.19
(D)  1.32

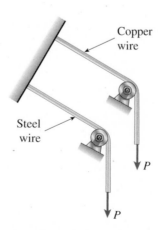

**A-2.2:** A plane truss with span length $L = 4.5$ m is constructed using cast iron pipes ($E = 170$ GPa) with a cross-sectional area of 4500 mm$^2$. The displacement of joint $B$ cannot exceed 2.7 mm. The maximum value of loads $P$ is approximately:

(A) 340 kN
(B) 460 kN
(C) 510 kN
(D) 600 kN

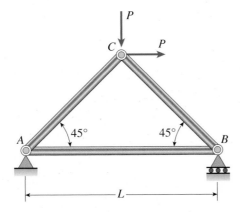

**A-2.3:** A brass rod ($E = 110$ GPa) with a cross-sectional area of 250 mm$^2$ is loaded by forces $P_1 = 15$ kN, $P_2 = 10$ kN, and $P_3 = 8$ kN. Segment lengths of the bar are $a = 2.0$ m, $b = 0.75$ m, and $c = 1.2$ m. The change in length of the bar is approximately:

(A) 0.9 mm
(B) 1.6 mm
(C) 2.1 mm
(D) 3.4 mm

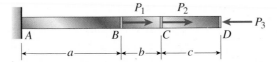

**A-2.4:** A brass bar ($E = 110$ MPa) of length $L = 2.5$ m has diameter $d_1 = 18$ mm over one-half of its length and diameter $d_2 = 12$ mm over the other half. Compare this nonprismatic bar to a prismatic bar of the same volume of material with constant diameter $d$ and length $L$. The elongation of the prismatic bar under the same load $P = 25$ kN is approximately:

(A) 3 mm
(B) 4 mm
(C) 5 mm
(D) 6 mm

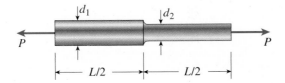

**A-2.5:** A nonprismatic cantilever bar has an internal cylindrical hole of diameter $d/2$ from 0 to $x$, so the net area of the cross section for Segment 1 is $(3/4)A$. Load $P$ is applied at $x$, and load $-P/2$ is applied at $x = L$. Assume that $E$ is constant. The length of the hollow segment, $x$, required to obtain axial displacement $\delta = PL/EA$ at the free end is:

(A) $x = L/5$
(B) $x = L/4$
(C) $x = L/3$
(D) $x = 3L/5$

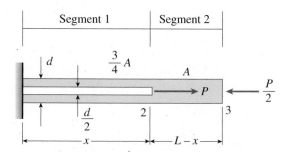

**A-2.6:** A nylon bar ($E = 2.1$ GPa) with diameter 12 mm, length 4.5 m, and weight 5.6 N hangs vertically under its own weight. The elongation of the bar at its free end is approximately:

(A) 0.05 mm
(B) 0.07 mm
(C) 0.11 mm
(D) 0.17 mm

**A-2.7:** A monel shell ($E_m = 170$ GPa, $d_3 = 12$ mm, $d_2 = 8$ mm) encloses a brass core ($E_b = 96$ GPa, $d_1 = 6$ mm). Initially, both shell and core are a length of 100 mm. A load $P$ is applied to both shell and core through a cap plate. The load $P$ required to compress both shell and core by 0.10 mm is approximately:

(A) 10.2 kN
(B) 13.4 kN
(C) 18.5 kN
(D) 21.0 kN

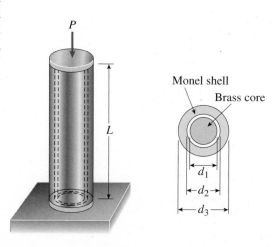

A-2.8: A steel rod ($E_s$ = 210 GPa, $d_r$ = 12 mm, $cte_s$ = 12 × $10^{-6}$/°C) is held stress-free between rigid walls by a clevis and pin ($d_p$ = 15 mm) assembly at each end. If the allowable shear stress in the pin is 45 MPa and the allowable normal stress in the rod is 70 MPa, the maximum permissible temperature drop $\Delta T$ is approximately:

(A)  14 °C
(B)  20 °C
(C)  28 °C
(D)  40 °C

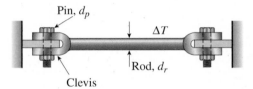

Pin, $d_p$

$\Delta T$

Rod, $d_r$

Clevis

A-2.9:    A    threaded    steel    rod    ($E_s$ = 210 GPa, $d_r$ = 15 mm, $cte_s$ = 12 × $10^{-6}$/°C.) is held stress-free between rigid walls by a nut and washer ($d_w$ = 22 mm) assembly at each end. If the allowable bearing stress between the washer and wall is 55 MPa and the allowable normal stress in the rod is 90 MPa, the maximum permissible temperature drop $\Delta T$ is approximately:

(A)  25 °C
(B)  30 °C
(C)  38 °C
(D)  46 °C

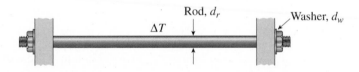

Rod, $d_r$

Washer, $d_w$

$\Delta T$

A-2.10: A steel bolt (area = 130 mm², $E_s$ = 210 GPa) is enclosed by a copper tube (length = 0.5 m, area = 400 mm², $E_c$ = 110 GPa) and the end nut is turned until it is just snug. The pitch of the bolt threads is 1.25 mm. The bolt is now tightened by a quarter turn of the nut. The resulting stress in the bolt is approximately:

(A)  56 MPa
(B)  62 MPa
(C)  74 MPa
(D)  81 MPa

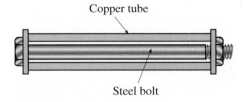

Copper tube

Steel bolt

**A-2.11:** A steel bar of rectangular cross section ($a = 38$ mm, $b = 50$ mm) carries a tensile load $P$. The allowable stresses in tension and shear are 100 MPa and 48 MPa, respectively. The maximum permissible load $P_{max}$ is approximately:

(A) 56 kN
(B) 62 kN
(C) 74 kN
(D) 91 kN

**A-2.12:** A brass wire ($d = 2.0$ mm, $E = 110$ GPa) is pretensioned to $T = 85$ N. The coefficient of thermal expansion for the wire is $19.5 \times 10^{-6}/°C$. The temperature change at which the wire goes slack is approximately:

(A) $+5.7 °C$
(B) $-12.6 °C$
(C) $+12.6 °C$
(D) $-18.2 °C$

**A-2.13:** A copper bar ($d = 10$ mm, $E = 110$ GPa) is loaded by tensile load $P = 11.5$ kN. The maximum shear stress in the bar is approximately:

(A) 73 MPa
(B) 87 MPa
(C) 145 MPa
(D) 150 MPa

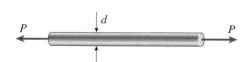

**A-2.14:** A steel plane truss is loaded at $B$ and $C$ by forces $P = 200$ kN. The cross-sectional area of each member is $A = 3970$ mm². Truss dimensions are $H = 3$ m and $L = 4$ m. The maximum shear stress in bar $AB$ is approximately:

(A) 27 MPa
(B) 33 MPa
(C) 50 MPa
(D) 69 MPa

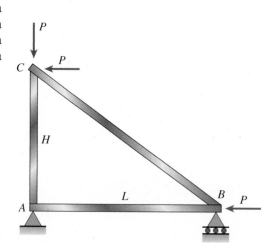

**A-2.15:** A plane stress element on a bar in uniaxial stress has a tensile stress of $\sigma_\theta = 78$ MPa (see fig.). The maximum shear stress in the bar is approximately:

(A) 29 MPa
(B) 37 MPa
(C) 50 MPa
(D) 59 MPa

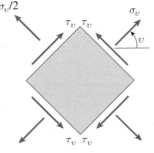

**A-2.16:** A prismatic bar (diameter $d_0 = 18$ mm) is loaded by force $P_1$. A stepped bar (diameters $d_1 = 20$ mm and $d_2 = 25$ mm with radius of fillets $R = 2$ mm) is loaded by force $P_2$. The allowable axial stress in the material is 75 MPa. The ratio $P_1/P_2$ of the maximum permissible loads that can be applied to the bars, considering stress concentration effects in the stepped bar, is:

(A) 0.9
(B) 1.2
(C) 1.4
(D) 2.1

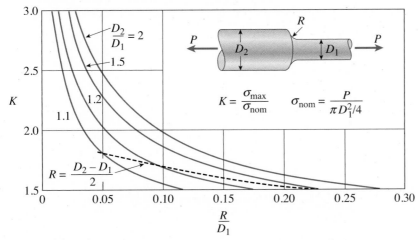

**FIG. 2-66** Stress-concentration factor $K$ for round bars with shoulder fillets. The dashed line is for a full quarter-circular fillet.

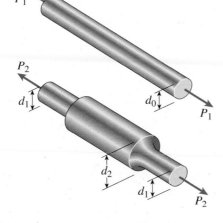

**A-3.1:** A brass rod of length $L = 0.75$ m is twisted by torques $T$ until the angle of rotation between the ends of the rod is 3.5°. The allowable shear strain in the copper is 0.0005 rad. The maximum permissible diameter of the rod is approximately:

(A) 6.5 mm
(B) 8.6 mm
(C) 9.7 mm
(D) 12.3 mm

**A-3.2:** The angle of rotation between the ends of a nylon bar is 3.5°. The bar diameter is 70 mm and the allowable shear strain is 0.014 rad. The minimum permissible length of the bar is approximately:

(A) 0.15 m
(B) 0.27 m
(C) 0.40 m
(D) 0.55 m

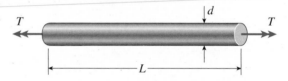

**A-3.3:** A brass bar twisted by torques $T$ acting at the ends has the following properties: $L = 2.1$ m, $d = 38$ mm, and $G = 41$ GPa. The torsional stiffness of the bar is approximately:

(A) 1200 N·m
(B) 2600 N·m
(C) 4000 N·m
(D) 4800 N·m

**A-3.4:** A brass pipe is twisted by torques $T = 800$ N·m acting at the ends causing an angle of twist of 3.5°. The pipe has the following properties: $L = 2.1$ m, $d_1 = 38$ mm, and $d_2 = 56$ mm. The shear modulus of elasticity $G$ of the pipe is approximately:

(A) 36.1 GPa
(B) 37.3 GPa
(C) 38.7 GPa
(D) 40.6 GPa

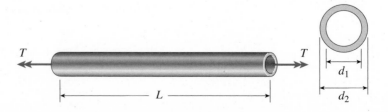

**A-3.5:** An aluminum bar of diameter $d = 52$ mm is twisted by torques $T_1$ at the ends. The allowable shear stress is 65 MPa. The maximum permissible torque $T_1$ is approximately:

(A)  1450 N·m
(B)  1675 N·m
(C)  1710 N·m
(D)  1800 N·m

**A-3.6:** A steel tube with diameters $d_2 = 86$ mm and $d_1 = 52$ mm is twisted by torques at the ends. The diameter of a solid steel shaft that resists the same torque at the same maximum shear stress is approximately:

(A)  56 mm
(B)  62 mm
(C)  75 mm
(D)  82 mm

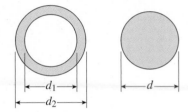

**A-3.7:** A stepped steel shaft with diameters $d_1 = 56$ mm and $d_2 = 52$ mm is twisted by torques $T_1 = 3.5$ kN·m and $T_2 = 1.5$ kN·m acting in opposite directions. The maximum shear stress is approximately:

(A)  54 MPa
(B)  58 MPa
(C)  62 MPa
(D)  79 MPa

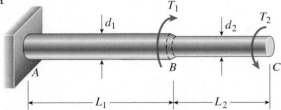

**A-3.8:** A stepped steel shaft ($G = 75$ GPa) with diameters $d_1 = 36$ mm and $d_2 = 32$ mm is twisted by torques $T$ at each end. Segment lengths are $L_1 = 0.9$ m and $L_2 = 0.75$ m. If the allowable shear stress is 28 MPa and maximum allowable twist is 1.8°, the maximum permissible torque is approximately:

(A)  142 N·m
(B)  180 N·m
(C)  185 N·m
(D)  257 N·m

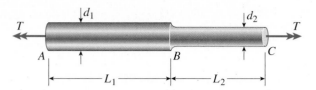

A-3.9: A gear shaft transmits torques $T_A = 975 \text{ N} \cdot \text{m}$, $T_B = 1500 \text{ N} \cdot \text{m}$, $T_C = 650 \text{ N} \cdot \text{m}$, and $T_D = 825 \text{ N} \cdot \text{m}$. If the allowable shear stress is 50 MPa, the required shaft diameter is approximately:

(A) 38 mm
(B) 44 mm
(C) 46 mm
(D) 48 mm

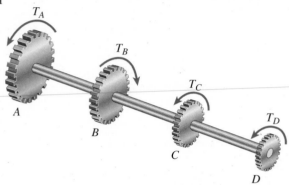

A-3.10: A hollow aluminum shaft ($G = 27 \text{ GPa}$, $d_2 = 96 \text{ mm}$, and $d_1 = 52 \text{ mm}$) has an angle of twist per unit length of 1.8°/m due to torques $T$. The resulting maximum tensile stress in the shaft is approximately:

(A) 38 MPa
(B) 41 MPa
(C) 49 MPa
(D) 58 MPa

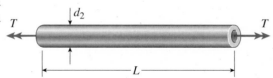

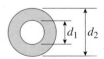

A-3.11: Torques $T = 5.7 \text{ kN} \cdot \text{m}$ are applied to a hollow aluminum shaft ($G = 27 \text{ GPa}$ and $d_1 = 52 \text{ mm}$). The allowable shear stress is 45 MPa and the allowable normal strain is $8.0 \times 10^{-4}$. The required outside diameter $d_2$ of the shaft is approximately:

(A) 38 mm
(B) 56 mm
(C) 87 mm
(D) 91 mm

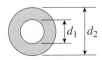

**A-3.12:** A motor drives a shaft with diameter $d = 46$ mm at $f = 5.25$ Hz and delivers $P = 25$ kW of power. The maximum shear stress in the shaft is approximately:

(A) 32 MPa
(B) 40 MPa
(C) 83 MPa
(D) 91 MPa

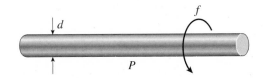

**A-3.13:** A motor drives a shaft at $f = 10$ Hz and delivers $P = 35$ kW of power. The allowable shear stress in the shaft is 45 MPa. The minimum diameter of the shaft is approximately:

(A) 35 mm
(B) 40 mm
(C) 47 mm
(D) 61 mm

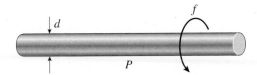

**A-3.14:** A drive shaft running at 2500 rpm has outer diameter 60 mm and inner diameter 40 mm. The allowable shear stress in the shaft is 35 MPa. The maximum power that can be transmitted is approximately:

(A) 220 kW
(B) 240 kW
(C) 288 kW
(D) 312 kW

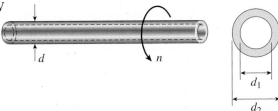

**A-3.15:** A prismatic shaft (diameter $d_0 = 19$ mm) is loaded by torque $T_1$. A stepped shaft (diameters $d_1 = 20$ mm and $d_2 = 25$ mm with radius of fillets $R = 2$ mm) is loaded by torque $T_2$. The allowable shear stress in the material is 42 MPa. The ratio $T_1/T_2$ of the maximum permissible torques that can be applied to the shafts, considering stress concentration effects in the stepped shaft is:

(A) 0.9
(B) 1.2
(C) 1.4
(D) 2.1

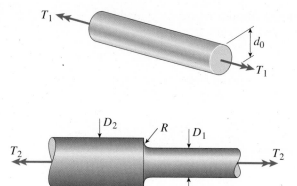

**FIG. 3-59** Stress-concentration factor $K$ for a stepped shaft in torsion. (The dashed line is for a full quarter-circular fillet.)

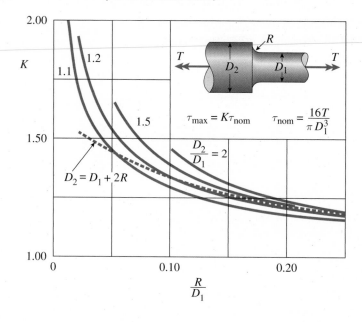

$$\tau_{max} = K\tau_{nom} \qquad \tau_{nom} = \frac{16T}{\pi D_1^3}$$

$$\frac{D_2}{D_1} = 2$$

$$D_2 = D_1 + 2R$$

**A-4.1:** A simply supported beam with proportional loading ($P = 4.1$ kN) has span length $L = 5$ m. Load $P$ is 1.2 m from support $A$ and load $2P$ is 1.5 m from support $B$. The bending moment just left of load $2P$ is approximately:

(A)  5.7 kN·m
(B)  6.2 kN·m
(C)  9.1 kN·m
(D)  10.1 kN·m

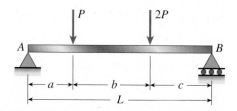

**A-4.2:** A simply supported beam is loaded as shown in the figure. The bending moment at point *C* is approximately:

(A)  5.7 kN·m
(B)  6.1 kN·m
(C)  6.8 kN·m
(D)  9.7 kN·m

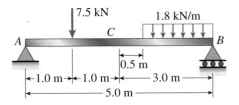

**A-4.3:** A cantilever beam is loaded as shown in the figure. The bending moment at 0.5 m from the support is approximately:

(A)  12.7 kN·m
(B)  14.2 kN·m
(C)  16.1 kN·m
(D)  18.5 kN·m

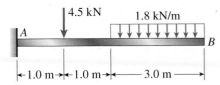

**A-4.4:** An L-shaped beam is loaded as shown in the figure. The bending moment at the midpoint of span *AB* is approximately:

(A)  6.8 kN·m
(B)  10.1 kN·m
(C)  12.3 kN·m
(D)  15.5 kN·m

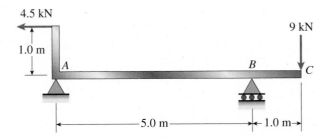

**A-4.5:** A T-shaped simple beam has a cable with force *P* anchored at *B* and passing over a pulley at *E*, as shown in the figure. The bending moment just left of *C* is 1.25 kN·m. The cable force *P* is approximately:

(A)  2.7 kN
(B)  3.9 kN
(C)  4.5 kN
(D)  6.2 kN

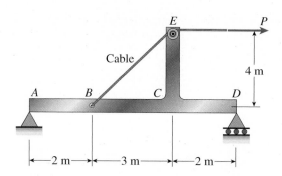

**A-4.6:** A simple beam ($L = 9$ m) with attached bracket *BDE* has force $P = 5$ kN applied downward at *E*. The bending moment just right of *B* is approximately:

(A) 6 kN·m
(B) 10 kN·m
(C) 19 kN·m
(D) 22 kN·m

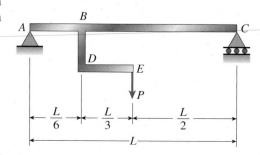

**A-4.7:** A simple beam *AB* with an overhang *BC* is loaded as shown in the figure. The bending moment at the midspan of *AB* is approximately:

(A) 8 kN·m
(B) 12 kN·m
(C) 17 kN·m
(D) 21 kN·m

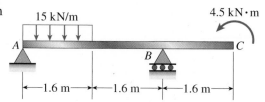

**A-5.1:** A copper wire ($d = 1.5$ mm) is bent around a tube of radius $R = 0.6$ m. The maximum normal strain in the wire is approximately:

(A) $1.25 \times 10^{-3}$
(B) $1.55 \times 10^{-3}$
(C) $1.76 \times 10^{-3}$
(D) $1.92 \times 10^{-3}$

**A-5.2:** A simply supported wood beam ($L = 5$ m) with rectangular cross section ($b = 200$ mm, $h = 280$ mm) carries uniform load $q = 6.5$ kN/m includes the weight of the beam. The maximum flexural stress is approximately:

(A) 8.7 MPa
(B) 10.1 MPa
(C) 11.4 MPa
(D) 14.3 MPa

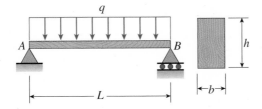

**A-5.3:** A cast iron pipe ($L = 12$ m, weight density $= 72$ kN/m$^3$, $d_2 = 100$ mm, and $d_1 = 75$ mm) is lifted by a hoist. The lift points are 6 m apart. The maximum bending stress in the pipe is approximately:

(A) 28 MPa
(B) 33 MPa
(C) 47 MPa
(D) 59 MPa

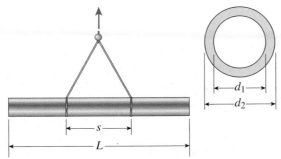

**A-5.4:** A beam with an overhang is loaded by a uniform load of 3 kN/m over its entire length. Moment of inertia $I_z = 3.36 \times 10^6$ mm$^4$ and distances to top and bottom of the beam cross section are 20 mm and 66.4 mm, respectively. It is known that reactions at $A$ and $B$ are 4.5 kN and 13.5 kN, respectively. The maximum bending stress in the beam is approximately:

(A) 36 MPa
(B) 67 MPa
(C) 102 MPa
(D) 119 MPa

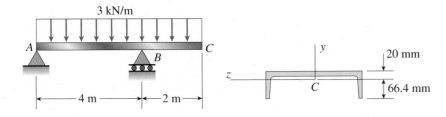

**A-5.5:** A steel hanger with solid cross section has horizontal force $P = 5.5\,kN$ applied at free end $D$. Dimension variable $b = 175\,mm$ and allowable normal stress is 150 MPa. Neglect self-weight of the hanger. The required diameter of the hanger is approximately:

(A) 5 cm
(B) 7 cm
(C) 10 cm
(D) 13 cm

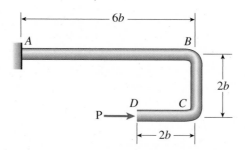

**A-5.6:** A cantilever wood pole carries force $P = 300\,N$ applied at its free end, as well as its own weight (weight density $= 6\,kN/m^3$). The length of the pole is $L = 0.75\,m$ and the allowable bending stress is 14 MPa. The required diameter of the pole is approximately:

(A) 4.2 cm
(B) 5.5 cm
(C) 6.1 cm
(D) 8.5 cm

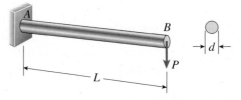

**A-5.7:** A simply supported steel beam of length $L = 1.5\,m$ and rectangular cross section ($h = 75\,mm$, $b = 20\,mm$) carries a uniform load of $q = 48\,kN/m$ that includes its own weight. The maximum transverse shear stress on the cross section at 0.25 m from the left support is approximately:

(A) 20 MPa
(B) 24 MPa
(C) 30 MPa
(D) 36 MPa

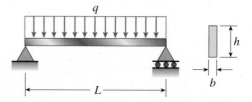

**A-5.8:** A simply supported laminated beam of length $L = 0.5$ m and square cross section weighs 4.8 N. Three strips are glued together to form the beam, with the allowable shear stress in the glued joint equal to 0.3 MPa. Considering also the weight of the beam, the maximum load $P$ that can be applied at $L/3$ from the left support is approximately:

(A) 240 N
(B) 360 N
(C) 434 N
(D) 510 N

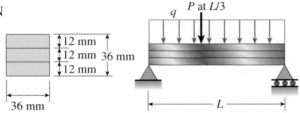

**A-5.9:** An aluminum cantilever beam of length $L = 0.65$ m carries a distributed load, which includes its own weight, of intensity $q/2$ at $A$ and $q$ at $B$. The beam cross section has a width of 50 mm and height of 170 mm. Allowable bending stress is 95 MPa and allowable shear stress is 12 MPa. The permissible value of load intensity $q$ is approximately:

(A) 110 kN/m
(B) 122 kN/m
(C) 130 kN/m
(D) 139 kN/m

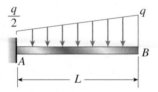

**A-5.10:** An aluminum light pole weighs 4300 N and supports an arm of weight 700 N, with the arm center of gravity at 1.2 m left of the centroidal axis of the pole. A wind force of 1500 N acts to the right at 7.5 m above the base. The pole cross section at the base has an outside diameter of 235 mm and thickness of 20 mm. The maximum compressive stress at the base is approximately:

(A) 16 MPa
(B) 18 MPa
(C) 21 MPa
(D) 24 MPa

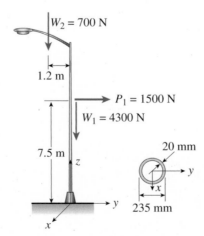

**A-5.11:** Two thin cables, each having a diameter of $d = t/6$ and carrying tensile loads $P$, are bolted to the top of a rectangular steel block with cross-sectional dimensions $b \times t$. The ratio of the maximum tensile to compressive stress in the block due to loads P is:

(A) 1.5
(B) 1.8
(C) 2.0
(D) 2.5

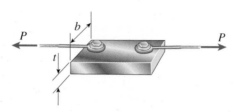

**A-5.12:** A rectangular beam with semicircular notches has dimensions $h = 160$ mm and $h_1 = 140$ mm. The maximum allowable bending stress in the plastic beam is $\sigma_{max} = 6.5$ MPa, and the bending moment is $M = 185$ N·m. The minimum permissible width of the beam is:

(A) 12 mm
(B) 20 mm
(C) 28 mm
(D) 32 mm

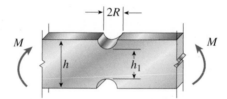

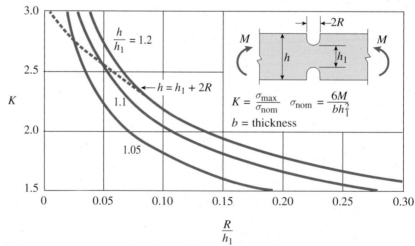

FIG. 5-50 Stress-concentration factor $K$ for a notched beam of rectangular cross section in pure bending ($h$ = height of beam; $b$ = thickness of beam, perpendicular to the plane of the figure). The dashed line is for semicircular notches ($h = h_1 + 2R$).

**A-6.1:** A composite beam is made up of a 200 mm × 300 mm core ($E_c$ = 14 GPa) and an exterior cover sheet (300 mm × 12 mm, $E_e$ = 100 GPa) on each side. Allowable stresses in core and exterior sheets are 9.5 MPa and 140 MPa, respectively. The ratio of the maximum permissible bending moment about the z axis to that about the y axis is most nearly:

(A) 0.5
(B) 0.7
(C) 1.2
(D) 1.5

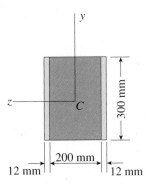

**A-6.2:** A composite beam is made up of a 90 mm × 160 mm wood beam ($E_w$ = 11 GPa) and a steel bottom cover plate (90 mm × 8 mm, $E_s$ = 190 GPa). Allowable stresses in wood and steel are 6.5 MPa and 110 MPa, respectively. The allowable bending moment about the z axis of the composite beam is most nearly:

(A) 2.9 kN·m
(B) 3.5 kN·m
(C) 4.3 kN·m
(D) 9.9 kN·m

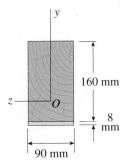

**A-6.3:** A steel pipe ($d_3$ = 104 mm, $d_2$ = 96 mm) has a plastic liner with inner diameter $d_1$ = 82 mm. The modulus of elasticity of the steel is 75 times that of the modulus of the plastic. Allowable stresses in steel and plastic are 40 MPa and 550 kPa, respectively. The allowable bending moment for the composite pipe is approximately:

(A) 1100 N·m
(B) 1230 N·m
(C) 1370 N·m
(D) 1460 N·m

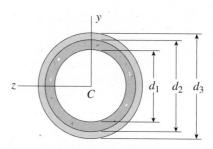

**A-6.4:** A bimetallic beam of aluminum ($E_a$ = 70 GPa) and copper ($E_c$ = 110 GPa) strips has a width of $b$ = 25 mm; each strip has a thickness $t$ = 1.5 mm. A bending moment of 1.75 N·m is applied about the $z$ axis. The ratio of the maximum stress in the aluminum to that in the copper is approximately:

(A) 0.6
(B) 0.8
(C) 1.0
(D) 1.5

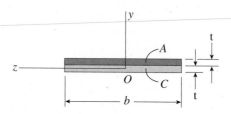

**A-6.5:** A composite beam of aluminum ($E_a$ = 72 GPa) and steel ($E_s$ = 190 GPa) has a width $b$ = 25 mm and heights $h_a$ = 42 mm and $h_s$ = 68 mm, respectively. A bending moment is applied about the $z$ axis resulting in a maximum stress in the aluminum of 55 MPa. The maximum stress in the steel is approximately:

(A) 86 MPa
(B) 90 MPa
(C) 94 MPa
(D) 98 MPa

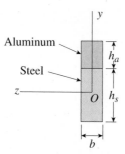

**A-7.1:** A rectangular plate ($a$ = 120 mm, $b$ = 160 mm) is subjected to compressive stress $\sigma_x$ = −4.5 MPa and tensile stress $\sigma_y$ = 15 MPa. The ratio of the normal stress acting perpendicular to the weld to the shear stress acting along the weld is approximately:

(A) 0.27
(B) 0.54
(C) 0.85
(D) 1.22

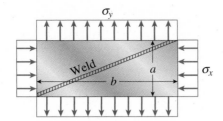

**A-7.2:** A rectangular plate in plane stress is subjected to normal stresses $\sigma_x$ and $\sigma_y$ and shear stress $\tau_{xy}$. Stress $\sigma_x$ is known to be 15 MPa, but $\sigma_y$ and $\tau_{xy}$ are unknown. However, the normal stress is known to be 33 MPa at counterclockwise angles of 35° and 75° from the $x$ axis. Based on this, the normal stress $\sigma_y$ on the element in the figure is approximately:

(A)  14 MPa
(B)  21 MPa
(C)  26 MPa
(D)  43 MPa

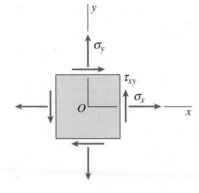

**A-7.3:** A rectangular plate in plane stress is subjected to normal stresses $\sigma_x = 35$ MPa, $\sigma_y = 26$ MPa, and shear stress $\tau_{xy} = 14$ MPa. The ratio of the magnitudes of the principal stresses $(\sigma_1/\sigma_2)$ is approximately:

(A)  0.8
(B)  1.5
(C)  2.1
(D)  2.9

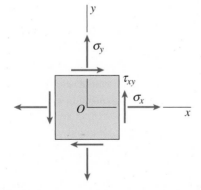

**A-7.4:** A drive shaft resists torsional shear stress of 45 MPa and axial compressive stress of 100 MPa. The ratio of the magnitudes of the principal stresses $(\sigma_1/\sigma_2)$ is approximately:

(A)  0.15
(B)  0.55
(C)  1.2
(D)  1.9

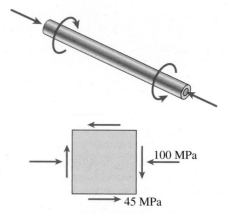

A-7.5: A drive shaft resists torsional shear stress of 45 MPa and axial compressive stress of 100 MPa. The maximum shear stress is approximately:

(A) 42 MPa
(B) 67 MPa
(C) 71 MPa
(D) 93 MPa

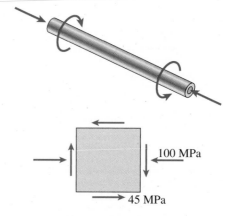

A-7.6: A drive shaft resists torsional shear stress of $\tau_{xy} = 40$ MPa and axial compressive stress $\sigma_x = -70$ MPa. One principal normal stress is known to be 38 MPa (tensile). The stress $\sigma_y$ is approximately:

(A) 23 MPa
(B) 35 MPa
(C) 62 MPa
(D) 75 MPa

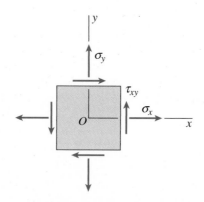

**A-7.7:** A cantilever beam with rectangular cross section ($b = 95$ mm, $h = 300$ mm) supports load $P = 160$ kN at its free end. The ratio of the magnitudes of the principal stresses ($\sigma_1/\sigma_2$) at point $A$ (at distance $c = 0.8$ m from the free end and distance $d = 200$ mm up from the bottom) is approximately:

(A) 5
(B) 12
(C) 18
(D) 25

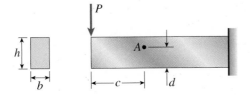

**A-7.8:** A simply supported beam ($L = 4.5$ m) with rectangular cross section ($b = 95$ mm, $h = 280$ mm) supports uniform load $q = 25$ kN/m. The ratio of the magnitudes of the principal stresses ($\sigma_1/\sigma_2$) at a point $a = 1.0$ m from the left support and distance $d = 100$ mm up from the bottom of the beam is approximately:

(A) 9
(B) 17
(C) 31
(D) 41

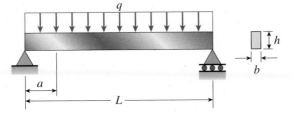

**A-8.1:** A thin-walled spherical tank with a diameter of 1.5 m and wall thickness of 65 mm has an internal pressure of 20 MPa. The maximum shear stress in the wall of the tank is approximately:

(A) 58 MPa
(B) 67 MPa
(C) 115 MPa
(D) 127 MPa

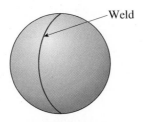

**A-8.2:** A thin-walled spherical tank has a diameter of 0.75 m and an internal pressure of 20 MPa. The yield stress in tension is 920 MPa, the yield stress in shear is 475 MPa, and the factor of safety is 2.5. The modulus of elasticity is 210 GPa, Poisson's ratio is 0.28, and maximum normal strain is $1220 \times 10^{-6}$. The minimum permissible thickness of the tank is approximately:

(A) 8.6 mm
(B) 9.9 mm
(C) 10.5 mm
(D) 11.1 mm

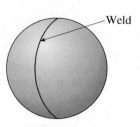

Weld

**A-8.3:** A thin-walled cylindrical tank with a diameter of 200 mm has an internal pressure of 11 MPa. The yield stress in tension is 250 MPa, the yield stress in shear is 140 MPa, and the factor of safety is 2.5. The minimum permissible thickness of the tank is approximately:

(A)  8.2 mm
(B)  9.1 mm
(C)  9.8 mm
(D)  11.0 mm

**A-8.4:** A thin-walled cylindrical tank with a diameter of 2.0 m and wall thickness of 18 mm is open at the top. The height $h$ of water (weight density = 9.81 kN/m³) in the tank at which the circumferential stress reaches 10 MPa in the tank wall is approximately:

(A)  14 m
(B)  18 m
(C)  20 m
(D)  24 m

**A-8.5:** The pressure relief valve is opened on a thin-walled cylindrical tank, with the radius-to-wall thickness ratio of 128, thereby decreasing the longitudinal strain by $150 \times 10^{-6}$. Assume $E = 73$ GPa and $v = 0.33$. The original internal pressure in the tank was approximately:

(A)  370 kPa
(B)  450 kPa
(C)  500 kPa
(D)  590 kPa

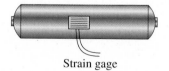

Strain gage

**A-8.6:** A cylindrical tank is assembled by welding steel sections circumferentially. Tank diameter is 1.5 m, thickness is 20 mm, and internal pressure is 2.0 MPa. The maximum stress in the heads of the tank is approximately:

(A)  38 MPa
(B)  45 MPa
(C)  50 MPa
(D)  59 MPa

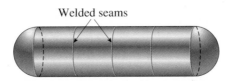

Welded seams

**A-8.7:** A cylindrical tank is assembled by welding steel sections circumferentially. Tank diameter is 1.5 m, thickness is 20 mm, and internal pressure is 2.0 MPa. The maximum tensile stress in the cylindrical part of the tank is approximately:

(A)  45 MPa
(B)  57 MPa
(C)  62 MPa
(D)  75 MPa

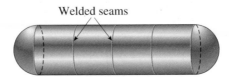

Welded seams

**A-8.8:** A cylindrical tank is assembled by welding steel sections circumferentially. Tank diameter is 1.5 m, thickness is 20 mm, and internal pressure is 2.0 MPa. The maximum tensile stress perpendicular to the welds is approximately:

(A)  22 MPa
(B)  29 MPa
(C)  33 MPa
(D)  37 MPa

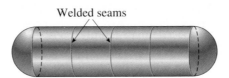

Welded seams

**A-8.9:** A cylindrical tank is assembled by welding steel sections circumferentially. Tank diameter is 1.5 m, thickness is 20 mm, and internal pressure is 2.0 MPa. The maximum shear stress in the heads is approximately:

(A)  19 MPa
(B)  23 MPa
(C)  33 MPa
(D)  35 MPa

**A-9.1:** An aluminum beam ($E = 72$ GPa) with a square cross section and span length $L = 2.5$ m is subjected to uniform load $q = 1.5$ kN/m. The allowable bending stress is 60 MPa. The maximum deflection of the beam is approximately:

(A) 10 mm
(B) 16 mm
(C) 22 mm
(D) 26 mm

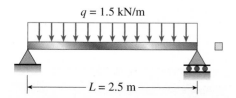

**A-9.2:** An aluminum cantilever beam ($E = 72$ GPa) with a square cross section and span length $L = 2.5$ m is subjected to uniform load $q = 1.5$ kN/m. The allowable bending stress is 55 MPa. The maximum deflection of the beam is approximately:

(A) 10 mm
(B) 20 mm
(C) 30 mm
(D) 40 mm

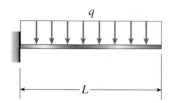

**A-9.3:** A steel beam ($E = 210$ GPa) with $I = 119 \times 10^6$ mm$^4$ and span length $L = 3.5$ m is subjected to uniform load $q = 9.5$ kN/m. The maximum deflection of the beam is approximately:

(A) 10 mm
(B) 13 mm
(C) 17 mm
(D) 19 mm

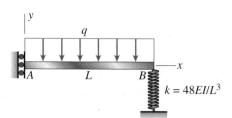

**A-9.4:** A steel bracket $ABC$ ($EI = 4.2 \times 10^6$ N $\cdot$ m$^2$) with span length $L = 4.5$ m and height $H = 2$ m is subjected to load $P = 15$ kN at $C$. The maximum rotation of joint $B$ is approximately:

(A) 0.1°
(B) 0.3°
(C) 0.6°
(D) 0.9°

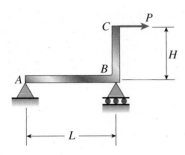

**A-9.5:** A steel bracket $ABC$ ($EI = 4.2 \times 10^6 \text{ N·m}^2$) with span length $L = 4.5$ m and height $H = 2$ m is subjected to load $P = 15$ kN at $C$. The maximum horizontal displacement of joint $C$ is approximately:

(A) 22 mm
(B) 31 mm
(C) 38 mm
(D) 40 mm

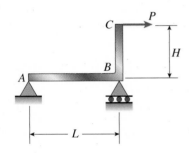

**A-9.6:** A nonprismatic cantilever beam of one material is subjected to load $P$ at its free end. Moment of inertia $I_2 = 2I_1$. The ratio $r$ of the deflection $\delta_B$ to the deflection $\delta_1$ at the free end of a prismatic cantilever with moment of inertia $I_1$ carrying the same load is approximately:

(A) 0.25
(B) 0.40
(C) 0.56
(D) 0.78

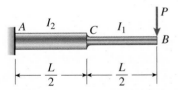

**A-9.7:** A steel bracket $ABCD$ ($EI = 4.2 \times 10^6 \text{ N·m}^2$), with span length $L = 4.5$ m and dimension $a = 2$ m, is subjected to load $P = 10$ kN at $D$. The maximum deflection at $B$ is approximately:

(A) 10 mm
(B) 14 mm
(C) 19 mm
(D) 24 mm

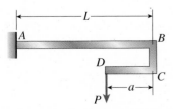

**A-10.1:** Propped cantilever beam $AB$ has moment $M_1$ applied at joint $B$. Framework $ABC$ has moment $M_2$ applied at $C$. Both structures have constant flexural rigidity $EI$. If the ratio of the *applied moments* $M_1/M_2 = 3/2$, the ratio of the *reactive moments* $M_{A1}/M_{A2}$ at clamped support $A$ is approximately:

(A) 1
(B) 3/2
(C) 2
(D) 5/2

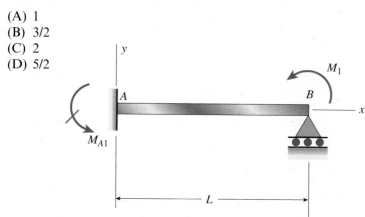

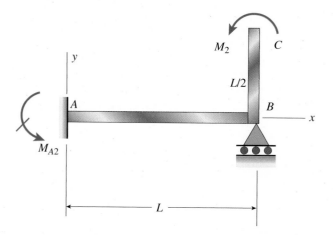

**A-10.2:** Propped cantilever beam $AB$ has moment $M_1$ applied at joint $B$. Framework $ABC$ has moment $M_2$ applied at $C$. Both structures have constant flexural rigidity $EI$. If the ratio of the *applied moments* $M_1/M_2 = 3/2$, the ratio of the *joint rotations at B*, $\theta_{B1}/\theta_{B2}$, is approximately:

(A) 1
(B) 3/2
(C) 2
(D) 5/2

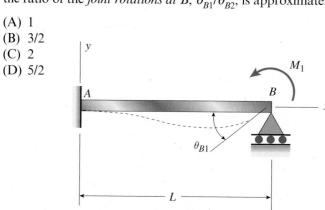

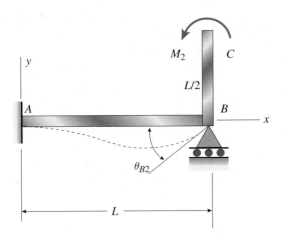

**A-10.3:** Structure 1 with member $BC$ of length $L/2$ has force $P_1$ applied at joint $C$. Structure 2 with member $BC$ of length $L$ has force $P_2$ applied at $C$. Both structures have constant flexural rigidity $EI$. If the ratio of the *applied forces* $P_1/P_2 = 5/2$, the ratio of the *joint B rotations* $\theta_{B1}/\theta_{B2}$ is approximately:

(A) 1
(B) 5/4
(C) 3/2
(D) 2

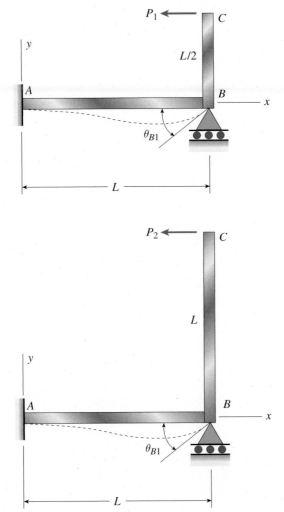

**A-10.4:** Structure 1 with member $BC$ of length $L/2$ has force $P_1$ applied at joint $C$. Structure 2 with member $BC$ of length $L$ has force $P_2$ applied at $C$. Both structures have constant flexural rigidity $EI$. The *required* ratio of the *applied forces* $P_1/P_2$ so that *joint B rotations* $\theta_{B1}$ and $\theta_{B2}$ are *equal* is approximately:

(A) 1
(B) 5/4
(C) 3/2
(D) 2

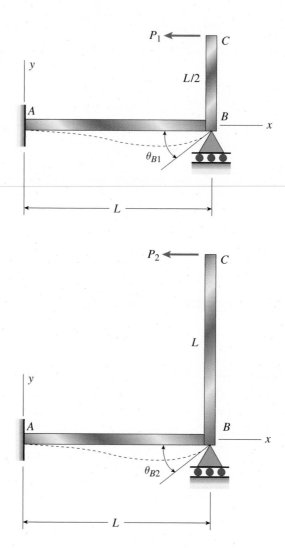

**A-10.5:** Structure 1 with member $BC$ of length $L/2$ has force $P_1$ applied at joint $C$. Structure 2 with member $BC$ of length $L$ has force $P_2$ applied at $C$. Both structures have constant flexural rigidity $EI$. If the ratio of the *applied forces* $P_1/P_2 = 5/2$, the ratio of the *joint B reactions* $R_{B1}/R_{B2}$ is approximately:

(A) 1
(B) 5/4
(C) 3/2
(D) 2

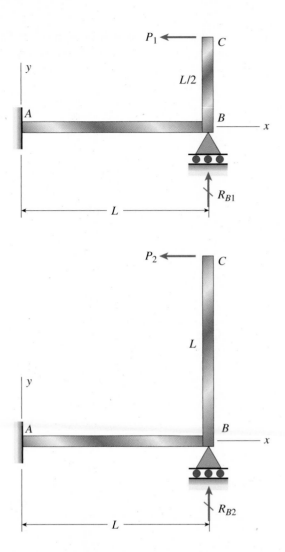

**A-10.6:** Structure 1 with member $BC$ of length $L/2$ has force $P_1$ applied at joint $C$. Structure 2 with member $BC$ of length $L$ has force $P_2$ applied at $C$. Both structures have constant flexural rigidity $EI$. The *required* ratio of the *applied forces* $P_1/P_2$ so that *joint B reactions* $R_{B1}$ and $R_{B2}$ are *equal* is approximately:

(A) 1
(B) 5/4
(C) 3/2
(D) 2

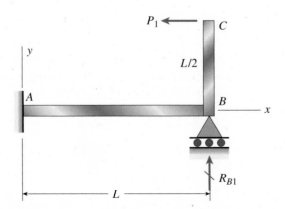

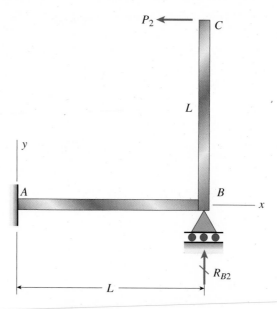

**A-10.7:** Structure 1 with member $BC$ of length $L/2$ has force $P_1$ applied at joint $C$. Structure 2 with member $BC$ of length $L$ has force $P_2$ applied at $C$. Both structures have constant flexural rigidity $EI$. If the ratio of the *applied forces* $P_1/P_2 = 5/2$, the ratio of the *joint C lateral deflections* $\delta_{C1}/\delta_{C2}$ is approximately:

(A) 1/2
(B) 4/5
(C) 3/2
(D) 2

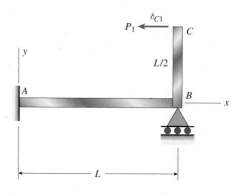

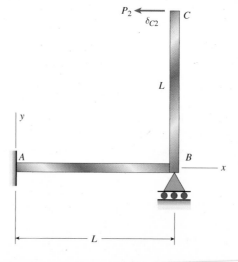

A-11.1: Beam $ACB$ has a sliding support at $A$ and is supported at $C$ by a pinned-end steel column with square cross section ($E = 200$ GPa, $b = 40$ mm) and height $L = 3.75$ m. The column must resist a load $Q$ at $B$ with a factor of safety of 2.0 with respect to the critical load. The maximum permissible value of $Q$ is approximately:

(A) 10.5 kN
(B) 11.8 kN
(C) 13.2 kN
(D) 15.0 kN

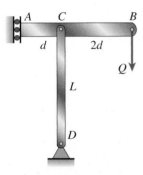

A-11.2: Beam $ACB$ has a pin support at $A$ and is supported at $C$ by a steel column with a square cross section ($E = 190$ GPa, $b = 42$ mm) and height $L = 5.25$ m. The column is pinned at $C$ and fixed at $D$. The column must resist a load $Q$ at $B$ with a factor of safety of 2.0 with respect to the critical load. The maximum permissible value of $Q$ is approximately:

(A) 3.0 kN
(B) 6.0 kN
(C) 9.4 kN
(D) 10.1 kN

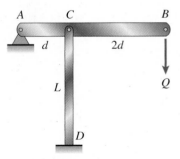

A-11.3: A steel pipe column ($E = 190$ GPa, $\alpha = 14 \times 10^{-6}/°C$, $d_2 = 82$ mm, and $d_1 = 70$ mm) of length $L = 4.25$ m is subjected to a temperature increase $\Delta T$. The column is pinned at the top and fixed at the bottom. The temperature increase at which the column will buckle is approximately:

(A) 36 °C
(B) 42 °C
(C) 54 °C
(D) 58 °C

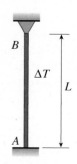

**A-11.4:** A steel pipe ($E = 190$ GPa, $\alpha = 14 \times 10^{-6}/°C$, $d_2 = 82$ mm, and $d_1 = 70$ mm) of length $L = 4.25$ m hangs from a rigid surface and is subjected to a temperature increase $\Delta T = 50\ °C$. The column is fixed at the top and has a small gap at the bottom. To avoid buckling, the minimum clearance at the bottom should be approximately:

(A) 2.55 mm
(B) 3.24 mm
(C) 4.17 mm
(D) 5.23 mm

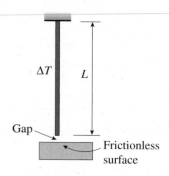

**A-11.5:** A pinned-end copper strut ($E = 110$ GPa) with length $L = 1.6$ m is constructed of circular tubing with outside diameter $d = 38$ mm. The strut must resist an axial load $P = 14$ kN with a factor of safety of 2.0 with respect to the critical load. The required thickness $t$ of the tube is approximately:

(A) 2.75 mm
(B) 3.15 mm
(C) 3.89 mm
(D) 4.33 mm

**A-11.6:** A plane truss composed of two steel pipes ($E = 210$ GPa, $d = 100$ mm, and wall thickness $= 6.5$ mm) is subjected to vertical load $W$ at joint $B$. Joints $A$ and $C$ are $L = 7$ m apart. The critical value of load $W$ for buckling in the plane of the truss is nearly:

(A) 138 kN
(B) 146 kN
(C) 153 kN
(D) 164 kN

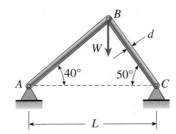

**A-11.7:** A beam is pin-connected to the tops of two identical pipe columns, each of height $h$, in a frame. The frame is restrained against sidesway at the top of column 1. Only buckling of columns 1 and 2 in the plane of the frame is of interest here. The ratio $(a/L)$ defining the placement of load $Q_{cr}$, which causes both columns to buckle simultaneously, is approximately:

(A) 0.25
(B) 0.33
(C) 0.67
(D) 0.75

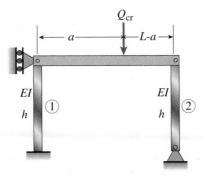

**A-11.8:** A steel pipe column ($E = 210$ GPa) with length $L = 4.25$ m is constructed of circular tubing with outside diameter $d_2 = 90$ mm and inner diameter $d_1 = 64$ mm. The pipe column is fixed at the base and pinned at the top and may buckle in any direction. The Euler buckling load of the column is most nearly:

(A) 303 kN
(B) 560 kN
(C) 690 kN
(D) 720 kN

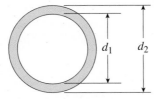

**A-11.9:** An aluminum tube ($E = 72$ GPa) $AB$ of circular cross section has a pinned support at the base and is pin-connected at the top to a horizontal beam supporting a load $Q = 600$ kN. The outside diameter of the tube is 200 mm and the desired factor of safety with respect to Euler buckling is 3.0. The required thickness $t$ of the tube is most nearly:

(A) 8 mm
(B) 10 mm
(C) 12 mm
(D) 14 mm

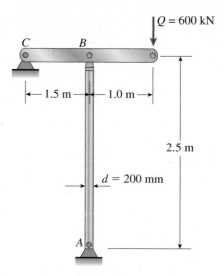

**A-11.10:** Two pipe columns are required to have the same Euler buckling load $P_{cr}$. Column 1 has flexural rigidity $EI$ and height $L_1$; column 2 has flexural rigidity $(4/3)EI$ and height $L_2$. The ratio $(L_2/L_1)$ at which both columns will buckle under the same load is approximately:

(A)  0.55
(B)  0.72
(C)  0.81
(D)  1.10

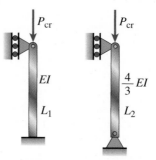

**A-11.11:** Two pipe columns are required to have the same Euler buckling load $P_{cr}$. Column 1 has flexural rigidity $EI_1$ and height $L$; column 2 has flexural rigidity $(2/3)EI_2$ and height $L$. The ratio $(I_2/I_1)$ at which both columns will buckle under the same load is approximately:

(A)  0.8
(B)  1.0
(C)  2.2
(D)  3.1

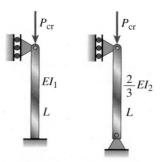

# Systems of Units and Conversion Factors

## B.1 SYSTEMS OF UNITS

Measurement systems have been a necessity since people first began to build and barter, and every ancient culture developed some sort of measurement system to serve its needs. Standardization of units took place gradually over the centuries, often through royal edicts. Development of the **British Imperial System** from earlier measurement standards began in the 13th century and was well established by the 18th century. The British system spread to many parts of the world, including the United States, through commerce and colonization. In the United States the system gradually evolved into the **U.S. Customary System (USCS)** that is in common use today.

The concept of the **metric system** originated in France about 300 years ago and was formalized in the 1790s, at the time of the French Revolution. France mandated the use of the metric system in 1840, and since then many other countries have done the same. In 1866, the United States Congress legalized the metric system without making it compulsory.

A new system of units was created when the metric system underwent a major revision in the 1950s. Officially adopted in 1960 and named the **International System of Units** (Système International d'Unités), this newer system is commonly referred to as **SI**. Although some SI units are the same as in the old metric system, SI has many new features and simplifications. Thus, SI is an improved metric system.

Length, time, mass, and force are the basic concepts of mechanics for which units of measurement are needed. However, only three of these quantities are independent since all four of them are related by Newton's second law of motion:

$$F = ma \tag{B-1}$$

in which $F$ is the force acting on a particle, $m$ is the mass of the particle, and $a$ is its acceleration. Since acceleration has units of length divided by time squared, all four quantities are involved in the second law.

The International System of Units, like the metric system, is based upon length, time, and mass as fundamental quantities. In these systems, force is derived from Newton's second law. Therefore, the unit of force is expressed in terms of the basic units of length, time, and mass, as shown in the next section.

SI is classified as an **absolute system of units** because measurements of the three fundamental quantities are independent of the locations at which the measurements are made; that is, the measurements do not depend upon the effects of gravity. Therefore, the SI units for length, time, and mass may be used anywhere on earth, in space, on the moon, or even on another planet. This is one of the reasons why the metric system has always been preferred for scientific work.

The British Imperial System and the U.S. Customary System are based upon length, time, and force as the fundamental quantities with mass being derived from the second law. Therefore, in these systems the unit of mass is expressed in terms of the units of length, time, and force. The unit of force is defined as the force required to give a certain standard mass an acceleration equal to the acceleration of gravity, which means that the unit of force varies with location and altitude. For this reason, these systems are called **gravitational systems of units**. Such systems were the first to evolve, probably because weight is such a readily discernible property and because variations in gravitational attraction were not noticeable. It is clear, however, that in the modern technological world an absolute system is preferable.

# B.2 SI UNITS

The International System of Units has seven **base units** from which all other units are derived. The base units of importance in mechanics are the meter (m) for length, second (s) for time, and kilogram (kg) for mass. Other SI base units pertain to temperature, electric current, amount of substance, and luminous intensity.

The **meter** was originally defined as one ten-millionth of the distance from the North Pole to the equator. Later, this distance was converted to a physical standard, and for many years the standard for the meter was the distance between two marks on a platinum-iridium bar stored at the headquarters of the International Bureau of Weights and Measures (Bureau International des Poids et Mesures) in Sèvres, a suburb on the western edge of Paris, France.

Because of the inaccuracies inherent in the use of a physical bar as a standard, the definition of the meter was changed in 1983 to the length of the path traveled by light in a vacuum during a time interval of 1/299,792,458 of a second.* The advantages of this "natural" standard are that it is not subject to physical damage and is reproducible at laboratories anywhere in the world.

The **second** was originally defined as 1/86,400 of a mean solar day (24 hours equals 86,400 seconds). However, since 1967 a highly accurate atomic clock has set the standard, and a second is now defined to be the duration of 9,192,631,770 periods of the radiation corresponding to the transition between the two hyperfine levels of the ground state of the cesium-133 atom. (Most engineers would probably prefer the original definition over the new one, which hasn't noticeably changed the second, but which is necessary because the earth's rotation rate is gradually slowing down.)

Of the seven base units in SI, the **kilogram** is the only one that is still defined by a physical object. Since the mass of an object can only be

---

*Taking the reciprocal of this number gives the speed of light in a vacuum (299,792,458 meters per second).

determined by comparing it experimentally with the mass of some other object, a physical standard is needed. For this purpose, a one-kilogram cylinder of platinum-iridium, called the International Prototype Kilogram (IPK), is kept by the International Bureau of Weights and Measures at Sèvres. (At the present time, attempts are being made to define the kilogram in terms of a fundamental constant, such as the Avogadro number, thus removing the need for a physical object.)

Other units used in mechanics, called **derived units**, are expressed in terms of the base units of meter, second, and kilogram. For instance, the unit of **force** is the **newton**, which is defined as the force required to impart an acceleration of one meter per second squared to a mass of one kilogram.* From Newton's second law ($F = ma$), we can derive the unit of force in terms of base units:

$$1 \text{ newton} = (1 \text{ kilogram})(1 \text{ meter per second squared})$$

Thus, the newton (N) is given in terms of base units by the formula

$$1 \text{ N} = 1 \text{ kg} \cdot \text{m/s}^2 \qquad \textbf{(B-2)}$$

To provide a point of reference, we note that a small apple weighs approximately one newton.

The unit of **work** and **energy** is the **joule**, defined as the work done when the point of application of a force of one newton is displaced a distance of one meter in the direction of the force.** Therefore,

$$1 \text{ joule} = (1 \text{ newton})(1 \text{ meter}) = 1 \text{ newton meter}$$

or
$$1 \text{ J} = 1 \text{ N} \cdot \text{m} \qquad \textbf{(B-3)}$$

When you raise this book from desktop to eye level, you do about 1 joule of work, and when you walk up one flight of stairs, you do about 200 joules of work.

The names, symbols, and formulas for SI units of importance in mechanics are listed in **Table B-1**. Some of the derived units have special names, such as newton, joule, hertz, watt, and pascal. These units are named for notable persons in science and engineering and have symbols (N, J, Hz, W, and Pa) that are capitalized, although the unit names themselves are written in lowercase letters. Other derived units have no special names (for example, the units of acceleration, area, and density) and must be expressed in terms of base units and other derived units.

The relationships between various SI units and some commonly used metric units are given in **Table B-2**. Metric units such as dyne, erg, gal, and micron are no longer recommended for engineering or scientific use.

The **weight** of an object is the **force of gravity** acting on that object, and therefore, weight is measured in newtons. Since the force of gravity depends upon altitude and position on the earth, weight is not an invariant property of a body. Furthermore, the weight of a body as measured by a spring scale is affected not only by the gravitational pull of the earth, but also by the centrifugal effects associated with the rotation of the earth.

---

*Sir Isaac Newton (1642–1727) was an English mathematician, physicist, and astronomer. He invented calculus and discovered the laws of motion and gravitation.

**James Prescott Joule (1818–1889) was an English physicist who developed a method for determining the mechanical equivalent of heat. His last name is pronounced "jool."

As a consequence, we must recognize two kinds of weight, **absolute weight** and **apparent weight**. The former is based upon the force of gravity alone, and the latter includes the effects of rotation. Thus, apparent weight is always less than absolute weight (except at the poles). Apparent weight, which is the weight of an object as measured with a spring scale,

## Table B-1

### Principal Units Used in Mechanics

| Quantity | International System (SI) | | | U.S. Customary System (USCS) | | |
|---|---|---|---|---|---|---|
| | Unit | Symbol | Formula | Unit | Symbol | Formula |
| Acceleration (angular) | radian per second squared | | $\text{rad/s}^2$ | radian per second squared | | $\text{rad/s}^2$ |
| Acceleration (linear) | meter per second squared | | $\text{m/s}^2$ | foot per second squared | | $\text{ft/s}^2$ |
| Area | square meter | | $\text{m}^2$ | square foot | | $\text{ft}^2$ |
| Density (mass) (Specific mass) | kilogram per cubic meter | | $\text{kg/m}^3$ | slug per cubic foot | | $\text{slug/ft}^3$ |
| Density (weight) (Specific weight) | newton per cubic meter | | $\text{N/m}^3$ | pound per cubic foot | pcf | $\text{lb/ft}^3$ |
| Energy; work | joule | J | $\text{N·m}$ | foot-pound | | ft-lb |
| Force | newton | N | $\text{kg·m/s}^2$ | pound | lb | (base unit) |
| Force per unit length (Intensity of force) | newton per meter | | $\text{N/m}$ | pound per foot | | lb/ft |
| Frequency | hertz | Hz | $\text{s}^{-1}$ | hertz | Hz | $\text{s}^{-1}$ |
| Length | meter | m | (base unit) | foot | ft | (base unit) |
| Mass | kilogram | kg | (base unit) | slug | | $\text{lb-s}^2/\text{ft}$ |
| Moment of a force; torque | newton meter | | $\text{N·m}$ | pound-foot | | lb-ft |
| Moment of inertia (area) | meter to fourth power | | $\text{m}^4$ | inch to fourth power | | $\text{in.}^4$ |
| Moment of inertia (mass) | kilogram meter squared | | $\text{kg·m}^2$ | slug foot squared | | $\text{slug-ft}^2$ |
| Power | watt | W | $\text{J/s}$ $(\text{N·m/s})$ | foot-pound per second | | ft-lb/s |
| Pressure | pascal | Pa | $\text{N/m}^2$ | pound per square foot | psf | $\text{lb/ft}^2$ |
| Section modulus | meter to third power | | $\text{m}^3$ | inch to third power | | $\text{in.}^3$ |
| Stress | pascal | Pa | $\text{N/m}^2$ | pound per square inch | psi | $\text{lb/in.}^2$ |
| Time | second | s | (base unit) | second | s | (base unit) |
| Velocity (angular) | radian per second | | $\text{rad/s}$ | radian per second | | $\text{rad/s}$ |
| Velocity (linear) | meter per second | | $\text{m/s}$ | foot per second | fps | $\text{ft/s}$ |
| Volume (liquids) | liter | L | $10^{-3}\,\text{m}^3$ | gallon | gal. | $231\ \text{in.}^3$ |
| Volume (solids) | cubic meter | | $\text{m}^3$ | cubic foot | cf | $\text{ft}^3$ |

*Notes:* 1 joule (J) = 1 newton meter (N · m) = 1 watt second (W · s)
1 hertz (Hz) = 1 cycle per second (cps) or 1 revolution per second (rev/s)
1 watt (W) = 1 joule per second (J/s) = 1 newton meter per second (N · m/s)
1 pascal (Pa) = 1 newton per meter squared (N/m²)
1 liter (L) = 0.001 cubic meter (m³) = 1000 cubic centimeters (cm³)

is the weight we customarily use in business and everyday life; absolute weight is used in astroengineering and certain kinds of scientific work. In this book, the term "weight" will always mean "apparent weight."

The **acceleration of gravity**, denoted by the letter $g$, is directly proportional to the force of gravity, and therefore it too depends upon position. In contrast, **mass** is a measure of the amount of material in a body and does not change with location.

The fundamental relationship between weight, mass, and acceleration of gravity can be obtained from Newton's second law ($F = ma$), which in this case becomes

$$W = mg \qquad \textbf{(B-4)}$$

In this equation, $W$ is the weight in newtons (N), $m$ is the mass in kilograms (kg), and $g$ is the acceleration of gravity in meters per second squared (m/s²). Equation (B-4) shows that *a body having a mass of one kilogram has a weight in newtons numerically equal to g.* The values of the weight $W$ and the acceleration $g$ depend upon many factors, including latitude and elevation. However, for scientific calculations a standard international value of $g$ has been established as

$$g = 9.806650 \text{ m/s}^2 \qquad \textbf{(B-5)}$$

## Table B-2

### Additional Units in Common Use

| SI and Metric Units | |
|---|---|
| 1 gal. = 1 centimeter per second squared (cm/s²) for example, g ≈ 981 gals) | 1 centimeter (cm) = $10^{-2}$ meters (m) |
| | 1 cubic centimeter (cm³) = 1 milliliter (mL) |
| 1 are (a) = 100 square meters (m²) | 1 micron = 1 micrometer ($\mu$m) = $10^{-6}$ meters (m) |
| 1 hectare (ha) = 10,000 square meters (m²) | 1 gram (g) = $10^{-3}$ kilograms (kg) |
| 1 erg = $10^{-7}$ joules (J) | 1 metric ton (t) = 1 megagram (Mg) = 1000 kilograms (kg) |
| 1 kilowatt-hour (kWh) = 3.6 megajoules (MJ) | 1 watt (W) = $10^7$ ergs per second (erg/s) |
| 1 dyne = $10^{-5}$ newtons (N) | 1 dyne per square centimeter (dyne/cm²) = $10^{-1}$ pascals (Pa) |
| 1 kilogram-force (kgf) = 1 kilopond (kp) | 1 bar = $10^5$ pascals (Pa) |
| = 9.80665 newtons (N) | 1 stere = 1 cubic meter (m³) |

| USCS and Imperial Units | |
|---|---|
| 1 kilowatt-hour (kWh) = 2,655,220 foot-pounds (ft-lb) | 1 kilowatt (kW) = 737.562 foot-pounds per second (ft-lb/s) = 1.34102 horsepower (hp) |
| 1 British thermal unit (Btu) = 778.171 foot-pounds (ft-lb) | |
| 1 kip (k) = 1000 pounds (lb) | 1 pound per square inch (psi) = 144 pounds per square foot (psf) |
| 1 ounce (oz) = 1/16 pound (lb) | |
| 1 ton = 2000 pounds (lb) | 1 revolution per minute (rpm) = $2\pi/60$ radians per second (rad/s) |
| 1 Imperial ton (or long ton) = 2240 pounds (lb) | |
| 1 poundal (pdl) = 0.0310810 pounds (lb) = 0.138255 newtons (N) | 1 mile per hour (mph) = 22/15 feet per second (fps) |
| 1 inch (in.) = 1/12 foot (ft) | 1 gallon (gal.) = 231 cubic inches (in.³) |
| 1 mil = 0.001 inch (in.) | 1 quart (qt) = 2 pints = 1/4 gallon (gal.) |
| 1 yard (yd) = 3 feet (ft) | 1 cubic foot (cf) = 576/77 gallons = 7.48052 gallons (gal.) |
| 1 mile = 5280 feet (ft) | |
| 1 horsepower (hp) = 550 foot-pounds per second (ft-lb/s) | 1 Imperial gallon = 277.420 cubic inches (in.³) |

This value is intended for use under standard conditions of elevation and latitude (sea level at a latitude of approximately 45°). The recommended value of $g$ for ordinary engineering purposes on or near the surface of the earth is

$$g = 9.81 \text{ m/s}^2 \qquad \textbf{(B-6)}$$

Thus, a body having a mass of one kilogram has a weight of 9.81 newtons.

**Atmospheric pressure** varies considerably with weather conditions, location, altitude, and other factors. Consequently, a standard international value for the pressure at the earth's surface has been defined:

$$1 \text{ standard atmosphere} = 101.325 \text{ kilopascals} \qquad \textbf{(B-7)}$$

The following simplified value is recommended for ordinary engineering work:

$$1 \text{ standard atmosphere} = 101 \text{ kPa} \qquad \textbf{(B-8)}$$

Of course, the values given in Eqs. (B-7) and (B-8) are intended for use in calculations and do not represent the actual ambient pressure at any given location.

A basic concept in mechanics is **moment** or **torque**, especially the moment of a force and the moment of a couple. Moment is expressed in units of force times length, or newton meters (N·m). Other important concepts in mechanics are **work** and **energy**, both of which are expressed in joules, a derived unit that happens to have the same units (newton meters) as the units of moment. However, moment is a distinctly different quantity from work or energy, and the joule should *never* be used for moment or torque.

**Frequency** is measured in units of **hertz** (Hz), a derived unit equal to the reciprocal of seconds (1/s or $s^{-1}$). The hertz is defined as the frequency of a periodic phenomenon for which the period is one second; thus, it is equivalent to one cycle per second (cps) or one revolution per second (rev/s). It is customarily used for mechanical vibrations, sound waves, and electromagnetic waves, and occasionally it is used for rotational frequency instead of the traditional units of revolution per minute (rpm) and revolution per second (rev/s).*

Two other derived units that have special names in SI are the **watt** (W) and the **pascal** (Pa). The watt is the unit of power, which is work per unit of time, and one watt is equal to one joule per second (J/s) or one newton meter per second (N · m/s). The pascal is the unit of pressure and stress, or force per unit area, and is equal to one newton per square meter (N/m²).**

The **liter** is not an accepted SI unit, yet it is so commonly used that it cannot be discarded easily. Therefore, SI permits its use under limited conditions for volumetric capacity, dry measure, and liquid measure. Both uppercase L and lowercase l are permitted as symbols for liter in SI, but in the United States only L is permitted (to avoid confusion with the numeral 1). The only prefixes permitted with liter are milli and micro.

**Loads on structures**, whether due to gravity or other actions, are usually expressed in force units, such as newtons, newtons per meter, or pascals

---

*Heinrich Rudolf Hertz (1857–1894) was a German physicist who discovered electromagnetic waves and showed that light waves and electromagnetic waves are identical.

**James Watt (1736–1819) was a Scottish inventor and engineer who developed a practical steam engine and discovered the composition of water. Watt also originated the term "horsepower." Blaise Pascal (1623–1662) was a French mathematician and philosopher. He founded probability theory, constructed the first calculating machine, and proved experimentally that atmospheric pressure varies with altitude.

(newtons per square meter). Examples of such loads are a concentrated load of 25 kN acting on an axle, a uniformly distributed load of intensity 800 N/m acting on a small beam, and air pressure of intensity 2.1 kPa acting on an airplane wing.

However, there is one circumstance in SI in which it is permissible to express a load in mass units. If the load acting on a structure is produced by gravity acting on a mass, then that load may be expressed in mass units (kilograms, kilograms per meter, or kilograms per square meter). The usual procedure in such cases is to convert the load to force units by multiplying by the acceleration of gravity ($g = 9.81$ m/s$^2$).

## SI Prefixes

Multiples and submultiples of SI units (both base units and derived units) are created by attaching prefixes to the units (see **Table B-3** for a list of prefixes). The use of a prefix avoids unusually large or small numbers. The general rule is that prefixes should be used to keep numbers in the range 0.1 to 1000.

All of the recommended prefixes change the size of the quantity by a multiple or submultiple of three. Similarly, when powers of 10 are used as multipliers, the exponents of 10 should be multiples of three (for example, $40 \times 10^3$ N is satisfactory, but $400 \times 10^2$ N is not). Also, the exponent on a unit with a prefix refers to the entire unit; for instance, the symbol mm$^2$ means (mm)$^2$ and not m(m)$^2$.

## Styles for Writing SI Units

Rules for writing SI units have been established by international agreement, and some of the most pertinent ones are described here. Examples of the rules are shown in parentheses.

1. Units are always written as symbols (kg) in equations and numerical calculations. In text, units are written as words (kilograms) unless numerical values are being reported, in which case either words or symbols may be used (12 kg or 12 kilograms).

2. Multiplication is shown in a compound unit by a raised dot (kN·m). When the unit is written in words, no dot is required (kilonewton meter).

| Prefix | Symbol | Multiplication factor | |
|--------|--------|------------------------|---|
| tera   | T   | $10^{12}$ | = 1,000,000,000,000 |
| giga   | G   | $10^{9}$  | = 1,000,000,000 |
| mega   | M   | $10^{6}$  | = 1,000,000 |
| kilo   | k   | $10^{3}$  | = 1000 |
| hecto  | h   | $10^{2}$  | = 100 |
| deka   | da  | $10^{1}$  | = 10 |
| deci   | d   | $10^{-1}$ | = 0.1 |
| centi  | c   | $10^{-2}$ | = 0.01 |
| milli  | m   | $10^{-3}$ | = 0.001 |
| micro  | $\mu$ | $10^{-6}$ | = 0.000 001 |
| nano   | n   | $10^{-9}$ | = 0.000 000 001 |
| pico   | p   | $10^{-12}$ | = 0.000 000 000 001 |

*Note:* The use of the prefixes hecto, deka, deci, and centi is not recommended in SI.

**Table B-3**

SI Prefixes

3. Division is shown in a compound unit by a slash (or *solidus*) or by multiplication using a negative exponent (m/s or m·s⁻¹). When the unit is written in words, the slash is always replaced by "per" (meter per second).

4. A space is always used between a number and its units (200 Pa or 200 pascals) with the exception of the degree symbol (either angle or temperature), where no space is used between the number and the symbol (45°, 20°C).

5. Units and their prefixes are always printed in roman type (that is, upright or vertical type) and never in italic type (slanted type), even when the surrounding text is in italic type.

6. When written as words, units are not capitalized (newton) except at the beginning of a sentence or in capitalized material such as a title. When written as a symbol, units are capitalized when they are derived from the name of a person (N). An exception is the symbol for liter, which may be either L or l, but the use of uppercase L is preferred to avoid confusion with the numeral 1. Also, some prefixes are written with capital letters when used in symbols (MPa) but not when used in words (megapascal).

7. When written as words, units are singular or plural as appropriate to the context (1 kilometer, 20 kilometers, 6 seconds). When written as symbols, units are always singular (1 km, 20 km, 6 s). The plural of hertz is hertz; the plurals of other units are formed in the customary manner (newtons, watts).

8. Prefixes are not used in the denominator of a compound unit. An exception is the kilogram (kg), which is a base unit and therefore, the letter "k" is not considered as a prefix. For example, we can write kN/m but not N/mm, and we can write J/kg, but not mJ/g.

## Pronunciation of SI Prefixes and Units

A guide to the pronunciation of a few SI names that are sometimes mispronounced is given in **Table B-4**. For instance, kilometer is pronounced *kill-oh-meter*, not *kil-om-eter*. The only prefix that generates arguments is giga—the official pronunciation is *jig-uh*, but many people say *gig-uh*.

# B.3 U.S. CUSTOMARY UNITS

The units of measurement traditionally used in the United States have never been made mandatory by the government; hence, for lack of a better name they are called the "customary" units. In this system the **base units** of relevance to mechanics are the foot (ft) for length, second (s) for time, and pound (lb) for force. The **foot** is defined as

$$1 \text{ ft} = 0.3048 \text{ m (exactly)} \qquad \textbf{(B-9)}$$

The **second** is the same as in SI and is described in the preceding section.

The **pound** is defined as the **force** that will give to a certain standard mass an acceleration equal to the acceleration of gravity. In other words, the

| Prefix | Pronunciation |
|--------|---------------|
| tera | same as *terra*, as in *terra firma* |
| giga | pronounced *jig-uh*; with *a* pronounced as in *about* (Alternate pronunciation: *gig-uh*) |
| mega | same as *mega* in *megaphone* |
| kilo | pronounced *kill-oh*; rhymes with *pillow* |
| milli | pronounced *mill-eh*, as in *military* |
| micro | same as *micro* in *microphone* |
| nano | pronounced *nan-oh*; rhymes with *man-oh* |
| pico | pronounced *pea-ko* |
|      | *Note:* The first syllable of every prefix is accented. |

| Unit | Pronunciation |
|------|---------------|
| joule | pronounced *jool*; rhymes with *cool* and *pool* |
| kilogram | pronounced *kill-oh-gram* |
| kilometer | pronounced *kill-oh-meter* |
| pascal | pronounced *pas-kal*, with the accent on *kal* |

### Table B-4

Pronunciation of SI Prefixes and Units

pound is the weight of the standard mass, which is defined as 0.45359237 kg (exactly). The weight of this amount of mass [see Eq. (B-4)] is

$$W = (0.45359237 \text{ kg})(9.806650 \text{ m/s}^2) = 4.448222 \text{ N}$$

in which the standard international value of $g$ is used [see Eq. (B-5)]. Thus, the pound is defined as follows:

$$1 \text{ lb} = 4.448222 \text{ N} \qquad \textbf{(B-10)}$$

which shows that the pound (like the foot) is actually defined in terms of SI units.

The unit of **mass** in USCS, called the **slug**, is a derived unit defined as the mass that will be accelerated one foot per second squared when acted upon by a force of one pound. Writing Newton's second law in the form $m = F/a$, we get

$$1 \text{ slug} = \frac{1 \text{ pound}}{1 \text{ ft/s}^2}$$

which shows that the slug is expressed in terms of base units by the formula

$$1 \text{ slug} = 1 \text{ lb-s}^2/\text{ft} \qquad \textbf{(B-11)}$$

To obtain the mass of an object of known weight, we use the second law in the form

$$m = \frac{W}{g} \qquad \textbf{(B-12)}$$

where $m$ is the mass in slugs, $W$ is the weight in pounds, and $g$ is the acceleration of gravity in feet per second squared.

As discussed previously, the value of $g$ depends upon the location, but in calculations where location is not relevant, the standard international value of $g$ may be used:

$$g = 32.1740 \text{ ft/s}^2 \qquad \textbf{(B-13)}$$

For ordinary purposes, the recommended value is

$$g = 32.2 \text{ ft/s}^2 \qquad \textbf{(B-14)}$$

From the preceding equations we conclude that an object having a mass of 1 slug will weigh 32.2 pounds at the earth's surface.

Another unit of mass in USCS is the pound-mass (lbm), which is the mass of an object weighing 1 pound, that is, 1 lbm = 1/32.2 slug.

As mentioned previously, **atmospheric pressure** varies considerably with local conditions; however, for many purposes the standard international value may be used:

$$\text{1 standard atmosphere} = 14.6959 \text{ pounds per square inch} \qquad \textbf{(B-15)}$$

or, for ordinary engineering work:

$$\text{1 standard atmosphere} = 14.7 \text{ psi} \qquad \textbf{(B-16)}$$

These values are intended for use in calculations and obviously do not represent the actual atmospheric pressure.

The unit of **work** and **energy** in USCS is the **foot-pound** (ft-lb), defined as the work done when the point of application of a force of one pound is displaced a distance of one foot in the direction of the force. The unit of **moment** or **torque** is the **pound-foot** (lb-ft), which comes from the fact that moment is expressed in units of force times length. Although in reality the same units apply to work, energy, and moment, it is common practice to use the pound-foot for moment and the foot-pound for work and energy.

The symbols and formulas for the most important USCS units used in mechanics are listed in **Table B-1**.

Many additional units from the U. S. Customary and Imperial systems appear in the mechanics literature; a few of these units are listed in the lower part of **Table B-2**.

# B.4 TEMPERATURE UNITS

Temperature is measured in SI by a unit called the kelvin (K), and the corresponding scale is the **Kelvin temperature scale**. The Kelvin scale is an absolute scale, which means that its origin (zero kelvins, or 0 K) is at absolute zero temperature, a theoretical temperature characterized by the complete absence of heat. On the Kelvin scale, water freezes at approximately 273 K and boils at approximately 373 K.

For nonscientific purposes the **Celsius temperature scale** is normally used. The corresponding unit of temperature is the degree Celsius (°C), which is equal to one kelvin. On this scale, water freezes at approximately zero degrees (0°C) and boils at approximately 100 degrees (100°C) under certain standard conditions. The Celsius scale is also known as the *centigrade temperature scale*.

The relationship between Kelvin temperature and Celsius temperature is given by the following equations:

$$\text{Temperature in degrees Celsius} = \text{temperature in kelvins} - 273.15$$

or
$$T(°C) = T(K) - 273.15 \qquad \textbf{(B-17)}$$

where $T$ denotes the temperature. When working with *changes* in temperature, or *temperature intervals*, as is usually the case in mechanics, either unit can be used because the intervals are the same.*

---

*Lord Kelvin (1824–1907), William Thomson, was a British physicist who made many scientific discoveries, developed theories of heat, and proposed the absolute scale of temperature. Anders Celsius (1701–1744) was a Swedish scientist and astronomer. In 1742 he developed the temperature scale in which 0 and 100 correspond, respectively, to the freezing and boiling points of water.

The U.S. Customary unit for temperature is the degree Fahrenheit (°F). On the **Fahrenheit temperature scale**, water freezes at approximately 32 degrees (32°F) and boils at approximately 212 degrees (212°F). Each Fahrenheit degree is exactly 5/9 of one kelvin or one degree Celsius. The corresponding absolute scale is the **Rankine temperature scale**, related to the Fahrenheit scale by the equation

$$T(°F) = T(°R) - 459.67 \qquad \textbf{(B-18)}$$

Thus, absolute zero corresponds to −459.67°F.*

The **conversion formulas** between the Fahrenheit and Celsius scales are as follows:

$$T(°C) = \frac{5}{9}[T(°F) - 32] \quad T(°F) = \frac{9}{5}T(°C) + 32 \qquad \textbf{(B-19a,b)}$$

As before, $T$ denotes the temperature on the indicated scale.

# B.5 CONVERSIONS BETWEEN UNITS

Quantities given in either USCS or SI units can be converted quickly to the other system by using the **conversion factors** listed in Table B-5.

If the given quantity is expressed in USCS units, it can be converted to SI units by *multiplying* by the conversion factor. To illustrate this process, assume that the stress in a beam is given as 10,600 psi and we wish to convert this quantity to SI units. From **Table B-5** we see that a stress of 1 psi converts to 6894.76 Pa. Therefore, the conversion of the given value is performed in the following manner:

$$(10,600 \text{ psi})(6894.76) = 73,100,000 \text{ Pa} = 73.1 \text{ MPa}$$

Because the original value is given to three significant digits, we have rounded the final result to three significant digits also (see Appendix C for a discussion of significant digits). Note that the conversion factor of 6894.76 has units of pascals divided by pounds per square inch, and therefore, the equation is dimensionally correct.

To reverse the conversion process (that is, to convert from SI units to USCS units), the quantity in SI units is *divided* by the conversion factor. For instance, suppose that the moment of inertia of the cross-sectional area of a beam is given as $94.73 \times 10^6$ mm$^4$. Then the moment of inertia in USCS units is

$$\frac{94.73 \times 10^6 \text{ mm}^4}{416,231} = 228 \text{ in.}^4$$

in which the term 416,231 is the conversion factor for moment of inertia.

---

*William John Macquorn Rankine (1820–1872) was a Scottish engineer and physicist. He made important contributions in such diverse fields as thermodynamics, light, sound, stress analysis, and bridge engineering. Gabriel Daniel Fahrenheit (1686–1736) was a German physicist who experimented with thermometers and made them more accurate by using mercury in the tube. He set the origin (0°) of his temperature scale at the freezing point of a mixture of ice, salt, and water.

## Table B-5

Conversions Between U.S. Customary Units and SI Units

| U.S. Customary unit | | Times conversion factor | | Equals SI unit | |
|---|---|---|---|---|---|
| | | **Accurate** | **Practical** | | |
| Acceleration (linear) | | | | | |
| foot per second squared | ft/s² | 0.3048* | 0.305 | meter per second squared | m/s² |
| inch per second squared | in./s² | 0.0254* | 0.0254 | meter per second squared | m/s² |
| Area | | | | | |
| square foot | ft² | 0.09290304* | 0.0929 | square meter | m² |
| square inch | in.² | 645.16* | 645 | square millimeter | mm² |
| Density (mass) | | | | | |
| slug per cubic foot | slug/ft³ | 515.379 | 515 | kilogram per cubic meter | kg/m³ |
| Density (weight) | | | | | |
| pound per cubic foot | lb/ft³ | 157.087 | 157 | newton per cubic meter | N/m³ |
| pound per cubic inch | lb/in.³ | 271.447 | 271 | kilonewton per cubic meter | kN/m³ |
| Energy; work | | | | | |
| foot-pound | ft-lb | 1.35582 | 1.36 | joule (N·m) | J |
| inch-pound | in.-lb | 0.112985 | 0.113 | joule | J |
| kilowatt-hour | kWh | 3.6* | 3.6 | megajoule | MJ |
| British thermal unit | Btu | 1055.06 | 1055 | joule | J |
| Force | | | | | |
| pound | lb | 4.44822 | 4.45 | newton (kg·m/s²) | N |
| kip (1000 pounds) | k | 4.44822 | 4.45 | kilonewton | kN |
| Force per unit length | | | | | |
| pound per foot | lb/ft | 14.5939 | 14.6 | newton per meter | N/m |
| pound per inch | lb/in. | 175.127 | 175 | newton per meter | N/m |
| kip per foot | k/ft | 14.5939 | 14.6 | kilonewton per meter | kN/m |
| kip per inch | k/in. | 175.127 | 175 | kilonewton per meter | kN/m |
| Length | | | | | |
| foot | ft | 0.3048* | 0.305 | meter | m |
| inch | in. | 25.4* | 25.4 | millimeter | mm |
| mile | mi | 1.609344* | 1.61 | kilometer | km |
| Mass | | | | | |
| slug | lb-s²/ft | 14.5939 | 14.6 | kilogram | kg |
| Moment of a force; torque | | | | | |
| pound-foot | lb-ft | 1.35582 | 1.36 | newton meter | N·m |
| pound-inch | lb-in. | 0.112985 | 0.113 | newton meter | N·m |
| kip-foot | k-ft | 1.35582 | 1.36 | kilonewton meter | kN·m |
| kip-inch | k-in. | 0.112985 | 0.113 | kilonewton meter | kN·m |
| Moment of inertia (area) | | | | | |
| inch to fourth power | in.⁴ | 416,231 | 416,000 | millimeter to fourth power | mm⁴ |
| inch to fourth power | in.⁴ | $0.416,231 \times 10^{-6}$ | $0.416 \times 10^{-6}$ | meter to fourth power | m⁴ |
| Moment of inertia (mass) | | | | | |
| slug foot squared | slug-ft² | 1.35582 | 1.36 | kilogram meter squared | kg·m² |

*An asterisk denotes an *exact* conversion factor.

**Note:** To convert from SI units to USCS units, *divide* by the conversion factor.

(*Continued*)

## Table B-5 (Continued)

| U.S. Customary unit | | Times conversion factor | | Equals SI unit | |
| --- | --- | --- | --- | --- | --- |
| | | Accurate | Practical | | |
| **Power** | | | | | |
| foot-pound per second | ft-lb/s | 1.35582 | 1.36 | watt (J/s or N·m/s) | W |
| foot-pound per minute | ft-lb/min | 0.0225970 | 0.0226 | watt | W |
| horsepower (550 ft-lb/s) | hp | 745.701 | 746 | watt | W |
| **Pressure; stress** | | | | | |
| pound per square foot | psf | 47.8803 | 47.9 | pascal (N/m$^2$) | Pa |
| pound per square inch | psi | 6894.76 | 6890 | pascal | Pa |
| kip per square foot | ksf | 47.8803 | 47.9 | kilopascal | kPa |
| kip per square inch | ksi | 6.89476 | 6.89 | megapascal | MPa |
| **Section modulus** | | | | | |
| inch to third power | in.$^3$ | 16,387.1 | 16,400 | millimeter to third power | mm$^3$ |
| inch to third power | in.$^3$ | $16.3871 \times 10^{-6}$ | $16.4 \times 10^{-6}$ | meter to third power | m$^3$ |
| **Velocity (linear)** | | | | | |
| foot per second | ft/s | 0.3048* | 0.305 | meter per second | m/s |
| inch per second | in./s | 0.0254* | 0.0254 | meter per second | m/s |
| mile per hour | mph | 0.44704* | 0.447 | meter per second | m/s |
| mile per hour | mph | 1.609344* | 1.61 | kilometer per hour | km/h |
| **Volume** | | | | | |
| cubic foot | ft$^3$ | 0.0283168 | 0.0283 | cubic meter | m$^3$ |
| cubic inch | in.$^3$ | $16.3871 \times 10^{-6}$ | $16.4 \times 10^{-6}$ | cubic meter | m$^3$ |
| cubic inch | in.$^3$ | 16.3871 | 16.4 | cubic centimeter (cc) | cm$^3$ |
| gallon (231 in.$^3$) | gal. | 3.78541 | 3.79 | liter | L |
| gallon (231 in.$^3$) | gal. | 0.00378541 | 0.00379 | cubic meter | m$^3$ |

*An asterisk denotes an *exact* conversion factor.

*Note:* To convert from SI units to USCS units, *divide* by the conversion factor.

# Problem Solving

## C.1 TYPES OF PROBLEMS

The study of mechanics of materials divides naturally into two parts: first, *understanding* the general concepts and principles, and second, *applying* those concepts and principles to physical situations. An understanding of the general concepts is obtained by studying the discussions and derivations presented in books such as this one. Skill in applying the concepts is accomplished by solving problems on your own. Of course, these two aspects of mechanics are closely related, and many experts in mechanics will argue that you do not really understand the concepts if you can not apply them. It is easy to recite the principles, but applying them to real situations requires an in-depth understanding. That is why teachers of mechanics place so much emphasis on problems. Problem solving gives meaning to the concepts and also provides an opportunity to gain experience and develop judgment.

Some of the homework problems in this book require symbolic solutions and others require numerical solutions. In the case of **symbolic problems** (also called *analytical*, *algebraic*, or *literal problems*), the data are supplied in the form of symbols for the various quantities, such as $P$ for load, $L$ for length, and $E$ for modulus of elasticity. Such problems are solved in terms of algebraic variables, and the results are expressed as formulas or mathematical expressions. Symbolic problems usually do not involve numerical calculations, except when numerical data are substituted into the final symbolic result in order to obtain a numerical value. However, this final substitution of numerical data should not obscure the fact that the problem was solved in symbolic terms.

In contrast, **numerical problems** are those in which the data are given in the form of numbers (with appropriate units); for example, a load might be given as 12 kN, a length as 3 m, and a dimension as 150 mm. The solution of a numerical problem is carried out by performing calculations from the beginning, and the results, both intermediate and final, are in the form of numbers.

An advantage of a numerical problem is that the magnitudes of all quantities are evident at every stage of the solution, thereby providing an opportunity to observe whether the calculations are producing reasonable results. Also, a numerical solution makes it possible to keep the magnitudes of quantities within prescribed limits. For instance, suppose the stress at a particular point in a beam must not exceed a certain allowable value. If this stress is calculated as an intermediate step in the numerical solution, you can verify immediately whether or not it exceeds the limit.

Symbolic problems have several advantages too. Because the results are algebraic formulas or expressions, you can see immediately how the variables affect the answers. For instance, if a load appears to the first power in the numerator of the final result, you know that doubling the load will double the result. Equally important is the fact that a symbolic solution shows what variables do *not* affect the result. For instance, a certain quantity may cancel out of the solution, a fact that might not even be noticed in a numerical solution. Furthermore, a symbolic solution makes it convenient to check the dimensional homogeneity of all terms in the solution. And most important, a symbolic solution provides a general formula that is applicable to many different problems, each with a different set of numerical data. In contrast, a numerical solution is good for only one set of circumstances, and a complete new solution is required if the data are changed. Of course, symbolic solutions are not feasible when the formulas become too complex to manipulate; when that happens, a numerical solution is required.

In more advanced work in mechanics, problem solving requires the use of **numerical methods**. This term refers to a wide variety of computational methods, including standard mathematical procedures (such as numerical integration and numerical solution of differential equations) and advanced methods of analysis (such as the finite-element method). Computer programs for these methods are readily available. More specialized computer programs are also available for performing routine tasks, such as finding deflections of beams and finding principal stresses. However, when studying mechanics of materials, we concentrate on the concepts rather than on the use of particular computer programs.

## C.2 STEPS IN SOLVING PROBLEMS

The procedures used in solving problems will vary among individuals and will vary according to the type of problem. Nevertheless, the following suggestions will help in reducing mistakes.

1.  Make a clear statement of the problem and draw a figure portraying the mechanical or structural system to be investigated. An important part of this step is identifying what is known and what is to be found.

2.  Simplify the mechanical or structural system by making assumptions about its physical nature. This step is called *modeling*, because it involves creating (on paper) an idealized model of the real system. The objective is to create a model that represents the real system to a sufficient degree of accuracy that the results obtained from the model can be applied to the real system.

    Here are a few examples of idealizations used in modeling mechanical systems: (a) Finite objects are sometimes modeled as particles, as when determining the forces acting on a joint of a truss.

(b) Deformable bodies are sometimes represented as rigid bodies, as when finding the reactions of a statically determinate beam or the forces in the members of a statically determinate truss. (c) The geometry and shapes of objects may be simplified, as when we consider the earth to be a sphere or a beam to be perfectly straight. (d) Distributed forces acting on machines and structures may be represented by equivalent concentrated forces. (e) Forces that are small compared to other forces, or forces that are known to have only a minor effect on the results, may be disregarded (friction forces are sometimes in this category). (f) Supports of structures often may be considered as immovable.

3. Draw large and clear sketches as you solve problems. Sketches always aid in understanding the physical situation and often bring out aspects of the problem that would otherwise be overlooked.

4. Apply the principles of mechanics to the idealized model to obtain the governing equations. In statics, the equations usually are equations of equilibrium obtained from Newton's first law; in dynamics, they usually are equations of motion obtained from Newton's second law. In mechanics of materials, the equations are associated with stresses, strains, deformations, and displacements.

5. Use mathematical and computational techniques to solve the equations and obtain results, either in the form of mathematical formulas or numerical values.

6. Interpret the results in terms of the physical behavior of the mechanical or structural system; that is, give meaning or significance to the results, and draw conclusions about the behavior of the system.

7. Check the results in as many ways as you can. Because errors can be disastrous and expensive, engineers should never rely on a single solution.

8. Finally, present your solution in clear, neat fashion so that it can be easily reviewed and checked by others.

# C.3 DIMENSIONAL HOMOGENEITY

The basic concepts in mechanics are length, time, mass, and force. Each of these physical quantities has a **dimension**, that is, a generalized unit of measurement. For example, consider the concept of length. There are many units of length, such as the meter, kilometer, yard, foot, and inch, yet all of these units have something in common—each one represents a distinct length and not some other quantity such as volume or force. Therefore, we can refer to the *dimension of length* without being specific as to the particular unit of measurement. Similar comments can be made for the dimensions of time, mass, and force. These four dimensions are customarily denoted by the symbols L, T, M, and F, respectively.

Every equation, whether in numeric form or symbolic form, must be **dimensionally homogeneous**, that is, the dimensions of all terms in the equation must be the same. To check the dimensional correctness of an equation, we disregard numerical magnitudes and write only the dimensions of each quantity in the equation. The resulting equation must have identical dimensions in all terms.

As an example, consider the following equation for the deflection $\delta$ at the midpoint of a simple beam with a uniformly distributed load:

$$\delta = \frac{5qL^4}{384EI}$$

The corresponding dimensional equation is obtained by replacing each quantity by its dimensions; thus, the deflection $\delta$ is replaced by the dimension L, the intensity of uniform load $q$ is replaced by F/L (force per unit of length), the length $L$ of the beam is replaced by the dimension L, the modulus of elasticity $E$ is replaced by F/L$^2$ (force per unit of area), and the moment of inertia $I$ is replaced by L$^4$. Therefore, the dimensional equation is

$$L = \frac{(F/L)L^4}{(F/L^2)L^4}$$

When simplified, this equation reduces to the dimensional equation L = L, as expected.

Dimensional equations can be written either in generalized terms using the LTMF notation or in terms of the actual units being used in the problem. For instance, if we are making calculations for the preceding beam deflection using USCS units, we can write the dimensional equation as follows:

$$\text{in.} = \frac{(\text{lb/in.})\text{in.}^4}{(\text{lb/in.}^2)\text{in.}^4}$$

which reduces to in. = in. and is dimensionally correct. Frequent checks for dimensional homogeneity (or *consistency of units*) help to eliminate errors when performing derivations and calculations.

## C.4 SIGNIFICANT DIGITS

**Engineering calculations** are performed by calculators and computers that operate with great precision. For instance, some computers routinely perform calculations with more than 25 digits in every numerical value, and output values with 10 or more digits are available in even the most inexpensive hand-held calculators. Under these conditions it is important to realize that the accuracy of the results obtained from an engineering analysis is determined not only by the calculations, but also by factors such as the accuracy of the given data, the approximations inherent in the analytical models, and the validity of the assumptions used in the theories. In many engineering situations, these considerations mean that the results are valid to only two or three significant digits.

As an example, suppose that a computation yields the result $R = 6287.46$ lb for the reaction of a statically indeterminate beam. To state the result in this manner is misleading, because it implies that the reaction is known to the nearest 1/100 of a pound even though its magnitude is over 6000 pounds. Thus, it implies an accuracy of approximately 1/600,000 and a precision of 0.01 lb, neither of which is justified. Instead, the accuracy of the calculated reaction depends upon matters such as the following: (1) how accurately the loads, dimensions, and other data used in the analysis are known, and (2) the approximations inherent in the theories of beam behavior. Most likely, the reaction $R$ in this example would be known only to the nearest 10 pounds, or perhaps only to the nearest 100 pounds. Consequently, the result of the computation should be stated as either $R = 6290$ lb or $R = 6300$ lb.

To make clear the accuracy of a given numerical value, it is common practice to use **significant digits**. A significant digit is a digit from 1 to 9 or any zero not used to show the position of the decimal point; for instance, the numbers 417, 8.29, 7.30, and 0.00254 each have three significant digits. However, the number of significant digits in a number such as 29,000 is not apparent. It may have two significant digits, with the three zeros serving only to locate the decimal point, or it may have three, four, or five significant digits if one or more of the zeros is valid. By using powers of ten, the accuracy of a number such as 29,000 can be made clearer. When written as $29 \times 10^3$ or $0.029 \times 10^6$, the number is understood to have two significant digits; when written as $29.0 \times 10^3$ or $0.0290 \times 10^6$, it has three significant digits.

When a number is obtained by calculation, its accuracy depends upon the accuracy of the numbers used in performing the calculations. A rule of thumb that serves for **multiplication and division** is the following: The number of significant digits in the calculated result is the same as the least number of significant digits in any of the numbers used in the calculation. As an illustration, consider the product of 2339.3 and 35.4. The calculated result is 82,811.220 when recorded to eight digits. However, stating the result in this manner is misleading because it implies much greater accuracy than is warranted by either of the original numbers. Inasmuch as the number 35.4 has only three significant digits, the proper way to write the result is $82.8 \times 10^3$.

For calculations involving **addition or subtraction** of a column of numbers, the last significant digit in the result is found in the last column of digits that has significant digits in all of the numbers being added or subtracted. To make this notion clearer, consider the following three examples:

|  | 459.637 | 838.49 | 856,400 |
|---|---|---|---|
|  | +7.2 | −7 | −847,900 |
| Result from calculator: | 466.837 | 831.49 | 8500 |
| Write the result as: | 466.8 | 831 | 8500 |

In the first example, the number 459.637 has six significant digits and the number 7.2 has two. When added, the result has four significant digits because all digits in the result to the right of the column containing the 2 are meaningless. In the second example, the number 7 is accurate to one significant digit (that is, it is not an exact number). Therefore, the final result is accurate only as far as the column containing the 7, which means it has three significant digits and is recorded as 831. In the third example, the numbers 856,400 and 847,900 are assumed to be accurate to four significant digits, but the result of the subtraction is accurate to only two significant digits since none of the zeros is significant. In general, subtraction results in reduced accuracy.

These three examples show that numbers obtained by calculation may contain superfluous digits having no physical meaning. Therefore, when reporting such numbers as final results, you should give only those digits that are significant.

In mechanics of materials, the data for problems are usually accurate to about 1%, or perhaps 0.1% in some cases, and therefore, the final results should be reported to a comparable accuracy. When greater accuracy is warranted, it will be obvious from the statement of the problem.

Although the use of significant digits provides a handy way to deal with the matter of **numerical accuracy**, it should be recognized that significant digits are not valid indicators of accuracy. To illustrate this fact, consider the numbers 999 and 101. Three significant digits in the number 999 correspond to an accuracy of 1/999, or 0.1%, whereas the same number of significant digits in the number 101 corresponds to an accuracy of only 1/101, or 1.0%. This disparity in accuracy can be reduced by always using one additional significant digit for numbers beginning with the digit 1. Thus, four significant digits in the number 101.1 gives about the same accuracy as three significant digits in the number 999.

In this book we generally will follow the rule that *final* numerical results beginning with the digits 2 through 9 should be recorded to three significant digits and those beginning with the digit 1 should be recorded to four significant digits. However, to preserve numerical accuracy and avoid round-off errors during the calculation process, the results of *intermediate* calculations will usually be recorded with additional digits.

Many of the numbers entering into our calculations are exact, for example, the number $\pi$, fractions such as 1/2, and integers such as the number 48 in the formula $PL^3/48EI$ for a beam deflection. Exact numbers are significant to an infinite number of digits and, therefore, have no role in determining the accuracy of a calculated result.

## C.5 ROUNDING OF NUMBERS

The process of discarding the insignificant digits and keeping only the significant ones is called *rounding*. To illustrate the process, assume that a number is to be rounded to three significant digits. Then the following rules apply:

(a) If the fourth digit is less than 5, the first three digits are left unchanged and all succeeding digits are dropped or replaced by zeros. For example, 37.44 rounds to 37.4 and 673,289 rounds to 673,000.

(b) If the fourth digit is greater than 5, or if the fourth digit is 5 and is followed by at least one digit other than zero, then the third digit is increased by 1 and all following digits are dropped or replaced by zeros. For example, 26.37 rounds to 26.4 and 3.245002 rounds to 3.25.

(c) Finally, if the fourth digit is 5 and all following digits (if any) are zeros, then the third digit is unchanged if it is an even number and increased by 1 if it is an odd number, and the 5 is replaced by a zero. (Trailing and leading zeros are retained only if they are needed to locate the decimal point.) This process is usually described as "rounding to the even digit." Since the occurrence of even and odd digits is more or less random, the use of this rule means that numbers are rounded upward about as often as downward, thereby reducing the chances of accumulating round-off errors.

The rules described in the preceding paragraphs for rounding to three significant digits apply in the same general manner when rounding to any other number of significant digits.

# Mathematical Formulas

## MATHEMATICAL CONSTANTS

$\pi = 3.14159\ldots \qquad e = 2.71828\ldots \qquad 2\pi$ radians $= 360$ degrees

1 radian $= \dfrac{180}{\pi}$ degrees $= 57.2958°$

1 degree $= \dfrac{\pi}{180}$ radians $= 0.0174533$ rad

Conversions: Multiply degrees by $\dfrac{\pi}{180}$ to obtain radians

Multiply radians by $\dfrac{180}{\pi}$ to obtain degrees

## EXPONENTS

$$A^n A^m = A^{n+m} \qquad \frac{A^m}{A^n} = A^{m-n} \qquad (A^m)^n = A^{mn} \qquad A^{-m} = \frac{1}{A^m}$$

$$(AB)^n = A^n B^n \qquad \left(\frac{A}{B}\right)^n = \frac{A^n}{B^n} \qquad A^{m/n} = \sqrt[n]{A^m} \qquad A^0 = 1 \ (A \neq 0)$$

## LOGARITHMS

$\log \equiv$ common logarithm (logarithm to the base 10)

$10^x = y \qquad \log y = x$

$\ln \equiv$ natural logarithm (logarithm to the base $e$) $\qquad e^x = y \qquad \ln y = x$

$e^{\ln A} = A \qquad 10^{\log A} = A \qquad \ln e^A = A \qquad \log 10^A = A$

$\log AB = \log A + \log B \qquad \log \dfrac{A}{B} = \log A - \log B \qquad \log \dfrac{1}{A} = -\log A$

$\log A^n = n \log A \qquad \log 1 = \ln 1 = 0 \qquad \log 10 = 1 \qquad \ln e = 1$

$\ln A = (\ln 10)(\log A) = 2.30259 \log A$

$\log A = (\log e)(\ln A) = 0.434294 \ln A$

## TRIGONOMETRIC FUNCTIONS

$$\tan x = \frac{\sin x}{\cos x} \quad \cot x = \frac{\cos x}{\sin x} \quad \sec x = \frac{1}{\cos x} \quad \csc x = \frac{1}{\sin x}$$

$$\sin^2 x + \cos^2 x = 1 \quad \tan^2 x + 1 = \sec^2 x \quad \cot^2 x + 1 = \csc^2 x$$

$$\sin(-x) = -\sin x \quad \cos(-x) = \cos x \quad \tan(-x) = -\tan x$$

$$\sin(x \pm y) = \sin x \cos y \pm \cos x \sin y$$

$$\cos(x \pm y) = \cos x \cos y \mp \sin x \sin y$$

$$\sin 2x = 2 \sin x \cos x \quad \cos 2x = \cos^2 x - \sin^2 x$$

$$\tan 2x = \frac{2 \tan x}{1 - \tan^2 x}$$

$$\tan x = \frac{1 - \cos 2x}{\sin 2x} = \frac{\sin 2x}{1 + \cos 2x}$$

$$\sin^2 x = \frac{1}{2}(1 - \cos 2x) \quad \cos^2 x = \frac{1}{2}(1 + \cos 2x)$$

For any triangle with sides $a$, $b$, $c$ and opposite angles $A$, $B$, $C$:

Law of sines:  $\dfrac{a}{\sin A} = \dfrac{b}{\sin B} = \dfrac{c}{\sin C}$

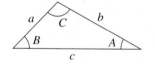

Law of cosines:  $c^2 = a^2 + b^2 - 2ab \cos C$

## QUADRATIC EQUATION AND QUADRATIC FORMULA

$$ax^2 + bx + c = 0 \qquad x = \frac{-b \pm \sqrt{b^2 - 4ac}}{2a}$$

## INFINITE SERIES

$$\frac{1}{1 + x} = 1 - x + x^2 - x^3 + \cdots \quad (-1 < x < 1)$$

$$\sqrt{1 + x} = 1 + \frac{x}{2} - \frac{x^2}{8} + \frac{x^3}{16} - \cdots \quad (-1 < x < 1)$$

$$\frac{1}{\sqrt{1 + x}} = 1 - \frac{x}{2} + \frac{3x^2}{8} - \frac{5x^3}{16} + \cdots \quad (-1 < x < 1)$$

$$e^x = 1 + x + \frac{x^2}{2!} + \frac{x^3}{3!} + \cdots \quad (-\infty < x < \infty)$$

$$\sin x = x - \frac{x^3}{3!} + \frac{x^5}{5!} - \frac{x^7}{7!} + \cdots \quad (-\infty < x < \infty)$$

$$\cos x = 1 - \frac{x^2}{2!} + \frac{x^4}{4!} - \frac{x^6}{6!} + \cdots \quad (-\infty < x < \infty)$$

*Note:* If $x$ is very small compared to 1, only the first few terms in the series are needed.

## DERIVATIVES

$$\frac{d}{dx}(ax) = a \qquad \frac{d}{dx}(x^n) = nx^{n-1} \qquad \frac{d}{dx}(au) = a\frac{du}{dx}$$

$$\frac{d}{dx}(uv) = u\frac{dv}{dx} + v\frac{du}{dx} \qquad \frac{d}{dx}\left(\frac{u}{v}\right) = \frac{v(du/dx) - u(dv/dx)}{v^2}$$

$$\frac{d}{dx}(u^n) = nu^{n-1}\frac{du}{dx} \qquad \frac{dy}{dx} = \frac{dy}{du}\frac{du}{dx} \qquad \frac{du}{dx} = \frac{1}{dx/du}$$

$$\frac{d}{dx}(\sin u) = \cos u\frac{du}{dx} \qquad \frac{d}{dx}(\cos u) = -\sin u\frac{du}{dx}$$

$$\frac{d}{dx}(\tan u) = \sec^2 u\frac{du}{dx} \qquad \frac{d}{dx}(\cot u) = -\csc^2 u\frac{du}{dx}$$

$$\frac{d}{dx}(\sec u) = \sec u \tan u\frac{du}{dx} \qquad \frac{d}{dx}(\csc u) = -\csc u \cot u\frac{du}{dx}$$

$$\frac{d}{dx}(\arctan u) = \frac{1}{1 + u^2}\frac{du}{dx} \qquad \frac{d}{dx}(\log u) = \frac{\log e}{u}\frac{du}{dx} \qquad \frac{d}{dx}(\ln u) = \frac{1}{u}\frac{du}{dx}$$

$$\frac{d}{dx}(a^u) = a^u \ln a\frac{du}{dx} \qquad \frac{d}{dx}(e^u) = e^u\frac{du}{dx}$$

## INDEFINITE INTEGRALS

*Note:* A constant must be added to the result of every integration.

$$\int a\, dx = ax \qquad \int u\, dv = uv - \int v\, du \quad \text{(integration by parts)}$$

$$\int x^n\, dx = \frac{x^{n+1}}{n+1} \quad (n \neq -1) \qquad \int \frac{dx}{x} = \ln |x| \quad (x \neq 0)$$

$$\int \frac{dx}{x^n} = \frac{x^{1-n}}{1-n} \quad (n \neq 1) \qquad \int (a + bx)^n\, dx = \frac{(a + bx)^{n+1}}{b(n+1)} \quad (n \neq -1)$$

$$\int \frac{dx}{a + bx} = \frac{1}{b} \ln (a + bx) \qquad \int \frac{dx}{(a + bx)^2} = \frac{1}{b(a + bx)}$$

$$\int \frac{dx}{(a + bx)^n} = -\frac{1}{(n-1)(b)(a + bx)^{n-1}} \quad (n \neq 1)$$

$$\int \frac{dx}{a^2 + b^2 x^2} = \frac{1}{ab} \tan^{-1} \frac{bx}{a} \quad (x \text{ in radians}) \quad (a > 0, b > 0)$$

$$\int \frac{dx}{a^2 - b^2 x^2} = \frac{1}{2ab} \ln \left( \frac{a + bx}{a - bx} \right) \quad (x \text{ in radians}) \quad (a > 0, b > 0)$$

$$\int \frac{x \, dx}{a + bx} = \frac{1}{b^2} [bx - a \ln (a + bx)]$$

$$\int \frac{x \, dx}{(a + bx)^2} = \frac{1}{b^2} \left[ \frac{a}{a + bx} + \ln (a + bx) \right]$$

$$\int \frac{x \, dx}{(a + bx)^3} = -\frac{a + 2bx}{2b^2(a + bx)^2} \qquad \int \frac{x \, dx}{(a + bx)^4} = -\frac{a + 3bx}{6b^2(a + bx)^3}$$

$$\int \frac{x^2 \, dx}{a + bx} = \frac{1}{2b^3} [(a + bx)(-3a + bx) + 2a^2 \ln (a + bx)]$$

$$\int \frac{x^2 \, dx}{(a + bx)^2} = \frac{1}{b^3} \left[ \frac{bx(2a + bx)}{a + bx} - 2a \ln (a + bx) \right]$$

$$\int \frac{x^2 \, dx}{(a + bx)^3} = \frac{1}{b^3} \left[ \frac{a(3a + 4bx)}{2(a + bx)^2} + \ln (a + bx) \right]$$

$$\int \frac{x^2 \, dx}{(a + bx)^4} = -\frac{a^2 + 3abx + 3b^2 x^2}{3b^3(a + bx)^3}$$

$$\int \sin ax \, dx = -\frac{\cos ax}{a} \qquad \int \cos ax \, dx = \frac{\sin ax}{a}$$

$$\int \tan ax \, dx = \frac{1}{a} \ln (\sec ax) \qquad \int \cot ax \, dx = \frac{1}{a} \ln (\sin ax)$$

$$\int \sec ax \, dx = \frac{1}{a} \ln (\sec ax + \tan ax)$$

$$\int \csc ax \, dx = \frac{1}{a} \ln (\csc ax - \cot ax)$$

$$\int \sin^2 ax \, dx = \frac{x}{2} - \frac{\sin 2ax}{4a}$$

$$\int \cos^2 ax \, dx = \frac{x}{2} + \frac{\sin 2ax}{4a} \quad (x \text{ in radians})$$

$$\int x \sin ax \, dx = \frac{\sin ax}{a^2} - \frac{x \cos ax}{a} \quad (x \text{ in radians})$$

$$\int x \cos ax \, dx = \frac{\cos ax}{a^2} + \frac{x \sin ax}{a} \quad (x \text{ in radians})$$

$$\int e^{ax} \, dx = \frac{e^{ax}}{a} \quad \int x e^{ax} \, dx = \frac{e^{ax}}{a^2} (ax - 1)$$

$$\int \ln ax \, dx = x(\ln ax - 1)$$

$$\int \frac{dx}{1 + \sin ax} = -\frac{1}{a} \tan \left( \frac{\pi}{4} - \frac{ax}{2} \right)$$

$$\int \sqrt{a + bx} \, dx = \frac{2}{3b} (a + bx)^{3/2}$$

$$\int \sqrt{a^2 + b^2 x^2} \, dx = \frac{x}{2} \sqrt{a^2 + b^2 x^2} + \frac{a^2}{2b} \ln \left( \frac{bx}{a} + \sqrt{1 + \frac{b^2 x^2}{a^2}} \right)$$

$$\int \frac{dx}{\sqrt{a^2 + b^2 x^2}} = \frac{1}{b} \ln \left( \frac{bx}{a} + \sqrt{1 + \frac{b^2 x^2}{a^2}} \right)$$

$$\int \sqrt{a^2 - b^2 x^2} \, dx = \frac{x}{2} \sqrt{a^2 - b^2 x^2} + \frac{a^2}{2b} \sin^{-1} \frac{bx}{a}$$

## DEFINITE INTEGRALS

$$\int_a^b f(x) \, dx = -\int_b^a f(x) \, dx \qquad \int_a^b f(x) \, dx = \int_a^c f(x) \, dx + \int_c^b f(x) \, dx$$

# Properties of Plane Areas

Notation: $A$ = area

$\overline{x}, \overline{y}$ = distances to centroid $C$

$I_x, I_y$ = moments of inertia with respect to the $x$ and $y$ axes, respectively

$I_{xy}$ = product of inertia with respect to the $x$ and $y$ axes

$I_P = I_x + I_y$ = polar moment of inertia with respect to the origin of the $x$ and $y$ axes

$I_{BB}$ = moment of inertia with respect to axis B–B

---

**1**

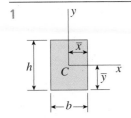

**Rectangle** (Origin of axes at centroid)

$$A = bh \quad \overline{x} = \frac{b}{2} \quad \overline{y} = \frac{h}{2}$$

$$I_x = \frac{bh^3}{12} \quad I_y = \frac{hb^3}{12} \quad I_{xy} = 0 \quad I_P = \frac{bh}{12}(h^2 + b^2)$$

---

**2**

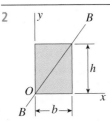

**Rectangle** (Origin of axes at corner)

$$I_x = \frac{bh^3}{3} \quad I_y = \frac{hb^3}{3} \quad I_{xy} = \frac{b^2h^2}{4} \quad I_P\frac{bh}{3}(h^2 + b^2)$$

$$I_{BB} = \frac{b^3h^3}{6(b^2 + h^2)}$$

---

**3**

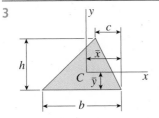

**Triangle** (Origin of axes at centroid)

$$A = \frac{bh}{2} \quad \overline{x} = \frac{b + c}{3} \quad \overline{y} = \frac{h}{3}$$

$$I_x = \frac{bh^3}{36} \quad I_y = \frac{bh}{36}(b^2 - bc + c^2)$$

$$I_{xy} = \frac{bh^2}{72}(b - 2c) \quad I_P = \frac{bh}{36}(h^2 + b^2 - bc + c^2)$$

**4**

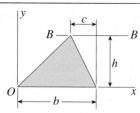

**Triangle** (Origin of axes at vertex)

$$I_x = \frac{bh^3}{12} \qquad I_y = \frac{bh}{12}(3b^2 - 3bc + c^2)$$

$$I_{xy} = \frac{bh^2}{24}(3b - 2c) \qquad I_{BB} = \frac{bh^3}{4}$$

**5**

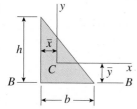

**Isosceles triangle** (Origin of axes at centroid)

$$A = \frac{bh}{2} \qquad \bar{x} = \frac{b}{2} \qquad \bar{y} = \frac{h}{3}$$

$$I_x = \frac{bh^3}{36} \qquad I_y = \frac{hb^3}{48} \qquad I_{xy} = 0$$

$$I_P = \frac{bh}{144}(4h^2 + 3b^2) \qquad I_{BB} = \frac{bh^3}{12}$$

(*Note:* For an equilateral triangle, $h = \sqrt{3}\ b/2$.)

**6**

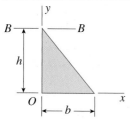

**Right triangle** (Origin of axes at centroid)

$$A = \frac{bh}{2} \qquad \bar{x} = \frac{b}{3} \qquad \bar{y} = \frac{h}{3}$$

$$I_x = \frac{bh^3}{36} \qquad I_y = \frac{hb^3}{36} \qquad I_{xy} = -\frac{b^2h^2}{72}$$

$$I_P = \frac{bh}{36}(h^2 + b^2) \qquad I_{BB} = \frac{bh^3}{12}$$

**7**

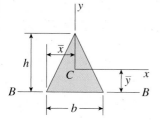

**Right triangle** (Origin of axes at vertex)

$$I_x = \frac{bh^3}{12} \qquad I_y = \frac{hb^3}{12} \qquad I_{xy} = \frac{b^2h^2}{24}$$

$$I_P = \frac{bh}{12}(h^2 + b^2) \qquad I_{BB} = \frac{bh^3}{4}$$

**8**

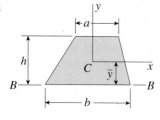

**Trapezoid** (Origin of axes at centroid)

$$A = \frac{h(a + b)}{2} \qquad \bar{y} = \frac{h(2a + b)}{3(a + b)}$$

$$I_x = \frac{h^3(a^2 + 4ab + b^2)}{36(a + b)} \qquad I_{BB} = \frac{h^3(3a + b)}{12}$$

9

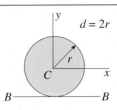

**Circle** (Origin of axes at center)

$$A = \pi r^2 = \frac{\pi d^2}{4} \quad I_x = I_y = \frac{\pi r^4}{4} = \frac{\pi d^4}{64}$$

$$I_{xy} = 0 \quad I_P = \frac{\pi r^4}{2} = \frac{\pi d^4}{32} \quad I_{BB} = \frac{5\pi r^4}{4} = \frac{5\pi d^4}{64}$$

10

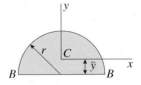

**Semicircle** (Origin of axes at centroid)

$$A = \frac{\pi r^2}{2} \quad \bar{y} = \frac{4r}{3\pi}$$

$$I_x = \frac{(9\pi^2 - 64)r^4}{72\pi} \approx 0.1098r^4 \quad I_y = \frac{\pi r^4}{8} \quad I_{xy} = 0 \quad I_{BB} = \frac{\pi r^4}{8}$$

11

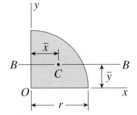

**Quarter circle** (Origin of axes at center of circle)

$$A = \frac{\pi r^2}{4} \quad \bar{x} = \bar{y} = \frac{4r}{3\pi}$$

$$I_x = I_y = \frac{\pi r^4}{16} \quad I_{xy} = \frac{r^4}{8} \quad I_{BB} = \frac{(9\pi^2 - 64)r^4}{144\pi} \approx 0.05488r^4$$

12

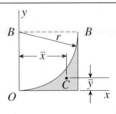

**Quarter-circular spandrel** (Origin of axes at point of tangency)

$$A = \left(1 - \frac{\pi}{4}\right)r^2 \quad \bar{x} = \frac{2r}{3(4 - \pi)} \approx 0.7766r \quad \bar{y} = \frac{(10 - 3\pi)r}{3(4 - \pi)} \approx 0.2234r$$

$$I_x = \left(1 - \frac{5\pi}{16}\right)r^4 \approx 0.01825r^4 \quad I_y = I_{BB} = \left(\frac{1}{3} - \frac{\pi}{16}\right)r^4 \approx 0.1370r^4$$

13

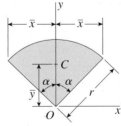

**Circular sector** (Origin of axes at center of circle)

$$\alpha = \text{angle in radians} \quad (\alpha \leq \pi/2)$$

$$A = \alpha r^2 \quad \bar{x} = r \sin \alpha \quad \bar{y} = \frac{2r \sin \alpha}{3\alpha}$$

$$I_x = \frac{r^4}{4}(\alpha + \sin \alpha \cos \alpha) \quad I_y = \frac{r^4}{4}(\alpha - \sin \alpha \cos \alpha)$$

$$I_{xy} = 0 \quad I_P = \frac{\alpha r^4}{2}$$

14

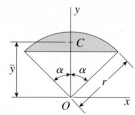

**Circular segment** (Origin of axes at center of circle)

$\alpha$ = angle in radians    $(\alpha \le \pi/2)$

$$A = r^2(\alpha - \sin\alpha\cos\alpha) \qquad \bar{y} = \frac{2r}{3}\left(\frac{\sin^3\alpha}{\alpha - \sin\alpha\cos\alpha}\right)$$

$$I_x = \frac{r^4}{4}(\alpha - \sin\alpha\cos\alpha + 2\sin^3\alpha\cos\alpha) \qquad I_{xy} = 0$$

$$I_y = \frac{r^4}{12}(3\alpha - 3\sin\alpha\cos\alpha - 2\sin^3\alpha\cos\alpha)$$

15

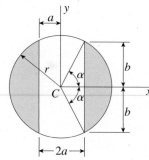

**Circle with core removed** (Origin of axes at center of circle)

$\alpha$ = angle in radians    $(\alpha \le \pi/2)$

$$\alpha = \arccos\frac{a}{r} \qquad b = \sqrt{r^2 - a^2} \qquad A = 2r^2\left(\alpha - \frac{ab}{r^2}\right)$$

$$I_x = \frac{r^4}{6}\left(3\alpha - \frac{3ab}{r^2} - \frac{2ab^3}{r^4}\right) \qquad I_y = \frac{r^4}{2}\left(\alpha - \frac{ab}{r^2} + \frac{2ab^3}{r^4}\right) \qquad I_{xy} = 0$$

16

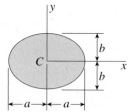

**Ellipse** (Origin of axes at centroid)

$$A = \pi ab \qquad I_x = \frac{\pi ab^3}{4} \qquad I_y = \frac{\pi ba^3}{4}$$

$$I_{xy} = 0 \qquad I_P = \frac{\pi ab}{4}(b^2 + a^2)$$

$$\text{Circumference} \approx \pi[1.5(a + b) - \sqrt{ab}] \qquad (a/3 \le b \le a)$$

$$\approx 4.17b^2/a + 4a \qquad (0 \le b \le a/3)$$

17

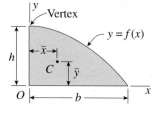

**Parabolic semisegment** (Origin of axes at corner)

$$y = f(x) = h\left(1 - \frac{x^2}{b^2}\right)$$

$$A = \frac{2bh}{3} \qquad \bar{x} = \frac{3b}{8} \qquad \bar{y} = \frac{2h}{5}$$

$$I_x = \frac{16bh^3}{105} \qquad I_y = \frac{2hb^3}{15} \qquad I_{xy} = \frac{b^2h^2}{12}$$

**18**

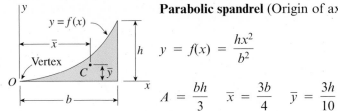

**Parabolic spandrel** (Origin of axes at vertex)

$$y = f(x) = \frac{hx^2}{b^2}$$

$$A = \frac{bh}{3} \qquad \bar{x} = \frac{3b}{4} \qquad \bar{y} = \frac{3h}{10}$$

$$I_x = \frac{bh^3}{21} \qquad I_y = \frac{hb^3}{5} \qquad I_{xy} = \frac{b^2h^2}{12}$$

**19**

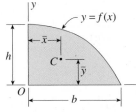

**Semisegment of $n$th degree** (Origin of axes at corner)

$$y = f(x) = h\left(1 - \frac{x^n}{b^n}\right) \qquad (n > 0)$$

$$A = bh\left(\frac{n}{n+1}\right) \qquad \bar{x} = \frac{b(n+1)}{2(n+2)} \qquad \bar{y} = \frac{hn}{2n+1}$$

$$I_x = \frac{2bh^3n^3}{(n+1)(2n+1)(3n+1)} \qquad I_y = \frac{hb^3n}{3(n+3)}$$

$$I_{xy} = \frac{b^2h^2n^2}{4(n+1)(n+2)}$$

**20**

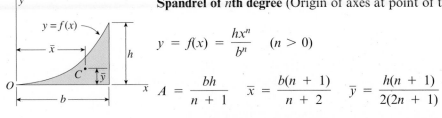

**Spandrel of $n$th degree** (Origin of axes at point of tangency)

$$y = f(x) = \frac{hx^n}{b^n} \qquad (n > 0)$$

$$A = \frac{bh}{n+1} \qquad \bar{x} = \frac{b(n+1)}{n+2} \qquad \bar{y} = \frac{h(n+1)}{2(2n+1)}$$

$$I_x = \frac{bh^3}{3(3n+1)} \qquad I_y = \frac{hb^3}{n+3} \qquad I_{xy} = \frac{b^2h^2}{4(n+1)}$$

**21**

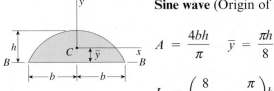

**Sine wave** (Origin of axes at centroid)

$$A = \frac{4bh}{\pi} \qquad \bar{y} = \frac{\pi h}{8}$$

$$I_x = \left(\frac{8}{9\pi} - \frac{\pi}{16}\right)bh^3 \approx 0.08659bh^3 \qquad I_y = \left(\frac{4}{\pi} - \frac{32}{\pi^3}\right)hb^3 \approx 0.2412hb^3$$

$$I_{xy} = 0 \qquad I_{BB} = \frac{8bh^3}{9\pi}$$

22

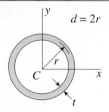

**Thin circular ring** (Origin of axes at center) Approximate formulas for case when $t$ is small

$$A = 2\pi rt = \pi dt \quad I_x = I_y = \pi r^3 t = \frac{\pi d^3 t}{8}$$

$$I_{xy} = 0 \quad I_P = 2\pi r^3 t = \frac{\pi d^3 t}{4}$$

23

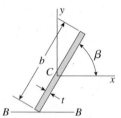

**Thin circular arc** (Origin of axes at center of circle) Approximate formulas for case when $t$ is small

$\beta$ = angle in radians   (*Note*: For a semicircular arc, $\beta = \pi/2$.)

$$A = 2\beta rt \quad \bar{y} = \frac{r \sin \beta}{\beta}$$

$$I_x = r^3 t(\beta + \sin \beta \cos \beta) \quad I_y = r^3 t(\beta - \sin \beta \cos \beta)$$

$$I_{xy} = 0 \quad I_{BB} = r^3 t \left( \frac{2\beta + \sin 2\beta}{2} - \frac{1 - \cos 2\beta}{\beta} \right)$$

24

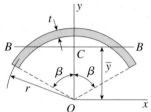

**Thin rectangle** (Origin of axes at centroid) Approximate formulas for case when $t$ is small

$$A = bt$$

$$I_x = \frac{tb^3}{12} \sin^2 \beta \quad I_y = \frac{tb^3}{12} \cos^2 \beta \quad I_{BB} = \frac{tb^3}{3} \sin^2 \beta$$

25

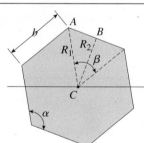

**Regular polygon with $n$ sides** (Origin of axes at centroid)

$C$ = centroid (at center of polygon)

$n$ = number of sides $(n \geq 3)$   $b$ = length of a side

$\beta$ = central angle for a side   $\alpha$ = interior angle (or vertex angle)

$$\beta = \frac{360°}{n} \quad \alpha = \left( \frac{n-2}{n} \right) 180° \quad \alpha + \beta = 180°$$

$R_1$ = radius of circumscribed circle (line $CA$)

$R_2$ = radius of inscribed circle (line $CB$)

$$R_1 = \frac{b}{2} \csc \frac{\beta}{2} \quad R_2 = \frac{b}{2} \cot \frac{\beta}{2} \quad A = \frac{nb^2}{4} \cot \frac{\beta}{2}$$

$I_c$ = moment of inertia about any axis through C (The centroid C is a principal point and every axis through C is a principal axis.)

$$I_c = \frac{nb^4}{192} \left( \cot \frac{\beta}{2} \right) \left( 3 \cot^2 \frac{\beta}{2} + 1 \right) \quad I_P = 2I_c$$

# Properties of Structural-Steel Shapes

In the following tables, the properties of a few structural-steel shapes are presented as an aid to the reader in solving problems in the text. These tables were compiled from the extensive tables in the *Manual of Steel Construction*, published by the American Institute of Steel Construction, Inc. (Ref. 5-4).

Notation:

$I$ = moment of inertia

$S$ = section modulus

$r = \sqrt{I/A}$ = radius of gyration

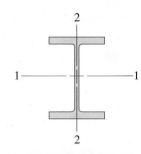

## Table F-1(a)

### Properties of Wide-Flange Sections (W Shapes) – USCS Units (Abridged List)

| Designation | Weight per foot | Area | Depth | Web thickness | Flange Width | Flange Thickness | Axis 1–1 I | Axis 1–1 S | Axis 1–1 r | Axis 2–2 I | Axis 2–2 S | Axis 2–2 r |
|---|---|---|---|---|---|---|---|---|---|---|---|---|
| | lb | in.² | in. | in. | in. | in. | in.⁴ | in.³ | in. | in.⁴ | in.³ | in. |
| W 30 × 211 | 211 | 62.2 | 30.9 | 0.775 | 15.1 | 1.32 | 10,300 | 665 | 12.9 | 757 | 100 | 3.49 |
| W 30 × 132 | 132 | 38.9 | 30.3 | 0.615 | 10.5 | 1.00 | 5770 | 380 | 12.2 | 196 | 37.2 | 2.25 |
| W 24 × 162 | 162 | 47.7 | 25.0 | 0.705 | 13.0 | 1.22 | 5170 | 414 | 10.4 | 443 | 68.4 | 3.05 |
| W 24 × 94 | 94.0 | 27.7 | 24.3 | 0.515 | 9.07 | 0.875 | 2700 | 222 | 9.87 | 109 | 24.0 | 1.98 |
| W 18 × 119 | 119 | 35.1 | 19.0 | 0.655 | 11.3 | 1.06 | 2190 | 231 | 7.90 | 253 | 44.9 | 2.69 |
| W 18 × 71 | 71.0 | 20.8 | 18.5 | 0.495 | 7.64 | 0.810 | 1170 | 127 | 7.50 | 60.3 | 15.8 | 1.70 |
| W 16 × 100 | 100 | 29.5 | 17.0 | 0.585 | 10.4 | 0.985 | 1490 | 175 | 7.10 | 186 | 35.7 | 2.51 |
| W 16 × 77 | 77.0 | 22.6 | 16.5 | 0.455 | 10.3 | 0.760 | 1110 | 134 | 7.00 | 138 | 26.9 | 2.47 |
| W 16 × 57 | 57.0 | 16.8 | 16.4 | 0.430 | 7.12 | 0.715 | 758 | 92.2 | 6.72 | 43.1 | 12.1 | 1.60 |
| W 16 × 31 | 31.0 | 9.13 | 15.9 | 0.275 | 5.53 | 0.440 | 375 | 47.2 | 6.41 | 12.4 | 4.49 | 1.17 |
| W 14 × 120 | 120 | 35.3 | 14.5 | 0.590 | 14.7 | 0.940 | 1380 | 190 | 6.24 | 495 | 67.5 | 3.74 |
| W 14 × 82 | 82.0 | 24.0 | 14.3 | 0.510 | 10.1 | 0.855 | 881 | 123 | 6.05 | 148 | 29.3 | 2.48 |
| W 14 × 53 | 53.0 | 15.6 | 13.9 | 0.370 | 8.06 | 0.660 | 541 | 77.8 | 5.89 | 57.7 | 14.3 | 1.92 |
| W 14 × 26 | 26.0 | 7.69 | 13.9 | 0.255 | 5.03 | 0.420 | 245 | 35.3 | 5.65 | 8.91 | 3.55 | 1.08 |
| W 12 × 87 | 87.0 | 25.6 | 12.5 | 0.515 | 12.1 | 0.810 | 740 | 118 | 5.38 | 241 | 39.7 | 3.07 |
| W 12 × 50 | 50.0 | 14.6 | 12.2 | 0.370 | 8.08 | 0.640 | 391 | 64.2 | 5.18 | 56.3 | 13.9 | 1.96 |
| W 12 × 35 | 35.0 | 10.3 | 12.5 | 0.300 | 6.56 | 0.520 | 285 | 45.6 | 5.25 | 24.5 | 7.47 | 1.54 |
| W 12 × 14 | 14.0 | 4.16 | 11.9 | 0.200 | 3.97 | 0.225 | 88.6 | 14.9 | 4.62 | 2.36 | 1.19 | 0.753 |
| W 10 × 60 | 60.0 | 17.6 | 10.2 | 0.420 | 10.1 | 0.680 | 341 | 66.7 | 4.39 | 116 | 23.0 | 2.57 |
| W 10 × 45 | 45.0 | 13.3 | 10.1 | 0.350 | 8.02 | 0.620 | 248 | 49.1 | 4.32 | 53.4 | 13.3 | 2.01 |
| W 10 × 30 | 30.0 | 8.84 | 10.5 | 0.300 | 5.81 | 0.510 | 170 | 32.4 | 4.38 | 16.7 | 5.75 | 1.37 |
| W 10 × 12 | 12.0 | 3.54 | 9.87 | 0.190 | 3.96 | 0.210 | 53.8 | 10.9 | 3.90 | 2.18 | 1.10 | 0.785 |
| W 8 × 35 | 35.0 | 10.3 | 8.12 | 0.310 | 8.02 | 0.495 | 127 | 31.2 | 3.51 | 42.6 | 10.6 | 2.03 |
| W 8 × 28 | 28.0 | 8.24 | 8.06 | 0.285 | 6.54 | 0.465 | 98.0 | 24.3 | 3.45 | 21.7 | 6.63 | 1.62 |
| W 8 × 21 | 21.0 | 6.16 | 8.28 | 0.250 | 5.27 | 0.400 | 75.3 | 18.2 | 3.49 | 9.77 | 3.71 | 1.26 |
| W 8 × 15 | 15.0 | 4.44 | 8.11 | 0.245 | 4.01 | 0.315 | 48.0 | 11.8 | 3.29 | 3.41 | 1.70 | 0.876 |

*Note:* Axes 1–1 and 2–2 are principal centroidal axes.

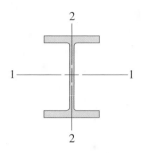

## Table F-1(b)

Properties of Wide-Flange Sections (W Shapes) – SI Units (Abridged List)

| Designation | Mass per meter | Area | Depth | Web thickness | Flange Width | Flange Thickness | Axis 1–1 $I$ | Axis 1–1 $S$ | Axis 1–1 $r$ | Axis 2–2 $I$ | Axis 2–2 $S$ | Axis 2–2 $r$ |
|---|---|---|---|---|---|---|---|---|---|---|---|---|
| | kg | mm² | mm | mm | mm | mm | $\times 10^6$ mm⁴ | $\times 10^3$ mm³ | mm | $\times 10^6$ mm⁴ | $\times 10^3$ mm³ | mm |
| W 760 × 314 | 314 | 40,100 | 785 | 19.7 | 384 | 33.5 | 4290 | 10,900 | 328 | 315 | 1640 | 88.6 |
| W 760 × 196 | 196 | 25,100 | 770 | 15.6 | 267 | 25.4 | 2400 | 6230 | 310 | 81.6 | 610 | 57.2 |
| W 610 × 241 | 241 | 30,800 | 635 | 17.9 | 330 | 31.0 | 2150 | 6780 | 264 | 184 | 1120 | 77.5 |
| W 610 × 140 | 140 | 17,900 | 617 | 13.1 | 230 | 22.2 | 1120 | 3640 | 251 | 45.4 | 393 | 50.3 |
| W 460 × 177 | 177 | 22,600 | 483 | 16.6 | 287 | 26.9 | 912 | 3790 | 201 | 105 | 736 | 68.3 |
| W 460 × 106 | 106 | 13,400 | 470 | 12.6 | 194 | 20.6 | 487 | 2080 | 191 | 25.1 | 259 | 43.2 |
| W 410 × 149 | 149 | 19,000 | 432 | 14.9 | 264 | 25.0 | 620 | 2870 | 180 | 77.4 | 585 | 63.8 |
| W 410 × 114 | 114 | 14,600 | 419 | 11.6 | 262 | 19.3 | 462 | 2200 | 178 | 57.4 | 441 | 62.7 |
| W 410 × 85 | 85.0 | 10,800 | 417 | 10.9 | 181 | 18.2 | 316 | 1510 | 171 | 17.9 | 198 | 40.6 |
| W 410 × 46.1 | 46.1 | 5890 | 404 | 6.99 | 140 | 11.2 | 156 | 773 | 163 | 5.16 | 73.6 | 29.7 |
| W 360 × 179 | 179 | 22,800 | 368 | 15.0 | 373 | 23.9 | 574 | 3110 | 158 | 206 | 1110 | 95.0 |
| W 360 × 122 | 122 | 15,500 | 363 | 13.0 | 257 | 21.7 | 367 | 2020 | 154 | 61.6 | 480 | 63.0 |
| W 360 × 79 | 79.0 | 10,100 | 353 | 9.40 | 205 | 16.8 | 225 | 1270 | 150 | 24.0 | 234 | 48.8 |
| W 360 × 39 | 39.0 | 4960 | 353 | 6.48 | 128 | 10.7 | 102 | 578 | 144 | 3.71 | 58.2 | 27.4 |
| W 310 × 129 | 129 | 16,500 | 318 | 13.1 | 307 | 20.6 | 308 | 1930 | 137 | 100 | 651 | 78.0 |
| W 310 × 74 | 74.0 | 9420 | 310 | 9.40 | 205 | 16.3 | 163 | 1050 | 132 | 23.4 | 228 | 49.8 |
| W 310 × 52 | 52.0 | 6650 | 318 | 7.62 | 167 | 13.2 | 119 | 747 | 133 | 10.2 | 122 | 39.1 |
| W 310 × 21 | 21.0 | 2680 | 302 | 5.08 | 101 | 5.72 | 36.9 | 244 | 117 | 0.982 | 19.5 | 19.1 |
| W 250 × 89 | 89.0 | 11,400 | 259 | 10.7 | 257 | 17.3 | 142 | 1090 | 112 | 48.3 | 377 | 65.3 |
| W 250 × 67 | 67.0 | 8580 | 257 | 8.89 | 204 | 15.7 | 103 | 805 | 110 | 22.2 | 218 | 51.1 |
| W 250 × 44.8 | 44.8 | 5700 | 267 | 7.62 | 148 | 13.0 | 70.8 | 531 | 111 | 6.95 | 94.2 | 34.8 |
| W 250 × 17.9 | 17.9 | 2280 | 251 | 4.83 | 101 | 5.33 | 22.4 | 179 | 99.1 | 0.907 | 18.0 | 19.9 |
| W 200 × 52 | 52.0 | 6650 | 206 | 7.87 | 204 | 12.6 | 52.9 | 511 | 89.2 | 17.7 | 174 | 51.6 |
| W 200 × 41.7 | 41.7 | 5320 | 205 | 7.24 | 166 | 11.8 | 40.8 | 398 | 87.6 | 9.03 | 109 | 41.1 |
| W 200 × 31.3 | 31.3 | 3970 | 210 | 6.35 | 134 | 10.2 | 31.3 | 298 | 88.6 | 4.07 | 60.8 | 32.0 |
| W 200 × 22.5 | 22.5 | 2860 | 206 | 6.22 | 102 | 8.00 | 20.0 | 193 | 83.6 | 1.42 | 27.9 | 22.3 |

*Note:* Axes 1–1 and 2–2 are principal centroidal axes.

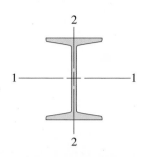

## Table F-2(a)

Properties of I-Beam Sections (S Shapes) – USCS Units (Abridged List)

| Designation | Weight per foot | Area | Depth | Web thickness | Flange Width | Flange Average thickness | Axis 1–1 $I$ | Axis 1–1 $S$ | Axis 1–1 $r$ | Axis 2–2 $I$ | Axis 2–2 $S$ | Axis 2–2 $r$ |
|---|---|---|---|---|---|---|---|---|---|---|---|---|
| | lb | in.² | in. | in. | in. | in. | in.⁴ | in.³ | in. | in.⁴ | in.³ | in. |
| S 24 × 100 | 100 | 29.3 | 24.0 | 0.745 | 7.25 | 0.870 | 2380 | 199 | 9.01 | 47.4 | 13.1 | 1.27 |
| S 24 × 80 | 80.0 | 23.5 | 24.0 | 0.500 | 7.00 | 0.870 | 2100 | 175 | 9.47 | 42.0 | 12.0 | 1.34 |
| S 20 × 96 | 96.0 | 28.2 | 20.3 | 0.800 | 7.20 | 0.920 | 1670 | 165 | 7.71 | 49.9 | 13.9 | 1.33 |
| S 20 × 75 | 75.0 | 22.0 | 20.0 | 0.635 | 6.39 | 0.795 | 1280 | 128 | 7.62 | 29.5 | 9.25 | 1.16 |
| S 18 × 70 | 70.0 | 20.5 | 18.0 | 0.711 | 6.25 | 0.691 | 923 | 103 | 6.70 | 24.0 | 7.69 | 1.08 |
| S 18 × 54.7 | 54.7 | 16.0 | 18.0 | 0.461 | 6.00 | 0.691 | 801 | 89.0 | 7.07 | 20.7 | 6.91 | 1.14 |
| S 15 × 50 | 50.0 | 14.7 | 15.0 | 0.550 | 5.64 | 0.622 | 485 | 64.7 | 5.75 | 15.6 | 5.53 | 1.03 |
| S 15 × 42.9 | 42.9 | 12.6 | 15.0 | 0.411 | 5.50 | 0.622 | 446 | 59.4 | 5.95 | 14.3 | 5.19 | 1.06 |
| S 12 × 50 | 50.0 | 14.6 | 12.0 | 0.687 | 5.48 | 0.659 | 303 | 50.6 | 4.55 | 15.6 | 5.69 | 1.03 |
| S 12 × 35 | 35.0 | 10.2 | 12.0 | 0.428 | 5.08 | 0.544 | 228 | 38.1 | 4.72 | 9.84 | 3.88 | 0.980 |
| S 10 × 35 | 35.0 | 10.3 | 10.0 | 0.594 | 4.94 | 0.491 | 147 | 29.4 | 3.78 | 8.30 | 3.36 | 0.899 |
| S 10 × 25.4 | 25.4 | 7.45 | 10.0 | 0.311 | 4.66 | 0.491 | 123 | 24.6 | 4.07 | 6.73 | 2.89 | 0.950 |
| S 8 × 23 | 23.0 | 6.76 | 8.00 | 0.441 | 4.17 | 0.425 | 64.7 | 16.2 | 3.09 | 4.27 | 2.05 | 0.795 |
| S 8 × 18.4 | 18.4 | 5.40 | 8.00 | 0.271 | 4.00 | 0.425 | 57.5 | 14.4 | 3.26 | 3.69 | 1.84 | 0.827 |
| S 6 × 17.2 | 17.3 | 5.06 | 6.00 | 0.465 | 3.57 | 0.359 | 26.2 | 8.74 | 2.28 | 2.29 | 1.28 | 0.673 |
| S 6 × 12.5 | 12.5 | 3.66 | 6.00 | 0.232 | 3.33 | 0.359 | 22.0 | 7.34 | 2.45 | 1.80 | 1.08 | 0.702 |
| S 4 × 9.5 | 9.50 | 2.79 | 4.00 | 0.326 | 2.80 | 0.293 | 6.76 | 3.38 | 1.56 | 0.887 | 0.635 | 0.564 |
| S 4 × 7.7 | 7.70 | 2.26 | 4.00 | 0.193 | 2.66 | 0.293 | 6.05 | 3.03 | 1.64 | 0.748 | 0.562 | 0.576 |

*Note:* Axes 1–1 and 2–2 are principal centroidal axes.

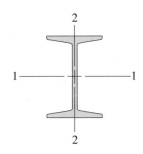

## Table F-2(b)

Properties of I-Beam Sections (S Shapes) – SI Units (Abridged List)

| Designation | Mass per meter | Area | Depth | Web thickness | Flange | | Axis 1–1 | | | Axis 2–2 | | |
|---|---|---|---|---|---|---|---|---|---|---|---|---|
| | | | | | Width | Average Thickness | $I$ | $S$ | $r$ | $I$ | $S$ | $r$ |
| | kg | mm² | mm | mm | mm | mm | $\times 10^6$ mm⁴ | $\times 10^3$ mm³ | mm | $\times 10^6$ mm⁴ | $\times 10^3$ mm³ | mm |
| S 610 × 149 | 149 | 18,900 | 610 | 18.9 | 184 | 22.1 | 991 | 3260 | 229 | 19.7 | 215 | 32.3 |
| S 610 × 119 | 119 | 15,200 | 610 | 12.7 | 178 | 22.1 | 874 | 2870 | 241 | 17.5 | 197 | 34.0 |
| S 510 × 143 | 143 | 18,200 | 516 | 20.3 | 183 | 23.4 | 695 | 2700 | 196 | 20.8 | 228 | 33.8 |
| S 510 × 112 | 112 | 14,200 | 508 | 16.1 | 162 | 20.2 | 533 | 2100 | 194 | 12.3 | 152 | 29.5 |
| S 460 × 104 | 104 | 13,200 | 457 | 18.1 | 159 | 17.6 | 384 | 1690 | 170 | 10.0 | 126 | 27.4 |
| S 460 × 81.4 | 81.4 | 10,300 | 457 | 11.7 | 152 | 17.6 | 333 | 1460 | 180 | 8.62 | 113 | 29.0 |
| S 380 × 74 | 74.0 | 9480 | 381 | 14.0 | 143 | 15.8 | 202 | 1060 | 146 | 6.49 | 90.6 | 26.2 |
| S 380 × 64 | 64.0 | 8130 | 381 | 10.4 | 140 | 15.8 | 186 | 973 | 151 | 5.95 | 85.0 | 26.9 |
| S 310 × 74 | 74.0 | 9420 | 305 | 17.4 | 139 | 16.7 | 126 | 829 | 116 | 6.49 | 93.2 | 26.2 |
| S 310 × 52 | 52.0 | 6580 | 305 | 10.9 | 129 | 13.8 | 94.9 | 624 | 120 | 4.10 | 63.6 | 24.9 |
| S 250 × 52 | 52.0 | 6650 | 254 | 15.1 | 125 | 12.5 | 61.2 | 482 | 96.0 | 3.45 | 55.1 | 22.8 |
| S 250 × 37.8 | 37.8 | 4810 | 254 | 7.90 | 118 | 12.5 | 51.2 | 403 | 103 | 2.80 | 47.4 | 24.1 |
| S 200 × 34 | 34.0 | 4360 | 203 | 11.2 | 106 | 10.8 | 26.9 | 265 | 78.5 | 1.78 | 33.6 | 20.2 |
| S 200 × 27.4 | 27.4 | 3480 | 203 | 6.88 | 102 | 10.8 | 23.9 | 236 | 82.8 | 1.54 | 30.2 | 21.0 |
| S 150 × 25.7 | 25.7 | 3260 | 152 | 11.8 | 90.7 | 9.12 | 10.9 | 143 | 57.9 | 0.953 | 21.0 | 17.1 |
| S 150 × 18.6 | 18.6 | 2360 | 152 | 5.89 | 84.6 | 9.12 | 9.16 | 120 | 62.2 | 0.749 | 17.7 | 17.8 |
| S 100 × 14.1 | 14.1 | 1800 | 102 | 8.28 | 71.1 | 7.44 | 2.81 | 55.4 | 39.6 | 0.369 | 10.4 | 14.3 |
| S 100 × 11.5 | 11.5 | 1460 | 102 | 4.90 | 67.6 | 7.44 | 2.52 | 49.7 | 41.7 | 0.311 | 9.21 | 14.6 |

*Note:* Axes 1–1 and 2–2 are principal centroidal axes.

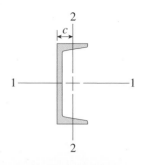

## Table F-3(a)

Properties of Channel Sections (C Shapes) – USCS Units (Abridged List)

| Designation | Weight per foot | Area | Depth | Web thickness | Flange | | Axis 1–1 | | | Axis 2–2 | | | |
|---|---|---|---|---|---|---|---|---|---|---|---|---|---|
| | | | | | Width | Average thickness | $I$ | $S$ | $r$ | $I$ | $S$ | $r$ | $c$ |
| | lb | in.² | in. | in. | in. | in. | in.⁴ | in.³ | in. | in.⁴ | in.³ | in. | in. |
| C 15 × 50 | 50.0 | 14.7 | 15.0 | 0.716 | 3.72 | 0.650 | 404 | 53.8 | 5.24 | 11.0 | 3.77 | 0.865 | 0.799 |
| C 15 × 40 | 40.0 | 11.8 | 15.0 | 0.520 | 3.52 | 0.650 | 348 | 46.5 | 5.45 | 9.17 | 3.34 | 0.883 | 0.778 |
| C 15 × 33.9 | 33.9 | 10.0 | 15.0 | 0.400 | 3.40 | 0.650 | 315 | 42.0 | 5.62 | 8.07 | 3.09 | 0.901 | 0.788 |
| C 12 × 30 | 30.0 | 8.81 | 12.0 | 0.510 | 3.17 | 0.501 | 162 | 27.0 | 4.29 | 5.12 | 2.05 | 0.762 | 0.674 |
| C 12 × 25 | 25.0 | 7.34 | 12.0 | 0.387 | 3.05 | 0.501 | 144 | 24.0 | 4.43 | 4.45 | 1.87 | 0.779 | 0.674 |
| C 12 × 20.7 | 20.7 | 6.08 | 12.0 | 0.282 | 2.94 | 0.501 | 129 | 21.5 | 4.61 | 3.86 | 1.72 | 0.797 | 0.698 |
| C 10 × 30 | 30.0 | 8.81 | 10.0 | 0.673 | 3.03 | 0.436 | 103 | 20.7 | 3.42 | 3.93 | 1.65 | 0.668 | 0.649 |
| C 10 × 25 | 25.0 | 7.34 | 10.0 | 0.526 | 2.89 | 0.436 | 91.1 | 18.2 | 3.52 | 3.34 | 1.47 | 0.675 | 0.617 |
| C 10 × 20 | 20.0 | 5.87 | 10.0 | 0.379 | 2.74 | 0.436 | 78.9 | 15.8 | 3.66 | 2.80 | 1.31 | 0.690 | 0.606 |
| C 10 × 15.3 | 15.3 | 4.48 | 10.0 | 0.240 | 2.60 | 0.436 | 67.3 | 13.5 | 3.87 | 2.27 | 1.15 | 0.711 | 0.634 |
| C 8 × 18.7 | 18.7 | 5.51 | 8.00 | 0.487 | 2.53 | 0.390 | 43.9 | 11.0 | 2.82 | 1.97 | 1.01 | 0.598 | 0.565 |
| C 8 × 13.7 | 13.7 | 4.04 | 8.00 | 0.303 | 2.34 | 0.390 | 36.1 | 9.02 | 2.99 | 1.52 | 0.848 | 0.613 | 0.554 |
| C 8 × 11.5 | 11.5 | 3.37 | 8.00 | 0.220 | 2.26 | 0.390 | 32.5 | 8.14 | 3.11 | 1.31 | 0.775 | 0.623 | 0.572 |
| C 6 × 13 | 13.0 | 3.81 | 6.00 | 0.437 | 2.16 | 0.343 | 17.3 | 5.78 | 2.13 | 1.05 | 0.638 | 0.524 | 0.514 |
| C 6 × 10.5 | 10.5 | 3.08 | 6.00 | 0.314 | 2.03 | 0.343 | 15.1 | 5.04 | 2.22 | 0.860 | 0.561 | 0.529 | 0.500 |
| C 6 × 8.2 | 8.20 | 2.39 | 6.00 | 0.200 | 1.92 | 0.343 | 13.1 | 4.35 | 2.34 | 0.687 | 0.488 | 0.536 | 0.512 |
| C 4 × 7.2 | 7.20 | 2.13 | 4.00 | 0.321 | 1.72 | 0.296 | 4.58 | 2.29 | 1.47 | 0.425 | 0.337 | 0.447 | 0.459 |
| C 4 × 5.4 | 5.40 | 1.58 | 4.00 | 0.184 | 1.58 | 0.296 | 3.85 | 1.92 | 1.56 | 0.312 | 0.277 | 0.444 | 0.457 |

*Notes:* 1. Axes 1–1 and 2–2 are principal centroidal axes.
   2. The distance $c$ is measured from the centroid to the back of the web.
   3. For axis 2–2, the tabulated value of $S$ is the smaller of the two section moduli for this axis.

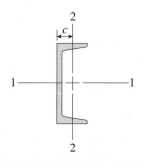

## Table F-3(b)

Properties of Channel Sections (C Shapes) – SI Units (Abridged List)

| Designation | Mass per meter | Area | Depth | Web thickness | Flange Width | Flange Average Thickness | Axis 1–1 I | Axis 1–1 S | Axis 1–1 r | Axis 2–2 I | Axis 2–2 S | Axis 2–2 r | c |
|---|---|---|---|---|---|---|---|---|---|---|---|---|---|
| | kg | mm² | mm | mm | mm | mm | × 10⁶ mm⁴ | × 10³ mm³ | mm | × 10⁶ mm⁴ | × 10³ mm³ | mm | mm |
| C 380 × 74 | 74.0 | 9480 | 381 | 18.2 | 94.5 | 16.5 | 168 | 882 | 133 | 4.58 | 61.8 | 22.0 | 20.3 |
| C 380 × 60 | 60.0 | 7610 | 381 | 13.2 | 89.4 | 16.5 | 145 | 762 | 138 | 3.82 | 54.7 | 22.4 | 19.8 |
| C 380 × 50.4 | 50.4 | 6450 | 381 | 10.2 | 86.4 | 16.5 | 131 | 688 | 143 | 3.36 | 50.6 | 22.9 | 20.0 |
| C 310 × 45 | 45.0 | 5680 | 305 | 13.0 | 80.5 | 12.7 | 67.4 | 442 | 109 | 2.13 | 33.6 | 19.4 | 17.1 |
| C 310 × 37 | 37.0 | 4740 | 305 | 9.83 | 77.5 | 12.7 | 59.9 | 393 | 113 | 1.85 | 30.6 | 19.8 | 17.1 |
| C 310 × 30.8 | 30.8 | 3920 | 305 | 7.16 | 74.7 | 12.7 | 53.7 | 352 | 117 | 1.61 | 28.2 | 20.2 | 17.7 |
| C 250 × 45 | 45.0 | 5680 | 254 | 17.1 | 77.0 | 11.1 | 42.9 | 339 | 86.9 | 1.64 | 27.0 | 17.0 | 16.5 |
| C 250 × 37 | 37.0 | 4740 | 254 | 13.4 | 73.4 | 11.1 | 37.9 | 298 | 89.4 | 1.39 | 24.1 | 17.1 | 15.7 |
| C 250 × 30 | 30.0 | 3790 | 254 | 9.63 | 69.6 | 11.1 | 32.8 | 259 | 93.0 | 1.17 | 21.5 | 17.5 | 15.4 |
| C 250 × 22.8 | 22.8 | 2890 | 254 | 6.10 | 66.0 | 11.1 | 28.0 | 221 | 98.3 | 0.945 | 18.8 | 18.1 | 16.1 |
| C 200 × 27.9 | 27.9 | 3550 | 203 | 12.4 | 64.3 | 9.91 | 18.3 | 180 | 71.6 | 0.820 | 16.6 | 15.2 | 14.4 |
| C 200 × 20.5 | 20.5 | 2610 | 203 | 7.70 | 59.4 | 9.91 | 15.0 | 148 | 75.9 | 0.633 | 13.9 | 15.6 | 14.1 |
| C 200 × 17.1 | 17.1 | 2170 | 203 | 5.59 | 57.4 | 9.91 | 13.5 | 133 | 79.0 | 0.545 | 12.7 | 15.8 | 14.5 |
| C 150 × 19.3 | 19.3 | 2460 | 152 | 11.1 | 54.9 | 8.71 | 7.20 | 94.7 | 54.1 | 0.437 | 10.5 | 13.3 | 13.1 |
| C 150 × 15.6 | 15.6 | 1990 | 152 | 7.98 | 51.6 | 8.71 | 6.29 | 82.6 | 56.4 | 0.358 | 9.19 | 13.4 | 12.7 |
| C 150 × 12.2 | 12.2 | 1540 | 152 | 5.08 | 48.8 | 8.71 | 5.45 | 71.3 | 59.4 | 0.286 | 8.00 | 13.6 | 13.0 |
| C 100 × 10.8 | 10.8 | 1370 | 102 | 8.15 | 43.7 | 7.52 | 1.91 | 37.5 | 37.3 | 0.177 | 5.52 | 11.4 | 11.7 |
| C 100 × 8 | 8.00 | 1020 | 102 | 4.67 | 40.1 | 7.52 | 1.60 | 31.5 | 39.6 | 0.130 | 4.54 | 11.3 | 11.6 |

*Notes:* 1. Axes 1–1 and 2–2 are principal centroidal axes.
2. The distance $c$ is measured from the centroid to the back of the web.
3. For axis 2–2, the tabulated value of $S$ is the smaller of the two section moduli for this axis.

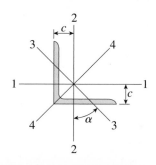

## Table F-4(a)

### Properties of Angle Sections with Equal Legs (L Shapes) – USCS Units (Abridged List)

| Designation | Weight per foot | Area | Axis 1–1 and Axis 2–2 | | | | Axis 3–3 |
|---|---|---|---|---|---|---|---|
| | | | $I$ | $S$ | $r$ | $c$ | $r_{min}$ |
| in. | lb | in.$^2$ | in.$^4$ | in.$^3$ | in. | in. | in. |
| L 8 × 8 × 1 | 51.0 | 15.0 | 89.1 | 15.8 | 2.43 | 2.36 | 1.56 |
| L 8 × 8 × 3/4 | 38.9 | 11.4 | 69.9 | 12.2 | 2.46 | 2.26 | 1.57 |
| L 8 × 8 × 1/2 | 26.4 | 7.75 | 48.8 | 8.36 | 2.49 | 2.17 | 1.59 |
| L 6 × 6 × 1 | 37.4 | 11.0 | 35.4 | 8.55 | 1.79 | 1.86 | 1.17 |
| L 6 × 6 × 3/4 | 28.7 | 8.46 | 28.1 | 6.64 | 1.82 | 1.77 | 1.17 |
| L 6 × 6 × 1/2 | 19.6 | 5.77 | 19.9 | 4.59 | 1.86 | 1.67 | 1.18 |
| L 5 × 5 × 7/8 | 27.2 | 7.98 | 17.8 | 5.16 | 1.49 | 1.56 | 0.971 |
| L 5 × 5 × 1/2 | 16.2 | 4.75 | 11.3 | 3.15 | 1.53 | 1.42 | 0.980 |
| L 5 × 5 × 3/8 | 12.3 | 3.61 | 8.76 | 2.41 | 1.55 | 1.37 | 0.986 |
| L 4 × 4 × 3/4 | 18.5 | 5.44 | 7.62 | 2.79 | 1.18 | 1.27 | 0.774 |
| L 4 × 4 × 1/2 | 12.8 | 3.75 | 5.52 | 1.96 | 1.21 | 1.18 | 0.776 |
| L 4 × 4 × 3/8 | 9.80 | 2.86 | 4.32 | 1.50 | 1.23 | 1.13 | 0.779 |
| L 3-1/2 × 3-1/2 × 3/8 | 8.50 | 2.48 | 2.86 | 1.15 | 1.07 | 1.00 | 0.683 |
| L 3-1/2 × 3-1/2 × 1/4 | 5.80 | 1.69 | 2.00 | 0.787 | 1.09 | 0.954 | 0.688 |
| L 3 × 3 × 1/2 | 9.40 | 2.75 | 2.20 | 1.06 | 0.895 | 0.929 | 0.580 |
| L 3 × 3 × 1/4 | 4.90 | 1.44 | 1.23 | 0.569 | 0.926 | 0.836 | 0.585 |

*Notes:* 1. Axes 1–1 and 2–2 are centroidal axes parallel to the legs.

2. The distance $c$ is measured from the centroid to the back of the legs.

3. For axes 1–1 and 2–2, the tabulated value of $S$ is the smaller of the two section moduli for those axes.

4. Axes 3–3 and 4–4 are principal centroidal axes.

5. The moment of inertia for axis 3–3, which is the smaller of the two principal moments of inertia, can be found from the equation $I_{33} = A r_{min}^2$.

6. The moment of inertia for axis 4–4, which is the larger of the two principal moments of inertia, can be found from the equation $I_{44} + I_{33} = I_{11} + I_{22}$.

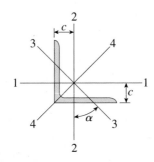

## Table F-4(b)

Properties of Angle Sections with Equal Legs (L Shapes) – SI Units (Abridged List)

| Designation | Mass per meter | Area | Axis 1–1 and Axis 2–2 | | | | Axis 3–3 |
|---|---|---|---|---|---|---|---|
| | | | $I$ | $S$ | $r$ | $c$ | $r_{min}$ |
| mm | kg | mm² | $\times 10^6$ mm⁴ | $\times 10^3$ mm³ | mm | mm | mm |
| L 203 × 203 × 25.4 | 75.9 | 9680 | 37.1 | 259 | 61.7 | 59.9 | 39.6 |
| L 203 × 203 × 19 | 57.9 | 7350 | 29.1 | 200 | 62.5 | 57.4 | 39.9 |
| L 203 × 203 × 12.7 | 39.3 | 5000 | 20.3 | 137 | 63.2 | 55.1 | 40.4 |
| L 152 × 152 × 25.4 | 55.7 | 7100 | 14.7 | 140 | 45.5 | 47.2 | 29.7 |
| L 152 × 152 × 19 | 42.7 | 5460 | 11.7 | 109 | 46.2 | 45.0 | 29.7 |
| L 152 × 152 × 12.7 | 29.2 | 3720 | 8.28 | 75.2 | 47.2 | 42.4 | 30.0 |
| L 127 × 127 × 22.2 | 40.5 | 5150 | 7.41 | 84.6 | 37.8 | 39.6 | 24.7 |
| L 127 × 127 × 12.7 | 24.1 | 3060 | 4.70 | 51.6 | 38.9 | 36.1 | 24.9 |
| L 127 × 127 × 9.5 | 18.3 | 2330 | 3.65 | 39.5 | 39.4 | 34.8 | 25.0 |
| L 102 × 102 × 19 | 27.5 | 3510 | 3.17 | 45.7 | 30.0 | 32.3 | 19.7 |
| L 102 × 102 × 12.7 | 19.0 | 2420 | 2.30 | 32.1 | 30.7 | 30.0 | 19.7 |
| L 102 × 102 × 9.5 | 14.6 | 1850 | 1.80 | 24.6 | 31.2 | 28.7 | 19.8 |
| L 89 × 89 × 9.5 | 12.6 | 1600 | 1.19 | 18.8 | 27.2 | 25.4 | 17.3 |
| L 89 × 89 × 6.4 | 8.60 | 1090 | 0.832 | 12.9 | 27.7 | 24.2 | 17.5 |
| L 76 × 76 × 12.7 | 14.0 | 1770 | 0.916 | 17.4 | 22.7 | 23.6 | 14.7 |
| L 76 × 76 × 6.4 | 7.30 | 929 | 0.512 | 9.32 | 23.5 | 21.2 | 14.9 |

*Notes:* 1. Axes 1–1 and 2–2 are centroidal axes parallel to the legs.

2. The distance $c$ is measured from the centroid to the back of the legs.

3. For axes 1–1 and 2–2, the tabulated value of $S$ is the smaller of the two section moduli for those axes.

4. Axes 3–3 and 4–4 are principal centroidal axes.

5. The moment of inertia for axis 3–3, which is the smaller of the two principal moments of inertia, can be found from the equation $I_{33} = Ar_{min}^2$.

6. The moment of inertia for axis 4–4, which is the larger of the two principal moments of inertia, can be found from the equation $I_{44} + I_{33} = I_{11} + I_{22}$.

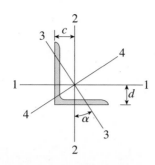

## Table F-5(a)

### Properties of Angle Sections with Unequal Legs (L Shapes) – USCS Units (Abridged List)

| Designation | Weight per foot | Area | Axis 1–1 | | | | Axis 2–2 | | | | Axis 3–3 | |
|---|---|---|---|---|---|---|---|---|---|---|---|---|
| | | | $I$ | $S$ | $r$ | $d$ | $I$ | $S$ | $r$ | $c$ | $r_{min}$ | $\tan \alpha$ |
| in. | lb | in.² | in.⁴ | in.³ | in. | in. | in.⁴ | in.³ | in. | in. | in. | |
| L 8 × 6 × 1 | 44.2 | 13.0 | 80.9 | 15.1 | 2.49 | 2.65 | 38.8 | 8.92 | 1.72 | 1.65 | 1.28 | 0.542 |
| L 8 × 6 × 1/2 | 23.0 | 6.75 | 44.4 | 8.01 | 2.55 | 2.46 | 21.7 | 4.79 | 1.79 | 1.46 | 1.30 | 0.557 |
| L 7 × 4 × 3/4 | 26.2 | 7.69 | 37.8 | 8.39 | 2.21 | 2.50 | 9.00 | 3.01 | 1.08 | 1.00 | 0.855 | 0.324 |
| L 7 × 4 × 1/2 | 17.9 | 5.25 | 26.6 | 5.79 | 2.25 | 2.40 | 6.48 | 2.10 | 1.11 | 0.910 | 0.866 | 0.334 |
| L 6 × 4 × 3/4 | 23.6 | 6.94 | 24.5 | 6.23 | 1.88 | 2.07 | 8.63 | 2.95 | 1.12 | 1.07 | 0.856 | 0.428 |
| L 6 × 4 × 1/2 | 16.2 | 4.75 | 17.3 | 4.31 | 1.91 | 1.98 | 6.22 | 2.06 | 1.14 | 0.981 | 0.864 | 0.440 |
| L 5 × 3-1/2 × 3/4 | 19.8 | 5.81 | 13.9 | 4.26 | 1.55 | 1.74 | 5.52 | 2.20 | 0.974 | 0.993 | 0.744 | 0.464 |
| L 5 × 3-1/2 × 1/2 | 13.6 | 4.00 | 10.0 | 2.97 | 1.58 | 1.65 | 4.02 | 1.55 | 1.00 | 0.901 | 0.750 | 0.479 |
| L 5 × 3 × 1/2 | 12.8 | 3.75 | 9.43 | 2.89 | 1.58 | 1.74 | 2.55 | 1.13 | 0.824 | 0.746 | 0.642 | 0.357 |
| L 5 × 3 × 1/4 | 6.60 | 1.94 | 5.09 | 1.51 | 1.62 | 1.64 | 1.41 | 0.600 | 0.853 | 0.648 | 0.652 | 0.371 |
| L 4 × 3-1/2 × 1/2 | 11.9 | 3.50 | 5.30 | 1.92 | 1.23 | 1.24 | 3.76 | 1.50 | 1.04 | 0.994 | 0.716 | 0.750 |
| L 4 × 3-1/2 × 1/4 | 6.20 | 1.81 | 2.89 | 1.01 | 1.26 | 1.14 | 2.07 | 0.794 | 1.07 | 0.897 | 0.723 | 0.759 |
| L 4 × 3 × 1/2 | 11.1 | 3.25 | 5.02 | 1.87 | 1.24 | 1.32 | 2.40 | 1.10 | 0.858 | 0.822 | 0.633 | 0.542 |
| L 4 × 3 × 3/8 | 8.50 | 2.48 | 3.94 | 1.44 | 1.26 | 1.27 | 1.89 | 0.851 | 0.873 | 0.775 | 0.636 | 0.551 |
| L 4 × 3 × 1/4 | 5.80 | 1.69 | 2.75 | 0.988 | 1.27 | 1.22 | 1.33 | 0.585 | 0.887 | 0.725 | 0.639 | 0.558 |

*Notes:* 1. Axes 1–1 and 2–2 are centroidal axes parallel to the legs.
2. The distances $c$ and $d$ are measured from the centroid to the backs of the legs.
3. For axes 1–1 and 2–2, the tabulated value of $S$ is the smaller of the two section moduli for those axes.
4. Axes 3–3 and 4–4 are principal centroidal axes.
5. The moment of inertia for axis 3–3, which is the smaller of the two principal moments of inertia, can be found from the equation $I_{33} = Ar_{min}^2$.
6. The moment of inertia for axis 4–4, which is the larger of the two principal moments of inertia, can be found from the equation $I_{44} + I_{33} = I_{11} + I_{22}$.

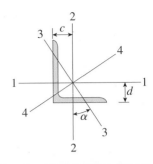

## Table F-5(b)

### Properties of Angle Sections with Unequal Legs (L Shapes) – SI Units (Abridged List)

| Designation | Mass per meter | Area | Axis 1–1 | | | | Axis 2–2 | | | | Axis 3–3 | |
|---|---|---|---|---|---|---|---|---|---|---|---|---|
| | | | $I$ | $S$ | $r$ | $d$ | $I$ | $S$ | $r$ | $c$ | $r_{min}$ | $\tan \alpha$ |
| mm | kg | mm$^2$ | $\times 10^6$ mm$^4$ | $\times 10^3$ mm$^3$ | mm | mm | $\times 10^6$ mm$^4$ | $\times 10^3$ mm$^3$ | mm | mm | mm | |
| L 203 × 152 × 25.4 | 65.5 | 8390 | 33.7 | 247 | 63.2 | 67.3 | 16.1 | 146 | 43.7 | 41.9 | 32.5 | 0.542 |
| L 203 × 152 × 12.7 | 34.1 | 4350 | 18.5 | 131 | 64.8 | 62.5 | 9.03 | 78.5 | 45.5 | 37.1 | 33.0 | 0.557 |
| L 178 × 102 × 19 | 38.8 | 4960 | 15.7 | 137 | 56.1 | 63.5 | 3.75 | 49.3 | 27.4 | 25.4 | 21.7 | 0.324 |
| L 178 × 102 × 12.7 | 26.5 | 3390 | 11.1 | 94.9 | 57.2 | 61.0 | 2.70 | 34.4 | 28.2 | 23.1 | 22.0 | 0.334 |
| L 152 × 102 × 19 | 35.0 | 4480 | 10.2 | 102 | 47.8 | 52.6 | 3.59 | 48.3 | 28.4 | 27.2 | 21.7 | 0.428 |
| L 152 × 102 × 12.7 | 24.0 | 3060 | 7.20 | 70.6 | 48.5 | 50.3 | 2.59 | 33.8 | 29.0 | 24.9 | 21.9 | 0.440 |
| L 127 × 89 × 19 | 29.3 | 3750 | 5.79 | 69.8 | 39.4 | 44.2 | 2.30 | 36.1 | 24.7 | 25.2 | 18.9 | 0.464 |
| L 127 × 89 × 12.7 | 20.2 | 2580 | 4.15 | 48.7 | 40.1 | 41.9 | 1.67 | 25.4 | 25.4 | 22.9 | 19.1 | 0.479 |
| L 127 × 76 × 12.7 | 19.0 | 2420 | 3.93 | 47.4 | 40.1 | 44.2 | 1.06 | 18.5 | 20.9 | 18.9 | 16.3 | 0.357 |
| L 127 × 76 × 6.4 | 9.80 | 1250 | 2.12 | 24.7 | 41.1 | 41.7 | 0.587 | 9.83 | 21.7 | 16.5 | 16.6 | 0.371 |
| L 102 × 89 × 12.7 | 17.6 | 2260 | 2.21 | 31.5 | 31.2 | 31.5 | 1.57 | 24.6 | 26.4 | 25.2 | 18.2 | 0.750 |
| L 102 × 89 × 6.4 | 9.20 | 1170 | 1.20 | 16.6 | 32.0 | 29.0 | 0.862 | 13.0 | 27.2 | 22.8 | 18.4 | 0.759 |
| L 102 × 76 × 12.7 | 16.4 | 2100 | 2.09 | 30.6 | 31.5 | 33.5 | 0.999 | 18.0 | 21.8 | 20.9 | 16.1 | 0.542 |
| L 102 × 76 × 9.5 | 12.6 | 1600 | 1.64 | 23.6 | 32.0 | 32.3 | 0.787 | 13.9 | 22.2 | 19.7 | 16.2 | 0.551 |
| L 102 × 76 × 6.4 | 8.60 | 1090 | 1.14 | 16.2 | 32.3 | 31.0 | 0.554 | 9.59 | 22.5 | 18.4 | 16.2 | 0.558 |

*Notes:* 1. Axes 1–1 and 2–2 are centroidal axes parallel to the legs.
2. The distances $c$ and $d$ are measured from the centroid to the backs of the legs.
3. For axes 1–1 and 2–2, the tabulated value of $S$ is the smaller of the two section moduli for those axes.
4. Axes 3–3 and 4–4 are principal centroidal axes.
5. The moment of inertia for axis 3–3, which is the smaller of the two principal moments of inertia, can be found from the equation $I_{33} = Ar_{min}^2$.
6. The moment of inertia for axis 4–4, which is the larger of the two principal moments of inertia, can be found from the equation $I_{44} + I_{33} = I_{11} + I_{22}$.

# Properties of Structural Lumber

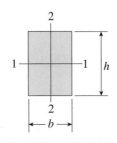

## Properties of Surfaced Lumber (Abridged List)

| Nominal dimensions $b \times h$ | Net dimensions $b \times h$ | Area $A = bh$ | Axis 1–1 Moment of inertia $I_1 = \dfrac{bh^3}{12}$ | Axis 1–1 Section modulus $S_1 = \dfrac{bh^2}{6}$ | Axis 2–2 Moment of inertia $I_2 = \dfrac{hb^3}{12}$ | Axis 2–2 Section modulus $S_2 = \dfrac{hb^2}{6}$ | Weight per linear foot (weight density = 35 lb/ft³) |
|---|---|---|---|---|---|---|---|
| in. | in. | in.² | in.⁴ | in.³ | in.⁴ | in.³ | lb |
| 2 × 4 | 1.5 × 3.5 | 5.25 | 5.36 | 3.06 | 0.98 | 1.31 | 1.3 |
| 2 × 6 | 1.5 × 5.5 | 8.25 | 20.80 | 7.56 | 1.55 | 2.06 | 2.0 |
| 2 × 8 | 1.5 × 7.25 | 10.88 | 47.63 | 13.14 | 2.04 | 2.72 | 2.6 |
| 2 × 10 | 1.5 × 9.25 | 13.88 | 98.93 | 21.39 | 2.60 | 3.47 | 3.4 |
| 2 × 12 | 1.5 × 11.25 | 16.88 | 177.98 | 31.64 | 3.16 | 4.22 | 4.1 |
| 3 × 4 | 2.5 × 3.5 | 8.75 | 8.93 | 5.10 | 4.56 | 3.65 | 2.1 |
| 3 × 6 | 2.5 × 5.5 | 13.75 | 34.66 | 12.60 | 7.16 | 5.73 | 3.3 |
| 3 × 8 | 2.5 × 7.25 | 18.13 | 79.39 | 21.90 | 9.44 | 7.55 | 4.4 |
| 3 × 10 | 2.5 × 9.25 | 23.13 | 164.89 | 35.65 | 12.04 | 9.64 | 5.6 |
| 3 × 12 | 2.5 × 11.25 | 28.13 | 296.63 | 52.73 | 14.65 | 11.72 | 6.8 |
| 4 × 4 | 3.5 × 3.5 | 12.25 | 12.51 | 7.15 | 12.51 | 7.15 | 3.0 |
| 4 × 6 | 3.5 × 5.5 | 19.25 | 48.53 | 17.65 | 19.65 | 11.23 | 4.7 |
| 4 × 8 | 3.5 × 7.25 | 25.38 | 111.15 | 30.66 | 25.90 | 14.80 | 6.2 |
| 4 × 10 | 3.5 × 9.25 | 32.38 | 230.84 | 49.91 | 33.05 | 18.89 | 7.9 |
| 4 × 12 | 3.5 × 11.25 | 39.38 | 415.28 | 73.83 | 40.20 | 22.97 | 9.6 |
| 6 × 6 | 5.5 × 5.5 | 30.25 | 76.3 | 27.7 | 76.3 | 27.7 | 7.4 |
| 6 × 8 | 5.5 × 7.5 | 41.25 | 193.4 | 51.6 | 104.0 | 37.8 | 10.0 |
| 6 × 10 | 5.5 × 9.5 | 52.25 | 393.0 | 82.7 | 131.7 | 47.9 | 12.7 |
| 6 × 12 | 5.5 × 11.5 | 63.25 | 697.1 | 121.2 | 159.4 | 58.0 | 15.4 |
| 8 × 8 | 7.5 × 7.5 | 56.25 | 263.7 | 70.3 | 263.7 | 70.3 | 13.7 |
| 8 × 10 | 7.5 × 9.5 | 71.25 | 535.9 | 112.8 | 334.0 | 89.1 | 17.3 |
| 8 × 12 | 7.5 × 11.5 | 86.25 | 950.5 | 165.3 | 404.3 | 107.8 | 21.0 |

*Note:* Axes 1–1 and 2–2 are principal centroidal axes.

# Deflections and Slopes of Beams

## Table H-1

Deflections and Slopes of Cantilever Beams

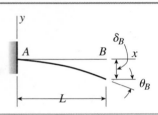

$v =$ deflection in the $y$ direction (positive upward)

$v' = dv/dx =$ slope of the deflection curve

$\delta_B = -v(L) =$ deflection at end $B$ of the beam (positive downward)

$\theta_B = -v'(L) =$ angle of rotation at end $B$ of the beam (positive clockwise)

$EI =$ constant

**1**

$$v = -\frac{qx^2}{24EI}(6L^2 - 4Lx + x^2) \qquad v' = -\frac{qx}{6EI}(3L^2 - 3Lx + x^2)$$

$$\delta_B = \frac{qL^4}{8EI} \qquad \theta_B = \frac{qL^3}{6EI}$$

**2**

$$v = -\frac{qx^2}{24EI}(6a^2 - 4ax + x^2) \qquad (0 \le x \le a)$$

$$v' = -\frac{qx}{6EI}(3a^2 - 3ax + x^2) \qquad (0 \le x \le a)$$

$$v = -\frac{qa^3}{24EI}(4x - a) \qquad v' = -\frac{qa^3}{6EI} \qquad (a \le x \le L)$$

$$\text{At } x = a: \; v = -\frac{qa^4}{8EI} \qquad v' = -\frac{qa^3}{6EI}$$

$$\delta_B = \frac{qa^3}{24EI}(4L - a) \qquad \theta_B = \frac{qa^3}{6EI}$$

(Continued)

## Table H-1 (Continued)

**3**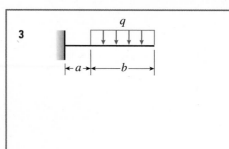

$$v = -\frac{qbx^2}{12EI}(3L + 3a - 2x) \quad (0 \le x \le a)$$

$$v' = -\frac{qbx}{2EI}(L + a - x) \quad (0 \le x \le a)$$

$$v = -\frac{q}{24EI}(x^4 - 4Lx^3 + 6L^2x^2 - 4a^3x + a^4) \quad (a \le x \le L)$$

$$v' = -\frac{q}{6EI}(x^3 - 3Lx^2 + 3L^2x - a^3) \quad (a \le x \le L)$$

At $x = a$: $v = -\dfrac{qa^2b}{12EI}(3L + a)$ $\quad v' = -\dfrac{qabL}{2EI}$

$$\delta_B = \frac{q}{24EI}(3L^4 - 4a^3L + a^4) \qquad \theta_B = \frac{q}{6EI}(L^3 - a^3)$$

**4**

$$v = -\frac{Px^2}{6EI}(3L - x) \qquad v' = -\frac{Px}{2EI}(2L - x)$$

$$\delta_B = \frac{PL^3}{3EI} \qquad \theta_B = \frac{PL^2}{2EI}$$

**5**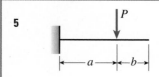

$$v = -\frac{Px^2}{6EI}(3a - x) \qquad v' = -\frac{Px}{2EI}(2a - x) \quad (0 \le x \le a)$$

$$v = -\frac{Pa^2}{6EI}(3x - a) \qquad v' = -\frac{Pa^2}{2EI} \quad (a \le x \le L)$$

At $x = a$: $v = -\dfrac{Pa^3}{3EI}$ $\qquad v' = -\dfrac{Pa^2}{2EI}$

$$\delta_B = \frac{Pa^2}{6EI}(3L - a) \qquad \theta_B = \frac{Pa^2}{2EI}$$

**6**

$$v = -\frac{M_0x^2}{2EI} \qquad v' = -\frac{M_0x}{EI}$$

$$\delta_B = \frac{M_0L^2}{2EI} \qquad \theta_B = \frac{M_0L}{EI}$$

## Table H-1 (Continued)

**7**

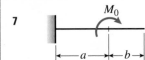

$$v = -\frac{M_0 x^2}{2EI} \qquad v' = -\frac{M_0 x}{EI} \qquad (0 \le x \le a)$$

$$v = -\frac{M_0 a}{2EI}(2x - a) \qquad v' = -\frac{M_0 a}{EI} \qquad (a \le x \le L)$$

$$At \ x = a: \qquad v = -\frac{M_0 a^2}{2EI} \qquad v' = -\frac{M_0 a}{EI}$$

$$\delta_B = \frac{M_0 a}{2EI}(2L - a) \qquad \theta_B = \frac{M_0 a}{EI}$$

**8**

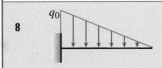

$$v = -\frac{q_0 x^2}{120 LEI}(10L^3 - 10L^2 x + 5Lx^2 - x^3)$$

$$v' = -\frac{q_0 x}{24 LEI}(4L^3 - 6L^2 x + 4Lx^2 - x^3)$$

$$\delta_B = \frac{q_0 L^4}{30EI} \qquad \theta_B = \frac{q_0 L^3}{24EI}$$

**9**

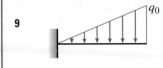

$$v = -\frac{q_0 x^2}{120 LEI}(20L^3 - 10L^2 x + x^3)$$

$$v' = -\frac{q_0 x}{24 LEI}(8L^3 - 6L^2 x + x^3)$$

$$\delta_B = \frac{11 q_0 L^4}{120EI} \qquad \theta_B = \frac{q_0 L^3}{8EI}$$

**10**

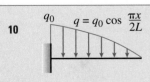
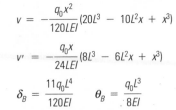

$$v = -\frac{q_0 L}{3\pi^4 EI}\left(48L^3 \cos\frac{\pi x}{2L} - 48L^3 + 3\pi^3 Lx^2 - \pi^3 x^3\right)$$

$$v' = -\frac{q_0 L}{\pi^3 EI}\left(2\pi^2 Lx - \pi^2 x^2 - 8L^2 \sin\frac{\pi x}{2L}\right)$$

$$\delta_B = \frac{2q_0 L^4}{3\pi^4 EI}(\pi^3 - 24) \qquad \theta_B = \frac{q_0 L^3}{\pi^3 EI}(\pi^2 - 8)$$

## Table H-2

### Deflections and Slopes of Simple Beams

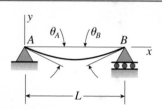

$v = $ deflection in the $y$ direction (positive upward)

$v' = dv/dx = $ slope of the deflection curve

$\delta_C = -v(L/2) = $ deflection at midpoint $C$ of the beam (positive downward)

$x_1 = $ distance from support $A$ to point of maximum deflection

$\delta_{max} = -v_{max} = $ maximum deflection (positive downward)

$\theta_A = -v'(0) = $ angle of rotation at left-hand end of the beam (positive clockwise)

$\theta_B = v'(L) = $ angle of rotation at right-hand end of the beam (positive counterclockwise)

$EI = $ constant

---

**1**

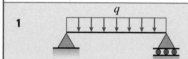

$$v = -\frac{qx}{24EI}(L^3 - 2Lx^2 + x^3)$$

$$v' = -\frac{q}{24EI}(L^3 - 6Lx^2 + 4x^3)$$

$$\delta_C = \delta_{max} = \frac{5qL^4}{384EI} \qquad \theta_A = \theta_B = \frac{qL^3}{24EI}$$

---

**2**

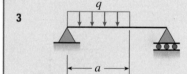

$$v = -\frac{qx}{384EI}(9L^3 - 24Lx^2 + 16x^3) \qquad \left(0 \le x \le \frac{L}{2}\right)$$

$$v' = -\frac{q}{384EI}(9L^3 - 72Lx^2 + 64x^3) \qquad \left(0 \le x \le \frac{L}{2}\right)$$

$$v = -\frac{qL}{384EI}(8x^3 - 24Lx^2 + 17L^2x - L^3) \qquad \left(\frac{L}{2} \le x \le L\right)$$

$$v' = -\frac{qL}{384EI}(24x^2 - 48Lx + 17L^2) \qquad \left(\frac{L}{2} \le x \le L\right)$$

$$\delta_C = \frac{5qL^4}{768EI} \qquad \theta_A = \frac{3qL^3}{128EI} \qquad \theta_B = \frac{7qL^3}{384EI}$$

---

**3**

$$v = -\frac{qx}{24LEI}(a^4 - 4a^3L + 4a^2L^2 + 2a^2x^2 - 4aLx^2 + Lx^3) \qquad (0 \le x \le a)$$

$$v' = -\frac{q}{24LEI}(a^4 - 4a^3L + 4a^2L^2 + 6a^2x^2 - 12aLx^2 + 4Lx^3) \qquad (0 \le x \le a)$$

$$v = -\frac{qa^2}{24LEI}(-a^2L + 4L^2x + a^2x - 6Lx^2 + 2x^3) \qquad (a \le x \le L)$$

$$v' = -\frac{qa^2}{24LEI}(4L^2 + a^2 - 12Lx + 6x^2) \qquad (a \le x \le L)$$

$$\theta_A = \frac{qa^2}{24LEI}(2L - a)^2 \qquad \theta_B = \frac{qa^2}{24LEI}(2L^2 - a^2)$$

## Table H-2 (Continued)

**4**

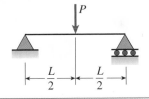

$$v = -\frac{Px}{48EI}(3L^2 - 4x^2) \quad v' = -\frac{P}{16EI}(L^2 - 4x^2) \quad \left(0 \le x \le \frac{L}{2}\right)$$

$$\delta_C = \delta_{max} = \frac{PL^3}{48EI} \quad \theta_A = \theta_B = \frac{PL^2}{16EI}$$

**5**

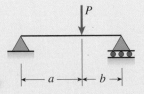

$$v = -\frac{Pbx}{6LEI}(L^2 - b^2 - x^2) \quad v' = -\frac{Pb}{6LEI}(L^2 - b^2 - 3x^2) \quad (0 \le x \le a)$$

$$\theta_A = \frac{Pab(L + b)}{6LEI} \quad \theta_B = \frac{Pab(L + a)}{6LEI}$$

If $a \ge b$, $\delta_C = \dfrac{Pb(3L^2 - 4b^2)}{48EI}$   If $a \le b$, $\delta_C = \dfrac{Pa(3L^2 - 4a^2)}{48EI}$

If $a \ge b$, $x_1 = \sqrt{\dfrac{L^2 - b^2}{3}}$   and   $\delta_{max} = \dfrac{Pb(L^2 - b^2)^{3/2}}{9\sqrt{3}LEI}$

**6**

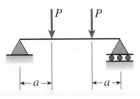

$$v = -\frac{Px}{6EI}(3aL - 3a^2 - x^2) \quad v' = -\frac{P}{2EI}(aL - a^2 - x^2) \quad (0 \le x \le a)$$

$$v = -\frac{Pa}{6EI}(3Lx - 3x^2 - a^2) \quad v' = -\frac{Pa}{2EI}(L - 2x) \quad (a \le x \le L - a)$$

$$\delta_C = \delta_{max} = \frac{Pa}{24EI}(3L^2 - 4a^2) \quad \theta_A = \theta_B = \frac{Pa(L - a)}{2EI}$$

**7**

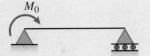

$$v = -\frac{M_0 x}{6LEI}(2L^2 - 3Lx + x^2) \quad v' = -\frac{M_0}{6LEI}(2L^2 - 6Lx + 3x^2)$$

$$\delta_C = \frac{M_0 L^2}{16EI} \quad \theta_A = \frac{M_0 L}{3EI} \quad \theta_B = \frac{M_0 L}{6EI}$$

$$x_1 = L\left(1 - \frac{\sqrt{3}}{3}\right) \quad \text{and} \quad \delta_{max} = \frac{M_0 L^2}{9\sqrt{3}EI}$$

**8**

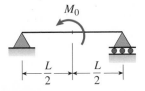

$$v = -\frac{M_0 x}{24LEI}(L^2 - 4x^2) \quad v' = -\frac{M_0}{24LEI}(L^2 - 12x^2) \quad \left(0 \le x \le \frac{L}{2}\right)$$

$$\delta_C = 0 \quad \theta_A = \frac{M_0 L}{24EI} \quad \theta_B = -\frac{M_0 L}{24EI}$$

**9**

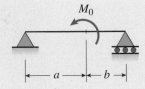

$$v = -\frac{M_0 x}{6LEI}(6aL - 3a^2 - 2L^2 - x^2) \quad (0 \le x \le a)$$

$$v' = -\frac{M_0}{6LEI}(6aL - 3a^2 - 2L^2 - 3x^2) \quad (0 \le x \le a)$$

At $x = a$: $v = \dfrac{M_0 ab}{3LEI}(2a - L) \quad v' = -\dfrac{M_0}{3LEI}(3aL - 3a^2 - L^2)$

$$\theta_A = \frac{M_0}{6LEI}(6aL - 3a^2 - 2L^2) \quad \theta_B = \frac{M_0}{6LEI}(3a^2 - L^2)$$

(Continued)

## Table H-2 (Continued)

**10**

$$v = -\frac{M_0 x}{2EI}(L - x) \qquad v' = -\frac{M_0}{2EI}(L - 2x)$$

$$\delta_C = \delta_{max} = \frac{M_0 L^2}{8EI} \qquad \theta_A = \theta_B = \frac{M_0 L}{2EI}$$

**11**

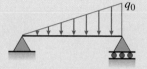

$$v = -\frac{q_0 x}{360LEI}(7L^4 - 10L^2 x^2 + 3x^4)$$

$$v' = -\frac{q_0}{360LEI}(7L^4 - 30L^2 x^2 + 15x^4)$$

$$\delta_C = \frac{5q_0 L^4}{768EI} \qquad \theta_A = \frac{7q_0 L^3}{360EI} \qquad \theta_B = \frac{q_0 L^3}{45EI}$$

$$x_1 = 0.5193L \qquad \delta_{max} = 0.00652\frac{q_0 L^4}{EI}$$

**12**

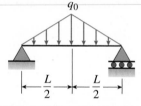

$$v = -\frac{q_0 x}{960LEI}(5L^2 - 4x^2)^2 \qquad \left(0 \le x \le \frac{L}{2}\right)$$

$$v' = -\frac{q_0}{192LEI}(5L^2 - 4x^2)(L^2 - 4x^2) \qquad \left(0 \le x \le \frac{L}{2}\right)$$

$$\delta_C = \delta_{max} = \frac{q_0 L^4}{120EI} \qquad \theta_A = \theta_B = \frac{5q_0 L^3}{192EI}$$

**13**

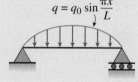

$$v = -\frac{q_0 L^4}{\pi^4 EI}\sin\frac{\pi x}{L} \qquad v' = -\frac{q_0 L^3}{\pi^3 EI}\cos\frac{\pi x}{L}$$

$$\delta_C = \delta_{max} = \frac{q_0 L^4}{\pi^4 EI} \qquad \theta_A = \theta_B = \frac{q_0 L^3}{\pi^3 EI}$$

# Properties of Materials

## Notes:

1. Properties of materials vary greatly depending upon manufacturing processes, chemical composition, internal defects, temperature, previous loading history, age, dimensions of test specimens, and other factors. The tabulated values are typical but should never be used for specific engineering or design purposes. Manufacturers and materials suppliers should be consulted for information about a particular product.

2. Except when compression or bending is indicated, the modulus of elasticity $E$, yield stress $\sigma_Y$, and ultimate stress $\sigma_U$ are for materials in tension.

## Table I-1

### Weights and Mass Densities

| Material | Weight density $\gamma$ | | Mass density $\rho$ | |
|---|---|---|---|---|
| | lb/ft³ | kN/m³ | slugs/ft³ | kg/m³ |
| Aluminum alloys | 160–180 | 26–28 | 5.2–5.4 | 2,600–2,800 |
| 2014-T6, 7075-T6 | 175 | 28 | 5.4 | 2,800 |
| 6061-T6 | 170 | 26 | 5.2 | 2,700 |
| Brass | 520–540 | 82–85 | 16–17 | 8,400–8,600 |
| Bronze | 510–550 | 80–86 | 16–17 | 8,200–8,800 |
| Cast iron | 435–460 | 68–72 | 13–14 | 7,000–7,400 |
| Concrete | | | | |
| Plain | 145 | 23 | 4.5 | 2,300 |
| Reinforced | 150 | 24 | 4.7 | 2,400 |
| Lightweight | 70–115 | 11–18 | 2.2–3.6 | 1,100–1,800 |
| Copper | 556 | 87 | 17 | 8,900 |
| Glass | 150–180 | 24–28 | 4.7–5.4 | 2,400–2,800 |
| Magnesium alloys | 110–114 | 17–18 | 3.4–3.5 | 1,760–1,830 |
| Monel (67% Ni, 30% Cu) | 550 | 87 | 17 | 8,800 |
| Nickel | 550 | 87 | 17 | 8,800 |
| Plastics | | | | |
| Nylon | 55–70 | 8.6–11 | 1.7–2.2 | 880–1,100 |
| Polyethylene | 60–90 | 9.4–14 | 1.9–2.8 | 960–1,400 |
| Rock | | | | |
| Granite, marble, quartz | 165–180 | 26–28 | 5.1–5.6 | 2,600–2,900 |
| Limestone, sandstone | 125–180 | 20–28 | 3.9–5.6 | 2,000–2,900 |
| Rubber | 60–80 | 9–13 | 1.9–2.5 | 960–1,300 |
| Sand, soil, gravel | 75–135 | 12–21 | 2.3–4.2 | 1,200–2,200 |
| Steel | 490 | 77.0 | 15.2 | 7,850 |
| Titanium | 280 | 44 | 8.7 | 4,500 |
| Tungsten | 1,200 | 190 | 37 | 1,900 |
| Water, fresh | 62.4 | 9.81 | 1.94 | 1,000 |
| sea | 63.8 | 10.0 | 1.98 | 1,020 |
| Wood (air dry) | | | | |
| Douglas fir | 30–35 | 4.7–5.5 | 0.9–1.1 | 480–560 |
| Oak | 40–45 | 6.3–7.1 | 1.2–1.4 | 640–720 |
| Southern pine | 35–40 | 5.5–6.3 | 1.1–1.2 | 560–640 |

## Table I-2

Moduli of Elasticity and Poisson's Ratios

| Material | Modulus of elasticity $E$ | | Shear modulus of elasticity $G$ | | Poisson's ratio $\nu$ |
|---|---|---|---|---|---|
| | ksi | GPa | ksi | GPa | |
| Aluminum alloys | 10,000–11,400 | 70–79 | 3,800–4,300 | 26–30 | 0.33 |
| 2014-T6 | 10,600 | 73 | 4,000 | 28 | 0.33 |
| 6061-T6 | 10,000 | 70 | 3,800 | 26 | 0.33 |
| 7075-T6 | 10,400 | 72 | 3,900 | 27 | 0.33 |
| Brass | 14,000–16,000 | 96–110 | 5,200–6,000 | 36–41 | 0.34 |
| Bronze | 14,000–17,000 | 96–120 | 5,200–6,300 | 36–44 | 0.34 |
| Cast iron | 12,000–25,000 | 83–170 | 4,600–10,000 | 32–69 | 0.2–0.3 |
| Concrete (compression) | 2,500–4,500 | 17–31 | | | 0.1–0.2 |
| Copper and copper alloys | 16,000–18,000 | 110–120 | 5,800–6,800 | 40–47 | 0.33–0.36 |
| Glass | 7,000–12,000 | 48–83 | 2,700–5,100 | 19–35 | 0.17–0.27 |
| Magnesium alloys | 6,000–6,500 | 41–45 | 2,200–2,400 | 15–17 | 0.35 |
| Monel (67% Ni, 30% Cu) | 25,000 | 170 | 9,500 | 66 | 0.32 |
| Nickel | 30,000 | 210 | 11,400 | 80 | 0.31 |
| Plastics | | | | | |
| Nylon | 300–500 | 2.1–3.4 | | | 0.4 |
| Polyethylene | 100–200 | 0.7–1.4 | | | 0.4 |
| Rock (compression) | | | | | |
| Granite, marble, quartz | 6,000–14,000 | 40–100 | | | 0.2–0.3 |
| Limestone, sandstone | 3,000–10,000 | 20–70 | | | 0.2–0.3 |
| Rubber | 0.1–0.6 | 0.0007–0.004 | 0.03–0.2 | 0.0002–0.001 | 0.45–0.50 |
| Steel | 28,000–30,000 | 190–210 | 10,800–11,800 | 75–80 | 0.27–0.30 |
| Titanium alloys | 15,000–17,000 | 100–120 | 5,600–6,400 | 39–44 | 0.33 |
| Tungsten | 50,000–55,000 | 340–380 | 21,000–23,000 | 140–160 | 0.2 |
| Wood (bending) | | | | | |
| Douglas fir | 1,600–1,900 | 11–13 | | | |
| Oak | 1,600–1,800 | 11–12 | | | |
| Southern pine | 1,600–2,000 | 11–14 | | | |

## Table I-3

Mechanical Properties

| Material | Yield stress $\sigma_Y$ | | Ultimate stress $\sigma_U$ | | Percent elongation (2 in. gage length) |
|---|---|---|---|---|---|
| | ksi | MPa | ksi | MPa | |
| Aluminum alloys | 5–70 | 35–500 | 15–80 | 100–550 | 1–45 |
| 2014-T6 | 60 | 410 | 70 | 480 | 13 |
| 6061-T6 | 40 | 270 | 45 | 310 | 17 |
| 7075-T6 | 70 | 480 | 80 | 550 | 11 |
| Brass | 10–80 | 70–550 | 30–90 | 200–620 | 4–60 |
| Bronze | 12–100 | 82–690 | 30–120 | 200–830 | 5–60 |
| Cast iron (tension) | 17–42 | 120–290 | 10–70 | 69–480 | 0–1 |
| Cast iron (compression) | | | 50–200 | 340–1,400 | |
| Concrete (compression) | | | 1.5–10 | 10–70 | |
| Copper and copper alloys | 8–110 | 55–760 | 33–120 | 230–830 | 4–50 |
| Glass | | | 5–150 | 30–1,000 | 0 |
| Plate glass | | | 10 | 70 | |
| Glass fibers | | | 1,000–3,000 | 7,000–20,000 | |
| Magnesium alloys | 12–40 | 80–280 | 20–50 | 140–340 | 2–20 |
| Monel (67% Ni, 30% Cu) | 25–160 | 170–1,100 | 65–170 | 450–1,200 | 2–50 |
| Nickel | 15–90 | 100–620 | 45–110 | 310–760 | 2–50 |
| Plastics | | | | | |
| Nylon | | | 6–12 | 40–80 | 20–100 |
| Polyethylene | | | 1–4 | 7–28 | 15–300 |
| Rock (compression) | | | | | |
| Granite, marble, quartz | | | 8–40 | 50–280 | |
| Limestone, sandstone | | | 3–30 | 20–200 | |
| Rubber | 0.2–1.0 | 1–7 | 1–3 | 7–20 | 100–800 |
| Steel | | | | | |
| High-strength | 50–150 | 340–1,000 | 80–180 | 550–1,200 | 5–25 |
| Machine | 50–100 | 340–700 | 80–125 | 550–860 | 5–25 |
| Spring | 60–240 | 400–1,600 | 100–270 | 700–1,900 | 3–15 |
| Stainless | 40–100 | 280–700 | 60–150 | 400–1,000 | 5–40 |
| Tool | 75 | 520 | 130 | 900 | 8 |
| Steel, structural | 30–100 | 200–700 | 50–120 | 340–830 | 10–40 |
| ASTM-A36 | 36 | 250 | 60 | 400 | 30 |
| ASTM-A572 | 50 | 340 | 70 | 500 | 20 |
| ASTM-A514 | 100 | 700 | 120 | 830 | 15 |

## Table I-3 (Continued)

| Material | Yield stress $\sigma_Y$ | | Ultimate stress $\sigma_U$ | | Percent elongation (2 in. gage length) |
|---|---|---|---|---|---|
| | ksi | MPa | ksi | MPa | |
| Steel wire | 40–150 | 280–1,000 | 80–200 | 550–1,400 | 5–40 |
| Titanium alloys | 110–150 | 760–1,000 | 130–170 | 900–1,200 | 10 |
| Tungsten | | | 200–600 | 1,400–4,000 | 0–4 |
| Wood (bending)<br>  Douglas fir<br>  Oak<br>  Southern pine | <br>5–8<br>6–9<br>6–9 | <br>30–50<br>40–60<br>40–60 | <br>8–12<br>8–14<br>8–14 | <br>50–80<br>50–100<br>50–100 | |
| Wood (compression parallel to grain)<br>  Douglas fir<br>  Oak<br>  Southern pine | <br>4–8<br>4–6<br>4–8 | <br>30–50<br>30–40<br>30–50 | <br>6–10<br>5–8<br>6–10 | <br>40–70<br>30–50<br>40–70 | |

## Table I-4

Coefficients of Thermal Expansion

| Material | Coefficient of thermal expansion $\alpha$ | | Material | Coefficient of thermal expansion $\alpha$ | |
|---|---|---|---|---|---|
| | $10^{-6}/°F$ | $10^{-6}/°C$ | | $10^{-6}/°F$ | $10^{-6}/°C$ |
| Aluminum alloys | 13 | 23 | Plastics<br>  Nylon<br>  Polyethylene | <br>40–80<br>80–160 | <br>70–140<br>140–290 |
| Brass | 10.6–11.8 | 19.1–21.2 | | | |
| Bronze | 9.9–11.6 | 18–21 | Rock | 3–5 | 5–9 |
| Cast iron | 5.5–6.6 | 9.9–12 | | | |
| Concrete | 4–8 | 7–14 | Rubber | 70–110 | 130–200 |
| Copper and copper alloys | 9.2–9.8 | 16.6–17.6 | Steel<br>  High-strength<br>  Stainless<br>  Structural | 5.5–9.9<br>8.0<br>9.6<br>6.5 | 10–18<br>14<br>17<br>12 |
| Glass | 3–6 | 5–11 | | | |
| Magnesium alloys | 14.5–16.0 | 26.1–28.8 | | | |
| Monel (67% Ni, 30% Cu) | 7.7 | 14 | Titanium alloys | 4.5–6.0 | 8.1–11 |
| Nickel | 7.2 | 13 | Tungsten | 2.4 | 4.3 |

# Answers to Problems

## CHAPTER 1

1.2-1 (a) $A_y = 5$ lb, $B_y = -5$ lb, $C_x = 50$ lb, $C_y = 0$; (b) $N = 50$ lb, $V = 5$ lb, $M = 75$ ft-lb

1.2-2 (a) $M_A = 0$, $C_y = 236$ N, $D_y = -75.6$ N; (b) $N = 0$, $V = -70$ N, $M = -36.7$ N·m; (c) $M_A = 0$, $C_y = 236$ N, $D_y = -75.6$ N; $N = 0$, $V = -70$ N, $M = -36.7$ N·m

1.2-3 (a) $A_x = 12.55$ lb, $A_y = -15$ lb, $C_y = 104.3$ lb, $D_x = 11.45$ lb, $D_y = -19.83$ lb; (b) Resultant$_B = 19.56$ lb; (c) $A_x = 42.7$ lb, $A_y = 37.2$ lb, $M_A = 522$ lb-ft, $D_x = -18.67$ lb, $D_y = 32.3$ lb, Resultant$_B = 56.6$ lb

1.2-4 (a) $R_{3x} = 40$ N, $R_{3y} = -25$ N, $R_{5x} = 20$ N; (b) $F_{11} = 0$, $F_{13} = 28.3$ N

1.2-5 (a) $A_x = 0$, $A_y = 1.0$ kip, $E_y = 5$ kips; (b) $F_{FE} = 1.898$ kips

1.2-6 (a) $F_x = 0$, $F_y = 12.0$ kN, $D_y = 6.0$ kN; (b) $F_{FE} = 0$

1.2-7 (a) $B_x = -0.8\ P$, $B_z = 2.0\ P$, $O_z = -1.25\ P$; (b) $F_{AC} = 0.960\ P$

1.2-8 (a) $A_x = -1.25P$, $B_y = 0$, $B_z = -P$; (b) $F_{AB} = 1.601P$

1.2-9 (a) $A_y = 4.67P$, $A_z = -4.0P$; (b) $F_{AB} = -8.33P$

1.2-10 (a) $A_z = 0$, $B_x = -3.75$ kN; (b) $F_{AB} = 6.73$ kN

1.2-11 (a) $T_A = 11,000$ lb-in.; (b) $T(L_1/2) = -T_A = -11,000$ lb-in., $T(L_1 + L_2/2) = T_2 = 10,000$ lb-in.

1.2-12 (a) $T_A = -1225$ N·m; (b) $T(L_1/2) = 62.5$ N·m, $T(L_1 + L_2/2) = -T_2 = -1100$ N·m

1.2-13 (a) $A_x = -540$ lb, $A_y = -55.6$ lb, $M_A = 4320$ lb-ft, $C_y = 55.6$ lb; (b) $N = 55.6$ lb, $V = 506$ lb, $M = -2374$ lb-ft

1.2-14 (a) $A_x = 280$ N, $A_y = 8.89$ N, $M_A = -1120$ N·m, $D_y = 151.1$ N; (b) Resultant$_B = 280$ N

1.2-15 (a) $A_x = 30$ lb, $A_y = 140$ lb, $C_x = -30$ lb, $C_y = 60$ lb; (b) $N = -23.3$ lb, $V = -20$ lb, $M = 33.3$ lb-ft

1.2-16 (a) $A_x = 10.98$ kN, $A_y = 29.0$ kN, $E_x = -8.05$ kN, $E_y = -22$ kN; (b) Resultant$_C = 23.4$ kN

1.2-17 (a) $A_y = -1250$ lb, $E_x = 0$, $E_y = 1750$ lb; (b) $N = 1750$ lb, $V = 500$ lb, $M = 575$ lb-ft

1.2-18 (a) $A_x = 320$ N, $A_y = -240$ N, $C_y = 192$ N, $E_y = -192$ N; (b) $N = -312$ N, $V = -57.9$ N, $M = 289$ N·m; (c) Resultant$_C = 400$ N

1.2-19 (a) $A_x = -28.9$ lb, $A_y = 50.0$ lb, $B_x = -65.0$ lb; (b) $F_{cable} = 71.6$ lb

1.2-20 (a) $A_x = -10$ kN, $A_y = -2.17$ kN, $C_y = 9.83$ kN, $E_y = 1.333$ kN; (b) Resultant$_D = 12.68$ kN

1.2-21 (a) $O_x = -48.3$ lb, $O_y = 40$ lb, $O_z = 12.94$ lb, $M_{Ox} = 331$ lb-in., $M_{Oy} = 690$ lb-in., $M_{Oz} = -338$ lb-in.; (b) $N = -40$ lb, $V = 50$ lb, $T = -690$ lb-in., $M = 473$ lb-in.

1.2-22 (a) $A_y = -120$ N, $A_z = -60$ N, $M_{Ax} = -70$ N·m, $M_{Ay} = -142.5$ N·m, $M_{Az} = -180$ N·m, $D_x = -60$ N, $D_y = 120$ N, $D_z = 30$ N; (b) $N = 120$ N, $V = 41.3$ N, $T = 142.5$ N·m, $M = 180.7$ N·m

1.2-23 (a) $A_x = 5.77$ lb, $A_y = 47.3$ lb, $A_z = -2.31$ lb, $M_{Az} = 200$ lb-in.; (b) $T_{DC} = 3.81$ lb, $T_{EC} = 6.79$ lb

1.2-24 $C_x = 120$ N, $C_y = -160$ N, $C_z = 506$ N, $D_z = 466$ N, $H_y = 320$ N, $H_z = 499$ N

1.2-25 $A_y = 57.2$ lb, $B_x = 44.2$ lb (to the left),
$B_y = 112.4$ lb, $C_x = 28.8$ lb, $C_y = 5.88$ lb

1.2-26 (a) $H_B = -104.6$ N, $V_B = 516$ N,
$V_F = 336$ N; (b) $N = -646$ N, $V = 176.8$ N,
$M = 44.9$ kN·m

1.3-1 (a) $\sigma_{AB} = 1443$ psi; (b) $P_2 = 1487.5$ lbs;
(c) $t_{BC} = 0.5$ in.

1.3-2 (a) $\sigma = 130.2$ MPa; (b) $\varepsilon = 4.652 \times 10^{-4}$

1.3-3 (a) $R_B =$ lb (cantilever), 191.3 lb (V-brakes);
$\sigma_C = 144$ psi (cantilever), 306 psi (V-brakes);
(b) $\sigma_{cable} = 26,946$ psi (both)

1.3-4 (a) $\varepsilon_s = 3.101 \times 10^{-4}$; (b) $\delta = 0.1526$ mm;
(c) $P_{max} = 89.5$ kN

1.3-5 (a) $\sigma_C = 2.46$ ksi; (b) $x_C = 19.56$ in.,
$y_C = 19.56$ in.

1.3-6 (a) $\sigma_t = 132.7$ MPa; (b) $\alpha_{max} = 34.4°$

1.3-7 (a) $\sigma_1 = 34.4$ ksi, $\sigma_2 = 30.6$ ksi;
(b) $d_{1new} = 3.18 \times 10^{-2}$ in.;
(c) $\sigma_1 = 19.6$ ksi, $\sigma_2 = 18.78$ ksi,
$\sigma_3 = 22.7$ ksi

1.3-8 $\sigma_C = 5.21$ MPa

1.3-9 (a) $T = 184$ lb, $\sigma = 10.8$ ksi;
(b) $\varepsilon_{cable} = 5 \times 10^{-4}$

1.3-10 (a) $T = 819$ N, $\sigma = 74.5$ MPa;
(b) $\varepsilon_{cable} = 4.923 \times 10^{-4}$

1.3-11 (a) $T = \begin{pmatrix} 5877 \\ 4679 \\ 7159 \end{pmatrix}$ lb; (b) $\sigma = \begin{pmatrix} 48,975 \\ 38,992 \\ 59,658 \end{pmatrix}$ psi;

(c) $T = \begin{pmatrix} 4278 \\ 6461 \\ 3341 \\ 4278 \end{pmatrix}$ lb, $\sigma = \begin{pmatrix} 35,650 \\ 53,842 \\ 27,842 \\ 35,650 \end{pmatrix}$ psi

1.3-12 (a) $\sigma_x = \gamma\omega^2(L^2 - x^2)/2g$;
(b) $\sigma_{max} = \gamma\omega^2 L^2/2g$

1.3-13 (a) $T_{AB} = 1620$ lb, $T_{BC} = 1536$ lb,
$T_{CD} = 1640$ lb; (b) $\sigma_{AB} = 13,501$ psi,
$\sigma_{BC} = 12,799$ psi, $\sigma_{CD} = 13,667$ psi

1.3-14 (a) $T_{AQ} = T_{BQ} = 50.5$ kN; (b) $\sigma = 166$ MPa

1.4-1 (a) $L_{max} = 11,800$ ft; (b) $L_{max} = 13,500$ ft

1.4-2 (a) $L_{max} = 7143$ m; (b) $L_{max} = 8209$ m

1.4-3 % elongation = 6.5, 24.0, 39.0;
% reduction = 8.1, 37.9, 74.9;
Brittle, ductile, ductile

1.4-4 $11.9 \times 10^3$ m; $12.7 \times 10^3$ m; $6.1 \times 10^3$ m;
$6.5 \times 10^3$ m; $23.9 \times 10^3$ m

1.4-5 $\sigma = 52.3$ ksi

1.4-6 $\sigma_{pl} \approx 47$ MPa, Slope $\approx 2.4$ GPa,
$\sigma_\gamma \approx 53$ MPa; Brittle

1.4-7 $\sigma_{pl} \approx 65,000$ psi, Slope $\approx 30 \times 10^6$ psi,
$\sigma_Y \approx 69,000$ psi, $\sigma_U \approx 113,000$ psi;
Elongation = 6%, Reduction = 31%

1.5-1 0.13 in. longer

1.5-2 4.0 mm longer

1.5-3 (a) $\delta_{pset} = 1.596$ in.; (b) $\sigma_B = 30$ ksi

1.5-4 (a) $\delta_{pset} = 4.28$ mm; (b) $\sigma_B = 65.6$ MPa

1.5-5 (b) 0.71 in.; (c) 0.58 in.; (d) 49 ksi

1.6-1 $P_{max} = 157$ k

1.6-2 $P = 27.4$ kN (tension)

1.6-3 $P = -15.708$ kips

1.6-4 (a) $P = 74.1$ kN; (b) $\delta = \varepsilon \cdot L = 0.469$ mm
shortening; $\dfrac{A_f - A}{A} = -0.081\%$,
$\Delta V_1 = V_{1f} - Vol_1 = -207$ mm$^3$;
(c) $d_3 = 65.4$ mm

1.6-5 $\Delta d = -1.56 \times 10^{-4}$ in., $P = 2.154$ kips

1.6-6 (a) $E = 104$ GPa; (b) $v = 0.34$

1.6-7 (a) $\Delta d_{BCinner} = 8 \times 10^{-4}$ in.;
(b) $v_{brass} = 0.34$; (c) $\Delta t_{AB} = 2.732 \times 10^{-4}$ in.,
$\Delta d_{ABinner} = 1.366 \times 10^{-4}$ in.

1.6-8 (a) $\Delta L_1 = 12.66$ mm; $\Delta L_2 = 5.06$ mm;
$\Delta L_3 = 3.8$ mm; (b) $\Delta Vol_1 = 21,548$ mm$^3$;
$\Delta Vol_2 = 21,601$ mm$^3$; $\Delta Vol_3 = 21,610$ mm$^3$

1.7-1 $\sigma_b = 7.04$ ksi, $\tau_{ave} = 10.756$ ksi

1.7-2 $\sigma_b = 139.86$ MPa; $P_{ult} = 144.45$ kN

1.7-3 (a) $\tau = 12.732$ ksi; (b) $\sigma_{bf} = 20$ ksi,
$\sigma_{bg} = 26.667$ ksi

1.7-4 (a) $B_x = -252.8$ N, $A_x = -B_x$,
$A_y = 1150.1$ N; (b) $A_{resultant} = 1178$ N;
(c) $\tau = 5.86$ MPa, $\sigma_{bshoe} = 7.36$ MPa

1.7-5 (a) $\tau_{max} = 2979$ psi; (b) $\sigma_{bmax} = 936$ psi

1.7-6 $T_1 = 13.176$ kN, $T_2 = 10.772$ kN,
$\tau_{1ave} = 25.888$ MPa, $\tau_{2ave} = 21.166$ MPa,
$\sigma_{b1} = 9.15$ MPa, $\sigma_{b2} = 7.48$ MPa

1.7-7 (a) Resultant = 1097 lb; (b) $\sigma_b = 4999$ psi;
(c) $\tau_{nut} = 2793$ psi, $\tau_{pl} = 609$ psi

1.7-8 $G = 2.5$ MPa

1.7-9 (a) $\gamma_{aver} = 0.004$; (b) $V = 89.6$ k

1.7-10 (a) $\gamma_{aver} = 0.50$; (b) $\delta = 4.92$ mm

1.7-11 (a) $\sigma_b = 69.5$ ksi, $\sigma_{brg} = 39.1$ ksi, $\tau_f = 21$ ksi;
(b) $\sigma_b = 60.4$ ksi, $\sigma_{brg} = 34$ ksi, $\tau_f = 18.3$ ksi

**1.7-12**  $\tau_{aver} = 42.9$ MPa

**1.7-13**  (a) $A_x = 0$, $A_y = 170$ lb, $M_A = 4585$ in.-lb;
(b) $B_x = 253.6$ lb, $B_y = 160$ lb,
$B_{res} = 299.8$ lb, $C_x = -B_x$; (c) $\tau_B = 3054$ psi,
$\tau_C = 1653$ psi; (d) $\sigma_{bB} = 4797$ psi,
$\sigma_{bC} = 3246$ psi

**1.7-14**  For a bicycle wih $L/R = 1.8$: (a) $T = 1440$ N;
(b) $\tau_{aver} = 147$ MPa

**1.7-15**  (a) $\tau = \dfrac{P}{2\pi rh}$; (b) $\delta = \dfrac{P}{2\pi hG} \ln \dfrac{b}{d}$

**1.7-16**  (a) $\tau_1 = 2.95$ MPa , $\tau_4 = 0$;
(b) $\sigma_{b1} = 1.985$ MPa, $\sigma_{b4} = 0$;
(c) $\sigma_{b4} = 41$ MPa; (d) $\tau = 10.62$ MPa;
(e) $\sigma_3 = 75.1$ MPa

**1.7-17**  (a) $O_x = 12.68$ lb, $O_y = 1.294$ lb,
$O_{res} = 12.74$ lb; (b) $\tau_O = 519$ psi,
$\sigma_{bO} = 816$ psi; (c) $\tau = 362$ psi

**1.7-18**  (a) $F_x = 153.9$ N, $\sigma = 3.06$ MPa;
(b) $\tau_{ave} = 1.96$ MPa; (c) $\sigma_b = 1.924$ MPa

**1.7-19**  (a) $P = 395$ lb; (b) $C_x = 374$ lb,
$C_y = -237$ lb, $C_{res} = 443$ lb; (c) $\tau = 18.04$
ksi, $\sigma_{bC} = 4.72$ ksi

**1.8-1**  $P_{allow} = 3140$ lb

**1.8-2**  $T_{max} = 33.4$ kN.m

**1.8-3**  $P_{allow} = 607$ lb

**1.8-4**  (a) $P_{allow} = 8.74$ kN; (b) $P_{allow} = 8.69$ kN;
(c) $P_{allow} = 21.2$ kN, $P_{allow} = 8.69$ kN
(shear controls)

**1.8-5**  $P = 294$ k

**1.8-6**  (a) $F = 1.171$ kN; (b) Shear: $F_a = 2.86$ kN

**1.8-7**  $W_{max} = 5110$ lb

**1.8-8**  (a) $F_A = \sqrt{2}T$, $F_B = 2T$, $F_C = T$;
(b) Shear at $A$: $W_{max} = 66.5$ kN

**1.8-9**  $P_a = 10.21$ kips

**1.8-10**  $C_{ult} = 5739$ N: $P_{max} = 445$ N

**1.8-11**  $W_{max} = 0.305$ kips

**1.8-12**  Shear in rivets in $CG$ & $CD$ controls:
$P_{allow} = 45.8$ kN

**1.8-13**  (a) $P_a = \sigma_a (0.587d^2)$; (b) $P_a = 21.6$ kips

**1.8-14**  $P_{allow} = 96.5$ kN

**1.8-15**  $P_{max} = 11.98$ psf

**1.8-16**  (a) $P_{allow} = \sigma_c (\pi d^2/4) \sqrt{1 - (R/L)^2}$;
(b) $P_{allow} = 9.77$ kN

**1.9-1**  (a) $d_{min} = 3.75$ in; (b) $d_{min} = 4.01$ in.

**1.9-2**  (a) $d_{min} = 164.6$ mm; (b) $d_{min} = 170.9$ mm

**1.9-3**  (a) $d_{min} = 0.704$ in.; (b) $d_{min} = 0.711$ in.

**1.9-4**  $d_{min} = 63.3$ mm

**1.9-5**  $d_{pin} = 1.029$ in.

**1.9-6**  (b) $A_{min} = 435$ mm$^2$

**1.9-7**  $d_{min} = 0.372$ in.

**1.9-8**  $d_{min} = 5.96$ mm

**1.9-9**  $n = 11.6$, or 12 bolts

**1.9-10**  $(d_2)_{min} = 131$ mm

**1.9-11**  $A_c = 1.189$ in.$^2$

**1.9-12**  (a) $t_{min} = 18.8$ mm, use $t = 20$ mm;
(b) $D_{min} = 297$ mm

**1.9-13**  (a) $\sigma_{DF} = 10.38$ ksi $< \sigma_{allow}$,
$\sigma_{bF} = 378$ psi $< \sigma_{ba}$; (b) new $\sigma_{BC} = 25$ ksi, so
increase rod $BC$ to $\frac{1}{4}$-in. diameter; required
diameter of washer $= 1 - \dfrac{5}{16}$ in. $= 1.312$ in.

**1.9-14**  (a) $d_m = 24.7$ mm; (b) $P_{max} = 49.4$ kN

**1.9-15**  $\theta = \arccos 1/\sqrt{3} = 54.7°$

# CHAPTER 2

**2.2-1**  (a) $\delta = \dfrac{6W}{5k}$; (b) $\delta = \dfrac{4W}{5k}$

**2.2-2**  (a) $\delta = 12.5$ mm; (b) $n = 5.8$

**2.2-3**  (a) $\dfrac{\delta_a}{\delta_s} = \dfrac{E_s}{E_a} = \dfrac{30}{11}$; (b) $\dfrac{d_a}{d_s} = \sqrt{\dfrac{E_s}{E_a}} = 1.651$;

(c) $\dfrac{L_a}{L_s} = 1.5 \dfrac{E_a}{E_s} = 0.55$;

(d) $E_1 = \dfrac{E_s}{1.7} = 17,647$ ksi (cast iron or

copper alloy) (see App. I)

**2.2-4**  $h = 13.4$ mm

**2.2-5**  $h = L - \pi\rho_{max}d^2/4k$

**2.2-6**  (a) $x = 102.6$ mm; (b) $x = 205$ mm;
(c) $P_{max} = 12.51$ N; (d) $\theta_{init} = 1.325°$;
(e) $P = 20.4$ N

**2.2-7**  (a) $\delta_4 = \dfrac{26P}{3k}$;

(b) $\delta_4 = \dfrac{104P}{45k}$, ratio $= \dfrac{15}{4} = 3.75$

**2.2-8**  (a) $\delta_B = 1.827$ mm; (b) $P_{max} = 390$ kN;
(c) $\delta_{Bx} = 6.71$ mm, $P_{max} = 106.1$ kN

**2.2-9**  $P_{max} = 72.3$ lb

**2.2-10**  (a) $x = 134.7$ mm; (b) $k_1 = 0.204$ N/mm;
(c) $b = 74.1$ mm; (d) $k_3 = 0.638$ N/mm

**2.2-11** (a) $t_{c,\min} = 0.021$ in.; (b) $\delta_r = 0.031$ in.;
(c) $h_{\min} = 0.051$ in.

**2.2-12** $\delta_A = 0.200$ mm, $\delta_D = 0.880$ mm

**2.2-13** (a) $\delta_D = \dfrac{P}{16}(28f_2 - 9f_1)$; (b) $\dfrac{L_1}{L_2} = \dfrac{27}{16}$

(c) $\dfrac{d_1}{d_2} = 1.225$; (d) $x = \dfrac{365L}{236}$

**2.2-14** (a) $\theta = 35.1°$, $\delta = 44.6$ mm, $R_A = 25$ N,
$R_C = 25$ N; (b) $\theta = 43.3°$, $\delta = 8.19$ mm,
$R_A = 31.5$ N, $R_C = 18.5$ N,
$M_A = 1.882$ N·m

**2.2-15** (a) $\theta = 35.1°$, $\delta = 1.782$ in., $R_A = 5$ lb,
$R_C = 5$ lb; (b) $\theta = 43.3°$, $\delta = 0.327$ in.,
$R_A = 6.3$ lb, $R_C = 3.71$ lb ,
$M_A = 1.252$ lb-ft

**2.3-1** (a) $\delta = 0.0276$ in.; (b) $d_B = 1.074$ in.

**2.3-2** (a) $\delta = 0.675$ mm; (b) $P_{\max} = 267$ kN

**2.3-3** (a) $\delta = 0.01125$ in. (elongation); (b) So new
value of $P_3$ is 1690 lb, an increase of 390
lb. (c) $A_{AB} = 0.78$ in.$^2$

**2.3-4** (a) $\delta = \dfrac{7PL}{6Ebt}$; (b) $\delta = 0.5$ mm;
(c) $L_{\text{slot}} = 244$ mm

**2.3-5** (a) $\delta = \dfrac{7PL}{6Ebt}$; (b) $\delta = 0.021$ in.;
(c) $L_{\text{slot}} = 10$ in.

**2.3-6** (a) $\delta_{AC} = 3.72$ mm; (b) $P_0 = 44.2$ kN

**2.3-7** (a) $\delta_a = 0.0589$ in.; (b) $\delta_b = 0.0501$ in.;
(c) $\dfrac{\delta_c}{\delta_a} = 0.58$, $\dfrac{\delta_c}{\delta_b} = 0.681$

**2.3-8** (a) $d_{\max} = 23.9$ mm; (b) $b = 4.16$ mm;
(c) $x = 183.3$ mm

**2.3-9** (a) $\delta = \dfrac{PL}{2EA}$; (b) $\sigma(y) = \dfrac{P}{A}\left(\dfrac{y}{L}\right)$;
(c) $\delta = \dfrac{PL}{EA}\left(\dfrac{2}{3}\right)$, $\sigma(y) = \dfrac{P}{A}\left[\dfrac{y}{L}\left(2 - \dfrac{y}{L}\right)\right]$

**2.3-10** (a) $\delta_{2\text{-}4} = 0.024$ mm; (b) $P_{\max} = 8.15$ kN;
(c) $L_2 = 9.16$ mm

**2.3-11** (a) $R_1 = -3P/2$; (b) $N_1 = 3P/2$ (tension),
$N_2 = P/2$ (tension); (c) $x = L/3$;
(d) $\delta_2 = 2PL/3EA$; (e) $\beta = 1/11$

**2.3-12** (a) $\delta_c = W(L^2 - h^2)/2EAL$;
(b) $\delta_B = WL/2EA$; (c) $\beta = 3$;

(d) $\delta = \dfrac{WL}{2EA} = 359$ mm (in sea water);
$\delta = \dfrac{WL}{2EA} = 412$ mm (in air)

**2.3-13** (b) $\delta = 0.010$ in.

**2.3-14** $\delta = 2PH/3Eb^2$

**2.3-15** $\delta = 2WL/\pi d^2E$

**2.3-16** (a) $\delta = 2.18$ mm; (b) $\delta = 6.74$ mm

**2.3-17** (b) $\delta = 11.14$ ft

**2.4-1** (a) $P = 1330$ lb; (b) $P_{\text{allow}} = 1300$ lb

**2.4-2** (a) $P = 104$ kN; (b) $P_{\max} = 116$ kN

**2.4-3** (a) $P_B/P = 3/11$; (b) $\sigma_B/\sigma_A = 1/2$;
(c) Ratio $= 1$

**2.4-4** (a) If $x \leq L/2$, $R_A = (-3PL)/(2(x + 3L))$,
$R_B = -P(2x + 3L)/(2(x + 3L))$. If
$x \geq L/2$, $R_A = (-P(x + L))/(x + 3L)$,
$R_B = (-2PL)/(x + 3L)$. (b) If $x \leq L/2$,
$\delta = PL(2x + 3L)/[(x + 3L)E\pi d^2]$. If
$x \geq L/2$, $\delta = 8PL(x + L)/[3(x + 3L)E\pi d^2]$.
(c) $x = 3L/10$ or $x = 2L/3$;
(d) $R_B = \rho g \pi d^2 L/8$, $R_A = 3\,\rho g \pi d^2 L/32$

**2.4-5** (a) 41.7%; (b) $\sigma_M = 32.7$ ksi, $\sigma_O = 51.4$ ksi

**2.4-6** (a) $\delta = 1.91$ mm; (b) $\delta = 1.36$ mm;
(c) $\delta = 2.74$ mm

**2.4-7** (a) $R_A = 2P/3$, $R_E = -5P/3$;
(b) $\delta_B = -\dfrac{LP}{6EA}$, $\delta_C = \dfrac{LP}{6EA}$, $\delta_D = \dfrac{5LP}{6EA}$;
(c) $\delta_{\max} = \dfrac{5LP}{6EA}$ (to the right), $\delta_A = \delta_E = 0$
(d) $P_{\max} = 12.37$ kip

**2.4-8** (a) $R_A = 10.5$ kN (to the left), $R_D = 2.0$ kN
(to the right); (b) $F_{BC} = 15.0$ kN
(compression)

**2.4-9** (b) $\sigma_a = 1610$ psi (compression),
$\sigma_s = 9350$ psi (tension)

**2.4-10** (a) $P = 13.73$ kN, $R_1 = 9.07$ kN,
$R_2 = 4.66$ kN, $\sigma_2 = 7$ MPa;
(b) $\delta_{\text{cap}} = 190.9$ mm, Axial Force Diagram:
$N(x) = -R_2$ if $x \leq L_2$, $N(x) = R_1$
if $x > L_2$; Axial Displacement Diagram:
$\delta(x) = \left[\dfrac{-R_2}{EA_2}(x)\right]$ if $x \leq L_2$,
$\delta(x) = \left[\dfrac{-R_2L_2}{EA_2} + \dfrac{R_1}{EA_1}(x - L_2)\right]$ if $x > L_2$;
(c) $q = 1.552$ kN/m

2.4-11  (a) $P_1 = PE_1/(E_1 + E_2)$;
(b) $e = b(E_2 - E_1)/[2(E_2 + E_1)]$;
(c) $\sigma_1/\sigma_2 = E_1/E_2$

2.4-12  (a) $P_{\text{allow}} = 1504$ N; (b) $P_{\text{allow}} = 820$ N;
(c) $P_{\text{allow}} = 703$ N

2.4-13  $d_2 = 0.338$ in., $L_2 = 48.0$ in.

2.4-14  (a) $A_x = -41.2$ kN, $A_y = -71.4$ kN,
$B_x = -329$ kN, $B_y = 256$ kN;
(b) $P_{\text{max}} = 233$ kN

2.4-15  (a) $\sigma_c = 10{,}000$ psi, $\sigma_D = 12{,}500$ psi;
(b) $\delta_B = 0.0198$ in.

2.4-16  $P_{\text{max}} = 1800$ N

2.4-17  $\sigma_s = 3.22$ ksi, $\sigma_b = 1.716$ ksi, $\sigma_c = 1.93$ ksi

2.5-1  $\sigma = 11{,}700$ psi

2.5-2  $T = 40.3°C$

2.5-3  $\Delta T = 185°F$

2.5-4  (a) $\Delta T = 24°C$, $\sigma_{\text{rod}} = 57.6$ MPa;
(b) Clevis: $\sigma_{bc} = 42.4$ MPa, Washer:
$\sigma_{bw} = 74.1$ MPa; (c) $d_b = 10.68$ mm

2.5-5  (a) $\sigma_c = E\alpha(\Delta T_B)/4$;
(b) $\sigma_c = E\alpha(\Delta T_B)/[4(EA/kL + 1)]$

2.5-6  (a) $N = 51.8$ kN, max. $\sigma_c = 26.4$ MPa,
$\delta_C = -0.314$ mm; (b) $N = 31.2$ kN,
max. $\sigma_c = 15.91$ MPa, $\delta_C = -0.546$ mm

2.5-7  $\delta = 0.123$ in.

2.5-8  $\Delta T = 34°C$

2.5-9  $\tau = 15.0$ ksi

2.5-10  $P_{\text{allow}} = 39.5$ kN

2.5-11  (a) $T_A = 400$ lb, $T_B = 200$ lb;
(b) $T_A = 454$ lb, $T_B = 92$ lb;
(c) $\Delta T = 153°F$

2.5-12  (a) $\sigma = 98$ MPa; (b) T = 35° C

2.5-13  (a) $\sigma = -957$ psi; (b) $F_k = 3006$ lbs (C);
(c) $\sigma = -2560$ psi

2.5-14  $s = PL/6EA$

2.5-15  (a) $P_1 = 231$ k, $R_A = -55.2$ k, $R_B = 55.2$ k;
(b) $P_2 = 145.1$ k, $R_A = -55.2$ k, $R_B = 55.2$ k;
(c) For $P_1$, $\tau_{\text{max}} = 13.39$ ksi, for $P_2$,
$\tau_{\text{max}} = 19.44$ ksi; (d) $\Delta T = 65.8°F$, $R_A = 0$,
$R_B = 0$; (e) $R_A = -55.2$ k, $R_B = 55.2$ k

2.5-16  (a) $R_A = [-s + \alpha\Delta T(L_1 + L_2)]/[(L_1/EA_1)$
$+ (L_2/EA_2) + (L/k_3)]$, $R_D = -R_A$;
(b) $\delta_B = \alpha\Delta T(L_1) - R_A(L_1/EA_1)$,
$\delta_C = \alpha\Delta T(L_1 + L_2) - R_A[(L_1/EA_1)$
$+ L_2/EA_2)]$

2.5-17  $T_B = 660$ lb, $T_C = 780$ lb

2.5-18  $P_{\text{allow}} = 1.8$ MN

2.5-19  (a) $\sigma_p = -0.196$ ksi, $\sigma_r = 3.42$ ksi;
(b) $\sigma_b = 2.74$ ksi, $\tau_c = 0.285$ ksi

2.5-20  $\sigma_p = 25.0$ MPa

2.5-21  $\sigma_p = 2400$ psi

2.5-22  (a) $P_B = 25.4$ kN, $P_s = -P_B$;
(b) $S_{\text{reqd}} = 25.7$ mm; (c) $\delta_{\text{final}} = 0.35$ mm

2.5-23  (a) $F_x = -0.174$ k; (b) $F_1 = 0.174$ k;
(c) $L_f = 12.01$ in.; (d) $\Delta T = 141.9°F$

2.5-24  $\sigma_a = 500$ MPa (tension), $\sigma_c = 10$ MPa
(compression)

2.5-25  (a) $F_k = 0.174$ k; (b) $F_t = -0.174$ k;
(c) $L_f = 11.99$ in.; (d) $\Delta T = -141.6°F$

2.6-1  $P_{\text{max}} = 42{,}600$ lb

2.6-2  $d_{\text{min}} = 6.81$ mm

2.6-3  $P_{\text{max}} = 24{,}000$ lb

2.6-4  (a) $\Delta T_{\text{max}} = -46°C$; (b) $\Delta T = +9.93°C$

2.6-5  (a) $\tau_{\text{max}} = 10{,}800$ psi; (b) $\Delta T_{\text{max}} = -49.9°F$;
(c) $\Delta T = +75.9°F$

2.6-6  (a) $\sigma_x = 84$ MPa; (b) $\tau_{\text{max}} = 42$ MPa;
(c) On rotated $x$ face: $\sigma_{x1} = 42$ MPa,
$\tau_{x1y1} = 42$ MPa; On rotated $y$ face:
$\sigma_{y1} = 42$ MPa, (d) On rotated $x$ face:
$\sigma_{x1} = 71.7$ MPa, $\tau_{x1y1} = -29.7$ MPa
On rotated $y$ face: $\sigma_{y1} = 12.3$ MPa

2.6-7  (a) $\sigma_{\text{max}} = 18{,}000$ psi; (b) $\tau_{\text{max}} = 9000$ psi

2.6-8  (a) Element $A$: $\sigma_x = 105$ MPa (compression), Element $B$: $\tau_{\text{max}} = 52.5$ MPa;
(b) $\theta = 33.1°$

2.6-9  (a) $\tau_{\text{max}AC} = \dfrac{\sigma_{AC}}{2} = 1.859$ ksi,

$\tau_{\text{max}AB} = \dfrac{\sigma_{AB}}{2} = 7.42$ ksi,

$\tau_{\text{max}DC} = \dfrac{\sigma_{BC}}{2} = -9.41$ ksi;

(b) $P_{\text{max}} = 36.5$ kip

2.6-10  (a) (1) $\sigma_x = -945$ kPa; (2) $\sigma_\theta = -807$ kPa,
$\tau_\theta = 334$ kPa; (3) $\sigma_\theta = -472$ kPa,
$\tau_\theta = 472$ kPa, $\sigma_{\text{max}} = -945$ kPa,
$\tau_{\text{max}} = -472$ kPa; (b) $\sigma_{\text{max}} = -378$ kPa,
$\tau_{\text{max}} = -189$ kPa

2.6-11  (a) $\tau_{pq} = 11.54$ psi; (b) $\sigma_{pq} = -1700$ psi,
$\sigma(pq + \pi/2) = -784$ psi; (c) $P_{\text{max}} = 14{,}688$ lb

2.6-12  (a) $\Delta T_{\text{max}} = 31.3°C$; (b) $\sigma_{pq} = -21.0$ MPa
(compression), $\tau_{pq} = 30$ MPa (CCW);
(c) $\beta = 0.62$

2.6-13    $N_{AC} = 5.77$ kips; $d_{min} = 1.08$ in.

2.6-14    (a) $\sigma_\theta = 0.57$ MPa, $\tau_\theta = -1.58$ MPa;
(b) $\alpha = 33.3°$; (c) $\alpha = 26.6°$

2.6-15    (a) $\theta = 35.26°$, $\tau_0 = -7070$ psi;
(b) $\sigma_{max} = 15,000$ psi, $\tau_{max} = 7500$ psi

2.6-16    $\sigma_{\theta 1} = 54.9$ MPa, $\sigma_{\theta 2} = 18.3$ MPa,
$\tau_\theta = -31.7$ MPa

2.6-17    $\sigma_{max} = 10,000$ psi, $\tau_{max} = 5000$ psi

2.6-18    (a) $\theta = 30.96°$; (b) $P_{max} = 1.53$ kN

2.6-19    (a) $\tau_\theta = 348$ psi, $\theta = 20.1°$;
(b) $\sigma_{x1} = -950$ psi, $\sigma_{y1} = -127.6$ psi;
(c) $k_{max} = 15,625$ lb/in.; (d) $L_{max} = 1.736$ ft;
(e) $\Delta T_{max} = 92.8°F$

2.7-1    (a) $U = 23P^2L/12EA$; (b) $U = 125$ in.-lb

2.7-2    (a) $U = 5P^2L/4\pi Ed^2$; (b) $U = 1.036$ J

2.7-3    $U = 5040$ in.-lb

2.7-4    (c) $U = P^2L/2EA + PQL/2EA + Q^2L/4EA$

2.7-5    Aluminum: 171 psi, 1740 in.

2.7-6    (a) $U = P^2L/EA$; (b) $\delta_B = 2PL/EA$

2.7-7    (a) $U_1 = 0.0375$ in.-lb; (b) $U_2 = 2.57$ in.-lb;
(c) $U_3 = 2.22$ in.-lb

2.7-8    (a) $U = 5k\delta^2$; (b) $\delta = W/10k$;
(c) $F_1 = 3W/10$, $F_2 = 3W/20$, $F_3 = W/10$

2.7-9    (a) $U = \dfrac{P^2L}{2Et(b_2 - b_1)} \ln \dfrac{b_2}{b_1}$;

        (b) $\delta = \dfrac{PL}{Et(b_2 - b_1)} \ln \dfrac{b_2}{b_1}$

2.7-10    (a) $P_1 = 270$ kN; (b) $\delta = 1.321$ mm;
(c) $U = 243$ J

2.7-11    (a) $x = 2s$, $P = 2(k_1 + k_2)s$;
(b) $U_1 = (2k_1 + k_2)s^2$

2.7-12    (a) $U = 6.55$ J; (b) $\delta_C = 168.8$ mm

2.8-1    (a) $\delta_{max} = 0.0361$ in.; (b) $\sigma_{max} = 22,600$ psi;
(c) Impact factor = 113

2.8-2    (a) $\delta_{max} = 6.33$ mm; (b) $\sigma_{max} = 359$ MPa;
(c) Impact factor = 160

2.8-3    (a) $\delta_{max} = 0.0312$ in.; (b) $\sigma_{max} = 26,000$ psi;
(c) Impact factor = 130

2.8-4    (a) $\delta_{max} = 215$ mm; (b) Impact factor = 3.9

2.8-5    (a) $\delta_{max} = 9.21$ in.; (b) Impact factor = 4.6

2.8-6    $v = 13.1$ m/s

2.8-7    $h_{max} = 8.55$ in.

2.8-8    $L_{min} = 9.25$ m

2.8-9    $L_{min} = 500$ in.

2.8-10    $V_{max} = 5.40$ m/s

2.8-11    $\delta_{max} = 11.0$ in.

2.8-12    $L = 25.5$ m

2.8-13    (a) Impact factor $= 1 + (1 + 2EA/W)^{1/2}$;
(b) 10

2.8-14    $\sigma_{max} = 33.3$ MPa

2.10-1    (a) $\sigma_{max} \approx 6.2$ ksi and 6.9 ksi;
(b) $\sigma_{max} \approx 11.0$ ksi and 9.0 ksi

2.10-2    (a) $\sigma_{max} \approx 26$ MPa and 29 MPa;
(b) $\sigma_{max} \approx 25$ MPa and 22 MPa

2.10-3    $P_{max} = \sigma_t bt/3$

2.10-4    $\sigma_{max} \approx 46$ MPa

2.10-5    $\sigma_{max} \approx 6100$ psi

2.10-6    (a) No, it makes it weaker: $P_1 = 25.1$ kN,
$P_2 \approx 14.4$ kN; (b) $d_0 \approx 15.1$ mm

2.10-7    $d_{max} \approx 0.51$ in.

2.11-2    (a) $\delta_C = 1.67$ mm; (b) $\delta_C = 5.13$ mm;
(c) $\delta_C = 11.88$ mm

2.11-3    (b) $P = 17.7$ k

2.11-4    For $P = 30$ kN: $\delta = 6.2$ mm; for $P = 40$ kN:
$\delta = 12.0$ mm

2.11-5    For $P = 24$ k: $\delta = 0.18$ in.; for $P = 40$ k:
$\delta = 0.68$ in.

2.11-6    For $P = 3.2$ kN: $\delta_B = 4.85$ mm; for
$P = 4.8$ kN: $\delta_B = 17.3$ mm

2.12-1    $P_Y = P_P = 2\sigma_Y A \sin \theta$

2.12-2    $P_P = 201$ kN

2.12-3    (a) $P_P = 5\sigma_Y A$

2.12-4    $P_P = 2\sigma_Y A(1 + \sin \alpha)$

2.12-5    $P_P = 47.9$ k

2.12-6    $P_P = 82.5$ kN

2.12-7    $P_P = 20.4$ k

2.12-8    (a) $P_Y = \sigma_Y A$, $\delta_Y = 3\sigma_Y L/2E$;
(b) $P_P = 4\sigma_Y A/3$, $\delta_P = 3\sigma_Y L/E$

2.12-9    (a) $P_Y = \sigma_Y A$, $\delta_Y = \sigma_Y L/E$;
(b) $P_P = 5\sigma_Y A/4$, $\delta_P = 2\sigma_Y L/E$

2.12-10    (a) $W_Y = 28.8$ kN, $\delta_Y = 125$ mm;
(b) $W_P = 48$ kN, $\delta_P = 225$ mm

2.12-11    (a) $P_Y = 70.1$ k, $\delta_Y = 0.01862$ in.;
(b) $P_P = 104.3$ k, $\delta_P = 0.0286$ in.

# CHAPTER 3

3.2-1    (a) $d_{max} = 0.413$ in.; (b) $L_{min} = 21.8$ in.

3.2-2    (a) $L_{min} = 162.9$ mm; (b) $d_{max} = 68.8$ mm

3.2-3    (a) $\gamma_1 = 267 \times 10^{-6}$ rad; (b) $r_{2, min} = 2.2$ in.

**3.2-4**  (a) $\gamma_1 = 393 \times 10^{-6}$ rad; (b) $r_{2,\,max} = 50.9$ mm

**3.2-5**  (a) $\gamma_1 = 195 \times 10^{-6}$ rad; (b) $r_{2,\,max} = 2.57$ in.

**3.3-1**  (a) $\tau_{max} = 8344$ psi; (b) $d_{min} = 0.651$ in.

**3.3-2**  (a) $\tau_{max} = 23.8$ MPa; (b) $T_{max} = 0.402$ N·m; (c) $\theta = 9.12°/$m

**3.3-3**  (a) $\tau_{max} = 18{,}300$ psi; (b) $\phi = 3.32°$

**3.3-4**  (a) $k_T = 2059$ N·m; (b) $\tau_{max} = 27.9$ MPa, $\gamma_{max} = 997 \times 10^{-6}$ radians;

(c) $\dfrac{k_{T\,hollow}}{k_{T\,solid}} = 0.938,\quad \dfrac{\tau_{maxH}}{\tau_{maxS}} = 1.067$;

(d) $d_2 = 32.5$ mm

**3.3-5**  (a) $L_{min} = 38.0$ in.; (b) $L_{min} = 40.9$ in.

**3.3-6**  $T_{max} = 6.03$ N·m, $\phi = 2.20°$

**3.3-7**  (a) $\tau_{max} = 7965$ psi; $\gamma_{max} = 0.00255$ radians, $G = 3.13 \times 10^6$ psi; (b) $T_{max} = 5096$ lb·in.

**3.3-8**  (a) $T_{max} = 9164$ N·m; (b) $T_{max} = 7765$ N·m;

**3.3-9**  $\tau_{max} = 4840$ psi

**3.3-10**  (a) $d_{min} = 63.3$ mm; (b) $d_{min} = 66$ mm (4.2% increase in diameter)

**3.3-11**  (a) $\tau_2 = 5170$ psi; (b) $\tau_1 = 3880$ psi; (c) $\theta = 0.00898°/$in.

**3.3-12**  (a) $\tau_2 = 30.1$ MPa; (b) $\tau_1 = 20.1$ MPa; (c) $\theta = 0.306°/$m

**3.3-13**  (a) $d_{min} = 2.50$ in.; (b) $k_T = 2941\ \dfrac{\text{in·kip}}{\text{rad}}$

(c) $d_{min} = 1.996$ in.

**3.3-14**  (a) $d_{min} = 64.4$ mm; (b) $k_T = 134.9$ kN·m/rad; (c) $d_{min} \doteq 50$ mm

**3.3-15**  (a) $T_{1,max} = 4.60$ in.-k; (b) $T_{1,max} = 4.31$ in.-k; (c) Torque: 6.25%, Weight: 25%

**3.3-16**  (a) $\phi = 5.19°$; (b) $d = 88.4$ mm; (c) Ratio = 0.524

**3.3-17**  (a) $r_2 = 1.399$ in. (b) $P_{max} = 1387$ lb

**3.4-1**  (a) $\tau_{max} = \tau_{BC} = 7602$ psi, $\phi_C = 0.16°$; (b) $d_{BC} = 1.966$ in., $\phi_C = -0.177°$

**3.4-2**  (a) $\tau_{bar} = 79.6$ MPa, $\tau_{tube} = 32.3$ MPa; (b) $\phi_A = 9.43°$

**3.4-3**  (a) $\tau_{max} = \tau_{BC} = 4653$ psi, $\phi_D = 0.978°$; (b) $d_{AB} = 3.25$ in., $d_{BC} = 2.75$ in., $d_{CD} = 2.16$ in., $\phi_D = 1.303°$

**3.4-4**  $T_{allow} = 439$ N·m

**3.4-5**  $d_1 = 0.818$ in.

**3.4-6**  (a) $d = 77.5$ mm; (b) $d = 71.5$ mm

**3.4-7**  (a) $d = 1.78$ in.; (b) $d = 1.83$ in.

**3.4-8**  (b) $d_B/d_A = 1.45$

**3.4-9**  Minimum $d_A = 2.52$ in.

**3.4-10**  Minimum $d_B = 48.6$ mm

**3.4-11**  (a) $R_1 = -3T/2$; (b) $T_1 = 1.5T$, $T_2 = 0.5T$; (c) $x = 7L/17$; (d) $\phi_2 = (12/17)(TL/GI_p)$

**3.4-12**  $\phi = 3TL/2\pi Gt d_A^3$

**3.4-13**  (a) $\phi = 2.79°$; (b) $\phi = 2.21°$

**3.4-14**  (a) $R_1 = \dfrac{-T}{2}$; (b) $\varphi_3 = \dfrac{19}{8} \cdot \dfrac{TL}{\pi Gt d^3}$

**3.4-15**  $\phi_D = \left| \dfrac{4Fd}{\pi G} \right| \dfrac{L_1}{t_{01} d_{01}^{\,3}}$

$+ \displaystyle\int_0^{L_2} \dfrac{L_2^A}{(d_{01}L_2 - d_{01}x + d_{03}x)^3 (t_{01}L_2 - t_{01}x + t_{01}x)}$

$dx + \dfrac{L_3}{t_{03} d_{03}^{\,3}}$,

$\phi_D = 0.142°$

**3.4-16**  (a) $\tau_{max} = 16tL/\pi d^3$; (b) $\phi = 16tL^2/\pi Gd^4$

**3.4-17**  (a) $\tau_{max} = 8t_A L/\pi d^3$; (b) $\phi = 16t_A L^2/3\pi Gd^4$

**3.4-18**  (a) $R_A = \dfrac{T_0}{6}$;

(b) $T_{AB}(x) = \left( \dfrac{T_0}{6} - \dfrac{x^2}{L^2}\, T_0 \right) 0 \le x \le \dfrac{L}{2}$,

$T_{BC}(x) = -\left[ \left( \dfrac{x-L}{L} \right)^2 \cdot \dfrac{T_0}{3} \right]$

$\dfrac{L}{2} \le x \le L$; (c) $\phi_c = \dfrac{T_0 L}{144 GI_p}$;

(d) $\tau_{max} = \dfrac{8}{3\pi} \cdot \dfrac{T_0}{d_{AB}^{\,3}}$

**3.4-19**  (a) $L_{max} = 4.42$ m; (b) $\phi = 170°$

**3.4-20**  (a) $T_{max} = 875$ N·m; (b) $\tau_{max} = 25.3$ MPa

**3.5-1**  (a) $\sigma_{max} = 6280$ psi; (b) $T = 74{,}000$ lb-in.

**3.5-2**  (a) $\varepsilon_{max} = 320 \times 10^{-6}$; (b) $\sigma_{max} = 51.2$ MPa; (c) $T = 20.0$ kN·m

**3.5-3**  (a) $d_1 = 2.40$ in.; (b) $\phi = 2.20°$; (c) $\gamma_{max} = 1600 \times 10^{-6}$ rad

**3.5-4**  $G = 30.0$ GPa

**3.5-5**  $T = 4200$ lb-in.

**3.5-6**  (a) $d_{min} = 37.7$ mm; (b) $T_{max} = 431$ N·m

**3.5-7**  (a) $d_1 = 0.6$ in.; (b) $d_{1max} = 0.661$ in.

**3.5-8**  (a) $d_2 = 79.3$ mm; (b) $d_2 = 74.4$ mm

**3.5-9**  (a) $\tau_{max} = 5090$ psi: (b) $\gamma_{max} = 432 \times 10^{-6}$ rad

**3.5-10**  (a) $\tau_{max} = 23.9$ MPa: (b) $\gamma_{max} = 884 \times 10^{-6}$ rad

**3.5-11** (a) $T_{1allow} = 17.84$ k-in., $T_{2allow} = 13.48$ k-in.;
(b) $L_{mid} = 18.54$ in.; (c) $d_{3new} = 2.58$ in.;
(d) $T_{max1} = 17.41$ k-in., $T_{max2} = 13.15$ k-in.,
$\varphi_{max1} = 1.487°$, $\varphi_{max2} = 1.245°$

**3.7-1** (a) $\tau_{max} = 4950$ psi; (b) $d_{min} = 3.22$ in.

**3.7-2** (a) $\tau_{max} = 50.0$ MPa; (b) $d_{min} = 32.3$ mm

**3.7-3** (a) $H = 6560$ hp: (b) Shear stress is halved.

**3.7-4** (a) $\tau_{max} = 16.8$ MPa; (b) $P_{max} = 267$ kW

**3.7-5** $d_{min} = 4.28$ in.

**3.7-6** $d_{min} = 110$ mm

**3.7-7** Minimum $d_1 = 1.221d$

**3.7-8** $P_{max} = 91.0$ kW

**3.7-9** $d = 2.75$ in.

**3.7-10** $d = 53.4$ mm

**3.8-1** (a) $\phi_{max} = 3T_0L/5GI_P$; (b) $\varphi_{max} = \dfrac{9LT_0}{25GI_p}$

**3.8-2** (a) $x = L/4$; (b) $\phi_{max} = T_0L/8GI_P$

**3.8-3** $\phi_{max} = 2b\tau_{allow}/Gd$

**3.8-4** $P_{allow} = 2710$ N

**3.8-5** (a) $T_{0,max} = 3678$ lb-in.;
(b) $T_{0,max} = 3898$ lb-in.

**3.8-6** (a) $T_{0,max} = 150$ N·m;
(b) $T_{0,max} = 140$ N·m

**3.8-7** (a) $a/L = d_A/(d_A + d_B)$:
(b) $a/L = d_A^4/(d_A^4 + d_B^4)$

**3.8-8** (a) $T_A = \dfrac{Lt_0}{6}$, $T_B = \dfrac{Lt_0}{3}$,

(b) $\phi_{max} = \phi\left(\dfrac{L}{\sqrt{3}}\right) = -\dfrac{\sqrt{3}L^2t_0}{27GI_P}$

**3.8-9** (a) $x = 30.12$ in.; (b) $\phi_{max} = -1°$
(at $x = 30.12$ in.)

**3.8-10** (a) $\tau_1 = 32.7$ MPa, $\tau_2 = 49.0$ MPa;
(b) $\phi = 1.030°$; (c) $k_T = 22.3$ kN·m

**3.8-11** (a) $\tau_1 = 1790$ psi, $\tau_2 = 2690$ psi;
(b) $\phi = 0.354°$; (c) $k_T = 809$ k-in.

**3.8-12** (a) $T_{max} = 1.521$ kN·m; (b) $d_2 = 56.9$ mm

**3.8-13** (a) $T_{max} = 9.13$ k-in.; (b) $d_2 = 2.27$ in.

**3.8-14** (a) $T_{1,allow} = 7.14$ kN·m;
(b) $T_{2,allow} = 6.35$ kN·m; (c) $T_{3,allow} = 7.41$ kN·m; (d) $T_{max} = 6.35$ kN·m;

**3.8-15** (a) $T_A = 15{,}292$ in.-lb, $T_B = 24{,}708$ in.-lb;
(b) $T_A = 8734$ in.-lb, $T_B = 31{,}266$ in.-lb

**3.8-16** (a) $R_1 = -0.77T$, $R_2 = -0.23T$;
(b) $T_{max} = 2.79$ kN·m; (c) $\phi_{max} = 7.51°$;

(d) $T_{max} = 2.48$ kN·m (shear in flange plate bolts controls); (e) $R_2 = \dfrac{\beta}{f_{T1} + f_{T2}}$,

$R_1 = -R_2$, with $f_{T1} = \dfrac{L_1}{G_1I_{p1}}$, $f_{T2} = \dfrac{L_2}{G_2I_{p2}}$;

(f) $\beta_{max} = 29.1°$

**3.9-1** (a) $U = 32.0$ in.-lb; (b) $\phi = 0.775°$

**3.9-2** (a) $U = 5.36$ J; (b) $\phi = 1.53°$

**3.9-3** $U = 22.6$ in.-lb

**3.9-4** $U = 1.84$ J

**3.9-5** (c) $U_3 = T^2L/2GI_P + TtL^2/2GI_P + t^2L^3/6GI_P$

**3.9-6** $U = 19T_0^2L/32GI_P$

**3.9-7** $\phi = T_0L_AL_B/[G(L_BI_{PA} + L_A I_{PB})]$

**3.9-8** $U = t_0^2L^3/40GI_P$

**3.9-9** (a) $U = \dfrac{T^2L(d_A + d_B)}{\pi Gtd_A^2d_B^2}$;

(b) $\phi = \dfrac{2TL(d_A + d_B)}{\pi Gtd_A^2d_B^2}$

**3.9-10** $U = \dfrac{\beta^2GI_{PA}I_{PB}}{2L(I_{PA} + I_{PB})}$

**3.9-11** $\phi = \dfrac{2n}{15d^2}\sqrt{\dfrac{2\pi I_mL}{G}}$; $\tau_{max} = \dfrac{n}{15d}\sqrt{\dfrac{2\pi GJ_m}{L}}$

**3.11-1** (a) $\tau_{approx} = 6310$ psi; (b) $\tau_{exact} = 6830$ psi

**3.11-2** $t_{min} = \pi d/64$

**3.11-3** (a) $\tau = 1250$ psi; (b) $\phi = 0.373°$

**3.11-4** (a) $\tau = 9.17$ MPa; (b) $\phi = 0.140°$

**3.11-5** $U_1/U_2 = 2$

**3.11-6** $\tau = 35.0$ MPa, $\phi = 0.570°$

**3.11-7** $\tau = 2390$ psi, $\theta = 0.00480°/$in.

**3.11-8** $\tau = T\sqrt{3}/9b^2t$, $\theta = 2T/9Gb^3t$

**3.11-9** (a) $\phi_1/\phi_2 = 1 + 1/4\beta^2$

**3.11-10** $\tau = 2T(1 + \beta)^2/tL_m^2\beta$

**3.11-11** $t_{min} = 0.140$ in.

**3.11-12** (a) $t = 6.66$ mm; (b) $t = 7.02$ mm

**3.12-1** $T_{max} \approx 6200$ lb-in.

**3.12-2** $R_{min} \approx 4.0$ mm

**3.12-3** For $D_1 = 0.8$ in.: $\tau_{max} \approx 6400$ psi

**3.12-4** $D_2 \approx 115$ mm; lower limit

**3.12-5** $D_1 \approx 1.31$ in.

# CHAPTER 4

**4.3-1**  $V = 333$ lb, $M = 50,667$ lb-in.

**4.3-2**  $V = -0.938$ kN, $M = 4.12$ kN·m

**4.3-3**  $V = 0$, $M = 0$

**4.3-4**  $V = 7.0$ kN, $M = -9.5$ kN·m

**4.3-5**  (a) $V = -190$ lb, M $= 16,580$ ft-lb;
(b) $q = 370.4$ lb/ft (upward)

**4.3-6**  (a) $V = -1.0$ kN, $M = -7$ kN·m;
(b) $P_2 = 6$ kN; (c) $P_1 = -8$ kN
(acts to right)

**4.3-7**  $b/L = 1/2$

**4.3-8**  $M = 108$ N·m

**4.3-9**  $N = P \sin \theta$, $V = P \cos \theta$, $M = Pr \sin \theta$

**4.3-10**  $V = -6.04$ kN, $M = 15.45$ kN·m

**4.3-11**  (a) $P = 1200$ lb; (b) $P = 133.3$ lb

**4.3-12**  $V = -4.17$ kN, $M = 75$ kN·m

**4.3-13**  (a) $V_B = 6000$ lb, $M_B = 9000$ lb-ft;
(b) $V_m = 0$, $M_m = 21,000$ lb-ft

**4.3-14**  (a) $N = 21.6$ kN (compression), $V = 7.2$ kN,
$M = 50.4$ kN·m; (b) $N = 21.6$ kN
(compression), $V = -5.4$ kN, $M = 0$ (at
moment release)

**4.3-15**  $V_{max} = 91wL^2\alpha/30g$, $M_{max} = 229wL^3\alpha/75g$

**4.5-1**  $V_{max} = P$, $M_{max} = Pa$

**4.5-2**  $V_{max} = M_0/L$, $M_{max} = M_0a/L$

**4.5-3**  $V_{max} = qL/2$, $M_{max} = -3qL^2/8$

**4.5-4**  $V_{max} = P$, $M_{max} = PL/4$

**4.5-5**  $V_{max} = -2P/3$, $M_{max} = PL/9$

**4.5-6**  $V_{max} = 2M_1/L$, $M_{max} = 7M_1/3$

**4.5-7**  (a) $V_{max} = \dfrac{P}{2}$ (on $AB$),

$M_{max} = R_C\left(\dfrac{3L}{4}\right) = \dfrac{3LP}{8}$ (just right of $B$);

(b) $N_{max} = P$ (tension on $AB$), $V_{max} = \dfrac{P}{5}$,

$M_{max} = \dfrac{-P}{5} \cdot \left(\dfrac{3L}{4}\right) = -\dfrac{3LP}{20}$ (just right of $B$)

**4.5-8**  (a) $V_{max} = P$, $M_{max} = -Pa$;
(b) $M = 3Pa$ (CCW); $V_{max} = 2P$,
$M_{max} = 2Pa$

**4.5-9**  $V_{max} = qL/2$, $M_{max} = 5qL^2/72$

**4.5-10**  (a) $V_{max} = -q_0L/2$, $M_{max} = -q_0L^2/6$;

(b) $V_{max} = -\dfrac{2Lq_0}{3}$, $M_{max} = -\dfrac{4L^2q_0}{15}$ (at $B$)

**4.5-11**  $R_B = 207$ lb, $R_A = 73.3$ lb,
$V_{max} = -207$ lb, $M_{max} = 2933$ lb-in.

**4.5-12**  $V_{max} = 1200$ N, $M_{max} = 960$ N.m

**4.5-13**  $V_{max} = 200$ lb, $M_{max} = -1600$ lb-ft

**4.5-14**  $V_{max} = 4.5$ kN, $M_{max} = -11.33$ kN·m

**4.5-15**  $V_{max} = -1300$ lb, $M_{max} = -28,800$ lb-in.

**4.5-16**  $V_{max} = 15.34$ kN, $M_{max} = 9.80$ kN·m

**4.5-17**  The first case has the larger maximum

moment: $\left(\dfrac{6}{5}PL\right)$

**4.5-18**  The third case has the larger maximum

moment: $\left(\dfrac{6}{5}PL\right)$

**4.5-19**  $V_{max} = 900$ lb, $M_{max} = -900$ lb-ft

**4.5-20**  $V_{max} = -10.0$ kN, $M_{max} = 16.0$ kN·m

**4.5-21**  Two cases have the same maximum
moment: $(PL)$.

**4.5-22**  $V_{max} = 33.0$ kN, $M_{max} = -61.2$ kN·m

**4.5-23**  (a) $V_{max} = -4P/9$, $M_{max} = 8PL/3$;
(b) $V_{max} = -4P/5$, $M_{max} = -4.8PL$

**4.5-24**  $M_{Az} = -PL$(clockwise), $A_x = 0$, $A_y = 0$,

$C_y = \dfrac{1}{12}P$ (upward), $D_y = \dfrac{1}{6}P$ (upward),

$V_{max} = P/12$, $M_{max} = PL$

**4.5-25**  (a) $R_A = 1.25$ k, $R_B = 13.75$ k;
(b) $P = 8$ k (upward)

**4.5-26**  $V_{max} = 4.6$ kN, $M_{max} = -6.24$ kN·m

**4.5-27**  (a) $R_A = 197.1$ lb, $R_B = 433$ lb;
(b) $a = 4.624$ ft; (c) $a = 3.143$ ft

**4.5-28**  $V_{max} = -2.8$ kN, $M_{max} = 1.450$ kN·m

**4.5-29**  $a = 0.5858L$, $V_{max} = 0.2929qL$,
$M_{max} = 0.02145qL^2$

**4.5-30**  $V_{max} = 2.5$ kN, $M_{max} = 5.0$ kN·m

**4.5-31**  (a) $V_{max} = -R_B = -\dfrac{Lq_0}{2}$,

$M_{max} = -M_A = \dfrac{L^2q_0}{6}$;

(b) $V_{max} = -R_B = -\dfrac{2Lq_0}{3}$,

$M_{max} = -M_A = \dfrac{4L^2q_0}{15}$

**4.5-32**  $M_{max} = 10$ kN·m

**4.5-33**  $M_{max} = M_{pos} = 897.6$ lb-ft (at $x = 9.6$ ft);
$M_{neg} = -600$ lb-ft (at $x = 20$ ft)

**4.5-34**  $V_{max} = -w_0L/3$, $M_{max} = -woL^2/12$

**4.5-35** $M_A = -\dfrac{w_0}{30}L^2$ (clockwise),

$A_x = -3w_0L/10$ (leftward), $A_y = -3w_0L/20$
(downward), $C_y = w_0L/12$ (upward),
$D_y = w_0L/6$ (upward), $V_{max} = w_0L/4$,
$M_{max} = -w_0L^2/24$ at **B**

**4.5-36** (a) $x = 9.6$ m, $V_{max} = 28$ kN; (b) $x = 4.0$ m,
$M_{max} = 78.4$ kN·m

**4.5-37** (a) $A_x = 50.38$ lb (right), $A_y = 210$ lb
(upward), $B_x = -50.38$ lb (left),
$N_{max} = -214.8$ lb, $V_{max} = -47.5$ lb,
$M_{max} = 270$ lb-ft; (b) $A_x = 0$,
$A_y = 67.5$ lb, $B_x = 0$, $B_y = 142.5$ lb,
$N_{max} = 134.4$ lb, $V_{max} = -47.5$ lb,
$M_{max} = 270$ lb-ft

**4.5-38** (a) $A_x = -q_0L/2$ (leftward), $A_y = 17q_0L/18$
(upward), $D_x = -q_0L/2$ (leftward),
$D_y = -4q_0L/9$ (downward), $M_D = 0$,
$N_{max} = q_0L^2$, $V_{max} = 17q_0L/18$,
$M_{max} = q_0L^2$; (b) $B_x = q_0L/2$ (rightward),
$B_y = -q_0L/2 + 5q_0L/3 = 7q_0L/6$ (upward);
$D_x = q_0L/2$ (rightward), $D_y = -5q_0L/3$
(downward), $M_D = 0$, $N_{max} = 5q_0L/3$
$V_{max} = 5q_0L/3$, $M_{max} q_0L^2$

**4.5-39** (a) $M_A = 0$, $R_{Ax} = 0$, $R_{Ay} = q_0L/6$
(upward), $R_{Cy} = q_0L/3$; $N_{max} = -q_0L/6$,
$V_{max} = -q_0L/3$, $M_{max} = 0.06415q_0L^2$;
(b)$M_A = (16/15)q_0L^2$, $R_{Ax} = -4q_0L/3$,
$R_{Ay} = q_0L/6$ (upward), $R_{Cy} = q_0L/3$;
$N_{max} = -q_0L/6$, $V_{max} = 4q_0L/3$ (in
column), $V_{max} = -q_0L/3$ (in beam),
$M_{max} = -(16/15)q_0L^2$ (in column),
$M_{max} = 0.06415q_0L^2$ (in beam)

**4.5-40** $M_A = 0$, $A_x = 0$, $A_y = -18.41$ kN (downward), $M_D = 0$, $D_x = -63.0$ kN (leftward),
$D_y = 62.1$ kN (upward), $N_{max} = -62.1$ kN,
$V_{max} = 63.0$ kN, $M_{max} = 756$ kN·m

# CHAPTER 5

**5.4-1** (a) $\varepsilon_{max} = 8.67 \times 10^{-4}$; (b) $R_{min} = 9.35$ in.;
(c) $d_{max} = 0.24$ in.

**5.4-2** (a) $L_{min} = 5.24$ m; (b) $d_{max} = 4.38$ mm

**5.4-3** (a) $\varepsilon_{max} = 5.98 \times 10^{-3}$; (b) $d_{max} = 4.85$ in.;
(c) $L_{min} = 51$ ft

**5.4-4** (a) $\rho = 85$ m, $\kappa = 0.0118\,\dfrac{1}{m}$, $\delta = 23.5$ mm;

(b) $h_{max} = 136$ mm; (c) $\delta = 75.3$ mm

**5.4-5** (a) $\varepsilon = 9.14 \times 10^{-4}$; (b) $t_{max} = 0.241$ in.;
(c) $\delta = 0.744$ in.; (d) $L_{max} = 37.1$ in.

**5.4-6** (a) $\varepsilon = 4.57 \times 10^{-4}$; (b) $L_{max} = 2$ m

**5.5-1** (a) $\sigma_{max} = 52.4$ ksi; (b) 33.3%;
(c) $L_{new} = 120$ in.

**5.5-2** (a) $\sigma_{max} = 250$ MPa; (b) $-19.98\%$; (c) $+25\%$

**5.5-3** (a) $\sigma_{max} = 38.2$ ksi; (b) $+10\%$; (c) $+10\%$

**5.5-4** (a) $\sigma_{max} = 8.63$ MPa; (b) $\sigma_{max} = 6.49$ MPa

**5.5-5** $\sigma_{max} = 21.6$ ksi

**5.5-6** $\sigma_{max} = 203$ MPa

**5.5-7** $\sigma_{max} = 3420$ psi

**5.5-8** $\sigma_{max} = 101$ MPa

**5.5-9** $\sigma_{max} = 10.82$ ksi

**5.5-10** $\sigma_{max} = 7.0$ MPa

**5.5-11** (a) $\sigma_{max} = 432$ psi; (b) $s = 0.58579L$,
$\sigma_{min} = 153.7$ psi; (c) $s = 0$ or $L$,
$\sigma_{max} = 896$ psi

**5.5-12** $\sigma_{max} = 2.10$ MPa

**5.5-13** (a) $\sigma_t = 30.93M/d^3$; (b) $\sigma_t = 360M/(73bh^2)$;
(c) $\sigma_t = 85.24M/d^3$

**5.5-14** $\sigma_{max} = 10.965M/d^3$

**5.5-15** (a) $\sigma_{max} = 21.4$ ksi; (b) $L = 20.9$ ft;
(c) $d = 8.56$ ft

**5.5-16** (a) $\sigma_t = 35.4$ MPa, $\sigma_c = 61$ MPa;

(b) $d_{max} = \dfrac{L}{2}$, $\sigma_t = 37.1$ MPa, $\sigma_c = 64.1$ MPa

**5.5-17** (a) $\sigma_t = 4.34$ ksi, $\sigma_c = 15.96$ ksi;
(b) $P_{max} = 214$ lb; (c) 4.28 ft

**5.5-18** (a) $\sigma_c = 1.456$ MPa, $\sigma_t = 1.514$ MPa;
(b) $\sigma_c = 1.666$ MPa $(+14\%)$,
$\sigma_t = 1.381$ MPa $(-9\%)$; (c) $\sigma_c = 0.728$ MPa
$(-50\%)$, $\sigma_t = 0.757$ MPa $(-50\%)$

**5.5-19** (a) $\sigma_t = 7810$ psi, $\sigma_c = 13{,}885$ psi;
(b) $a = 12.73$ ft

**5.5-20** $\sigma_{max} = 3pL^2a_0/t$

**5.5-21** (a) $\sigma_t = 20{,}360$ psi, $\sigma_c = 13{,}188$ psi;
(b) $h = 3.20$ in.; (c) $q = 97.2$ lb/ft,
$P = 675$ lb

**5.5-22** $\sigma = 25.1$ MPa, 17.8 MPa, $-23.5$ MPa

**5.5-23** $d = 3$ ft, $\sigma_{max} = 171$ psi, $d = 6$ ft,
$\sigma_{max} = 830$ psi

**5.5-24** (a) $c_1 = 91.7$ mm, $c_2 = 108.3$ mm,
$I_z = 7.969 \times 10^7$ mm$^4$; (b) $\sigma_t = 4659$ kPa
(top of beam at $C$), $\sigma_c = 5506$ kPa (bottom
of beam at $C$)

5.5-25  (a) $F_{res} = 104.8$ lb; (b) $\sigma_{max} = 36.0$ ksi (compression at base); (c) $\sigma_{max} = 32.4$ ksi (tension at base)

5.6-1   $d_{min} = 4.00$ in.

5.6-2   (a) $d_{min} = 12.62$ mm; (b) $P_{max} = 39.8$ N

5.6-3   (a) C $15 \times 33.9$; (b) S $8 \times 18.4$; (c) W $8 \times 35$

5.6-4   (a) W $360 \times 39$; (b) W $250 \times 89$

5.6-5   (a) S $10 \times 35$; (b) $P_{max} = 3152$ lb

5.6-6   (a) $b_{min} = 161.6$ mm; (b) $b_{min} = 141.2$ mm, area$_{(b)}$/area$_{(a)} = 1.145$

5.6-7   (a) $2 \times 12$; (b) $w_{max} = 137.3$ lb/ft$^2$

5.6-8   (a) $s_{max} = 429$ mm; (b) $h_{min} = 214$ mm

5.6-9   (a) $q_{0,allow} = 424$ lb/ft; (b) $q_{0,allow} = 268$ lb/ft

5.6-10  $h_{min} = 30.6$ mm

5.6-11  (a) $S_{reqd} = 15.37$ in.$^3$; (b) S $8 \times 23$

5.6-12  (a) $d_{min} = 37.6$ mm; (b) $d_{min} = 45.2$ mm, area$_{(b)}$/area$_{(a)} = 0.635$

5.6-13  (a) $4 \times 12$; (b) $q_{max} = 14.2$ lb/ft

5.6-14  $b = 152$ mm, $h = 202$ mm

5.6-15  $b = 10.25$ in.

5.6-16  $t = 13.61$ mm

5.6-17  $W_1:W_2:W_3:W_4 = 1:1.260:1.408:0.888$

5.6-18  (a) $q_{max} = 6.61$ kN/m; (b) $q_{max} = 9.37$ kN/m

5.6-19  6.57%

5.6-20  (a) $b_{min} = 11.91$ mm; (b) $b_{min} = 11.92$ mm

5.6-21  (a) $s_{max} = 49.2$ in.; (b) $d = 12.65$ in.

5.6-22  (a) $\beta = 1/9$; (b) 5.35%

5.6-23  Increase when $d/h > 0.6861$; decrease when $d/h < 0.6861$

5.7-1   (a) $x = L/4$, $\sigma_{max} = 4PL/9h_A^3$, $\sigma_{max}/\sigma_B = 2$; (b) $x = 0.209L$, $\sigma_{max} = 0.394PL/h_A^3$, $\sigma_{max}/\sigma_B = 3.54$

5.7-2   (a) $x = 4$ m, $\sigma_{max} = 37.7$ MPa, $\sigma_{max}/\sigma_B = 9/8$; (b) $x = 2$ m, $\sigma_{max} = 25.2$ MPa, $\sigma_{max}/\sigma_m = 4/3$

5.7-3   (a) $x = 8$ in., $\sigma_{max} = 1250$ psi, $\sigma_{max}/\sigma_B = 1.042$; (b) $x = 4.64$ in., $\sigma_{max} = 1235$ psi, $\sigma_{max}/\sigma_m = 1.215$

5.7-4   (a) $\sigma_A = 210$ MPa; (b) $\sigma_B = 221$ MPa; (c) $x = 0.625$ m; (d) $\sigma_{max} = 231$ MPa; (e) $\sigma_{max} = 214$ MPa

5.7-5   (a) $1 \le d_B/d_A \le 1.5$; (b) $\sigma_{max} = \sigma_B = 32PL/\pi d_B^3$

5.7-6   $h_x = h_B x/L$

5.7-7   $b_y = 2b_B x/L$

5.7-8   $h_x = h_B \sqrt{x/L}$

5.8-2   (a) $\tau_{max} = 731$ kPa, $\sigma_{max} = 4.75$ MPa; (b) $\tau_{max} = 1462$ kPa, $\sigma_{max} = 19.01$ MPa

5.8-3   (a) $M_{max} = 25.4$ k-ft; (b) $M_{max} = 4.95$ k-ft

5.8-4   $\tau_{max} = 500$ kPa

5.8-5   $\tau_{max} = 2400$ psi

5.8-6   (a) $L_0 = h(\sigma_{allow}/\tau_{allow})$; (b) $L_0 = (h/2)(\sigma_{allow}/\tau_{allow})$

5.8-7   (a) $P_{max} = 1.914$ kip; (b) $P_{max} = 2.05$ kip

5.8-8   (a) $M_{max} = 72.2$ N·m; (b) $M_{max} = 9.01$ N·m

5.8-9   (a) $8 \times 12$-in. beam; (b) $8 \times 12$-in. beam

5.8-10  (a) $P = 38.0$ kN; (b) $P = 35.6$ kN

5.8-11  (a) $w_1 = 121$ lb/ft$^2$; (b) $w_2 = 324$ lb/ft$^2$; (c) $w_{allow} = 121$ lb/ft$^2$

5.8-12  (a) $b = 89.3$ mm; (b) $b = 87.8$ mm

5.9-1   $d_{min} = 5.70$ in.

5.9-2   (a) $W = 28.6$ kN; (b) $W = 38.7$ kN

5.9-3   (a) $d = 10.52$ in.; (b) $d = 2.56$ in.

5.9-4   (a) $q_{0,max} = 55.7$ kN/m; (b) $L_{max} = 2.51$ m

5.10-1  (a) $\tau_{max} = 5795$ psi; (b) $\tau_{min} = 4555$ psi; (c) $\tau_{aver} = 5714$ psi; (d) $V_{web} = 28.25$ k

5.10-2  (a) $\tau_{max} = 28.43$ MPa; (b) $\tau_{min} = 21.86$ MPa; (c) $\tau_{aver} = 27.41$ MPa; (d) $V_{web} = 119.7$ kN

5.10-3  (a) $\tau_{max} = 4861$ psi; (b) $\tau_{min} = 4202$ psi; (c) $\tau_{aver} = 4921$ psi; (d) $V_{web} = 9.432$ k

5.10-4  (a) $\tau_{max} = 32.28$ MPa; (b) $\tau_{min} = 21.45$ MPa; (c) $\tau_{aver} = 29.24$ MPa; (d) $V_{web} = 196.1$ kN

5.10-5  (a) $\tau_{max} = 2634$ psi; (b) $\tau_{min} = 1993$ psi; (c) $\tau_{aver} = 2518$ psi; (d) $V_{web} = 20.19$ k

5.10-6  (a) $\tau_{max} = 28.40$ MPa; (b) $\tau_{min} = 19.35$ MPa; (c) $\tau_{aver} = 25.97$ MPa; (d) $V_{web} = 58.63$ kN

5.10-7  $q_{max} = 1270$ lb/ft

5.10-8  (a) $q_{max} = 184.7$ kN/m; (b) $q_{max} = 247$ kN/m

5.10-9  S $8 \times 23$

5.10-10 $V = 273$ kN

5.10-11 $\tau_{max} = 1.42$ ksi, $\tau_{min} = 1.03$ ksi

5.10-12 $\tau_{max} = 19.7$ MPa

5.10-13 $\tau_{max} = 2221$ psi

5.11-1  $V_{max} = 676$ lb

5.11-2  $V_{max} = 1.924$ MN

5.11-3  $F = 1994$ lb/in.

5.11-4  $V_{max} = 10.7$ kN

5.11-5    (a) $s_{max} = 5.08$ in.; (b) $s_{max} = 4.63$ in.

5.11-6    (a) $s_A = 78.3$ mm; (b) $s_B = 97.9$ mm

5.11-7    (a) $s_{max} = 2.77$ in.; (b) $s_{max} = 1.85$ in.

5.11-8    $s_{max} = 92.3$ mm

5.11-9    $V_{max} = 18.30$ k

5.11-10    $s_{max} = 236$ mm

5.11-11    (a) Case (1); (b) Case (3); (c) Case (1);
(d) Case (3)

5.11-12    $s_{max} = 180$ mm

5.12-1    $\sigma_t = 14,660$ psi, $\sigma_c = -14,990$ psi

5.12-2    $\sigma_t = 5770$ kPa, $\sigma_c = -6668$ kPa

5.12-3    $t_{min} = 0.477$ in.

5.12-4    $\sigma_t = -11.83$ MPa, $\sigma_c = -12.33$ MPa,
$t_{min} = 12.38$ mm

5.12-5    $\sigma_t = 302$ psi, $\sigma_c = -314$ psi

5.12-6    $T_{max} = 108.6$ kN

5.12-7    $\alpha = \arctan [(d_2^2 + d_1^2)/(4hd_2)]$

5.12-8    (a) $d_{min} = 8.46$ cm; (b) $d_{min} = 8.91$ cm

5.12-9    $H_{max} = 32.2$ ft

5.12-10    $W = 33.3$ kN

5.12-11    (a) $\sigma_t = 87.6$ psi, $\sigma_c = -99.6$ psi;
(b) $d_{max} = 28.9$ in.

5.12-12    (a) $b = \pi - d/6$; (b) $b = \pi - d/3$;
(c) Rectangular post

5.12-13    (a) $\sigma_t = 1900$ psi, $\sigma_c = -1100$ psi;
(b) Both stresses increase in magnitude.

5.12-14    (a) $\sigma_t = 8P/b^2$, $\sigma_c = -4P/b^2$;
(b) $\sigma_t = 9.11P/b^2$, $\sigma_c = 6.36P/b^2$

5.12-15    (a) $\sigma_t = 857$ psi, $\sigma_c = -5711$ psi;
(b) $y_0 = -4.62$ in.; (c) $\sigma_t = 453$ psi,
$\sigma_c = -2951$ psi, $y_0 = -6.33$ in.

5.12-16    (a) $\sigma_t = 3.27$ MPa, $\sigma_c = -24.2$ MPa;
(b) $y_0 = -76.2$ mm; (c) $\sigma_t = 1.587$ MPa,
$\sigma_c = -20.3$ MPa, $y_0 = -100.8$ mm

5.12-17    (a) $\sigma_t = 15.48$ ksi; (b) $\sigma_t = 2.91$ ksi

5.12-18    (a) $y_0 = -21.5$ mm; (b) $P = 67.3$ kN;
(c) $y_0 = 148.3$ mm, $P = 149.6$ kN

5.13-1    (a) $d = 0.50$ in., $\sigma_{max} = 15,500$ psi;
(b) $R = 0.10$ in., $\sigma_{max} \approx 49,000$ psi

5.13-2    (a) $d = 16$ mm, $\sigma_{max} = 81$ MPa;
(b) $R = 4$mm, $\sigma_{max} \approx 200$ MPa

5.13-3    $b_{min} \approx 0.24$ in.

5.13-4    $b_{min} \approx 0.33$ mm

5.13-5    (a) $R_{min} \approx 0.45$ in.; (b) $d_{max} = 4.13$ in.

# CHAPTER 6

6.2-1    $\sigma_{face} = \pm 1980$ psi, $\sigma_{core} = \pm531$ psi

6.2-2    (a) $M_{max} = 58.7$ kN·m;
(b) $M_{max} = 90.9$ kN·m; (c) $t = 7.08$ mm

6.2-3    (a) $M_{max} = 172$ k-in.; (b) $M_{max} = 96$ k-in.

6.2-4    (a) $M_{allow,steel} = \dfrac{\pi\sigma_s(E_B d_1^4 - E_s d_1^4 + E_s d_2^4)}{32E_s d_2}$,

$M_{allow,brass} = \dfrac{\pi\sigma_B(E_B d_1^4 - E_s d_1^4 + E_s d_2^4)}{32E_s d_1}$;

(b) $M_{max,brass} = 1235$ N·m; (c) $d_1 = 33.3$ mm

6.2-5    (a) $\sigma_w = 666$ psi, $\sigma_s = 13,897$ psi;
(b) $q_{max} = 665$ lb/ft; (c) $M_{0,max} = 486$ lb-ft

6.2-6    (a) $M_{allow} = 768$ N·m; (b) $\sigma_{sa} = 47.9$ MPa,
$M_{max} = 1051$ N·m

6.2-7    (a) $\sigma_{face} = 3610$ psi, $\sigma_{core} = 4$ psi;
(b) $\sigma_{face} = 3630$ psi, $\sigma_{core} = 0$

6.2-8    (a) $\sigma_{face} = 14.1$ MPa, $\sigma_{core} = 0.214$ MPa;
(b) $\sigma_{face} = 14.9$ MPa, $\sigma_{core} = 0$

6.2-9    $\sigma_a = 4120$ psi, $\sigma_c = 5230$ psi

6.2-10    (a) $\sigma_w = 5.1$ MPa (compression),
$\sigma_s = 37.6$ MPa (tension); (b) $t_s = 3.09$ mm

6.2-11    (a) $\sigma_{plywood} = 1131$ psi, $\sigma_{pine} = 969$ psi;
(b) $q_{max} = 95.5$ lb/ft

6.2-12    $Q_{0.max} = 15.53$ kN/m

6.3-1    (a) $M_{max} = 442$ k-in.; (b) $M_{max} = 189$ k-in.

6.3-2    $t_{min} = 15.0$ mm

6.3-3    (a) $q_{allow} = 454$ lb/ft; (b) $\sigma_{wood} = 277$ psi,
$\sigma_{steel} = 11,782$ psi

6.3-4    (a) $\sigma_B = 60.3$ MPa, $\sigma_w = 7.09$ MPa;
(b) $t_B = 25.1$ mm, $M_{max} = 80$ kN·m

6.3-5    $\sigma_a = 1860$ psi, $\sigma_P = 72$ psi

6.3-6    $\sigma_a = 12.14$ MPa, $\sigma_P = 0.47$ MPa

6.3-7    (a) $q_{allow} = 264$ lb/ft; (b) $q_{allow} = 280$ lb/ft

6.3-8    (a) $\sigma_s = 93.5$ MPa; (b) $h_s = 5.08$ mm,
$h_a = 114.92$ mm

6.3-9    $M_{max} = 81.1$ k-in.

6.3-10    $S_A = 50.6$ mm³; Metal $A$

6.3-11    $\sigma_s = 13,400$ psi (tension), $\sigma_c = 812$ psi
(compression).

6.3-12    (a) $\sigma_c = 8.51$ MPa, $\sigma_s = 118.3$ MPa;
(b) $M_{max} = M_c = 172.9$ kN·m;
(c) $A_s = 2254$ mm², $M_{allow} = 167.8$ kN·m

6.3-13    (a) $\sigma_c = 649$ psi, $\sigma_s = 15,246$ psi;
(b) $M_{allow} = M_s = 207$ kip-ft

6.3-14  (a) $M_{max} = M_s = 10.59$ kN·m;
(b) $A_s = 1262$ mm$^2$, $M_{allow}$ 15.79 kN·m

6.3-15  $M_{allow} = M_W = 12.58$ kip-ft

6.4-1  $\tan \beta = h/b$, so $NA$ lies along other diagonal

6.4-2  $\beta = 51.8°$, $\sigma_{max} = 17.5$ MPa

6.4-3  $\beta = 42.8°$, $\sigma_{max} = 1036$ psi

6.4-4  $\beta = 78.9°$, $\sigma_A = -\sigma_E = 102$ MPa, $\sigma_B = -\sigma_D = -48$ MPa

6.4-5  $\beta = 72.6°$, $\sigma_A = -\sigma_E = 14{,}554$ psi, $\sigma_B = -\sigma_D = -4953$ psi

6.4-6  $\beta = -79.3°$, $\sigma_{max} = 8.87$ MPa

6.4-7  $\beta = -78.8°$, $\sigma_{max} = 1660$ psi

6.4-8  $\beta = -81.8°$, $\sigma_{max} = 69.4$ MPa

6.4-9  $\beta = 72.9°$, $\sigma_{max} = 8600$ psi

6.4-10  $\beta = 60.6°$, $\sigma_{max} = 20.8$ MPa

6.4-11  (a) $\sigma_A = 45{,}420 \sin \alpha + 3629 \cos \alpha$ (psi);
(b) $\tan \beta = 37.54 \tan \alpha$

6.4-12  $\beta = 79.0°$, $\sigma_{max} = 16.6$ MPa

6.4-13  (a) $\beta = -76.2°$, $\sigma_{max} = 8469$ psi;
(b) $\beta = -79.4°$, $\sigma_{max} = 8704$ psi

6.5-1  $\beta = 83.1°$, $\sigma_t = 5{,}060$ psi, $\sigma_c = -10{,}420$ psi

6.5-2  $\beta = 83.4°$, $\sigma_t = 10.5$ MPa, $\sigma_c = -23.1$ MPa

6.5-3  $\beta = 75.6°$, $\sigma_t = 3{,}080$ psi, $\sigma_c = -3450$ psi

6.5-4  $\beta = 75.8°$, $\sigma_t = 31.7$ MPa, $\sigma_c = -39.5$ MPa

6.5-5  (a) $\beta = -28.7°$, $\sigma_t = 4263$ psi, $\sigma_c = -4903$ psi; (b) $\beta = -38.5°$, $\sigma_t = 5756$ psi, $\sigma_c = -4868$ psi

6.5-6  $\beta = 78.1°$, $\sigma_t = 40.7$ MPa, $\sigma_c = -40.7$ MPa

6.5-7  $\beta = 82.3°$, $\sigma_t = 1397$ psi, $\sigma_c = -1157$ psi

6.5-8  $\beta = 2.93°$, $\sigma_t = 6.56$ MPa, $\sigma_c = -6.54$ MPa

6.5-9  For $\theta = 0$: $\sigma_t = -\sigma_c = 2.546 M/r^3$; for $\theta = 45°$: $\sigma_t = 4.535 M/r^3$, $\sigma_c = -3.955 M/r^3$; for $\theta = 90°$: $\sigma_t = 3.867 M/r^3$, $\sigma_c = -5.244 M/r^3$;

6.5-10  $\beta = -78.9°$, $\sigma_t = 131.1$ MPa, $\sigma_t = -148.5$ MPa

6.5-11  $\beta = -11.7°$, $\sigma_t = 28.0$ ksi, $\sigma_c = -24.2$ ksi

6.5-12  $\beta = -56.5°$, $\sigma_t = 31.0$ MPa, $\sigma_c = -29.0$ MPa

6.8-1  (a) $\tau_{max} = 3584$ psi; (b) $\tau_B = 430$ psi

6.8-2  (a) $\tau_{max} = 29.7$ MPa; (b) $\tau_B = 4.65$ MPa

6.8-3  (a) $\tau_{max} = 3448$ psi; (b) $\tau_{max} = 3446$ psi

6.8-4  (a) $\tau_{max} = 27.04$ MPa; (b) $\tau_{max} = 27.02$ MPa

6.9-1  $e = 1.027$ in.

6.9-2  $e = 22.1$ mm

6.9-6  (b) $e = \dfrac{63 \pi r}{24\pi + 38} = 1.745r$

6.9-8  (a) $e = \dfrac{b}{2}\left(\dfrac{2h + 3b}{h + 3b}\right)$;
(b) $e = \dfrac{b}{2}\left(\dfrac{43h + 48b}{23h + 48b}\right)$

6.10-1  $f = 2(2b_1 + b_2)/(3b_1 + b_2)$

6.10-2  (a) $f = 16\tau_2(r_2^3 - r_1^3)/3\pi(r_2^4 - r_1^4)$;
(b) $f = 4/\pi$

6.10-3  $q = 1000$ lb/in.

6.10-4  (a) 56.7%: (b) $M = 12.3$ kN·m

6.10-5  $f = 1.12$

6.10-6  $f = 1.15$

6.10-7  $Z = 16.98$ in.$^3$, $f = 1.14$

6.10-8  $Z = 1.209 \times 10^6$ mm$^3$, $f = 1.11$

6.10-9  $M_Y = 525$ k-ft, $M_P = 591$ k-ft, $f = 1.13$

6.10-10  $M_Y = 378$ kN·m, $M_P = 427$ kN·m, $f = 1.13$

6.10-11  $M_Y = 4320$ k-in., $M_P = 5450$ k-in., $f = 1.26$

6.10-12  $M_Y = 672$ kN·m, $M_P = 878$ kN·m, $f = 1.31$

6.10-13  $M_Y = 1619$ k-in., $M_P = 1951$ k-in., $f = 1.21$

6.10-14  $M_Y = 122$ kN·m, $M_P = 147$ kN·m, $f = 1.20$

6.10-15  (a) $M = 5977$ k-in.; (b) 22.4%

6.10-16  (a) $M = 524$ kN·m; (b) 36%

6.10-17  (a) $M = 2551$ k-in.; (b) 7.7%

6.10-18  $Z = 136 \times 10^3$ mm$^3$, $f = 1.79$

6.10-19  $M_P = 1120$ k-in.

6.10-20  $M_P = 295$ kN·m

# CHAPTER 7

7.2-1  For $\theta = 55°$: $\sigma_{x1} = 4221$ psi, $\sigma_{y1} = 4704$ psi, $\tau_{x1y1} = -3411$ psi

7.2-2  For $\theta = 40°$: $\sigma_{x1} = 117.2$ MPa, $\sigma_{y1} = 62.8$ MPa, $\tau_{x1y1} = -10.43$ MPa

7.2-3  For $\theta = 30°$: $\sigma_{x1} = -3041$ psi, $\sigma_{y1} = -8959$ psi, $\tau_{x1y1} = -12{,}725$ psi

7.2-4  For $\theta = 52°$: $\sigma_{x1} = -136.6$ MPa, $\sigma_{y1} = 16.6$ MPa, $\tau_{x1y1} = -84$ MPa

7.2-5　For $\theta = 50°$: $\sigma_{x1} = -1243$ psi, $\sigma_{y1} = -6757$ psi, $\tau_{x1y1} = 1240$ psi

7.2-6　For $\theta = -40°$: $\sigma_{x1} = -5.5$ MPa, $\sigma_{y1} = -27$ MPa, $\tau_{x1y1} = -28.1$ MPa

7.2-7　For $\theta = 38°$: $\sigma_{x1} = -13,359$ psi, $\sigma_{y1} = -3671$ psi, $\tau_{x1y1} = 4960$ psi

7.2-8　For $\theta = -40°$: $\sigma_{x1} = -66.5$ MPa, $\sigma_{y1} = -6.52$ MPa, $\tau_{x1y1} = -14.52$ MPa

7.2-9　Normal stress on seam, 187 psi tension. Shear stress, 163 psi (clockwise)

7.2-10　Normal stress on seam, 1440 kPa tension. Shear stress, 1030 kPa (clockwise)

7.2-11　$\sigma_w = -125$ psi, $\tau_w = 375$ psi

7.2-12　$\sigma_w = 10.0$ MPa, $\tau_w = -5.0$ MPa

7.2-13　$\theta = 51.3°$, $\sigma_{y1} = 500$ psi, $\tau_{x1y1} = -1122$ psi

7.2-14　$\theta = 36.6°$, $\sigma_{y1} = -9$ MPa, $\tau_{x1y1} = -14.83$ MPa

7.2-15　For $\theta = -36°$: $\sigma_{x1} = -12,068$ psi, $\sigma_{y1} = -4732$ psi, $\tau_{x1y1} = -4171$ psi

7.2-16　For $\theta = -50°$: $\sigma_{x1} = 51.4$ MPa, $\sigma_{y1} = -14.4$ MPa, $\tau_{x1y1} = -31.3$ MPa

7.2-17　$\sigma_y = 3673$ psi, $\tau_{xy} = 1405$ psi

7.2-18　$\sigma_y = -77.7$ MPa, $\tau_{xy} = -27.5$ MPa

7.2-19　$\sigma_b = -4700$ psi, $\tau_b = 2655$ psi, $\theta_1 = 48.04°$

7.3-1　$\sigma_1 = 5868$ psi, $\sigma_2 = 982$ psi, $\theta_{p1} = 8.94°$

7.3-2　$\sigma_1 = 119.2$ MPa, $\sigma_2 = 60.8$ MPa, $\theta_{p1} = 29.52°$

7.3-3　$\sigma_1 = -6333$ psi, $\sigma_2 = -1167$ psi, $\theta_{p1} = -23.68°$

7.3-4　$\sigma_1 = 53.6$ MPa, $\theta_{p1} = -14.2°$

7.3-5　$\sigma_1 = 5771$ psi, $\sigma_2 = 18,029$ psi,

$$\tau_{max} = \frac{\sigma_1 - \sigma_2}{2} = -6129 \text{psi}$$

$\theta_{p1} = -14.12°$

7.3-6　$\tau_{max} = 24.2$ MPa, $\sigma_{x1} = -14.25$ MPa, $\sigma_{y1} = -14.25$ MPa, $\theta_{s1} = 60.53°$

7.3-7　$\tau_{max} = 6851$ psi, $\theta_{s1} = 61.8°$

7.3-8　$\tau_{max} = 26.7$ MPa, $\theta_{s1} = 19.08°$

7.3-9　(a) $\sigma_1 = 180$ psi, $\theta_{p1} = -20.56°$; (b) $\tau_{max} = 730$ psi, $\theta_{s1} = -65.56°$

7.3-10　(a) $\sigma_1 = 25$ MPa, $\sigma_2 = -130$ MPa; (b) $\tau_{max} = 77.5$ MPa, $\sigma_{ave} = -52.5$ MPa

7.3-11　(a) $\sigma_1 = 2693$ psi, $\sigma_2 = 732$ psi; (b) $\tau_{max} = 980$ psi, $\sigma_{ave} = 1713$ psi

7.3-12　(a) $\sigma_1 = 2262$ kPa, $\theta_{p1} = -13.70°$; (b) $\tau_{max} = 1000$ kPa, $\theta_{s1} = -58.7°$

7.3-13　(a) $\sigma_1 = 14,764$ psi, $\theta_{p1} = 7.90°$; (b) $\tau_{max} = 6979$ psi, $\theta_{s1} = -37.1°$

7.3-14　(a) $\sigma_1 = 29.2$ MPa, $\theta_{p1} = -17.98°$; (b) $\tau_{max} = 66.4$ MPa, $\theta_{s1} = -63.0°$

7.3-15　(a) $\sigma_1 = -1228$ psi, $\theta_{p1} = 24.7°$; (b) $\tau_{max} = 5922$ psi, $\theta_{s1} = -20.3°$

7.3-16　(a) $\sigma_1 = 76.3$ MPa, $\theta_{p1} = 107.5°$; (b) $\tau_{max} = 101.3$ MPa, $\theta_{s1} = -62.5°$

7.3-17　$3030$ psi $\le \sigma_y \le 9470$ psi

7.3-18　$18.5$ MPa $\le \sigma_y \le 85.5$ MPa

7.3-19　(a) $\sigma_y = 3961$ psi; (b) $\theta_{p1} = -38.93°$, $\sigma_1 = 6375$ psi, $\theta_{p2} = 51.07°$, $\sigma_2 = 2386$ psi

7.3-20　(a) $\sigma_y = 23.3$ MPa; (b) $\theta_{p1} = 65.6°$, $\sigma_1 = 41$ MPa, $\theta_{p2} = -24.4°$, $\sigma_2 = -62.7$ MPa

7.4-1　(a) $\sigma_{x1} = 10,901$ psi, $\sigma_{y1} = 3349$ psi, $\tau_{x1y1} = -6042$ psi; (b) $\tau_{max} = 7125$ psi, $\sigma_{ave} = 7125$ psi

7.4-2　(a) $\sigma_{x1} = 40.1$ MPa, $\sigma_{y1} = 16.91$ MPa, $\tau_{x1y1} = 26$ MPa; (b) $\tau_{max} = 28.5$ MPa, $\sigma_{ave} = 28.5$ MPa

7.4-3　(a) $\sigma_{x1} = -5400$ psi, $\sigma_{y1} = -1350$ psi, $\tau_{x1y1} = 2700$ psi; (b) $\tau_{max} = -3375$ psi, $\sigma_{aver} = -3375$ psi

7.4-4　For $\theta = 25°$: (a) $\sigma_{x1} = -36.0$ MPa, $\tau_{x1y1} = 25.7$ MPa; (b) $\tau_{max} = 33.5$ MPa, $\theta_{s1} = 45.0°$

7.4-5　For $\theta = 55°$: (a) $\sigma_{x1} = 882$ psi, $\tau_{x1y1} = -3759$ psi; $\sigma_{y1} = 3618$ psi (b) $\tau_{max} = 4000$ psi, $\theta_{x1} = -45.0°$; $\sigma_{aver} = 2250$ psi

7.4-6　For $\theta = 21.80°$: (a) $\sigma_{x1} = -17.1$ MPa, $\tau_{x1y1} = 29.7$ MPa; (b) $\tau_{max} = 43.0$ MPa, $0_{x1} = 45.0°$

7.4-7　For $\theta = 52°$: (a) $\sigma_{x1} = 2620$ psi, $\tau_{x1y1} = -653$ psi; (b) $\sigma_1 = 2700$ psi, $\theta_{p1} = 45.0°$

7.4-8　(a) $\sigma_{x1} = -60.8$ MPa, $\sigma_{y1} = 128.8$ MPa, $\tau_{x1y1} = -46.7$ MPa; (b) $\sigma_1 = 139.6$ MPa, $\sigma_2 = -71.6$ MPa, $\tau_{max} = 105.6$ MPa

7.4-9　For $\theta = 36.87°$: (a) $\sigma_{x1} = 3600$ psi, $\tau_{x1y1} = 1050$ psi; (b) $\sigma_1 = 3750$ psi, $\theta_{p1} = 45.0°$

7.4-10　For $\theta = 40°$: $\sigma_{x1} = 27.5$ MPa, $\tau_{x1y1} = -5.36$ MPa

**7.4-11** For $\theta = -51°$: $\sigma_{x1} = 11,982$ psi
$\tau_{x1y1} = -3569$ psi

**7.4-12** For $\theta = -33°$: $\sigma_{x1} = 61.7$ MPa,
$\tau_{x1y1} = -51.7$ MPa, $\sigma_{y1} = -171.3$ MPa

**7.4-13** For $\theta = 14°$: $\sigma_{x1} = -1509$ psi,
$\tau_{x1y1} = 527$ psi, $\sigma_{y1} = -891$ psi

**7.4-14** For $\theta = 35°$: $\sigma_{x1} = 46.4$ MPa,
$\tau_{x1y1} = -9.81$ MPa

**7.4-15** For $\theta = 65°$: $\sigma_{x1} = -1846$ psi,
$\tau_{x1y1} = 3897$ psi

**7.4-16** (a) $\sigma_1 = 10,865$ kPa, $\theta_{p1} = 115.2°$;
(b) $\tau_{max} = 4865$ kPa, $\theta_{s1} = 70.2°$

**7.4-17** (a) $\sigma_1 = 2565$ psi, $\theta_{p1} = 31.3°$;
(b) $\tau_{max} = 3265$ psi, $\theta_{s1} = -13.70°$

**7.4-18** (a) $\sigma_1 = 18.2$ MPa, $\theta_{p1} = 123.3°$;
(b) $\tau_{max} = 15.4$ MPa, $\theta_{s1} = 78.3°$

**7.4-19** (a) $\sigma_1 = -6923$ psi, $\theta_{p1} = -32.4°$;
(b) $\tau_{max} = 7952$ psi, $\theta_{s1} = 102.6°$

**7.4-20** (a) $\sigma_1 = 40.0$ MPa, $\theta_{p1} = 68.8°$;
(b) $\tau_{max} = 40.0$ MPa, $\theta_{s1} = 23.8°$

**7.4-21** (a) $\sigma_1 = 7490$ psi, $\theta_{p1} = 63.2°$;
(b) $\tau_{max} = 3415$ psi, $\theta_{s1} = -18.20°$

**7.4-22** (a) $\sigma_1 = 3.43$ MPa, $\theta_{p1} = -19.68°$:
(b) $\tau_{max} = 15.13$ MPa, $\theta_{s1} = -64.7°$

**7.4-23** (a) $\sigma_1 = 7525$ psi, $\theta_{p1} = 9.80°$;
(b) $\tau_{max} = 3875$ psi, $\theta_{s1} = -35.2°$

**7.5-1** $\sigma_x = 25,385$ psi, $\sigma_y = 19,615$ psi,
$\Delta t = -2.81 \times 10^{-4}$ in.

**7.5-2** $\sigma_x = 102.6$ MPa, $\sigma_y = -11.21$ MPa,
$\Delta t = -1.646 \times 10^{-3}$ mm

**7.5-3** (a) $\varepsilon_z = -v(\varepsilon_x + \varepsilon_y)/(1 - v)$;
(b) $e = (1 - 2v)(\varepsilon_x + \varepsilon_y)/(1 - v)$

**7.5-4** $v = 0.24$, $E = 112.1$ GPa

**7.5-5** $v = 0.3$, $E = 29,560$ ksi

**7.5-6** (a) $\gamma_{max} = 5.85 \times 10^{-4}$;
(b) $\Delta t = -1.32 \times 10^{-3}$ mm;
(c) $\Delta V = 387$ mm$^3$

**7.5-7** (a) $\gamma_{max} = 1900 \times 10^{-6}$;
(b) $\Delta t = -141 \times 10^{-6}$ in. (decrease);
(c) $\Delta V = 0.0874$ in.$^3$ (increase)

**7.5-8** (a) $\Delta V_b = -49.2$ mm$^3$, $U_b = 3.52$ J;
(b) $\Delta V_a = -71.5$ mm$^3$, $U_a = 4.82$ J

**7.5-9** $\Delta V = -0.0377$ in.$^3$, $U = 55.6$ in.-lb

**7.5-10** (a) $\Delta V = 2766$ mm$^3$, $U = 56$ J;
(b) $t_{max} = 36.1$ mm; (c) $b_{min} = 640$ mm

**7.5-11** (a) $\Delta V = 0.0385$ in.$^3$, $U = 574$ lb-in.;
(b) $t_{max} = 0.673$ in.; (c) $b_{min} = 10.26$ in.

**7.5-12** (a) $\Delta ac = \varepsilon_x d = 0.1296$ mm (increase);
(b) $\Delta bc = \varepsilon_y d = -0.074$ mm (decrease);
(c) $\Delta t = \varepsilon_z t = -2.86 \times 10^{-3}$ mm
(decrease); (d) $\Delta V = eV_0 = 430$ mm$^3$;
(e) $U = uV_0 = 71.2$ N·m; (f) $t_{max} = 22.0$ mm;
(g) $\sigma_{xmax} = 63.9$ MPa

**7.6-1** (a) $\tau_{max} = \dfrac{\sigma_1 - \sigma_3}{2} = 8750$ psi;
(b) $\Delta a = a\varepsilon_x = 7.73 \times 10^{-3}$ in.,
$\Delta b = \varepsilon_y b = -3.75 \times 10^{-3}$ in.,
$\Delta c = \varepsilon_z c = -1.3 \times 10^{-3}$ in.;
(c) $\Delta V = eV_0 = 0.0173$ in.$^3$;
(d) $U = uV_0 = 964$ in.-lb;
(e) $\sigma_{xmax} = 12,824$ psi;
(f) $\sigma_{xmax} = 11,967$ psi

**7.6-2** (a) $\tau_{max} = \dfrac{\sigma_1 - \sigma_3}{2} = 8.5$ MPa;
(b) $\Delta a = a\varepsilon_x = -0.0525$ mm,
$\Delta b = \varepsilon_y b = -9.67 \times 10^{-3}$ mm,
$\Delta c = \varepsilon_z c = -9.67 \times 10^{-3}$ mm;
(c) $\Delta V = eV_0 = -2.052 \times 10^3$ mm$^3$;
(d) $U = uV_0 = 56.2$ N·m;
(e) $\sigma_{xmax} = -50$ MPa;
(f) $\sigma_{xmax} = -65.1$ MPa

**7.6-3** (a) $\sigma_x = -4200$ psi, $\sigma_y = -2100$ psi,
$\sigma_z = -2100$ psi;
(b) $\tau_{max} = \dfrac{\sigma_1 - \sigma_3}{2} = 1050$ psi;
(c) $\Delta V = eV_0 = -0.0192$ in.$^3$;
(d) $U = uV_0 = 35.3$ in.-lb;
(e) $\sigma_{xmax} = -3864$ psi
(f) $\varepsilon_{xmax} = -235 \cdot (10^{-6})$

**7.6-4** (a) $\sigma_x = -82.6$ MPa, $\sigma_y = -54.7$ MPa,
$\sigma_z = -54.7$ MPa;
(b) $\tau_{max} = \dfrac{\sigma_1 - \sigma_3}{2} = 13.92$ MPa;
(c) $\Delta V = eV_0 = -846$ mm$^3$;
(d) $U = uV_0 = 29.9$ N·m;
(e) $\sigma_{xmax} = -73$ MPa
(f) $\varepsilon_{xmax} = -741 \cdot (10^{-6})$

**7.6-5** (a) $K_{Al} = 1 \times 10^7$ psi; (b) $E = 6139$ ksi,
$v = 0.35$

**7.6-6** (a) $K = 4.95$ GPa; (b) $E = 1.297$ GPa,
$v = 0.40$

**7.6-7** (a) $p = vF/[A(1 - v)]$;
(b) $\delta = FL(1 + v)(1 - 2v)/EA(1 - v)$

**7.6-8** (a) $p = vp_0$; (b) $e = -p_0(1 + v)(1 - 2v)/E$;
(c) $u = p_0^2(1 - v^2)/2E$

**7.6-9** (a) $\Delta d = 1.472 \times 10^{-3}$ in., $\Delta V = 0.187$ in.$^3$, $U = 332$ in.-lb; (b) $h = 5282$ ft

**7.6-10** (a) $p = 700$ MPa; (b) $K = 175$ GPa; (c) $U = 2470$ J

**7.6-11** $\varepsilon_0 = 276 \times 10^{-6}$, $e = 828 \times 10^{-6}$, $u = 4.97$ psi

**7.7-1** (a) $\Delta d = 1.878 \times 10^{-3}$ in.; (b) $\Delta\phi = -\alpha = 1.425 \times 10^{-4}$ (decrease, radians); (c) $\Delta\psi = -\alpha = 1.425 \times 10^{-4}$ (increase, radians)

**7.7-2** (a) $\Delta d = \varepsilon_{x1} L_d = 0.062$ mm; (b) $\Delta\phi = -\alpha = 1.89 \times 10^{-4}$ (decrease, radians); (c) $\Delta\psi = -\alpha = 1.89 \times 10^{-4}$ (increase, radians)

**7.7-3** (a) $\Delta d = 0.00458$ in. (increase); (b) $\Delta\phi = 157 \times 10^{-6}$ rad (decrease); (c) $\gamma = -314 \times 10^{-6}$ rad (angle $ced$ increases)

**7.7-4** (a) $\Delta d = 0.168$ mm (increase); (b) $\Delta\phi = 317 \times 10^{-6}$ rad (decrease); (c) $\gamma = -634 \times 10^{-6}$ rad (angle $ced$ increases)

**7.7-5** $\varepsilon_{x1} = 3.97 \times 10^{-4}$, $\varepsilon_{y1} = 3.03 \times 10^{-4}$, $\gamma_{x1y1} = 1.829 \times 10^{-4}$

**7.7-6** $\varepsilon_{x1} = 9.53 \times 10^{-5}$, $\varepsilon_{y1} = -1.353 \times 10^{-4}$, $\gamma_{x1y1} = -3.86 \times 10^{-4}$

**7.7-7** $\varepsilon_1 = 554 \times 10^{-6}$, $\theta_{p1} = -22.9°$, $\gamma_{max} = 488 \times 10^{-6}$

**7.7-8** $\varepsilon_1 = 172 \times 10^{-6}$, $\theta_{p1} = 163.9°$, $\gamma_{max} = 674 \times 10^{-6}$

**7.7-9** For $\theta = 75°$: (a) $\varepsilon_{x1} = 202 \times 10^{-6}$, $\gamma_{x1y1} = -569 \times 10^{-6}$; (b) $\varepsilon_1 = 568 \times 10^{-6}$, $\theta_{P1} = 22.8°$; (c) $\gamma_{max} = 587 \times 10^{-6}$

**7.7-10** For $\theta = 45°$: (a) $\varepsilon_{x1} = -385 \times 10^{-6}$, $\gamma_{x1y1} = 690 \times 10^{-6}$; (b) $\varepsilon_1 = -254 \times 10^{-6}$, $\theta_{p1} = 65.7°$; (c) $\gamma_{max} = 1041 \times 10^{-6}$

**7.7-11** $\tau_{maxxy} = \dfrac{\sigma_x - \sigma_y}{2} = 4076$ psi,

$$\gamma_{xymax} = 2\sqrt{\left(\frac{\varepsilon_x - \varepsilon_y}{2}\right)^2 + \left(\frac{\gamma_{xy}}{2}\right)^2}$$
$$= 6.83 \times 10^{-4},$$

$$\gamma_{xzmax} = 2\sqrt{\left(\frac{\varepsilon_x - \varepsilon_z}{2}\right)^2 + \gamma_{xz}^2} = 8.96 \times 10^{-4},$$

$$\gamma_{yzmax} = 2\sqrt{\left(\frac{\varepsilon_y - \varepsilon_z}{2}\right)^2 + \gamma_{yz}^2} = 2.13 \times 10^{-4}$$

**7.7-12** $\tau_{maxxy} = \dfrac{\sigma_x - \sigma_y}{2} = 33.7$ MPa,

$$\gamma_{xymax} = 2\sqrt{\left(\frac{\varepsilon_x - \varepsilon_y}{2}\right)^2 + \left(\frac{\gamma_{xy}}{2}\right)^2}$$
$$= 1.244 \times 10^{-3},$$

$$\gamma_{xzmax} = 2\sqrt{\left(\frac{\varepsilon_x - \varepsilon_z}{2}\right)^2 + \gamma_{xz}^2}$$
$$= 1.459 \times 10^{-3},$$

$$\gamma_{yzmax} = 2\sqrt{\left(\frac{\varepsilon_y - \varepsilon_z}{2}\right)^2 + \gamma_{yz}^2}$$
$$= 2.15 \times 10^{-4}$$

**7.7-13** For $\theta = 30°$: (a) $\varepsilon_{x1} = -756 \times 10^{-6}$, $\gamma_{x1y1} = 868 \times 10^{-6}$; (b) $\varepsilon_1 = 426 \times 10^{-6}$, $\theta_{p1} = 99.8°$; (c) $\gamma_{max} = 1342 \times 10^{-6}$

**7.7-14** For $\theta = 50°$: (a) $\varepsilon_{x1} = -1469 \times 10^{-6}$, $\gamma_{x1y1} = -717 \times 10^{-6}$; (b) $\varepsilon_1 = -732 \times 10^{-6}$, $\theta_{p1} = 166.0°$; (c) $\gamma_{max} = 911 \times 10^{-6}$

**7.7-15** $\varepsilon_1 = 551 \times 10^{-6}$, $\theta_{p1} = 12.5°$, $\gamma_{max} = 662 \times 10^{-6}$

**7.7-16** $\varepsilon_1 = 332 \times 10^{-6}$, $\theta_{p1} = 12.0°$, $\gamma_{max} = 515 \times 10^{-6}$

**7.7-17** (a) $P = 5154$ lb, $T = -978$ in.-lb; (b) $\gamma_{max} = 2.84 \times 10^{-4}$, $\tau_{max} = 3304$ psi

**7.7-18** $P = 121.4$ kN, $\alpha = 56.7°$

**7.7-19** $P = 9726$ lb, $\alpha = 75.2°$

**7.7-20** $\varepsilon_x = \varepsilon_a$, $\varepsilon_y = (2\varepsilon_b + 2\varepsilon_c - \varepsilon_a)/3$, $\gamma_{xy} = 2(\varepsilon_b - \varepsilon_c)/\sqrt{3}$

**7.7-21** For $\theta_{p1} = 30°$: $\varepsilon_1 = 1550 \times 10^{-6}$, $\varepsilon_2 = -250 \times 10^{-6}$, $\sigma_1 = 10,000$ psi, $\sigma_2 = 2000$ psi

**7.7-22** $\sigma_x = 91.6$ Mpa

**7.7-23** $\varepsilon_{x1} = 3.97 \times 10^{-4}$, $\varepsilon_{y1} = 3.03 \times 10^{-4}$, $\gamma_{x1y1} = 1.829 \times 10^{-4}$

**7.7-24** $\varepsilon_{x1} = 9.53 \times 10^{-5}$, $\varepsilon_{y1} = -1.353 \times 10^{-4}$, $\gamma_{x1y1} = -3.86 \times 10^{-4}$

**7.7-25** $\varepsilon_1 = 554 \times 10^{-6}$, $\theta_{p1} = 157.1°$, $\gamma_{max} = 488 \times 10^{-6}$

**7.7-26** $\varepsilon_1 = 172 \times 10^{-6}$, $\theta_{p1} = 163.9°$, $\gamma_{max} = 674 \times 10^{-6}$

**7.7-27** For $\theta = 75°$: (a) $\varepsilon_{x1} = 202 \times 10^{-6}$, $\gamma_{x1y1} = -569 \times 10^{-6}$: (b) $\varepsilon_1 = 568 \times 10^{-6}$, $\theta_{p1} = 22.8°$; (c) $\gamma_{max} = 587 \times 10^{-6}$

**7.7-28** For $\theta = 45°$: (a) $\varepsilon_{x1} = -385 \times 10^{-6}$, $\gamma_{x1y1} = 690 \times 10^{-6}$; (b) $\varepsilon_1 = -254 \times 10^{-6}$, $\theta_{p1} = 65.7°$; (c) $\gamma_{max} = 1041 \times 10^{-6}$

# CHAPTER 8

8.2-1 (a) Use $t = 2.5$ in. (b) $p_{max} = 381$ psi

8.2-2 (a) Use $t = 98$ mm. (b) $p_{max} = 3.34$ MPa

8.2-3 (a) $F = 1073$ lb, $\sigma = 255$ psi;
(b) $d_b = 0.286$ in.; (c) $r = 7.35$ in.

8.2-4 (a) $\sigma_{max} = 3.12$ MPa, $\varepsilon_{max} = 0.438$;
(b) $t_{reqd} = 1.29$ mm

8.2-5 (a) $\sigma_{max} = 425$ psi, $\varepsilon_{max} = 1.105$;
(b) $p_{max} = 7.77$ psi

8.2-6 (a) $p_{max} = 3.51$ MPa; (b) $p_{max} = 2.93$ MPa

8.2-7 (a) $f = 26.4$ k/in.; (b) $\tau_{max} = 7543$ psi;
(c) $\varepsilon_{max} = 3.57 \times 10^{-4}$

8.2-8 (a) $f = 5.5$ MN/m; (b) $\tau_{max} = 57.3$ MPa;
(c) $\varepsilon_{max} = 3.87 \times 10^{-4}$

8.2-9 (a) $t_{min} = 0.291$ in.; (b) $p = 1904$ psi

8.2-10 (a) $t_{min} = 7.17$ mm; (b) $p = 19.25$ MPa

8.2-11 $D_0 = 90$ ft

8.3-1 (a) $t_{min} = 0.289$ in.; (b) $p_{max} = 2286$ psi

8.3-2 (a) $h = 22.2$ m; (b) zero

8.3-3 $n = 2.25$

8.3-4 (a) $F = 3\pi p r^2$; (b) $t_{reqd} = 10.91$ mm

8.3-5 (a) $p = 55$ psi; (b) $\varepsilon_r = 9.18 \times 10^{-4}$

8.3-6 (a) $\varepsilon_{max} = 6.67 \times 10^{-5}$; (b) $\varepsilon_r = 2.83 \times 10^{-4}$

8.3-7 $t_{min} = 0.113$ in.

8.3-8 $t_{min} = 3.71$ mm

8.3-9 (a) $h = 25$ ft; (b) $\sigma_1 \approx 125$ psi

8.3-10 (a) $\sigma_h = 24.9$ MPa; (b) $\sigma_c = 49.7$ MPa;
(c) $\sigma_w = 24.9$ MPa; (d) $\tau_h = 12.43$ MPa;
(e) $\tau_c = 24.9$ MPa

8.3-11 (a) $t_{min} = 0.675$ in.; (b) $t_{min} = 0.338$ in.

8.3-12 (a) $\sigma_1 = 93.3$ MPa, $\sigma_2 = 46.7$ MPa;
(b) $\tau_1 = 23.2$ MPa, $\tau_2 = 46.7$ MPa;
(c) $\varepsilon_1 = 3.97 \times 10^{-4}$, $\varepsilon_2 = 9.33 \times 10^{-5}$;
(d) $\theta = 35°$, $\sigma_{x_1} = 62.0$ MPa,
$\sigma_{y_1} = 78.0$ MPa, $\tau_{x_1 y_1} = 21.9$ MPa

8.3-13 (a) $\sigma_1 = 7015$ psi, $\sigma_2 = 3508$ psi;
(b) $\tau_1 = 1754$ psi, $\tau_2 = 3508$ psi;
(c) $\varepsilon_1 = 1.988 \times 10^{-4}$, $\varepsilon_2 = 4.68 \times 10^{-5}$;
(d) $\theta = 28°$, $\sigma_{x1} = 4281$ psi, $\sigma_{y1} = 6242$ psi,
$\tau_{x_1 y_1} = 1454$ psi

8.4-1 (a) $\sigma_y = 0$, $\sigma_x = \dfrac{-M\left(d - \dfrac{h}{2}\right)}{I} = -5$ ksi,

$\tau_{xy} = \dfrac{VQ}{Ib} = 1.111$ ksi, $\sigma_1 = -5.24$ ksi,

$\theta_{p1} = -11.98°$, $\sigma_2 = 0.236$ ksi, $\theta_{p2} = 78.02°$;
$\tau_{max} = 2.74 \cdot$ ksi

(b) $\sigma_y = 0$,

$$\sigma_x = \frac{-N}{A} - \frac{M\left(d - \dfrac{h}{2}\right)}{I}$$

$$= -5.95 \text{ ksi}, \quad \tau_{xy} = \frac{VQ}{Ib} = 1.11 \text{ ksi},$$

$\sigma_1 = -6.15$ ksi, $\theta_{p1} = -10.24°$,
$\sigma_2 = 0.201$ ksi, $\theta_{p2} = 79.76°$; $\tau_{max} = 3.18$ ksi

8.4-2 (a) $\sigma_y = 0$,

$$\sigma_x = \frac{-M\left(d - \dfrac{h}{2}\right)}{I} = -37.4 \text{ MPa},$$

$$\tau_{xy} = \frac{VQ}{Ib} = 7.49 \text{ MPa},$$

$\sigma_1 = -38.9$ MPa, $\theta_{p1} = -10.9°$,
$\sigma_2 = 1.442$ MPa, $\theta_{p2} = 79.1°$,
$\tau_{max} = 20.2$ MPa;

(b) $\sigma_y = 0$,

$$\sigma_x = \frac{-N}{A} - \frac{M\left(d - \dfrac{h}{2}\right)}{I} = -44.4 \text{ MPa},$$

$$\tau_{xy} = \frac{VQ}{Ib} = 7.49 \text{ MPa},$$

$\sigma_1 = -45.7$ MPa, $\theta_{p1} = -9.3°$,
$\sigma_2 = 1.227$ MPa, $\theta_{p2} = 80.7°$,
$\tau_{max} = 23.4$ MPa

8.4-3 (a) $\sigma_1 = 219$ psi, $\sigma_2 = -219$ psi,
$\tau_{max} = 219$ psi; (b) $\sigma_1 = 35.9$ psi,
$\sigma_2 = -749$ psi, $\tau_{max} = 392$ psi;
(c) $\sigma_1 = 0$ psi, $\sigma_2 = -2139$ psi,
$\tau_{max} = 1069$ psi

8.4-4 $P = 20$ kN

8.4-5 $P = 2.91$ k

8.4-6 (b) $\sigma_1 = 4.5$ MPa, $\sigma_2 = -76.1$ MPa,
$\tau_{max} = 40.3$ MPa

8.4-7 (b) $\sigma_1 = 14{,}100$ psi, $\sigma_2 = -220$ psi,
$\tau_{max} = 7160$ psi

8.4-8 (b) $\sigma_1 = 8.27$ MPa, $\sigma_2 = -64.3$ MPa,
$\tau_{max} = 36.3$ MPa

8.4-9 (b) $\sigma_1 = 159.8$ psi, $\sigma_2 = -3393$ psi,
$\tau_{max} = 1777$ psi

8.4-10 $\sigma_1 = 17.86$ MPa, $\sigma_2 = -0.145$ MPa,
$\tau_{max} = 9.00$ MPa

8.4-11 $\dfrac{\sigma_1}{\sigma_2} = -184$

**8.4-12** $\dfrac{\sigma_1}{\sigma_2} = -663$

**8.5-1** $t_{min} = 0.125$ in.

**8.5-2** $p_{max} = 9.60$ MPa

**8.5-3** (a) $\sigma_{max} = \sigma_1 = 11.09$ ksi, $\tau_{max} = 3.21$ ksi; (b) $T_{max} = 178$ k-ft; (c) $t_{min} = 0.519$ in.

**8.5-4** (a) $P_{max} = 52.7$ kN; (b) $p_{max} = 6$ MPa

**8.5-5** $\sigma_t = 10,680$ psi: No compressive stresses. $\tau_{max} = 5340$ psi

**8.5-6** $\phi_{max} = 0.552$ rad $= 31.6°$

**8.5-7** $\sigma_t = 3963$ psi, $\sigma_c = -8791$ psi, $\tau_{max} = 6377$ psi

**8.5-8** $\sigma_t = 16.93$ MPa, $\sigma_c = -41.4$ MPa, $\tau_{max} = 28.9$ MPa

**8.5-9** $P = 194.2$ k

**8.5-10** (a) $\sigma_{max} = \sigma_1 = 35.8$ MPa, $\tau_{max} = 18.05$ MPa; (b) $P_{max} = 6.73$ kN

**8.5-11** (a) $\sigma_x = 0$ psi, $\sigma_y = 6145$ psi, $\tau_{xy} = 345$ psi; (b) $\sigma_1 = 6164$ psi, $\sigma_2 = -19.30$ psi, $\tau_{max} = 3092$ psi

**8.5-12** $\tau_A = 76.0$ MPa, $\tau_B = 19.94$ MPa, $\tau_C = 23.7$ MPa

**8.5-13** $\sigma_1 = 1094$ psi (max. tensile stress at base of pole), $\sigma_2 = -7184$ psi (max. compressive stress at base of pole), $\tau_{max} = 3731$ psi (max. shear stress at base of pole)

**8.5-14** $d_{min} = 48.4$ mm

**8.5-15** $\sigma_t = 39,950$ psi, $\sigma_c = -2226$ psi, $\tau_{max} = 21,090$ psi

**8.5-16** (a) $\sigma_t = 29.15\dfrac{qR^2}{d^3}$, $\sigma_c = -8.78\dfrac{qR^2}{d^3}$, $\tau_{max} = 18.97\dfrac{qR^2}{d^3}$; (b) $\sigma_t = 14.04\dfrac{qR^2}{d^3}$, $\sigma_c = -2.41\dfrac{qR^2}{d^3}$, $\tau_{max} = 8.22\dfrac{qR^2}{d^3}$

**8.5-17** $\sigma_t = 4320$ psi, $\sigma_c = -1870$ psi, $\tau_{max} = 3100$ psi

**8.5-18** Pure shear: $\tau_{max} = 0.804$ MPa

**8.5-19** (a) $d_{min} = 1.65$ in.; (b) $P_{max} = 19.25$ lb

**8.5-20** (a) $\sigma_1 = 29.3$ MPa, $\sigma_2 = -175.9$ MPa, $\tau_{max} = 102.6$ MPa; (b) $\sigma_1 = 156.1$ MPa, $\sigma_2 = -33$ MPa, $\tau_{max} = 94.5$ MPa

**8.5-21** (a) $\sigma_1 = 0$ psi, $\sigma_2 = -20,730$ psi, $\tau_{max} = 10,365$ psi; (b) $\sigma_1 = 988$ psi, $\sigma_2 = -21,719$ psi, $\tau_{max} = 11,354$ psi

**8.5-22** Maximum: $\sigma_t = 18.35$ MPa, $\sigma_C = -18.35$ MPa, $\tau_{max} = 9.42$ MPa

**8.5-23** Top of beam: $\sigma_1 = 8591$ psi, $\sigma_2 = 0$ psi, $\tau_{max} = 4295$ psi

**8.5-24** (a) $d_{Al} = 26.3$ mm; (b) $d_{Ti} = 21.4$ mm

**8.5-25** $\sigma_y = 0$, $\sigma_x = \dfrac{(FL)\dfrac{d_2}{2}}{\dfrac{I_p}{2}} = 1943$ psi,

$\tau_{xy} = \dfrac{-T\dfrac{d_2}{2}}{I_p} = -547$ psi, $\sigma_1 = 2087$ psi,

$\sigma_2 = -143.2$ psi, $\tau_{max} = 1115$ psi

**8.5-26** (a) $\sigma_1 = 0$, $\sigma_2 = \sigma_x = -108.4$ MPa,

$\tau_{max} = \dfrac{\sigma_x}{2} = -54.2$ MPa;

(b) $\sigma_1 = 0.703$ MPa, $\sigma_2 = -1.153$ MPa, $\tau_{max} = 0.928$ MPa; (c) $P_{max} = 348$ N

**8.5-27** $\sigma_x = -18.6$ ksi, $\sigma_y = 0$, $\tau_{xy} = 4.45$ ksi, $\sigma_1 = 1.012$ ksi, $\sigma_2 = -19.62$ ksi, $\tau_{max} = 10.31$ ksi

# CHAPTER 9

**9.2-1** $q = q_0 x/L$; Triangular load, acting downward

**9.2-2** (a) $q = q_0 \sin \pi x/L$, Sinusoidal load; (b) $R_A = R_B = q_0 L/\pi$; (c) $M_{max} = q_o L^2/\pi^2$

**9.2-3** $q = q_0(1 - x/L)$; Triangular load, acting downward

**9.2-4** (a) $q = q_0(L^2 - x^2)/L^2$; Parabolic load, acting downward; (b) $R_A = 2q_0 L/3$, $M_A = -q_0 L^2/A$

**9.3-1** $\delta_{max} = 0.182$ in., $\theta = 0.199°$

**9.3-2** $h = 96$ mm

**9.3-3** $L = 120$ in. $= 10$ ft

**9.3-4** $\delta_{max} = 15.4$ mm

**9.3-5** $\delta/L = 1/400$

**9.3-6** $E_g = 80.0$ GPa

**9.3-7** Let $\beta = a/L$: $\dfrac{\delta_C}{\delta_{max}}$

$= 3\sqrt{3}(-1 + 8\beta - 4\beta^2)$

The deflection at the midpoint is close to the maximum deflection. The maximum difference is only 2.6%.

**9.3-11** $v = -mx^2(3L - x)/6EI$, $\delta_B = mL^2/3EI$, $\theta_B = mL^2/2EI$

**9.3-12** $v(x) = -\dfrac{q}{48EI}(2x^4 - 12x^2L^2 + 11L^4)$,

$\delta_B = \dfrac{qL^4}{48EI}$

**9.3-13** See Table H-2, Case 9.

**9.3-14** See Table H-1, Case 2.

**9.3-15** $v(x) = \dfrac{q_0L}{24\,EI}(x^3 - 2Lx^2)$ for $0 \le x \le \dfrac{L}{2}$,

$v(x) = \dfrac{-q_0}{960LEI}(-160L^2x^3 + 160L^3x^2 +$

$80Lx^4 - 16x^5 - 25L^4x + 3L^5)$ for

$\dfrac{L}{2} \le x \le L, \delta_B = \dfrac{7}{160}\dfrac{q_0L^4}{EI}, \delta_C = \dfrac{1}{64}\dfrac{q_0L^4}{EI}$

**9.3-16** $v(x) = \dfrac{q_0x}{5760LEI}(200x^2L^2 - 240x^3L$

$+96x^4 - 53L^4)$ for $0 \le x \le \dfrac{L}{2}$,

$v(x) = \dfrac{-q_0L}{5760EI}(40x^3 - 120Lx^2 +$

$83L^2x - 3L^3)$ for $\dfrac{L}{2} \le x \le L$,

$\delta_C = \dfrac{3q_0L^4}{1280EI}$

**9.3-17** $v(x) = -\dfrac{PL}{10{,}368EI}(-4104x^2 + 3565L^2)$

for $0 \le x \le \dfrac{L}{3}$, $v(x) = -\dfrac{P}{1152EI}$

$(-648Lx^2 + 192x^3 + 64L^2x + 389L^3)$

for $\dfrac{L}{3} \le x \le \dfrac{L}{2}$, $v(x) = -\dfrac{P}{144EIL}$

$(-72L^2x^2 + 12Lx^3 + 6x^4 + 5L^3x + 49L^4)$

for $\dfrac{L}{2} \le x \le L$, $\delta_A = \dfrac{3565PL^3}{10{,}368EI}$,

$\delta_C = \dfrac{3109PL^3}{10{,}368EI}$

**9.4-3** $v = -M_0x(L - x)^2/2LEI$,
$\delta_{max} = 2M_0L^2/27EI$ (downward)

**9.4-4** $v(x) = -\dfrac{q}{48EI}(2x^4 - 12x^2L^2 + 11L^4)$,

$\theta_B = \dfrac{-qL^3}{3EI}$

**9.4-5** See Table H-1, Case 10.

**9.4-6** $v = -q_0x^2(45L^4 - 40L^3x + 15L^2x^2 - x^4)/$
$360L^2EI$, $\delta_B = 19q_0L^4/360EI$, $\theta_B = q_0L^3/15EI$

**9.4-7** $v = -q_0x(3L^5 - 5L^3x^2 + 3Lx^4 - x^5)/90L^2EI$,
$\delta_{max} = 61q_0L^4/5760EI$

**9.4-8** $v(x) = \dfrac{q_0}{120EIL}$

$(x^5 - 5Lx^4 + 20L^3x^2 - 16L^5)$,

$\delta_{max} = \dfrac{2q_0L^4}{15EI}$

**9.4-9** $v(x) = -\dfrac{qL^2}{16EI}(x^2 - L^2)$ for $0 \le x \le L$,

$v(x) = -\dfrac{q}{48EI}(-20L^3x + 27L^2x^2 -$

$12Lx^3 + 2x^4 + 3L^4)$ for $L \le x \le \dfrac{3L}{2}$,

$\delta_C = \dfrac{9qL^4}{128EI}$, $\theta_C = \dfrac{7qL^3}{48EI}$

**9.4-10** $v(x) = -\dfrac{q_0L^2}{480EI}(-20x^2 + 19L^2)$ for

$0 \le x \le \dfrac{L}{2}$,

$v(x) = -\dfrac{q_0}{960EIL}(80Lx^4 - 16x^5 -$

$120L^2x^3 + 40L^3x^2 - 25L^4x + 41L^5)$

for $\dfrac{L}{2} \le x \le L$,

$\delta_A = \dfrac{19q_0L^4}{480EI}, \theta_B = -\dfrac{13q_0L^3}{192EI}, \delta_C = \dfrac{7q_0L^4}{240EI}$

**9.5-1** $\theta_B = 7PL^2/9EI$, $\delta_B = 5PL^3/9EI$

**9.5-2** (a) $\delta_1 = 11PL^3/144EI$;
(b) $\delta_2 = 25PL^3/384EI$;
(c) $\delta_1/\delta_2 = 88/75 = 1.173$

**9.5-3** (a) $a/L = 2/3$; (b) $a/L = 1/2$

**9.5-4** (a) $\delta_C = 6.25$ mm (upward)
(b) $\delta_C = 18.36$ mm (downward)

**9.5-5** $y = Px^2(L - x)^2/3LEI$

**9.5-6** $\theta_B = 7qL^3/162EI$, $\delta_B = 23qL^4/648EI$

**9.5-7** $\delta_C = 0.0905$ in., $\delta_B = 0.293$ in.

**9.5-8** (a) $M = PL/2$; (b) $M = 5PL/24$,
$\theta_B = PL^2/12EI$; (c) $M = PL/8$,
$\delta_B = -PL^3/24EI$

**9.5-9** $M = (19/180)q_0L^2$

**9.5-10** (a) $\delta_A = PL^2(10L - 9a)/324EI$ (positive upward); (b) Upward when $a/L < 10/9$, downward when $a/L > 10/9$

**9.5-11** (a) $\delta_C = PH^2(L + H)/3EI$;
(b) $\delta_{max} = PHL^2/9\sqrt{3}EI$

9.5-12  $\delta_C = 3.5$ mm

9.5-13  $\theta_B = q_0 L^3/10EI$, $\delta_B = 13q_0 L^4/180EI$

9.5-14  $\theta_A = q(L^3 - 6La^2 + 4a^3)/24EI$,
$\delta_{max} = q(5L^4 - 24L^2 a^2 + 16a^4)/384EI$

9.5-15  (a) $P/Q = 9a/4L$; (b) $P/Q = 8a(3L + a)/9L^2$;
(c) $P/qa = 9a/8L$ for $\delta_B = 0$,
$P/qa = a(4L + a)/3L^2$ for $\delta_D = 0$

9.5-16  $\delta = 19WL^3/31{,}104EI$

9.5-17  $k = 3.33$ lb/in.

9.5-18  $M_1 = 7800$ N·m, $M_2 = 4200$ N·m

9.5-19  $\delta = \dfrac{6Pb^3}{EI}$

9.5-20  $\delta_E = \dfrac{47Pb^3}{12EI}$

9.5-21  $\delta_C = 0.120$ in.

9.5-22  $q = 16cEI/7L^4$

9.5-23  $\delta_h = Pcb^2/2EI$, $\delta_v = Pc^2(c + 3b)/3EI$

9.5-24  $\delta = PL^2(2L + 3a)/3EI$

9.5-25  (a) $H_B = 0$, $V_B = \dfrac{M}{L}$, $V_C = -V_B$;

(b) $\theta_A = \dfrac{5ML}{6EI}$, $\theta_B = \dfrac{ML}{3EI}$, $\theta_C = \dfrac{-ML}{6EI}$,
$\theta_D = \theta_C$; (c) $\delta_A = (7/24)ML^2/EI$ (to the left),
$\delta_D = (1/12)ML^2/EI$ (to the right);

(d) $L_{CD} = \dfrac{\sqrt{14}}{2} L = 1.871L$

9.5-26  (a) $H_B = 0$, $V_B = \dfrac{P}{3}$, $V_C = \dfrac{2P}{3}$;

(b) $\theta_A = \left(\dfrac{-4}{81}\right)\dfrac{ML}{EI}$, $\theta_B = \theta_A$,

$\theta_C = \left(\dfrac{5}{81}\right)\dfrac{ML}{EI}$, $\theta_D = \theta_C$;

(c) $\delta_A = -\theta_B\left(\dfrac{L}{2}\right) = \dfrac{2L^2 M}{81EI}$ (to the right),

$\delta_D = \theta_C\left(\dfrac{L}{2}\right) = \dfrac{5L^2 M}{162EI}$ (to the left);

(d) $L_{CD} = \dfrac{2\sqrt5 L}{5} = 0.894L$

9.5-27  (a) $b/L = 0.403$; (b) $\delta_C = 0.00287qL^4/EI$

9.5-28  $\alpha = 22.5°, 112.5°, -67.5°,$ or $-157.5°$

9.6-4  $\theta_B = 7qL^3/162EI$, $\delta_B = 23qL^4/648EI$

9.6-5  $\delta_B = 0.443$ in., $\delta_C = 0.137$ in.

9.6-6  $\delta_B = 11.8$ mm, $\delta_C = 4.10$ mm

9.6-8  $P = 64$ kN

9.6-9  $\theta_A = M_0 L/6EI$, $\theta_B = M_0 L/3EI$,
$\delta = M_0 L^2/16EI$

9.6-10  $\theta_A = Pa(L - a)(L - 2a)/6LEI$,
$\delta_1 = Pa^2(L - 2a)^2/6LEI$, $\delta_2 = 0$

9.6-11  $\theta_A = M_0 L/6EI$, $\theta_B = 0$, $\delta = M_0 L^2/27EI$
(downward)

9.7-1  (a) $\delta_B = PL^3(1 + 7I_1/I_2)/24EI_1$;
(b) $r = (1 + 7I_1/I_2)/8$

9.7-2  (a) $\delta_B = qL^4(1 + 15I_1/I_2)/128EI_1$;
(b) $r = (1 + 15I_1/I_2)/16$

9.7-3  (a) $\delta_C = 0.31$ in. (upward);
(b) $\delta_C = 0.75$ in. (downward)

9.7-4  $v = -qx(21L^3 - 64Lx^2 + 32x^3)/768EI$ for
$0 \le x \le L/4$,
$v = -q(13L^4 + 256L^3 x - 512Lx^3$
$+ 256x^4)/12{,}288EI$ for $L/4 \le x \le L/2$,
$\theta_A = 7qL^3/256EI$, $\delta_{max} = 31qL^4/4096EI$

9.7-5  $\theta_A = 8PL^2/243EI$, $\delta_B = 8PL^3/729EI$,
$\delta_{max} = 0.01363PL^3/EI$

9.7-6  $v = -2Px(19L^2 - 27x^2)/729EI$ for
$0 \le x \le L/3$,
$v = P(13L^3 - 175L^2 x + 243Lx^2 - 81x^3)/$
$1458EI$ for $L/3 \le x \le L$,
$\theta_A = 38PL^2/729EI$, $\theta_C = 34PL^2/729EI$,
$\delta_B = 32PL^3/2187EI$

9.7-7  $v = \dfrac{PL^3}{EI_A}\left[\dfrac{L}{2(L + x)} - \dfrac{3x}{8L} + \dfrac{1}{8}\right.$

$\left. + \ln\left(\dfrac{L + x}{2L}\right)\right]$,

$\delta_A = \dfrac{PL^3}{8EI_A}(8 \ln 2 - 5)$

9.7-8  $v = \dfrac{PL^3}{24EI_A}\left[7 - \dfrac{4L(2L + 3x)}{(L + x)^2} - \dfrac{2x}{L}\right]$,

$\delta_A = \dfrac{PL^3}{24EI_A}$

9.7-9  $v = \dfrac{8PL^3}{EI_A}\left[\dfrac{L}{2L + x} - \dfrac{2x}{9L} - \dfrac{1}{9}\right.$

$\left. + \ln\left(\dfrac{2L + x}{3L}\right)\right]$,

$\delta_A = \dfrac{8PL^3}{EI_A}\left(\ln\dfrac{3}{2} - \dfrac{7}{18}\right)$

9.7-10 $v(x) = \dfrac{19{,}683PL^3}{2000EI_A}\left(\dfrac{81L}{81L + 40x} + 2\ln\right)$

$\left(\dfrac{81}{121} + \dfrac{40x}{121L}\right) - \left(\dfrac{6440x}{14{,}641L} - \dfrac{3361}{14{,}641}\right),$

$\delta_A = \dfrac{19{,}683PL^3}{7{,}320{,}500EI_A}$

$\left(-2820 + 14{,}641\ln\left(\dfrac{11}{9}\right)\right)$

9.7-11 $v(x) = -\dfrac{19{,}683PL^3}{2000EI_A}$

$\left(\dfrac{81L}{81L + 40x} + 2\ln\left(1 + \dfrac{40x}{81L}\right)\right.$

$\left. -\dfrac{6440x}{14{,}641L} - 1\right),$

$\delta_B = \dfrac{19{,}683PL^3}{7{,}320{,}500EI_A}$

$\left(-2820 + 14{,}641\ln\left(\dfrac{11}{9}\right)\right)$

9.7-12 (a) $v' = -\dfrac{qL^3}{16EI_A}\left[1 - \dfrac{8Lx^2}{(L + x)^3}\right]$

for $0 \le x \le L,$

$v = -\dfrac{qL^4}{2EI_A}\left[\dfrac{(9L^2 + 14Lx + x^2)x}{8L(L + x)^2}\right.$

$\left. - \ln\left(1 + \dfrac{x}{L}\right)\right]$ for $0 \le x \le L;$

(b) $\theta_A = \dfrac{qL^3}{16EI_A}, \delta_C = \dfrac{qL^4(3 - 4\ln 2)}{8EI_A}$

9.8-1 $U = 4bhL\sigma_{max}^2/45E$

9.8-2 (a) and (b) $U = P^2L^3/96EI$;
(c) $\delta = PL^3/48EI$

9.8-3 (a) and (b) $U = \dfrac{q^2L^3}{15EI}$

9.8-4 (a) $U = 32EI\delta^2/L^3$; (b) $U = \pi^4EI\delta^2/4L^3$

9.8-5 (a) $U = P^2a^2(L + a)/6EI$;
(b) $\delta_C = Pa^2(L + a)/3EI$;
(c) $U = 241$ in.-lb, $\delta_C = 0.133$ in.

9.8-6 $U = \dfrac{L}{15{,}360EI}(17L^4q^2 + 280qL^2M_0$
$+ 2560M_0^2)$

9.8-7 $\delta_B = 2PL^3/3EI + 8\sqrt{2}PL/EA$

9.9-2 $\delta_D = Pa^2b^2/3LEI$

9.9-3 $\delta_C = Pa^2(L + a)/3EI$

9.9-6 $\delta_C = L^3(2P_1 + 5P_2)/48EI,$
$\delta_B = L^3(5P_1 + 16P_2)/48EI$

9.9-7 $\theta_A = 7qL^3/48EI$

9.9-8 $\delta_C = Pb^2(b + 3h)/3EI,$
$\theta_C = Pb(b + 2h)/2EI$

9.9-9 $\delta_C = 31qL^4/4096EI$

9.9-10 $\theta_A = M_A(L + 3a)/3EI,$
$\delta_A = M_Aa(2L + 3a)/6EI$

9.9-11 $\delta_C = Pa^2(L + a)/3EI + P(L + a)^2/kL^2$

9.9-12 $\delta_D = 37qL^4/6144EI$ (upward)

9.10-1 $\sigma_{max} = \sigma_{st}[1 + (1 + 2h/\delta_{st})^{1/2}]$

9.10-2 $\sigma_{max} = \sqrt{18WEh/AL}$

9.10-3 $\delta_{max} = 0.302$ in., $\sigma_{max} = 21{,}700$ psi

9.10-4 $d = 281$ mm

9.10-5 W $14 \times 53$

9.10-6 $h = 360$ mm

9.10-7 $R = \sqrt{3EII_m\omega^2/L^3}$

9.11-1 $v = -\alpha(T_2 - T_1)(x)(L - x)/2h$
(pos. upward),
$\theta_A = \alpha L(T_2 - T_1)/2h$ (clockwise),
$\delta_{max} = \alpha L^2(T_2 - T_1)/8h$ (downward)

9.11-2 $v = -\alpha(T_2 - T_1)(x^2)/2h$ (upward),
$\theta_B = \alpha L(T_2 - T_1)/h$ (counterclockwise),
$\delta_B = \alpha L^2(T_2 - T_1)/2h$ (upward)

9.11-3 $v(x) = \dfrac{\alpha(T_2 - T_1)(x^2 - L^2)}{2h},$

$\theta_C = \dfrac{\alpha(T_2 - T_1)(L + a)}{h}$
(counterclockwise),

$\delta_C = \dfrac{\alpha(T_2 - T_1)(2La + a^2)}{2h}$ (upward)

9.11-4 (a) $\delta_{max} = \dfrac{\alpha T_0L^3}{9\sqrt{3}h}$ (downward);

(b) $\delta_{max} = \dfrac{\alpha T_0L^4(2\sqrt{2} - 1)}{48h}$ (downward)

9.11-5 (a) $\delta_{max} = \dfrac{\alpha T_0L^3}{6h}$ (downward);

(b) $\delta_{max} = \dfrac{\alpha T_0L^4}{12h}$ (downward);

(c) $\delta_{max} = \dfrac{\alpha T_0L^3}{6h}$ (downward),

$\delta_{max} = \dfrac{\alpha T_0L^4}{12h}$ (downward)

# CHAPTER 10

**10.3-1** $R_A = -R_B = 3M_0/2L$, $M_A = M_0/2$,
$v = -M_0x^2(L-x)/4LEI$

**10.3-2** $R_A = R_B = qL/2$, $M_A = M_B = qL^2/12$,
$v = -qx^2(L-x)^2/24EI$

**10.3-3** $R_A = R_B = 3EI\delta_B/L^3$, $M_A = 3EI\delta_B/L^2$,
$v = -\delta_Bx^2(3L-x)/2L^3$

**10.3-4** $\theta_B = \dfrac{qL^3}{6(k_RL - EI)}$,

$\delta_B = -\dfrac{1}{8}qL^4 + \dfrac{k_RqL^5}{12(k_RL - EI)}$

**10.3-5** $R_A = V(0) = \dfrac{9}{40}q_0L$,

$R_B = -V(L) = \dfrac{11}{40}q_0L$,

$M_A = \dfrac{7}{120}q_0L^2$

**10.3-6** (a) $R_A = V(0) = \dfrac{7}{60}q_0L$,

$R_B = -V(L) = \dfrac{13}{60}q_0L$,

$M_A = \dfrac{1}{30}q_0L^2$,

$v = \dfrac{q_0}{360L^2EI}(-x^6 + 7L^3x^3 - 6q_0L^4x^2)$;

(b) $R_A = V(0) = 0.31q_0L$

$= \left(\dfrac{2}{\pi} - 6\dfrac{\pi^2 - 4\pi + 8}{\pi^4}\right)q_0L$,

$R_B = -V(L) = 0.327q_0L$

$= \left(6\dfrac{\pi^2 - 4\pi + 8}{\pi^4}\right)q_0L$,

$M_A = -2q_0L^2\dfrac{\pi^2 - 12\pi + 24}{\pi^4}$,

$v = \dfrac{1}{EI}$

$\left[-q_0\left(\dfrac{2L}{\pi}\right)^4\sin\left(\dfrac{\pi x}{2L}\right) - 6q_0L\dfrac{\pi^2 - 4\pi + 8}{\pi^4}\dfrac{x^3}{6}\right.$

$\left.+ 2q_0L^2\dfrac{\pi^2 - 12\pi + 24}{\pi^4}\dfrac{x^2}{2} + q_0\left(\dfrac{2L}{\pi}\right)^3x\right]$

**10.3-7** (a) $R_A = \dfrac{61Lq_0}{120}$,

$M_A = \dfrac{11L^2q_0}{120}$,

$R_B = \dfrac{19Lq_0}{120}$,

$v(x) = -\dfrac{q_0x^2(33L^4 - 61L^3x + 30L^2x^2 - 2x^4)}{720EIL^2}$;

(b) $R_A = \dfrac{48Lq_0}{\pi^4}$,

$R_B = \displaystyle\int_0^L q(x)\,dx - R_A = \dfrac{2Lq_0}{\pi} - \dfrac{48Lq_0}{\pi^4}$,

$M_A = \displaystyle\int_0^L q(x)x\,dx - R_BL = \dfrac{2L^2q_0(\pi - 2)}{\pi^2}$

$\quad - L\left(\dfrac{2Lq_0}{\pi} - \dfrac{48Lq_0}{\pi^4}\right)$,

$v(x) = \dfrac{16L^4q_0 - 24L^2q_0x^2 + 8Lq_0x^3 - 16L^4q_0\cos\left(\frac{\pi x}{2L}\right)}{\pi^4EI}$

**10.3-8** (a) $R_A = V(0) = \dfrac{24}{\pi^4}q_0L$,

$R_B = -V(L) = -\dfrac{24}{\pi^4}q_0L$,

$M_A = \left(\dfrac{12}{\pi^4} - \dfrac{1}{\pi^2}\right)q_0L^2$

$\quad\quad\quad\quad\quad\quad$ (counterclockwise),

$M_B = \left(\dfrac{12}{\pi^4} - \dfrac{1}{\pi^2}\right)q_0L^2$

$\quad\quad\quad\quad\quad\quad$ (counterclockwise),

$v = \dfrac{1}{\pi^4EI}\left[-q_0L^4\cos\left(\dfrac{\pi x}{L}\right)\right.$

$\left.+ 4q_0Lx^3 - 6q_0L^2x^2 + q_0L^4\right]$;

(b) $R_A = R_B = q_0L/\pi$, $M_A = M_B = 2q_0L^2/\pi^3$,
$v = -q_0L^2(L^2\sin \pi x/L + \pi x^2 - \pi Lx)/\pi^4EI$

**10.3-9** (a) $R_A = V(0) = \dfrac{48(4 - \pi)}{\pi^4}q_0L$,

$R_B = -V(L) = \left(\dfrac{2}{\pi} - \dfrac{48(4 - \pi)}{\pi^4}\right)q_0L$,

$M_A = -q_0\left(\dfrac{2L}{\pi}\right)^2 + \dfrac{16(6 - \pi)}{\pi^4}q_0L^2$,

$M_B = -\dfrac{32(\pi - 3)}{\pi^4}q_0L^2$,

$v = \dfrac{1}{\pi^4EI}\left[-16q_0L^4\cos\left(\dfrac{\pi x}{2L}\right)\right.$

$\quad + 8(4 - \pi)q_0Lx^3$

$\left.- 8(6 - \pi)q_0L^2x^2 + 16q_0L^4\right]$;

(b) $R_A = V(0) = \dfrac{13}{30} q_0 L$,

$R_B = -V(L) = \dfrac{7}{30} q_0 L$,

$M_A = \dfrac{1}{15} q_0 L^2$ (counterclockwise),

$M_B = -\dfrac{1}{20} q_0 L^2$ (counterclockwise),

$v = \dfrac{q_0}{360 L^2 EI} [x^6 - 15 L^2 x^4 + 26 L^3 x^3 - 12 L^4 x^2]$

**10.3-10** $R_A = V(0) = \dfrac{3}{20} q_0 L$,

$R_B = -V(L) = \dfrac{7}{20} q_0 L$,

$M_A = \dfrac{1}{30} q_0 L^2$,

$v = \dfrac{1}{120 L EI}(-q_0 x^5 + 3 q_0 L x^3 - 2 q_0 L^2 x^2)$

**10.3-11** $R_A = -R_B = 3 M_0 / 2L$, $M_A = -M_B = M_0/4$,
$v = -M_0 x^2 (L - 2x)/8 LEI$ for $0 \le x \le L/2$

**10.3-12** $R_B = -\dfrac{9}{8} \dfrac{M_0}{L}$,

$R_A = \dfrac{9}{8} \dfrac{M_0}{L}$,

$M_A = \dfrac{1}{8} \dfrac{M_0}{L}$,

$v = \dfrac{1}{EI}\left(\dfrac{9 M_0}{48 L} x^3 - \dfrac{M_0}{16} x^2\right)\left(0 \le x \le \dfrac{L}{2}\right)$,

$v = \dfrac{1}{EI}$
$\left(\dfrac{9 M_0}{48 L} x^3 - \dfrac{9 M_0}{16} x^2 + \dfrac{M_0 L}{2} x - \dfrac{M_0 L^2}{8}\right)$
$\left(\dfrac{L}{2} \le x \le L\right)$

**10.4-1** $R_A = Pb(3L^2 - b^2)/2L^3$,
$R_B = Pa^2(3L - a)/2L^3$, $M_A = Pab(L + b)/2L^2$

**10.4-2** $R_A = qL$, $M_A = \dfrac{qL^2}{3}$, $M_B = \dfrac{qL^2}{6}$

**10.4-3** $R_A = -\dfrac{1}{8} qL$, $R_B = \dfrac{17}{8} qL$, $M_A = -\dfrac{1}{8} qL^2$

**10.4-4** (a) $R_A = M_0/3L$, $H_A = -4M_0/3L$,
$R_B = -R_A$, $R_C = -H_A$;
(b) $\theta_A = -M_0 L/18 EI$, $\theta_B = M_0 L/9 EI$,
$\theta_C = \theta_A$; (c) $L_{BC} = 2L$

**10.4-5** (a) $R_A = \dfrac{4 M_0}{3L}$, $H_A = \dfrac{2 M_0}{3L}$, $R_B = -\dfrac{4 M_0}{3L}$,

$R_C = -\dfrac{2 M_0}{3L}$;

(b) $\theta_A = \dfrac{5}{18} \dfrac{M_0 L}{EI}$, $\theta_B = \dfrac{-M_0 L}{18 EI}$,

$\theta_C = \dfrac{M_0 L}{36 EI}$;

(c) $L_{AB} = 2.088 L$

**10.4-6** (a) $R_A = \dfrac{M_0}{L} + \dfrac{M_0 k_R}{2(3 EI + L k_R)}$,

$R_B = -R_A$,

$M_B = \dfrac{L M_0 k_R}{6 EI + 2 L k_R}$ (CCW);

(b) $\theta_A = \dfrac{L M_0}{4 EI} + \dfrac{L M_0}{4(3 EI + L k_R)}$

For $k_R$ goes to zero:

$\theta_A = \dfrac{L M_0}{4 EI} + \dfrac{L M_0}{4(3 EI)} = \dfrac{L M_0}{3 EI}$

For $k_R$ goes to infinity: $\theta_A = \dfrac{M_0 L}{4 EI}$

For $k_R$ goes to $6 EI/L$: $\theta_A = \dfrac{L M_0}{4 EI}$

$+ \dfrac{L M_0}{4\left[3 EI + L\left(\dfrac{6 EI}{L}\right)\right]} = \dfrac{5 L M_0}{18 EI}$

**10.4-7** (a) $H_A = \dfrac{3}{2} \dfrac{M_0}{L}$,

$H_B = 0$, $V_B = 0$, $V_C = 0$, $H_D = -H_A$;

(b) $\theta_A = \dfrac{-M_0 L}{16 EI}$, $\theta_D = -\theta_A$,

$\theta_B = \dfrac{M_0 L}{8 EI}$, $\theta_C = -\theta_B$;

(c) $H_A = \dfrac{M_0}{L}$, $H_B = -2\dfrac{M_0}{L}$,

$V_B = \dfrac{M_0}{L}$, $V_C = \dfrac{-M_0}{L}$, $H_D = H_A$,

$\theta_A = \dfrac{-M_0 L}{24 EI}$, $\theta_D = \theta_A$, $\theta_B = \dfrac{M_0 L}{12 EI}$,

$\theta_C = \theta_B$

**10.4-8** $t_{AB}/t_{CD} = L_{AB}/L_{CD}$

**10.4-9** $R_A = \dfrac{7}{12}qL$, $R_B = \dfrac{17}{12}qL$, $M_A = \dfrac{7}{12}qL^2$

**10.4-10** $R_A = 2qL$, $M_B = \dfrac{7}{12}qL^2$

**10.4-11** $R_A = R_B = q_0L/4$, $M_A = M_B = 5q_0L^2/96$

**10.4-12** $R_A = qL/8$, $R_B = 33qL/16$,
$R_C = 13qL/16$

**10.4-13** $R_A = 1100$ lb (downward), $R_B = 2800$ lb (upward), $M_A = 30,000$ lb-in. (clockwise)

**10.4-14** $R_B = 6.44$ kN

**10.4-15** (a) The tension force in the tie rod $=$
$R_D = 604$ lb; (b) $R_A = 796$ lb,
$M_A = 1308$ lb-ft $= 1.567 \times 10^4$ lb-in.

**10.4-16** $R_A = 31qL/48$, $R_B = 17qL/48$,
$M_A = 7qL^2/48$

**10.4-17** (a) $R_A = -23P/17$, $R_D = R_E = 20P/17$,
$M_A = 3PL/17$; (b) $M_{max} = PL/2$

**10.4-18** $R_A = R_D = 2qL/5$, $R_B = R_C = 11qL/10$

**10.4-19** $M_B(q) = (-800 \cdot q)$ lb-in. for $q < 250$ lb/in.,
$M_B(q) = (-200 \cdot q - 150,000)$ lb-in. for
$q \geq 250$ lb/in.

**10.4-20** $R_A = -R_B = 6M_0ab/L^3$,
$M_A = M_0b(3a - L)/L^2$,
$M_B = -M_0a(3b - L)/L^2$

**10.4-21** $\sigma = 509$ psi

**10.4-22** $(M_{AB})_{max} = 121qL^2/2048 = 6.05$ kN·m;
$(M_{CD})_{max} = 5qL^2/64 = 8.0$ kN·m

**10.4-23** $F = 3160$ lb, $M_{AB} = 18,960$ lb-ft,
$M_{DE} = 7320$ lb-ft

**10.4-24** $k = 48EI(6 + 5\sqrt{2})/7L^3 = 89.63EI/L^3$

**10.4-25** (a) $V_A = V_C = 3P/32$, $H_A = P$,
$M_A = 13PL/32$; (b) $M_{max} = 13PL/32$

**10.4-26** $H_A = -\dfrac{35}{64}P$, $H_C = -\dfrac{29}{64}P$,

$M_{max} = \dfrac{35}{128}PL$

**10.4-27** $R_A = R_B = 3000$ lb, $R_C = 0$

**10.4-28** (a) $M_A = M_B = qb(3L^2 - b^2)/24L$;
(b) $b/L = 1.0$, $M_A = qL^2/12$;
(c) For $a = b = L/3$, $(M_{max})_{pos} = 19qL^2/648$

**10.4-29** (a) $d_2/d_1 = \sqrt[4]{8} = 1.682$;
(b) $M_{max} = qL^2(3 - 2\sqrt{2})/2 = 0.08579qL^2$;
(c) Point $C$ is below points $A$ and $B$ by the amount $0.01307qL^4/EI$

**10.4-30** $M_{max} = 19q_0L^2/256$, $\sigma_{max} = 13.4$ MPa,
$\sigma_{max} = 19q_0L^4/7680EI = 0.00891$ mm

**10.5-1** $S = \dfrac{243E_SE_WIAH\alpha(\Delta T)}{4AL^3E_S + 243IHE_W}$

**10.5-2** (a) $R_A = -\dfrac{\alpha(T_2 - T_1)L^2}{2h}$
$\cdot \left(\dfrac{3EI \cdot k}{3EI + L^3 \cdot k}\right)$,

$R_B = \dfrac{\alpha(T_2 - T_1)L^2}{2h} \cdot \left(\dfrac{3EI \cdot k}{3EI + L^3 \cdot k}\right)$,

$M_A = R_BL$
$= \dfrac{\alpha(T_2 - T_1)L^3}{2h} \cdot \left(\dfrac{3EI \cdot k}{3EI + L^3 \cdot k}\right)$;

(b) $R_A = -R_B = -\dfrac{3EI\alpha(T_2 - T_1)}{2hL}$
(upward),

$R_B = \dfrac{3EI\alpha(T_2 - T_1)}{2hL}$ (downward),

$M_A = R_BL = \dfrac{3EI\alpha(T_2 - T_1)}{2h}$
(counterclockwise)

**10.5-3** $R_A = -R_B = -\dfrac{\alpha(T_2 - T_1)L^2}{2h}$
$\cdot \left(\dfrac{3EI \cdot k}{3EI + L^3 \cdot k}\right)$ (upward),

$R_B = \dfrac{\alpha(T_2 - T_1)L^2}{2h} \cdot \left(\dfrac{3EI \cdot k}{3EI + L^3 \cdot k}\right)$
(downward),

$M_A = R_BL = \dfrac{\alpha(T_2 - T_1)L^3}{2h}$
$\cdot \left(\dfrac{3EI \cdot k}{3EI + L^3 \cdot k}\right)$ (counterclockwise)

**10.5-4** (a) $R_B = -\dfrac{\alpha(T_1 - T_2)L^2}{h}$
$\left(\dfrac{6EI \cdot k}{36EI + L^3 \cdot k}\right)$ (downward),

$R_A = -\dfrac{1}{4}R_B = \dfrac{\alpha(T_1 - T_2)L^3}{2h}$
$\left(\dfrac{3EI \cdot k}{36EI + L^3 \cdot k}\right)$ (upward),

$R_C = -\dfrac{3}{4}R_B = \dfrac{\alpha(T_1 - T_2)L^2}{2h}$
$\left(\dfrac{9EI \cdot k}{36EI + L^3 \cdot k}\right)$ (upward);

(b) $R_B = -\dfrac{6EI\alpha(T_1 - T_2)}{Lh}$ (downward),

$R_A = \dfrac{3EI\alpha(T_1 - T_2)}{2Lh}$ (upward),

$R_C = \dfrac{9EI\alpha(T_1 - T_2)}{2Lh}$ (upward)

10.5-5    $R_B = -\dfrac{\alpha(T_1 - T_2)L^2}{h}\left(\dfrac{6EI\cdot k}{36EI + L^3\cdot k}\right)$

(downward),

$R_A = -\dfrac{1}{4}R_B = \dfrac{\alpha(T_1 - T_2)L^2}{2h}$

$\left(\dfrac{3EI\cdot k}{36EI + L^3\cdot k}\right)$ (upward),

$R_C = -\dfrac{3}{4}R_B = \dfrac{\alpha(T_1 - T_2)L^2}{2k}$

$\left(\dfrac{9EI\cdot k}{36EI + L^3\cdot k}\right)$ (upward)

10.6-1    (a) $H = \pi^2 EA\delta^2/4L^2$, $\sigma_t = \pi^2 E\delta^2/4L^2$;
(b) $\sigma_t = 617$, 154, and 69 psi

10.6-2    (a) $\lambda = 17q^2 L^7/40{,}320E^2 I^2$, $\sigma_b = qhL^2/16I$;
(b) $\sigma_t = 17q^2 L^6/40{,}320EI^2$;
(c) $\lambda = 0.01112$ mm, $\sigma_b = 117.2$ MPa,
$\sigma_t = 0.741$ MPa

# CHAPTER 11

11.2-1    $P_{cr} = \beta_R/L$

11.2-2    (a) $P_{cr} = \dfrac{\beta a^2 + \beta_R}{L}$; (b) $P_{cr} = \dfrac{\beta a^2 + 2\beta_R}{L}$

11.2-3    $P_{cr} = 6\beta_R/L$

11.2-4    (a) $P_{cr} = \dfrac{(L - a)(\beta a^2 + \beta_R)}{aL}$;

(b) $P_{cr} = \dfrac{\beta L^2 + 20\beta_R}{4L}$

11.2-5    $P_{cr} = \dfrac{3\beta_R}{L}$

11.2-6    $P_{cr} = \dfrac{3}{5}\beta L$

11.2-7    $P_{cr} = \dfrac{7}{4}\beta L$

11.3-1    (a) $P_{cr} = 453$ k; (b) $P_{cr} = 152$ k

11.3-2    (a) $P_{cr} = 2803$ kN; (b) $P_{cr} = 953$ kN

11.3-3    (a) $P_{cr} = 650$ k; (b) $P_{cr} = 140$ k

11.3-4    $M_{allow} = 1143$ kN·m

11.3-5    $Q_{allow} = 23.8$ k

11.3-6    (a) $Q_{cr} = \dfrac{\pi^2 EI}{L^2}$; (b) $Q_{cr} = \dfrac{2\pi^2 EI}{9L^2}$

11.3-7    (a) $Q_{cr} = \dfrac{2\pi^2 EI}{L^2}$; (b) $M_{cr} = \dfrac{3d\pi^2 EI}{L^2}$

11.3-8    $\Delta T = \pi^2 I/\alpha AL^2$

11.3-9    $h/b = 2$

11.3-10    (a) $P_{cr} = 3\pi^3 Er^4/4L^2$; (b) $P_{cr} = 11\pi^3 Er^4/4L^2$

11.3-11    $P_1 : P_2 : P_3 = 1.000 : 1.047 : 1.209$

11.3-12    $P_{allow} = 604$ kN

11.3-13    $F_{allow} = 54.4$ k

11.3-14    $W_{max} = 124$ kN

11.3-15    $t_{min} = 0.165$ in.

11.3-16    $P_{cr} = 497$ kN

11.3-17    $W_{cr} = 51.9$ k

11.3-18    $\theta = \arctan 0.5 = 26.57°$

11.3-19    (a) $q_{max} = 142.4$ lb/ft; (b) $I_{b,min} = 38.5$ in.$^4$;
(c) $s = 0.264$ ft, 2.42 ft

11.3-20    $P_{cr} = 3.56$ kN

11.3-21    $P_{cr} = 16.28$ k

11.4-1    $P_{cr} = 235$ k, 58.7 k, 480 k, 939 k

11.4-2    $P_{cr} = 62.2$ kN, 15.6 kN, 127 kN, 249 kN

11.4-3    $P_{allow} = 253$ k, 63.2 k, 517 k, 1011 k

11.4-4    $P_{allow} = 678$ kN, 169.5 kN, 1387 kN, 2712 kN

11.4-5    $P_{cr} = 229$ k

11.4-6    $T_{allow} = 18.1$ kN

11.4-7    (a) $Q_{cr} = 4575$ lb; (b) $Q_{cr} = 10{,}065$ lb,
$a = 0$ in.

11.4-8    $P_{cr} = 447$ kN, 875 kN, 54.7 kN, 219 kN

11.4-9    $P_{cr} = 4\pi^2 EI/L^2$, $v = \delta(1-\cos 2\pi x/L)/2$

11.4-10    $t_{min} = 10.0$ mm

11.4-11    (b) $P_{cr} = 413.89EI/L^2$

11.5-1    $\delta = 0.112$ in., $M_{max} = 1710$ lb-in.

11.5-2    $\delta = 8.87$ mm, $M_{max} = 2.03$ kN·m

11.5-3    For $P = 0.3P_{cr}$:
$M/Pe = 1.162$ (sin 1.721 $x/L$) + cos 1.721$x/L$

11.5-4    $P = 583.33\{\arccos [5/(5 + \delta)]\}^2$, in which
$P =$ kN and $\delta =$ mm; $P = 884$ kN when
$\delta = 10$ mm

**11.5-5** $P = 125.58\{\text{arccos } [0.2/(0.2 + \delta)]\}^2$, in which $P = $ k and $\delta = $ in.; $P = 190$ k when $\delta = 0.4$ in.

**11.5-6** $P_{\text{allow}} = 49.9$ kN

**11.5-7** $L_{\text{max}} = 150.5$ in. $= 12.5$ ft

**11.5-8** $L_{\text{max}} = 3.14$ m

**11.5-9** $\delta = e(\sec kL - 1)$, $M_{\text{max}} = Pe \sec kL$

**11.5-10** $L_{\text{max}} = 2.21$ m

**11.5-11** $L_{\text{max}} = 130.3$ in. $= 10.9$ ft

**11.5-12** $T_{\text{max}} = 8.29$ kN

**11.5-13** (a) $q_0 = 2230$ lb/ft $= 186$ lb/in.; (b) $M_{\text{max}} = 37.7$ k·in., ratio $= 0.47$

**11.6-1** (a) $\sigma_{\text{max}} = 17.3$ ksi; (b) $L_{\text{max}} = 46.2$ in.

**11.6-2** $P_{\text{allow}} = 37.2$ kN

**11.6-3** $b_{\text{min}} = 4.10$ in.

**11.6-4** (a) $\sigma_{\text{max}} = 38.8$ MPa; (b) $L_{\text{max}} = 5.03$ m

**11.6-5** (a) $\sigma_{\text{max}} = 9.65$ ksi; (b) $P_{\text{allow}} = 3.59$ k

**11.6-6** $d_2 = 131$ mm

**11.6-7** (a) $\sigma_{\text{max}} = 10.9$ ksi; (b) $P_{\text{allow}} = 160$ k

**11.6-8** (a) $\sigma_{\text{max}} = 104.5$ MPa; (b) $L_{\text{max}} = 3.66$ m

**11.6-9** (a) $\sigma_{\text{max}} = 9.60$ ksi; (b) $P_{\text{allow}} = 53.6$ k

**11.6-10** (a) $\sigma_{\text{max}} = 47.6$ MPa; (b) $n = 2.30$

**11.6-11** (a) $\sigma_{\text{max}} = 13.4$ ksi; (b) $n = 2.61$

**11.6-12** (a) $\sigma_{\text{max}} = 120.4$ MPa; (b) $P_2 = 387$ kN

**11.6-13** (a) $\sigma_{\text{max}} = 17.6$ ksi; (b) $n = 1.89$

**11.6-14** (a) $\sigma_{\text{max}} = 115.2$ MPa; (b) $P_2 = 193$ kN

**11.9-1** $P_{\text{allow}} = \begin{pmatrix} 254 \\ 177 \\ 97 \\ 55 \end{pmatrix}$ k for $L = \begin{pmatrix} 8 \\ 16 \\ 24 \\ 32 \end{pmatrix}$ ft

**11.9-2** $P_{\text{allow}} = \begin{pmatrix} 3019 \\ 2193 \\ 1285 \\ 723 \end{pmatrix}$ kN for $\begin{pmatrix} 3\text{ m} \\ 6\text{ m} \\ 9\text{ m} \\ 12\text{ m} \end{pmatrix}$

**11.9-3** $P_{\text{allow}} = \begin{pmatrix} 338 \\ 240 \\ 135 \\ 76 \end{pmatrix}$ k for $L = \begin{pmatrix} 10 \\ 20 \\ 30 \\ 40 \end{pmatrix}$ ft

**11.9-4** W $250 \times 67$

**11.9-5** W $12 \times 87$

**11.9-6** W $360 \times 122$

**11.9-7** $P_{\text{allow}} = \begin{pmatrix} 60.7 \\ 42.4 \\ 23.3 \\ 13.1 \end{pmatrix}$ k for $L = \begin{pmatrix} 6 \\ 12 \\ 18 \\ 24 \end{pmatrix}$ ft

**11.9-8** $P_{\text{allow}} = \begin{pmatrix} 1104 \\ 919 \\ 678 \\ 441 \end{pmatrix}$ kN for $L = \begin{pmatrix} 2.5 \\ 5.0 \\ 7.5 \\ 10.0 \end{pmatrix}$ m

**11.9-9** $P_{\text{allow}} = \begin{pmatrix} 96.9 \\ 73.9 \\ 50.6 \\ 32.6 \end{pmatrix}$ k for $L = \begin{pmatrix} 6.0 \\ 9.0 \\ 12.0 \\ 15.0 \end{pmatrix}$ ft

**11.9-10** $P_{\text{allow}} = \begin{pmatrix} 229 \\ 207 \\ 185 \\ 164 \end{pmatrix}$ kN for $L = \begin{pmatrix} 2.6 \\ 2.8 \\ 3.0 \\ 3.2 \end{pmatrix}$ m

**11.9-11** $L_{\text{max}} = 5.13$ ft

**11.9-12** $L_{\text{max}} = 3.52$ m

**11.9-13** $L_{\text{max}} = 13.9$ ft

**11.9-14** $P_{\text{allow}} = 5520$ kN

**11.9-15** (a) $L_{\text{max}} = 21.25$ ft; (b) $L_{\text{max}} = 14.10$ ft

**11.9-16** (a) $L_{\text{max}} = 6.44$ m; (b) $L_{\text{max}} = 4.68$ m

**11.9-17** $d = 4.88$ in.

**11.9-18** $d = 99.8$ mm

**11.9-19** $d = 5.25$ in.

**11.9-20** $d = 190$ mm

**11.9-21** $P_{\text{allow}} = \begin{pmatrix} 143.5 \\ 124.0 \\ 85.5 \\ 59.4 \end{pmatrix}$ k for $L = \begin{pmatrix} 6.0 \\ 8.0 \\ 10.0 \\ 12.0 \end{pmatrix}$ ft

**11.9-22** $P_{\text{allow}} = \begin{pmatrix} 315 \\ 245 \\ 127 \\ 71 \end{pmatrix}$ kN for $L = \begin{pmatrix} 1.0 \\ 2.0 \\ 3.0 \\ 4.0 \end{pmatrix}$ m

**11.9-23** $P_{\text{allow}} = \begin{pmatrix} 18.1 \\ 14.8 \\ 8.4 \\ 5.4 \end{pmatrix}$ k for $L = \begin{pmatrix} 2.0 \\ 3.0 \\ 4.0 \\ 5.0 \end{pmatrix}$ ft

**11.9-24** $P_{\text{allow}} = \begin{pmatrix} 96.5 \\ 84.1 \\ 61.0 \\ 42.4 \end{pmatrix}$ kN for $L = \begin{pmatrix} 0.6 \\ 0.8 \\ 1.0 \\ 1.2 \end{pmatrix}$ m

11.9-25 (a) $L_{max} = 25.2$ in.; (b) $d_{min} = 2.11$ in.

11.9-26 (a) $L_{max} = 468$ mm; (b) $d_{min} = 42.8$ mm

11.9-27 (a) $L_{max} = 14.9$ in.; (b) $d_{min} = 1.12$ in.

11.9-28 (a) $L_{max} = 473$ mm; (b) $d_{min} = 33.4$ mm

11.9-29 $P_{allow} = \begin{pmatrix} 34.6 \\ 28.0 \\ 19.9 \end{pmatrix}$ k for $L = \begin{pmatrix} 5.0 \\ 7.5 \\ 10.0 \end{pmatrix}$ ft

11.9-30 $P_{allow} = \begin{pmatrix} 193.8 \\ 177.3 \\ 153.5 \end{pmatrix}$ kN for $L = \begin{pmatrix} 1.5 \\ 2.0 \\ 2.5 \end{pmatrix}$ m

11.9-31 $P_{allow} = \begin{pmatrix} 22.8 \\ 20.2 \\ 16.7 \end{pmatrix}$ k for $L = \begin{pmatrix} 6.0 \\ 8.0 \\ 10.0 \end{pmatrix}$ ft

11.9-32 $P_{allow} = \begin{pmatrix} 310 \\ 255 \\ 190 \end{pmatrix}$ kN for $L = \begin{pmatrix} 2.5 \\ 3.5 \\ 4.5 \end{pmatrix}$ m

11.9-33 (a) $L_{max} = 10.37$ ft; (b) $b_{min} = 5.59$ in.

11.9-34 (a) $L_{max} = 3.45$ m; (b) $b_{min} = 154.9$ mm

11.9-35 (a) $L_{max} = 10.25$ ft; (b) $b_{min} = 3.47$ in.

11.9-36 (a) $L_{max} = 2.50$ m; (b) $b_{min} = 134.8$ mm

# CHAPTER 12

12.3-2   $\bar{x} = \bar{y} = 5a/12$

12.3-3   $\bar{y} = 1.10$ in.

12.3-4   $2c^2 = ab$

12.3-5   $\bar{y} = 13.94$ in.

12.3-6   $\bar{y} = 52.5$ mm

12.3-7   $\bar{x} = 0.99$ in., $\bar{y} = 1.99$ in.

12.3-8   $\bar{x} = 137$ mm., $\bar{y} = 132$ mm

12.4-6   $I_x = 518 \times 10^3$ mm$^4$

12.4-7   $I_x = 36.1$ in.$^4$, $I_y = 10.9$ in.$^4$

12.4-8   $I_x = I_y = 194.6 \times 10^6$ mm$^4$, $r_x = r_y = 80.1$ mm

12.4-9   $I_1 = 1480$ in.$^4$, $I_2 = 186$ in.$^4$, $r_1 = 7.10$ in., $r_2 = 2.52$ in.

12.5-1   $I_b = 940$ in.$^4$

12.5-2   $I_c = 11a^4/192$

12.5-3   $I_{x_c} = 7.23$ in.$^4$

12.5-4   $I_2 = 405 \times 10^3$ mm$^4$

12.5-5   $I_{x_c} = 6050$ in.$^4$, $I_{y_c} = 475$ in.$^4$

12.5-6   $I_{x_c} = 106 \times 10^6$ mm$^4$

12.5-7   $I_{x_c} = 17.40$ in.$^4$, $I_{y_c} = 6.27$ in.$^4$

12.5-8   $b = 250$ mm

12.6-1   $I_P = bh(b^2 + 12h^2)/48$

12.6-2   $(I_P)_C = r^4(9\alpha^2 - 8 \sin^2 \alpha)/18\alpha$

12.6-3   $I_P = 233$ in.$^4$

12.6-4   $I_P = bh(b^2 + h^2)/24$

12.6-5   $(I_P)_C = r^4(176 - 84\pi + 9\pi^2)/[72(4 - \pi)]$

12.7-2   $I_{xy} = r^4/24$

12.7-3   $b = 2r$

12.7-4   $I_{xy} = t^2(2b^2 - t^2)/4$

12.7-5   $I_{12} = -20.5$ in.$^4$

12.7-6   $I_{xy} = 24.3 \times 10^6$ mm$^4$

12.7-7   $I_{x_c y_c} = -6.079$ in.$^4$

12.8-1   $I_{x_1} = I_{y_1} = b^4/12$, $I_{x_1 y_1} = 0$

12.8-2   $I_{x_1} = \dfrac{b^3 h^3}{6(b^2 + h^2)}, I_{y_1} = \dfrac{bh(b^4 + h^4)}{12(b^2 + h^2)},$

$I_{x_1 y_1} = \dfrac{b^2 h^2 (h^2 - b^2)}{12(b^2 + h^2)}$

12.8-3   $I_d = 159$ in.$^4$

12.8-4   $I_{x_1} = 12.44 \times 10^6$ mm$^4$, $I_{y_1} = 9.68 \times 10^6$ mm$^4$, $I_{x_1 y_1} = 6.03 \times 10^6$ mm$^4$

12.8-5   $I_{x_1} = 13.50$ in.$^4$, $I_{y_1} = 3.84$ in.$^4$, $I_{x_1 y_1} = 4.76$ in.$^4$

12.8-6   $I_{x_1} = 8.75 \times 10^6$ mm$^4$, $I_{y_1} = 1.02 \times 10^6$ mm$^4$, $I_{x_1 y_1} = -0.356 \times 10^6$ mm$^4$

12.9-1   (a) $c = \sqrt{a^2 - b^2}/2$; (b) $a/b = \sqrt{5}$; (c) $1 \leq a/b < \sqrt{5}$

12.9-2   Shows that two different sets of principal axes exist at each point.

12.9-3   $\theta_{p1} = -29.87°$, $\theta_{p2} = 60.13°$, $I_1 = 311.1$ in.$^4$, $I_2 = 88.9$ in.$^4$

12.9-4   $\theta_{p1} = -8.54°$, $\theta_{p2} = 81.46°$, $I_1 = 17.24 \times 10^6$ mm$^4$, $I_2 = 4.88 \times 10^6$ mm$^4$

12.9-5   $\theta_{p1} = 37.73°$, $\theta_{p2} = 127.73°$, $I_1 = 15.45$ in.$^4$, $I_2 = 1.89$ in.$^4$

**12.9-6**  $\theta_{p1} = 32.63°$, $\theta_{p2} = 122.63°$,
$I_1 = 8.76 \times 10^6$ mm$^4$, $I_2 = 1.00 \times 10^6$ mm$^4$

**12.9-7**  $\theta_{p1} = 16.85°$, $\theta_{p2} = 106.85°$, $I_1 = 0.2390b^4$,
$I_2 = 0.0387b^4$

**12.9-8**  $\theta_{p1} = 74.08°$, $\theta_{p2} = -15.92°$,
$I_1 = 8.29 \times 10^6$ mm$^4$, $I_2 = 1.00 \times 10^6$ mm$^4$

**12.9-9**  $\theta_{p1} = 75.73°$, $\theta_{p2} = -14.27°$, $I_1 = 20.07$ in.$^4$,
$I_2 = 2.12$ in.$^4$

# NAME INDEX

## A

Andrews, E. S., 992

## B

Bernoulli, Daniel, 991
Bernoulli, Jacob, 40, 416, 735, 987–993
Bredt, R., 321, 989
Budynas, R., 989

## C

Castigliano, C. A. P., 781, 992
Cauchy, Augustin Louis, 601, 991
Celsius, Anders, 1045
Clapeyron, Benoit Paul Emile,
    177, 988
Clebsch, A., 992
Considère, A. G., 915, 994
Coulomb, Charles Augustin de, 265,
    989–990, 992
Culmann, Karl, 687, 991

## D

da Vinci, Leonardo, 4
De Salvio, Alfonso, 989
Didion, I., 988
Duleau, Alphonse, 265, 989

## E

Elzevir, Louis, 989
Engesser, F., 916, 994
Euler, Leonhard, 4, 140, 417, 735, 882, 889,
    915, 988, 990, 993

## F

Fahrenheit, Daniel, 1047
Fazekas, G. A., 989
Föppl, A., 993

## G

Galilei, Galileo, 4, 417, 989–990
Goodier, James Norman, 988

## H

Hauff, E., 992
Hertz, Heinrich Rudolf, 1042
Hoff, N. J., 916, 994
Hooke, Robert, 52, 987

## J

Jasinsky, F. S., 916, 994
Johnston, B. G., 916, 994
Joule, James Prescott, 1039
Jourawski, D. J., 444, 990

## K

Keller, J. B., 994
Kelvin, William Thomas Lord, 1046
Kuenzi, E. W., 990

## L

L'Hôpital, G. F. A. de, 992
Lamarle, A. H. E., 915, 994
Love, A. E. H., 987
Lüders, W., 170, 988

## M

Macaulay, W. H., 992
Maki, A. C., 990
Mariotte, Edme, 417, 990
Maxwell, James Clerk, 990
McLean, L., 988
Mohr, Otto Christian, 991
Morin, A. J., 988

## N

Navier, Louis Marie Henri,
    140, 988–990
Newton, Isaac Sir, 987, 1039

## O

Oravas, G. A., 988, 992
Osgood, W. R., 989

## P

Parent, Antoine, 417, 990
Pascal, Blaise, 1042
Pearson, K., 987
Pilkey, W. D., 993
Piobert, G., 170, 988
Poisson, Siméon Denis, 54, 988, 991
Poncelet, Jean Victor, 40, 191, 987, 989

## R

Ramberg, W. A., 989
Rankine, William John Macquorn, 601,
    991, 1047
Roark, R. J., 989

## S

Saint-Venant, Barré de, 198, 265, 417,
    601, 989–992
Shanley, F. R., 915–916, 994

## T

Timoshenko, S. P., 555, 987–988, 991, 994
Todhunter, I., 987

## V

Van den Broek, J. A., 994
von Kármán, Theodore, 916, 994

## W

Watt, James, 1042
Williot, Joseph Victor, 991

## Y

Young, Donovan, 994
Young, Thomas, 53, 265, 987–989
Young, W. C., 989

## Z

Zaslavsky, A., 989, 990, 993

## A

Absolute weight, 1040
Acceleration of gravity, 187, 1038,
    1041–1044
Allowable loads, 68–73
Allowable Stress Design, 918
Allowable stresses, 68–73
Alternating (reverse) loads, 195–196
Aluminum, 42–45, 121, 196, 427, 920, 924
    alloys, 42
    column design formulations, 920
    column limitations, 924
    structural sections, 427
Aluminum Association, 427, 920, 990
American Forest and Paper Association,
    427, 921, 990
American Institute of Steel Construction
    (AISC), 427, 918, 990, 1069
Analysis of stress and strain, *see* Plane
    stress
Angle of twist, 255, 257–259, 263, 272–274,
    308, 321, 989, 1006, 1008
    bar of linearly elastic material, 263
    and shear stress distribution, 308
    thin-walled tubes, 321
Angle section, 552
Angles of rotation
    beams, 732
Angular speed, 291, 292
Apparent weight, 1040
Applications of plane stress, 670–714.
        *See also* Plane stress
Areas, 3, 28, 121–123, 592, 685, 917–920,
    956–971
    composite, 959–965
    plane, 965–971
Atmospheric pressure, 1042
Axial force, 27, 31–32
Axial loads, 3–4, 27, 118–217, 462–467
    bars, 130
    beams limitations, 464
    changes in lengths, 120–129
    changes in lengths under nonuniform
        conditions, 130–137
    elastoplastic analysis, 210–215
    impact loading, 187–194
    nonlinear behavior, 205–209
    repeated loading and fatigue,
        195–196
    statically indeterminate structures,
        138–148
    stress concentrations, 197–204
    stresses in beams, 462–467
    stresses on inclined sections,
        164–175
    thermal effects, misfits, and prestrains,
        149–163
Axial rigidity, 122
Axis of symmetry, 971

## B

Bars, 2–3, 70, 119–123, 130–132, 169,
    179–180, 188–190, 206–217,
    259, 263, 265, 274
    angle of twist, 263
    changes in lengths, 206–207
    consisting of prismatic segments, 131
    continuously varying loads or
        dimensions, 131
    with intermediate axial loads, 130
    limitations of circular cross
        section, 265
    linearly elastic material, 263
    made of linearly elastic materials, 274
    maximum elongation, 188–189
    maximum stress, 189–190
    nonuniform, 179–180
    shear strains, 259
    in tension, 169
Beams, 14, 350, 354–360, 402–475, 506–569,
    685–693, 728–799, 853–854,
    1083–1088
    axial loads limitations, 464
    axial load stresses, 462–467
    concentrated load, 374
    constant strength, 435
    cross sections, 408, 417
    curvature shortening, 853
    deflection, 728–799
    deflection curve, 732, 753
    deflections and slopes, 1083–1088
    elastoplastic material, 558, 561
    finding reactions, 356
    fully stressed, 435
    glulam, 458
    hole at neutral axis, 468
    horizontal reactions, 854
    linearly elastic material limitations, 444
    maximum stresses, 685–693
    moment-curvature relationship, 518
    normal stresses, 518
    overhang, 350
    plastic modulus and shape factor, 561
    plastic moment and neutral axis, 558
    principle of superposition, 754
    properties, 417
    rectangular cross section, 468, 561
    shapes and relative efficiency, 428
    shear flow, 458
    shear forces and bending moments,
        354–360
    shear stresses, 543–545
    small angles of rotation, 732
    stresses, 402–475, 506–568
    structure, 14
    thin-walled open cross sections,
        543–545
    transformed, 517–525
    types, 354–360

wide-flange shape, 563
wide-flange shear stresses, 546–549
yield moment, 558
Bearing stresses, 47
Bending, 352–389, 404, 468–472, 511,
    533–540, 774–778
    pure and nonuniform, 404
    sandwich beams, 511
    strain energy, 774–778
    stress concentration in, 468–472
    unsymmetric beams, 533–540
Bending moment, 352–389, 526,
    879, 902
    column with eccentric axial
        loads, 902
    diagrams, 375–387
    distributed load, 373
    double symmetric beam, 526
    equation and column, 879
    relationships between loads and shear
        forces, 371–374
    shear-force, 375–387
    types of beams, loads and reactions,
        354–360
Biaxial stress, 595, 600, 624, 631–632,
    674–675, 679
    Hooke's law, 624
    spherical pressure vessel, 674–675
    strain energy density, 631–632
Bifurcation point, 873
Bolts, 161
Box beam, 458
British Imperial System, 1037
Brittle, 43
Buckling, 70, 74, 210, 321, 428, 672,
    870–879, 882, 915
    angle of twist, 321
    columns, 870–877
    critical load, 871–873, 882
    differential equation, 879
    elastoplastic analysis, 210
    Euler, 882, 915
    spherical pressure vessels, 672
    and stability, 870–877
Built-up beams, 403, 458–461
    glulam, 458
    plate girder, 458
    shear flow, 458–459
    stresses, 458–461
    wide flange, 460
    wood box, 460

## C

Cables, 122
Cantilever beam, 354–362, 463, 541, 687,
    769, 779, 823–835
    eccentric axial loads, 463
    propped, 359, 823, 826–833
    shaft center, 541

Castigliano's theorem, 779–790
    application, 781–782
    deflection of beams, 779–790
    integral sign, 784–785
    modified, 784
Celsius temperature scale, 1046
Center of flexure, 541
Centroids, 954–979
    centroids of composite areas, 959–979
    centroids of plane areas, 956–958
    composite areas, 959–961
    and moments of inertia, 954–979
    moments of inertia of plane areas,
        962–964
    parallel-axis theorem for moments of
        inertia, 965–968
    plane areas, 956–958
    polar moments of inertia, 969–970
    principal axes and principal moments of
        inertia, 976–979
    products of inertia, 971–973
    rotation of axes, 974–975
Channel section, 550–552
Circular bar, 257–271
    linearly elastic materials, 260–271
    torsion, 257–259
Circular beam, 429, 448–449, 543
Circular members, 259–264
    cross section, 265, 448–450
    limitations of bars, 265
    shear stress, 448–450
    transmission of power, 291–295
Circumferential stress
    cylindrical pressure vessels, 678
Coefficient of thermal expansion, 149
Coil spring, 120
Columns, 868–935
    bending-moment equation, 879
    both ends fixed against rotation, 893
    buckling and differential
        equation, 879
    buckling and stability, 870–877
    critical loads, 871, 892
    critical stress, 894
    deflection curve, 891
    design formulas for columns, 916–933
    eccentric axial loads, 899–903
    effective lengths, 892
    effects of large deflections, 895
    elastic column behavior, 909–910
    fixed at base and free at top, 889
    fixed at base and pinned
        at top, 894
    imperfections, 895
    inelastic behavior, 895
    inelastic column behavior, 909–910
    limitations, 896
    maximum bending moment, 902
    maximum deflection, 900
    neutral equilibrium, 873
    optimum shapes, 886
    other support conditions, 889–898
    pinned ends, 878–888
    reduced-modulus theory, 913
    secant formula, 904–908

Combined loading, 694–711
    method illustration, 695
    method of analysis, 694
    plane stress, 694–711
Compatibility equations, 832
Composite areas, 959–961, 963
    areas and first moments of, 959
    moment of inertia, 963
    review of centroids and moments of
        inertia, 959–961
Composite beams 508, 510–512
    limitations, 512
    moment-curvature relationships, 510
    neutral axis, 510
    normal stresses, 511
    strains and stresses, 508
Compression, 3–82, 911
    allowable stresses and allowable
        loads, 68–73
    design for axial loads and direct
        shear, 74–79
    elasticity, plasticity and creep, 45–51
    Hooke's law, 52–53
    ideal column, 911
    linear elasticity, 52–56
    mechanical properties of materials,
        37–45
    mechanics of materials, 4–6
    normal stress and strain, 27–36
    Poisson's ratio, 53–54
    shear stress and strain, 57–60
    strain, 30
    stresses, 28
    stress-strain diagram, 911
    tension, compression, and shear,
        44–45
    test, 38
Concentrated load, 356, 374–375, 775
    beam element, 374
    deflection curve, 775
    shear-force and bending-moment
        diagrams, 375
Continuous beam, 825
Conversions between units, 1047–1049
Coordinate axes, 404
Couple, 7, 60–61, 256, 354–364, 371
    bending moment, 354–360
    shear force, 371
Creep, 44–51
    tension, compression, and shear, 45–51
Critical loads, 5, 867, 871–888
    columns, 871
    pinned end column, 882
Critical stress, 894
Cross sections, 27, 310, 408, 414–415,
        439–450, 542–545
    beam, 408
    circular shear stress, 448–450
    equilateral triangular, 310
    hollow circular, 449
    maximum stresses, 415
    moment inertia, 414
    rectangular shear stresses, 439–447
    shear center, 542
    shear stresses in beams, 543–545

Curvature, 402–417, 510–511, 732, 774,
        845, 853–856
    beam, 405–407
    determine, 853
    moments, 510–511
    shortening, 853–856
    stresses in beams, 405–406
    temperature effects, 845
Cylindrical pressure vessels, 678–684
    circumferential stress, 678
    longitudinal stress, 679
    plane stress, 678–684
    stresses at inner surface, 680
    stresses at outer surface, 679

## D

Definite integrals, 1061
Deflection, 404, 728–799, 825–831, 891, 900
    analysis by differential equations,
        825–831
    beams, 728–799
    bending-moment equation, 735–745
    Castigliano's theorem, 779–790
    caused by single load, 775
    column, 891
    column with eccentric axial loads, 900
    concentrated load, 775
    curves, 404, 733–734, 782–784, 795,
        825–831, 891
    differential equations of deflection
        curve, 730–734
    differentiation under integral sign, 784
    distributed loads, 753
    expression for curvature, 734
    integration, 735–751
    load equation, 746–751
    method of superposition, 752–759
    moment-area method, 760–768
    nonprismatic beams, 733–734, 769–773
    prismatic beams, 734
    produced by impact, 791–793
    shear-force equation, 746–751
    small angles of rotation, 732
    statically indeterminate beams, 825–831
    strain energy of bending, 774–778
    tables of beam deflections, 753
    temperature effects, 793–797
    use of fictitious load, 782–784
Derivatives, 1059
Design, 74–79, 426–434, 916–933
    axial loads and direct shear, 74–79
    bending stresses, 426–434
    formulas for columns, 916–933
Differential equation of deflection,
        730–734, 825–831, 847–848, 879
    column buckling, 879
    deflection of beams, 730–734
    fixed-end beam, 847–848
    statically indeterminate beams,
        825–831
Direct shear, 59
Displacement method, 138
Displacements caused by single
       load, 180

Distributed loads, 7–8, 356, 371–373,
384–386, 443, 733, 753, 1043
   bending moment, 373
   deflection curve, 753
   shear force, 371
   shear stresses, 443
Double shear, 58
Doubly symmetric beams, 526–532, 542
   bending moments, 526
   inclined loads, 526–532
   load inclination, 527
   neutral axis, 527
   normal stresses, 526
   shear center, 542
Doubly symmetric shapes, 416
Ductility, 42
Dummy-load method, 784
Dynamic load, 187
Dynamic test, 38

**E**

Eccentric axial loads, 463, 899–903
   cantilever beam, 463
   columns, 899–903
   maximum bending moment, 902
Effects of shear strains
   rectangular beam, 445
Elastic column behavior
   columns, 909–910
Elastic core, 558
Elasticity, 3, 40, 45–51, 62, 121–123,
290–291, 511–512
   composite beam, 511–512
   shear modulus, 62
   tension, compression, and shear, 45–51
Elastoplastic analysis, 210–215
Elastoplastic bending, 558–565
Elastoplastic material, 206, 210, 409,
558, 561
   beam, 558, 561
Electrical-resistance strain gage, 641
Elliptical cross section, 309
Elongation, 30, 37, 40–42, 120–129,
178–179
   axially loaded members, 120–129
   linearly elastic behavior, 178–179
Energy, 1042
Equations of compatibility, 832
Equations of equilibrium, 832
Equilateral triangular cross section, 310
Equilibrium, 4–6, 8, 14, 26, 60, 123, 140,
210–211, 273, 358–359, 832,
869–876
   columns, 869–876
   equations, 6, 123, 140, 832
   forces, 358
   moment, 359
   static, 4, 6, 8, 14, 26
   tension, compression, and shear, 6
Euler buckling, 892, 915, 923
   coefficient for wood columns, 923
Exponents, 1057
Expression for curvature, 734
Extensometer, 37

**F**

Factors of safety, 3, 68–69
Fahrenheit temperature, 1047
Fatigue, 195–196
FE exam review problems, 995–1036
Fictitious load, 782–784
Filament-reinforced material, 44
First moment, 443, 760, 959
   area theorem, 760
   composite areas, 959
Fixed-end beam, 824, 847–848
   differential equation, 847–848
Flexibility, 120–121, 138, 140, 263, 785, 835
   bars, 263
   method, 138, 140, 785, 835
Flexure formula, 403, 414–417, 429, 435,
439–444, 506–511, 526–528,
685, 904
   beam, 417, 685
   columns, 904
   composite beam, 506–511
   doubly symmetric beams, 526–528
   nonprismatic beam, 435
   rectangular cross section, 439–444
Force, 7–8, 138, 832
   displacement relations, 138, 832
   method, 138
Free-body diagrams, 8–13
Frequency, 1042
Fully stressed beams, 435

**G**

Gage length, 38–40
Gage pressure, 672
Glass, 43–44, 201, 1090, 1092–1093
   fibers, 44
Glulam beam, 458

**H**

Hertz, 292, 1039–1040, 1042, 1044
Hole at neutral axis of beams, 468
Hollow circular cross section, 449
Homogeneous material, 54, 121
Hooke's law, 52–53, 62, 178, 260, 413,
623–628, 630
   biaxial stress, 624
   compression, 52–53, 62
   plane stress, 623–628
   pure shear, 624
   shear, 52–53, 62, 260
   strain, 623–628
   tension, 52–53, 62
   triaxial stress, 630
   uniaxial stress, 624
   volume change, 624–625
Horizontal reactions, 854
Horsepower, 292–293, 1041, 1049
Hydrostatic stress, 589, 633

**I**

I-beams, 307, 311, 426–427, 458, 508
Ideal column, 878, 911

Impact factor, 190
Impact loads, 187–194, 989
Imperial Units, 1041
Inclined sections, 119, 164–175, 255,
589–590, 592
   stresses, 164–175, 592
Indefinite integrals, 1059–1061
Inelastic buckling, 909, 914
   Shanley theory, 914
Inelastic column behavior, 909–910
Inertia, 262, 264–266, 273, 301, 320, 527,
543, 549, 695, 769, 956, 965–966,
969–970, 975
   cross sections, 262, 264–266, 527,
543, 695
   moments and products, 965–966, 975
   polar moments, 969–970
Infinite series, 1058–1059
In-plane principal stresses, 601
In-plane shear stresses, 604
Internal forces (stress resultants)
   tension, compression, and shear,
13–14
International System (SI), 1038–1044
   and metric units in common, 1041
   prefixes, 1043
   principal units used in mechanics, 1040
   pronunciation, 1044
   styles for writing, 1043–1044
   units, 1038–1044
Isotropic, 54

**J**

Joule, 177, 181, 292, 1039–1040, 1048

**K**

Kelvin temperature, 1046
Kinetic energy, 187–191, 194

**L**

Lateral contraction, 41
Limitations, 191
Linear elastic material, 52–56, 178, 263,
274, 444
   angle of twist, 263
   behavior, 178
   limitations of beams, 444
   tension, compression, and shear,
52–56
Line of action, 31–32
Loads, 63–73, 75, 176, 190–191,
195–196, 354–360, 371–375,
527, 1042
   allowable, 63–73
   axially loaded members, 195–196
   displacement diagram, 176
   double symmetric beam, 527
   and fatigue, 195–196
   form of couples, 375
   inclination, 527
   shear forces and bending moments,
354–360

shear forces and bending moments relations, 371–374

structures, 1042

suddenly applied, 190–191

Logarithms, 1057

Longitudinal displacements, 853–856

Longitudinal strains, 407–411

Longitudinal stress, 679

Lower flange, 549

# M

Margin of safety, 59

Materials, 1089–1094

Mathematical constants, 1057

Mathematical formulas, 1057–1061

Mechanics of materials, 3–80, 955–977

tension, compression, and shear, 4–6, 37–45

Metric system, 1037

Misfits, 149–163

axially loaded members, 149–163

Modified Castigliano's theorem, 784

Modulus of elasticity, 40

Modulus of rigidity, 62

Mohr's circle, 61, 607–622, 640, 975

construction, 609

equations, 607

maximum shear stresses, 613

moments and products of inertia, 975

plane stress, 607–622

principal strains, 640

principal stresses, 612

strain, 607–622

stresses on inclined element, 61

two forms, 608

Moment-area method, 760–768

Moment-area theorem, 762–764

Moment-curvature relationship, 413, 510, 518, 1042

composite beams, 510

transformed beam, 518

Moment reaction as redundant, 834–835

Moments of inertia, 256, 414, 954–979

centroids of composite areas, 959–961

centroids of plane areas, 956–958

composite area, 963

cross-sectional area, 414

formulas, 963

Mohr's circle, 975

moments of inertia of plane areas, 962–964

parallel-axis theorem, 965–969

parallel-axis theorem for moments of inertia, 965–968

plane areas, 962–964

polar, 969–970

polar moments of inertia, 969–970

principal, 976–979

principal axes and principal moments of inertia, 976–979

products of inertia, 971–973

rotation of axes, 974–975

transformation equations, 975

twisting, 256

# N

Necking, 41, 43–44

Neutral axis, 412–413, 510, 517, 527, 533, 558

beam of elastoplastic material, 558

composite beams, 510

double symmetric beam, 527

location, 412–413

transformed section, 517

unsymmetric beam, 533

Neutral equilibrium, 873

Nominal strain, 40

Nominal test, 39

Noncircular prismatic shafts, 307–315

Nonlinear behavior, 205–209

Nonlinear stress-strain curves, 205–206

Nonprismatic beams, 435–438, 733–734, 769–773

deflection curve, 733–734

deflection of beams, 769–773

stresses in beams, 435–438

Nonuniform bars, 179–180

Nonuniform bending, 404

Nonuniform torsion, 272–282, 301

Normal strain, 27–36

tension, compression, and shear, 27–36

Normal stresses, 27–36, 412–425, 511, 518, 526

composite beams, 511

double symmetric beam, 526

stresses in beams, 412–425

tension, compression, and shear, 27–36

transformed beam, 518

Notches in rectangular beams, 468–475

# O

Optimization, 75

Optimum shapes and columns, 886

Out-of-plane shear stresses, 604

# P

Parallel-axis theorem, 965–969, 971

moments of inertia, 965–968

polar moments of inertia, 969

Pascal, 1042

Percent elongation, 43

Percent reduction, 43

Perfectly plastic, 40

Perpendicular planes, 60–61

Pinned-end columns, 892, 895, 911

critical loads, 892

effects of large deflections, 895

tangent-modulus theory, 911

Plane areas, 956–958, 962–964, 1063–1068

centroids and moments of inertia, 956–958

moments of inertia, 962–964

properties, 1063–1068

Plane frame structure, 14

Plane of bending, 404

Plane strain, 633, 635–638

*versus* plane stress, 633

transformation equations, 635–638

Plane stress, 588–714

applications, 670–714

combined loading, 694–711

cylindrical pressure vessels, 678–684

Hooke's law, 623–628

Hooke's law for plane stress, 623–628

maximum stresses in beams, 685–693

Mohr's circle for plane stress, 607–622

plane strain, 633–647

principal stresses and maximum shear stresses, 598–606

special cases, 594

spherical pressure vessels, 672–677

strain-energy density, 626

transformation equations, 592

triaxial stress, 629–632

Plasticity, 45–51

Plastic modulus and shape factor, 561

Plastic moment and neutral axis, 558

Plastics, 44

Poisson's ratio, 53–55, 62, 409, 623

Polar moments of inertia, 969–970

parallel-axis theorem, 969

Pressure vessels, 672

Prestrains, 149–163

axially loaded members, 149–163

Principal angles, 600

Principal axes, 976–979

Principal moments of inertia, 976–979

Principal planes, 600

Principal points, 977

Principal strains, 639–640

Mohr's circle, 640

Principal stresses, 598–606, 612, 674, 686, 689

in-plane, 601

Mohr's circle, 612

plane stress, 598–606

principal angles, 600

rectangular beam, 686

spherical pressure vessel, 674

wide-flange beam, 689

Prismatic bars, 27, 121, 131

segments, 131

Prismatic beams, 734

Prismatic shafts, 307–315

Problem solving, 1051–1057

dimensional homogeneity, 1053–1054

rounding of numbers, 1056

significant digits, 1054–1056

steps, 1052–1053

types, 1051–1052

Products of inertia, 971–973, 975

Mohr's circle, 975

review of centroids and moments of inertia, 971–973

Propped cantilever beam, 822, 832–835

moment reaction as redundant, 834–835

simple support redundant, 832–834

Pure bending, 404

Pure shear, 283–289, 300–306, 624

Hooke's law, 624

strain-energy density in, 302

torsion, 300–306

## Q

Quadratic equation, 1058
Quadratic formula, 1058

## R

Radius of curvature, 405–406, 409, 508, 731, 774
Radius of gyration, 884, 904, 922, 963
Ramberg-Osgood stress-strain law, 208
Rankine temperature scale, 1047
Reactions, 75, 354–360
    bending moments, 354–360
Rectangular beam, 310, 443, 445, 468–469, 555, 686–688
    cross section, 310, 468, 561
    distribution of shear stresses, 443
    effects of shear strains, 445
    maximum shear stresses, 687
    notches, 469
    principal stresses, 686
    shear centers, 555
    stress contour, 688
    stresses, 685–687
    stress trajectories, 687
Rectangular cross section, 439–447
Reduced modulus theory, 913
Redundant reaction, 833
Reinforced concrete beams, 512
Released structure, 833
Reloading of material, 46–47
Repeated loading and fatigue, 195–196
Resilience, 181
Review of centroids and moments of inertia, 954–980
Rotation of axes, 974–975

## S

Saint-Venant's principle, 197–198
Sandwich beams, 508, 511
    bending, 511
Secant formula, 904–908
Second moment of area, *see* Moments of inertia
Section moduli, 415
Shafts, 256, 291–295, 307–315
    torsion, 307–315
    transmission of power by circular, 291–295
Shanley theory, 914–915
    inelastic buckling, 914
Shape factor, 561
Shear, 3–82, 283–289, 302, 440
    allowable stresses and allowable loads, 68–73
    design for axial loads and direct shear, 74–79
    elasticity, plasticity and creep, 45–51
    formula derivation, 440
    Hooke's law, 52–53, 62
    linear elasticity, 52–56
    mechanical properties of materials, 37–45
    mechanics of materials, 4–6
    normal stress and strain, 27–36
    Poisson's ratio, 53–54

shear stress and strain, 57–60
    strain-energy density, 302
    torsion, 283–289
Shear-center concept, 541–542
Shear centers, 541–542, 550–557
    channel section, 550
    double symmetric cross section, 542
    stresses in beams, 550–557
    symmetric cross section, 542
    thin-walled open cross sections, 542
    thin-walled open sections, 550–557
    two intersecting narrow rectangles, 555
    unsymmetric cross section, 542
Shear flow, 316, 458–461, 545
    beam, 458
    formula, 459
    stresses in beams, 458–461
    thin-walled open cross sections, 545
Shear force, 352–389, 453
    and bending moments, 352–401
    diagrams, 375–387
        concentrated load, 375
    distributed load, 371
    loads and bending moments relations, 371–374
    shear-force and bending-moment diagrams, 375–387
    types of beams, loads and reactions, 354–360
    web of wide-flange beam, 453
Shear modulus of elasticity, 62
Shear strains, 57–61, 257, 259
    bar, 259
    outer surface, 257
    tension, compression, and shear, 57–61
Shear stresses, 57–60, 168, 308, 316, 439–457, 543–549, 598–606, 613, 629, 687
    beams of thin-walled open cross sections, 543–545
    circular cross section, 448–450
    distribution and angle of twist, 308
    in-plane, 604
    lower flange, 549
    Mohr's circle, 613
    out-of-plane, 604
    perpendicular planes, 60–61
    plane stress, 598–606
    principal planes, 600
    rectangular beam, 687
    rectangular cross section, 439–447
    and shear flow, 316
    tension, compression, and shear, 57–60
    triaxial stress, 629
    upper flange, 547
    vertical and horizontal, 439
    web of wide-flange beam, 451, 453, 548
    webs of beams with flanges, 451–457
    wide-flange beams, 546–549
Shell structures, 672
Simple support redundant, 832–834
Slenderness ratio, 894
Solid circular cross section, 309
Spherical pressure vessels, 672–677
    plane stress, 672–677
    principal stresses, 674
    stresses at inner surface, 674–675

stresses at outer surface, 674
    tensile stresses, 673
Spherical stress, 632
Springs, 120
Stability and columns, 870–877
Statically determinate structures, 4, 6, 138, 151, 160, 206, 356, 360, 735, 822, 825, 1053
    deflections, 735, 825
    reactions, 356, 360, 1053
Statically indeterminate structures, 138–148, 208, 210–211, 296–299, 820–858
    analysis by differential equations, 825–831
    axially loaded members, 138–148
    deflection curves, 825–831
    longitudinal displacements at ends of beam, 853–856
    method of superposition, 832–844
    temperature effects, 845–852
    torsion, 296–299
    torsional members, 296–299
    types, 822–824
Static load, 187
Static test, 38
Steel, 427, 458, 918–919, 924, 1068–1080
    column design formulas, 918–919
    column imitations, 924
    plate girder, 458
    sections, 427
    shapes, 1069–1080
Stepped shaft, 14
Stiffness method, 138
Strain, 30, 40, 508, 588–669
    calculation, 642
    composite beams, 508
    compression, 30
    Hooke's law for plane stress, 623–628
    measurements, 641
    Mohr's circle for plane stress, 607–622
    plane strain, 633–647
    plane stress, 590–597
    principal stresses and maximum shear stresses, 598–606
    pure shear, 283–289
    torsion, 283–289
    triaxial stress, 629–632
    true, 40
Strain energy, 177–178, 181, 300–306, 319, 626, 631, 774–778
    bending, 774–778
    deflection of beams, 774–778
    density, 181, 302, 626, 631
    elastic and inelastic, 177–178
    plane stress, 626
    pure shear, 302
    thin-walled tubes, 319
    torsion, 300–306
    triaxial stress, 631
Strength, 41
Stress, 28, 31–32, 39, 61, 68–73, 164–175, 197–204, 283–289, 402–569, 642, 674–675, 679, 685–693
    allowable, 68–73
    axial loads, 462–467
    axially loaded members, 164–175, 197–204

beams, 402–569
  bending of unsymmetric beams, 533–540
  built-up beams and shear flow, 458–461
  calculation, 642
  circular cross section, 448–450
  composite beams, 508–516
  compression, 28
  concentration in bending, 468–472
  contour of rectangular beam, 688
  cross section, 415
  curvature of beam, 405–406
  cylindrical pressure vessels, 679
  design for bending stresses, 426–434
  designing, 201
  distribution, 31–32
  doubly symmetric beams, 526–532
  elastoplastic bending, 558–565
  elements, 165
  inclined element, 61
  inclined planes, 283
  inclined sections, 164–175, 592
  inner surface, 674–675
  line of action of axial forces, 31–32
  longitudinal strains, 407–411
  Mohr's circle, 61
  nonprismatic beams, 435–438
  nonuniform bending, 404
  normal stress, 412–425
  outer surface, 674, 679
  plane stress, 685–693
  pure bending, 404
  pure shears, 283–289
  rectangular beam, 685–687
  rectangular cross section, 439–447
  shear-center concept, 541–542
  spherical pressure vessel, 674–675
  thin-walled open cross sections, 543–545
  thin-walled open sections, 550–557
  torsion, 283–289
  trajectories, 687
  transformed-section method, 517–525
  true, 39
  webs of beams with flanges, 451–457
  wide-flange beams, 546–549
Stress concentration, 197–204, 324–327,
    468–472
  bending, 468–472
  factors, 198–200
  torsion, 324–327
Stress-strain diagram, 40, 911
  ideal column, 911
Superposition, 752–759, 832–844
  deflection of beams, 752–759
  principle, 754
  statically indeterminate beams, 832–844
Symmetric cross section
  shear center, 542
Systems of units and conversion factors,
    1037–1051

**T**

Tangent-modulus theory, 911
Temperature, 150, 793–797, 845–852,
    1046–1047
  deflection of beams, 793–797

displacement relation, 150
  effects, 793–797, 845–852
  statically indeterminate beams, 845–852
  units, 1046–1047
Tensile strain, 30
Tensile stresses, 28, 673
  spherical pressure vessel, 673
Tensile-test machine, 37
Tension, 3–83, 169
  allowable loads, 70
  allowable stresses, 69
  allowable stresses and allowable loads,
      68–73
  applied forces, 7–8
  bar, 169
  compression, 44–45
  design for axial loads and direct
      shear, 74–79
  elasticity, plasticity and creep, 45–51
  equality of shear stresses on
      perpendicular planes, 60–61
  equilibrium equations, 6
  free-body diagrams, 8–13
  Hooke's law, 52–53, 62
  internal forces, 13–14
  limitations, 29–30, 54
  linear elasticity, 52–56
  line of action of axial forces, 31–32
  mechanical properties of materials, 37–45
  mechanics of materials, 4–6
  normal strain, 30–31
  normal stress and strain, 27–36
  Poisson's ratio, 53–54
  reloading of material, 46–47
  safety factors, 68–69
  shear strain, 61
  shear stress and strain, 57–60
  sign conventions for shear stresses and
      strains, 61–62
  stress resultants, 13–14
  tables of mechanical properties, 45
  uniaxial stress and strain, 31
  uniform stress distribution, 31–32
Thermal effects, 149–163
  axially loaded members, 149–163
  strains, 149
  stress, 149
Thermal expansion, 149
Thin-walled open cross sections, 311, 542,
    545, 550–557
  shear center, 542
  shear flow, 545
  stresses in beams, 550–557
Thin-walled tubes, 316–323
  angle of twist, 321
  limitations, 321
  strain energy, 319
  torsion, 316–323
Third principal stress, 601
Torque, 256, 1042
Torsion, 254–329
  circular bars of linearly elastic
      materials, 260–271
  constant and thin-walled tubes, 319
  formula, 261–262, 317
  noncircular prismatic shafts, 307–315

nonuniform, 272–282, 301
  nonuniform torsion, 272–282
  statically indeterminate torsional
      members, 296–299
  strain energy in torsion and pure shear,
      300–306
  stress concentrations in torsion,
      324–327
  stresses and strains in pure shears,
      283–289
  thin-walled tubes, 316–323
  torsional deformations of circular bar,
      257–259
  torsion of noncircular prismatic shafts,
      307–315
  transmission of power by circular
      shafts, 291–295
Toughness, 181
Trajectories, 687
Transformation equations, 592,
    635–638, 975
  application, 635
  moments and products of inertia, 975
  plane strain, 635–638
  plane stress, 592
Transformed beam, 517–525
  moment-curvature relationship, 518
  neutral axis, 517
  stresses, 518
Transmission of power by circular shafts,
    291–295
Trial-and-error procedure, 917
Triaxial stress, 629–632
  Hooke's law, 630
  plane stress, 629–632
  shear stresses, 629
  strain-energy density, 631
  volume change, 631
Trigonometric functions, 1058
True strain, 40
True stress, 39
Truss structure, 14
Tubes, 316–323
Turnbuckles, 161
Twisting moments, 256
Two intersecting narrow rectangles, 555

**U**

Unbalanced I-beam, 541
Uniaxial stress, 31, 170, 624
  Hooke's law, 624
  tension, compression, and shear, 31
Uniform load, 377
Uniform stress distribution, 31–32
United States Customary System
    (USCS), 1037
  Imperial Units in common,
      1044–1046
  principal units used in mechanics, 1040
  SI units conversions, 1048–1049
  units, 1044–1046
Unsymmetric beams, 533–540
  bending, 533–540
  neutral axis, 533
  procedure for analyzing, 535

Unsymmetric cross section, 542
Upper flange, 547

## V

Volume change, 624–625

## W

Watt, 1042
Web of wide-flange beam, 451–457, 548
    limitations, 454
    shear force, 453
    shear stresses, 451–457, 548
Welded steel plate girder, 460

Wide-flange beams, 546–549, 688–690
    calculating first moment, 460
    maximum shear stresses, 689
    principal stresses, 689
    reduced modulus, 913
    shape, 563
    shear stresses, 546–549
Wire rope, 122
Wood, 427, 460, 921, 923–924
    beams, 427
    box beam, 460
    columns, 923–924
    Euler buckling coefficient, 923
    limitations, 924
    structural members design, 921
Work, 1042

## Y

Yield displacement, 210, 212
Yielding, 40–44, 52, 210
    elastoplastic analysis, 210
    linear elasticity, 52
Yield load, 210–215
Yield moment, 558–564
    beam of elastoplastic material, 558
Yield point, 40–42, 558
Young's modulus, 53

## Z

Z-section, 555

## PRINCIPAL UNITS USED IN MECHANICS

| Quantity | International System (SI) | | | U.S. Customary System (USCS) | | |
|---|---|---|---|---|---|---|
| | Unit | Symbol | Formula | Unit | Symbol | Formula |
| Acceleration (angular) | radian per second squared | | $\text{rad/s}^2$ | radian per second squared | | $\text{rad/s}^2$ |
| Acceleration (linear) | meter per second squared | | $\text{m/s}^2$ | foot per second squared | | $\text{ft/s}^2$ |
| Area | square meter | | $\text{m}^2$ | square foot | | $\text{ft}^2$ |
| Density (mass) (Specific mass) | kilogram per cubic meter | | $\text{kg/m}^3$ | slug per cubic foot | | $\text{slug/ft}^3$ |
| Density (weight) (Specific weight) | newton per cubic meter | | $\text{N/m}^3$ | pound per cubic foot | pcf | $\text{lb/ft}^3$ |
| Energy; work | joule | J | N·m | foot-pound | | ft-lb |
| Force | newton | N | $\text{kg·m/s}^2$ | pound | lb | (base unit) |
| Force per unit length (Intensity of force) | newton per meter | | N/m | pound per foot | | lb/ft |
| Frequency | hertz | Hz | $\text{s}^{-1}$ | hertz | Hz | $\text{s}^{-1}$ |
| Length | meter | m | (base unit) | foot | ft | (base unit) |
| Mass | kilogram | kg | (base unit) | slug | | $\text{lb-s}^2/\text{ft}$ |
| Moment of a force; torque | newton meter | | N·m | pound-foot | | lb-ft |
| Moment of inertia (area) | meter to fourth power | | $\text{m}^4$ | inch to fourth power | | $\text{in.}^4$ |
| Moment of inertia (mass) | kilogram meter squared | | $\text{kg·m}^2$ | slug foot squared | | $\text{slug-ft}^2$ |
| Power | watt | W | J/s (N·m/s) | foot-pound per second | | ft-lb/s |
| Pressure | pascal | Pa | $\text{N/m}^2$ | pound per square foot | psf | $\text{lb/ft}^2$ |
| Section modulus | meter to third power | | $\text{m}^3$ | inch to third power | | $\text{in.}^3$ |
| Stress | pascal | Pa | $\text{N/m}^2$ | pound per square inch | psi | $\text{lb/in.}^2$ |
| Time | second | s | (base unit) | second | s | (base unit) |
| Velocity (angular) | radian per second | | rad/s | radian per second | | rad/s |
| Velocity (linear) | meter per second | | m/s | foot per second | fps | ft/s |
| Volume (liquids) | liter | L | $10^{-3}\ \text{m}^3$ | gallon | gal. | $231\ \text{in.}^3$ |
| Volume (solids) | cubic meter | | $\text{m}^3$ | cubic foot | cf | $\text{ft}^3$ |